ENDOCRINE SYSTEM
Acts by means of hormones secreted into the blood to regulate processes that require duration rather than speed—e.g., metabolic activities and water and electrolyte balance.
See Chapter 17.

HOMEOSTASIS
A dynamic steady state of the constituents in the internal fluid environment that surrounds and exchanges materials with the cells.
See Chapter 1.
Factors homeostatically maintained are:
- Concentration of nutrient molecules
  See Chapters 15, 16, and 17.
- Concentration of O_2 and CO_2
  See Chapter 12.
- Concentration of waste products
  See Chapter 13.
- pH   *See Chapter 14.*
- Concentration of water, salts, and other electrolytes
  See Chapters 13, 14, and 17.
- Volume and pressure
  See Chapters 10 and 14.
- Temperature   *See Chapter 16.*

INTEGUMENTARY SYSTEM
Serves as a protective barrier between the external environment and the remainder of the body; the sweat glands and adjustments in skin blood flow are important in temperature regulation.
See Chapters 11 and 16.

Keeps internal fluids in

Keeps foreign material out

IMMUNE SYSTEM
Defends against foreign invaders and cancer cells; paves way for tissue repair.
See Chapter 11.

Protects against foreign invaders

MUSCULAR AND SKELETAL SYSTEMS
Support and protect body parts and allow body movement; heat-generating muscle contractions are important in temperature regulation; calcium is stored in the bone.
See Chapters 8, 16, and 17.

Enables the body to interact with the external environment

Exchanges with all other systems.

CELLS
Need homeostasis for their own survival and for performing specialized functions essential for survival of the whole body.
See Chapters 1, 2, and 3.
Need a continual supply of nutrients and O_2 and ongoing elimination of acid-forming CO_2 to generate the energy needed to power life-sustaining cellular activities as follows:
$$Food + O_2 \rightarrow CO_2 + H_2O + energy$$
See Chapter 15.

Cells make up body systems

Fundamentals of Physiology
A HUMAN PERSPECTIVE

THIRD EDITION

Lauralee Sherwood
*Department of Physiology and
Pharmacology
School of Medicine
West Virginia University*

THOMSON

BROOKS/COLE

Australia • Canada • Mexico • Singapore • Spain
United Kingdom • United States

THOMSON

BROOKS/COLE

Executive Editors, Biology: Peter Adams, Nedah Rose
Editor-in-Chief: Michelle Julet
Development Editor: Mary Arbogast
Assistant Editor: Kari Hopperstead
Editorial Assistant: Kristin Lenore, Sarah Lowe
Technology Project Manager: Travis Metz
Marketing Manager: Ann Caven
Marketing Assistant: Leyla Jowza
Advertising Project Manager: Kelley McAllister
Project Manager, Editorial Production: Teri Hyde
Art Director: Rob Hugel
Print/Media Buyer: Barbara Britton

Permissions Editor: Joohee Lee
Production Service: Martha Emry
Text Designer: Jeanne Calabrese
Photo Researcher: Myrna Engler
Copy Editor: Linda Purrington
Illustrator: Precision Graphics
Cover Designer: Jeanne Calabrese
Cover Image: Mike Powell, Getty Images
Cover Printer: Phoenix Color Corp
Compositor: Thompson Type
Printer: Courier/Kendallville

For more information about our products, contact us at:
Thomson Learning Academic Resource Center
1-800-423-0563
For permission to use material from this text or product, submit a request online at http://www.thomsonrights.com. Any additional questions about permissions can be submitted by email to thomsonrights@thomson.com.

ExamView® and ExamView Pro® are registered trademarks of FSCreations, Inc. Windows is a registered trademark of the Microsoft Corporation used herein under license. Macintosh and Power Macintosh are registered trademarks of Apple Computer, Inc. Used herein under license.

Library of Congress Control Number: 2004116551

Student Edition: ISBN 0-534-46697-4

Instructor's Edition: ISBN 0-534-46699-0

Thomson Higher Education
10 Davis Drive
Belmont, CA 94002-3098
USA

Asia (including India)
Thomson Learning
5 Shenton Way
#01-01 UIC Building
Singapore 068808

Australia/New Zealand
Thomson Learning Australia
102 Dodds Street
Southbank, Victoria 3006
Australia

Canada
Thomson Nelson
1120 Birchmount Road
Toronto, Ontario M1K 5G4
Canada

UK/Europe/Middle East/Africa
Thomson Learning
High Holborn House
50/51 Bedford Row
London WC1R 4LR
United Kingdom

Latin America
Thomson Learning
Seneca, 53
Colonia Polanco
11560 Mexico
D.F. Mexico

Spain (includes Portugal)
Thomson Paraninfo
Calle Magallanes, 25
28015 Madrid, Spain

To my family,
in celebration of the continuity of life,
with memories of the past, joys of the present,
and expectations of the future:

My parents,
Larry (in memoriam) and Lee Sherwood

My husband,
Peter Marshall

My daughters and sons-in-law,
Melinda and Mark Marple
Allison and Hany Tadros

My grandchildren,
Lindsay Marple
Emily Marple
"Baby boy" Tadros, to be born the same time this book is to be published

Brief Contents

Contents

Chapter 7
The Peripheral Nervous System: Efferent Division   184

Chapter 8
Muscle Physiology   202

Chapter 9
Cardiac Physiology **240**

Chapter 12
The Respiratory System 364

Chapter 14
Fluid and Acid–Base Balance   442

Chapter 15
The Digestive System   464

Chapter 16

Energy Balance and Temperature Regulation   510

Chapter 17

The Endocrine System   528

Chapter 18
The Reproductive System   582

GOALS, PHILOSOPHY, AND THEME

My goal in writing physiology textbooks is to help students not only learn about how the body works but also to share my enthusiasm for the subject matter. I have been teaching physiology since the mid-1960s and remain awestruck at the miraculous intricacies and efficiency of body function. No machine can take over even a portion of natural body function as effectively. Most of us, even infants, have a natural curiosity about how our bodies work. When a baby first discovers it can control its own hands, it is fascinated and spends many hours manipulating them in front of its face. By capitalizing on students' natural curiosity about themselves, I try to make physiology a subject they can enjoy learning.

Even the most tantalizing subject matter, however, can be difficult to comprehend if not effectively presented. Therefore, this book is not cluttered with unnecessary details and has a logical, understandable format with an emphasis on how each concept is an integral part of the whole subject matter. Too often, students view the components of a physiology course as isolated entities; by understanding how each component depends on the others, a student can appreciate the integrated functioning of the human body. The text focuses on the mechanisms of body function from cells to systems and is organized around the central theme of homeostasis—how the body meets changing demands while maintaining the internal constancy necessary for all cells and organs to function.

Fundamentals of Physiology: A Human Perspective is a carefully condensed version of my *Human Physiology: From Cells to Systems,* which is now in its fifth edition (Brooks/Cole 2004). The parent book is designed primarily for upper-level undergraduates who have some math and science background and need an in-depth coverage of physiology. This new third edition of *Fundamentals of Physiology* continues the tradition of being a briefer book suitable for lower-level physiology courses. It is of manageable length and depth while still providing scientifically sound coverage of the fundamental concepts of physiology and incorporating pedagogical features that make it interesting and comprehensible.

This text is especially geared toward beginning health professional students who have limited science and math background yet need clear, current, concise, clinically oriented coverage of physiology. Its approach and depth are appropriate for other undergraduates as well. Because it is intended to serve as an introductory text and, for most students, may be their only exposure to a formal physiology text, all aspects of physiology receive broad coverage. Materials were selected for inclusion on a "need to know" basis, not just because a given fact happens to be known. In other words, content is restricted to relevant information needed to understand basic physiologic concepts and to serve as a foundation for future careers in the health professions.

No assumptions regarding prerequisite courses have been made. Biochemical and quantitative details have been minimized because of the scientifically limited background of the students for whom the book is intended. Also, because anatomy is not a prerequisite course, enough relevant anatomy is integrated within the text to make the inseparable relation between structure and function meaningful. Furthermore, new pedagogical features have been added to make this third edition even more approachable by students with limited background.

In consideration of the clinical orientation of most students, research methodologies and data are not emphasized, although the material is based on up-to-date evidence. New information on recent discoveries has been included in all chapters. Students can be assured of the timeliness and accuracy of the material presented. Some controversial ideas and hypotheses are presented to illustrate that physiology is a dynamic, changing discipline.

To keep pace with today's rapid advances in the health sciences, students in the health professions must be able to draw on their conceptual understanding of physiology instead of merely recalling isolated facts that soon may be outdated. Therefore, this text is designed to promote understanding of the basic principles and concepts of physiology rather than memorization of details.

A comfortable, "down to earth" writing style that is neither terse nor too informal has been used so that students would not feel overwhelmed by language that "goes over their heads" or insulted by language that "talks down to them." The text is written in simple, straightforward language, and every effort has been made to assure smooth reading through good transitions, logical reasoning, and integration of ideas throughout the text.

ORGANIZATION

There is no ideal organization of physiologic processes into a logical sequence. With the sequence I chose for this book, most chapters build on material presented in immediately preceding chapters, yet each chapter is designed to stand on its own, allowing the instructor flexibility in curriculum design. This flexibility is facilitated by cross-references to related material in other chapters. The cross-references let students quickly refresh their memory of material already learned or to proceed if desired to a more in-depth coverage of a particular topic.

The general flow is from introductory background information to cells to excitable tissue to organ systems. I tried to provide logical transitions from one chapter to the next. For example, Chapter 8, "Muscle Physiology," ends with a discussion of cardiac muscle, which is carried forward in Chapter 9, "Cardiac Physiology." Even topics that seem unrelated in sequence, such as Chapter 11, "Blood and Body Defenses," and Chapter 12, "The Respiratory System," are linked together, in this case by ending Chapter 11 with a discussion of respiratory defense mechanisms.

Several organizational features warrant specific mention. The most difficult decision in organizing this text was placement of the endocrine material. For the following reasons, I put the general properties of hormones in Chapter 4 and specific actions of endocrine glands in Chapter 17. Intermediary metabolism of absorbed nutrient molecules is largely under endocrine control, providing a link from digestion (Chapter 15) and energy balance (Chapter 16) to the chapter focusing on the endocrine system (Chapter 17). There is merit in placing the chapters on the nervous and endocrine systems in close proximity because of these systems' roles as the body's major regulatory systems. Placing the endocrine system chapter earlier, immediately after the discussion of the nervous system (Chapters 5 through 7), however, would have created two problems. First, it would have disrupted the logical flow of material related to excitable tissue. Second, the endocrine system could not have been covered at the level of depth its importance warrants if it had been discussed before the students were provided the background essential to understanding this system's roles in maintaining homeostasis.

My solution to this dilemma was the development of a new comparative chapter, "Principles of Neural and Hormonal Communication" (Chapter 4). This chapter introduces the underlying mechanisms of neural and hormonal action early,

before the nervous system and specific hormones are mentioned in later chapters. Topics in this chapter include intercellular communication and signal transduction; graded potentials, action potentials, synapses, and neuronal integration; and the molecular/biochemical/cellular features of hormonal action. Thus this chapter brings together the similarities and differences in how nerve cells and endocrine cells communicate with other cells in carrying out their regulatory actions. Building on the different modes of action of nerve and endocrine cells, the last section of this new chapter compares in a general way how the nervous and endocrine systems differ as regulatory systems. Chapter 5 then begins with the nervous system, providing a good link between Chapters 4 and 5.

Specific hormones are subsequently introduced in appropriate chapters, such as vasopressin and aldosterone in the chapters on kidney and fluid balance. The endocrine system chapter (Chapter 17) pulls together in one chapter the source, functions, and control of specific endocrine secretions and serves as a summarizing/unifying capstone chapter. Finally, building on the gonadotropic hormones introduced in Chapter 17, the final chapter diverges from the theme of homeostasis to focus on reproductive physiology (Chapter 18).

In addition to the novel handling of hormones and the endocrine system, other organizational features are unique to this book. For example, unlike other physiology texts, the skin is covered in the chapter on body defenses, in consideration of the skin's newly recognized immune functions. Bone is also covered more extensively in the endocrine chapter than in most undergraduate physiology texts, especially with regard to hormonal control of bone growth and bone's dynamic role in calcium metabolism. Departure from traditional groupings of material in several important instances has permitted more independent and more extensive coverage of topics that are frequently omitted or buried within chapters concerned with other subject matter. For example, a separate chapter (Chapter 14) is devoted to fluid balance and acid–base regulation, topics often tucked within the kidney chapter. Another example is the grouping of energy balance and temperature regulation into an independent chapter.

New to this book, the chapter on the peripheral nervous system from the previous edition is now divided into two chapters, to give instructors more flexibility in assigning material. Chapter 6 focuses on the afferent division of the peripheral nervous system, ending with the special senses. With this organization, the special senses can easily be skipped (by instructors who wish to do so) without omitting the middle of a chapter, as was the case with the last edition. Chapter 7 focuses on the efferent division of the peripheral nervous system, grouping together coverage of the autonomic nervous system and motor neurons and ending with the neuromuscular junction, providing a good link to Chapter 8 on muscle physiology.

Although there is a rationale for covering the various aspects of physiology in the order given here, it is by no means the only logical way of presenting the topics. Each chapter is able to stand on its own, especially with the cross-references provided, so that the sequence of presentation can be varied at the instructor's discretion. Some chapters may even be omitted, depending on the students' needs and interests and the time constraints of the course.

TEXT FEATURES AND LEARNING AIDS

▌ Implementing the homeostasis theme

 A unique, easy-to-follow, pictorial homeostatic model showing the relationship among cells, systems, and homeostasis is developed in the introductory chapter and presented on the inside front cover as a quick reference. Each chapter begins with a specially tailored version of this model, accompanied by a brief written introduction, emphasizing how the body system considered in the chapter functionally fits in with the body as a whole. This opening feature is designed to orient students to the homeostatic aspects of the material that follows. Then, at the close of each chapter, **Chapter in Perspective: Focus on Homeostasis** helps students put into perspective how the part of the body just discussed contributes to homeostasis. This capstone feature, the opening homeostatic model, and the introductory comments are designed to work together to facilitate students' comprehension of the interactions and interdependency of body systems, even though each system is discussed separately.

▌ Analogies

Many analogies and frequent references to everyday experiences are included to help students relate to the physiology concepts presented. These useful tools have been drawn in large part from my four decades of teaching experience. Knowing which areas are likely to give students the most difficulty, I have tried to develop links that help them relate the new material to something with which they are already familiar.

▌ Pathophysiology and clinical coverage

Clinical Note Another effective way to keep students' interest is to help them realize they are learning worthwhile and applicable material. Because most students using this text will have health-related careers, frequent references to pathophysiology and clinical physiology demonstrate the content's relevance to their professional goals. For the first time, *Clinical Note* icons flag clinically relevant material, which is integrated throughout the text.

▌ Boxed features

Boxed features integrated within the chapter expose students to high-interest, tangentially relevant information on such diverse topics as exercise physiology, stem-cell research, acupuncture, strokes, historical perspectives, and body responses to new environments such as those encountered in space flight and deep-sea diving.

▌ Pedagogical illustrations

The anatomic illustrations, schematic representations, photographs, tables, and graphs are designed to complement and reinforce the written material. Now more three-dimensional and realistic, most of the cellular art and much of the anatomic art are new to this edition. Also new, numerous process-oriented figures incorporating step-by-step descriptions allow visually oriented students to review processes through figures.

Flow diagrams are used extensively to help students integrate the written information. In the flow diagrams, lighter and darker shades of the same color are used to denote a decrease or an increase in a controlled variable, such as blood pressure or the concentration of blood glucose. Furthermore, the corners of all physical entities, such as body structures or chemicals, are rounded to distinguish them from the square corners of all actions.

Integrated color-coded figure/table combinations help students better visualize what part of the body is responsible for what activities. For example, anatomic depiction of the brain is integrated with a table of the functions of the major brain components, with each component shown in the same color in the figure and the table.

A unique feature of this book is that people depicted in the various illustrations are realistic representatives of a cross section of humanity (they were drawn from photographs of real people). Sensitivity to various races, sexes, and ages should enable all students to identify with the material being presented.

▌ Feedforward statements as subsection titles

Instead of traditional short topic titles for each major subsection (for example, "Heart Valves"), feedforward statements alert students to the main point of the subsection to come (for example, "Pressure-operated heart valves ensure that blood flows in the right direction through the heart"). New to this edition, more of these headings have been added to present large concepts in smaller, more manageable pieces for the student. As an added bonus, the additional headings make it easier to locate specific information at a glance.

▌ Key terms and word derivations

Key terms are defined as they appear in the text. Because physiology is laden with a myriad of new vocabulary words, many of which are rather intimidating at first glance, word derivations are provided to enhance understanding of new words.

▌ Integrated text references to the accompanying CD-ROM

To help students effectively coordinate all of their learning assets, they are directed throughout the text to appropriate media exercises and animated tutorials in the accompanying PhysioEdge 2 CD-ROM.

▌End-of-Chapter learning and review

The **Chapter Summary** presents the major points of each chapter in concise, section-by-section bulleted lists, including cross-references for page numbers, figures, and tables. With this new summary design students can review more efficiently by using both written and visual information to focus on the main concepts before moving on.

The **Review Exercises** at the end of each chapter include a variety of question formats for students to self-test their knowledge and application of the facts and concepts presented. A **Points to Ponder** section features thought-provoking problems that encourage students to analyze what they have learned, and the **Clinical Consideration**, a mini case history, challenges them to apply their knowledge to a patient's specific symptoms. Answers and explanations for all of these questions are found in Appendix F. Finally, the **PhysioEdge Resources** section directs students to the wealth of study aids and ideas for further reading through the human physiology website and PhysioEdge CD-ROM, described in more detail below.

▌Appendixes and Glossary

The appendixes are designed for the most part to help students who need to brush up on some foundation materials or who need additional resources.

- *Appendix A,* **The Metric System,** is a conversion table between metric measures and their English equivalents.
- Most undergraduate physiology texts have a chapter on chemistry, yet physiology instructors rarely teach basic chemistry concepts. Knowledge of chemistry beyond that introduced in secondary schools is not required for understanding this text. Therefore, I decided to reserve valuable text space for physiology concepts, and to provide instead *Appendix B,* **A Review of Chemical Principles,** as a handy reference for students who need an introduction or a brief review of basic chemistry concepts that apply to physiology.
- Likewise, *Appendix C,* **Storage, Replication, and Expression of Genetic Information,** serves as a reference for students or as assigned material if the instructor deems it appropriate. It includes a discussion of DNA and chromosomes, protein synthesis, cell division, and mutations.
- The chemical details of acid–base balance are omitted from the text proper but are included in *Appendix D,* **The Chemistry of Acid–Base Balance,** for those students who have the background and need for a more chemically oriented approach to this topic.
- New to this edition, *Appendix E,* **Text References to Exercise Physiology,** provides an index of all relevant content on this topic.
- *Appendix F,* **Answers to End-of-Chapter Objective Questions, Points to Ponder, and Clinical Considerations,** provides answers to all objective learning activities and explanations for the Points to Ponder and Clinical Consideration.

- The **Glossary,** which offers a way to review the meaning of key terminology, includes phonetic pronunciations of the entries.

ANCILLARIES FOR STUDENTS

▌PhysioEdge 2 CD-ROM

A complimentary text-correlated interactive CD-ROM, PhysioEdge 2, accompanies each new copy of the text. This new learning companion illustrates concepts that are harder conceptually to understand from the text alone. PhysioEdge 2's dynamic representation of concepts through animations and graphic depictions helps students learn the concepts. In addition, they can gauge their understanding by taking the quizzes on the CD because the program's strong diagnostic component gives immediate feedback on answers. When a student does well on the quizzes, the PhysioEdge quizzing program is prompted to introduce the next level of difficulty. It's like having a personal tutor available on demand. PhysioEdge 2 is completely integrated with the text. Icons in the text indicate which figures have complementary CD activities and which concepts can be explored in greater detail through animations, tutorials, and media exercises.

▌InfoTrac College Edition®

Available exclusively from Brooks/Cole, this online library offers students unlimited access, at any time of the day, to full-length research articles—updated daily and spanning four years. With InfoTrac, a student can search for complete articles from thousands of scholarly and popular periodicals, such as *Physiological Reviews, Science, American Journal of Sports Medicine, Science News,* and *Discover.* A four-month subscription to this password-protected site is offered free to students with each new text purchase. An online student guide correlates each chapter in this text to Info-Trac articles. This student guide can be accessed free at the following website:

http://infotrac.thomsonlearning.com

▌Book Companion Website

Through this book's content-rich website at:

http://biology.brookscole.com/hpfundamentals3

students have access to helpful study aids such as **Flash Cards,** a **Glossary, Web Links, Quizzing, Internet Exercises, InfoTrac College Edition Exercises, Chapter Outlines, Chapter Objectives,** and **Chapter Summaries.** This website also features a **Visual Learning Resource,** with enlargements and informative animations of selected figures from the text. These animations help bring to life some of the physiologic processes most difficult to visualize, enhancing the understanding of complex sequences of events. Moreover, with this edition the website offers valuable **Case Histories** that introduce the clinical aspects facilitating the instruction and learning of human physiology.

Study Guide

Each chapter of this student-oriented supplementary manual, which is correlated with the corresponding chapter in *Fundamentals of Physiology: A Human Perspective,* Third Edition, contains a chapter overview, a detailed chapter outline, a list of Key Terms, and Review Exercises (multiple choice, true/false, fill-in-the-blanks, and matching). This learning resource also offers Points to Ponder, questions that stimulate use of material in the chapter as a starting point to critical thinking and further learning. Clinical Perspectives, common applications of the physiology under consideration; Experiments of the Day; and simple hands-on activities enhance the learning process. Answers to the Review Exercises are provided at the back of the Study Guide.

ISBN: 0-495-03202-6

Photo Atlas for Anatomy and Physiology

This full-color atlas (with more than 600 photographs) depicts structures in the same colors as they would appear in real life or in a slide. Labels as well as color differentiations within each structure are employed to facilitate identification of the structure's various components. The atlas includes photographs of tissue and organ slides, the human skeleton, commonly used models, cat dissections, cadavers, some fetal pig dissections, and some physiology materials.

ISBN: 0534-51716-1

Fundamentals of Physiology Laboratory Manual

This manual, which may be required by the instructor in courses that have a laboratory component, contains a variety of exercises that reinforce concepts covered in *Fundamentals of Physiology: A Human Perspective,* Third Edition. These laboratory experiences increase students' understanding of the subject matter in a straightforward manner, with thorough directions to guide them through the process and relevant questions for reviewing, explaining, and applying results.

ISBN: 0-495-03203-4

ACKNOWLEDGMENTS

I gratefully acknowledge the many people who helped with the first two editions of the *Fundamentals of Physiology* textbook as well as with the five editions of the Sherwood, *Human Physiology: From Cells to Systems* textbook on which the *Fundamentals* version is based.

In addition to the 116 reviewers who carefully evaluated the forerunner books for accuracy, clarity, and relevance, I express sincere appreciation to the following individuals who served as reviewers for this book: William Blaker, Furman University; Maureen Burton, University of South Dakota; Linda Curtis, High Point University; Mary Sue Evans, Johnson County Community College; Michael Finkler, Indiana University Kokomo; Nicholas Geist, Sonoma State University; Richard Gunasekera, University of Houston—Victoria; Pamela Gunter-Smith, Spelman College; Patricia Gwirtz, University of North Texas Health Science Center; Ronald Harris, Marymount College; Mark Harrison, University of Indianapolis; Kenneth Kaloustian, Quinnipiac University; Joan Esterline Lafuze, Indiana University East; V. Patteson Lombardi, University of Oregon; Suguru Nakamura, Murray State University; Onkar S. Phalora, Anderson University; Jeffrey Plochocki, Pennsylvania State University, Altoona; David Quadagno, Florida State University; Gary Ritchison, Eastern Kentucky University; Laurel Roberts, University of Pittsburgh; Gail Sabbadini, San Diego State University; Philip Stephens, Villanova University; and David Stepp, Medical College of Georgia.

I have been very fortunate to work with a highly competent, dedicated team from Brooks/Cole, along with other very capable external suppliers selected by the publishing company. I would like to acknowledge all of their contributions, which collectively made this book possible. It has been a source of comfort and inspiration to know that so many people have been working so diligently in so many ways to make this book the best it could possibly be.

Two editors from Brooks/Cole guided this book to fruition. Nedah Rose launched this edition and helped shape the new directions it would take, then Peter Adams took over midstream and helped bring the book to conclusion without missing a beat. Both deserve warm thanks for helping this belated book become a reality after an 11-year hiatus since the last edition. Thanks also to Editorial Assistants Sarah Lowe and Kristin Lenore, who trafficked paperwork and coordinated many tasks for Nedah and Peter during the development process. Furthermore, I appreciate the efforts of Senior Developmental Editor Mary Arbogast for facilitating and offering valuable insights during development and production. I can always count on Mary. She's a valuable resource because she "knows the ropes" and is always thinking of ways to make the book better and the process smoother.

Assistant Editor Kari Hopperstead oversaw the development of multiple components of the ancillary package that accompanies *Fundamentals of Physiology: A Human Perspective,* Third Edition, making sure the package was a cohesive whole. The technology enhanced learning tools were developed under the guidance of Technology Project Manager Travis Metz. These include the new interactive CD-ROM and expanded Web-based complementary activities. A hearty note of gratitude is extended to both of them for these welcome additions to the multimedia package that accompanies this edition.

On the production side, Project Manager Teri Hyde is a highly capable, responsive, caring person who always takes time to respond to my queries and concerns and closely monitors every step of the production process, even though she simultaneously oversees the complex production process of multiple books. I felt confident knowing that she was in the background, making sure that everything was going according to plan. I am grateful for the creative insight of Art Directors Rob Hugel and Lee Friedman, who ensured that the visual aspects of the text are aesthetically pleasing and meaningful. I also thank Permissions Editor Joohee Lee for tracking down permissions for a myriad of art and other copyrighted materials incorporated in the text—a tedious but absolutely essential task.

With everything finally coming together, Print Buyer Barbara Britton oversaw the manufacturing process, coordinating the actual printing of the book. No matter how well a book is conceived, produced, and printed, it would not reach its full potential as an educational tool without being efficiently and effectively marketed. Marketing Manager Ann Caven, Marketing Assistant Leyla Jowza, and Advertising Project Manager Kelley McAllister played the lead roles in marketing this text, for which I am most appreciative.

Brooks/Cole also did an outstanding job in selecting highly skilled vendors to carry out particular production tasks. First and foremost, it has been my personal and professional pleasure to work with a very capable production service, Martha Emry, who coordinated the day-to-day management of production. In her competent hands lay the responsibility of seeing that all art, typesetting, page layout, and other related details got done right and in a timely fashion. Martha is a great facilitator. The production of this book was among the smoothest I have ever experienced. Thanks to Martha, there were no snags or delays. Furthermore, Martha was always understanding and supportive when my schedule got especially hectic, especially when final exams, Christmas preparations, and production deadlines all hit at the same time.

My appreciation is also extended to freelance Photo Researcher Myrna Engler. She always finds multiple versions from which I can choose, so I don't have to "settle" for something less than desirable.

Designer Jeanne Calabrese deserves thanks for the fresh and attractive, yet space-conscious appearance of the book's interior, and for envisioning the book's visually appealing exterior.

I also want to extend a hearty note of gratitude to Compositor Thompson Type for their accurate typesetting, execution of most of the art revisions, and attractive, logical layout design. They seemed to work magic in making up the pages, because the page breaks and chapter endings always seemed to come out "just right," and not by chance alone.

Finally, my love and gratitude go to my family for another year and a half of sacrifice in my family life as this third edition was being developed and produced. I want to thank them for their patience and understanding during the times I was working on the book instead of being there with them or for them. My husband, Peter Marshall, deserves special appreciation and recognition for taking over my share of household responsibilities, freeing up time for me to work on the text. I could not have done this, or any of the preceding books, without his help, support, and encouragement.

Thanks to all!

Fundamentals of Physiology

A HUMAN PERSPECTIVE

THIRD EDITION

During the minute that it will take you to read this page:

Your eyes will convert the image from this page into electrical signals (nerve impulses) that will transmit the information to your brain for processing.

Your heart will beat 70 times, pumping 5 liters (about 5 quarts) of blood to your lungs and another 5 liters to the rest of your body.

More than 1 liter of blood will flow through your kidneys, which will act on the blood to conserve the "wanted" materials and eliminate the "unwanted" materials in the urine. Your kidneys will produce 1 ml (about a thimbleful) of urine.

Your digestive system will be processing your last meal for transfer into your bloodstream for delivery to your cells.

Besides receiving and processing information such as visual input, your brain will also provide output to your muscles to help maintain your posture, move your eyes across the page as you read, and turn the page as needed. Chemical messengers will carry signals between your nerves and muscles to trigger appropriate muscle contraction.

You will breathe in and out about 12 times, exchanging 6 liters of air between the atmosphere and your lungs.

Your cells will consume 250 ml (about a cup) of oxygen and produce 200 ml of carbon dioxide.

You will use about 2 calories of energy derived from food to support your body's "cost of living," and your contracting muscles will burn additional calories.

Homeostasis: The Foundation of Physiology

INTRODUCTION TO PHYSIOLOGY

The activities described on the preceding page are a sampling of the processes that occur in our bodies all the time just to keep us alive. We usually take these life-sustaining activities for granted and don't really think about "what makes us tick," but that's what physiology is about. **Physiology** is the study of the functions of living things. Specifically, we will focus on how the human body works. Physiologists view the body as a machine whose mechanisms of action can be explained in terms of cause-and-effect sequences of physical and chemical processes—the same types of processes that occur in other components of the universe.

Physiology is closely interrelated with **anatomy**, the study of the structure of the body. Physiological mechanisms are made possible by the structural design and relationships of the various body parts that carry out each of these functions. Just as the functioning of an automobile depends on the shapes, organization, and interactions of its various parts, the structure and function of the human body are inseparable. Therefore, as we tell the story of how the body works we will provide sufficient anatomic background for you to understand the function of the body part being discussed.

LEVELS OF ORGANIZATION IN THE BODY

We now turn our attention to how the body is structurally organized into a total functional unit, from the chemical level to the whole body (● Figure 1-1). These levels of organization make possible life as we know it.

▮ **The chemical level: Various atoms and molecules make up the body.**

Like all matter on this planet, the human body is a combination of specific chemicals. *Atoms* are the smallest building blocks of all nonliving and living

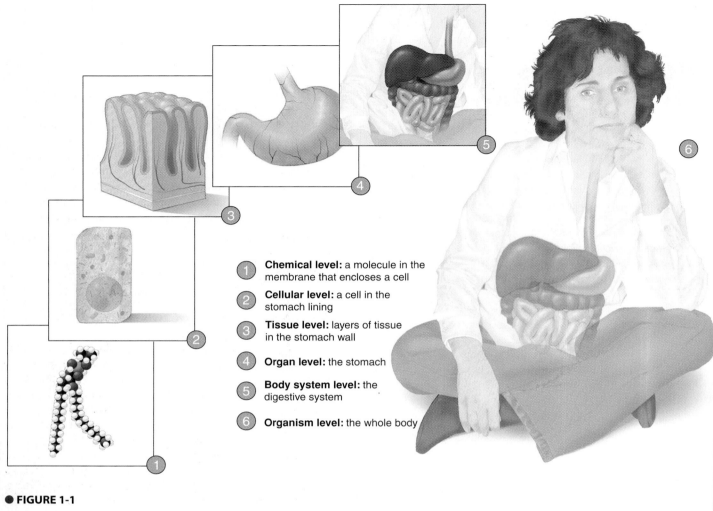

● FIGURE 1-1

Levels of organization in the body

1. **Chemical level:** a molecule in the membrane that encloses a cell
2. **Cellular level:** a cell in the stomach lining
3. **Tissue level:** layers of tissue in the stomach wall
4. **Organ level:** the stomach
5. **Body system level:** the digestive system
6. **Organism level:** the whole body

matter. The most common atoms in the body—oxygen, carbon, hydrogen, and nitrogen—make up approximately 96% of the total body chemistry. These common atoms and a few others combine to form the *molecules* of life, such as proteins, carbohydrates, fats, and nucleic acids (genetic material, such as deoxyribonucleic acid, or DNA). These important atoms and molecules are the inanimate raw ingredients from which all living things arise. (See Appendix B for a review of this chemical level.)

▮ The cellular level: Cells are the basic units of life.

The mere presence of a particular collection of atoms and molecules does not confer the unique characteristics of life. Instead, these nonliving chemical components must be arranged and packaged in very precise ways to form a living entity. The **cell**, the basic or fundamental unit of both structure and function in a living being, is the smallest unit capable of carrying out the processes associated with life. Cellular physiology is the focus of Chapter 2.

An extremely thin, oily barrier, the *plasma membrane,* encloses the contents of each cell, separating these chemicals from those outside the cell. Because the plasma membrane can control movement of materials into and out of the cell, the cell's interior contains a combination of atoms and molecules that differs from the mixture of chemicals in the environment surrounding the cell. Because the plasma membrane and its associated functions are so crucial for carrying out life processes, Chapter 3 is devoted entirely to this structure.

Organisms are independent living entities. The simplest forms of independent life are single-celled organisms such as bacteria and amoebas. Complex multicellular organisms, such as trees and humans, are structural and functional aggregates of trillions of cells (*multi* means "many"). In multicellular organisms, cells are the living building blocks. In the simpler multicellular forms of life—for example, a sponge—the cells of the organism are all similar. However, more complex organisms, such as humans, have many different kinds of cells, such as muscle cells, nerve cells, and gland cells.

Each human organism begins when an egg and sperm unite to form a single cell that starts to multiply and form a growing mass through myriad cell divisions. If cell multiplication were the only process involved in development, all the body cells would be essentially identical, as in the simplest multicellular life forms. During development of complex multicellular organisms such as humans, however, each cell *differentiates*, or becomes specialized to carry out a particular function. As a result of **cell differentiation**, your body is made up of many different specialized types of cells.

BASIC CELL FUNCTIONS

All cells, whether they exist as solitary cells or as part of a multicellular organism, perform certain basic functions essential for survival of the cell. These basic cell functions include the following:

1. Obtaining food (nutrients) and oxygen (O_2) from the environment surrounding the cell.
2. Performing chemical reactions that use nutrients and O_2 to provide energy for the cells, as follows:

$$\text{Food} + O_2 \rightarrow CO_2 + H_2O + \text{energy}$$

3. Eliminating to the cell's surrounding environment carbon dioxide (CO_2) and other by-products, or wastes, produced during these chemical reactions.
4. Synthesizing proteins and other components needed for cell structure, for growth, and for carrying out particular cell functions.
5. Controlling to a large extent the exchange of materials between the cell and its surrounding environment.
6. Moving materials from one part of the cell to another in carrying out cell activities, with some cells even being able to move in entirety through their surrounding environment.
7. Being sensitive and responsive to changes in the surrounding environment.
8. In the case of most cells, reproducing. Some body cells, most notably nerve cells and muscle cells, lose the ability to reproduce after they are formed during early development. This is why strokes, which result in lost nerve cells in the brain, and heart attacks, which bring about death of heart muscle cells, can be so devastating.

Cells are remarkably similar in the ways they carry out these basic functions. Thus all cells share many common characteristics.

SPECIALIZED CELL FUNCTIONS

In multicellular organisms, each cell also performs a specialized function, which is usually a modification or elaboration of a basic cell function. For example, by taking special advantage of their protein-synthesizing ability, the gland cells of the digestive system secrete digestive enzymes that break down ingested food; these enzymes are all proteins.

Each cell performs these specialized activities in addition to carrying on the unceasing, fundamental activities required of all cells. The basic cell functions are essential for survival of each individual cell, whereas the specialized contributions and interactions among the cells of a multicellular organism are essential for survival of the whole body.

▌ The tissue level: Tissues are groups of cells of similar specialization.

Just as a machine does not function unless all its parts are properly assembled, the cells of the body must be specifically organized to carry out the life-sustaining processes of the body as a whole, such as digestion, respiration, and circulation. Cells are progressively organized into tissues, organs, body systems, and finally the whole body.

Cells of similar structure and specialized function combine to form **tissues**, of which there are four *primary types*: muscle, nervous, epithelial, and connective tissue. Each tissue consists of cells of a single specialized type, along with varying amounts of extracellular ("outside the cell") material.

- **Muscle tissue** consists of cells specialized for contracting and generating force. There are three types of muscle tissue: *skeletal muscle,* which moves the skeleton; *cardiac muscle,* which pumps blood out of the heart; and *smooth muscle,* which encloses and controls movement of contents through hollow tubes and organs, such as movement of food through the digestive tract.
- **Nervous tissue** consists of cells specialized for initiating and transmitting electrical impulses, sometimes over long distances. These electrical impulses act as signals that relay information from one part of the body to another. Nervous tissue is found in (1) the brain; (2) the spinal cord; (3) the nerves that signal information about the external environment and about the status of various internal factors in the body that are regulated, such as blood pressure; and (4) the nerves that influence muscle contraction or gland secretion.
- **Epithelial tissue** consists of cells specialized for exchanging materials between the cell and its environment. Any substance that enters or leaves the body proper must cross an epithelial barrier. Epithelial tissue is organized into two general types of structures: epithelial sheets and secretory glands. Epithelial cells join together very tightly to form sheets of tissue that cover and line various parts of the body. For example, the outer layer of the skin is epithelial tissue, as is the lining of the digestive tract. In general, these epithelial sheets serve as boundaries that separate the body from the external environment and from the contents of cavities that open to the external environment, such as the digestive tract lumen. (A **lumen** is the cavity within a hollow organ or tube.) Only selective transfer of materials is possible between regions separated by an epithelial barrier. The type and extent of controlled exchange vary, depending on the location and function of the epithelial tissue. For example, the skin can exchange very little between the body and external environment, whereas the epithelial cells lining the digestive tract are specialized for absorbing nutrients.

Glands are epithelial tissue derivatives that are specialized for secreting. **Secretion** is the release from a cell, in response to appropriate stimulation, of specific products that have been produced by the cell. There are two categories of glands: *exocrine* and *endocrine* (● Figure 1-2). **Exocrine glands** secrete through ducts to the outside of the body (or

● FIGURE 1-2

Exocrine and endocrine gland formation. (a) Exocrine gland cells release their secretory product through a duct to the outside of the body (or to a cavity in communication with the outside). (b) Endocrine glands release their secretory product into the blood.

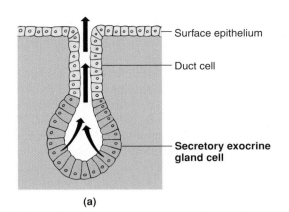
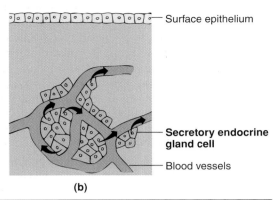

Surface epithelium

Duct cell

Secretory exocrine gland cell

(a)

Surface epithelium

Secretory endocrine gland cell

Blood vessels

(b)

into a cavity that opens to the outside) (*exo* means "external"; *crine* means "secretion"). Examples are sweat glands and glands that secrete digestive juices. **Endocrine glands** lack ducts and release their secretory products, known as *hormones*, into the blood internally, within the body (*endo* means "internal"). For example, the pancreas secretes insulin into the blood, which transports this hormone to its sites of action throughout the body. Most cell types depend on insulin for taking up glucose (sugar).

- **Connective tissue** is distinguished by having relatively few cells dispersed within an abundance of extracellular material. As its name implies, connective tissue connects, supports, and anchors various body parts. It includes such diverse structures as the loose connective tissue that attaches epithelial tissue to underlying structures; tendons, which attach skeletal muscles to bones; bone, which gives the body shape, support, and protection; and blood, which transports materials from one part of the body to another. Except for

● FIGURE 1-3

Components of the body systems

(*Source:* Adapted from Cecie Starr and Ralph Taggart, *Biology: The Unity and Diversity of Life,* Eighth Edition, Fig. 33.11, pp. 552–553. Copyright ©1998 Wadsworth Publishing Company.)

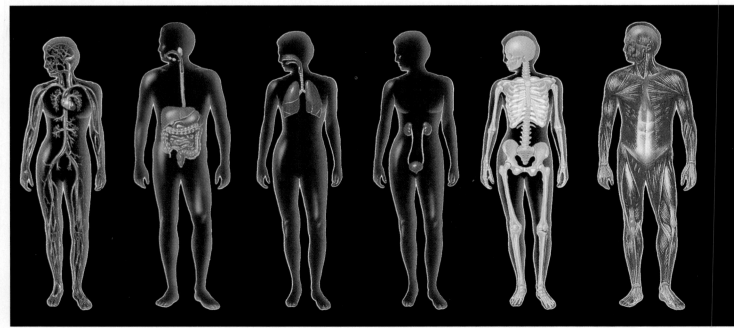

Circulatory system
heart, blood vessels, blood

Digestive system
mouth, pharynx, esophagus, stomach, small intestine, large intestine, salivary glands, exocrine pancreas, liver, gallbladder

Respiratory system
nose, pharynx, larynx, trachea, bronchi, lungs

Urinary system
kidneys, ureters, urinary bladder, urethra

Skeletal system
bones, cartilage, joints

Muscular system
skeletal muscles

blood, the cells within connective tissue produce specific structural molecules that they release into the extracellular spaces between the cells. One such molecule is the rubber band–like protein fiber *elastin*, whose presence facilitates the stretching and recoiling of structures such as the lungs, which alternately inflate and deflate during breathing.

Muscle, nervous, epithelial, and connective tissue are the primary tissues in a classical sense; that is, each is an integrated collection of cells of the same specialized structure and function. The term *tissue* is also often used, as in clinical medicine, to mean the aggregate of various cellular and extracellular components that make up a particular organ (for example, lung tissue or liver tissue).

▌ The organ level: An organ is a unit made up of several tissue types.

Organs consist of two or more types of primary tissue organized together to perform a particular function or functions. The stomach is an example of an organ made up of all four primary tissue types. The tissues that make up the stomach function collectively to store ingested food, move it forward into the rest of the digestive tract, and begin the digestion of protein. The stomach is lined with epithelial tissue that restricts the transfer of harsh digestive chemicals and undigested food from the stomach lumen into the blood. Epithelial gland cells in the stomach include exocrine cells, which

secrete protein-digesting juices into the lumen, and endocrine cells, which secrete a hormone that helps regulate the stomach's exocrine secretion and muscle contraction. The wall of the stomach contains smooth muscle tissue, whose contractions mix ingested food with the digestive juices and push the mixture out of the stomach and into the intestine. The stomach wall also contains nervous tissue, which, along with hormones, controls muscle contraction and gland secretion. Connective tissue binds together all these various tissues.

▌ The body system level: A body system is a collection of related organs.

Groups of organs are further organized into **body systems**. Each system is a collection of organs that perform related functions and interact to accomplish a common activity that is essential for survival of the whole body. For example, the digestive system consists of the mouth, salivary glands, pharynx (throat), esophagus, stomach, pancreas, liver, gallbladder, small intestine, and large intestine. These digestive organs cooperate to break food down into small nutrient molecules that can be absorbed into the blood for distribution to all cells.

The human body has 11 systems: the circulatory, digestive, respiratory, urinary, skeletal, muscular, integumentary, immune, nervous, endocrine, and reproductive systems (● Figure 1-3). Chapters 4 through 18 cover the details of these systems.

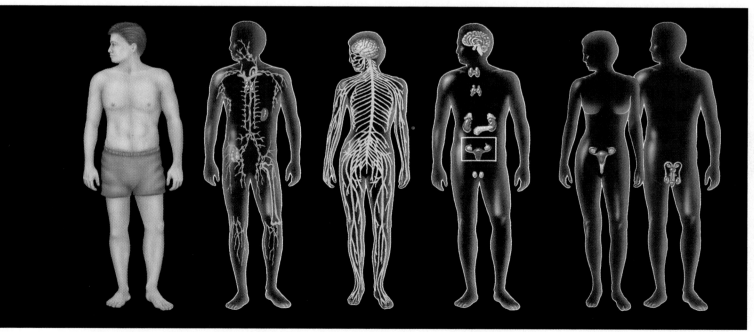

Integumentary system
skin, hair, nails

Immune system
lymph nodes, thymus, bone marrow, tonsils, adenoids, spleen, appendix, and, not shown, white blood cells and gut-associated lymphoid tissue

Nervous system
brain, spinal cord, peripheral nerves, and, not shown, special sense organs

Endocrine system
all hormone-secreting tissues, including hypothalamus, pituitary, thyroid, adrenals, endocrine pancreas, gonads, kidneys, pineal, thymus, and, not shown, parathyroids, intestine, heart, and skin

Reproductive system
Male: testes, penis, prostate gland, seminal vesicles, bulbourethral glands, and associated ducts

Female: ovaries, oviducts, uterus, vagina, breasts

Stem-Cell Science and Tissue Engineering: The Quest to Make Defective Body Parts Like New Again

Liver failure, stroke, paralyzing spinal-cord injury, diabetes mellitus, damaged heart muscle, arthritis, extensive burns, surgical removal of a cancerous breast, an arm mangled in an accident. Although our bodies are remarkable and normally serve us well, sometimes a body part is defective, injured beyond repair, or lost in such situations. In addition to the loss of quality of life for the affected individuals, billions of dollars are spent to treat patients with lost, permanently damaged, or failing organs, accounting for about half of the total health-care costs in the United States. Ideally, when the body suffers an irreparable loss, new, permanent replacement parts could be substituted to restore normal function and appearance. Fortunately, this possibility is moving rapidly from the realm of science fiction to the reality of scientific progress.

The Medical Promise of Stem Cells

The recent isolation of stem cells offers incredibly exciting medical promise for repairing or replacing tissues or organs that are diseased, damaged, or worn out. Two categories of stem cells are being explored for their potential in alleviating many conditions associated with tissue or organ failure: embryonic stem cells and tissue-specific stem cells harvested from adults. **Embryonic stem cells** are the "mother cells" resulting from the early divisions of a fertilized egg. These undifferentiated cells ultimately give rise to all the mature, specialized cells of the body, while at the same time renewing themselves. Embryonic stem cells are totipotent (which means "having total potential"), because they have the potential of generating any cell type in the body if given the appropriate cues. As the embryonic stem cells divide over the course of development, they branch off into various specialty tracks under the guidance of particular genetically controlled chemical signals.

More specifically, the undifferentiated embryonic stem cells give rise to many partially differentiated **tissue-specific stem cells,** each of which becomes committed to generating the highly differentiated, specialized cell types that comprise a particular tissue. For example, tissue-specific muscle stem cells are forced along a lifelong career path of becoming specialized muscle cells. Some tissue-specific stem cells remain in adult tissues and serve as a continual source of new specialized cells in that particular tissue or organ. For example, various partially differentiated stem cells found in the bone marrow generate the different types of blood cells, which after being fully differentiated, are released into the blood as needed. Tissue-specific stem cells have even been found in adult brain and muscle tissue. Even though mature nerve and muscle cells cannot reproduce themselves, researchers recently discovered that, to a limited extent, adult brains and muscles can grow new cells throughout life by means of these persisting stem cells. However, in nerve and muscle tissue this process is too slow to keep pace with major losses, as in a stroke or heart attack. Some investigators are searching for drugs that might spur a person's own tissue-specific stem cells into increased action to make up for damage or loss of that tissue—a feat that is not currently feasible.

More hope is pinned on nurturing stem cells outside of the body for possible transplant into the body. In 1998, for the first time, researchers succeeded in isolating embryonic stem cells and maintaining them indefinitely in an undifferentiated state in culture. With **cell culture,** cells isolated from a living organism continue to thrive and reproduce in laboratory dishes when supplied with appropriate nutrients and supportive materials. Many scientists believe that research involving cultured embryonic stem cells will lead to groundbreaking treatments for a wide variety of disorders.

The medical promise of embryonic stem cells lies in their potential to serve as an all-purpose material that can be coaxed into whatever cell types are needed to patch up the body. Early experiments suggest that these cells have the ability to differentiate into particular cells when exposed to the appropriate chemical signals. Further stem cell research has far-reaching implications that could revolutionize the practice of medicine in the 21st century. As scientists gradually learn to prepare the right cocktail of chemical signals to direct the undifferentiated cells into the desired cell types, they will have the potential to fill in deficits in damaged or dead tissue with healthy cells. Scientists even foresee the ability to grow customized tissues and eventually whole, made-to-order replacement organs, such as livers, hearts, and kidneys.

The Medical Promise of Tissue Engineering

Tissue engineering, another exciting frontier in clinical research, is focused on growing new tissues and even whole, complex organs in the laboratory that can be implanted to serve as permanent replacements for body parts that cannot be repaired. The era of tissue engineering is being ushered in by advances in cell biology, plastic manufacturing, and computer graphics. Using computer-aided designs, very pure, dissolvable plastics are shaped into three-dimensional molds or scaffoldings that mimic the structure of a particular tissue or organ. The plastic mold is then "seeded" with the desired cell types, which are coaxed, by applying appropriate nourishing and stimulatory chemicals, into multiplying and assembling into the desired tissue. After the biodegradable plastic scaffolding dissolves, only the newly generated tissue remains, ready to be implanted into a patient as a permanent, living replacement part.

What about the source of cells to seed the plastic mold? The immune system is pro-

---

▊ The organism level:
The body systems are packaged together into a functional whole body.

Each body system depends on the proper functioning of other systems to carry out its specific responsibilities. The whole body of a multicellular organism—a single, independently living individual—consists of the various body systems structurally and functionally linked together as an entity that is separate from the external (outside the body) environment.

Thus the body is made up of living cells organized into life-sustaining systems.

Currently researchers are hotly pursuing several approaches to repairing or replacing tissues or organs that can no longer adequately perform vital functions because of disease, trauma, or age-related changes. (See the accompanying boxed feature, ▶ Beyond the Basics. Each chapter has similar boxed features ("boxes") that explore in greater depth high-interest, tangential information on such diverse topics as environmental impact on the body, aging, ethical issues, exercise physiology, new discoveries regarding common diseases, historical perspectives, and so on.)

grammed to attack foreign cells such as bacterial invaders. This system also launches an attack on foreign cells transplanted into the body from another individual. Such an attack brings about rejection of transplanted organs, tissues, or cells unless the transplant recipient is treated with *immunosuppressive drugs* (drugs that suppress the immune system's attack on the transplanted material). An unfortunate side effect of immunosuppressive drugs is the reduced ability of the patient's immune system to defend against potential disease-causing bacteria and viruses. To prevent rejection by the immune system and to avoid the necessity of lifelong immunosuppressive drugs, the plastic mold used for tissue engineering could be seeded, if possible, with appropriate cells harvested from the recipient. Often, however, because of the very need for a replacement part, the patient does not have any of the appropriate cells to seed the synthetic scaffold. This is what makes the recent isolation of embryonic stem cells so exciting. Through genetic engineering, these stem cells could be converted into "universal" seed cells that would be immunologically acceptable to any recipient; that is, they could be genetically programmed to not be rejected by any body. Thus the vision of tissue engineers is to create universal replacement parts that could be put into any patient who needs them, without fear of transplant rejection or use of troublesome immunosuppressive drugs.

Here are some of the tissue engineers' early accomplishments and future predictions:

- Engineered skin patches have already been used to treat victims of severe burns, and cartilage substitutes are already available to patients.
- Considerable progress has been made on building artificial bone, teeth, and bladders.
- Tissue-engineered scaffolding to promote nerve regeneration is currently undergoing testing in animals.

- Progress has been made on growing two complicated organs, the liver and the pancreas.
- Ultimately, complex body parts such as arms and hands will be produced in the laboratory for attachment as needed.

Tissue engineering thus holds the promise that damaged body parts can be replaced with the best alternative, a laboratory-grown version of "the real thing."

Ethical Concerns and Political Issues

Despite this potential, embryonic stem-cell research is fraught with ethical controversy because of the source of these cells. These embryonic stem cells were isolated from embryos from an abortion clinic and from unused embryos from an in-vitro fertility clinic. Opponents of using embryonic stem cells are morally and ethically concerned because embryos are destroyed in the process of harvesting these cells. Proponents argue that these embryos were destined to be destroyed anyway—a decision already made by the parents of the embryos—and that these stem cells have great potential for alleviating much human suffering. Thus embryonic stem-cell science has become inextricably linked with stem cell politics.

Because federal policy currently prohibits use of public funding to support research involving human embryos, the scientists who isolated these embryonic stem cells relied on private money. Public policy makers, scientists, and bioethicists now face balancing a host of ethical issues against the tremendous potential clinical application of embryonic stem-cell research. Such research will proceed at a much faster pace if federal money is available to support it. In a controversial decision, President George W. Bush issued guidelines in August 2001 permitting the use of government funds to support studies using established lines of human embryonic stem cells but not research aimed at deriving new cell lines. This decision

was based on the premise that investigations using already established embryonic stem-cell lines would not involve destroying more embryos for scientific purposes.

As a possible alternative to using the controversial totipotent embryonic stem cells, other researchers are exploring the possibility of using tissue-specific stem cells harvested from various adult tissues. Until recently, most investigators believed these adult stem cells could only give rise to the specialized cells of a particular tissue. Although these partially differentiated adult stem cells do not have the complete developmental potential of embryonic stem cells, the adult cells have been coaxed into producing a wider variety of cells than originally thought possible. To name a few examples, provided the right supportive environment, stem cells from the brain have given rise to blood cells, bone-marrow stem cells to liver and nerve cells, and fat-tissue stem cells to bone, cartilage, and muscle cells. Thus researchers may be able to tap into the more limited but still versatile developmental potential of specialized stem cells in the adult human body. Although embryonic stem cells hold greater potential for developing treatments for a broader range of diseases, adult stem cells are more accessible than embryonic stem cells, and their use is not controversial. It may even be possible to harvest stem cells from a patient's own body and manipulate them for use in treating the individual, thus avoiding the issue of transplant rejection. For example, researchers dream of being able to take fat stem cells from a person and transform them into a needed replacement knee joint.

Whatever the source, stem cell research promises to revolutionize medicine. According to the Center for Disease Control's National Center for Health Statistics, an estimated 3000 Americans die every day from conditions that may in the future be treatable with stem cell derivatives.

We next focus on how the different body systems normally work together to maintain the internal conditions necessary for life.

CONCEPT OF HOMEOSTASIS

If each cell has basic survival skills, why can't the body cells live without performing specialized tasks and being organized according to specialization into systems that accomplish functions essential for the whole body's survival? The cells in a multicellular organism must contribute to survival of the organism as a whole and cannot live and function

without contributions from the other body cells, because the vast majority of cells are not in direct contact with the external environment in which the organism lives. A single-celled organism such as an amoeba can directly obtain nutrients and O_2 from its immediate external surroundings and eliminate wastes back into those surroundings. A muscle cell or any other cell in a multicellular organism has the same need for life-supporting nutrient and O_2 uptake and waste elimination, yet the muscle cell cannot directly make these exchanges with the environment surrounding the body, because the cell is isolated from this external environment. How can a muscle cell make vital exchanges with the external environment with which it has no contact? The key is the presence of a watery **internal environment** with which the body cells are in direct contact and make life-sustaining exchanges.

Body cells are in contact with a privately maintained internal environment.

The fluid collectively contained within all body cells is called **intracellular fluid (ICF)** and the fluid outside the cells is called **extracellular fluid (ECF)** (*intra* means "within"; *extra* means "outside of"). The extracellular fluid is the internal environment of the body. It is the fluid environment in which the cells live. Note that the internal environment is outside the cells but inside the body. By contrast, the intracellular fluid is inside the cells and the external environment is outside the body. You live in the external environment; your cells live within the body's internal environment.

The extracellular fluid (internal environment) is made up of two components: the **plasma**, the fluid portion of the blood; and the **interstitial fluid**, which surrounds and bathes the cells (*inter* means "between"; *stitial* means "that which stands") (● Figure 1-4).

No matter how remote a cell is from the external environment, it can make life-sustaining exchanges with its own surrounding internal environment. In turn, particular body systems accomplish the transfer of materials between the external environment and the internal environment so that the composition of the internal environment is appropriately maintained to support the life and functioning of the cells. For example, the digestive system transfers the nutrients required by all body cells from the external environment into the plasma. Likewise, the respiratory system transfers O_2 from the external environment into the plasma. The circulatory system distributes these nutrients and O_2 throughout the body. Materials are thoroughly mixed and exchanged between the plasma and the interstitial fluid across the capillaries, the

Components of the extracellular fluid (internal environment)

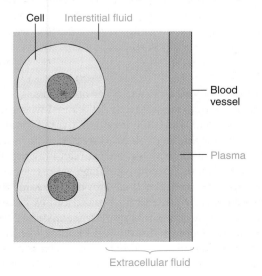

smallest and thinnest of the blood vessels. As a result, the nutrients and O_2 originally obtained from the external environment are delivered to the interstitial fluid surrounding the cells. The body cells, in turn, pick up these needed supplies from the interstitial fluid. Similarly, wastes produced by the cells are extruded into the interstitial fluid, picked up by the plasma, and transported to the organs that specialize in eliminating these wastes from the internal environment to the external environment. The lungs remove CO_2 from the plasma, and the kidneys remove other wastes for elimination in the urine.

Thus a body cell takes in essential nutrients from its watery surroundings and eliminates wastes into these same surroundings, just as an amoeba does. The main difference is that each body cell must help maintain the composition of the internal environment so that this fluid continuously remains suitable to support the existence of all the body cells. In contrast, an amoeba does nothing to regulate its surroundings.

Body systems maintain homeostasis, a dynamic steady state in the internal environment.

The body cells can live and function only when the extracellular fluid is compatible with their survival; thus the chemical composition and physical state of this internal environment must be maintained within narrow limits. As cells take up nutrients and O_2 from the internal environment, these essential materials must constantly be replenished. Likewise, wastes must constantly be removed from the internal environment so they do not reach toxic levels. Other aspects of the internal environment that are important for maintaining life, such as temperature, also must be kept relatively constant. Maintenance of a relatively stable internal environment is termed **homeostasis** (*homeo* means "the same"; *stasis* means "to stand or stay").

The functions performed by each body system contribute to homeostasis, thereby maintaining within the body the environment required for the survival and function of all the cells. Cells, in turn, make up body systems. This is the central theme of physiology and of this book: *Homeostasis is essential for the survival of each cell, and each cell, through its specialized activities, contributes as part of a body system to the maintenance of the internal environment shared by all cells* (● Figure 1-5).

The fact that the internal environment must be kept relatively stable does not mean that its composition, temperature, and other characteristics are absolutely unchanging. Both external and internal factors continuously threaten to disrupt homeostasis. When any factor starts to move the internal environment away from optimal conditions, the body systems initiate appropriate counterreactions to minimize the change. For example, exposure to a cold environmental temperature (an external factor) tends to reduce the body's internal temperature. In response, the temperature control center in the brain initiates compensatory measures, such as shivering, to raise body temperature to normal. By contrast, production of extra heat by working muscles during exercise (an internal factor) tends to raise the body's internal temperature. In re-

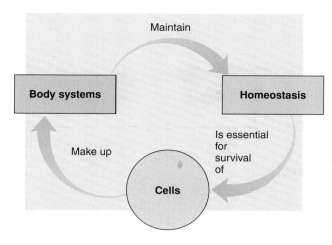

● **FIGURE 1-5**

Interdependent relationship of cells, body systems, and homeostasis. The depicted interdependent relationship serves as the foundation for modern-day physiology: *homeostasis is essential for the survival of cells, body systems maintain homeostasis, and cells make up body systems.*

sponse, the temperature control center brings about sweating and other compensatory measures to reduce body temperature to normal.

Thus homeostasis is not a rigid, fixed state but a dynamic steady state in which the changes that do occur are minimized by compensatory physiological responses. The term *dynamic* refers to the fact that each homeostatically regulated factor is marked by continuous change, whereas *steady state* implies that these changes do not deviate far from a constant, or steady, level. This situation is comparable to the minor steering adjustments you make as you steer a car along a straight course down the highway. Small fluctuations around the optimal level for each factor in the internal environment are normally kept by carefully regulated mechanisms, within the narrow limits compatible with life.

FACTORS HOMEOSTATICALLY REGULATED

Many factors of the internal environment must be homeostatically maintained. They include the following (● Figure 1-6):

1. *Concentration of nutrient molecules.* Cells need a constant supply of nutrient molecules for energy production. Energy, in turn, is needed to support life-sustaining and specialized cell activities.

2. *Concentration of O_2 and CO_2.* Cells need O_2 to carry out energy-yielding chemical reactions. The CO_2 produced during these reactions must be removed so that acid-forming CO_2 does not increase the acidity of the internal environment.

3. *Concentration of waste products.* Some chemical reactions produce end products that exert a toxic effect on the body's cells if these wastes are allowed to accumulate.

4. *pH.* Changes in the pH (relative amount of acid) adversely affect nerve cell function and wreak havoc with the enzyme activity of all cells.

5. *Concentration of water, salt, and other electrolytes.* Because the relative concentrations of salt (NaCl) and water in the

extracellular fluid (internal environment) influence how much water enters or leaves the cells, these concentrations are carefully regulated to maintain the proper volume of the cells. Cells do not function normally when they are swollen or shrunken. Other electrolytes perform a variety of vital functions. For example, the rhythmic beating of the heart depends on a relatively constant concentration of potassium (K^+) in the extracellular fluid.

6. *Volume and pressure.* The circulating component of the internal environment, the plasma, must be maintained at adequate volume and blood pressure to ensure body-wide distribution of this important link between the external environment and the cells.

7. *Temperature.* Body cells function best within a narrow temperature range. If cells are too cold, their functions slow down too much, and worse yet, if they get too hot their structural and enzymatic proteins are impaired or destroyed.

CONTRIBUTIONS OF THE BODY SYSTEMS TO HOMEOSTASIS

The 11 body systems contribute to homeostasis in the following important ways (● Figure 1-6):

1. The *circulatory system* is the transport system that carries materials such as nutrients, O_2, CO_2, wastes, electrolytes, and hormones from one part of the body to another. It includes the heart, blood vessels, and blood. Also, the lymph vessels serve as an accessory drainage pathway from the tissues to the heart.

2. The *digestive system* breaks down dietary food into small nutrient molecules that can be absorbed into the plasma for distribution to the body cells. It also transfers water and electrolytes from the external environment into the internal environment. It eliminates undigested food residues to the external environment in the feces.

3. The *respiratory system,* consisting of the lungs and major airways, gets O_2 from and eliminates CO_2 to the external environment. By adjusting the rate of removal of acid-forming CO_2, the respiratory system is also important in maintaining the proper pH of the internal environment.

4. The *urinary system* removes excess water, salt, acid, and other electrolytes from the plasma and eliminates them in the urine, along with waste products other than CO_2. This system includes the kidneys and associated "plumbing."

5. The *skeletal system* (bones, joints) provides support and protection for the soft tissues and organs. It also serves as a storage reservoir for calcium (Ca^{2+}), an electrolyte whose plasma concentration must be maintained within very narrow limits. Together with the muscular system, the skeletal system also enables movement of the body and its parts. Furthermore, the bone marrow—the soft interior portion of some types of bone—is the ultimate source of all blood cells.

6. The *muscular system* (skeletal muscles) moves the bones to which the skeletal muscles are attached. From a purely homeostatic view, this system enables an individual to move toward food or away from harm. Furthermore, the heat generated by muscle contraction is important in temperature regulation. In addition, because skeletal muscles are under

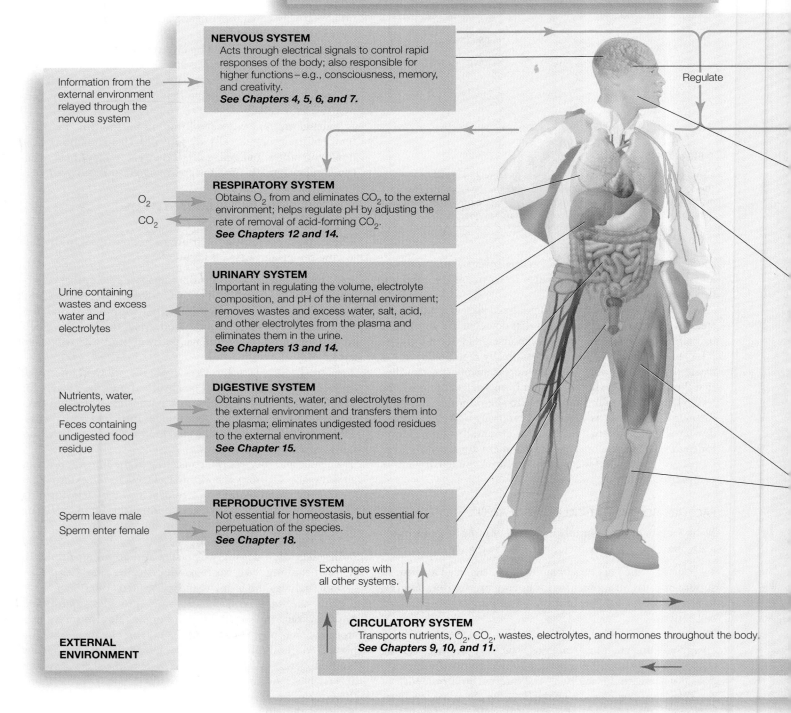

BODY SYSTEMS
Made up of cells organized according to specialization to maintain homeostasis.
See Chapter 1.

NERVOUS SYSTEM
Acts through electrical signals to control rapid responses of the body; also responsible for higher functions—e.g., consciousness, memory, and creativity.
See Chapters 4, 5, 6, and 7.

Information from the external environment relayed through the nervous system

Regulate

O_2
CO_2

RESPIRATORY SYSTEM
Obtains O_2 from and eliminates CO_2 to the external environment; helps regulate pH by adjusting the rate of removal of acid-forming CO_2.
See Chapters 12 and 14.

Urine containing wastes and excess water and electrolytes

URINARY SYSTEM
Important in regulating the volume, electrolyte composition, and pH of the internal environment; removes wastes and excess water, salt, acid, and other electrolytes from the plasma and eliminates them in the urine.
See Chapters 13 and 14.

Nutrients, water, electrolytes

Feces containing undigested food residue

DIGESTIVE SYSTEM
Obtains nutrients, water, and electrolytes from the external environment and transfers them into the plasma; eliminates undigested food residues to the external environment.
See Chapter 15.

Sperm leave male
Sperm enter female

REPRODUCTIVE SYSTEM
Not essential for homeostasis, but essential for perpetuation of the species.
See Chapter 18.

Exchanges with all other systems.

EXTERNAL ENVIRONMENT

CIRCULATORY SYSTEM
Transports nutrients, O_2, CO_2, wastes, electrolytes, and hormones throughout the body.
See Chapters 9, 10, and 11.

● **FIGURE 1-6**

Role of the body systems in maintaining homeostasis

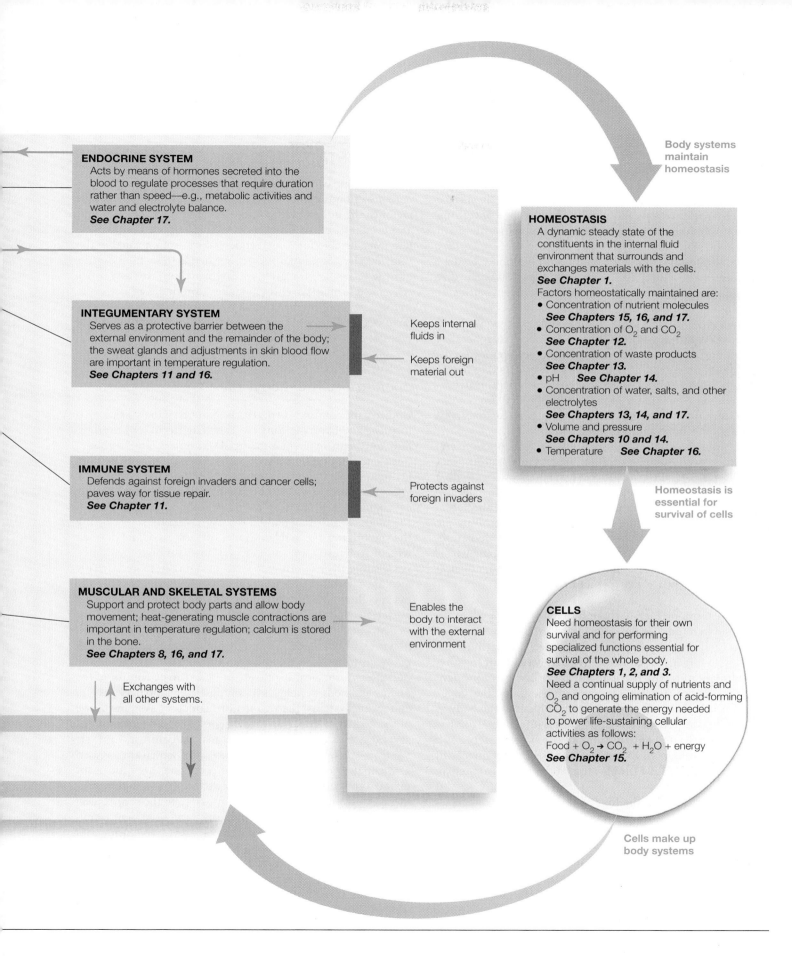

ENDOCRINE SYSTEM
Acts by means of hormones secreted into the blood to regulate processes that require duration rather than speed—e.g., metabolic activities and water and electrolyte balance.
See Chapter 17.

INTEGUMENTARY SYSTEM
Serves as a protective barrier between the external environment and the remainder of the body; the sweat glands and adjustments in skin blood flow are important in temperature regulation.
See Chapters 11 and 16.

Keeps internal fluids in

Keeps foreign material out

IMMUNE SYSTEM
Defends against foreign invaders and cancer cells; paves way for tissue repair.
See Chapter 11.

Protects against foreign invaders

MUSCULAR AND SKELETAL SYSTEMS
Support and protect body parts and allow body movement; heat-generating muscle contractions are important in temperature regulation; calcium is stored in the bone.
See Chapters 8, 16, and 17.

Enables the body to interact with the external environment

Exchanges with all other systems.

Body systems maintain homeostasis

HOMEOSTASIS
A dynamic steady state of the constituents in the internal fluid environment that surrounds and exchanges materials with the cells.
See Chapter 1.
Factors homeostatically maintained are:
• Concentration of nutrient molecules
  See Chapters 15, 16, and 17.
• Concentration of O_2 and CO_2
  See Chapter 12.
• Concentration of waste products
  See Chapter 13.
• pH   *See Chapter 14.*
• Concentration of water, salts, and other electrolytes
  See Chapters 13, 14, and 17.
• Volume and pressure
  See Chapters 10 and 14.
• Temperature   *See Chapter 16.*

Homeostasis is essential for survival of cells

CELLS
Need homeostasis for their own survival and for performing specialized functions essential for survival of the whole body.
See Chapters 1, 2, and 3.
Need a continual supply of nutrients and O_2 and ongoing elimination of acid-forming CO_2 to generate the energy needed to power life-sustaining cellular activities as follows:
Food + O_2 → CO_2 + H_2O + energy
See Chapter 15.

Cells make up body systems

voluntary control, a person can use them to accomplish myriad other movements of his or her own choice. These movements, which range from the fine motor skills required for delicate needlework to the powerful movements involved in weight lifting, are not necessarily directed toward maintaining homeostasis.

7. The *integumentary system* (skin and related structures) serves as an outer protective barrier that prevents internal fluid from being lost from the body and foreign micro-organisms from entering. This system is also important in regulating body temperature. The amount of heat lost from the body surface to the external environment can be adjusted by controlling sweat production and by regulating the flow of warm blood through the skin.

8. The *immune system* (white blood cells, lymphoid organs) defends against foreign invaders and against body cells that have become cancerous. It also paves the way for repairing or replacing injured or worn-out cells.

9. The *nervous system* (brain, spinal cord, nerves) is one of the two major regulatory systems of the body. In general, it controls and coordinates bodily activities that require swift responses. It is especially important in detecting and initiating reactions to changes in the external environment. Furthermore, it is responsible for higher functions that are not entirely directed toward maintaining homeostasis, such as consciousness, memory, and creativity.

10. The *endocrine system* is the other major regulatory system. In contrast to the nervous system, in general the hormone-secreting glands of the endocrine system regulate activities that require duration rather than speed, such as growth. This system is especially important in controlling the concentration of nutrients and, by adjusting kidney function, controlling the internal environment's volume and electrolyte composition.

11. The *reproductive system* is not essential for homeostasis and therefore is not essential for survival of the individual. It is essential, however, for perpetuating the species.

As we examine each of these systems in greater detail, always keep in mind that the body is a coordinated whole even though each system provides its own special contributions. It is easy to forget that all the body parts actually fit together into a functioning, interdependent whole body. Accordingly, each chapter begins with a figure and discussion that focus on how the body system to be described fits into the body as a whole. In addition, each chapter ends with a brief overview of the homeostatic contributions of the body system. As a further tool to help you keep track of how all the pieces fit together, ● Figure 1-6 is duplicated on the inside front cover as a handy reference.

Also be aware that the functioning whole is greater than the sum of its separate parts. Through specialization, cooperation, and interdependence, cells combine to form an integrated, unique, single living organism with more diverse and complex capabilities than are possessed by any of the cells that make it up. For humans, these capabilities go far beyond the processes needed to maintain life. A cell, or even a random combination of cells, obviously cannot create an artistic masterpiece or design a spacecraft, but body cells working together permit those capabilities in an individual.

You have now learned what homeostasis is and how the functions of different body systems maintain it. Next let's look at the regulatory mechanisms by which the body reacts to changes and controls the internal environment.

HOMEOSTATIC CONTROL SYSTEMS

A **homeostatic control system** is a functionally interconnected network of body components that operate to maintain a given factor in the internal environment relatively constant around an optimal level. To maintain homeostasis, the control system must be able to (1) detect deviations from normal in the internal environmental factor that needs to be held within narrow limits; (2) integrate this information with any other relevant information; and (3) make appropriate adjustments in the activity of the body parts responsible for restoring this factor to its desired value. We next turn our attention to the types of control systems in the body.

▌ **Homeostatic control systems may operate locally or bodywide.**

Homeostatic control systems can be grouped into two classes—intrinsic and extrinsic controls. **Intrinsic (local) controls** are built into or are inherent in an organ (*intrinsic* means "within"). For example, as an exercising skeletal muscle rapidly uses up O_2 to generate energy to support its contractile activity, the O_2 concentration within the muscle falls. This local chemical change acts directly on the smooth muscle in the walls of the blood vessels that supply the exercising muscle, causing the smooth muscle to relax so that the vessels dilate, or open widely. As a result, increased blood flows through the dilated vessels into the exercising muscle, bringing in more O_2. This local mechanism helps maintain an optimal level of O_2 in the internal fluid environment immediately around the exercising muscle's cells.

Most factors in the internal environment are maintained, however, by **extrinsic controls,** which are regulatory mechanisms initiated outside an organ to alter the activity of the organ (*extrinsic* means "outside of"). Extrinsic control of the organs and body systems is accomplished by the nervous and endocrine systems, the two major regulatory systems of the body. Extrinsic control permits coordinated regulation of several organs toward a common goal; in contrast, intrinsic controls are self-serving for the organ in which they occur. Coordinated, overall regulatory mechanisms are crucial for maintaining the dynamic steady state in the internal environment as a whole. To restore blood pressure to the proper level when it falls too low, for example, the nervous system simultaneously acts on the heart and the blood vessels throughout the body to increase the blood pressure to normal.

To stabilize the physiological factor being regulated, homeostatic control systems must be able to detect and resist change. The term **feedback** refers to responses made after a change has been detected; the term **feedforward** is used for responses made in anticipation of a change. Let's take a look at these mechanisms in more detail.

▌ Negative feedback opposes an initial change and is widely used to maintain homeostasis.

Homeostatic control mechanisms primarily operate on the principle of negative feedback. In **negative feedback,** a change in a homeostatically controlled factor triggers a response that seeks to restore the factor to normal by moving the factor in the opposite direction of its initial change. That is, a corrective adjustment opposes the original deviation from the normal desired level.

A common example of negative feedback is control of room temperature. Room temperature is a **controlled variable,** a factor that can vary but is held within a narrow range by a control system. In our example, the control system includes a thermostatic device, a furnace, and all their electrical connections. The room temperature is determined by the activity of the furnace, a heat source that can be turned on or off. To switch on or off appropriately, the control system as a whole must "know" what the *actual* room temperature is, "compare" it with the *desired* room temperature, and "adjust" the output of the heat source to bring the actual temperature to the desired level. A thermometer in the thermostat provides information about the actual room temperature. The thermometer is the **sensor,** which monitors the magnitude of the controlled variable. The sensor typically converts the original information regarding a change into a form of "language" the control system can "understand." For example, the thermometer converts the magnitude of the air temperature into electrical impulses. This message serves as the input into the control system. The thermostat setting provides the desired temperature level, or **set point.** The thermostat acts as an **integrator,** or **control center:** It compares the sensor's input with the set point and adjusts the heat output of the furnace to bring about the appropriate effect, or response, to oppose a deviation from the set point. The furnace is the **effector,** the component of the control system commanded to bring about the desired effect. These general components of a negative-feedback control system are summarized in ● Figure 1-7a. Carefully examine this figure and the guidelines in the footnote, because the symbols and conventions introduced here are used in comparable flow diagrams throughout the text.

Let's look at a typical negative-feedback loop. For example, if in cold weather the room temperature falls below the set point, the thermostat, through connecting circuitry, activates the furnace, which produces heat to raise the room temperature (● Figure 1-7b). Once the room temperature reaches the set point, the thermometer no longer detects a deviation from that point. As a result, the activating mechanism in the thermostat and the furnace are switched off. Thus the heat from the furnace counteracts or is "negative" to the original fall in temperature. If the heat-generating pathway were not shut off once the target temperature was reached, heat production would continue and the room would get hotter and hotter. Overshooting beyond the set point does not occur,

because the heat "feeds back" to shut off the thermostat that triggered its output. Thus a negative-feedback control system detects a change in a controlled variable away from the ideal value, initiates mechanisms to correct the situation, then shuts itself off. In this way, the controlled variable does not drift too far below or too far above the set point.

What if the original deviation is a rise in room temperature above the set point because it is hot outside? A heat-producing furnace is of no use in returning the room temperature to the desired level. In this case, the thermostat, through connecting circuitry, can activate the air conditioner, which cools the room air, the opposite effect from that of the furnace. In negative-feedback fashion, once the set point is reached, the air conditioner is turned off to prevent the room from becoming too cold. Note that if the controlled variable can be deliberately adjusted to oppose a change in one direction only, the variable can move in uncontrolled fashion in the opposite direction. For example, if the house is equipped only with a furnace that produces heat to oppose a fall in room temperature, no mechanism is available to prevent the house from getting too hot in warm weather. However, the room temperature can be kept relatively constant through two opposing mechanisms, one that heats and one that cools the room, despite wide variations in the temperature of the external environment.

Homeostatic negative-feedback systems in the human body operate in the same way. For example, when temperature-monitoring nerve cells detect a decrease in body temperature below the desired level, these sensors signal the temperature control center, which begins a sequence of events that ends in shivering, among other responses, to generate heat and increase the temperature to the proper level (● Figure 1-7c). When the body temperature rises to the set point, the temperature-monitoring nerve cells turn off the stimulatory signal to the skeletal muscles. As a result, the body temperature doesn't go on rising above the set point. Conversely, when the temperature-monitoring nerve cells detect a rise in body temperature above normal, cooling mechanisms such as sweating are called into play to reduce the temperature to normal. When the temperature reaches the set point, the cooling mechanisms are shut off. As with body temperature, opposing mechanisms can move most homeostatically controlled variables in either direction as needed.

▌ Positive feedback amplifies an initial change.

In negative feedback, a control system's output is regulated to resist change, so that the controlled variable is kept at a relatively steady set point. With **positive feedback,** by contrast, the output enhances or amplifies a change so the controlled variable continues to move in the direction of the initial change. Such action is comparable to the heat generated by a furnace triggering the thermostat to call for even *more* heat output from the furnace, so that the room temperature would continuously rise.

Because the major goal in the body is to maintain stable, homeostatic conditions, positive feedback does not occur

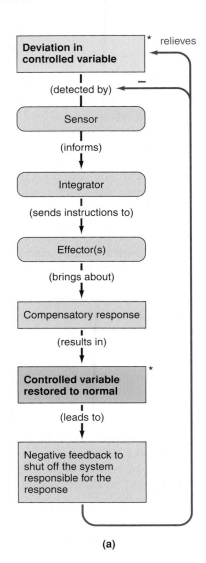

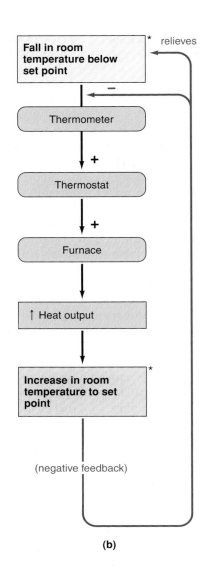

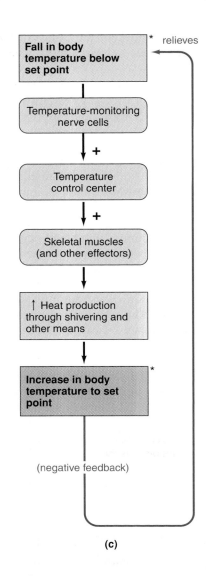

(a)   (b)   (c)

For flow diagrams throughout the text:

+ = Stimulates or activates

− = Inhibits or shuts off

⬭ = Physical entity, such as body structure or a chemical

▭ = Actions

❙ = Compensatory pathway

❙ = Turning off of compensatory pathway (negative feedback)

* Note that lighter and darker shades of the same color are used to denote respectively a decrease or an increase in a controlled variable.

● **FIGURE 1-7**

Negative feedback. (a) Components of a negative-feedback control system. (b) Negative-feedback control of room temperature. (c) Negative-feedback control of body temperature.

nearly as often as negative feedback. Positive feedback does play an important role in certain instances, however, as in the birth of a baby. The hormone oxytocin causes powerful contractions of the uterus. As uterine contractions push the baby against the cervix (the exit from the uterus), the resultant stretching of the cervix triggers a sequence of events that brings about the release of even more oxytocin, which causes even stronger uterine contractions, triggering the release of more oxytocin, and so on. This positive-feedback cycle does not stop until the baby is finally born. Likewise, all other nor-

mal instances of positive feedback cycles in the body include some mechanism for stopping the cycle.

❙ Feedforward mechanisms initiate responses in anticipation of a change.

In addition to feedback mechanisms, which bring about a reaction to a change in a regulated variable, the body less frequently employs feedforward mechanisms, which respond in anticipation of a change in a regulated variable. For example, when a meal is still in the digestive tract, a feedforward mechanism increases secretion of a hormone that will promote the cellular uptake and storage of ingested nutrients after they have been absorbed from the digestive tract. This anticipatory response helps limit the rise in blood nutrient concentration after nutrients have been absorbed.

▌ Disruptions in homeostasis can lead to illness and death.

Clinical Note Despite control mechanisms, when one or more of the body's systems malfunction, homeostasis is disrupted, and all the cells suffer, because they no longer have an optimal environment in which to live and function. Various pathophysiological states ensue, depending on the type and extent of homeostatic disruption. The term **pathophysiology** refers to the abnormal functioning of the body (altered physiology) associated with disease. When a homeostatic disruption becomes so severe that it is no longer compatible with survival, death results.

CHAPTER IN PERSPECTIVE: FOCUS ON HOMEOSTASIS

In this chapter, you have learned what homeostasis is: a dynamic steady state of the constituents in the internal fluid environment (the extracellular fluid) that surrounds and exchanges materials with the cells. Maintenance of homeostasis is essential for the survival and normal functioning of cells. Each cell, through its specialized activities, contributes as part of a body system to the maintenance of homeostasis.

This relationship is the foundation of physiology and the central theme of this book. We have described how cells are organized according to specialization into body systems. How homeostasis is essential for cell survival and how body systems maintain this internal constancy are the topics covered in the rest of this book. Each chapter concludes with this capstone feature to facilitate your understanding of how the system under discussion contributes to homeostasis, as well as of the interactions and interdependency of the body systems.

CHAPTER SUMMARY

Introduction to Physiology (p. 1)
▌ Physiology is the study of body functions.
▌ Physiologists explain body function in terms of the mechanisms of action involving cause-and-effect sequences of physical and chemical processes.
▌ Physiology and anatomy are closely interrelated because body functions are highly dependent on the structure of the body parts that carry them out.

Levels of Organization in the Body (pp. 1–7)
▌ The human body is a complex combination of specific atoms and molecules.
▌ These nonliving chemical components are organized in a precise way to form cells, the smallest entities capable of carrying out the processes associated with life. Cells are the basic structural and functional living building blocks of the body. *(Review Figure 1-1.)*
▌ The basic functions performed by each cell for its own survival include (1) obtaining O_2 and nutrients, (2) performing energy-generating chemical reactions, (3) eliminating wastes, (4) synthesizing proteins and other cell components, (5) controlling movement of materials between the cell and its environment, (6) moving materials throughout the cell, (7) responding to the environment, and (8) reproducing.
▌ In addition to its basic cell functions, each cell in a multicellular organism performs a specialized function.
▌ Combinations of cells of similar structure and specialized function form the four primary tissues of the body: muscle, nervous, epithelial, and connective tissue.
▌ Glands are derived from epithelial tissue and specialized for secretion. Exocrine glands secrete through ducts to the body surface or a cavity that communicates with the outside; endocrine glands secrete hormones into the blood. *(Review Figure 1-2.)*

▌ Organs are combinations of two or more types of tissues that act together to perform one or more functions. An example is the stomach.
▌ Body systems are collections of organs that perform related functions and interact to accomplish a common activity essential for survival of the whole body. An example is the digestive system. *(Review Figure 1-3.)*
▌ Organ systems combine to form the organism, or whole body.

Concept of Homeostasis (pp. 7–12)
▌ The fluid inside the cells of the body is intracellular fluid, and the fluid outside the cells is extracellular fluid.
▌ Because most body cells are not in direct contact with the external environment, cell survival depends on maintaining a relatively stable internal fluid environment with which the cells directly make life-sustaining exchanges.
▌ The extracellular fluid serves as the internal environment. It consists of the plasma and interstitial fluid. *(Review Figure 1-4.)*
▌ Homeostasis is the maintenance of a dynamic steady state in the internal environment.
▌ The factors of the internal environment that must be homeostatically maintained are its (1) concentration of nutrient molecules; (2) concentration of O_2 and CO_2; (3) concentration of waste products; (4) pH; (5) concentration of water, salt, and other electrolytes; (6) volume and pressure; and (7) temperature. *(Review Figure 1-6.)*
▌ The functions performed by the body systems are directed toward maintaining homeostasis. The body systems' functions ultimately depend on the specialized activities of the cells that make up the system. Thus, homeostasis is essential for each cell's survival, and each cell contributes to homeostasis. *(Review Figures 1-5 and 1-6.)*

Homeostatic Control Systems (pp. 12–15)

▮ A homeostatic control system is a network of body components working together to maintain a given factor in the internal environment relatively constant near an optimal set level.

▮ Homeostatic control systems can be classified as (1) intrinsic (local) controls, which are inherent compensatory responses of an organ to a change; and (2) extrinsic controls, which are responses of an organ that are triggered by factors external to the organ, namely, by the nervous and endocrine systems.

▮ Both intrinsic and extrinsic control systems generally operate on the principle of negative feedback: A change in a controlled variable triggers a response that drives the variable in the opposite direction of the initial change, thus opposing the change. *(Review Figure 1-7.)*

▮ In positive feedback, a change in a controlled variable triggers a response that drives the variable in the same direction as the initial change, thus amplifying the change.

▮ Feedforward mechanisms are compensatory responses that occur in anticipation of a change.

REVIEW EXERCISES

Objective Questions (Answers on p. A-40)

1. Which of the following activities is *not* carried out by every cell in the body?
   a. obtaining O_2 and nutrients
   b. performing chemical reactions to acquire energy for the cell's use
   c. eliminating wastes
   d. controlling to a large extent exchange of materials between the cell and its external environment
   e. reproducing

2. Which of the following is the proper progression of the levels of organization in the body?
   a. chemicals, cells, organs, tissues, body systems, whole body
   b. chemicals, cells, tissues, organs, body systems, whole body
   c. cells, chemicals, tissues, organs, whole body, body systems
   d. cells, chemicals, organs, tissues, whole body, body systems
   e. chemicals, cells, tissues, body systems, organs, whole body

3. The term *tissue* can apply either to one of the four primary tissue types or to a particular organ's aggregate of cellular and extracellular components. *(True or false?)*

4. Cells in a multicellular organism have specialized to such an extent that they have little in common with single-celled organisms. *(True or false?)*

5. Cell specializations are usually a modification or elaboration of one of the basic cell functions *(True or false?)*

6. The four primary types of tissue are _____, _____, _____, and _____.

7. The term _____ refers to the release from a cell, in response to appropriate stimulation, of specific products that have in large part been synthesized by the cell.

8. _____ glands secrete through ducts to the outside of the body, whereas _____ glands release their secretory products, known as _____, internally into the blood.

9. _____ controls are inherent to an organ, whereas _____ controls are regulatory mechanisms initiated outside of an organ that alter the activity of the organ.

10. Match the following:
    ____ 1. circulatory system
    ____ 2. digestive system
    ____ 3. respiratory system
    ____ 4. urinary system
    ____ 5. muscular and skeletal systems
    ____ 6. integumentary system
    ____ 7. immune system
    ____ 8. nervous system
    ____ 9. endocrine system
    ____ 10. reproductive system

    (a) obtains O_2 and eliminates CO_2
    (b) support, protect, and move body parts
    (c) controls, via hormones it secretes, processes that require duration
    (d) transport system
    (e) removes wastes and excess water, salt, and other electrolytes
    (f) perpetuates the species
    (g) obtains nutrients, water, and electrolytes
    (h) defends against foreign invaders and cancer
    (i) acts through electrical signals to control body's rapid responses
    (j) serves as outer protective barrier

Essay Questions

1. Define *physiology*.
2. What are the basic cell functions?
3. Distinguish between the external environment and the internal environment. What constitutes the internal environment?
4. What fluid compartments make up the internal environment?
5. Define *homeostasis*.
6. Describe the interrelationships among cells, body systems, and homeostasis.
7. What factors must be homeostatically maintained?
8. Define and describe the components of a homeostatic control system.
9. Compare negative and positive feedback.

POINTS TO PONDER

(Explanations on p. A-40)

1. Considering the nature of negative-feedback control and the function of the respiratory system, what effect do you predict that a decrease in CO_2 in the internal environment would have on how rapidly and deeply a person breathes?

2. Would the O_2 levels in the blood be (a) normal, (b) below normal, or (c) elevated in a patient with severe pneumonia resulting in impaired exchange of O_2 and CO_2 between the air and blood in the lungs? Would the CO_2 levels in the same patient's blood be (a) normal, (b) below normal, or (c) elevated? Because CO_2 reacts with H_2O to form carbonic acid (H_2CO_3), would the patient's blood (a) have a normal pH, (b) be too acidic, or (c) not be acidic enough (that is, be too alkaline), assuming that other compensatory measures have not yet had time to act?

3. The hormone insulin enhances the transport of glucose (sugar) from the blood into most body cells. Its secretion is controlled by a negative-feedback system between the concentration of glucose in the blood and the insulin-secreting cells. Therefore, which of the following statements is correct?

    a. A decrease in blood glucose concentration stimulates insulin secretion, which in turn further lowers blood glucose concentration.

    b. An increase in blood glucose concentration stimulates insulin secretion, which in turn lowers blood glucose concentration.

    c. A decrease in blood glucose concentration stimulates insulin secretion, which in turn increases blood glucose concentration.

    d. An increase in blood glucose concentration stimulates insulin secretion, which in turn further increases blood glucose concentration.

    e. None of the preceding are correct.

4. Given that most AIDS victims die from overwhelming infections or rare types of cancer, what body system do you think HIV (the AIDS virus) impairs?

5. Body temperature is homeostatically regulated around a set point. Given your knowledge of negative feedback and homeostatic control systems, predict whether narrowing or widening of the blood vessels of the skin will occur when a person exercises strenuously. (*Hints:* Muscle contraction generates heat. Narrowing of the vessels supplying an organ decreases blood flow through the organ, whereas vessel widening increases blood flow through the organ. The more warm blood flowing through the skin, the greater the loss of heat from the skin to the surrounding environment.)

CLINICAL CONSIDERATION

(Explanation on p. A-40)

Jennifer R. has the "stomach flu" that is going around campus and has been vomiting profusely for the past 24 hours. Not only has she been unable to keep down fluids or food, but she has also lost the acidic digestive juices secreted by the stomach that are normally reabsorbed back into the blood farther down the digestive tract. In what ways might this condition threaten to disrupt the homeostasis in Jennifer's internal environment? That is, what homeostatically maintained factors are moved away from normal by her profuse vomiting? What body systems respond to resist these changes?

PHYSIOEDGE RESOURCES

PhysioEdge CD-ROM

PhysioEdge, the CD-ROM packaged with your text, focuses on the concepts students find most difficult to learn. Figures marked with this icon have associated activities on the CD. A diagnostic quiz component allows you to get immediate feedback on your understanding of the concept and to advance through various levels of difficulty.

PhysioEdge Website

The website for this book contains a wealth of helpful study aids, as well as many ideas for further reading and research. Log on to:
http://www.brookscole.com/hpfundamentals3
Select a chapter from the drop-down menu, or click on one of the following resource areas:

▪ **Case Histories** introduce the clinical aspects of human physiology.

▪ For two- and three-dimensional (2-D and 3-D) graphic illustrations and animations of physiological concepts, visit our **Visual Learning Resource.**

▪ Review the **Chapter Outline, Learning Objectives,** and **Chapter Summary.** To test your mastery of important terminology for this chapter, you can use the **Glossary** or the electronic **Flash Cards,** which you can sort by definition or by term.

▪ For testing your knowledge and preparing for in-class examinations, look for **Fill in the Blanks, Matching, Multiple Choice,** and **True–False Quizzes** based on each chapter.

▪ Clicking on **Web Links** takes you to an extensive list of current links to Internet sites with news, research, and images related to individual topics in the chapter.

▪ **Internet Exercises** are critical thinking questions that involve research on the Internet with starter URLs (Web addresses) provided.

▪ For **InfoTrac Exercises,** projects that use InfoTrac College Edition® as a research tool, go to **InfoTrac College Edition/Research.**

 For Suggested Readings, consult **InfoTrac College Edition/ Research** on the PhysioEdge website or go directly to InfoTrac College Edition, your online research library, at:
http://infotrac.thomsonlearning.com

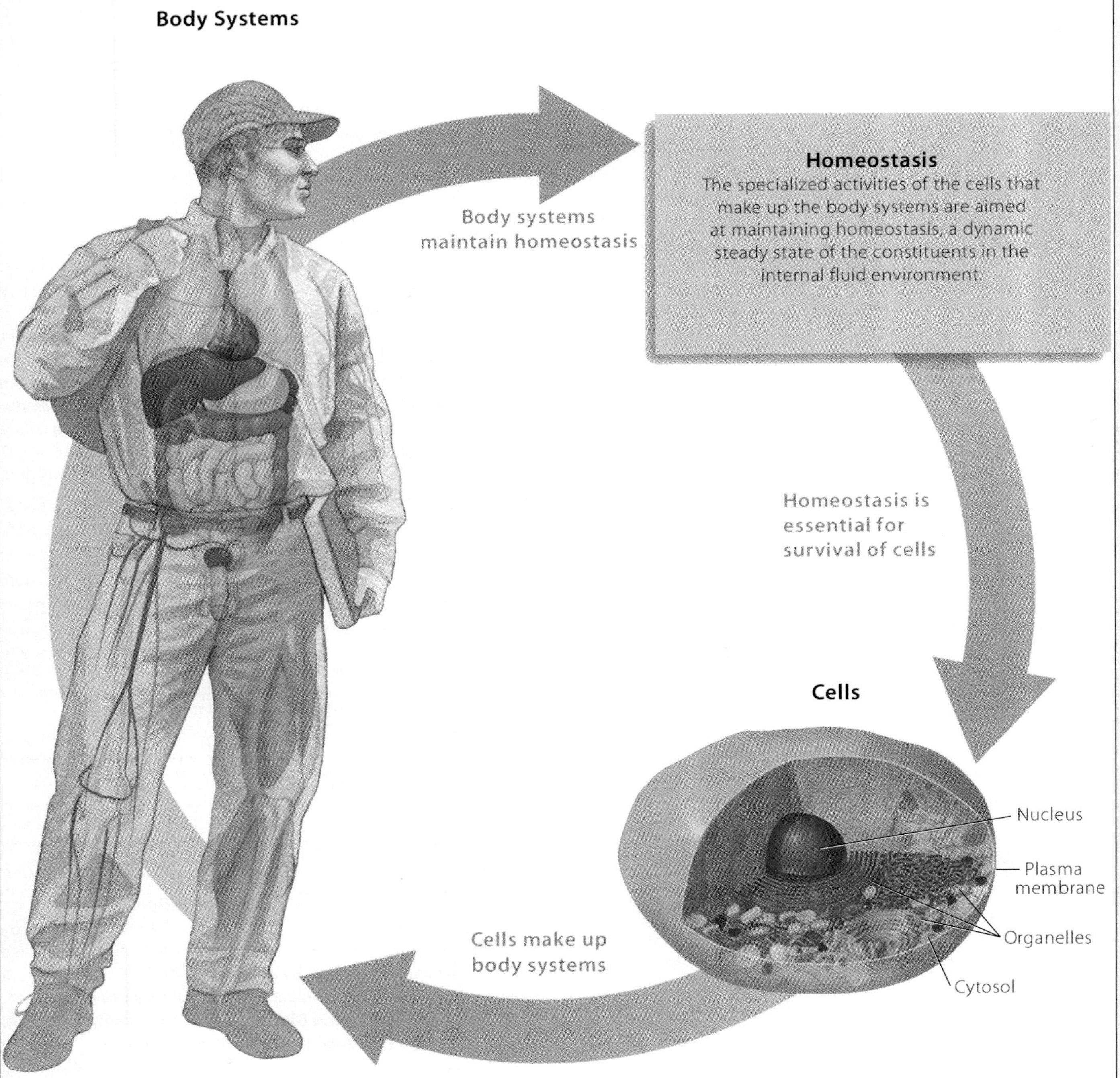

Body Systems

Body systems maintain homeostasis

Homeostasis
The specialized activities of the cells that make up the body systems are aimed at maintaining homeostasis, a dynamic steady state of the constituents in the internal fluid environment.

Homeostasis is essential for survival of cells

Cells

Nucleus

Plasma membrane

Organelles

Cytosol

Cells make up body systems

Cells are the body's living building blocks. Just as the body as a whole is highly organized, so too is a cell's interior. A cell has three major parts: a **plasma membrane** that encloses the cell; the **nucleus,** which houses the cell's genetic material; and the **cytoplasm,** which is organized into discrete highly specialized *organelles* dispersed throughout a gel-like liquid, the *cytosol*. The cytosol is pervaded by a protein scaffolding, the *cytoskeleton,* that serves as the "bone and muscle" of the cell.

Through the coordinated action of each of these cell components, every cell can perform certain basic functions essential to its own survival and a specialized task that helps maintain homeostasis. Cells are organized according to their specialization into body systems that maintain the stable internal environment essential for the whole body's survival. All body functions ultimately depend on the activities of the individual cells that compose the body.

Cell Physiology

 Click on the Tutorials menu of the CD-ROM for a tutorial on Transport across Membranes.

Cells consist of the same chemicals found in nonliving objects on our planet. Even though researchers have analyzed the chemicals of which cells are made, they have not been able to organize these molecules into a living cell in a laboratory. Life stems from the complex organization and interaction of these molecules within the cell. Groups of inanimate chemical molecules are structurally organized and function together in unique ways to form a cell, the smallest living entity. Cells, in turn, serve as the living building blocks for the immensely complicated whole body. Thus cells are the bridge between molecules and humans (and all other living organisms). By probing deeper into the molecular structure and organization of the cells that make up the body, modern physiologists are unraveling many of the broader mysteries of how the body works.

OBSERVATIONS OF CELLS

The cells that make up the human body are so small they cannot be seen by the unaided eye. The smallest visible particle is about 5 to 10 times larger than a typical human cell, which averages about 10 to 20 micrometers (μm) in diameter. About 100 average-sized cells lined up side by side would stretch a distance of only 1 mm (one μm = 1 millionth of a meter; 1 mm = 1 thousandth of a meter; 1 m = 39.37 inches. (See Appendix A for a comparison of metric units and their English equivalents. This appendix also provides a visual comparison of the size of cells in relation to other selected structures.)

Until the microscope was invented in the middle of the 17th century, scientists did not even know that cells existed. In the early part of the 19th century, with the development of better light microscopes, researchers learned that all plant and animal tissues consist of individual cells. The cells of a hummingbird, a human, and a whale are all about the same size. Larger species have a greater number of cells, not larger cells. These early investigators also discovered that cells are filled with a fluid, which, given the microscopic capabilities of the time, appeared to be a rather uniform, soupy mixture believed to be the elusive "stuff of life." When in the 1940s scientists first employed the technique of electron microscopy to observe living matter, they began to understand the great diversity and complexity of the internal structure of cells. (Electron microscopes are about 100 times more powerful than light microscopes.) Now that scientists have even more sophisticated microscopes, biochemical techniques, cell culture technology, and genetic engineering, the concept of the cell as a microscopic bag of formless fluid has given way to the current understanding of the cell as a complex, highly organized, compartmentalized structure.

AN OVERVIEW OF CELL STRUCTURE

The trillions of cells in a human body are classified into about 200 different cell types based on specific variations in structure and function. Even though there is no such thing as a "typical" cell, because of these diverse structural and functional specializations, different cells share many common features. Most cells have three major subdivisions: the *plasma membrane,* which encloses the cells; the *nucleus,* which contains the cell's genetic material; and the *cytoplasm,* the portion of the cell's interior not occupied by the nucleus. For now, we will provide an overview of each subdivision, then will describe each in detail later.

▎ The plasma membrane bounds the cell.

The **plasma membrane,** or **cell membrane,** is a very thin membranous structure that encloses each cell. This oily barrier separates the cell's contents from its surroundings; it keeps the *intracellular fluid (ICF)* within the cells from mingling with the *extracellular fluid (ECF)* outside of the cells. The plasma membrane is not simply a mechanical barrier to hold in the contents of the cells; it also has the ability to selectively control movement of molecules between the ICF and ECF. The plasma membrane can be likened to the gated walls that enclosed ancient cities. Through this structure, the cell can control the entry of food and other needed supplies and the export of products manufactured within, while at the same time guarding against unwanted traffic into or out of the cell. The plasma membrane is discussed thoroughly in Chapter 3.

▎ The nucleus contains the DNA.

The two major parts of the cell's interior are the nucleus and the cytoplasm. The **nucleus,** which is typically the largest single organized cell component, can be seen as a distinct spherical or oval structure, usually located near the center of the cell. It is surrounded by a double-layered membrane, which separates the nucleus from the rest of the cell. The nuclear membrane is pierced by many **nuclear pores** that allow necessary traffic to move between the nucleus and the cytoplasm.

The nucleus houses the cell's genetic material, **deoxyribonucleic acid (DNA),** which has two important functions: (1) directing protein synthesis and (2) serving as a genetic blueprint during cell replication. DNA provides codes, or "instructions," for directing synthesis of specific structural and enzymatic proteins within the cell. By directing the kinds and amounts of various enzymes and other proteins that are produced, the nucleus indirectly governs most cell activities and serves as the cell's control center.

Three types of **ribonucleic acid (RNA)** play a role in this protein synthesis. First, DNA's genetic code for a particular protein is transcribed into a **messenger RNA** molecule, which exits the nucleus through the nuclear pores. Within the cytoplasm, messenger RNA delivers the coded message to **ribosomal RNA,** which "reads" the code and translates it into the appropriate amino acid sequence for the designated protein being synthesized. Finally, **transfer RNA** transfers the appropriate amino acids within the cytoplasm to their designated site in the protein under construction.

In addition to providing codes for protein synthesis, DNA also serves as a genetic blueprint during cell replication to ensure that the cell produces additional cells just like itself, thus continuing the identical type of cell line within the body. Furthermore, in the reproductive cells, the DNA blueprint serves to pass on genetic characteristics to future generations. (See Appendix C for further details of DNA and RNA function and protein synthesis.)

▎ The cytoplasm consists of various organelles and the cytosol.

The **cytoplasm** is that portion of the cell interior not occupied by the nucleus. It contains a number of distinct, highly organized, membrane-enclosed structures—the *organelles* ("little organs")—dispersed within the *cytosol,* which is a complex, gel-like liquid.

On average, nearly half of the total cell volume is occupied by organelles. Each organelle is a separate compartment within the cell that is enclosed by a membrane similar to the plasma membrane. Thus the contents of an organelle are separated from the surrounding cytosol and from the contents of the other organelles. Nearly all cells contain six main types of organelles—the *endoplasmic reticulum, Golgi complex, lysosomes, peroxisomes, mitochondria,* and *vaults* (● Figure 2-1). These organelles are similar in all cells, although there are some variations depending on the specialized capabilities of each cell type. Organelles are like intracellular "specialty shops." Each is a separate internal compartment that contains a specific set of chemicals for carrying out a particular cellular function. This compartmentalization is advantageous because it permits chemical activities that would not be compatible with each other to occur simultaneously within the cell. For example, the enzymes that destroy unwanted proteins in the cell do so within the protective confines of the lysosomes without the risk of destroying essential cell proteins. Just as organs each play a role that is essential for survival of the whole body, organelles each perform a specialized activity necessary for survival of the whole cell.

The remainder of the cytoplasm not occupied by organelles consists of the **cytosol** ("cell liquid"). The cytosol is made up of a semiliquid, gel-like mass laced with an elaborate protein network known as the *cytoskeleton.* Many of the chemical reactions that are compatible with each other are

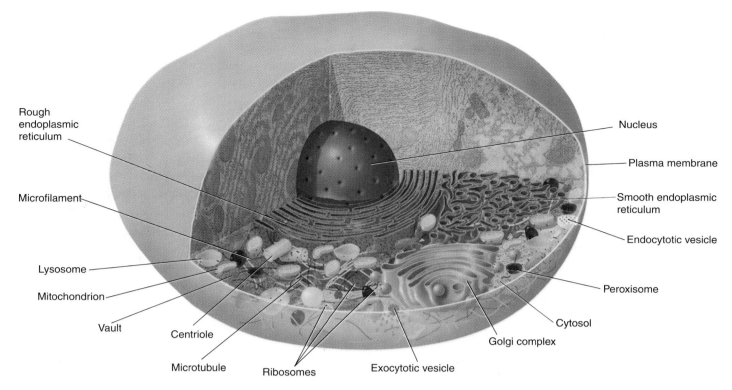

Rough endoplasmic reticulum

Microfilament

Lysosome

Mitochondrion

Vault

Centriole

Microtubule

Ribosomes

Exocytotic vesicle

Golgi complex

Cytosol

Peroxisome

Endocytotic vesicle

Smooth endoplasmic reticulum

Plasma membrane

Nucleus

● **FIGURE 2-1**

Schematic three-dimensional diagram of cell structures visible under an electron microscope

For an interaction related to this figure, see Media Exercise 2.1: Anatomy of a Generic Cell on the CD-ROM.

conducted in the cytosol. The cytoskeletal network gives the cell its shape, provides for its internal organization, and regulates its various movements. (For clarification, the intracellular fluid includes all the fluid inside the cell, including that within the cytosol, the organelles, and the nucleus.)

In this chapter, we will examine each of the cytoplasmic components in more detail, concentrating first on the six types of organelles.

ENDOPLASMIC RETICULUM AND SEGREGATED SYNTHESIS

The **endoplasmic reticulum (ER)** is an elaborate fluid-filled membranous system distributed extensively throughout the cytosol. It is primarily a protein- and lipid-manufacturing factory. Two distinct types of endoplasmic reticulum—the smooth ER and the rough ER—can be distinguished. The **smooth ER** is a meshwork of tiny interconnected tubules, whereas the **rough ER** projects outward from the smooth ER as stacks of relatively flattened sacs (● Figure 2-2). Even though these two regions differ considerably in appearance and function, they are connected to each other. In other words, the ER is one continuous organelle with many interconnected channels. The relative amount of smooth and rough ER varies between cells, depending on the activity of the cell.

▌ The rough endoplasmic reticulum synthesizes proteins for secretion and membrane construction.

The outer surface of the rough ER membrane is studded with small, dark-staining particles that give it a "rough" or granular appearance under a light microscope. These particles are **ribosomes**, which are ribosomal RNA-protein complexes that synthesize proteins under the direction of nuclear DNA. Messenger RNA carries the genetic message from the nucleus to the ribosome "workbench," where protein synthesis takes place (see p. A-24). Not all ribosomes in the cell are attached to the rough ER. Unattached or "free" ribosomes are dispersed throughout the cytosol.

The rough ER, in association with its ribosomes, synthesizes and releases a variety of new proteins into the ER lumen, the fluid-filled space enclosed by the ER membrane. These proteins serve one of two purposes: (1) Some proteins are destined for export to the cell's exterior as secretory products, such as protein hormones or enzymes. (All enzymes are proteins.) (2) Other proteins are transported to sites within the cell for use in constructing new cellular membrane (either new plasma membrane or new organelle membrane) or other protein components of organelles. Cellular membranes consist predominantly of proteins and lipids (fats). The membranous wall of the ER also contains enzymes essential for the synthe-

Endoplasmic reticulum (ER). (a) Schematic three-dimensional representation of the relationship between the smooth ER, which is a meshwork of tiny interconnected tubules, and the rough ER, which is studded with ribosomes and projects outward from the smooth ER as stacks of relatively flattened sacs. (b) Electron micrograph of the rough ER. Note the layers of flattened sacs studded with small, dark-staining ribosomes. (c) Electron micrograph of the smooth ER.

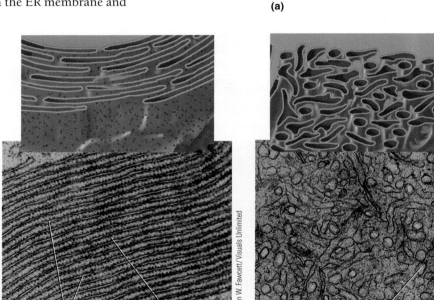

ER lumen   **Rough ER**   Ribosomes   **Smooth ER**

(a)

sis of nearly all the lipids needed to produce new membranes. These newly synthesized lipids enter the ER lumen along with the proteins. Predictably, the rough ER is most abundant in cells specialized for protein secretion (for example, cells that secrete digestive enzymes), or in cells that require extensive membrane synthesis (for example, rapidly growing cells such as immature egg cells).

After being synthesized and released into the ER lumen, a new protein cannot pass out through the ER membrane and therefore becomes permanently separated from the cytosol as soon as it has been synthesized. In contrast to the rough ER ribosomes, the free ribosomes synthesize proteins that are used intracellularly within the cytosol. In this way, newly produced molecules that are destined for export out of the cell or for synthesis of new cell components (those synthesized by the ER) are physically separated from those that belong in the cytosol (those produced by the free ribosomes).

How do the newly synthesized molecules within the ER lumen get to their destinations at other sites inside the cell or to the outside of the cell, if they cannot pass out through the ER membrane? They do so through the action of the smooth endoplasmic reticulum.

Rough ER lumen   Ribosomes

(b)

Smooth ER lumen

(c)

Don W. Fawcett / Visuals Unlimited

▌ The smooth endoplasmic reticulum packages new proteins in transport vesicles.

The smooth ER does not contain ribosomes, so it is "smooth." Lacking ribosomes, it is not involved in protein synthesis.

In the majority of cells, the smooth ER is rather sparse and serves primarily as a central packaging and discharge site for molecules that are to be transported from the ER. Newly synthesized proteins and lipids pass from the rough ER to gather in the smooth ER. Portions of the smooth ER then "bud off" (that is, balloon outward on the surface, then are pinched off), forming **transport vesicles** that contain the new molecules enclosed in a spherical capsule of membrane derived from the smooth ER membrane (● Figure 2-3). (A **vesicle** is a fluid-filled, membrane-enclosed intracellular cargo container.) Newly synthesized membrane components are rapidly incor-

porated into the ER membrane itself to replace the membrane that was used to "wrap up" the molecules in the transport vesicle. Transport vesicles move to the Golgi complex, described in the next section, for further processing of their cargo.

In contrast to the sparseness of the smooth ER in most cells, some specialized types of cells have an extensive smooth ER, which has additional responsibilities. For example, muscle cells have an elaborate, modified smooth ER known as the *sarcoplasmic reticulum,* which stores calcium that plays an important role in the process of muscle contraction (see p. 210).

GOLGI COMPLEX AND EXOCYTOSIS

Closely associated with the endoplasmic reticulum is the **Golgi complex.** Each Golgi complex consists of a stack of flattened, slightly curved, membrane-enclosed sacs (● Figure 2-4). The

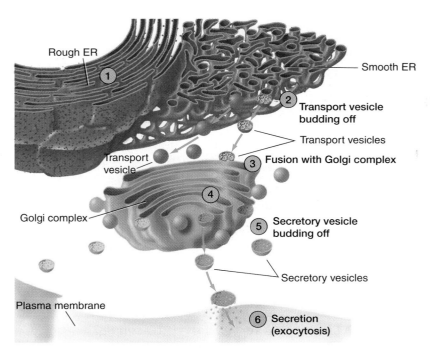

Rough ER ① Smooth ER

② Transport vesicle budding off

Transport vesicles

Transport vesicle

③ Fusion with Golgi complex

Golgi complex

④

⑤ Secretory vesicle budding off

Secretory vesicles

Plasma membrane

⑥ Secretion (exocytosis)

① The rough ER synthesizes proteins to be secreted to the exterior or to be incorporated into the cellular membrane.

② The smooth ER packages the secretory product into transport vesicles, which bud off and move to the Golgi complex.

③ The transport vesicle fuses with the Golgi complex, opens up, and empties its contents into the closest Golgi sac.

④ As the newly synthesized proteins from the ER travel by vesicular transport through the layers of the Golgi complex, this complex modifies the raw proteins into final form and sorts and directs the finished products to their final destination by varying their wrappers.

⑤ Secretory vesicles containing the finished protein product bud off the Golgi complex and remain in the cytosol, storing the product until signaled to empty.

⑥ On appropriate stimulation, the secretory vesicles fuse with the plasma membrane, open, and empty their contents to the cell's exterior. Secretion has occurred by exocytosis, with the secretory product never having come into contact with the cytosol.

● **FIGURE 2-3**

Overview of the secretion process for proteins synthesized by the endoplasmic reticulum

PhysioEdge For an animation of this figure, click the Exocytosis tab in the Transport across Membranes tutorial on the CD-ROM.

sacs within each Golgi stack are not physically connected to each other. Note that the flattened sacs are thin in the middle but have dilated, or bulging, edges. The number of Golgi complexes varies, depending on the cell type. Some cells have only one Golgi stack, whereas cells highly specialized for protein secretion may have hundreds of stacks.

▌ Transport vesicles carry their cargo to the Golgi complex for further processing.

The majority of the newly synthesized molecules that have just budded off from the smooth ER enter a Golgi stack. When a transport vesicle carrying its newly synthesized cargo reaches a Golgi stack, the vesicle membrane fuses with the membrane of the sac closest to the center of the cell. The vesicle membrane opens up and becomes integrated into the Golgi membrane, and the contents of the vesicle are released to the interior of the sac (● Figure 2-3).

These newly synthesized raw materials from the ER travel by means of vesicle formation through the layers of the Golgi stack, from the innermost sac closest to the ER to the outermost sac near the plasma membrane. During this transit, two important, interrelated functions take place:

1. *Processing the raw materials into finished products.* Within the Golgi complex, the "raw" proteins from the ER are modified into their final form, for example, by having sugar attached to them.

2. *Sorting and directing the finished products to their final destinations.* The Golgi complex is responsible for sorting and segregating different types of products according to their function and destination, such as products to be secreted to the cell's exterior or to be used for constructing new plasma membrane.

● **FIGURE 2-4**

Golgi complex. (a) Schematic three-dimensional diagram of a Golgi complex. (b) Electron micrograph of a Golgi complex. The vesicles at the dilated edges of the sacs contain finished protein products packaged for distribution to their final destination.

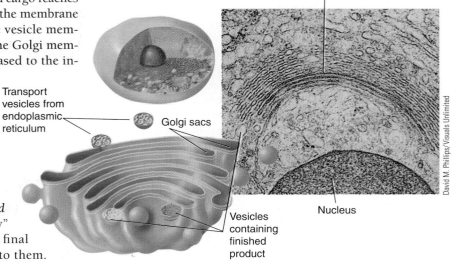

Golgi complex

Transport vesicles from endoplasmic reticulum

Golgi sacs

Nucleus

Vesicles containing finished product

David M. Phillips/Visuals Unlimited

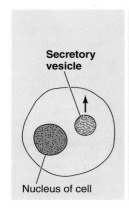

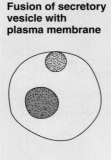

Secretory vesicle

Fusion of secretory vesicle with plasma membrane

Secretion of vesicle contents (exocytosis)

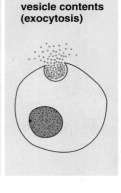

Nucleus of cell

(a)

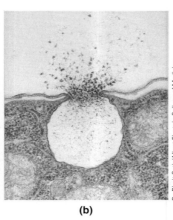

(b)

● **FIGURE 2-5**

Exocytosis of secretory product. (a) Schematic overview of the process of exocytosis. (b) Transmission electron micrograph of exocytosis.

▌ The Golgi complex packages secretory vesicles for release by exocytosis.

How does the Golgi complex sort and direct finished proteins to the proper destinations? Finished products are collected within the dilated edges of the Golgi complex's sacs. The dilated edge of the outermost sac then pinches off to form a membrane-enclosed vesicle that contains the finished product. For each type of product to reach its appropriate site of function, each distinct type of vesicle takes up a specific product before budding off. Vesicles with their selected cargo destined for different sites are wrapped in membranes containing different surface protein molecules. Each different surface protein marker serves as a specific **docking marker** (like an address on an envelope). Each vesicle can "dock" and "unload" its cargo only at the appropriate "**docking-marker acceptor**," a protein located only at the proper destination within the cell (like a house address). Thus, Golgi products are sorted and delivered like addressed envelopes containing particular pieces of mail being delivered only to the appropriate house addresses.

In secretory cells, numerous large **secretory vesicles**, which contain proteins to be secreted, bud off from the Golgi stacks. Specialized secretory cells include endocrine cells that secrete protein hormones and digestive gland cells that secrete digestive enzymes. The secretory proteins remain stored within the secretory vesicles until the cell is stimulated by a specific signal that indicates a need for release of that particular secretory product. On appropriate stimulation, the vesicles move to the cell's periphery. Vesicular contents are quickly released to the cell's exterior as the vesicle fuses with the plasma membrane, opens, and empties its contents to the outside (● Figures 2-3 and 2-5). This mechanism—extrusion to the exterior of substances originating within the cell—is referred to as **exocytosis** (*exo* means "out of"; *cyto* means "cell"). Release of the contents of a secretory vesicle by means of exocytosis constitutes the process of **secretion**. Secretory vesicles fuse only with the plasma membrane and not with any of the internal membranes that bound organelles, thereby preventing fruitless or even dangerous discharge of secretory products into the organelles.

LYSOSOMES AND ENDOCYTOSIS

On the average, a cell contains about 300 lysosomes.

▌ Lysosomes serve as the intracellular digestive system.

Lysosomes are membrane-enclosed sacs containing powerful **hydrolytic enzymes**, which catalyze *hydrolysis reactions* (see p. A-17). These reactions break down the organic molecules that make up cell debris and foreign material, such as bacteria that have been brought into the cell. (*Lys* means "breakdown"; *some* means "body." Lysosomes are small bodies within the cell that break down organic molecules.) The lysosomal enzymes are similar to the hydrolytic enzymes that the digestive system secretes to digest food. Thus lysosomes serve as the intracellular "digestive system." Instead of having a uniform structure, as is characteristic of all other organelles, lysosomes vary in size and shape, depending on the contents they are digesting. Most commonly, lysosomes are small oval or spherical bodies (● Figure 2-6).

● **FIGURE 2-6**

Lysosomes and peroxisomes. Electron micrograph showing both lysosomes, which contain hydrolytic enzymes, and peroxisomes, which contain oxidative enzymes.

Lysosome

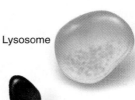

Peroxisome

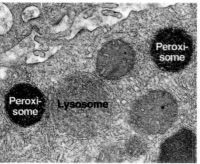

Peroxisome

Peroxisome

Lysosome

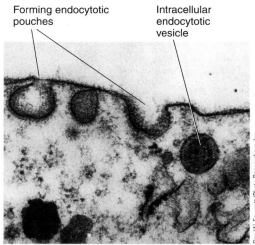

Forming endocytotic pouches

Intracellular endocytotic vesicle

Don W. Fawcett/Photo Researchers, Inc.

(a) Pinocytosis

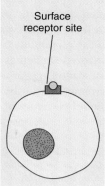
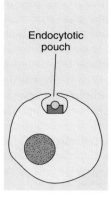
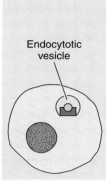

Particular protein

Nucleus of cell

Surface receptor site

Endocytotic pouch

Endocytotic vesicle

(b) Receptor-mediated endocytosis

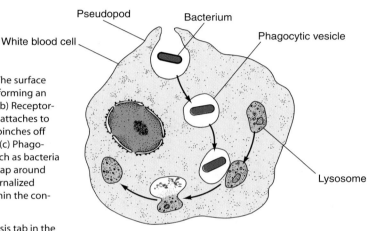

Pseudopod

Bacterium

White blood cell

Phagocytic vesicle

Lysosome

(c) Phagocytosis

● **FIGURE 2-7**

Forms of endocytosis. (a) Electron micrograph of pinocytosis. The surface membrane dips inward to form a pouch, then seals the surface, forming an intracellular vesicle that nonselectively internalizes a bit of ECF. (b) Receptor-mediated endocytosis. When a large molecule such as a protein attaches to a specific surface receptor site, the membrane dips inward and pinches off to selectively internalize the molecule in an intracellular vesicle. (c) Phagocytosis. White blood cells internalize multimolecular particles such as bacteria by extending surface projections known as pseudopods that wrap around and seal in the targeted material. A lysosome fuses with the internalized vesicle, releasing enzymes that attack the engulfed material within the confines of the vesicle.

 For an animation of this figure, click the Endocytosis tab in the Transport across Membranes tutorial on the CD-ROM.

▌ Extracellular material is brought into the cell by phagocytosis for attack by lysosomal enzymes.

Extracellular material to be attacked by lysosomal enzymes is brought inside the cell through the process of **endocytosis** (*endo* means "within"). Endocytosis can be accomplished in three ways—*pinocytosis, receptor-mediated endocytosis,* and *phagocytosis*—depending on the contents of the internalized material.

PINOCYTOSIS

With **pinocytosis** ("cell drinking"), a small droplet of extracellular fluid is internalized. First, the plasma membrane dips inward, forming a pouch that contains a small bit of ECF (● Figure 2-7a). The plasma membrane then seals at the surface of the pouch, trapping the contents in a small, intracellular **endocytotic vesicle.** Besides bringing ECF into a cell, pinocytosis provides a means to retrieve extra plasma membrane that has been added to the cell surface during exocytosis.

RECEPTOR-MEDIATED ENDOCYTOSIS

Unlike pinocytosis, which involves the nonselective uptake of the surrounding fluid, **receptor-mediated endocytosis** is a highly selective process that enables cells to import specific

large molecules that it needs from its environment. Receptor-mediated endocytosis is triggered by the binding of a specific molecule such as a protein to a surface membrane receptor site specific for that protein. This binding causes the plasma membrane at that site to sink in, then seal at the surface, trapping the protein inside the cell (● Figure 2-7b). Cholesterol complexes, vitamin B_{12}, the hormone insulin, and iron are examples of substances selectively taken into cells by receptor-mediated endocytosis.

Unfortunately, some viruses can sneak into cells by exploiting this mechanism. For instance, flu viruses and HIV, the virus that causes AIDS (see p. 347), gain entry to cells via receptor-mediated endocytosis. They do so by binding with membrane receptor sites normally designed to trigger the internalization of a needed molecule.

PHAGOCYTOSIS

During **phagocytosis** ("cell eating"), large multimolecular particles are internalized. Most body cells perform pinocytosis, many carry out receptor-mediated endocytosis, but only a few specialized cells are capable of phagocytosis. The latter are the "professional" phagocytes, the most notable being certain types of white blood cells that play an important role in the body's defense mechanisms. When a white blood cell encounters a large multimolecular particle, such as a bacterium or tissue debris, it extends surface projections known as **pseudopods** ("false feet") that completely surround or engulf the particle and trap it within an internalized vesicle (● Figure 2-7c). A lysosome fuses with the membrane of the internalized vesicle and releases its hydrolytic enzymes into the vesicle, where they safely attack the bacterium or other trapped material without damaging the remainder of the cell. The enzymes largely break down the engulfed material into reusable raw ingredients, such as amino acids, glucose, and fatty acids, that the cell can use.

▌ Lysosomes remove useless but not useful parts of the cell.

Lysosomes can also fuse with aged or damaged organelles to remove these useless parts of the cell. This selective self-digestion makes way for new replacement parts. In most cells all the organelles are renewable.

Clinical Note   Some individuals lack the ability to synthesize one or more of the lysosomal enzymes. The result is massive accumulation within the lysosomes of the compound normally digested by the missing enzyme. Clinical manifestations often accompany such disorders, because the engorged lysosomes interfere with normal cell activity. The nature and severity of the symptoms depend on the type of substance accumulating, which in turn depends on what lysosomal enzyme is missing. Among these so-called *storage diseases* is **Tay-Sachs disease**, which is characterized by abnormal accumulation of complex molecules found in nerve cells. As the accumulation continues, profound symptoms of progressive nervous system degeneration result.

PEROXISOMES AND DETOXIFICATION

Typically, several hundred small peroxisomes about one third to one half the average size of lysosomes are present in a cell (● Figure 2-6).

▌ Peroxisomes house oxidative enzymes that detoxify various wastes.

Peroxisomes are similar to lysosomes in that they are membrane-enclosed sacs containing enzymes, but unlike the lysosomes, which contain hydrolytic enzymes, peroxisomes house several powerful oxidative enzymes. **Oxidative enzymes**, as the name implies, use oxygen (O_2), in this case to strip hydrogen from certain organic molecules. This oxidation helps detoxify various wastes produced within the cell or foreign toxic compounds that have entered the cell, such as alcohol consumed in beverages.

MITOCHONDRIA AND ATP PRODUCTION

A single cell may contain as few as a hundred or as many as several thousand mitochondria.

▌ Mitochondria, the energy organelles, are enclosed by a double membrane.

Mitochondria are the energy organelles, or "power plants," of the cell; they extract energy from the nutrients in food and transform it into a usable form for cell activities. Mitochondria generate about 90% of the energy that cells—and, accordingly, the whole body—need to survive and function. The number of mitochondria per cell varies greatly, depending on the energy needs of each particular cell type. In some cell types, the mitochondria are densely compacted in cell regions that use most of the cell's energy. For example, mitochondria are packed between the contractile units in the muscle cells of the heart.

Mitochondria are rod-shaped or oval structures. Each mitochondrion is enclosed by a double membrane—a smooth outer membrane that surrounds the mitochondrion itself, and an inner membrane that forms a series of infoldings or shelves called **cristae**, which project into an inner cavity filled with a gel-like solution known as the **matrix** (● Figure 2-8). These cristae contain crucial proteins (the electron transport proteins, to be described shortly) that ultimately are responsible for converting much of the energy in food into a usable form. The generous folds of the inner membrane greatly increase the surface area available for housing these important proteins. The matrix consists of a concentrated mixture of hundreds of different dissolved enzymes (the citric acid cycle

● **FIGURE 2-8**

Mitochondrion. (a) Schematic representation of a mitochondrion. The electron transport proteins embedded in the cristae folds of the mitochondrial inner membrane are ultimately responsible for converting much of the energy of food into a usable form. (b) Electron micrograph of a mitochondrion.

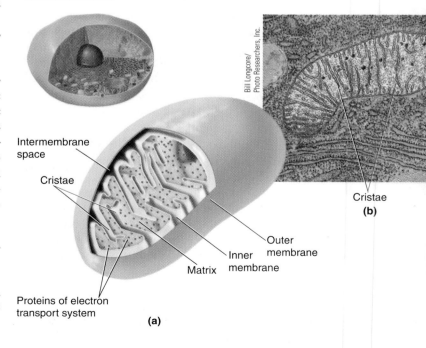

Bill Longcore/Photo Researchers, Inc.

Intermembrane space

Cristae

Cristae
(b)

Outer membrane

Inner membrane

Matrix

Proteins of electron transport system

(a)

enzymes, soon to be described) that prepare nutrient molecules for the final extraction of usable energy by the cristae proteins.

■ Mitochondria play a major role in generating ATP.

The source of energy for the body is the chemical energy stored in the carbon bonds of ingested food. Body cells are not equipped to use this energy directly, however. Instead, they must extract energy from food nutrients and convert it into an energy form that they can use—namely, the high-energy phosphate bonds of **adenosine triphosphate (ATP)**, which consists of adenosine with three phosphate groups attached (*tri* means "three") (see p. A-18). When a high-energy bond such as that binding the terminal phosphate to adenosine is split, a substantial amount of energy is released. Adenosine triphosphate is the universal energy carrier—the common energy "currency" of the body. Cells can "cash in" ATP to pay the energy "price" for running the cell machinery. To obtain immediate usable energy, cells split the terminal phosphate bond of ATP, which yields **adenosine diphosphate (ADP)**—adenosine with two phosphate groups attached (*di* means "two")—plus inorganic phosphate (P_i) plus energy:

$$\text{ATP} \xrightarrow{\text{splitting}} \text{ADP} + P_i + \text{energy for use by the cell}$$

In this energy scheme, food can be thought of as the "crude fuel," whereas ATP is the "refined fuel" for operating the body's machinery. Let us elaborate on this fuel conversion process. Dietary food is digested, or broken down, by the digestive system into smaller absorbable units that can be transferred from the digestive tract lumen into the blood (Chapter 15). For example, dietary carbohydrates are broken down primarily into glucose, which is absorbed into the blood. No usable energy is released during the digestion of food. When delivered to the cells by the blood, the nutrient molecules are transported across the plasma membrane into the cytosol. (Details of how materials cross the membrane are covered in Chapter 3.)

We are now going to turn our attention to the steps involved in ATP production within the cell and the role of the mitochondria in these steps. ATP is generated in most cells from the sequential dismantling of absorbed nutrient molecules in three different steps: *glycolysis*, the *citric acid cycle*, and the *electron transport chain*. (Muscle cells use an additional cytosolic pathway for immediately generating energy at the onset of exercise; see p. 219.) We will use glucose as an example to describe these three steps.

GLYCOLYSIS

Among the thousands of enzymes within the cytosol are those responsible for **glycolysis**, a chemical process involving 10 separate se-

quential reactions that break down the simple six-carbon sugar molecule, glucose, into two pyruvic acid molecules, each of which contains three carbons (*glyc-* means "sweet;" *lysis* means "breakdown"). During glycolysis, some of the energy from the broken chemical bonds of glucose is used to convert ADP into ATP (● Figure 2-9). However, glycolysis is not very efficient in terms of energy extraction: the net yield is only two molecules of ATP per glucose molecule processed. Much of the energy originally contained in the glucose molecule is still locked in the chemical bonds of the pyruvic acid molecules. The low-energy yield of glycolysis is not enough to support the body's demand for ATP. This is where the mitochondria come into play.

CITRIC ACID CYCLE (Krebs)

The pyruvic acid produced by glycolysis in the cytosol can be selectively transported into the mitochondrial matrix. Here it is further broken down into a two-carbon molecule, acetic acid, by enzymatic removal of one of the carbons in the form of carbon dioxide (CO_2), which eventually is eliminated from the body as an end product, or waste (● Figure 2-10). During this breakdown process, a carbon–hydrogen bond is disrupted, so a hydrogen atom is also released. This hydrogen atom is held by a hydrogen carrier molecule, the function of which will be discussed shortly. The acetic acid thus formed combines with coenzyme A, a derivative of pantothenic acid (a B vitamin), producing the compound acetyl coenzyme A (acetyl CoA).

Acetyl CoA then enters the **citric acid cycle**, which consists of a cyclical series of eight separate biochemical reactions that are directed by the enzymes of the mitochondrial matrix. This cycle of reactions can be compared to one revolution around a Ferris wheel. Keep in mind that ● Figure 2-10 is highly schematic. It depicts a cyclic series of biochemical reactions. The molecules themselves are not physically moved around in a cycle. On the top of the Ferris wheel, acetyl CoA, a two-carbon molecule, enters a seat already occupied by oxaloacetic acid, a four-carbon molecule. These two molecules link together to form a six-carbon citric acid molecule, and the trip around the citric acid cycle begins. (This cycle is alternatively known as the **Krebs cycle**, in honor of its principal discoverer, or as the **tricarboxylic acid cycle**, because cit-

---

● **FIGURE 2-9**

A simplified summary of glycolysis. Glycolysis involves the breakdown of glucose into two pyruvic acid molecules, with a net yield of two molecules of ATP for every glucose molecule processed.

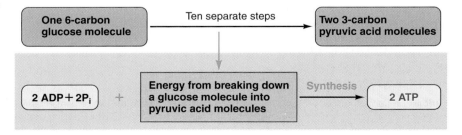

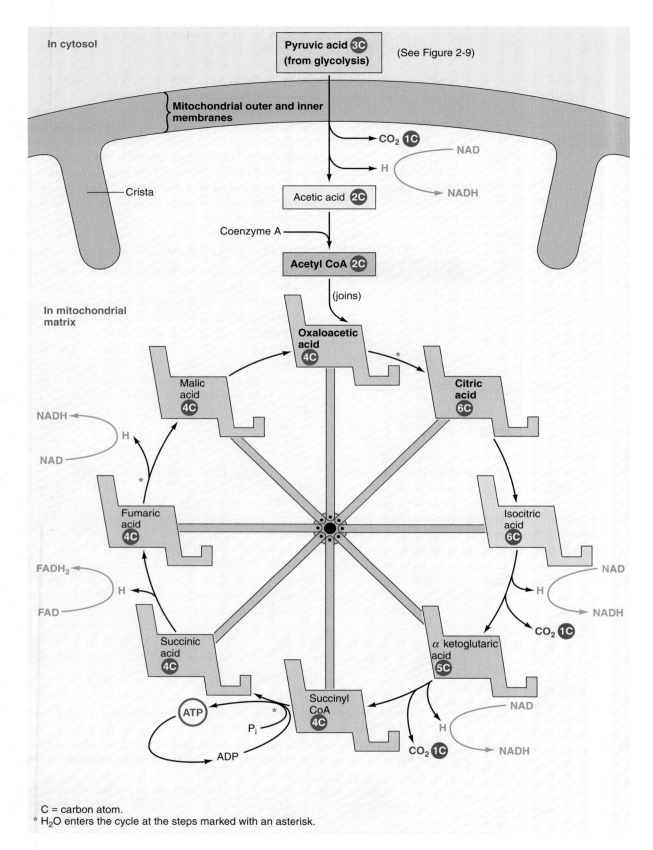

● FIGURE 2-10

Citric acid cycle. A simplified version of the citric acid cycle, showing how the two carbons entering the cycle by means of acetyl CoA are eventually converted to CO_2, with oxaloacetic acid, which accepts acetyl CoA, being regenerated at the end of the cyclical pathway. Also denoted is the release of hydrogen atoms at specific points along the pathway. These hydrogens bind to the hydrogen carrier molecules NAD and FAD for further processing by the electron transport chain. One molecule of ATP is generated for each molecule of acetyl CoA that enters the citric acid cycle, for a total of two molecules of ATP for each molecule of processed glucose.

ric acid contains three carboxylic acid groups.) As the seat moves around the cycle, at each new position, matrix enzymes modify the passenger molecule to form a slightly different molecule. These molecular alterations have the following important consequences:

1. Two carbons are sequentially "kicked off the ride" as they are removed from the six-carbon citric acid molecule, converting it back into the four-carbon oxaloacetic acid, which is now available at the top of the cycle to pick up another acetyl CoA for another revolution through the cycle.

2. The released carbon atoms, which were originally present in the acetyl CoA that entered the cycle, are converted into two molecules of CO_2. This CO_2, as well as the CO_2 produced during the formation of acetic acid from pyruvic acid, passes out of the mitochondrial matrix and subsequently out of the cell to enter the blood. In turn, the blood carries the CO_2 to the lungs, where it is finally eliminated into the atmosphere through the process of breathing. The oxygen used to make CO_2 from these released carbon atoms is derived from the molecules that were involved in the reactions, not from free molecular oxygen supplied by breathing.

3. Hydrogen atoms are also "bumped off" during the cycle at four of the chemical conversion steps. The key purpose of the citric acid cycle is to produce these hydrogens for entry into the electron transport chain. These hydrogens are "caught" by two other compounds that act as hydrogen carrier molecules—**nicotinamide adenine dinucleotide (NAD)**, a derivative of the B vitamin niacin, and **flavine adenine dinucleotide (FAD)**, a derivative of the B vitamin riboflavin. The transfer of hydrogen converts these compounds to NADH and $FADH_2$, respectively.

4. One more molecule of ATP is produced for each molecule of acetyl CoA processed. Because each glucose molecule is converted into two acetic acid molecules, thus permitting two turns of the citric acid cycle, two more ATP molecules are produced from each glucose molecule.

So far, the cell still does not have much of an energy profit. However, the citric acid cycle is important in preparing the hydrogen carrier molecules for their entry into the next step, the electron transport chain, which produces far more energy than the sparse amount of ATP produced by the cycle itself.

ELECTRON TRANSPORT CHAIN

Considerable untapped energy is still stored in the released hydrogen atoms, which contain electrons at high energy levels. The "big payoff" comes when NADH and $FADH_2$ enter the **electron transport chain**, which consists of electron carrier molecules located in the inner mitochondrial membrane lining the cristae (● Figure 2-11). The high-energy electrons are extracted from the hydrogens held in NADH and $FADH_2$ and are transferred through a series of steps from one electron-carrier molecule to another, within the cristae membrane, in a kind of assembly line. As a result of giving up hydrogen and electrons within the electron transport chain, NADH and $FADH_2$ are converted back to NAD and FAD. These molecules are now free to pick up more hydrogen atoms released during glycolysis and the citric acid cycle. Thus NAD and FAD serve

as the link between the citric acid cycle and the electron transport chain.

The electron carriers are arranged in a specifically ordered fashion on the inner membrane so that the high-energy electrons are progressively transferred through a chain of reactions, with the electrons falling to successively lower energy levels with each step. Ultimately, the electrons are passed to molecular oxygen (O_2) derived from the air we breathe. Electrons bound to O_2 are in their lowest energy state. Oxygen breathed in from the atmosphere enters the mitochondria to serve as the final electron acceptor of the electron transport chain. This negatively charged oxygen (negative because it has acquired additional electrons) then combines with the positively charged hydrogen ions (positive because they have donated the electrons at the beginning of the electron transport chain) to form water (H_2O).

As the electrons move through this chain of reactions, they release energy. Part of the released energy is lost as heat, but some is harnessed by the mitochondrion to synthesize ATP. Energy released by the electrons as they move to ever lower energy levels activates through a sequence of steps the enzyme **ATP synthase**, which is located on the surface of the inner membrane. On activation, ATP synthase converts ADP + P_i to ATP, providing a rich yield of 32 more ATP molecules for each glucose molecule thus processed. ATP is subsequently transported out of the mitochondrion into the cytosol for use as the cell's energy source.

The harnessing of energy into a useful form as the electrons tumble from a high-energy state to a low-energy state can be likened to a power plant converting the energy of water tumbling down a waterfall into electricity. Because O_2 is used in these final steps of energy conversion when a phosphate is added to form ATP, this process is called **oxidative phosphorylation**. The electron transport chain is also called the *respiratory chain* because it is crucial to **cellular respiration**, a term that refers to the intracellular oxidation of nutrient derivatives.

The series of steps that lead to oxidative phosphorylation might at first seem an unnecessary complication. Why not just directly oxidize, or "burn," food molecules to release their energy? When this process is carried out outside the body, all the energy stored in the food molecule is released explosively in the form of heat (● Figure 2-12). In the body, oxidation of food molecules occurs in many small, controlled steps so that the food molecule's chemical energy is gradually made available for convenient packaging in a storage form that is useful to the cell. The cell, by means of its mitochondria, can more efficiently capture the energy from the food molecules within ATP bonds when it is released in small quantities. In this way, much less of the energy is converted to heat. The heat that is produced is not completely wasted energy; it is used to help maintain body temperature, with any excess heat being eliminated to the environment.

❚ **The cell generates more energy in aerobic than in anaerobic conditions.**

The cell is a much more efficient energy converter when oxygen is available (● Figure 2-13). In an **anaerobic** ("lack of air,"

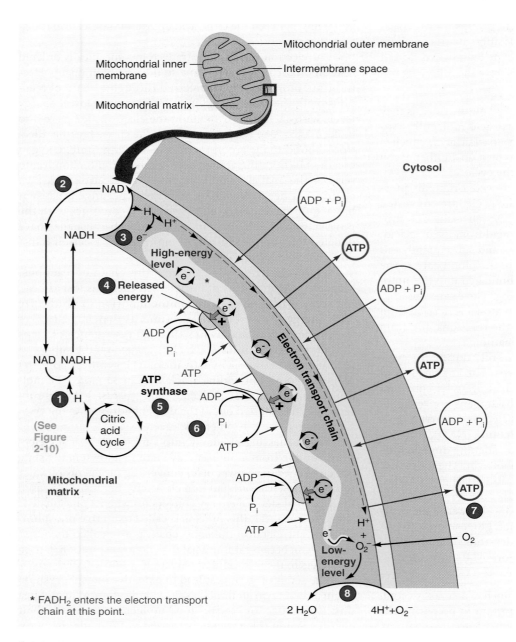

Mitochondrial outer membrane
Mitochondrial inner membrane
Intermembrane space
Mitochondrial matrix

Cytosol

NAD
NADH
H
H⁺
e⁻
High-energy level
e⁻ *
Released energy
ADP + P_i
ATP
ADP
P_i
ATP
ATP synthase
ADP
ADP + P_i
P_i
ATP
ATP
NAD NADH
H
Citric acid cycle
ADP + P_i
ATP
ATP
Electron transport chain
Mitochondrial matrix
ADP
P_i
ATP
e⁻
e⁻
e⁻
e⁻
e⁻
e⁻
e⁻
Low-energy level
H⁺ + O_2^-
e⁻
O_2
ATP

(See Figure 2-10)

* FADH₂ enters the electron transport chain at this point.

2 H_2O 4H⁺+O_2^-

1 Hydrogen (H) that is released during the degradation of carbon-containing nutrient molecules by the citric acid cycle in the mitochondrial matrix is carried to the mitochondrial inner membrane by hydrogen carriers such as NADH.

2 After releasing hydrogen at the inner membrane, NAD shuttles back to pick up more hydrogen generated by the citric acid cycle in the matrix.

3 Meanwhile, high-energy electrons extracted from the hydrogen are passed through the electron transport chain located on the mitochondrial inner membrane.

4 Energy is gradually released as the electrons fall to successively lower energy levels by moving through the electron transport chain of reactions.

5 The released energy triggers a sequence of steps that ultimately results in the activation of the enzyme ATP synthase within the mitochondrial inner membrane.

6 Activated ATP synthase brings about the formation of ATP from ADP and P_i.

7 ATP is transported out of the mitochondrion for use by the cell.

8 Molecular oxygen, after serving as the final electron acceptor, combines with the hydrogen ions (H⁺) generated from hydrogen on extraction of high-energy electrons to produce water.

● **FIGURE 2-11**

ATP synthesis by the mitochondrial inner membrane. ATP synthesis resulting from the activation of ATP synthase during passage of high-energy electrons through the mitochondrial electron-transport chain.

specifically "lack of O_2") condition, the degradation of glucose cannot proceed beyond glycolysis. Recall that glycolysis takes place in the cytosol and involves the breakdown of glucose into pyruvic acid, producing a low yield of two molecules of ATP per molecule of glucose. The untapped energy of the glucose molecule remains locked in the bonds of the pyruvic acid molecules, which are eventually converted to lactic acid if they do not enter the pathway that ultimately leads to oxidative phosphorylation.

When sufficient O_2 is present—an **aerobic** ("with air" or "with O_2") condition—mitochondrial processing (that is, the citric acid cycle in the matrix and the electron transport chain

on the inner membrane) harnesses sufficient energy to generate 34 more molecules of ATP, for a total net yield of 36 ATPs per molecule of glucose processed. (For a description of aerobic exercise, see the boxed feature, ❱ Beyond the Basics, on p. 32.) The overall reaction for the oxidation of food molecules to yield energy is as follows:

Food + O_2 → CO_2 + H_2O + ATP
(necessary for oxidative phosphorylation) (produced primarily by the citric acid cycle) (produced by the electron transport chain) (produced primarily by the electron transport chain)

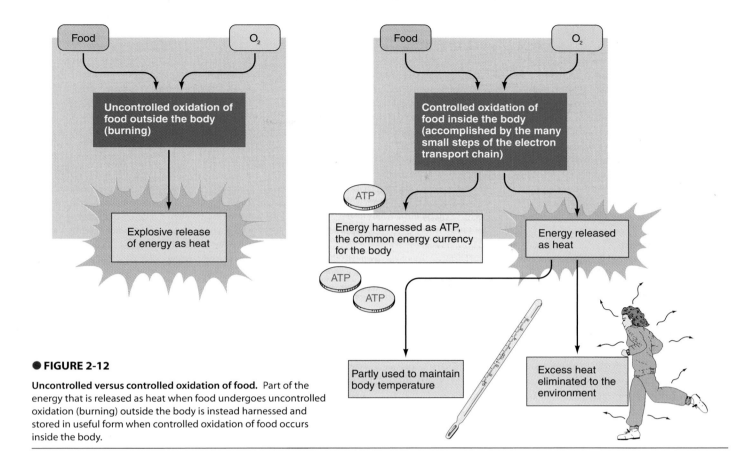

● FIGURE 2-12

Uncontrolled versus controlled oxidation of food. Part of the energy that is released as heat when food undergoes uncontrolled oxidation (burning) outside the body is instead harnessed and stored in useful form when controlled oxidation of food occurs inside the body.

Glucose, the principal nutrient derived from dietary carbohydrates, is the fuel preference of most cells. However, nutrient molecules derived from fats (fatty acids) and, if necessary, from protein (amino acids) can also participate at specific points in this overall chemical reaction to eventually produce energy. Amino acids are usually used for protein synthesis instead of energy production, but they can be used as fuel if insufficient glucose and fat are available (Chapter 17).

● FIGURE 2-13

Comparison of energy yield and products under anaerobic and aerobic conditions. In anaerobic conditions, only 2 ATPs are produced for every glucose molecule processed, but in aerobic conditions a total of 36 ATPs are produced per glucose molecule.

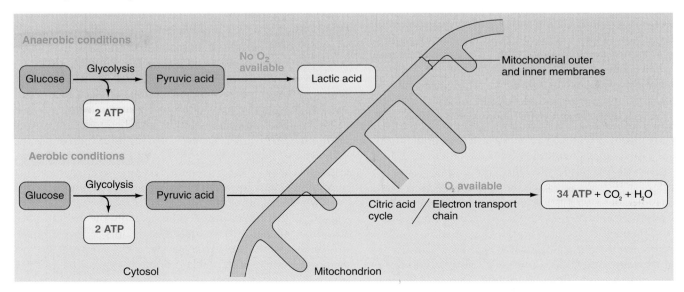

Aerobic Exercise: What For and How Much?

Aerobic ("with O$_2$") **exercise** involves large muscle groups and is performed at a low-enough intensity and for a long-enough period of time that fuel sources can be converted to ATP by using the citric acid cycle and electron transport chain as the predominant metabolic pathway. Aerobic exercise can be sustained for from 15 to 20 minutes to several hours at a time. Short-duration, high-intensity activities such as weight training and the 100-meter dash, which last for a matter of seconds and rely solely on energy stored in the muscles and on glycolysis, are forms of **anaerobic** ("without O$_2$") **exercise.**

Inactivity is associated with increased risk of developing both hypertension (high blood pressure) and coronary artery disease (block-age of the arteries that supply the heart). To reduce the risk of hypertension and coronary artery disease and to improve physical work capacity, the American College of Sports Medicine recommends that an individual participate in aerobic exercise a minimum of three times per week for 20 to 60 minutes. Recent studies have shown the same health benefits are derived whether the exercise is accomplished in one long stretch or is broken down into multiple shorter stints. This is good news, because many individuals find it easier to stick with brief bouts of exercise sprinkled throughout the day.

The intensity of the exercise should be based on a percentage of the individual's maximal capacity to work. The easiest way to estab-lish the proper intensity of exercise and to monitor intensity levels is by checking the heart rate. The estimated maximal heart rate is determined by subtracting the person's age from 220. Significant benefits can be derived from aerobic exercise performed between 70% and 80% of maximal heart rate. For example, the estimated maximal heart rate for a 20-year-old is 200 beats per minute. If this person exercised three times per week for 20 to 60 minutes at an intensity that increased the heart rate to 140 to 160 beats per minute, the participant should significantly improve his or her aerobic work capacity and reduce the risk of cardiovascular disease.

Note that the oxidative reactions within the mitochondria generate energy, unlike the oxidative reactions controlled by the peroxisome enzymes. Both organelles use O$_2$, but for different purposes.

▌ The energy stored within ATP is used for synthesis, transport, and mechanical work.

Once formed, ATP is transported out of the mitochondria and is then available as an energy source as needed within the cell. Cell activities that require energy expenditure fall into three main categories:

1. *Synthesis of new chemical compounds,* such as protein synthesis by the endoplasmic reticulum. Some cells, especially cells with a high rate of secretion and cells in the growth phase, use up to 75% of the ATP they generate just to synthesize new chemical compounds.
2. *Membrane transport,* such as the selective transport of molecules across the kidney tubules during the process of urine formation. Kidney cells can expend as much as 80% of their ATP currency to operate their selective membrane-transport mechanisms.
3. *Mechanical work,* such as contraction of the heart muscle to pump blood or contraction of skeletal muscles to lift an object. These activities require tremendous quantities of ATP.

As a result of cell energy expenditure to support these various activities, large quantities of ADP are produced. These energy-depleted ADP molecules enter the mitochondria for "recharging" and then cycle back into the cytosol as energy-rich ATP molecules after participating in oxidative phosphorylation. In this recharging–expenditure cycle, a single ADP/ATP molecule may shuttle back and forth thousands of times per day between the mitochondria and cytosol.

The high demands for ATP make glycolysis alone an insufficient as well as inefficient supplier of power for most cells. Were it not for the mitochondria, which house the metabolic machinery for oxidative phosphorylation, the body's energy capability would be very limited. However, glycolysis does provide cells with a sustenance mechanism that can produce at least some ATP under anaerobic conditions. Skeletal muscle cells in particular take advantage of this ability during short bursts of strenuous exercise, when energy demands for contractile activity outstrip the body's ability to bring adequate O$_2$ to the exercising muscles to support oxidative phosphorylation.

VAULTS AS CELLULAR TRUCKS

In addition to the five well-documented organelles, in the early 1990s researchers identified a sixth type of organelle—vaults.

▌ Vaults may serve as cellular transport vehicles.

Vaults, which are three times as large as ribosomes, are shaped like octagonal barrels (● Figure 2-14a and c). Their name comes from their multiple arches, which reminded their discoverers of vaulted or cathedral ceilings. Just like barrels, vaults have a hollow interior. Sometimes vaults are seen in an open state, appearing like pairs of unfolded flowers with each half of the vault bearing eight "petals" attached to a central ring (● Figure 2-14b). A cell may contain thousands of vaults. Why would the presence of these numerous, relatively large organelles have been elusive until recently? The reason is that they do not show up with ordinary staining techniques.

Two clues to the function of vaults may be their octagon shape and their hollow interior. Intriguingly, the nuclear pores are also octagonal and the same size as vaults, leading to speculation that vaults may be cellular "trucks." According to this proposal, vaults would dock at or enter nuclear pores, pick up molecules synthesized in the nucleus, and deliver their

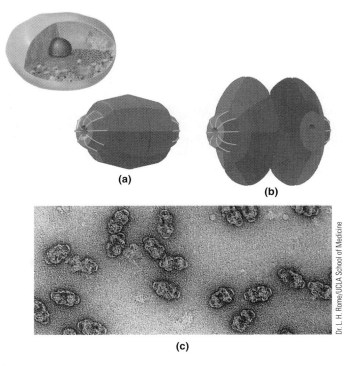

● FIGURE 2-14

Vaults. (a) Schematic 3-D representation of a vault, an octagonal barrel-shaped organelle believed to transport either messenger RNA or the ribosomal subunits from the nucleus to the cytoplasmic ribosomes. (b) Schematic representation of an opened vault, showing its hollow interior. (c) Electron micrograph of vaults.

Dr. L. H. Rome/UCLA School of Medicine

cargo elsewhere in the cell. Ongoing research supports vaults' role in nucleus-to-cytoplasm transport, but what cargo they are carrying has not been determined. One possibility is that vaults may be carrying messenger RNA from the nucleus to the ribosomal sites of protein synthesis within the cytoplasm. Another possibility is that vaults' unknown cargo may be the two subunits that make up ribosomes (see ● Figure C-6). These two subunits are produced in the nucleus, then exit through the nuclear pores by unknown means to reach their sites of action. Of interest, the interior of a vault is the right size to accommodate these ribosomal subunits.

CYTOSOL: CELL GEL

Occupying about 55% of the total cell volume, the **cytosol** is the semiliquid portion of the cytoplasm that surrounds the organelles. Its nondescript appearance under an electron microscope gives the mistaken impression that the cytosol is a liquid mixture of uniform consistency, but it is actually a highly organized, gel-like mass with differences in composition and consistency from one part of the cell to another. Furthermore, dispersed throughout the cytosol is a *cytoskeleton,* a protein scaffolding that gives shape to the cell, provides an intracellular organizational framework, and is responsible for various cell movements. For now we will concentrate on the gelatinous portion of the cytosol, then turn our attention to its cytoskeletal component in the next section.

▌ The cytosol is important in intermediary metabolism, ribosomal protein synthesis, and nutrient storage.

Three general categories of activities are associated with the gelatinous portion of the cytosol: (1) enzymatic regulation of intermediary metabolism, (2) ribosomal protein synthesis, and (3) storage of fat, carbohydrate, and secretory vesicles.

ENZYMATIC REGULATION OF INTERMEDIARY METABOLISM

The term **intermediary metabolism** refers collectively to the large set of chemical reactions inside the cell that involve the degradation, synthesis, and transformation of small organic molecules such as simple sugars, amino acids, and fatty acids. These reactions are critical for ultimately capturing energy to be used for cell activities and for providing the raw materials needed for maintaining the cell's structure and function and for the cell's growth. All intermediary metabolism occurs in the cytoplasm, with most of it being accomplished in the cytosol. The cytosol contains thousands of enzymes involved in glycolysis and other intermediary biochemical reactions.

RIBOSOMAL PROTEIN SYNTHESIS

Also dispersed throughout the cytosol are the free ribosomes, which synthesize proteins for use in the cytosol itself. In contrast, recall that the rough ER ribosomes synthesize proteins for secretion and for constructing new cell components.

STORAGE OF FAT, GLYCOGEN, AND SECRETORY VESICLES

Excess nutrients not immediately used for ATP production are converted in the cytosol into storage forms that are readily visible under a light microscope. Such nonpermanent masses of stored material are known as **inclusions.** Inclusions are not surrounded by membrane, and they may or may not be present, depending on the type of cell and the circumstances. The largest and most important storage product is fat. Small fat droplets are present within the cytosol in various cells. In **adipose tissue**, the tissue specialized for fat storage, the stored fat molecules can occupy almost the entire cytosol, where they merge to form one large fat droplet (● Figure 2-15a). The other visible storage product is **glycogen,** the storage form of glucose, which appears as aggregates or clusters dispersed throughout the cell (● Figure 2-15b). Cells vary in their ability to store glycogen, with liver and muscle cells having the greatest stores. When food is not available to provide fuel for the citric acid cycle and electron transport chain, stored glycogen and fat are broken down to release glucose and fatty acids, respectively, which can feed the mitochondrial energy-producing machinery. An average adult human has enough glycogen stored to provide sufficient energy for about a day of normal activities, and typically has enough fat stored to provide energy for two months.

Secretory vesicles that have been processed and packaged by the endoplasmic reticulum and Golgi complex also remain in the cytosol, where they are stored until signaled to empty their contents to the outside. In addition, transport and endocytotic vesicles move through the cytosol.

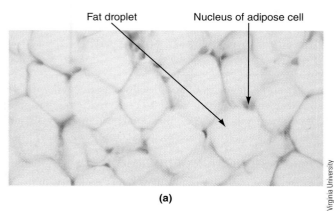

Fat droplet   Nucleus of adipose cell

(a)

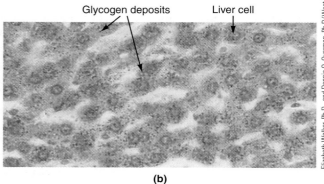

Glycogen deposits   Liver cell

(b)

Elizabeth Walker, Ph.D., and Dennis O. Overman, Ph.D./West Virginia University

● **FIGURE 2-15**

Inclusions. (a) Light micrograph showing fat storage in an adipose cell. Note that the fat droplet occupies almost the entire cytosol. (b) Light micrograph showing glycogen storage in a liver cell. The red-staining granules throughout the liver cell's cytosol are glycogen deposits.

CYTOSKELETON: CELL "BONE AND MUSCLE"

Different cells in the body have distinct shapes, structural complexities, and functional specializations. Maintenance of the unique characteristics of each cell type necessitates intracellular scaffolding to support and organize the cell components into an appropriate arrangement and to control their movements. These functions are performed by the **cytoskeleton**, the complex protein network portion of the cytosol that acts as the "bone and muscle" of the cell.

This elaborate network has three distinct elements: (1) *microtubules*, (2) *microfilaments*, and (3) *intermediate filaments*. The different parts of the cytoskeleton are structurally linked and functionally coordinated to provide certain integrated functions for the cell. Because of the complexity of this network and the variety of functions it serves, we address its elements separately. These functions, along with the functions of all other cell structures, are summarized in ▲ Table 2-1, with an emphasis on the components of the cytoplasm.

▌ Microtubules help maintain asymmetric cell shapes and play a role in complex cell movements.

The **microtubules** are the largest of the cytoskeletal elements. They are very slender, long, hollow, unbranched tubes composed primarily of **tubulin**, a small, globular, protein molecule (● Figure 2-16a). Microtubules are essential for maintaining an asymmetric cell shape, such as that of a nerve cell, whose elongated axon may extend up to a meter in length from where the cell body originates in the spinal cord to where the axon ends at a muscle (see ● Figure 4-9, p. 79). Microtubules, along with specialized intermediate filaments, stabilize this asymmetric axonal extension.

Microtubules also play an important role in coordinating numerous complex cell movements, including (1) transport of secretory vesicles or other materials from one region of the cell to another by forming microtubular "highways," (2) distribution of chromosomes during cell division through formation of a mitotic spindle, and (3) movement of specialized cell projections such as cilia and flagella.

● **FIGURE 2-16**

Components of the cytoskeleton. (a) Microtubules, the largest of the cytoskeletal elements, are long, hollow tubes formed by two slightly different variants of globular-shaped tubulin molecules. (b) Most microfilaments, the smallest of the cytoskeletal elements, consist of two chains of actin molecules wrapped around each other. (c) The intermediate filament keratin found in skin is made up of three polypeptide strands wound around each other. The composition of intermediate filaments, which are intermediate in size between the microtubules and microfilaments, varies among different cell types.

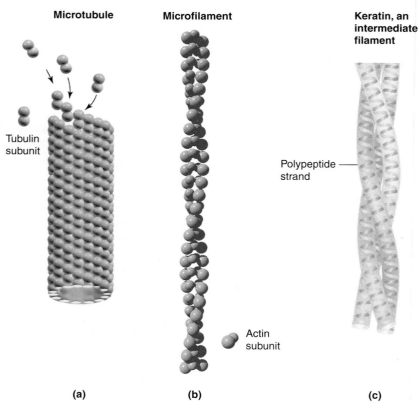

Microtubule   Microfilament   Keratin, an intermediate filament

Tubulin subunit

Polypeptide strand

Actin subunit

(a)   **(b)**   **(c)**

TABLE 2-1

Summary of Cell Structures and Functions

CELL PART	NUMBER PER CELL	STRUCTURE	FUNCTION
Plasma Membrane	1	Lipid bilayer studded with proteins and small amounts of carbohydrate	Acts as selective barrier between cell contents and extracellular fluid; controls traffic in and out of the cell
Nucleus	1	DNA and specialized proteins enclosed by a double-layered membrane	Acts as control center of the cell, providing storage of genetic information; nuclear DNA provides codes for the synthesis of structural and enzymatic proteins and serves as blueprint for cell replication
Cytoplasm			
Organelles			
Endoplasmic reticulum	1	Extensive, continuous membranous network of fluid-filled tubules and flat sacs, partially studded with ribosomes	Forms new cell membrane and other cell components and manufactures products for secretion
Golgi complex	1 to several hundred	Sets of stacked, flat membranous sacs	Modifies, packages, and distributes newly synthesized proteins
Lysosomes	300 average	Membranous sacs containing hydrolytic enzymes	Serve as digestive system of the cell, destroying foreign substances and cellular debris
Peroxisomes	200 average	Membranous sacs containing oxidative enzymes	Perform detoxification activities
Mitochondria	100–2,000	Rod- or oval-shaped bodies enclosed by two membranes, with the inner membrane folded into cristae that project into an interior matrix	Act as energy organelles; major site of ATP production; contain enzymes for citric acid cycle and electron transport chain
Vaults	Thousands	Shaped like hollow octagonal barrels	Serve as cellular trucks for transport from nucleus to cytoplasm
Cytosol: gel–like portion			
Intermediary metabolism enzymes	Many	Dispersed within the cytosol	Facilitate intracellular reactions involving the degradation, synthesis, and transformation of small organic molecules
Ribosomes	Many	Granules of RNA and proteins—some attached to rough endoplasmic reticulum, some free in the cytoplasm	Serve as workbenches for protein synthesis
Inclusions	Varies	Glycogen granules, fat droplets	Store excess nutrients
Transport, secretory, and endocytotic vesicles	Varies	Transiently formed membrane-enclosed products synthesized within or engulfed by the cell	Transport and/or store products being moved within, out of, or into the cell, respectively
Cytosol: cytoskeleton portion			As an integrated whole, serves as the cell's "bone and muscle"
Microtubules	Many	Long, slender, hollow tubes composed of tubulin molecules	Maintain asymmetric cell shapes and coordinate complex cell movements
Microfilaments	Many	Intertwined helical chains of actin molecules; microfilaments composed of myosin molecules also present in muscle cells	Play a vital role in various cell contractile systems; serve as a mechanical stiffener for microvilli
Intermediate filaments	Many	Irregular, threadlike proteins	Help resist mechanical stress

Cilia (meaning "eyelashes") are numerous tiny, hairlike protrusions, whereas a **flagellum** (meaning "whip") is a single, long, whiplike appendage. Even though they project from the surface of the cell, cilia and a flagellum are both intracellular structures—they are covered by the plasma membrane. Cilia beat or stroke in the same direction, much like the coordinated efforts of a rowing team. In humans, ciliated cells are found in the stationary cells that line the respiratory tract, the oviduct of the female reproductive tract, and the fluid-filled ventricles (chambers) of the brain. The coordinated stroking of the thousands of respiratory cilia help keep foreign particles out of the lungs by sweeping outward dust and other inspired (breathed-in) particles (● Figure 2-17). In the female reproductive tract, the sweeping action of the cilia that line the oviduct draws the egg (ovum) released from the ovary during ovulation into the oviduct and then guides it toward the uterus (womb). In the brain, the ciliated cells that line the ventricles produce cerebrospinal fluid, which flows through the ventricles and around the brain and spinal cord, cushioning and bathing these fragile neural structures. Beating of the cilia helps promote flow of this supportive fluid.

The only human cells that have a flagellum are sperm (see ● Figure 18-6, p. 593). The whiplike motion of the flagellum or "tail" enables a sperm to move through its environment, which is crucial for maneuvering into position to fertilize the female ovum.

❚ Microfilaments are important to cell contractile systems and as mechanical stiffeners.

The **microfilaments** are the smallest proven elements of the cytoskeleton. The most obvious microfilaments in most cells are those composed of **actin**, a protein molecule that has a globular shape similar to tubulin. Unlike tubulin, which forms a hollow tube, actin is assembled into two strands twisted around each other to form a microfilament (● Figure 2-16b). In muscle cells, another protein called **myosin** forms a different kind of microfilament. In most cells, myosin is not as abundant and does not form such distinct filaments.

Microfilaments serve two functions: (1) They play a vital role in various cell contractile systems, and (2) they act as mechanical stiffeners for several specific cell projections.

MICROFILAMENTS IN CELL CONTRACTILE SYSTEMS

The most obvious, best organized, and most clearly understood cell contractile system is that found in muscle. Muscle contains an abundance of actin and myosin microfilaments, which accomplish muscle contraction by sliding past each other, using ATP as an energy source. This ATP-powered microfilament sliding and force development is triggered by a complex sequence of electrical, biochemical, and mechanical events initiated when the muscle cell is stimulated to contract (see Chapter 8 for details).

Nonmuscle cells may also contain "musclelike" assemblies. For example, white blood cells are mobile because of **amoeboid movement**, a cell-crawling process that depends on the activity of their actin filaments, in a mechanism similar to that used by amoebas to maneuver through their environment. When crawling, the motile cell forms the fingerlike extensions known as *pseudopods* at the "front" or leading edge of the cell in the direction of the target (● Figure 2-18). For example, the target that triggers amoeboid movement might be the proximity of food in the case of an amoeba or a bacterium in the case of a white blood cell. Pseudopods are formed as a result of the organized assembly and disassembly of actin networks. During amoeboid movement, actin filaments continuously grow at the cell's leading edge through the addition of actin molecules at the front of the actin chain. This filament growth pushes that portion of the cell forward as a pseudopod extension. Simultaneously, actin molecules at the rear of the filament are being disassembled and transferred to the front of the line. Thus the filament does not get any longer; it stays the same length but moves forward through the continuous transfer of actin molecules from the rear to the front of the filament in what is termed *treadmilling* fashion. The cell progressively moves forward through repeated cycles of pseudopod formation at the leading edge.

● **FIGURE 2-17**

Cilia in the respiratory tract. Scanning electron micrograph of cilia on cells lining the human respiratory tract. The respiratory airways are lined by goblet cells, which secrete a sticky mucus that traps inspired particles, and epithelial cells that bear numerous hairlike cilia. The cilia all beat in the same direction to sweep inspired particles up and out of the airways.

Cilia   Goblet cell

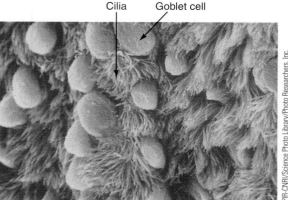

PIR-CNRI/Science Photo Library/Photo Researchers, Inc.

● **FIGURE 2-18**

An amoeba undergoing amoeboid movement

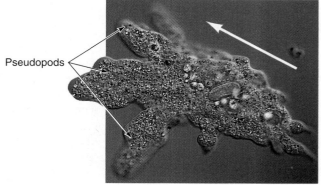

Pseudopods

M. Abbey/Visuals Unlimited

White blood cells are the most active crawlers in the body. These cells leave the circulatory system and travel by amoeboid movement to areas of infection or inflammation, where they engulf and destroy micro-organisms and cell debris. Amazingly, it is estimated that the total distance traveled collectively per day by all your white blood cells while they roam the tissues in their search-and-destroy tactic would circle Earth twice!

MICROFILAMENTS AS MECHANICAL STIFFENERS

Besides their role in cell contractile systems, the actin filaments' second major function is to serve as mechanical supports or stiffeners for several cell extensions, of which the most common are microvilli. **Microvilli** are microscopic, nonmotile, hairlike projections from the surface of epithelial cells lining the small intestine and kidney tubules (● Figure 2-19). Their presence greatly increases the surface area available for transferring material across the plasma membrane. In the case of the small intestine, the microvilli increase the area available for absorbing digested nutrients. In the kidney tubules, microvilli enlarge the absorptive surface that salvages useful substances passing through the kidney so that these materials are saved for the body instead of being eliminated in the urine. Within each microvillus, a core consisting of parallel actin filaments linked together forms a rigid mechanical stiffener that keeps these valuable surface projections intact.

▌Intermediate filaments are important in cell regions subject to mechanical stress.

The **intermediate filaments** are intermediate in size between the microtubules and the microfilaments—hence their name. The proteins that compose the intermediate filaments vary between cell types, but in general they appear as irregular, threadlike molecules. These proteins form tough, durable fibers that play a central role in maintaining the structural integrity of a cell and in resisting mechanical stresses externally applied to a cell.

Different types of intermediate filaments are tailored to suit their structural or tension-bearing role in specific cell types. For example, skin cells contain irregular networks of intermediate filaments made of the protein **keratin** (see ● Figure 2-16c). These intracellular filaments interconnect with extracellular filaments that tie adjacent cells together, thereby creating a continuous filamentous network that extends throughout the skin and gives it strength. When the surface skin cells die, their tough keratin skeletons persist to form a protective, waterproof outer layer. Hair and nails are also keratin structures.

▌The cytoskeleton functions as an integrated whole and links other parts of the cell together.

With high-voltage electron microscopy, which provides a three-dimensional view of the internal organization of the cell, a meshwork of exceedingly fine, interlinked filaments called the **microtrabecular lattice** can be seen extending throughout the cytoplasm and connecting to the inner layer of the plasma membrane. Some cell biologists believe this meshwork is an artifact that arises during preparation of the specimen, but others think this lattice constitutes intricate interconnections between the cytoskeletal structures as well as various organelles (● Figure 2-20). Collectively, the cyto-

● **FIGURE 2-19**

Microvilli in the small intestine. Scanning electron micrograph showing microvilli on the surface of a small-intestine epithelial cell.
(*Source:* © Dr. R. G. Kessel and Dr. R. H. Kardon: *Tissues and Organs: A Text-Atlas of Scanning Electron Microscopy.* W.H. Freeman, 1979, all rights reserved.)

Microvilli

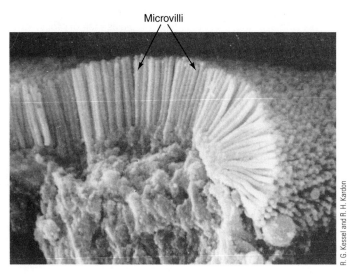

R. G. Kessel and R. H. Kardon

● **FIGURE 2-20**

Interconnections between cytoskeletal structures and organelles

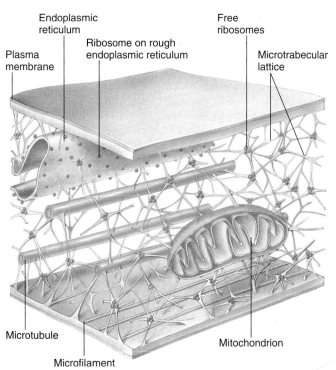

skeletal elements and their interconnections support the plasma membrane and are responsible for the particular shape, rigidity, and spatial geometry of each different cell type. This internal framework thus acts as the cell's "skeleton."

Furthermore, as you have learned, the coordinated action of the cytoskeletal elements is responsible for directing intracellular transport and for regulating numerous cell movements and thereby also serves as the cell's "muscle."

 PhysioEdge Click on the Media Exercises menu of the CD-ROM, and work Media Exercise 2.1: Anatomy of a Generic Cell and Media Exercise 2.2: Basic Functions of Organelles to test your understanding of the previous sections.

CHAPTER IN PERSPECTIVE: FOCUS ON HOMEOSTASIS

The ability of cells to perform functions essential for their own survival as well as specialized tasks that help maintain homeostasis within the body ultimately depends on the successful, cooperative operation of the intracellular components. For example, to support life-sustaining activities, all cells must generate energy, in a usable form, from nutrient molecules. Energy is generated intracellularly by chemical reactions that take place within the cytosol and mitochondria.

In addition to being essential for basic cell survival, the organelles and cytoskeleton also participate in many cells' specialized tasks that contribute to homeostasis. Here are several examples:

• Nerve and endocrine cells both release protein chemical messengers that are important in regulatory activities aimed at maintaining homeostasis—for example, chemical messengers released from nerve cells stimulate the respiratory muscles, which accomplish life-sustaining exchanges of O_2 and CO_2 between the body and atmosphere through breathing. These protein chemical messengers (neurotransmitters in nerve cells and hormones in endocrine cells) are all produced by the endoplasmic reticulum and Golgi complex and released by exocytosis from the cell when needed.

• The ability of muscle cells to contract depends on their highly developed cytoskeletal microfilaments sliding past each other. Muscle contraction is responsible for many homeostatic activities, including (1) contracting the heart muscle, which pumps life-supporting blood throughout the body; (2) contracting the muscles attached to bones, which enables the body to procure food; and (3) contracting the muscle in the walls of the stomach and intestine, which moves the food along the digestive tract so that ingested nutrients can be progressively broken down into a form that can be absorbed into the blood for delivery to the cells.

• White blood cells help the body resist infection by making extensive use of lysosomal destruction of engulfed particles as they police the body for microbial invaders. These white blood cells are able to roam the body by means of amoeboid movement, a cell-crawling process accomplished by coordinated assembly and disassembly of actin, one of their cytoskeletal components.

As we begin to examine the various organs and systems, keep in mind that proper cell functioning is the foundation of all organ activities.

CHAPTER SUMMARY

❚ The complex organization and interaction of the chemicals within a cell confer the unique characteristics of life.
❚ Cells are the living building blocks of the body.

Observations of Cells (pp. 19–20)
❚ Cells are too small for the unaided eye to see.
❚ Using early microscopes, investigators learned that all plant and animal tissues consist of individual cells.
❚ Through more sophisticated techniques, scientists now know that a cell is a complex, highly organized, compartmentalized structure.

An Overview of Cell Structure (pp. 20–21)
❚ Cells have three major subdivisions: the plasma membrane, the nucleus, and the cytoplasm. (Review Figure 2-1 and Table 2-1, p. 35.)
❚ The plasma membrane encloses the cell and separates the intracellular and extracellular fluid.
❚ The nucleus contains the deoxyribonucleic acid (DNA), the cell's genetic material.
❚ Three types of RNA play a role in the protein synthesis coded by DNA—messenger RNA, ribosomal RNA, and transfer RNA.
❚ The cytoplasm consists of cytosol, a complex gel-like mass laced with a cytoskeleton, and organelles, which are highly organized, membrane-enclosed structures dispersed within the cytosol.

❚ The six types of organelles are the endoplasmic reticulum, Golgi complex, lysosomes, peroxisomes, mitochondria, and vaults. (Review Figure 2-1.)

Endoplasmic Reticulum and Segregated Synthesis (pp. 21–22)
❚ The endoplasmic reticulum (ER) is a single, complex membranous network that encloses a fluid-filled lumen.
❚ The primary function of the ER is to serve as a factory for synthesizing proteins and lipids to be used for (1) secreting special products such as enzymes and hormones to the exterior of the cell and (2) producing new cell components, particularly cellular membranes.
❚ The two types of endoplasmic reticulum are rough endoplasmic reticulum, which is studded with ribosomes, and smooth endoplasmic reticulum, which lacks ribosomes. (Review Figure 2-2.)
❚ The rough endoplasmic reticular ribosomes synthesize proteins, which are released into the ER lumen so that they are separated from the cytosol. Also entering the lumen are lipids produced within the membranous walls of the ER.
❚ Synthesized products move from the rough ER to the smooth ER, where they are packaged and discharged as transport vesicles.
❚ Transport vesicles are formed as a portion of the smooth ER "buds off," containing a collection of newly synthesized proteins and lipids wrapped in smooth ER membrane. (Review Figure 2-3.)

Golgi Complex and Exocytosis (pp. 22–24)

- Transport vesicles move to and fuse with the Golgi complex, which consists of stacks of flattened, membrane-enclosed sacs. (*Review Figure 2-4.*)
- The Golgi complex serves a twofold function: (1) to act as a refining plant for modifying into a finished product the newly synthesized molecules delivered to it in crude form from the endoplasmic reticular factory; and (2) to sort, package, and direct molecular traffic to appropriate intracellular and extracellular destinations.
- Before budding off of the Golgi complex, a vesicle takes up a specific product that has been processed within the Golgi sacs. The membrane that wraps the vesicle contains docking markers, which ensure that the vesicle docks and unloads its captured cargo only at the appropriate destination within the cell.
- The Golgi complex of secretory cells packages proteins to be exported out of the cell in secretory vesicles that are released by exocytosis on appropriate stimulation. (*Review Figures 2-3 and 2-5.*)

Lysosomes and Endocytosis (pp. 24–26)

- Lysosomes are membrane-enclosed sacs that contain powerful hydrolytic (digestive) enzymes. (*Review Figure 2-6.*)
- Serving as the intracellular digestive system, lysosomes destroy foreign material such as bacteria that has been internalized by the cell and demolish worn-out cell parts to make way for new replacement parts.
- Extracellular material is brought into the cell by endocytosis for attack by lysosomal enzymes.
- The three forms of endocytosis are pinocytosis, receptor-mediated endocytosis, and phagocytosis. (*Review Figure 2-7.*)

Peroxisomes and Detoxification (p. 26)

- Peroxisomes are small membrane-enclosed sacs containing powerful oxidative enzymes. (*Review Figure 2-6.*)
- They are specialized for carrying out particular oxidative reactions, including detoxifying various wastes and toxic foreign compounds that have entered the cell.

Mitochondria and ATP Production (pp. 26–32)

- The rod-shaped mitochondria are the energy organelles of the cell. They house the enzymes of the citric acid cycle (in the mitochondrial matrix) and electron transport chain (on the cristae of the mitochondrial inner membrane). (*Review Figure 2-8.*) Together, these two biochemical steps efficiently convert the energy in food molecules to the usable energy stored in ATP molecules. (*Review Figures 2-10 and 2-11.*)
- During this process, which is known as *oxidative phosphorylation*, the mitochondria use molecular O_2 and produce CO_2 and H_2O as by-products.
- A cell is more efficient at converting food energy into ATP when O_2 is available. In an anaerobic (without O_2) condition, a cell can produce only 2 molecules of ATP for every glucose molecule processed by means of glycolysis, which takes place in the cytosol. (*Review Figure 2-9.*) In an aerobic (with O_2) condition, the mitochondria further degrade the products of glycolysis to yield another 34 molecules of ATP for every glucose molecule processed. (*Review Figure 2-13.*)
- Cells use ATP as an energy source for synthesis of new chemical compounds, for membrane transport, and for mechanical work.

Vaults as Cellular Trucks (pp. 32–33)

- Vaults are recently discovered structures shaped like hollow octagonal barrels. (*Review Figure 2-14.*)
- They are the same shape and size as the nuclear pores. Researchers speculate that vaults are cellular trucks that dock at the nuclear pores and pick up cargo for transport from the nucleus.
- The leading proposals are that vaults may transport messenger RNA or the ribosomal subunits from the nucleus to the cytoplasmic sites of protein synthesis.

Cytosol: Cell Gel (p. 33)

- The cytosol contains the enzymes involved in intermediary metabolism and the ribosomal machinery essential for synthesis of these enzymes as well as other cytosolic proteins.
- Many cells store unused nutrients within the cytosol in the form of glycogen granules or fat droplets. (*Review Figure 2-15.*)
- Also present in the cytosol are various secretory, transport, and endocytotic vesicles.

Cytoskeleton: Cell "Bone and Muscle" (pp. 34–38)

- Extending throughout the cytosol is the cytoskeleton, which serves as the "bone and muscle" of the cell.
- The three types of cytoskeletal elements—microtubules, microfilaments, and intermediate filaments—each consist of different proteins and perform various roles. (*Review Figure 2-16.*)
- Collectively, the cytoskeletal elements give the cell shape and support, enable it to organize and move its internal structures as needed, and, in some cells, allow movement between the cell and its environment.

REVIEW EXERCISES

Objective Questions (Answers on p. A-40)

1. The barrier that separates and controls movement between the cell contents and the extracellular fluid is the _____.
2. The chemical that directs protein synthesis and serves as a genetic blueprint is _____, which is found in the _____ of the cell.
3. The cytoplasm consists of _____, which are specialized, membrane-enclosed intracellular compartments, and a gel-like mass known as _____, which contains an elaborate protein network called the _____.
4. Transport vesicles from the _____ fuse with and enter the _____ for modification and sorting.
5. The (*what kind of*) _____ enzymes within the peroxisomes primarily detoxify various wastes produced within the cell or foreign compounds that have entered the cell.
6. The universal energy carrier of the body is _____.
7. Amoeboid movement is accomplished by the coordinated assembly and disassembly of microtubules. (*True or false?*)
8. Using the following answer code, indicate which type of ribosome is being described:
   (a) free ribosome
   (b) rough ER-bound ribosome
   _____ 1. synthesizes proteins used to construct new cell membrane
   _____ 2. synthesizes proteins used intracellularly within the cytosol
   _____ 3. synthesizes secretory proteins such as enzymes or hormones

9. Using the following answer code, indicate which form of energy production is being described:
    (a) glycolysis
    (b) citric acid cycle
    (c) electron transport chain
    ____ 1. takes place in the mitochondrial matrix
    ____ 2. produces H_2O as a by-product
    ____ 3. rich yield of ATP
    ____ 4. takes place in the cytosol
    ____ 5. processes acetyl CoA
    ____ 6. located in the mitochondrial inner-membrane cristae
    ____ 7. converts glucose into two pyruvic acid molecules
    ____ 8. uses molecular oxygen

Essay Questions
1. What are a cell's three major subdivisions?
2. State an advantage of organelle compartmentalization.
3. List the six types of organelles.
4. Describe the structure of the endoplasmic reticulum, distinguishing between the rough and the smooth ER. What is the function of each?
5. Compare exocytosis and endocytosis. Define *secretion, pinocytosis, receptor-mediated endocytosis,* and *phagocytosis.*
7. Which organelles serve as the intracellular digestive system? What type of enzymes do they contain? What functions do these organelles serve?
8. Compare lysosomes with peroxisomes.
9. Describe the structure of mitochondria, and explain their role in oxidative phosphorylation.
10. Distinguish between the oxidative enzymes found in peroxisomes and those found in mitochondria.
11. Cells expend energy on what three categories of activities?
12. List and describe the functions of each of the components of the cytoskeleton.

POINTS TO PONDER

(Explanations on p. A-40)

1. Let's consider how much ATP you synthesize in a day. Assume that you consume 1 mole of O_2 per hour or 24 moles/day (a mole is the number of grams of a chemical equal to its molecular weight). About 6 moles of ATP are produced per mole of O_2 consumed. The molecular weight of ATP is 507. How many grams of ATP do you produce per day at this rate? Given that 1000 g equal 2.2 pounds, how many pounds of ATP do you produce per day at this rate? (This is under relatively inactive conditions!)

2. The stomach has two types of exocrine secretory cells: *chief cells* that secrete an inactive form of a protein-digesting enzyme, *pepsinogen,* and *parietal cells* that secrete *hydrochloric acid (HCl),* which activates pepsinogen. Both these cell types have an abundance of mitochondria for ATP production—the chief cells for the energy needed to synthesize pepsinogen, the parietal cells for the energy needed to transport H^+ and Cl^- from the blood into the stomach lumen. Only one of these cell types also has an extensive rough endoplasmic reticulum and abundant Golgi stacks. Would this type be the chief cells or the parietal cells? Why?

3. The poison *cyanide* acts by binding irreversibly to one component of the electron transport chain, blocking its action. As a result, the entire electron-transport process comes to a screeching halt, and the cells lose over 94% of their ATP-producing capacity. Considering the types of cell activities that depend on energy expenditure, what would be the consequences of cyanide poisoning?

4. Why do you think a person can only briefly perform anaerobic exercise (such as lifting and holding a heavy weight) but can sustain aerobic exercise (such as walking or swimming) for long periods? (*Hint:* Muscles have limited energy stores.)

5. One type of the affliction *epidermolysis bullosa* is caused by a genetic defect that results in production of abnormally weak keratin. Based on your knowledge of the role of keratin, what part of the body do you think would be affected by this condition?

CLINICAL CONSIDERATION

(Explanation on p. A-41)

Kevin S. and his wife have been trying to have a baby for the past three years. On seeking the help of a fertility specialist, Kevin learned that he has a hereditary form of male sterility involving nonmotile sperm. His condition can be traced to defects in the cytoskeletal components of the sperm's flagella. Based on this finding, the physician suspected that Kevin also has a long history of recurrent respiratory tract disease. Kevin confirmed that indeed he has had colds, bronchitis, and influenza more frequently than his friends. Why would the physician suspect that Kevin probably had a history of frequent respiratory disease based on his diagnosis of sterility due to nonmotile sperm?

PHYSIOEDGE RESOURCES

 PhysioEdge CD-ROM

PhysioEdge, the CD-ROM packaged with your text, focuses on the concepts students find most difficult to learn. Figures marked with this icon have associated activities on the CD. For a visual review of concepts in this chapter, check out the following:

Tutorial: Transport across Membranes

Media Exercise 2.1: Anatomy of a Generic Cell

Media Exercise 2.2: Basic Functions of Organelles

PhysioEdge Website

The website for this book contains a wealth of helpful study aids, as well as many ideas for further reading and research. Log on to: **http://www.brookscole.com/hpfundamentals3** Select Chapter 2 from the drop-down menu or click on one of the many resource areas.

For Suggested Readings, consult **InfoTrac College Edition/ Research** on the PhysioEdge website or go directly to InfoTrac College Edition, your online research library, at: **http://infotrac.thomsonlearning.com**

Body Systems

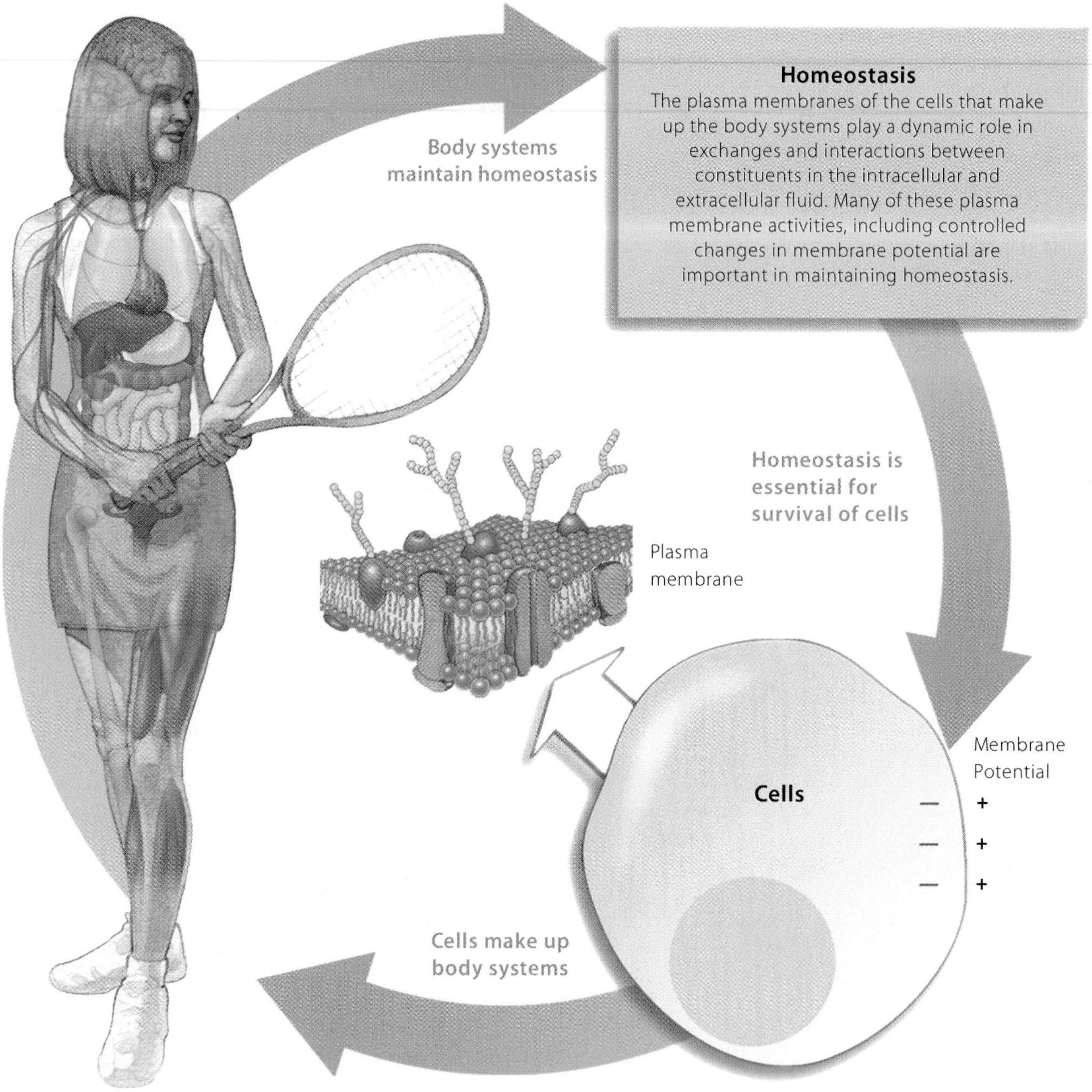

Homeostasis
The plasma membranes of the cells that make up the body systems play a dynamic role in exchanges and interactions between constituents in the intracellular and extracellular fluid. Many of these plasma membrane activities, including controlled changes in membrane potential are important in maintaining homeostasis.

Body systems maintain homeostasis

Homeostasis is essential for survival of cells

Plasma membrane

Cells

Membrane Potential

−   +
−   +
−   +

Cells make up body systems

All cells are enveloped by a **plasma membrane,** a thin, flexible, lipid barrier that separates the contents of the cell from its surroundings. To carry on life-sustaining and specialized activities, each cell must exchange materials across this membrane with the homeostatically maintained internal fluid environment that surrounds it. This discriminating barrier contains specific proteins, some of which enable selective passage of materials. Other membrane proteins serve as receptor sites for interaction with specific chemical messengers in the cell's environment. These messengers control many cell activities crucial to homeostasis.

Cells have a membrane potential, a slight excess of negative charges lined up along the inside of the membrane and a slight excess of positive charges on the outside. The specialization of nerve and muscle cells depends on the ability of these cells to alter their potential on appropriate stimulation.

The Plasma Membrane and Membrane Potential

PhysioEdge Click on the Tutorials menu of the CD-ROM for a tutorial on Membrane Potential.

The survival of every cell depends on the maintenance of intracellular contents unique for that cell type despite the remarkably different composition of the extracellular fluid surrounding it. This difference in fluid composition inside and outside a cell is maintained by the **plasma membrane,** an extremely thin layer of lipids and proteins that forms the outer boundary of every cell and encloses the intracellular contents. In addition to serving as a mechanical barrier that traps needed molecules within the cell, the plasma membrane plays an active role in determining the composition of the cell by selectively permitting specific substances to pass between the cell and its environment. Besides controlling the entry of nutrient molecules and the exit of secretory and waste products, the plasma membrane maintains differences in ion concentrations between the cell's interior and exterior. These ionic differences, as you will learn, are important in the electrical activity of the plasma membrane. Also, the plasma membrane participates in the joining of cells together to form tissues and organs. Furthermore, it plays a key role in the ability of a cell to respond to changes, or signals, in the cell's environment. This ability is important in communication between cells. No matter what the cell type, these common membrane functions are crucial to the cell's survival, to its ability to perform specialized homeostatic activities, and to its ability to function cooperatively and in a coordinated fashion with other cells. Many of the functional differences between cell types are due to subtle variations in the composition of their plasma membranes, which in turn enable different cells to interact in different ways with essentially the same extracellular fluid environment.

MEMBRANE STRUCTURE AND COMPOSITION

The plasma membrane is too thin to be seen under an ordinary light microscope, but with an electron microscope it appears as a **trilaminar structure** consisting of two dark layers separated by a light middle

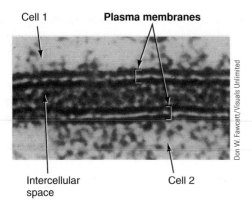

Cell 1

Plasma membranes

Intercellular space

Cell 2

Don W. Fawcett/Visuals Unlimited

● **FIGURE 3-1**

Trilaminar appearance of a plasma membrane in an electron micrograph. Depicted are the plasma membranes of two adjacent cells. Note the trilaminar structure (that is, two dark layers separated by a light middle layer) of each membrane.

layer (● Figure 3-1) (*tri* means "three;" *lamina* means "layer"). The specific arrangement of the molecules that make up the plasma membrane is responsible for this three-layered "sandwich" appearance.

■ The plasma membrane is a fluid lipid bilayer embedded with proteins.

The plasma membrane of every cell consists mostly of lipids (fats) and proteins plus small amounts of carbohydrate. The most abundant membrane lipids are phospholipids, with lesser amounts of cholesterol. An estimated billion phospholipid molecules are present in the plasma membrane of a typical human cell. **Phospholipids** have a polar (electrically charged; see p. A-13) head containing a negatively charged phosphate group and two nonpolar (electrically neutral) fatty acid tails (● Figure 3-2a). The polar end is hydrophilic ("water loving") because it can interact with water molecules, which are also polar; the nonpolar end is hydrophobic ("water fearing") and will not mix with water. Such two-sided molecules self-assemble into a **lipid bilayer**, a double layer of lipid molecules, when in contact with water (● Figure 3-2b) (*bi* means "two"). The hydrophobic tails bury themselves in the center away from the water, whereas the hydrophilic heads line up on both sides in contact with the water. The outer surface of the layer is exposed to extracellular fluid (ECF), whereas the inner surface is in contact with the intracellular fluid (ICF) (● Figure 3-2c).

Choline

Phosphate

Glycerol

Head (polar, hydrophilic)

Fatty acid

Tails (nonpolar, hydrophobic)

(a)

Polar heads (hydrophilic)

ECF (water)

Nonpolar tails (hydrophobic)

Lipid bilayer

Polar heads (hydrophilic)

ICF (water)

(b)

⊖ = Negative charge on phosphate group

● **FIGURE 3-2**

Structure and organization of phospholipid molecules in a lipid bilayer. (a) Phospholipid molecule. (b) When in contact with water, phospholipid molecules organize themselves into a lipid bilayer with the polar heads interacting with the polar water molecules at each surface and the nonpolar tails all facing the interior of the bilayer. (c) An exaggerated view of the plasma membrane enclosing a cell, separating the ICF from the ECF. (*Source:* Part c adapted from Cecie Starr and Ralph Taggart, *Biology: The Unity and Diversity of Life,* Eighth Edition, Fig. 4-26, p. 56. ©1998 Wadsworth Publishing Company.)

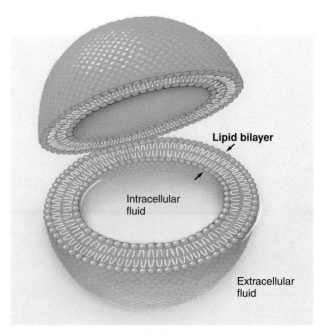

Lipid bilayer

Intracellular fluid

Extracellular fluid

(c)

The lipid bilayer is not a rigid structure but instead is fluid in nature, with a consistency more like liquid cooking oil than like solid shortening. The phospholipids, which are not held together by strong chemical bonds, can twirl around rapidly as well as move about within their own half of the layer, accounting in large part for membrane fluidity.

Cholesterol also contributes to the fluidity as well as the stability of the membrane. The cholesterol molecules are tucked in between the phospholipid molecules, where they prevent the fatty acid chains from packing together and crystallizing, a process that would drastically reduce membrane fluidity.

Because of its fluid nature, the plasma membrane has structural integrity but at the same time is flexible, enabling the cell to change shape. For example, muscle cells change shape as they contract, and red blood cells must change shape considerably as they squeeze their way single file through the capillaries, the tiniest of blood vessels.

The **membrane proteins** are attached to or inserted within the lipid bilayer (● Figure 3-3). Some of these proteins extend through the entire thickness of the membrane. Other proteins stud only the outer or inner surface. The fluidity of the lipid bilayer enables many membrane proteins to float freely like "icebergs" in a moving "sea" of lipid. This view of membrane structure is known as the **fluid mosaic model**, in reference to the membrane fluidity and the ever-changing mosaic pattern of the proteins embedded within the lipid bilayer. (A mosaic is a surface decoration made by inlaying small pieces of variously colored tiles to form patterns or pictures.)

The small amount of **membrane carbohydrate** is located only at the outer surface. Thus your cells are "sugar coated." Short-chain carbohydrates protrude like tiny antennas from the outer surface, bound primarily to membrane proteins and, to a lesser extent, to lipids (● Figure 3-3).

This proposed structure can account for the trilaminar appearance of the plasma membrane. When stains are used to help visualize the plasma membrane under an electron microscope, the two dark lines represent the hydrophilic polar regions of the lipid and protein molecules that have taken up the stain. The light space between corresponds to the poorly stained hydrophobic core formed by the nonpolar regions of these molecules.

● FIGURE 3-3

Fluid mosaic model of plasma membrane structure. The plasma membrane is composed of a lipid bilayer embedded with proteins. Some of these proteins extend through the thickness of the membrane, some are partially submerged in the membrane, and others are loosely attached to the surface of the membrane. Short carbohydrate chains are attached to proteins or lipids on the outer surface only.

 For an interaction related to this figure, see Media Exercise 3.1: The Plasma Membrane and Cell–Cell Connections on the CD-ROM.

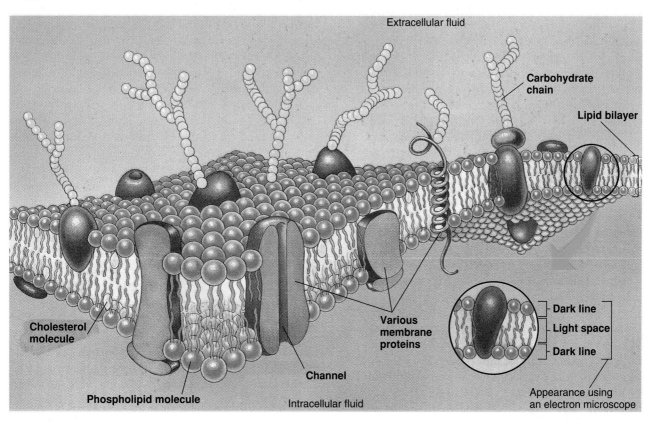

The different components of the plasma membrane carry out a variety of functions. As a quick preview, the lipid bilayer forms the primary barrier to diffusion, the proteins perform most of the specific membrane functions, and the carbohydrates play an important role in "self"-recognition processes and cell-to-cell interactions. We will now examine the functions of each of these membrane components in more detail.

▌The lipid bilayer forms the basic structural barrier that encloses the cell.

The lipid bilayer serves three important functions:

1. It forms the basic structure of the membrane. The phospholipids can be visualized as the "pickets" that form the "fence" around the cell.
2. Its hydrophobic interior serves as a barrier to passage of water-soluble substances between the ICF and ECF. Water-soluble substances cannot dissolve in and pass through the lipid bilayer. By means of this barrier, the cell can maintain different mixtures and concentrations of solutes inside and outside the cell.
3. It is responsible for the fluidity of the membrane.

▌The membrane proteins perform a variety of specific membrane functions.

Different types of membrane proteins serve the following specialized functions:

1. Some proteins span the membrane to form water-filled pathways, or **channels**, across the lipid bilayer. Their presence enables water-soluble substances that are small enough to enter a channel to pass through the membrane without coming into direct contact with the hydrophobic lipid interior (● Figure 3-3). Channels are highly selective. Their small diameter precludes passage of particles greater than 0.8 nm (40 billionths of an inch) in diameter. Only small ions can fit through channels. Furthermore, a given channel can selectively attract or repel particular ions. For example, only sodium (Na^+) can pass through Na^+ channels, whereas only potassium (K^+) can pass through K^+ channels. This selectivity is due to specific arrangements of chemical groups on the interior surfaces of the proteins that form the channel walls. Cells vary in the number, kind, and activity of channels they have. It is even possible for a given channel to be open or closed to its specific ion as a result of changes in channel shape in response to a controlling mechanism. This is a good example of function depending on structural details. (To learn how a specific channel defect can lead to a devastating disease, see the accompanying boxed feature, ▶ Beyond the Basics.)
2. Another group of proteins that span the membrane serve as **carrier molecules**, which transfer specific substances across the membrane that are unable to cross on their own. The means by which carriers accomplish this transport will be described later. (Thus, channels and carrier molecules are both important in the transport of substances between the ECF and ICF.) Each carrier can transport only a particular molecule or closely related molecules. Cells of different types have different kinds of carriers. As a result, they vary as to which substances they can selectively transport across their membranes. For example, thyroid gland cells are the only cells in the body to use iodine. Appropriately, only the plasma membranes of thyroid gland cells have carriers for iodine, so these cells alone can transport this element from the blood into the cell interior.
3. Other proteins located on the inner membrane surface serve as **docking-marker acceptors** that bind lock-and-key fashion with the docking markers of secretory vesicles (see p. 24). Secretion is initiated as stimulatory signals trigger the fusion of the secretory vesicle membrane with the inner surface of the plasma membrane through interactions between these matching labels. The secretory vesicle subsequently opens up and empties its contents to the outside by exocytosis.
4. Another group of surface-located proteins function as **membrane-bound enzymes** that control specific chemical reactions at either the inner or the outer cell surface. Cells are specialized in the types of enzymes embedded within their plasma membranes. For example, the outer layer of the plasma membrane of skeletal muscle cells contains an enzyme that destroys the chemical messenger that triggers muscle contraction, thus enabling the muscle to relax.
5. Many proteins on the outer surface serve as **receptor sites** (or, simply, **receptors**) that "recognize" and bind with specific molecules in the cell's environment. This binding initiates a series of membrane and intracellular events (to be described later) that alter the activity of the particular cell. In this way, chemical messengers in the blood, such as water-soluble hormones, can influence only the specific cells that have receptors for a given messenger. Even though every cell is exposed to the same messenger via its widespread distribution by the blood, a given messenger has no effect on other cells that do not have receptors for this specific messenger. To illustrate, the anterior pituitary gland secretes into the blood thyroid-stimulating hormone (TSH), which can attach only to the surface of thyroid gland cells to stimulate secretion of thyroid hormone. No other cells have receptors for TSH, so only thyroid cells are influenced by TSH despite its widespread distribution.
6. Still other proteins serve as **cell adhesion molecules (CAMS)**. Many CAMs protrude from the outer membrane surface and form loops or hooks that the cells use to grip each other and to grasp the connective tissue fibers that interlace between cells. Thus these molecules help hold the cells within tissues and organs together.
7. Finally, still other proteins on the outer membrane surface, especially in conjunction with carbohydrates, are important in the cells' ability to recognize "self" (that is, cells of the same type) and in cell-to-cell interactions. *MHC proteins*

▌The membrane carbohydrates serve as self-identity markers.

The short sugar chains on the outer membrane surface serve as self-identity markers that let cells identify and interact with each other in the following ways:

1. Different cell types have different markers. The unique combination of sugar chains projecting from the surface mem-

Cystic Fibrosis: A Fatal Defect in Membrane Transport

Cystic fibrosis (CF), the most common fatal genetic disease in the United States, strikes 1 in every 2000 Caucasian children. It is characterized by the production of an abnormally thick, sticky mucus. Most dramatically affected are the respiratory airways and the pancreas.

Respiratory Problems

The presence of the thick, sticky mucus in the respiratory airways makes it difficult to get adequate air in and out of the lungs. Also, because bacteria thrive in the accumulated mucus, CF patients suffer from repeated respiratory infections. They are especially susceptible to *Pseudomonas aeruginosa,* an "opportunistic" bacterium that is frequently present in the environment but usually causes infection only when some underlying problem handicaps the body's defenses. Gradually, the involved lung tissue becomes scarred (fibrotic), making the lungs harder to inflate. This complication increases the work of breathing beyond the extra work required to move air through the clogged airways.

Underlying Cause

During the last decade, researchers found that cystic fibrosis is caused by any one of several different genetic defects that lead to production of a flawed version of a protein known as *cystic fibrosis transmembrane conductance regulator (CFTR).* CFTR normally helps form and regulate the chloride (Cl^-) channels in the plasma membrane. With CF, the defective CFTR "gets stuck" in the endoplasmic reticulum/Golgi system, which normally manufactures and processes this product and ships it to the plasma membrane (see p. 23). That is, in CF patients the mutated version of CFTR is only partially processed and never makes it to the cell surface. The resultant absence of CFTR protein in the plasma membrane's Cl^- channels leads to membrane impermeability to Cl^-. Because Cl^- transport across the membrane is closely linked to Na^+ transport, cells lining the respiratory airways cannot absorb salt (NaCl) properly. As a result, salt accumulates in the fluid lining the airways.

What has puzzled researchers is how this Cl^- channel defect and resultant salt accumulation lead to the excess mucus problem. Two recent discoveries have perhaps provided an answer, although these proposals remain to be proven and research into other possible mechanisms continues to be pursued. One group of investigators found that the airway cells produce a natural antibiotic, *defensin,* which normally kills most of the inhaled airborne bacteria. It turns out that defensin cannot function properly in a salty environment. Bathed in the excess salt associated with CF, the disabled antibiotic cannot rid the lungs of inhaled bacteria. This leads to repeated infections. One of the outcomes of the body's response to these infections is excess mucus production. In turn, this mucus serves as a breeding ground for more bacterial growth. The cycle continues as the lung-clogging mucus accumulation and frequency of lung infections grows ever worse. To make matters worse, the excess mucus is especially thick and sticky, making it difficult for the normal ciliary defense mechanisms of the lung to sweep up the bacteria-laden mucus from the lungs (see p. 36 and p. 357). The mucus is thick and sticky because it is underhydrated (has too little water), a problem believed to be linked to the defective salt transport.

The second new study found an additional complicating factor in the CF story. These researchers demonstrated that CFTR appears to serve a dual role as a Cl^- channel and as a membrane receptor that binds with *Pseudomonas aeruginosa* (and perhaps other bacteria). CFTR subsequently destroys the captured bacteria. In the absence of CFTR in the airway cell membranes of CF patients, *P. aeruginosa* is not cleared from the airways as usual. In a double onslaught, these bacteria were shown to trigger the airway cells to produce unusually large amounts of an abnormal, thick, sticky mucus. This mucus promotes more bacterial growth, as the vicious cycle continues.

Pancreatic Problems

Furthermore, in CF patients the pancreatic duct, which carries secretions from the pancreas to the small intestine, becomes plugged with thick mucus. Because the pancreas produces enzymes important in the digestion of food, malnourishment eventually results. In addition, as the pancreatic digestive secretions accumulate behind the blocked pancreatic duct, fluid-filled cysts form in the pancreas, with the affected pancreatic tissue gradually degenerating and becoming fibrotic. The name "cystic fibrosis" aptly describes long-term changes that occur in the pancreas and lungs as the result of a single genetic flaw in CFTR.

Treatment

Treatment consists of physical therapy to help clear the airways of the excess mucus and antibiotic therapy to combat respiratory infections, plus special diets and administration of supplemental pancreatic enzymes to maintain adequate nutrition. Despite this supportive treatment, most CF victims do not survive beyond their early 20s, with most dying from lung complications.

With the recent discovery of the genetic defect responsible for the majority of CF cases, investigators are hopeful of developing a means to correct or compensate for the defective gene. Another potential cure being studied is development of drugs that induce the mutated CFTR to be "finished off" and inserted in the plasma membrane. Furthermore, several promising new drug therapy approaches, such as a mucus-thinning aerosol drug that can be inhaled, offer hope of reducing the number of lung infections and extending the life span of CF victims until a cure can be found.

---

brane proteins serves as the "trademark" of a particular cell type, enabling a cell to recognize others of its own kind. Thus these carbohydrate chains play an important role in recognition of "self" and in cell-to-cell interactions. Cells can recognize other cells of the same type and join together to form tissues. This is especially important during embryonic development. If cultures of embryonic cells of two different types, such as nerve cells and muscle cells, are mixed together, the cells sort themselves into separate aggregates of nerve cells and muscle cells.

2.   Carbohydrate-containing surface markers are also involved in tissue growth, which is normally held within certain limits of cell density. Cells do not "trespass" across the boundaries of neighboring tissues; that is, they do not overgrow their own territory. The exception is the uncontrolled spread of cancer cells, which have been shown to bear abnormal surface carbohydrate markers.

CELL-TO-CELL ADHESIONS

In multicellular organisms such as humans, plasma membranes not only serve as the outer boundaries of all cells but also participate in cell-to-cell adhesions. These adhesions bind groups of cells together into tissues and package them further into organs. The life-sustaining activities of the body systems depend not only on the functions of the individual cells of which they are made but also on how these cells live and work together in tissue and organ communities.

Organization of cells into appropriate groupings is at least partially attributable to the carbohydrate chains on the membrane surface. Once arranged, cells are held together by three different means: (1) the extracellular matrix, (2) cell adhesion molecules in the cells' plasma membranes, and (3) specialized cell junctions.

▌The extracellular matrix serves as the biological "glue."

Tissues are not made up solely of cells, and many cells within a tissue are not in direct physical contact with neighboring cells. Instead, they are held together by the **extracellular matrix (ECM)**, an intricate meshwork of fibrous proteins embedded in a watery, gel-like substance composed of complex carbohydrates. The ECM serves as the biological "glue." The watery gel provides a pathway for diffusion of nutrients, wastes, and other water-soluble traffic between the blood and tissue cells. It is usually called the *interstitial fluid* (see p. 8). Interwoven within this gel are three major types of protein fibers: collagen, elastin, and fibronectin.

1. **Collagen** forms cablelike fibers or sheets that provide tensile strength (resistance to longitudinal stress).
2. **Elastin** is a rubberlike protein fiber most abundant in tissues that must be capable of easily stretching and then recoiling after the stretching force is removed. It is found, for example, in the lungs, which stretch and recoil as air moves in and out.
3. **Fibronectin** promotes cell adhesion and holds cells in position. Reduced amounts of this protein have been found within certain types of cancerous tissue, possibly accounting for the fact that cancer cells do not adhere well to each other but tend to break loose and metastasize (spread elsewhere in the body).

The ECM is secreted by local cells present in the matrix. The relative amount of ECM compared to cells varies greatly among tissues. For example, the ECM is scant in epithelial tissue but is the predominant component of connective tissue. Most of this abundant matrix in connective tissue is secreted by **fibroblasts** ("fiber formers"). The exact composition of ECM components varies for different tissues, thus providing distinct local environments for the various cell types in the body. In some tissues, the matrix becomes highly specialized to form such structures as cartilage or tendons or, on appropriate calcification, the hardened structures of bones and teeth.

▌Some cells are directly linked together by specialized cell junctions.

In tissues where the cells lie in close proximity to each other, some tissue cohesion is provided by the cell adhesion molecules, or CAMs. As you just learned, these special loop- and hook-shaped surface membrane proteins "Velcro" adjacent cells to each other. In addition, some cells within given types of tissues are directly linked together by one of three types of specialized cell junctions: (1) desmosomes (adhering junctions), (2) tight junctions (impermeable junctions), or (3) gap junctions (communicating junctions).

DESMOSOMES

Desmosomes act like "spot rivets" that anchor together two closely adjacent but nontouching cells. A desmosome consists of two components: (1) a pair of dense, buttonlike cytoplasmic thickenings known as *plaque* located on the inner surface of each of the two adjacent cells; and (2) strong filaments that extend across the space between the two cells and attach to the plaque on both sides (● Figure 3-4). These intercellular filaments bind adjacent plasma membranes together so that they resist being pulled apart. Thus desmosomes are adhering junctions. They are most abundant in tissues that are subject to considerable stretching, such as those found in the skin, the heart, and the uterus.

TIGHT JUNCTIONS

At **tight junctions**, adjacent cells firmly bind with each other at points of direct contact to seal off passageway between the two cells. They are found primarily in sheets of epithelial tissue. Epithelial tissue covers the surface of the body and lines

---

● FIGURE 3-4

Desmosome. Desmosomes are adhering junctions that spot-rivet cells, anchoring them together in tissues subject to considerable stretching.

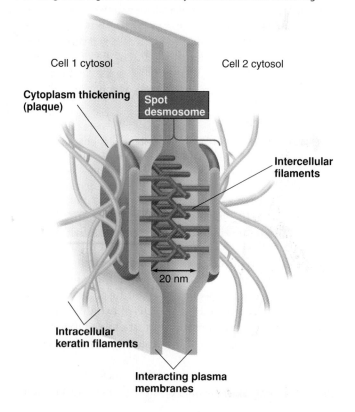

Cell 1 cytosol   Cell 2 cytosol

Cytoplasm thickening (plaque)

Spot desmosome

Intercellular filaments

20 nm

Intracellular keratin filaments

Interacting plasma membranes

all its internal cavities. All epithelial sheets serve as highly selective barriers between two compartments that have considerably different chemical compositions. For example, the epithelial sheet lining the digestive tract separates the food and potent digestive juices within the inner cavity (lumen) from the blood vessels that lie on the other side. It is important that only completely digested food particles and not undigested food particles or digestive juices move across the epithelial sheet from the lumen to the blood. Accordingly, the lateral (side) edges of the adjacent cells in the epithelial sheet are joined together in a tight seal near their luminal border by "kiss sites," sites of direct fusion of *junctional proteins* on the outer surfaces of the two interacting plasma membranes (● Figure 3-5). These tight junctions are impermeable and thus prevent materials from passing between the cells. Passage across the epithelial barrier, therefore, must take place *through* the cells, not *between* them. This traffic across the cell is regulated by means of the channels and carriers present. If the cells were not joined by tight junctions, uncontrolled exchange of molecules could take place between the compartments by unpoliced traffic through the spaces between adjacent cells. Tight junctions thus prevent undesirable leaks within epithelial sheets.

GAP JUNCTIONS

At a **gap junction**, as the name implies, a gap exists between adjacent cells, which are linked by small, connecting tunnels formed by connexons. A **connexon** is made up of six protein subunits arranged in a hollow tubelike structure. Two connexons extend outward, one from each of the plasma membranes of two adjacent cells, and join together end-to-end to form the connecting tunnel between the two cells (● Figure 3-6). Gap junctions are communicating junctions. The small diameter of the tunnels permits small, water-soluble particles to pass between the connected cells but precludes passage of large molecules such as vital intracellular proteins. Through these specialized anatomic arrangements, ions (electrically charged particles) and small molecules can be directly exchanged between interacting cells without ever entering the ECF.

Gap junctions are especially abundant in cardiac muscle and smooth muscle. In these tissues, movement of ions between cells through gap junctions transmits electrical activity throughout an entire muscle mass. Because this electrical activity brings about contraction, the presence of gap junctions enables synchronized contraction of a whole muscle mass, such as the pumping chamber of the heart.

Gap junctions are also found in some nonmuscle tissues, where they permit unrestricted passage of small nutrient molecules between cells. For example, glucose, amino acids and other nutrients pass through gap junctions to a developing egg cell from cells surrounding the egg within the ovary, thus helping the egg stockpile these essential nutrients.

Gap junctions also serve as avenues for the direct transfer of small signaling molecules from one cell to the next. Such transfer permits the cells connected by gap junctions to directly communicate with each other. This communication provides one possible mechanism by which cooperative cell activity may be coordinated. In the next chapter we will examine other means by which cells "talk to each other."

We are now going to turn our attention to the topic of membrane transport, focusing on how the plasma membrane can selectively control what enters and exits the cell.

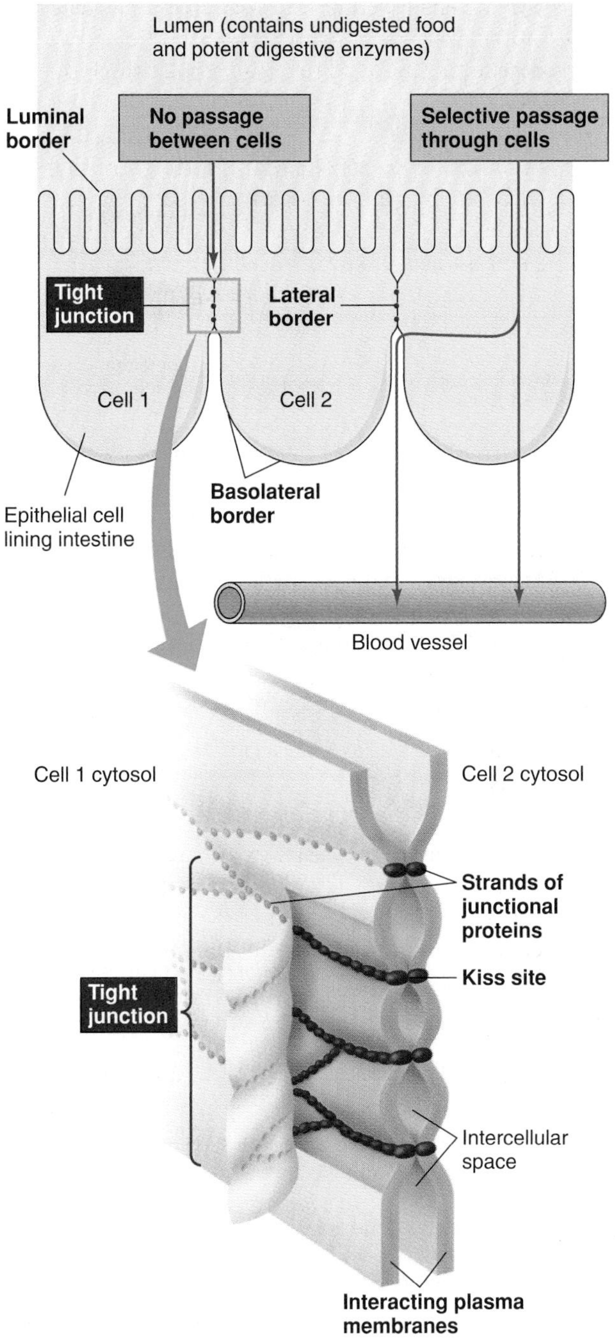

● FIGURE 3-5

Tight junction. Tight junctions are impermeable junctions that join the lateral edges of epithelial cells near their luminal borders, thus preventing materials from passing *between* the cells. Only regulated passage of materials can occur *through* these cells, which form highly selective barriers that separate two compartments of highly different chemical composition.

For an interaction related to this figure, see Media Exercise 3.1: The Plasma Membrane and Cell–Cell Connections on the CD-ROM.

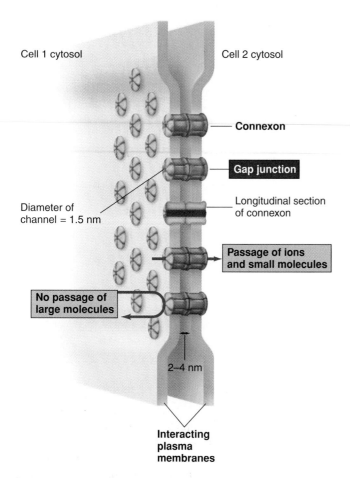

Cell 1 cytosol

Cell 2 cytosol

Connexon

Gap junction

Longitudinal section
of connexon

Diameter of
channel = 1.5 nm

Passage of ions
and small molecules

No passage of
large molecules

2–4 nm

Interacting
plasma
membranes

● **FIGURE 3-6**

Gap junction. Gap junctions are communicating junctions consisting of small connecting tunnels made up of connexons that permit movement of charge-carrying ions and small molecules between two adjacent cells.

 For an interaction related to this figure, see Media Exercise 3.1: The Plasma Membrane and Cell–Cell Connections on the CD-ROM.

 Click on the Media Exercises menu of the CD-ROM and work Media Exercise 3.1: The Plasma Membrane and Cell–Cell Connections to test your understanding of the previous section.

OVERVIEW OF MEMBRANE TRANSPORT

Anything that passes between a cell and the surrounding extracellular fluid must be able to penetrate the plasma membrane. If a substance can cross the membrane, the membrane is said to be **permeable** to that substance; if a substance cannot pass, the membrane is **impermeable** to it. The plasma membrane is **selectively permeable** in that it permits some particles to pass through while excluding others.

Two properties of particles influence whether they can permeate the plasma membrane without any assistance: (1) the relative solubility of the particle in lipid and (2) the size of the particle. Highly lipid-soluble particles can dissolve in the lipid bilayer and pass through the membrane. Uncharged or nonpolar molecules (such as O_2, CO_2, and fatty acids) are highly lipid soluble and readily permeate the membrane.

Charged particles (ions such as Na^+ and K^+) and polar molecules (such as glucose and proteins) have low lipid solubility but are very soluble in water. The lipid bilayer serves as an impermeable barrier to particles poorly soluble in lipid. For water-soluble (and thus lipid-insoluble) ions less than 0.8 nm in diameter, the protein channels serve as an alternate route for passage across the membrane. Only ions for which specific channels are available and open can permeate the membrane.

Particles that have low lipid solubility and are too large for channels cannot permeate the membrane on their own. Yet some of these particles must cross the membrane for the cell to survive and function. Glucose is an example of a large, poorly lipid soluble particle that must gain entry to the cell but cannot permeate by dissolving in the lipid bilayer or passing through a channel. Cells have several means of assisted transport to move particles across the membrane that must enter or leave the cell but cannot do so unaided, as you will learn shortly.

Even if a particle can permeate the membrane by virtue of its lipid solubility or its ability to fit through a channel, some force is needed to produce its movement across the membrane. Two general types of forces are involved in accomplishing transport across the membrane: (1) forces that do not require the cell to expend energy to produce movement (**passive forces**) and (2) forces requiring cell energy (ATP) expenditure to transport a substance across the membrane (**active forces**).

We will now examine the various methods of membrane transport, indicating whether each is an unassisted or assisted means of transport and whether each is a passive or active transport mechanism.

UNASSISTED MEMBRANE TRANSPORT

Molecules (or ions) that can penetrate the plasma membrane on their own are passively driven across the membrane by two forces: diffusion down a concentration gradient and/or movement along an electrical gradient. We will first examine diffusion down a concentration gradient.

❚ **Particles that can permeate the membrane passively diffuse down their concentration gradient.**

All molecules (or ions) are in continuous random motion at temperatures above absolute zero as a result of heat energy. This motion is most evident in liquids and gases, where the individual molecules have more room to move before colliding with another molecule. Each molecule moves separately and randomly in any direction. As a consequence of this haphazard movement, the molecules frequently collide, bouncing off each other in different directions like billiard balls striking each other. The greater the molecular concentration of a substance in a solution, the greater the likelihood of collisions. Consequently, molecules within a particular space tend to become evenly distributed over time. Such uniform spreading out of molecules due to their random intermingling is

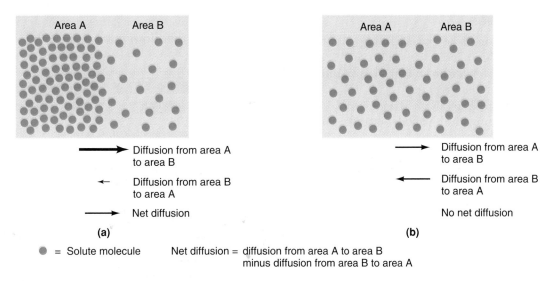

Diffusion from area A to area B

← Diffusion from area B to area A

Net diffusion

(a)

Diffusion from area A to area B

← Diffusion from area B to area A

No net diffusion

(b)

● = Solute molecule   Net diffusion = diffusion from area A to area B minus diffusion from area B to area A

● **FIGURE 3-7**

Diffusion. (a) Diffusion down a concentration gradient. (b) Steady state, with no net diffusion occurring.

 PhysioEdge For an animation of this figure, click the Diffusion and Membrane Transport tab in the Membrane Potential tutorial on the CD-ROM.

known as **diffusion** (*diffusere* means "to spread out"). To illustrate, in ● Figure 3-7a, the concentration differs between area A and area B in a solution. Such a difference in concentration between two adjacent areas is called a **concentration gradient** (or **chemical gradient**). Random molecular collisions will occur more frequently in area A because of its greater concentration of molecules. For this reason, more molecules will bounce from area A into area B than in the opposite direction. In both areas, the individual molecules will move randomly and in all directions, but the net movement of molecules by diffusion will be from the area of higher concentration to the area of lower concentration.

The term **net diffusion** refers to the difference between two opposing movements. If 10 molecules move from area A to area B while 2 molecules simultaneously move from B to A, the net diffusion is 8 molecules moving from A to B. Molecules will spread in this way until the substance is uniformly distributed between the two areas and a concentration gradient no longer exists (● Figure 3-7b). At this point, even though movement is still taking place, no net diffusion is occurring because the opposing movements exactly counterbalance each other; that is, are in equilibrium. Movement of molecules from area A to area B will be exactly matched by movement of molecules from B to A. This situation is known as a **steady state**.

What happens if a plasma membrane separates different concentrations of a substance? If the substance can permeate the membrane, net diffusion of the substance will occur through the membrane down its concentration gradient from the area of high con-

centration to the area of low concentration (● Figure 3-8a). No energy is required for this movement, so it is a passive mechanism of membrane transport. As an example, O_2 is transferred across the lung membrane by this means. The blood carried to the lungs is low in O_2, having given up O_2 to the body tissues for cell metabolism. The air in the lungs, in contrast, is high in O_2 because it is continuously exchanged with fresh air by the process of breathing. Because of this concentration gradient, net diffusion of O_2 occurs from the lungs into the blood as blood flows through the lungs. Thus as blood leaves the lungs for delivery to the tissues, it is high in O_2.

● **FIGURE 3-8**

Diffusion through a membrane. (a) Net diffusion across the membrane down a concentration gradient. (b) No diffusion through the membrane despite the presence of a concentration gradient.

 PhysioEdge For interactions related to this figure, see the Diffusion and Membrane Transport tab in the Membrane Potential tutorial and Media Exercise 3.2: Means of Transmembrane Exchange on the CD-ROM.

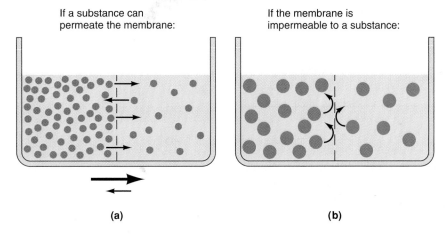

If a substance can permeate the membrane:

If the membrane is impermeable to a substance:

(a)   **(b)**

FICK'S LAW OF DIFFUSION

Several factors in addition to the concentration gradient influence the rate of net diffusion across a membrane. The effects of these factors collectively make up **Fick's law of diffusion** (▲ Table 3-1). Note that the larger the surface area available, the greater the rate of diffusion it can accommodate. Various strategies are used throughout the body for increasing the membrane surface area across which diffusion and other types of transport take place. For example, absorption of nutrients in the small intestine is enhanced by the presence of microvilli, which greatly increase the available absorptive surface in contact with the nutrient-rich contents of the small intestine lumen (see p. 37). Conversely, abnormal loss of membrane surface area decreases the rate of net diffusion. For example, in *emphysema* O_2 and CO_2 exchange between the air and blood in the lungs is reduced. In this condition, the walls of the air sacs break down, resulting in less surface area available for diffusion of these gases.

Also note that the greater the distance, the slower the rate of diffusion. Accordingly, membranes across which diffusing particles must travel are normally relatively thin, such as the membranes separating the air and blood in the lungs. Thickening of this air–blood interface (as in *pneumonia*, for example) slows down exchange of O_2 and CO_2. Furthermore, diffusion is efficient only for short distances between cells and their surroundings. It becomes an inappropriately slow process for distances of more than a few centimeters. To illustrate, it would take months or even years for O_2 to diffuse from the surface of the body to the cells in the interior. Instead, the circulatory system provides a network of tiny vessels that deliver and pick up materials at every "block" of a few cells, with diffusion accomplishing short local exchanges between the blood and surrounding cells.

Ions that can permeate the membrane also passively move along their electrical gradient.

Movement of ions (electrically charged particles that have either lost or gained an electron) is also affected by their electrical charge. Like charges (those with the same kind of charge) repel each other, and opposite charges attract each other. If a relative difference in charge exists between two adjacent areas, the positively charged ions (*cations*) tend to move toward the more negatively charged area, whereas the negatively charged ions (*anions*) tend to move toward the more positively charged area. A difference in charge between two adjacent areas thus produces an **electrical gradient** that promotes the movement of ions toward the area of opposite charge. Because the cell does not have to expend energy for ions to move into or out of the cell along an electrical gradient, this method of membrane transport is passive. When an electrical gradient exists between the ICF and ECF, only ions that can permeate the plasma membrane can move along this gradient.

Both an electrical and a concentration (chemical) gradient may be acting on a particular ion at the same time. The net effect of simultaneous electrical and concentration gradients on this ion is called an **electrochemical gradient**. Later in this chapter you will learn how electrochemical gradients contribute to the electrical properties of the plasma membrane.

Osmosis is the net diffusion of water down its own concentration gradient.

Water can readily permeate the plasma membrane. It is small enough to slip between the phospholipid molecules within the lipid bilayer. Also, in some cell types, membrane proteins form **aquaporins**, which are channels specific for the passage of water (*aqua* means "water"). The driving force for diffusion of water across the membrane is the same as for any other diffusing molecule, namely its concentration gradient. Usually the term *concentration* refers to the density of the solute (dissolved substance) in a given volume of water. It is important to recognize, however, that adding a solute to pure water in essence decreases the water concentration. In general, one molecule of a solute displaces one molecule of water.

Compare the water and solute concentrations in the two containers in ● Figure 3-9. The container in part (a) of the figure is full of pure water, so the water concentration is 100% and the solute concentration is 0%. In part (b), solute has replaced 10% of the water molecules. The water concentration is now 90%, and the solute concentration is 10%—a lower water concentration and a higher solute concentration than in part (a). Note that as the solute concentration increases, the water concentration decreases correspondingly.

If solutions of unequal solute concentration (and hence unequal water concentration) are separated by a membrane that permits passage of water, such as the plasma membrane, water will diffuse passively down its own concentration gradi-

---

▲**TABLE 3-1**

Factors Influencing the Rate of Net Diffusion of a Substance across a Membrane (Fick's Law of Diffusion)

FACTOR	EFFECT ON RATE OF NET DIFFUSION
↑ Concentration gradient of substance (ΔC)	↑
↑ Permeability of membrane to substance (P)	↑
↑ Surface area of membrane (A)	↑
↑ Molecular weight of substance (MW)	↓
↑ Distance (thickness) (ΔX)	↓

Modified Fick's equation:

$$\text{Net rate of diffusion } (Q) = \frac{\Delta C \cdot P \cdot A}{MW \cdot \Delta X}$$

$$\left[\frac{P}{\sqrt{MW}} = \text{diffusion coefficient } (D) \right]$$

$$\text{Restated } Q \propto \frac{\Delta C \cdot A \cdot D}{\Delta X}$$

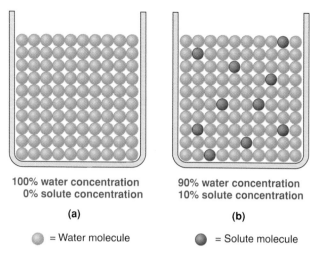

100% water concentration
0% solute concentration
(a)

90% water concentration
10% solute concentration
(b)

⬤ = Water molecule   ⬤ = Solute molecule

⬤ **FIGURE 3-9**

Relationship between solute and water concentration in a solution.
(a) Pure water. (b) Solution.

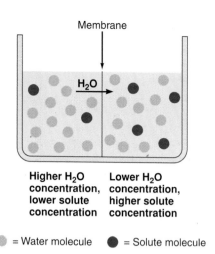

Higher H₂O | **Lower H₂O**
concentration, | **concentration,**
lower solute | **higher solute**
concentration | **concentration**

⬤ = Water molecule   ⬤ = Solute molecule

⬤ **FIGURE 3-10**

Osmosis. Osmosis is the net diffusion of water down its own concentration gradient (to the area of higher solute concentration).

ent from the area of higher water concentration (lower solute concentration) to the area of lower water concentration (higher solute concentration) (⬤ Figure 3-10). This net diffusion of water is known as **osmosis**. Because solutions are always referred to in terms of concentration of solute, *water moves by osmosis to the area of higher solute concentration*. Despite the impression that the solutes are "drawing," or attracting, water, osmosis is nothing more than diffusion of water down its own concentration gradient across the membrane.

Thus far in our discussion of osmosis, we have ignored any solute movement. Let us compare the results of osmosis when the solute can and cannot permeate the membrane:

OSMOSIS WHEN A MEMBRANE SEPARATES UNEQUAL SOLUTIONS OF A PENETRATING SOLUTE

Assume that solutions of unequal solute concentration are separated by a membrane that permits passage of both water and the solute. Because the membrane is permeable to the solute as well as to water, the solute can move down its own concentration gradient in the opposite direction of the net water movement (⬤ Figure 3-11). This movement continues until both the solute and water are evenly distributed across the membrane. With all concentration gradients gone, osmosis ceases. The final volume of the compartments when the steady state is achieved is the same as at the onset. Water and solute molecules have merely exchanged places between the two compartments until their distributions have equalized; that is, an equal number of water molecules have moved from side 1 to side 2 as solute molecules have moved from side 2 to side 1.

OSMOSIS WHEN A MEMBRANE SEPARATES UNEQUAL SOLUTIONS OF A NONPENETRATING SOLUTE

Now assume that solutions of unequal solute concentration are separated by a membrane that is permeable to water but excludes passage of the solute. Because the membrane is impermeable to the solute, the solute cannot cross the membrane down its concentration gradient (⬤ Figure 3-12). At first the

⬤ **FIGURE 3-11**

Movement of water and a penetrating solute unequally distributed across a membrane

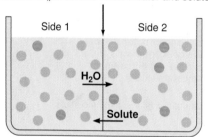

Higher H₂O concentration, | Lower H₂O concentration,
lower solute concentration | higher solute concentration

H₂O moves from side 1 to side 2 down its concentration gradient

Solute moves from side 2 to side 1 down its concentration gradient

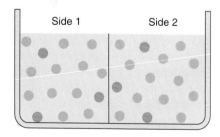

• Water concentrations equal
• Solute concentrations equal
• No further net diffusion
• Steady state exists

⬤ = Water molecule   ⬤ = Solute molecule

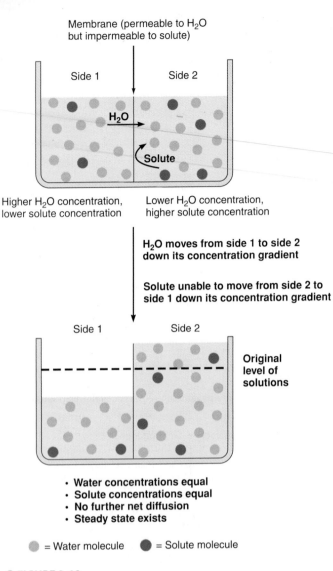

Membrane (permeable to H₂O but impermeable to solute)

Side 1   Side 2

H₂O →

Solute

Higher H₂O concentration, lower solute concentration | Lower H₂O concentration, higher solute concentration

H₂O moves from side 1 to side 2 down its concentration gradient

Solute unable to move from side 2 to side 1 down its concentration gradient

Side 1   Side 2

---- Original level of solutions

- Water concentrations equal
- Solute concentrations equal
- No further net diffusion
- Steady state exists

○ = Water molecule   ● = Solute molecule

● **FIGURE 3-12**

Osmosis in the presence of an unequally distributed nonpenetrating solute

Membrane (permeable to H₂O but impermeable to solute)

Side 1   Side 2

H₂O →

Solute

Pure water | Lower H₂O concentration, higher solute concentration

H₂O moves from side 1 to side 2 down its concentration gradient

Solute unable to move from side 2 to side 1 down its concentration gradient

Side 1   Side 2

Hydrostatic (fluid) pressure difference

---- Original level of solutions

Osmosis →
← Hydrostatic pressure

- Water concentrations not equal
- Solute concentrations not equal
- Tendency for water to diffuse by osmosis into side 2 is exactly balanced by opposing tendency for hydrostatic pressure difference to push water into side 1
- Osmosis ceases
- Opposing pressure necessary to completely stop osmosis is equal to osmotic pressure of solution

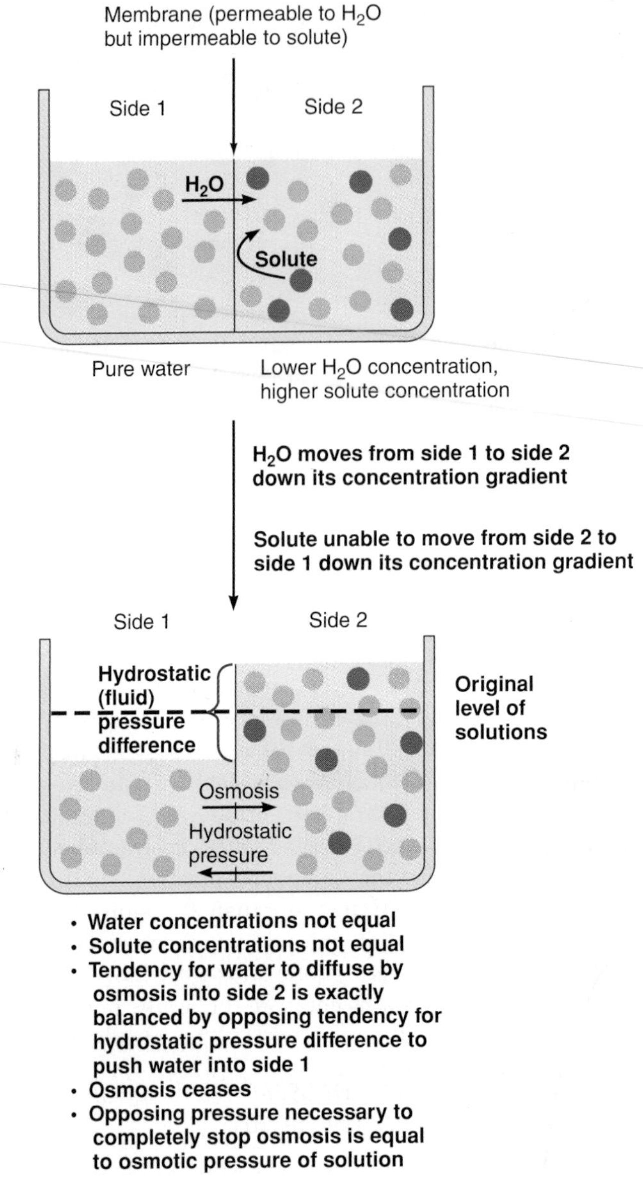

● **FIGURE 3-13**

Osmosis when pure water is separated from a solution containing a nonpenetrating solute

concentration gradients are identical to those in the previous example. However, even though net diffusion of water takes place from side 1 to side 2, the solute cannot move. As a result of water movement alone, the volume of side 2 increases while the volume of side 1 correspondingly decreases. Loss of water from side 1 increases the solute concentration on side 1, whereas addition of water to side 2 reduces the solute concentration on that side. Eventually the concentrations of water and solute on the two sides of the membrane become equal, and net diffusion of water ceases. Unlike the situation in which the solute can also permeate, diffusion of water alone has resulted in a change in the final volumes of the two compartments. The side originally containing the greater solute concentration has a larger volume, having gained water.

OSMOSIS WHEN A MEMBRANE SEPARATES PURE WATER FROM A SOLUTION OF A NONPENETRATING SOLUTE

What will happen if a nonpenetrating solute is present on side 2 and pure water is present on side 1 (● Figure 3-13)? Osmo-

sis occurs from side 1 to side 2, but the concentrations between the two compartments can never become equal. No matter how dilute side 2 becomes because of water diffusing into it, it can never become pure water, nor can side 1 ever acquire any solute. Because equilibrium is impossible to achieve, does net diffusion of water (osmosis) continue until all the water has left side 1? No. As the volume expands in compartment 2, a difference in hydrostatic pressure between the two compartments is created, and it opposes osmosis. **Hydrostatic (fluid) pressure is** the pressure exerted by a standing, or stationary, fluid on an object—in this case the plasma membrane (*hydro* means "fluid"; *static* means "standing"). The hydrostatic pressure exerted by the larger volume of fluid on side 2 is greater than the hydrostatic pressure exerted on side 1. This hydrostatic pressure difference tends to push fluid from side 2 to side 1.

The **osmotic pressure** of a solution is a measure of the tendency for water to move into that solution because of its relative concentration of nonpenetrating solutes and water. Net movement of water continues until the opposing hydrostatic pressure exactly counterbalances the osmotic pressure. The magnitude of the osmotic pressure is equal to the magnitude of the opposing hydrostatic pressure necessary to completely stop osmosis. The greater the concentration of nonpenetrating solute → the lower the concentration of water → the greater the drive for water to move by osmosis from pure water into the solution → the greater the opposing pressure required to stop the osmotic flow → the greater the osmotic pressure of the solution. Therefore, a solution with a high concentration of nonpenetrating solute exerts greater osmotic pressure than a solution with a lower concentration of nonpenetrating solute does.

TONICITY

The **tonicity** of a solution is the effect the solution has on cell volume—whether the cell remains the same size, swells, or shrinks—when the solution surrounds the cell. The tonicity of a solution is determined by its concentration of nonpenetrating solutes. Solutes that can penetrate the plasma membrane quickly become equally distributed between the ECF and ICF, so they do not contribute to osmotic differences. An **isotonic solution** (*iso* means "same") has the same concentration of nonpenetrating solutes as normal body cells. When a cell is bathed in an isotonic solution, no water enters or leaves the cell by osmosis, so cell volume remains constant. For this reason, the ECF is normally kept isotonic so that no net diffusion of water occurs across the plasma membranes of body cells. This is important because cells, especially brain cells, do not function properly if they are swollen or shrunken.

Any change in the concentration of nonpenetrating solutes in the ECF produces a corresponding change in the water concentration difference across the plasma membrane. The resultant osmotic movement of water brings about changes in cell volume. The easiest way to demonstrate this phenomenon is to place red blood cells in solutions with varying concentrations of nonpenetrating solutes. Normally the plasma in which red blood cells are suspended has the same osmotic activity as the fluid inside these cells, so that the cells maintain a constant volume. If red blood cells are placed in a dilute or **hypotonic solution** (*hypo* means "below"), a solution with a below-normal concentration of nonpenetrating solutes (and therefore a higher concentration of water), water enters the cells by osmosis. Net gain of water by the cells causes them to swell, perhaps to the point of rupturing. If, in contrast, red blood cells are placed in a concentrated or **hypertonic solution** (*hyper* means "above"), a solution with an above-normal concentration of nonpenetrating solutes (and therefore a lower concentration of water), the cells shrink as they lose water by osmosis. Thus it is crucial that the concentration of nonpenetrating solutes in the ECF quickly be restored to normal should the ECF become hypotonic (as with ingesting too much water) or hypertonic (as with losing too much water through severe diarrhea).

ASSISTED MEMBRANE TRANSPORT

All the kinds of transport we have discussed thus far—diffusion down concentration gradients, movement along electrical gradients, and osmosis—produce net movement of molecules capable of permeating the plasma membrane by virtue of their lipid solubility or small size. Large, poorly lipid-soluble molecules such as proteins, glucose, and amino acids cannot cross the plasma membrane on their own no matter what forces are acting on them. This impermeability ensures that the large polar intracellular proteins cannot escape from the cell. Thus these proteins stay in the cell where they belong and can carry out their life-sustaining functions, for example, serving as metabolic enzymes.

However, because large, poorly lipid-soluble molecules cannot cross the plasma membrane on their own, the cell must provide mechanisms for deliberately transporting these types of molecules into or out of the cell as needed. For example, the cell must usher into the cell essential nutrients, such as glucose for energy and amino acids for the synthesis of proteins, and transport out of the cell metabolic wastes and secretory products, such as water-soluble protein hormones. Furthermore, passive diffusion alone cannot always account for the movement of small ions. Cells use two different mechanisms to accomplish these selective transport processes: *carrier-mediated transport* for transfer of small water-soluble substances across the membrane and *vesicular transport* for movement of large molecules and multimolecular particles between the ECF and ICF. We will examine each of these methods of assisted membrane transport in turn.

▌ Carrier-mediated transport is accomplished by a membrane carrier flipping its shape.

Carrier proteins span the thickness of the plasma membrane and can reverse shape so that specific binding sites can alternately be exposed at either side of the membrane. That is, the carrier "flip-flops" so that binding sites located in the interior of the carrier are alternately exposed to the ECF and the ICF. ● Figure 3-14 is a schematic representation of this **carrier-mediated transport**. As the molecule to be transported attaches to a binding site within the interior of the carrier on one side of the membrane (step 1), this binding causes the carrier to flip its shape so that the same site is now exposed to the other side of the membrane (step 2). Then, having been moved in this way from one side of the membrane to the other, the bound molecule detaches from the carrier (step 3). After the passenger detaches, the carrier reverts to its original shape (step 4).

Carrier-mediated transport systems display three important characteristics that determine the kind and amount of material that can be transferred across the membrane: *specificity, saturation,* and *competition.*

1. **Specificity.** Each carrier protein is specialized to transport a specific substance, or at most a few closely related chemical compounds. For example, amino acids cannot bind to glucose carriers, although several similar amino acids may be able to

Schematic representation of carrier-mediated transport: facilitated diffusion

For interactions related to this figure, see the Diffusion and Membrane Transport tab in the Membrane Potential tutorial and Media Exercise 3.2: Means of Transmembrane Exchange on the CD-ROM.

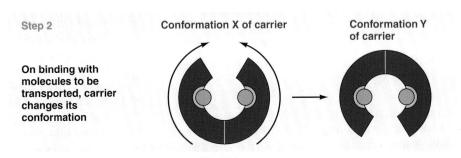

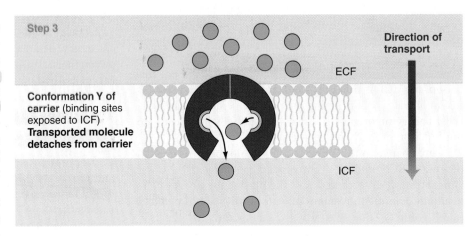

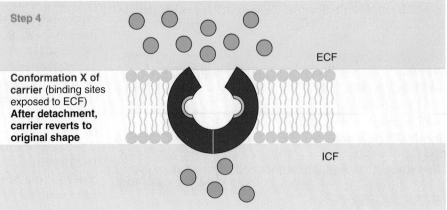

use the same carrier. Cells vary in the types of carriers they have, thus permitting transport selectivity among cells.

2. **Saturation.** A limited number of carrier binding sites are available within a particular plasma membrane for a specific substance. Therefore, there is a limit to the amount of a substance a carrier can transport across the membrane in a given time. This limit is known as the **transport maximum (T_m)**. Until the T_m is reached, the number of carrier binding sites occupied by a substance and, accordingly, the substance's rate of transport across the membrane are directly related to its concentration. The more of a substance available for transport, the more will be transported. When the T_m is reached, the carrier is saturated (all binding sites are occupied), and the rate of the substance's transport across the membrane is maximal. Further increases in the substance's concentration are not accompanied by corresponding increases in the rate of transport (● Figure 3-15).

As an analogy, consider a ferry boat that can maximally carry 100 people across a river during one trip in an hour. If 25 people are on hand to board the ferry, 25 will be transported that hour. Doubling the number of people on hand to 50 will double the rate of transport to 50 people that hour. Such a direct relationship will exist between the number of people waiting to board (the concentration) and the rate of transport until the ferry is fully occupied (its T_m is reached). The ferry can maximally transport 100 people per hour. Even if 150 people are waiting to board, still only 100 will be transported per hour.

Saturation of carriers is a critical rate-limiting factor in the transport of selected substances across the kidney membranes during urine formation and across the intestinal membranes during absorption of digested foods. Furthermore, it is sometimes possible to regulate (for example, by hormones) the rate of carrier-mediated transport by varying the affinity (attraction) of the binding site for its passenger or by varying

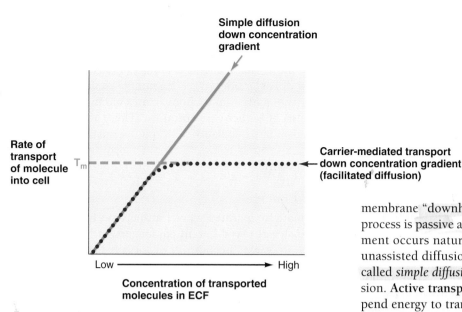

Simple diffusion down concentration gradient

Rate of transport of molecule into cell

T_m

Carrier-mediated transport down concentration gradient (facilitated diffusion)

Low ⟶ High

Concentration of transported molecules in ECF

● **FIGURE 3-15**

Comparison of carrier-mediated transport and simple diffusion down a concentration gradient. With simple diffusion of a molecule down its concentration gradient, the rate of transport of the molecule into the cell is directly proportional to the extracellular concentration of the molecule. With carrier-mediated transport of a molecule down its concentration gradient, the rate of transport of the molecule into the cell is directly proportional to the extracellular concentration of the molecule until the carrier is saturated, at which time the rate of transport reaches a maximal value (transport maximum, or T_m). The rate of transport does not increase with further increases in the ECF concentration of the molecule.

PhysioEdge For an interaction related to this figure, see Media Exercise 3.2: Means of Transmembrane Exchange on the CD-ROM.

the number of binding sites. For example, the hormone insulin greatly increases the carrier-mediated transport of glucose into most cells of the body by promoting an increase in the number of glucose carriers in the cell's plasma membrane. Deficient insulin action (*diabetes mellitus*) drastically impairs the body's ability to take up and use glucose as the primary energy source.

3. **Competition.** Several closely related compounds may compete for a ride across the membrane on the same carrier. If a given binding site can be occupied by more than one type of molecule, the rate of transport of each substance is less when both molecules are present than when either is present by itself. To illustrate, assume the ferry has 100 seats (binding sites) that can be occupied by either men or women. If only men are waiting to board, up to 100 men can be transported during each trip; the same holds true if only women are waiting to board. If, however, both men and women are waiting to board, they will compete for the available seats so that fewer men and fewer women will be transported than when either group is present alone. Fifty of each might make the trip, although the total number of people transported will still be the same, 100 people. In other words, when a carrier can transport two closely related substances, such as the amino acids glycine and alanine, the presence of both diminishes the rate of transfer for either.

Carrier-mediated transport may be passive or active.

Carrier-mediated transport takes two forms, depending on whether energy must be supplied to complete the process: *facilitated diffusion* (not requiring energy) and *active transport* (requiring energy). **Facilitated diffusion** uses a carrier to facilitate (assist) the transfer of a particular substance across the membrane "downhill" from high to low concentration. This process is passive and does not require energy because movement occurs naturally down a concentration gradient. The unassisted diffusion that we described earlier is sometimes called *simple diffusion*, to distinguish it from facilitated diffusion. **Active transport**, in contrast, requires the carrier to expend energy to transfer its passenger "uphill" against a concentration gradient, from an area of lower concentration to an area of higher concentration. An analogous situation is a car on a hill. To move the car downhill requires no energy; it will coast from the top down. Driving the car uphill, however, requires the use of energy (gasoline).

FACILITATED DIFFUSION

The most notable example of facilitated diffusion is the transport of glucose into cells. Glucose is in higher concentration in the blood than in the tissues. Fresh supplies of this nutrient are regularly added to the blood by eating and from reserve energy stores in the body. Simultaneously, the cells metabolize glucose almost as rapidly as it enters the cells from the blood. As a result, a continuous gradient exists for net diffusion of glucose into the cells. However, glucose cannot cross cell membranes on its own. Being polar, it is not lipid soluble, and it is too large to fit through a channel. Without the glucose carrier molecules to facilitate membrane transport of glucose, the cells would be deprived of glucose, their preferred source of fuel.

The carrier-binding sites involved in facilitated diffusion can bind with their passenger molecules when exposed to either side of the membrane (● Figure 3-14). Passenger binding triggers the carrier to flip its conformation and drop off the passenger on the opposite side of the membrane. Because passengers are more likely to bind with the carrier on the high-concentration side than on the low-concentration side, the net movement always proceeds down the concentration gradient from higher to lower concentration. As is characteristic of mediated transport, the rate of facilitated diffusion is limited by saturation of the carrier binding sites, unlike the rate of simple diffusion, which is always directly proportional to the concentration gradient (● Figure 3-15).

ACTIVE TRANSPORT

Active transport also involves the use of a protein carrier to transfer a specific substance across the membrane, but in this case the carrier transports the substance against its concentration gradient. For example, the uptake of iodine by thyroid gland cells necessitates active transport because 99% of the iodine in the body is concentrated in the thyroid. To move iodine from the blood, where its concentration is low, into the thyroid, where its concentration is high, requires expenditure of energy to drive the carrier. Specifically, energy in the form of ATP is required in active transport to vary the affinity of the binding site when exposed on opposite sides of the plasma membrane. In contrast, the affinity of the binding site in facilitated diffusion is the same when exposed to either the outside or the inside of the cell.

With active transport, the binding site has a greater affinity for its passenger on the low-concentration side as a result of *phosphorylation* of the carrier on this side (● Figure 3-16, step 1). The carrier exhibits ATPase activity in that it splits the terminal phosphate from an ATP molecule to yield ADP plus a free inorganic phosphate (see p. 27). The phosphate group is then attached to the carrier. This phosphorylation and the binding of the passenger on the low-concentration side cause the carrier protein to flip its conformation so that the passenger is now exposed to the high-concentration side of the membrane (● Figure 3-16, step 2). The change in carrier shape is accompanied by *dephosphorylation*; that is, the phosphate group detaches from the carrier. Removal of phosphate reduces the affinity of the binding site for the passenger,

so the passenger is released on the high-concentration side. The carrier then returns to its original conformation. Thus ATP energy is used in the phosphorylation–dephosphorylation cycle of the carrier. It alters the affinity of the carrier's binding sites on opposite sides of the membrane so that transported particles are moved uphill from an area of low concentration to an area of higher concentration. These active transport mechanisms are frequently called **pumps**, analogous to water pumps that require energy to lift water against the downward pull of gravity.

Na⁺–K⁺ PUMP

Other more complicated active-transport mechanisms involve the transfer of two different passengers, either simultaneously in the same direction or sequentially in opposite directions. For example, the plasma membrane of all cells contains a sequentially active **Na⁺–K⁺ ATPase pump** (**Na⁺–K⁺ pump** for short). This carrier transports Na⁺ out of the cell, concentrating it in the ECF, and picks up K⁺ from the outside, concentrating it in the ICF (● Figure 3-17). Splitting of ATP through ATPase activity and the subsequent phosphorylation of the carrier on the intracellular side increases the carrier's affinity for Na⁺ and induces a change in carrier shape,

---

● **FIGURE 3-16**

Active transport. The energy of ATP is required in the phosphorylation–dephosphorylation cycle of the carrier to transport the molecule uphill from a region of low concentration to a region of high concentration.

 For interactions related to this figure, see the Diffusion and Membrane Transport tab in the Membrane Potential tutorial and Media Exercise 3.2: Means of Transmembrane Exchange on the CD-ROM.

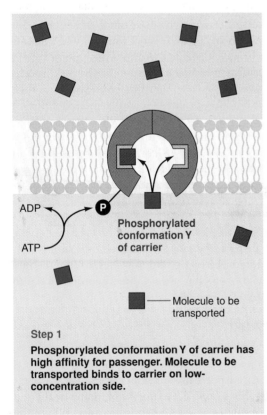

Step 1

Phosphorylated conformation Y of carrier has high affinity for passenger. Molecule to be transported binds to carrier on low-concentration side.

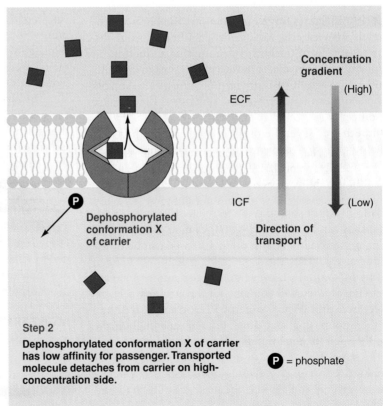

Step 2

Dephosphorylated conformation X of carrier has low affinity for passenger. Transported molecule detaches from carrier on high-concentration side.

Ⓟ = phosphate

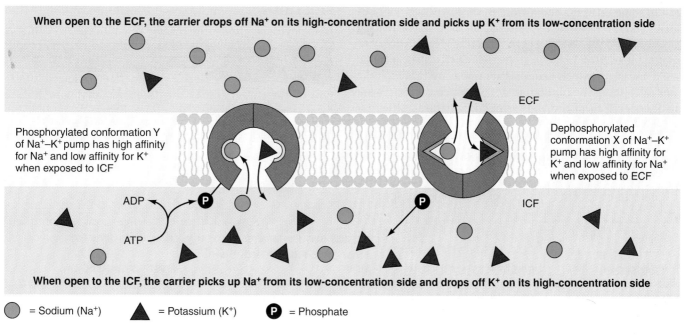

When open to the ECF, the carrier drops off Na⁺ on its high-concentration side and picks up K⁺ from its low-concentration side

ECF

Phosphorylated conformation Y of Na⁺–K⁺ pump has high affinity for Na⁺ and low affinity for K⁺ when exposed to ICF

Dephosphorylated conformation X of Na⁺–K⁺ pump has high affinity for K⁺ and low affinity for Na⁺ when exposed to ECF

ADP

ATP

ICF

When open to the ICF, the carrier picks up Na⁺ from its low-concentration side and drops off K⁺ on its high-concentration side

◯ = Sodium (Na⁺)   ▲ = Potassium (K⁺)   Ⓟ = Phosphate

● **FIGURE 3-17**

Na⁺–K⁺ ATPase pump. The plasma membrane of all cells contains an active transport carrier, the Na⁺–K⁺ ATPase pump, which uses energy in the carrier's phosphorylation–dephosphorylation cycle to sequentially transport Na⁺ out of the cell and K⁺ into the cell against these ions' concentration gradients. This pump moves three Na⁺ out and two K⁺ in for each ATP split.

leading to the drop-off of Na⁺ on the exterior. The subsequent dephosphorylation of the carrier increases its affinity for K⁺ on the extracellular side and restores the original carrier conformation, thereby transferring K⁺ into the cytoplasm. There is not a direct exchange of Na⁺ for K⁺, however; the Na⁺–K⁺ pump moves three Na⁺ out of the cell for every two K⁺ it pumps in. (To appreciate the magnitude of active Na⁺–K⁺ pumping that takes place, consider that a single nerve cell membrane contains perhaps 1 million Na⁺–K⁺ pumps capable of transporting about 200 million ions per second.)

The two most important roles of the Na⁺–K⁺ pump are as follow:

1. It establishes Na⁺ and K⁺ concentration gradients across the plasma membrane of all cells; these gradients are critically important in the ability of nerve and muscle cells to generate electrical signals essential to their functioning (a topic discussed more thoroughly later).

2. It helps regulate cell volume by controlling the concentrations of solutes inside the cell and thus minimizing osmotic effects that would induce swelling or shrinking of the cell.

■ **With vesicular transport, material is moved into or out of the cell wrapped in membrane.**

The special carrier-mediated transport systems embedded in the plasma membrane can selectively transport ions and small polar molecules. But what about large polar molecules or even multimolecular materials that must leave or enter the cell, such as during secretion of protein hormones (large polar molecules) by endocrine cells or during ingestion of invading bacteria (multimolecular particles) by white blood cells? These materials are unable to cross the plasma membrane, even with assistance: They are too large for channels, and no carriers exist for them (they would not even fit into a carrier molecule). These large particles are transferred between the ICF and ECF not by crossing the membrane but by being wrapped in a membrane-enclosed vesicle, a process known as **vesicular transport**. Vesicular transport requires energy expenditure by the cell, so this is an active method of membrane transport. Energy is needed to accomplish vesicle formation and vesicle movement within the cell. Transport into the cell in this manner is termed *endocytosis*, whereas transport out of the cell is called *exocytosis*.

ENDOCYTOSIS

To review, in **endocytosis** the plasma membrane surrounds the substance to be ingested, then fuses over the surface, pinching off a membrane-enclosed vesicle so that the engulfed material is trapped within the cell (see p. 25). Recall that there are three forms of endocytosis, depending on the nature of the material internalized: pinocytosis (nonselective uptake of ECF), receptor-mediated endocytosis (selective uptake of a large molecule), and phagocytosis (selective uptake of a multimolecular particle).

Once inside the cell, an engulfed vesicle has two possible destinies:

1. In most instances, lysosomes fuse with the vesicle to degrade and release its contents into the intracellular fluid.

2. In some cells, the endocytotic vesicle bypasses the lysosomes and travels to the opposite side of the cell, where it releases its contents by exocytosis. This provides a pathway to shuttle intact particles through the cell. Such vesicular traffic is one means by which materials are transferred through the thin cells lining the capillaries, across which exchanges are made between the blood and surrounding tissues.

EXOCYTOSIS

In **exocytosis**, almost the reverse of endocytosis occurs. A membrane-enclosed vesicle formed within the cell fuses with the plasma membrane, then opens up and releases its contents to the exterior (see p. 24). Materials packaged for export by the endoplasmic reticulum and Golgi complex are externalized by exocytosis.

▲ **TABLE 3-2**

Characteristics of the Methods of Membrane Transport

METHODS OF TRANSPORT	SUBSTANCES INVOLVED	ENERGY REQUIREMENTS AND FORCE-PRODUCING MOVEMENT	LIMIT TO TRANSPORT
Diffusion			
Through lipid bilayer	Nonpolar molecules of any size (e.g., O_2, CO_2, fatty acids)	Passive; molecules move down concentration gradient (from high to low concentration)	Continues until gradient is abolished (steady state with no net diffusion)
Through protein channel	Specific small ions (e.g., Na^+, K^+, Ca^{2+}, Cl^-)	Passive; ions move through open channels down electrochemical gradient (from high to low concentration and attraction of ion to area of opposite charge)	Continues until there is no net movement and a steady state is established
Special case of osmosis	Water only	Passive; water moves down its own concentration gradient (water moves to area of lower water concentration, i.e., higher solute concentration)	Continues until concentration difference is abolished or until stopped by an opposing hydrostatic pressure or until cell is destroyed
Carrier-Mediated Transport			
Facilitated diffusion	Specific polar molecules for which a carrier is available (e.g., glucose)	Passive; molecules move down concentration gradient (from high to low concentration)	Shows a transport maximum (T_m); carrier can become saturated
Active transport	Specific ions or polar molecules for which carriers are available (e.g., Na^+, K^+, amino acids)	Active; ions or molecules move against concentration gradient (from low to high concentration); requires ATP	Shows a transport maximum; carrier can become saturated
Vesicular Transport			
Endocytosis			
Pinocytosis	Small volume of ECF	Active; plasma membrane dips inward and pinches off at surface, forming internalized vesicle	Control poorly understood
Receptor-mediated endocytosis	Specific large polar molecule (e.g., protein)	Active; plasma membrane dips inward and pinches off at surface, forming internalized vesicle	Necessitates binding to specific receptor site on membrane surface
Phagocytosis	Multimolecular particles (e.g., bacteria and cell debris)	Active; cell extends pseudopods that surround particle, forming internalized vesicle	Necessitates binding to specific receptor site on membrane surface
Exocytosis	Secretory products (e.g., hormones and enzymes) as well as large molecules passing through cell intact	Active; secretory vesicle formed within cell fuses with plasma membrane, opens up and releases contents to outside	Secretion triggered by specific neural or hormonal stimuli

Exocytosis serves two different purposes:

1. It provides a mechanism for secreting large polar molecules, such as protein hormones and enzymes that are unable to cross the plasma membrane. In this case, the vesicular contents are highly specific and are released only on receipt of appropriate signals.
2. It enables the cell to add specific components to the membrane, such as selected carriers, channels, or receptors, depending on the cell's needs. In such cases, the composition of the membrane surrounding the vesicle is important, and the contents may be merely a sampling of ICF.

Our discussion of membrane transport is now complete; ▲ Table 3-2 summarizes the pathways by which materials can pass between the ECF and ICF. Cells are differentially selective in what enters or leaves because they have varying numbers and kinds of channels, carriers, and mechanisms for vesicular transport. Large polar molecules (too large for channels and not lipid soluble) for which there are no special transport mechanisms are unable to permeate.

The selective transport of K^+ and Na^+ is responsible for the electrical properties of cells. We turn our attention to this topic next.

 Click on the Media Exercises menu of the CD-ROM and work Media Exercise 3.2: Means of Transmembrane Exchange to test your understanding of the previous section.

MEMBRANE POTENTIAL

The plasma membranes of all living cells have a membrane potential, or are polarized electrically.

▌ Membrane potential is a separation of opposite charges across the plasma membrane.

The term **membrane potential** refers to a separation of charges across the membrane or to a difference in the relative number of cations and anions in the ICF and ECF. Recall that opposite charges tend to attract each other and like charges tend to repel each other. Work must be performed (energy expended) to separate opposite charges after they have come together. Conversely, when oppositely charged particles have been separated, the electrical force of attraction between them can be harnessed to perform work when the charges are permitted to come together again. This is the basic principle underlying electrically powered devices. Because separated charges have the "potential" to do work, a separation of charges across the membrane is called a membrane potential. Potential is measured in units of volts (the same unit used for the voltage in electrical devices), but because the membrane potential is relatively low, the unit used is the **millivolt (mV)** (1 mV = 1/1,000 volt).

Because the concept of potential is fundamental to understanding much of physiology, especially nerve and muscle physiology, it is important to understand clearly what this term means. The membrane in ● Figure 3-18a is electrically neutral. An equal number of positive (+) and negative (−)

charges are on each side of the membrane, so no membrane potential exists. In ● Figure 3-18b, some of the positive charges from the right side have been moved to the left. Now the left has an excess of positive charges, leaving an excess of negative charges on the right. In other words, there is a separation of opposite charges across the membrane, or a difference in the relative number of positive and negative charges between the two sides. That is, now a membrane potential exists. The attractive force between these separated charges causes them to accumulate in a thin layer along the outer and inner surfaces of the plasma membrane (● Figure 3-18c). These separated charges represent only a small fraction of the total number of charged particles (ions) present in the ICF and ECF. The vast majority of the fluid inside and outside the cells is electrically neutral (● Figure 3-18d). The electrically balanced ions can be ignored, because they do not contribute to membrane potential. Thus an almost insignificant fraction of the total number of charged particles present in the body fluids is responsible for the membrane potential. Note that the membrane itself is not charged. The term *membrane potential* refers to the difference in charge between the wafer-thin regions of ICF and ECF lying next to the inside and outside of the membrane, respectively.

The magnitude of the potential depends on the degree of separation of the opposite charges: The greater the number of charges separated, the larger the potential. Therefore, in ● Figure 3-18e membrane B has more potential than A and less potential than C.

▌ Membrane potential is due to differences in the concentration and permeability of key ions.

All cells have membrane potential. The cells of *excitable tissues*—namely nerve cells and muscle cells—have the ability to produce rapid, transient changes in their membrane potential when excited. These brief fluctuations in potential serve as electrical signals. The constant membrane potential present in the cells of nonexcitable tissues and those of excitable tissues when they are at rest—that is, when they are not producing electrical signals—is known as the **resting membrane potential**. We will concentrate now on the generation and maintenance of the resting membrane potential and will examine in later chapters the changes that take place in excitable tissues during electrical signaling.

The unequal distribution of a few key ions between the ICF and ECF and their selective movement through the plasma membrane are responsible for the electrical properties of the membrane. In the body, electrical charges are carried by ions. The ions primarily responsible for the generation of the resting membrane potential are Na^+, K^+, and A^-. The last refers to the large, negatively charged (anionic) intracellular proteins. Other ions (calcium, chloride, and bicarbonate, to name a few) do not make a direct contribution to the resting electrical properties of the plasma membrane in most cells, even though they play other important roles in the body.

The concentrations and relative permeabilities of the ions critical to membrane electrical activity are compared in

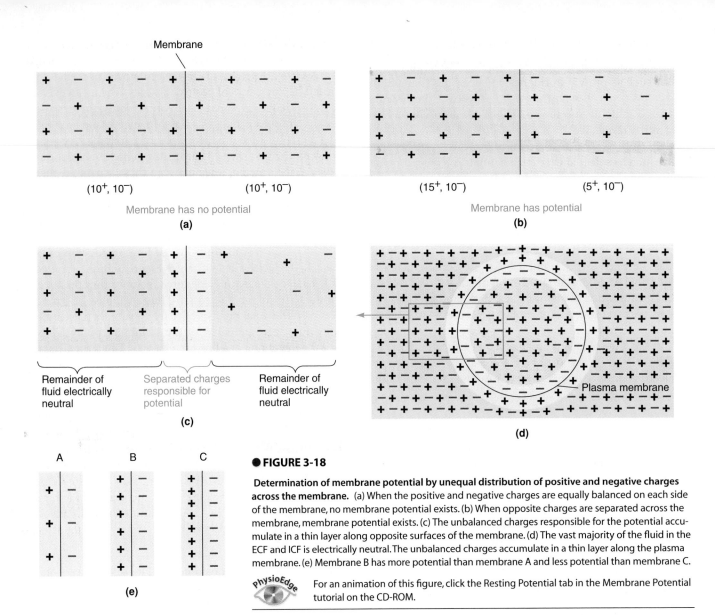

● FIGURE 3-18

Determination of membrane potential by unequal distribution of positive and negative charges across the membrane. (a) When the positive and negative charges are equally balanced on each side of the membrane, no membrane potential exists. (b) When opposite charges are separated across the membrane, membrane potential exists. (c) The unbalanced charges responsible for the potential accumulate in a thin layer along opposite surfaces of the membrane. (d) The vast majority of the fluid in the ECF and ICF is electrically neutral. The unbalanced charges accumulate in a thin layer along the plasma membrane. (e) Membrane B has more potential than membrane A and less potential than membrane C.

 For an animation of this figure, click the Resting Potential tab in the Membrane Potential tutorial on the CD-ROM.

▲ Table 3-3. Note that Na^+ is in greater concentration in the extracellular fluid and K^+ is in much higher concentration in the intracellular fluid. These concentration differences are maintained by the Na^+–K^+ pump at the expense of energy. Because the plasma membrane is virtually impermeable to A^-, these large, negatively charged proteins are found only inside the cell. After they have been synthesized from amino acids transported into the cell, they remain trapped within the cell.

In addition to the active carrier mechanism, Na^+ and K^+ can passively cross the membrane through protein channels specific for them. It is usually much easier for K^+ than for Na^+ to get through the membrane, because typically the membrane has many more channels open for passive K^+ traffic than for passive Na^+ traffic across the membrane. At resting potential in a nerve cell, the membrane is about 50 to 75 times more permeable to K^+ than to Na^+.

Armed with a knowledge of the relative concentrations and permeabilities of these ions, we can now analyze the forces acting across the plasma membrane. This analysis will be

▲ **TABLE 3-3**

Concentration and Permeability of Ions Responsible for Membrane Potential in a Resting Nerve Cell

ION	CONCENTRATION (millimoles/liter)		RELATIVE PERMEABILITY
	Extracellular	**Intracellular**	
Na⁺	150	15	1
K⁺	5	150	50–75
A⁻	0	65	0

PhysioEdge For an animation related to this table, click the Ion Concentration and Permeability tab in the Membrane Potential tutorial on the CD-ROM.

broken down as follows: We will consider first the direct contributions of the Na^+–K^+ pump to membrane potential; second, the effect that the movement of K^+ alone would have on membrane potential; third, the effect of Na^+ alone; and finally, the situation that exists in the cells when both K^+ and Na^+ effects are taking place concurrently. Remember throughout this discussion that *the concentration gradient for K^+ will always be outward* and *the concentration gradient for Na^+ will always be inward*, because the Na^+–K^+ pump maintains a higher concentration of K^+ inside the cell and a higher concentration of Na^+ outside the cell. Also, note that because K^+ and Na^+ are both cations (positively charged), *the electrical gradient for both of these ions will always be toward the negatively charged side of the membrane.*

EFFECT OF THE SODIUM–POTASSIUM PUMP ON MEMBRANE POTENTIAL

About 20% of the membrane potential is directly generated by the Na^+–K^+ pump. This active transport mechanism pumps three Na^+ out for every two K^+ it transports in. Because Na^+ and K^+ are both positive ions, this unequal transport generates a membrane potential, with the outside becoming relatively more positive than the inside as more positive ions are transported out than in. However, most of the membrane potential—the remaining 80%—is caused by the passive diffusion of K^+ and Na^+ down concentration gradients. Thus most of the Na^+–K^+ pump's role in producing membrane potential is indirect, through its critical contribution to maintaining the concentration gradients directly responsible for the ion movements that generate most of the potential.

EFFECT OF THE MOVEMENT OF POTASSIUM ALONE ON MEMBRANE POTENTIAL: K^+ EQUILIBRIUM POTENTIAL

Let's consider a hypothetical situation characterized by (1) the concentrations that exist for K^+ and A^- across the plasma membrane, (2) free permeability of the membrane to K^+ but not to A^-, and (3) no potential as yet present. The concentration gradient for K^+ would tend to move this ion out of the cell (● Figure 3-19). Because the membrane is permeable to K^+, this ion would readily pass through. As potassium ions moved to the outside, they would carry their positive charge with them, so more positive charges would be on the outside. At the same time, negative charges in the form of A^- would be left behind on the inside, similar to the situation shown in ● Figure 3-18b. (Remember that the large protein anions cannot diffuse out, despite a tremendous concentration gradient.) A membrane potential would now exist. Because an electrical gradient would also be present, K^+, being a positively charged ion, would be attracted toward the negatively charged interior and repelled by the positively charged exterior. Thus two opposing forces would now be acting on K^+: the concentration gradient tending to move K^+ out of the cell and the electrical gradient tending to move these same ions into the cell.

Initially the concentration gradient would be stronger than the electrical gradient, so net diffusion of K^+ out of the cell would continue, and the membrane potential would increase. As more and more K^+ moved down its concentration gradient and out of the cell, however, the opposing electrical gradient would also become greater as the outside became increasingly more positive and the inside more negative. Net outward diffusion would gradually be reduced as the strength

● **FIGURE 3-19**

Equilibrium potential for K^+

 For an animation of this figure, click the Resting Potential tab in the Membrane Potential tutorial on the CD-ROM.

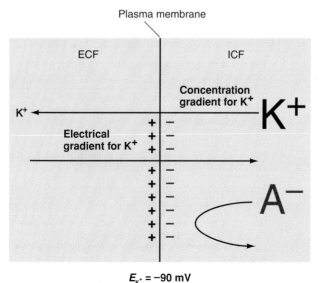

① The concentration gradient for K^+ tends to push this ion out of the cell.

② The outside of the cell becomes more + as the positively charged K^+ ions move to the outside down their concentration gradient.

③ The membrane is impermeable to the large intracellular protein anion (A^-). The inside of the cell becomes more − as the positively charged K^+ ions move out, leaving behind the negatively charged A^-.

④ The resulting electrical gradient tends to move K^+ into the cell.

⑤ No further net movement of K^+ occurs when the inward electrical gradient exactly counterbalances the outward concentration gradient. The membrane potential at this equilibrium point is the equilibrium potential for K^+ (E_{K^+}) at −90mV.

of the electrical gradient approached that of the concentration gradient. Finally, when these two forces exactly balanced each other (that is, when they were in equilibrium), no further net movement of K^+ would occur. The potential that would exist at this equilibrium is known as the **K^+ equilibrium potential (E_{K^+})**. At this point, a large concentration gradient for K^+ would still exist, but no more net movement of K^+ would occur out of the cell down this concentration gradient because of the exactly equal opposing electrical gradient (● Figure 3-19).

The membrane potential at E_{K^+} is −90 mV. By convention, *the sign always designates the polarity of the excess charge on the inside of the membrane.* A membrane potential of −90 mV means that the potential is of a magnitude of 90 mV, with the inside being negative relative to the outside. A potential of +90 mV would have the same strength, but in this case the inside would be more positive than the outside.

EFFECT OF MOVEMENT OF SODIUM ALONE ON MEMBRANE POTENTIAL: Na^+ EQUILIBRIUM POTENTIAL

A similar hypothetical situation could be developed for Na^+ alone (● Figure 3-20). The concentration gradient for Na^+ would move this ion into the cell, producing a buildup of positive charges on the interior of the membrane and leaving negative charges unbalanced outside (primarily in the form of chloride, Cl^-; Na^+ and Cl^-—that is, salt—are the predominant ECF ions). Net diffusion inward would continue until equilibrium was established by the development of an opposing electrical gradient that exactly counterbalanced the concentration gradient. At this point, given the concentrations for Na^+, the **Na^+ equilibrium potential (E_{Na^+})** would be +60 mV. In this case the inside of the cell would be positive,

in contrast to the equilibrium potential for K^+. The magnitude of E_{Na^+} is somewhat less than for E_{K^+} (60 mV compared to 90 mV) because the concentration gradient for Na^+ is not as large (▲ Table 3-3); thus, the opposing electrical gradient (membrane potential) is not as great at equilibrium.

CONCURRENT POTASSIUM AND SODIUM EFFECTS ON MEMBRANE POTENTIAL

Neither K^+ nor Na^+ exists alone in the body fluids, so equilibrium potentials are not present in the body cells. They exist only in hypothetical or experimental conditions. In a living cell, the effects of both K^+ and Na^+ must be taken into account. *The greater the permeability of the plasma membrane for a given ion, the greater the tendency for that ion to drive the membrane potential toward the ion's own equilibrium potential.* Because the membrane at rest is 50 to 75 times more permeable to K^+ than to Na^+, K^+ passes through more readily than Na^+; thus K^+ influences the resting membrane potential to a much greater extent than Na^+ does. Recall that K^+ acting alone would establish an equilibrium potential of −90 mV. The membrane is somewhat permeable to Na^+, however, so some Na^+ enters the cell in a limited attempt to reach its equilibrium potential. This Na^+ influx neutralizes, or cancels, some of the potential produced by K^+ alone.

To better understand this concept, assume that each separated pair of charges in ● Figure 3-21 represents 10 mV of potential. (This is not technically correct, because in reality many separated charges must be present to account for a potential of 10 mV.) In this simplified example, nine separated pluses and minuses, with the minuses on the inside, would represent the E_{K^+} of −90 mV. Superimposing the slight influence of Na^+ on this K^+-dominated membrane, assume

---

● FIGURE 3-20

Equilibrium potential for Na^+

 PhysioEdge For an animation of this figure, click the Resting Potential tab in the Membrane Potential tutorial on the CD-ROM.

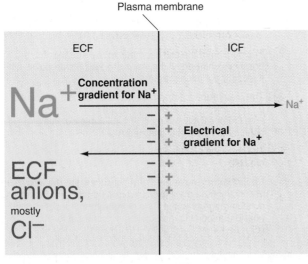

1. The concentration gradient for Na^+ tends to push this ion into the cell.

2. The inside of the cell becomes more + as the positively charged Na^+ ions move to the inside down their concentration gradient.

3. The outside becomes more − as the positively charged Na^+ ions move in, leaving behind in the ECF unbalanced negatively charged ions, mostly Cl^-.

4. The resulting electrical gradient tends to move Na^+ out of the cell.

5. No further net movement of Na^+ occurs when the outward electrical gradient exactly counterbalances the inward concentration gradient. The membrane potential at this equilibrium point is the equilibrium potential for Na^+ (E_{Na^+}) at +60 mV.

E_{Na^+} = +60 mV

1. The Na+–K+ pump actively transports Na+ out of and K+ into the cell, keeping the concentration of Na+ high in the ECF and the concentration of K+ high in the ICF.

2. Given the concentration gradients that exist across the plasma membrane, K+ tends to drive the membrane potential to K+'s equilibrium potential (−90 mV), whereas Na+ tends to drive the membrane potential to Na+'s equilibrium potential (+60 mV).

3. However, K+ exerts the dominant effect on the resting membrane potential because the membrane is more permeable to K+. As a result, the resting potential (−70 mV) is much closer to E_{K^+} than to E_{Na^+}.

4. During the establishment of resting potential, the relatively large net diffusion of K+ outward does not produce a potential of −90 mV because the resting membrane is slightly permeable to Na+ and the relatively small net diffusion of Na+ inward neutralizes (in gray shading) some of the potential that would be created by K+ alone, bringing the resting potential to −70 mV, slightly less than E_{K^+}.

5. The negatively charged intracellular proteins (A−) that cannot permeate the membrane remain unbalanced inside the cell during the net outward movement of the positively charged ions, so the inside of the cell is more negative than the outside.

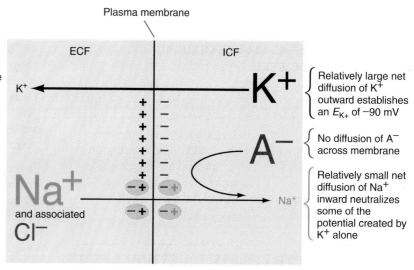

Resting membrane potential = −70 mV

(A− = Large intracellular anionic proteins)

● **FIGURE 3-21**

Effect of concurrent K+ and Na+ movement on establishing the resting membrane potential

 For an animation of this figure, click the Resting Potential tab in the Membrane Potential tutorial on the CD-ROM.

---

that two sodium ions enter the cell down the Na+ concentration and electrical gradients. (Note that the electrical gradient for Na+ is now inward in contrast to the outward electrical gradient for Na+ at E_{Na^+}. At E_{Na^+}, the inside of the cell is positive as a result of the inward movement of Na+ down its concentration gradient. In a resting nerve cell, however, the inside is negative because of the dominant influence of K+ on membrane potential. Thus both the concentration and electrical gradients now favor the inward movement of Na+.) The inward movement of these two positively charged sodium ions neutralizes some of the potential established by K+, so now only seven pairs of charges are separated, and the potential is −70 mV. This is the resting membrane potential of a typical nerve cell. The resting potential is much closer to E_{K^+} than to E_{Na^+} because of the greater permeability of the membrane to K+, but it is slightly less than E_{K^+} (−70 mV is a lower potential than −90 mV) because of the weak influence of Na+.

BALANCE OF PASSIVE LEAKS AND ACTIVE PUMPING AT RESTING MEMBRANE POTENTIAL

At resting potential, neither K+ nor Na+ is at equilibrium. A potential of −70 mV does not exactly counterbalance the concentration gradient for K+; it takes a potential of −90 mV to do that. Thus there is a continual tendency for K+ to passively exit through its leak channels. In the case of Na+, the concentration and electrical gradients do not even oppose each other; they both favor the inward movement of Na+.

Therefore, Na+ continually leaks inward down its electrochemical gradient, but only slowly, because of its low permeability; that is, because of the scarcity of Na+ leak channels.

Because such leaking goes on all the time, why doesn't the intracellular concentration of K+ continue to fall and the concentration of Na+ inside the cell progressively increase? Because of the Na+–K+ pump, this does not happen. This active transport mechanism counterbalances the rate of leakage. At resting potential, the pump transports back into the cell essentially the same number of potassium ions that have leaked out and simultaneously transports to the outside the sodium ions that have leaked in. Because the pump offsets the leaks, the concentration gradients for K+ and Na+ remain constant across the membrane. Thus not only is the Na+–K+ pump initially responsible for the Na+ and K+ concentration differences across the membrane, but it also maintains these differences.

As just discussed, the magnitude of these concentration gradients, together with the difference in permeability of the membrane to these ions, accounts for the magnitude of the membrane potential. Because the concentration gradients and permeabilities for Na+ and K+ remain constant in the resting state, the resting membrane potential established by these forces remains constant. At this point, no net movement of any ions takes place, because all passive leaks are exactly balanced by active pumping. A steady state exists, even though there is still a strong concentration gradient for both K+ and Na+ in opposite directions, as well as a slight excess of positive charges in the ECF accompanied by a corresponding slight excess of negative charges in the ICF (enough to account for a potential of the magnitude of 70 mV).

SPECIALIZED USE OF MEMBRANE POTENTIAL IN NERVE AND MUSCLE CELLS

Nerve and muscle cells have developed a specialized use for membrane potential. They can rapidly and transiently alter their membrane permeabilities to the involved ions in response to appropriate stimulation, thereby bringing about fluctuations in membrane potential. The rapid fluctuations in potential are responsible for producing nerve impulses in nerve cells and for triggering contraction in muscle cells. These activities are the major focus of the next five chapters.

CHAPTER IN PERSPECTIVE: FOCUS ON HOMEOSTASIS

All cells of the body must obtain vital materials such as nutrients and O_2 from the surrounding ECF and must transfer to the ECF wastes to be eliminated as well as secretory products such as chemical messengers and digestive enzymes. Thus transport of materials across the plasma membrane between the ECF and ICF is essential for cell survival, and the constituents in the ECF must be homeostatically maintained in order to support these life-sustaining exchanges.

Many cell types use membrane transport to carry out their specialized activities geared toward maintaining homeostasis. Here are several examples:

1. Absorption of nutrients from the digestive tract lumen involves the transport of these energy-giving molecules across the membranes of the cells lining the tract.

2. Exchange of O_2 and CO_2 between the air and blood in the lungs involves the transport of these gases across the membranes of the cells lining the lungs' air sacs and blood vessels.

3. Urine formation is accomplished by the selective transfer of materials between the blood and the fluid within the kidney tubules across the membranes of the cells lining the tubules.

4. The beating of the heart is triggered by cyclic changes in the transport of Na^+, K^+, and Ca^{2+} across the heart cells' membranes.

5. Secretion of chemical messengers such as neurotransmitters from nerve cells and hormones from endocrine cells involves the transport of these regulatory products to the ECF on appropriate stimulation.

In addition to providing selective transport of materials between the ECF and ICF, the plasma membrane contains receptors for binding with specific chemical messengers that regulate various cell activities, many of which are specialized activities aimed toward maintaining homeostasis. For example, the hormone vasopressin, which is secreted in response to a water deficit in the body, binds with receptors in the plasma membrane of a specific type of kidney cell. This binding triggers these cells to conserve water during urine formation, thus helping alleviate the water deficit that initiated the response.

All living cells have a membrane potential, with the cell's interior being slightly more negative than the fluid surrounding the cell when the cell is electrically at rest. The specialized activities of nerve and muscle cells depend on these cells' ability to change their membrane potential rapidly on appropriate stimulation. These transient, rapid changes in potential in nerve cells serve as electrical signals or nerve impulses, which provide a means to transmit information along nerve pathways. This information is used to accomplish homeostatic adjustments, such as restoring blood pressure to normal when signaled that it has fallen too low.

Rapid changes in membrane potential in muscle cells trigger muscle contraction, the specialized activity of muscle. Muscle contraction contributes to homeostasis in many ways, including the pumping of blood by the heart and moving food through the digestive tract.

CHAPTER SUMMARY

Membrane Structure and Composition (pp. 43–47)

- All cells are bounded by a plasma membrane, a thin lipid bilayer in which proteins are interspersed and to which carbohydrates are attached on the outer surface.

- The electron microscopic appearance of the plasma membrane as a trilaminar structure (two dark lines separated by a light interspace) is caused by the arrangement of the molecules composing it (Review Figure 3-1). The phospholipids orient themselves to form a bilayer with a hydrophobic interior (light interspace) sandwiched between the hydrophilic outer and inner surfaces (dark lines). (Review Figure 3-2.)

- This lipid bilayer forms the structural boundary of the cell, serving as a barrier for water-soluble substances and being responsible for the fluid nature of the membrane.

- Cholesterol molecules tucked between the phospholipids contribute to the fluidity and stability of the membrane.

- According to the fluid mosaic model of membrane structure, the lipid bilayer is embedded with proteins (Review Figure 3-3). Membrane proteins, which vary in type and distribution among cells, serve as (1) channels for passage of small ions across the membrane; (2) carriers for transport of specific substances in or out of the cell; (3) docking-marker acceptors for fusion with and subsequent exocytosis of secretory vesicles; (4) membrane-bound enzymes that govern specific chemical reactions; (5) receptors for detecting and responding to chemical messengers that alter cell function; and (6) cell adhesion molecules that help hold cells together.

- The membrane carbohydrates, short sugar chains that project from the outer surface only, serve as self-identity markers (Review Figure 3-3). They are important in recognition of "self" in cell-to-cell interactions such as tissue formation and tissue growth.

Cell-to-Cell Adhesions (pp. 47–50)

- Special cells locally secrete a complex extracellular matrix, which serves as a biological "glue" between the cells of a tissue.
- The extracellular matrix consists of a watery, gel-like substance interspersed with three major types of protein fibers: collagen, elastin, and fibronectin.
- Many cells are further joined by specialized cell junctions, of which there are three types: desmosomes, tight junctions, and gap junctions.
- Desmosomes serve as adhering junctions to hold cells together mechanically and are especially important in tissues subject to a great deal of stretching. (Review Figure 3-4.)
- Tight junctions actually fuse cells together to seal off passage between cells, thereby permitting only regulated passage of materials through the cells. These impermeable junctions are found in the epithelial sheets that separate compartments with very different chemical compositions. (Review Figure 3-5.)
- Gap junctions are communicating junctions between two adjacent but not touching cells. Cells joined by gap junctions are connected by small tunnels that permit exchange of ions and small molecules between the cells. Such movement of ions plays a key role in the spread of electrical activity to synchronize contraction in heart and smooth muscle. (Review Figure 3-6.)

Membrane Transport (pp. 50–61)

- Materials can pass between the ECF and ICF by unassisted and assisted means.
- Transport mechanisms may also be passive (the particle moves across the membrane without the cell expending energy) or active (the cell expends energy to move the particle across the membrane). (Review Table 3-2, p. 60.)
- Lipid-soluble particles and ions can cross the membrane unassisted. Nonpolar (lipid-soluble) molecules of any size can dissolve in and passively pass through the lipid bilayer down concentration gradients (Review Figures 3-7 and 3-8). Small ions traverse the membrane passively down electrochemical gradients through open protein channels specific for the ion.

- Osmosis is a special case of water passively moving down its own concentration gradient to an area of higher solute concentration. (Review Figures 3-9 through 3-13.)
- Carrier mechanisms are important for the assisted transfer of small polar molecules and for selected movement of ions across the membrane.
- In carrier-mediated transport, the particle is transported across by specific membrane carrier proteins. Carrier-mediated transport may be passive and move the particle down its concentration gradient (facilitated diffusion), or active and move the particle against its concentration gradient (active transport). (Review Figures 3-14 through 3-17.)
- Large polar molecules and multimolecular particles can leave or enter the cell by being wrapped in a piece of membrane to form vesicles that can be internalized (endocytosis) or externalized (exocytosis). (Review Figures 2-5 and 2-7.)

Membrane Potential (pp. 61–66)

- All cells have a membrane potential, which is a separation of opposite charges across the plasma membrane. (Review Figure 3-18)
- The Na^+–K^+ pump makes a small direct contribution to membrane potential through its unequal transport of positive ions; it transports more Na^+ ions out than K^+ ions in. (Review Figure 3-17.)
- The primary role of the Na^+–K^+ pump, however, is to actively maintain a greater concentration of Na^+ outside the cell and a greater concentration of K^+ inside the cell. These concentration gradients tend to passively move K^+ out of the cell and Na^+ into the cell. (Review Table 3-3 and Figures 3-19 and 3-20.)
- Because the resting membrane is much more permeable to K^+ than to Na^+, substantially more K^+ leaves the cell than Na^+ enters. This results in an excess of positive charges outside the cell and leaves an unbalanced excess of negative charges inside in the form of large protein anions (A^-) that are trapped within the cell. (Review Table 3-3 and Figure 3-21.) When the resting membrane potential of -70 mV is achieved, no further net movement of K^+ and Na^+ takes place, because any further leaking of these ions down their concentration gradients is quickly reversed by the Na^+–K^+ pump.

REVIEW EXERCISES

Objective Questions (answers on p. A-41)

1. The nonpolar tails of the phospholipid molecules bury themselves in the interior of the plasma membrane. (True or false?)
2. The hydrophobic regions of the molecules composing the plasma membrane correspond to the two dark layers of this structure visible under an electron microscope. (True or false?)
3. Through its unequal pumping, the Na^+–K^+ pump is directly responsible for separating sufficient charges to establish a resting membrane potential of -70 mV. (True or false?)
4. At resting membrane potential, there is a slight excess of _____ (positive/negative) charges on the inside of the membrane, with a corresponding slight excess of _____ charges on the outside.
5. Using the following answer code, indicate which membrane component is responsible for the function in question:
    (a) lipid bilayer
    (b) proteins
    (c) carbohydrates

____ 1. channel formation
____ 2. barrier to passage of water-soluble substances
____ 3. receptor sites
____ 4. membrane fluidity
____ 5. recognition of "self"
____ 6. membrane-bound enzymes
____ 7. structural boundary
____ 8. carriers

6. Using the following answer code, indicate the direction of net movement in each case:
    (a) movement from high to low concentration
    (b) movement from low to high concentration

____ 1. simple passive diffusion
____ 2. facilitated diffusion
____ 3. active transport
____ 4. water with regard to the water concentration gradient during osmosis
____ 5. water with regard to the solute concentration gradient during osmosis

7. Using the following answer code, indicate the type of cell junction described:
    (a) gap junction
    (b) tight junction
    (c) desmosome

    _____ 1. adhering junction
    _____ 2. impermeable junction
    _____ 3. communicating junction
    _____ 4. made up of connexons, which permit passage of ions and small molecules between cells
    _____ 5. consists of interconnecting fibers, which spot-rivet adjacent cells
    _____ 6. formed by an actual fusion of proteins on the outer surfaces of two interacting cells
    _____ 7. important in tissues subject to mechanical stretching
    _____ 8. important in synchronizing contractions within heart and smooth muscle by allowing spread of electrical activity between the cells composing the muscle mass
    _____ 9. important in preventing passage between cells in epithelial sheets that separate compartments of two different chemical compositions

Essay Questions

1. Describe the fluid mosaic model of membrane structure.
2. What are the functions of the three major types of protein fibers in the extracellular matrix?
3. What two properties of a particle influence whether it can permeate the plasma membrane?
4. List and describe the methods of membrane transport. Indicate what types of substances are transported by each method, and state whether each is a passive or active means of transport.
5. Explain what effect an increase in surface area would have on the rate of net diffusion across a membrane. Explain what effect an increase in thickness would have on the rate of net diffusion across a membrane.
6. State two important roles of the Na^+–K^+ pump.
7. Describe the contribution of each of the following to establishing and maintaining membrane potential: (a) the Na^+–K^+ pump; (b) passive movement of K^+ across the membrane; (c) passive movement of Na^+ across the membrane; and (d) the large intracellular anions.

POINTS TO PONDER

(Explanations on p. A-41)

1. Assume that a membrane permeable to Na^+ but not to Cl^- separates two solutions. The concentration of sodium chloride on side 1 is much higher than on side 2. Which of the following ionic movements would occur?
    a. Na^+ would move until its concentration gradient is dissipated (until the concentration of Na^+ on side 2 is the same as the concentration of Na^+ on side 1).
    b. Cl^- would move down its concentration gradient from side 1 to side 2.
    c. A membrane potential, negative on side 1, would develop.
    d. A membrane potential, positive on side 1, would develop.
    e. None of the above are correct.

2. Compared to resting potential, would the membrane potential become more negative or more positive if the membrane were more permeable to Na^+ than to K^+?

3. Which of the following methods of transport is being used to transfer the substance into the cell in the accompanying graph?

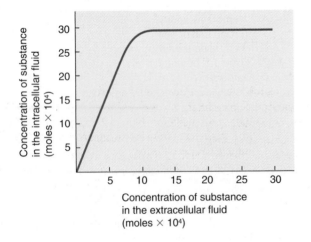

    a. diffusion down a concentration gradient
    b. osmosis
    c. facilitated diffusion
    d. active transport
    e. vesicular transport
    f. It is impossible to tell with the information provided.

4. Colostrum, the first milk that a mother produces, contains an abundance of antibodies, large protein molecules. These maternal antibodies help protect breast-fed infants from infections until the babies are capable of producing their own antibodies. By what means would you suspect these maternal antibodies are transported across the cells lining a newborn's digestive tract into the bloodstream?

5. The rate at which the Na^+–K^+ pump operates is not constant but is controlled by a combined effect of changes in ICF Na^+ concentration and ECF K^+ concentration. Do you think an increase in both ICF Na^+ and ECF K^+ concentrations would accelerate or slow down the Na^+–K^+ pump? What would be the benefit of this response? Before you reply, consider the following additional information about Na^+ and K^+ movement across the membrane. Not only do Na^+ and K^+ slowly and passively leak through their channels in a resting cell, but during an electrical impulse, known as an *action potential*, Na^+ rapidly and passively enters the cell; this movement is followed by a rapid, passive outflow of K^+. (These ion movements, which result from rapid changes in membrane permeability, bring about rapid, pronounced changes in membrane potential. This sequence of rapid potential changes—an action potential—serves as an electrical signal for conveying information along a nerve pathway.)

CLINICAL CONSIDERATION

(Explanation on p. A-41)

When William H. was helping victims following a devastating earthquake in a region that was not prepared to swiftly set up adequate temporary shelter, he developed severe diarrhea. He was diagnosed as having *cholera,* a disease transmitted through unsanitary water supplies that have been contaminated by fecal material from infected individuals. In this condition, the toxin produced by the cholera bacteria leads to opening of the Cl^- channels in the luminal membranes of the intestinal cells, thereby increasing the secretion of Cl^- from the cells into the intestinal tract lumen. By what mechanisms would Na^+ and water be secreted into the lumen in accompaniment with Cl^- secretion? How does this secretory response account for the severe diarrhea that is characteristic of cholera?

PHYSIOEDGE RESOURCES

 PhysioEdge CD-ROM

PhysioEdge, the CD-ROM packaged with your text, focuses on the concepts students find most difficult to learn. Figures marked with this icon have associated activities on the CD. For a visual review of concepts in this chapter, check out the following:

Tutorial: Membrane Potential

Media Exercise 3.1: The Plasma Membrane and Cell–Cell Connections

Media Exercise 3.2: Means of Transmembrane Exchange

PhysioEdge Website

The website for this book contains a wealth of helpful study aids, as well as many ideas for further reading and research. Log on to:
http://www.brookscole.com/hpfundamentals3
Select Chapter 3 from the drop-down menu, or click on one of the many resource areas.

For Suggested Readings, consult **InfoTrac College Edition/ Research** on the PhysioEdge website or go directly to InfoTrac College Edition, your online research library, at:
http://infotrac.thomsonlearning.com

Nervous and Endocrine Systems

Female

Body systems maintain homeostasis

Homeostasis
The nervous and endocrine systems, as the body's two major regulatory systems, regulate many body activities aimed at maintaining a stable internal fluid environment.

Homeostasis is essential for survival of cells

Cells

Cells make up body systems

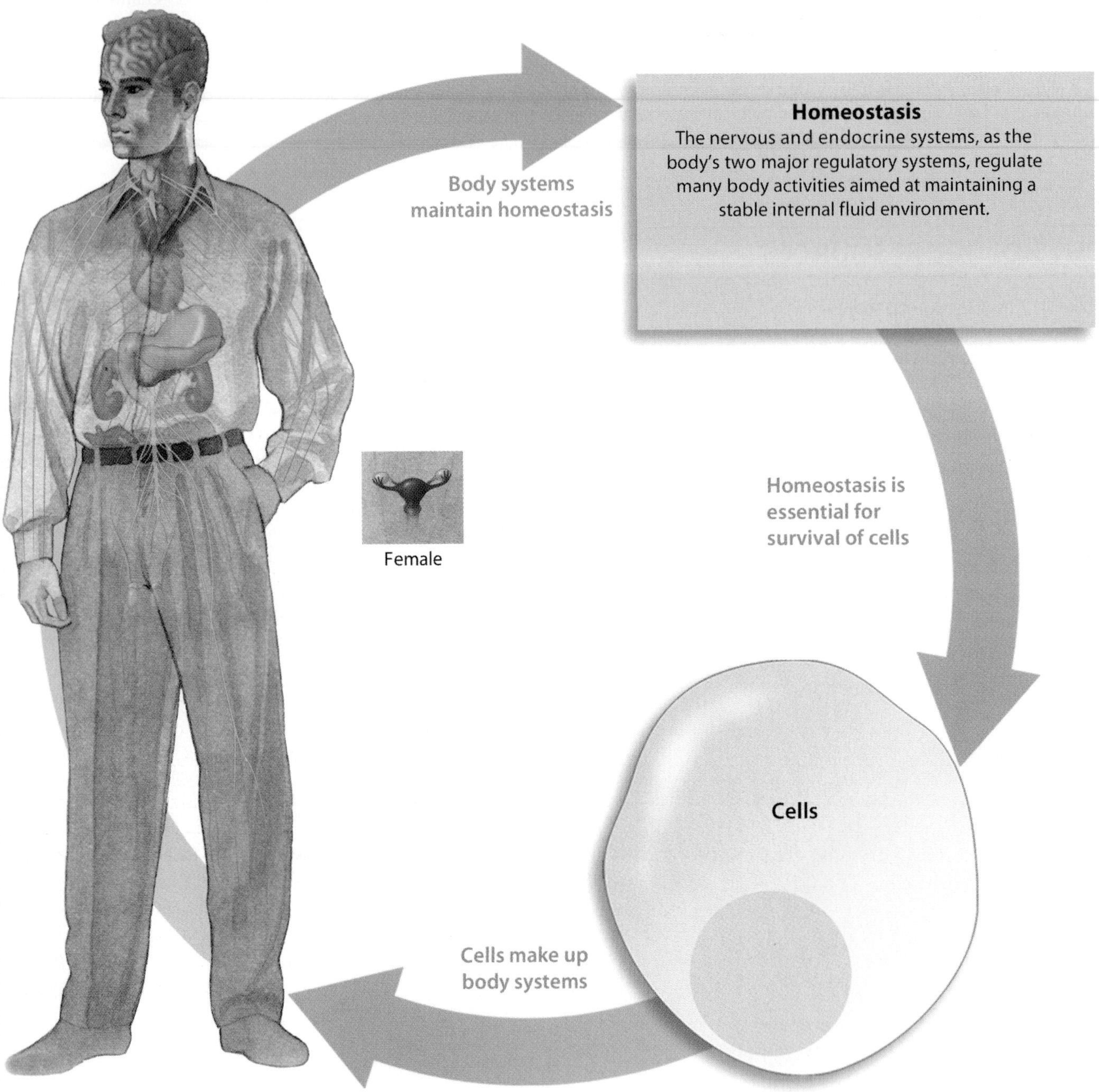

To maintain homeostasis, cells must work together in a co-ordinated fashion toward common goals. The two major regulatory systems of the body that help ensure life-sustaining coordinated responses are the nervous and endocrine systems. **Neural communication** is accomplished by means of nerve cells, or neurons, which are specialized for rapid electrical signaling and for secreting neurotransmitters, short-distance chemical messengers that act on nearby target organs. The nervous system exerts rapid control over most of the body's muscular and glandular activities. **Hormonal communication** is accomplished by hormones, which are long-distance chemical messengers secreted by the endo-crine glands into the blood. The blood carries the hormones to distant target sites, where they regulate processes that require duration rather than speed, such as metabolic activi-ties, water and electrolyte balance, and growth.

Principles of Neural and Hormonal Communication

CONTENTS AT A GLANCE

 Click on the Tutorials menu of the CD-ROM for a tutorial on Neuronal Physiology and Hormonal Communication.

Communication is critical for the survival of the society of cells that collectively compose the body. The ability of cells to communicate with each other is essential for coordination of their diverse activities to maintain homeostasis as well as to control growth and development of the body as a whole. In this chapter, we will consider the molecular and cellular means by which the two major regulatory systems of the body—the nervous and endocrine systems—communicate with the cells/tissues/organs/systems whose activities they control. We will begin with neural communication, then turn our attention to hormonal communication, and conclude with a general comparison of the action modes of the nervous and endocrine systems.

INTRODUCTION TO NEURAL COMMUNICATION

All body cells display a membrane potential, which is a separation of positive and negative charges across the membrane, as discussed in the preceding chapter. This potential is related to the uneven distribution of Na^+, K^+, and large intracellular protein anions between the intracellular fluid (ICF) and extracellular fluid (ECF), and to the differential permeability of the plasma membrane to these ions (see pp. 61–66).

▌ Nerve and muscle are excitable tissues.

Two types of cells, *nerve cells* and *muscle cells,* have developed a specialized use for this membrane potential. They can undergo transient, rapid changes in their membrane potentials. These fluctuations in potential serve as electrical signals. The constant membrane potential that exists when a nerve or muscle cell is not displaying rapid changes in potential is referred to as the *resting potential*. In Chapter 3 you learned that the resting potential of a typical nerve cell is −70 mV.

Nerve and muscle are considered **excitable tissues** because when excited they change their resting potential to produce electrical signals. Nerve cells, which are known as *neurons,* use these electrical signals to receive, process, initiate, and transmit messages. In muscle cells, these electrical signals initiate contraction. Thus electrical signals are critical to the function of the nervous system as well as all muscles. In this chapter, we will consider how neurons undergo changes in potential to accomplish their function. Muscle cells are discussed in later chapters.

▌Membrane potential decreases during depolarization and increases during hyperpolarization.

Before you can understand what electrical signals are and how they are created, it will help to become familiar with the following terms, used to describe changes in potential, as graphically represented on ● Figure 4-1:

1. **Polarization:** Charges are separated across the plasma membrane, so that the membrane has potential. Any time the value of the membrane potential is other than 0 mV, in either the positive or negative direction, the membrane is in a state of polarization. Recall that the magnitude of the potential is directly proportional to the number of positive and negative charges separated by the membrane and that the sign of the potential (+ or −) always designates whether excess positive or excess negative charges are present, respectively, on the inside of the membrane.
2. **Depolarization:** A change in potential that makes the membrane less polarized (less negative) than at resting potential. Depolarization decreases membrane potential, moving it closer to 0 mV (for example, a change from −70 mV to −60 mV); fewer charges are separated than at resting potential.
3. **Repolarization:** The membrane returns to resting potential after having been depolarized.
4. **Hyperpolarization:** A change in potential that makes the membrane more polarized (more negative) than at resting potential. Hyperpolarization increases membrane potential, moving it even farther from 0 mV (for instance, a change from

−70 mV to −80 mV); more charges are separated than at resting potential.

One possibly confusing point should be clarified. On the device used for recording rapid changes in potential, a *decrease* in potential (that is, the inside being less negative than at resting) is represented as an upward deflection, whereas an *increase* in potential (that is, the inside being more negative than at resting) is represented by a *downward* deflection.

▌Electrical signals are produced by changes in ion movement across the plasma membrane.

Changes in membrane potential are brought about by changes in ion movement across the membrane. For example, if the net inward flow of positively charged ions increases compared to the resting state, the membrane becomes depolarized (less negative inside). By contrast, if the net outward flow of positively charged ions increases compared to the resting state, the membrane becomes hyperpolarized (more negative inside).

Changes in ion movement in turn are brought about by changes in membrane permeability in response to *triggering events.* Depending on the type of electrical signal, a triggering event might be (1) a change in the electrical field in the vicinity of an excitable membrane; (2) an interaction of a chemical messenger with a surface receptor on a nerve or muscle cell membrane; (3) a stimulus, such as sound waves stimulating specialized nerve cells in your ear; or (4) a spontaneous change of potential caused by inherent imbalances in the leak–pump cycle. (You will learn more about the nature of these various triggering events as our discussion of electrical signals continues.)

Because the water-soluble ions responsible for carrying charge cannot penetrate the plasma membrane's lipid bilayer, these charges can only cross the membrane through channels specific for them. Membrane channels may be either *leak channels* or *gated channels.* **Leak channels** are open all the time, thus permitting unregulated leakage of their chosen ion across the membrane through the channels. **Gated channels,** in contrast, have gates that can alternately be open, permitting ion passage through the channel, or closed, preventing ion passage through the channels. Gate opening and closing results from a change in the three-dimensional conformation (shape) of the protein that forms the gated channel. There are four kinds of gated channels, depending on the factor that induces the change in channel conformation: (1) **voltage-gated channels,** which open or close in response to changes in membrane potential; (2) **chemically gated channels,** which change conformation in response to the binding of a specific chemical messenger with a membrane receptor in close association with the channel; (3) **mechanically gated channels,** which respond to stretching or other mechanical deformation; and (4) **thermally gated channels,** which respond to local changes in temperature (heat or cold).

Thus, triggering events alter membrane permeability and consequently alter ion flow across the membrane by opening or closing the gates guarding particular ion channels. These

● FIGURE 4-1

Types of changes in membrane potential

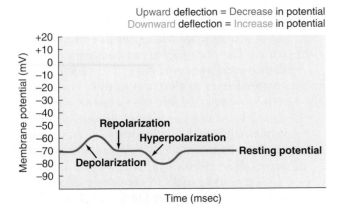

Upward **deflection** = Decrease in potential
Downward **deflection** = Increase in potential

ion movements redistribute charge across the membrane, causing membrane potential to fluctuate.

There are two basic forms of electrical signals: (1) *graded potentials,* which serve as short-distance signals; and (2) *action potentials,* which signal over long distances. We are now going to examine these types of signals in more detail, beginning with graded potentials, and then will explore how nerve cells use these signals to convey messages.

GRADED POTENTIALS

Graded potentials are local changes in membrane potential that occur in varying grades or degrees of magnitude or strength. For example, membrane potential could change from -70 mV to -60 mV (a 10-mV graded potential) or from -70 mV to -50 mV (a 20-mV graded potential).

▌ The stronger a triggering event, the larger the resultant graded potential.

Graded potentials are usually produced by a specific triggering event that causes gated ion channels to open in a specialized region of the excitable cell membrane. Most commonly, gated Na^+ channels open, leading to the inward movement of Na^+ down its concentration and electrical gradients. The resultant depolarization—the graded potential—is confined to this small, specialized region of the total plasma membrane.

The magnitude of this initial graded potential (that is, the difference between the new potential and the resting potential) is related to the magnitude of the triggering event: *The stronger the triggering event, the more gated channels that open, the greater the positive charge entering the cell, and the larger the depolarizing graded potential at the point of origin. Also, the longer the duration of the triggering event, the longer the duration of the graded potential.*

▌ Graded potentials spread by passive current flow.

When a graded potential occurs locally in a nerve or muscle cell membrane, the remainder of the membrane is still at resting potential. The temporarily depolarized region is called an *active area.* Note from ● Figure 4-2 that inside the cell, the active area is relatively more positive than the neighboring *inactive areas* that are still at resting potential. Outside the cell, the active area is relatively less positive than these adjacent areas. Because of this difference in potential, electrical charges, in this case carried by ions, passively flow between the active and adjacent resting regions on both the inside and outside of the membrane. Any flow of electrical charges is called a **current.** By convention, the direction of current flow is always designated by the direction in which the positive charges are moving (● Figure 4-2c). On the inside, positive charges flow through the ICF away from the relatively more positive depolarized active region toward the more negative adjacent resting regions. Similarly, outside the cell positive

charges flow through the ECF from the more positive adjacent inactive regions toward the relatively more negative active region. Ion movement (that is, current) is occurring *along* the membrane between regions next to each other on the same side of the membrane. This flow is in contrast to ion movement *across* the membrane through ion channels.

As a result of local current flow between an active depolarized area and an adjacent inactive area, the potential changes in the previously inactive area. Positive charges have flowed into this adjacent area on the inside, while simultaneously positive charges have flowed out of this area on the outside. Thus at this adjacent site the inside is more positive (or less negative), and the outside is less positive (or more negative) than before (● Figure 4-2c). Stated differently, the previously inactive adjacent region has been depolarized, so the graded potential has spread. This area's potential now differs from that of the inactive region immediately next to it on the other side, inducing further current flow at this new site, and so on. In this manner, current spreads in both directions away from the initial site of the potential change.

The amount of current that flows between two areas depends on the difference in potential between the areas and on the resistance of the material through which the charges are moving. **Resistance** is the hindrance to electrical charge movement. The greater the difference in potential, the greater the current flow. The lower the resistance, the greater the current flow. *Conductors* have low resistance, providing little hindrance to current flow. Electrical wires and the ICF and ECF are all good conductors, so current readily flows through them. *Insulators* have high resistance and greatly hinder movement of charge. The plastic surrounding electrical wires has high resistance, as do body lipids. Thus current does not flow across the plasma membrane's lipid bilayer. Current, carried by ions, can move across the membrane only through ion channels.

▌ Graded potentials die out over short distances.

The passive current flow between active and adjacent inactive areas is similar to the means by which current is carried through electrical wires. We know from experience that current leaks out of an electrical wire with dangerous results unless the wire is covered with an insulating material such as plastic. (People can get an electric shock if they touch a bare wire.) Likewise, current is lost across the plasma membrane as charge-carrying ions leak through the "uninsulated" parts of the membrane, that is, through open channels. Because of this current loss, the magnitude of the local current progressively diminishes with increasing distance from the initial site of origin (● Figure 4-3). Thus the magnitude of the graded potential continues to decrease the farther it moves away from the initial active area. Another way of saying this is that the spread of a graded potential is *decremental* (gradually decreases) (● Figure 4-4). Note that in ● Figure 4-3, the magnitude of the initial change in potential is 15 mV (a change from -70 mV to -55 mV), then decreases as it moves along the membrane to a change in potential of 10 mV (from -70 mV

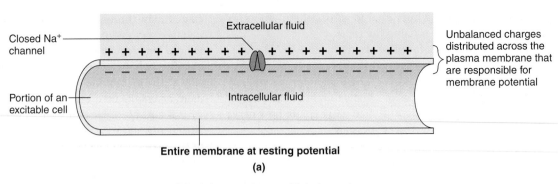

Extracellular fluid

Closed Na⁺ channel

Unbalanced charges distributed across the plasma membrane that are responsible for membrane potential

Portion of an excitable cell

Intracellular fluid

Entire membrane at resting potential

(a)

FIGURE 4-2

Current flow during a graded potential. (a) The membrane of an excitable cell at resting potential. (b) A triggering event opens Na⁺ channels, leading to the Na⁺ entry that brings about depolarization. The adjacent inactive areas are still at resting potential. (c) Local current flow occurs between the active and adjacent inactive areas. This local current flow results in depolarization of the previously inactive areas. In this way, the depolarization spreads away from its point of origin.

 PhysioEdge For an interaction related to this figure, see Media Exercise 4.2: Graded Potentials and Action Potentials on the CD-ROM.

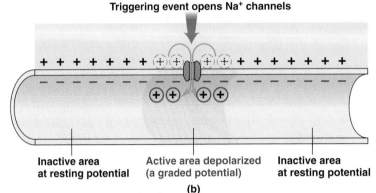

Triggering event opens Na⁺ channels

Inactive area at resting potential | Active area depolarized (a graded potential) | Inactive area at resting potential

(b)

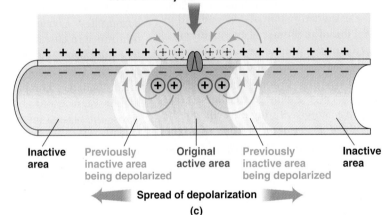

Local current flow occurs between the active and adjacent inactive areas

Inactive area | Previously inactive area being depolarized | Original active area | Previously inactive area being depolarized | Inactive area

Spread of depolarization

(c)

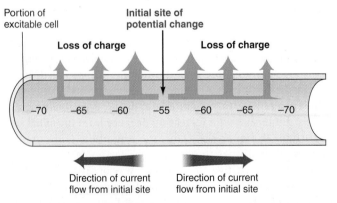

Portion of excitable cell

Initial site of potential change

Loss of charge   **Loss of charge**

−70   −65   −60   −55   −60   −65   −70

Direction of current flow from initial site   Direction of current flow from initial site

* Numbers refer to the local potential in mV at various points along the membrane.

FIGURE 4-3

Current loss across the plasma membrane. Leakage of charge-carrying ions across the plasma membrane results in progressive loss of current with increasing distance from the initial site of potential change.

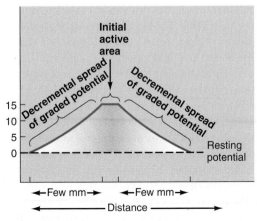

Change in membrane potential in mV relative to resting potential— i.e., magnitude of electrical signal

Initial active area

Decremental spread of graded potential

Decremental spread of graded potential

15
10
5
0 — — — Resting potential

←Few mm→   ←Few mm→

Distance

FIGURE 4-4

Decremental spread of graded potentials. Because of leaks in current, the magnitude of a graded potential continues to decrease as it passively spreads from the initial active area. The potential dies out altogether within a few millimeters of its site of initiation.

to −60 mV), and continues to diminish the farther it moves away from the initial active area, until there is no longer a change in potential. In this way, these local currents die out within a few millimeters from the initial site of change in potential and consequently can function as signals for only very short distances.

Although graded potentials have limited signaling distance, they are critically important to the body's function, as explained in later chapters. The following are all graded potentials: *postsynaptic potentials, receptor potentials, end-plate potentials, pacemaker potentials,* and *slow-wave potentials.* These terms are unfamiliar to you now, but you will become well acquainted with them as we continue discussing nerve and muscle physiology. We are including this list here because it is the only place all these graded potentials will be grouped together. For now it's enough to say that for the most part, excitable cells produce one of these types of graded potentials in response to a triggering event. In turn, graded potentials can initiate *action potentials,* the long-distance signals, in an excitable cell.

ACTION POTENTIALS

Action potentials are brief, rapid, large (100 mV) changes in membrane potential during which the potential actually reverses, so that the inside of the excitable cell transiently becomes more positive than the outside. As with a graded potential, a single action potential involves only a small portion of the total excitable cell membrane. Unlike graded potentials, however, action potentials are conducted, or propagated, throughout the entire membrane in *nondecremental* fashion; that is, they do not diminish in strength as they travel from their site of initiation throughout the remainder of the cell membrane. Thus action potentials can serve as faithful long-distance signals. Think about the nerve cell that brings about contraction of muscle cells in your big toe. If you want to wiggle your big toe, commands are sent from your brain down your spinal cord to initiate an action potential at the beginning of this nerve cell, which is located in the spinal cord. This action potential travels in undiminishing fashion all the way down the nerve cell's long axon, which runs through your leg to terminate on your big-toe muscle cells. The signal has not weakened or died off, being instead preserved at full strength from beginning to end.

Let's now consider the changes in potential during an action potential and the permeability and ion movements responsible for generating this change in potential, before we turn our attention to the means by which action potentials spread throughout the cell membrane in undiminishing fashion.

▌ During an action potential, the membrane potential rapidly, transiently reverses.

If of sufficient magnitude, a graded potential can initiate an action potential before the graded potential dies off. (Later you will discover the means by which this initiation is accomplished for the various types of graded potentials.) Typi-

cally, the portion of the excitable membrane where graded potentials are produced in response to a triggering event does not undergo action potentials. Instead, the graded potential, by electrical or chemical means, brings about depolarization of adjacent portions of the membrane where action potentials can take place. For convenience in this discussion, we will now jump from the triggering event to the depolarization of the membrane portion that is to undergo an action potential, without considering the involvement of the intervening graded potential.

To initiate an action potential, a triggering event causes the membrane to depolarize from the resting potential of −70 mV (● Figure 4-5). Depolarization proceeds slowly at first, until it reaches a critical level known as **threshold potential,** typically between −50 and −55 mV. At threshold potential, an explosive depolarization takes place. A recording of the potential at this time shows a sharp upward deflection to +30 mV as the potential rapidly reverses itself so that the inside of the cell becomes positive compared to the outside. Just as rapidly, the membrane repolarizes, dropping back to resting potential.

The entire rapid change in potential from threshold to peak and then back to resting is called the *action potential.* Unlike the variable duration of a graded potential, the duration of an action potential is always the same in a given excitable cell. In a nerve cell, an action potential lasts for only 1 msec (0.001 sec). It lasts longer in muscle, with the duration depending on the muscle type. Often an action potential is referred to as a **spike,** because of its spikelike recorded appearance. Alternatively, when an excitable membrane is trig-

● **FIGURE 4-5**

Changes in membrane potential during an action potential

 PhysioEdge For an animation related to this figure, click the Action Potential tab in the Neuronal Physiology and Hormonal Communication tutorial on the CD-ROM.

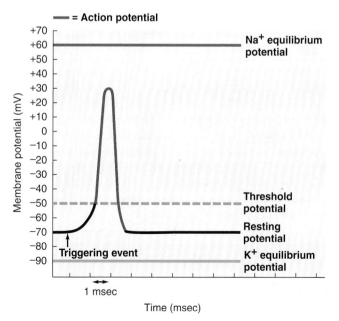

Voltage-Gated Sodium Channel

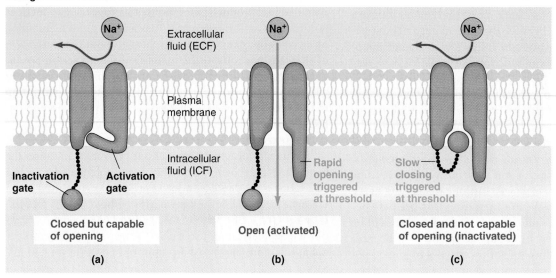

Extracellular fluid (ECF)

Plasma membrane

Intracellular fluid (ICF)

Inactivation gate   **Activation gate**

Rapid opening triggered at threshold

Slow closing triggered at threshold

Closed but capable of opening	Open (activated)	Closed and not capable of opening (inactivated)
(a)	**(b)**	**(c)**

Voltage-Gated Potassium Channel

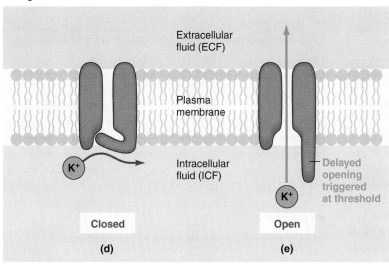

Extracellular fluid (ECF)

Plasma membrane

Intracellular fluid (ICF)

Delayed opening triggered at threshold

Closed	Open
(d)	**(e)**

● **FIGURE 4-6**

Conformations of voltage-gated sodium and potassium channels

 For an animation of this figure, click the Voltage-gated Ion Channels tab in the Neuronal Physiology and Hormonal Communication tutorial on the CD-ROM.

gered to undergo an action potential, it is said to **fire.** Thus the terms *action potential, spike,* and *firing* all refer to the same phenomenon of rapid reversal of membrane potential.

If threshold potential is not reached by the initial triggered depolarization, no action potential takes place. Thus threshold is a critical all-or-none point. Either the membrane is depolarized to threshold and an action potential takes place, or threshold is not reached in response to the depolarizing event and no action potential occurs.

■ **Marked changes in membrane permeability and ion movement lead to an action potential.**

How is the membrane potential, which is usually maintained at a constant resting level, thrown out of balance to such an

extent as to produce an action potential? Recall that K^+ makes the greatest contribution to the establishment of the resting potential, because the membrane at rest is considerably more permeable to K^+ than to Na^+ (see p. 64). During an action potential, marked changes in membrane permeability to Na^+ and K^+ take place, permitting rapid fluxes of these ions down their electrochemical gradients. These ion movements carry the current responsible for the potential changes that occur during an action potential. Action potentials take place as a result of the triggered opening and subsequent closing of two specific types of channels: voltage-gated Na^+ channels and voltage-gated K^+ channels.

VOLTAGE-GATED Na$^+$ AND K$^+$ CHANNELS

Voltage-gated membrane channels consist of proteins that have a number of charged groups. The electric field (potential) surrounding the channels can exert a distorting force on the channel structure as charged portions of the channel proteins are electrically attracted or repelled by charges in the fluids surrounding the membrane. Unlike the majority of membrane proteins, which remain stable despite fluctuations in membrane potential, the voltage-gated channel proteins are especially sensitive to voltage changes. Small distortions in channel shape induced by potential changes can cause them to flip to another conformation. Here again is an example of how subtle changes in structure can profoundly influence function.

The voltage-gated Na^+ channel has two gates: an *activation gate* and an *inactivation gate* (● Figure 4-6). The activation gate guards the channel by opening and closing like a hinged door. The inactivation gate consists of a ball-and-chain–like sequence of amino acids. This gate is open when the ball is dangling free on its chain and closed when the ball binds to its receptor located at the channel opening, thus blocking the opening. Both gates must be open to permit passage of Na^+ through the channel, and closure of either gate prevents passage. This voltage-gated Na^+ channel can exist in three different conformations: (1) *closed but capable of opening* (activation gate closed, inactivation gate open, ● Figure 4-6a); (2) *open,* or *activated* (both gates open, ● Figure 4-6b); and (3) *closed and not capable of opening* (activation gate open, inactivation gate closed, ● Figure 4-6c).

The voltage-gated K^+ channel is simpler. It has only one gate, which can be either open or closed (● Figure 4-6d and e). These voltage-gated Na^+ and K^+ channels exist in addition to the Na^+–K^+ pump and the leak channels for these ions (described in Chapter 3).

CHANGES IN PERMEABILITY AND ION MOVEMENT DURING AN ACTION POTENTIAL

At resting potential (-70 mV), all the voltage-gated Na^+ and K^+ channels are closed, with the Na^+ channels' activation gates being closed and their inactivation gates being open; that is, the voltage-gated Na^+ channels are in their "closed but capable of opening" conformation. Therefore, passage of Na^+ and K^+ does not occur through these voltage-gated channels at resting potential. However, because of the presence of many K^+ leak channels and very few Na^+ leak channels, the resting membrane is 50 to 75 times more permeable to K^+ than to Na^+.

When a membrane starts to depolarize toward threshold as a result of a triggering event, the activation gates of some of its voltage-gated Na^+ channels open. Now both gates of these activated channels are open. Because both the concentration and electrical gradients for Na^+ favor its movement into the cell, Na^+ starts to move in. The inward movement of positively charged Na^+ depolarizes the membrane further, thereby opening even more voltage-gated Na^+ channels and allowing more Na^+ to enter, and so on, in a positive-feedback cycle (● Figure 4-7).

At threshold potential, there is an explosive increase in Na^+ permeability, which is symbolized as P_{Na^+}, as the membrane swiftly becomes 600 times more permeable to Na^+ than to K^+. Each individual channel is either closed or open and cannot be partially open. However, the delicately poised gating mechanisms of the various voltage-gated Na^+ channels are jolted open by slightly different voltage changes. During the early depolarizing phase, more and more of the Na^+ channels open as the potential progressively decreases. At threshold, enough Na^+ gates have opened to set off the positive feedback cycle that rapidly causes the remaining Na^+ gates to swing open. Now Na^+ permeability dominates the membrane, in contrast to the K^+ domination at resting potential. Thus at threshold Na^+ rushes into the cell, rapidly eliminating the internal negativity and even making the inside of the cell more positive than the outside as the membrane potential is driven toward the Na^+ equilibrium potential (which is $+60$ mV; see p. 64). The potential reaches $+30$ mV, close to the Na^+ equilibrium potential. The potential does not become any more positive, because at the peak of the action potential the Na^+ channels start to close to the inactivated state and P_{Na^+} starts to fall to its low resting value.

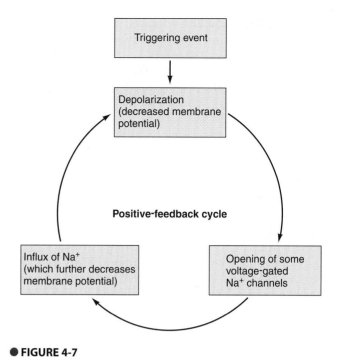

● **FIGURE 4-7**

Positive-feedback cycle responsible for opening Na^+ channels at threshold

What causes the Na^+ channels to close? When the membrane potential reaches threshold, two closely related events take place in the gates of each Na^+ channel. First the activation gates are triggered to *open rapidly,* in response to the depolarization, converting the channel to its open (activated) conformation (● Figure 4-6b). Surprisingly, this channel opening initiates the process of channel closing. The conformational change that opens the channel also allows the inactivation gate's ball to bind to its receptor at the channel opening, thereby physically blocking the mouth of the channel. However, this closure process takes time, so the inactivation gate *closes slowly* compared to the rapidity of channel opening. Meanwhile, during the 0.5-msec delay after the activation gate opens and before the inactivation gate closes, both gates are open, and Na^+ rushes into the cell through these open channels, bringing the action potential to its peak. Then the inactivation gate closes, membrane permeability to Na^+ plummets to its low resting value, and further Na^+ entry is prevented. The channel remains in this inactivated conformation until the membrane potential has been restored to its resting value.

Simultaneous with inactivation of Na^+ channels, the voltage-gated K^+ channels start to slowly open, with maximum opening occurring at the peak of the action potential. Opening of the K^+ channel gate is a delayed voltage-gated response triggered by the initial depolarization to threshold. Thus three action-potential–related events occur at threshold: (1) the rapid opening of the Na^+ activation gates, which permits Na^+ to enter, moving the potential from threshold to its positive peak; (2) the slow closing of the Na^+ inactivation gates, which halts further Na^+ entry after a brief time delay, thus keeping the potential from rising any further; and

(3) the slow opening of the K^+ gates, which is responsible for the potential plummeting from its peak back to resting.

Opening of the voltage-gated K^+ channels greatly increases K^+ permeability (designated P_{K^+}) to about 300 times the resting P_{Na^+} at the peak of the action potential. This marked increase in P_{K^+} causes K^+ to rush out of the cell down its concentration and electrical gradients, carrying positive charges back to the outside. Note that at the peak of the action potential, the positive potential inside the cell tends to repel the positive K^+ ions, so the electrical gradient for K^+ is outward, unlike at resting potential. The outward movement of K^+ rapidly restores the negative resting potential.

To review (● Figure 4-8), *the rising phase of the action potential* (from threshold to +30 mV) *is due to Na^+ influx* (Na^+ entering the cell) induced by an explosive increase in P_{Na^+} at threshold. The *falling phase* (from +30 mV to resting potential) *is brought about largely by K^+ efflux* (K^+ leaving the cell) caused by the marked increase in P_{K^+} occurring simultaneously with the inactivation of the Na^+ channels at the peak of the action potential.

As the potential returns to resting, the changing voltage shifts the Na^+ channels to their "closed but capable of opening" conformation, with the activation gate closed and the inactivation gate open. Now the channel is reset, ready to respond to another triggering event. The newly opened voltage-gated K^+ channels also close, so the membrane returns to the resting number of open K^+ leak channels. The membrane remains at resting potential until another triggering event alters the gated Na^+ and K^+ channels.

▌ The Na^+–K^+ pump gradually restores the concentration gradients disrupted by action potentials.

At the completion of an action potential, the membrane potential has been restored to its resting condition, but the ion distribution has been altered slightly. Sodium has entered the cell during the rising phase, and a comparable amount of K^+ has left during the falling phase. The Na^+–K^+ pump restores these ions to their original locations in the long run, but not after each action potential.

The active pumping process takes much longer to restore Na^+ and K^+ to their original locations than it takes for the passive fluxes of these ions during an action potential. However, the membrane does not need to wait until the Na^+–K^+ pump slowly restores the concentration gradients before it can undergo another action potential. Actually, the movement of only relatively few of the total number of Na^+ and K^+ ions present causes the large swings in potential that occur during an action potential. Only about 1 out of 100,000 K^+ ions present in the cell leaves during an action potential, while a comparable number of Na^+ ions enters from the ECF. The movement of this extremely small proportion of the total Na^+ and K^+ during a single action potential produces dramatic 100-mV changes in potential (between −70 mV and +30 mV), but only infinitesimal changes in the ICF and ECF concentrations of these ions. Much more K^+ is still inside the cell than outside, and Na^+ is still predominantly an extracellular cation. Consequently, the Na^+ and K^+ concentration gradients still exist, so repeated action potentials can occur without the pump having to keep pace to restore the gradients.

Were it not for the pump, of course, even tiny fluxes accompanying repeated action potentials would eventually "run down" the concentration gradients so that further action potentials would be impossible. If the concentrations of Na^+ and K^+ were equal between the ECF and ICF, changes in permeability to these ions would not bring about ion fluxes, so no change in potential would occur. Thus the Na^+–K^+ pump is critical to maintaining the concentration gradients in the long run. However, it does not have to perform its role between action potentials, nor is it directly involved in the ion fluxes or potential changes that occur during an action potential.

▌ Action potentials are propagated from the axon hillock to the axon terminals.

A single action potential involves only a small patch of the total surface membrane of an excitable cell. But if action potentials are to serve as long-distance signals, they cannot be merely isolated events occurring in a limited area of a nerve or muscle cell membrane. Mechanisms must exist to conduct or spread the action potential throughout the entire cell membrane. Furthermore, the signal must be transmitted from one cell to the next cell (for example, along specific nerve pathways). To explain how these mechanisms are accomplished, we will first begin with a brief look at neuronal structure. Then we will examine how an action potential (nerve impulse) is conducted throughout a nerve cell, before we turn our attention to how the signal is passed to another cell.

A single nerve cell, or **neuron**, typically consists of three basic parts: the *cell body,* the *dendrites,* and the *axon,* although there are variations in structure, depending on the location and function of the neuron. The nucleus and organelles are

● **FIGURE 4-8**

Permeability changes and ion fluxes during an action potential

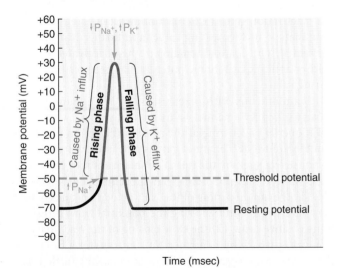

housed in the **cell body**, from which numerous extensions known as **dendrites** typically project like antennae to increase the surface area available for receiving signals from other nerve cells (● Figure 4-9). Some neurons have up to 400,000 of these elongated surface extensions. Dendrites carry signals *toward* the cell body. In most neurons the plasma membrane of the dendrites and cell body contains protein receptors for binding chemical messengers from other neurons. Therefore, the dendrites and cell body are the neuron's *input zone*, because these components receive and integrate incoming signals. This is the region where graded potentials are produced in response to triggering events, in this case, incoming chemical messengers.

The **axon**, or **nerve fiber**, is a single, elongated, tubular extension that conducts action potentials *away from* the cell body and eventually terminates at other cells. The axon frequently gives off side branches along its course. The first portion of the axon plus the region of the cell body from which the axon leaves is known as the **axon hillock**. The axon hillock is the neuron's *trigger zone*, because it is the site where action potentials are triggered, or initiated, by the graded potential if it is of sufficient magnitude. The action potentials

---

● **FIGURE 4-9**

Anatomy of the most abundant structural type of neuron (nerve cell). (a) Most but not all neurons consist of the basic parts represented in the figure. (b) An electron micrograph highlighting the cell body, dendrites, and part of the axon of a neuron within the central nervous system.

 For an animation of this figure, click the Conduction of the Action Potential tab in the Neural and Hormonal Communication tutorial on the CD-ROM. Also review Media Exercise 4.1: Basics of a Neuron.

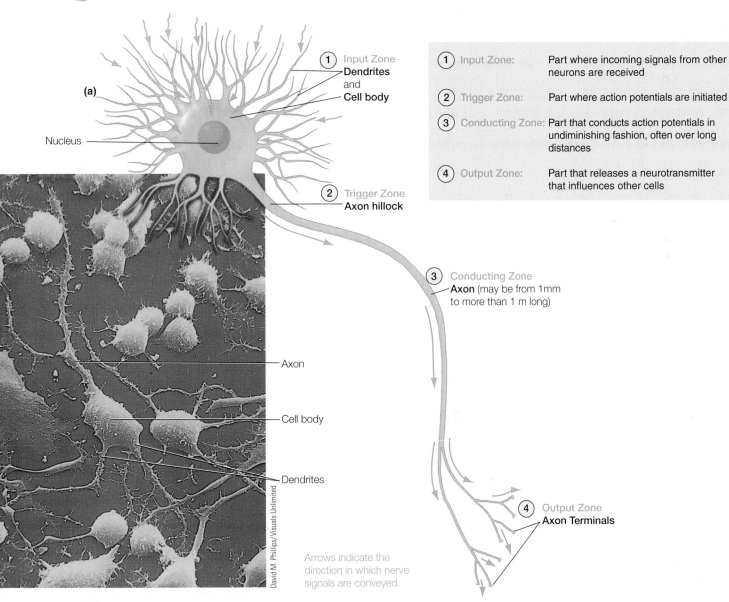

① Input Zone:	Part where incoming signals from other neurons are received
② Trigger Zone:	Part where action potentials are initiated
③ Conducting Zone:	Part that conducts action potentials in undiminishing fashion, often over long distances
④ Output Zone:	Part that releases a neurotransmitter that influences other cells

(a)

Nucleus

① Input Zone
Dendrites and **Cell body**

② Trigger Zone
Axon hillock

③ Conducting Zone
Axon (may be from 1mm to more than 1 m long)

④ Output Zone
Axon Terminals

Axon

Cell body

Dendrites

David M. Phillips/Visuals Unlimited

Arrows indicate the direction in which nerve signals are conveyed.

(b)

are then conducted along the axon from the axon hillock to the typically highly branched ending at the **axon terminals.** These terminals release chemical messengers that simultaneously influence numerous other cells with which they come into close association. Functionally, therefore, the axon is the *conducting zone* of the neuron, and the axon terminals constitute its *output zone.* (The major exception to this typical neuronal structure and functional organization is neurons specialized to carry sensory information, a topic described in a later chapter.)

Axons vary in length from less than a millimeter in neurons that communicate only with neighboring cells to longer than a meter in neurons that communicate with distant parts of the nervous system or with peripheral organs. For example, the axon of the nerve cell innervating your big toe must traverse the distance from the origin of its cell body within the spinal cord in the lower region of your back all the way down your leg to your toe.

Action potentials can be initiated only in portions of the membrane that have an abundance of voltage-gated Na^+ channels that can be triggered to open by a depolarizing event. Typically, regions of excitable cells where graded potentials take place do not undergo action potentials, because voltage-gated Na^+ channels are sparse there. Therefore, sites specialized for graded potentials do not undergo action potentials, even though they might be considerably depolarized. However, graded potentials can, before dying out, trigger action potentials in adjacent portions of the membrane by bringing these more sensitive regions to threshold through local current flow spreading from the site of the graded potential. In a typical neuron, for example, graded potentials are generated in the dendrites and cell body in response to incoming signals. If these graded potentials have sufficient magnitude by the time they have spread to the axon hillock, they initiate an action potential at this triggering zone.

▌ Once initiated, action potentials are conducted throughout a nerve fiber.

Once an action potential is initiated at the axon hillock, no further triggering event is necessary to activate the remainder of the nerve fiber. The impulse is automatically conducted throughout the neuron without further stimulation by one of two methods of propagation: *contiguous conduction* or *saltatory conduction.*

Contiguous conduction involves the spread of the action potential along every patch of membrane down the length of the axon (*contiguous* means "touching" or "next to in sequence.") This process is illustrated in ● Figure 4-10. You are viewing a schematic representation of a longitudinal section of the axon hillock and the portion of the axon immediately beyond it. The membrane at the axon hillock is at the peak of an action potential. The inside of the cell is positive in this active area, because Na^+ has already rushed into the nerve cell at this point. The remainder of the axon, still at resting potential and negative inside, is considered inactive. For the action potential to spread from the active to the inactive areas, the inactive areas must somehow be depolarized to

threshold before they can undergo an action potential. This depolarization is accomplished by local current flow between the area already undergoing an action potential and the adjacent inactive area, similar to the current flow responsible for the spread of graded potentials. Because opposite charges attract, current can flow locally between the active area and the neighboring inactive area on both the inside and the outside of the membrane. This local current flow in effect neutralizes or eliminates some of the unbalanced charges in the inactive area; that is, it reduces the number of opposite charges separated across the membrane, reducing the potential in this area. This depolarizing effect quickly brings the involved inactive area to threshold, at which time the voltage-gated Na^+ channels in this region of the membrane are all thrown open, leading to an action potential in this previously inactive area. Meanwhile, the original active area returns to resting potential as a result of K^+ efflux.

In turn, beyond the new active area is another inactive area, so the same thing happens again. This cycle repeats itself in a chain reaction until the action potential has spread to the end of the axon. *Once an action potential is initiated in one part of a nerve cell membrane, a self-perpetuating cycle is initiated so that the action potential is propagated along the rest of the fiber automatically.* In this way, the axon is like a firecracker fuse that needs to be lit at only one end. Once ignited, the fire spreads down the fuse; it is not necessary to hold a match to every separate section of the fuse.

Note that the original action potential does not travel along the membrane. Instead, it triggers an identical new action potential in the adjacent area of the membrane, with this process being repeated along the axon's length. An analogy is the "wave" at a stadium. Each section of spectators stands up (the rising phase of an action potential), then sits down (the falling phase) in sequence one after another as the wave moves around the stadium. The wave, not individual spectators, travels around the stadium. Similarly, new action potentials arise sequentially down the axon. Each new action potential in the conduction process is a fresh local event that depends on the induced permeability changes and electrochemical gradients, which are virtually identical down the length of the axon. Therefore, the last action potential at the end of the axon is identical to the original one, no matter how long the axon. Thus an action potential is spread along the axon in undiminished fashion. In this way, action potentials can serve as long-distance signals without attenuation or distortion.

This nondecremental propagation of an action potential contrasts with the decremental spread of a graded potential, which dies out over a very short distance because it cannot regenerate itself. ▲ Table 4-1 summarizes the differences between graded potentials and action potentials, some of which are yet to be discussed.

▌ The refractory period ensures one-way propagation of the action potential.

What ensures the one-way propagation of an action potential away from the initial site of activation? Note from ● Figure 4-11 that once the action potential has been regenerated

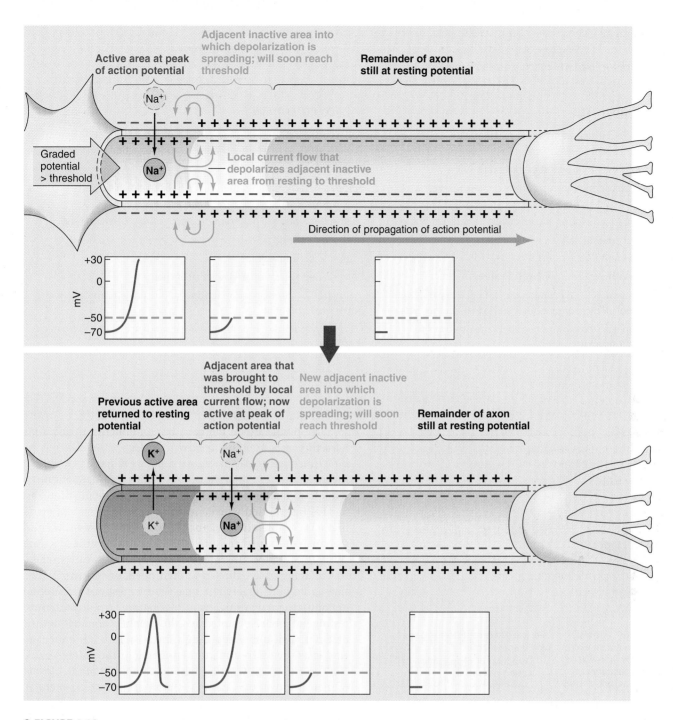

● FIGURE 4-10

Contiguous conduction. Local current flow between the active area at the peak of an action potential and the adjacent inactive area still at resting potential reduces the potential in this contiguous inactive area to threshold, which triggers an action potential in the previously inactive area. The original active area returns to resting potential, and the new active area induces an action potential in the next adjacent inactive area by local current flow as the cycle repeats itself down the length of the axon.

 For an animation of this figure, click the Conduction of the Action Potential tab in the Neural and Hormonal Communication tutorial on the CD-ROM.

at a new neighboring site (now positive inside) and the original active area has returned to resting (once again negative inside), the close proximity of opposite charges between these two areas is conducive to local current flow taking place in the backward direction, as well as in the forward direction into as yet unexcited portions of the membrane. If such backward current flow were able to bring the just inactivated area to threshold, another action potential would be initiated here, which would spread both forward and backward, initiating still other action potentials, and so on. But if action poten-

TABLE 4-1

Comparison of Graded Potentials and Action Potentials

GRADED POTENTIALS	ACTION POTENTIALS
Graded potential change; magnitude varies with magnitude of triggering event	All-or-none membrane response; magnitude of triggering event coded in frequency rather than amplitude of action potentials
Duration varies with duration of triggering event	Constant duration
Decremental conduction; magnitude diminishes with distance from initial site	Propagated throughout membrane in undiminishing fashion
Passive spread to neighboring inactive areas of membrane	Self-regeneration in neighboring inactive areas of membrane
No refractory period	Refractory period
Can be summed	Summation impossible
Can be depolarization or hyperpolarization	Always depolarization and reversal of charges
Triggered by stimulus, by combination of neurotransmitter with receptor, or by spontaneous shifts in leak–pump cycle	Triggered by depolarization to threshold, usually through spread of graded potential
Occurs in specialized regions of membrane designed to respond to triggering event	Occurs in regions of membrane with abundance of voltage-gated Na^+ channels

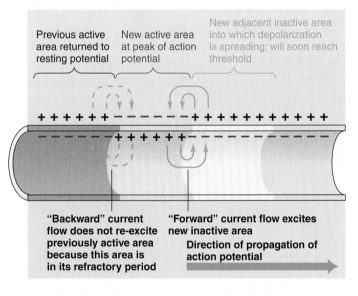

Previous active area returned to resting potential   New active area at peak of action potential   New adjacent inactive area into which depolarization is spreading; will soon reach threshold

"Backward" current flow does not re-excite previously active area because this area is in its refractory period   "Forward" current flow excites new inactive area   Direction of propagation of action potential

● **FIGURE 4-11**

Value of the refractory period. "Backward" current flow is prevented by the refractory period. During an action potential and slightly beyond, an area cannot be restimulated by normal events to undergo another action potential. Thus the refractory period ensures that an action potential can be propagated only in the forward direction along the axon.

tials were to move in both directions, the situation would be chaotic, with numerous action potentials bouncing back and forth along the axon until the nerve cell eventually fatigued. Fortunately, neurons are saved from this fate of oscillating action potentials by the **refractory period**, during which a new action potential cannot be initiated by normal events in a region that has just undergone an action potential.

The refractory period has two components: the *absolute refractory period* and the *relative refractory period*. During the time that a particular patch of axonal membrane is undergoing an action potential, it cannot initiate another action potential, no matter how strongly a triggering event stimulates it. This time period when a recently activated patch of membrane is completely refractory (meaning "stubborn," or unresponsive) to further stimulation is known as the **absolute refractory period** (● Figure 4-12). Once the voltage-gated Na^+ channels have flipped to their open, or activated, state, they cannot be triggered to open again in response to another depolarizing triggering event, no matter how strong, until resting potential is restored and the channels are reset to their original positions. Accordingly, the absolute refractory period lasts the entire time from opening of the voltage-gated Na^+ channels' activation gates at threshold, through closure of their inactivation gates at the peak of the action potential, until the return to resting potential when the channels' activation gates close and inactivation gates open once again; that is, until the channels are in their "closed but capable of opening" conformation. Only then can they respond to another depolarization with an explosive increase in P_{Na^+} to initiate another action potential. Because of this absolute refractory period, one action potential must be over before another can be initiated at the same site. Action potentials cannot overlap or be added one on top of another "piggyback fashion."

Following the absolute refractory period is a **relative refractory period**, during which a second action potential can be produced only by a triggering event considerably stronger than is usually necessary. The relative refractory period occurs during the time when the voltage-gated K^+ channels that opened at the peak of the action potential are in the process of closing. During the relative refractory period, Na^+

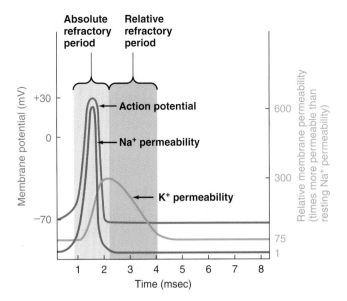

● FIGURE 4-12

Absolute and relative refractory periods. During the absolute refractory period, the portion of the membrane that has just undergone an action potential cannot be restimulated. This period corresponds to the time during which the Na$^+$ gates are not in their resting conformation. During the relative refractory period, the membrane can be restimulated only by a stronger stimulus than is usually necessary. This period corresponds to the time during which the K$^+$ gates opened during the action potential have not yet closed.

entry in response to another triggering event is opposed by a persistent outward leak of K$^+$ through its not-yet-closed channels, and thus a greater-than-normal depolarizing triggering event is needed to bring the membrane to threshold during the relative refractory period.

By the time the original site has recovered from its refractory period and is capable of being restimulated by normal current flow, the action potential has been rapidly propagated in the forward direction only and is so far away that it can no longer influence the original site. Thus, *the refractory period ensures the one-way propagation of the action potential down the axon away from the initial site of activation.*

▌ Action potentials occur in all-or-none fashion.

If any portion of the neuronal membrane is depolarized to threshold, an action potential is initiated and relayed along the membrane in undiminished fashion. Furthermore, once threshold has been reached the resultant action potential always goes to maximal height. The reason for this effect is that the changes in voltage during an action potential result from ion movements down concentration and electrical gradients, and these gradients are not affected by the strength of the depolarizing triggering event. A triggering event stronger than one necessary to bring the membrane to threshold does not produce a larger action potential. However, a triggering event that fails to depolarize the membrane to threshold does not trigger an action potential at all. Thus *an excitable membrane either responds to a triggering event with a maximal action potential that spreads nondecrementally throughout the membrane,* or *it does not respond with an action potential at all.* This property is called the **all-or-none law.**

This all-or-none concept is analogous to firing a gun. Either the trigger is not pulled sufficiently to fire the bullet (threshold is not reached), or it is pulled hard enough to elicit the full firing response of the gun (threshold is reached). Squeezing the trigger harder does not produce a greater explosion. Just as it is not possible to fire a gun halfway, it is not possible to cause a halfway action potential.

The threshold phenomenon allows some discrimination between important and unimportant stimuli or other triggering events. Stimuli too weak to bring the membrane to threshold do not initiate action potentials and therefore do not clutter up the nervous system by transmitting insignificant signals.

▌ The strength of a stimulus is coded by the frequency of action potentials.

How is it possible to differentiate between two stimuli of varying strengths when both stimuli bring the membrane to threshold and generate action potentials of the same magnitude? For example, how can one distinguish between touching a warm object or touching a very hot object if both trigger identical action potentials in a nerve fiber relaying information about skin temperature to the central nervous system? The answer lies in the *frequency* with which the action potentials are generated. A stronger stimulus does not produce a larger action potential, but it does trigger a greater *number* of action potentials per second. In addition, a stronger stimulus in a region causes more neurons to reach threshold, increasing the total information sent to the central nervous system.

Once initiated, the velocity, or speed, with which an action potential travels down the axon depends on whether the fiber is myelinated. Contiguous conduction occurs in unmyelinated fibers. In this case, as you just learned, each individual action potential initiates an identical new action potential in the next contiguous (bordering) segment of the axon membrane so that every portion of the membrane undergoes an action potential as this electrical signal is conducted from the beginning to the end of the axon. A faster method of propagation, *saltatory conduction,* takes place in myelinated fibers. We are next going to see how a myelinated fiber compares with an unmyelinated fiber, then see how saltatory conduction compares with contiguous conduction.

▌ Myelination increases the speed of conduction of action potentials.

Myelinated fibers, as the name implies, are covered with myelin at regular intervals along the length of the axon (● Figure 4-13a). **Myelin** is composed primarily of lipids. Because the water-soluble ions responsible for carrying current across the membrane cannot permeate this thick lipid barrier, the myelin coating acts as an insulator, just like plastic around an electrical wire, to prevent current leakage across the myelinated portion of the membrane. Myelin is not actually a part of the nerve cell but consists of separate myelin-forming cells that wrap themselves around the axon in jelly-roll fashion

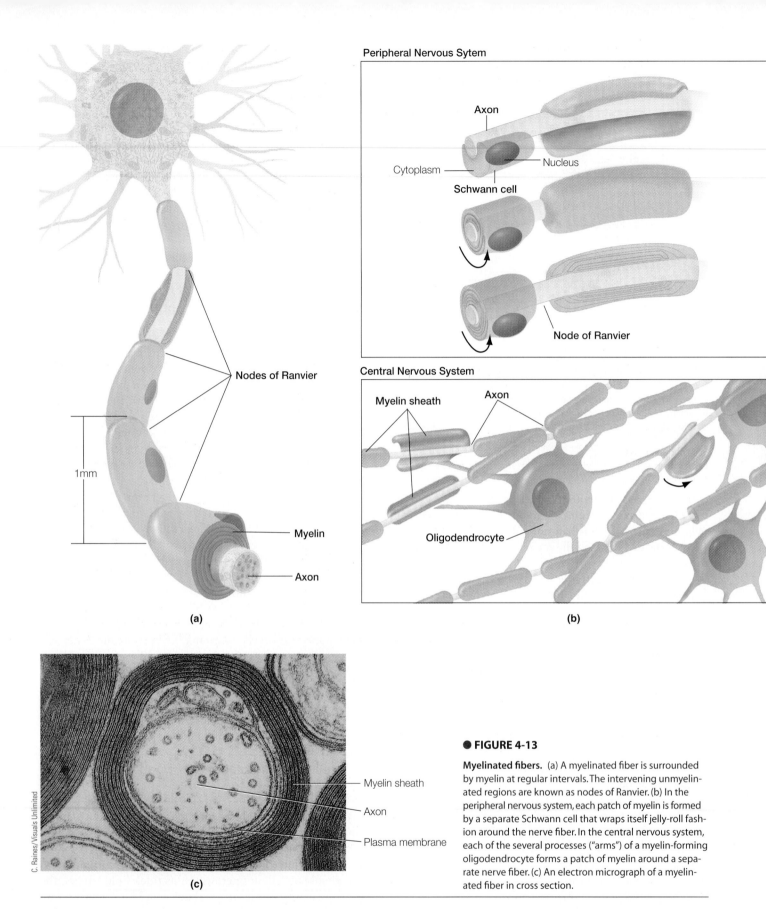

Peripheral Nervous Sytem

Axon

Cytoplasm

Nucleus

Schwann cell

Node of Ranvier

Nodes of Ranvier

1mm

Myelin

Axon

(a)

Central Nervous System

Myelin sheath

Axon

Oligodendrocyte

(b)

Myelin sheath

Axon

Plasma membrane

C. Raines/Visuals Unlimited

(c)

● **FIGURE 4-13**

Myelinated fibers. (a) A myelinated fiber is surrounded by myelin at regular intervals. The intervening unmyelinated regions are known as nodes of Ranvier. (b) In the peripheral nervous system, each patch of myelin is formed by a separate Schwann cell that wraps itself jelly-roll fashion around the nerve fiber. In the central nervous system, each of the several processes ("arms") of a myelin-forming oligodendrocyte forms a patch of myelin around a separate nerve fiber. (c) An electron micrograph of a myelinated fiber in cross section.

(● Figure 4-13b and c). These myelin-forming cells are **oligodendrocytes** in the central nervous system (the brain and spinal cord) and **Schwann cells** in the peripheral nervous system (the nerves running between the central nervous system and the various regions of the body). The lipid composition of myelin is due to the presence of layer on layer of the lipid bilayer that composes the plasma membrane of these myelin-forming cells. Between the myelinated regions, at the **nodes of Ranvier**, the axonal membrane is bare and exposed to the ECF. Only at these bare spaces can current flow across the membrane to produce action potentials. Voltage-gated Na$^+$ channels are concentrated at the nodes, whereas the myelin-covered regions are almost devoid of these special passageways. By contrast, an unmyelinated fiber has a high density of voltage-gated Na$^+$ channels throughout its entire length. As you now know, action potentials can be generated only at portions of the membrane furnished with an abundance of these channels.

The nodes are usually about 1 mm apart, short enough that local current from an active node can reach an adjacent node before dying off. When an action potential occurs at one node, opposite charges attract from the adjacent inactive node, reducing its potential to threshold so that it undergoes an action potential, and so on. Consequently, in a myelinated fiber, the impulse "jumps" from node to node, skipping over the myelinated sections of the axon; this process is called **saltatory conduction** (*saltere* means "to jump or leap"). Saltatory conduction propagates action potentials more rapidly than does contiguous conduction, because the action potential does not have to be regenerated at myelinated sections but must be regenerated within every section of an unmyelinated axonal membrane from beginning to end. Myelinated fibers conduct impulses about 50 times faster than unmyelinated fibers of comparable size. Thus the most urgent types of information are transmitted via myelinated fibers, whereas nervous pathways carrying less urgent information are unmyelinated.

Clinical Note **Multiple sclerosis (MS)** is a pathophysiological condition in which nerve fibers in various locations throughout the nervous system become demyelinated (lose their myelin). MS is an autoimmune disease, in which the body's defense system erroneously attacks the myelin sheath surrounding myelinated nerve fibers (*auto* means "self"; *immune* means "defense against"). Loss of myelin slows transmission of impulses in the affected neurons. A hardened scar known as a *sclerosis* (meaning "hardness") forms at the multiple sites of myelin damage. These scars further interfere with and can eventually block action potential propagation in the underlying axons. The symptoms of MS vary considerably, depending on the extent and location of the myelin damage.

You have now seen how an action potential is propagated along the axon. But what happens when an action potential reaches the end of the axon? We are going to now turn our attention to this topic.

 Click on the Media Exercises menu of the CD-ROM and work Media Exercises 4.1: Basics of a Neuron and 4.2: Graded Potentials and Action Potentials to test your understanding of the previous sections.

SYNAPSES AND NEURONAL INTEGRATION

When the action potential reaches the axon terminals, they release a chemical messenger that alters the activity of the cells on which the neuron terminates. A neuron may terminate on one of three structures: a muscle, a gland, or another neuron. Therefore, depending on where a neuron terminates, it can cause a muscle cell to contract, a gland cell to secrete, another neuron to convey an electrical message along a nerve pathway, or some other function. When a neuron terminates on a muscle or a gland, the neuron is said to **innervate**, or supply, the structure. The junctions between nerves and the muscles and glands that they innervate will be described later. For now we will concentrate on the junction between two neurons—a **synapse**. (Sometimes the term *synapse* is used to describe a junction between any two excitable cells, but we will reserve this term for the junction between two neurons.)

▌ Synapses are junctions between presynaptic and postsynaptic neurons.

Typically, a synapse involves a junction between an axon terminal of one neuron, known as the *presynaptic neuron,* and the dendrites or cell body of a second neuron, known as the *postsynaptic neuron.* (*Pre* means "before" and *post* means "after"; the presynaptic neuron lies before the synapse and the postsynaptic neuron lies after the synapse.) The dendrites and to a lesser extent the cell body of most neurons receive thousands of synaptic inputs, which are axon terminals from many other neurons. It has been estimated that some neurons within the central nervous system receive as many as 100,000 synaptic inputs (● Figure 4-14).

The anatomy of one of these thousands of synapses is shown in ● Figure 4-15a. The axon terminal of the **presynaptic neuron,** which conducts its action potentials *toward* the synapse, ends in a slight swelling, the **synaptic knob.** The synaptic knob contains **synaptic vesicles,** which store a specific chemical messenger, a **neurotransmitter** that has been synthesized and packaged by the presynaptic neuron. The synaptic knob comes into close proximity to, but does not actually directly touch, the **postsynaptic neuron,** the neuron whose action potentials are propagated *away* from the synapse. The space between the presynaptic and postsynaptic neurons, the **synaptic cleft,** is too wide for the direct spread of current from one cell to the other and therefore prevents action potentials from electrically passing between the neurons. The portion of the postsynaptic membrane immediately underlying the synaptic knob is referred to as the **subsynaptic membrane** (*sub* means "under").

Synapses operate in one direction only; that is, the presynaptic neuron brings about changes in membrane potential of the postsynaptic neuron, but the postsynaptic neuron does not directly influence the potential of the presynaptic neuron. The reason for this becomes readily apparent when you examine the events that occur at a synapse.

▌ A neurotransmitter carries the signal across a synapse.

Here are the events that occur at a synapse (● Figure 4-15):

1. When an action potential in a presynaptic neuron has been propagated to the axon terminal (step **1** in ● Fig-

ure 4-15), this local change in potential triggers the opening of voltage-gated Ca^{2+} channels in the synaptic knob.

2. Because Ca^{2+} is much more highly concentrated in the ECF and its electrical gradient is inward, this ion flows into the synaptic knob through the opened channels (step 2).

3. Ca^{2+} induces the release of a neurotransmitter from some of the synaptic vesicles into the synaptic cleft (step 3). The release is accomplished by exocytosis (see p. 24).

4. The released neurotransmitter diffuses across the cleft and binds with specific protein receptor sites on the subsynaptic membrane (step 4).

5. This binding triggers the opening of specific ion channels in the subsynaptic membrane, changing the ion permeability of the postsynaptic neuron (step 5). These are chemically gated channels, in contrast to the voltage-gated channels responsible for the action potential and for the Ca^{2+} influx into the synaptic knob.

Because the presynaptic terminal releases the neurotransmitter and the subsynaptic membrane of the postsynaptic neuron has receptor sites for the neurotransmitter, the synapse can operate only in the direction from presynaptic to postsynaptic neuron.

▌Some synapses excite whereas others inhibit the postsynaptic neuron.

Each presynaptic neuron typically releases only one neurotransmitter; however, different neurons vary in the neurotransmitter they release. On binding with their subsynaptic receptor sites, different neurotransmitters cause different ion permeability changes. There are two types of synapses, depending on the permeability changes induced in the postsynaptic neuron by the combination of a specific neurotransmitter with its receptor sites: *excitatory synapses* and *inhibitory synapses*.

EXCITATORY SYNAPSES

At an **excitatory synapse**, the response to the binding of a neurotransmitter to the receptor is the opening of nonspecific cation channels within the subsynaptic membrane that permit simultaneous passage of Na^+ and K^+ through them. (These are a different type of channel from those you have encountered before.) Thus permeability to both these ions is increased at the same time. How much of each ion diffuses through an open cation channel depends on their electrochemical gradients. At resting potential, both the concentration and electrical gradients for Na^+ favor its movement into the postsynaptic neuron, whereas only the concentration gradient for K^+ favors its movement outward. Therefore, the permeability change induced at an excitatory synapse results in the movement of a few K^+ ions out of the postsynaptic neuron, while a relatively larger number of Na^+ ions simultaneously enter this neuron. The result is a net movement of positive ions into the cell. This makes the inside of the membrane slightly less negative than at resting potential, thus producing a *small depolarization* of the postsynaptic neuron.

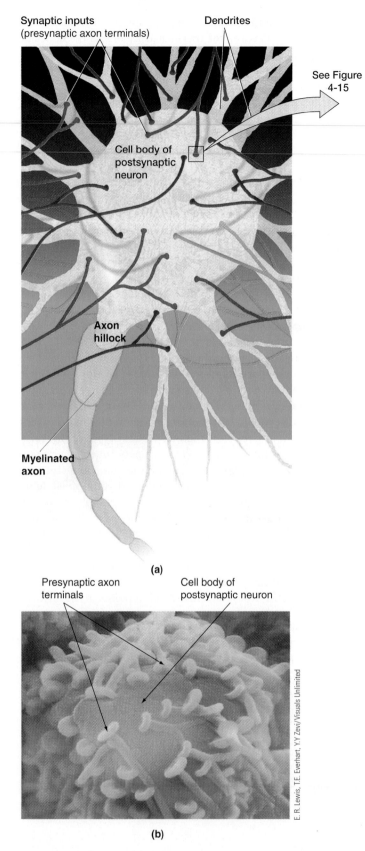

Synaptic inputs (presynaptic axon terminals)

Dendrites

See Figure 4-15

Cell body of postsynaptic neuron

Axon hillock

Myelinated axon

(a)

Presynaptic axon terminals

Cell body of postsynaptic neuron

E. R. Lewis, T.E. Everhart, Y.Y Zevi/Visuals Unlimited

(b)

● **FIGURE 4-14**

Synaptic inputs to a postsynaptic neuron. (a) Schematic representation of synaptic inputs (presynaptic axon terminals) to the dendrites and cell body of a single postsynaptic neuron. (b) Electron micrograph showing multiple presynaptic axon terminals to a single postsynaptic cell body.

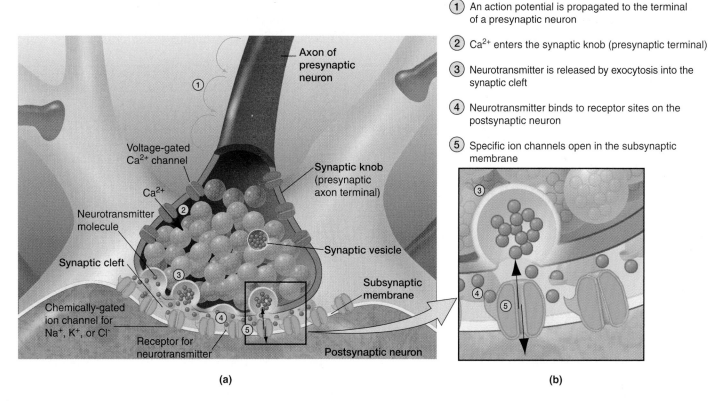

1. An action potential is propagated to the terminal of a presynaptic neuron

2. Ca²⁺ enters the synaptic knob (presynaptic terminal)

3. Neurotransmitter is released by exocytosis into the synaptic cleft

4. Neurotransmitter binds to receptor sites on the postsynaptic neuron

5. Specific ion channels open in the subsynaptic membrane

(a)

(b)

● **FIGURE 4-15**

Synaptic structure and function. (a) Schematic representation of the structure of a single synapse. The circled numbers designate the sequence of events that take place at a synapse. (b) A blow-up depicting the release by exocytosis of neurotransmitter from the presynaptic axon terminal and its subsequent binding with receptor sites specific for it on the subsynaptic membrane of the postsynaptic neuron.

 For an animation of this figure, click the Synapses tab (page 2) in the Neural and Hormonal Communication tutorial on the CD-ROM.

Activation of one excitatory synapse can rarely depolarize the postsynaptic neuron sufficiently to bring it to threshold. Too few channels are involved at a single subsynaptic membrane to permit adequate ion flow to reduce the potential to threshold. This small depolarization, however, does bring the membrane of the postsynaptic neuron closer to threshold, increasing the likelihood that threshold will be reached (in response to further excitatory input) and an action potential will occur. That is, the membrane is now more excitable (easier to bring to threshold) than when at rest. Accordingly, this postsynaptic potential change occurring at an excitatory synapse is called an **excitatory postsynaptic potential**, or **EPSP** (● Figure 4-16a).

INHIBITORY SYNAPSES

At an **inhibitory synapse**, the binding of a different released neurotransmitter with its receptor sites increases the permeability of the subsynaptic membrane to either K^+ or Cl^-. In either case, the resulting ion movements bring about a *small hyperpolarization* of the postsynaptic neuron—that is, greater internal negativity. In the case of increased P_{K^+}, more positive charges leave the cell via K^+ efflux, leaving more negative charges behind on the inside; in the case of

increased P_{Cl^-}, negative charges enter the cell in the form of Cl^- ions, because Cl^- concentration is higher outside the cell. This small hyperpolarization moves the membrane potential even farther away from threshold (● Figure 4-16b), lessening the likelihood that the postsynaptic neuron will reach threshold and undergo an action potential. That is, the membrane is now less excitable (harder to bring to threshold by excitatory input) than when it is at resting potential. The membrane is said to be inhibited under these circumstances, and the small hyperpolarization of the postsynaptic cell is called an **inhibitory postsynaptic potential**, or **IPSP**.

SYNAPTIC DELAY

This conversion of the electrical signal in the presynaptic neuron (an action potential) to an electrical signal in the postsynaptic neuron (either an EPSP or IPSP) by chemical means (via the neurotransmitter–receptor combination) takes time. This **synaptic delay** is usually about 0.5 to 1 msec. In a neural pathway, chains of neurons often must be traversed. The more complex the pathway, the more synaptic delays, and the longer the *total reaction time* (the time required to respond to a particular event).

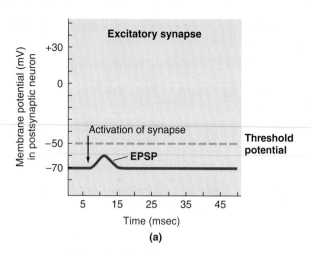

(a)

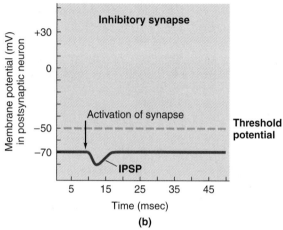

(b)

● **FIGURE 4-16**

Postsynaptic potentials. (a) Excitatory synapse. An excitatory postsynaptic potential (EPSP) brought about by activation of an excitatory presynaptic input brings the postsynaptic neuron closer to threshold potential. (b) Inhibitory synapse. An inhibitory postsynaptic potential (IPSP) brought about by activation of an inhibitory presynaptic input moves the postsynaptic neuron farther from threshold potential.

 For an animation of this figure, click the Synapses tab (page 4) in the Neural and Hormonal Communication tutorial on the CD-ROM.

Each synapse is either always excitatory or always inhibitory.

Many different chemicals serve as neurotransmitters (▲ Table 4-2). Even though neurotransmitters vary from synapse to synapse, the same neurotransmitter is always released at a particular synapse. Furthermore, at a given synapse, binding of a neurotransmitter with its appropriate subsynaptic receptors always leads to the same change in permeability and resultant change in potential of the postsynaptic membrane. That is, the response to a given neurotransmitter–receptor combination is always constant. *Each synapse is either always excitatory or always inhibitory.* It does not give rise to an EPSP under one circumstance and produce an IPSP at another time. As examples, *glutamate* is a common excitatory neurotrans-

▲ **TABLE 4-2**	
Some Common Neurotransmitters	
Acetylcholine	Histamine
Dopamine	Glycine
Norepinephrine	Glutamate
Epinephrine	Aspartate
Serotonin	Gamma-aminobutyric acid (GABA)

mitter and *gamma-aminobutyric acid (GABA)* is a common inhibitory neurotransmitter in the central nervous system.

Neurotransmitters are quickly removed from the synaptic cleft.

As long as the neurotransmitter remains bound to the receptor sites, the alteration in membrane permeability responsible for the EPSP or IPSP continues. The neurotransmitter must be inactivated or removed after it has produced the appropriate response in the postsynaptic neuron, however, so that the postsynaptic "slate" is "wiped clean," leaving it ready to receive additional messages from the same or other presynaptic inputs. Thus, after combining with the postsynaptic receptor, chemical transmitters are removed and the response is terminated. Several mechanisms can remove the neurotransmitter: It may diffuse away from the synaptic cleft, be inactivated by specific enzymes within the subsynaptic membrane, or be actively taken back up into the axon terminal by transport mechanisms in the presynaptic membrane. The method employed depends on the particular synapse.

The grand postsynaptic potential depends on the sum of the activities of all presynaptic inputs.

The events that occur at a single synapse result in either an EPSP or an IPSP at the postsynaptic neuron. But if a single EPSP is inadequate to bring the postsynaptic neuron to threshold and an IPSP moves it even farther from threshold, how can an action potential be initiated in the postsynaptic neuron? The answer lies in the thousands of presynaptic inputs that a typical neuronal cell body receives from many other neurons. Some of these presynaptic inputs may be carrying sensory information from the environment; some may be signaling internal changes in homeostatic balance; others may be transmitting signals from control centers in the brain; and still others may arrive carrying other bits of information. At any given time, any number of these presynaptic neurons (probably hundreds) may be firing and thus influencing the postsynaptic neuron's level of activity. The total potential in the postsynaptic neuron, the **grand postsynaptic potential (GPSP)**, is a composite of all EPSPs and IPSPs occurring at approximately the same time.

The postsynaptic neuron can be brought to threshold in two ways: (1) *temporal summation* and (2) *spatial summation.* To illustrate these methods of summation, we will examine the possible interactions of three presynaptic inputs—two excitatory inputs (Ex1 and Ex2) and one inhibitory input (In1)— on a hypothetical postsynaptic neuron (● Figure 4-17). The recording shown in the figure represents the potential in the postsynaptic cell. Bear in mind during our discussion of this simplified version that many thousands of synapses are actually interacting in the same way on a single cell body and its dendrites.

TEMPORAL SUMMATION

Suppose that Ex1 has an action potential that causes an EPSP in the postsynaptic neuron. If another action potential occurs later in Ex1, an EPSP of the same magnitude takes place (panel A in ● Figure 4-17). Next suppose that Ex1 has two action potentials in close succession (panel B). The first action potential in Ex1 produces an EPSP in the postsynaptic membrane. While the postsynaptic membrane is still partially depolarized from this first EPSP, the second action potential in Ex1 produces a second EPSP. The second EPSP adds on to the first EPSP, bringing the membrane to threshold, so an action potential occurs in the postsynaptic neuron. Graded potentials do not have a refractory period, so this additive effect is possible.

The summing of several EPSPs occurring very close together in time because of successive firing of a single presynaptic neuron is known as **temporal summation** (*tempus* means "time"). In reality, the situation is much more complex than just described. The sum of up to 50 EPSPs might be needed to bring the postsynaptic membrane to threshold. Each action potential in a single presynaptic neuron triggers the emptying of a certain number of synaptic vesicles. The amount of neurotransmitter released and the resultant magnitude of the change in postsynaptic potential are thus directly related to the frequency of presynaptic action potentials. One way, then,

---

● FIGURE 4-17

Determination of the grand postsynaptic potential by the sum of activity in the presynaptic inputs.
Two excitatory (Ex1 and Ex2) and one inhibitory (In1) presynaptic inputs terminate on this hypothetical postsynaptic neuron. The potential of the postsynaptic neuron is being recorded.

 For an animation of this figure, click the Synapses and Neural Integration tab in the Neuronal Physiology and Hormonal Communication tutorial on the CD-ROM.

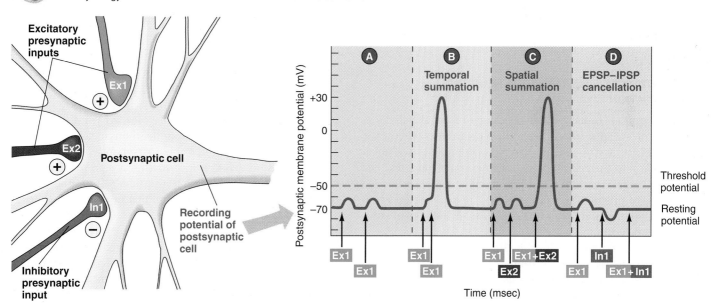

Panel **A** If an excitatory presynaptic input (Ex1) is stimulated a second time after the first EPSP in the postsynaptic cell has died off, a second EPSP of the same magnitude will occur.

Panel **B** If, however, Ex1 is stimulated a second time before the first EPSP has died off, the second EPSP will add onto, or sum with, the first EPSP, resulting in *temporal summation*, which may bring the postsynaptic cell to threshold.

Panel **C** The postsynaptic cell may also be brought to threshold by *spatial summation* of EPSPs that are initiated by simultaneous activation of two (Ex1 and Ex2) or more excitatory presynaptic inputs.

Panel **D** Simultaneous activation of an excitatory (Ex1) and inhibitory (In1) presynaptic input does not change the postsynaptic potential, because the resultant EPSP and IPSP cancel each other out.

in which the postsynaptic membrane can be brought to threshold is through rapid, repetitive excitation from a single persistent input.

SPATIAL SUMMATION

Let us now see what happens in the postsynaptic neuron if both excitatory inputs are stimulated simultaneously (panel C). An action potential in either Ex1 or Ex2 will produce an EPSP in the postsynaptic neuron; however, neither of these alone brings the membrane to threshold to elicit a postsynaptic action potential. But simultaneous action potentials in Ex1 and Ex2 produce EPSPs that add to each other, bringing the postsynaptic membrane to threshold, so an action potential does occur. The summation of EPSPs originating simultaneously from several different presynaptic inputs (that is, from different points in "space") is known as **spatial summation.** A second way, therefore, to elicit an action potential in a postsynaptic cell is through concurrent activation of several excitatory inputs. Again, in reality up to 50 simultaneous EPSPs are required to bring the postsynaptic membrane to threshold.

Similarly, IPSPs can undergo temporal and spatial summation. As IPSPs add together, however, they progressively move the potential further from threshold.

CANCELLATION OF CONCURRENT EPSPS AND IPSPS

If an excitatory and an inhibitory input are simultaneously activated, the concurrent EPSP and IPSP more or less cancel each other out. The extent of cancellation depends on their respective magnitudes. In most cases, the postsynaptic membrane potential remains close to resting (panel D).

IMPORTANCE OF POSTSYNAPTIC NEURONAL INTEGRATION

The magnitude of the GPSP depends on the sum of activity in all the presynaptic inputs and in turn determines whether or not the neuron will undergo an action potential to pass information on to the cells on which the neuron terminates. The following oversimplified real-life example demonstrates the benefits of this neuronal integration. The explanation is not completely accurate technically, but the principles of summation are accurate.

Assume for simplicity's sake that urination is controlled by a postsynaptic neuron that innervates the urinary bladder. When this neuron fires, the bladder contracts. (Actually, voluntary control of urination is accomplished by postsynaptic integration at the neuron controlling the external urethral sphincter rather than the bladder itself.) As the bladder starts to fill with urine and becomes stretched, a reflex is initiated that ultimately produces EPSPs in the postsynaptic neuron responsible for causing bladder contraction. Partial filling of the bladder does not cause enough excitation to bring the neuron to threshold, so urination does not take place (panel A of ● Figure 4-17). As the bladder becomes progressively filled, the frequency of action potentials progressively increases in the presynaptic neuron that signals the postsynaptic neuron of the extent of bladder filling (Ex1 in panel B of ● Fig-

ure 4-17). When the frequency becomes great enough that the EPSPs are temporally summed to threshold, the postsynaptic neuron undergoes an action potential that stimulates bladder contraction.

What if the time is inopportune for urination to take place? IPSPs can be produced at the bladder postsynaptic neuron by presynaptic inputs originating in higher levels of the brain responsible for voluntary control (In1 panel D of ● Figure 4-17). These "voluntary" IPSPs in effect cancel out the "reflex" EPSPs triggered by stretching of the bladder. Thus the postsynaptic neuron remains at resting potential and does not have an action potential, so the bladder is prevented from contracting and emptying even though it is full.

What if a person's bladder is only partially filled, so that the presynaptic input originating from this source is insufficient to bring the postsynaptic neuron to threshold to cause bladder contraction, and yet he or she needs to supply a urine specimen for laboratory analysis? The person can voluntarily activate an excitatory presynaptic neuron (Ex2 in panel C of ● Figure 4-17). The EPSPs originating from this neuron and the EPSPs of the reflex-activated presynaptic neuron (Ex1) are spatially summed to bring the postsynaptic neuron to threshold. This achieves the action potential necessary to stimulate bladder contraction, even though the bladder is not full.

This example illustrates the importance of postsynaptic neuronal integration. Each postsynaptic neuron in a sense "computes" all the input it receives and makes a "decision" about whether to pass the information on (that is, whether threshold is reached and an action potential is transmitted down the axon). In this way, neurons serve as complex computational devices, or integrators. The dendrites function as the primary processors of incoming information. They receive and tally the signals coming in from all the presynaptic neurons. Each neuron's output in the form of frequency of action potentials to other cells (muscle cells, gland cells, or other neurons) reflects the balance of activity in the inputs it receives via EPSPs or IPSPs from the thousands of other neurons that terminate on it. Each postsynaptic neuron filters out and does not pass on information it receives that is not significant enough to bring it to threshold. If every action potential in every presynaptic neuron that impinges on a particular postsynaptic neuron were to cause an action potential in the postsynaptic neuron, the neuronal pathways would be overwhelmed with trivia. Only if an excitatory presynaptic signal is reinforced by other supporting signals through summation will the information be passed on. Furthermore, interaction of postsynaptic potentials provides a way for one set of signals to offset another set (IPSPs negating EPSPs). This allows a fine degree of discrimination and control in determining what information will be passed on.

Let us now see why action potentials are initiated at the axon hillock.

▌ Action potentials are initiated at the axon hillock because it has the lowest threshold.

Threshold potential is not uniform throughout the postsynaptic neuron. The lowest threshold is present at the axon hillock, because this region has a much greater density of voltage-gated Na^+ channels than anywhere else in the neuron. This greater density of these voltage-sensitive channels makes the axon hillock considerably more responsive to changes in potential than the dendrites or remainder of the cell body. The latter regions have a significantly higher threshold than the axon hillock. Because of local current flow, changes in membrane potential (EPSPs or IPSPs) occurring anywhere on the dendrites or cell body spread throughout the dendrites, cell body, and axon hillock. When summation of EPSPs takes place, the lower threshold of the axon hillock is reached first, whereas the dendrites and cell body at the same potential are still considerably below their own, much higher thresholds. Therefore, an action potential originates in the axon hillock and is propagated from there to the end of the axon.

▍ Neuropeptides act primarily as neuromodulators.

Researchers recently discovered that in addition to the classical neurotransmitters just described, some neurons also release neuropeptides. **Neuropeptides** are larger than classical neurotransmitters and are cosecreted along with the neurotransmitter. Most neuropeptides function as neuromodulators. **Neuromodulators** are chemical messengers that do not cause the formation of EPSPs or IPSPs, but rather bring about long-term changes that subtly *modulate*—depress or enhance—the action of the synapse. They bind to neuronal receptors at nonsynaptic sites—that is, not at the subsynaptic membrane. Neuromodulators may act at either presynaptic or postsynaptic sites. For example, a neuromodulator may influence the level of an enzyme involved in the synthesis of a neurotransmitter by a presynaptic neuron, or it may alter the sensitivity of the postsynaptic neuron to a particular neurotransmitter by causing long-term changes in the number of subsynaptic receptor sites for the neurotransmitter. Thus neuromodulators delicately fine-tune the synaptic response. The effect may last for days or even months or years. Whereas neurotransmitters are involved in rapid communication between neurons, neuromodulators are involved with more long-lasting events, such as learning and motivation.

▍ Drugs and diseases can modify synaptic transmission.

Clinical Note The vast majority of drugs that influence the nervous system perform their function by altering synaptic mechanisms. Synaptic drugs may block an undesirable effect or enhance a desirable effect. Possible drug actions include (1) altering the synthesis, storage, or release of a neurotransmitter; (2) modifying neurotransmitter interaction with the postsynaptic receptor; (3) influencing neurotransmitter reuptake or destruction; and (4) replacing a deficient neurotransmitter with a substitute transmitter.

For example, the socially abused drug **cocaine** blocks the reuptake of the neurotransmitter *dopamine* at presynaptic terminals. It does so by binding competitively with the dopamine reuptake transporter, which is a protein molecule that picks up released dopamine from the synaptic cleft and shuttles it back to the axon terminal. With cocaine occupying the dopamine transporter, dopamine remains in the synaptic cleft longer than usual and continues to interact with its postsynaptic receptor sites. The result is prolonged activation of neural pathways that use this chemical as a neurotransmitter. Among these pathways are those that play a role in emotional responses, especially feelings of pleasure. In essence, when cocaine is present the neural switches in the pleasure pathway are locked in the "on" position.

Cocaine is addictive because it causes long-term molecular adaptations of the involved neurons such that they cannot transmit normally across synapses without increasingly higher doses of the drug. Because the postsynaptic cells are incessantly stimulated for an extended time, they become accustomed or adapt to "expecting" this high level of stimulation; that is, they are "hooked" on the drug. The term **tolerance** refers to this *desensitization* to an addictive drug so that the user needs greater quantities of the drug to achieve the same effect. Specifically, with prolonged use of cocaine, the number of dopamine receptors in the brain is reduced in response to the glut of the abused substance. As a result of this desensitization, the user must steadily increase the dosage of the drug to get the same "high," or sensation of pleasure. When the cocaine molecules diffuse away, the sense of pleasure evaporates, because the normal level of dopamine activity does not sufficiently "satisfy" the overly needy demands of the postsynaptic cells for stimulation. Cocaine users reaching this low become frantic and profoundly depressed. Only more cocaine makes them feel good again. But repeated use of cocaine modifies responsiveness to the drug. Over the course of abuse, the user often no longer can derive pleasure from the drug but suffers unpleasant *withdrawal symptoms* once the effect of the drug has worn off. Furthermore, the amount of cocaine needed to overcome the devastating crashes progressively increases. The user typically becomes **addicted** to the drug, compulsively seeking out and taking the drug at all costs, first to experience the pleasurable sensations and later to avoid the negative withdrawal symptoms, even when the drug no longer provides pleasure. Cocaine is abused by millions who have become addicted to its mind-altering properties, with devastating social and economic effects.

Whereas cocaine abuse leads to excessive dopamine activity, **Parkinson's disease** is attributable to a deficiency of dopamine in the *basal nuclei,* a region of the brain involved in controlling complex movements. This movement disorder is characterized by muscle rigidity and involuntary tremors at rest. The standard treatment for Parkinson's disease is the administration of *levodopa* (L-dopa), a precursor of dopamine. Dopamine itself cannot be administered because it is unable to cross the blood–brain barrier (discussed in the following chapter), but L-dopa can enter the brain from the blood. Once inside the brain, L-dopa is converted into dopamine, thus substituting for the deficient neurotransmitter. This therapy greatly alleviates the symptoms associated with the deficit in most patients. You will learn more about this condition when we discuss the basal nuclei.

Synaptic transmission is also vulnerable to neural toxins, which may cause nervous system disorders by acting at either presynaptic or postsynaptic sites. For example, **tetanus toxin** prevents the release of the neurotransmitter GABA from inhibitory presynaptic inputs terminating at neurons that supply skeletal muscles. Unchecked excitatory inputs to these neurons result in uncontrolled muscle spasms. These spasms occur especially in the jaw muscles early in the disease, giving rise to the common name of *lockjaw* for this condition. Later they progress to the muscles responsible for breathing, at which point death occurs.

Other drugs and diseases that influence synaptic transmission are too numerous to mention, but as these examples illustrate, any site along the synaptic pathway is vulnerable to interference.

▌ Neurons are linked through complex converging and diverging pathways.

Two important relationships exist between neurons: convergence and divergence. A given neuron may have many other neurons synapsing on it. Such a relationship is known as **convergence** (● Figure 4-18). Through this converging input, a single cell is influenced by thousands of other cells. This single cell, in turn, influences the level of activity in many other cells by divergence of output. The term **divergence** refers to the branching of axon terminals so that a single cell synapses with and influences many other cells.

Note that a particular neuron is postsynaptic to the neurons converging on it but presynaptic to the other cells at which it terminates. Thus the terms *presynaptic* and *postsynaptic* refer only to a single synapse. Most neurons are presynaptic to one group of neurons and postsynaptic to another group.

There are an estimated 100 billion neurons and 10^{14} (100 quadrillion) synapses in the brain alone! When you consider the vast and intricate interconnections possible between these neurons through converging and diverging pathways, you can begin to imagine how complex the wiring mechanism of our nervous system really is. Even the most sophisticated computers are far less complex than the human brain. The "language" of the nervous system—that is, all communication between neurons—is in the form of graded potentials, action potentials, neurotransmitter signaling across synapses, and other nonsynaptic forms of chemical chatter. All activities for which the nervous system is responsible—every sensation you feel, every command to move a muscle, every thought, every emotion, every memory, every spark of creativity—all depend on the patterns of electrical and chemical signaling between neurons along these complexly wired neural pathways.

A neuron communicates with the cells it influences by releasing a neurotransmitter, but this is only one means of intercellular ("between cell") communication. We will now consider all the ways by which cells can "talk with each other."

Click on the Media Exercises menu of the CD-ROM and work Media Exercise 4.3: Synapses and Neuronal Integration to test your understanding of the previous section.

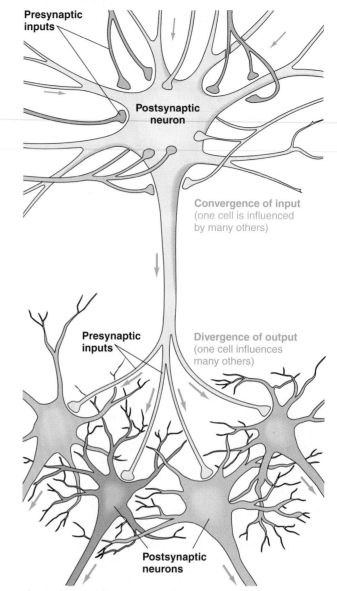

Arrows indicate direction in which information is being conveyed.

● **FIGURE 4-18**

Convergence and divergence

INTERCELLULAR COMMUNICATION AND SIGNAL TRANSDUCTION

Coordination of the diverse activities of cells throughout the body to accomplish life-sustaining and other desired activities depends on the ability of cells to communicate with each other.

▌ Communication between cells is largely orchestrated by extracellular chemical messengers.

There are three types of intercellular ("between cell") communication (● Figure 4-19):

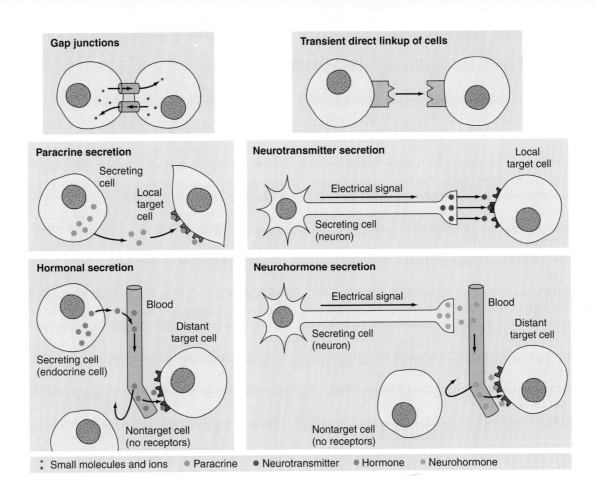

● FIGURE 4-19

Types of intercellular communication. Gap junctions and transient direct linkup of cells are both means of direct communication between cells. Paracrines, neurotransmitters, hormones, and neurohormones are all extracellular chemical messengers that accomplish indirect communication between cells. These chemical messengers differ in their source and the distance they travel to reach their target cells.

1. The most intimate means of intercellular communication is through gap junctions, which are minute tunnels that bridge the cytoplasm of neighboring cells in some types of tissues. Through these specialized anatomic arrangements, small ions and molecules are directly exchanged between interacting cells without ever entering the extracellular fluid (see p. 49).

2. The presence of identifying markers on the surface membrane of some cells permits them to directly link up transiently and interact with certain other cells in a specialized way. This is the means by which the phagocytes of the body's defense system specifically recognize and selectively destroy only undesirable cells, such as cancer cells, while leaving the body's own healthy cells alone.

3. The most common means by which cells communicate with each other is through **extracellular chemical messengers**, of which there are four types: *paracrines, neurotransmitters, hormones,* and *neurohormones*. In each case, a specific chemical messenger is synthesized by specialized cells to serve a designated purpose. On being released into the ECF by appropriate stimulation, these signaling agents act on other particular cells, the messenger's **target cells**, in a pre-scribed manner. To exert its effect, an extracellular chemical messenger must bind with target cell receptors specific for it.

The four types of chemical messengers differ in their source and the distance and means by which they get to their site of action as follows:

*   **Paracrines** are local chemical messengers whose effect is exerted only on neighboring cells in the immediate environment of their site of secretion. Because paracrines are distributed by simple diffusion, their action is restricted to short distances. They do not gain entry to the blood in any significant quantity because they are rapidly inactivated by locally existing enzymes. One example of a paracrine is *histamine*, which is released from a specific type of connective tissue cell during an inflammatory response within an invaded or injured tissue (see p. 331). Among other things, histamine dilates (opens more widely) the blood vessels in the vicinity to increase blood flow to the tissue. This action brings additional blood-borne combat supplies into the affected area.

Paracrines must be distinguished from chemicals that influence neighboring cells after being nonspecifically released during the course of cellular activity. For example, an increased local concentration of CO_2 in an exercising muscle is among the factors that promote local dilation of the blood vessels supplying the muscle. The resultant increased blood flow helps to meet the more active tissue's increased metabolic demands. However, CO_2 is produced by all cells and is not specifically released to accomplish this particular response, so it and similar nonspecifically released chemicals are not considered paracrines.

• As you just learned, neurons communicate directly with the cells they innervate (their target cells) by releasing **neurotransmitters,** which are very short-range chemical messengers, in response to electrical signals (action potentials). Like paracrines, neurotransmitters diffuse from their site of release across a narrow extracellular space to act locally on only an adjoining target cell, which may be another neuron, a muscle, or a gland.

• **Hormones** are long-range chemical messengers that are specifically secreted into the blood by endocrine glands in response to an appropriate signal. The blood carries the messengers to other sites in the body, where they exert their effects on their target cells some distance away from their site of release. Only the target cells of a particular hormone have membrane receptors for binding with this hormone. Nontarget cells are not influenced by any blood-borne hormones that reach them.

• **Neurohormones** are hormones released into the blood by *neurosecretory neurons.* Like ordinary neurons, neurosecretory neurons can respond to and conduct electrical signals. Instead of directly innervating target cells, however, a neurosecretory neuron releases its chemical messenger, a neurohormone, into the blood on appropriate stimulation. The neurohormone is then distributed through the blood to distant target cells. Thus, like endocrine cells, neurosecretory neurons release blood-borne chemical messengers, whereas ordinary neurons secrete short-range neurotransmitters into a confined space. In the future, the general term "hormone" will tacitly include both blood-borne hormonal and neurohormonal messengers.

In every case, extracellular chemical messengers are released from one cell type and interact with other target cells to bring about a desired effect in the target cells. We now turn our attention to how these chemical messengers bring about the desired cell response.

▌ Extracellular chemical messengers bring about cell responses primarily by signal transduction.

The term **signal transduction** refers to the process by which incoming signals (instructions from extracellular chemical messengers) are conveyed to the target cell's interior for execution. (A *transducer* is a device that receives energy from one system and transmits it in a different form to another system. For example, your radio receives radio waves sent out from the broadcast station and transmits these signals in the form of sound waves that can be detected by your ears.) Lipid-soluble extracellular chemical messengers, such as cholesterol-derived steroid hormones, can gain entry into the cell by dissolving in and passing through the lipid bilayer of the target cell's plasma membrane. Thus these extracellular chemical messengers bind to receptors inside the target cell to initiate the desired intracellular response themselves. By contrast, extracellular chemical messengers that are water soluble cannot gain entry to the target cell because they are poorly soluble in lipid and cannot dissolve in the plasma membrane. Protein hormones delivered by the blood and neurotransmitters released from nerve endings are the major water-soluble extracellular messengers. These messengers signal the cell to perform a given response by first binding with surface membrane receptors specific for that given messenger. These receptors are specialized proteins within the plasma membrane (see p. 46). The combination of extracellular messenger with a surface membrane receptor triggers a sequence of intracellular events that ultimately controls a particular cellular activity, such as membrane transport, secretion, metabolism, or contraction.

Despite the wide range of possible responses, binding the receptor with an extracellular messenger (the **first messenger**) brings about the desired intracellular response by only two general means: (1) by opening or closing channels or (2) by activating second-messenger systems. Because of the universal nature of these events, let's examine each more closely.

▌ Some extracellular chemical messengers open chemically gated channels.

Some extracellular messengers accomplish the desired intracellular response by opening or closing specific chemically gated channels in the membrane to regulate the movement of particular ions into or out of the cell. An example is the opening of chemically gated channels in the subsynaptic membrane in response to neurotransmitter binding. The resultant small, short-lived movement of given charge-carrying ions across the membrane through these open channels generates electrical signals, in this example EPSPs and IPSPs.

Stimulation of muscle cells to bring about contraction likewise occurs when chemically gated channels in the muscle cells open in response to binding of neurotransmitter released from the neurons supplying the muscle. You will learn more about this mechanism in the muscle chapter (Chapter 8). Thus control of chemically gated channels by extracellular messengers is an important regulatory mechanism in both nerve and muscle physiology.

On completion of the response, the extracellular messenger is removed from the receptor site, and the chemically gated channels close once again. The ions that moved across the membrane through opened channels to trigger the response are returned to their original location by special membrane carriers.

▌ Many extracellular chemical messengers activate second-messenger pathways.

Many extracellular chemical messengers that cannot actually enter their target cells bring about the desired intracellular response by another means than opening chemically gated channels. These first messengers issue their orders by triggering a "Psst, pass it on" process. Binding of the first messenger to a membrane receptor serves as a signal for activating an intracellular **second messenger.** The second messenger ultimately relays the orders through a series of biochemical intermediaries to particular intracellular proteins that carry out the dictated response, such as changes in cellular metabolism or secretory activity. The intracellular pathways activated by a second messenger in response to binding of the first messenger to a surface receptor are remarkably similar among different cells despite the diversity of ultimate responses to that signal. The variability in response depends on the specialization of the cell, not on the mechanism used.

Second-messenger systems are widely used throughout the body, including being one of the key means by which most water-soluble hormones ultimately bring about their effects. Let's now turn our attention to hormonal communication, where we will examine second-messenger systems in more detail.

PRINCIPLES OF HORMONAL COMMUNICATION

Endocrinology is the study of homeostatic chemical adjustments and other activities accomplished by hormones, which are secreted into the blood by endocrine glands. The nervous system and the endocrine system are the body's two major regulatory systems. The first part of this chapter described the underlying molecular and cellular mechanisms that serve as the basis for how the whole nervous system operates—electrical signaling within neurons and chemical transmission of signals between neurons. We are now going to focus on the molecular and cellular features of hormonal action and will compare the similarities and differences in how nerve cells and endocrine cells communicate with other cells in carrying out their regulatory actions. Finally, building on the different modes of action at the molecular and cellular level, the last section of the chapter will compare in a general way how the nervous and endocrine systems differ as regulatory systems.

▌ Hormones are chemically classified as being hydrophilic or lipophilic.

Hormones are not all similar chemically but instead fall into two distinct groups based on their solubility properties: hydrophilic or lipophilic hormones. Hormones within each group are further classified according to their biochemical structure and/or source as follows:

1. **Hydrophilic** ("water loving") **hormones** are highly water soluble and have low lipid solubility. Most hydrophilic hormones are peptide or protein hormones consisting of specific amino acids arranged in a chain of varying length. The shorter chains are peptides and the longer ones are proteins. For convenience, in the future we will refer to this entire category as *peptides.* An example is insulin from the pancreas. Another

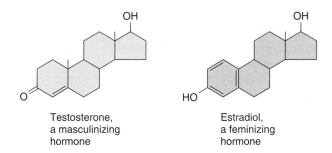

● **FIGURE 4-20**

Comparison of two steroid hormones, testosterone and estradiol

group of hydrophilic hormones are the *catecholamines,* which are derived from the amino acid tyrosine and are specifically secreted by the adrenal medulla. The adrenal gland consists of an inner adrenal medulla surrounded by an outer adrenal cortex. (You will learn more about the location and structure of the endocrine glands in later chapters.) Epinephrine is the major catecholamine.

2. **Lipophilic** ("lipid-loving") **hormones** have high lipid solubility and are poorly soluble in water. Lipophilic hormones include *thyroid hormone* and the *steroid hormones.* Steroids are neutral lipids derived from cholesterol. The hormones secreted by the adrenal cortex, such as cortisol, and the sex hormones (testosterone in males and estrogen in females) secreted by the reproductive organs are all steroids.

Minor differences in chemical structure between hormones within each category often result in profound differences in biologic response. Comparing two steroid hormones in ● Figure 4-20, for example, note the subtle difference between testosterone, the male sex hormone responsible for inducing the development of masculine characteristics, and estradiol, a form of estrogen, which is the feminizing female sex hormone.

The solubility properties of a hormone determine the means by which (1) the hormone is processed by the endocrine cell, (2) the way the hormone is transported in the blood, and (3) the mechanism by which the hormone exerts its effects at the target cell. We are first going to consider the different ways in which these hormone types are processed at their site of origin, the endocrine cell, before comparing their means of transport and their mechanisms of action.

▌ The mechanisms of synthesis, storage, and secretion of hormones vary according to their chemical differences.

Because of their chemical differences, the means by which the various types of hormones are synthesized, stored, and secreted differ as follows:

PROCESSING OF HYDROPHILIC PEPTIDE HORMONES

Peptide hormones are synthesized and secreted by the same steps used for manufacturing any protein that is exported from the cell (see ● Figure 2-3, p. 23). From the time peptide hormones are synthesized until they are secreted, they are always segregated from intracellular proteins by being contained within membrane-enclosed compartments. Here is a brief overview of these steps:

1.   Large precursor proteins, or **preprohormones**, are synthesized by ribosomes on the rough endoplasmic reticulum (ER). They then migrate to the Golgi complex in membrane-enclosed vesicles that pinch off from the smooth ER.
2.   During their journey through the ER and Golgi complex, the large preprohormone precursor molecules are pruned to active hormones.
3.   The Golgi complex then packages the finished hormones into secretory vesicles that are pinched off and stored in the cytoplasm until an appropriate signal triggers their secretion.
4.   On appropriate stimulation, the secretory vesicles fuse with the plasma membrane and release their contents to the outside by exocytosis (see p. 24). Such secretion usually does not go on continuously; it is triggered only by specific stimuli. The blood then picks up the secreted hormone for distribution.

PROCESSING OF LIPOPHILIC STEROID HORMONES

All steroidogenic (steroid-producing) cells perform the following steps to produce and release their hormonal product:

1.   Cholesterol is the common precursor for all steroid hormones.
2.   Synthesis of the various steroid hormones from cholesterol requires a series of enzymatic reactions that modify the basic cholesterol molecule—for example, by varying the type and position of side groups attached to the cholesterol framework (● Figure 4-21). Each conversion from cholesterol to a specific steroid hormone requires the help of a number of enzymes that are limited to certain steroidogenic organs. Accordingly, each steroidogenic organ can produce only the steroid hormone or hormones for which it has a complete set of appropriate enzymes. For example, a key enzyme

● **FIGURE 4-21**

Steroidogenic pathways for the major steroid hormones. All steroid hormones are produced through a series of enzymatic reactions that modify cholesterol molecules, such as by varying the side groups attached to them. Each steroidogenic organ can produce only those steroid hormones for which it has a complete set of the enzymes needed to appropriately modify cholesterol. For example, the testes have the enzymes necessary to convert cholesterol into testosterone (male sex hormone), whereas the ovaries have the enzymes needed to yield progesterone and the various estrogens (female sex hormones).

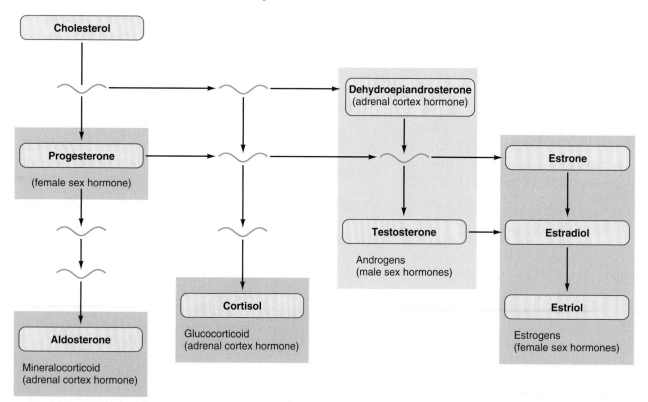

= Intermediates not biologically active in humans

necessary for producing cortisol is found only in the adrenal cortex, so no other steroidogenic organ can produce this hormone.

3. Unlike peptide hormones, steroid hormones are not stored. Once formed, the lipid-soluble steroid hormones immediately diffuse through the steroidogenic cell's lipid plasma membrane to enter the blood. Only the hormone precursor cholesterol is stored in significant quantities within steroidogenic cells. Accordingly, the rate of steroid hormone secretion is controlled entirely by the rate of hormone synthesis. In contrast, peptide hormone secretion is controlled primarily by regulating the release of presynthesized stored hormone.

The adrenomedullary catecholamines and thyroid hormone have unique synthetic and secretory pathways that will be described when addressing each of these hormones specifically in the endocrine chapter, Chapter 17.

Hydrophilic hormones dissolve in the plasma; lipophilic hormones are transported by plasma proteins.

All hormones are carried by the blood, but they are not all transported in the same manner:

• The hydrophilic peptide hormones are transported simply dissolved in the plasma.
• Lipophilic steroids and thyroid hormone, which are poorly soluble in water, cannot dissolve to any extent in the watery plasma. Instead, the majority of the lipophilic hormones circulate in the blood to their target cells reversibly bound to plasma proteins. Some are bound to specific plasma proteins designed to carry only one type of hormone, whereas other plasma proteins, such as albumin, indiscriminately pick up any "hitchhiking" hormone.

Only the small, unbound, freely dissolved fraction of a lipophilic hormone is biologically active (that is, free to cross capillary walls and bind with target cell receptors to exert an effect). The bound form of steroid and thyroid hormones provides a large reserve of these lipophilic hormones that can be called on to replenish the active free pool. To maintain normal endocrine function, the magnitude of the small free, effective pool, rather than the total plasma concentration of a particular lipophilic hormone, is monitored and adjusted.

Clinical Note The chemical properties of a hormone dictate not only the means by which blood transports it but also how it can be artificially introduced into the blood for therapeutic purposes. Because the digestive system does not secrete enzymes that can digest steroid and thyroid hormones, when taken orally these hormones, such as the sex steroids contained in birth control pills, can be absorbed intact from the digestive tract into the blood. No other type of hormones can be taken orally, because protein-digesting enzymes would attack and convert them into inactive fragments. Therefore, these hormones must be administered by nonoral routes; for example, insulin deficiency is treated with daily injections of insulin.

We will now examine how the hydrophilic and lipophilic hormones vary in their mechanisms of action at their target cells.

Hormones generally produce their effect by altering intracellular proteins.

To induce their effect, hormones must bind with target cell receptors specific for them. Each interaction between a particular hormone and a target cell receptor produces a highly characteristic response that differs among hormones and among different target cells influenced by the same hormone. Both the location of the receptors within the target cell, and the mechanism by which binding of the hormone with the receptors induces a response, vary depending on the hormone's solubility characteristics.

LOCATION OF RECEPTORS FOR HYDROPHILIC AND LIPOPHILIC HORMONES

Hormones can be grouped into two categories based on the location of their receptors:

1. The hydrophilic peptides and catecholamines, which are poorly soluble in lipid, cannot pass through the lipid membrane barriers of their target cells. Instead, they bind with specific receptors located on the *outer plasma membrane surface* of the target cell.
2. The lipophilic steroids and thyroid hormone easily pass through the surface membrane to bind with specific receptors located inside the target cell.

GENERAL MEANS OF HYDROPHILIC AND LIPOPHILIC HORMONE ACTION

Even though hormones elicit a wide variety of biologic responses, all hormones ultimately influence their target cells by altering the cell's proteins through three general means:

1. A few hydrophilic hormones, on binding with a target cell's surface receptors change the cell's permeability (either opening or closing channels to one or more ions) by *altering the conformation (shape) of adjacent channel-forming proteins already in the membrane.*
2. Most surface-binding hydrophilic hormones function by activating second-messenger systems within the target cell. This activation directly *alters the activity of pre-existing intracellular proteins, usually enzymes, to produce the desired effect.*
3. All lipophilic hormones function by *activating specific genes in the target cell to cause formation of new intracellular proteins,* which in turn produce the desired effect. The new proteins may be enzymatic or structural.

Let us examine the two major mechanisms of hormonal action (activation of second-messenger systems and activation of genes) in more detail.

▌Hydrophilic hormones alter pre-existing proteins via second-messenger systems.

Most hydrophilic hormones (peptides and catecholamines) bind to surface membrane receptors and produce their effects in their target cells by acting through a second-messenger system to alter the activity of pre-existing proteins. There are two major second-messenger pathways: One uses **cyclic adenosine monophosphate** (**cyclic AMP**, or **cAMP**) as a second messenger, and the other employs Ca^{2+} in this role. Let's examine the cAMP pathway in more detail to illustrate how second-messenger systems function.

CYCLIC AMP SECOND-MESSENGER PATHWAY

In the following description of the cAMP pathway, the numbered steps correlate to the numbered steps in ● Figure 4-22.

1. Binding of an appropriate extracellular messenger (a first messenger) to its surface membrane receptor eventually acti-

vates the enzyme **adenylyl cyclase** (step ①), which is located on the cytoplasmic side of the plasma membrane. A membrane-bound "middleman," a **G protein**, acts as an intermediary between the receptor and adenylyl cyclase. G proteins are found on the inner surface of the plasma membrane. An unactivated G protein consists of a complex of alpha (α), beta (β), and gamma (γ) subunits. A number of different G proteins with varying α subunits have been identified. The different G proteins are activated in response to binding of various first messengers to surface receptors. When a first messenger binds with its receptor, the receptor attaches to the appropriate G protein, resulting in activation of the α subunit. Once activated, the α subunit breaks away from the G protein complex and moves along the inner surface of the plasma membrane until it reaches an effector protein. An effector protein is either an ion channel or an enzyme within the membrane. The α subunit links up with the effector protein and alters its activity. In the cAMP pathway, adenylyl cyclase is the effector protein activated. Researchers have identified more than 300 different receptors that convey instructions of extracellular

---

● FIGURE 4-22

Mechanism of action of hydrophilic hormones via activation of the cyclic AMP second-messenger system

 For an animation of this figure, click the Cellular Communication tab in the Neural and Hormonal Communication tutorial on the CD-ROM.

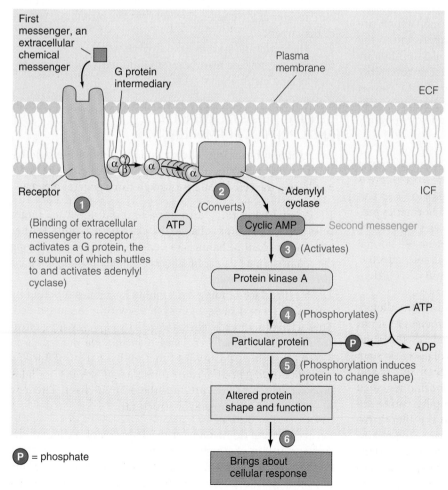

1. Binding of an extracellular messenger, the *first messenger*, to a surface membrane receptor activates by means of a G protein intermediary the membrane-bound enzyme adenylyl cyclase.

2. Adenylyl cyclase converts intracellular ATP into cyclic AMP.

3. Cyclic AMP acts as an intracellular *second messenger*, triggering the desired cellular response by activating protein kinase A.

4. Protein kinase A in turn phosphorylates a particular intracellular protein.

5. Phosphorylation induces a change in the shape and function of the protein.

6. The altered protein then accomplishes the cellular response dictated by the extracellular messenger.

messengers through the membrane to effector proteins by means of G proteins.

2. Adenylyl cyclase induces the conversion of intracellular ATP to cAMP by cleaving off two of the phosphates (step ②). (This is the same ATP used as the common energy currency in the body.)

3. Acting as the intracellular second messenger, cAMP triggers a preprogrammed series of biochemical steps within the cell to bring about the response dictated by the first messenger. To begin, cyclic AMP activates a specific intracellular enzyme, **protein kinase A** (step ③).

4. Protein kinase A in turn phosphorylates (attaches a phosphate group from ATP to) a specific pre-existing intracellular protein (step ④), such as an enzyme important in a particular metabolic pathway.

5. Phosphorylation causes the protein to change its shape and function (either activating or inhibiting it) (step ⑤).

6. This altered protein brings about a change in cell function (step ⑥). The resultant change is the target cell's ultimate physiologic response to the first messenger. For example, the activity of a particular enzymatic protein that regulates a specific metabolic event may be increased or decreased.

After the response is accomplished and the first messenger is removed, the α subunit rejoins the α and γ subunits to restore the inactive G-protein complex. Cyclic AMP and the other participating chemicals are inactivated so that the intracellular message is "erased" and the response can be terminated. Otherwise, once triggered the response would go on indefinitely until the cell ran out of necessary supplies.

Note that in this signal transduction pathway, the steps involving the extracellular first messenger, the receptor, the G protein complex, and the effector protein occur *in the plasma membrane* and lead to activation of the second messenger. The extracellular messenger cannot gain entry into the cell to "personally" deliver its message to the proteins that carry out the desired response. Instead, it initiates membrane events that activate an intracellular messenger, cAMP. The second messenger then triggers a chain reaction of biochemical events *inside the cell* that leads to the cellular response.

Different types of cells have different pre-existing proteins available for phosphorylation and modification by protein kinase A. Therefore, a *common second messenger, cAMP, can induce widely differing responses in different cells,* depending on what proteins are modified. Cyclic AMP can be thought of as a commonly used molecular "switch" that can "turn on" (or "turn off") different cell events, depending on the kinds of protein activity ultimately modified in the various target cells. The type of proteins altered by a second messenger depends on the unique specialization of a particular cell type. This can be likened to being able to either illuminate or cool off a room depending on whether the wall switch you flip on is wired to a device specialized to light up (a chandelier) or one specialized to create air movement (a ceiling fan). In the body, the variable responsiveness once the switch is turned on is due to the genetically programmed differences in the sets of proteins within different cells. For example, activating the cAMP system brings about modification of heart rate

in the heart, stimulation of the formation of female sex hormones in the ovaries, breakdown of stored glucose in the liver, control of water conservation during urine formation in the kidneys, creation of some simple memory traces in the brain, and perception of a sweet taste by a taste bud. (See the accompanying boxed feature, ❙ Beyond the Basics, for a description of a surprising signal-transduction pathway—one that causes a cell to kill itself.)

Many hydrophilic hormones use cAMP as their second messenger. A few use intracellular Ca^{2+} in this role; for others, the second messenger is still unknown.

Recognize that activation of second messengers is a universal mechanism employed by a variety of extracellular messengers in addition to hydrophilic hormones. Other chemical messengers that use the cAMP second-messenger system to accomplish their effects include neuromodulators and molecules that give rise to a number of different taste or smell sensations.

AMPLIFICATION BY A SECOND-MESSENGER PATHWAY

Several remaining points about receptor activation and the ensuing events merit attention. First, considering the number of steps involved in a second-messenger relay chain, you might wonder why so many cell types use the same complex system to accomplish such a wide range of functions. The multiple steps of a second-messenger pathway are actually advantageous, because a **cascading** (multiplying) **effect** of these pathways greatly amplifies the initial signal. Amplification means that the magnitude of the output of a system is much greater than the input. Binding of one extracellular chemical-messenger molecule to a receptor activates a number of adenyl cyclase molecules (let us arbitrarily say 10), each of which activates many (in our hypothetical example, let's say 100) cAMP molecules. Each cAMP molecule then acts on a single protein kinase A, which phosphorylates and thereby influences many (again, let us say 100) specific proteins, such as enzymes. Each enzyme, in turn, is responsible for producing many (perhaps 100) molecules of a particular product, such as a secretory product. The result of this cascade of events, with one event triggering the next event in sequence, is a tremendous amplification of the initial signal. In our hypothetical example, one chemical-messenger molecule has been responsible for inducing a yield of 10 million molecules of a secretory product. In this way, very low concentrations of hormones and other chemical messengers can trigger pronounced cell responses.

MODIFICATIONS OF SECOND-MESSENGER PATHWAYS

Although membrane receptors serve as links between extracellular first messengers and intracellular second messengers in the regulation of specific cellular activities, the receptors themselves are also frequently subject to regulation. In many instances, the number and affinity (attraction of a receptor for its chemical messenger) can be altered, depending on the circumstances.

Many disease processes can be linked to malfunctioning receptors or defects in signal transduction pathways. For ex-

Programmed Cell Suicide: A Surprising Example of a Signal Transduction Pathway

In the vast majority of cases, the signal transduction pathways triggered by the binding of an extracellular chemical messenger to a cell's surface membrane receptor are aimed at promoting proper functioning, growth, survival, or reproduction of the cell. By contrast, every cell has an unusual built-in pathway that, if triggered, causes the cell to commit suicide by activating intracellular protein-snipping enzymes, which slice the cell into small, disposable pieces. Such intentional programmed cell death is called **apoptosis.** (This term means "dropping off," in reference to the dropping off of cells that are no longer useful, much as autumn leaves drop off trees.) Apoptosis is a normal part of life: Individual cells that have become superfluous or disordered are triggered to self-destruct for the greater good of maintaining the whole body's health.

Roles of Apoptosis
Here are examples of the vital roles played by this intrinsic sacrificial program:

- *Predictable self-elimination of selected cells is a normal part of development.* Certain unwanted cells produced during development are programmed to kill themselves as the body is sculpted into its final form. During the development of a female, for example, apoptosis deliberately prunes the embryonic ducts capable of forming a male reproductive tract Likewise, apoptosis carves fingers from a mitten-shaped devel-

oping hand by eliminating the weblike membranes between them.
- *Apoptosis is important in tissue turnover in the adult body.* Optimal functioning of most tissues depends on a balance between controlled production of new cells and regulated cell self-destruction. This balance maintains the proper number of cells in a given tissue while ensuring a controlled supply of fresh cells that are at their peak of performance.
- *Programmed cell death plays an important role in the immune system.* Apoptosis provides a means to remove cells infected with harmful viruses. Furthermore, infection-fighting white blood cells that have finished their prescribed function and are no longer needed, execute themselves.
- *Undesirable cells that threaten homeostasis are typically culled from the body by apoptosis.* Included in this hit list are aged cells, cells that have suffered irreparable damage by exposure to radiation or other poisons, and cells that have somehow gone awry. Many mutated cells are eliminated by this means before they become fully cancerous.

Comparison of Apoptosis and Necrosis
Apoptosis is not the only means by which a cell can die, but it is the neatest way. Apoptosis is a controlled, intentional, tidy way of removing individual cells that are no longer needed or that pose a threat to the body. The other

form of cell death, **necrosis** (meaning "making dead"), is uncontrolled, accidental, messy murder of useful cells that have been severely injured by an agent external to the cell, as by a physical blow, O_2 deprivation, or disease. For example, heart muscle cells deprived of their O_2 supply by complete blockage of the blood vessels supplying them during a heart attack die as a result of necrosis (see p. 267).

Even though necrosis and apoptosis both result in cell death, the steps involved are very different. In necrosis the dying cells are passive victims, whereas in apoptosis the cells actively participate in their own deaths. In necrosis, the injured cell cannot pump out Na^+ as usual. As a result, water streams in by osmosis, causing the cell to swell and rupture. Typically, in necrosis the insult that prompted cell death injures many cells in the vicinity, so many neighboring cells swell and rupture together. Release of intracellular contents into the surrounding tissues initiates an inflammatory response at the damaged site (see p. 330). Unfortunately, the inflammatory response can potentially harm healthy neighboring cells.

By contrast, apoptosis targets individual cells for destruction, leaving the surrounding cells intact. A cell signaled to commit suicide detaches itself from its neighbors, then shrinks instead of swelling and bursting. As its lethal weapon, the suicidal cell activates a cascade of normally inactive intracellular protein-cutting enzymes, the **caspases,** which kill the cell from

---

Clinical Note ample, in Laron dwarfism, the person is abnormally short despite having normal levels of growth hormone because the tissues cannot respond normally to growth-promoting factors. This is in contrast to the more usual type of dwarfism in which the person is abnormally short because of growth hormone deficiency.

 Click on the Media Exercises menu of the CD-ROM and work Media Exercise 3.3: Signaling at Cell Membranes and Membrane Potential to test your understanding of the previous section.

By stimulating genes, lipophilic hormones promote synthesis of new proteins.

All lipophilic hormones (steroids and thyroid hormone) bind with intracellular receptors and produce their effects in their target cells by activating specific genes to cause the synthesis of new enzymatic or structural proteins. The following steps in this process correlate with those numbered in ● Figure 4-23:

1. Free lipophilic hormone (hormone not bound with its carrier) diffuses through the plasma membrane of the target cell and binds with its specific receptor inside the cell (step ①). Most lipophilic hormone receptors are located in the nucleus.
2. Each receptor has a specific region for binding with its hormone and another region for binding with DNA. The receptor cannot bind with DNA unless it first binds with the hormone. Once the hormone is bound to the receptor, the hormone receptor complex binds with DNA at a specific attachment site on DNA known as the **hormone response element (HRE)** (step ②). Different steroid hormones and thyroid hormone, once bound with their respective receptors, attach at different HREs on DNA. For example, the estrogen receptor complex binds at DNA's estrogen response element.

● **FIGURE 4-23**

Mechanism of action of lipophilic hormones via activation of genes
(*Source:* Adapted with permission from George A. Hedge, Howard D. Colby, and Robert L. Goodman, *Clinical Endocrine Physiology.* Philadelphia: W.B. Saunders Company, 1987, Figure 1-9, p. 20.)

within. Once unleashed, the caspases act like molecular scissors to systematically dismantle the cell. Snipping protein after protein, they chop up the nucleus, disassembling its life-essential DNA, then break down the internal shape-holding cytoskeleton, and finally fragment the cell itself into disposable membrane-enclosed packets. Importantly, the contents of the dying cell remain wrapped by plasma membrane throughout the entire self-execution process, thus avoiding the spewing of potentially harmful intracellular contents characteristic of necrosis. No inflammatory response is triggered, so no neighboring cells are harmed. Instead, cells in the vicinity swiftly engulf and destroy the apoptotic cell fragments by phagocytosis. The breakdown products are recycled for other purposes as needed. Meanwhile, the tissue as a whole has continued to function normally, while the targeted cell has unobtrusively killed itself.

Control of Apoptosis
If every cell contains the potent self-destructive caspases, what normally keeps these powerful enzymes under control (that is, in inactive form) in cells that are useful to the body and deserve to live? Likewise, what activates the death-wielding caspase cascade in unwanted cells destined to eliminate themselves? Given the importance of these life-or-death decisions, it is not surprising that multiple pathways tightly control whether a cell is "to be or not to be."

A cell normally receives a constant stream of "survival signals," which reassure the cell that it is useful to the body, that all is right in the internal environment surrounding the cell, and that everything is in good working order within the cell. These signals include tissue-specific growth factors, certain hormones, and appropriate contact with neighboring cells and the extracellular matrix. These extracellular survival signals trigger intracellular pathways that block activation of the caspase cascade, thus restraining the cell's death machinery. Most cells are programmed to commit suicide if they do not receive their normal reassuring survival signals. The usual safeguards are removed, and the lethal protein-snipping enzymes are unleashed. For example, withdrawal of growth factors or detachment from the extracellular matrix causes a cell to promptly execute itself.

Furthermore, cells display "death receptors" in their plasma membrane that receive specific extracellular "death signals," such as a particular hormone or a specific chemical messenger from white blood cells. Activation of death pathways by these signals can override the life-saving pathways triggered by the survival signals. The death-signal transduction pathway swiftly ignites the internal apoptotic machinery, driving the cell to its own demise. Likewise, the self-execution machinery is set in motion when a cell suffers irreparable intracellular damage. Thus some signals block apoptosis, whereas others promote it. Whether a

cell lives or dies depends on which of these competing signals dominates at any given time. Although all cells have the same death machinery, they vary in the specific signals that induce them to commit suicide.

Considering that every cell's life hangs in delicate balance at all times, it is not surprising that faulty control of apoptosis—resulting in either too much or too little cell suicide—appears to participate in many major diseases. Excessive apoptotic activity is believed to contribute to the brain cell death associated with Alzheimer's disease, Parkinson's disease, and stroke as well as to the premature demise of important infection-fighting cells in AIDS. Conversely, too little apoptosis most likely plays a role in cancer. Evidence suggests that cancer cells fail to respond to the normal extracellular signals that promote cell death. Because these cells neglect to die on cue, they grow in unchecked fashion, forming a chaotic, out-of-control mass.

Apoptosis is currently one of the hottest topics of investigation. Researchers are scrambling to sort out the multiple factors involved in the signal transduction pathways controlling this process. Their hope is to find ways to tinker with the apoptotic machinery to find badly needed new therapies to treat a variety of big killers.

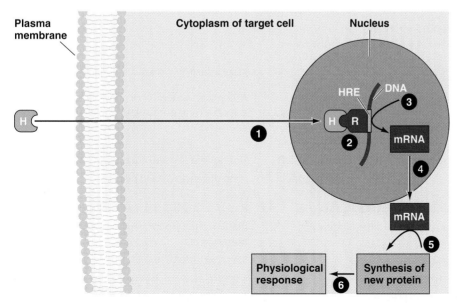

1. A lipophilic hormone diffuses through the plasma and nuclear membranes of its target cells and binds with a nuclear receptor specific for it.

2. The hormone receptor complex in turn binds with the hormone response element, a segment of DNA specific for the hormone receptor complex.

3. DNA binding activates specific genes, which produce complementary messenger RNA.

4. Messenger RNA leaves the nucleus.

5. In the cytoplasm, messenger RNA directs the synthesis of new proteins.

6. These new proteins, either enzymatic or structural, accomplish the target cell's ultimate physiologic response to the hormone.

H = Free lipophilic hormone
R = Lipophilic hormone receptor

HRE = Hormone response element
mRNA = Messenger RNA

3. Binding of the hormone receptor complex with DNA ultimately "turns on" a specific gene within the target cell. This gene contains a code for synthesizing a given protein. The code of the activated gene is transcribed into complementary messenger RNA (step ③).

4. The new messenger RNA leaves the nucleus and enters the cytoplasm (step ④).

5. In the cytoplasm, messenger RNA binds to a ribosome, the "workbench" that mediates the assembly of new proteins (p. A-24). Here, messenger RNA directs the synthesis of the designated new proteins according to the DNA code in the activated genes (step ⑤).

6. The newly synthesized protein, either enzymatic or structural, produces the target cell's ultimate physiologic response to the hormone (step ⑥).

By means of this mechanism, different genes are activated by different lipophilic hormones, resulting in different biologic effects.

▌ Hormonal responses are slower and longer than neural responses.

Compared to neural responses that are brought about within milliseconds, hormone action is relatively slow and prolonged, taking minutes to hours after the hormone binds to its receptor, for the response to take place. The variability in time of onset for hormonal responses depends on the mechanism employed. Hormones that act through a second-messenger system to alter a pre-existing enzyme's activity elicit full action within a few minutes. In contrast, hormonal responses that require the synthesis of new protein may take up to several hours before any action is initiated.

Also in contrast to neural responses that are quickly terminated once the triggering signal ceases, hormonal responses persist for a period of time after the hormone is no longer bound to its receptor. Once an enzyme is activated in response to hydrophilic hormonal input, it no longer depends on the presence of the hormone. Thus the response lasts until the enzyme is inactivated. Likewise, once a new protein is synthesized in response to lipophilic hormonal input, it continues to function until it is degraded. As a result, a hormone's effect usually lasts for some time after its withdrawal. Predictably, the responses that depend on protein synthesis last longer than do those stemming from enzyme activation.

Let's now compare further the similarities and differences between neural and hormonal responses at the system level.

COMPARISON OF THE NERVOUS AND ENDOCRINE SYSTEMS

As you know, the nervous and endocrine systems are the two main regulatory systems of the body. The **nervous system** swiftly transmits electrical impulses to the skeletal muscles and the exocrine glands that it innervates. The **endocrine system** secretes hormones into the blood for delivery to distant sites of action. Although these two systems differ in many respects, they have much in common (▲ Table 4-3). They both ultimately alter their target cells (their sites of action) by releasing chemical messengers (neurotransmitters in the case of nerve cells, hormones in the case of endocrine cells), which interact in particular ways with specific receptors of the target cells. Let's examine the anatomic distinctions between these two systems and the different ways in which they accomplish specificity of action.

▲ **TABLE 4-3**

Comparison of the Nervous System and the Endocrine System

PROPERTY	NERVOUS SYSTEM	ENDOCRINE SYSTEM
Anatomic Arrangement	A "wired" system; specific structural arrangement between neurons and their target cells; structural continuity in the system	A "wireless" system; endocrine glands widely dispersed and not structurally related to one another or to their target cells
Type of Chemical Messenger	Neurotransmitters released into synaptic cleft	Hormones released into blood
Distance of Action of Chemical Messenger	Very short distance (diffuses across synaptic cleft)	Long distance (carried by blood)
Means of Specificity of Action on Target Cell	Dependent on close anatomic relationship between nerve cells and their target cells	Dependent on specificity of target cell binding and responsiveness to a particular hormone
Speed of Response	Rapid (milliseconds)	Slow (minutes to hours)
Duration of Action	Brief (milliseconds)	Long (minutes to days or longer)
Major Functions	Coordinates rapid, precise responses	Controls activities that require long duration rather than speed

Anatomically, the nervous and endocrine systems are quite different. In the nervous system, each nerve cell terminates directly on its specific target cells; that is, the nervous system is "wired" in a very specific way into highly organized, distinct anatomic pathways for transmission of signals from one part of the body to another. Information is carried along chains of neurons to the desired destination through action potential propagation coupled with synaptic transmission. In contrast, the endocrine system is a "wireless" system in that the endocrine glands are not anatomically linked with their target cells. Instead, the endocrine chemical messengers are secreted into the blood and delivered to distant target sites. In fact, the components of the endocrine system itself are not anatomically interconnected; the endocrine glands are scattered throughout the body (see ● Figure 17-1, p. 530). These glands constitute a system in a functional sense, however, because they all secrete hormones and many interactions take place between various endocrine glands.

■ Neural specificity is due to anatomic proximity and endocrine specificity to receptor specialization.

As a result of their anatomic differences, the nervous and endocrine systems accomplish specificity of action by distinctly different means. Specificity of neural communication depends on nerve cells having a close anatomic relationship with their target cells, so that each neuron has a very narrow range of influence. A neurotransmitter is released for restricted distribution only to specific adjacent target cells, then is swiftly inactivated or removed before it can gain access to the blood. The target cells for a particular neuron have receptors for the neurotransmitter, but so do many other cells in other locations, and they could respond to this same mediator if it were delivered to them. For example, the entire system of nerve cells supplying all your body's skeletal muscles (motor neurons) use the same neurotransmitter, *acetylcholine (ACh)*, and all your skeletal muscles bear complementary ACh receptors (Chapter 7). Yet you can specifically wiggle your big toe without influencing any of your other muscles, because ACh can be discretely released from the motor neurons that are specifically wired to the muscles controlling your toe. If ACh were indiscriminately released into the blood, as are the hormones of the wireless endocrine system, all the skeletal muscles would simultaneously respond by contracting, because they all have identical receptors for ACh. This does not happen, of course, because of the precise wiring patterns that provide direct lines of communication between motor neurons and their target cells.

This specificity sharply contrasts to the way specificity of communication is built into the endocrine system. Because hormones travel in the blood, they can reach virtually all tissues. Yet despite this ubiquitous distribution, only specific target cells can respond to each hormone. Specificity of hormonal action depends on specialization of target cell receptors. For a hormone to exert its effect, the hormone must first bind with receptors specific for it that are located only on or in the hormone's target cells. Target cell receptors are highly discerning in their binding function. They will recognize and bind only a certain hormone, even though they are exposed simultaneously to many other blood-borne hormones, some of which are structurally very similar to the one that they discriminately bind. A receptor recognizes a specific hormone because the conformation of a portion of the receptor molecule matches a unique portion of its binding hormone in "lock-and-key" fashion. Binding of a hormone with target cell receptors initiates a reaction that culminates in the hormone's final effect. The hormone cannot influence any other cells, because they lack the right binding receptors.

■ The nervous and endocrine systems have their own realms of authority but interact functionally.

The nervous and endocrine systems are specialized for controlling different types of activities. In general, the nervous system governs the coordination of rapid, precise responses. It is especially important in the body's interactions with the external environment. Neural signals in the form of action potentials are rapidly propagated along nerve cell fibers, resulting in the release at the nerve terminal of a neurotransmitter that has to diffuse only a microscopic distance to its target cell before a response is effected. A neurally mediated response is not only rapid but brief; the action is quickly halted as the neurotransmitter is swiftly removed from the target site. This permits either ending the response, almost immediately repeating the response, or rapidly initiating an alternate response, as circumstances demand (for example, the swift changes in commands to muscle groups needed to coordinate walking). This mode of action makes neural communication extremely rapid and precise. The target tissues of the nervous system are the muscles and glands, especially exocrine glands, of the body.

The endocrine system, in contrast, is specialized to control activities that require duration rather than speed, such as regulating organic metabolism and water and electrolyte balance; promoting smooth, sequential growth and development; and controlling reproduction. The endocrine system responds more slowly to its triggering stimuli than does the nervous system for several reasons. First, the endocrine system must depend on blood flow to convey its hormonal messengers over long distances. Second, hormones' mechanism of action at their target cells is more complex than that of neurotransmitters and thus requires more time before a response occurs. The ultimate effect of some hormones cannot be detected until a few hours after they bind with target cell receptors. Also, because of the receptors' high affinity for their respective hormone, hormones often remain bound to receptors for some time, thus prolonging their biological effectiveness. Furthermore, unlike the brief, neurally induced responses that stop almost immediately after the neurotransmitter is removed, endocrine effects usually last for some time after the hormone's withdrawal. Neural responses to a single burst of neurotransmitter release usually last only milliseconds to seconds, whereas the alterations that hormones induce in target cells range from minutes to days or, in the case of growth-promoting effects, even a lifetime. Thus hormonal action is relatively slow and prolonged, making endocrine control particularly suitable for regulating metabolic activities that require long-term stability.

Although the endocrine and nervous systems have their own areas of specialization, they are intimately interconnected functionally. Some nerve cells do not release neurotransmitters at synapses but instead end at blood vessels and release their chemical messengers (neurohormones) into the blood, where these chemicals act as hormones. A given messenger may even be a neurotransmitter when released from a nerve ending and a hormone when secreted by an endocrine cell (see p. 558). The nervous system directly or indirectly controls the secretion of many hormones (see Chapter 17). At the same time, many hormones act as neuromodulators, altering synaptic effectiveness and thereby influencing the excitability of the nervous system. The presence of certain key hormones is even essential for the proper development and maturation of the brain during fetal life. Furthermore, in many instances the nervous and endocrine systems both influence the same target cells in supplementary fashion. For example, these two major regulatory systems both help regulate the circulatory and digestive systems. Thus many important regulatory interfaces exist between the nervous and endocrine systems.

In the next three chapters, we will concentrate on the nervous system and will examine the endocrine system in more detail in later chapters. Throughout the text we will continue to point out the numerous ways these two regulatory systems interact, so that the body is a coordinated whole, even though each system has its own realm of authority.

CHAPTER IN PERSPECTIVE: FOCUS ON HOMEOSTASIS

To maintain homeostasis, cells must communicate so that they work together to accomplish life-sustaining activities. To bring about desired responses, the two major regulatory systems of the body, the nervous system and the endocrine system, in particular must communicate with the target cells they are controlling. Neural and hormonal communication are therefore critical

in maintaining a stable internal environment as well as in coordinating nonhomeostatic activities.

Nerve cells are specialized to receive, process, encode, and rapidly transmit information from one part of the body to another. The information is transmitted over intricate nerve pathways by propagation of action potentials along the nerve cell's length as well as by chemical transmission of the signal from neuron to neuron at synapses and from neuron to muscles and glands through other neurotransmitter–receptor interactions at these junctions.

Collectively, the nerve cells make up the nervous system. Many of the activities controlled by the nervous system are geared toward maintaining homeostasis. Some neuronal electrical signals convey information about changes to which the body must rapidly respond in order to maintain homeostasis—for example, information about a fall in blood pressure. Other neuronal electrical signals swiftly convey messages to muscles and glands to stimulate appropriate responses to counteract these changes—for example, adjustments in heart and blood vessel activity to restore blood pressure to normal when it starts to fall.

The endocrine system secretes hormones into the blood, which carries these chemical messengers to distant target cells where they bring about their effect by changing the activity of enzymatic or structural proteins within these cells. Water-soluble hormones largely alter pre-existing intracellular proteins by activating second-messenger systems. Lipid-soluble hormones activate genes to promote the synthesis of new intracellular proteins. The resultant changes in activity of specific intracellular proteins accomplish the physiologic response directed by the hormonal messenger. Through its relatively slow acting hormonal messengers, the endocrine system generally regulates activities that require duration rather than speed. Most of these activities are directed toward maintaining homeostasis. For example, hormones help maintain the proper concentration of nutrients in the internal environment by directing chemical reactions involved in the cellular uptake, storage, release and use of these molecules. Also, hormones help maintain the proper water and electrolyte balance in the internal environment. Unrelated to homeostasis, hormones direct growth and control most aspects of the reproductive system.

Together the nervous and endocrine systems orchestrate a wide range of adjustments that help the body maintain homeostasis in response to stress. Likewise, these systems work in concert to control the circulatory and digestive systems, which in turn carry out important homeostatic activities.

CHAPTER SUMMARY

Introduction to Neural Communication (pp. 71–73)

▪ Nerve and muscle cells are known as *excitable tissues* because they can rapidly alter their membrane permeabilities and thus undergo transient membrane potential changes when excited. These rapid changes in potential serve as electrical signals.

▪ Compared to resting potential, a membrane becomes depolarized when its potential is reduced (becomes less negative) and

hyperpolarized when its potential is increased (becomes more negative). *(Review Figure 4-1.)*

▪ Changes in potential are brought about by triggering events that alter membrane permeability, in turn leading to changes in ion movement across the membrane.

▪ There are two kinds of potential change: (1) graded potentials, which serve as short-distance signals, and (2) action potentials, the long-distance signals. *(Review Table 4-1, p. 82.)*

Graded Potentials (pp. 73–75)

▮ Graded potentials occur in a small, specialized region of an excitable cell membrane. *(Review Figure 4-2.)*

▮ The magnitude of a graded potential varies directly with the magnitude of the triggering event.

▮ Graded potentials passively spread decrementally by local current flow and die out over a short distance. *(Review Figures 4-3 and 4-4.)*

Action Potentials (pp. 75–85)

▮ During an action potential, depolarization of the membrane to threshold potential triggers sequential changes in permeability caused by conformational changes in voltage-gated Na^+ and K^+ channels. *(Review Figures 4-5 and 4-6.)*

▮ These permeability changes bring about a brief reversal of membrane potential, with Na^+ influx causing the rising phase (from -70 mV to $+30$ mV), followed by K^+ efflux causing the falling phase (from peak back to resting potential). *(Review Figure 4-8.)*

▮ Before an action potential returns to resting, it regenerates an identical new action potential in the area next to it by means of current flow that brings the previously inactive area to threshold. This self-perpetuating cycle continues until the action potential has spread throughout the cell membrane in undiminished fashion.

▮ There are two types of action potential propagation: (1) contiguous conduction in unmyelinated fibers, in which the action potential spreads along every portion of the membrane; and (2) the more rapid, saltatory conduction in myelinated fibers, where the impulse jumps over the sections of the fiber covered with insulating myelin. *(Review Figures 4-10 and 4-13.)*

▮ The Na^+–K^+ pump gradually restores the ions that moved during propagation of the action potential to their original location, to maintain the concentration gradients.

▮ It is impossible to restimulate the portion of the membrane where the impulse has just passed until it has recovered from its refractory period. The refractory period ensures the one-way propagation of action potentials away from the original site of activation. *(Review Figures 4-11 and 4-12.)*

▮ Action potentials occur either maximally in response to stimulation or not at all.

▮ Variable strengths of stimuli are coded by varying the frequency of action potentials, not their magnitude.

Synapses and Neuronal Integration (pp. 85–92)

▮ The primary means by which one neuron directly interacts with another neuron is through a synapse. *(Review Figure 4-14.)*

▮ Most neurons have four different functional parts:

1. The dendrite/cell body region is specialized to serve as the postsynaptic component that binds with and responds to neurotransmitters released from other neurons.
2. The axon hillock is specialized for initiation of action potentials in response to graded potential changes induced by binding of a neurotransmitter with receptors on the dendrite/cell body region
3. The axon, or nerve fiber, is specialized to conduct action potentials in undiminished fashion from the axon hillock to the axon terminals.
4. The axon terminal is specialized to serve as the presynaptic component, which releases a neurotransmitter that influences other postsynaptic cells in response to action potential propagation down the axon. *(Review Figure 4-9.)*

▮ Released neurotransmitter combines with receptor sites on the postsynaptic neuron with which the presynaptic axon terminal interacts. This combination opens chemically gated channels in the postsynaptic neuron. *(Review Figure 4-15.)*

1. If nonspecific cation channels that permit passage of both Na^+ and K^+ are opened, the resultant ionic fluxes cause an EPSP, a small depolarization that brings the postsynaptic cell closer to threshold. *(Review Figure 4-16.)*
2. However, the likelihood that the postsynaptic neuron will reach threshold is diminished when an IPSP, a small hyperpolarization, is produced as a result of the opening of either K^+ or Cl^- channels, or both. *(Review Figure 4-16.)*

▮ Even though there are a number of different neurotransmitters, each synapse always releases the same neurotransmitter to produce a given response when combined with a particular receptor.

▮ The interconnecting synaptic pathways between various neurons are incredibly complex, due to convergence of neuronal input and divergence of its output. Usually, many presynaptic inputs converge on a single neuron and jointly control its level of excitability. This same neuron, in turn, diverges to synapse with and influence the excitability of many other cells. Each neuron thus has the task of computing an output to numerous other cells from a complex set of inputs to itself. *(Review Figure 4-18.)*

▮ If the dominant activity is in its excitatory inputs, the postsynaptic cell is likely to be brought to threshold and have an action potential. This can be accomplished by either (1) temporal summation (EPSPs from a single, repetitively firing, presynaptic input occurring so close together in time that they add together) or (2) spatial summation (adding of EPSPs occurring simultaneously from several different presynaptic inputs). *(Review Figure 4-17.)*

▮ If inhibitory inputs dominate, the postsynaptic potential is brought farther than usual away from threshold.

▮ If excitatory and inhibitory activity to the postsynaptic neuron is balanced, the membrane remains close to resting.

Intercellular Communication and Signal Transduction (pp. 92–95)

▮ Intercellular communication is accomplished by (1) gap junctions, (2) transient direct link up and interaction between cells, and (3) extracellular chemical messengers. *(Review Figure 4-19.)*

▮ Cells communicate with each other to carry out various coordinated activities largely by dispatching extracellular chemical messengers, which act on particular target cells to bring about the desired response.

▮ There are four types of extracellular chemical messengers, depending on their source and the distance and means by which they get to their site of action: (1) paracrines (local chemical messengers); (2) neurotransmitters (very short-range chemical messengers released by neurons); (3) hormones (long-range chemical messengers secreted into the blood by endocrine glands; and (4) neurohormones (long-range chemical messengers secreted into the blood by neurosecretory neurons). *(Review Figure 4-19.)*

▮ Transfer of the signal carried by the extracellular messenger into the cell for execution is known as *signal transduction*.

▮ Attachment of an extracellular chemical messenger that cannot gain entry to the cell such as a protein hormone (the first messenger) to a membrane triggers cellular responses by two major methods: (1) opening or closing specific channels or (2) activating an intracellular messenger (the second messenger). *(Review Figure 4-22.)*

Principles of Hormonal Communication (pp. 95–102)

▮ Hormones are long-distance chemical messengers secreted by the endocrine glands into the blood, which transports the hormones to specific target sites where they control a particular function by altering protein activity within the target cells.

- Hormones are grouped into two categories based on their solubility differences: (1) hydrophilic (water-soluble) hormones, which include peptides (the majority of hormones) and catecholamines (secreted by the adrenal medulla) and (2) lipophilic (lipid-soluble) hormones, which include thyroid hormone and steroid hormones (the sex hormones and those secreted by the adrenal cortex).
- Hydrophilic peptide hormones are synthesized and packaged for export by the endoplasmic reticulum/Golgi complex, stored in secretory vesicles, and released by exocytosis on appropriate stimulation.
- Hydrophilic hormones dissolve freely in the plasma for transport to their target cells.
- At their target cells, hydrophilic hormones bind with surface membrane receptors. On binding, a hydrophilic hormone triggers a chain of intracellular events by means of a second-messenger system that ultimately alters pre-existing cell proteins, usually enzymes, which exert the effect leading to the target cell's response to the hormone. (Review Figure 4-22.)
- Steroids are synthesized by modifications of stored cholesterol by means of enzymes specific for each steroidogenic tissue. (Review Figure 4-21.)
- Steroids are not stored in the endocrine cells. Being lipophilic, they diffuse out through the lipid membrane barrier as soon as they are synthesized. Control of steroids is directed at their synthesis.
- Lipophilic steroids and thyroid hormone are both transported in the blood largely bound to carrier plasma proteins, with only free, unbound hormone being biologically active.
- Lipophilic hormones readily enter through the lipid membrane barriers of their target cells and bind with nuclear receptors. Hormonal binding activates the synthesis of new enzymatic or structural intracellular proteins that carry out the hormone's effect on the target cell. (Review Figure 4-23.)

Comparison of the Nervous and Endocrine Systems (pp. 102–104)

- The nervous and endocrine systems are the two main regulatory systems of the body. (Review Table 4-3.)
- The nervous system is anatomically "wired" to its target organs, whereas the "wireless" endocrine system secretes blood-borne hormones that reach distant target organs.
- Specificity of neural action depends on the anatomic proximity of the neurotransmitter-releasing neuronal terminal to its target organ. Specificity of endocrine action depends on specialization of target-cell receptors for a specific circulating hormone.
- In general, the nervous system coordinates rapid responses, whereas the endocrine system regulates activities that require duration rather than speed.

REVIEW EXERCISES

Objective Questions (Answers on p. A-41)

1. Conformational changes in channel proteins brought about by voltage changes are responsible for opening and closing Na^+ and K^+ gates during the generation of an action potential. (True or false?)
2. The Na^+–K^+ pump restores the membrane to resting potential after it reaches the peak of an action potential. (True or false?)
3. Following an action potential, there is more K^+ outside the cell than inside because of the efflux of K^+ during the falling phase. (True or false?)
4. Postsynaptic neurons can either excite or inhibit presynaptic neurons. (True or false?)
5. Second-messenger systems ultimately bring about the desired cell response by inducing a change in the shape and function of particular intracellular proteins. (True or false?)
6. Each steroidogenic organ has all the enzymes necessary to produce any steroid hormone. (True or false?)
7. The one-way propagation of action potentials away from the original site of activation is ensured by the _____.
8. The _____ is the site of action potential initiation in most neurons because it has the lowest threshold.
9. A junction in which electrical activity in one neuron influences the electrical activity in another neuron by means of a neurotransmitter is called a _____.
10. Summing of EPSPs occurring very close together in time as a result of repetitive firing of a single presynaptic input is known as _____.
11. Summing of EPSPs occurring simultaneously from several different presynaptic inputs is known as _____.
12. The neuronal relationship where synapses from many presynaptic inputs act on a single postsynaptic cell is called _____, whereas the relationship in which a single presynaptic neuron synapses with and thereby influences the activity of many postsynaptic cells is known as _____.
13. A common membrane-bound intermediary between the receptor and the effector protein within the plasma membrane is the _____.
14. Using the following answer code, indicate which potential is being described:
    - (a) graded potential
    - (b) action potential

    _____ 1. behaves in all-or-none fashion
    _____ 2. magnitude of the potential change varies with the magnitude of the triggering response
    _____ 3. decremental spread away from the original site
    _____ 4. spreads throughout the membrane in nondiminishing fashion
    _____ 5. serves as a long-distance signal
    _____ 6. serves as a short-distance signal
15. Using the following answer code, indicate which characteristics apply to peptide and steroid hormones:
    - (a) peptide hormones
    - (b) steroid hormones
    - (c) both peptide and steroid hormones
    - (d) neither peptide or steroid hormones

    _____ 1. are hydrophilic
    _____ 2. are lipophilic
    _____ 3. are synthesized by the ER
    _____ 4. are synthesized by modifying cholesterol
    _____ 5. includes epinephrine from the adrenal medulla
    _____ 6. includes cortisol from the adrenal cortex
    _____ 7. bind to plasma proteins
    _____ 8. bind to nuclear receptors
    _____ 9. bind to surface membrane receptors
    _____ 10. activate genes to promote synthesis of new proteins
    _____ 11. act via second-messenger to alter pre-existing proteins
    _____ 12. are secreted into blood by endocrine glands and carried to distant target sites

Essay Questions

1. What are the two types of excitable tissue?
2. Define the following terms: *polarization, depolarization, hyper-polarization, repolarization, resting membrane potential, threshold potential, action potential, refractory period, all-or-none law.*
3. Describe the permeability changes and ion fluxes that occur during an action potential.
4. Compare contiguous conduction and saltatory conduction.
5. Compare the events that occur at excitatory and inhibitory synapses.
6. Compare the four kinds of gated channels in terms of the factor that opens or closes them.
7. List and describe the types of intercellular communication.
8. Discuss the sequence of events in the cAMP second-messenger pathway.
9. Compare the nervous and endocrine systems.

POINTS TO PONDER

(Explanations on p. A-41)

1. Which of the following would occur if a neuron were experimentally stimulated simultaneously at both ends?
   a. The action potentials would pass in the middle and travel to the opposite ends.
   b. The action potentials would meet in the middle and then be propagated back to their starting positions.
   c. The action potentials would stop as they met in the middle.
   d. The stronger action potential would override the weaker action potential.
   e. Summation would occur when the action potentials met in the middle, resulting in a larger action potential.

2. Compare the expected changes in membrane potential of a neuron stimulated with a *subthreshold stimulus* (a stimulus not sufficient to bring a membrane to threshold), a *threshold stimulus* (a stimulus just sufficient to bring the membrane to threshold), and a *suprathreshold stimulus* (a stimulus larger than that necessary to bring the membrane to threshold).

3. Assume you touched a hot stove with your finger. Contraction of the biceps muscle causes flexion (bending) of the elbow, whereas contraction of the triceps muscle causes extension (straightening) of the elbow. What pattern of postsynaptic potentials (EPSPs and IPSPs) would you expect to be initiated as a reflex in the cell bodies of the neurons controlling these muscles to pull your hand away from the painful stimulus?

   Now assume your finger is being pricked to obtain a blood sample. The same *withdrawal reflex* would be initiated. What pattern of postsynaptic potentials would you voluntarily produce in the neurons controlling the biceps and triceps to keep your arm extended in spite of the painful stimulus?

4. Researchers believe *schizophrenia* is caused by excessive dopamine activity in a particular region of the brain. Explain why symptoms of schizophrenia sometimes occur as a side effect in patients being treated for Parkinson's disease.

5. Assume presynaptic excitatory neuron A terminates on a postsynaptic cell near the axon hillock and presynaptic excitatory neuron B terminates on the same postsynaptic cell on a dendrite located on the side of the cell body opposite the axon hillock. Explain why rapid firing of presynaptic neuron A could bring the postsynaptic neuron to threshold through temporal summation, thus initiating an action potential, whereas firing of presynaptic neuron B at the same frequency and the same magnitude of EPSPs may not bring the postsynaptic neuron to threshold.

CLINICAL CONSIDERATION

(Explanation on p. A-42)

Becky N. was apprehensive as she sat in the dentist's chair awaiting the placement of her first silver amalgam (the "filling" in a cavity in a tooth). Before preparing the tooth for the amalgam by drilling away the decayed portion of the tooth, the dentist injected a local anesthetic in the nerve pathway supplying the region. As a result, Becky, much to her relief, did not feel any pain during the drilling and filling procedure. Local anesthetics block Na^+ channels. Explain how this action prevents the transmission of pain impulses to the brain.

PHYSIOEDGE RESOURCES

PhysioEdge CD-ROM

PhysioEdge, the CD-ROM packaged with your text, focuses on the concepts students find most difficult to learn. Figures marked with this icon have associated activities on the CD. For a visual review of concepts in this chapter, check out the following:

Tutorial: Neural and Hormonal Communication

Media Exercise 4.1: Basics of a Neuron

Media Exercise 4.2: Graded Potentials and Action Potentials

Media Exercise 4.3: Synapses and Neural Integration

Media Exercise 3.3: Signaling at Cell Membranes and Membrane Potential

PhysioEdge Website

The website for this book contains a wealth of helpful study aids, as well as many ideas for further reading and research. Log on to: **http://www.brookscole.com/hpfundamentals3**

Select Chapter 4 from the drop-down menu, or click on one of the many resource areas, including **Case Histories,** which introduce clinical aspects of human physiology. For this chapter check out: Case History 15: A Stiff Baby, and Case History 16: And a Limp Baby.

 For Suggested Readings, consult **InfoTrac College Edition/ Research** on the PhysioEdge website or go directly to InfoTrac College Edition, your online research library, at: **http://infotrac.thomsonlearning.com**

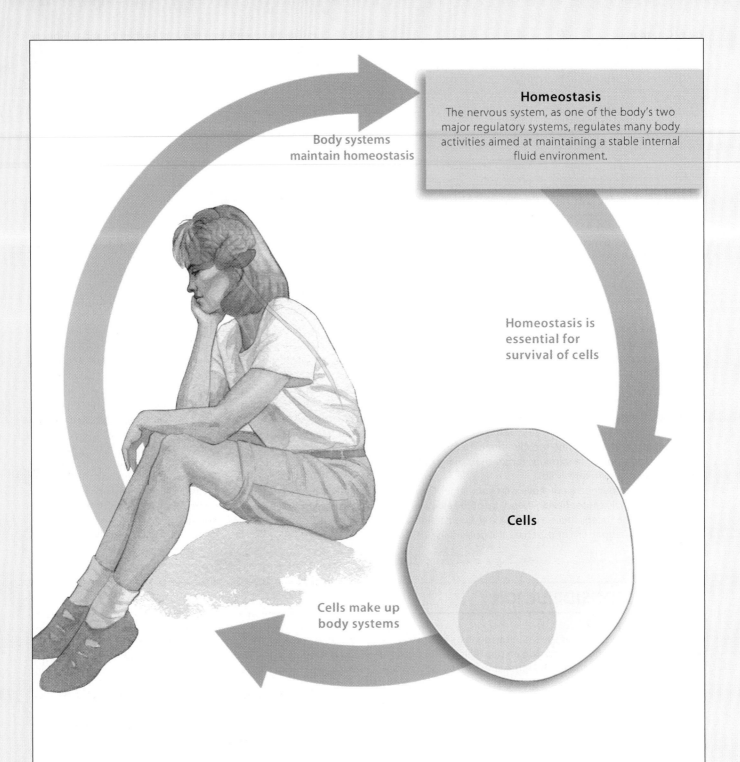

Homeostasis
The nervous system, as one of the body's two major regulatory systems, regulates many body activities aimed at maintaining a stable internal fluid environment.

Body systems maintain homeostasis

Homeostasis is essential for survival of cells

Cells

Cells make up body systems

The **nervous system** is one of the two major regulatory systems of the body, the other being the endocrine system. A complex interactive network of three basic functional types of nerve cells—afferent neurons, efferent neurons, and interneurons—constitutes the excitable cells of the nervous system. The **central nervous system (CNS)** consists of the brain and spinal cord, which receive input about the external and internal environment from the afferent neurons. The CNS sorts and processes this input, then initiates appropriate directions in the efferent neurons, which carry the instructions to glands or muscles to bring about the desired response—some type of secretion or movement. Many of these neurally controlled activities are directed toward maintaining homeostasis. In general, the nervous system acts by means of its electrical signals (action potentials) to control the rapid responses of the body.

The Central Nervous System

The way humans act and react depends on complex, organized, discrete neuronal processing. Many of the basic life-supporting neuronal patterns, such as those controlling respiration and circulation, are similar in all individuals. However, there must be subtle differences in neuronal integration between someone who is a talented composer and someone who cannot carry a tune, or between someone who is a math wizard and someone who struggles with long division. Some differences in the nervous systems of individuals are genetically endowed. The rest, however, are due to environmental encounters and experiences. When the immature nervous system develops according to its genetic plan, an overabundance of neurons and synapses is formed. Depending on external stimuli and the extent these pathways are used, some are retained, firmly established, and even enhanced, whereas others are eliminated.

Clinical Note   A case in point is **amblyopia** (lazy eye), in which the weaker of the two eyes is not used for vision. A lazy eye that does not get appropriate visual stimulation during a critical developmental period will almost completely and permanently lose the power of vision. The functionally blind eye itself is completely normal; the defect lies in the lost neuronal connections in the brain's visual pathways. If, however, the weak eye is forced to work, by covering the stronger eye with a patch during the sensitive developmental period, the weaker eye will retain full vision.

The maturation of the nervous system involves many instances of "use it or lose it." Once the nervous system has matured, ongoing modifications still occur as we continue to learn from our unique set of experiences. For example, the act of reading this page is somehow altering the neuronal activity of your brain as you (it is hoped) tuck the information away in your memory.

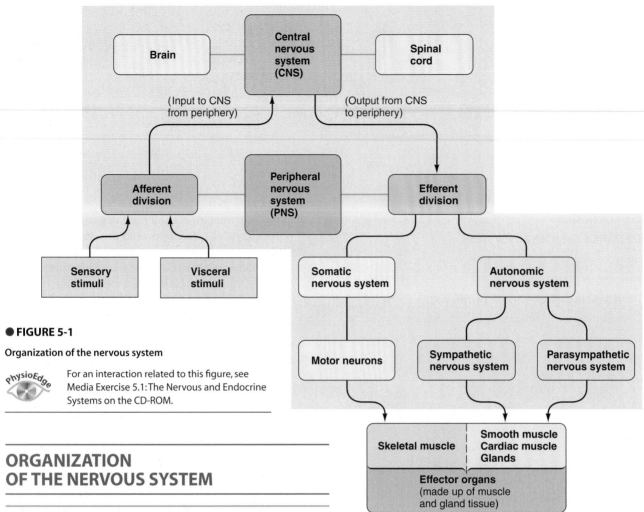

● FIGURE 5-1

Organization of the nervous system

PhysioEdge For an interaction related to this figure, see Media Exercise 5.1: The Nervous and Endocrine Systems on the CD-ROM.

ORGANIZATION OF THE NERVOUS SYSTEM

❚ The nervous system is organized into the central nervous system and the peripheral nervous system.

The nervous system is organized into the **central nervous system (CNS)**, consisting of the brain and spinal cord, and the **peripheral nervous system (PNS)**, consisting of nerve fibers that carry information between the CNS and other parts of the body (the periphery) (● Figure 5-1). The PNS is further subdivided into afferent and efferent divisions. The **afferent division** carries information *to* the CNS, apprising it of the external environment and providing status reports on internal activities being regulated by the nervous system (*a* is from *ad,* meaning "toward," as in *advance; ferent* means "carrying"; thus *afferent* means "carrying toward"). Instructions *from* the CNS are transmitted via the **efferent division** to **effector organs**—the muscles or glands that carry out the orders to bring about the desired effect (*e* is from *ex,* meaning "from," as in *exit;* thus *efferent* means "carrying from"). The efferent nervous system is divided into the **somatic nervous system**, which consists of the fibers of the motor neurons that supply the skeletal muscles, and the **autonomic nervous system** fibers, which innervate smooth muscle, cardiac muscle, and glands. The latter system is further subdivided into the **sympathetic nervous system** and the **parasympathetic ner-**

vous system, both of which innervate most of the organs supplied by the autonomic system.

It is important to recognize that all these "nervous systems" are really subdivisions of a single, integrated nervous system. They are arbitrary divisions based on differences in the structure, location, and functions of the various diverse parts of the whole nervous system.

❚ The three functional classes of neurons are afferent neurons, efferent neurons, and interneurons.

Three functional classes of neurons make up the nervous system: *afferent neurons, efferent neurons,* and *interneurons.* The afferent division of the peripheral nervous system consists of **afferent neurons,** which are shaped differently from efferent neurons and interneurons (● Figure 5-2). At its peripheral ending, a typical afferent neuron has a **sensory receptor** that generates action potentials in response to a particular type of stimulus. (This stimulus-sensitive afferent neuronal receptor should not be confused with the special protein receptors that bind chemical messengers and are found in the plasma membrane of all cells.) The afferent neuron cell body, which

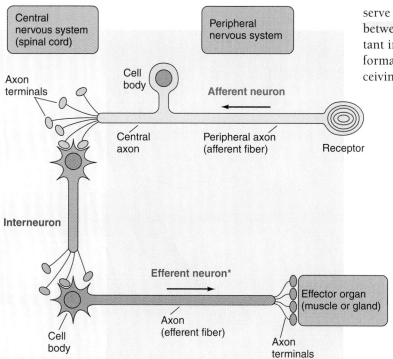

* Efferent autonomic nerve pathways consist of a two-neuron chain between the CNS and the effector organ.

● **FIGURE 5-2**

Structure and location of the three functional classes of neurons

PhysioEdge For an interaction related to this figure, see Media Exercise 5.1: The Nervous and Endocrine Systems on the CD-ROM.

is devoid of dendrites and presynaptic inputs, is located adjacent to the spinal cord. A long *peripheral axon,* commonly called the *afferent fiber,* extends from the receptor to the cell body, and a short central axon passes from the cell body into the spinal cord. Action potentials are initiated at the receptor end of the peripheral axon in response to a stimulus and are propagated along the peripheral axon and central axon toward the spinal cord. The terminals of the central axon diverge and synapse with other neurons within the spinal cord, thus disseminating information about the stimulus. Afferent neurons lie primarily within the peripheral nervous system. Only a small portion of their central axon endings project into the spinal cord to relay peripheral signals.

Efferent neurons also lie primarily in the peripheral nervous system (● Figure 5-2). Efferent neuron cell bodies originate in the CNS, where many centrally located presynaptic inputs converge on them to influence their outputs to the effector organs. Efferent axons *(efferent fibers)* leave the CNS to course their way to the muscles or glands they innervate, conveying their integrated output for the effector organs to put into effect. (An autonomic nerve pathway consists of a two-neuron chain between the CNS and the effector organ.)

Interneurons lie entirely within the CNS. About 99% of all neurons belong to this category. The human CNS is estimated to have over 100 billion interneurons! These neurons

serve two main roles. First, as their name implies, they lie between the afferent and efferent neurons and are important in integrating peripheral responses to peripheral information (*inter* means "between"). For example, on receiving information through afferent neurons that you are touching a hot object, appropriate interneurons signal efferent neurons that transmit to your hand and arm muscles the message, "Pull the hand away from the hot object!" The more complex the required action, the greater the number of interneurons interposed between the afferent message and efferent response. Second, interconnections between interneurons themselves are responsible for the abstract phenomena associated with the "mind," such as thoughts, emotions, memory, creativity, intellect, and motivation. These activities are the least understood functions of the nervous system.

With this brief introduction to the types of neurons and their location in the various divisions of the nervous system, we will now turn our attention to the central nervous system, followed in the next two chapters by a discussion of the two divisions of the peripheral nervous system.

PhysioEdge Click on the Media Exercises menu of the CD-ROM and work Media Exercise 5.1: The Nervous and Endocrine Systems to test your understanding of the previous section.

PROTECTION AND NOURISHMENT OF THE BRAIN

About 90% of the cells within the CNS are not neurons but **glial cells** or **neuroglia**. Despite their large numbers, the glial cells occupy only about half the volume of the brain, because they do not branch as extensively as neurons do.

▮ Glial cells support the interneurons physically, metabolically, and functionally.

Unlike neurons, glial cells do not initiate or conduct nerve impulses. However, they do communicate with neurons and among themselves by means of chemical signals. For much of the time since their discovery in the 19th century, scientists thought glial cells were passive "mortar" that physically supported the functionally important neurons. In the last decade, however, the varied and important roles of these dynamic cells have become apparent. Glial cells serve as the connective tissue of the CNS and as such help support the neurons both physically and metabolically. They homeostatically maintain the composition of the specialized extracellular environment surrounding the neurons within the narrow limits optimal for normal neuronal function. Furthermore, they actively modulate synaptic function and are now considered nearly as important as neurons to learning and memory. The specific roles of the four major types of glial cells in the CNS—*astrocytes, oligodendrocytes, microglia,* and *ependymal cells*—are as follows (● Figure 5-3).

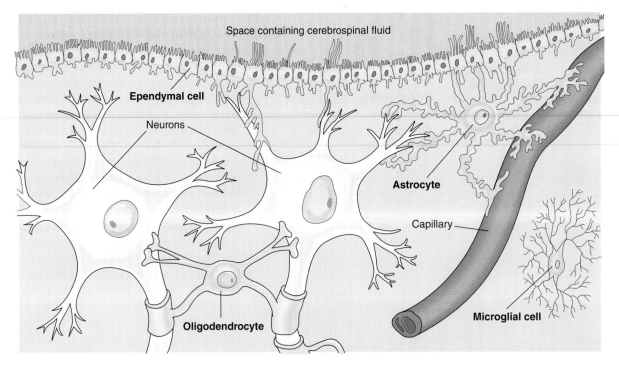

● **FIGURE 5-3**

Glial cells of the central nervous system. The glial cells include the astrocytes, oligodendrocytes, microglia, and ependymal cells.

ASTROCYTES

Named for their starlike shape (*astro* means "star," *cyte* means "cell") (● Figure 5-4), **astrocytes** are the most abundant glial cells. They fill a number of critical functions:

1. As the main "glue" (*glia* means "glue") of the CNS, astrocytes hold the neurons together in proper spatial relationships.
2. Astrocytes serve as a scaffold to guide neurons to their proper final destination during fetal brain development.
3. These glial cells induce the small blood vessels of the brain to undergo the anatomic and functional changes that are responsible for establishing the blood–brain barrier, a highly selective barricade between the blood and brain that we will soon describe in greater detail.
4. Astrocytes are important in the repair of brain injuries and in neural scar formation.
5. They play a role in neurotransmitter activity. Astrocytes take up and degrade glutamate and gamma-aminobutyric acid (GABA), excitatory and inhibitory neurotransmitters, respectively, thus bringing the actions of these chemical messengers to a halt.
6. Astrocytes take up excess K^+ from the brain ECF when high action-potential activity outpaces the ability of the Na^+–K^+ pump to return the effluxed K^+ to the neurons. (Recall that K^+ leaves a neuron during the falling phase of an action potential; see p. 78.) By taking up excess K^+, the astrocytes help maintain the proper brain ECF ion concentration to sustain normal neural excitability.
7. Recent discoveries have shown that astrocytes communicate with neurons by means of chemical signals passing locally in both directions between these cells. Evidence suggests

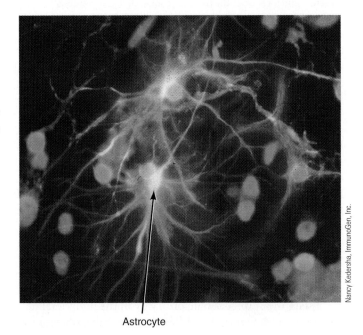

Nancy Kedersha, ImmunoGen, Inc.

Astrocyte

● **FIGURE 5-4**

Astrocytes. Note the starlike shape of these astrocytes, which have been grown in tissue culture.

this two-directional extracellular signaling plays an important role in synaptic transmission and the brain's processing of information. Furthermore, these glial cells converse with each other by chemical signaling, further influencing how well the brain performs.

OLIGODENDROCYTES

Oligodendrocytes form the insulative myelin sheaths around axons in the CNS. An oligodendrocyte has several elongated projections, each of which is wrapped jelly-roll fashion around a section of an interneuronal axon to form a patch of myelin (see p. 84). Late in fetal life, oligodendrocytes begin secreting *nerve-growth–inhibiting proteins,* such as the recently identified chemical messenger dubbed *Nogo.* Scientists speculate that nerve-growth inhibitors normally serve as "guardrails" to keep new nerve endings from straying outside their proper paths. The growth-inhibiting action of oligodendrocytes may thus serve to stabilize the enormously complex structure of the CNS.

Clinical Note Growth inhibition is a disadvantage, however, when central axons need to be mended, as when the spinal cord has been severed accidentally. Damaged neuronal fibers in the brain and spinal cord never regenerate, leaving the patient with permanent deficits, such as paralysis and loss of sensation below the level of spinal cord severance. Researchers are currently exploring promising ways to spur repair of central axonal pathways, with the goal of enabling victims to walk again.

By contrast, Schwann cells, the myelin-forming cells of the peripheral nervous system, form a *regeneration tube* and secrete *nerve-growth–enhancing proteins* that respectively guide and promote regrowth of damaged peripheral axons, as long as the cell body and dendrites remain intact (that is, as long as the neuron is still alive). Successful fiber regeneration permits the return of sensation and movement at some time after peripheral nerve injuries, although regeneration is not always successful.

MICROGLIA

Microglia are the immune defense cells of the CNS. They are phagocytes delivered by the blood to the CNS, where they remain stationary until activated by an infection or injury. When trouble occurs in the CNS, microglia become highly mobile, moving toward the affected area to remove any foreign invaders or tissue debris. Activated microglia release destructive chemicals for assault against their target.

Clinical Note Researchers increasingly suspect that excessive release of these chemicals from overzealous microglia may damage the neurons they are meant to protect, thus contributing to the insidious neuronal damage seen in stroke, Alzheimer's disease, multiple sclerosis, the dementia (mental failing) of AIDS, and other *neurodegenerative diseases.*

EPENDYMAL CELLS

Ependymal cells line the internal, fluid-filled cavities of the CNS—the ventricles of the brain and the central canal of the spinal cord. The **ventricles** consist of four interconnected chambers within the interior of the brain that are continuous with the narrow, hollow **central canal** that tunnels through the middle of the spinal cord. The ependymal cells lining the ventricles help form cerebrospinal fluid, a topic we discuss shortly.

Importantly, exciting new research has identified a totally different role for the ependymal cells: They serve as neural stem cells with the potential of forming not only other glial cells but new neurons as well (see p. 6). The traditional view has long held that new neurons are not produced in the mature brain, thus neurons are considered irreplaceable. But the discovery that ependymal cells are precursors for new neurons suggests that the adult brain has more potential for repairing damaged regions than previously assumed. Currently there is no evidence that the brain spontaneously repairs itself following neuron-losing insults such as head trauma, strokes, and neurodegenerative disorders. Apparently most brain regions cannot activate this mechanism for replenishing neurons, probably because the appropriate "cocktail" of supportive chemicals is not present. Researchers hope that probing into why these ependymal cells are dormant and how they might be activated will lead to the possibility of unlocking the brain's latent capacity for self-repair.

▌ The delicate central nervous tissue is well protected.

Central nervous tissue is very delicate. This characteristic, coupled with the fact that damaged nerve cells cannot be replaced, makes it imperative that this fragile, irreplaceable tissue be well protected. Four major features help protect the CNS from injury:

1. It is enclosed by hard, bony structures. The **cranium (skull)** encases the brain, and the **vertebral column** surrounds the spinal cord.
2. Three protective and nourishing membranes, the **meninges,** lie between the bony covering and the nervous tissue.
3. The brain "floats" in a special cushioning fluid, the **cerebrospinal fluid (CSF).** CSF circulates throughout the ventricles and central canal within the interior of the brain and spinal cord as well as over the entire surface of the brain and spinal cord in a space between the meningeal layers. CSF has about the same density as the brain itself, so the brain is essentially suspended in this special fluid environment. The major function of CSF is to serve as a shock-absorbing fluid to prevent the brain from bumping against the interior of the hard skull when the head is subjected to sudden, jarring movements.
4. A highly selective **blood–brain barrier (BBB)** limits access of blood-borne materials into the vulnerable brain tissue. Throughout the body, exchange of materials between the blood and the surrounding interstitial fluid can take place only across the walls of capillaries, the smallest of blood vessels. Capillary walls are formed by a single layer of cells. The holes or pores usually present between the cells making up a capillary wall permit rather free exchange across capillaries elsewhere. However, the cells that form the walls of a brain capillary are joined by tight junctions (see p. 48). These impermeable junctions completely seal the capillary wall so that nothing can be exchanged across the wall by passing between the cells. The only permissible exchanges occur through the capillary cells themselves. Only selected, carefully regulated exchanges can be made across this barrier. Thus, transport

Strokes: A Deadly Domino Effect

The most common cause of brain damage is *cerebrovascular accidents (strokes)*. When a brain (cerebral) blood vessel is blocked by a clot or ruptures, the brain tissue supplied by that vessel loses its vital O_2 and glucose supply. The result is damage and usually death of the deprived tissue. Recently, researchers have learned that neural damage (and the subsequent loss of neural function) extends well beyond the blood-deprived area as a result of a neurotoxic effect that leads to the death of additional nearby cells. Whereas the initial blood-deprived cells die by necrosis (unintentional cell death), the doomed neighbors undergo apoptosis (deliberate cell suicide; see p. 100). The initial O_2-starved cells release excessive amounts of glutamate, a common excitatory neurotransmitter. Glutamate or other neurotransmitters are normally released in small amounts from neurons as a means of chemical communication between brain cells. The excitatory overdose of glutamate from the damaged brain cells binds with and over-excites surrounding neurons. Specifically, glutamate binds with excitatory receptors known as NMDA receptors, which function as Ca^{2+} channels. As a result of toxic activation of these receptor-channels, they remain open for too long, permitting too much Ca^{2+} to rush into the affected neighboring neurons. This elevated intracellular Ca^{2+} triggers these cells to self-destruct. Cell-damaging free radicals (see p. 536) are produced during this process. Adding to the injury, researchers speculate that the Ca^{2+} apoptotic signal may spread from these dying cells to abutting healthy cells through gap junctions, cell-to-cell conduits that allow Ca^{2+} and other small ions to diffuse freely between cells. This action kills even more neuronal victims. Thus the majority of neurons that die following a stroke are originally unharmed cells that commit suicide in response to the chain of reactions unleashed by the toxic release of glutamate from the initial site of O_2 deprivation.

Until the last decade, physicians could do nothing to halt the inevitable neuronal loss following a stroke, leaving patients with an unpredictable mix of neural deficits. Treatment was limited to rehabilitative therapy after the damage was already complete. In recent years, armed with the new knowledge about the underlying factors in stroke-related neuronal death, the medical community has been seeking ways to halt the cell-killing domino effect. The goal, of course, is to limit the extent of neuronal damage and thus minimize or even prevent clinical symptoms such as paralysis. In the early 1990s, doctors started administering clot-dissolving drugs within the first three hours after the onset of a stroke to restore blood flow through blocked cerebral vessels. Clot busters were the first drugs used to treat strokes, but they are only the beginning of new stroke therapies. Other methods are currently under investigation to prevent adjacent nerve cells from succumbing to the neurotoxic release of glutamate. These include blocking the NMDA receptors that initiate the death-wielding chain of events in response to glutamate, halting the apoptosis pathway that results in self-execution, and blocking the gap junctions that permit the Ca^{2+} death messenger to spread to adjacent cells. These tactics hold much promise for treating strokes, which are the most prevalent cause of adult disability and the third leading cause of death in the United States. However, to date no new neuroprotective drugs have been found that do not cause serious side effects.

---

across brain capillary walls *between* the wall-forming cells is *anatomically prevented* and transport *through* the cells is *physiologically restricted*. Together, these mechanisms constitute the BBB. By strictly limiting exchange between the blood and the brain, the BBB protects the delicate brain from chemical fluctuations in the blood. For example, even if the K^+ level in the blood is doubled, little change occurs in the K^+ concentration of the fluid bathing the central neurons. This is beneficial because alterations in interstitial fluid K^+ would be detrimental to neuronal function. Also, the BBB minimizes the possibility that potentially harmful blood-borne substances might reach the central neural tissue. It further prevents certain circulating hormones that could also act as neurotransmitters from reaching the brain, where they could produce uncontrolled nervous activity. On the negative side, the BBB limits the use of drugs for the treatment of brain and spinal cord disorders, because many drugs cannot penetrate this barrier.

▌ The brain depends on constant delivery of oxygen and glucose by the blood.

Even though many substances in the blood never actually come in contact with the brain tissue, the brain, more than any other tissue, is highly dependent on a constant blood supply. Unlike most tissues, which can resort to anaerobic metabolism to produce ATP in the absence of O_2 for at least short periods (see p. 29), the brain cannot produce ATP in the absence of O_2. Furthermore, in contrast to most tissues, which can use other sources of fuel for energy production in lieu of glucose, the brain normally uses only glucose but does not store any of this nutrient. Therefore, the brain absolutely depends on a continuous, adequate blood supply of O_2 and glucose. Brain damage results if this organ is deprived of its critical O_2 supply for more than 4 to 5 minutes or if its glucose supply is cut off for more than 10 to 15 minutes. The most common cause of inadequate blood supply to the brain is a stroke. (See the accompanying boxed feature, ▎ Beyond the Basics, for details.)

OVERVIEW OF THE CENTRAL NERVOUS SYSTEM

The central nervous system consists of the brain and spinal cord. The brain has an estimated 100 billion neurons, which are assembled into complex networks that enable you to (1) subconsciously regulate your internal environment by neural means; (2) experience emotions; (3) voluntarily control your movements; (4) perceive (be consciously aware of) your own body and your surroundings; and (5) engage in other higher cognitive processes such as thought and memory. The term **cognition** refers to the act or process of "knowing," including both awareness and judgment.

No part of the brain acts in isolation from other brain regions, because networks of neurons are anatomically linked by synapses, and neurons throughout the brain communicate extensively with each other by electrical and chemical means. However, neurons that work together to ultimately accomplish a given function tend to be organized within a discrete location. Therefore, even though the brain is a functional whole, it is organized into different regions. The parts of the brain can be arbitrarily grouped in various ways based on anatomic distinctions, functional specialization, and evolutionary development. We will use the following grouping (▲ Table 5-1, pp. 116–117):

1. Brain stem
2. Cerebellum
3. Forebrain
    a. Diencephalon
        (1) Hypothalamus
        (2) Thalamus
    b. Cerebrum
        (1) Basal nuclei
        (2) Cerebral cortex

The order in which these components are listed generally represents both their anatomic location (from bottom to top) and their complexity and sophistication of function (from the least specialized, oldest level to the newest, most specialized level).

A primitive nervous system consists of comparatively few interneurons interspersed between afferent and efferent neurons. During evolutionary development, the interneuronal component progressively expanded, formed more complex interconnections, and became localized at the head end of the nervous system, forming the brain. Newer, more sophisticated layers of the brain were added on to the older, more primitive layers. The human brain represents the present peak of development.

The *brain stem,* the oldest region of the brain, is continuous with the spinal cord (▲ Table 5-1 and ● Figure 5-5b). It consists of the midbrain, pons, and medulla. The brain stem controls many of the life-sustaining processes, such as respiration, circulation, and digestion, which are common to many of the lower vertebrate forms. These processes are often referred to as "vegetative" functions because, with the loss of higher brain functions, these lower brain levels, in accompaniment with appropriate supportive therapy such as provision of adequate nourishment, can still sustain the functions essential for survival. Because the person has no awareness or control of that life, however, someone in that condition is sometimes described as "being a vegetable."

Attached at the top rear portion of the brain stem is the *cerebellum,* which is concerned with maintaining proper position of the body in space and subconscious coordination of motor activity (movement). The cerebellum also plays a key role in learning skilled motor tasks, such as a dance routine. On top of the brain stem, tucked within the interior of the cerebrum, is the *diencephalon.* It houses two brain components: the *hypothalamus,* which controls many homeostatic functions important in maintaining stability of the internal environment, and the *thalamus,* which performs some primitive sensory processing.

● **FIGURE 5-5**

Brain of human cadaver. (a) Dorsal view looking down on the top of the brain. Note that the deep longitudinal fissure divides the cerebrum into the right and left cerebral hemispheres. (b) Sagittal view of the right half of the brain. All major brain regions are visible from this midline interior view. The corpus callosum serves as a neural bridge between the two cerebral hemispheres.

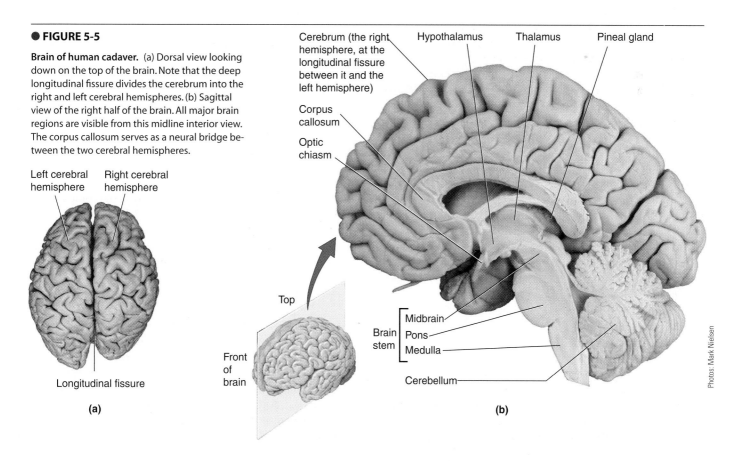

▲ TABLE 5-1

Overview of Structures and Functions of the Major Components of the Brain

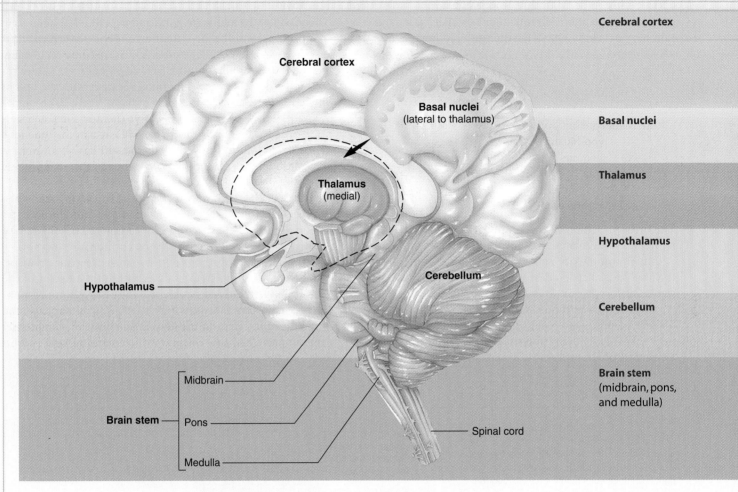

BRAIN COMPONENT

Cerebral cortex

Basal nuclei

Thalamus

Hypothalamus

Cerebellum

Brain stem
(midbrain, pons, and medulla)

On top of this "cone" of lower brain regions is the *cerebrum,* whose "scoop" gets progressively larger and more highly convoluted (that is, has tortuous ridges delineated by deep grooves or folds) the more advanced the vertebrate species is. The cerebrum is most highly developed in humans, where it constitutes about 80% of the total brain weight. The outer layer of the cerebrum is the highly convoluted *cerebral cortex,* which caps an inner core that houses the *basal nuclei.* The myriad convolutions of the human cerebral cortex give it the appearance of a much-folded walnut (● Figure 5-5a). In many lower mammals, the cortex is perfectly smooth. Without these surface wrinkles, the human cortex would take up to three times the area it does, and, accordingly, would not fit like a cover over the underlying structures. The increased neural circuitry housed in the extra cerebral cortical area not found in lower species is responsible for much of our unique human abilities. The cerebral cortex plays a key role in the most sophisticated neural functions, such as voluntary initiation

of movement, final sensory perception, conscious thought, language, personality traits, and other factors we associate with the mind or intellect. It is the highest, most complex, integrating area of the brain.

Each of these regions of the CNS will be discussed in turn, starting with the highest level, the cerebral cortex, and moving down to the lowest level, the spinal cord.

CEREBRAL CORTEX

The **cerebrum,** by far the largest portion of the human brain, is divided into two halves, the right and left **cerebral hemispheres** (● Figure 5-5a). They are connected to each other by the **corpus callosum,** a thick band consisting of an estimated 300 million neuronal axons traveling between the two hemispheres (● Figure 5-5b; also see ● Figure 5-12, p. 124, and ● Figure 5-13, p. 125). The corpus callosum is the body's "information superhighway." The two hemispheres commu-

MAJOR FUNCTIONS

1. Sensory perception
2. Voluntary control of movement
3. Language
4. Personality traits
5. Sophisticated mental events, such as thinking, memory, decision making, creativity, and self-consciousness

1. Inhibition of muscle tone
2. Coordination of slow, sustained movements
3. Suppression of useless patterns of movement

1. Relay station for all synaptic input
2. Crude awareness of sensation
3. Some degree of consciousness
4. Role in motor control

1. Regulation of many homeostatic functions, such as temperature control, thirst, urine output, and food intake
2. Important link between nervous and endocrine systems
3. Extensive involvement with emotion and basic behavioral patterns

1. Maintenance of balance
2. Enhancement of muscle tone
3. Coordination and planning of skilled voluntary muscle activity

1. Origin of majority of peripheral cranial nerves
2. Cardiovascular, respiratory, and digestive control centers
3. Regulation of muscle reflexes involved with equilibrium and posture
4. Reception and integration of all synaptic input from spinal cord; arousal and activation of cerebral cortex
5. Role in sleep–wake cycle

nicate and cooperate with each other by means of constant information exchange through this neural connection.

The cerebral cortex is an outer shell of gray matter covering an inner core of white matter.

Each hemisphere is composed of a thin outer shell of *gray matter,* the **cerebral cortex,** covering a thick central core of *white matter* (see ● Figure 5-13, p. 125). Another region of gray matter, the basal nuclei, is located deep within the white matter. Throughout the entire CNS, **gray matter** consists predominantly of densely packaged neuronal cell bodies and their dendrites as well as most glial cells. Bundles or tracts of myelinated nerve fibers (axons) constitute the **white matter;** its white appearance is due to the lipid composition of the myelin. The gray matter can be viewed as the "computers" of the CNS and the white matter as the "wires" that connect the computers to each other. Integration of neural input and ini-

tiation of neural output take place at synapses within the gray matter. The fiber tracts in the white matter transmit signals from one part of the cerebral cortex to another or between the cortex and other regions of the CNS. Such communication between different areas of the cortex and elsewhere facilitates integration of their activity. This integration is essential for even a relatively simple task such as picking a flower. Vision of the flower is received by one area of the cortex, reception of its fragrance takes place in another area, and movement is initiated by still another area. More subtle neuronal responses, such as appreciation of the flower's beauty and the urge to pick it, are poorly understood but undoubtedly extensively involve interconnecting fibers between different cortical regions.

The four pairs of lobes in the cerebral cortex are specialized for different activities.

We are now going to consider the locations of the major functional areas of the cerebral cortex. Throughout this discussion, keep in mind that even though a discrete activity is ultimately attributed to a particular region of the brain, no part of the brain functions in isolation. Each part depends on complex interplay among numerous other regions for both incoming and outgoing messages.

The anatomic landmarks used in cortical mapping are certain deep folds that divide each half of the cortex into four major lobes: the *occipital, temporal, parietal,* and *frontal lobes* (● Figure 5-6). Look at the basic functional map of the cortex in ● Figure 5-7a during the following discussion of the major activities attributed to various regions of these lobes.

● **FIGURE 5-6**

Cortical lobes. Each half of the cerebral cortex is divided into the occipital, temporal, parietal, and frontal lobes, as depicted in this schematic lateral view of the brain.

 For an interaction related to this figure, see Media Exercise 5.2: The Cerebral Cortex on the CD-ROM.

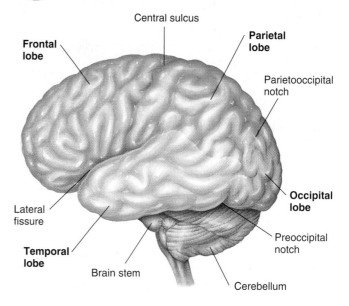

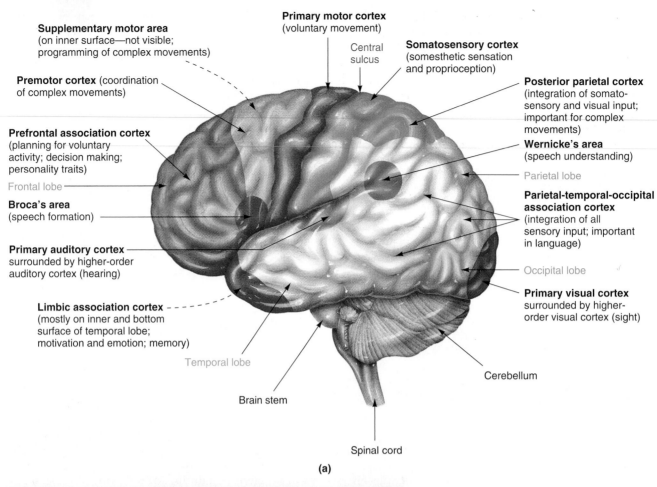

Supplementary motor area (on inner surface—not visible; programming of complex movements)

Premotor cortex (coordination of complex movements)

Prefrontal association cortex (planning for voluntary activity; decision making; personality traits)

Frontal lobe

Broca's area (speech formation)

Primary auditory cortex surrounded by higher-order auditory cortex (hearing)

Limbic association cortex (mostly on inner and bottom surface of temporal lobe; motivation and emotion; memory)

Temporal lobe

Brain stem

Primary motor cortex (voluntary movement)

Central sulcus

Somatosensory cortex (somesthetic sensation and proprioception)

Posterior parietal cortex (integration of somatosensory and visual input; important for complex movements)

Wernicke's area (speech understanding)

Parietal lobe

Parietal-temporal-occipital association cortex (integration of all sensory input; important in language)

Occipital lobe

Primary visual cortex surrounded by higher-order visual cortex (sight)

Cerebellum

Spinal cord

(a)

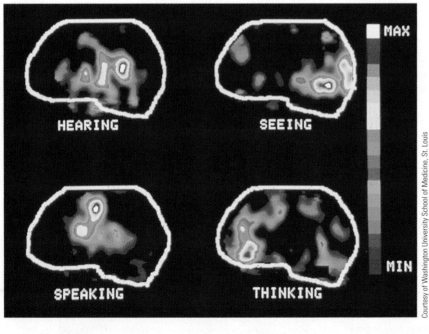

HEARING

SEEING

SPEAKING

THINKING

MAX

MIN

Courtesy of Washington University School of Medicine, St. Louis

(b)

● **FIGURE 5-7**

Functional areas of the cerebral cortex. (a) Various regions of the cerebral cortex are primarily responsible for various aspects of neural processing, as indicated in this schematic lateral view of the brain. (b) Different areas of the brain "light up" on positron emission tomography (PET) scans as a person performs different tasks. PET scans detect the magnitude of blood flow in various regions of the brain. Because more blood flows into a particular region of the brain when it is more active, neuroscientists can use PET scans to "take pictures" of the brain at work on various tasks.

 PhysioEdge

For an interaction related to this figure, see Media Exercise 5.2: The Cerebral Cortex on the CD-ROM.

The **occipital lobes**, which are located posteriorly (at the back of the head), do the initial processing of visual input. Sound sensation is initially received by the **temporal lobes**, located laterally (on the sides of the head) (● Figure 5-7a and

b). You will learn more about the functions of these regions in Chapter 6 when we discuss vision and hearing.

The parietal lobes and frontal lobes, located on the top of the head, are separated by a deep infolding, the **central sul-**

118 Chapter 5

cus, which runs roughly down the middle of the lateral surface of each hemisphere. The **parietal lobes** lie to the rear of the central sulcus on each side, and the **frontal lobes** lie in front of it. The parietal lobes are primarily responsible for receiving and processing sensory input. The frontal lobes are responsible for three main functions: (1) voluntary motor activity, (2) speaking ability, and (3) elaboration of thought. We are next going to examine the role of the parietal lobes in sensory perception, then turn our attention to the functions of the frontal lobes in more detail.

▌The parietal lobes accomplish somatosensory processing.

Sensations from the surface of the body, such as touch, pressure, heat, cold, and pain are collectively known as **somesthetic sensations** (*somesthetic* means "body feelings"). The means by which afferent neurons detect and relay information to the CNS regarding these sensations will be covered in Chapter 6 when we explore the afferent division of the peripheral nervous system in detail. Within the CNS, this information is "projected" (transmitted along specific neural pathways to higher brain levels) to the **somatosensory cortex.** The somatosensory cortex is located in the front portion of each parietal lobe immediately behind the central sulcus (● Figures 5-7a and 5-8a). It is the site for initial cortical processing and perception of somesthetic input as well as proprioceptive input. **Proprioception** is the awareness of body position.

Each region within the somatosensory cortex receives somesthetic and proprioceptive input from a specific area of the body. This distribution of cortical sensory processing is depicted in ● Figure 5-8b. Note that on this so-called **sensory homunculus** (*homunculus* means "little man"), the body is represented upside down on the somatosensory cortex and, more importantly, different parts of the body are not equally represented. The size of each body part in this homunculus indicates the relative proportion of the somatosensory cortex devoted to that area. The exaggerated size of the face, tongue, hands, and genitalia indicates the high degree of sensory perception associated with these body parts.

The somatosensory cortex on each side of the brain for the most part receives sensory input from the opposite side of the body, because most of the ascending pathways carrying sensory information up the spinal cord cross over to the opposite side before eventually terminating in the cortex. Thus damage to the somatosensory cortex in the left hemisphere produces sensory deficits on the right side of the body, whereas sensory losses on the left side are associated with damage to the right half of the cortex.

Simple awareness of touch, pressure, temperature, or pain is detected by the thalamus, a lower level of the brain, but the somatosensory cortex goes beyond pure recognition of sensations, to fuller sensory perception. The thalamus makes you aware that something hot versus something cold is touching your body, but it does not tell you where or of what intensity. The somatosensory cortex localizes the source of sensory input and perceives the level of intensity of the stimulus. It also is

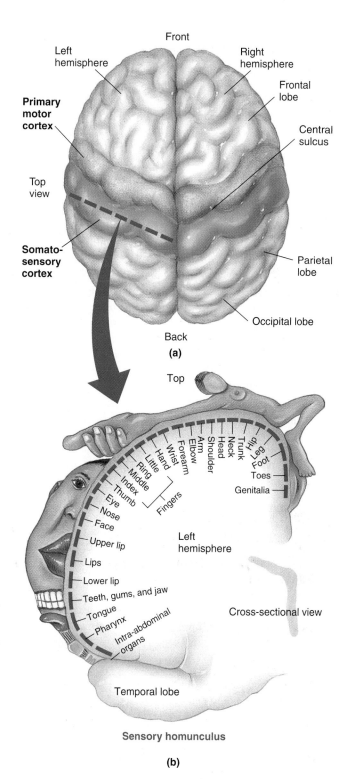

(a)

(b)

Sensory homunculus

● **FIGURE 5-8**

Somatotopic map of the somatosensory cortex. (a) Top view of cerebral hemispheres. (b) Sensory homunculus showing the distribution of sensory input to the somatosensory cortex from different parts of the body. The distorted graphic representation of the body parts indicate the relative proportion of the somatosensory cortex devoted to reception of sensory input from each area.

capable of spatial discrimination, so it can discern shapes of objects being held and can distinguish subtle differences in similar objects that come into contact with the skin.

The somatosensory cortex, in turn, projects this sensory input via white matter fibers to adjacent higher sensory areas for even further elaboration, analysis, and integration of sensory information. These higher areas are important in perceiving complex patterns of somatosensory stimulation—for example, simultaneous appreciation of the texture, firmness, temperature, shape, position, and location of an object you are holding.

▌ The primary motor cortex is located in the frontal lobes.

The area in the rear portion of the frontal lobe immediately in front of the central sulcus and next to the somatosensory cortex is the **primary motor cortex** (● Figures 5-7a and 5-9a). It confers voluntary control over movement produced by skeletal muscles. As in sensory processing, the motor cortex on each side of the brain primarily controls muscles on the opposite side of the body. Neuronal tracts originating in the motor cortex of the left hemisphere cross over before passing down the spinal cord to terminate on efferent motor neurons that trigger skeletal muscle contraction on the right side of the body. Accordingly, damage to the motor cortex on the left side of the brain produces paralysis on the right side of the body, and the converse is also true.

Stimulation of different areas of the primary motor cortex brings about movement in different regions of the body. Like the sensory homunculus for the somatosensory cortex, the **motor homunculus**, which depicts the location and relative amount of motor cortex devoted to output to the muscles of each body part, is upside down and distorted (● Figure 5-9b). The fingers, thumbs, and muscles important in speech, especially those of the lips and tongue, are grossly exaggerated, indicating the fine degree of motor control these body parts have. Compare this to how little brain tissue is devoted to the trunk, arms, and lower extremities, which are not capable of such complex movements. Thus the extent of representation in the motor cortex is proportional to the precision and complexity of motor skills required of the respective part.

▌ Other brain regions besides the primary motor cortex are important in motor control.

Even though signals from the primary motor cortex terminate on the efferent neurons that trigger voluntary skeletal muscle contraction, the motor cortex is not the only region of the brain involved with motor control. First, lower brain regions and the spinal cord control involuntary skeletal muscle activity, such as in maintaining posture. Some of these same regions also play an important role in monitoring and coordinating voluntary motor activity that the primary motor cortex has set in motion. Second, although fibers originating from the motor cortex can activate motor neurons to bring about muscle contraction, the motor cortex itself does not *initiate* voluntary movement. The motor cortex is activated by a widespread pattern of neuronal discharge, the **readiness**

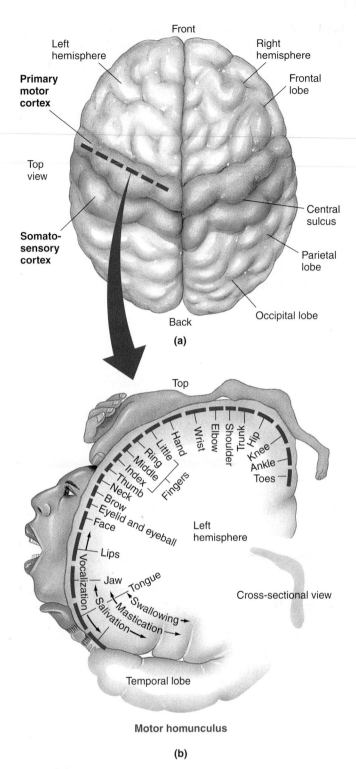

(a)

(b)

Motor homunculus

● **FIGURE 5-9**

Somatotopic map of the primary motor cortex. (a) Top view of cerebral hemispheres. (b) Motor homunculus showing the distribution of motor output from the primary motor cortex to different parts of the body. The distorted graphic representation of the body parts indicates the relative proportion of the primary motor cortex devoted to controlling skeletal muscles in each area.

potential, which occurs about 750 msec before specific electrical activity is detectable in the motor cortex. The higher motor areas of the brain that researchers believe are involved in this voluntary decision-making period include the *supplementary motor area,* the *premotor cortex,* and the *posterior parietal cortex* (● Figure 5-7a). These higher areas all command the primary motor cortex. Furthermore, a subcortical region of the brain, the *cerebellum,* plays an important role, by sending input to the motor areas of the cortex, in planning, initiating, and timing certain kinds of movement.

Each of these other four regions of the brain carry out different, related functions that are all important in programming and coordinating complex movements involving simultaneous contraction of many muscles. Even though electrical stimulation of the primary motor cortex brings about contraction of particular muscles, no purposeful coordinated movement can be elicited, just as pulling on isolated strings of a puppet does not produce any meaningful movement. A puppet displays purposeful movements only when a skilled puppeteer manipulates the strings in a coordinated manner. In the same way, these four regions (and perhaps other areas as yet undetermined) develop a **motor program** for the specific voluntary task and then "pull" the appropriate pattern of "strings" in the primary motor cortex to bring about the sequenced contraction of appropriate muscles to accomplish the desired complex movement. When one of the higher motor areas is damaged, the person cannot process complex sensory information to accomplish purposeful movement in a spatial context. Such patients, for example, cannot successfully manipulate eating utensils.

Even though these higher motor areas command the primary motor cortex and are important in preparing for the execution of deliberate, meaningful movement, researchers cannot say that voluntary movement is actually initiated by these areas. This pushes the question of how and where voluntary activity is initiated one step further. Probably no single area is responsible; undoubtedly, numerous pathways can ultimately bring about deliberate movement.

For example, think about the neural systems called into play during the simple act of picking up an apple to eat. Your memory tells you the fruit is in a bowl on the kitchen counter. Sensory systems, coupled with your knowledge based on past experience, enable you to distinguish the apple from the other kinds of fruit in the bowl. On receiving this integrated sensory information, motor systems issue commands to the exact muscles of the body in the proper sequence to enable you to move to the fruit bowl and pick up the targeted apple. During execution of this act, minor adjustments in the motor command are made as needed, based on continual updating provided by sensory input about the position of your body relative to the goal. Then there is the issue of motivation and behavior. Why are you reaching for the apple in the first place? Is it because you are hungry (detected by a neural system in the hypothalamus) or because of a more complex behavioral scenario unrelated to a basic hunger drive, such as the fact that you started to think about food because you just saw someone eating on television? Why did you choose an apple rather than a banana when both are in the fruit bowl and you like the taste of both, and so on? Thus initiating and executing purposeful voluntary movement actually include a complex neuronal interplay involving output from the motor regions guided by integrated sensory information and ultimately depending on motivational systems and elaboration of thought. All this plays against a background of memory stores from which you can make meaningful decisions about desirable movements.

Because of its plasticity, the brain can be remodeled in response to varying demands.

The brain displays a degree of **plasticity**, that is, an ability to change or be functionally remodeled in response to the demands placed on it. The term *plasticity* is used to describe this ability because plastics can be manipulated into any desired shape to serve a particular purpose. The ability of the brain to modify as needed is more pronounced in the early developmental years, but even adults retain some plasticity. When an area of the brain associated with a particular activity is destroyed, other areas of the brain may gradually assume some or all of the functions of the damaged region. Researchers are only beginning to unravel the underlying molecular mechanisms responsible for the brain's plasticity. Current evidence suggests that the formation of new neural pathways (not new neurons, but new connections between existing neurons) in response to changes in experience are mediated in part by alterations in dendritic shape resulting from modifications in certain cytoskeletal elements (see p. 34). As its dendrites become more branched and elongated, a neuron becomes able to receive and integrate more signals from other neurons. Thus the precise synaptic connections between neurons are not fixed but can be modified by experience. The gradual modification of each person's brain by a unique set of experiences provides a biological basis for individuality.

Different regions of the cortex control different aspects of language.

Unlike the sensory and motor regions of the cortex, which are present in both hemispheres, in the vast majority of people the areas of the brain responsible for language ability are found in only one hemisphere—the left hemisphere. **Language** is a complex form of communication in which written or spoken words symbolize objects and convey ideas. It involves the integration of two distinct capabilities—namely, *expression* (speaking ability) and *comprehension*—each of which is related to a specific area of the cortex. The primary areas of cortical specialization for language are Broca's area and Wernicke's area. **Broca's area,** which governs speaking ability, is located in the left frontal lobe in close association with the motor areas of the cortex that control the muscles necessary for articulation (● Figure 5-7a and b). **Wernicke's area,** located in the left cortex at the juncture of the parietal, temporal, and occipital lobes, is concerned with language comprehension. It plays a critical role in understanding both spoken and written messages. Furthermore, it is responsible for formulating coherent patterns of speech that are transferred via a bundle of fibers to Broca's area, which in turn controls articulation of this speech. Wernicke's area receives input from the visual cortex in the occipital lobe, a pathway important in reading comprehension and in describing objects seen, as well as from the auditory cortex in the temporal lobe, a pathway essential for understanding spoken words.

Clinical Note Because various aspects of language are localized in different regions of the cortex, damage to specific regions of the brain can result in selective disturbances of language. Damage to Broca's area results in a failure of word formation, although the patient can still understand the spoken and written word. Such people know what they want to say but cannot express themselves. Even though they can move their lips and tongue, they cannot establish the proper motor command to articulate the desired words. In contrast, patients with a lesion in Wernicke's area cannot understand words they see or hear. They can speak fluently, even though their perfectly articulated words make no sense. They cannot attach meaning to words or choose appropriate words to convey their thoughts. Such language disorders caused by damage to specific cortical areas are known as **aphasias**, most of which result from strokes. Aphasias should not be confused with **speech impediments**, which are caused by a defect in the mechanical aspect of speech, such as weakness or incoordination of the muscles controlling the vocal apparatus.

Dyslexia, another language disorder, is a difficulty in learning to read because of inappropriate interpretation of words. The problem arises from developmental abnormalities in connections between the visual and language areas of the cortex or within the language areas themselves; that is, the person is born with "faulty wiring" within the language-processing system. The condition is in no way related to intellectual ability.

The association areas of the cortex are involved in many higher functions.

The motor, sensory, and language areas account for only about half of the total cerebral cortex. The remaining areas, called **association areas**, are involved in higher functions. There are three association areas: (1) the *prefrontal association cortex,* (2) the *parietal-temporal-occipital association cortex,* and (3) the *limbic association cortex* (● Figure 5-7a). At one time the association areas were called "silent" areas, because stimulation does not produce any observable motor response or sensory perception. (During brain surgery, typically the patient remains awake and only local anesthetic is used along the cut scalp. This is possible because the brain itself is insensitive to pain. Before cutting into this precious, nonregenerative tissue, the neurosurgeon explores the exposed region with a tiny stimulating electrode. The patient is asked to describe what happens with each stimulation—the flick of a finger, a prickly feeling on the bottom of the foot, nothing? In this way, the surgeon can ascertain the appropriate landmarks on the neural map before making an incision.)

The **prefrontal association cortex** is the front portion of the frontal lobe just anterior to the premotor cortex. The roles attributed to this region are (1) planning for voluntary activity, (2) decision making (that is, weighing consequences of future actions and choosing between different options for various social or physical situations) (● Figure 5-7b), (3) creativity, and (4) personality traits. To carry out these highest of neural functions, the prefrontal association cortex is the site of operation of *working memory* where the brain temporarily stores and actively manipulates information used in reasoning and planning. You will learn more about working memory later. Stimulating the prefrontal association cortex does not produce any observable effects, but deficits in this area change personality and social behavior.

The **parietal-temporal-occipital association cortex** lies at the interface of the three lobes for which it is named. In this strategic location, it pools and integrates somatic, auditory, and visual sensations projected from these three lobes for complex perceptual processing. It enables you to "get the complete picture" of the relationship of various parts of your body with the external world. For example, it integrates visual information with proprioceptive input to let you place what you are seeing in proper perspective, such as realizing that a bottle is in an upright position despite the angle from which you view it (that is, whether you are standing up, lying down, or hanging upside down from a tree branch).

The **limbic association cortex** is located mostly on the bottom and adjoining inner portion of each temporal lobe. This area is concerned primarily with motivation and emotion and is extensively involved in memory.

The cortical association areas are all interconnected by bundles of fibers within the cerebral white matter. Collectively, the association areas integrate diverse information for purposeful action. An oversimplified basic sequence of linkage between the various functional areas of the cortex is schematically represented in ● Figure 5-10.

The cerebral hemispheres have some degree of specialization.

The cortical areas described thus far appear to be equally distributed in both the right and left hemispheres, except for the language areas, which are found only on one side, usually the left. The left side is also most commonly the dominant hemisphere for fine motor control. Thus most people are right-handed, because the left side of the brain controls the right side of the body. Furthermore, each hemisphere is somewhat specialized in the types of mental activities it carries out best. The **left cerebral hemisphere** excels in the performance of logical, analytic, sequential, and verbal tasks, such as math, language forms, and philosophy. In contrast, the **right cerebral hemisphere** excels in nonlanguage skills, especially spatial perception and artistic and musical talents. Whereas the left hemisphere tends to process information in a fine-detail, fragmentary way, the right hemisphere views the world in a big-picture, holistic way. Normally, much sharing of information occurs between the two hemispheres so that they complement each other, but in many individuals the skills associated with one hemisphere seem more strongly developed. Left cerebral hemisphere dominance tends to be associated with "thinkers," whereas the right-hemisphere skills dominate in "creators."

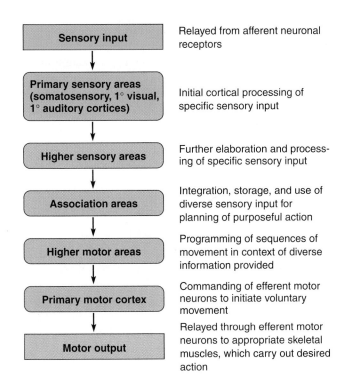

Sensory input	Relayed from afferent neuronal receptors
Primary sensory areas (somatosensory, 1° visual, 1° auditory cortices)	Initial cortical processing of specific sensory input
Higher sensory areas	Further elaboration and processing of specific sensory input
Association areas	Integration, storage, and use of diverse sensory input for planning of purposeful action
Higher motor areas	Programming of sequences of movement in context of diverse information provided
Primary motor cortex	Commanding of efferent motor neurons to initiate voluntary movement
Motor output	Relayed through efferent motor neurons to appropriate skeletal muscles, which carry out desired action

For simplicity, a number of interconnections have been omitted.

● **FIGURE 5-10**

Schematic linking of various regions of the cortex

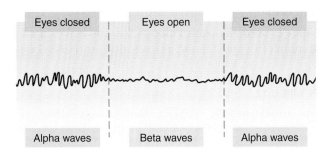

● **FIGURE 5-11**

Replacement of an alpha rhythm on an EEG with a beta rhythm when the eyes are opened

▌ An electroencephalogram is a record of postsynaptic activity in cortical neurons.

Extracellular current flow arising from electrical activity within the cerebral cortex can be detected by placing recording electrodes on the scalp to produce a graphic record known as an **electroencephalogram** or **EEG**. These "brain waves" for the most part are not due to action potentials but instead represent the momentary collective postsynaptic potential activity (that is, EPSPs and IPSPs; see p. 88) in the cell bodies and dendrites located in the cortical layers under the recording electrode.

Electrical activity can always be recorded from the living brain, even during sleep and unconscious states, but the waveforms vary, depending on the degree of activity of the cerebral cortex. Often the waveforms appear irregular, but sometimes distinct patterns in the wave's amplitude and frequency can be observed. A dramatic example of this is illustrated in ● Figure 5-11, in which the EEG waveform recorded over the occipital (visual) cortex changes markedly in response to simply opening and closing the eyes.

The EEG has three major uses:

1. It is used to *distinguish various stages of sleep* as described later in this chapter.

 2. The EEG is often used as a *clinical tool in the diagnosis of cerebral dysfunction*. Diseased or damaged cortical tissue often gives rise to altered EEG patterns. One of the most common neurologic diseases accompanied by a distinctively abnormal EEG is **epilepsy**. Epileptic seizures occur when a large collection of neurons abnormally undergo synchronous action potentials that produce stereotypical, involuntary spasms and alterations in behavior.

3. The EEG finds further use in the *legal determination of brain death*. Even though a person may have stopped breathing and the heart may have stopped pumping blood, it is often possible to restore and maintain respiratory and circulatory activity if resuscitative measures are instituted soon enough. Yet because the brain is susceptible to O_2 deprivation, irreversible brain damage may have already occurred before lung and heart function have been re-established, resulting in the paradoxical situation of a dead brain in a living body. The determination of whether a comatose patient being maintained by artificial respiration and other supportive measures is alive or dead has important medical, legal, and social implications. The need for viable organs for modern transplant surgery has made the timeliness of such life/death determinations of utmost importance. Physicians, lawyers, and the American public in general have accepted the notion of brain death—that is, a brain that is not functioning, with no possibility of recovery—as the determinant of death under such circumstances. The most widely accepted indication of brain death is *electrocerebral silence*—an essentially flat EEG.

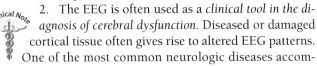 Click on the Media Exercises menu of the CD-ROM and work Media Exercise 5.2: The Cerebral Cortex to test your understanding of the previous section.

BASAL NUCLEI, THALAMUS, AND HYPOTHALAMUS

The **basal nuclei** (also known as **basal ganglia**) consist of several masses of gray matter located deep within the cerebral white matter (see ▲ Table 5-1 and ● Figure 5-12). In the CNS, a **nucleus** (plural, **nuclei**) is a functioning group of neuron cell bodies.

▌ The basal nuclei play an important inhibitory role in motor control.

The basal nuclei play a complex role in controlling movement in addition to having nonmotor functions that are less understood. In particular, the basal nuclei are important in

(1) inhibiting muscle tone throughout the body (proper muscle tone is normally maintained by a balance of excitatory and inhibitory inputs to the neurons that innervate skeletal muscles); (2) selecting and maintaining purposeful motor activity while suppressing useless or unwanted patterns of movement; and (3) helping monitor and coordinate slow, sus-

● **FIGURE 5-12**

Frontal section of the brain. (a) Schematic frontal section of the brain. The cerebral cortex, an outer shell of gray matter, surrounds an inner core of white matter. Deep within the cerebral white matter are several masses of gray matter, the basal nuclei. The ventricles are cavities in the brain through which the cerebrospinal fluid flows. The thalamus forms the walls of the third ventricle. (b) Photograph of a frontal section of the brain of a cadaver.

For an interaction related to this figure, see Media Exercise 5.3: Subcortical Structures on the CD-ROM.

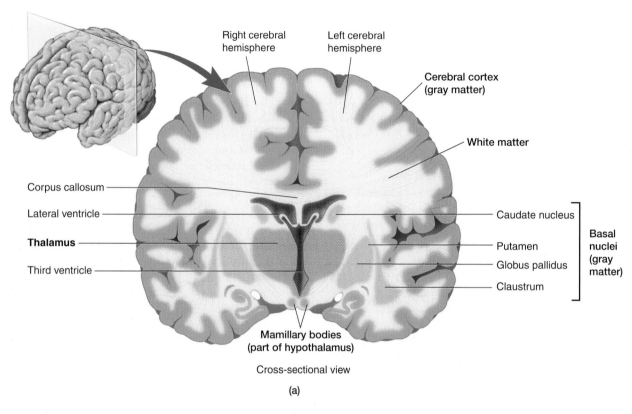

(a)

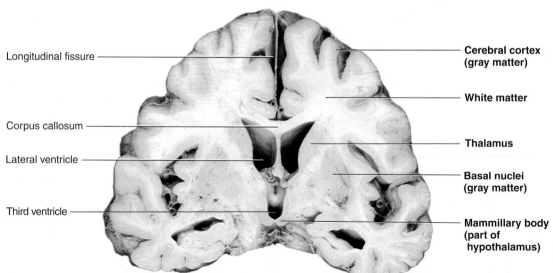

(b)

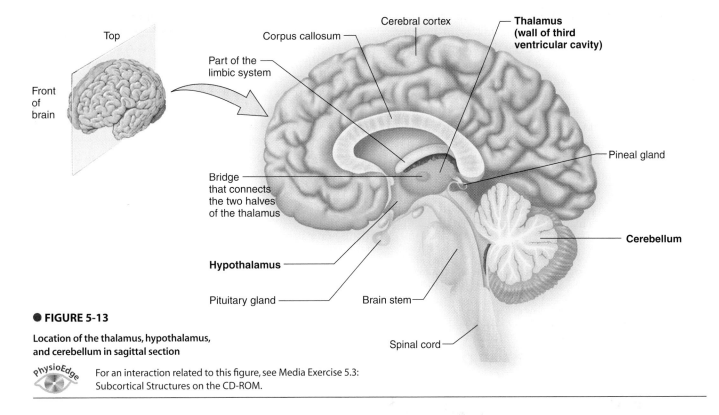

Top

Front of brain

Corpus callosum

Cerebral cortex

Thalamus (wall of third ventricular cavity)

Part of the limbic system

Pineal gland

Bridge that connects the two halves of the thalamus

Cerebellum

Hypothalamus

Pituitary gland

Brain stem

Spinal cord

● **FIGURE 5-13**

Location of the thalamus, hypothalamus, and cerebellum in sagittal section

PhysioEdge For an interaction related to this figure, see Media Exercise 5.3: Subcortical Structures on the CD-ROM.

tained contractions, especially those related to posture and support. The basal nuclei do not directly influence the efferent motor neurons that bring about muscle contraction but act instead by modifying ongoing activity in motor pathways.

Clinical Note The importance of the basal nuclei in motor control is evident in diseases involving this region, the most common of which is **Parkinson's disease (PD)**. This condition is associated with a deficiency of dopamine, an important neurotransmitter in the basal nuclei (see p. 91). Because the basal nuclei lack enough dopamine to exert their normal roles, three types of motor disturbances characterize PD: (1) increased muscle tone, or rigidity; (2) involuntary, useless, or unwanted movements, such as *resting tremors* (for example, hands rhythmically shaking, making it difficult or impossible to hold a cup of coffee); and (3) slowness in initiating and carrying out different motor behaviors. People with PD find it difficult to stop ongoing activities. If sitting down, they tend to remain seated, and if they get up, they do so very slowly.

The thalamus is a sensory relay station and is important in motor control.

Deep within the brain near the basal nuclei is the **diencephalon,** a midline structure that forms the walls of the third ventricular cavity, one of the spaces through which cerebrospinal fluid flows. The diencephalon consists of two main parts, the *thalamus* and the *hypothalamus* (see ▲ Table 5-1 and ● Figures 5-5b, 5-12, and 5-13).

The **thalamus** serves as a "relay station" and synaptic integrating center for preliminary processing of all sensory input on its way to the cortex. It screens out insignificant sig-

nals and routes the important sensory impulses to appropriate areas of the somatosensory cortex, as well as to other regions of the brain. Along with the brain stem and cortical association areas, the thalamus is important in the ability to direct attention to stimuli of interest. For example, parents can sleep soundly through the noise of outdoor traffic but become instantly aware of their baby's slightest whimper. The thalamus is also capable of crude awareness of various types of sensation but cannot distinguish their location or intensity. Some degree of consciousness resides here as well. The thalamus also plays an important role in motor control by positively reinforcing voluntary motor behavior initiated by the cortex.

The hypothalamus regulates many homeostatic functions.

The **hypothalamus** is a collection of specific nuclei and associated fibers that lie beneath the thalamus. It is an integrating center for many important homeostatic functions and serves as an important link between the autonomic nervous system and the endocrine system. Specifically, the hypothalamus (1) controls body temperature; (2) controls thirst and urine output; (3) controls food intake; (4) controls anterior pituitary hormone secretion; (5) produces posterior pituitary hormones; (6) controls uterine contractions and milk ejection; (7) serves as a major autonomic nervous system coordinating center, which in turn affects all smooth muscle, cardiac muscle, and exocrine glands; (8) plays a role in emotional and behavioral patterns; and (9) participates in the sleep–wake cycle.

The hypothalamus is the brain area most involved in directly regulating the internal environment. For example, when the body is cold, the hypothalamus initiates internal responses to increase heat production (such as shivering) and to decrease heat loss (such as constricting the skin blood vessels to reduce the flow of warm blood to the body surface, where heat could be lost to the external environment). Other areas of the brain, such as the cerebral cortex, act more indirectly to regulate the internal environment. For example, a person who feels cold is motivated to voluntarily put on warmer clothing, close the window, turn up the thermostat, and so on. Even these voluntary behavioral activities are strongly influenced by the hypothalamus, which, as a part of the limbic system, functions together with the cortex in controlling emotions and motivated behavior. We are now going to turn our attention to the limbic system and its functional relations with the higher cortex.

 Click on the Media Exercises menu of the CD-ROM and work Media Exercise 5.3: Subcortical Structures to test your understanding of the previous section.

THE LIMBIC SYSTEM AND ITS FUNCTIONAL RELATIONS WITH THE HIGHER CORTEX

The **limbic system** is not a separate structure but a ring of forebrain structures that surround the brain stem and are interconnected by intricate neuron pathways (● Figure 5-14). It includes portions of each of the following: the lobes of the cerebral cortex (especially the limbic association cortex), the basal nuclei, the thalamus, and the hypothalamus. This complex interacting network is associated with emotions, basic

● **FIGURE 5-14**

Limbic system. This partially transparent view of the brain reveals the structures composing the limbic system.

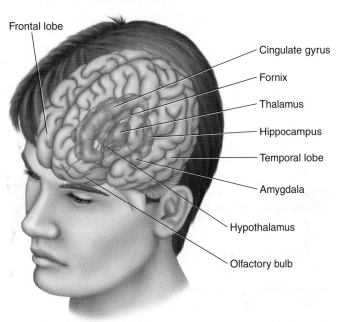

Frontal lobe

Cingulate gyrus
Fornix
Thalamus
Hippocampus
Temporal lobe
Amygdala
Hypothalamus
Olfactory bulb

survival and sociosexual behavioral patterns, motivation, and learning. Let's examine each of these brain functions further.

▌ The limbic system plays a key role in emotion.

The concept of **emotion** encompasses subjective emotional feelings and moods (such as anger, fear, and happiness) plus the overt physical responses associated with these feelings. These responses include specific behavioral patterns (for example, preparing for attack or defense when angered by an adversary) and observable emotional expressions (for example, laughing, crying, or blushing). Evidence points to a central role for the limbic system in all aspects of emotion. Stimulating specific regions within the limbic system of humans during brain surgery produces various vague subjective sensations that the patient describes as joy, satisfaction, or pleasure in one region and discouragement, fear, or anxiety in another. For example, the **amygdala**, on the interior underside of the temporal lobe (● Figure 5-14), is an especially important region for processing inputs that give rise to the sensation of fear. In humans and to an undetermined extent in other species, higher levels of the cortex are also crucial for conscious awareness of emotional feelings.

▌ The limbic system and higher cortex participate in controlling basic behavioral patterns.

Basic behavioral patterns controlled at least in part by the limbic system include those aimed at individual survival (attack, searching for food) and those directed toward perpetuating the species (sociosexual behaviors conducive to mating). In experimental animals, stimulating the limbic system brings about complex and even bizarre behaviors. For example, stimulation in one area can elicit responses of anger and rage in a normally docile animal, whereas stimulation in another area results in placidity and tameness, even in an otherwise vicious animal. Stimulation in yet another limbic area can induce sexual behaviors such as copulatory movements.

ROLE OF THE HYPOTHALAMUS IN BASIC BEHAVIORAL PATTERNS

The relationships among the hypothalamus, limbic system, and higher cortical regions regarding emotions and behavior are still not well understood. Apparently the extensive involvement of the hypothalamus in the limbic system governs the involuntary internal responses of various body systems in preparation for appropriate action to accompany a particular emotional state. For example, the hypothalamus controls the increase of heart rate and respiratory rate, elevation of blood pressure, and diversion of blood to skeletal muscles that occur in anticipation of attack or when angered. These preparatory changes in internal state require no conscious control.

ROLE OF THE HIGHER CORTEX IN BASIC BEHAVIORAL PATTERNS

In executing complex behavioral activities such as attack, flight, or mating, the individual (animal or human) must interact with the external environment. Higher cortical mechanisms are called into play to connect the limbic system and hypothalamus with the outer world so that appropriate overt behaviors are manifested. At the simplest level, the cortex provides the neural mechanisms necessary for implementing the appropriate skeletal muscle activity required to approach or avoid an adversary, participate in sexual activity, or display emotional expression.

Higher cortical levels also can reinforce, modify, or suppress basic behavioral responses so that actions can be guided by planning, strategy, and judgment based on an understanding of the situation. Even if you were angry at someone and your body was internally preparing for attack, you probably would judge that an attack would be inappropriate and could consciously suppress the external manifestation of this basic emotional behavior. Thus the higher levels of the cortex, particularly the prefrontal and limbic association areas, are important in conscious learned control of innate behavioral patterns. Using fear as an example, exposure to an aversive experience calls two parallel tracks into play for processing this emotional stimulus: a fast track in which the lower-level amygdala plays a key role and a slower track mediated primarily by the higher-level prefrontal cortex. The fast track permits a rapid, rather crude, instinctive response ("gut reaction") and is essential for the "feeling" of being afraid. The slower track involving the prefrontal cortex permits a more refined response to the aversive stimulus based on a rational analysis of the current situation compared to stored past experiences. The prefrontal cortex formulates plans and guides behavior, suppressing amygdala-induced responses that may be inappropriate for the situation at hand.

REWARD AND PUNISHMENT CENTERS

An individual tends to reinforce behaviors that have proved gratifying and to suppress behaviors that have been associated with unpleasant experiences. Certain regions of the limbic system have been designated as "**reward**" and "**punishment**" **centers**, because stimulation in these respective areas gives rise to pleasant or unpleasant sensations. When a self-stimulating device is implanted in a reward center, an experimental animal will self-deliver up to 5000 stimulations per hour and will even shun food when starving, in preference for the pleasure derived from self-stimulation. In contrast, when the device is implanted in a punishment center, animals will avoid stimulation at all costs. Reward centers are found most abundantly in regions involved in mediating the highly motivated behavioral activities of eating, drinking, and sexual activity.

▌ Motivated behaviors are goal directed.

Motivation is the ability to direct behavior toward specific goals. Some goal-directed behaviors are aimed at satisfying specific identifiable physical needs related to homeostasis. **Homeostatic drives** represent the subjective urges associated with specific bodily needs that motivate appropriate behavior to satisfy those needs. As an example, the sensation of thirst accompanying a water deficit in the body drives an individual to drink to satisfy the homeostatic need for water. However, whether water, a soft drink, or another beverage is chosen as the thirst quencher is unrelated to homeostasis. Much human behavior does not depend on purely homeostatic drives related to simple tissue deficits such as thirst. Human behavior is influenced by experience, learning, and habit, shaped in a complex framework of unique personal gratifications blended with cultural expectations.

▌ Norepinephrine, dopamine, and serotonin are neurotransmitters in pathways for emotions and behavior.

The underlying neurophysiological mechanisms responsible for the psychological observations of emotions and motivated behavior largely remain a mystery, although the neurotransmitters *norepinephrine, dopamine,* and *serotonin* all have been implicated. Norepinephrine and dopamine, both chemically classified as *catecholamines,* are known transmitters in the regions that elicit the highest rates of self-stimulation in animals equipped with do-it-yourself devices. A number of **psychoactive drugs** affect moods in humans, and some of these drugs have also been shown to influence self-stimulation in experimental animals. For example, increased self-stimulation is observed after the administration of drugs that increase catecholamine synaptic activity, such as *amphetamine,* an "upper" drug.

Clinical Note Although most psychoactive drugs are used therapeutically to treat various mental disorders, others, unfortunately, are abused. Many abused drugs act by enhancing the effectiveness of dopamine in the "pleasure" pathways, thus initially giving rise to an intense sensation of pleasure. As you have already learned, an example is cocaine, which blocks the reuptake of dopamine at synapses (see p. 91).

Depression is among the mental disorders associated with defects in limbic system neurotransmitters. A functional deficiency of serotonin or norepinephrine or both is implicated in this disorder, which is characterized by a pervasive negative mood accompanied by a generalized loss of interests, an inability to experience pleasure, and suicidal tendencies. All effective antidepressant drugs increase the available concentration of these neurotransmitters in the CNS. Prozac, the most widely prescribed drug in American psychiatry, is illustrative. It blocks the reuptake of released serotonin, thus prolonging serotonin activity at synapses. Serotonin and norepinephrine are synaptic messengers in the limbic regions of the brain involved in pleasure and motivation, suggesting that the pervasive sadness and lack of interest (no motivation) in depressed patients are related at least in part to disruption of these regions by deficiencies or decreased effectiveness of these neurotransmitters.

Researchers are optimistic that as understanding of the molecular mechanisms of mental disorders is expanded in the future, many psychiatric problems can be corrected or better managed through drug intervention, a hope of great medical significance.

Learning is the acquisition of knowledge as a result of experiences.

Learning is the acquisition of knowledge or skills as a consequence of experience, instruction, or both. It is widely believed that rewards and punishments are integral parts of many types of learning. If an animal is rewarded on responding in a particular way to a stimulus, the likelihood increases that the animal will respond in the same way again to the same stimulus as a consequence of this experience. Conversely, if a particular response is accompanied by punishment, the animal is less likely to repeat the same response to the same stimulus. When behavioral responses that give rise to pleasure are reinforced or those accompanied by punishment are avoided, learning has taken place. Housebreaking a puppy is an example. If the puppy is praised when it urinates outdoors but scolded when it wets the carpet, it will soon learn the acceptable place to empty its bladder. Thus learning is a change in behavior that occurs as a result of experiences. It is highly dependent on the organism's interaction with its environment. The only limits to the effects that environmental influences can have on learning are the biological constraints imposed by species-specific and individual genetic endowments.

Memory is laid down in stages.

Memory is the storage of acquired knowledge for later recall. Learning and memory form the basis by which individuals adapt their behavior to their particular external circumstances. Without these mechanisms, planning for successful interactions and intentional avoidance of predictably disagreeable circumstances would be impossible.

The neural change responsible for retention or storage of knowledge is known as the **memory trace**. Generally, concepts, not verbatim information, are stored. As you read this page, you are storing the concept discussed, not the specific words. Later, when you retrieve the concept from memory, you will convert it into your own words. It is possible, however, to memorize bits of information word by word.

Storage of acquired information is accomplished in at least two stages: short-term memory and long-term memory (▲ Table 5-2). **Short-term memory** lasts for seconds to hours, whereas **long-term memory** is retained for days to years. The process of transferring and fixing short-term memory traces into long-term memory stores is known as **consolidation**. Stored knowledge is of no use unless it can be retrieved and used to influence current or future behavior.

A recently developed concept is that of **working memory**, or what has been called "the erasable blackboard of the mind." Working memory temporarily holds and interrelates various pieces of information that are relevant to a current mental task. Through your working memory, you briefly hold and process data for immediate use—both newly acquired information and related, previously stored knowledge that is transiently brought forth into working memory—so that you can evaluate the incoming data in context. This integrative function is crucial to your ability to reason, plan, and make

▲ **TABLE 5-2**

Comparison of Short-Term and Long-Term Memory

CHARACTERISTIC	SHORT-TERM MEMORY	LONG-TERM MEMORY
Time of Storage after Acquisition of New Information	Immediate	Later; must be transferred from short-term to long-term memory through consolidation; enhanced by practice or recycling of information through short-term mode
Duration	Lasts for seconds to hours	Retained for days to years
Capacity of Storage	Limited	Very large
Retrieval Time (remembering)	Rapid retrieval	Slower retrieval, except for thoroughly ingrained memories, which are rapidly retrieved
Inability to Retrieve (forgetting)	Permanently forgotten; memory fades quickly unless consolidated into long-term memory	Usually only transiently unable to access; relatively stable memory trace
Mechanism of Storage	Involves transient changes in functions of pre-existing synapses, such as altering amount of neurotransmitter released	Involves relatively permanent functional or structural changes between existing neurons, such as formation of new synapses; synthesis of new proteins plays key role

judgments. By comparing and manipulating new and old information within your working memory, you can comprehend what you are reading, carry on a conversation, calculate a restaurant tip in your head, find your way home, and know that you should put on warm clothing if you see snow outside. In short, working memory enables people to string thoughts together in a logical sequence and plan for future action.

COMPARISON OF SHORT-TERM AND LONG-TERM MEMORY

Newly acquired information is initially deposited in short-term memory, which has a limited capacity for storage. Information in short-term memory has one of two eventual fates. Either it is soon forgotten (for example, forgetting a telephone number after you have looked it up and finished dialing), or it is transferred into the more permanent long-term memory mode through *active practice* or *rehearsal*. The recycling of newly acquired information through short-term memory increases the likelihood that the information will be consolidated into long-term memory. (Therefore, when you cram for an exam your long-term retention of the information is poor!) This relationship can be likened to developing photographic film. The originally developed image (short-term memory) will rapidly fade unless it is chemically fixed (consolidated) to provide a more enduring image (long-term memory). Sometimes only parts of memories are fixed, whereas others fade away. Information of interest or importance to the individual is more likely to be recycled and fixed in long-term stores, whereas less important information is quickly erased.

The storage capacity of the long-term memory bank is much larger than the capacity of short-term memory. Different informational aspects of long-term memory traces seem to be processed and codified, then stored with other memories of the same type; for example, visual memories are stored separately from auditory memories. This organization facilitates future searching of memory stores to retrieve desired information. For example, in remembering a woman you once met you may use various recall cues from different storage pools, such as her name, her appearance, the fragrance she wore, an incisive point she made, or the song playing in the background.

Because long-term memory stores are larger, it often takes longer to retrieve information from long-term memory than from short-term memory. *Remembering* is the process of retrieving specific information from memory stores; *forgetting* is the inability to retrieve stored information. Information lost from short-term memory is permanently forgotten, but information in long-term storage is frequently forgotten only transiently. Often you are only temporarily unable to access the information—for example, being unable to remember an acquaintance's name, then having it suddenly "come to you" later.

Some forms of long-term memory involving information or skills used on a daily basis are essentially never forgotten and are rapidly accessible, such as knowing your own name or being able to write. Even though long-term memories are relatively stable, stored information may be gradually lost or modified over time unless it is thoroughly ingrained as a result of years of practice.

AMNESIA

Clinical Note Occasionally, individuals suffer from a lack of memory that involves whole portions of time rather than isolated bits of information. This condition, known as **amnesia**, occurs in two forms. The most common form, *retrograde* ("going backward") *amnesia,* is the inability to recall recent past events. It usually follows a traumatic event that interferes with electrical activity of the brain, such as a concussion or stroke. If a person is knocked unconscious, the content of short-term memory is essentially erased, resulting in loss of memory about activities that occurred within about the last half hour before the event. Severe trauma may interfere with access to recently acquired information in long-term stores as well.

Anterograde ("going forward") *amnesia,* conversely, is the inability to store memory in long-term storage for later retrieval. It is usually associated with lesions of the medial portions of the temporal lobes, which are generally considered critical regions for memory consolidation. People suffering from this condition may be able to recall things they learned before the onset of their problem, but they cannot establish new permanent memories. New information is lost as quickly as it fades from short-term memory. In one case study, the person could not remember where the bathroom was in his new home but still had total recall of his old home.

❚ Memory traces are present in multiple regions of the brain.

What parts of the brain are responsible for memory? There is no single "memory center" in the brain. Instead, the neurons involved in memory traces are widely distributed throughout the subcortical and cortical regions of the brain. The regions of the brain most extensively implicated in memory include the hippocampus and associated structures of the medial (inner) temporal lobes, the limbic system, the cerebellum, the prefrontal cortex, and other regions of the cerebral cortex.

THE HIPPOCAMPUS AND DECLARATIVE MEMORIES

The **hippocampus**, the elongated, medial portion of the temporal lobe that is part of the limbic system (● Figure 5-14), plays a vital role in short-term memory involving the integration of various related stimuli and is also crucial for consolidation into long-term memory. The hippocampus is believed to store new long-term memories only temporarily and then transfer them to other cortical sites for more permanent storage. The sites for long-term storage of various types of memories are only beginning to be identified by neuroscientists.

The hippocampus and surrounding regions play an especially important role in **declarative memories**—the "what" memories of specific people, places, objects, facts, and events that often result after only one experience and that can be declared in a statement such as "I saw the Statue of Liberty last summer" or conjured up in a mental image. Declarative memories require conscious recall. The hippocampus and associated temporal/limbic structures are especially important in maintaining a durable record of the everyday episodic events in our lives. People with hippocampal damage are profoundly forgetful of facts critical to daily functioning. Declarative memories typically are the first to be lost. Interestingly, extensive damage in the hippocampus region is evident in patients with Alzheimer's disease during autopsy.

THE CEREBELLUM AND PROCEDURAL MEMORIES

In contrast to the role of the hippocampus and surrounding temporal/limbic regions in declarative memories, the cerebellum and relevant cortical regions play an essential role in the "how to" **procedural memories** involving motor skills gained through repetitive training, such as memorizing a particular dance routine. The cortical areas important for a given procedural memory are the specific motor or sensory systems engaged in performing the routine. In contrast to declarative memories, which are consciously recollected from previous experiences, procedural memories can be brought forth without conscious effort. For example, an ice skater during a competition typically performs best by "letting the body take over" the routine instead of thinking about exactly what needs to be done next. The distinct localization in different parts of the brain of these two types of memory is apparent in people who have temporal/limbic lesions. They can perform a skill, such as playing a piano, but the next day they have no recollection that they did so.

THE PREFRONTAL CORTEX AND WORKING MEMORY

The major orchestrator of the complex reasoning skills associated with *working memory* is the prefrontal association cortex. The prefrontal cortex not only serves as a temporary storage site for holding relevant data online but also is largely responsible for the so-called executive functions involving manipulation and integration of information for planning, juggling competing priorities, solving problems, and organizing activities. Researchers have identified different storage bins in the prefrontal cortex, depending on the nature of the current relevant data. For example, working memory involving spatial cues is in a prefrontal location distinct from working memory involving verbal cues or cues about an object's appearance. One recent fascinating proposal suggests that how intelligent a person is may be determined by the capacity of his or her working memory to temporarily hold and relate a variety of relevant data.

▌ Short-term and long-term memory involve different molecular mechanisms.

Another question besides the "where" of memory is the "how" of memory. Despite a vast amount of psychological data, only a few tantalizing scraps of physiologic evidence concerning the cellular basis of memory traces are available. Obviously, some change must take place within the neural circuitry of the brain to account for the altered behavior that follows learning. A single memory does not reside in a single neuron but rather in changes in the pattern of signals transmitted across synapses within a vast neuronal network.

Different mechanisms are responsible for short-term and long-term memory, as follows:

1. Short-term memory involves transient modifications in the function of pre-existing synapses, such as a temporary change in the amount of neurotransmitter released in response to stimulation or temporary increased responsiveness of the postsynaptic cell to the neurotransmitter within affected nerve pathways.

2. Long-term memory storage, in contrast, requires the activation of specific genes that control synthesis of proteins, needed for lasting structural or functional changes at specific synapses. Thus, long-term memory storage involves rather permanent physical changes in the brain. Examples of such changes include the formation of new synaptic connections or permanent changes in pre- or postsynaptic membranes.

CEREBELLUM

The **cerebellum**, which is attached to the back of the upper portion of the brain stem, lies underneath the occipital lobe of the cortex (see ▲ Table 5-1 and ● Figures 5-5b and 5-13).

▌ The cerebellum is important in balance and in planning and executing voluntary movement.

The cerebellum consists of three functionally distinct parts with different roles that collectively are largely concerned with subconscious control of motor activity (● Figure 5-15). Specifically, the different parts of the cerebellum perform the following functions:

1. The **vestibulocerebellum** is important for maintaining balance and controls eye movements.

2. The **spinocerebellum** enhances muscle tone and coordinates skilled, voluntary movements. This brain region is especially important in ensuring the accurate timing of various muscle contractions to coordinate movements involving multiple joints. For example, the movements of your shoulder, elbow, and wrist joints must be synchronized even during the simple act of reaching for a pencil. When cortical motor areas send messages to muscles for executing a particular movement, the spinocerebellum is informed of the intended motor command. This region also receives input from peripheral receptors that inform it about the body movements and positions that are actually taking place. The spinocerebellum essentially acts as "middle management," comparing the "intentions" or "orders" of the higher centers with the "performance" of the muscles and then correcting any "errors" or deviations from the intended movement. The spinocerebellum even seems able to predict the position of a body part in the next, future, fraction of a second during a complex movement and to make adjustments accordingly. If you are reaching for a pencil, for example, this region "puts on the

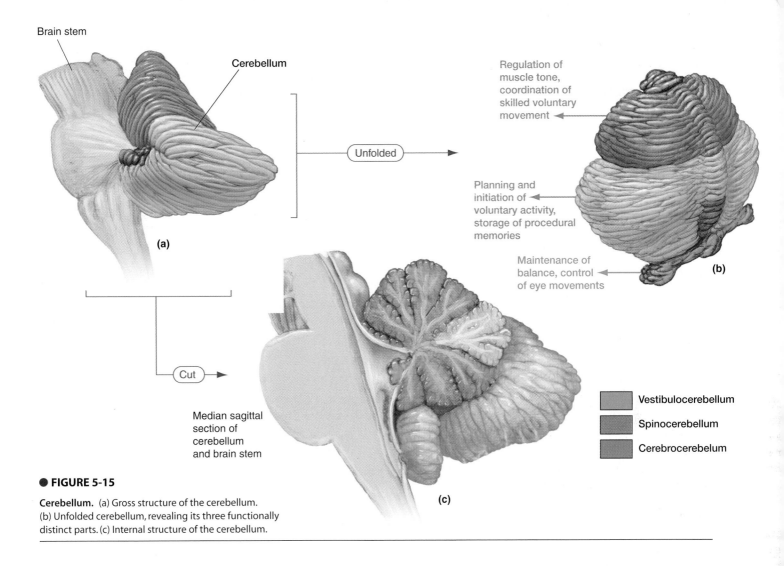

Brain stem

Cerebellum

Unfolded

Regulation of muscle tone, coordination of skilled voluntary movement

Planning and initiation of voluntary activity, storage of procedural memories

Maintenance of balance, control of eye movements

(a)

(b)

Cut

Median sagittal section of cerebellum and brain stem

(c)

Vestibulocerebellum

Spinocerebellum

Cerebrocerebelum

● **FIGURE 5-15**

Cerebellum. (a) Gross structure of the cerebellum.
(b) Unfolded cerebellum, revealing its three functionally
distinct parts. (c) Internal structure of the cerebellum.

brakes" soon enough to stop the forward movement of your hand at the intended location rather than letting you overshoot your target. These ongoing adjustments, which ensure smooth, precise, directed movement, are especially important for rapidly changing (phasic) activities such as typing, playing the piano, or running.

3. The **cerebrocerebellum** plays a role in planning and initiating voluntary activity by providing input to the cortical motor areas. This is also the cerebellar region that stores procedural memories.

Clinical Note All the following symptoms of cerebellar disease can be referred to a loss of these functions: poor balance, reduced muscle tone but no paralysis, inability to perform rapid movements smoothly, and inability to stop and start skeletal muscle action quickly. The latter gives rise to an *intention tremor* characterized by oscillating to-and-fro movements of a limb as it approaches its intended destination. As a person with cerebellar damage tries to pick up a pencil, he or she may overshoot the pencil and then rebound excessively, repeating this to-and-fro process until success is finally achieved. No tremor is observed except in performing intentional activity, in contrast to the resting tremor associated with disease of the basal nuclei.

The cerebellum and basal nuclei both monitor and adjust motor activity commanded from the motor cortex, and like the basal nuclei, the cerebellum does not directly influence the efferent motor neurons. Although they perform different roles (for example, the cerebellum enhances muscle tone, whereas the basal nuclei inhibits it), both function indirectly by modifying the output of major motor systems in the brain. The motor command for a particular voluntary activity arises from the motor cortex, but the actual execution of that activity is coordinated subconsciously by these subcortical regions. To illustrate, you can voluntarily decide that you want to walk, but you do not have to consciously think about the specific sequence of movements that you will have to perform to accomplish this intentional act. Accordingly, much voluntary activity is actually involuntarily regulated.

You will learn more about motor control when we discuss skeletal muscle physiology in Chapter 8. For now, we are going to move on to the remaining part of the brain, the brain stem.

BRAIN STEM

The **brain stem** consists of the **medulla, pons,** and **midbrain** (see ▲ Table 5-1 and ● Figure 5-5b).

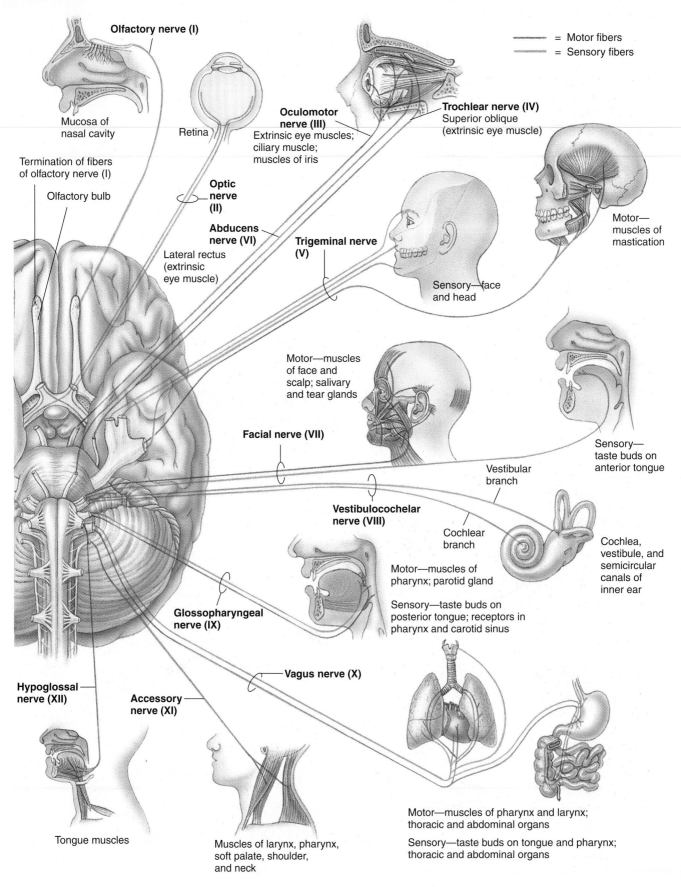

= Motor fibers
= Sensory fibers

Olfactory nerve (I)

Mucosa of nasal cavity

Retina

Oculomotor nerve (III)
Extrinsic eye muscles; ciliary muscle; muscles of iris

Trochlear nerve (IV)
Superior oblique (extrinsic eye muscle)

Termination of fibers of olfactory nerve (I)

Olfactory bulb

Optic nerve (II)

Abducens nerve (VI)

Lateral rectus (extrinsic eye muscle)

Trigeminal nerve (V)

Motor— muscles of mastication

Sensory—face and head

Motor—muscles of face and scalp; salivary and tear glands

Facial nerve (VII)

Sensory— taste buds on anterior tongue

Vestibular branch

Vestibulocochelar nerve (VIII)

Cochlear branch

Cochlea, vestibule, and semicircular canals of inner ear

Motor—muscles of pharynx; parotid gland

Sensory—taste buds on posterior tongue; receptors in pharynx and carotid sinus

Glossopharyngeal nerve (IX)

Vagus nerve (X)

Hypoglossal nerve (XII)

Accessory nerve (XI)

Tongue muscles

Muscles of larynx, pharynx, soft palate, shoulder, and neck

Motor—muscles of pharynx and larynx; thoracic and abdominal organs

Sensory—taste buds on tongue and pharynx; thoracic and abdominal organs

● **FIGURE 5-16**

Cranial nerves. Inferior (underside) view of the brain, showing the attachments of the 12 pairs of cranial nerves to the brain and many of the structures innervated by those nerves.

■ The brain stem is a vital link between the spinal cord and higher brain regions.

All incoming and outgoing fibers traversing between the periphery and higher brain centers must pass through the brain stem, with incoming fibers relaying sensory information to the brain and outgoing fibers carrying command signals from the brain for efferent output. A few fibers merely pass through, but most synapse within the brain stem for important processing. Thus the brain stem is a critical connecting link between the rest of the brain and the spinal cord.

The functions of the brain stem include the following:

1. The majority of the 12 pairs of **cranial nerves** arise from the brain stem (● Figure 5-16). With one major exception, these nerves supply structures in the head and neck with both sensory and motor fibers. They are important in sight, hearing, taste, smell, sensation of the face and scalp, eye movement, chewing, swallowing, facial expressions, and salivation. The major exception is cranial nerve X, the **vagus nerve.** Instead of innervating regions in the head, most branches of the vagus nerve supply organs in the thoracic and abdominal cavities. The vagus is the major nerve of the parasympathetic nervous system.
2. Collected within the brain stem are neuronal clusters, or "centers," that control heart and blood vessel function, respiration, and many digestive activities.
3. The brain stem plays a role in regulating muscle reflexes involved in equilibrium and posture.
4. A widespread network of interconnected neurons called the **reticular formation** runs throughout the entire brain stem and into the thalamus. This network receives and integrates all incoming sensory synaptic input. Ascending fibers originating in the reticular formation carry signals upward to arouse and activate the cerebral cortex (● Figure 5-17). These fibers compose the **reticular activating system (RAS),** which controls the overall degree of cortical alertness and is important in the ability to direct attention. In turn, fibers descending from the cortex, especially its motor areas, can activate the RAS.
5. The centers that govern sleep traditionally have been considered to be housed within the brain stem, although recent evidence suggests that the center that promotes slow-wave sleep lies in the hypothalamus.

We are now going to examine sleep and the other states of consciousness in further detail.

■ Sleep is an active process consisting of alternating periods of slow-wave and paradoxical sleep.

The term **consciousness** refers to subjective awareness of the external world and self, including awareness of the private inner world of one's own mind—that is, awareness of thoughts, perceptions, dreams, and so on. Even though the final level of awareness resides in the cerebral cortex and a crude sense of awareness is detected by the thalamus, conscious experience

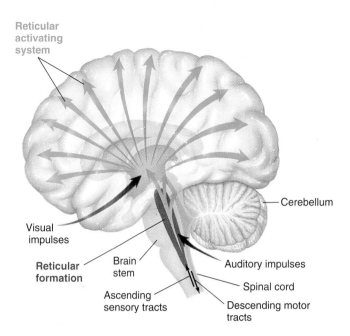

● **FIGURE 5-17**

The reticular activating system. The reticular formation, a widespread network of neurons within the brain stem (*in red*), receives and integrates all synaptic input. The reticular activating system, which promotes cortical alertness and helps direct attention toward specific events, consists of ascending fibers (*in blue*) that originate in the reticular formation and carry signals upward to arouse and activate the cerebral cortex.

depends on the integrated functioning of many parts of the nervous system.

The following states of consciousness are listed in decreasing order of arousal level, based on extent of interaction between peripheral stimuli and the brain:

*   maximum alertness
*   wakefulness
*   sleep (several different types)
*   coma

Maximum alertness depends on attention-getting sensory input that "energizes" the RAS and subsequently the activity level of the CNS as a whole. At the other extreme, coma is the total unresponsiveness of a living person to external stimuli, caused either by brain stem damage that interferes with the RAS or by widespread depression of the cerebral cortex, such as accompanies O_2 deprivation.

The **sleep–wake cycle** is a normal cyclic variation in awareness of surroundings. In contrast to being awake, sleeping people are not consciously aware of the external world, but they do have inward conscious experiences such as dreams. Furthermore, they can be aroused by external stimuli, such as an alarm going off.

Sleep is an active process, not just the absence of wakefulness. The brain's overall level of activity is not reduced during sleep. During certain stages of sleep, O_2 uptake by the brain is even increased above normal waking levels.

There are two types of sleep, characterized by different EEG patterns and different behaviors: *slow-wave sleep* and *paradoxical,* or *REM, sleep* (▲ Table 5-3).

Comparison of Slow-Wave and Paradoxical Sleep

| | TYPE OF SLEEP | |
CHARACTERISTIC	Slow-wave sleep	Paradoxical sleep
EEG	Displays slow waves	Similar to EEG of alert, awake person
Motor Activity	Considerable muscle tone; frequent shifting	Abrupt inhibition of muscle tone; no movement
Heart Rate, Respiratory Rate, Blood Pressure	Minor reductions	Irregular
Dreaming	Rare (mental activity is extension of waking-time thoughts)	Common
Percentage of Sleeping Time	80%	20%
Other Important Characteristics	Has four stages; sleeper must pass through this type of sleep first	Rapid eye movements

EEG PATTERNS DURING SLEEP

Slow-wave sleep occurs in four stages, each displaying progressively slower EEG waves of higher amplitude (hence "slow-wave" sleep) (● Figure 5-18). At the onset of sleep, you move from the light sleep of stage 1 to the deep sleep of stage 4 of slow-wave sleep during a period of 30 to 45 minutes, then you reverse through the same stages in the same amount of time. A 10- to 15-minute episode of **paradoxical sleep** punctuates the end of each slow-wave sleep cycle. Paradoxically, your EEG pattern during this time abruptly becomes similar to that of a wide-awake, alert individual, even though you are still asleep (hence, "paradoxical" sleep) (● Figure 5-18). After the paradoxical episode, the stages of slow-wave sleep repeat again.

A person cyclically alternates between the two types of sleep throughout the night. In a normal sleep cycle, you always pass through slow-wave sleep before entering paradoxical sleep. On average, paradoxical sleep occupies 20% of total sleeping time throughout adolescence and most of adulthood. Infants spend considerably more time in paradoxical sleep. In contrast, paradoxical as well as deep stage 4 slow-wave sleep declines in the elderly. People who require less total sleeping time than normal spend proportionately more time in paradoxical and deep stage 4 slow-wave sleep and less time in the lighter stages of slow-wave sleep.

BEHAVIORAL PATTERNS DURING SLEEP

In addition to distinctive EEG patterns, the two types of sleep are distinguished by behavioral differences. It is difficult to pinpoint exactly when an individual drifts from drowsiness into slow-wave sleep. In this type of sleep, the person still has considerable muscle tone and frequently shifts body po-

sition. Only minor reductions in respiratory rate, heart rate, and blood pressure occur. During this time the sleeper can be easily awakened and rarely dreams. The mental activity

● **FIGURE 5-18**

EEG patterns during different types of sleep. Note that the EEG pattern during paradoxical sleep is similar to that of an alert, awake person, whereas the pattern during slow-wave sleep displays distinctly different waves.

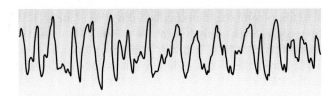

Slow-wave sleep, stage 4

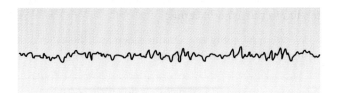

Paradoxical sleep

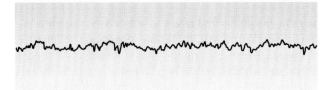

Awake, eyes open

associated with slow-wave sleep is less visual than dreaming. It is more conceptual and plausible—like an extension of waking-time thoughts concerned with everyday events—and it is less likely to be recalled. The major exception is nightmares, which occur during stages 3 and 4. People who walk and talk in their sleep do so during slow-wave sleep.

The behavioral pattern accompanying paradoxical sleep is marked by abrupt inhibition of muscle tone throughout the body. The muscles are completely relaxed, with no movement taking place. Paradoxical sleep is further characterized by *rapid eye movements,* hence the alternative name REM sleep. Heart rate and respiratory rate become irregular and blood pressure may fluctuate. Another characteristic of REM sleep is *dreaming.* The rapid eye movements are not related to "watching" the dream imagery. The eye movements are driven in a locked, oscillating pattern uninfluenced by dream content.

Brain imaging of volunteers during REM sleep shows heightened activity in the higher-level visual processing areas and limbic system (the seat of emotions), coupled with reduced activity in the prefrontal cortex (the seat of reasoning). This pattern of activity lays the groundwork for the characteristics of dreaming: internally generated visual imagery reflecting activation of the person's "emotional memory bank" with little guidance or interpretation from the complex thinking areas. As a result, dreams are often charged with intense emotions, a distorted sense of time, and bizarre content that is uncritically accepted as real, with little reflection about all the strange happenings.

▮ The sleep–wake cycle is controlled by interactions among three neural systems.

The sleep–wake cycle as well as the various stages of sleep are due to the cyclic interplay of three different neural systems: (1) an **arousal system**, which is part of the reticular activating system originating in the brain stem, (2) a **slow-wave sleep center** in the hypothalamus that contains *sleep-on neurons* believed to induce sleep; and (3) a **paradoxical sleep center** in the brain stem that houses *REM sleep-on neurons* that become very active during REM sleep. The patterns of interaction among these three neural regions, which bring about the fairly predictable cyclical sequence between being awake and passing alternately between the two types of sleep, are the subject of intense investigation. Nevertheless, the molecular mechanisms controlling the sleep–wake cycle remain poorly understood.

The normal cycle can easily be interrupted, with the arousal system more readily overriding the sleep systems than vice versa; that is, it is easier to stay awake when you are sleepy than to fall asleep when you are wide awake. The arousal system can be activated by afferent sensory input (for example, a person has difficulty falling asleep when it is noisy) or by input descending to the brain stem from higher brain regions. Intense concentration or strong emotional states, such as anxiety or excitement, can keep a person from falling asleep, just as motor activity, such as getting up and walking around, can arouse a drowsy person.

▮ The function of sleep is unclear.

Even though humans spend about a third of their lives sleeping, why sleep is needed largely remains a mystery. Although still speculative, recent studies suggest that slow-wave sleep and REM sleep serve different purposes. One widely accepted proposal holds that sleep provides "catch-up" time for the brain to restore biochemical or physiological processes that have progressively degraded during wakefulness. These restoration and recovery activities are believed to take place during slow-wave sleep. Another leading proposal is that sleep, especially paradoxical sleep, is necessary to allow the brain to "shift gears," to accomplish the long-term structural and chemical adjustments necessary for learning and memory, especially consolidation of procedural memories. This theory might explain why infants need so much sleep. Their highly plastic brains are rapidly undergoing profound synaptic modifications in response to environmental stimulation. In contrast, mature individuals, in whom neural changes are less dramatic, sleep less.

We have now completed our discussion of the brain and are going to shift our attention to the other component of the central nervous system, the spinal cord.

SPINAL CORD

The **spinal cord** is a long, slender cylinder of nerve tissue that extends from the brain stem. It is about 45 cm long (18 inches) and 2 cm in diameter (about the size of your thumb).

▮ The spinal cord extends through the vertebral canal and is connected to the spinal nerves.

Exiting through a large hole in the base of the skull, the spinal cord is enclosed by the protective vertebral column as it descends through the vertebral canal. Paired **spinal nerves** emerge from the spinal cord through spaces formed between the bony, winglike arches of adjacent vertebrae. The spinal nerves are named according to the region of the vertebral column from which they emerge (● Figure 5-19): there are 8 pairs of *cervical (neck) nerves* (namely C1–C8), 12 *thoracic (chest) nerves,* 5 *lumbar (abdominal) nerves,* 5 *sacral (pelvic) nerves,* and 1 *coccygeal (tailbone) nerve.*

During development, the vertebral column grows about 25 cm longer than the spinal cord. Because of this differential growth, segments of the spinal cord that give rise to various spinal nerves are not aligned with the corresponding intervertebral spaces. Most spinal nerve roots must descend along the cord before emerging from the vertebral column at the corresponding space. The spinal cord itself extends only to the level of the first or second lumbar vertebra (about waist level), so the nerve roots of the remaining nerves are greatly elongated, to exit the vertebral column at their appropriate space. The thick bundle of elongated nerve roots within the lower vertebral canal is called the **cauda equina** ("horse's tail") because of its appearance (● Figure 5-19b).

The white matter of the spinal cord is organized into tracts.

Although there are some slight regional variations, the cross-sectional anatomy of the spinal cord is generally the same throughout its length (● Figure 5-20). In contrast to the gray matter in the brain, the gray matter in the spinal cord forms a butterfly-shaped region on the inside and is surrounded by the outer white matter. As in the brain, the cord gray matter consists primarily of neuronal cell bodies and their dendrites, short interneurons, and glial cells. The white matter is organized into tracts, which are bundles of nerve fibers (axons of long interneurons) with a similar function. The bundles are grouped into columns that extend the length of the cord. Each of these tracts begins or ends within a particular area of the brain, and each transmits a specific type of information. Some are **ascending** (cord to brain) **tracts** that transmit to the brain signals derived from afferent input. Others are **descending**

(brain to cord) **tracts** that relay messages from the brain to efferent neurons. Because various types of signals are carried in different tracts within the spinal cord, damage to particular areas of the cord can interfere with some functions, whereas other functions remain intact.

Spinal nerves carry both afferent and efferent fibers.

Spinal nerves connect with each side of the spinal cord by a **dorsal root** and a **ventral root** (● Figure 5-20). Afferent fibers carrying incoming signals from peripheral receptors enter the spinal cord through the dorsal root. The cell bodies for the afferent neurons at each level are clustered together in a **dorsal root ganglion.** (A collection of neuronal cell bodies located outside of the CNS is called a *ganglion*, whereas a functional collection of cell bodies within the CNS is referred to as a *center* or a *nucleus*.) The cell bodies for the efferent neu-

● **FIGURE 5-19**

Spinal nerves. There are 31 pairs of spinal nerves named according to the region of the vertebral column from which they emerge. Because the spinal cord is shorter than the vertebral column, spinal nerve roots must descend along the cord before emerging from the vertebral column at the corresponding intervertebral space, especially those beyond the level of the first lumbar vertebra (L1). Collectively these rootlets are called the cauda equina, literally "horse's tail." (a) Posterior view of the brain, spinal cord, and spinal nerves (on the right side only). (b) Lateral view of the spinal cord and spinal nerves emerging from the vertebral column.

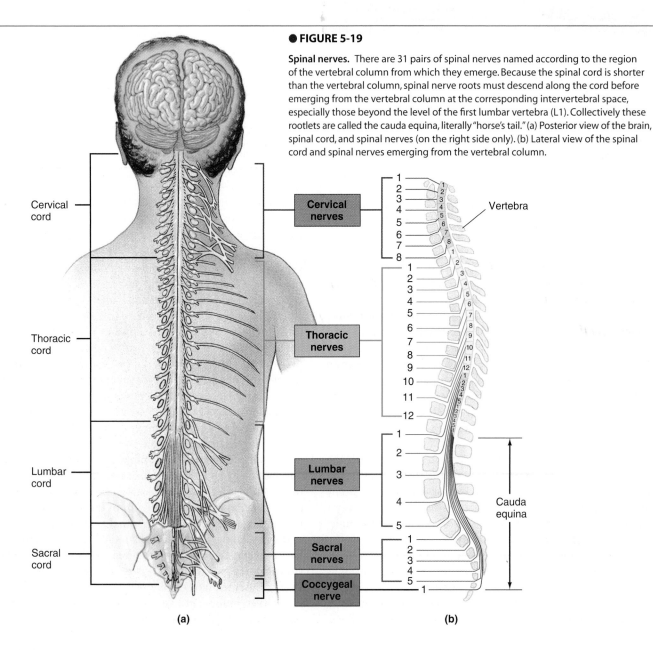

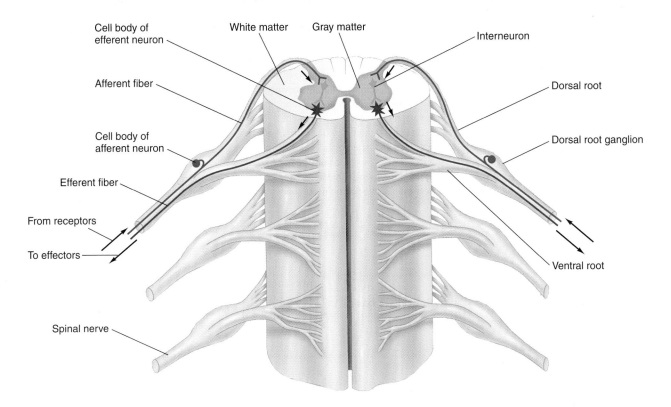

● **FIGURE 5-20**

Spinal cord in cross section. Schematic representation of the spinal cord in cross section showing the relationship between the spinal cord and spinal nerves. The afferent fibers enter through the dorsal root, and the efferent fibers exit through the ventral root. Afferent and efferent fibers are enclosed together within a spinal nerve.

rons originate in the gray matter and send axons out through the ventral root. Therefore, efferent fibers carrying outgoing signals to muscles and glands exit through the ventral root.

The dorsal and ventral roots at each level join to form a spinal nerve that emerges from the vertebral column (● Figure 5-20). A spinal nerve contains both afferent and efferent fibers traversing between a particular region of the body and the spinal cord. Note the relationship between a *nerve* and a *neuron*. A **nerve** is a bundle of peripheral neuronal axons, some afferent and some efferent, enclosed by a connective tissue covering and following the same pathway. A nerve does not contain a complete nerve cell, only the axonal portions of many neurons. (By this definition, there are no nerves in the CNS! Bundles of axons in the CNS are called *tracts.*) The individual fibers within a nerve generally do not have any direct influence on each other. They travel together for convenience, just as many individual telephone lines are carried within a telephone cable, yet any particular phone connection can be private without interference or influence from other lines in the cable.

The 31 pairs of spinal nerves, along with the 12 pairs of cranial nerves that arise from the brain, constitute the *peripheral nervous system.* After they emerge, the spinal nerves progressively branch to form a vast network of peripheral nerves that supply the tissues. Each segment of the spinal cord gives rise to a pair of spinal nerves that ultimately supply a particular region of the body with both afferent and efferent fibers.

Thus the location and extent of sensory and motor deficits associated with spinal cord injuries can be clinically important in determining the level and extent of the cord injury.

▌ The spinal cord is responsible for the integration of many basic reflexes.

The spinal cord is strategically located between the brain and afferent and efferent fibers of the peripheral nervous system; this location enables the spinal cord to fulfill its two primary functions: (1) serving as a link for transmission of information between the brain and the remainder of the body and (2) integrating reflex activity between afferent input and efferent output without involving the brain. This type of reflex activity is called a *spinal reflex.*

A **reflex** is any response that occurs automatically without conscious effort. There are two types of reflexes: (1) **simple,** or **basic, reflexes,** which are built-in, unlearned responses, such as pulling the hand away from a burning hot object; and (2) **acquired,** or **conditioned, reflexes,** which are a result of practice and learning, such as a pianist striking a particular key on seeing a given note on the music staff. The musician reads music and plays it automatically, but only after considerable conscious training effort. (For a discussion of the role of acquired reflexes in many sports skills, see the boxed feature, ❯ Beyond the Basics, on p. 138.)

Swan Dive or Belly Flop: It's a Matter of CNS Control

Sport skills must be learned. Much of the time, strong basic reflexes must be overridden in order to perform the skill. Learning to dive into water, for example, is very difficult initially. Strong head-righting reflexes controlled by sensory organs in the neck and ears initiate a straightening of the neck and head before the beginning diver enters the water, causing what is commonly known as a "belly flop." In a backward dive, the head-righting reflex causes the beginner to land on his or her back or even in a sitting position. To perform any motor skill that involves body inversions, somersaults, back flips, or other abnormal postural movements, the person must learn to consciously inhibit basic postural reflexes. This is accomplished by having the person concentrate on specific body positions during the movement. For example, to perform a somersault, the person must concentrate on keeping the chin tucked and grabbing the knees. After the skill is performed repeatedly, new synaptic patterns are formed in the CNS, and the new or conditioned response substitutes for the natural reflex responses. Sport skills must be practiced until the movement becomes automatic; then the athlete is free during competition to think about strategy or the next move to be performed in a routine.

REFLEX ARC

The neural pathway involved in accomplishing reflex activity is known as a **reflex arc**, which typically includes five basic components:

1. receptor
2. afferent pathway
3. integrating center
4. efferent pathway
5. effector

The **receptor** responds to a **stimulus**, which is a detectable physical or chemical change in the environment of the receptor. In response to the stimulus, the receptor produces an action potential that is relayed by the **afferent pathway** to the **integrating center** for processing. Usually the integrating center is the CNS. The spinal cord and brain stem integrate basic reflexes, whereas higher brain levels usually process acquired reflexes. The integrating center processes all information available to it from this receptor as well as from all other inputs, then "makes a decision" about the appropriate response. The instructions from the integrating center are transmitted via the **efferent pathway** to the **effector**—a muscle or gland—which carries out the desired response. Unlike conscious behavior, in which any one of a number of responses is possible, a reflex response is predictable, because the pathway between the receptor and effector is always the same.

WITHDRAWAL REFLEX

A basic **spinal reflex** is one integrated by the spinal cord; that is, all components necessary for linking afferent input to efferent response are present within the spinal cord. The **withdrawal reflex** can serve to illustrate a basic spinal reflex (● Figure 5-21). When a person touches a hot stove (or receives another painful stimulus), a withdrawal reflex is initiated to pull the hand away from the stove (to withdraw from the painful stimulus). The skin has different receptors for warmth, cold, light touch, pressure, and pain. Even though all information is sent to the CNS by way of action potentials, the CNS can distinguish between various stimuli because different receptors and consequently different afferent pathways are activated by different stimuli. When a receptor is stimulated enough to reach threshold, an action potential is generated in the afferent sensory neuron. The stronger the stimulus, the greater the frequency of action potentials generated and propagated to the CNS. Once the afferent neuron enters the spinal cord, it diverges to synapse with the following different interneurons (the numbers correspond to those on ● Figure 5-21).

1. An excited afferent neuron stimulates excitatory interneurons that in turn stimulate the efferent motor neurons supplying the biceps, the muscle in the arm that flexes (bends) the elbow joint. By contracting, the biceps pulls the hand away from the hot stove.

2. The afferent neuron also stimulates inhibitory interneurons that in turn inhibit the efferent neurons supplying the triceps to prevent it from contracting. The triceps is the muscle in the arm that extends (straightens out) the elbow joint. When the biceps is contracting to flex the elbow, it would be counterproductive for the triceps to be contracting. Therefore, built into the withdrawal reflex is inhibition of the muscle that antagonizes (opposes) the desired response. This type of neuronal connection involving stimulation of the nerve supply to one muscle and simultaneous inhibition of the nerves to its antagonistic muscle is known as **reciprocal innervation**.

3. The afferent neuron stimulates still other interneurons that carry the signal up the spinal cord to the brain via an ascending pathway. Only when the impulse reaches the sensory area of the cortex is the person aware of the pain, its location, and the type of stimulus. Also, when the impulse reaches the brain, the information can be stored as memory, and the person can start thinking about the situation—how it happened, what to do about it, and so on. All this activity at the conscious level is above and beyond the basic reflex.

As is characteristic of all spinal reflexes, the brain can modify the withdrawal reflex. Impulses may be sent down descending pathways to the efferent motor neurons supplying the involved muscles to override the input from the receptors, actually preventing the biceps from contracting in spite of the painful stimulus. When your finger is being pricked to obtain a blood sample, pain receptors are stimulated, initiating the withdrawal reflex. However, knowing that you must be brave and not pull your hand away, you can consciously override the reflex by sending IPSPs via descending pathways to the motor neurons supplying the biceps and EPSPs to those supplying the triceps. The activity in these efferent neurons

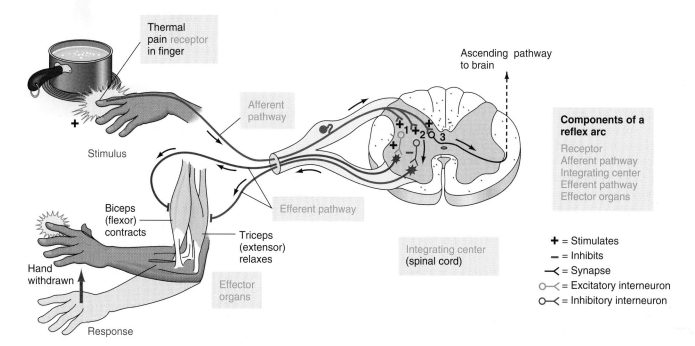

● **FIGURE 5-21**

The withdrawal reflex. When a painful stimulus activates a receptor in the finger, action potentials are generated in the corresponding afferent pathway, which propagates the electrical signals to the CNS. Once the afferent neuron enters the spinal cord, it diverges and terminates on three different types of interneurons (only one of each type is depicted): (1) excitatory interneurons, which in turn stimulate the efferent motor neurons to the biceps, causing the arm to flex and pull the hand away from the painful stimulus; (2) inhibitory interneurons, which inhibit the efferent motor neurons to the triceps, thus preventing counterproductive contraction of this antagonistic muscle; and (3) interneurons that carry the signal up the spinal cord via an ascending pathway to the brain for awareness of pain, memory storage, and so on.

depends on the sum of activity of all their synaptic inputs. Because the neurons supplying the biceps are now receiving more IPSPs from the brain (voluntary) than EPSPs from the afferent pain pathway (reflex), these neurons are inhibited and do not reach threshold. Therefore, the biceps is not stimulated to contract and withdraw the hand. Simultaneously, the neurons to the triceps are receiving more EPSPs from the brain than IPSPs via the reflex arc, so they reach threshold, fire, and consequently stimulate the triceps to contract. Thus the arm is kept extended despite the painful stimulus. In this way, the withdrawal reflex has been voluntarily overridden.

STRETCH REFLEX

Only one reflex is simpler than the withdrawal reflex: the **stretch reflex**, in which an afferent neuron originating at a stretch-detecting receptor in a skeletal muscle terminates directly on the efferent neuron supplying the same skeletal muscle to cause it to contract and counteract the stretch. (You will learn more about the role of this reflex in Chapter 8). The stretch reflex is a **monosynaptic** ("one synapse") **reflex**, because the only synapse in the reflex arc is the one between the afferent neuron and the efferent neuron. The withdrawal

reflex and all other reflexes are **polysynaptic** ("many synapses"), because interneurons are interposed in the reflex pathway and, therefore, a number of synapses are involved.

OTHER REFLEX ACTIVITY

Besides protective reflexes such as the withdrawal reflex and simple postural reflexes, basic spinal reflexes also mediate emptying of pelvic organs (for example, urination, defecation, and expulsion of semen). All spinal reflexes can be voluntarily overridden at least temporarily by higher brain centers.

Not all reflex activity involves a clear-cut reflex arc, although the basic principles of a reflex (that is, an automatic response to a detectable change) are still present. Pathways for unconscious responsiveness digress from the typical reflex arc in two general ways:

1. *Responses mediated at least in part by hormones.* A particular reflex may be mediated solely by either neurons or hormones or may involve a pathway using both.
2. *Local responses that do not involve either nerves or hormones.* For example, the blood vessels in an exercising muscle dilate because of local metabolic changes, thereby increasing blood flow to match the active muscle's metabolic needs.

CHAPTER IN PERSPECTIVE: FOCUS ON HOMEOSTASIS

To interact in appropriate ways with the external environment to sustain the body's viability, such as in acquiring food, and to make the internal adjustments necessary to maintain homeostasis, the body must be informed about any changes taking place in the external and internal environment and must be able to process this information and send messages to various muscles and glands to accomplish the desired results. The nervous system, one of the body's two major regulatory systems, plays a central role in this life-sustaining communication. The central nervous system (CNS), which consists of the brain and spinal cord, receives information about the external and internal environment by means of the afferent peripheral nerves. After sorting, processing, and integrating this input, the CNS sends directions, by means of efferent peripheral nerves, to bring about appropriate muscular contractions and glandular secretions.

With its swift electrical signaling system, the nervous system is especially important in controlling the rapid responses of the body. Many neurally controlled muscular and glandular activities are aimed toward maintaining homeostasis. The CNS is the main site of integration between the afferent input and efferent output. It links the appropriate response to a particular input so that conditions compatible with life are maintained in the body. For example, when informed by the afferent nervous system that blood pressure has fallen, the CNS sends appropriate commands to the heart and blood vessels to increase the blood pressure to normal. Likewise, when informed that the body is overheated the CNS promotes secretion of sweat by the sweat glands. Evaporation of sweat helps cool the body to normal temperature. Were it not for this processing and integrating ability of the CNS, maintaining homeostasis in an organism as complex as a human would be impossible.

At the simplest level, the spinal cord integrates many basic protective and evacuative reflexes that do not require conscious participation, such as withdrawing from a painful stimulus and emptying of the urinary bladder. In addition to serving as a more complex integrating link between afferent input and efferent output, the brain is also responsible for the initiation of all voluntary movement, complex perceptual awareness of the external environment and self, language, and abstract neural phenomena such as thinking, learning, remembering, consciousness, emotions, and personality traits. All neural activity—from the most private thoughts to commands for motor activity, from enjoying a concert to retrieving memories from the distant past—is ultimately attributable to propagation of action potentials along individual nerve cells and chemical transmission between cells.

During evolutionary development, the nervous system has become progressively more complex. Newer, more complicated, and more sophisticated layers of the brain have been piled on top of older, more primitive regions. Mechanisms for governing many basic activities necessary for survival are built into the older parts of the brain. The newer, higher levels progressively modify, enhance, or nullify actions coordinated by lower levels in a hierarchy of command, and they also add new capabilities. Many of these higher neural activities are not aimed toward maintaining life, but they add immeasurably to the quality of being alive.

CHAPTER SUMMARY

Organization of the Nervous System (pp. 110–111)

- The nervous system consists of the central nervous system (CNS), which includes the brain and spinal cord, and the peripheral nervous system, which includes the nerve fibers carrying information to (afferent division) and from (efferent division) the CNS. *(Review Figure 5-1.)*

- Three functional classes of neurons—afferent neurons, efferent neurons, and interneurons—compose the excitable cells of the nervous system. *(Review Figure 5-2.)* (1) Afferent neurons inform the CNS about conditions in both the external and internal environment. (2) Efferent neurons carry instructions from the CNS to effector organs, namely muscles and glands. (3) Interneurons are responsible for integrating afferent information and formulating an efferent response, as well as for all higher mental functions associated with the "mind."

Protection and Nourishment of the Brain (pp. 111–114)

- Glial cells form the connective tissue within the CNS and physically, metabolically, and functionally support the neurons.

- The four types of glial cells are the astrocytes, oligodendrocytes, microglia, and ependymal cells. *(Review Figures 5-3 and 5-4.)*

- The brain is provided with several protective devices, which is important because neurons cannot divide to replace damaged cells. (1) The brain is wrapped in three layers of protective membranes—the meninges—and is further surrounded by a hard, bony covering. (2) Cerebrospinal fluid flows within and around the brain to cushion it against physical jarring. (3) Protection against chemical injury is conferred by a blood–brain barrier that limits access of blood-borne substances to the brain.

- The brain depends on a constant blood supply for delivery of O_2 and glucose because it cannot generate ATP in the absence of either of these substances.

Overview of the Central Nervous System (pp. 114–116)

- Even though no part of the brain acts in isolation of other brain regions, the brain is organized into networks of neurons within discrete locations that are ultimately responsible for carrying out particular tasks.

- The parts of the brain from the lowest, most primitive level to the highest, most sophisticated level are the brain stem, cerebellum, hypothalamus, thalamus, basal nuclei, and cerebral cortex. *(Review Table 5-1 and Figure 5-5.)*

Cerebral Cortex (pp. 116–123)

▪ The cerebral cortex is the outer shell of gray matter that caps an underlying core of white matter. The white matter consists of bundles of nerve fibers that interconnect various cortical regions with other areas. The cortex itself consists primarily of neuronal cell bodies, dendrites, and glial cells.

▪ Ultimate responsibility for many discrete functions is localized in particular regions of the cortex as follows: (1) the occipital lobes house the visual cortex; (2) the auditory cortex is in the temporal lobes; (3) the parietal lobes are responsible for reception and perceptual processing of somatosensory (somesthetic and proprioceptive) input; and (4) voluntary motor movement is set into motion by the frontal lobe, where the primary motor cortex and higher motor areas are located. *(Review Figures 5-6 through 5-9.)*

▪ Language ability depends on the integrated activity of two primary language areas—Broca's area and Wernicke's area—typically located only in the left cerebral hemisphere. *(Review Figure 5-7.)*

▪ The association areas are regions of the cortex not specifically assigned to processing sensory input or commanding motor output or language ability. These areas provide an integrative link between diverse sensory information and purposeful action; they also play a key role in higher brain functions such as memory and decision making. The association areas include the prefrontal association cortex, the parietal-temporal-occipital association cortex, and the limbic association cortex. *(Review Figures 5-7 and 5-10.)*

Basal Nuclei, Thalamus, and Hypothalamus (pp. 123–126)

▪ The subcortical brain structures, which include the basal nuclei, thalamus, and hypothalamus, interact extensively with the cortex in performing their functions. *(Review Figures 5-12 and 5-13.)*

▪ The basal nuclei inhibit muscle tone; coordinate slow, sustained postural contractions; and suppress useless patterns of movement.

▪ The thalamus serves as a relay station for preliminary processing of sensory input on its way to the cortex. It also accomplishes a crude awareness of sensation and some degree of consciousness.

▪ The hypothalamus regulates many homeostatic functions, in part through its extensive control of the autonomic nervous system and endocrine system.

The Limbic System and Its Functional Relations with the Higher Cortex (pp. 126–130)

▪ The limbic system, which includes portions of the hypothalamus and other forebrain structures that encircle the brain stem, is responsible for emotion as well as basic, inborn behavioral patterns related to survival and perpetuation of the species. It also plays an important role in motivation and learning. *(Review Figure 5-14.)*

▪ There are two types of memory: (1) a short-term memory with limited capacity and brief retention, coded by modification of activity at pre-existing synapses, and (2) a long-term memory with large storage capacity and enduring memory traces, involving relatively permanent structural or functional changes, such as the formation of new synapses, between existing neurons. Enhanced protein synthesis underlies these long-term changes. *(Review Table 5-2.)*

▪ The hippocampus and associated structures are especially important in declarative or "what" memories of specific objects, facts, and events.

▪ The cerebellum and associated structures are especially important in procedural or "how to" memories of motor skills gained through repetitive training.

▪ The prefrontal association cortex is the site of working memory, which temporarily holds currently relevant data—both new information and knowledge retrieved from memory stores—and manipulates and relates them to accomplish the higher-reasoning processes of the brain.

Cerebellum (pp. 130–131)

▪ The vestibulocerebellum helps maintain balance and controls eye movements.

▪ The spinocerebellum enhances muscle tone and helps coordinate voluntary movement, especially fast, phasic motor activities.

▪ The cerebrocerebellum plays a role in initiating voluntary movement and in storing procedural memories. *(Review Figure 5-15.)*

Brain Stem (pp. 131–135)

▪ The brain stem is an important link between the spinal cord and higher brain levels.

▪ The brain stem is the origin of the cranial nerves. *(Review Figure 5-16.)* It also contains centers that control cardiovascular, respiratory, and digestive function; regulates postural muscle reflexes; controls the overall degree of cortical alertness; and plays a key role in the sleep–wake cycle.

▪ The prevailing state of consciousness depends on the cyclical interplay between an arousal system (the reticular activating system) originating in the brain stem along with a slow-wave sleep center in the hypothalamus and a paradoxical sleep center in the brain stem. *(Review Figure 5-17.)*

▪ Sleep is an active process, not just the absence of wakefulness. While sleeping, a person cyclically alternates between slow-wave sleep and paradoxical sleep. *(Review Table 5-3.)*

▪ Slow-wave sleep is characterized by slow waves on the EEG and little change in behavior pattern from the waking state except for not being consciously aware of the external world. *(Review Figure 5-18.)*

▪ Paradoxical, or REM, sleep is characterized by an EEG pattern similar to that of an alert, awake individual. Rapid eye movements, dreaming, and abrupt changes in behavior pattern occur. *(Review Figure 5-18.)*

Spinal Cord (pp. 135–139)

▪ The spinal cord has two vital functions. First, it serves as the neuronal link between the brain and the peripheral nervous system. All communication up and down the spinal cord is located in ascending and descending tracts in the cord's outer white matter. Second, it is the integrating center for spinal reflexes, including some of the basic protective and postural reflexes and those involved with the emptying of the pelvic organs.

▪ The basic reflex arc includes a receptor, an afferent pathway, an integrating center, an efferent pathway, and an effector. *(Review Figure 5-21.)*

▪ The centrally located gray matter of the spinal cord contains the interneurons interposed between the afferent input and efferent output as well as the cell bodies of efferent neurons. *(Review Figure 5-20.)*

▪ Afferent and efferent fibers, which carry signals to and from the spinal cord, respectively, are bundled together into spinal nerves. These nerves supply specific body regions and are attached to the spinal cord in paired fashion throughout its length. *(Review Figures 5-19 and 5-20.)*

REVIEW EXERCISES

Objective Questions (Answers on p. A-42)

1. The major function of the CSF is to nourish the brain. *(True or false?)*

2. In emergencies when O_2 supplies are low, the brain can perform anaerobic metabolism. *(True or false?)*

3. Damage to the left cerebral hemisphere brings about paralysis and loss of sensation on the left side of the body. *(True or false?)*

4. The hands and structures associated with the mouth have a disproportionately large share of representation in both the sensory and motor cortexes. *(True or false?)*

5. The left cerebral hemisphere specializes in artistic and musical ability, whereas the right side excels in verbal and analytical skills. *(True or false?)*

6. The process of transferring and fixing short-term memory traces into long-term memory stores is known as _____.

7. Afferent fibers enter through the _____ root of the spinal cord, and efferent fibers leave through the _____ root.

8. Match the following:

   ___ 1. consists of nerves carrying information between the periphery and the CNS
   ___ 2. consists of the brain and spinal cord
   ___ 3. division of the peripheral nervous system that transmits signals to the CNS
   ___ 4. division of the peripheral nervous system that transmits signals from the CNS
   ___ 5. supplies skeletal muscles
   ___ 6. supplies smooth muscle, cardiac muscle, and glands

   (a) somatic nervous system
   (b) autonomic nervous system
   (c) central nervous system
   (d) peripheral nervous system
   (e) efferent division
   (f) afferent division

9. Using the following answer code, indicate which neurons are being described (a characteristic may apply to more than one class of neurons):

   (a) afferent neurons
   (b) efferent neurons
   (c) interneurons

   1. have receptors at peripheral endings
   2. lie entirely within the CNS
   3. lie primarily within the peripheral nervous system
   4. innervate muscles and glands
   5. cell bodies are devoid of presynaptic inputs
   6. predominant type of neuron
   7. responsible for thoughts, emotions, memory, and so forth

Essay Questions

1. Discuss the function of each of the following: astrocytes, oligodendrocytes, ependymal cells, microglia, cerebrospinal fluid, and blood–brain barrier.
2. Compare the composition of white and gray matter.
3. Draw and label the major functional areas of the cerebral cortex, indicating the functions attributable to each area.
4. Discuss the function of each of the following parts of the brain: thalamus, hypothalamus, basal nuclei, limbic system, cerebellum, and brain stem.
5. Define *somesthetic sensations* and *proprioception.*
6. What is an electroencephalogram?
7. Discuss the roles of Broca's area and Wernicke's area in language.
8. Compare short-term and long-term memory.
9. What is the reticular activating system?
10. Compare slow-wave and paradoxical sleep.
11. Draw and label a cross section of the spinal cord.
12. List the five components of a basic reflex arc.
13. Distinguish between a monosynaptic and a polysynaptic reflex.

POINTS TO PONDER

(Explanations on p. A-42)

1. Special studies designed to assess the specialized capacities of each cerebral hemisphere have been performed on "split-brain" patients. In these people the corpus callosum—the bundle of fibers that links the two halves of the brain together—has been surgically cut to prevent the spread of epileptic seizures from one hemisphere to the other. Even though no overt changes in behavior, intellect, or personality occur in these patients, because both hemispheres individually receive the same information, deficits are observable with tests designed to restrict information to one brain hemisphere at a time. One such test involves limiting a visual stimulus to only half of the brain. Because of a crossover in the nerve pathways from the eyes to the occipital cortex, the visual information to the right of a midline point is transmitted to only the left half of the brain, whereas visual information to the left of this point is received by only the right half of the brain. A split-brain patient presented with a visual stimulus that reaches only the left hemisphere accurately describes the object seen, but when a visual stimulus is presented to only the right hemisphere, the patient denies having seen anything. The right hemisphere does receive the visual input, however, as demonstrated by nonverbal tests. Even though a split-brain patient denies having seen anything after an object is presented to the right hemisphere, he or she can correctly match the object by picking it out from among a number of objects, usually to the patient's surprise. What is your explanation of this finding?

2. Which of the following symptoms are most likely to occur as the result of a severe blow to the back of the head?
   a. paralysis
   b. hearing impairment
   c. visual disturbances
   d. burning sensations
   e. personality disorders

3.   The hormone insulin enhances the carrier-mediated transport of glucose into most of the body's cells but not into brain cells. The uptake of glucose from the blood by neurons is not dependent on insulin. Knowing the brain's need for a continuous supply of blood-borne glucose, what effect do you predict that insulin excess would have on the brain?

4.   Give examples of conditioned reflexes you have acquired.
5.   Under what circumstances might it be inadvisable to administer a clot-dissolving drug to a stroke victim?

CLINICAL CONSIDERATION

(Explanation on p. A-42)

Julio D., who had recently retired, was enjoying an afternoon of playing golf when suddenly he experienced a severe headache and dizziness. These symptoms were quickly followed by numbness and partial paralysis on the upper right side of his body, accompanied by an inability to speak. After being rushed to the emergency room, Julio was diagnosed as having suffered a stroke. Given the observed neurologic impairment, what areas of his brain were affected?

PHYSIOEDGE RESOURCES

 PhysioEdge CD-ROM

PhysioEdge, the CD-ROM packaged with your text, focuses on the concepts students find most difficult to learn. Figures marked with this icon have associated activities on the CD. For a visual review of concepts in this chapter, check out the following:

Tutorial: Skeletal Muscle Contraction—Muscle Spindles tab

Media Exercise 5.1: The Nervous and Endocrine Systems

Media Exercise 5.2: The Cerebral Cortex

Media Exercise 5.3: Subcortical Structures

PhysioEdge Website

The website for this book contains a wealth of helpful study aids, as well as many ideas for further reading and research. Log on to: http://www.brookscole.com/hpfundamentals3

Select Chapter 5 from the drop-down menu or click on one of the many resource areas, including **Case Histories,** which introduce clinical aspects of human physiology. For this chapter check out Case History 13: A Critical Twenty-four Hours and Case History 14: One Disease, Two Outcomes.

For Suggested Readings, consult **InfoTrac College Edition/ Research** on the PhysioEdge website or go directly to InfoTrac College Edition, your online research library, at: **http://infotrac.thomsonlearning.com**

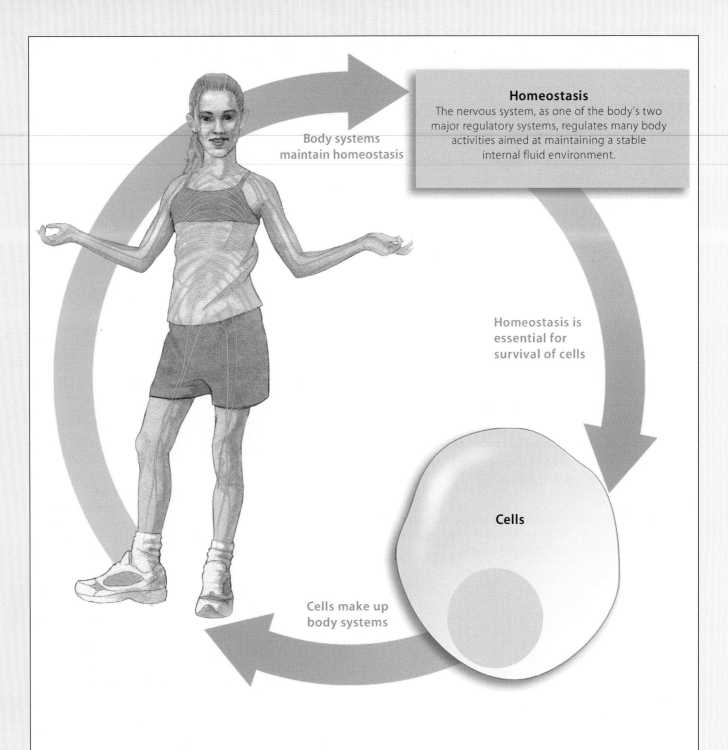

Homeostasis
The nervous system, as one of the body's two major regulatory systems, regulates many body activities aimed at maintaining a stable internal fluid environment.

Body systems maintain homeostasis

Homeostasis is essential for survival of cells

Cells

Cells make up body systems

The nervous system, one of the two major regulatory systems of the body, consists of the central nervous system (CNS), composed of the brain and spinal cord, and the **peripheral nervous system (PNS),** composed of the afferent and efferent fibers that relay signals between the CNS and periphery (other parts of the body).

The **afferent division** of the peripheral nervous system detects, encodes, and transmits peripheral signals to the central nervous system, thus informing the CNS about the internal and external environment. This afferent input to the controlling centers of the CNS is essential in maintaining homeostasis. To make appropriate adjustments in effector organs via efferent output, the CNS has to "know" what is going on. Afferent input is also used to plan for voluntary actions unrelated to homeostasis.

CHAPTER 6

The Peripheral Nervous System: Afferent Division; Special Senses

Click on the Tutorials menu of the CD-ROM for a tutorial on Special Senses.

INTRODUCTION

The peripheral nervous system consists of nerve fibers that carry information between the CNS and other parts of the body. The afferent division of the peripheral nervous system sends information about the internal and external environment to the CNS.

❚ Visceral afferents carry subconscious input whereas sensory afferents carry conscious input.

Afferent information about the internal environment, such as the blood pressure and the concentration of CO_2 in the body fluids, never reaches the level of conscious awareness, but this input is essential for determining the appropriate efferent output to maintain homeostasis. The incoming pathway for subconscious information derived from the internal viscera (organs) is called a **visceral afferent**. Afferent input that does reach the level of conscious awareness is known as *sensory information,* and the incoming pathway is considered a **sensory afferent**. Sensory information is categorized as either (1) **somatic** (body sense) **sensation** arising from the body surface, including *somesthetic sensation* from the skin and *proprioception* from the muscles, joints, skin, and inner ear (see p. 119) or (2) **special senses**, including *vision, hearing, taste,* and *smell.* Final processing of sensory input by the CNS not only is essential for interaction with the environment for basic survival (for example, food procurement and defense from danger) but also adds immeasurably to the richness of life.

❚ Perception is the conscious awareness of surroundings derived from interpretation of sensory input.

Perception is our conscious interpretation of the external world as created by the brain from a pattern of nerve impulses delivered to it from sensory receptors.

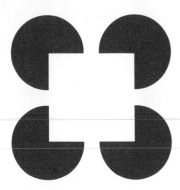

● **FIGURE 6-1**

Do you "see" a white square that is not really there?

● **FIGURE 6-2**

Variable perceptions from the same visual input. Do you see two faces in profile, or a wineglass?

Is the world as we perceive it, reality? The answer is a resounding no. Our perception is different from what is really "out there," for several reasons. First, humans have receptors that detect only a limited number of existing energy forms. We perceive sounds, colors, shapes, textures, smells, tastes, and temperature but are not informed of magnetic forces, polarized light waves, radio waves, or x-rays because we do not have receptors to respond to the latter energy forms. What is not detected by receptors, the brain will never know. Our response range is limited even for the energy forms for which we do have receptors. For example, dogs can hear a whistle whose pitch is above our level of detection. Second, the information channels to our brains are not high-fidelity recorders. During precortical processing of sensory input, some features of stimuli are accentuated and others are suppressed or ignored. Third, the cerebral cortex further manipulates the data, comparing the sensory input with other incoming information as well as with memories of past experiences to extract the significant features—for example, sifting out a friend's words from the hubbub of sound in a school cafeteria. In the process, the cortex often fills in or distorts the information to abstract a logical perception; that is, it "completes the picture." As a simple example, you "see" a white square in ● Figure 6-1 even though there is no white square but merely right-angle wedges taken out of four red circles. Optical illusions illustrate how the brain interprets reality according to its own rules. Do you see two faces in profile or a wineglass in ● Figure 6-2? You can alternately see one or the other out of identical visual input. Thus our perceptions do not replicate reality. Other species, equipped with different types of receptors and sensitivities and with different neural processing, perceive a markedly different world from what we perceive.

RECEPTOR PHYSIOLOGY

A **stimulus** is a change detectable by the body. Stimuli exist in a variety of energy forms, or **modalities**, such as heat, light, sound, pressure, and chemical changes. Afferent neurons have **receptors** at their peripheral endings that respond to stimuli in both the external world and internal environment. Because the only way afferent neurons can transmit information to the CNS about these stimuli is via action potential propagation, receptors must convert these other forms of energy into electrical signals (action potentials). This energy conversion process is known as **transduction.**

▌ **Receptors have differential sensitivities to various stimuli.**

Each type of receptor is specialized to respond more readily to one type of stimulus, its **adequate stimulus**, than to other stimuli. For example, receptors in the eye are most sensitive to light, receptors in the ear to sound waves, and warmth receptors in the skin to heat energy. Because of this differential sensitivity of receptors, we cannot "see" with our ears or "hear" with our eyes. Some receptors can respond weakly to stimuli other than their adequate stimulus, but even when activated by a different stimulus a receptor still gives rise to the sensation usually detected by that receptor type. As an example, the adequate stimulus for eye receptors (photoreceptors) is light, to which they are exquisitely sensitive, but these receptors can also be activated to a lesser degree by mechanical stimulation. When hit in the eye, a person often "sees stars," because the mechanical pressure stimulates the photoreceptors. Thus the sensation perceived depends on the type of receptor stimulated rather than on the type of stimulus. However, because receptors typically are activated by their adequate stimulus, the sensation usually corresponds to the stimulus modality.

TYPES OF RECEPTORS ACCORDING TO THEIR ADEQUATE STIMULUS

Depending on the type of energy to which they ordinarily respond, receptors are categorized as follows:

• **Photoreceptors** are responsive to visible wavelengths of light.

• **Mechanoreceptors** are sensitive to mechanical energy. Examples include skeletal muscle receptors sensitive to stretch, the receptors in the ear containing fine hair cells that are bent as a result of sound waves, and blood pressure–monitoring baroreceptors.

• **Thermoreceptors** are sensitive to heat and cold.

• **Osmoreceptors** detect changes in the concentration of solutes in the body fluids and the resultant changes in osmotic activity (see p. 53).

• **Chemoreceptors** are sensitive to specific chemicals. Chemoreceptors include the receptors for smell and taste, as well as those located deeper within the body that detect O_2 and CO_2 concentrations in the blood or the chemical content of the digestive tract.

• **Nociceptors,** or **pain receptors,** are sensitive to tissue damage such as pinching or burning or to distortion of tissue. Intense stimulation of any receptor is also perceived as painful.

Some sensations are compound sensations in that their perception arises from central integration of several simultaneously activated primary sensory inputs. For example, the perception of wetness comes from touch, pressure, and thermal receptor input; there is no such thing as a "wetness receptor."

USES FOR INFORMATION DETECTED BY RECEPTORS

The information detected by receptors is conveyed via afferent neurons to the CNS, where it is used for a variety of purposes:

- First, afferent input is essential for the control of efferent output, both for regulating motor behavior in accordance with external circumstances and for coordinating internal activities directed at maintaining homeostasis. At the most basic level, afferent input provides information (of which the person may or may not be consciously aware) for the CNS to use in directing activities necessary for survival. On a broader level, we could not interact successfully with our environment or with each other without sensory input.
- Second, processing of sensory input by the reticular activating system in the brain stem is critical for cortical arousal and consciousness (see p. 133).
- Third, central processing of sensory information gives rise to our perceptions of the world around us.
- Fourth, selected information delivered to the CNS may be stored for future reference.
- Finally, sensory stimuli can have a profound impact on our emotions. The smell of just-baked apple pie, the sensuous feel of silk, the sight of a loved one, hearing bad news—sensory input can gladden, sadden, arouse, calm, anger, frighten or evoke any other range of emotions.

We will next examine how adequate stimuli initiate action potentials that ultimately are used for these purposes.

▌ A stimulus alters the receptor's permeability, leading to a graded receptor potential.

A receptor may be either (1) a specialized ending of the afferent neuron or (2) a separate cell closely associated with the peripheral ending of the neuron. Stimulation of a receptor alters its membrane permeability, usually by causing a nonselective opening of all small ion channels. The means by which this permeability change takes place is individualized for each receptor type. Because the electrochemical driving force is greater for Na^+ than for other small ions at resting potential, the predominant effect is an inward flux of Na^+, which depolarizes the receptor membrane (see p. 72). This local depolarizing change in potential is known as a **receptor potential.** The receptor potential is a graded potential whose amplitude and duration can vary, depending on the strength and the rate of application or removal of the stimulus (see p. 73). The stronger the stimulus, the greater the permeability change and the larger the receptor potential. As is true of all graded potentials, receptor potentials have no refractory period, so summation in response to rapidly successive stimuli is possible. Because the receptor region has a very high threshold, action potentials do not take place at the receptor itself. For

long-distance transmission, the receptor potential must be converted into action potentials that can be propagated along the afferent fiber.

▌ Receptor potentials may initiate action potentials in the afferent neuron.

A receptor potential may initiate an action potential in the afferent neuron membrane next to the receptor by triggering Na^+ channels in this region to open. If the resulting ionic flux is big enough to bring the adjacent membrane to threshold, the action potential that is initiated self-propagates along the afferent fiber to the CNS.

Note that the initiation site of action potentials in an afferent neuron differs from the site in an efferent neuron or interneuron. In the latter two types of neurons, action potentials are initiated at the axon hillock located at the start of the axon next to the cell body (see p. 79). By contrast, action potentials are initiated at the peripheral end of an afferent nerve fiber next to the receptor, a long distance from the cell body (● Figure 6-3).

The intensity of the stimulus is reflected by the magnitude of the receptor potential. In turn, the larger the receptor potential, the greater the frequency of action potentials generated in the afferent neuron. A larger receptor potential cannot bring about a larger action potential (because of the all-or-none law), but it can induce more rapid firing of action potentials (see p. 83). Stimulus strength is also reflected by the size of the area stimulated. Stronger stimuli usually affect larger areas, so correspondingly more receptors respond. For example, a light touch does not activate as many pressure receptors in the skin as does a more forceful touch applied to

---

● **FIGURE 6-3**

Comparison of the initiation site of an action potential in the three types of neurons

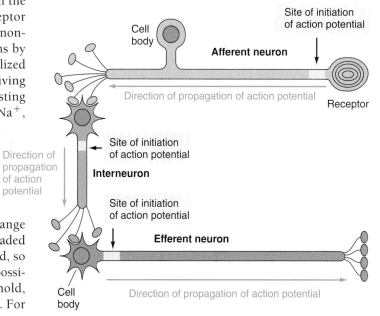

the same area. Stimulus intensity is therefore distinguished both by the frequency of action potentials generated in the afferent neuron and by the number of receptors activated within the area.

Receptors may adapt slowly or rapidly to sustained stimulation.

Stimuli of the same intensity do not always bring about receptor potentials of the same magnitude from the same receptor. Some receptors can diminish the extent of their depolarization despite sustained stimulus strength, a phenomenon called **adaptation**. Subsequently, the frequency of action potentials generated in the afferent neuron decreases. That is, the receptor "adapts" to the stimulus by no longer responding to it to the same degree.

TYPES OF RECEPTORS ACCORDING TO THEIR SPEED OF ADAPTATION

There are two types of receptors—*tonic receptors* and *phasic receptors*—based on their speed of adaptation. **Tonic receptors** do not adapt at all or adapt slowly (● Figure 6-4a). These receptors are important in situations where it is valuable to maintain information about a stimulus. Examples of tonic receptors are muscle stretch receptors, which monitor muscle length, and joint proprioceptors, which measure the degree of joint flexion. To maintain posture and balance, the CNS must continually get information about the degree of muscle length and joint position. It is important, therefore, that these receptors do *not* adapt to a stimulus but continue to generate action potentials to relay this information to the CNS.

    Phasic receptors, in contrast, are rapidly adapting receptors. The receptor rapidly adapts by no longer responding to a maintained stimulus, but when the stimulus is removed the receptor typically responds with a slight depolarization called the **off response** (● Figure 6-4b). Phasic receptors are useful in situations where it is important to signal a change in stimulus intensity rather than to relay status quo information. Rapidly adapting receptors include *tactile (touch) receptors* in the skin that signal changes in pressure on the skin surface. Because these receptors adapt rapidly, you are not continually conscious of wearing your watch, rings, and clothing. When you put something on, you soon become accustomed to it, because of these receptors' rapid adaptation. When you take the item off, you are aware of its removal, because of the off response.

Each somatosensory pathway is "labeled" according to modality and location.

On reaching the spinal cord, afferent information has two possible destinies: (1) it may become part of a reflex arc, bringing about an appropriate effector response, or (2) it may be relayed upward to the brain via ascending pathways for further processing and possible conscious awareness. Pathways conveying conscious somatic sensation, the **somatosensory pathways**, consist of discrete chains of neurons synaptically interconnected in a particular sequence to accomplish progressively

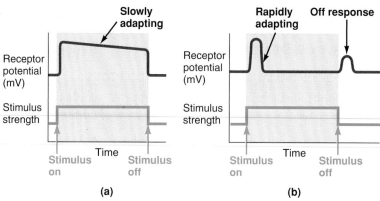

● FIGURE 6-4

Tonic and phasic receptors. (a) *Tonic receptor*. This receptor type does not adapt at all or adapts slowly to a sustained stimulus and thus provides continuous information about the stimulus. (b) *Phasic receptor*. This receptor type adapts rapidly to a sustained stimulus and frequently exhibits an off response when the stimulus is removed. Thus the receptor signals changes in stimulus intensity rather than relaying status quo information.

more sophisticated processing of the sensory information. A particular sensory modality detected by a specialized receptor type is sent over a specific afferent and ascending pathway (a neural pathway committed to that modality) to excite a defined area in the somatosensory cortex. That is, a particular sensory input is **projected** to a specific region of the cortex. Thus different types of incoming information are kept separated within specific **labeled lines** between the periphery and the cortex. In this way, even though all information is propagated to the CNS via the same type of signal (action potentials), the brain can decode the type and location of the stimulus. ▲ Table 6-1 summarizes how the CNS is informed of the type (what?), location (where?), and intensity (how much?) of a stimulus.

Acuity is influenced by receptive field size.

Each somatosensory neuron responds to stimulus information only within a circumscribed region of the skin surface surrounding it; this region is called its **receptive field**. The size of a receptive field varies inversely with the density of receptors in the region; the more closely receptors of a particular type are spaced, the smaller the area of skin each monitors. The smaller the receptive field in a region, the greater its **acuity** or **discriminative ability**. Compare the tactile discrimination in your fingertips with that in your elbow by "feeling" the same object with both. You can sense more precise information about the object with your richly innervated fingertips because the receptive fields there are small; as a result, each neuron signals information about small, discrete portions of the object's surface. An estimated 17,000 tactile mechanoreceptors are present in the fingertips and palm of each hand. In contrast, the skin over the elbow is served by relatively few sensory endings with larger receptive fields. Subtle differences within each large receptive field cannot be detected (● Figure 6-5). The distorted cortical representation of vari-

TABLE 6-1
Coding of Sensory Information

STIMULUS PROPERTY	MECHANISM OF CODING
Type of Stimulus (stimulus modality)	Distinguished by type of receptor activated and specific pathway over which this information is transmitted to a particular area of cerebral cortex
Location of Stimulus	Distinguished by location of activated receptor field and pathway subsequently activated to transmit this information to the area of the somatosensory cortex representing that particular location
Intensity of Stimulus (stimulus strength)	Distinguished by frequency of action potentials initiated in an activated afferent neuron and by number of receptors (and afferent neurons) activated

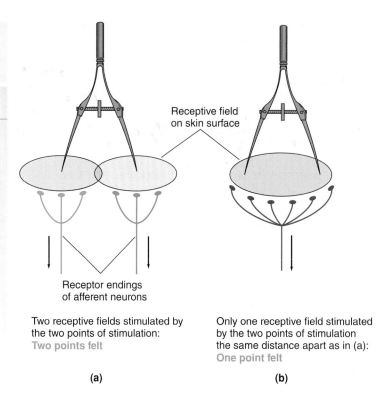

Two receptive fields stimulated by the two points of stimulation:
Two points felt

Only one receptive field stimulated by the two points of stimulation the same distance apart as in (a):
One point felt

(a)   (b)

● **FIGURE 6-5**

Comparison of discriminative ability of regions with small versus large receptive fields. The relative tactile acuity of a given region can be determined by the *two-point threshold-of-discrimination test*. If the two points of a pair of calipers applied to the surface of the skin stimulate two different receptive fields, two separate points are felt. If the two points touch the same receptive field, they are perceived as only one point. By adjusting the distance between the caliper points, one can determine the minimal distance at which the two points can be recognized as two rather than one, which reflects the size of the receptive fields in the region. With this technique, it is possible to plot the discriminative ability of the body surface. The two-point threshold ranges from 2 mm in the fingertip (enabling a person to read Braille, where the raised dots are spaced 2.5 mm apart) to 48 mm in the poorly discriminative skin of the calf of the leg. (a) Region with small receptive fields. (b) Region with large receptive fields.

ous body parts in the sensory homunculus (see p. 119) corresponds precisely with the innervation density; more cortical space is allotted for sensory reception from areas with smaller receptive fields and, accordingly, greater tactile discriminative ability.

Having completed our general discussion of receptor physiology, we are now going to examine one important somatic sensation in greater detail—pain.

PAIN

Pain is primarily a protective mechanism meant to bring to conscious awareness the fact that tissue damage is occurring or is about to occur.

▌ Stimulation of nociceptors elicits the perception of pain plus motivational and emotional responses.

Unlike other somatosensory modalities, the sensation of pain is accompanied by motivated behavioral responses (such as withdrawal or defense) as well as emotional reactions (such as crying or fear). Also, unlike other sensations, the subjective perception of pain can be influenced by other past or present experiences (for example, heightened pain perception accompanying fear of the dentist or lowered pain perception in an injured athlete during a competitive event).

There are three categories of pain receptors, or nociceptors: **Mechanical nociceptors** respond to mechanical damage such as cutting, crushing, or pinching; **thermal nociceptors**

respond to temperature extremes, especially heat; and **polymodal nociceptors** respond equally to all kinds of damaging stimuli, including irritating chemicals released from injured tissues. None of the nociceptors have specialized receptor structures; they are all naked nerve endings. Because of their value to survival, nociceptors do not adapt to sustained or repetitive stimulation.

All nociceptors can be sensitized by the presence of *prostaglandins,* which greatly enhance the receptor response to noxious stimuli (that is, it hurts more when prostaglandins are present). Prostaglandins are a special group of fatty acid derivatives that are cleaved from the lipid bilayer of the plasma membrane and act locally on being released (see p. 595). Tissue injury, among other things, can lead to the local release of prostaglandins. These chemicals act on the nociceptors' peripheral endings to lower their threshold for activation. Aspirin-like drugs inhibit the synthesis of prostaglandins, accounting at least in part for the **analgesic** (pain-relieving) properties of these drugs.

FAST AND SLOW AFFERENT PAIN FIBERS

Pain impulses originating at nociceptors are transmitted to the CNS via one of two types of afferent fibers. Signals arising from mechanical and thermal nociceptors are transmitted over small, myelinated fibers at rates of up to 30 meters/sec (the **fast pain pathway**). Impulses from polymodal nociceptors are carried by small, unmyelinated fibers at a much slower rate of 12 meters/sec (the **slow pain pathway**). Think about the last time you cut or burned your finger. You undoubtedly felt a sharp twinge of pain at first, with a more diffuse, disagreeable pain commencing shortly thereafter. Pain typically is perceived initially as a brief, sharp, prickling sensation that is easily localized; this is the fast pain pathway originating from specific mechanical or heat nociceptors. This feeling is followed by a dull, aching, poorly localized sensation that persists for a longer time and is more unpleasant; this is the slow pain pathway, which is activated by chemicals released into the ECF from damaged tissue. The persistence of these chemicals might explain the long-lasting, aching pain that continues after removal of the mechanical or thermal stimulus that caused the tissue damage.

HIGHER-LEVEL PROCESSING OF PAIN INPUT

The afferent pain fibers synapse with specific interneurons in the spinal cord from which the signal is transmitted to the brain for perceptual processing. One of the neurotransmitters released from these afferent pain terminals is **substance P**, which is believed to be unique to pain fibers.

Clinical Note Chronic pain, sometimes excruciating, occasionally occurs in the absence of tissue injury. In contrast to the pain accompanying peripheral injury, which serves as a normal protective mechanism to warn of impending or actual damage to the body, abnormal chronic pain states result from damage within the pain pathways in the peripheral nerves or the CNS. That is, pain is perceived because of abnormal signaling within the pain pathways in the absence of peripheral injury or typical painful stimuli. For example, strokes that damage ascending pathways can lead to an abnormal, persistent sensation of pain.

▌ The brain has a built-in analgesic system.

In addition to the chain of neurons connecting peripheral nociceptors with higher CNS structures for pain perception, the CNS also contains a built-in pain-suppressing or **analgesic system** that suppresses transmission in the pain pathways as they enter the spinal cord. This analgesic system suppresses pain by blocking the release of substance P from afferent pain-fiber terminals

The analgesic system depends on the presence of **opiate receptors**. People have long known that **morphine**, a component of the opium poppy, is a powerful analgesic. Researchers considered it very unlikely that the body has been endowed with opiate receptors only to interact with chemicals derived from a flower! They therefore began to search for the substances that normally bind with these opiate receptors. The result was the discovery of **endogenous opiates** (morphine-like substances)—the **endorphins**, **enkephalins**, and **dynor-**

phin—which are important in the body's natural analgesic system. These endogenous opiates serve as analgesic neurotransmitters; they are released from the descending analgesic pathway and bind with opiate receptors on the afferent pain-fiber terminal. This binding suppresses the release of substance P, blocking further transmission of the pain signal. Morphine binds to these same opiate receptors, which accounts for its analgesic properties.

Usually the endogenous opiate system is dormant. It is unclear how this natural pain-suppressing mechanism is activated. Factors known to modulate pain include exercise, stress, and acupuncture. Researchers believe that endorphins are released during prolonged exercise and presumably produce the "runner's high." Some types of stress also induce analgesia. It is sometimes disadvantageous for a stressed organism to display the normal reaction to pain. For example, when two male lions are fighting for dominance of the group, withdrawing, escaping, or resting when injured would mean certain defeat. (See the accompanying boxed feature, ▶ Beyond the Basics, for an examination of how acupuncture relieves pain.)

We have now completed our coverage of somatic sensation. As you are now aware, somatic sensation is detected by widely distributed receptors that provide information about the body's interactions with the environment in general. In contrast, each of the special senses has highly localized, extensively specialized receptors that respond to unique environmental stimuli. The special senses include *vision, hearing, taste,* and *smell,* to which we now turn our attention, starting with vision.

 PhysioEdge Click on the Media Exercises menu of the CD-ROM and work Media Exercise 6.1: Pain to test your understanding of the previous section.

EYE: VISION

For **vision**, the eyes capture the patterns of illumination in the environment as an "optical picture" on a layer of light-sensitive cells, the retina, much as a camera captures an image on film. Just as film can be developed into a visual likeness of the original image, the coded image on the retina is transmitted through a series of progressively more complex steps of visual processing until it is finally consciously perceived as a visual likeness of the original image. Before considering the steps involved in the process of vision, we are first going to examine how the eyes are protected from injury.

▌ Protective mechanisms help prevent eye injuries.

Several mechanisms help protect the eyes from injury. Except for its anterior (front) portion, the eyeball is sheltered by the bony socket in which it is positioned. The **eyelids** act like shutters to protect the anterior portion of the eye from environmental insults. They close reflexly to cover the eye under threatening circumstances, such as rapidly approaching objects, dazzling light, and instances when the exposed sur-

Acupuncture: Is It for Real?

It sounds like science fiction. How can a needle inserted in the hand relieve a toothache? **Acupuncture analgesia (AA),** the technique of relieving pain by inserting and manipulating threadlike needles at key points, has been practiced in China for over 2000 years but is relatively new to Western medicine and still remains controversial in the United States.

Traditional Chinese teaching holds that disease can occur when the normal patterns of flow of healthful energy (called *Qi;* pronounced "chee") become disrupted, with acupuncture correcting this imbalance and restoring health. Many Western scientists have been skeptical because, until recently, the phenomenon could not be explained on the basis of any known, logical, physiologic principles, although a tremendous body of anecdotal evidence in support of the effectiveness of AA existed in China. In Western medicine, the success of acupuncture was considered a placebo effect. The term *placebo effect* refers to a chemical or technique that brings about a desired response through the power of suggestion or distraction rather than through any direct action.

Because the Chinese were content with anecdotal evidence for the success of AA, this phenomenon did not come under close scientific scrutiny until the last several decades, when European and American scientists started studying it. As a result of these efforts, an impressive body of rigorous scientific investigation supports the contention that AA really works (that is, by a physiological rather than a placebo/psychological effect). Furthermore, its mechanisms of action have become apparent. Indeed, more is known about the underlying physiological mechanisms of AA than about those of many conventional medical techniques, such as gas anesthesia.

The overwhelming body of evidence supports the *acupuncture endorphin hypothesis* as the primary mechanism of AA's action. According to this hypothesis, acupuncture needles activate specific afferent nerve fibers, which send impulses to the central nervous system. Here the incoming impulses activate three centers (a spinal cord center, a midbrain center, and a hormonal center, the hypothalamus-anterior pituitary unit) to cause analgesia. Researchers have shown that all three centers block pain transmission through use of endorphins and closely related compounds. Several other neurotransmitters, such as serotonin and norepinephrine, as well as cortisol, the major hormone released during stress, are implicated as well. (Pain relief in placebo controls is believed to occur as a result of placebo responders subconsciously activating their own built-in analgesic system.)

AA has been proven effective in treating chronic pain and to exert a real physical effect in that it is more effective than placebos. In fact, AA compares favorably with morphine for treating chronic pain. In controlled clinical studies, 55 to 85% of patients were helped by AA. (By comparison, 70% of patients benefit from morphine therapy.) Pain relief was reported by only 30 to 35% of placebo controls (people who thought they were receiving proper AA treatment, but in whom needles were inserted in the wrong places or not deep enough).

In the United States, AA has not been used in mainstream medicine, even by physicians who have been convinced by scientific evidence that the technique is valid. AA methodology has traditionally not been taught in U.S. medical colleges, and the techniques take time to learn. Also, AA is much more time consuming than using drugs. Western physicians who have been trained to use drugs to solve most pain problems are generally reluctant to scrap their known methods for an unfamiliar, time-consuming technique. However, acupuncture is gaining favor as an alternative treatment for relief of chronic pain, especially because analgesic drugs can have troublesome side effects. After decades of being spurned by the vast majority of the U.S. medical community, acupuncture started gaining respectability following a 1997 report issued by an expert panel convened by the National Institutes of Health (NIH). This report, based on an evaluation of published scientific studies, concluded that acupuncture is effective as an alternative or adjunct to conventional treatment for many kinds of pain and nausea. Now that acupuncture has been sanctioned by NIH, some medical insurers have taken the lead in paying for this now scientifically legitimate treatment, and some of the nation's medical schools are beginning to incorporate the technique into their curricula. There are also 40 nationally accredited acupuncture schools for nonphysicians. Of the 13,000 licensed acupuncture practitioners in the United States, only 3000 are physicians.

---

face of the eye or eyelashes are touched. Frequent spontaneous blinking of the eyelids helps disperse the lubricating, cleansing, bactericidal ("germ-killing") **tears**. Tears are produced continuously by the **lacrimal gland** in the upper lateral corner under the eyelid. This eye-washing fluid flows across the anterior surface of the eye and drains into tiny canals in the corner of each eye (● Figure 6-6a), eventually emptying into the back of the nasal passageway. This drainage system cannot handle the profuse tear production during crying, so the tears overflow from the eyes. The eyes are also equipped with protective **eyelashes**, which trap fine, airborne debris such as dust before it can fall into the eye.

▌ The eye is a fluid-filled sphere enclosed by three specialized tissue layers.

Each **eye** is a spherical, fluid-filled structure enclosed by three layers. From outermost to innermost, these are (1) the *sclera/cornea*; (2) the *choroid/ciliary body/iris*; and (3) the *retina* (● Figure 6-6b). Most of the eyeball is covered by a tough outer layer of connective tissue, the **sclera**, which forms the visible white part of the eye (● Figure 6-6a). Anteriorly, the outer layer consists of the transparent **cornea**, through which light rays pass into the interior of the eye. The middle layer underneath the sclera is the highly pigmented **choroid**, which contains many blood vessels that nourish the retina. The choroid layer becomes specialized anteriorly to form the *ciliary body* and *iris*, which will be described shortly. The innermost coat under the choroid is the **retina**, which consists of an outer pigmented layer and an inner nervous-tissue layer. The latter contains the **rods** and **cones**, the photoreceptors that convert light energy into nerve impulses. Like the black walls of a photographic studio, the pigment in the choroid and retina absorbs light after it strikes the retina to prevent reflection or scattering of light within the eye.

The interior of the eye consists of two fluid-filled cavities, separated by an elliptical **lens**, all of which are transparent to permit light to pass through the eye from the cornea

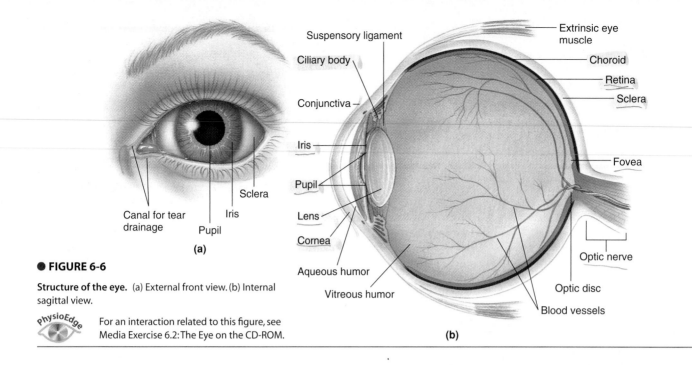

● FIGURE 6-6

Structure of the eye. (a) External front view. (b) Internal sagittal view.

For an interaction related to this figure, see Media Exercise 6.2: The Eye on the CD-ROM.

(Figure labels, part a: Canal for tear drainage, Pupil, Iris, Sclera)

(Figure labels, part b: Suspensory ligament, Ciliary body, Conjunctiva, Iris, Pupil, Lens, Cornea, Aqueous humor, Vitreous humor, Extrinsic eye muscle, Choroid, Retina, Sclera, Fovea, Optic nerve, Optic disc, Blood vessels)

to the retina. The larger posterior (rear) cavity between the lens and retina contains a semifluid, jellylike substance, the **vitreous humor.** The vitreous humor is important in maintaining the spherical shape of the eyeball. The anterior cavity between the cornea and lens contains a clear, watery fluid, the **aqueous humor.** The aqueous humor carries nutrients for the cornea and lens, both of which lack a blood supply. Blood vessels in these structures would impede the passage of light to the photoreceptors.

Aqueous humor is produced by a capillary network within the **ciliary body,** a specialized anterior derivative of the choroid layer. This fluid drains into a canal at the edge of the cornea and eventually enters the blood (● Figure 6-7).

Clinical Note If aqueous humor is not drained as rapidly as it forms (for example, because of a blockage in the drainage canal), the excess will accumulate in the anterior cavity, causing the pressure to rise within the eye. This condition is known as **glaucoma.** The excess aqueous humor pushes the lens backward into the vitreous humor, which in turn is pushed against the inner neural layer of the retina. This compression causes retinal and optic nerve damage that can lead to blindness if the condition is not treated.

▌ The amount of light entering the eye is controlled by the iris.

Not all the light passing through the cornea reaches the light-sensitive photoreceptors, because of the presence of the iris, a thin, pigmented smooth muscle that forms a visible ring-like structure within the aqueous humor (● Figure 6-6a and b). The pigment in the iris is responsible for eye color. The varied flecks, lines, and other nuances of the iris are unique for each individual, making the iris the basis of the latest identification technology. Recognition of iris patterns by a video

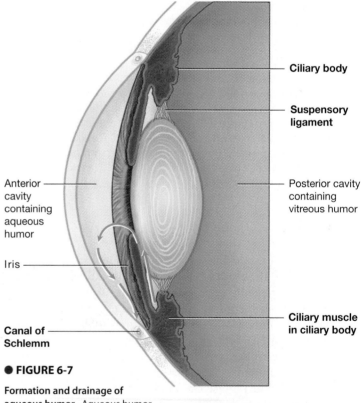

● FIGURE 6-7

Formation and drainage of aqueous humor. Aqueous humor is formed by a capillary network in the ciliary body, then drains into the canal of Schlemm, and eventually enters the blood.

PhysioEdge For an interaction related to this figure, see Media Exercise 6.2: The Eye on the CD-ROM.

(Figure labels: Ciliary body, Suspensory ligament, Posterior cavity containing vitreous humor, Ciliary muscle in ciliary body, Anterior cavity containing aqueous humor, Iris, Canal of Schlemm)

camera that captures iris images and translates the landmarks into a computerized code is more foolproof than fingerprinting or even DNA testing.

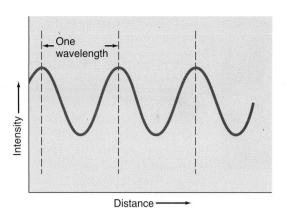

● FIGURE 6-8

Properties of an electromagnetic wave. A *wavelength* is the distance between two wave peaks. The *intensity* is the amplitude of the wave.

The round opening in the center of the iris through which light enters the interior portions of the eye is the **pupil**. The size of this opening can be adjusted by variable contraction of the iris muscles to admit more or less light as needed, much as the shutter controls the amount of light entering a camera. Iris muscles, and thus pupillary size, are controlled by the autonomic nervous system. Parasympathetic stimulation causes pupillary constriction, whereas sympathetic stimulation causes pupillary dilation.

The eye refracts the entering light to focus the image on the retina.

Light energy is a form of electromagnetic radiation that travels in wavelike fashion. The distance between two wave peaks is known as the *wavelength* (● Figure 6-8). The wavelengths in the electromagnetic spectrum range from 10^{-14} m (quadrillionths of a meter, as in the extremely short cosmic rays) to 10^4 m (10 kilometers, as in long radio waves) (● Figure 6-9). The photoreceptors in the eye are sensitive only to wavelengths between 400 and 700 nanometers (nm; billionths of a meter). Thus **visible light** is only a small portion of the total electromagnetic spectrum. Light of different wavelengths in this visible band is perceived as different color sensations. The shorter visible wavelengths are sensed as violet and blue; the longer wavelengths are interpreted as orange and red.

In addition to having variable wavelengths, light energy also varies in intensity; that is, the amplitude, or height, of the wave (● Figure 6-8). Dimming a bright red light does not change its color; it just becomes less intense or less bright.

Light waves *diverge* (radiate outward) in all directions from every point of a light source. The forward movement of a light wave in a particular direction is known as a **light ray**. Divergent light rays reaching the eye must be bent inward to be focused back into a point (the **focal point**) on the light-sensitive retina to provide an accurate image of the light source (● Figure 6-10).

PROCESS OF REFRACTION

Light travels faster through air than through other transparent media such as water and glass. When a light ray enters a medium of greater density, it is slowed down (the converse is also true). The course of direction of the ray changes if it strikes the surface of the new medium at any angle other than perpendicular (● Figure 6-11). The bending of a light ray is known as **refraction**. With a curved surface such as a lens, the greater the curvature the greater the degree of bending and the stronger the lens. When a light ray strikes the

● FIGURE 6-9

Electromagnetic spectrum. The wavelengths in the electromagnetic spectrum range from less than 10^{-14} m to 10^4 m. The visible spectrum includes wavelengths ranging from 400 to 700 nanometers (nm).

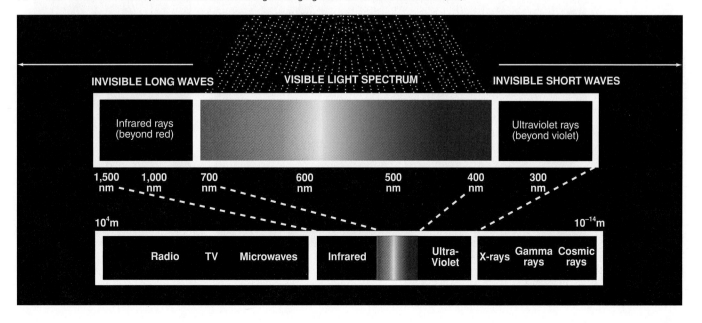

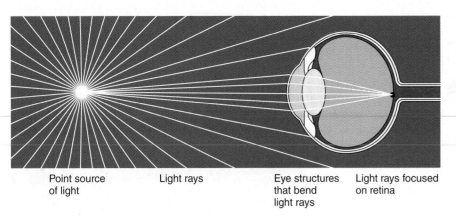

Point source of light   Light rays   Eye structures that bend light rays   Light rays focused on retina

● **FIGURE 6-10**

Focusing of diverging light rays. Diverging light rays must be bent inward to be focused.

certain refractive errors of the eye, such as nearsightedness.

EYE'S REFRACTIVE STRUCTURES

The two structures most important in the eye's refractive ability are the *cornea* and the *lens*. The curved corneal surface, the first structure light passes through as it enters the eye, contributes most extensively to the eye's total refractive ability because the difference in density at the air–cornea interface is much greater than the differences in density between the lens and the fluids surrounding it. In **astigmatism** the curvature of the cornea is uneven, so light rays are unequally refracted. The refractive ability of a person's cornea remains constant, because the curvature of the cornea never changes. In contrast, the refractive ability of the lens can be adjusted by changing its curvature as needed for near or far vision.

Rays from light sources more than 20 feet away are considered parallel by the time they reach the eye. By contrast, light rays originating from near objects are still diverging when they reach the eye. For a given refractive ability of the eye, it takes a greater distance behind the lens to bring the divergent rays of a near source to a focal point than to bring the parallel

curved surface of any object of greater density, the direction of refraction depends on the angle of the curvature (● Figure 6-12). A **convex surface** curves outward (like the outer surface of a ball), whereas a **concave surface** curves inward (like a cave). Convex surfaces converge light rays, bringing them closer together. Because convergence is essential for bringing an image to a focal point, refractive surfaces of the eye are convex. Concave surfaces diverge light rays (spread them farther apart). A concave lens is useful for correcting

● **FIGURE 6-11**

Refraction. A light ray is bent (refracted) when it strikes the surface of a medium of different density from the one in which it had been traveling (for example, moving from air into glass) at any angle other than perpendicular to the new medium's surface. (b) The pencil in the glass of water *appears* to bend. What is happening, though, is that the light rays coming to the camera (or your eyes) are bent as they pass through the water, then the glass, and then the air. Consequently the pencil appears distorted.

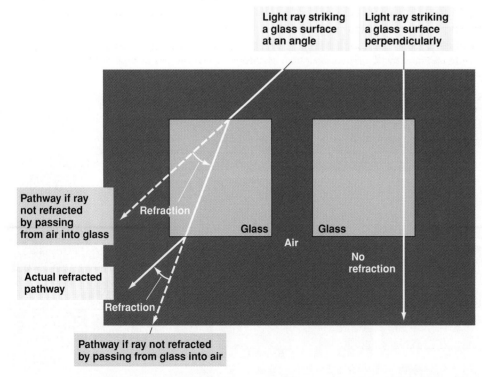

Light ray striking a glass surface at an angle

Light ray striking a glass surface perpendicularly

Pathway if ray not refracted by passing from air into glass

Refraction

Actual refracted pathway

Glass   Glass

Air

No refraction

Refraction

Pathway if ray not refracted by passing from glass into air

(a)

Bill Beatty/Visuals Unlimited

(b)

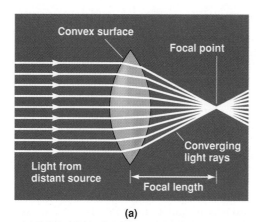

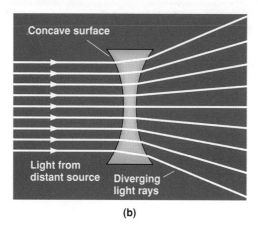

● FIGURE 6-12

Refraction by convex and concave lenses. (a) A lens with a convex surface, which converges the rays (brings them closer together). (b) A lens with a concave surface, which diverges the rays (spreads them farther apart).

rays of a far source to a focal point (● Figure 6-13 a and b). However, in a particular eye the distance between the lens and the retina always remains the same. Therefore, a greater distance beyond the lens is not available for bringing near objects into focus. Yet for clear vision the refractive structures of the eye must bring both near and far light sources into focus on the retina. If an image is focused before it reaches the retina or is not yet focused when it reaches the retina, it will be blurred (● Figure 6-14). To bring both near and far light sources into focus on the retina (that is, in the same distance), a stronger lens must be used for the near source (● Figure 6-13c). Let's see how the strength of the lens can be adjusted as needed.

▌ Accommodation increases the strength of the lens for near vision.

The ability to adjust the strength of the lens is known as **accommodation.** The strength of the lens depends on its shape, which in turn is regulated by the ciliary muscle.

● FIGURE 6-14

Comparison of images that do and do not come into focus on the retina

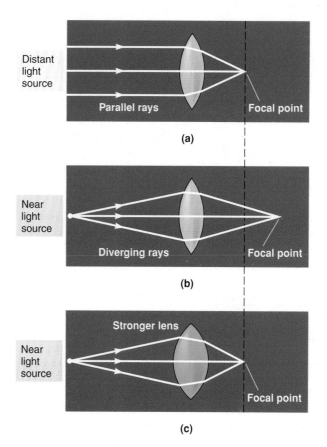

(a)

(b)

(c)

● FIGURE 6-13

Focusing of distant and near sources of light. The rays from a distant (far) light source (more than 20 feet from the eye) are parallel by the time the rays reach the eye. (b) The rays from a near light source (less than 20 feet from the eye) are still diverging when they reach the eye. A longer distance is required for a lens of a given strength to bend the diverging rays from a near light source into focus compared to the parallel rays from a distant light source. (c) To focus both a distant and a near light source in the same distance (the distance between the lens and retina), a stronger lens must be used for the near source.

PhysioEdge For an animation of this figure, click the Sight tab (page 3) in the Special Senses tutorial on the CD-ROM.

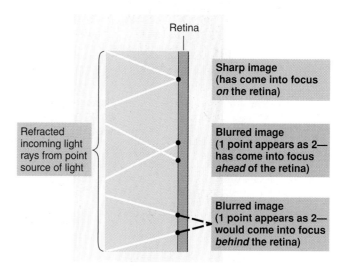

● = Points of stimulation of the retina

The **ciliary muscle** is part of the ciliary body, an anterior specialization of the choroid layer. The ciliary body has two major components: the ciliary muscle and the capillary network that produces aqueous humor (see ● Figure 6-7). The ciliary muscle is a circular ring of smooth muscle attached to the lens by **suspensory ligaments** (● Figure 6-15).

When the ciliary muscle is relaxed, the suspensory ligaments are taut, and they pull the lens into a flattened, weakly refractive shape (● Figure 6-15c). As the muscle contracts, its circumference decreases, slackening the tension in the suspensory ligaments (● Figure 6-15d). When the suspensory ligaments subject the lens to less tension, it becomes more spherical, because of its inherent elasticity. The greater curvature of the more rounded lens increases its strength, further bending light rays. In the normal eye, the ciliary muscle is relaxed and the lens is flat for far vision, but the muscle contracts to let the lens become more convex and stronger for near vision. The ciliary muscle is controlled by the autonomic nervous system, with sympathetic stimulation causing its relaxation and parasympathetic stimulation causing its contraction.

Clinical Note Throughout life, only cells at the outer edges of the lens are replaced. Cells in the center of the lens are in double jeopardy. Not only are they oldest, but they also are the farthest away from the aqueous humor, the lens's nutrient source. With advancing age, these nonrenewable central cells die and become stiff. With loss of elasticity, the lens can no longer assume the spherical shape required to accommodate for near vision. This age-related reduction in accommodative ability, **presbyopia**, affects most people by middle age (45 to 50), requiring them to resort to corrective lenses for near vision (reading).

● **FIGURE 6-15**

Mechanism of accommodation. (a) Schematic representation of suspensory ligaments extending from the ciliary muscle to the outer edge of the lens. (b) Scanning electron micrograph showing the suspensory ligaments attached to the lens. (c) When the ciliary muscle is relaxed, the suspensory ligaments are taut, putting tension on the lens so that it is flat and weak. (d) When the ciliary muscle is contracted, the suspensory ligaments become slack, reducing the tension on the lens. The lens can then assume a stronger, rounder shape because of its elasticity.

 PhysioEdge For an animation of this figure, click the Sight tab (page 4) in the Special Senses tutorial on the CD-ROM.

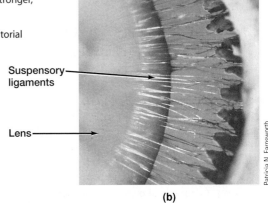

Suspensory ligaments

Lens

Patricia N. Farnsworth

(b)

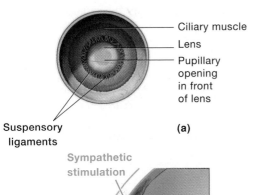

Ciliary muscle
Lens
Pupillary opening in front of lens

Suspensory ligaments

(a)

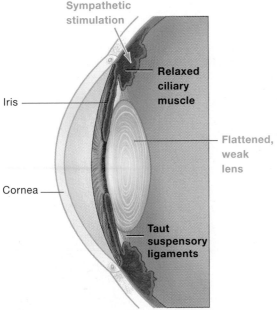

Sympathetic stimulation

Relaxed ciliary muscle

Iris

Flattened, weak lens

Cornea

Taut suspensory ligaments

(c)

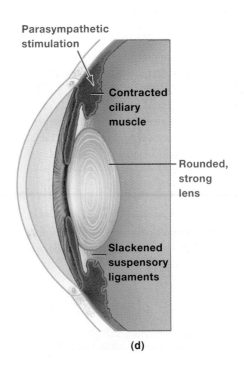

Parasympathetic stimulation

Contracted ciliary muscle

Rounded, strong lens

Slackened suspensory ligaments

(d)

The elastic fibers in the lens are normally transparent. These fibers occasionally become opaque so that light rays cannot pass through, a condition known as a **cataract**. The defective lens can usually be surgically removed and vision restored by an implanted artificial lens or by compensating eyeglasses.

Other common vision disorders are *nearsightedness (myopia)* and *farsightedness (hyperopia)*. In a normal eye (**emmetropia**) (● Figure 6-16a), a far light source is focused on the retina without accommodation, whereas the strength of the lens is increased by accommodation to bring a near source

● **FIGURE 6-16**

Emmetropia, myopia, and hyperopia. This figure compares far vision and near vision in the (a) normal eye with (b) nearsightedness and (c) farsightedness in both their (1) uncorrected and (2) corrected states. The vertical dashed line represents the normal distance of the retina from the cornea; that is, the site at which an image is brought into focus by the refractive structures in a normal eye.

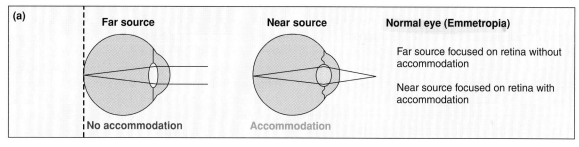

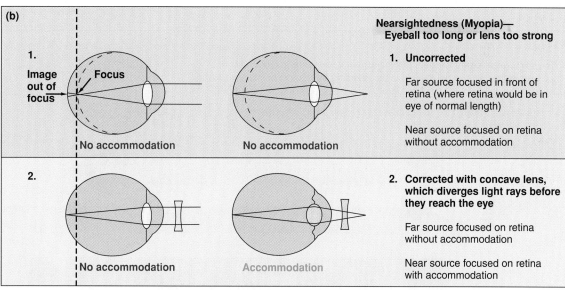

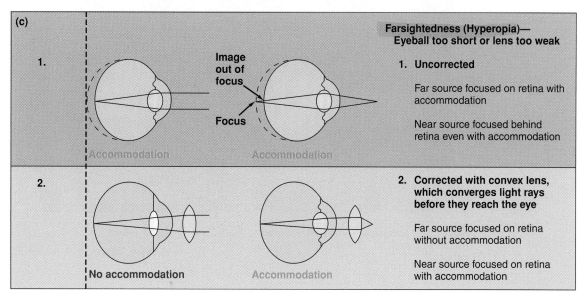

into focus. In **myopia** (● Figure 6-16b1), because the eyeball is too long or the lens is too strong, a near light source is brought into focus on the retina without accommodation (even though accommodation is normally used for near vision), whereas a far light source is focused in front of the retina and is blurry. Thus a myopic individual has better near vision than far vision, a condition that can be corrected by a concave lens (● Figure 6-16b2). With **hyperopia** (● Figure 6-16c1), either the eyeball is too short or the lens is too weak. Far objects are focused on the retina only with accommodation, whereas near objects are focused behind the retina even with accommodation and, accordingly, are blurry. Thus a hyperopic individual has better far vision than near vision, a condition that can be corrected by a convex lens (● Figure 6-16c2). Such vision tends to get worse as the person gets older because of loss of accommodative ability with the onset of presbyopia.

❚ **Light must pass through several retinal layers before reaching the photoreceptors.**

The major function of the eye is to focus light rays from the environment on the rods and cones, the photoreceptor cells of the retina. The photoreceptors then transform the light energy into electrical signals for transmission to the CNS.

The receptor-containing portion of the retina is actually an extension of the CNS and not a separate peripheral organ. During embryonic development, the retinal cells "back out" of the nervous system, so the retinal layers, surprisingly, are facing backward! The neural portion of the retina consists of three layers of excitable cells (● Figure 6-17): (1) the outermost layer (closest to the choroid) containing the **rods** and **cones,** whose light-sensitive ends face the choroid (away from the incoming light); (2) a middle layer of **bipolar cells;** and (3) an inner layer of **ganglion cells.** Axons of the ganglion cells join to form the **optic nerve,** which leaves the retina slightly off center. The point on the retina at which the optic nerve leaves and through which blood vessels pass is the **optic disc** (● Figure 6-6b). This region is often called the **blind spot**; no image can be detected in this area because it has no rods and cones (● Figure 6-18). We are normally not aware of the blind spot, because central processing somehow "fills in" the missing spot. You can discover the existence of your own blind spot by a simple demonstration (● Figure 6-19).

● **FIGURE 6-17**

Retinal layers. The retinal visual pathway extends from the photoreceptor cells (rods and cones, whose light-sensitive ends face the choroid away from the incoming light) to the bipolar cells to the ganglion cells. The horizontal and amacrine cells act locally for retinal processing of visual input.

 For an interaction related to this figure, see Media Exercise 6.2: The Eye on the CD-ROM.

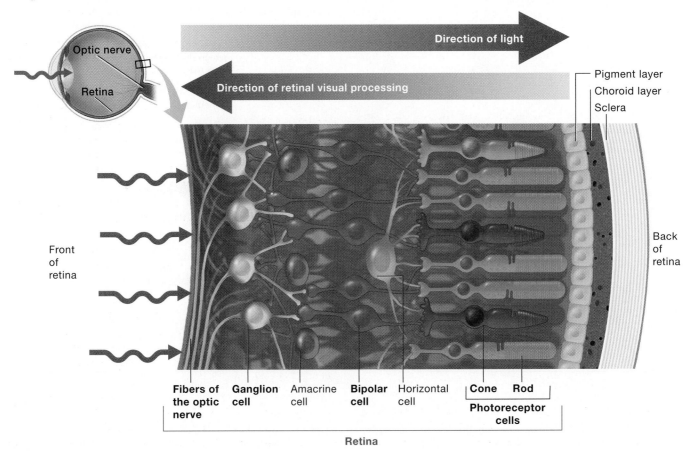

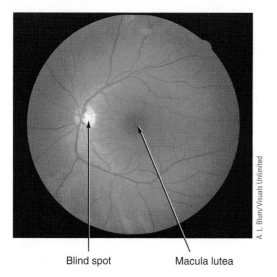

Blind spot   Macula lutea

A. L. Blum/Visuals Unlimited

● **FIGURE 6-18**

View of the retina seen through an ophthalmoscope. With an ophthalmoscope, a lighted viewing instrument, it is possible to view the optic disc (blind spot) and macula lutea within the retina at the rear of the eye.

● **FIGURE 6-19**

Demonstration of the blind spot. Find the blind spot in your left eye by closing your right eye and holding the book about 4 inches from your face. While focusing on the cross, gradually move the book away from you until the circle vanishes from view. At this time, the image of the circle is striking the blind spot of your left eye. You can similarly locate the blind spot in your right eye by closing your left eye and focusing on the circle. The cross will disappear when its image strikes the blind spot of your right eye.

Light must pass through the ganglion and bipolar layers before reaching the photoreceptors in all areas of the retina except the fovea. In the **fovea**, which is a pinhead-sized depression located in the exact center of the retina, the bipolar and ganglion cell layers are pulled aside so that light strikes the photoreceptors directly (● Figure 6-6b). This feature, coupled with the fact that only cones (which have greater acuity or discriminative ability than the rods) are found here, makes the fovea the point of most distinct vision. Thus we turn our eyes so that the image of the object at which we are looking is focused on the fovea. The area immediately surrounding the fovea, the **macula lutea**, also has a high concentration of cones and fairly high acuity (● Figure 6-18). Macular acuity, however, is less than that of the fovea, because of the overlying ganglion and bipolar cells in the macula.

Clinical Note **Macular degeneration** is the leading cause of blindness in the western hemisphere. This condition is characterized by loss of photoreceptors in the macula lutea in association with advancing age. Its victims have "donut" vision. They suffer a loss in the middle of their visual field, which normally has the highest acuity, and are left with only the less distinct peripheral vision.

▌ **Phototransduction by retinal cells converts light stimuli into neural signals.**

Photoreceptors (rod and cone cells) consist of three parts (● Figure 6-20a):

1. An *outer segment,* which lies closest to the eye's exterior, facing the choroid. It detects the light stimulus.
2. An *inner segment,* which lies in the middle of the photoreceptor's length. It contains the metabolic machinery of the cell.
3. A *synaptic terminal,* which lies closest to the eye's interior, facing the bipolar cells. It transmits the signal generated in the photoreceptor on light stimulation to these next cells in the visual pathway.

The outer segment, which is rod shaped in rods and cone shaped in cones (● Figure 6-20a), consists of stacked, flattened, membranous discs containing an abundance of light-sensitive *photopigment* molecules. Each retina has about 150 million photoreceptors, and over a billion photopigment molecules may be packed into the outer segment of each photoreceptor.

Photopigments undergo chemical alterations when activated by light. Through a series of steps, this light-induced change and subsequent activation of the photopigment bring about a receptor potential that ultimately leads to the generation of action potentials, which transmit this information to the brain for visual processing. A photopigment consists of two components: **opsin**, a protein that is an integral part of the disc membrane; and **retinene**, a derivative of vitamin A that is bound within the interior of the opsin molecule (● Figure 6-20b). Retinene is the light-absorbing part of the photopigment. There are four different photopigments, one in the rods and one in each of three types of cones. These four photopigments differentially absorb various wavelengths of light. **Rhodopsin**, the rod photopigment, absorbs all visible wavelengths. Using visual input from the rods, the brain cannot discriminate between various wavelengths in the visible spectrum. Therefore, rods provide vision only in shades of gray by detecting different intensities, not different colors. The photopigments in the three types of cones—**red, green,** and **blue cones**—respond selectively to various wavelengths of light, making color vision possible.

Phototransduction, the process of converting light stimuli into electrical signals, is basically the same for all photoreceptors. When retinene absorbs light, it changes shape (● Figure 6-20b). This change in conformation activates the photopigment, producing a receptor potential that passively spreads from the outer segment to the synaptic terminal of the photoreceptor (● Figure 6-21). Here the potential change alters the release of neurotransmitter from the synaptic terminal, graded according to light intensity.

Photoreceptors synapse with bipolar cells, the next layer of excitable cells in the retina. The potential of bipolar cells is influenced by the neurotransmitter released from the photoreceptors. Bipolar cells in turn terminate on the ganglion cells, whose axons form the optic nerve for transmission of signals to the brain. Bipolar cells display graded potentials similar to the photoreceptors. Action potentials originate for the first time in the visual pathway in the ganglion cells, the first neurons in the chain that must propagate the visual message over long distance to the visual cortex in the occipital lobe of the brain (see p. 118). Neural messages are sent to the visual cortex only from photoreceptors that are "turned on" sufficiently by light to bring to threshold the ganglion cells to which they are "wired." Thus the resulting image perceived by the brain depends on the pattern of light striking the photoreceptors.

The altered photopigments are restored by enzyme action to their original conformation in the dark. Subsequently, the membrane potential and rate of neurotransmitter release of the photoreceptor are returned to their unexcited state, and no further action potentials are transmitted to the visual cortex.

▮ Rods provide indistinct gray vision at night, whereas cones provide sharp color vision during the day.

The retina contains more than 30 times more rods than cones (100 million rods compared to 3 million cones per eye). Cones are most abundant in the macula lutea in the center of the retina. From this point outward, the concentration of cones decreases and the concentration of rods increases. Rods are most abundant in the periphery. We have examined the similar way in which phototransduction takes place in rods and cones. Now we will focus on the differences between these photoreceptors. You already know that rods provide vision only in shades of gray, whereas cones provide color vision. The capabilities of the rods and cones also differ in other respects because of a difference in the "wiring patterns" between these photoreceptor types and other retinal neuronal layers (▲ Table 6-2). Cones have low sensitivity to light, being "turned on" only by bright daylight, but they have high acuity (sharpness; ability to distinguish between two nearby points). Thus cones provide sharp vision with high resolution for fine detail. Humans use cones for day vision, which is in color and distinct. Rods, in contrast, have low acuity but high sensitivity, so they

● **FIGURE 6-20**

Photoreceptors. (a) Schematic representation of the three parts of the rods and cones, the eye's photoreceptors. Note in the outer segment of the rod and cone the stacked, flattened, membranous discs, which contain an abundance of photopigment molecules. (b) A photopigment, such as rhodopsin, depicted here and found in rods, consists of the membrane protein opsin and the vitamin-A derivative retinene. In the dark, retinene is bound within the interior of opsin and the photopigment is inactivated. In the light, retinene changes shape and activates the photopigment.

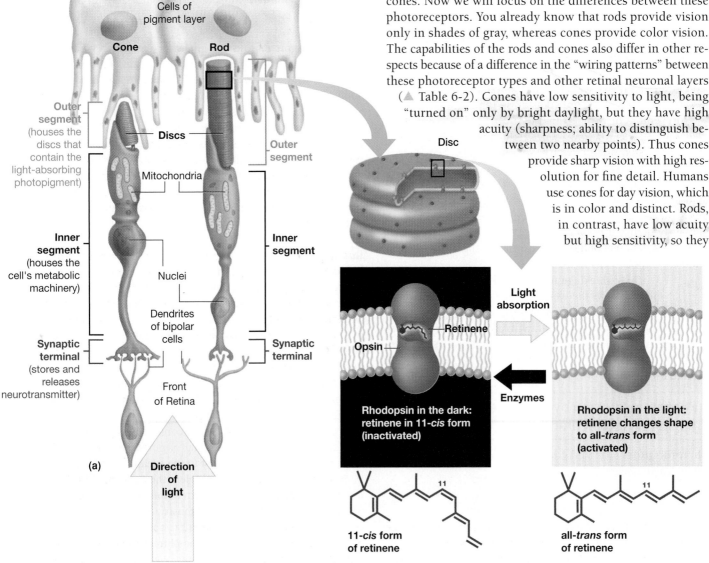

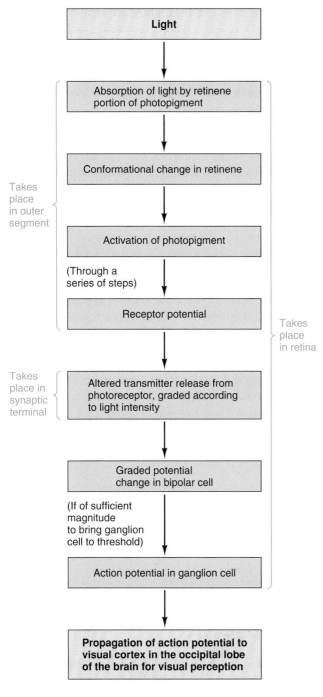

● FIGURE 6-21

Phototransduction

Light
↓
Absorption of light by retinene portion of photopigment
↓
Conformational change in retinene
↓
Activation of photopigment
↓
Receptor potential
↓
Altered transmitter release from photoreceptor, graded according to light intensity
↓
Graded potential change in bipolar cell
↓
Action potential in ganglion cell
↓
Propagation of action potential to visual cortex in the occipital lobe of the brain for visual perception

Takes place in outer segment

(Through a series of steps)

Takes place in retina

Takes place in synaptic terminal

(If of sufficient magnitude to bring ganglion cell to threshold)

TABLE 6-2

Properties of Rod Vision and Cone Vision

RODS	CONES
100 million per retina	3 million per retina
Vision in shades of gray	Color vision
High sensitivity	Low sensitivity
Low acuity	High acuity
Night vision	Day vision
More numerous in periphery	Concentrated in fovea

respond to the dim light of night. You can see at night with your rods but at the expense of color and distinctness.

▌The sensitivity of the eyes can vary markedly through dark and light adaptation.

The eyes' sensitivity to light depends on the amount of photopigment present in the rods and cones. When you go from bright sunlight into darkened surroundings, you cannot see anything at first, but gradually you begin to distinguish ob-

jects as a result of the process of **dark adaptation.** Breakdown of photopigments during exposure to sunlight tremendously decreases photoreceptor sensitivity. For example, a reduction in rhodopsin content of only 0.6% from its maximum value decreases rod sensitivity approximately 3000 times. In the dark, the photopigments broken down during light exposure are gradually regenerated. As a result, the sensitivity of your eyes gradually increases so you can begin to see in the darkened surroundings. However, only the highly sensitive, rejuvenated rods are "turned on" by the dim light.

Conversely, when you move from the dark to the light (for example, leaving a movie theater and entering the bright sunlight), at first your eyes are very sensitive to the dazzling light. With little contrast between lighter and darker parts, the entire image appears bleached. As some of the photopigments are rapidly broken down by the intense light, the sensitivity of the eyes decreases and normal contrasts can once again be detected, a process known as **light adaptation.** The rods are so sensitive to light that enough rhodopsin is broken down in bright light to essentially "burn out" the rods; that is, after the rod photopigments have already been broken down by the bright light, they no longer can respond to the light. Furthermore, a central neural adaptive mechanism switches the eye from the rod system to the cone system on exposure to bright light. Therefore, only the less sensitive cones are used for day vision.

Researchers estimate that our eyes' sensitivity can change as much as 1 million times as they adjust to various levels of illumination through dark and light adaptation. These adaptive measures are also enhanced by pupillary reflexes that adjust the amount of available light permitted to enter the eye.

Clinical Note Because retinene, one of the photopigment components, is a derivative of vitamin A, adequate amounts of this nutrient must be available for the ongoing resynthesis of photopigments. **Night blindness** occurs as a result of dietary deficiencies of vitamin A. Although photopigment concentrations in both rods and cones are reduced in this condition, there is still enough cone photopigment to respond to the intense stimulation of bright light, except in the most severe cases. However, even modest reductions in rhodopsin content can decrease the sensitivity of rods so much that they cannot respond to dim light. The person can see in the day using cones but cannot see at night because the rods are no longer functional. Thus carrots are "good for your eyes" because they are rich in vitamin A.

Color vision depends on the ratios of stimulation of the three cone types.

Vision depends on stimulation of retinal photoreceptors by light. Certain objects in the environment such as the sun, fire, and lightbulbs, emit light. But how do you see objects such as chairs, trees, and people, which do not emit light? The pigments in various objects selectively absorb particular wavelengths of light transmitted to them from light-emitting sources, and the unabsorbed wavelengths are reflected from the objects' surfaces. These reflected light rays enable you to see the objects. An object perceived as blue absorbs the longer red and green wavelengths of light and reflects the shorter blue wavelengths, which can be absorbed by the photopigment in the eyes' blue cones, thereby activating them.

Each cone type is most effectively activated by a particular wavelength of light in the range of color indicated by its name—blue, green, or red. However, cones also respond in varying degrees to other wavelengths (● Figure 6-22). **Color vision**, the perception of the many colors of the world, depends on the three cone types' various *ratios of stimulation* in response to different wavelengths. A wavelength perceived as blue does not stimulate red or green cones at all but excites blue cones maximally (the percentage of maximal stimulation for red, green, and blue cones, respectively, is 0:0:100).

The sensation of yellow, in comparison, arises from a stimulation ratio of 83:83:0, red and green cones each being stimulated 83% of maximum, while blue cones are not excited at all. The ratio for green is 31:67:36, and so on, with various combinations giving rise to the sensation of all the different colors. White is a mixture of all wavelengths of light, whereas black is the absence of light.

The extent to which each of the cone types is excited is coded and transmitted in separate parallel pathways to the brain. A distinct color vision center in the primary visual cortex combines and processes these inputs to generate the perception of color, taking into consideration the object in comparison with its background. The concept of color is therefore in the mind of the beholder. Most of us agree on what color we see because we have the same types of cones and use similar neural pathways for comparing their output. Occasionally, however, individuals lack a particular cone type, so their color vision is a product of the differential sensitivity of only two types of cones, a condition known as **color blindness**. Not only do color-defective individuals perceive certain colors differently, but they are also unable to distinguish as many varieties of colors (● Figure 6-23). For example, people with certain color defects cannot distinguish between red and green. At a traffic light they can tell which light is "on" by its intensity, but they must rely on the position of the bright light to know whether to stop or go.

Visual information is separated before reaching the visual cortex.

The field of view that can be seen without moving the head is known as the **visual field**. Because of the pattern of wiring between the eyes and the visual cortex, the left half of the cortex receives information only from the right half of the vi-

● **FIGURE 6-22**

Sensitivity of the three types of cones to different wavelengths. The ratios of stimulation of the three cone types are shown for three sample colors.

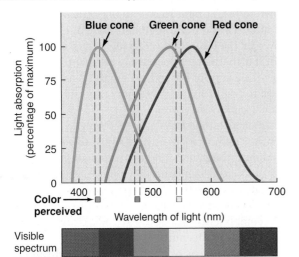

Color perceived	Percent of maximum stimulation		
	Red cones	Green cones	Blue cones
	0	0	100
	31	67	36
	83	83	0

● **FIGURE 6-23**

Color blindness chart. People with red–green color blindness cannot detect the number 29 in this chart.

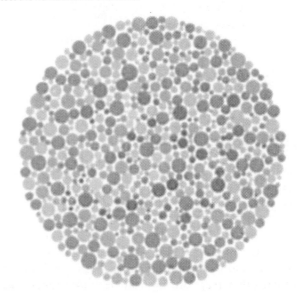

sual field as detected by both eyes, and the right half receives input only from the left half of the visual field of both eyes.

As light enters the eyes, light rays from the left half of the visual field fall on the right half of the retina of both eyes (the medial or inner half of the left retina and the lateral or outer half of the right retina) (● Figure 6-24a). Similarly, rays from the right half of the visual field reach the left half of each retina (the lateral half of the left retina and the medial half of the right retina). Each optic nerve exiting the retina carries information from both halves of the retina it serves. This information is separated as the optic nerves meet at the **optic chiasm** located underneath the hypothalamus (*chiasm* means "cross") (see ● Figure 5-5b, p. 115). Within the optic chiasm, the fibers from the medial half of each retina cross to the opposite side, but those from the lateral half remain on the original side. The reorganized bundles of fibers leaving the optic chiasm are known as **optic tracts**. Each optic tract carries information from the lateral half of one retina and the medial half of the other retina. Therefore, this partial crossover brings together from the two eyes fibers that carry information from the same half of the visual field. Each optic tract, in turn, delivers to the half of the brain on its same side information about the opposite half of the visual field. A knowledge of these pathways can facilitate diagnosis of visual defects arising from interruption of the visual pathway at various points (● Figure 6-24b).

Before we move on to how the brain processes visual information, take a look at ▲ Table 6-3, which summarizes the functions of the various components of the eyes.

▌The thalamus and visual cortexes elaborate the visual message.

The first stop in the brain for information in the visual pathway is the thalamus (● Figure 6-24a). It separates information received from the eyes and relays it via fiber bundles known as **optic radiations** to different zones in the cortex, each of which processes different aspects of the visual stimulus (for example, color, form, depth, movement). This sorting process is no small task, because each optic nerve contains more than a million fibers carrying information from the photoreceptors in one retina. This is more than all the afferent fibers carrying somatosensory input from all the other regions of the body! Researchers estimate that hundreds of millions of neurons occupying about 30% of the cortex participate in visual processing, compared to 8% devoted to touch perception and 3% to hearing.

DEPTH PERCEPTION

Although each half of the visual cortex receives information simultaneously from the same part of the visual field as received by both eyes, the messages from the two eyes are not identical. Each eye views an object from a slightly different vantage point, even though the overlap is tremendous. The overlapping area seen by both eyes at the same time is known as the **binocular** ("two-eyed") field of vision, which is important for **depth perception**. The brain uses the slight disparity

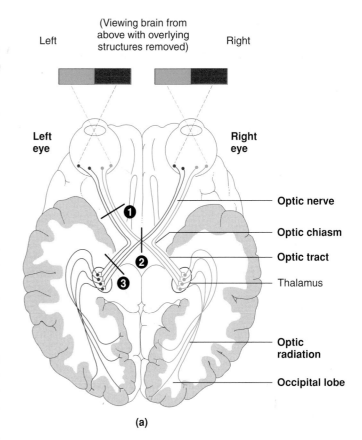

(Viewing brain from above with overlying structures removed)

Left   Right

Left eye

Right eye

— Optic nerve
— Optic chiasm
— Optic tract
— Thalamus

— Optic radiation

— Occipital lobe

(a)

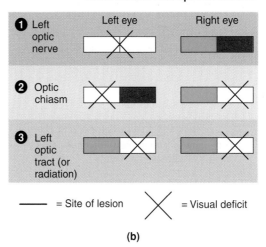

Visual deficits with specific lesions

❶ Left optic nerve

❷ Optic chiasm

❸ Left optic tract (or radiation)

—— = Site of lesion   ✕ = Visual deficit

(b)

● **FIGURE 6-24**

The visual pathway and visual deficits associated with lesions in the pathway. (a) Visual pathway. Note that the left half of the visual cortex in the occipital lobe receives information from the right half of the visual field of both eyes (in blue), and the right half of the cortex receives information from the left half of the visual field of both eyes (in red). (b) Visual deficits with specific lesions in the visual pathway. Each visual deficit illustrated is associated with a lesion at the corresponding numbered point in the visual pathway in part (a).

---

in the information received from the two eyes to estimate distance, allowing you to perceive three-dimensional objects in spatial depth. Some depth perception is possible using only one eye, based on experience and comparison with other

Functions of the Major Components of the Eye

STRUCTURE (in alphabetical order)	LOCATION	FUNCTION
Aqueous Humor	Anterior cavity between cornea and lens	Clear watery fluid that is continually formed and carries nutrients to the cornea and lens
Bipolar Cells	Middle layer of nerve cells in retina	Important in retinal processing of light stimulus
Blind Spot	Point slightly off-center on retina where optic nerve exits; lacks photoreceptors (also known as *optic disc*)	Route for passage of optic nerve and blood vessels
Choroid	Middle layer of eye	Pigmented to prevent scattering of light rays in eye; contains blood vessels that nourish retina; anteriorly specialized to form ciliary body and iris
Ciliary Body	Specialized anterior derivative of the choroid layer; forms a ring around the outer edge of the lens	Produces aqueous humor and contains ciliary muscle
Ciliary Muscle	Circular muscular component of ciliary body; attached to lens by means of suspensory ligaments	Important in accommodation
Cones	Photoreceptors in outermost layer of retina	Responsible for high acuity, color, and day vision
Cornea	Anterior clear outermost layer of eye	Contributes most extensively to eye's refractive ability
Fovea	Exact center of retina	Region with greatest acuity
Ganglion Cells	Inner layer of nerve cells in retina	Important in retinal processing of light stimulus; form optic nerve
Iris	Visible pigmented ring of muscle within aqueous humor	Varies size of pupil by variable contraction; responsible for eye color
Lens	Between aqueous humor and vitreous humor; attaches to ciliary muscle by suspensory ligaments	Provides variable refractive ability during accommodation
Macula Lutea	Area immediately surrounding the fovea	Has high acuity because of abundance of cones
Optic Disc	(see entry for blind spot)	
Optic Nerve	Leaves each eye at optic disc (blind spot)	First part of visual pathway to the brain
Pupil	Anterior round opening in middle of iris	Permits variable amounts of light to enter eye
Retina	Innermost layer of eye	Contains the photoreceptors (rods and cones)
Rods	Photoreceptors in outermost layer of retina	Responsible for high-sensitivity, black-and-white, and night vision
Sclera	Tough outer layer of eye	Protective connective tissue coat; forms visible white part of eye; anteriorly specialized to form cornea
Suspensory Ligaments	Suspended between ciliary muscle and lens	Important in accommodation
Vitreous Humor	Between lens and retina	Semifluid, jellylike substance that helps maintain spherical shape of eye

cues. For example, if your one-eyed view includes a car and a building and the car is much larger, you correctly interpret that the car must be closer to you than the building is.

HIERARCHY OF VISUAL CORTICAL PROCESSING

Within the cortex, visual information is first processed in the primary visual cortex, then is sent to higher-level visual areas for even more complex processing and abstraction. Each level of cortical visual neurons has increasingly greater capacity for abstraction of information built up from the increasing convergence of input from lower-level neurons. In this way, the cortex transforms the dotlike pattern of photoreceptors stimulated to varying degrees by varying light intensities in the retinal image, into information about depth, position, orientation, movement, contour, and length. Other aspects of this information, such as color perception, are processed simultaneously. How and where the entire image is finally put together is still unresolved. Only when these separate bits of processed information are integrated by higher visual regions is a reassembled picture of the visual scene perceived. This is similar to the blobs of paint on an artist's palette versus the finished portrait; the separate pigments do not represent a portrait of a face until they are appropriately integrated on a canvas.

▌ Visual input goes to other areas of the brain not involved in vision perception.

Not all fibers in the visual pathway terminate in the visual cortexes. Some are projected to other regions of the brain for purposes other than direct vision perception. Examples of nonsight activities dependent on input from the rods and cones include (1) contribution to cortical alertness and attention, (2) control of pupil size, and (3) control of eye movements. Each eye is equipped with a set of six **external eye muscles** that position and move the eye so that it can better locate, see, and track objects. Eye movements are among the fastest, most discretely controlled movements of the body.

Now let's shift attention from the eyes to the ears.

 Click on the Media Exercises menu of the CD-ROM and work Media Exercise 6.2: The Eye to test your understanding of the previous section.

EAR: HEARING AND EQUILIBRIUM

Each **ear** consists of three parts: the *external,* the *middle,* and the *inner ear* (● Figure 6-25). The external and middle portions of the ear transmit airborne sound waves to the fluid-filled inner ear, amplifying the sound energy in the process. The inner ear houses two different sensory systems: the cochlea, which contains the receptors for conversion of sound waves into nerve impulses, making hearing possible; and the vestibular apparatus, which is necessary for the sense of equilibrium.

● **FIGURE 6-25**

Anatomy of the ear

 For an interaction related to this figure, see Media Exercise 6.3: The Ear on the CD-ROM.

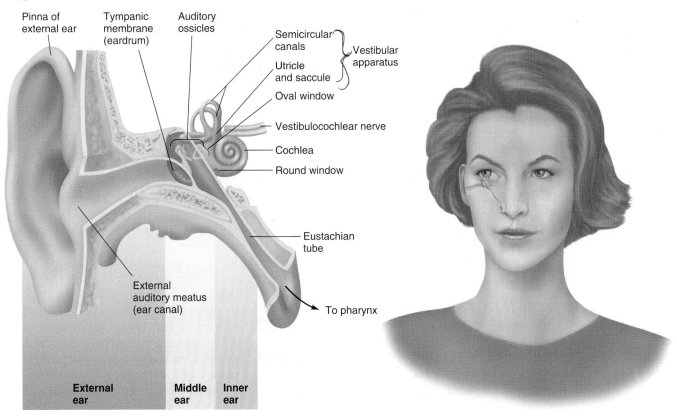

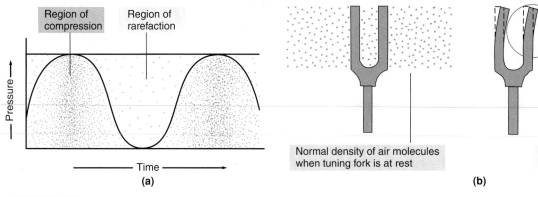

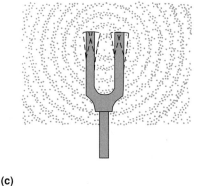

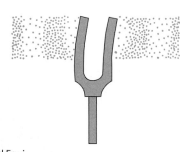

● FIGURE 6-26

Formation of sound waves. (a) Sound waves are alternating regions of compression and rarefaction of air molecules. (b) A vibrating tuning fork sets up sound waves as the air molecules ahead of the advancing arm of the tuning fork are compressed while the molecules behind the arm are rarefied. (c) Disturbed air molecules bump into molecules beyond them, setting up new regions of air disturbance more distant from the original source of sound. In this way, sound waves travel progressively farther from the source, even though each individual air molecule travels only a short distance when it is disturbed. The sound wave dies out when the last region of air disturbance is too weak to disturb the region beyond it.

For an animation of this figure, click the Hearing and Equilibrium tab in the Special Senses tutorial on the CD-ROM.

(c)

■ Sound waves consist of alternate regions of compression and rarefaction of air molecules.

Hearing is the neural perception of sound energy. Hearing involves two aspects: identification of the sounds ("what") and their localization ("where"). We will first examine the characteristics of sound waves, then how the ears and brain process sound input to accomplish hearing.

Sound waves are traveling vibrations of air that consist of regions of high pressure caused by compression of air molecules alternating with regions of low pressure caused by rarefaction of the molecules (● Figure 6-26a). Any device capable of producing such a disturbance pattern in air molecules is a source of sound. A simple example is a tuning fork. When a tuning fork is struck, its prongs vibrate. As a prong of the fork moves in one direction (● Figure 6-26b), air molecules ahead of it are pushed closer together, or compressed, increasing the pressure in this area. Simultaneously, as the prong moves forward the air molecules behind the prong spread out or are rarefied, lowering the pressure in that region. As the prong moves in the opposite direction, an opposite wave of compression and rarefaction is created. Even though individual molecules are moved only short distances as the tuning fork vibrates, alternating waves of compression and rarefaction spread out considerable distances in a rippling fashion. Disturbed air molecules disturb other molecules in adjacent regions, setting up new regions of compression and rarefaction, and so on (● Figure 6-26c). Sound energy is gradually dissipated as sound waves travel farther from the original

sound source. The intensity of the sound decreases, until it finally dies out when the last sound wave is too weak to disturb the air molecules around it.

Sound waves can also travel through media other than air, such as water. They do so less efficiently, however; greater pressures are required to cause movements of fluid than movements of air because of the fluid's greater inertia (resistance to change).

Sound is characterized by its pitch (tone), intensity (loudness), and timbre (quality) (● Figure 6-27):

• The **pitch**, or **tone**, of a sound (for example, whether it is a *C* or a *G* note) is determined by the *frequency* of vibrations. The greater the frequency of vibration, the higher the pitch. Human ears can detect sound waves with frequencies from 20 to 20,000 cycles per second but are most sensitive to frequencies between 1000 and 4000 cycles per second.

• The **intensity**, or **loudness**, of a sound depends on the *amplitude* of the sound waves, or the pressure differences between a high-pressure region of compression and a low-pressure region of rarefaction. Within the hearing range, the greater the amplitude, the louder the sound. Human ears can detect a wide range of sound intensities, from the slightest whisper to the painfully loud takeoff of a jet. Loudness is measured in **decibels (dB)**, which are a logarithmic measure of intensity compared with the faintest sound that can be heard—the **hearing threshold**. Because of the logarithmic relationship, every 10 decibels indicates a 10fold increase in loudness. A few examples of common sounds illustrate the magnitude of these increases (▲ Table 6-4). Note that the rustle of leaves at 10 dB

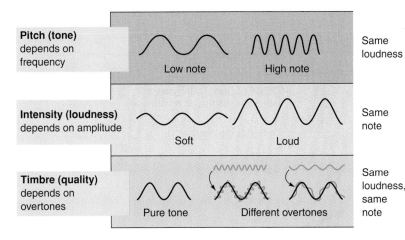

Pitch (tone) depends on frequency	Low note High note — Same loudness
Intensity (loudness) depends on amplitude	Soft Loud — Same note
Timbre (quality) depends on overtones	Pure tone Different overtones — Same loudness, same note

● **FIGURE 6-27**

Properties of sound waves

is 10 times louder than hearing threshold, but the sound of a jet taking off at 150 dB is a quadrillion (a million billion) times, not 150 times, louder than the faintest audible sound. Sounds greater than 100 dB can permanently damage the sensitive sensory apparatus in the cochlea.

- The **timbre**, or **quality**, of a sound depends on its *overtones,* which are additional frequencies superimposed on top of the fundamental pitch or tone. A tuning fork has a pure tone, but most sounds lack purity. For example, complex mixtures of overtones impart different sounds to different instruments playing the same note (a *C* note on a trumpet sounds

▲ **TABLE 6-4**

Relative Magnitude of Common Sounds

SOUND	LOUDNESS IN DECIBELS (dB)	COMPARISON TO FAINTEST AUDIBLE SOUND (hearing threshold)
Rustle of Leaves	10 dB	10 times louder
Ticking of Watch	20 dB	100 times louder
Hush of Library	30 dB	1 thousand times louder
Normal Conversation	60 dB	1 million times louder
Food Blender	90 dB	1 billion times louder
Loud Rock Concert	120 dB	1 trillion times louder
Takeoff of Jet Plane	150 dB	1 quadrillion times louder

different from *C* on a piano). Overtones are likewise responsible for characteristic differences in voices. Timbre enables the listener to distinguish the source of sound waves, because each source produces a different pattern of overtones. Thanks to timbre, you can tell whether it is your mother or girlfriend calling on the telephone before you say the wrong thing.

▌ The external ear plays a role in sound localization.

The specialized receptors for sound are located in the fluid-filled inner ear. Airborne sound waves must therefore be channeled toward and transferred into the inner ear, compensating in the process for the loss in sound energy that naturally occurs as sound waves pass from air into water. This function is performed by the external ear and the middle ear.

The **external ear** (● Figure 6-25) consists of the *pinna* (ear), *external auditory meatus* (ear canal), and *tympanic membrane* (eardrum). The **pinna**, a prominent skin-covered flap of cartilage, collects sound waves and channels them down the external ear canal. Many species (dogs, for example) can cock their ears in the direction of sound to collect more sound waves, but human ears are relatively immobile. Because of its shape, the pinna partially shields sound waves that approach the ear from the rear and thus helps a person distinguish whether a sound is coming from directly in front or behind.

Sound localization for sounds approaching from the right or left is determined by two cues. First, the sound wave reaches the ear closer to the sound source slightly before it arrives at the farther ear. Second, the sound is less intense as it reaches the farther ear, because the head acts as a sound barrier that partially disrupts the propagation of sound waves. The auditory cortex integrates all these cues to determine the location of the sound source. It is difficult to localize sound with only one ear.

The entrance to the **ear canal** is guarded by fine hairs. The skin lining the canal contains modified sweat glands that produce **cerumen** (earwax), a sticky secretion that traps fine foreign particles. Together the hairs and earwax help prevent airborne particles from reaching the inner portions of the ear canal, where they could accumulate or injure the tympanic membrane and interfere with hearing.

▌ The tympanic membrane vibrates in unison with sound waves in the external ear.

The **tympanic membrane**, which is stretched across the entrance to the middle ear, vibrates when struck by sound waves. The alternating higher- and lower-pressure regions of a sound wave cause the exquisitely sensitive eardrum to bow inward and outward in unison with the wave's frequency.

For the membrane to be free to move as sound waves strike it, the resting air pressure on both sides of the tympanic membrane must be equal. The outside of the eardrum is exposed to atmospheric pressure that reaches it through the ear

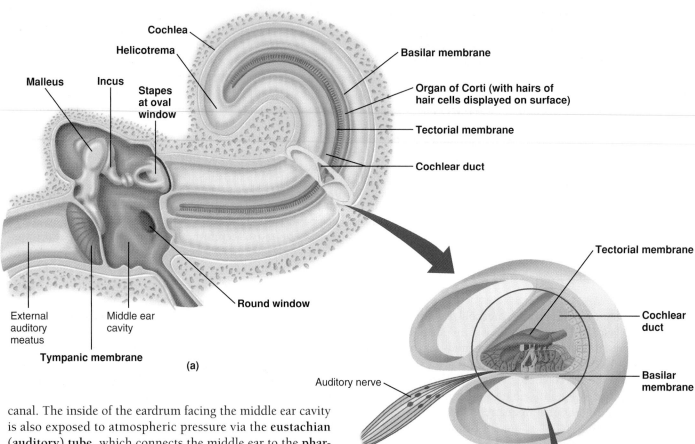

Cochlea

Helicotrema

Basilar membrane

Malleus **Incus**

Stapes at oval window

Organ of Corti (with hairs of hair cells displayed on surface)

Tectorial membrane

Cochlear duct

External auditory meatus

Middle ear cavity

Round window

Tympanic membrane

(a)

Auditory nerve

Tectorial membrane

Cochlear duct

Basilar membrane

(b)

Hairs

Hair cells

Tectorial membrane

Supporting cell

Nerve fibers

Basilar membrane

(c)

● **FIGURE 6-28**

Middle ear and cochlea. (a) Gross anatomy of the middle ear and cochlea, with the cochlea "unrolled." (b) Cross section of the cochlea. (c) Enlargement of the organ of Corti.

canal. The inside of the eardrum facing the middle ear cavity is also exposed to atmospheric pressure via the **eustachian (auditory) tube**, which connects the middle ear to the **pharynx** (back of the throat) (● Figure 6-25). The eustachian tube is normally closed, but it can be pulled open by yawning, chewing, and swallowing. Such opening permits air pressure within the middle ear to equilibrate with atmospheric pressure so that pressures on both sides of the tympanic membrane are equal. During rapid external pressure changes (for example, during air flight), the eardrum bulges painfully as the pressure outside the ear changes, while the pressure in the middle ear remains unchanged. Opening the eustachian tube by yawning allows the pressure on both sides of the tympanic membrane to equalize, relieving the pressure distortion as the eardrum "pops" back into place. Infections originating in the throat sometimes spread through the eustachian tube to the middle ear. The resulting fluid accumulation in the middle ear not only is painful but also interferes with sound conduction across the middle ear.

▌ **The middle ear bones convert tympanic membrane vibrations into fluid movements in the inner ear.**

The **middle ear** transfers the vibratory movements of the tympanic membrane to the fluid of the inner ear. This transfer is facilitated by a movable chain of three small bones or **ossicles** (the **malleus, incus,** and **stapes**) that extend across the middle ear (● Figure 6-28a). The first bone, the malleus, is attached to the tympanic membrane, and the last bone, the stapes, is attached to the oval window, the entrance into the fluid-filled cochlea. As the tympanic membrane vibrates in response to sound waves, the chain of bones is set into mo-

tion at the same frequency, transmitting this frequency of movement from the tympanic membrane to the oval window. The resulting pressure on the oval window with each vibration produces wavelike movements in the inner ear fluid at the same frequency as the original sound waves. However, as noted earlier, greater pressure is required to set fluid in motion. Two mechanisms related to the ossicular system amplify the pressure of the airborne sound waves to set up fluid vibrations in the cochlea. First, because the surface area of the tympanic membrane is much larger than that of the oval window, pressure is increased as force exerted on the tympanic membrane is conveyed to the oval window (pressure = force/unit area). Second, the lever action of the ossicles provides an additional mechanical advantage. Together, these mechanisms increase the force exerted on the oval window by 20 times what it would be if the sound wave struck the oval window directly. This additional pressure is sufficient to set the cochlear fluid in motion.

The cochlea contains the organ of Corti, the sense organ for hearing.

The pea-sized, snail-shaped **cochlea**, the "hearing" portion of the inner ear, is a coiled tubular system lying deep within the temporal bone (● Figure 6-25). It is easier to understand the functional components of the cochlea by "unrolling" it, as shown in ● Figure 6-28a. The cochlea is divided throughout most of its length into three fluid-filled longitudinal compartments. A blind-ended **cochlear duct**, which constitutes the middle compartment, tunnels lengthwise through the center of the cochlea, almost but not quite reaching its end. The upper compartment is sealed from the middle ear cavity by the oval window, to which the stapes is attached. Another small membrane-covered opening, the **round window**, seals the lower compartment from the middle ear. The region beyond the tip of the cochlear duct where the fluid in the upper and lower compartments is continuous is called the **helicotrema**. The **basilar membrane** forms the floor of the cochlear duct, separating it from the lower compartment. The basilar membrane is especially important because it bears the **organ of Corti**, the sense organ for hearing.

Hair cells in the organ of Corti transduce fluid movements into neural signals.

The organ of Corti, which rests on top of the basilar membrane throughout its full length, contains **hair cells** that are the receptors for sound. (● Figure 6-28c). Hair cells generate neural signals when their surface hairs are mechanically deformed in association with fluid movements in the inner ear. These hairs are mechanically embedded in the **tectorial membrane**, an awninglike projection overhanging the organ of Corti throughout its length (● Figure 6-28b and c).

The pistonlike action of the stapes against the oval window sets up pressure waves in the upper compartment. Because fluid is incompressible, pressure is dissipated in two ways as the stapes causes the oval window to bulge inward: (1) displacement of the round window and (2) deflection of the basilar membrane (● Figure 6-29a). In the first of these pathways, the pressure wave pushes the fluid forward in the upper compartment, then around the helicotrema, and into the lower compartment, where it causes the round window to bulge outward into the middle ear cavity to compensate for the pressure increase. As the stapes rocks backward and pulls

the oval window outward toward the middle ear, the fluid shifts in the opposite direction, displacing the round window inward. This pathway does not result in sound reception; it just dissipates pressure. Pressure waves of frequencies associated with sound reception take a "shortcut." Pressure waves in the upper compartment are transferred into the cochlear duct and then through the basilar membrane into the lower compartment, where they cause the round window to alternately bulge outward and inward. The main difference in this pathway is that transmission of pressure waves through the basilar membrane causes this membrane to move up and down, or vibrate, in synchrony with the pressure wave. Because the organ of Corti rides on the basilar membrane, the hair cells also move up and down as the basilar membrane oscillates. Because the hairs of these receptor cells are embedded in the stiff, stationary tectorial membrane, they are bent back and forth when the oscillating basilar membrane shifts their position in relationship to the tectorial membrane (● Figure 6-30). This back-and-forth mechanical deformation of the hairs alternately opens and closes mechanically gated ion channels (see p. 72) in the hair cell, resulting in alternating depolarizing and hyperpolarizing potential changes—the receptor potential—at the same frequency as the original sound stimulus (● Figure 6-31). This receptor potential in the hair cells is converted into action potentials in the afferent nerve fibers that make up the **auditory (cochlear) nerve**, which conducts the impulses to the brain. The neural pathway between the organ of Corti and the auditory cortex in the temporal lobe of the brain (see p. 118) involves several synapses en route, the most notable of which are in the brain stem and the thalamus. The brain stem uses the auditory input for alertness and arousal. The thalamus sorts and relays the signals upward.

Unlike the visual pathways, auditory signals from each ear are transmitted to both temporal lobes because the fibers partially cross over in the brain stem. For this reason, a disruption of the auditory pathways on one side beyond the brain stem does not affect hearing in either ear to any extent.

The primary auditory cortex appears to perceive discrete sounds, whereas the surrounding higher-order auditory cortex integrates the separate sounds into a coherent, meaningful pattern. Think about the complexity of the task accomplished by your auditory system. When you are at a concert, your organ of Corti responds to the simultaneous mixture of the instruments, the applause and hushed talking of the audience, and the background noises in the theater. You can distinguish these separate parts of the many sound waves reaching your ears and can pay attention to those of importance to you.

Pitch discrimination depends on the region of the basilar membrane that vibrates.

Pitch discrimination (that is, the ability to distinguish between various frequencies of incoming sound waves) depends on the shape and properties of the basilar membrane, which is narrow and stiff at its oval window end and wide and flexible at its helicotrema end (● Figure 6-29b). Different regions of the basilar membrane naturally vibrate maxi-

● **FIGURE 6-29**

Transmission of sound waves. (a) Fluid movement within the cochlea set up by vibration of the oval window follows two pathways, one dissipating sound energy and the other initiating the receptor potential. (b) Different regions of the basilar membrane vibrate maximally at different frequencies. (c) The narrow, stiff end of the basilar membrane nearest the oval window vibrates best with high-frequency pitches. The wide, flexible end of the basilar membrane near the helicotrema vibrates best with low-frequency pitches.

PhysioEdge For an animation of this figure, click the Hearing and Equilibrium tab in the Special Senses tutorial on the CD-ROM.

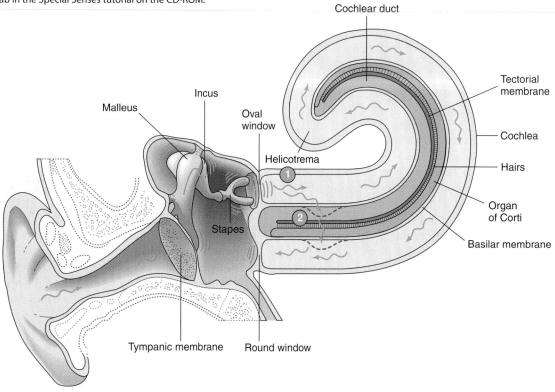

Fluid movement within the cochlea set up by vibration of the oval window follows two pathways:

Pathway ① : Through the upper compartment, around the heliocotrema, and through the lower compartment, causing the round window to vibrate. This pathway just dissipates sound energy.

Pathway ② : A "shortcut" from the upper compartment through the basilar membrane to the lower compartment. This pathway triggers activation of the receptors for sound by bending the hairs of hair cells as the organ of Corti on top of the vibrating basilar membrane is displaced in relation to the overlying tectorial membrane.

(a)

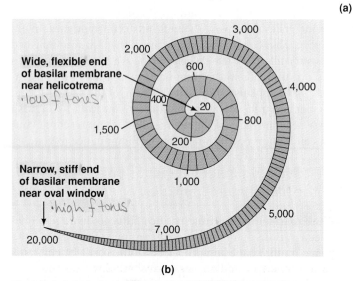

Wide, flexible end of basilar membrane near helicotrema
·low f tones·

Narrow, stiff end of basilar membrane near oval window
·high f tones·

The numbers indicate the frequencies in cycles per second with which different regions of the basilar membrane maximally vibrate.

(b)

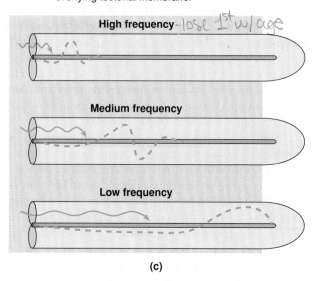

High frequency - lose 1st w/age

Medium frequency

Low frequency

(c)

The hairs from the hair cells of the basilar membrane are embedded in the overlying tectorial membrane. These hairs are bent when the basilar membrane is deflected in relation to the stationary tectorial membrane. This bending results in a receptor potential.

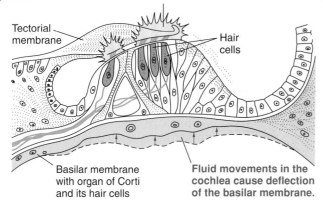

Tectorial membrane

Hair cells

Basilar membrane with organ of Corti and its hair cells

Fluid movements in the cochlea cause deflection of the basilar membrane.

● **FIGURE 6-30**

Bending of hairs on deflection of the basilar membrane

 PhysioEdge For an animation of this figure, click the Hearing and Equilibrium tab in the Special Senses tutorial on the CD-ROM.

mally at different frequencies; that is, each frequency displays peak vibration at a different position along the membrane. The narrow end nearest the oval window vibrates best with high-frequency pitches, whereas the wide end nearest the helicotrema vibrates maximally with low-frequency tones (● Figure 6-29c). The pitches in between are sorted out precisely along the length of the membrane from higher to lower frequency. As a sound wave of a particular frequency is set up in the cochlea by oscillation of the stapes, the wave travels to the region of the basilar membrane that naturally responds maximally to that frequency. The energy of the pressure wave is dissipated with this vigorous membrane oscillation, so the wave dies out at the region of maximal displacement. This information is propagated to the CNS, which interprets the pattern of hair cell stimulation as a sound of a particular frequency. Modern techniques have determined that the basilar membrane is so fine-tuned that the peak membrane response to a single pitch probably extends no more than the width of a few hair cells.

Overtones of varying frequencies cause many points along the basilar membrane to vibrate simultaneously but less intensely than the fundamental tone, enabling the CNS to distinguish the timbre of the sound (**timbre discrimination**).

▌ Loudness discrimination depends on the amplitude of vibration.

Intensity (loudness) discrimination depends on the amplitude of vibration. As sound waves originating from louder sound sources strike the eardrum, they cause it to vibrate more vigorously (that is, bulge in and out to a greater extent) but at the same frequency as a softer sound of the same pitch. The greater tympanic membrane deflection is converted into a greater amplitude of basilar membrane movement in the re-

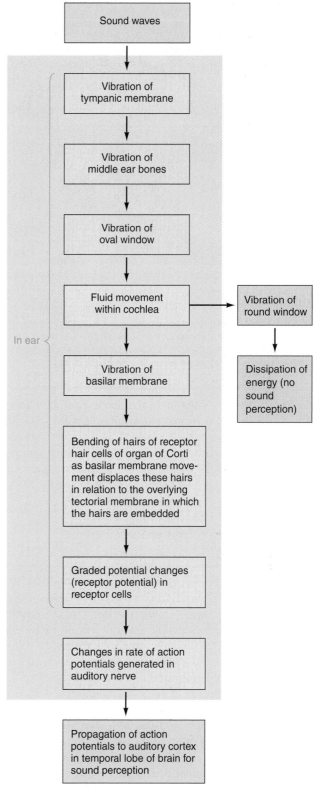

Sound waves

Vibration of tympanic membrane

Vibration of middle ear bones

Vibration of oval window

Fluid movement within cochlea → Vibration of round window → Dissipation of energy (no sound perception)

In ear

Vibration of basilar membrane

Bending of hairs of receptor hair cells of organ of Corti as basilar membrane movement displaces these hairs in relation to the overlying tectorial membrane in which the hairs are embedded

Graded potential changes (receptor potential) in receptor cells

Changes in rate of action potentials generated in auditory nerve

Propagation of action potentials to auditory cortex in temporal lobe of brain for sound perception

● **FIGURE 6-31**

Sound transduction

gion of peak responsiveness. The CNS interprets greater basilar membrane oscillation as a louder sound.

The auditory system is so sensitive and can detect sounds so faint that the distance of basilar membrane deflection is

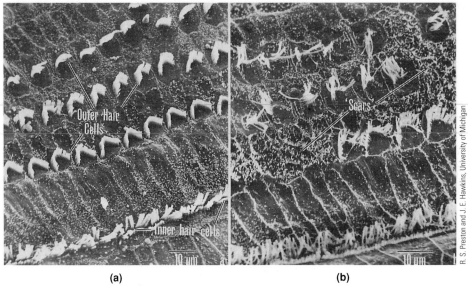

(a)　　　　　　　　　　　　　　　　　(b)

● **FIGURE 6-32**

Loss of hair cells caused by loud noises. Injury and loss of hair cells caused by intense noise. Portions of the organ of Corti, with its three rows of outer hair cells and one row of inner hair cells, from the inner ear of (a) a normal guinea pig and (b) a guinea pig after a 24-hour exposure to noise at 120 decibels SPL (sound pressure level), a level approached by loud rock music.

comparable to only a fraction of the diameter of a hydrogen atom, the smallest of atoms. No wonder very loud sounds (for example, the sounds of a typical rock concert) can set up such violent vibrations of the basilar membrane that irreplaceable hair cells are actually sheared off or permanently distorted, leading to partial hearing loss (● Figure 6-32).

▌ Deafness is caused by defects either in conduction or neural processing of sound waves.

Clinical Note   Loss of hearing, or **deafness**, may be temporary or permanent, partial or complete. Deafness is classified into two types—*conductive deafness* and *sensorineural deafness*—depending on the part of the hearing mechanism that fails to function adequately. **Conductive deafness** occurs when sound waves are not adequately conducted through the external and middle portions of the ear to set the fluids in the inner ear in motion. Possible causes include physical blockage of the ear canal with earwax, rupture of the eardrum, middle ear infections with accompanying fluid accumulation, or restriction of the ossicular movement because of bony adhesions between the stapes and the oval window. In **sensorineural deafness**, the sound waves are transmitted to the inner ear, but they are not translated into nerve signals that are interpreted by the brain as sound sensations. The defect can lie in the organ of Corti or the auditory nerves or, rarely, in the ascending auditory pathways or auditory cortex.

One of the most common causes of partial hearing loss, **neural presbycusis**, is a degenerative, age-related process that occurs as hair cells "wear out" with use. Over time, exposure to even ordinary modern-day sounds eventually damages hair cells, so that on average adults have lost more than 40% of their cochlear hair cells by age 65. The hair cells that process high-frequency sounds are the most vulnerable to destruction.

Hearing aids are helpful in conductive deafness but are less beneficial for sensorineural deafness. These devices increase the intensity of airborne sounds and may modify the sound spectrum and tailor it to the patient's particular pattern of hearing loss at higher or lower frequencies. For the sound to be perceived, however, the receptor cell–neural pathway system must still be intact.

In recent years, **cochlear implants** have become available. These electronic devices, which are surgically implanted, transduce sound signals into electrical signals that can directly stimulate the auditory nerve, thus bypassing a defective cochlear system. Cochlear implants cannot restore normal hearing, but they do permit recipients to recognize sounds. Success ranges from an ability to "hear" a phone ringing to being able to carry on a conversation over the telephone.

▌ The vestibular apparatus is important for equilibrium by detecting position and motion of the head.

In addition to its cochlear-dependent role in hearing, the inner ear has another specialized component, the **vestibular apparatus**, which provides information essential for the sense of equilibrium and for coordinating head movements with eye and postural movements. The vestibular apparatus consists of two sets of structures lying within a tunneled-out region of the temporal bone near the cochlea—the *semicircular canals* and the *otolith organs*, namely the *utricle* and *saccule* (● Figure 6-33a).

The vestibular apparatus detects changes in position and motion of the head. As in the cochlea, all components of the vestibular apparatus contain fluid. Also, similar to the organ of Corti, the vestibular components each contain hair cells that respond to mechanical deformation triggered by specific movements of the fluid. Unlike the auditory system, much of the information provided by the vestibular apparatus does not reach the level of conscious awareness.

ROLE OF THE SEMICIRCULAR CANALS

The **semicircular canals** detect rotational or angular acceleration or deceleration of the head, such as when starting or stopping spinning, somersaulting, or turning the head. Each ear contains three semicircular canals arranged three-

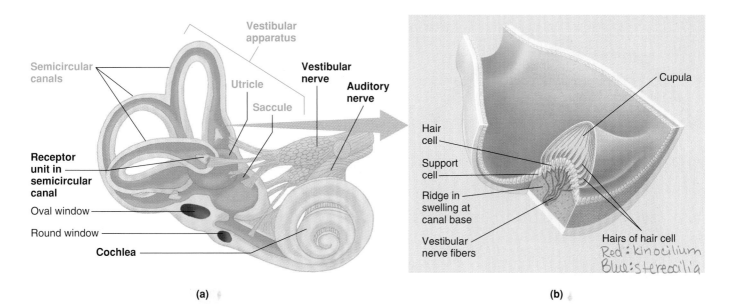

(handwritten: Red: kinocilium / Blue: stereocilia)

(a)

(b)

(vertical text, left margin: Dean Hillman, New York University Medical School)

(handwritten labels: Kinocilium / stereocilia)

Cluster of hairs of hair cell within vestibular apparatus

(c)

● **FIGURE 6-33**

Vestibular apparatus. (a) Gross anatomy of the vestibular apparatus. (b) Receptor unit in semicircular canal. (c) Scanning electron micrograph of the hairs on the hair cells within the vestibular apparatus.

(*Source:* Figure 6-33b adapted from Cecie Starr and Ralph Taggart, *Biology: The Unity and Diversity of Life*, Eighth Edition, Fig. 36.10b, p. 595. Copyright 1998 Wadsworth Publishing Company.)

dimensionally in planes that lie at right angles to each other. The receptive hair cells of each semicircular canal are situated on top of a ridge located in a swelling at the base of the canal (● Figures 6-33a and b). The hairs are embedded in an overlying, caplike, gelatinous layer, the **cupula**, which protrudes into the fluid within this swelling. The cupula sways in the direction of fluid movement, much like seaweed leaning in the direction of the prevailing tide.

Acceleration or deceleration during rotation of the head in any direction causes fluid movement in at least one of the semicircular canals, because of their three-dimensional arrangement. As you start to move your head, the bony canal and the ridge of hair cells embedded in the cupula move with your head. Initially, however, the fluid within the canal, not being attached to your skull, does not move in the direction of the rotation but lags behind because of its inertia. (Because of inertia, a resting object remains at rest, and a moving object continues to move in the same direction unless the object is acted on by some external force that induces change.) When the fluid is left behind as you start to rotate your head, the fluid that is in the same plane as the head movement is in effect shifted in the opposite direction from the movement (similar to your body tilting to the right as the car in which you are riding suddenly turns to the left) (● Figure 6-34). This fluid movement causes the cupula to lean in the opposite direction from the head movement, bending the sensory hairs embedded in it. If your head movement continues at the same rate in the same direction, the fluid catches up and moves in unison with your head so that the hairs return to their unbent position. When your head slows down and stops, the reverse situation occurs. The fluid briefly continues to move in the direction of the rotation while your head decelerates to a stop. As a result, the cupula and its hairs are transiently bent in the direction of the preceding spin, which is opposite to the way they were bent during acceleration. Bending the hairs in one direction increases the rate of firing in afferent fibers within the **vestibular nerve**, whereas bending in the opposite direction decreases the frequency of action potentials in these afferent fibers. When the fluid gradually comes to a halt, the hairs straighten again. Thus the semicircular canals detect changes in the rate of rotational movement of your head. They do not respond when your head is motionless or when it is moving in a circle at a constant speed.

ROLE OF THE OTOLITH ORGANS

The **otolith organs** provide information about the position of the head relative to gravity and also detect changes in the rate of linear motion (moving in a straight line regardless of direction). The otolith organs, the **utricle** and the **saccule**,

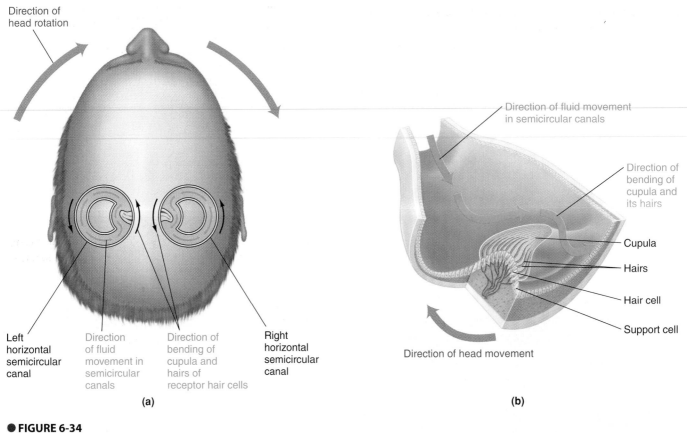

● **FIGURE 6-34**

Activation of the hair cells in the semicircular canals

are saclike structures housed within a bony chamber situated between the semicircular canals and the cochlea (● Figure 6-33a). The hairs of the receptive hair cells in these sense organs also protrude into an overlying gelatinous sheet, whose movement displaces the hairs and results in changes in hair cell potential. Many tiny crystals of calcium carbonate—the **otoliths** ("ear stones")—are suspended within the gelatinous layer, making it heavier and giving it more inertia than the surrounding fluid (● Figure 6-35a). When a person is in an upright position, the hairs within the utricles are oriented vertically and the saccule hairs are lined up horizontally.

Let's look at the *utricle* as an example. Its otolith-embedded, gelatinous mass shifts positions and bends the hairs in two ways:

1. When you tilt your head in any direction other than vertical (that is, other than straight up and down), the hairs are bent in the direction of the tilt because of the gravitational force exerted on the top-heavy gelatinous layer (● Figure 6-35b). This bending produces depolarizing or hyperpolarizing receptor potentials depending on the tilt of your head. The CNS thus receives different patterns of neural activity depending on head position with respect to gravity.
2. The utricle hairs are also displaced by any change in horizontal linear motion (such as moving straight forward, backward, or to the side). As you start to walk forward (● Figure 6-35c), the top-heavy otolith membrane at first lags

behind the fluid and hair cells because of its greater inertia. The hairs are thus bent to the rear, in the opposite direction of the forward movement of your head. If you maintain your walking pace, the gelatinous layer soon catches up and moves at the same rate as your head so that the hairs are no longer bent. When you stop walking, the otolith sheet continues to move forward briefly as your head slows and stops, bending the hairs toward the front. Thus the hair cells of the utricle detect horizontally directed linear acceleration and deceleration, but they do not provide information about movement in a straight line at constant speed.

The saccule functions similarly to the utricle, except that it responds selectively to tilting of the head away from a horizontal position (such as getting up from bed) and to vertically directed linear acceleration and deceleration (such as jumping up and down or riding in an elevator).

Signals arising from the various components of the vestibular apparatus are carried through the vestibulocochlear nerve to the **vestibular nuclei**, a cluster of neuronal cell bodies in the brain stem, and to the cerebellum. Here the vestibular information is integrated with input from the skin surface, eyes, joints, and muscles for (1) maintaining balance and desired posture; (2) controlling the external eye muscles so that the eyes remain fixed on the same point, despite movement of the head; and (3) perceiving motion and orientation. Some people, for poorly understood reasons, are especially sensitive to particular motions that activate the vestibular apparatus and cause symptoms of dizziness and nausea; this sensitivity is called **motion sickness**.

△ Table 6-5 summarizes the functions of the major components of the ear.

 Click on the Media Exercises menu of the CD-ROM and work Media Exercise 6.3: The Ear to test your understanding of the previous section.

● **FIGURE 6-35**

Utricle. (a) Receptor unit in utricle. (b) Activation of the utricle by a change in head position. (c) Activation of the utricle by horizontal linear acceleration.

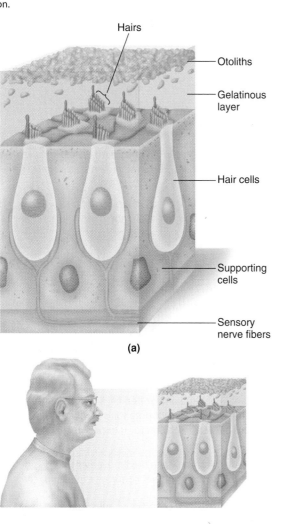

Hairs

Otoliths

Gelatinous layer

Hair cells

Supporting cells

Sensory nerve fibers

(a)

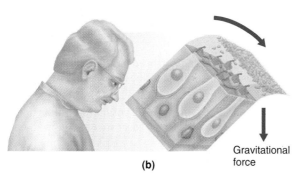

Gravitational force

(b)

CHEMICAL SENSES: TASTE AND SMELL

Unlike the eyes' photoreceptors and the ears' mechanoreceptors, the receptors for taste and smell are chemoreceptors, which generate neural signals on binding with particular chemicals in their environment. The sensations of taste and smell in association with food intake influence the flow of digestive juices and affect appetite. Furthermore, stimulation of taste or smell receptors induces pleasurable or objectionable sensations and signals the presence of something to seek (a nutritionally useful, good-tasting food) or to avoid (a potentially toxic, bad-tasting substance). Thus the chemical senses provide a "quality control" checkpoint for substances available for ingestion. In lower animals, smell also plays a major role in finding direction, in seeking prey or avoiding predators, and in sexual attraction to a mate. The sense of smell is less sensitive in humans and much less important in influencing our behavior (although millions of dollars are spent annually on perfumes and deodorants to make us smell better and thereby be more socially attractive). We will first examine the mechanism of taste (**gustation**) and then turn our attention to smell (**olfaction**).

▌ **Taste receptor cells are located primarily within tongue taste buds.**

The chemoreceptors for taste sensation are packaged in taste buds, about 10,000 of which are present in the oral cavity and throat, with the greatest percentage on the upper surface of the tongue (● Figure 6-36). A **taste bud** consists of about 50 *taste receptor cells* packaged with *supporting cells* in an arrangement like slices of an orange. Each taste bud has a

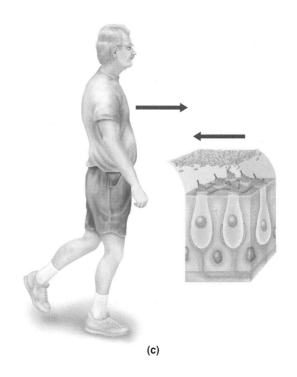

(c)

Functions of the Major Components of the Ear

STRUCTURE	LOCATION	FUNCTION
External ear		**Collects and transfers sound waves to middle ear**
Pinna (ear)	Skin-covered flap of cartilage located on each side of head	Collects sound waves and channels them down ear canal; contributes to sound localization
External auditory meatus (ear canal)	Tunnels from exterior through temporal bone to tympanic membrane	Directs sound waves to tympanic membrane; contains filtering hairs and secretes earwax to trap foreign particles
Tympanic membrane (eardrum)	Thin membrane that separates external ear and middle ear	Vibrates in synchrony with sound waves that strike it, setting middle ear bones in motion
Middle ear		**Transfers vibrations of tympanic membrane to fluid in cochlea**
Malleus, incus, stapes	Movable chain of bones that extends across middle ear cavity; malleus attaches to tympanic membrane, and stapes attaches to oval window	Oscillate in synchrony with tympanic membrane vibrations and set up wavelike movements in cochlea fluid at the same frequency
Inner ear: Cochlea		**Houses sensory system for hearing**
Oval window	Thin membrane at entrance to cochlea; separates middle ear from upper compartment of cochlea	Vibrates in unison with movement of stapes, to which it is attached; oval window movement sets cochlea fluid in motion
Upper and lower compartments of cochlea	Snail-shaped tubular system that lies deep within the temporal bone	Contains fluid that is set in motion by oval window movement driven by oscillation of middle ear bones
Cochlear duct	Blind-ended tubular compartment that tunnels through center of cochlea between the upper and lower compartments	Houses basilar membrane and organ of Corti
Basilar membrane	Forms floor of cochlear duct	Vibrates in unison with fluid movements; bears the organ of Corti, the sense organ for hearing
Organ of Corti	Rests on top of basilar membrane throughout its length	Contains hair cells, the receptors for sound, which undergo receptor potentials when their hairs are bent as a result of fluid movement in cochlea
Tectorial membrane	Stationary membrane that overhangs organ of Corti and within which the surface hairs of the receptor hair cells are embedded	Site at which the embedded hairs of receptor cells are bent and undergo receptor potentials as vibrating basilar membrane moves in relation to stationary tectorial membrane
Round window	Thin membrane that separates lower compartment of cochlea from middle ear	Vibrates in unison with fluid movements to dissipate pressure in cochlea; does not contribute to sound reception
Inner Ear: Vestibular Apparatus		**Houses sensory systems for equilibrium, and provides input essential for maintaining posture and balance**
Semicircular canals	Three semicircular canals arranged three-dimensionally in planes at right angles to each other near cochlea	Detect rotational or angular acceleration or deceleration
Utricle	Saclike structure in a bony chamber between cochlea and semicircular canals	Detects (1) changes in head position away from vertical and (2) horizontally directed linear acceleration and deceleration
Saccule	Lies next to utricle	Detects (1) changes in head position away from horizontal and (2) vertically directed linear acceleration and deceleration

small opening, the **taste pore**, through which fluids in the mouth come into contact with the surface of its receptor cells. **Taste receptor cells** are modified epithelial cells with many surface folds, or microvilli, that protrude slightly through the taste pore, greatly increasing the surface area exposed to the oral contents (see p. 37). The plasma membrane of the microvilli contains receptor sites that bind selectively with chemical molecules in the environment. Only chemicals in solution—either ingested liquids or solids that have been dissolved in saliva—can attach to receptor cells and evoke the sensation of taste.

Most receptors are carefully sheltered from direct exposure to the environment, but the taste receptor cells, by virtue of their task, frequently come into contact with potent chemicals. Unlike the eye or ear receptors, which are irreplaceable, taste receptors have a life span of about 10 days. Epithelial cells surrounding the taste bud differentiate first into supporting cells and then into receptor cells to constantly renew the taste bud components.

Binding of a taste-provoking chemical, a **tastant**, with a receptor cell produces a receptor potential that in turn initiates action potentials in afferent nerve fibers. These signals are conveyed via synaptic stops in the brain stem and thalamus to the **cortical gustatory area**, a region in the parietal lobe adjacent to the "tongue" area of the somatosensory cortex, where the taste is perceived. Taste signals are also sent to the hypothalamus and limbic system to add affective dimensions, such as whether the taste is pleasant or unpleasant, and to process behavioral aspects associated with taste and smell.

▌ Taste discrimination is coded by patterns of activity in various taste bud receptors.

We can discriminate among thousands of different taste sensations, yet all tastes are varying combinations of four **primary tastes**: *salty, sour, sweet,* and *bitter.* A growing number of sensory physiologists suggest a fifth primary taste—*umami,* a meaty or savory taste—should be added to the list. These primary taste sensations are elicited by the following stimuli:

- **Salt taste** is stimulated by chemical salts, especially NaCl (table salt).
- **Sour taste** is caused by acids. The citric acid content of lemons, for example, accounts for their distinctly sour taste.
- **Sweet taste** is evoked by the particular configuration of glucose. From an evolutionary perspective, we crave sweet foods because they supply necessary calories in a readily usable form. However, other organic molecules with similar structures but no calories, such as saccharin, aspartame, and other artificial sweeteners, can also interact with "sweet" receptor binding sites.
- **Bitter taste** is elicited by a more chemically diverse group of tastants than the other taste sensations. For example, al-

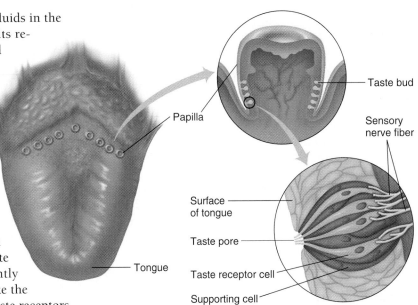

● **FIGURE 6-36**

Location and structure of the taste buds. Taste buds are located primarily along the edges of moundlike papillae on the upper surface of the tongue. The receptor cells and supporting cells of a taste bud are arranged like slices of an orange.

 For an interaction related to this figure, see Media Exercise 6.4: Taste on the CD-ROM.

kaloids (such as caffeine, nicotine, strychnine, morphine, and other toxic plant derivatives), as well as poisonous substances, all taste bitter, presumably as a protective mechanism to discourage ingestion of these potentially dangerous compounds.

- **Umami taste**, which was first identified and named by a Japanese researcher, is triggered by amino acids, especially glutamate. The presence of amino acids, as found in meat for example, serves as a marker for a desirable, nutritionally protein-rich food.

Each receptor cell responds in varying degrees to all four (or perhaps five) primary tastes but is generally preferentially responsive to one of the taste modalities. The richness of fine taste discrimination beyond the primary tastes depends on subtle differences in the stimulation patterns of all the taste buds in response to various substances, similar to the variable stimulation of the three cone types that gives rise to the range of color sensations.

Taste perception is also influenced by information derived from other receptors, especially odor. When you temporarily lose your sense of smell because of swollen nasal passageways during a cold, your sense of taste is also markedly reduced, even though your taste receptors are unaffected by the cold. Other factors affecting taste include temperature and texture of the food as well as psychological factors associated with past experiences with the food. How the cortex accomplishes the complex perceptual processing of taste sensation is currently not known.

▌ The olfactory receptors in the nose are specialized endings of renewable afferent neurons.

The **olfactory** (smell) **mucosa**, a 3-cm² patch of mucosa located in the ceiling of the nasal cavity, contains three cell types: *olfactory receptor cells, supporting cells,* and *basal cells* (● Figure 6-37). The supporting cells secrete mucus, which coats the nasal passages. The basal cells are precursors for new olfactory receptor cells, which are replaced about every two months. An **olfactory receptor cell** is an afferent neuron that has a receptor portion that lies in the olfactory mucosa of the nose and whose afferent axon traverses into the brain. The axons of the olfactory receptor cells collectively form the **olfactory nerve**.

The receptor portion of an olfactory receptor cell consists of an enlarged knob bearing several long cilia that extend like a tassel to the surface of the mucosa. These cilia contain the binding sites for attachment of **odorants**, molecules that can be smelled. During quiet breathing, odorants typically reach the sensitive receptors only by diffusion because the olfactory mucosa is above the normal path of air flow. The act of sniffing enhances this process by drawing the air currents upward within the nasal cavity so that a greater percentage of the odoriferous molecules in the air come into contact with the olfactory mucosa. Odorants also reach the olfactory mucosa during eating by wafting up to the nose from the mouth through the pharynx (back of the throat).

To be smelled, a substance must be (1) sufficiently volatile (easily vaporized) that some of its molecules can enter the nose in the inspired air and (2) sufficiently water soluble that it can dissolve in the mucus coating the olfactory mucosa. As with taste receptors, to be detected by olfactory receptors molecules must be dissolved.

▌ Various parts of an odor are detected by different olfactory receptors and sorted into "smell files."

The human nose contains 5 million olfactory receptors, of which there are 1000 different types. During smell detection, an odor is "dissected" into various components. Each receptor responds to only one discrete component of an odor rather than to the whole odorant molecule. Accordingly, each of the various parts of an odor is detected by one of the thousand different receptors, and a given receptor can respond to a particular odor component shared in common by different scents. Compare this to the three cone types for coding color vision and the taste buds that respond differentially to only four (or five) primary tastes to accomplish coding for taste discrimination.

Binding of an appropriate scent signal to an olfactory receptor brings about a receptor potential that generates action potentials in the afferent fiber. The frequency of the action potentials depends on the concentration of the stimulating chemical molecules.

● **FIGURE 6-37**

Location and structure of the olfactory receptors

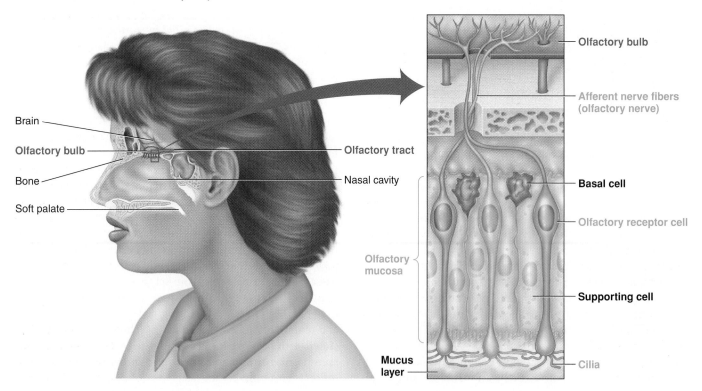

Brain

Olfactory bulb

Bone

Soft palate

Olfactory tract

Nasal cavity

Olfactory bulb

Afferent nerve fibers (olfactory nerve)

Basal cell

Olfactory receptor cell

Olfactory mucosa

Supporting cell

Mucus layer

Cilia

The afferent fibers arising from the receptor endings in the nose pass through tiny holes in the flat bone plate separating the olfactory mucosa from the overlying brain tissue (● Figure 6-37). They immediately synapse in the **olfactory bulb**, a complex neural structure containing several different layers of cells that are functionally similar to the retinal layers of the eye. Each olfactory bulb is lined by small ball-like neural junctions known as **glomeruli** ("little balls") (● Figure 6-38). Within each glomerulus, the terminals of receptor cells carrying information about a particular scent component synapse with the next cells in the olfactory pathway, the **mitral cells**. Because each glomerulus receives signals only from receptors that detect a particular odor component, the glomeruli serve as "smell files." The separate components of an odor are sorted into different glomeruli, one component per file. Thus the glomeruli, which serve as the first relay station in the brain for processing olfactory information, play a key role in organizing scent perception.

The mitral cells on which the olfactory receptors terminate in the glomeruli refine the smell signals and relay them to the brain for further processing. Fibers leaving the olfactory bulb travel in two different routes:

1. A subcortical route going primarily to regions of the limbic system, especially the lower medial sides of the temporal lobes (considered the **primary olfactory cortex**). This route, which includes hypothalamic involvement, permits close coordination between smell and behavioral reactions associated with feeding, mating, and direction orienting.
2. A route through the thalamus to the cortex. As with the other senses, the cortical route is important for conscious perception and fine discrimination of smell.

Odor discrimination is coded by patterns of activity in the olfactory bulb glomeruli.

Because each given odorant activates multiple receptors and glomeruli in response to its various odor components, odor discrimination is based on different patterns of glomeruli activated by various scents. In this way, the cortex can distinguish over 10,000 different scents. This mechanism for sorting out and distinguishing different odors is very effective. A noteworthy example is our ability to detect methyl mercaptan (garlic odor) at a concentration of 1 molecule per 50 billion molecules in the air! This substance is added to odorless natural gas to enable us to detect potentially lethal gas leaks. Despite this impressive sensitivity, humans have a poor sense of smell compared to other species. By comparison, dogs'

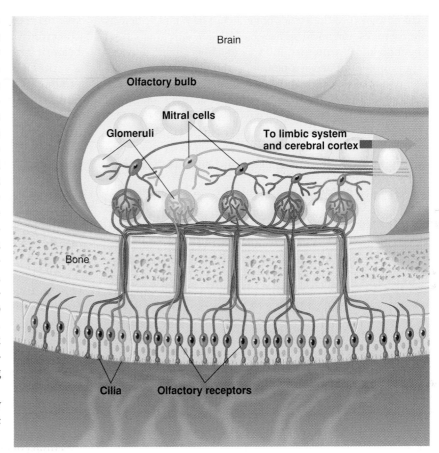

● FIGURE 6-38

Processing of scents in the olfactory bulb. Each of the glomeruli lining the olfactory bulb receives synaptic input from only one type of olfactory receptor, which, in turn, responds to only one discrete component of an odorant. Thus the glomeruli sort and file the various components of an odoriferous molecule before relaying the smell signal to the mitral cells and higher brain levels for further processing.

sense of smell is hundreds of times more sensitive than that of humans.

The olfactory system adapts quickly.

Although the olfactory system is sensitive and highly discriminating, it is also quickly adaptive. Sensitivity to a new odor diminishes rapidly after a short period of exposure to it, even though the odor source continues to be present. This reduced sensitivity does not involve receptor adaptation, as researchers thought for years; actually, the olfactory receptors themselves adapt slowly. It apparently involves some sort of adaptation process in the CNS. Adaptation is specific for a particular odor, and responsiveness to other odors remains unchanged.

The vomeronasal organ detects pheromones.

In addition to the olfactory mucosa, the nose contains another sense organ, the **vomeronasal organ (VNO)**, which is common in animals but until recently was thought nonexistent in humans. The VNO is located about half an inch inside the human nose next to the vomer bone, hence its name. It detects **pheromones**, chemical signals passed subconsciously from one individual to another. Binding of a pheromone to its receptor on the surface of a neuron in the VNO triggers an action potential that travels through nonolfactory pathways to the limbic system, the brain region that governs emotional responses and sociosexual behaviors. These signals never reach the higher levels of conscious awareness. In animals, the VNO is known as the "sexual nose" for its role in governing reproductive and social behaviors, such as identifying and attracting a mate.

Scientists have now proved the existence of pheromones in humans. Although the role of the VNO in human behavior has not been validated, researchers suspect that it is responsible for spontaneous "feelings" between people, either "good chemistry," such as "love at first sight," or "bad chemistry," such as "getting bad vibes" from someone you just met. They speculate that pheromones in humans subtly influence sexual activity, compatibility with others, or group behavior, similar to the role they play in animals. Because messages conveyed by the VNO seem to bypass cortical consciousness, the response to the largely odorless pheromones is not a distinct, discrete perception, such as smelling a favorite fragrance, but more like an inexplicable impression.

 Click on the Media Exercises menu of the CD-ROM and work Media Exercise 6.4: Taste to test your understanding of the previous section.

CHAPTER IN PERSPECTIVE: FOCUS ON HOMEOSTASIS

To maintain a life-sustaining stable internal environment, adjustments must constantly be made to compensate for a myriad external and internal factors that continuously threaten to disrupt homeostasis, such as external exposure to cold or internal acid production. Many of these adjustments are directed by the nervous system, one of the body's two major regulatory systems. The central nervous system (CNS), the integrating and decision-making component of the nervous system, must continuously be informed of "what's happening" in both the internal and the external environment so that it can command appropriate re-

sponses in the organ systems to maintain the body's viability. In other words, the CNS must know what changes are taking place before it can respond to these changes.

The afferent division of the peripheral nervous system is the communication link by which the CNS is informed about the internal and external environment. The afferent division detects, encodes, and transmits peripheral signals to the CNS for processing. Afferent input is necessary for arousal, perception, and determination of efferent output.

Afferent information about the internal environment, such as the CO_2 level in the blood, never reaches the level of conscious awareness, but this input to the controlling centers of the CNS is essential for maintaining homeostasis. Afferent input that reaches the level of conscious awareness, called *sensory information*, includes somesthetic and proprioceptive sensation (body sense) and special senses (vision, hearing, taste, and smell).

The body sense receptors are distributed over the entire body surface as well as throughout the joints and muscles. Afferent signals from these receptors provide information about what's happening directly to each specific body part in relation to the external environment (that is, the "what," "where," and "how much" of stimulatory inputs to the body's surface and the momentary position of the body in space). In contrast, each special sense organ is restricted to a single site in the body. Rather than providing information about a specific body part, a special sense organ provides a specific type of information about the external environment that is useful to the body as a whole. For example, through their ability to detect, extensively analyze, and integrate patterns of illumination in the external environment, the eyes and visual processing system enable you to "see" your surroundings. The same integrative effect could not be achieved if photoreceptors were scattered over your entire body surface, as are touch receptors.

Sensory input (both body sense and special senses) enables a complex multicellular organism such as a human to interact in meaningful ways with the external environment in procuring food, defending against danger, and engaging in other behavioral actions geared toward maintaining homeostasis. In addition to providing information essential for interactions with the external environment for basic survival, the perceptual processing of sensory input adds immeasurably to the richness of life, such as enjoyment of a good book, concert, or meal.

CHAPTER SUMMARY

Introduction (pp. 145–146)
- The afferent division of the peripheral nervous system carries information about the internal and external environment to the CNS.
- Sensory information, afferent information that reaches the level of conscious awareness, includes (1) somatic sensation (somesthetic sensation and proprioception) and (2) special senses.
- Perception is the conscious interpretation of the external world that the brain creates from sensory input.

Receptor Physiology (pp. 146–149)
- Receptors are specialized peripheral endings of afferent neurons. *(Review Figure 6-3.)* Each type of receptor responds to its adequate stimulus (a change in the energy form or modality to which it is responsive), translating the energy form of the stimulus into electrical signals, a process known as *transduction*.
- A stimulus brings about a graded, depolarizing receptor potential by altering the receptor's membrane permeability. Recep-

tor potentials, if of sufficient magnitude, generate action potentials in the afferent neuronal membrane next the receptor. These action potentials self-propagate along the afferent neuron to the CNS.

- The strength and rate of change of the stimulus determine the magnitude of the receptor potential, which in turn determines the frequency of action potentials generated in the afferent neuron. (Review Table 6-1.)
- Receptor potential size is also influenced by the extent of receptor adaptation, which is a reduction in receptor potential despite sustained stimulation. (1) Tonic receptors adapt slowly or not at all and thus provide continuous information about the stimuli they monitor. (2) Phasic receptors adapt rapidly and frequently exhibit off responses, thereby providing information about changes in the energy form they monitor. (Review Figure 6-4.)
- Discrete labeled-line pathways lead from the receptors to the CNS so that information about the type and location of the stimuli can be deciphered by the CNS, even though all the information arrives in the form of action potentials. (Review Table 6-1.)

Pain (pp. 149–150)

- Painful experiences are elicited by noxious mechanical, thermal, or chemical stimuli and consist of two components: the perception of pain coupled with emotional and behavioral responses to it.
- Three categories of nociceptors, or pain receptors, respond to these stimuli: mechanical nociceptors, thermal nociceptors, and polymodal nociceptors.
- Pain signals are transmitted over two afferent pathways: a fast pathway that carries sharp, prickling pain signals; and a slow pathway that carries dull, aching, persistent pain signals.
- Afferent pain fibers terminate in the spinal cord on ascending pathways that transmit the signal to the brain for processing.
- Descending pathways from the brain use endogenous opiates to suppress the release of substance P, a pain-signaling neurotransmitter from the afferent pain-fiber terminal. Thus these descending pathways block further transmission of the pain signal and serve as a built-in analgesic system.

Eye: Vision (pp. 150–165)

- The eye is a specialized structure housing the light-sensitive receptors essential for vision perception—namely the rods and cones found in its retinal layer. (Review Table 6-3, p. 164, and Figure 6-6.)
- The iris controls the size of the pupil, thereby adjusting the amount of light permitted to enter the eye.
- The cornea and lens are the primary refractive structures that bend the incoming light rays to focus the image on the retina. The cornea contributes most to the total refractive ability of the eye. The strength of the lens can be adjusted through action of the ciliary muscle to accommodate for differences in near and far vision. (Review Figures 6-10 through 6-15.)
- The rod and the cone photoreceptors are activated when the photopigments they contain differentially absorb various wavelengths of light. Light absorption causes a biochemical change in the photopigment that is ultimately converted into a change in the rate of action potential propagation in the visual pathway leaving the retina. (Review Figures 6-17, 6-20, and 6-21.)
- The visual message is transmitted via a complex crossed and uncrossed pathway to the visual cortex in the occipital lobe of the brain for perceptual processing. (Review Figure 6-24.)
- Cones display high acuity but can be used only for day vision because of their low sensitivity to light. Different ratios of stimulation of three cone types by varying wavelengths of light lead to color vision. (Review Figure 6-22 and Table 6-2.)
- Rods provide only indistinct vision in shades of gray, but because they are very sensitive to light they can be used for night vision. (Review Table 6-2.)
- The eyes' sensitivity is increased during dark adaptation, by the regeneration of rod photopigments that had been broken down during preceding light exposure. Sensitivity is decreased during light adaptation, by the rapid breakdown of cone photopigments.

Ear: Hearing and Equilibrium (pp. 165–175)

- The ear performs two unrelated functions: (1) hearing, which involves the external ear, middle ear, and cochlea of the inner ear; and (2) sense of equilibrium, which involves the vestibular apparatus of the inner ear. In contrast to the photoreceptors of the eye, the ear receptors located in the inner ear—the hair cells in the cochlea and vestibular apparatus—are mechanoreceptors. (Review Table 6-5, p. 176, and Figure 6-25.)
- Hearing depends on the ear's ability to convert airborne sound waves into mechanical deformations of receptive hair cells, thereby initiating neural signals.
- Sound waves consist of high-pressure regions of compression alternating with low-pressure regions of rarefaction of air molecules. The pitch (tone) of a sound is determined by the frequency of its waves, the loudness (intensity) by the amplitude of the waves, and the timbre (quality) by its characteristic overtones. (Review Figures 6-26 and 6-27.)
- Sound waves are funneled through the external ear canal to the tympanic membrane, which vibrates in synchrony with the waves.
- Middle ear bones bridging the gap between the tympanic membrane and the inner ear amplify the tympanic movements and transmit them to the oval window, whose movement sets up traveling waves in the cochlear fluid. (Review Figure 6-28.)
- These waves, which are at the same frequency as the original sound waves, set the basilar membrane in motion. Various regions of this membrane selectively vibrate more vigorously in response to different frequencies of sound. (Review Figure 6-29.)
- On top of the basilar membrane are the receptive hair cells of the organ of Corti, whose hairs are bent as the basilar membrane is deflected up and down in relation to the overhanging stationary tectorial membrane in which the hairs are embedded. (Review Figures 6-28 and 6-30.)
- This mechanical deformation of specific hair cells in the region of maximal basilar membrane vibration is transduced into neural signals that are transmitted to the auditory cortex in the temporal lobe of the brain for sound perception. (Review Figure 6-31.)
- The vestibular apparatus in the inner ear consists of (1) the semicircular canals, which detect rotational acceleration or deceleration in any direction, and (2) the utricle and the saccule, which detect changes in the rate of linear movement in any direction and provide information important for determining head position in relation to gravity. (Review Figure 6-33.)
- Neural signals are generated in response to the mechanical deformation of hair cells by specific movement of fluid and related structures within these vestibular sense organs. This information is important for the sense of equilibrium and for maintaining posture. (Review Figures 6-34 and 6-35.)

Chemical Senses: Taste and Smell (pp. 175–180)

- Taste and smell are chemical senses. In both cases, attachment of specific dissolved molecules to binding sites on the receptor membrane causes receptor potentials that, in turn, set up neural impulses that signal the presence of the chemical.

- Taste receptors are housed in taste buds on the tongue; olfactory receptors are located in the olfactory mucosa in the upper part of the nasal cavity. Both sensory pathways include two routes: one to the limbic system for emotional and behavioral processing and one to the cortex for conscious perception and fine discrimination. (*Review Figures 6-36 and 6-37.*)
- Taste and olfactory receptors are continuously renewed, unlike visual and hearing receptors, which are irreplaceable.
- The four classic primary tastes are salt, sour, sweet, and bitter, along with a likely fifth primary taste, umami, a meaty, "amino-acid" taste. Taste discrimination beyond the primary tastes depends on the patterns of stimulation of the taste buds, each of which responds in varying degrees to the different primary tastes.
- There are 1000 different types of olfactory receptors, each of which responds to only one discrete component of an odor. The afferent signals arising from the olfactory receptors are sorted according to scent component by the glomeruli within the olfactory bulb. Odor discrimination depends on the patterns of activation of the glomeruli. (*Review Figure 6-38.*)

REVIEW EXERCISES

Objective Questions (Answers on p. A-42)

1. The sensation perceived depends on the type of receptor stimulated rather than on the type of stimulus. (*True or false?*)
2. All afferent information is sensory information. (*True or false?*)
3. During dark adaptation, rhodopsin is gradually regenerated to increase the sensitivity of the eyes. (*True or false?*)
4. An optic nerve carries information from the lateral and medial halves of the same eye, whereas an optic tract carries information from the lateral half of one eye and the medial half of the other. (*True or false?*)
5. Displacement of the round window generates neural impulses that are perceived as sound sensations. (*True or false?*)
6. Hair cells in different regions of the organ of Corti are activated by different tones. (*True or false?*)
7. Each taste receptor responds to just one of the primary tastes. (*True or false?*)
8. Rapid adaptation to odors results from adaptation of the olfactory receptors. (*True or false?*)
9. Conversion of the energy forms of stimuli into electrical energy by the receptors is known as _____.
10. The type of stimulus to which a particular receptor is most responsive is called its _____.
11. Match the following:

   _____ 1. layer that contains the photoreceptors
   _____ 2. point from which the optic nerve leaves the retina
   _____ 3. forms the white part of the eye
   _____ 4. colored diaphragm of muscle that controls the amount of light entering the eye
   _____ 5. contributes most to the eye's refractive ability
   _____ 6. supplies nutrients to the lens and cornea
   _____ 7. produces aqueous humor
   _____ 8. has adjustable refractive ability
   _____ 9. portion of the retina with greatest acuity

   (a) aqueous humor
   (b) fovea
   (c) cornea
   (d) retina
   (e) lens
   (f) optic disc; blind spot
   (g) iris
   (h) ciliary body
   (i) sclera

12. Using the following answer code, indicate which properties apply to taste and/or smell:

   (a) applies to taste
   (b) applies to smell
   (c) applies to both taste and smell

   _____ 1. Receptors are separate cells that synapse with terminal endings of afferent neurons.
   _____ 2. Receptors are specialized endings of afferent neurons.
   _____ 3. Receptors are regularly replaced.
   _____ 4. Specific chemicals in the environment attach to special binding sites on the receptor surface, leading to a depolarizing receptor potential.
   _____ 5. There are two processing pathways: a limbic system route and a cortical route.
   _____ 6. Four (or possibly five) different receptor types are used.
   _____ 7. A thousand different receptor types are used.
   _____ 8. Information from receptor cells is filed and sorted by neural junctions called *glomeruli.*

Essay Questions

1. List and describe the receptor types according to their adequate stimulus.
2. Compare tonic and phasic receptors.
3. Explain how acuity is influenced by receptive field size.
4. Compare the fast and slow pain pathways.
5. Describe the built-in analgesic system of the brain.
6. Describe the process of phototransduction.
7. Compare the functional characteristics of rods and cones.
8. What are sound waves? What is responsible for the pitch, intensity, and timbre of a sound?
9. Describe the function of each of the following parts of the ear: pinna, ear canal, tympanic membrane, ossicles, oval window, and the various parts of the cochlea. Include a discussion of how sound waves are transduced into action potentials.
10. Discuss the functions of the semicircular canals, the utricle, and the saccule.
11. Describe the location, structure, and activation of the receptors for taste and smell.
12. Compare the processes of color vision, hearing, taste, and smell discrimination.

POINTS TO PONDER

(Explanations on p. A-42)

1. Patients with certain nerve disorders are unable to feel pain. Why is this disadvantageous?

2. Ophthalmologists frequently instill eye drops in their patients' eyes to bring about pupillary dilation, which makes it easier for the physician to view the eye's interior. The iris contains two sets of smooth muscle, one *circular* (the muscle fibers run in a circle around the pupil) and the other *radial* (the fibers radiate outward from the pupil like bicycle spokes). Remember that a muscle fiber shortens when it contracts. Parasympathetic stimulation causes contraction of the circular muscle; sympathetic stimulation causes contraction of the radial muscle. In what way would the drug in the eye drops affect autonomic nervous system activity in the eye to cause the pupils to dilate?

3. A patient complains of not being able to see the right half of the visual field with either eye. At what point in the patient's visual pathway does the defect lie?

4. Explain how middle ear infections interfere with hearing. Of what value are the "tubes" that are sometimes surgically placed in the eardrums of patients with a history of repeated middle ear infections accompanied by chronic fluid accumulation?

5. Explain why your sense of smell is reduced when you have a cold, even though the cold virus does not directly adversely affect the olfactory receptor cells.

CLINICAL CONSIDERATION

(Explanation on p. A-42)

Suzanne J. complained to her physician of bouts of dizziness. The physician asked her whether by "dizziness" she meant a feeling of lightheadedness, as if she felt she were going to faint (a condition known as *syncope*), or a feeling that she or surrounding objects in the room were spinning around (a condition known as *vertigo*). Why is this distinction important in the differential diagnosis of her condition? What are some possible causes of each of these symptoms?

PHYSIOEDGE RESOURCES

PhysioEdge CD-ROM

PhysioEdge, the CD-ROM packaged with your text, focuses on the concepts students find most difficult to learn. Figures marked with this icon have associated activities on the CD. For a visual review of concepts in this chapter, check out the following:

Tutorial: Special Senses

Media Exercise 6.1: Pain

Media Exercise 6.2: The Eye

Media Exercise 6.3: The Ear

Media Exercise 6.4: Taste

PhysioEdge Website

The website for this book contains a wealth of helpful study aids, as well as many ideas for further reading and research. Log on to:
http://infotrac.thomsonlearning.com
Select Chapter 6 from the drop-down menu, or click on one of the many resource areas.

For Suggested Readings, consult **InfoTrac College Edition/ Research** on the PhysioEdge website or go directly to InfoTrac College Edition, your online research library, at:
http://infotrac.thomsonlearning.com

Nervous System
(Peripheral Nervous System)

Body systems maintain homeostasis

Homeostasis
The nervous system, as one of the body's two major control systems, regulates many body activities aimed at maintaining a stable internal fluid environment.

Homeostasis is essential for survival of cells

Cells

Cells make up body systems

The nervous system, one of the two major regulatory systems of the body, consists of the central nervous system (CNS), composed of the brain and spinal cord, and the **peripheral nervous system,** composed of the afferent and efferent fibers that relay signals between the CNS and periphery (other parts of the body).

Once informed by the afferent division of the peripheral nervous system that a change in the internal or external environment is threatening homeostasis, the CNS makes appropriate adjustments to maintain homeostasis. The CNS makes these adjustments by controlling the activities of effector organs (muscles and glands) by transmitting signals *from* the CNS to these organs through the **efferent division** of the peripheral nervous system.

The Peripheral Nervous System: Efferent Division

INTRODUCTION

The efferent division of the peripheral nervous system is the communication link by which the central nervous system controls the activities of muscles and glands, the effector organs that carry out the intended effects, or actions. The CNS regulates these effector organs by initiating action potentials in the cell bodies of efferent neurons whose axons terminate on these organs. Cardiac muscle, smooth muscle, most exocrine glands, and some endocrine glands are innervated by the **autonomic nervous system**, the involuntary branch of the peripheral efferent division. Skeletal muscle is innervated by the **somatic nervous system**, the branch of the efferent division subject to voluntary control. Efferent output typically influences either movement or secretion. Much of this efferent output is directed toward maintaining homeostasis. The efferent output to skeletal muscles is also directed toward voluntarily controlled nonhomeostatic activities, such as riding a bicycle. (It is important to realize that many effector organs are also subject to hormonal control and/or intrinsic control mechanisms; see p. 12).

How many different neurotransmitters would you guess are released from the various efferent neuronal terminals to elicit essentially all the neurally controlled effector organ responses? Only two—acetylcholine and norepinephrine! Acting independently, these neurotransmitters bring about such diverse effects as salivary secretion, bladder contraction, and voluntary motor movements. These effects are a prime example of how the same chemical messenger may elicit a multiplicity of responses from various tissues, depending on specialization of the effector organs.

AUTONOMIC NERVOUS SYSTEM

▌ An autonomic nerve pathway consists of a two-neuron chain.

Each autonomic nerve pathway extending from the CNS to an innervated organ is a two-neuron chain

(● Figure 7-1). The cell body of the first neuron in the series is located in the CNS. Its axon, the **preganglionic fiber**, synapses with the cell body of the second neuron, which lies within a ganglion. (Recall that a ganglion is a cluster of neuronal cell bodies outside the CNS.) The axon of the second neuron, the **postganglionic fiber**, innervates the effector organ.

The autonomic nervous system has two subdivisions—the **sympathetic** and the **parasympathetic nervous systems** (● Figure 7-2). Sympathetic nerve fibers originate in the thoracic and lumbar regions of the spinal cord (see p. 135). Most sympathetic preganglionic fibers are very short, synapsing with cell bodies of postganglionic neurons within ganglia that lie in a **sympathetic ganglion chain** (also called the **sympathetic trunk**) located along either side of the spinal cord. Long postganglionic fibers originating in the ganglion chain end on the effector organs. Some preganglionic fibers pass through the ganglion chain without synapsing. Instead, they end later in sympathetic **collateral ganglia** about halfway between the CNS and the innervated organs, with postganglionic fibers traveling the rest of the distance.

Parasympathetic preganglionic fibers arise from the cranial (brain) and sacral (lower spinal cord) areas of the CNS. These fibers are longer than sympathetic preganglionic fibers because they do not end until they reach **terminal ganglia** that lie in or near the effector organs. Very short postganglionic fibers end on the cells of an organ itself.

▮ Parasympathetic postganglionic fibers release acetylcholine; sympathetic ones release norepinephrine.

Sympathetic and parasympathetic preganglionic fibers release the same neurotransmitter, **acetylcholine (ACh)**, but the postganglionic endings of these two systems release different neurotransmitters (the neurotransmitters that influence the effector organs). Parasympathetic postganglionic fibers release acetylcholine. Accordingly, they, along with all autonomic preganglionic fibers, are called **cholinergic fibers**. Most sympathetic postganglionic fibers, in contrast, are called **adrenergic fibers** because they release **noradrenaline**, commonly known as **norepinephrine**.

Postganglionic autonomic fibers do not end in a single terminal swelling like a synaptic knob. Instead, the terminal branches of autonomic fibers have numerous swellings, or **varicosities**, that simultaneously release neurotransmitter over a large area of the innervated organ rather than on single cells (● Figure 7-1 and ● Figure 8-25, p. 234). This diffuse release of neurotransmitter, coupled with the fact that any resulting change in electrical activity is spread throughout a smooth or cardiac muscle mass via gap junctions (see p. 49), means that autonomic activity typically influences whole organs instead of discrete cells.

▮ The sympathetic and parasympathetic nervous systems dually innervate most visceral organs.

Afferent information coming from the viscera (internal organs) usually does not reach the conscious level. Likewise visceral activities such as circulation, digestion, sweating, and pupillary size are regulated outside the realm of consciousness and voluntary control via autonomic efferent output.

Most visceral organs are innervated by both sympathetic and parasympathetic nerve fibers (● Figure 7-3). ▲ Table 7-1 (p. 189) summarizes the major effects of these autonomic branches. Although the details of this wide array of autonomic responses are described more fully in later chapters that discuss the individual organs involved, you can consider several general concepts now. As you can see from the table, the sympathetic and parasympathetic nervous systems generally exert opposite effects in a particular organ. Sympathetic stimulation increases the heart rate, whereas parasympathetic stimulation decreases it; sympathetic stimulation slows down movement within the digestive tract, whereas parasympathetic stimulation enhances digestive motility. Note that both systems increase the activity of some organs and reduce the activity of others.

Rather than memorizing a list such as in ▲ Table 7-1, it is better to logically deduce the actions of the two systems by first understanding the circumstances under which each system dominates. Usually both systems are partially active; that is, normally some level of action potential activity exists in both

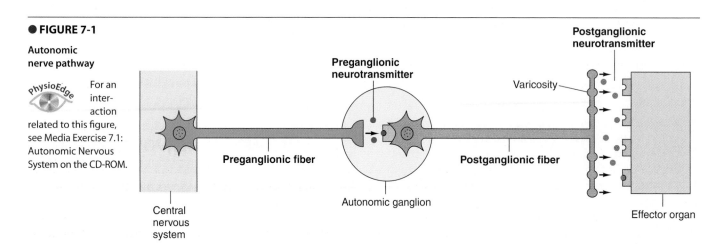

● **FIGURE 7-1**

Autonomic nerve pathway

PhysioEdge For an interaction related to this figure, see Media Exercise 7.1: Autonomic Nervous System on the CD-ROM.

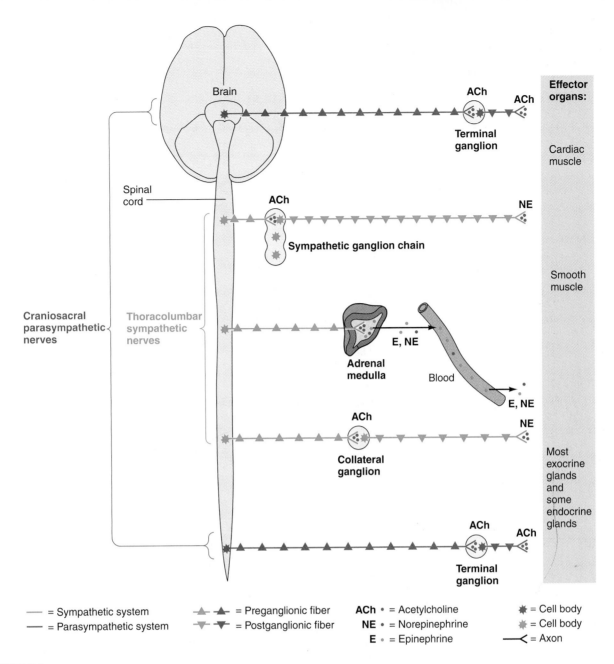

● FIGURE 7-2

Autonomic nervous system. The sympathetic nervous system, which originates in the thoracolumbar regions of the spinal cord, has short cholinergic (acetylcholine-releasing) preganglionic fibers and long adrenergic (norepinephrine-releasing) postganglionic fibers. The parasympathetic nervous system, which originates in the brain and sacral region of the spinal cord, has long cholinergic preganglionic fibers and short cholinergic postganglionic fibers. In most instances, sympathetic and parasympathetic postganglionic fibers both innervate the same effector organs. The adrenal medulla is a modified sympathetic ganglion, which releases epinephrine and norepinephrine into the blood.

the sympathetic and the parasympathetic fibers supplying a particular organ. This ongoing activity is called **sympathetic** or **parasympathetic tone** or **tonic activity.** Under given circumstances, activity of one division can dominate the other. *Sympathetic dominance* to a particular organ exists when the sympathetic fibers' rate of firing to that organ increases above tonic level, coupled with a simultaneous decrease below tonic

level in the parasympathetic fibers' frequency of action potentials to the same organ. The reverse situation is true for *parasympathetic dominance.* The balance between sympathetic and parasympathetic activity can be shifted separately for individual organs to meet specific demands, or a more generalized, widespread discharge of one autonomic system in favor of the other can be elicited to control bodywide functions.

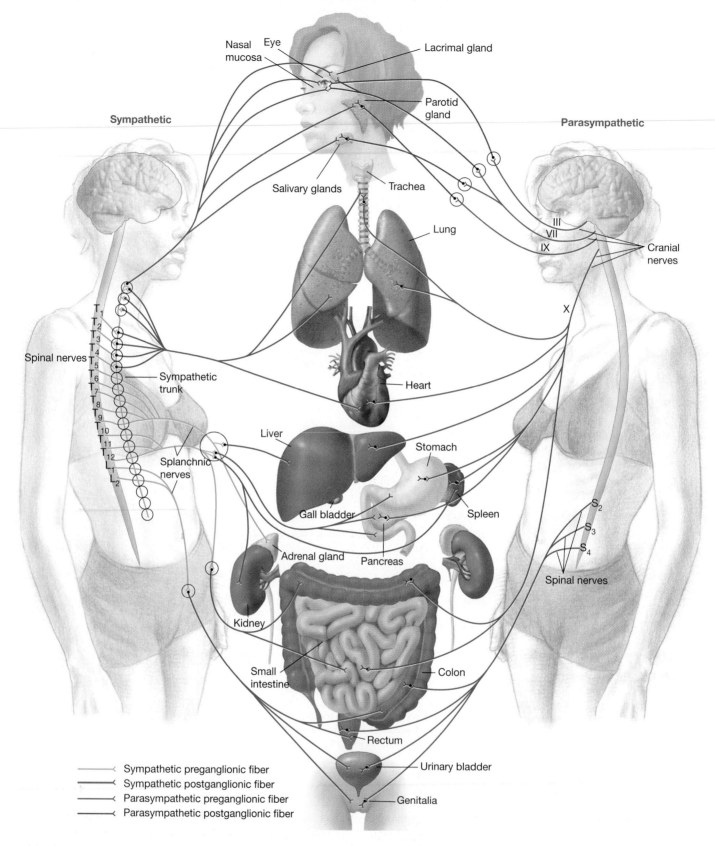

Sympathetic

Parasympathetic

Nasal mucosa · Eye · Lacrimal gland · Parotid gland · Salivary glands · Trachea · Lung · Heart · Liver · Stomach · Gall bladder · Spleen · Adrenal gland · Pancreas · Kidney · Small intestine · Colon · Rectum · Urinary bladder · Genitalia

Spinal nerves

Sympathetic trunk

Splanchnic nerves

III
VII
IX
X
Cranial nerves

S₂
S₃
S₄
Spinal nerves

————< Sympathetic preganglionic fiber
————< Sympathetic postganglionic fiber
————< Parasympathetic preganglionic fiber
————< Parasympathetic postganglionic fiber

● **FIGURE 7-3**

Schematic representation of the structures innervated by the sympathetic and parasympathetic nervous system

For an interaction related to this figure, see Media Exercise 7.1: Autonomic Nervous System on the CD-ROM.

Effects of Autonomic Nervous System on Various Organs

ORGAN	EFFECT OF SYMPATHETIC STIMULATION	EFFECT OF PARASYMPATHETIC STIMULATION
Heart	Increased rate, increased force of contraction (of whole heart)	Decreased rate, decreased force of contraction (of atria only)
Blood Vessels	Constriction	Dilation of vessels supplying the penis and clitoris only
Lungs	Dilation of bronchioles (airways)	Constriction of bronchioles
	Inhibition (?) of mucus secretion	Stimulation of mucus secretion
Digestive Tract	Decreased motility (movement)	Increased motility
	Contraction of sphincters (to prevent forward movement of contents)	Relaxation of sphincters (to permit forward movement of contents)
	Inhibition (?) of digestive secretions	Stimulation of digestive secretions
Urinary Bladder	Relaxation	Contraction (emptying)
Eye	Dilation of pupil	Constriction of pupil
	Adjustment of eye for far vision	Adjustment of eye for near vision
Liver (glycogen stores)	Glycogenolysis (glucose released)	None
Adipose Cells (fat stores)	Lipolysis (fatty acids released)	None
Exocrine Glands		
Exocrine pancreas	Inhibition of pancreatic exocrine secretion	Stimulation of pancreatic exocrine secretion (important for digestion)
Sweat glands	Stimulation of secretion by most sweat glands	Stimulation of secretion by some sweat glands
Salivary glands	Stimulation of small volume of thick saliva rich in mucus	Stimulation of large volume of watery saliva rich in enzymes
Endocrine Glands		
Adrenal medulla	Stimulation of epinephrine and norepinephrine secretion	None
Endocrine pancreas	Inhibition of insulin secretion; stimulation of glucagon secretion	Stimulation of insulin and glucagon secretion
Genitals	Ejaculation and orgasmic contractions (males); orgasmic contractions (females)	Erection (caused by dilation of blood vessels in penis [male] and clitoris [female])
Brain Activity	Increased alertness	None

Massive widespread discharges take place more frequently in the sympathetic system. The value of this potential for massive sympathetic discharge is clear, considering the circumstances during which this system usually dominates.

TIMES OF SYMPATHETIC DOMINANCE

The sympathetic system promotes responses that prepare the body for strenuous physical activity in emergency or stressful situations, such as a physical threat from the outside. This response is typically referred to as a **fight-or-flight response**, because the sympathetic system readies the body to fight against or flee from the threat. Think about the body resources needed in such circumstances. The heart beats more rapidly and more forcefully, blood pressure is elevated by generalized constriction of the blood vessels, respiratory airways open wide to permit maximal air flow, glycogen (stored sugar) and fat stores are broken down to release extra fuel into the blood, and blood vessels supplying skeletal muscles dilate (open more widely). All these responses are aimed at providing increased flow of oxygenated, nutrient-rich blood to the skeletal muscles in anticipation of strenuous physical activity. Furthermore, the pupils dilate and the eyes adjust for far vision, letting the person visually assess the entire threatening scene. Sweating is promoted in anticipation of excess heat production by the physical exertion. Because digestive and urinary activities are not essential in meeting the threat, the sympathetic system inhibits these activities.

TIMES OF PARASYMPATHETIC DOMINANCE

The parasympathetic system dominates in quiet, relaxed situations. Under such nonthreatening circumstances, the body can be concerned with its own "general housekeeping" activities, such as digestion. The parasympathetic system promotes these **"rest-and-digest"** types of bodily functions while slowing down those activities that are enhanced by the sympathetic system. There is no need, for example, to have the heart beating rapidly and forcefully when the person is in a tranquil setting.

ADVANTAGE OF DUAL AUTONOMIC INNERVATION

What is the advantage of dual innervation of organs with nerve fibers whose actions oppose each other? It enables precise control over an organ's activity, like having both an accelerator and a brake to control the speed of a car. If an animal suddenly darts across the road as you are driving, you could eventually stop if you just took your foot off the accelerator, but you might stop too slowly to avoid hitting the animal. If you simultaneously apply the brake as you lift up on the accelerator, however, you can come to a more rapid, controlled stop. In a similar manner, a sympathetically accelerated heart rate could gradually be reduced to normal following a stressful situation by decreasing the firing rate in the cardiac sympathetic nerve (letting up on the accelerator), but the heart rate can be reduced more rapidly by simultaneously increasing activity in the parasympathetic supply to the heart (applying the brake). Indeed, the two divisions of the autonomic nervous system are usually reciprocally controlled; increased activity in one division is accompanied by a corresponding decrease in the other.

We are now going to turn our attention to the adrenal medulla, a unique endocrine component of the sympathetic nervous system.

▌ The adrenal medulla is a modified part of the sympathetic nervous system.

There are two *adrenal glands,* one lying above the kidney on each side (*ad* means "next to"; *renal* means "kidney"). The adrenal glands are endocrine glands, each with an outer portion, the *adrenal cortex,* and an inner portion, the *adrenal medulla.* The **adrenal medulla** is considered a modified sympathetic ganglion that does not give rise to postganglionic fibers. Instead, on stimulation by the preganglionic fiber that originates in the CNS it secretes hormones into the blood (● Figures 7-2 and 7-4). Not surprisingly, the hormones are identical or similar to postganglionic sympathetic neurotransmitters.

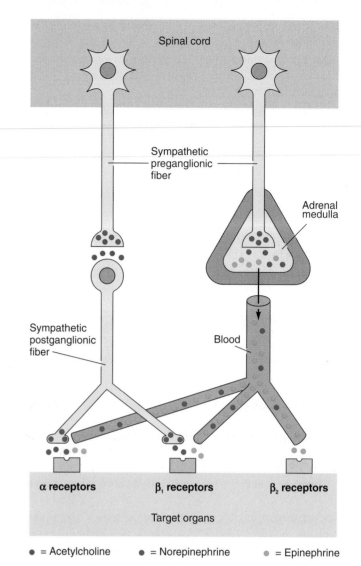

● **FIGURE 7-4**

Comparison of the release and binding to receptors of epinephrine and norepinephrine. Norepinephrine is released both as a neurotransmitter from sympathetic postganglionic fibers and as a hormone from the adrenal medulla. Beta-1 (β_1) receptors bind equally with both norepinephrine and epinephrine, whereas beta-2 (β_2) receptors bind primarily with epinephrine and alpha (α) receptors of both subtypes have a greater affinity for norepinephrine than for epinephrine.

About 20% of the adrenal medullary hormone output is norepinephrine, and the remaining 80% is the closely related substance **epinephrine (adrenaline).** These hormones, in general, reinforce activity of the sympathetic nervous system.

▌ Several different receptor types are available for each autonomic neurotransmitter.

Because each autonomic neurotransmitter and medullary hormone stimulate activity in some tissues but inhibit activity in others, the particular responses must depend on specialization of the tissue cells rather than on properties of the chemicals themselves. Responsive tissue cells have one or more of several different types of plasma membrane receptor proteins for these chemical messengers. Binding of a neurotransmitter to a receptor induces the tissue-specific response.

CHOLINERGIC RECEPTORS

Researchers have identified two types of acetylcholine (cholinergic) receptors—*nicotinic* and *muscarinic receptors*—on the basis of their response to particular drugs. Nicotinic receptors are activated by the tobacco plant derivative nicotine, whereas muscarinic receptors are activated by the mushroom poison muscarine. **Nicotinic receptors** are found on the postganglionic cell bodies in all autonomic ganglia. These receptors respond to acetylcholine released from both sympathetic and parasympathetic preganglionic fibers. **Muscarinic receptors** are found on effector cell membranes (smooth muscle, cardiac muscle, and glands). They bind with acetylcholine released from parasympathetic postganglionic fibers.

ADRENERGIC RECEPTORS

There are two major classes of adrenergic receptors for norepinephrine and epinephrine: *alpha* (α) and *beta* (β) *receptors*, which are further subclassified into α_1 and α_2 and β_1 and β_2 receptors. These various receptor types are distinctly distributed among the effector organs. Receptors of the β_2 type bind primarily with epinephrine, whereas β_1 receptors have about equal affinities for norepinephrine and epinephrine, and α receptors of both subtypes have a greater sensitivity to norepinephrine than to epinephrine (● Figure 7-4).

Activation of α_1 receptors usually brings about an excitatory response in the effector organ—for example, arteriolar constriction caused by increased contraction of smooth muscle in the walls of these blood vessels. Alpha-1 receptors are present in most sympathetic target tissues. Activation of α_2 receptors, in contrast, brings about an inhibitory response in the effector organ, such as decreased smooth-muscle contraction in the digestive tract. Stimulation of β_1 receptors, which are found primarily in the heart, causes an excitatory response, namely increased rate and force of cardiac contraction. The response to β_2 receptor activation is generally inhibitory, such as arteriolar or bronchiolar (respiratory airway) dilation caused by relaxation of the smooth muscle in the walls of these tubular structures.

AUTONOMIC AGONISTS AND ANTAGONISTS

Drugs are available that selectively enhance or depress autonomic responses at each of the receptor types. An **agonist** binds with a neurotransmitter's receptor and elicits an effect that mimics that of the neurotransmitter. An **antagonist** by contrast binds with a neurotransmitter's receptor and blocks the neurotransmitter's response. When an antagonist binds with a receptor, it does not elicit a response itself, but, by occupying the receptor, it prevents a released neurotransmitter from binding with its own receptor and bringing about its effect.

Some of these drugs are only of experimental interest, but others are very important therapeutically. For example, *atro-*

Clinical Note *pine* blocks the effect of acetylcholine at muscarinic receptors but does not affect nicotinic receptors. Because the acetylcholine released at both parasympathetic and sympathetic preganglionic fibers combines with nicotinic receptors, blockage at nicotinic synapses would knock out both these autonomic branches. By acting selectively to interfere with acetylcholine action only at muscarinic junctions, which are the sites of parasympathetic postganglionic action, atropine effectively blocks parasympathetic effects but does not influence sympathetic activity at all. Doctors use this principle to suppress salivary and bronchial secretions before surgery, to reduce the risk of a patient inhaling these secretions into the lungs.

Likewise, drugs that act selectively at α and β adrenergic receptor sites to either activate or block specific sympathetic effects are widely used. *Salbutamol* is an excellent example. It selectively activates β_2 adrenergic receptors at low doses, making it possible to dilate the bronchioles in the treatment of asthma without undesirably stimulating the heart (the heart has mostly β_1 receptors).

▌ Many regions of the central nervous system are involved in the control of autonomic activities.

Messages from the CNS are delivered to cardiac muscle, smooth muscle, and glands via the autonomic nerves, but what regions of the CNS regulate autonomic output?

- Some autonomic reflexes, such as urination, defecation, and erection, are integrated at the spinal cord level, but all these spinal reflexes are subject to control by higher levels of consciousness.
- The medulla within the brain stem is the region most directly responsible for autonomic output. Centers for controlling cardiovascular, respiratory, and digestive activity via the autonomic system are located there.
- The hypothalamus plays an important role in integrating the autonomic, somatic, and endocrine responses that automatically accompany various emotional and behavioral states. For example, the increased heart rate, blood pressure, and respiratory activity associated with anger or fear are brought about by the hypothalamus acting through the medulla.
- Autonomic activity can also be influenced by the prefrontal association cortex through its involvement with emotional expression characteristic of the individual's personality. An example is blushing when embarrassed, which is caused by dilation of blood vessels supplying the skin of the cheeks. Such responses are mediated through hypothalamic-medullary pathways.

▲ Table 7-2 summarizes the main distinguishing features of the sympathetic and parasympathetic nervous systems.

PhysioEdge Click on the Media Exercises menu of the CD-ROM and work Media Exercise 7.1: Autonomic Nervous System to test your understanding of the previous section.

FEATURE	SYMPATHETIC SYSTEM	PARASYMPATHETIC SYSTEM
Origin of Preganglionic Fiber	Thoracic and lumbar regions of spinal cord	Brain and sacral region of spinal cord
Origin of Postganglionic Fiber (location of ganglion)	Sympathetic ganglion chain (near spinal cord) or collateral ganglia (about halfway between spinal cord and effector organs)	Terminal ganglia (in or near effector organs)
Length and Type of Fiber	Short cholinergic preganglionic fibers	Long cholinergic preganglionic fibers
	Long adrenergic postganglionic fibers	Short cholinergic postganglionic fibers
Effector Organs Innervated	Cardiac muscle, almost all smooth muscle, most exocrine glands, and some endocrine glands	Cardiac muscle, most smooth muscle, most exocrine glands, and some endrocrine glands
Types of Receptors for Neurotransmitters	For preganglionic neurotransmitter: nicotinic For postganglionic neurotransmitter: $\alpha_1, \alpha_2, \beta_1, \beta_2$	For preganglionic neurotransmitter: nicotinic For postganglionic neurotransmitter: muscarinic
Dominance	Dominates in emergency "fight-or-flight" situations; prepares body for strenuous physical activity	Dominates in quiet, relaxed situations; promotes "general housekeeping" activities such as digestion

SOMATIC NERVOUS SYSTEM

▌ Motor neurons supply skeletal muscle.

Skeletal muscle is innervated by **motor neurons**, the axons of which constitute the somatic nervous system. The cell bodies of almost all motor neurons are within the spinal cord. The only exception: the cell bodies of motor neurons supplying muscles in the head are in the brain stem. Unlike the two-neuron chain of autonomic nerve fibers, the axon of a motor neuron is continuous from its origin in the CNS to its ending on skeletal muscle. Motor neuron axon terminals release acetylcholine, which brings about excitation and contraction of the innervated muscle cells. Motor neurons can only stimulate skeletal muscles, in contrast to autonomic fibers, which can either stimulate or inhibit their effector organs. Inhibition of skeletal muscle activity can be accomplished only within the CNS through inhibitory synaptic input to the dendrites and cell bodies of the motor neurons supplying that particular muscle.

▌ Motor neurons are the final common pathway.

Motor neuron dendrites and cell bodies are influenced by many converging presynaptic inputs, both excitatory and inhibitory. Some of these inputs are part of spinal reflex pathways originating with peripheral sensory receptors. Others are part of descending pathways originating within the brain. Areas of the brain that exert control over skeletal muscle movements include the motor regions of the cortex, the basal nuclei, the cerebellum, and the brain stem (see p. 120 and p. 223).

Motor neurons are considered the **final common pathway**, because the only way any other parts of the nervous system can influence skeletal muscle activity is by acting on these motor neurons. The level of activity in a motor neuron and its subsequent output to the skeletal muscle fibers it innervates depend on the relative balance of EPSPs and IPSPs (see p. 88) brought about by its presynaptic inputs originating from these diverse sites in the brain.

The somatic system is under voluntary control, but much of the skeletal muscle activity involving posture, balance, and stereotypical movements is subconsciously controlled. You may decide you want to start walking, but you do not have to consciously bring about the alternate contraction and relaxation of the involved muscles because these movements are involuntarily coordinated by lower brain centers.

Clinical Note The cell bodies of the crucial motor neurons may be selectively destroyed by **poliovirus**. The result is paralysis of the muscles innervated by the affected neurons.

Amyotrophic lateral sclerosis (ALS), also known as **Lou Gehrig's disease**, is the most common motor-neuron disease. This incurable condition is characterized by degeneration and eventually death of motor neurons. The result is gradual loss

Comparison of the Autonomic Nervous System and the Somatic Nervous System

FEATURE	AUTONOMIC NERVOUS SYSTEM	SOMATIC NERVOUS SYSTEM
Site of Origin	Brain or spinal cord	Spinal cord for most; those supplying muscles in head originate in brain
Number of Neurons from Origin in CNS to Effector Organ	Two-neuron chain (preganglionic and postganglionic)	Single neuron (motor neuron)
Organs Innervated	Cardiac muscle, smooth muscle, exocrine and some endocrine glands	Skeletal muscle
Type of Innervation	Most effector organs dually innervated by the two antagonistic branches of this system (sympathetic and parasympathetic)	Effector organs innervated only by motor neurons
Neurotransmitter at Effector Organs	May be acetylcholine (parasympathetic terminals) or norepinephrine (sympathetic terminals)	Only acetylcholine
Effects on Effector Organs	Either stimulation or inhibition (antagonistic actions of two branches)	Stimulation only (inhibition possible only centrally through IPSPs on dendrites and cell body of motor neuron)
Types of Control	Under involuntary control	Subject to voluntary control; much activity subconsciously coordinated
Higher Centers Involved in Control	Spinal cord, medulla, hypothalamus, prefrontal association cortex	Spinal cord, motor cortex, basal nuclei, cerebellum, brain stem

of motor control, progressive paralysis, and finally, death within one to five years of onset.

Before turning our attention to the junction between a motor neuron and the muscle cells it innervates, we are going to pull together in tabular form two groups of information we have been examining. ▲ Table 7-3 summarizes the features of the two divisions of the efferent division of the peripheral nervous system. ▲ Table 7-4 compares the three functional types of neurons.

▍ Motor neurons and skeletal muscle fibers are chemically linked at neuromuscular junctions.

An action potential in a motor neuron is rapidly propagated from the cell body within the CNS to the skeletal muscle along the large myelinated axon (efferent fiber) of the neuron. As the axon approaches a muscle, it divides into many terminal branches and loses its myelin sheath. Each of these axon terminals forms a special junction, a **neuromuscular junction**, with one of the many muscle cells that compose the whole

muscle (● Figure 7-5). A single muscle cell, called a **muscle fiber**, is long and cylindrical. The axon terminal is enlarged into a knoblike structure, the **terminal button**, which fits into a shallow depression, or groove, in the underlying muscle fiber (● Figure 7-6). Some scientists alternatively call the neuromuscular junction a "motor end plate." However, we will reserve the term **motor end plate** for the specialized portion of the muscle cell membrane immediately under the terminal button.

▍ Acetylcholine is the neuromuscular junction neurotransmitter.

Nerve and muscle cells do not actually come into direct contact at a neuromuscular junction. The space, or cleft, between these two structures is too large to permit electrical transmission of an impulse between them (that is, the action potential cannot "jump" that far). Therefore, just as at a neuronal synapse (see p. 85), a chemical messenger carries the signal between the neuron terminal and the muscle fiber. This neurotransmitter is acetylcholine (ACh.)

TABLE 7-4

Comparison of Neuron Types

FEATURE	AFFERENT NEURON	EFFERENT NEURON		INTERNEURON
		Autonomic Nervous System	Somatic Nervous System	
Origin, Structure, Location	Receptor at peripheral ending; elongated peripheral axon, which travels in peripheral nerve; cell body located in dorsal root ganglion; short central axon entering spinal cord	Two-neuron chain; first neuron (preganglionic fiber) originating in CNS and terminating on a ganglion; second neuron (postganglionic fiber) originating in the ganglion and terminating on effector organ	Cell body of motor neuron lying in spinal cord; long axon traveling in peripheral nerve and terminating on effector organ	Various shapes; lying entirely within CNS; some cell bodies originating in brain, with long axons traveling down spinal cord in descending pathways; some originating in spinal cord, with long axons traveling up cord to brain in ascending pathways; others forming short local connections
Termination	Interneurons*	Effector organs (cardiac muscle, smooth muscle, exocrine and some endocrine glands)	Effector organs (skeletal muscle)	Other interneurons and efferent neurons
Function	Carries information about external and internal environment to CNS	Carries instructions from CNS to effector organs	Carries instructions from CNS to effector organs	Processes and integrates afferent input; initiates and coordinates efferent output; responsible for thought and other higher mental functions
Convergence of Input on Cell Body	No (only input is through receptor)	Yes	Yes	Yes
Effect of Input to Neuron	Can only be excited (through receptor potential induced by stimulus; must reach threshold for action potential)	Can be excited or inhibited (through EPSPs and IPSPs at first neuron; must reach threshold for action potential)	Can be excited or inhibited (through EPSPs and IPSPs; must reach threshold for action potential)	Can be excited or inhibited (through EPSPs and IPSPs; must reach threshold for action potential)
Site of Action Potential Initiation	First excitable portion of membrane adjacent to receptor	Axon hillock	Axon hillock	Axon hillock
Divergence of Output	Yes	Yes	Yes	Yes
Effect of Output on Structure on Which It Terminates	Only excites	Postganglionic fiber either excites or inhibits	Only excites	Either excites or inhibits

*Except in the stretch reflex, where afferent neurons terminate on efferent neurons.

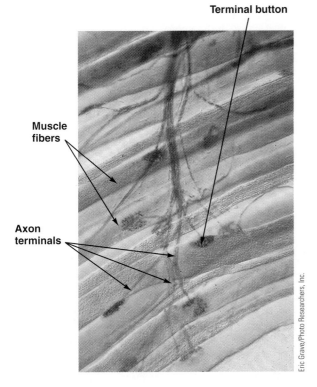

Terminal button

Muscle fibers

Axon terminals

Eric Grave/Photo Researchers, Inc.

● **FIGURE 7-5**

Motor neuron innervating skeletal muscle cells. When a motor neuron reaches a skeletal muscle, it divides into many terminal branches, each of which forms a neuromuscular junction with a single muscle cell (muscle fiber).

RELEASE OF ACh AT THE NEUROMUSCULAR JUNCTION

Each terminal button contains thousands of vesicles that store ACh. Propagation of an action potential to the axon terminal (step 1, ● Figure 7-6) triggers the opening of voltage-gated Ca^{2+} channels in the terminal button (see p. 72). Opening of Ca^{2+} channels permits Ca^{2+} to diffuse into the terminal button from its higher extracellular concentration (step 2), which in turn causes the release of ACh by exocytosis from several hundred of the vesicles into the cleft (step 3).

FORMATION OF AN END-PLATE POTENTIAL

The released ACh diffuses across the cleft and binds with specific receptor sites, which are specialized membrane proteins unique to the motor end-plate portion of the muscle fiber membrane (step 4). (These cholinergic receptors are of the nicotinic type.) Binding of ACh with these receptor sites induces the opening of chemically gated channels in the motor end plate. These channels permit a small amount of cation traffic through them (both Na^+ and K^+) but no anions (step 5). Because the permeability of the end-plate membrane to Na^+ and K^+ on opening of these channels is essentially equal, the relative movement of these ions through the channels depends on their electrochemical driving forces. Recall that at resting potential, the net driving force for Na^+ is much greater than

that for K^+, because the resting potential is much closer to the K^+ equilibrium potential. Both the concentration and electrical gradients for Na^+ are inward, whereas the outward concentration gradient for K^+ is almost, but not quite, balanced by the opposing inward electrical gradient. As a result, when ACh triggers the opening of these channels, considerably more Na^+ moves inward than K^+ outward, depolarizing the motor end plate. This potential change is called the **end-plate potential (EPP)**. It is a graded potential similar to an EPSP (excitatory postsynaptic potential; see p. 87), except that an EPP is much larger.

INITIATION OF AN ACTION POTENTIAL

The motor end-plate region itself does not have a threshold potential, so an action potential cannot be initiated at this site. However, an EPP brings about an action potential in the rest of the muscle fiber, as follows. The neuromuscular junction is usually in the middle of the long, cylindrical muscle fiber. When an EPP takes place, local current flow occurs between the depolarized end plate and the adjacent, resting cell membrane in both directions (step 6), opening voltage-gated Na^+ channels and thus reducing the potential to threshold in the adjacent areas (step 7). The subsequent action potential initiated at these sites propagates throughout the muscle fiber membrane by contiguous conduction (step 8) (see p. 80). The spread runs in both directions, away from the motor end plate toward both ends of the fiber. This electrical activity triggers contraction of the muscle fiber. Thus, by means of ACh, an action potential in a motor neuron brings about an action potential and subsequent contraction in the muscle fiber.

Unlike synaptic transmission, the magnitude of an EPP is normally enough to cause an action potential in the muscle cell. Therefore, one-to-one transmission of an action potential typically occurs at a neuromuscular junction; one action potential in a nerve cell triggers one action potential in a muscle cell that it innervates. Other comparisons of neuromuscular junctions with synapses can be found in ▲ Table 7-5.

▌ Acetylcholinesterase ends acetylcholine activity at the neuromuscular junction.

To ensure purposeful movement, electrical activity and the resultant contraction of the muscle cell turned on by motor neuron action potentials must be switched off promptly when there is no longer a signal from the motor neuron. The muscle cell's electrical response is turned off by an enzyme present in the motor end-plate membrane, **acetylcholinesterase (AChE)**, which inactivates ACh.

As a result of diffusion, many of the released ACh molecules come into contact with and bind to the receptor sites, which are on the surface of the motor end-plate membrane. However, some of the ACh molecules bind with AChE, which is also at the end-plate surface. Being quickly inactivated, this ACh never contributes to the end-plate potential. The acetylcholine that does bind with receptor sites does so very briefly

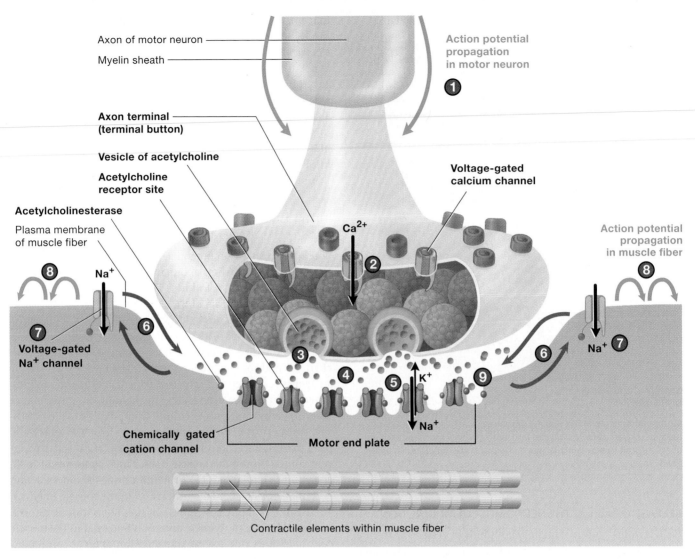

Axon of motor neuron

Myelin sheath

Action potential propagation in motor neuron

Axon terminal (terminal button)

Vesicle of acetylcholine

Acetylcholine receptor site

Acetylcholinesterase

Plasma membrane of muscle fiber

Voltage-gated calcium channel

Ca^{2+}

Action potential propagation in muscle fiber

Na$^+$

Voltage-gated Na$^+$ channel

Na$^+$

Chemically gated cation channel

K^+

Na$^+$

Motor end plate

Contractile elements within muscle fiber

① An action potential in a motor neuron is propagated to the axon terminal (terminal button).

② The presence of an action potential in the terminal button triggers the opening of voltage-gated Ca^{2+} channels and the subsequent entry of Ca^{2+} into the terminal button.

③ Ca^{2+} triggers the release of acetylcholine by exocytosis from a portion of the vesicles.

④ Acetylcholine diffuses across the space separating the nerve and muscle cells and binds with receptor sites specific for it on the motor end plate of the muscle cell membrane.

⑤ This binding brings about the opening of cation channels, leading to a relatively large movement of Na$^+$ into the muscle cell compared to a smaller movement of K$^+$ outward.

⑥ The result is an end-plate potential. Local current flow occurs between the depolarized end plate and adjacent membrane.

⑦ This local current flow opens voltage-gated Na$^+$ channels in the adjacent membrane.

⑧ The resultant Na$^+$ entry reduces the potential to threshold, initiating an action potential, which is propagated throughout the muscle fiber.

⑨ Acetylcholine is subsequently destroyed by acetylcholinesterase, an enzyme located on the motor end-plate membrane, terminating the muscle cell's response.

● **FIGURE 7-6**

Events at a neuromuscular junction

 For an animation of this figure, click the Nerve-to-Muscle Communication tab in the Skeletal Muscle Contraction tutorial on the CD-ROM.

Comparison of a Synapse and a Neuromuscular Junction

SIMILARITIES	DIFFERENCES
Both consist of two excitable cells separated by a narrow cleft that prevents direct transmission of electrical activity between them.	A synapse is a junction between two neurons. A neuromuscular junction exists between a motor neuron and a skeletal muscle fiber.
Axon terminals of both store chemical messengers (neurotransmitters) that are released by the Ca^{2+}-induced exocytosis of storage vesicles when an action potential reaches the terminal.	One-to-one transmission of action potentials occurs at a neuromuscular junction, whereas one action potential in a presynaptic neuron cannot by itself bring about an action potential in a postsynaptic neuron. An action potential in a postsynaptic neuron occurs only when summation of EPSPs brings the membrane to threshold.
In both, binding of the neurotransmitter with receptor sites in the membrane of the cell underlying the axon terminal opens specific channels in the membrane, permitting ionic movements that alter the membrane potential of the cell.	A neuromuscular junction is always excitatory (an EPP); a synapse may be either excitatory (an EPSP) or inhibitory (an IPSP).
The resultant change in membrane potential in both cases is a graded potential.	Inhibition of skeletal muscles cannot be accomplished at the neuromuscular junction; can occur only in CNS through IPSPs at dendrites and cell body of the motor neuron.

(for about 1 millionth of a second), then detaches. Some of the detached ACh molecules quickly rebind with receptor sites, keeping the end-plate channels open, but some randomly contact AChE instead this time around and are inactivated (step ⑨). As this process repeats, more and more ACh is inactivated until it has been virtually removed from the cleft within a few milliseconds after its release. Removing ACh ends the EPP, so the remainder of the muscle cell membrane returns to resting potential. Now the muscle cell can relax. Or, if sustained contraction is essential for the desired movement, another motor neuron action potential leads to the release of more ACh, which keeps the contractile process going. By removing contraction-inducing ACh from the motor end plate, AChE permits the choice of allowing relaxation to take place (no more ACh released) or keeping the contraction going (more ACh released), depending on the body's momentary needs.

▌ The neuromuscular junction is vulnerable to several chemical agents and diseases.

Clinical Note Several chemical agents and diseases are known to affect the neuromuscular junction by acting at different sites in the transmission process. Two well-known toxins—black widow spider venom and botulinum toxin—alter the release of ACh, but in opposite directions.

BLACK WIDOW SPIDER VENOM CAUSES EXPLOSIVE RELEASE OF ACh

The venom of black widow spiders exerts its deadly effect by triggering explosive release of ACh from the storage vesicles, not only at neuromuscular junctions but at all cholinergic sites. All cholinergic sites undergo prolonged depolarization, the most harmful result of which is respiratory failure. Breathing is accomplished by alternate contraction and relaxation of skeletal muscles, particularly the diaphragm. Respiratory paralysis occurs as a result of prolonged depolarization of the diaphragm. During this so-called *depolarization block,* the voltage-gated Na^+ channels are trapped in their inactivated state (see p. 76), prohibiting the initiation of new action potentials and resultant contraction of the diaphragm. Thus the victim cannot breathe.

BOTULINUM TOXIN BLOCKS RELEASE OF ACh

Botulinum toxin, in contrast, exerts its lethal blow by blocking the release of ACh from the terminal button in response to a motor neuron action potential. *Clostridium Botulinum* toxin causes botulism, a form of food poisoning. It most frequently results from improperly canned foods contaminated with clostridial bacteria that survive and multiply, producing their toxin in the process. When this toxin is consumed, it prevents muscles from responding to nerve impulses. Death is due to respiratory failure caused by inability to contract the diaphragm. Botulinum toxin is one of the most lethal poisons known; ingesting less than 0.0001 mg can kill an adult human. (See the accompanying boxed feature, ▌ Beyond the Basics, to learn about a surprising new wrinkle in the botulinum toxin story.)

CURARE BLOCKS ACTION OF ACh AT RECEPTOR SITES

Other chemicals interfere with neuromuscular junction activity by blocking the effect of released ACh. The best known example is the antagonist **curare**, which reversibly binds to the ACh receptor sites on the motor end plate. Unlike ACh,

Beyond the Basics

Botulinum Toxin's Reputation Gets a Facelift

The powerful toxin produced by *Clostridium botulinum* causes the deadly food poisoning, botulism. Yet this dreaded, highly lethal poison has been put to use as a treatment for alleviating specific movement disorders and, more recently, has been added to the list of tools that cosmetic surgeons use to fight wrinkles.

During the last decade, botulinum toxin, marketed in therapeutic doses as *Botox,* has offered welcome relief to people with a number of painful, disruptive neuromuscular diseases known categorically as **dystonias.** These conditions are characterized by spasms (excessive, sustained, involuntarily produced muscle contractions) that result in involuntary twisting or abnormal postures, depending on the body part affected. For example, painful neck spasms that twist the head to one side result from *spasmodic torticollis* (*tortus* means "twisted"; *collum* means "neck"), the most common dystonia. The problem is believed to arise from too little inhibitory compared to excitatory input to the motor neurons that supply the affected muscle. The reasons for this imbalance in motor neuron input are unknown. The end result of excessive motor neuron activation is sustained, disabling con-

traction of the muscle supplied by the overactive motor neurons. Fortunately, injecting minuscule amounts of botulinum toxin into the affected muscle causes a reversible, partial paralysis of the muscle. Botulinum toxin interferes with the release of muscle-contraction–causing acetylcholine from the overactive motor neurons at the neuromuscular junctions in the treated muscle. The goal is to inject just enough botulinum toxin to alleviate the troublesome spasmodic contractions but not enough to eliminate the normal contractions needed for ordinary movements. The therapeutic dose is considerably less than the amount of toxin needed to induce even mild symptoms of botulinum poisoning. Botulinum toxin is eventually cleared away, so its muscle-relaxing effects wear off after three to six months, at which time the treatment must be repeated.

The first dystonia for which Botox was approved as a treatment by the Food and Drug Administration (FDA) was *blepharospasm* (*blepharo* means "eyelid"). In this condition, sustained and involuntary contractions of the muscles around the eye nearly permanently close the eyelids.

Botulinum toxin's potential as a treatment option for cosmetic surgeons was accidentally discovered when physicians noted that injections used to counter abnormal eye muscle contractions also smoothed the appearance of wrinkles in the treated areas. It turns out that frown lines, crow's feet, and furrowed brows are caused by facial muscles that have become overactivated, or permanently contracted, as a result of years of performing certain repetitive facial expressions. By relaxing these muscles, botulinum toxin temporarily smoothes out these age-related wrinkles. Botox has recently received FDA approval as an antiwrinkle treatment. The agent is considered an excellent alternative to facelift surgery for combating lines and creases. However, as with its therapeutic use to treat dystonias, the costly injections of botulinum toxin must be repeated every three to six months to maintain the desired effect in appearance. Furthermore, Botox does not work against the fine, crinkly wrinkles associated with years of excessive sun exposure, because these wrinkles are caused by skin damage, not by contracted muscles.

however, curare does not alter membrane permeability, nor is it inactivated by AChE. When curare occupies ACh receptor sites, ACh cannot combine with these sites to open the channels that would permit the ionic movement responsible for an EPP. Consequently, because muscle action potentials cannot occur in response to nerve impulses to these muscles, paralysis ensues. When enough curare is present to block a significant number of ACh receptor sites, the person dies from respiratory paralysis caused by inability to contract the diaphragm. In the past some peoples used curare as a deadly arrowhead poison.

MYASTHENIA GRAVIS INACTIVATES ACh RECEPTOR SITES

Myasthenia gravis, a disease involving the neuromuscular junction, is characterized by extreme muscular weakness (*myasthenia* means "muscular weakness"; *gravis* means "severe"). It is an autoimmune condition (*autoimmune* means "immunity against self") in which the body erroneously produces antibodies against its own motor end-plate ACh receptors. Thus not all the released ACh molecules can find a functioning receptor site with which to bind. As a result, AChE destroys much of the ACh before it ever has a chance to interact with a receptor site and contribute to the EPP.

ORGANOPHOSPHATES PREVENT INACTIVATION OF ACh

Organophosphates are a group of chemicals that modify neuromuscular junction activity in yet another way—namely,

by irreversibly inhibiting AChE. Inhibition of AChE prevents the inactivation of released ACh. Death from organophosphates is also due to respiratory failure, because the diaphragm cannot repolarize and return to resting conditions, then contract again to bring in a fresh breath of air. These toxic agents are used in some pesticides and military nerve gases.

 Click on the Media Exercises menu of the CD-ROM and work Media Exercise 7.2: Neuromuscular Junction to test your understanding of the previous section.

CHAPTER IN PERSPECTIVE: FOCUS ON HOMEOSTASIS

The nervous system, along with the other major regulatory system, the endocrine system, controls most of the body's muscular and glandular activities. Whereas the afferent division of the peripheral nervous system detects and carries information to the central nervous system (CNS) for processing and decision making, the efferent division of the peripheral nervous system carries directives from the CNS to the effector organs (muscles and glands), which carry out the intended response. Much of this efferent output is directed toward maintaining homeostasis.

The autonomic nervous system, which is the efferent branch that innervates smooth muscle, cardiac muscle, and glands, plays a major role in the following range of homeostatic activities:

- regulating blood pressure
- controlling digestive juice secretion and digestive tract contractions that mix ingested food with the digestive juices
- controlling sweating to help maintain body temperature

The somatic nervous system, the efferent branch that innervates skeletal muscle, contributes to homeostasis by stimulating the following activities:

- Skeletal muscle contractions that enable the body to move in relation to the external environment contribute to homeostasis by moving the body toward food or away from harm.

- Skeletal muscle contraction also accomplishes breathing to maintain appropriate levels of O_2 and CO_2 in the body.
- Shivering is a skeletal muscle activity important in maintaining body temperature.

In addition, efferent output to skeletal muscles accomplishes many movements that are not aimed at maintaining a stable internal environment but nevertheless enrich our lives and enable us to engage in activities that contribute to society, such as dancing, building bridges, or performing surgery.

CHAPTER SUMMARY

Introduction (p. 185)

▮ The CNS controls muscles and glands by transmitting signals to these effector organs through the efferent division of the PNS.

▮ There are two types of efferent output: the autonomic nervous system, which is under involuntary control and supplies cardiac and smooth muscle as well as most exocrine and some endocrine glands; and the somatic nervous system, which is subject to voluntary control and supplies skeletal muscle. *(Review Table 7-3, p. 193.)*

Autonomic Nervous System (pp. 185–192)

▮ The autonomic nervous system consists of two subdivisions—the sympathetic and parasympathetic nervous systems. *(Review Figures 7-2 and 7-3 and Tables 7-1 and 7-2.)*

▮ An autonomic nerve pathway consists of a two-neuron chain. The preganglionic fiber originates in the CNS and synapses with the cell body of the postganglionic fiber in a ganglion outside the CNS. The postganglionic fiber ends on the effector organ. *(Review Figures 7-1 and 7-2 and Table 7-4.)*

▮ All preganglionic fibers and parasympathetic postganglionic fibers release acetylcholine. Sympathetic postganglionic fibers release norepinephrine. *(Review Figure 7-2 and Table 7-2.)*

▮ The same neurotransmitter elicits different responses in different tissues. Thus the response depends on specialization of the tissue cells, not on the properties of the messenger.

▮ Tissues innervated by the autonomic nervous system possess one or more of several different receptor types for the postganglionic chemical messengers. Cholinergic receptors include nicotinic and muscarinic receptors; adrenergic receptors include α_1, α_2, β_1, and β_2 receptors. *(Review Figure 7-4 and Table 7-2.)*

▮ A given autonomic fiber either excites or inhibits activity in the organ it innervates. *(Review Table 7-1.)*

▮ Most visceral organs are innervated by both sympathetic and parasympathetic fibers, which in general produce opposite effects in a particular organ. Dual innervation of organs by both branches of the autonomic nervous system permits precise control over an organ's activity. *(Review Figure 7-3 and Table 7-1.)*

▮ The sympathetic system dominates in emergency or stressful ("fight-or-flight") situations and promotes responses that prepare the body for strenuous physical activity. The parasympathetic system dominates in quiet, relaxed ("rest-and-digest") situations and promotes body maintenance activities such as digestion. *(Review Tables 7-1 and 7-2.)*

Somatic Nervous System (pp. 192–198)

▮ The somatic nervous system consists of the axons of motor neurons, which originate in the spinal cord or brain stem and end on skeletal muscle.

▮ Acetylcholine, the neurotransmitter released from a motor neuron, stimulates muscle contraction.

▮ Motor neurons are the final common pathway by which various regions of the CNS exert control over skeletal muscle activity.

▮ Each axon terminal of a motor neuron forms a neuromuscular junction with a single muscle cell (fiber). *(Review Figure 7-5.)*

▮ Because these structures do not make direct contact, signals are passed between the nerve terminal and muscle fiber by means of the chemical messenger acetylcholine (ACh). *(Review Figure 7-6.)*

▮ An action potential in the axon terminal causes the release of ACh from its storage vesicles. The released ACh diffuses across the space separating the nerve and muscle cell and binds to special receptor sites on the underlying motor end plate of the muscle cell membrane. This combination of ACh with the receptor sites triggers the opening of specific channels in the motor end plate. The subsequent ion movements depolarize the motor end plate, producing the end-plate potential (EPP).

▮ Local current flow between the depolarized end plate and adjacent muscle-cell membrane brings these adjacent areas to threshold, initiating an action potential that is propagated throughout the muscle fiber.

▮ This muscle action potential triggers muscle contraction.

▮ Acetylcholinesterase inactivates ACh, ending the EPP and, subsequently, the action potential and resultant contraction. *(Review Figure 7-6.)*

REVIEW EXERCISES

Objective Questions (Answers on p. A-42)

1. Sympathetic preganglionic fibers begin in the thoracic and lumbar segments of the spinal cord. *(True or false?)*

2. Action potentials are transmitted on a one-to-one basis at both a neuromuscular junction and a synapse. *(True or false?)*

3. The sympathetic nervous system
    a. is always excitatory.
    b. innervates only tissues concerned with protecting the body against challenges from the outside environment.
    c. has short preganglionic and long postganglionic fibers.
    d. is part of the afferent division of the peripheral nervous system.
    e. is part of the somatic nervous system.
4. Acetylcholinesterase
    a. is stored in vesicles in the terminal button.
    b. combines with receptor sites on the motor end plate to bring about an end-plate potential.
    c. is inhibited by organophosphates.
    d. is the chemical transmitter at the neuromuscular junction.
    e. paralyzes skeletal muscle by strongly binding with acetylcholine receptor sites.
5. The two divisions of the autonomic nervous system are the _____ nervous system, which dominates in fight-or-flight situations, and the _____ nervous system, which dominates in quiet, relaxed situations.
6. The _____ is a modified sympathetic ganglion that does not give rise to postganglionic fibers but instead secretes hormones similar or identical to sympathetic postganglionic neurotransmitters into the blood.
7. Using the following answer code, identify the autonomic transmitter being described:
    (a) acetylcholine
    (b) norepinephrine
    _____ 1. secreted by all preganglionic fibers
    _____ 2. secreted by sympathetic postganglionic fibers
    _____ 3. secreted by parasympathetic postganglionic fibers
    _____ 4. secreted by the adrenal medulla
    _____ 5. secreted by motor neurons
    _____ 6. binds to muscarinic or nicotinic receptors
    _____ 7. binds to α or β receptors
8. Using the following answer code, indicate which type of efferent output is being described:
    (a) characteristic of the somatic nervous system
    (b) characteristic of the autonomic nervous system
    _____ 1. composed of two-neuron chains
    _____ 2. innervates cardiac muscle, smooth muscle, and glands
    _____ 3. innervates skeletal muscle
    _____ 4. consists of the axons of motor neurons
    _____ 5. exerts either an excitatory or an inhibitory effect on its effector organs
    _____ 6. dually innervates its effector organs
    _____ 7. exerts only an excitatory effect on its effector organs

Essay Questions
1. Distinguish between preganglionic and postganglionic fibers.
2. What is the advantage of dual innervation of many organs by both branches of the autonomic nervous system?
3. Distinguish among the following types of receptors: nicotinic receptors, muscarinic receptors, α_1 receptors, α_2 receptors, β_1 receptors, and β_2 receptors.
4. What regions of the CNS regulate autonomic output?
5. Why are motor neurons called the "final common pathway"?
6. Describe the sequence of events that occurs at a neuromuscular junction.
7. Discuss the effect each of the following has at the neuromuscular junction: black widow spider venom, botulinum toxin, curare, myasthenia gravis, and organophosphates.

POINTS TO PONDER

(Explanations on p. 42)
1. Explain why epinephrine, which causes arteriolar constriction (narrowing) in most tissues, is frequently administered in conjunction with local anesthetics.
2. Would skeletal muscle activity be affected by atropine? Why or why not?
3. Considering that you can voluntarily control the emptying of your urinary bladder by contracting (preventing emptying) or relaxing (permitting emptying) your external urethral sphincter, a ring of muscle that guards the exit from the bladder, of what type of muscle is this sphincter composed and what branch of the nervous system supplies it?
4. The venom of certain poisonous snakes contains α bungarotoxin, which binds tenaciously to acetylcholine receptor sites on the motor end-plate membrane. What would the resultant symptoms be?
5. Explain how destruction of motor neurons by poliovirus or amyotrophic lateral sclerosis can be fatal.

CLINICAL CONSIDERATION

(Explanation on p. A-43)
Christopher K. experienced chest pains when he climbed the stairs to his fourth-floor office or played tennis but had no symptoms when not physically exerting himself. His condition was diagnosed as *angina pectoris* (*angina* means "pain"; *pectoris* means "chest"), heart pain that occurs whenever the blood supply to the heart muscle cannot meet the muscle's need for oxygen delivery. This condition usually is caused by narrowing of the blood vessels supplying the heart by cholesterol-containing deposits. Most people with this condition do not have any pain at rest but experience bouts of pain whenever the heart's need for oxygen increases, such as during exercise or emotionally stressful situations that increase sympathetic nervous activity. Christopher obtains immediate relief of angina attacks by promptly taking a vasodilator drug such as *nitroglycerin*, which relaxes the smooth muscle in the walls of his narrowed heart vessels. Consequently, the vessels open more widely and more blood can flow through them. For prolonged treatment, his doctor has indicated that Christopher will experience fewer and less severe angina attacks if he takes a β_1-blocker drug, such as *metoprolol*, on a regular basis. Explain why.

PHYSIOEDGE RESOURCES

 PhysioEdge CD-ROM

PhysioEdge, the CD-ROM packaged with your text, focuses on the concepts students find most difficult to learn. Figures marked with this icon have associated activities on the CD. For a visual review of concepts in this chapter, check out the following:

Tutorial: Skeletal Muscle Contraction—Nerve-to-Muscle Communication tab

Media Exercise 7.1: Autonomic Nervous System

Media Exercise 7.2: Neuromuscular Junction

PhysioEdge Website

The website for this book contains a wealth of helpful study aids, as well as many ideas for further reading and research. Log on to:
http://www.brookscole.com/hpfundamentals3
Select Chapter 7 from the drop-down menu or click on one of the many resource areas.

For Suggested Readings, consult InfoTrac College Edition/ Research on the PhysioEdge website or go directly to InfoTrac College Edition, your online research library, at:
http://infotrac.thomsonlearning.com

Muscular System

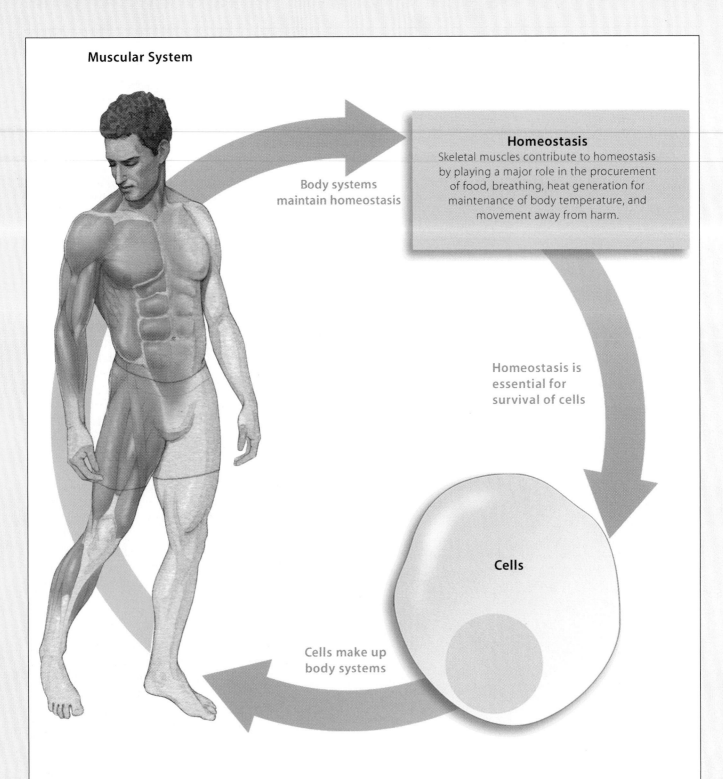

Homeostasis
Skeletal muscles contribute to homeostasis by playing a major role in the procurement of food, breathing, heat generation for maintenance of body temperature, and movement away from harm.

Body systems maintain homeostasis

Homeostasis is essential for survival of cells

Cells

Cells make up body systems

Muscles are the contraction specialists of the body. **Skeletal muscle** attaches to the skeleton. Contraction of skeletal muscles moves bones to which they are attached, allowing the body to perform a variety of motor activities. Skeletal muscles that support homeostasis include those important in acquiring, chewing, and swallowing food and those essential for breathing. Skeletal muscle contraction is also used to move the body away from harm. Heat-generating muscle contractions are important in regulating temperature. Skele-

tal muscles are also used for nonhomeostatic activities, such as dancing or operating a computer. **Smooth muscle** is found in the walls of hollow organs and tubes. Controlled contraction of smooth muscle regulates movement of blood through blood vessels, food through the digestive tract, air through respiratory airways, and urine to the exterior. **Cardiac muscle** is found only in the walls of the heart, whose contraction pumps life-sustaining blood throughout the body.

Muscle Physiology

Click on the Tutorials menu of the CD-ROM for a tutorial on Skeletal Muscle Contraction.

INTRODUCTION

By moving specialized intracellular components, muscle cells can develop tension and shorten, that is, contract. Recall that the three types of muscle are *skeletal muscle, cardiac muscle,* and *smooth muscle* (see p. 3). Through their highly developed ability to contract, groups of muscle cells working together within a muscle can produce movement and do work. Controlled contraction of muscles allows (1) purposeful movement of the whole body or parts of the body (such as walking or waving your hand), (2) manipulation of external objects (such as driving a car or moving a piece of furniture), (3) propulsion of contents through various hollow internal organs (such as circulation of blood or movement of materials through the digestive tract), and (4) emptying the contents of certain organs to the external environment (such as urination or giving birth).

Muscle comprises the largest group of tissues in the body, accounting for approximately half of the body's weight. Skeletal muscle alone makes up about 40% of body weight in men and 32% in women, with smooth and cardiac muscle making up another 10% of the total weight. Although the three muscle types are structurally and functionally distinct, they can be classified in two different ways according to their common characteristics (● Figure 8-1). First, muscles are categorized as *striated* (skeletal and cardiac muscle) or *unstriated* (smooth muscle), depending on whether alternating dark and light bands, or striations (stripes), can be seen when the muscle is viewed under a light microscope. Second, muscles are categorized as *voluntary* (skeletal muscle) or *involuntary* (cardiac and smooth muscle), depending respectively on whether they are innervated by the somatic nervous system and are subject to voluntary control, or are innervated by the autonomic nervous system and are not subject to voluntary control (see p. 185). Although skeletal muscle is categorized as voluntary, because it can be consciously controlled, much skeletal muscle activity is also subject to subconscious, involuntary regulation, such as that related to posture, balance, and stereotypical movements like walking.

Most of this chapter is a detailed examination of the most abundant and best understood muscle, skeletal muscle. Skeletal muscles make up the muscular system. We will begin with a discussion of skeletal muscle structure, then examine how it works from the molecular level through the cell level and finally to the whole muscle. The chapter concludes with a discussion of the unique properties of smooth and cardiac muscle in comparison to skeletal muscle. Smooth muscle appears throughout the body systems as components of hollow organs and tubes. Cardiac muscle is found only in the heart.

STRUCTURE OF SKELETAL MUSCLE

A single skeletal muscle cell, known as a **muscle fiber**, is relatively large, elongated, and cylinder-shaped. A skeletal muscle consists of a number of muscle fibers lying parallel to each other and bundled together by connective tissue (● Figure 8-2a). The fibers usually extend the entire length of the muscle.

▌ Skeletal muscle fibers are striated by a highly organized internal arrangement.

The predominant structural feature of a skeletal muscle fiber is numerous **myofibrils**. These specialized contractile elements, which constitute 80% of the volume of the muscle fiber, are cylindrical intracellular structures that extend the entire length of the muscle fiber (● Figure 8-2b), each myofibril consisting of a regular arrangement of highly organized cytoskeletal elements—the thick and the thin filaments (● Figure 8-2c). The **thick filaments** are special assemblies of the protein *myosin*, whereas the thin filaments are made up primarily of the protein *actin* (● Figure 8-2d). The levels of organization in a skeletal muscle can be summarized as follows:

$$\text{Whole muscle} \rightarrow \text{muscle fiber} \rightarrow \text{myofibril} \rightarrow \substack{\text{thick} \\ \text{and thin} \\ \text{filaments}} \rightarrow \substack{\text{myosin} \\ \text{and} \\ \text{actin}}$$

(an organ)	(a cell)	(a specialized intracellular structure)	(cytoskeletal elements)	(protein molecules)

Viewed with a light microscope, a myofibril displays alternating dark bands (the A bands) and light bands (the I bands) (● Figure 8-3a). The bands of all the myofibrils lined up parallel to each other collectively produce the striated or striped appearance of a skeletal muscle fiber (● Figure 8-3b). Alternate stacked sets of thick and thin filaments that slightly overlap each other are responsible for the A and I bands (● Figure 8-2c).

An **A band** is a stacked set of thick filaments along with the portions of the thin filaments that overlap on both ends of the thick filaments. The thick filaments lie only within the A band and extend its entire width; that is, the two ends of the thick filaments within a stack define the outer limits of a given A band. The lighter area within the middle of the A band, where the thin filaments do not reach, is the **H zone**. Only the central portions of the thick filaments are found in this

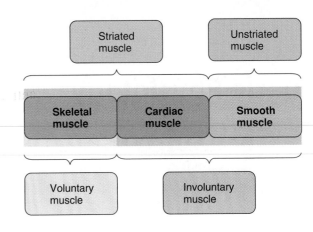

● **FIGURE 8-1**

Categorization of muscle

region. A system of supporting proteins holds the thick filaments together vertically within each stack. These proteins can be seen as the **M line**, which extends vertically down the middle of the A band within the center of the H zone.

The **I band** consists of the remaining portion of the thin filaments that do not project into the A band. Visible in the middle of each I band is a dense, vertical **Z line**. The area between two Z lines is called a **sarcomere**, which is the functional unit of skeletal muscle. A **functional unit** of any organ is the smallest component that can perform all the functions of that organ. Accordingly, a sarcomere is the smallest component of a muscle fiber that can contract. The Z line is a flat, cytoskeletal disc that connects the thin filaments of two adjoining sarcomeres. Thus the I band contains only thin filaments from two adjacent sarcomeres but not the entire length of these filaments.

Not visible under a light microscope, single strands of a giant, highly elastic protein known as **titin** extend in both directions from the M line along the length of the thick filament to the Z lines at opposite ends of the sarcomere. Titin is the largest protein in the body, being made up of nearly 30,000 amino acids. It serves two important roles: (1) along with the M-line proteins, titin helps stabilize the position of the thick filaments in relation to the thin filaments, and (2) by acting like a spring, it greatly augments a muscle's elasticity. That is, titin helps a muscle stretched by an external force passively recoil or spring back to its resting length when the stretching force is removed, much like a stretched spring.

With an electron microscope, fine **cross bridges** can be seen extending from each thick filament toward the surrounding thin filaments in the areas where the thick and thin filaments overlap (● Figure 8-2c). To give you an idea of the magnitude of these filaments, a single muscle fiber may contain an estimated 16 billion thick and 32 billion thin filaments, all arranged in a very precise pattern within the myofibrils.

▌ Myosin forms the thick filaments.

Each thick filament has several hundred myosin molecules packed together in a specific arrangement. A **myosin molecule**

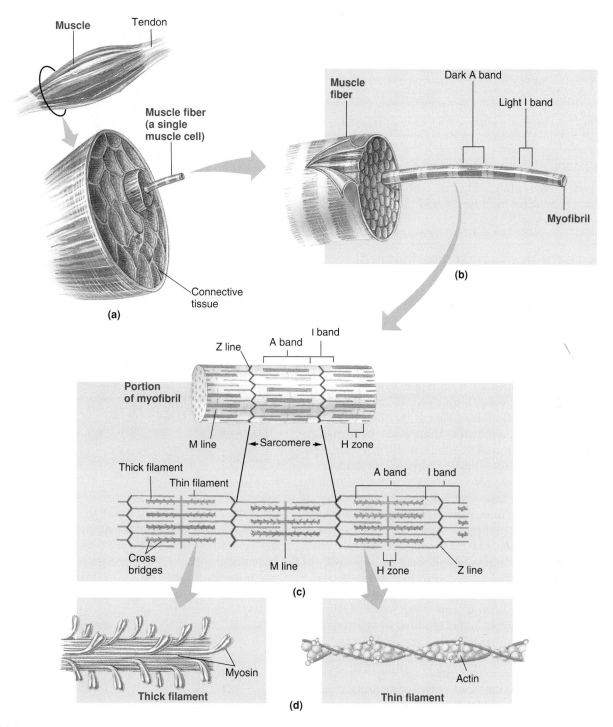

FIGURE 8-2

Levels of organization in a skeletal muscle. (a) Enlargement of a cross section of a whole muscle. (b) Enlargement of a myofibril within a muscle fiber. (c) Cytoskeletal components of a myofibril. (d) Protein components of thick and thin filaments.

 For interactions related to this figure, see the Anatomy Review tab in the Skeletal Muscle Contraction tutorial and Media Exercise 8.2: Skeletal Muscle Mechanics on the CD-ROM.

is a protein consisting of two identical subunits, each shaped somewhat like a golf club (● Figure 8-4a). The protein's tail ends are intertwined around each other like golf-club shafts twisted together, with the two globular heads projecting out at one end. The two halves of each thick filament are mirror im-

ages made up of myosin molecules lying lengthwise in a regular, staggered array, with their tails oriented toward the center of the filament and their globular heads protruding outward at regular intervals (● Figure 8-4b). These heads form the cross bridges between the thick and thin filaments. Each

Light-microscope view of skeletal muscle components. (a) High-power
light-microscope view of a myofibril. (b) Low-power light-microscope view
of skeletal muscle fibers. Note striated appearance.

(*Source:* Reprinted with permission from Sydney Schochet Jr., M.D., Professor, Department
of Pathology, School of Medicine, West Virginia University: *Diagnostic Pathology of
Skeletal Muscle and Nerve* (Stamford, Connecticut: Appleton & Lange, 1986), Figure 1-13.)

 For an interaction related to this figure, see Media Exercise 8.1:
Muscle Types on the CD-ROM.

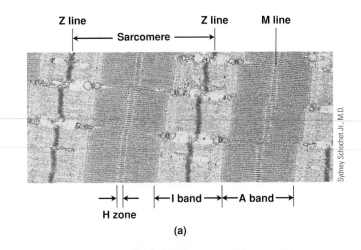

(a)

cross bridge has two important sites crucial to the contractile
process: (1) an actin-binding site and (2) a myosin ATPase
(ATP-splitting) site.

▌ Actin is the main structural component of the thin filaments.

Thin filaments consist of three proteins: *actin, tropomyosin,*
and *troponin* (● Figure 8-5). **Actin molecules,** the primary
structural proteins of the thin filament, are spherical. The
backbone of a thin filament is formed by actin molecules
joined into two strands and twisted together, like two intertwined strings of pearls. Each actin molecule has a special
binding site for attachment with a myosin cross bridge. By
a mechanism to be described shortly, binding of myosin
and actin molecules at the cross bridges results in energy-
consuming contraction of the muscle fiber. Accordingly,
myosin and actin are often called **contractile proteins,** even
though, as you will see, neither myosin nor actin actually
contracts. Myosin and actin are not unique to muscle cells,
but these proteins are more abundant and more highly organized in muscle cells.

In a relaxed muscle fiber, contraction does not take place;
actin cannot bind with cross bridges, because of the way the
two other types of protein—tropomyosin and troponin—are
positioned within the thin filament. **Tropomyosin molecules**
are threadlike proteins that lie end-to-end alongside the
groove of the actin spiral. In this position, tropomyosin
covers the actin sites that bind with the cross bridges,
blocking the interaction that leads to muscle contraction. The other thin filament component, **troponin,** is a
protein complex made of three polypeptide units: one
binds to tropomyosin, one binds to actin, and a third can
bind with Ca^{2+}.

When troponin is not bound to Ca^{2+}, this protein stabi-
lizes tropomyosin in its blocking position over actin's cross-
bridge binding sites (● Figure 8-6a). When Ca^{2+} binds to
troponin, the shape of this protein is changed in such a way
that tropomyosin slips away from its blocking position (● Fig-

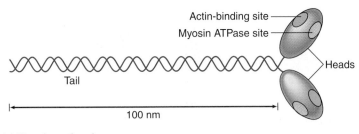

(a) **Myosin molecule**

● **FIGURE 8-4**

**Structure of myosin molecules and their organization
within a thick filament.** (a) *Myosin molecule.* Each myosin
molecule consists of two identical, golf club–shaped subunits
with their tails intertwined and their globular heads, each of
which contains an actin-binding site and a myosin ATPase site,
projecting out at one end. (b) *Thick filament.* A thick filament
is made up of myosin molecules lying lengthwise parallel
to each other. Half are oriented in one direction and half in
the opposite direction. The globular heads, which protrude at
regular intervals along the thick filament, form the cross bridges.

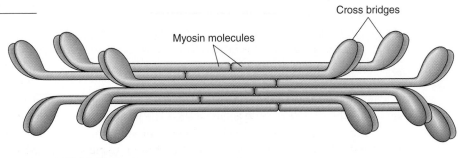

(b) **Thick filament**

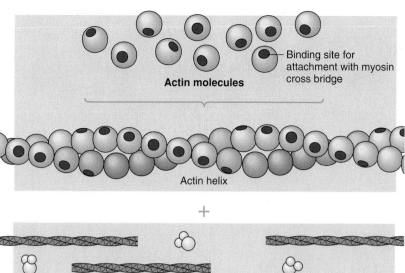

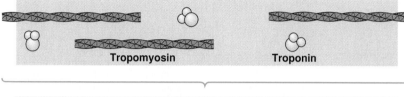

Actin molecules

Binding site for attachment with myosin cross bridge

Actin helix

+

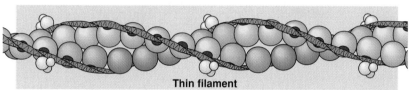

Tropomyosin

Troponin

Thin filament

● **FIGURE 8-5**

Composition of a thin filament. The main structural component of a thin filament is two chains of spherical actin molecules that are twisted together. Troponin molecules (which consist of three small, spherical subunits) and threadlike tropomyosin molecules are arranged to form a ribbon that lies alongside the groove of the actin helix and physically covers the binding sites on actin molecules for attachment with myosin cross bridges. (The thin filaments shown here are not drawn in proportion to the thick filaments in ● Figure 8-4. Thick filaments are two to three times larger in diameter than thin filaments.)

ure 8-6b). With tropomyosin out of the way, actin and myosin can bind and interact at the cross bridges, resulting in muscle contraction. Tropomyosin and troponin are often called **regulatory proteins** because of their role in covering (preventing contraction) or exposing (permitting contraction) the binding sites for cross-bridge interaction between actin and myosin.

Click on the Media Exercises menu of the CD-ROM and work Media Exercises 8.1: Muscle Types and 8.3: Skeletal Muscle Structure to test your understanding of the previous section.

MOLECULAR BASIS OF SKELETAL MUSCLE CONTRACTION

Several important links in the contractile process remain to be discussed. How does cross-bridge interaction between actin and myosin bring about muscle contraction? How does a muscle action potential trigger this contractile process? What is the source of the Ca^{2+} that physically repositions troponin and tropomyosin to permit cross-bridge binding? We will turn our attention to these topics in this section.

❚ **During contraction, cycles of cross-bridge binding and bending pull the thin filaments inward.**

Cross-bridge interaction between actin and myosin brings about muscle contraction by means of the sliding filament mechanism.

SLIDING FILAMENT MECHANISM

The thin filaments on each side of a sarcomere slide inward over the stationary thick filaments toward the A band's center during contraction (● Figure 8-7). As they slide inward, the thin filaments pull the Z lines to which they are attached closer together, so the sarcomere shortens. As all the sarcomeres throughout the muscle fiber's length shorten simultaneously, the entire fiber shortens. This is the **sliding filament mechanism** of muscle contraction. The H zone, in the center of the A band where the thin filaments do not reach, becomes smaller as the thin filaments approach each other when they slide more deeply inward. The I band, which consists of the portions of the thin filaments that do not overlap with the thick filaments, narrows as the thin filaments further overlap the thick filaments during their inward slide. The thin filaments themselves do not change length during muscle fiber shortening. The width of the A band remains unchanged during contraction, because its width is determined by the length of the thick filaments, and the thick filaments do not change length during the shortening process. Note that neither the thick nor thin filaments decrease in length to shorten the sarcomere. Instead, contraction is accomplished by the thin filaments from the opposite sides of each sarcomere sliding closer together between the thick filaments.

POWER STROKE

Cross-bridge activity pulls the thin filaments inward relative to the stationary thick filaments. During contraction, with the tropomyosin and troponin "chaperones" pulled out of the way

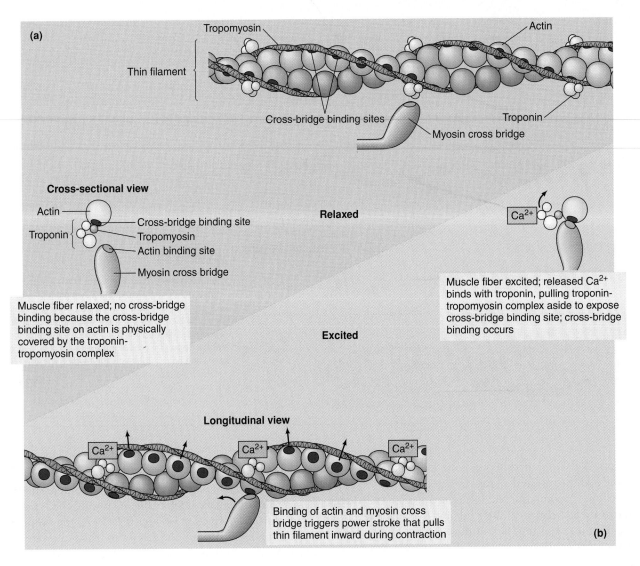

(a)

Tropomyosin

Actin

Thin filament

Cross-bridge binding sites

Troponin

Myosin cross bridge

Cross-sectional view

Actin

Cross-bridge binding site

Troponin

Tropomyosin

Actin binding site

Myosin cross bridge

Relaxed

Ca²⁺

Muscle fiber relaxed; no cross-bridge binding because the cross-bridge binding site on actin is physically covered by the troponin-tropomyosin complex

Muscle fiber excited; released Ca²⁺ binds with troponin, pulling troponin-tropomyosin complex aside to expose cross-bridge binding site; cross-bridge binding occurs

Excited

Longitudinal view

Ca²⁺ Ca²⁺ Ca²⁺

Binding of actin and myosin cross bridge triggers power stroke that pulls thin filament inward during contraction

(b)

● **FIGURE 8-6**

Role of calcium in turning on cross bridges

by Ca²⁺, the myosin cross bridges from a thick filament can bind with the actin molecules in the surrounding thin filaments. Let's concentrate on a single cross-bridge interaction (● Figure 8-8a). The two myosin heads of each myosin molecule act independently, with only one head attaching to actin at a given time. When myosin and actin make contact at a cross bridge, the bridge changes shape, bending inward as if it were on a hinge, "stroking" toward the center of the sarcomere, like the stroking of a boat oar. This so-called **power stroke** of a cross bridge pulls inward the thin filament to which it is attached. A single power stroke pulls the thin filament inward only a small percentage of the total shortening distance. Repeated cycles of cross-bridge binding and bending complete the shortening.

At the end of one cross-bridge cycle, the link between the myosin cross bridge and actin molecule breaks. The cross bridge returns to its original shape and binds to the next actin molecule behind its previous actin partner. The cross bridge bends once again to pull the thin filament in further, then de-

taches and repeats the cycle. Repeated cycles of cross-bridge power strokes successively pull in the thin filaments, much like pulling in a rope hand over hand.

Because of the way myosin molecules are oriented within a thick filament (● Figure 8-8b), all the cross bridges stroke toward the center of the sarcomere, so that all the surrounding thin filaments on each end of the sarcomere are pulled inward simultaneously. The cross bridges aligned with given thin filaments do not all stroke in unison, however. At any time during contraction, part of the cross bridges are attached to the thin filaments and are stroking, while others are returning to their original conformation in preparation for binding with another actin molecule. Thus some cross bridges are "holding on" to the thin filaments, whereas others "let go" to bind with new actin. Were it not for this asynchronous cycling of the cross bridges, the thin filaments would slip back toward their resting position between strokes.

How does muscle excitation switch on this cross-bridge cycling? The term **excitation–contraction coupling** refers to

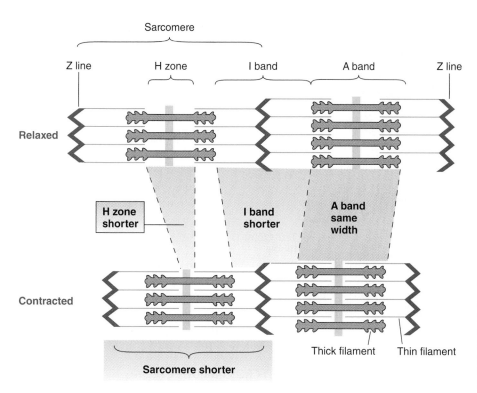

● **FIGURE 8-7**

Changes in banding pattern during shortening. During muscle contraction, each sarcomere shortens as the thin filaments slide closer together between the thick filaments so that the Z lines are pulled closer together. The width of the A bands does not change as a muscle fiber shortens, but the I bands and H zones become shorter.

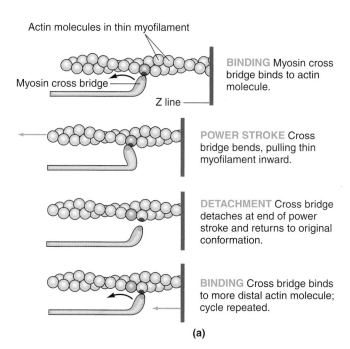

BINDING Myosin cross bridge binds to actin molecule.

POWER STROKE Cross bridge bends, pulling thin myofilament inward.

DETACHMENT Cross bridge detaches at end of power stroke and returns to original conformation.

BINDING Cross bridge binds to more distal actin molecule; cycle repeated.

(a)

the series of events linking muscle excitation (the presence of an action potential in a muscle fiber) to muscle contraction (cross-bridge activity that causes the thin filaments to slide closer together to produce sarcomere shortening). We will now turn our attention to excitation–contraction coupling.

▌ Calcium is the link between excitation and contraction.

Skeletal muscles are stimulated to contract by release of acetylcholine (ACh) at neuromuscular junctions between motor neuron terminals and muscle fibers. Recall that the binding of ACh with the motor end plate of a muscle fiber brings about permeability changes in the muscle fiber, resulting in an action potential that is conducted over the entire surface of the muscle cell membrane (see p. 195). Two membranous structures within the muscle fiber play an important role in linking this excitation to contraction—transverse tubules and the sarcoplasmic reticulum. Let's examine the structure and function of each.

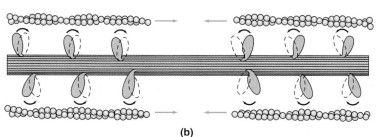

(b)

● **FIGURE 8-8**

Cross-bridge activity. (a) During each cross-bridge cycle, the cross bridge binds with an actin molecule, bends to pull the thin filament inward during the power stroke, then detaches and returns to its resting conformation, ready to repeat the cycle. (b) The power strokes of all cross bridges extending from a thick filament are directed toward the center of the thick filament.

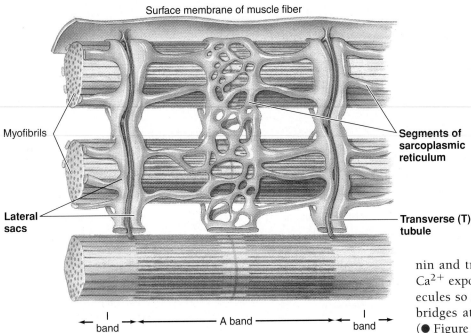

Surface membrane of muscle fiber

Myofibrils

Lateral
sacs

Segments of
sarcoplasmic
reticulum

Transverse (T)
tubule

I
band

A band

I
band

● FIGURE 8-9

The T tubules and sarcoplasmic reticulum in relationship to the myofibrils. The transverse (T) tubules are membranous, perpendicular extensions of the surface membrane that dip deep into the muscle fiber at the junctions between the A and I bands of the myofibrils. The sarcoplasmic reticulum is a fine, membranous network that runs longitudinally and surrounds each myofibril, with separate segments encircling each A band and I band. The ends of each segment are expanded to form lateral sacs that lie next to the adjacent T tubules.

 For an interaction related to this figure, see Media Exercise 8.2: Skeletal Muscle Mechanics on the CD-ROM.

SPREAD OF THE ACTION POTENTIAL DOWN THE T TUBULES

At each junction of an A band and I band, the surface membrane dips into the muscle fiber to form a **transverse tubule (T tubule)**, which runs perpendicularly from the surface of the muscle cell membrane into the central portions of the muscle fiber (● Figure 8-9). Because the T tubule membrane is continuous with the surface membrane, an action potential on the surface membrane also spreads down into the T tubule, rapidly transmitting the surface electric activity into the central portions of the fiber. The presence of a local action potential in the T tubules induces permeability changes in a separate membranous network within the muscle fiber, the sarcoplasmic reticulum.

RELEASE OF CALCIUM FROM THE SARCOPLASMIC RETICULUM

The **sarcoplasmic reticulum** is a modified endoplasmic reticulum (see p. 22) that consists of a fine network of interconnected compartments surrounding each myofibril like a mesh sleeve (● Figure 8-9). This membranous network encircles the myofibril throughout its length but is not continuous. Separate segments of sarcoplasmic reticulum are wrapped around each A band and each I band. The ends of each segment expand to form saclike regions, the **lateral sacs**, which are separated from the adjacent T tubules by a slight gap (● Figure 8-9). The sarcoplasmic reticulum's lateral sacs store Ca^{2+}. Spread of an action potential down a T tubule triggers the opening of Ca^{2+}-release channels that span the gap between the T tubule and adjacent lateral sacs on each side. Through these open Ca^{2+}-release channels, calcium is released into the cytosol from the lateral sacs. By slightly repositioning the troponin and tropomyosin molecules, this released Ca^{2+} exposes the binding sites on the actin molecules so they can link with the myosin cross bridges at their complementary binding sites (● Figure 8-10).

ATP-POWERED CROSS-BRIDGE CYCLING

Recall that a myosin cross bridge has two special sites, an actin-binding site and an ATPase site. The latter is an enzymatic site that can bind the energy carrier *adenosine triphosphate (ATP)* and split it into *adenosine diphosphate (ADP)* and *inorganic phosphate (P_i)*, yielding energy in the process. The breakdown of ATP occurs on the myosin cross bridge before the bridge ever links with an actin molecule (step ① in ● Figure 8-11). The ADP and P_i remain tightly bound to the myosin, and the generated energy is stored within the cross bridge to produce a high-energy form of myosin. To use an analogy, the cross bridge is "cocked" like a gun, ready to be fired when the trigger is pulled. When the muscle fiber is excited, Ca^{2+} pulls the troponin-tropomyosin complex out of its blocking position so that the energized (cocked) myosin cross bridge can bind with an actin molecule (step ②a). This contact between myosin and actin "pulls the trigger," causing the cross-bridge bending that produces the power stroke (step ③). Researchers have not found the mechanism by which the chemical energy released from ATP is stored within the myosin cross bridge and then translated into the mechanical energy of the power stroke. Inorganic phosphate is released from the cross bridge during the power stroke. After the power stroke is complete, ADP is released .

When the muscle is not excited and Ca^{2+} is not released, troponin and tropomyosin remain in their blocking position, so that actin and the myosin cross bridges do not bind and no power stroking takes place (step ②b).

When P_i and ADP are released from myosin following contact with actin and the subsequent power stroke, the myosin ATPase site is free for attachment of another ATP molecule. The actin and myosin remain linked together at the cross bridge until a fresh molecule of ATP attaches to myosin at the end of the power stroke. Attachment of the new ATP molecule permits detachment of the cross bridge, which returns

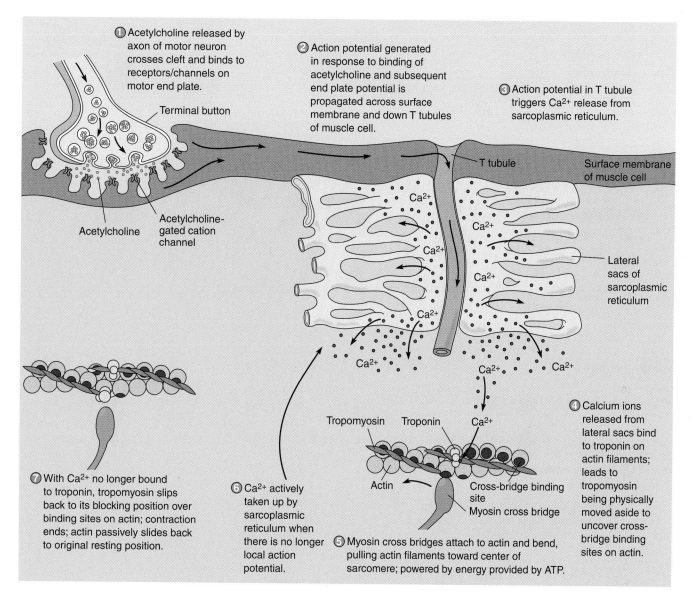

① Acetylcholine released by axon of motor neuron crosses cleft and binds to receptors/channels on motor end plate.

Terminal button

Acetylcholine

Acetylcholine-gated cation channel

② Action potential generated in response to binding of acetylcholine and subsequent end plate potential is propagated across surface membrane and down T tubules of muscle cell.

③ Action potential in T tubule triggers Ca²⁺ release from sarcoplasmic reticulum.

T tubule

Surface membrane of muscle cell

Ca^{2+}

Lateral sacs of sarcoplasmic reticulum

Tropomyosin Troponin Ca^{2+}

Actin

Cross-bridge binding site
Myosin cross bridge

④ Calcium ions released from lateral sacs bind to troponin on actin filaments; leads to tropomyosin being physically moved aside to uncover cross-bridge binding sites on actin.

⑦ With Ca²⁺ no longer bound to troponin, tropomyosin slips back to its blocking position over binding sites on actin; contraction ends; actin passively slides back to original resting position.

⑥ Ca²⁺ actively taken up by sarcoplasmic reticulum when there is no longer local action potential.

⑤ Myosin cross bridges attach to actin and bend, pulling actin filaments toward center of sarcomere; powered by energy provided by ATP.

● **FIGURE 8-10**

Calcium release in excitation–contraction coupling. Steps ① through ⑤ show the events that couple neurotransmitter release and subsequent electrical excitation of the muscle cell with muscle contraction. Steps ⑥ and ⑦ show events associated with muscle relaxation.

 For an animation of this figure, click the Contraction/Relaxation tab in the Skeletal Muscle Contraction tutorial on the CD-ROM.

to its unbent form, ready to start another cycle (step ④a). The newly attached ATP is then split by myosin ATPase, energizing the myosin cross bridge once again (step ①). On binding with another actin molecule, the energized cross bridge again bends, and so on, successively pulling the thin filament inward to accomplish contraction.

RIGOR MORTIS

Note that fresh ATP must attach to myosin to permit the cross-bridge link between myosin and actin to break at the end of a cycle, even though the ATP is not split during this dissociation process. The need for ATP in separating myosin and

actin is amply shown in **rigor mortis.** This "stiffness of death" is a generalized locking in place of the skeletal muscles that begins 3 to 4 hours after death and completes in about 12 hours. Following death, the cytosolic concentration of Ca^{2+} begins to rise, most likely because the inactive muscle-cell membrane cannot keep out extracellular Ca^{2+} and perhaps also because Ca^{2+} leaks out of the lateral sacs. This Ca^{2+} moves the regulatory proteins aside, letting actin bind with the myosin cross bridges, which were already charged with ATP before death. Dead cells cannot produce any more ATP, so actin and myosin, once bound, cannot detach, because they lack fresh ATP. The thick and thin filaments thus stay

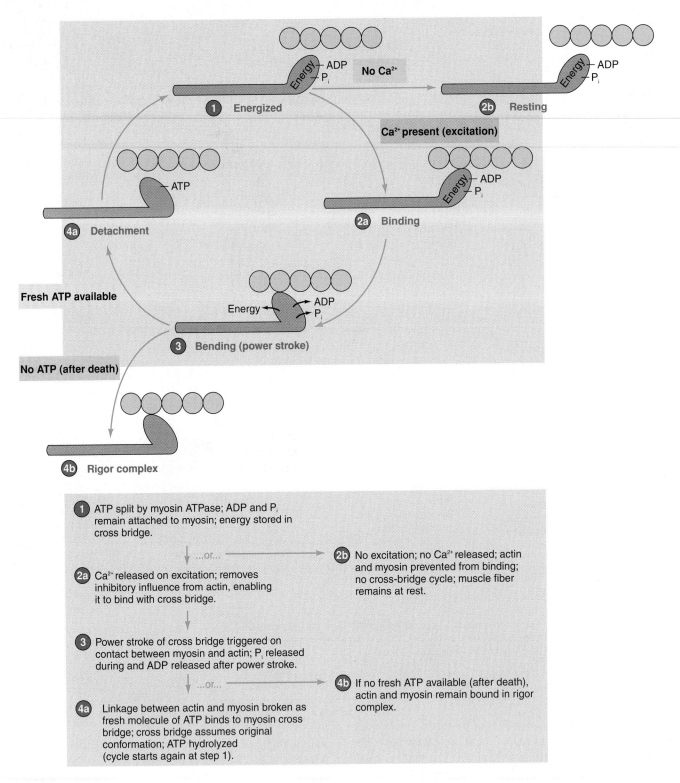

● FIGURE 8-11

Cross-bridge cycle

 For an animation of this figure, click the Contraction/Relaxation tab in the Skeletal Muscle Contraction tutorial on the CD-ROM.

linked together by the immobilized cross bridges, leaving dead muscles stiff (step 4b). During the next several days, rigor mortis gradually subsides as the proteins involved in the rigor complex begin to degrade.

RELAXATION

How is **relaxation** normally accomplished in a living muscle? Just as an action potential in a muscle fiber turns on the contractile process by triggering release of Ca^{2+} from the lateral sacs into the cytosol, the contractile process is turned off when Ca^{2+} is returned to the lateral sacs when local electrical activity stops. The sarcoplasmic reticulum has an energy-consuming carrier, a Ca^{2+}–ATPase pump, which actively transports Ca^{2+} from the cytosol and concentrates it in the lateral sacs. When acetylcholinesterase removes ACh from the neuro-muscular junction, the muscle-fiber action potential stops. When a local action potential is no longer in the T tubules to trigger release of Ca^{2+}, the ongoing activity of the sarcoplasmic reticulum's Ca^{2+} pump returns released Ca^{2+} back into its lateral sacs. Removing cytosolic Ca^{2+} lets the troponin-tropomyosin complex slip back into its blocking position, so actin and myosin can no longer bind at the cross bridges. The thin filaments, freed from cycles of cross-bridge attachment and pulling, return passively to their resting position. The muscle fiber has relaxed.

We are now going to compare the duration of contractile activity to the duration of excitation, and we then shift gears to discuss skeletal muscle mechanics.

Contractile activity far outlasts the electrical activity that initiated it.

A single action potential in a skeletal muscle fiber lasts only 1 to 2 msec. The onset of the resulting contractile response lags behind the action potential because the entire excitation–contraction coupling must occur before cross-bridge activity begins. In fact, the action potential is completed before the contractile apparatus even becomes operational. This time delay of a few milliseconds between stimulation and the onset of contraction is called the **latent period** (● Figure 8-12).

Time is also required for generating tension within the muscle fiber, produced by the sliding interactions between the thick and thin filaments through cross-bridge activity. The time from contraction onset until peak tension develops—**contraction time**—averages about 50 msec, although this time varies, depending on the type of muscle fiber. The contractile response does not end until the lateral sacs have taken up all the Ca^{2+} released in response to the action potential. This reuptake of Ca^{2+} is also time consuming. Even after Ca^{2+} is removed, it takes time for the filaments to return to their resting positions. The time from peak tension until relaxation is complete—the **relaxation time**—usually lasts another 50 msec or more. Consequently, the entire contractile response to a single action potential may last up to 100 msec or more; this is much longer than the duration of the action potential that initiated it (100 msec compared to 1 to 2 msec). This fact is important in the body's ability to produce muscle contractions of variable strength, as you will discover in the next section.

SKELETAL MUSCLE MECHANICS

Thus far we have described the contractile response in a single muscle fiber. In the body, groups of muscle fibers are organized into whole muscles. We will now turn our attention to contraction of whole muscles.

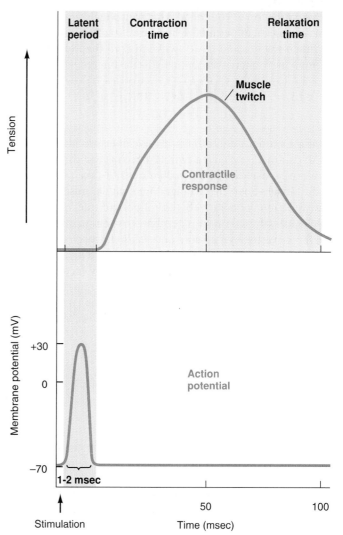

The duration of the action potential is not drawn to scale but is exaggerated.

● **FIGURE 8-12**

Relationship of an action potential to the resulting muscle twitch

Whole muscles are groups of muscle fibers bundled together and attached to bones.

Each person has about 600 skeletal muscles, which range in size from the delicate external eye muscles that control eye movements and contain only a few hundred fibers, to the large, powerful leg muscles that contain several hundred thousand fibers.

Each muscle is sheathed by connective tissue that penetrates from the surface into the muscle to envelop each individual fiber and divide the muscle into columns or bundles. The connective tissue extends beyond the ends of the muscle to form tough, collagenous **tendons** that attach the muscle to bones. A tendon may be quite long, attaching to a bone some distance from the fleshy part of the muscle. For example, some of the muscles involved in finger movement are in the forearm, with long tendons extending down to attach to the bones of the fingers. (You can readily see these tendons move on the top of your hand when you wiggle your fingers.) This arrangement permits greater dexterity; the fingers would be much thicker and more awkward if all the muscles used in finger movement were actually in the fingers.

Contractions of a whole muscle can be of varying strength.

A single action potential in a muscle fiber produces a brief, weak contraction called a **twitch**, which is too short and too weak to be useful and normally does not take place in the body. Muscle fibers are arranged into whole muscles, where they function cooperatively to produce contractions of variable grades of strength stronger than a twitch. In other words, the force exerted by the same muscle can be made to vary, depending on whether the person is picking up a piece of paper, a book, or a 50-pound weight. Two primary factors can be adjusted to accomplish gradation of whole-muscle tension: (1) *the number of muscle fibers contracting within a muscle* and (2) *the tension developed by each contracting fiber*. We will discuss each of these factors in turn.

The number of fibers contracting within a muscle depends on the extent of motor unit recruitment.

The greater the number of fibers contracting, the greater the total muscle tension. Therefore, larger muscles consisting of more muscle fibers obviously can generate more tension than can smaller muscles with fewer fibers.

Each whole muscle is innervated by a number of different motor neurons. When a motor neuron enters a muscle, it branches, with each axon terminal supplying a single muscle fiber (● Figure 8-13). One motor neuron innervates a number of muscle fibers, but each muscle fiber is supplied by only one motor neuron. When a motor neuron is activated, all the muscle fibers it supplies are stimulated to contract simultaneously. This team of concurrently activated components— one motor neuron plus all the muscle fibers it innervates—is called a **motor unit**. The muscle fibers that compose a motor unit are dispersed throughout the whole muscle; thus their simultaneous contraction results in an evenly distributed, although weak, contraction of the whole muscle. Each muscle consists of a number of intermingled motor units. For a weak contraction of the whole muscle, only one or a few of its motor units are activated. For stronger and stronger contractions, more and more motor units are recruited, or stimulated to contract, a phenomenon known as **motor unit recruitment**.

How much stronger the contraction will be with the recruitment of each additional motor unit depends on the size of the motor units (that is, the number of muscle fibers controlled by a single motor neuron). The number of muscle fibers per motor unit and the number of motor units per muscle vary widely, depending on the specific function of the muscle. For muscles that produce precise, delicate movements, such as external eye muscles and hand muscles, a single motor unit may contain as few as a dozen muscle fibers. Because so few muscle fibers are involved with each motor unit, recruit-

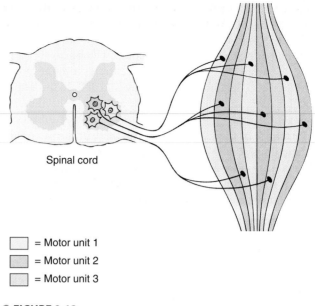

Spinal cord

☐ = Motor unit 1
▨ = Motor unit 2
▨ = Motor unit 3

● **FIGURE 8-13**

Schematic representation of motor units in a skeletal muscle

ment of each additional motor unit adds only a small increment in the whole muscle's strength of contraction. These small motor units allow very fine control over muscle tension. In contrast, in muscles designed for powerful, coarsely controlled movement, such as those of the legs, a single motor unit may contain 1500 to 2000 muscle fibers. Recruitment of motor units in these muscles results in large incremental increases in whole-muscle tension. More powerful contractions occur at the expense of less precisely controlled gradations. Thus the number of muscle fibers participating in the whole muscle's total contractile effort depends on the number of motor units recruited and the number of muscle fibers per motor unit in that muscle.

To delay or prevent **fatigue** (inability to maintain muscle tension at a given level) during a sustained contraction involving only a portion of a muscle's motor units, as is necessary in muscles supporting the weight of the body against the force of gravity, **asynchronous recruitment of motor units** takes place. The body alternates motor unit activity, like shifts at a factory, to give motor units that have been active an opportunity to rest while others take over. Changing of the shifts is carefully coordinated, so the sustained contraction is smooth rather than jerky. Asynchronous motor unit recruitment is possible only for submaximal contractions, during which only some of the motor units must maintain the desired level of tension. During maximal contractions, when all the muscle fibers must participate, it is impossible to alternate motor unit activity to prevent fatigue. This is one reason why you cannot support a heavy object as long as a light one.

The frequency of stimulation can influence the tension developed by each muscle fiber.

Whole-muscle tension depends not only on the number of muscle fibers contracting but also on the tension developed by each contracting fiber. Various factors influence the extent to which tension can be developed. These factors include the following:

1. Frequency of stimulation
2. Length of the fiber at the onset of contraction
3. Extent of fatigue
4. Thickness of the fiber

We will now examine the effect of frequency of stimulation. (The other factors are discussed in later sections.)

Even though a single action potential in a muscle fiber produces only a twitch, contractions with longer duration and greater tension can be achieved by repeated stimulation of the fiber. Let us see what happens when a second action potential occurs in a muscle fiber. If the muscle fiber has completely relaxed before the next action potential takes place, a second twitch of the same magnitude as the first occurs (● Figure 8-14a). The same excitation–contraction events take place each time, resulting in identical twitch responses. If, however, the muscle fiber is stimulated a second time before it has completely relaxed from the first twitch, a second action potential causes a second contractile response, which is added "piggyback" on top of the first twitch (● Figure 8-14b). The two twitches from the two action potentials add together, or sum, to produce greater tension in the fiber than that produced by a single action potential. This **twitch summation** is similar to temporal summation of EPSPs at the postsynaptic neuron (see p. 89).

Twitch summation is possible only because the duration of the action potential (1 to 2 msec) is much shorter than the duration of the resulting twitch (100 msec). Once an action potential has been initiated, a brief refractory period occurs during which another action potential cannot be initiated (see p. 82). It is therefore impossible to achieve summation of action potentials. The membrane must return to resting potential and recover from its refractory period before another action potential can occur. However, because the action potential and refractory period are over long before the resulting muscle twitch is completed, the muscle fiber may be restimulated while some contractile activity still exists, to produce summation of the mechanical response.

If the muscle fiber is stimulated so rapidly that it does not have a chance to relax at all between stimuli, a smooth, sustained contraction of maximal strength known as **tetanus** occurs (● Figure 8-14c). A tetanic contraction is usually three to four times stronger than a single twitch. (Don't confuse this normal physiologic tetanus with the disease tetanus; see p. 92.)

Twitch summation results from a sustained elevation in cytosolic calcium.

What is the mechanism of twitch summation and tetanus at the cell level? The tension produced by a contracting muscle fiber increases as a result of greater cross-bridge cycling. As the frequency of action potentials increases, the resulting tension development increases until a maximum tetanic contraction is achieved. Enough Ca^{2+} is released in response to a single action potential to interact with all the troponin within the cell. As a result, all the cross bridges are free to participate in the contractile response. How, then, can repetitive action potentials bring about a greater contractile response? The difference depends on how long enough Ca^{2+} is available.

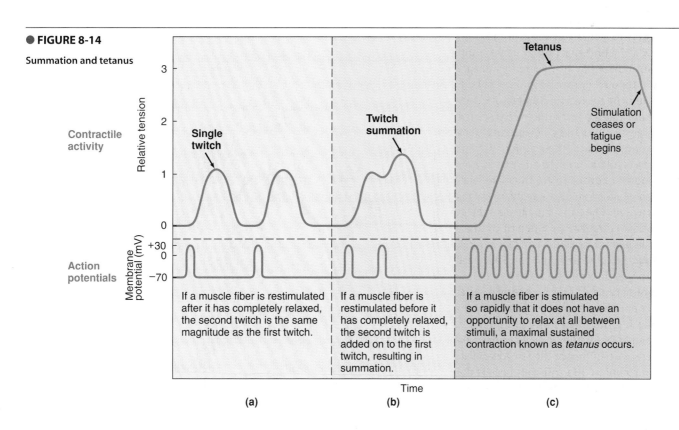

● **FIGURE 8-14**

Summation and tetanus

Single twitch

Twitch summation

Tetanus

Stimulation ceases or fatigue begins

If a muscle fiber is restimulated after it has completely relaxed, the second twitch is the same magnitude as the first twitch.

If a muscle fiber is restimulated before it has completely relaxed, the second twitch is added on to the first twitch, resulting in summation.

If a muscle fiber is stimulated so rapidly that it does not have an opportunity to relax at all between stimuli, a maximal sustained contraction known as *tetanus* occurs.

Time

(a) (b) (c)

The cross bridges remain active and continue to cycle as long as enough Ca^{2+} is present to keep the troponin-tropomyosin complex away from the cross-bridge binding sites on actin. Each troponin-tropomyosin complex spans a distance of seven actin molecules. Thus binding of Ca^{2+} to one troponin molecule leads to the uncovering of only seven cross-bridge binding sites on the thin filament.

As soon as Ca^{2+} is released in response to an action potential, the sarcoplasmic reticulum starts pumping Ca^{2+} back into the lateral sacs. As the cytosolic Ca^{2+} concentration declines with the reuptake of Ca^{2+} by the lateral sacs, less Ca^{2+} is present to bind with troponin, so some of the troponin-tropomyosin complexes slip back into their blocking positions. Consequently, not all the cross-bridge binding sites remain available to participate in the cycling process during a single twitch induced by a single action potential. Because not all the cross bridges find a binding site, the resulting contraction during a single twitch is not of maximal strength.

If action potentials and twitches occur far enough apart in time for all the released Ca^{2+} from the first contractile response to be pumped back into the lateral sacs between the action potentials, an identical twitch response will occur as a result of the second action potential. If, however, a second action potential occurs and more Ca^{2+} is released while the Ca^{2+} that was released in response to the first action potential is being taken back up, the cytosolic Ca^{2+} concentration remains elevated. This prolonged availability of Ca^{2+} in the cytosol permits more of the cross bridges to continue participating in the cycling process for a longer time. As a result, tension development increases correspondingly. As the frequency of action potentials increases, the duration of elevated cytosolic Ca^{2+} concentration increases, and contractile activity likewise increases until a maximum tetanic contraction is reached. With tetanus, the maximum number of cross-bridge binding sites remain uncovered so that cross-bridge cycling, and consequently tension development, are at their peak.

Because skeletal muscle must be stimulated by motor neurons to contract, the nervous system plays a key role in regulating contraction strength. The two main factors subject to control to accomplish gradation of contraction are the *number of motor units stimulated* and the *frequency of their stimulation*. The areas of the brain that direct motor activity combine tetanic contractions and precisely timed shifts of asynchronous motor unit recruitment to execute smooth rather than jerky contractions.

Additional factors not directly under nervous control also influence the tension developed during contraction. Among these is the length of the fiber at the onset of contraction, to which we now turn our attention.

There is an optimal muscle length at which maximal tension can be developed.

A relationship exists between the length of the muscle before the onset of contraction and the tetanic tension that each contracting fiber can subsequently develop at that length. For every muscle there is an **optimal length** (l_o) at which maximal force can be achieved on a subsequent tetanic contraction. More tension can be achieved during tetanus when beginning at the optimal muscle length than can be achieved when the contraction begins with the muscle less than or greater than its optimal length. This **length–tension relationship** can be explained by the sliding filament mechanism of muscle contraction.

AT OPTIMAL LENGTH (l_o)

At l_o, when maximum tension can be developed (point A in ● Figure 8-15), the thin filaments optimally overlap the regions of the thick filaments from which the cross bridges

● **FIGURE 8-15**

Length–tension relationship. Maximal tetanic contraction can be achieved when a muscle fiber is at its optimal length (l_o) before the onset of contraction, because this is the point of optimal overlap of thick-filament cross bridges and thin-filament cross-bridge binding sites (point A). The percentage of maximal tetanic contraction that can be achieved decreases when the muscle fiber is longer or shorter than l_o before contraction. When it is longer, fewer thin-filament binding sites are accessible for binding with thick-filament cross bridges, because the thin filaments are pulled out from between the thick filaments (points B and C). When the fiber is shorter, fewer thin-filament binding sites are exposed to thick-filament cross bridges because the thin filaments overlap (point D). Also, further shortening and tension development are impeded as the thick filaments become forced against the Z lines (point D). In the body, the resting muscle length is at l_o. Furthermore, because of restrictions imposed by skeletal attachments, muscles cannot vary beyond 30% of their l_o in either direction (the range screened in light green). At the outer limits of this range, muscles still can achieve about 50% of their maximal tetanic contraction.

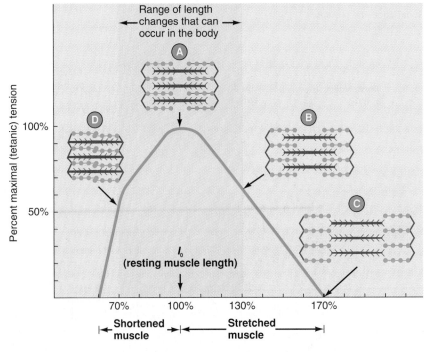

Muscle fiber length compared with resting length

project. At this length, a maximal number of cross-bridge sites are accessible to the actin molecules for binding and bending. The central region of thick filaments, where the thin filaments do not overlap at l_o, lacks cross bridges; only myosin tails are found here.

AT LENGTHS GREATER THAN l_o

At greater lengths, as when a muscle is passively stretched (point B), the thin filaments are pulled out from between the thick filaments, decreasing the number of actin sites available for cross-bridge binding; that is, some of the actin sites and cross bridges no longer "match up," so they "go unused." When less cross-bridge activity can occur, less tension can develop. In fact, when the muscle is stretched to about 70% longer than its l_o (point C) the thin filaments are completely pulled out from between the thick filaments, preventing cross-bridge activity, and consequently no contraction can occur.

AT LENGTHS LESS THAN l_o

If a muscle is shorter than l_o before contraction (point D), less tension can be developed for two main reasons:

1. The thin filaments from the opposite sides of the sarcomere become overlapped, which limits the opportunity for the cross bridges to interact with actin.
2. The ends of the thick filaments become forced against the Z lines, so further shortening is impeded.

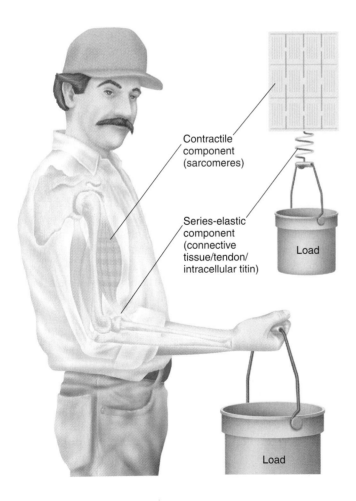

Contractile component (sarcomeres)

Series-elastic component (connective tissue/tendon/intracellular titin)

Load

Load

LIMITATIONS ON MUSCLE LENGTH

The extremes in muscle length that prevent development of tension occur only under experimental conditions, when a muscle is removed and stimulated at various lengths. In the body the muscles are so positioned that their relaxed length is approximately their optimal length; thus they can achieve near-maximal tetanic contraction most of the time. Because attachment to the skeleton imposes limitations, a muscle cannot be stretched or shortened more than 30% of its resting optimal length, and usually it deviates much less than 30% from normal length. Even at the outer limits (130% and 70% of l_o), the muscles still can generate half their maximum tension.

The factors we have discussed thus far that influence how much tension can be developed by a contracting muscle fiber—the frequency of stimulation and the muscle length at the onset of contraction—can vary from contraction to contraction. Other determinants of muscle fiber tension—the metabolic capability of the fiber relative to resistance to fatigue and the thickness of the fiber—do not vary from contraction to contraction but depend on the fiber type and can be modified over a period of time. After we complete our discussion of skeletal muscle mechanics, we will consider these other factors in the next section, on skeletal muscle metabolism and fiber types.

> ▌ Muscle tension is transmitted to bone as the contractile component tightens the series-elastic component.

Tension is produced internally within the sarcomeres, considered the **contractile component** of the muscle, as a result of cross-bridge activity and the resulting sliding of filaments. However, the sarcomeres are not attached directly to the bones. Instead, the tension generated by these contractile elements must be transmitted to the bone via the connective tissue and tendons before the bone can be moved. Connective tissue and tendon, as well as other components of the muscle, such as the intracellular titin, have a certain degree of passive elasticity. These noncontractile tissues are called the **series-elastic component** of the muscle; they behave like a stretchy spring placed between the internal tension-generating elements and the bone that is to be moved against an external load (● Figure 8-16). Shortening of the sarcomeres stretches the series-elastic component. Muscle tension is transmitted to the bone by this tightening of the series-elastic component. This force applied to the bone moves the bone against a load.

A muscle is typically attached to at least two different bones across a joint by means of tendons that extend from each end of the muscle (● Figure 8-17). When the muscle

● **FIGURE 8-16**

Relationship between the contractile component and the series-elastic component in transmitting muscle tension to bone. Muscle tension is transmitted to the bone by means of the stretching and tightening of the muscle's elastic connective tissue and tendon as a result of sarcomere shortening brought about by cross-bridge cycling.

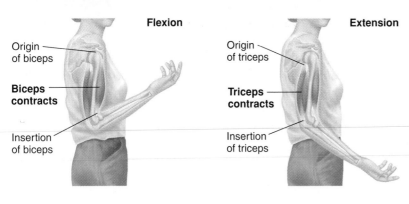

Flexion

Origin of biceps

Biceps contracts

Insertion of biceps

Extension

Origin of triceps

Triceps contracts

Insertion of triceps

● **FIGURE 8-17**

Flexion and extension of the elbow joint

shortens during contraction, the position of the joint is changed as one bone is moved in relation to the other—for example, *flexion* of the elbow joint by contraction of the biceps muscle and *extension* of the elbow by contraction of the triceps. The end of the muscle attached to the more stationary part of the skeleton is called the **origin**, and the end attached to the skeletal part that moves is the **insertion**.

Not all muscle contractions shorten muscles and move bones, however. For a muscle to shorten during contraction, the tension developed in the muscle must exceed the forces that oppose movement of the bone to which the muscle's insertion is attached. In the case of elbow flexion, the opposing force, or **load**, is the weight of an object being lifted. When you flex your elbow without lifting any external object, there is still a load, albeit a minimal one—the weight of your forearm being moved against the force of gravity.

▌ The two primary types of contraction are isotonic and isometric.

There are two primary types of contraction, depending on whether the muscle changes length during contraction. In an **isotonic contraction**, muscle tension remains constant as the muscle changes length. In an **isometric contraction**, the muscle is prevented from shortening, so tension develops at constant muscle length. The same internal events occur in both isotonic and isometric contractions: muscle excitation turns on the tension-generating contractile process; the cross bridges start cycling; and filament sliding shortens the sarcomeres, which stretches the series-elastic component to exert force on the bone at the site of the muscle's insertion.

Considering your biceps as an example, assume you are going to lift an object. When the tension developing in your biceps becomes great enough to overcome the weight of the object in your hand, you can lift the object, with the whole muscle shortening in the process. Because the weight of the object does not change as it is lifted, the muscle tension remains constant throughout the period of shortening. This is an *isotonic* (literally, "constant tension") *contraction*. Isotonic contractions are used for body movements and for moving external objects.

What happens if you try to lift an object too heavy for you (that is, if the tension you are capable of developing in your arm muscles is less than required to lift the load)? In this case, the muscle cannot shorten and lift the object but remains at constant length despite the development of tension, so an *isometric* ("constant length") *contraction* occurs. In addition to occurring when the load is too great, isometric contractions also take place when the tension developed in the muscle is deliberately less than needed to move the load. In this case, the goal is to keep the muscle at fixed length although it can develop more tension. These submaximal isometric contractions are important for maintaining posture (such as keeping the legs stiff while standing) and for supporting objects in a fixed position. During a given movement, a muscle may shift between isotonic and isometric contractions. For example, when you pick up a book to read, your biceps undergoes an isotonic contraction while you are lifting the book, but the contraction becomes isometric as you stop to hold the book in front of you.

CONCENTRIC AND ECCENTRIC ISOTONIC CONTRACTIONS

There are actually two types of isotonic contraction—*concentric* and *eccentric*. In both, the muscle changes length at constant tension. With **concentric contractions**, however, the muscle shortens, whereas with eccentric contractions the muscle lengthens, because it is being stretched by an external force while contracting. With an eccentric contraction, the contractile activity is resisting the stretch. An example is lowering a load to the ground. During this action, the muscle fibers in the biceps are lengthening but are still contracting in opposition to being stretched. This tension supports the weight of the object.

OTHER CONTRACTIONS

The body is not limited to pure isotonic and isometric contractions. Muscle length and tension frequently vary throughout a range of motion. Think about pulling back a bow and arrow. The tension of your biceps muscle continuously increases to overcome the progressively increasing resistance as you stretch the bow further. At the same time, the muscle progressively shortens as you draw the bow farther back. Such a contraction occurs at neither constant tension nor constant length.

Some skeletal muscles do not attach to bones at both ends but still produce movement. For example, the tongue muscles are not attached at the free end. Isotonic contractions of the tongue muscles maneuver the free, unattached portion of the tongue to facilitate speech and eating. A few skeletal muscles that are completely unattached to bone actually prevent movement. These are the voluntarily controlled rings of skeletal muscles, known as **sphincters**, that guard the exit of urine and feces from the body by isotonically contracting.

The load is also an important determinant of the **velocity**, or speed, of shortening during a concentric contraction. The greater the load, the lower the velocity at which a single muscle fiber (or a constant number of contracting fibers within a muscle) shortens during an isotonic tetanic contraction. The speed of shortening is maximal when there is no external load, progressively decreases with an increasing load, and falls to zero (no shortening—isometric contraction) when the load cannot be overcome by maximal tetanic tension. You have frequently experienced this load–velocity relationship. You can lift light objects requiring little muscle tension quickly, whereas you can lift very heavy objects only slowly, if at all. This relationship between load and shortening velocity is a fundamental property of muscle, presumably because it takes the cross bridges longer to stroke against a greater load.

Now let's shift attention from muscle mechanics to the metabolic means by which muscles power these movements.

 Click on the Media Exercises menu of the CD-ROM and work Media Exercise 8.2: Skeletal Muscle Mechanics to test your understanding of the previous section.

SKELETAL MUSCLE METABOLISM AND FIBER TYPES

Three different steps in the contraction–relaxation process require ATP:

1. Splitting of ATP by myosin ATPase provides the energy for the power stroke of the cross bridge.
2. Binding (but not splitting) of a fresh molecule of ATP to myosin lets the bridge detach from the actin filament at the end of a power stroke so that the cycle can be repeated. This ATP is later split to provide energy for the next stroke of the cross bridge.
3. The active transport of Ca^{2+} back into the sarcoplasmic reticulum during relaxation depends on energy derived from the breakdown of ATP.

▌ **Muscle fibers have alternate pathways for forming ATP.**

Because ATP is the only energy source that can be directly used for these activities, for contractile activity to continue ATP must constantly be supplied. Only limited stores of ATP are immediately available in muscle tissue, but three pathways supply additional ATP as needed during muscle contraction: (1) transfer of a high-energy phosphate from creatine phosphate to ADP, (2) oxidative phosphorylation (the citric acid cycle and electron transport system), and (3) glycolysis.

CREATINE PHOSPHATE

Creatine phosphate is the first energy storehouse tapped at the onset of contractile activity. Like ATP, creatine phosphate contains a high-energy phosphate group, which can be donated directly to ADP to form ATP. A rested muscle contains about five times as much creatine phosphate as ATP. Thus most energy is stored in muscle in creatine phosphate pools. At the onset of contraction, when the meager reserves of ATP are rapidly used, more ATP is quickly formed by the transfer of energy and phosphate from creatine phosphate to ADP. Because only one enzymatic reaction is involved in this energy transfer, ATP can be formed rapidly (within a fraction of a second) by using creatine phosphate.

Thus creatine phosphate is the first source for supplying additional ATP when exercise begins. Muscle ATP levels actually remain fairly constant early in contraction, but creatine phosphate stores become depleted. In fact, short bursts of high-intensity contractile effort, such as high jumps, sprints, or weight lifting, are supported primarily by ATP derived at the expense of creatine phosphate. Other energy systems do not have a chance to become operable before the activity is over. Creatine phosphate stores typically power the first minute or less of exercise.

OXIDATIVE PHOSPHORYLATION

If the energy-dependent contractile activity is to continue, the muscle shifts to the alternate pathways of oxidative phosphorylation and glycolysis to form ATP. These multistepped pathways require time to pick up their rates of ATP formation to match the increased demands for energy, time provided by the immediate supply of energy from the one-step creatine phosphate system.

Oxidative phosphorylation takes place within the muscle mitochondria if enough O_2 is present (see p. 29). Although it provides a rich yield of 36 ATP molecules for each glucose molecule processed, oxidative phosphorylation is relatively slow because of the number of steps involved.

During light exercise (such as walking) to moderate exercise (such as jogging or swimming), muscle cells can form enough ATP through oxidative phosphorylation to keep pace with the modest energy demands of the contractile machinery for prolonged periods of time. To sustain ongoing oxidative phosphorylation, the exercising muscles depend on delivery of adequate O_2 and nutrients to maintain their activity. Activity that can be supported in this way is **aerobic** ("with O_2") or **endurance-type exercise**.

GLYCOLYSIS

There are respiratory and cardiovascular limits to how much O_2 can be delivered to a muscle. That is, the lungs and heart can pick up and deliver just so much O_2 to exercising muscles. Furthermore, in near-maximal contractions, the powerful contraction compresses almost closed the blood vessels that course through the muscle, severely limiting the O_2 available to the muscle fibers. Even when O_2 is available, the relatively slow oxidative-phosphorylation system may not be able to produce ATP rapidly enough to meet the muscle's needs during intense activity. A skeletal muscle's energy consumption may increase up to 100-fold when going from rest to high-intensity exercise. When O_2 delivery or oxidative phosphorylation cannot keep pace with the demand for ATP formation as the intensity of exercise increases, the muscle fibers rely increasingly on glycolysis to generate ATP (see p. 27). The chemical reactions of **glycolysis** yield products for ultimate entry into the oxidative phosphorylation pathway, but glycolysis can also proceed alone in the absence of further processing of its products by oxidative phosphorylation. During glycolysis, a glucose molecule is broken down into two **pyruvic acid** molecules, yielding two ATP molecules in the process. Pyruvic acid can be further degraded by oxidative phosphorylation to extract more energy. However, glycolysis alone has two advantages over the oxidative phosphorylation pathway: (1) glycolysis can form ATP in the absence of O_2 (operating *anaerobically,* that is, "without O_2"), and (2) it can proceed more rapidly than oxidative phosphorylation. Although glycolysis extracts considerably fewer ATP molecules from each nutrient molecule processed, it can proceed so much more rapidly that it can outproduce oxidative phosphorylation over a given period of time if enough glucose is present. Activity that can be supported in this way is **anaerobic** or **high-intensity exercise.**

Even though anaerobic glycolysis provides a means of performing intense exercise when the O_2 delivery/oxidative phosphorylation capacity is exceeded, using this pathway has two consequences. First, large amounts of nutrient fuel must be processed, because glycolysis is much less efficient than oxidative phosphorylation in converting nutrient energy into the energy of ATP. (Glycolysis yields a net of 2 ATP molecules for each glucose molecule degraded, whereas the oxidative phosphorylation pathway can extract 36 molecules of ATP from each glucose molecule.) Muscle cells can store limited quantities of glucose in the form of glycogen, but anaerobic glycolysis rapidly depletes the muscle's glycogen supplies. Second, when the end product of anaerobic glycolysis, pyruvic acid, cannot be further processed by the oxidative phosphorylation pathway, it is converted to **lactic acid.** Lactic acid accumulation has been implicated in the muscle soreness that occurs during the time that intense exercise is actually taking place. (The delayed-onset pain and stiffness that begin the day after unaccustomed muscular exertion, however, are probably caused by reversible structural damage.) Furthermore, lactic acid picked up by the blood produces the metabolic acidosis accompanying intense exercise. Researchers believe that both depletion of energy reserves and the fall in muscle pH caused by lactic acid accumulation play a role in the onset of muscle fatigue, when an exercising muscle can no longer respond to stimulation with the same degree of contractile activity. Therefore, anaerobic high-intensity exercise can be sustained for only a short duration, in contrast to the body's prolonged ability to sustain aerobic endurance-type activities.

▌ Increased oxygen consumption is necessary to recover from exercise.

A person continues to breathe deeply and rapidly for a period of time after exercising. The necessity for the elevated O_2 up-take during recovery from exercise (**excess postexercise oxygen consumption,** or **EPOC**) is due to a variety of factors. The best known is repayment of an **oxygen deficit** that was incurred during exercise, when contractile activity was being supported by ATP derived from nonoxidative sources such as creatine phosphate and anaerobic glycolysis. During exercise, the creatine phosphate stores of active muscles are reduced, lactic acid may accumulate, and glycogen stores may be tapped; the extent of these effects depends on the intensity and duration of the activity. During the recovery period, the creatine phosphate system is restored, lactic acid is removed, and glycogen stores are at least partially replenished. The biochemical transformations that restore the energy systems all need O_2, which is provided by the sustained increase in respiratory activity after exercise has stopped.

Part of the EPOC is not directly related to repayment of energy stores but instead results from a general metabolic disturbance following exercise. For example, secretion of epinephrine, a hormone that increases O_2 consumption by the body, is elevated during exercise. Until the circulating level of epinephrine returns to its pre-exercise state, O_2 uptake is increased above normal. Furthermore, during exercise body temperature rises several degrees Fahrenheit. A rise in temperature speeds up O_2-consuming chemical reactions. Until body temperature returns to pre-exercise levels, the increased speed of these chemical reactions accounts in part for the EPOC.

We have been looking at the contractile and metabolic activities of skeletal muscle fibers in general. Yet not all skeletal muscle fibers use these mechanisms to the same extent. We are next going to examine the different types of muscle fibers based on their speed of contraction and how they are metabolically equipped to generate ATP.

▌ There are three types of skeletal muscle fibers, based on differences in ATP hydrolysis and synthesis.

Classified by their biochemical capacities, there are three major types of muscle fibers (▲ Table 8-1):

1. **Slow-oxidative (type I) fibers**
2. **Fast-oxidative (type IIa) fibers**
3. **Fast-glycolytic (type IIb) fibers**

As their names imply, the two main differences among these fiber types are their speed of contraction (slow or fast) and the type of enzymatic machinery they primarily use for ATP formation (oxidative or glycolytic).

FAST VERSUS SLOW FIBERS

Fast fibers have higher myosin ATPase (ATP-splitting) activity than do slow fibers. The higher the ATPase activity, the more rapidly ATP is split and the faster the rate at which energy is made available for cross-bridge cycling. The result is a fast twitch, compared to the slower twitches of those fibers that split ATP more slowly. Thus two factors determine the speed with which a muscle contracts: the load (load–velocity

TABLE 8-1

Characteristics of Skeletal Muscle Fibers

CHARACTERISTIC	Slow-oxidative (type I)	Fast-oxidative (type IIa)	Fast-glycolytic (type IIb)
	TYPE OF FIBER		
Myosin-ATPase Activity	Low	High	High
Speed of Contraction	Slow	Fast	Fast
Resistance to Fatigue	High	Intermediate	Low
Oxidative Phosphorylation Capacity	High	High	Low
Enzymes for Anaerobic Glycolysis	Low	Intermediate	High
Mitochondria	Many	Many	Few
Capillaries	Many	Many	Few
Myoglobin Content	High	High	Low
Color of Fiber	Red	Red	White
Glycogen Content	Low	Intermediate	High

relationship) and the myosin ATPase activity of the contracting fibers (fast or slow twitch).

OXIDATIVE VERSUS GLYCOLYTIC FIBERS

Fiber types also differ in ATP-synthesizing ability. Those with a greater capacity to form ATP are more resistant to fatigue. Some fibers are better equipped for oxidative phosphorylation, whereas others rely primarily on anaerobic glycolysis for synthesizing ATP. Because oxidative phosphorylation yields considerably more ATP from each nutrient molecule processed, it does not readily deplete energy stores. Furthermore, it does not result in lactic acid accumulation. Oxidative types of muscle fibers are therefore more resistant to fatigue than glycolytic fibers are.

Other related characteristics distinguishing these three fiber types are summarized in ▲ Table 8-1. As you would expect, the oxidative fibers, both slow and fast, contain an abundance of mitochondria, the organelles that house the enzymes involved in oxidative phosphorylation. Because adequate oxygenation is essential to support this pathway, these fibers are richly supplied with capillaries. Oxidative fibers also have a high myoglobin content. **Myoglobin**, which is similar to he-

moglobin, can store small amounts of O_2, but more importantly, it increases the rate of O_2 transfer from the blood into muscle fibers. Myoglobin not only helps support oxidative fibers' O_2 dependency, but it also gives them a red color, just as oxygenated hemoglobin produces the red color of arterial blood. Accordingly, these muscle fibers are called **red fibers**.

In contrast, the fast fibers specialized for glycolysis contain few mitochondria but have a high content of glycolytic enzymes instead. Also, to supply the large amounts of glucose needed for glycolysis, they contain a lot of stored glycogen. Because the glycolytic fibers need relatively less O_2 to function, they have only a meager capillary supply compared with the oxidative fibers. The glycolytic fibers contain very little myoglobin and therefore are pale in color, so they are sometimes called **white fibers**. (The most readily observable comparison between red and white fibers is the dark and white meat in poultry.)

GENETIC ENDOWMENT OF MUSCLE FIBER TYPES

In humans, most muscles contain a mixture of all three fiber types; the percentage of each type is largely determined by the type of activity for which the muscle is specialized. Accordingly, a high proportion of slow-oxidative fibers are found in muscles specialized for maintaining low-intensity contractions for long periods of time without fatigue, such as the muscles of the back and legs that support the body's weight against the force of gravity. A preponderance of fast-glycolytic fibers are found in the arm muscles, which are adapted for performing rapid, forceful movements such as lifting heavy objects.

The percentage of these various fibers not only differs between muscles within an individual but also varies considerably among individuals. Athletes genetically endowed with a higher percentage of the fast-glycolytic fibers are good candidates for power and sprint events, whereas those with a greater proportion of slow-oxidative fibers are more likely to succeed in endurance activities such as marathon races.

Of course, success in any event depends on many factors other than genetic endowment, such as the extent and type of training and the level of dedication. Indeed, the mechanical and metabolic capabilities of muscle fibers can change a lot in response to the patterns of demands placed on them. Let's see how.

❚ Muscle fibers adapt considerably in response to the demands placed on them.

Different types of exercise produce different patterns of neuronal discharge to the muscle involved. Depending on the pattern of neural activity, long-term adaptive changes occur in the muscle fibers, enabling them to respond most efficiently to the types of demands placed on the muscle. Two types of changes can be induced in muscle fibers: changes in their ATP-synthesizing capacity and changes in their diameter.

IMPROVEMENT IN OXIDATIVE CAPACITY

Regular aerobic endurance exercise, such as long-distance jogging or swimming, induces metabolic changes within the

Are Athletes Who Use Steroids to Gain Competitive Advantage Really Winners or Losers?

The testing of athletes for drugs, and the much publicized exclusion from competition of those found to be using substances outlawed by sports federations, have stirred considerable controversy. One such group of drugs are **anabolic androgenic steroids** (*anabolic* means "buildup of tissues," *androgenic* means "male producing," and *steroids* are a class of hormone). These agents are closely related to testosterone, the natural male sex hormone, which is responsible for promoting the increased muscle mass characteristic of males.

Although their use is outlawed (possessing anabolic steroids without a prescription became a federal offense in 1991), these agents are taken by many athletes who specialize in power events such as weight lifting and sprinting in the hopes of increasing muscle mass and, accordingly, muscle strength. Both male and female athletes have resorted to using these substances in an attempt to gain a competitive edge. Bodybuilders also take anabolic steroids. There are an estimated 1 million anabolic steroid abusers in the United States. Unfortunately, use of these agents has spread into our nation's high schools and even younger age groups. In a survey of 17,000 high school youths by the National Institute of Drug Abuse, 5% of the respondents reported current or past steroid use to "build muscles" or to "improve appearance." The manager of the steroid abuse hotline of the National Steroid Research Center reports having calls for help from abusers as young as 12 years old.

Studies have confirmed that steroids can increase muscle mass when used in large amounts and coupled with heavy exercise. One reputable study demonstrated an average 8.9-pound gain of lean muscle in bodybuilders who used steroids during a 10-week period. Anecdotal evidence suggests that some steroid users have added as much as 40 pounds of muscle in a year.

The adverse effects of these drugs, however, outweigh any benefits derived. In females, who normally lack potent androgenic hormones, anabolic steroid drugs not only promote "male-type" muscle mass and strength but also "masculinize" the users in other ways, such as by inducing growth of facial hair and by lowering the voice. More importantly, in both males and females, these agents adversely affect the reproductive and cardiovascular systems and the liver, may have an impact on behavior, and may be addictive.

Adverse Effects on the Reproductive System

In males, testosterone secretion and sperm production by the testes are normally controlled by hormones from the anterior pituitary gland. In negative-feedback fashion, testosterone inhibits secretion of these controlling hormones, so that a constant circulating level of testosterone is maintained. The anterior pituitary is similarly inhibited by androgenic steroids taken as a drug. As a result, because the testes do not receive their normal stimulatory input from the anterior pituitary, testosterone secretion and sperm production decrease and the testes shrink. This hormone abuse also may set the stage for cancer of the testes and prostate gland.

In females, inhibition of the anterior pituitary by the androgenic drugs represses the hormonal output that controls ovarian function. The result is failure to ovulate, menstrual irregularities, and decreased secretion of "feminizing" female sex hormones. Their decline diminishes breast size and other female characteristics.

Adverse Effects on the Cardiovascular System

Use of anabolic steroids induces several cardiovascular changes that increase the risk of developing atherosclerosis, which in turn is associated with an increased incidence of heart attacks and strokes (see p. 265). Among these adverse cardiovascular effects are (1) a reduction in high-density lipoproteins (HDL), the "good" cholesterol carriers that help remove cholesterol from the body, and (2) elevated blood pressure. Animal studies have also demonstrated damage to the heart muscle itself.

Adverse Effects on the Liver

Liver dysfunction is common with high steroid intake, because the liver, which normally inactivates steroid hormones and prepares them for urinary excretion, is overloaded by the excess steroid intake. The incidence of liver cancer is also increased.

Adverse Effects on Behavior

Although the evidence is still controversial, anabolic steroid use appears to promote aggressive, even hostile behavior—the so-called 'roid rages.

Addictive Effects

A troubling new concern is the addiction to anabolic steroids of some who abuse these drugs. In one study involving face-to-face interviews, 14% of steroid users were judged on the basis of their responses to be addicted. In another survey, using anonymous, self-administered questionnaires, 57% of steroid users qualified as being addicted. This apparent tendency to become chemically dependent on steroids is alarming, because the potential for adverse effects on health increases with long-term, heavy use, the kind of use that would be expected in someone hooked on the drug.

Thus, for health reasons, without even taking into account the legal and ethical issues, people should not use anabolic steroids. However, the problem appears to be worsening. Currently, the international black market for anabolic steroids is estimated at $1 billion per year. What should be done about this problem, in your opinion?

---

oxidative fibers, which are the ones primarily recruited during aerobic exercise. These changes enable the muscles to use O_2 more efficiently. For example, mitochondria increase in number in the oxidative fibers. Muscles so adapted can better endure prolonged activity without fatiguing, but they do not change in size.

MUSCLE HYPERTROPHY

The actual size of the muscles can be increased by regular bouts of anaerobic, short-duration, high-intensity resistance training, such as weight lifting. The resulting muscle enlargement comes primarily from an increase in diameter (**hypertrophy**) of the fast-glycolytic fibers that are called into play during such powerful contractions. Most of the fiber thickening results from increased synthesis of myosin and actin filaments, which permits a greater opportunity for cross-bridge interaction and consequently increases the muscle's contractile strength. The resultant bulging muscles are better adapted to activities that require intense strength for brief periods, but endurance has not been improved.

INFLUENCE OF TESTOSTERONE

Men's muscle fibers are thicker, and accordingly, their muscles are larger and stronger than those of women, even without weight training, because of the actions of testosterone, a steroid hormone secreted primarily in males. Testosterone promotes the synthesis and assembly of myosin and actin. This fact has led some athletes, both males and females, to the dangerous practice of taking this or closely related steroids to increase their athletic performance. (To explore this topic further, see the accompanying boxed feature, ▶ Beyond the Basics).

MUSCLE ATROPHY

Clinical Note At the other extreme, if a muscle is not used, its actin and myosin content decreases, its fibers become smaller, and the muscle accordingly **atrophies** (decreases in mass) and becomes weaker. Muscle atrophy can result in two ways. **Disuse atrophy** occurs when a muscle is not used for a long period of time even though the nerve supply is intact, as when a cast or brace must be worn or during prolonged bed confinement. **Denervation atrophy** occurs after the nerve supply to a muscle is lost. If the muscle is stimulated electrically until innervation can be re-established, such as during regeneration of a severed peripheral nerve, atrophy can be diminished but not entirely prevented. Contractile activity itself obviously plays an important role in preventing atrophy; however, poorly understood factors released from active nerve endings, perhaps packaged with the ACh vesicles, apparently contribute to the integrity and growth of muscle tissue. (For a discussion of how the weightlessness of space travel affects muscle mass, see the boxed feature, ▶ Beyond the Basics, on p. 224.)

We have now completed our discussion of all the determinants of whole-muscle tension in a skeletal muscle, which are summarized in ▲ Table 8-2. For the remaining section on skeletal muscle, we are going to examine the central and local mechanisms involved in regulating the motor activity performed by these muscles.

CONTROL OF MOTOR MOVEMENT

Particular patterns of motor unit output govern motor activity, ranging from maintenance of posture and balance to stereotypical locomotor movements, such as walking, to individual, highly skilled motor activity, such as gymnastics. Control of any motor movement, no matter what its level of complexity, depends on converging input to the motor neurons of specific motor units. The motor neurons in turn trigger contraction of the muscle fibers within their respective motor units by means of the events that occur at the neuromuscular junction.

▌ Multiple neural inputs influence motor unit output.

Three levels of input control motor neuron output:

1. *Input from afferent neurons,* usually through intervening interneurons, at the level of the spinal cord—that is, spinal reflexes (see p. 138).
2. *Input from the primary motor cortex.* Fibers originating from neuronal cell bodies known as **pyramidal cells** within the primary motor cortex (see p. 120) descend directly without synaptic interruption to terminate on motor neurons. These fibers make up the **corticospinal** (or **pyramidal**) **motor system**.
3. *Input from the brain stem as part of the multineuronal motor system.* The pathways composing the **multineuronal** (or **extrapyramidal**) **motor system** include a number of synapses that involve many regions of the brain (*extra* means "outside of";

▲ **TABLE 8-2**

Determinants of Whole-Muscle Tension in Skeletal Muscle

NUMBER OF FIBERS CONTRACTING	TENSION DEVELOPED BY EACH CONTRACTING FIBER
Number of motor units recruited*	Frequency of stimulation (twitch summation and tetanus)*
Number of muscle fibers per motor unit	Fiber length at onset of contraction (length–tension relationship)

Size of muscle (number of muscle fibers available to contract)	Extent of fatigue
	   Duration of activity
	   Amount of asynchronous recruitment of motor units
	   Type of fiber (fatigue-resistant oxidative or fatigue-prone glycolytic)

	Thickness of fiber
	   Pattern of neural activity (hypertrophy, atrophy)
	   Amount of testosterone (larger fibers in males than females)

*Factors subject to regulation.

Loss of Muscle Mass: A Plight of Space Flight

Skeletal muscles are a case of "use it or lose it." Stimulation of skeletal muscles by motor neurons is essential not only to induce the muscles to contract but also to maintain their size and strength. Muscles that are not routinely stimulated gradually atrophy, or diminish in size and strength.

Skeletal muscles are important in supporting upright posture in the face of gravitational forces in addition to moving body parts. When humans entered the weightlessness of space, it became apparent that the muscular system required the stress of work or gravity to maintain its size and strength. In 1991, the space shuttle *Columbia* was launched for a nine-day mission dedicated among other things to comprehensive research on physiologic changes brought on by weightlessness. The three female and four male astronauts aboard suffered a dramatic and significant 25% reduction of mass in their weight-bearing muscles. The effort required to move the body is remarkably

less in space than on Earth, and there is no need for active muscular opposition to gravity. Furthermore, the muscles used to move around the confines of a space capsule differ from those used for walking down the street. As a result, some muscles rapidly undergo what is known as *functional atrophy*.

The muscles most affected are those in the lower extremities, the gluteal muscles (buttocks), the extensor muscles of the neck and back, and the muscles of the trunk—that is, the muscles used for antigravity support on the ground. Changes include a decrease in muscle volume and mass, decrease in strength and endurance, increased breakdown of muscle protein, and loss of muscle nitrogen (an important component of muscle protein). The exact biological mechanisms that induce muscle atrophy are unknown, but most scientists believe the lack of customary forcefulness of contraction plays a major role. This atrophy presents no problem while within the space

capsule, but such loss of muscle mass must be restricted if astronauts are to perform heavy work during space walks and are to resume normal activities on their return to Earth.

Space programs in the United States and the former Soviet Union have employed intervention techniques that emphasize both diet and exercise in an attempt to prevent muscle atrophy. Faithfully doing vigorous, carefully designed physical exercise several hours daily has helped reduce the severity of functional atrophy. Studies of nitrogen and mineral balances, however, suggest that muscle atrophy continues to progress during exposure to weightlessness despite efforts to prevent it. Furthermore, only half of the muscle mass was restored in the *Columbia* crew after the astronauts had been back on the ground a length of time equal to their flight. These and other findings suggest that further muscle-preserving interventions will be necessary for extended stays in space.

---

pyramidal refers to the pyramidal system). The final link in multineuronal pathways is the brain stem, especially the reticular formation (see p. 133), which in turn is influenced by motor regions of the cortex, the cerebellum, and the basal nuclei. In addition, the motor cortex itself is interconnected with the thalamus as well as with premotor and supplementary motor areas, all part of the multineuronal system.

The only brain regions that directly influence motor neurons are the primary motor cortex and brain stem; the other involved brain regions indirectly regulate motor activity by adjusting motor output from the motor cortex and brain stem. A number of complex interactions take place between these various brain regions. (See Chapter 5 for further discussion of the specific roles and interactions of these brain regions.)

Spinal reflexes involving afferent neurons are important in maintaining posture and in executing basic protective movements, such as the withdrawal reflex. The corticospinal system primarily mediates performance of fine, discrete, voluntary movements of the hands and fingers, such as those required for doing intricate needlework. The multineuronal system, in contrast, primarily regulates overall body posture involving involuntary movements of large muscle groups of the trunk and limbs. The corticospinal and multineuronal systems show considerable complex interaction and overlapping of function. To voluntarily manipulate your fingers to do needlework, for example, you subconsciously assume a particular posture of your arms that lets you hold your work.

Some of the inputs converging on motor neurons are excitatory, whereas others are inhibitory. Coordinated movement depends on an appropriate balance of activity in these

inputs. The following types of motor abnormalities result from defective motor control:

Clinical Note • If an inhibitory system originating in the brain stem is disrupted, muscles become hyperactive because of the unopposed activity in excitatory inputs to motor neurons. This condition, characterized by increased muscle tone and augmented limb reflexes, is known as **spastic paralysis.**

• In contrast, loss of excitatory input, such as that accompanying destruction of descending excitatory pathways exiting the primary motor cortex, brings about **flaccid paralysis.** In this condition, the muscles are relaxed and the person cannot voluntarily contract muscles, although spinal reflex activity is still present. Damage to the primary motor cortex on one side of the brain, as with a stroke, leads to flaccid paralysis on the opposite half of the body (**hemiplegia**, or paralysis of one side of the body). Disruption of all descending pathways, as in traumatic severance of the spinal cord, produces flaccid paralysis below the level of the damaged region—**quadriplegia** (paralysis of all four limbs) in upper spinal cord damage and **paraplegia** (paralysis of the legs) in lower spinal cord injury.

• Destruction of motor neurons—either their cell bodies or efferent fibers—causes flaccid paralysis and lack of reflex responsiveness in the affected muscles.

• Damage to the cerebellum or basal nuclei does not result in paralysis but instead in uncoordinated, clumsy activity and inappropriate patterns of movement. These regions normally smooth out activity initiated voluntarily.

• Damage to higher cortical regions involved in planning motor activity results in the inability to establish appropriate motor commands to accomplish desired goals.

Muscle receptors provide afferent information needed to control skeletal muscle activity.

Coordinated, purposeful skeletal muscle activity depends on afferent input from a variety of sources. At a simple level, afferent signals indicating that your finger is touching a hot stove trigger reflex contractile activity in appropriate arm muscles to withdraw the hand from the injurious stimulus. At a more complex level, if you are going to catch a ball, the motor systems of your brain must program sequential motor commands that will move and position your body correctly for the catch, using predictions of the ball's direction and rate of movement provided by visual input. Many muscles acting simultaneously or alternately at different joints are called into play to shift your body's location and position rapidly, while maintaining your balance in the process. It is critical to have ongoing input about your body position with respect to the surrounding environment, as well as the position of your various body parts in relationship to each other. This information is necessary for establishing a neuronal pattern of activity to perform the desired movement. To appropriately program muscle activity, your CNS must know the starting position of your body. Further, it must be constantly informed about the progression of movement it has initiated, so that it can make adjustments as needed. Your brain receives this information, which is known as *proprioceptive input* (see p. 119), from receptors in your eyes, joints, vestibular apparatus, and skin, as well as from the muscles themselves.

You can demonstrate your joint and muscle proprioceptive receptors in action by closing your eyes and bringing the tips of your right and left index fingers together at any point in space. You can do so without seeing where your hands are, because your brain is informed of the position of your hands and other body parts at all times by afferent input from the joint and muscle receptors.

Two types of muscle receptors—*muscle spindles* and *Golgi tendon organs*—monitor changes in muscle length and tension. Muscle length is monitored by muscle spindles; changes in muscle tension are detected by Golgi tendon organs. Both these receptor types are activated by muscle stretch, but they convey different types of information. Let us see how.

MUSCLE SPINDLE STRUCTURE

Muscle spindles, which are distributed throughout the fleshy part of a skeletal muscle, consist of collections of specialized muscle fibers known as **intrafusal fibers**, which lie within spindle-shaped connective tissue capsules parallel to the "ordinary" **extrafusal fibers** (*fusus* means "spindle") (● Figure 8-18). Unlike an ordinary extrafusal skeletal muscle fiber, which contains contractile elements (myofibrils) throughout its entire length, an intrafusal fiber has a noncontractile central portion, with the contractile elements being limited to both ends.

Each muscle spindle has its own private efferent and afferent nerve supply. The efferent neuron that innervates a muscle spindle's intrafusal fibers is known as a **gamma motor neuron**, whereas the motor neurons that supply the extrafusal fibers are called **alpha motor neurons.** Two types of afferent sensory endings terminate on the intrafusal fibers and serve as muscle spindle receptors, both of which are activated by stretch. Together they detect changes in the length of the fibers during stretching as well as the speed with which it occurs. Muscle spindles play a key role in the stretch reflex.

STRETCH REFLEX

Whenever a whole muscle is passively stretched, its muscle spindle intrafusal fibers are likewise stretched, increasing the firing rate in the afferent nerve fibers whose sensory endings terminate on the stretched spindle fibers. The afferent neuron directly synapses on the alpha motor neuron that innervates the extrafusal fibers of the same muscle, resulting in contraction of that muscle (● Figure 8-19a, pathway 1 → 2). This **stretch reflex** serves as a local negative-feedback mechanism to resist any passive changes in muscle length so that optimal resting length can be maintained.

The classic example of the stretch reflex is the **patellar tendon**, or **knee-jerk**, **reflex** (● Figure 8-20). The extensor muscle of the knee is the *quadriceps femoris*, which forms the anterior (front) portion of the thigh and is attached just below the knee to the tibia (shinbone) by the *patellar tendon*. Tapping this tendon with a rubber mallet passively stretches the quadriceps muscle, activating its spindle receptors. The resulting stretch reflex brings about contraction of

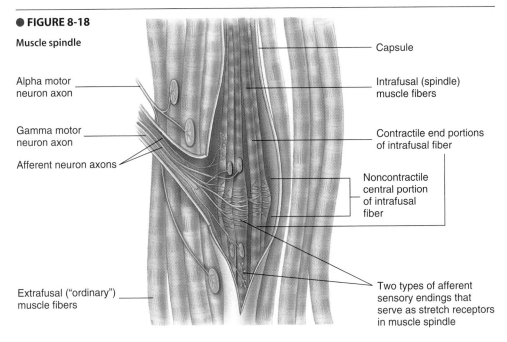

● **FIGURE 8-18**

Muscle spindle

Alpha motor neuron axon

Gamma motor neuron axon

Afferent neuron axons

Extrafusal ("ordinary") muscle fibers

Capsule

Intrafusal (spindle) muscle fibers

Contractile end portions of intrafusal fiber

Noncontractile central portion of intrafusal fiber

Two types of afferent sensory endings that serve as stretch receptors in muscle spindle

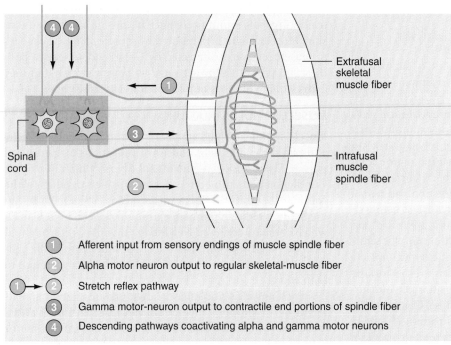

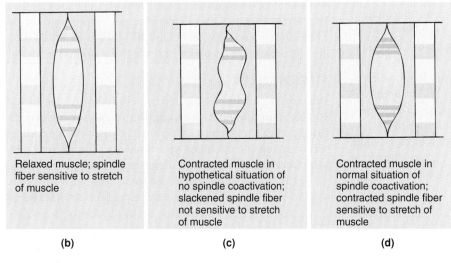

Extrafusal skeletal muscle fiber

Intrafusal muscle spindle fiber

Spinal cord

① Afferent input from sensory endings of muscle spindle fiber

② Alpha motor neuron output to regular skeletal-muscle fiber

①→② Stretch reflex pathway

③ Gamma motor-neuron output to contractile end portions of spindle fiber

④ Descending pathways coactivating alpha and gamma motor neurons

(a)

Relaxed muscle; spindle fiber sensitive to stretch of muscle

(b)

Contracted muscle in hypothetical situation of no spindle coactivation; slackened spindle fiber not sensitive to stretch of muscle

(c)

Contracted muscle in normal situation of spindle coactivation; contracted spindle fiber sensitive to stretch of muscle

(d)

● **FIGURE 8-19**

Muscle spindle function. (a) Pathways involved in the monosynaptic stretch reflex and coactivation of alpha and gamma motor neurons. (b) Status of a muscle spindle when the muscle is relaxed. (c) Status of a muscle spindle in the hypothetical situation of a muscle being contracted on alpha motor-neuron stimulation in the absence of spindle coactivation. (d) Status of a muscle spindle in the actual physiologic situation when both the muscle and muscle spindle are contracted on alpha and gamma motor-neuron coactivation.

 For an animation of this figure, click the Muscle Spindles tab (page 4) in the Skeletal Muscle Contraction tutorial on the CD-ROM.

It also indicates an appropriate balance of excitatory and inhibitory input to the motor neurons from higher brain levels. Muscle jerks may be absent or depressed with loss of higher-level excitatory inputs, or may be greatly exaggerated with loss of inhibitory input to the motor neurons from higher brain levels.

The primary purpose of the stretch reflex is to resist the tendency for the passive stretch of extensor muscles by gravitational forces when a person is standing upright. Whenever the knee joint tends to buckle because of gravity, the quadriceps muscle is stretched. The resulting enhanced contraction of this extensor muscle brought about by the stretch reflex quickly straightens out the knee, holding the limb extended so that the person remains standing.

COACTIVATION OF GAMMA AND ALPHA MOTOR NEURONS

Gamma motor neurons initiate contraction of the muscular end regions of intrafusal fibers (● Figure 8-19a, pathway ③). This contractile response is too weak to have any influence on whole-muscle tension, but it does have an important localized effect on the muscle spindle itself. If there were no compensating mechanisms, shortening of the whole muscle by alpha motor-neuron stimulation of extrafusal fibers would slacken the spindle fibers so that they would be less sensitive to stretch and therefore not as effective as muscle length detectors (● Figures 8-19b and c). **Coactivation** of the gamma motor-neuron system along with the alpha motor-neuron system during reflex and voluntary contractions (● Figure 8-19a, pathway ④) takes the slack out of the spindle fibers as the whole muscle shortens, letting these receptor structures maintain their high sensitivity to stretch over a wide range of muscle lengths. When gamma motor-neuron stimulation triggers simultaneous contraction of both end muscular portions of an intrafusal fiber, the noncontractile central portion is pulled in opposite directions, tightening this region and taking out the slack (● Figure 8-19d). Whereas the extent of alpha motor-neuron activation depends on the intended strength of the motor response, the extent of simultaneous gamma motor-neuron activity to the same muscle depends on the anticipated distance of shortening.

this extensor muscle, causing the knee to extend and raise the foreleg in the well-known knee-jerk fashion. This test is routinely done as a preliminary assessment of nervous system function. A normal knee jerk indicates that a number of neural and muscular components—muscle spindle, afferent input, motor neurons, efferent output, neuromuscular junctions, and the muscles themselves—are functioning normally.

	TYPE OF MUSCLE			
CHARACTERISTIC	**Skeletal**	**Multiunit Smooth**	**Single-unit Smooth**	**Cardiac**
Presence of T Tubules	Yes	No	No	Yes
Level of Development of Sarcoplasmic Reticulum	Well developed	Poorly developed	Poorly developed	Moderately developed
Cross Bridges Turned on by Ca^{2+}	Yes	Yes	Yes	Yes
Source of Increased Cytosolic Ca^{2+}	Sarcoplasmic reticulum	Extracellular fluid and sarcoplasmic reticulum	Extracellular fluid and sarcoplasmic reticulum	Extracellular fluid and sarcoplasmic reticulum
Site of Ca^{2+} Regulation	Troponin in thin filaments	Myosin in thick filaments	Myosin in thick filaments	Troponin in thin filaments
Mechanism of Ca^{2+} Action	Physically repositions troponin-tropomyosin complex to uncover actin cross-bridge binding sites	Chemically brings about phosphorylation of myosin cross bridges so they can bind with actin	Chemically brings about phosphorylation of myosin cross bridges so they can bind with actin	Physically repositions troponin-tropomyosin complex
Presence of Gap Junctions	No	Yes (very few)	Yes	Yes
ATP Used Directly by Contractile Apparatus	Yes	Yes	Yes	Yes
Myosin ATPase Activity; Speed of Contraction	Fast or slow, depending on type of fiber	Very slow	Very slow	Slow
Means by Which Gradation Accomplished	Varying number of motor units contracting (motor unit recruitment) and frequency at which they're stimulated (twitch summation)	Varying number of muscle fibers contracting and varying cytosolic Ca^{2+} concentration in each fiber by autonomic and hormonal influences	Varying cytosolic Ca^{2+} concentration through myogenic activity and influences of autonomic nervous system, hormones, mechanical stretch, and local metabolites	Varying length of fiber (depending on extent of filling of heart chambers) and varying cytosolic Ca^{2+} concentration through autonomic, hormonal, and local metabolite influence
Clear-cut Length–Tension Relationship	Yes	No	No	Yes

constituent found in Z lines (● Figure 8-21b). Dense bodies are positioned throughout the smooth muscle cell as well as being attached to the internal surface of the plasma membrane. The dense bodies are held in place by a scaffold of intermediate filaments. The actin filaments are anchored to the dense bodies. The thick- and thin-filament contractile units are oriented slightly diagonally from side to side within the smooth muscle cell in an elongated, diamond-shaped lattice, rather than running parallel with the long axis as myofibrils do in skeletal muscle (● Figure 8-22a). Relative sliding of the thin filaments past the thick filaments during contraction causes the filament lattice to shorten and expand from side to side. As a result, the whole cell shortens and bulges out between the points where the thin filaments are attached to the inner surface of the plasma membrane (● Figure 8-22b).

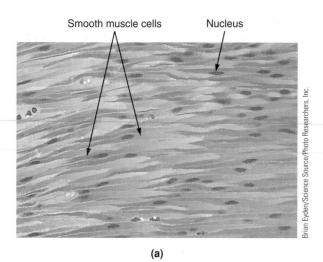

Smooth muscle cells   Nucleus

Brian Eyden/Science Source/Photo Researchers, Inc.

(a)

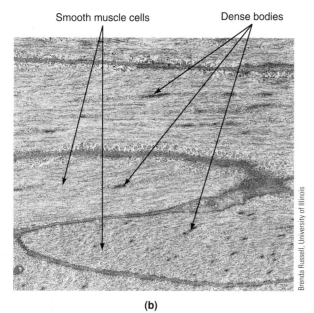

Smooth muscle cells   Dense bodies

Brenda Russell, University of Illinois

(b)

● **FIGURE 8-21**

Microscopic view of smooth muscle cells. (a) Low-power light micrograph of smooth muscle cells. Note the spindle shape and single, centrally located nucleus. (b) Electron micrograph of smooth muscle cells at 14,000× magnification. Note the presence of dense bodies and lack of banding.

For an interaction related to this figure, see Media Exercise 8.4: Smooth Muscle on the CD-ROM.

▌ Smooth muscle cells are turned on by Ca^{2+}-dependent phosphorylation of myosin.

The thin filaments of smooth muscle cells do not contain troponin, and tropomyosin does not block actin's cross-bridge binding sites. Then what prevents actin and myosin from binding at the cross bridges in the resting state, and how is cross-bridge activity switched on in the excited state? Smooth muscle myosin can interact with actin only when the myosin is *phosphorylated* (that is, has a phosphate group attached to it). During excitation, the increased cytosolic Ca^{2+} acts as an intracellular messenger, initiating a chain of biochemical

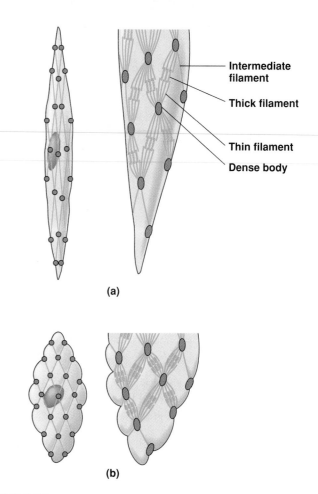

Intermediate filament

Thick filament

Thin filament

Dense body

(a)

(b)

● **FIGURE 8-22**

Schematic representation of the arrangement of thick and thin filaments in a smooth muscle cell in contracted and relaxed states. (a) Relaxed smooth muscle cell. (b) Contracted smooth muscle cell.

events that results in phosphorylation of myosin. Phosphorylated myosin then binds with actin so that cross-bridge cycling can begin. Thus smooth muscle is triggered to contract by a rise in cytosolic Ca^{2+}, similar to what happens in skeletal muscle. In smooth muscle, however, Ca^{2+} ultimately turns on the cross bridges by inducing a *chemical* change in myosin in the *thick* filaments, whereas in skeletal muscle it exerts its effects by invoking a *physical* change at the *thin* filaments (● Figure 8-23). Recall that in skeletal muscle Ca^{2+} moves troponin and tropomyosin from their blocking position, so actin and myosin are free to bind with each other.

The way excitation increases cytosolic Ca^{2+} concentration in smooth muscle cells also differs from that for skeletal muscle. A smooth muscle cell has no T tubules and a poorly developed sarcoplasmic reticulum. The increased cytosolic Ca^{2+} that triggers the contractile response comes from two sources. Most Ca^{2+} enters down its concentration gradient from the ECF on opening of surface-membrane Ca^{2+} channels. The entering Ca^{2+} triggers the opening of Ca^{2+} channels in the sarcoplasmic reticulum, so that small additional amounts of Ca^{2+} are released intracellularly from this meager source. Because smooth muscle cells are so much smaller in diame-

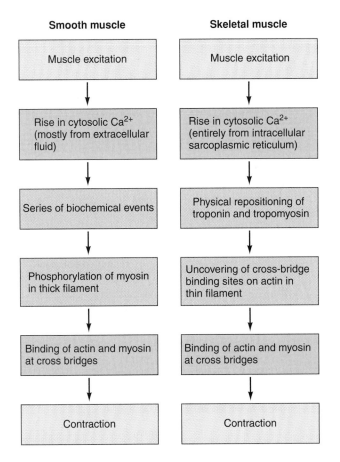

Smooth muscle	Skeletal muscle
Muscle excitation	Muscle excitation
Rise in cytosolic Ca^{2+} (mostly from extracellular fluid)	Rise in cytosolic Ca^{2+} (entirely from intracellular sarcoplasmic reticulum)
Series of biochemical events	Physical repositioning of troponin and tropomyosin
Phosphorylation of myosin in thick filament	Uncovering of cross-bridge binding sites on actin in thin filament
Binding of actin and myosin at cross bridges	Binding of actin and myosin at cross bridges
Contraction	Contraction

● FIGURE 8-23

Comparison of the role of calcium in bringing about contraction in smooth muscle and skeletal muscle

ter than skeletal muscle fibers, most Ca^{2+} entering from the ECF can influence cross-bridge activity, even in the central portions of the cell, without requiring an elaborate T tubule–sarcoplasmic reticulum mechanism.

Relaxation is accomplished by removal of Ca^{2+} as it is actively transported out across the plasma membrane and back into the sarcoplasmic reticulum. When Ca^{2+} is removed, myosin is dephosphorylated (the phosphate is removed) and can no longer interact with actin, so the muscle relaxes.

We still have not addressed the question of how smooth muscle becomes excited to contract; that is, what opens the Ca^{2+} channels in the plasma membrane? Smooth muscle is grouped into two categories—*multiunit* and *single-unit smooth muscle*—based on differences in how the muscle fibers become excited. Let's compare these two types of smooth muscle.

Multiunit smooth muscle is neurogenic.

Multiunit smooth muscle exhibits properties partway between skeletal muscle and single-unit smooth muscle. As the name implies, a multiunit smooth muscle consists of multiple discrete units that function independently of each other and must be separately stimulated by nerves to contract, similar to skeletal-muscle motor units. Thus contractile activity in both

skeletal muscle and multiunit smooth muscle is **neurogenic** ("nerve produced"). That is, contraction in these muscle types is initiated only in response to stimulation by the nerves supplying the muscle. Whereas skeletal muscle is innervated by the voluntary somatic nervous system (motor neurons), multiunit (as well as single-unit) smooth muscle is supplied by the involuntary autonomic nervous system.

Multiunit smooth muscle is found (1) in the walls of large blood vessels; (2) in large airways to the lungs; (3) in the muscle of the eye that adjusts the lens for near or far vision; (4) in the iris of the eye, which alters the pupil size to adjust the amount of light entering the eye; and (5) at the base of hair follicles, contraction of which causes "goose bumps."

Single-unit smooth muscle cells form functional syncytia.

Most smooth muscle is **single-unit smooth muscle**, alternately called **visceral smooth muscle**, because it is found in the walls of the hollow organs or viscera (for example, the digestive, reproductive, and urinary tracts and small blood vessels). The term "single-unit smooth muscle" derives from the fact that the muscle fibers that make up this type of muscle become excited and contract as a single unit. The muscle fibers in single-unit smooth muscle are electrically linked by gap junctions (see p. 49). When an action potential occurs anywhere within a sheet of single-unit smooth muscle, it is quickly propagated via these special points of electrical contact throughout the entire group of interconnected cells, which then contract as a single, coordinated unit. Such a group of interconnected muscle cells that function electrically and mechanically as a unit is known as a **functional syncytium** (plural, *syncytia; syn* means "together"; *cyt* means "cell").

Thinking about the role of the uterus during labor can help you appreciate the significance of this arrangement. Muscle cells composing the uterine wall act as a functional syncytium. They repetitively become excited and contract as a unit during labor, exerting a series of coordinated "pushes" that eventually deliver the baby. Independent, uncoordinated contractions of individual muscle cells in the uterine wall could not exert the uniformly applied pressure needed to expel the baby. Single-unit smooth muscle elsewhere in the body is arranged in similar functional syncytia.

Single-unit smooth muscle is myogenic.

Single-unit smooth muscle is **self-excitable** rather than requiring nervous stimulation for contraction. Clusters of specialized smooth muscle cells within a functional syncytium display spontaneous electrical activity; that is, they can undergo action potentials without any external stimulation. In contrast to the other excitable cells we have been discussing (such as neurons, skeletal muscle fibers, and multiunit smooth muscle), the self-excitable cells of single-unit smooth muscle do not maintain a constant resting potential. Instead, their membrane potential inherently fluctuates without any influence by factors external to the cell. Two major types of spontaneous depolarizations displayed by self-excitable cells are *pacemaker potentials* and *slow-wave potentials*.

PACEMAKER POTENTIALS

With **pacemaker potentials**, the membrane potential gradually depolarizes on its own because of shifts in passive ionic fluxes accompanying automatic changes in channel permeability (● Figure 8-24a). When the membrane has depolarized to threshold, an action potential is initiated. After repolarizing, the membrane potential once again depolarizes to threshold, cyclically continuing in this manner to repetitively self-generate action potentials.

SLOW-WAVE POTENTIALS

Slow-wave potentials are gradually alternating hyperpolarizing and depolarizing swings in potential caused by automatic cyclic changes in the rate at which sodium ions are actively transported across the membrane (● Figure 8-24b). The potential moves farther from threshold during each hyperpolarizing swing and closer to threshold during each depolarizing swing. If threshold is reached, a burst of action potentials occurs at the peak of a depolarizing swing. Threshold is not always reached, however, so oscillating slow-wave potentials can continue without generating action potentials. Whether threshold is reached depends on the starting point of the membrane potential at the onset of its depolarizing swing. The starting point, in turn, is influenced by neural and local factors. (We have now discussed all the means by which excitable tissues can be brought to threshold. ▲ Table 8-4 summarizes the different triggering events that can initiate action potentials in various excitable tissues.)

MYOGENIC ACTIVITY

Self-excitable smooth muscle cells are specialized to initiate action potentials, but they are not equipped to contract. Only a very few of all the cells in a functional syncytium are noncontractile, pacemaker cells. These cells are typically clustered together in a specific location. The vast majority of smooth muscle cells in a functional syncytium are specialized to contract but cannot self-initiate action potentials. However, once an action potential is initiated by a self-excitable smooth muscle cell, it is conducted to the remaining contractile, nonpacemaker cells of the functional syncytium via gap junctions, so the entire group of connected cells contracts as a unit without any nervous input. Such nerve-independent contractile activity initiated by the muscle itself is called **myogenic** ("muscle-produced") **activity**, in contrast to the neurogenic activity of skeletal muscle and multiunit smooth muscle.

▌ Gradation of single-unit smooth muscle contraction differs from that of skeletal muscle.

Single-unit smooth muscle differs from skeletal muscle in the way contraction is graded. Gradation of skeletal muscle contraction is entirely under neural control, primarily involving motor unit recruitment and twitch summation. In smooth muscle, the gap junctions ensure that an entire smooth-muscle

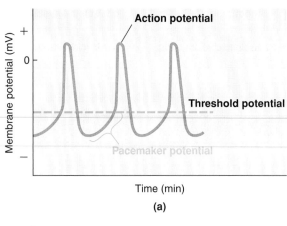

(a)

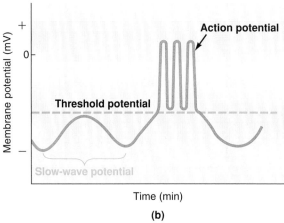

(b)

● **FIGURE 8-24**

Self-generated electrical activity in smooth muscle. (a) With pacemaker potentials, the membrane gradually depolarizes to threshold on a regular periodic basis without any nervous stimulation. These regular depolarizations cyclically trigger self-induced action potentials. (b) In slow-wave potentials, the membrane gradually undergoes self-induced hyperpolarizing and depolarizing swings in potential. A burst of action potentials occurs if a depolarizing swing brings the membrane to threshold.

mass contracts as a single unit, making it impossible to vary the number of muscle fibers contracting. Only the tension of the fibers can be modified to achieve varying strengths of contraction of the whole organ. The portion of cross bridges activated and the tension subsequently developed in single-unit smooth muscle can be graded by varying the cytosolic Ca^{2+} concentration. A single excitation in smooth muscle does not cause all the cross bridges to switch on, in contrast to skeletal muscles, where a single action potential triggers release of enough Ca^{2+} to permit all cross bridges to cycle. As Ca^{2+} concentration increases in smooth muscle, more cross bridges are brought into play, and greater tension develops.

SMOOTH MUSCLE TONE

Many single-unit smooth muscle cells have sufficient levels of cytosolic Ca^{2+} to maintain a low level of tension, or **tone**, even in the absence of action potentials. A sudden drastic change in Ca^{2+}, such as accompanies a myogenically induced action potential, brings about a contractile response superimposed on the ongoing tonic tension. Besides self-

Various Means of Initiating Action Potentials in Excitable Tissues

METHOD OF DEPOLARIZING THE MEMBRANE TO THRESHOLD POTENTIAL	TYPE OF EXCITABLE TISSUE INVOLVED	DESCRIPTION OF THIS TRIGGERING EVENT
Summation of Excitatory Postsynaptic Potentials (EPSPs) (see p. 87)	Efferent neurons, interneurons	Temporal or spatial summation (see p. 89) of slight depolarizations (EPSPs) of the dendrite/cell body end of the neuron brought about by changes in channel permeability in response to binding of excitatory neurotransmitter with surface membrane receptors
Receptor Potential (see p. 147)	Afferent neurons	Typically a depolarization of afferent neuron's receptor initiated by changes in channel permeability in response to neuron's adequate stimulus (see p. 146)
End-plate Potential (see p. 195)	Skeletal muscle	Depolarization of motor end plate (see p. 193) brought about by changes in channel permeability in response to binding of neurotransmitter acetylcholine with receptors on end-plate membrane
Pacemaker Potential (see p. 232)	Smooth muscle, cardiac muscle	Gradual depolarization of membrane on its own because of shifts in passive ionic fluxes accompanying automatic changes in channel permeability
Slow-wave Potential (see p. 232)	Smooth muscle (in digestive tract only)	Gradual alternating hyperpolarizing and depolarizing swings in potential caused by automatic cyclical changes in active ionic transport across membrane

induced action potentials, a number of other factors, including autonomic neurotransmitters, can influence contractile activity and the development of tension in smooth muscle cells by altering their cytosolic Ca^{2+} concentration.

MODIFICATION OF SMOOTH MUSCLE ACTIVITY BY THE AUTONOMIC NERVOUS SYSTEM

Smooth muscle is typically innervated by both branches of the autonomic nervous system. In single-unit smooth muscle, this nerve supply does not *initiate* contraction, but it can *modify* the rate and strength of contraction, either enhancing or retarding the inherent contractile activity of a given organ. Recall that the isolated motor end-plate region of a skeletal muscle fiber interacts with ACh released from a single axon terminal of a motor neuron. In contrast, the receptor proteins that bind with autonomic neurotransmitters are dispersed throughout the entire surface membrane of a smooth muscle cell. Smooth muscle cells are sensitive to varying degrees and in varying ways to autonomic neurotransmitters, depending on the cells' distribution of cholinergic and adrenergic receptors (see p. 191).

Each terminal branch of a postganglionic autonomic fiber travels across the surface of one or more smooth-muscle cells, releasing neurotransmitter from the vesicles within its multiple **varicosities** (bulges) as an action potential passes along the terminal (● Figure 8-25). The neurotransmitter diffuses to the many receptor sites specific for it on the cells underlying the terminal. Thus, in contrast to the discrete one-to-one relationship at motor end plates, a given smooth-muscle cell can be influenced by more than one type of neurotransmitter, and each autonomic terminal can influence more than one smooth-muscle cell.

OTHER FACTORS INFLUENCING SMOOTH MUSCLE ACTIVITY

Other factors (besides autonomic neurotransmitters) can influence the rate and strength of both multiunit and single-unit smooth muscle contraction, including mechanical stretch, certain hormones, local metabolites, and specific drugs. All these factors ultimately act by modifying the permeability of Ca^{2+} channels in the plasma membrane, the sarcoplasmic reticulum, or both, through a variety of mechanisms. Thus smooth muscle is subject to more external influences than skeletal muscle is, even though smooth muscle can contract on its own, whereas skeletal muscle cannot.

For now, as we look at the length–tension relationship in smooth muscle, we are going to discuss in more detail only the effect of mechanical stretch on smooth muscle contractility. We will examine the extracellular chemical influences on smooth muscle contractility in later chapters when we

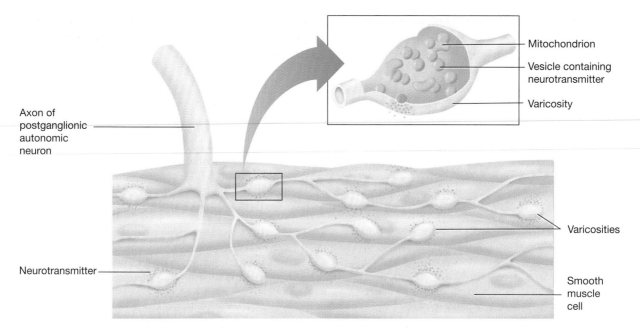

Axon of postganglionic autonomic neuron

Mitochondrion

Vesicle containing neurotransmitter

Varicosity

Varicosities

Neurotransmitter

Smooth muscle cell

● **FIGURE 8-25**

Schematic representation of innervation of smooth muscle by autonomic postganglionic nerve terminals

discuss regulation of the various organs that contain smooth muscle.

▌ Smooth muscle can still develop tension when stretched.

The relationship between the length of the muscle fibers before contraction and the tension that can be developed on a subsequent contraction is less closely linked in smooth muscle than in skeletal muscle. The range of lengths over which a smooth muscle fiber can develop near-maximal tension is much greater than for skeletal muscle. Smooth muscle can still develop considerable tension even when stretched up to 2.5 times its resting length, for two likely reasons. First, in contrast to skeletal muscle, in which the resting length is at l_o, in smooth muscle the resting (nonstretched) length is much shorter than the l_o. Therefore, smooth muscle can be stretched considerably before reaching its optimal length. Second, the thin filaments still overlap the much longer thick filaments even in the stretched-out position, so that cross-bridge interaction and tension development can still take place. In contrast, when skeletal muscle is stretched only three fourths longer than its resting length the thick and thin filaments are completely pulled apart and can no longer interact (see ● Figure 8-15, p. 216).

The ability of a considerably stretched smooth-muscle fiber to still develop tension is important, because the smooth muscle fibers within the wall of a hollow organ are progressively stretched as the volume of the organ's contents expands. Consider the urinary bladder as an example. Even though the muscle fibers in the urinary bladder are stretched as the blad-

der gradually fills with urine, they still maintain their tone and can even develop further tension in response to inputs that regulate bladder emptying. If considerable stretching prevented tension development, as in skeletal muscle, a filled bladder would not be able to contract to empty. Smooth muscle fibers can contract to half their normal length, enabling hollow organs to dramatically empty their contents on increased contractile activity; thus smooth-muscled viscera can easily accommodate large volumes but can empty to practically zero volume. This length range in which smooth muscle normally functions (anywhere from 0.5 to 2.5 times the normal length) is much greater than the limited length range within which skeletal muscle remains functional.

Smooth muscle contains a lot of connective tissue, which resists being stretched. Unlike skeletal muscle, in which the skeletal attachments restrict how far the muscle can be stretched, this connective tissue prevents smooth muscle from being overstretched and thus puts an upper limit on how much a smooth-muscled hollow organ can hold.

▌ Smooth muscle is slow and economical.

A smooth muscle contractile response proceeds more slowly than a skeletal muscle twitch. A single smooth-muscle contraction may last as long as 3 seconds (3000 msec), compared to the maximum of 100 msec required for a single contractile response in skeletal muscle. ATP splitting by myosin ATPase is much slower in smooth muscle, so cross-bridge activity and filament sliding occur more slowly. Smooth muscle also relaxes more slowly because of slower Ca^{2+} removal. Slowness should not be equated with weakness, however.

Smooth muscle can generate the same contractile tension per unit of cross-sectional area as skeletal muscle, but it does so more slowly and at considerably less energy expense. Because of slow cross-bridge cycling during smooth muscle contraction, cross bridges stay attached for more time during each cycle, compared with skeletal muscle; that is, the cross bridges "latch onto" the thin filaments for a longer time each cycle. This so-called **latch phenomenon** enables smooth muscle to maintain tension with comparatively less ATP consumption, because each cross-bridge cycle uses up one molecule of ATP. Smooth muscle is therefore an economical contractile tissue, making it well suited for long-term sustained contractions with little energy consumption and without fatigue. In contrast to the rapidly changing demands placed on your skeletal muscles as you maneuver through and manipulate your external environment, your smooth-muscle activities are geared for long-term duration and slower adjustments to change.

Because of its slowness and the less ordered arrangement of its filaments, smooth muscle has often been mistakenly viewed as a poorly developed version of skeletal muscle. Actually, smooth muscle is just as highly specialized for the demands placed on it—that is, being able to economically maintain tension for prolonged periods without fatigue and being able to vary considerably in volume of contents it holds, with little change in tension. It is an extremely adaptive, efficient tissue.

Cardiac muscle blends features of both skeletal and smooth muscle.

Cardiac muscle, found only in the heart, shares structural and functional characteristics with both skeletal and single-unit smooth muscle. Like skeletal muscle, cardiac muscle is striated, with its thick and thin filaments highly organized into a regular banding pattern. Cardiac thin filaments contain troponin and tropomyosin, which constitute the site of Ca^{2+} action in switching on cross-bridge activity, as in skeletal muscle. Also like skeletal muscle, cardiac muscle has a clear length–tension relationship. Like the oxidative skeletal-muscle fibers, cardiac muscle cells have lots of mitochondria and myoglobin. They also have T tubules and a moderately well-developed sarcoplasmic reticulum.

As in smooth muscle, Ca^{2+} enters the cytosol from both the ECF and the sarcoplasmic reticulum during cardiac excitation. Ca^{2+} entry from the ECF triggers release of Ca^{2+} intracellularly from the sarcoplasmic reticulum. Like single-unit smooth muscle, the heart displays pacemaker (but not slow-wave) activity, initiating its own action potentials without any external influence. Cardiac cells are interconnected by gap junctions that enhance the spread of action potentials throughout the heart, just as in single-unit smooth muscle. Also similarly, the heart is innervated by the autonomic nervous system, which, along with certain hormones and local factors, can modify the rate and strength of contraction.

Unique to cardiac muscle, the cardiac fibers are joined together in a branching network, and the action potentials of cardiac muscle last much longer before repolarizing. Further details and the importance of cardiac muscle's features are addressed in the next chapter.

 Click on the Media Exercises menu of the CD-ROM and work Media Exercise 8.4: Smooth Muscle to test your understanding of the previous section.

 ## CHAPTER IN PERSPECTIVE: FOCUS ON HOMEOSTASIS

The skeletal muscles comprise the muscular system itself. Cardiac and smooth muscle are part of the organs that make up other body systems. Cardiac muscle is found only in the heart, which is part of the circulatory system. Smooth muscle is found in the walls of hollow organs and tubes, including the blood vessels in the circulatory system, airways in the respiratory system, bladder in the urinary system, stomach and intestines in the digestive system, and uterus and ductus deferens (the duct through which sperm leave the testes) in the reproductive system.

Contraction of skeletal muscles accomplishes movement of the body parts in relation to each other and movement of the whole body in relation to the external environment. Thus these muscles permit you to move through and manipulate your external environment. At a very general level, some of these movements are aimed at maintaining homeostasis, such as moving the body toward food or away from harm. Examples of more specific homeostatic functions accomplished by skeletal muscles include chewing and swallowing food for further breakdown in the digestive system into usable energy-producing nutrient molecules (the mouth and throat muscles are all skeletal muscles), and breathing to obtain O_2 and get rid of CO_2 (the respiratory muscles are all skeletal muscles). Heat generation by contracting skeletal muscles also is the major source of heat production in maintaining body temperature. The skeletal muscles further accomplish many nonhomeostatic activities that enable us to work and play—for example, operating a computer or riding a bicycle—so that we can contribute to society and enjoy ourselves.

All the other systems of the body, except the immune (defense) system, depend on their nonskeletal muscle components to enable them to accomplish their homeostatic functions. For example, contraction of cardiac muscle in the heart pushes life-sustaining blood forward into the blood vessels, and contraction of smooth muscle in the stomach and intestines pushes the ingested food through the digestive tract at a rate appropriate for the digestive juices secreted along the route to break down the food into usable units.

CHAPTER SUMMARY

Introduction (pp. 203–204)

▪ Muscles are specialized for contraction.

▪ The three types of muscle—skeletal, cardiac, and smooth—are categorized in two different ways according to common characteristics. (1) Skeletal and cardiac muscle are *striated*, whereas smooth muscle is *unstriated*. (2) Skeletal muscle is *voluntary*, whereas cardiac and smooth muscle are *involuntary*. *(Review Figure 8-1 and Table 8-3, pp. 228–229.)*

Structure of Skeletal Muscle (pp. 204–207)

▪ Skeletal muscles are made up of bundles of long, cylindrical muscle cells known as *muscle fibers,* wrapped in connective tissue.

▪ Muscle fibers are packed with myofibrils, each myofibril consisting of alternating, slightly overlapping stacked sets of thick and thin filaments. This arrangement leads to a skeletal muscle fiber's striated microscopic appearance, which consists of alternating dark A bands and light I bands. *(Review Figures 8-2 and 8-3.)*

▪ Thick filaments consist of the protein myosin. Cross bridges made up of the myosin molecules' globular heads project from each thick filament toward the surrounding thin filaments. *(Review Figure 8-4.)*

▪ Thin filaments consist primarily of the protein actin, which can bind and interact with the myosin cross bridges to bring about contraction. Accordingly, myosin and actin are known as *contractile proteins*. However, in the resting state two other regulatory proteins, tropomyosin and troponin, lie across the surface of the thin filament to prevent this cross-bridge interaction. *(Review Figure 8-5.)*

Molecular Basis of Skeletal Muscle Contraction (pp. 207–213)

▪ Excitation of a skeletal muscle fiber by its motor neuron brings about contraction through a series of events that results in the thin filaments sliding closer together between the thick filaments. *(Review Figure 8-7.)*

▪ This sliding filament mechanism of muscle contraction is switched on by the release of Ca^{2+} from the lateral sacs of the sarcoplasmic reticulum. *(Review Figures 8-9 and 8-10.)*

▪ Calcium release occurs in response to the spread of a muscle-fiber action potential into the central portions of the fiber via the T tubules. *(Review Figures 8-9 and 8-10.)*

▪ Released Ca^{2+} binds to the troponin–tropomyosin complex of the thin filament, slightly repositioning the complex to uncover actin's cross-bridge binding sites. *(Review Figure 8-6.)*

▪ After the exposed actin attaches to a myosin cross bridge, molecular interaction between actin and myosin releases energy within the myosin head that was stored from prior splitting of ATP by the myosin ATPase site. This released energy powers cross-bridge stroking. *(Review Figure 8-11.)*

▪ During a power stroke, an activated cross bridge bends toward the center of the thick filament, "rowing" in the thin filament to which it is attached. *(Review Figure 8-8.)*

▪ With the addition of a fresh ATP molecule to the myosin cross bridge, myosin and actin detach, the cross bridge returns to its original shape, and the cycle is repeated. *(Review Figure 8-11.)*

▪ Repeated cycles of cross-bridge activity slide the thin filaments inward step by step.

▪ When there is no longer a local action potential, the lateral sacs actively take up the Ca^{2+}, troponin and tropomyosin slip back into their blocking position, and relaxation occurs. *(Review Figure 8-10.)*

▪ The entire contractile response lasts about 100 times longer than the action potential. *(Review Figure 8-12.)*

Skeletal Muscle Mechanics (pp. 213–219)

▪ Gradation of whole-muscle contraction can be accomplished by (1) varying the number of muscle fibers contracting within the muscle and (2) varying the tension developed by each contracting fiber. *(Review Table 8-2, p. 223.)*

▪ The greater the number of active muscle fibers, the greater the whole-muscle tension. The number of fibers contracting depends on (1) size of the muscle (number of muscle fibers present), (2) extent of motor unit recruitment (how many motor neurons supplying the muscle are active), and (3) size of each motor unit (how many muscle fibers are activated simultaneously by a single motor neuron). *(Review Figure 8-13 and Table 8-2, p. 223.)*

▪ Also, the greater the tension developed by each contracting fiber, the stronger the contraction of the whole muscle. Two readily variable factors that affect fiber tension are (1) frequency of stimulation, which determines the extent of twitch summation, and (2) length of the fiber before the onset of contraction. *(Review Table 8-2, p. 223.)*

▪ The term *twitch summation* refers to the increase in tension accompanying repetitive stimulation of the muscle fiber. After undergoing an action potential, the muscle cell membrane recovers from its refractory period and can be restimulated again while some contractile activity triggered by the first action potential still remains. As a result, the contractile responses (twitches) induced by the two rapidly successive action potentials can sum, increasing the tension developed by the fiber. If the muscle fiber is stimulated so rapidly that it does not have a chance to start relaxing between stimuli, a smooth, sustained maximal (maximal for the fiber at that length) contraction known as *tetanus* occurs. *(Review Figure 8-14.)*

▪ The tension developed on a tetanic contraction also depends on the length of the fiber at the onset of contraction. At the optimal length (l_o), there is maximal opportunity for cross-bridge interaction, because of optimal overlap of thick and thin filaments. Thus at l_o, which is the resting muscle length, the greatest tension can develop. At lengths shorter or longer than l_o, less tension can develop on contraction, primarily because some of the cross bridges cannot participate. *(Review Figure 8-15.)*

▪ The two primary types of muscle contraction—isometric (constant length) and isotonic (constant tension)—depend on the relationship between muscle tension and the load. The load is the force opposing contraction, that is, the weight of an object being lifted. (1) If tension is less than the load, the muscle cannot shorten and lift the object but remains at constant length, producing an isometric contraction. (2) In an isotonic contraction, the tension exceeds the load so the muscle can shorten and lift the object, maintaining constant tension throughout the period of shortening. *(Review Figure 8-16.)*

▪ The velocity, or speed, of shortening is inversely proportional to the load.

Skeletal Muscle Metabolism and Fiber Types (pp. 219–223)

▪ Three biochemical pathways furnish the ATP needed for muscle contraction: (1) the transfer of high-energy phosphates from stored creatine phosphate to ADP, providing the first source of ATP at the onset of exercise; (2) oxidative phosphorylation, which efficiently extracts large amounts of ATP from nutrient molecules if enough O_2 is available to support this system; and (3) glycolysis, which can synthesize ATP in the absence of O_2 but uses large amounts of stored glycogen and produces lactic acid in the process.

- The three types of muscle fibers are classified by the pathways they use for ATP synthesis (oxidative or glycolytic) and the rapidity with which they split ATP and subsequently contract (slow twitch or fast twitch): (1) slow-oxidative fibers, (2) fast-oxidative fibers, and (3) fast-glycolytic fibers. *(Review Table 8-1.)*

- Muscle fibers adapt in response to different demands placed on them. Regular endurance exercise promotes improved oxidative capacity in oxidative fibers, whereas high-intensity resistance training promotes hypertrophy of fast glycolytic fibers.

Control of Motor Movement (pp. 223–227)

- Control of any motor movement depends on activity level in the presynaptic inputs that converge on the motor neurons supplying various muscles. These inputs come from three sources: (1) spinal reflex pathways, which originate with afferent neurons; (2) the corticospinal (pyramidal) motor system, which originates at the pyramidal cells in the primary motor cortex and is concerned primarily with discrete, intricate movements of the hands; and (3) the multineuronal (extrapyramidal) motor system, which originates in the brain stem and is mostly involved with postural adjustments and involuntary movements of the trunk and limbs. The final motor output from the brain stem is influenced by the cerebellum, basal nuclei, and cerebral cortex.

- Establishment and adjustment of motor commands depend on continuous afferent input, especially feedback about changes in muscle length (monitored by muscle spindles) and muscle tension (monitored by Golgi tendon organs). *(Review Figure 8-18.)*

- When a whole muscle is passively stretched, the accompanying stretch of its muscle spindles triggers the stretch reflex, which results in reflex contraction of that muscle. This reflex resists any passive changes in muscle length. *(Review Figures 8-19 and 8-20.)*

Smooth and Cardiac Muscle (pp. 227–235) *(Review Table 8-3.)*

- The thick and thin filaments of smooth muscle are not arranged in an orderly pattern, so the fibers are not striated. *(Review Figures 8-21 and 8-22.)*

- In smooth muscle, cytosolic Ca^{2+}, which enters from the extracellular fluid as well as being released from sparse intracellular stores, activates cross-bridge cycling by initiating a series of biochemical reactions that result in phosphorylation of the myosin cross bridges to enable them to bind with actin. *(Review Figure 8-23.)*

- Multiunit smooth muscle is neurogenic, requiring stimulation of individual muscle fibers by its autonomic nerve supply to trigger contraction.

- Single-unit smooth muscle is myogenic; it can initiate its own contraction without any external influence, as a result of spontaneous depolarizations to threshold potential brought about by automatic shifts in ionic fluxes. Only a few of the smooth muscle cells in a functional syncytium are self-excitable. The two major types of spontaneous depolarizations displayed by self-excitable smooth muscle cells are pacemaker potentials and slow-wave potentials. *(Review Figure 8-24 and Table 8-4.)*

- Once an action potential is initiated within a self-excitable smooth muscle cell, this electrical activity spreads by means of gap junctions to the surrounding cells within the functional syncytium, so the entire sheet becomes excited and contracts as a unit.

- The level of tension in single-unit smooth muscle depends on the level of cytosolic Ca^{2+}. Many single-unit smooth muscle cells have sufficient cytosolic Ca^{2+} to maintain a low level of tension known as *tone*, even in the absence of action potentials.

- The autonomic nervous system as well as hormones and local metabolites can modify the rate and strength of the self-induced smooth-muscle contractions. All these factors influence smooth muscle activity by altering cytosolic Ca^{2+} concentration.

- Smooth muscle does not have a clear-cut length–tension relationship. It can develop tension when considerably stretched.

- Smooth muscle contractions are energy efficient, enabling this type of muscle to economically sustain long-term contractions without fatigue. This economy, coupled with the fact that single-unit smooth muscle can exist at a variety of lengths with little change in tension, makes single-unit smooth muscle ideally suited for its task of forming the walls of hollow organs that can distend.

- Cardiac muscle is found only in the heart. It has highly organized striated fibers, like skeletal muscle. Like single-unit smooth muscle, some cardiac muscle fibers can generate action potentials, which are spread throughout the heart with the aid of gap junctions.

REVIEW EXERCISES

Objective Questions (Answers on p. A-43)

1. On completion of an action potential in a muscle fiber, the contractile activity initiated by the action potential ceases. *(True or false?)*

2. The velocity at which a muscle shortens depends entirely on the ATPase activity of its fibers. *(True or false?)*

3. When a skeletal muscle is maximally stretched, it can develop maximal tension on contraction, because the actin filaments can slide in a maximal distance. *(True or false?)*

4. A pacemaker potential always initiates an action potential. *(True or false?)*

5. A slow-wave potential always initiates an action potential. *(True or false?)*

6. Smooth muscle can develop tension even when considerably stretched, because the thin filaments still overlap with the long thick filaments. *(True or false?)*

7. A(n) _____ contraction is an isotonic contraction in which the muscle shortens, whereas the muscle lengthens in a(n) _____ isotonic contraction.

8. _____ motor neurons supply extrafusal muscle fibers, whereas intrafusal fibers are innervated by _____ motor neurons.

9. The two types of atrophy are _____ and _____.

10. Which of the following provide(s) direct input to alpha motor neurons? *(Indicate all correct answers.)*
    a. primary motor cortex
    b. brain stem
    c. cerebellum
    d. basal nuclei
    e. spinal reflex pathways

11. Which of the following is *not* involved in bringing about muscle relaxation?
    a. reuptake of Ca^{2+} by the sarcoplasmic reticulum
    b. no more ATP
    c. no more action potential
    d. removal of ACh at the end plate by acetylcholinesterase
    e. filaments sliding back to their resting position

12. Match the following (with reference to skeletal muscle):
    ____ 1. Ca^{2+}
    ____ 2. T tubule
    ____ 3. ATP
    ____ 4. lateral sac of the sarcoplasmic reticulum
    ____ 5. myosin
    ____ 6. troponin-tropomyosin complex
    ____ 7. actin

    (a) cyclically binds with the myosin cross bridges during contraction
    (b) has ATPase activity
    (c) supplies energy for the power stroke of a cross bridge
    (d) rapidly transmits the action potential to the central portion of the muscle fiber
    (e) stores Ca^{2+}
    (f) pulls the troponin-tropomyosin complex out of its blocking position
    (g) prevents actin from interacting with myosin when the muscle fiber is not excited

13. Using the answer code at the right, indicate what happens in the banding pattern during contraction:
    ____ 1. thick myofilament
    ____ 2. thin myofilament
    ____ 3. A band
    ____ 4. I band
    ____ 5. H zone
    ____ 6. sarcomere

    (a) remains the same size during contraction
    (b) decreases in length (shortens) during contraction

Essay Questions

1. Describe the levels of organization in a skeletal muscle.
2. What produces the striated appearance of skeletal muscles? Describe the arrangement of thick and thin filaments that gives rise to the banding pattern.
3. What is the functional unit of skeletal muscle?
4. Describe the composition of thick and thin filaments.
5. Describe the sliding filament mechanism of muscle contraction. How do cross-bridge power strokes bring about shortening of the muscle fiber?
6. Compare the excitation–contraction coupling process in skeletal muscle with that in smooth muscle.
7. How can gradation of skeletal muscle contraction be accomplished?
8. What is a motor unit? Compare the size of motor units in finely controlled muscles with those specialized for coarse, powerful contractions. Describe motor unit recruitment.
9. Explain twitch summation and tetanus.
10. How does a skeletal muscle fiber's length at the onset of contraction affect the strength of the subsequent contraction?
11. Compare isotonic and isometric contractions.
12. Describe the role of each of the following in powering skeletal muscle contraction: ATP, creatine phosphate, oxidative phosphorylation, and glycolysis. Distinguish between aerobically and anaerobically supported exercise.
13. Compare the three types of skeletal muscle fibers.
14. What are the roles of the corticospinal system and multineuronal system in controlling motor movement?
15. Describe the structure and function of muscle spindles and Golgi tendon organs.
16. Distinguish between multiunit and single-unit smooth muscle.
17. Differentiate between neurogenic and myogenic muscle activity.
18. How can smooth muscle contraction be graded?
19. Compare the contractile speed and relative energy expenditure of skeletal muscle with that of smooth muscle.
20. In what ways is cardiac muscle functionally similar to skeletal muscle and to single-unit smooth muscle?

POINTS TO PONDER

(Explanations on p. A-43)

1. Why does regular aerobic exercise provide more cardiovascular benefit than weight training does? *(Hint:* The heart responds to the demands placed on it in a way similar to skeletal muscle.)

2. Put yourself in the position of the scientists who discovered the sliding filament mechanism of muscle contraction, by considering what molecular changes must be involved to account for the observed alterations in the banding pattern during contraction. If you were comparing a relaxed and contracted muscle fiber under a high-power light microscope (see ● Figure 8-3a, p. 206), how could you determine that the thin filaments do not change in length during muscle contraction? You cannot see or measure a single thin filament at this magnification. (*Hint:* What landmark in the banding pattern represents each end of the thin filament? If these landmarks are the same distance apart in a relaxed and contracted fiber, then the thin filaments must not change in length.)

3. What type of off-the-snow training would you recommend for a competitive downhill skier versus a competitive cross-country skier? What adaptive skeletal muscle changes would you hope to accomplish in the athletes in each case?

4. When the bladder is filled and the micturition (urination) reflex is initiated, the nervous supply to the bladder promotes contraction of the bladder and relaxation of the external urethral sphincter, a ring of muscle that guards the exit from the bladder. If the time is inopportune for bladder emptying when the micturition reflex is initiated, the external urethral sphincter can be voluntarily tightened to prevent urination even though the bladder is contracting. Using your knowledge of the muscle types and their innervation, of what types of muscle are the bladder and the external urethral sphincter composed, and what branch of the efferent division of the peripheral nervous system supplies each of these muscles?

5. Three-dimensionally, the thin filaments are arranged hexagonally around the thick filaments. Cross bridges project from each thick filament in all six directions toward the surrounding thin filaments. Each thin filament, in turn, is surrounded by three thick filaments. Sketch this geometric arrangement of thick and think filaments in cross section.

CLINICAL CONSIDERATION

(Explanation on p. A-43)
Jason W. is waiting impatiently for the doctor to finish removing the cast from his leg, which Jason broke the last day of school six weeks ago. Summer vacation is half over, and he hasn't been able to swim, play baseball, or participate in any of his favorite sports. When the cast is finally off, Jason's excitement gives way to concern when he sees that the injured limb is noticeably smaller in diameter than his normal leg. What explains this reduction in size? How can the leg be restored to its normal size and functional ability?

PHYSIOEDGE RESOURCES

 PhysioEdge CD-ROM
PhysioEdge, the CD-ROM packaged with your text, focuses on the concepts students find most difficult to learn. Figures marked with this icon have associated activities on the CD. For a visual review of concepts in this chapter, check out the following:

Tutorial: Skeletal Muscle Contraction

Media Exercise 8.1: Muscle Types

Media Exercise 8.2: Skeletal Muscle Mechanics

Media Exercise 8.3: Skeletal Muscle Structure

Media Exercise 8.4: Smooth Muscle

PhysioEdge Website
The website for this book contains a wealth of helpful study aids, as well as many ideas for further reading and research. Log on to:
http://www.brookscole.com/hpfundamentals3
Select Chapter 8 from the drop-down menu or click on one of the many resource areas.

For Suggested Readings, consult InfoTrac College Edition/ Research on the PhysioEdge website or go directly to InfoTrac College Edition, your online research library, at:
http://infotrac.thomsonlearning.com

Circulatory System (Heart)

Homeostasis
The circulatory system contributes to homeostasis by transporting O_2, CO_2, wastes, electrolytes and hormones from one part of the body to another.

Body systems maintain homeostasis

Homeostasis is essential for survival of cells

Cells
Cells need a constant supply of O_2 and nutrients delivered to them by the circulatory system, which also carries away CO_2 and other wastes, in order to power life-sustaining cellular activities by the chemical reaction:

$$Food + O_2 \rightarrow CO_2 + H_2O + Energy$$

Cells make up body systems

The maintenance of homeostasis depends on essential materials such as O_2 and nutrients being continually picked up from the external environment and delivered to the cells and on waste products being continually removed. Homeostasis also depends on the transfer of hormones, which are important regulatory chemical messengers, from their site of production to their site of action. The circulatory system, which contributes to homeostasis by serving as the body's transport system, consists of the heart, blood vessels, and blood.

All body tissues constantly depend on the life-supporting blood flow the heart provides them by contracting or beating. The heart drives blood through the blood vessels for delivery to the tissues in sufficient amounts, whether the body is at rest or engaging in vigorous exercise.

Cardiac Physiology

PhysioEdge Click on the Tutorials menu of the CD-ROM for a tutorial on Cardiovascular Physiology.

INTRODUCTION

From just a matter of days following conception until death, the beat goes on. In fact, throughout an average human life span, the heart contracts about 3 billion times, never stopping except for a fraction of a second between beats. Within about three weeks after conception, the heart of the developing embryo starts to function. Scientists believe it is the first organ to become functional. At this time the human embryo is only a few millimeters long, about the size of a capital letter on this page.

Why does the heart develop so early, and why is it so crucial throughout life? It is this important because the circulatory system is the body's transport system. A human embryo, having very little yolk available as food, depends on promptly establishing a circulatory system that can interact with the mother's circulation to pick up and distribute to the developing tissues the supplies so critical for survival and growth. Thus begins the story of the circulatory system, which continues throughout life to be a vital pipeline for transporting materials on which the cells of the body are absolutely dependent.

The **circulatory system** has three basic components:

1. The **heart** serves as the pump that imparts pressure to the blood to establish the pressure gradient needed for blood to flow to the tissues. Like all liquids, blood flows down a pressure gradient from an area of higher pressure to an area of lower pressure. This chapter focuses on cardiac physiology (*cardia* means "heart").
2. The **blood vessels** serve as the passageways through which blood is directed and distributed from the heart to all parts of the body and subsequently returned to the heart (see Chapter 10).
3. **Blood** is the transport medium within which materials being transported long distances in the body (such as O_2, CO_2, nutrients, wastes, electrolytes, and hormones) are dissolved or suspended (see Chapter 11).

Blood travels continuously through the circulatory system to and from the heart through two separate vascular (blood vessel) loops, both originating and terminating at the heart (● Figure 9-1). The **pulmonary circulation** consists of a closed loop of vessels carrying blood between the heart and lungs (*pulmo* means "lung). The **systemic circulation** is a circuit of vessels carrying blood between the heart and other body systems.

ANATOMY OF THE HEART

The heart is a hollow, muscular organ about the size of a clenched fist. It lies in the **thoracic** (chest) **cavity** about midline between the **sternum** (breastbone) anteriorly and the **vertebrae** (backbone) posteriorly. Place your hand over your heart. People usually put their hand on the left side of the chest, even though the heart is actually in the middle of the chest. The heart has a broad **base** at the top and tapers to a pointed tip, the **apex**, at the bottom. It is situated at an angle under the sternum so that its base lies predominantly to the right and the apex to the left of the sternum. When the heart beats forcefully, the apex actually thumps against the inside of the chest wall on the left side. Because we become aware of the beating heart through the apex beat on the left side of the chest, we tend to think—erroneously—that the entire heart is on the left.

Clinical Note The heart's position between two bony structures, the sternum and vertebrae, makes it possible to manually drive blood out of the heart when it is not pumping effectively, by rhythmically depressing the sternum. This maneuver compresses the heart between the sternum and vertebrae so that blood is squeezed out into the blood vessels, maintaining blood flow to the tissues. Often this *external cardiac compression*, which is part of **cardiopulmonary resuscitation (CPR)**, serves as a lifesaving measure until appropriate therapy can restore the heart to normal function.

▌ The heart is a dual pump.

Even though anatomically the heart is a single organ, the right and left sides of the heart function as two separate pumps. The heart is divided into right and left halves and has four chambers, an upper and a lower chamber within each half (● Figure 9-2a). The upper chambers, the **atria** (singular, **atrium**), receive blood returning to the heart and transfer it to the lower chambers, the **ventricles**, which pump blood from the heart. The vessels that return blood from the tissues to the atria are **veins**, and those that carry blood away from the ventricles to the tissues are **arteries**. The two halves of the heart are separated by the **septum**, a continuous muscular partition that prevents mixture of blood from the two sides of the heart. This separation is extremely important, because the right half of the heart is receiving and pumping O_2-poor blood, whereas the left side of the heart receives and pumps O_2-rich blood.

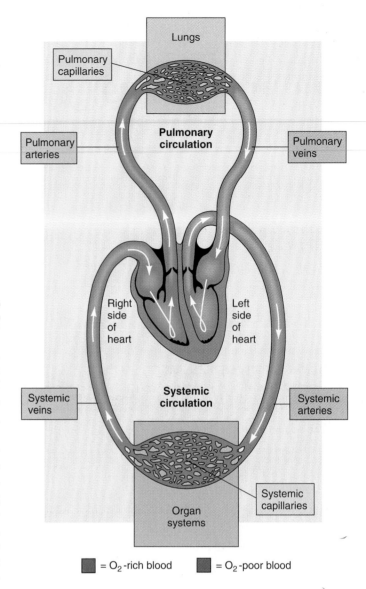

● **FIGURE 9-1**

Pulmonary and systemic circulation in relation to the heart. The circulatory system consists of two separate vascular loops: the pulmonary circulation, which carries blood between the heart and lungs; and the systemic circulation, which carries blood between the heart and organ systems.

 For an interaction related to this figure, see Media Exercise 9.2: The Heart: A Dual Pump on the CD-ROM.

THE COMPLETE CIRCUIT OF BLOOD FLOW

Let us look at how the heart functions as a dual pump, by tracing a drop of blood through one complete circuit (● Figure 9-2a and b). Blood returning from the systemic circulation enters the right atrium via two large veins, the **venae cavae,** one returning blood from above and the other returning blood from below heart level. The drop of blood entering the right atrium has returned from the body tissues, where O_2 has been taken from it and CO_2 has been added to it. This partially deoxygenated blood flows from the right atrium into the right ventricle, which pumps it out through the **pulmonary artery**, which immediately forms two branches, one

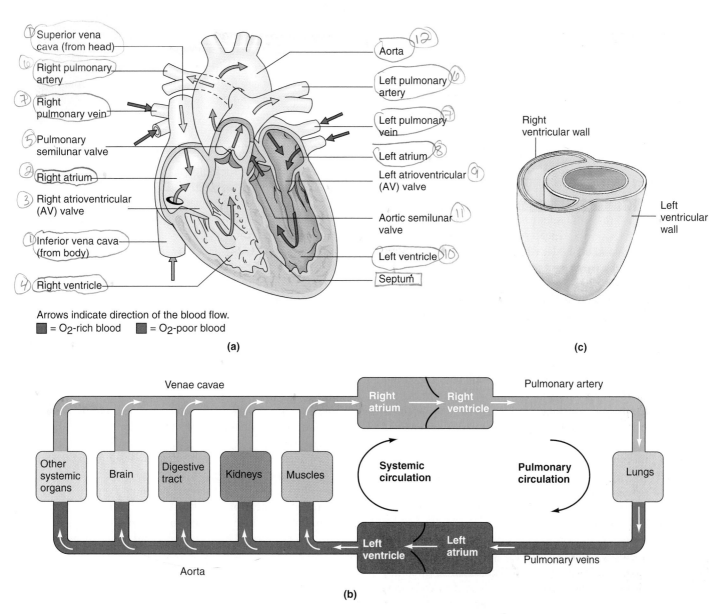

Figure key (a):

1. Superior vena cava (from head)
6. Right pulmonary artery
7. Right pulmonary vein
5. Pulmonary semilunar valve
2. Right atrium
3. Right atrioventricular (AV) valve
1. Inferior vena cava (from body)
4. Right ventricle

12. Aorta
6. Left pulmonary artery
7. Left pulmonary vein
8. Left atrium
9. Left atrioventricular (AV) valve
11. Aortic semilunar valve
10. Left ventricle
Septum

Arrows indicate direction of the blood flow.
■ = O₂-rich blood   ■ = O₂-poor blood

(a)

Right ventricular wall

Left ventricular wall

(c)

(b) — Venae cavae, Right atrium, Right ventricle, Pulmonary artery; Other systemic organs, Brain, Digestive tract, Kidneys, Muscles; Systemic circulation, Pulmonary circulation, Lungs; Left ventricle, Left atrium, Pulmonary veins; Aorta

● **FIGURE 9-2**

Blood flow through and pump action of the heart. (a) Blood flow through the heart. (b) Dual pump action of the heart. The right side of the heart receives O₂-poor blood from the systemic circulation and pumps it into the pulmonary circulation. The left side of the heart receives O₂-rich blood from the pulmonary circulation and pumps it into the systemic circulation. Note the parallel pathways of blood flow through the systemic organs. (The relative volume of blood flowing through each organ is not drawn to scale.) (c) Comparison of the thickness of the right and left ventricular walls. Note that the left ventricular wall is much thicker than the right wall.

---

going to each of the two lungs. Thus the *right side of the heart receives blood from the systemic circulation and pumps it into the pulmonary circulation.*

Within the lungs, the drop of blood loses its extra CO₂ and picks up a fresh supply of O₂ before being returned to the left atrium via the **pulmonary veins** coming from both lungs. This O₂-rich blood returning to the left atrium subsequently flows into the left ventricle, the pumping chamber that propels the blood to all body systems except the lungs; that is, *the left side of the heart receives blood from the pulmonary circulation and pumps it into the systemic circulation.*

The single large artery carrying blood away from the left ventricle is the **aorta**. Major arteries branch from the aorta to supply the various organs of the body.

In contrast to the pulmonary circulation, in which all the blood flows through the lungs, the systemic circulation may be viewed as a series of parallel pathways. Part of the blood pumped out by the left ventricle goes to the muscles, part to the kidneys, part to the brain, and so on. Thus the output of the left ventricle is distributed so that each part of the body receives a fresh blood supply; the same arterial blood does not pass from organ to organ. Accordingly, the drop of blood we are tracing goes to only one of the systemic organs. Tissue cells within the organ take O_2 from the blood and use it to oxidize nutrients for energy production; in the process, the tissue cells form CO_2 as a waste product that is added to the blood (see p. 3 and p. 30). The drop of blood, now partially depleted of O_2 content and increased in CO_2 content, returns to the right side of the heart, which once again will pump it to the lungs. One circuit is complete.

COMPARISON OF THE RIGHT AND LEFT PUMPS

Both sides of the heart simultaneously pump equal amounts of blood. The volume of O_2-poor blood being pumped to the lungs by the right side of the heart soon becomes the same volume of O_2-rich blood being delivered to the tissues by the left side of the heart. The pulmonary circulation is a low-pressure, low-resistance system, whereas the systemic circulation is a high-pressure, high-resistance system. Pressure is the force exerted on the vessel walls by the blood pumped into the vessels by the heart. Resistance is the opposition to blood flow, largely caused by friction between the flowing blood and the vessel wall. Even though the right and left sides of the heart pump the same amount of blood, the left side performs more work, because it pumps an equal volume of blood at a higher pressure into a higher-resistance and longer system. Accordingly, the heart muscle on the left side is much thicker than the muscle on the right side, making the left side a stronger pump (● Figure 9-2c).

▌ Pressure-operated heart valves ensure that blood flows in the right direction through the heart.

Blood flows through the heart in one fixed direction from veins to atria to ventricles to arteries. The presence of 4 one-way heart valves ensures this unidirectional flow of blood. The valves are positioned so that they open and close passively because of pressure differences, similar to a one-way door (● Figure 9-3). A forward pressure gradient (that is, a greater pressure behind the valve) forces the valve open, much as you open a door by pushing on one side of it, whereas a backward pressure gradient (that is, a greater pressure in front of the valve) forces the valve closed, just as you apply pressure to the opposite side of the door to close it. Note that a backward gradient can force the valve closed but cannot force it to swing open in the opposite direction; that is, heart valves are not like swinging, saloon-type doors.

AV VALVES BETWEEN THE ATRIA AND VENTRICLES

Two of the heart valves, the **right** and **left atrioventricular (AV) valves**, are positioned between the atrium and the ventricle on the right and left sides, respectively (● Figure 9-4a). These valves let blood flow from the atria into the ventricles during ventricular filling (when atrial pressure exceeds ventricular pressure) but prevent the backflow of blood from the ventricles into the atria during ventricular emptying (when ventricular pressure greatly exceeds atrial pressure). If the rising ventricular pressure did not force the AV valves to close as the ventricles contracted to empty, much of the blood would inefficiently be forced back into the atria and veins instead of being pumped into the arteries. The right AV valve is also called the **tricuspid valve** (*tri* means "three"), because it consists of three cusps or leaflets (● Figure 9-4b). Likewise, the left AV valve, which has two cusps, is often called the **bicuspid valve** (*bi* means "two") or, alternatively, the **mitral valve** (because of its physical resemblance to a mitre, or bishop's traditional hat).

The edges of the AV valve leaflets are fastened by tough, thin, fibrous cords of tendinous-type tissue, the **chordae tendineae**, which prevent the valves from being everted. That is, the chordae tendineae prevent the AV valve from being forced by the high ventricular pressure to open in the opposite direction into the atria. These cords extend from the edges of each cusp and attach to small, nipple-shaped **papillary muscles**, which protrude from the inner surface of the ventricular walls (*papilla* means "nipple"). When the ventricles contract, the papillary muscles also contract, pulling downward on the chordae tendineae. This pulling exerts tension on the closed AV valve cusps to hold them in position, much like tethering ropes hold down a hot-air balloon. This action helps keep the valve tightly sealed in the face of a strong backward pressure gradient (● Figure 9-4c).

SEMILUNAR VALVES BETWEEN THE VENTRICLES AND MAJOR ARTERIES

The two remaining heart valves, the **aortic** and **pulmonary valves**, lie at the juncture where the major arteries leave the ventricles (● Figure 9-4a). They are known as **semilunar valves** because they have three cusps, each resembling a shallow half-moon-shaped pocket (*semi* means "half"; *lunar* means "moon") (● Figure 9-4b). These valves are forced open when the left and right ventricular pressures exceed the pressure in the aorta and pulmonary artery, respectively, during ventricular contraction and emptying. Closure results when the ventricles relax and ventricular pressures fall below the aortic and pulmonary artery pressures. The closed valves prevent blood from flowing from the arteries back into the ventricles from which it has just been pumped.

● **FIGURE 9-3**

Mechanism of valve action

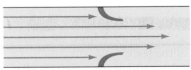

Valve opened

When pressure is greater behind the valve, it opens.

Valve closed; does not open in opposite direction

When pressure is greater in front of the valve, it closes. Note that when pressure is greater in front of the valve, it does not open in the opposite direction; that is, it is a one-way valve.

The semilunar valves are prevented from everting by the anatomic structure and positioning of the cusps. When on ventricular relaxation a backward pressure gradient is created, the back surge of blood fills the pocketlike cusps and sweeps them into a closed position, with their unattached upturned edges fitting together in a deep, leakproof seam (● Figure 9-4d).

NO VALVES BETWEEN THE ATRIA AND VEINS

Even though there are no valves between the atria and veins, backflow of blood from the atria into the veins usually is not a significant problem, for two reasons: (1) atrial pressures usually are not much higher than venous pressures, and (2) the sites where the venae cavae enter the atria are partially compressed during atrial contraction.

Now let's turn our attention to the part of the heart that actually generates the forces that produce blood flow, the cardiac muscle within the heart walls.

The heart walls consist primarily of spirally arranged cardiac muscle fibers.

The heart wall has three distinct layers:

• A thin inner layer, the **endothelium**, a unique type of epithelial tissue that lines the entire circulatory system

• A middle layer, the **myocardium**, which is composed of cardiac muscle and constitutes the bulk of the heart wall (*myo* means "muscle"; *cardia* means "heart")

• A thin external membrane, the **epicardium** that covers the heart (*epi* means "on")

The myocardium consists of interlacing bundles of cardiac muscle fibers arranged spirally around the circumference of the heart. As a result of this arrangement, when the ventricular muscle contracts and shortens it exerts a "wringing" effect, efficiently exerting pressure on the blood within the enclosed chambers and directing it upward toward the openings of the major arteries that exit at the base of the ventricles.

Cardiac muscle fibers are interconnected by intercalated discs and form functional syncytia.

The individual cardiac muscle cells interconnect to form branching fibers, with adjacent cells joined end to end at specialized structures called **intercalated discs**. Inside an inter-

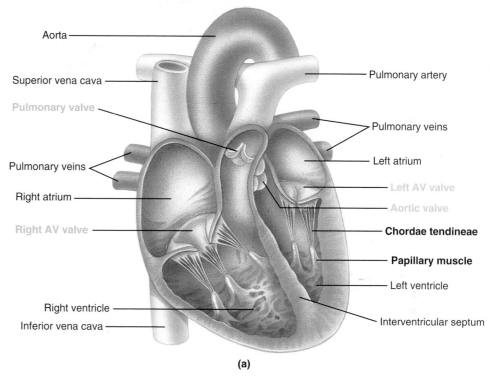

(a)

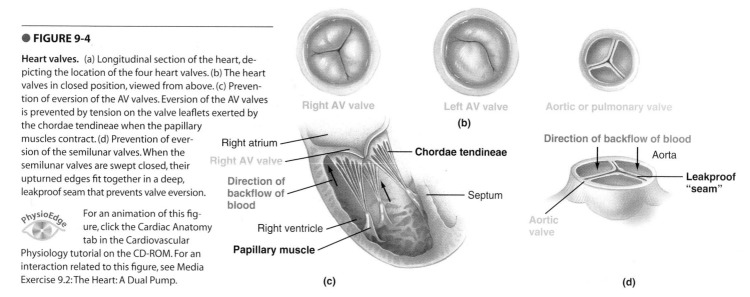

(b)

(c)

(d)

● **FIGURE 9-4**

Heart valves. (a) Longitudinal section of the heart, depicting the location of the four heart valves. (b) The heart valves in closed position, viewed from above. (c) Prevention of eversion of the AV valves. Eversion of the AV valves is prevented by tension on the valve leaflets exerted by the chordae tendineae when the papillary muscles contract. (d) Prevention of eversion of the semilunar valves. When the semilunar valves are swept closed, their upturned edges fit together in a deep, leakproof seam that prevents valve eversion.

PhysioEdge For an animation of this figure, click the Cardiac Anatomy tab in the Cardiovascular Physiology tutorial on the CD-ROM. For an interaction related to this figure, see Media Exercise 9.2: The Heart: A Dual Pump.

calated disc are two types of membrane junctions: desmosomes and gap junctions (● Figure 9-5). A *desmosome* is a type of adhering junction that "spot-rivets" adjacent cells, mechanically holding them together. Desmosomes are particularly abundant in tissues, such as the heart, that are subject to a lot of mechanical stress (see p. 48). Furthermore, at intervals along the intercalated disc, the opposing membranes approach each other very closely to form *gap junctions,* which are areas of low electrical resistance that let action potentials spread from one cardiac cell to adjacent cells (see p. 49). Some cardiac-muscle cells can generate action potentials without any nervous stimulation. When one of the cardiac cells spontaneously undergoes an action potential, the electrical impulse spreads to all the other cells that are joined by gap junctions in the surrounding muscle mass, so they become excited and contract as a single unit or functional syncytium (see p. 231). The atria and the ventricles each form a functional syncytium and contract as separate units. The synchronous contraction of the muscle cells that make up the walls of each of these chambers produces the force needed to eject the enclosed blood.

● **FIGURE 9-5**

Organization of cardiac muscle fibers. Adjacent cardiac muscle cells are joined end to end by intercalated discs, which contain two types of specialized junctions: desmosomes, which act as spot rivets mechanically holding the cells together; and gap junctions, which permit action potentials to spread from one cell to adjacent cells.

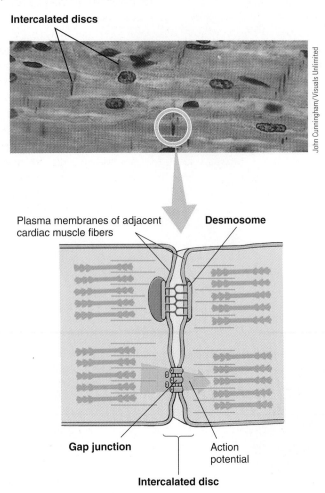

No gap junctions join the atrial and ventricular contractile cells, and furthermore, the atria and ventricles are separated by electrically nonconductive fibrous tissue that surrounds and supports the valves. However, an important, specialized conduction system facilitates and coordinates transmission of electrical excitation from the atria to the ventricles to ensure synchronization between atrial and ventricular pumping.

Because of both the syncytial nature of cardiac muscle and the conduction system between the atria and ventricles, an impulse spontaneously generated in one part of the heart spreads throughout the entire heart. Therefore, unlike skeletal muscle, where graded contractions can be produced by varying the number of muscle cells contracting within the muscle (recruitment of motor units), either all the cardiac muscle fibers contract or none do. A "half-hearted" contraction is not possible. Cardiac contraction is graded by varying the strength of contraction of all the cardiac muscle cells. You will learn more about this process in a later section.

Building on this foundation of heart structure, we are now going to explain how action potentials are initiated and spread throughout the heart, followed by a discussion of how this electrical activity brings about coordinated pumping of the heart.

 Click on the Media Exercises menu of the CD-ROM and work Media Exercise 9.2: The Heart: A Dual Pump to test your understanding of the previous section.

ELECTRICAL ACTIVITY OF THE HEART

Contraction of cardiac muscle cells to eject blood is triggered by action potentials sweeping across the muscle cell membranes. The heart contracts, or beats, rhythmically as a result of action potentials that it generates by itself, a property called **autorhythmicity** (*auto* means "self"). There are two specialized types of cardiac muscle cells:

1. **Contractile cells,** which are 99% of the cardiac muscle cells, do the mechanical work of pumping. These working cells normally do not initiate their own action potentials.
2. In contrast, the small but extremely important remainder of the cardiac cells, the **autorhythmic cells,** do not contract but instead are specialized for initiating and conducting the action potentials responsible for contraction of the working cells.

▌ Cardiac autorhythmic cells display pacemaker activity.

In contrast to nerve and skeletal muscle cells, in which the membrane remains at constant resting potential unless the cell is stimulated, the cardiac autorhythmic cells do not have a resting potential. Instead, they display *pacemaker activity;* that is, their membrane potential slowly depolarizes, or drifts, between action potentials until threshold is reached, at which time the membrane fires or has an action potential. An autorhythmic cell membrane's slow drift to threshold is called the

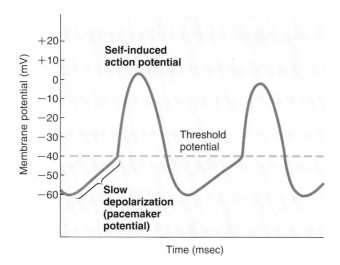

FIGURE 9-6

Pacemaker activity of cardiac autorhythmic cells

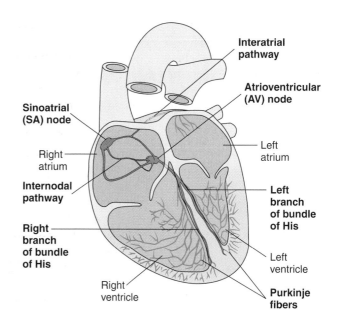

FIGURE 9-7

Specialized conduction system of the heart

For an animation of this figure, click the Cardiac Electrical Activity tab in the Cardiovascular Physiology tutorial on the CD-ROM. For an interaction related to this figure, see Media Exercise 9.2: The Heart: A Dual Pump.

pacemaker potential (● Figure 9-6; see also p. 232). This slow depolarization results from built-in, complex changes in passive ion movement across the membrane between action potentials. Through repeated cycles of drift and fire, these autorhythmic cells cyclically initiate action potentials, which then spread throughout the heart to trigger rhythmic beating without any nervous stimulation.

▍The sinoatrial node is the normal pacemaker of the heart.

The specialized noncontractile cardiac cells capable of autorhythmicity lie in the following specific sites (● Figure 9-7):

1. The **sinoatrial node (SA node)**, a small, specialized region in the right atrial wall near the opening of the superior vena cava.
2. The **atrioventricular node (AV node)**, a small bundle of specialized cardiac muscle cells located at the base of the right atrium near the septum, just above the junction of the atria and ventricles.
3. The **bundle of His (atrioventricular bundle)**, a tract of specialized cells that originates at the AV node and enters the interventricular septum. Here, it divides to form the right and left bundle branches that travel down the septum, curve around the tip of the ventricular chambers, and travel back toward the atria along the outer walls.
4. **Purkinje fibers**, small terminal fibers that extend from the bundle of His and spread throughout the ventricular myocardium much like small twigs of a tree branch.

NORMAL PACEMAKER ACTIVITY

Because these various autorhythmic cells have different rates of slow depolarization to threshold, the rates at which they are normally capable of generating action potentials also differ (▲ Table 9-1). The heart cells with the fastest rate of action potential initiation are localized in the SA node. Once an action potential occurs in any cardiac muscle cell, it is prop-

agated throughout the rest of the myocardium via gap junctions and the specialized conduction system. Therefore, the SA node, which normally has the fastest rate of autorhythmicity, at 70 to 80 action potentials per minute, drives the rest of the heart at this rate and thus is known as the **pacemaker** of the heart. That is, the entire heart becomes excited, triggering the contractile cells to contract and the heart to beat at the pace or rate set by SA node autorhythmicity, normally at 70 to 80 beats per minute. The other autorhythmic tissues cannot assume their own naturally slower rates, because they are activated by action potentials originating in the SA node before they can reach threshold at their own, slower rhythm.

The following analogy shows how the SA node drives the rest of the heart at its own pace. Suppose a train has 100

▲ TABLE 9-1

Normal Rate of Action Potential Discharge in Autorhythmic Tissues of the Heart

TISSUE	ACTION POTENTIALS PER MINUTE*
SA node (normal pacemaker)	70–80
AV node	40–60
Bundle of His and Purkinje fibers	20–40

*In the presence of parasympathetic tone; see p. 187 and p. 261.

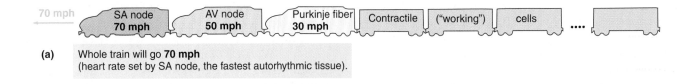

(a) Whole train will go **70 mph**
(heart rate set by SA node, the fastest autorhythmic tissue).

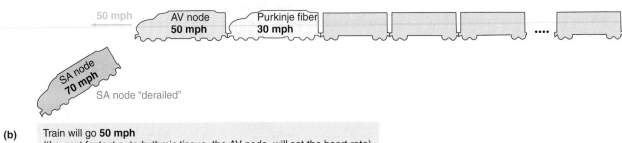

(b) Train will go **50 mph**
(the next fastest autorhythmic tissue, the AV node, will set the heart rate).

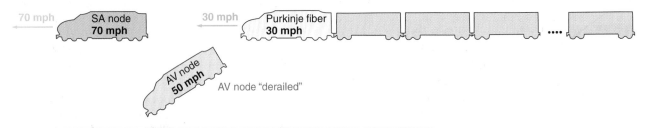

(c) First part of train will go **70 mph**; last part will go **30 mph**
(atria will be driven by SA node; ventricles will assume own, much slower rhythm).

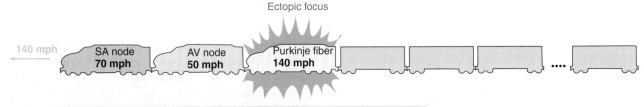

(d) Train will be driven by ectopic focus, which is now going faster than the SA node
(the whole heart will be driven more rapidly by an abnormal pacemaker).

● **FIGURE 9-8**

Analogy of pacemaker activity. (a) Normal pacemaker activity by the SA node. (b) Takeover of pacemaker activity by the AV node when the SA node is nonfunctional. (c) Takeover of ventricular rate by the slower ventricular autorhythmic tissue in the condition of heart block even though the SA node is still functioning. (d) Takeover of pacemaker activity by an ectopic focus.

cars, 3 of which are engines capable of moving on their own; the other 97 cars must be pulled (● Figure 9-8a). One engine (the SA node) can travel at 70 miles/hour (mph) on its own, another engine (the AV node) at 50 mph, and the last engine (the Purkinje fibers) at 30 mph. If all these cars are joined together, the engine that can travel at 70 mph will pull the rest of the cars at that speed. The engines that can travel at lower speeds on their own will be pulled at a faster speed by the fastest engine and therefore cannot assume their own slower rate as long as they are being driven by a faster engine. The

other 97 cars (nonautorhythmic, contractile cells), being unable to move on their own, will likewise travel at whatever speed the fastest engine pulls them.

ABNORMAL PACEMAKER ACTIVITY

Clinical Note If for some reason the fastest engine breaks down (SA node damage), the next fastest engine (AV node) takes over and the entire train travels at 50 mph; that is, if the SA node becomes nonfunctional, the AV node assumes pacemaker activity (● Figure 9-8b). The non-SA nodal autorhythmic tissues are **latent pacemakers** that can take over, although at a lower rate, if the normal pacemaker fails.

If impulse conduction becomes blocked between the atria and the ventricles, the atria continue at the typical rate of 70 beats per minute, and the ventricular tissue, not being driven by the faster SA nodal rate, assumes its own much slower autorhythmic rate of about 30 beats per minute, initiated by the ventricular autorhythmic cells (Purkinje fibers). This situation is like a breakdown of the second engine (AV node) so that the lead engine (SA node) becomes disconnected from the slow third engine (Purkinje fibers) and the rest of the cars (● Figure 9-8c). The lead engine continues at 70 mph while the rest of the train proceeds at 30 mph. This **complete heart block** occurs when the conducting tissue between the atria and ventricles is damaged and becomes nonfunctional. A ventricular rate of 30 beats per minute will support only a very sedentary existence; in fact, the patient usually becomes comatose.

When a person has an abnormally low heart rate, as in SA node failure or heart block, an **artificial pacemaker** can be used. Such an implanted device rhythmically generates impulses that spread throughout the heart to drive both the atria and ventricles at the typical rate of 70 beats per minute.

Occasionally an area of the heart, such as a Purkinje fiber, becomes overly excitable and depolarizes more rapidly than the SA node. (The slow engine suddenly goes faster than the lead engine; see ● Figure 9-8d). This abnormally excitable area, an **ectopic focus**, initiates a premature action potential that spreads throughout the rest of the heart before the SA node can initiate a normal action potential (*ectopic* means "out of place"). An occasional abnormal impulse from a ventricular ectopic focus produces a **premature ventricular contraction (PVC)**. If the ectopic focus continues to discharge at its more rapid rate, pacemaker activity shifts from the SA node to the ectopic focus. The heart rate abruptly becomes greatly accelerated and continues this rapid rate for a variable time period until the ectopic focus returns to normal. Such overly irritable areas may be associated with organic heart disease, but more frequently they occur in response to anxiety, lack of sleep, or excess caffeine, nicotine, or alcohol consumption.

We will now turn our attention to how an action potential, once initiated, is conducted throughout the heart.

▋ The spread of cardiac excitation is coordinated to ensure efficient pumping.

Once initiated in the SA node, an action potential spreads throughout the rest of the heart. For efficient cardiac function, the spread of excitation should satisfy three criteria:

1. *Atrial excitation and contraction should be complete before the onset of ventricular contraction.* Complete ventricular filling requires that atrial contraction precede ventricular contraction. During cardiac relaxation, the AV valves are open, so that venous blood entering the atria continues to flow directly into the ventricles. Almost 80% of ventricular filling occurs by this means before atrial contraction. When the atria do contract, more blood is squeezed into the ventricles to complete ventricular filling. Ventricular contraction then occurs to eject blood from the heart into the arteries.

If the atria and ventricles were to contract simultaneously, the AV valves would close immediately, because ventricular pressures would greatly exceed atrial pressures. The ventricles have much thicker walls and, accordingly, can generate more pressure. Atrial contraction would be unproductive, because the atria could not squeeze blood into the ventricles through closed valves. Therefore, to ensure complete filling of the ventricles—to obtain the remaining 20% of ventricular filling that occurs during atrial contraction—the atria must become excited and contract before ventricular excitation and contraction.

2. *Excitation of cardiac muscle fibers should be coordinated to ensure that each heart chamber contracts as a unit to pump efficiently.* If the muscle fibers in a heart chamber became excited and contracted randomly rather than contracting simultaneously in a coordinated fashion, they would be unable to eject blood. A smooth, uniform ventricular contraction is essential to squeeze out the blood. As an analogy, assume you have a basting syringe full of water. If you merely poke a finger here or there into the rubber bulb of the syringe, you will not eject much water. However, if you compress the bulb in a smooth, coordinated fashion, you can squeeze out the water.

Clinical Note — In a similar manner, contraction of isolated cardiac muscle fibers is not successful in pumping blood. Such random, uncoordinated excitation and contraction of the cardiac cells is known as **fibrillation**. Fibrillation of the ventricles is much more serious than atrial fibrillation. Ventricular fibrillation rapidly causes death, because the heart cannot pump blood into the arteries. This condition can often be corrected by **electrical defibrillation**, in which a very strong electrical current is applied on the chest wall. When this current reaches the heart, it stimulates (depolarizes) all parts of the heart simultaneously. Usually the first part of the heart to recover is the SA node, which takes over pacemaker activity, once again initiating impulses that trigger the synchronized contraction of the rest of the heart.

3. *The pair of atria and pair of ventricles should be functionally coordinated so that both members of the pair contract simultaneously.* This coordination permits synchronized pumping of blood into the pulmonary and systemic circulation.

The normal spread of cardiac excitation is carefully orchestrated to ensure that these criteria are met and the heart functions efficiently, as follows (● Figure 9-9).

ATRIAL EXCITATION

An action potential originating in the SA node first spreads throughout both atria, primarily from cell to cell via gap junctions. In addition, ~~several~~ poorly delineated, specialized conduction pathways speed up conduction of the impulse through the atria (● Figures 9-7 and 9-9).

- ① The **interatrial pathway** extends from the SA node within the right atrium to the left atrium. Because this pathway rapidly transmits the action potential from the SA node to the pathway's termination in the left atrium, a wave of excitation can spread across the gap junctions throughout the left atrium

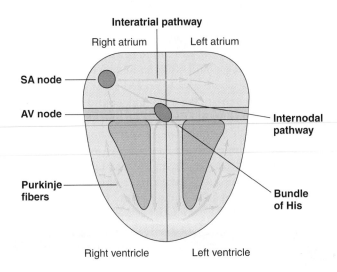

Interatrial pathway

Right atrium   Left atrium

SA node

AV node

Internodal pathway

Purkinje fibers

Bundle of His

Right ventricle   Left ventricle

● **FIGURE 9-9**

Spread of cardiac excitation. An action potential initiated at the SA node first spreads throughout both atria. Its spread is facilitated by two specialized atrial conduction pathways, the interatrial and internodal pathways. The AV node is the only point where an action potential can spread from the atria to the ventricles. From the AV node, the action potential spreads rapidly throughout the ventricles, hastened by a specialized ventricular conduction system consisting of the bundle of His and Purkinje fibers.

at the same time excitation is similarly spreading throughout the right atrium. This ensures that both atria become depolarized to contract simultaneously.

• ② The **internodal pathway** extends from the SA node to the AV node. The AV node is the only point of electrical contact between the atria and ventricles; in other words, because the atria and ventricles are structurally connected by electrically nonconductive fibrous tissue, the only way an action potential in the atria can spread to the ventricles is by passing through the AV node. The internodal conduction pathway directs the spread of an action potential originating at the SA node to the AV node to ensure sequential contraction of the ventricles following atrial contraction

CONDUCTION BETWEEN THE ATRIA AND THE VENTRICLES

The action potential is conducted relatively slowly through the AV node. This slowness is advantageous because it allows time for complete ventricular filling. The impulse is delayed about 100 msec (the **AV nodal delay**), which enables the atria to become completely depolarized and to contract, emptying their contents into the ventricles, before ventricular depolarization and contraction occur.

VENTRICULAR EXCITATION

After the AV nodal delay, the impulse travels rapidly down the septum via the right and left branches of the bundle of His and throughout the ventricular myocardium via the Purkinje fibers. The network of fibers in this ventricular conduction system is specialized for rapid propagation of action potentials. Its presence hastens and coordinates the spread of ventricular excitation to ensure that the ventricles contract as a unit.

Although this system carries the action potential rapidly to a large number of cardiac muscle cells, it does not terminate on every cell. The impulse quickly spreads from the excited cells to the rest of the ventricular muscle cells by means of gap junctions.

The ventricular conduction system is more highly organized and more important than the interatrial and internodal conduction pathways. Because the ventricular mass is so much larger than the atrial mass, it is crucial that a rapid conduction system be present to hasten the spread of excitation in the ventricles. Purkinje fibers can transmit an action potential six times faster than the ventricular syncytium of contractile cells could. If the entire ventricular depolarization process depended on cell-to-cell spread of the impulse via gap junctions, the ventricular tissue immediately next to the AV node would become excited and contract before the impulse had even passed to the heart apex. This, of course, would not allow efficient pumping. Rapid conduction of the action potential down the bundle of His and its swift, diffuse distribution throughout the Purkinje network lead to almost simultaneous activation of the ventricular myocardial cells in both ventricular chambers, which ensures a single, smooth, coordinated contraction that can efficiently eject blood into both the systemic and pulmonary circulations at the same time.

The action potential of cardiac contractile cells shows a characteristic plateau.

The action potential in cardiac contractile cells, although initiated by the nodal pacemaker cells, varies considerably in ionic mechanisms and shape from the SA node potential (compare ● Figures 9-6 and 9-10). Unlike autorhythmic cells, the membrane of contractile cells remains essentially at rest at about −90 mV until excited by electrical activity propagated from the pacemaker. Once the membrane of a ventricular myocardial contractile cell is excited, the membrane potential rapidly reverses to a positive value of +30 mV as a result of activation of voltage-gated Na$^+$ channels and Na$^+$ subsequently rapidly entering the cell, as it does in other excitable cells undergoing an action potential (see p. 77). Unique to the cardiac contractile cells, however, the membrane potential is maintained close to this peak positive level for several hundred milliseconds, producing a *plateau phase* of the action potential. In contrast, the short action potential of neurons and skeletal muscle cells lasts 1 to 2 msec. Whereas the rising phase of the action potential is brought about by activation of comparatively "fast" Na$^+$ channels, this plateau is maintained primarily by activation of relatively "slow" voltage-gated Ca^{2+} channels in the cardiac contractile cell membrane. These channels open in response to the sudden change in voltage during the rising phase of the action potential. Opening of these Ca^{2+} channels results in a slow, inward diffusion of Ca^{2+}, because Ca^{2+} is in greater concentration in the ECF. This continued influx of positively charged Ca^{2+} prolongs the positivity inside the cell and is primarily responsible for the plateau part of the action potential. The rapid falling phase of the action potential results from inactivation of the Ca^{2+} channels and delayed activation of voltage-gated K$^+$ chan-

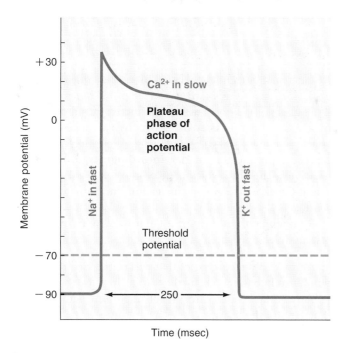

● FIGURE 9-10

Action potential in contractile cardiac muscle cells. The action potential in cardiac contractile cells differs considerably from the action potential in cardiac autorhythmic cells (compare with ● Figure 9-6).

nels. As in other excitable cells, the cell returns to resting potential as K^+ rapidly leaves the cell.

Let's now see how this action potential brings about contraction.

▮ Ca^{2+} entry from the ECF induces a more abundant Ca^{2+} release from the sarcoplasmic reticulum.

In cardiac contractile cells, the slow Ca^{2+} channels lie primarily in the T tubules. As you just learned, these voltage-gated channels open during a local action potential. Thus, unlike in skeletal muscle, Ca^{2+} diffuses into the cytosol from the ECF across the T tubule membrane during a cardiac action potential. This entering Ca^{2+} triggers the opening of nearby Ca^{2+}-release channels in the adjacent lateral sacs of the sarcoplasmic reticulum. In this way, a small amount Ca^{2+} entering the cytosol from the ECF induces a much larger release of Ca^{2+} into the cytosol from the intracellular stores (● Figure 9-11). The resultant increase in cytosolic Ca^{2+} turns on the contractile machinery. The extra supply of Ca^{2+} from the sarcoplasmic reticulum is responsible for the long period of cardiac contraction, which lasts about three times longer than a single contraction of a skeletal muscle fiber (300 msec compared to 100 msec). This increased contractile time ensures adequate time to eject the blood.

As in skeletal muscle, the role of Ca^{2+} within the cytosol is to bind with the troponin-tropomyosin complex and physically pull it aside to allow cross-bridge cycling and contraction (● Figure 9-11) (see p. 210). However, unlike skeletal muscle, in which sufficient Ca^{2+} is always released to turn

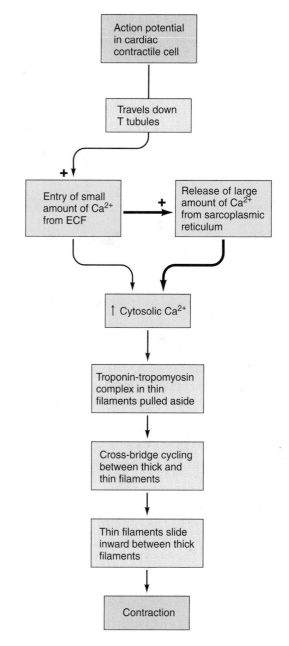

● FIGURE 9-11

Excitation–contraction coupling in cardiac contractile cells

on all the cross bridges, in cardiac muscle the extent of cross-bridge activity varies with the amount of cytosolic Ca^{2+}. As we will show, various regulatory factors can alter the amount of cytosolic Ca^{2+}.

Removal of Ca^{2+} from the cytosol by energy-dependent mechanisms in both the plasma membrane and sarcoplasmic reticulum restores the blocking action of troponin and tropomyosin, so contraction ceases and the heart muscle relaxes.

Clinical Note Some drugs alter cardiac function by influencing Ca^{2+} movement across the myocardial cell membranes. For example, Ca^{2+}-channel blocking agents, such as *verapamil*, block Ca^{2+} influx during an action potential, reducing the force of cardiac contraction. Other drugs, such as *digitalis,* increase cardiac contractility by inducing an accumulation of cytosolic Ca^{2+}.

▮ A long refractory period prevents tetanus of cardiac muscle.

Like other excitable tissues, cardiac muscle has a refractory period. During the refractory period, a second action potential cannot be triggered until an excitable membrane has recovered from the preceding action potential. In skeletal muscle, the refractory period is very short compared with the duration of the resulting contraction, so the fiber can be restimulated again before the first contraction is complete to produce summation of contractions. Rapidly repetitive stimulation that does not let the muscle fiber relax between stimulations results in a sustained, maximal contraction known as *tetanus* (see ● Figure 8-14, p. 215).

In contrast, cardiac muscle has a long refractory period that lasts about 250 msec because of the prolonged plateau phase of the action potential. This is almost as long as the period of contraction initiated by the action potential; a cardiac muscle fiber contraction averages about 300 msec (● Figure 9-12). Consequently, cardiac muscle cannot be restimulated until contraction is almost over, precluding summation of contractions and tetanus of cardiac muscle. This is a valuable protective mechanism, because the pumping of blood requires alternate periods of contraction (emptying) and relaxation (filling). A prolonged tetanic contraction would prove fatal. The heart chambers could not be filled and emptied again.

● **FIGURE 9-12**

Relationship of an action potential and the refractory period to the duration of the contractile response in cardiac muscle

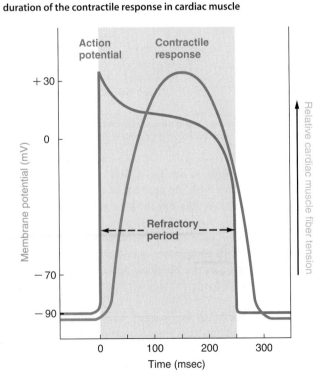

▮ The ECG is a record of the overall spread of electrical activity through the heart.

The electrical currents generated by cardiac muscle during depolarization and repolarization (see p. 72) spread into the tissues around the heart and are conducted through the body fluids. A small part of this electrical activity reaches the body surface, where it can be detected using recording electrodes. The record produced is an **electrocardiogram**, or **ECG**. (Alternatively, the abbreviation **EKG** is often used, from the ancient Greek word "*kardia*" instead of the Latin "*cardia*" for "heart".)

Remember three important points when considering what an ECG represents:

1. An ECG is a recording of that part of the electrical activity induced in body fluids by the cardiac impulse that reaches the body surface, not a direct recording of the actual electrical activity of the heart.
2. The ECG is a complex recording representing the *overall* spread of activity throughout the heart during depolarization and repolarization. It is not a recording of a *single* action potential in a single cell at a single point in time. The record at any given time represents the sum of electrical activity in all the cardiac muscle cells, some of which may be undergoing action potentials while others may not yet be activated. For example, immediately after the SA node fires, the atrial cells are undergoing action potentials while the ventricular cells are still at resting potential. At a later point, the electrical activity will have spread to the ventricular cells while the atrial cells will be repolarizing. Therefore, the overall pattern of cardiac electrical activity varies with time as the impulse passes throughout the heart.
3. The recording represents comparisons in voltage detected by electrodes at two different points on the body surface, not the actual potential. For example, the ECG does not record a potential at all when the ventricular muscle is either completely depolarized or completely repolarized; both electrodes are "viewing" the same potential, so no difference in potential between the two electrodes is recorded.

The exact pattern of electrical activity recorded from the body surface depends on the orientation of the recording electrodes. Electrodes may be loosely thought of as "eyes" that "see" electrical activity and translate it into a visible recording, the ECG record. Whether an upward deflection or downward deflection is recorded is determined by the way the electrodes are oriented with respect to the current flow in the heart. For example, the spread of excitation across the heart is "seen" differently from the right arm, from the left leg, or from a recording directly over the heart. Even though the same electrical events are occurring in the heart, different waveforms representing the same electrical activity result when this activity is recorded by electrodes at different points on the body.

To provide standard comparisons, ECG records routinely consist of 12 conventional electrode systems, or leads. When an electrocardiograph machine is connected between recording electrodes at two points on the body, the specific arrangement of each pair of connections is called a **lead**. The 12 different leads each record electrical activity in the heart from different locations—six different electrical arrangements from the limbs and six chest leads at various sites around the heart. To provide a common basis for comparison and for recognizing deviations from normal, the same 12 leads are routinely used in all ECG recordings (● Figure 9-13).

● FIGURE 9-13

Electrocardiogram leads. (a) Limb leads. The six limb leads include leads I, II, III, aVR, aVL, and aVF. Leads I, II, and III are bipolar leads because two recording electrodes are used. The tracing records the *difference* in potential between the two electrodes. For example, lead I records the difference in potential detected at the right arm and left arm. The electrode placed on the right leg serves as a ground and is not a recording electrode. The aVR, aVL, and aVF leads are unipolar leads. Even though two electrodes are used, only the actual potential under one electrode, the exploring electrode, is recorded. The other electrode is set at zero potential and serves as a neutral reference point. For example, the aVR lead records the potential reaching the right arm in comparison to the rest of the body. (b) Chest leads. The six chest leads, V1 through V6, are also unipolar leads. The exploring electrode mainly records the electrical potential of the cardiac musculature immediately beneath the electrode in six different locations surrounding the heart.

▌ Different parts of the ECG record can be correlated to specific cardiac events.

Interpretation of the wave configurations recorded from each lead depends on a thorough knowledge about the sequence of cardiac excitation spread and about the position of the heart relative to electrode placement. A normal ECG has three distinct waveforms: the P wave, the QRS complex, and the T wave (● Figure 9-14). (The letters only indicate the orderly sequence of the waves. The inventor of the technique just started in mid-alphabet when naming the waves.)

- The **P wave** represents atrial depolarization.
- The **QRS complex** represents ventricular depolarization.
- The **T wave** represents ventricular repolarization.

Because these shifting waves of depolarization and repolarization bring about the alternating contraction and relaxation of the heart respectively, the cyclic mechanical events of the heart lag slightly behind the rhythmic changes in electrical activity. The following points about the ECG record should also be noted:

1. Firing of the SA node does not generate enough electrical activity to reach the body surface, so no wave is recorded for SA nodal depolarization. Therefore, the first recorded wave, the P wave, occurs when the impulse or wave of depolarization spreads across the atria.

2. In a normal ECG, no separate wave for atrial repolarization is visible. The electrical activity associated with atrial repolarization normally occurs simultaneously with ventricular depolarization and is masked by the QRS complex.

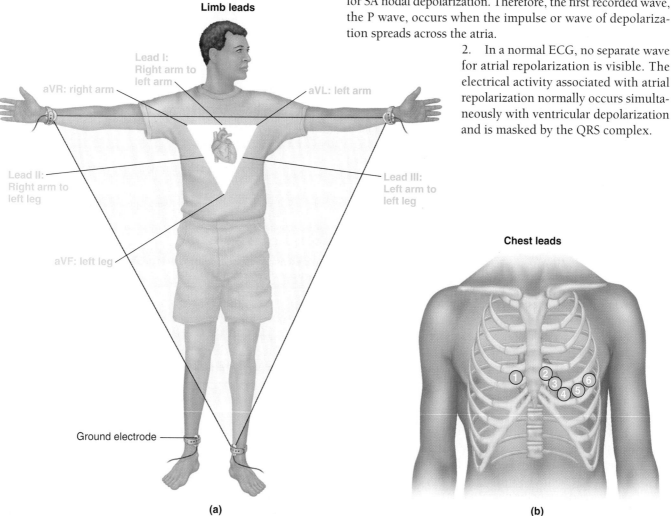

Limb leads

Lead I:
Right arm to left arm

aVR: right arm

aVL: left arm

Lead II:
Right arm to left leg

Lead III:
Left arm to left leg

aVF: left leg

Ground electrode

Chest leads

(a)

(b)

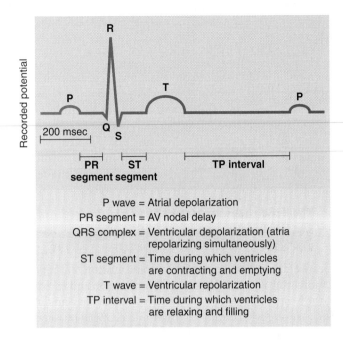

● **FIGURE 9-14**

Electrocardiogram waveforms in lead II

PhysioEdge For an interaction related to this figure, see Media Exercise 9.1: The Electrocardiogram on the CD-ROM.

3. The P wave is much smaller than the QRS complex, because the atria have a much smaller muscle mass than the ventricles and consequently generate less electrical activity.

4. At the following three points in time no net current flow is taking place in the heart musculature, so the ECG remains at baseline:

   a. *During the AV nodal delay.* This delay is represented by the interval of time between the end of P and the onset QRS; this segment of the ECG is known as the **PR segment.** (It is called the "PR segment" rather than the "PQ segment" because the Q deflection is small and sometimes absent, whereas the R deflection is the dominant wave of the complex.) Current is flowing through the AV node, but the magnitude is too small for the ECG electrodes to detect.

   b. *When the ventricles are completely depolarized and the cardiac contractile cells are undergoing the plateau phase of their action potential before they repolarize,* represented by the **ST segment.** This segment lies between QRS and T; it coincides with the time during which ventricular activation is complete and the ventricles are contracting and emptying. Note that the ST segment is *not* a record of cardiac contractile activity. The ECG is a measure of the electrical activity that triggers the subsequent mechanical activity.

   c. *When the heart muscle is completely repolarized and at rest and ventricular filling is taking place,* after the T wave and before the next P wave; this period is called the **TP interval.**

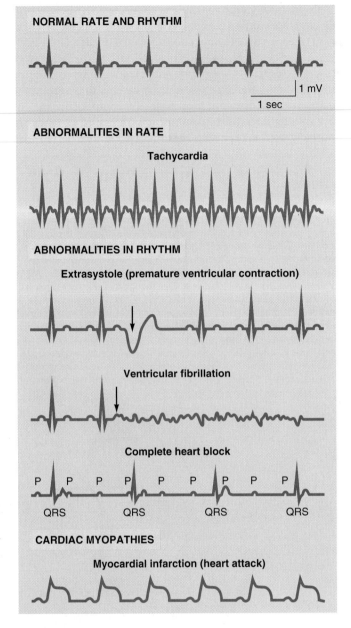

● **FIGURE 9-15**

Representative heart conditions detectable through electrocardiography

▍ **The ECG can be used to diagnose abnormal heart rates, arrhythmias, and damage of heart muscle.**

Clinical Note Because electrical activity triggers mechanical activity, abnormal electrical patterns are usually accompanied by abnormal contractile activity of the heart. Thus evaluation of ECG patterns can provide useful information about the status of the heart. The main deviations from normal that can be found through electrocardiography are (1) abnormalities in rate, (2) abnormalities in rhythm, and (3) cardiac myopathies (● Figure 9-15).

ABNORMALITIES IN RATE

The distance between two consecutive QRS complexes on an ECG record is calibrated to the beat-to-beat heart rate. A rapid heart rate of more than 100 beats per minute is called **tachycardia** (*tachy* means "fast"), whereas a slow heart rate of fewer than 60 beats per minutes is called **bradycardia** (*brady* means "slow").

ABNORMALITIES IN RHYTHM

The term *rhythm* refers to the regularity or spacing of the ECG waves. Any variation from the normal rhythm and sequence of excitation of the heart is termed an **arrhythmia**. It may result from ectopic foci, alterations in SA node pacemaker activity, or interference with conduction. For example, with *complete heart block* the SA node continues to govern atrial depolarization, but the ventricles generate their own impulses at a rate much slower than the atria. On the ECG, the P waves exhibit a normal rhythm. The QRS and T waves also occur regularly but much more slowly than the P waves and are completely independent of P wave rhythm. Because atrial and ventricular activity are not synchronized, waves for atrial repolarization may appear, no longer masked by the QRS complex.

CARDIAC MYOPATHIES

Abnormal ECG waves are also important in recognizing and assessing **cardiac myopathies** (damage of the heart muscle). **Myocardial ischemia** is inadequate delivery of oxygenated blood to the heart tissue. Actual death, or **necrosis**, of heart muscle cells occurs when a blood vessel supplying that area of the heart becomes blocked or ruptured. This condition is **acute myocardial infarction**, commonly called a **heart attack**. Abnormal QRS waveforms appear when part of the heart muscle becomes necrotic.

 Click on the Media Exercises menu of the CD-ROM and work Media Exercise 9.1: The Electrocardiogram to test your understanding of the previous section.

MECHANICAL EVENTS OF THE CARDIAC CYCLE

The mechanical events of the cardiac cycle—contraction, relaxation, and the resultant changes in blood flow through the heart—are brought about by the rhythmic changes in cardiac electrical activity.

▌ The heart alternately contracts to empty and relaxes to fill.

The cardiac cycle consists of alternate periods of **systole** (contraction and emptying) and **diastole** (relaxation and filling). Contraction results from the spread of excitation across the heart, whereas relaxation follows the subsequent repolarization of the cardiac musculature. The atria and ventricles go through separate cycles of systole and diastole. Unless quali-

fied, the terms *systole* and *diastole* refer to what's happening with the ventricles.

The following discussion and corresponding ● Figure 9-16 correlate various events that occur concurrently during the cardiac cycle, including ECG features, pressure changes, volume changes, valve activity, and heart sounds. Only the events on the left side of the heart are described, but keep in mind that identical events are occurring on the right side of the heart, except that the pressures are lower. To complete one full cardiac cycle, our discussion will begin and end with ventricular diastole.

EARLY VENTRICULAR DIASTOLE

During early ventricular diastole, the atrium is still also in diastole. This stage corresponds to the TP interval on the ECG—the interval after ventricular repolarization and before another atrial depolarization. Because of the continuous inflow of blood from the venous system into the atrium, atrial pressure slightly exceeds ventricular pressure even though both chambers are relaxed (point 1 in ● Figure 9-16). Because of this pressure differential, the AV valve is open, and blood flows directly from the atrium into the ventricle throughout ventricular diastole (heart A in ● Figure 9-16). As a result of this passive filling, the ventricular volume slowly continues to rise even before atrial contraction takes place (point 2).

LATE VENTRICULAR DIASTOLE

Late in ventricular diastole, the SA node reaches threshold and fires. The impulse spreads throughout the atria, which appears on the ECG as the P wave (point 3). Atrial depolarization brings about atrial contraction, raising the atrial pressure curve (point 4) and squeezing more blood into the ventricle. The corresponding rise in ventricular pressure (point 5) that occurs simultaneous to the rise in atrial pressure is due to the additional volume of blood added to the ventricle by atrial contraction (point 6 and heart B). Throughout atrial contraction, atrial pressure still slightly exceeds ventricular pressure, so the AV valve remains open.

END OF VENTRICULAR DIASTOLE

Ventricular diastole ends at the onset of ventricular contraction. By this time, atrial contraction and ventricular filling are completed. The volume of blood in the ventricle at the end of diastole (point 7) is known as the **end-diastolic volume (EDV)**, which averages about 135 ml. No more blood will be added to the ventricle during this cycle. Therefore, the end-diastolic volume is the maximum amount of blood that the ventricle will contain during this cycle.

VENTRICULAR EXCITATION AND ONSET OF VENTRICULAR SYSTOLE

After atrial excitation, the impulse travels through the AV node and specialized conduction system to excite the ventricle. Simultaneously, the atria are contracting. By the time ventricular activation is complete, atrial contraction is already over. The QRS complex represents this ventricular ex-

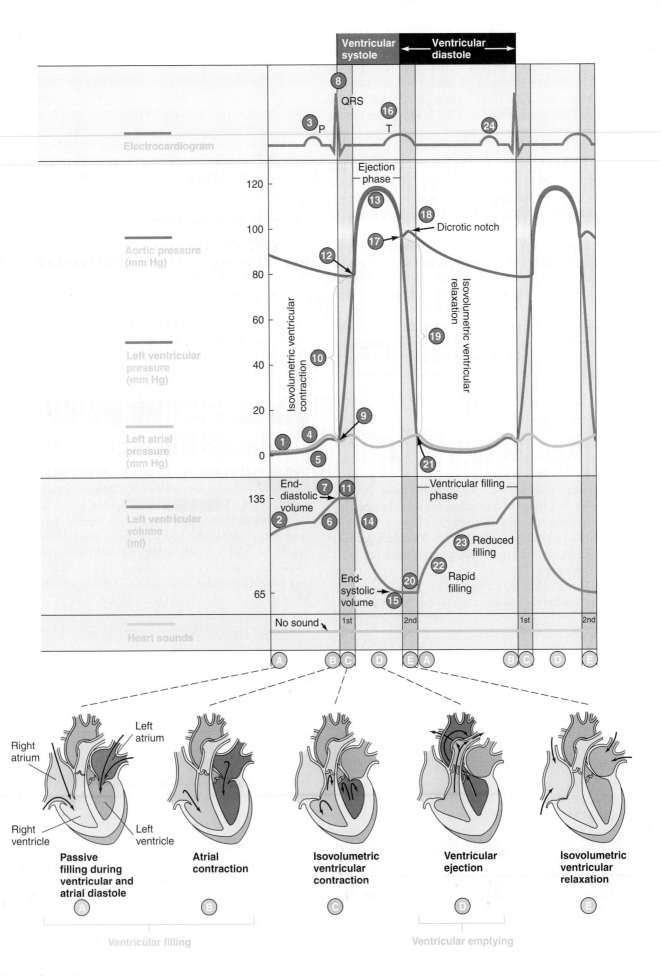

Ventricular systole

Ventricular diastole

Electrocardiogram

8
QRS

3 P

16 T

24

Ejection phase

120

13

18 → Dicrotic notch

Aortic pressure (mm Hg)

100

17

12

80

Left ventricular pressure (mm Hg)

Isovolumetric ventricular contraction

Isovolumetric ventricular relaxation

60

19

40

10

20

9

Left atrial pressure (mm Hg)

1 4

0

5

21

End-diastolic volume

7 11

135

Left ventricular volume (ml)

2 6

14

Ventricular filling phase

23 Reduced filling

22 Rapid filling

20

End-systolic volume

65

15

Heart sounds

No sound →   1st   2nd   1st   2nd

A   B C   D E A   B C   D E

Right atrium

Left atrium

Right ventricle

Left ventricle

Passive filling during ventricular and atrial diastole

A

Atrial contraction

B

Isovolumetric ventricular contraction

C

Ventricular ejection

D

Isovolumetric ventricular relaxation

E

Ventricular filling

Ventricular emptying

citation (point ⑧), which induces ventricular contraction. The ventricular pressure curve sharply increases shortly after the QRS complex, signaling the onset of ventricular systole (point ⑨). As ventricular contraction begins, ventricular pressure immediately exceeds atrial pressure. This backward pressure differential forces the AV valve closed (point ⑨).

ISOVOLUMETRIC VENTRICULAR CONTRACTION

After ventricular pressure exceeds atrial pressure and the AV valve has closed, to open the aortic valve the ventricular pressure must continue to increase until it exceeds aortic pressure. Therefore, after closing of the AV valve and before opening of the aortic valve is a brief period of time when the ventricle remains a closed chamber (point ⑩). Because all valves are closed, no blood can enter or leave the ventricle during this time. This interval is termed the period of **isovolumetric ventricular contraction** (*isovolumetric* means "constant volume and length") (heart Ⓒ). Because no blood enters or leaves the ventricle, the ventricular chamber stays at constant volume, and the muscle fibers stay at constant length. This isovolumetric condition is similar to an isometric contraction in skeletal muscle. During isovolumetric ventricular contraction, ventricular pressure continues to increase as the volume remains constant (point ⑪).

VENTRICULAR EJECTION

When ventricular pressure exceeds aortic pressure (point ⑫), the aortic valve is forced open and ejection of blood begins (heart Ⓓ). The amount of blood pumped out of each ventricle with each contraction is called the **stroke volume (SV)**. The aortic pressure curve rises as blood is forced into the aorta from the ventricle faster than blood is draining off into the smaller vessels at the other end (point ⑬). The ventricular volume decreases substantially as blood is rapidly pumped out (point ⑭). Ventricular systole includes both the period of isovolumetric contraction and the ventricular ejection phase.

END OF VENTRICULAR SYSTOLE

The ventricle does not empty completely during ejection. Normally, only about half the blood within the ventricle at the end of diastole is pumped out during the subsequent systole.

● **FIGURE 9-16**

Cardiac cycle. This graph depicts various events that occur concurrently during the cardiac cycle. Follow each horizontal strip across to see the changes that take place in the electrocardiogram; aortic, ventricular, and atrial pressures; ventricular volume; and heart sounds throughout the cycle. Late diastole, one full systole and diastole (one full cardiac cycle), and another systole are shown for the left side of the heart. Follow each vertical strip downward to see what happens simultaneously with each of these factors during each phase of the cardiac cycle. See the text (pp. 255 and 257) for a detailed explanation of the circled numbers. The sketches of the heart illustrate the flow of O₂-poor (*dark blue*) and O₂-rich (*bright red*) blood in and out of the ventricles during the cardiac cycle.

 For an animation of this figure, click the ECG and the Cardiac Cycle tab in the Cardiovascular Physiology tutorial on the CD-ROM.

The amount of blood left in the ventricle at the end of systole when ejection is complete is the **end-systolic volume (ESV)**, which averages about 65 ml (point ⑮). This is the least amount of blood that the ventricle will contain during this cycle.

The difference between the volume of blood in the ventricle before contraction and the volume after contraction is the amount of blood ejected during the contraction; that is, EDV − ESV = SV. In our example, the end-diastolic volume is 135 ml, the end-systolic volume is 65 ml, and the stroke volume is 70 ml.

VENTRICULAR REPOLARIZATION AND ONSET OF VENTRICULAR DIASTOLE

The T wave signifies ventricular repolarization at the end of ventricular systole (point ⑯). As the ventricle starts to relax, on repolarization, ventricular pressure falls below aortic pressure and the aortic valve closes (point ⑰). Closure of the aortic valve produces a disturbance or notch on the aortic pressure curve, the **dicrotic notch** (point ⑱). No more blood leaves the ventricle during this cycle, because the aortic valve has closed.

ISOVOLUMETRIC VENTRICULAR RELAXATION

When the aortic valve closes, the AV valve is not yet open, because ventricular pressure still exceeds atrial pressure, so no blood can enter the ventricle from the atrium. Therefore, all valves are once again closed for a brief period of time known as **isovolumetric ventricular relaxation** (point ⑲ and heart Ⓔ). The muscle fiber length and chamber volume (point ⑳) remain constant. No blood leaves or enters as the ventricle continues to relax and the pressure steadily falls.

VENTRICULAR FILLING

When ventricular pressure falls below atrial pressure, the AV valve opens (point ㉑), and ventricular filling occurs again (point ㉒). Ventricular diastole includes both the period of isovolumetric ventricular relaxation and the ventricular filling phase.

Atrial repolarization and ventricular depolarization occur simultaneously, so the atria are in diastole throughout ventricular systole. Blood continues to flow from the pulmonary veins into the left atrium. When the AV valve opens at the end of ventricular systole, blood that accumulated in the atrium during ventricular systole pours rapidly into the ventricle (heart Ⓐ again). Ventricular filling thus occurs rapidly at first (point ㉒). Then ventricular filling slows down (point ㉓) as the accumulated blood has already been delivered to the ventricle. During this period of reduced filling, blood continues to flow from the pulmonary veins into the left atrium and through the open AV valve into the left ventricle. During late ventricular diastole, when the ventricle is filling slowly, the SA node fires again (point ㉔), and the cardiac cycle starts over.

When the body is at rest, one complete cardiac cycle lasts 800 msec, with 300 msec devoted to ventricular systole and

500 msec taken up by ventricular diastole. Significantly, much of ventricular filling occurs early in diastole during the rapid-filling phase. During times of rapid heart rate, diastole length is shortened much more than systole length is. For example, if the heart rate increases from 75 to 180 beats per minute, the duration of diastole decreases about 75%, from 500 msec to 125 msec. This greatly reduces the time available for ventricular relaxation and filling. Importantly, because much ventricular filling is done during early diastole, filling is not seriously impaired during periods of increased heart rate, such as during tachycardia.

▌ The two heart sounds are associated with valve closures.

Two major heart sounds normally can be heard with a stethoscope during the cardiac cycle. The **first heart sound** is low-pitched, soft, and relatively long—often said to sound like "lub." The **second heart sound** has a higher pitch and is shorter and sharper—often said to sound like "dup." Thus, one normally hears "lub-dup-lub-dup-lub-dup. . . ." The first heart sound is associated with closure of the AV valves, whereas the second sound is associated with closure of the semilunar valves (● Figure 9-16). Opening of valves does not produce any sound.

Because the AV valves close at the onset of ventricular contraction, when ventricular pressure first exceeds atrial pressure, the first heart sound signals the onset of ventricular systole. The semilunar valves close at the onset of ventricular relaxation, as the left and right ventricular pressures fall below the aortic and pulmonary artery pressures, respectively. The second heart sound, therefore, signals the onset of ventricular diastole.

▌ Turbulent blood flow produces heart murmurs.

Clinical Note Abnormal heart sounds, or **murmurs**, are usually (but not always) associated with cardiac disease. Murmurs not involving heart pathology, so-called **functional murmurs**, are more common in young people.

Blood normally flows in a *laminar* fashion; that is, layers of the fluid slide smoothly over each other (*lamina* means "layer"). Laminar flow does not produce any sound. When blood flow becomes turbulent, however, a sound can be heard (● Figure 9-17). Such an abnormal sound is due to vibrations that the turbulent flow creates in the surrounding structures.

STENOTIC AND INSUFFICIENT VALVES

The most common cause of turbulence is valve malfunction, either a stenotic or an insufficient valve. A **stenotic valve** is a stiff, narrowed valve that does not open completely. Blood must be forced through the constricted opening at tremendous velocity, resulting in turbulence that produces an abnormal whistling sound similar to the sound produced when you force air rapidly through narrowed lips to whistle.

Laminar flow (does not create any sound)

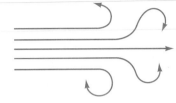

Turbulent flow (can be heard)

● **FIGURE 9-17**

Comparison of laminar and turbulent flow

An **insufficient** or **incompetent valve** is one that cannot close completely, usually because the valve edges are scarred and do not fit together properly. Turbulence is produced when blood flows backward through the insufficient valve and collides with blood moving in the opposite direction, creating a swishing or gurgling murmur. Such backflow of blood is known as **regurgitation**. An insufficient heart valve is often called a **leaky valve**, because it lets blood leak back through at a time when the valve should be closed.

TIMING OF MURMURS

The valve involved and the type of defect can usually be detected by the *location* and *timing* of the murmur. Each heart valve may be heard best at a specific location on the chest. Noting where a murmur is loudest helps the diagnostician tell which valve is involved.

The "timing" of the murmur refers to the part of the cardiac cycle during which the murmur is heard. Recall that the first heart sound signals the onset of ventricular systole, and the second heart sound signals the onset of ventricular diastole. Thus a murmur between the first and second heart sounds (lub-murmur-dup, lub-murmur-dup) is a **systolic murmur**. A **diastolic murmur**, in contrast, occurs between the second and first heart sound (lub-dup-murmur, lub-dup-murmur). The sound of the murmur characterizes it as either a stenotic (whistling) murmur or an insufficient (swishy) murmur. Armed with these facts, one can determine the cause of a valvular murmur (▲ Table 9-2). As an example, a whistling murmur (denoting a stenotic valve) occurring between the first and second heart sounds (denoting a systolic murmur) indicates stenosis in a valve that should be open during systole. It could be either the aortic or the pulmonary semilunar valve through which blood is being ejected. Identifying which of these valves is stenotic is done by finding where the murmur is best heard.

The main concern with heart murmurs, of course, is not the murmur itself but the harmful circulatory results of the defect.

Timing and Type of Murmur in Various Heart Valve Disorders

PATTERN HEARD WITH STETHOSCOPE	TYPE OF VALVE DEFECT	TIMING OF MURMUR	VALVE DISORDER	COMMENT
Lub-Whistle-Dup	Stenotic	Systolic	Stenotic semilunar valve	A whistling systolic murmur signifies that a valve that should be open during systole (a semilunar valve) does not open completely.
Lub-Dup-Whistle	Stenotic	Diastolic	Stenotic AV valve	A whistling diastolic murmur signifies that a valve that should be open during diastole (an AV valve) does not open completely.
Lub-Swish-Dup	Insufficient	Systolic	Insufficient AV valve	A swishy systolic murmur signifies that a valve that should be closed during systole (an AV valve) does not close completely.
Lub-Dup-Swish	Insufficient	Diastolic	Insufficient semilunar valve	A swishy diastolic murmur signifies that a valve that should be closed during diastole (a semilunar valve) does not close completely.

CARDIAC OUTPUT AND ITS CONTROL

Cardiac output (CO) is the volume of blood pumped by *each ventricle* per minute (not the total amount of blood pumped by the heart). During any period of time, the volume of blood flowing through the pulmonary circulation is the same as the volume flowing through the systemic circulation. Therefore, the cardiac output from each ventricle normally is the same, although on a beat-to-beat basis, minor variations may occur.

▌ Cardiac output depends on the heart rate and the stroke volume.

The two determinants of cardiac output are *heart rate* (beats per minute) and *stroke volume* (volume of blood pumped per beat or stroke). The average resting heart rate is 70 beats per minute, established by SA node rhythmicity, and the average resting stroke volume is 70 ml per beat, producing an average cardiac output of 4900 ml/min, or close to 5 liters/min:

$$
\begin{aligned}
\text{Cardiac output (CO)} &= \text{heart rate} \times \text{stroke volume} \\
&= 70 \text{ beats/min} \times 70 \text{ ml/beat} \\
&= 4900 \text{ ml/min} \approx 5 \text{ liters/min}
\end{aligned}
$$

Because the body's total blood volume averages 5 to 5.5 liters, each half of the heart pumps the equivalent of the entire blood volume each minute. In other words, each minute the right ventricle normally pumps 5 liters of blood through the lungs, and the left ventricle pumps 5 liters through the systemic circulation. At this rate, each half of the heart would pump about 2.5 million liters of blood in just one year. Yet this is only the resting cardiac output! During exercise cardiac output can increase to 20 to 25 liters per minute, and

outputs as high as 40 liters per minute have been recorded in trained athletes during heavy endurance-type exercise. The difference between the cardiac output at rest and the maximum volume of blood the heart can pump per minute is called the **cardiac reserve**.

How can cardiac output vary so tremendously, depending on the demands of the body? You can readily answer this question by thinking about how your own heart pounds rapidly (increased heart rate) and forcefully (increased stroke volume) when you engage in strenuous physical activities (need for increased cardiac output). Thus regulation of cardiac output depends on the control of both heart rate and stroke volume, topics that we discuss next.

▌ Heart rate is determined primarily by autonomic influences on the SA node.

The SA node is normally the pacemaker of the heart, because it has the fastest spontaneous rate of depolarization to threshold. When the SA node reaches threshold, an action potential is initiated that spreads throughout the heart, inducing the heart to contract or have a "heartbeat." This happens about 70 times per minute, setting the average heart rate at 70 beats per minute.

The heart is innervated by both divisions of the autonomic nervous system, which can modify the rate (as well as the strength) of contraction, even though nervous stimulation is not required to initiate contraction. The parasympathetic nerve to the heart, the **vagus nerve**, primarily supplies the atrium, especially the SA and AV nodes. Parasympathetic innervation of the ventricles is sparse. The cardiac sympathetic nerves also supply the atria, including the SA and AV

Effects of the Autonomic Nervous System on the Heart and Structures That Influence the Heart

AREA AFFECTED	EFFECT OF PARASYMPATHETIC STIMULATION	EFFECT OF SYMPATHETIC STIMULATION
SA Node	Decreases rate of depolarization to threshold; decreases heart rate	Increases rate of depolarization to threshold; increases heart rate
AV Node	Decreases excitability; increases AV nodal delay	Increases excitability; decreases AV nodal delay
Ventricular Conduction Pathway	No effect	Increases excitability; hastens conduction through bundle of His and Purkinje cells
Atrial Muscle	Decreases contractility; weakens contraction	Increases contractility; strengthens contraction
Ventricular Muscle	No effect	Increases contractility; strengthens contraction
Adrenal Medulla (an endocrine gland)	No effect	Promotes adrenomedullary secretion of epinephrine, a hormone that augments the sympathetic nervous system's actions on the heart
Veins	No effect	Increases venous return, which increases strength of cardiac contraction through Frank-Starling mechanism

nodes, and richly innervate the ventricles as well. Parasympathetic and sympathetic stimulation have the following specific effects on the heart (▲ Table 9-3).

EFFECT OF PARASYMPATHETIC STIMULATION ON THE HEART

• Parasympathetic stimulation decreases the SA node's rate of spontaneous depolarization, prolonging the time required to drift to threshold. Therefore, the SA node reaches threshold and fires less frequently, decreasing the heart rate (● Figure 9-18).

● **FIGURE 9-18**

Autonomic control of heart rate. Increased parasympathetic activity decreases the heart rate, whereas increased sympathetic activity increases the heart rate.

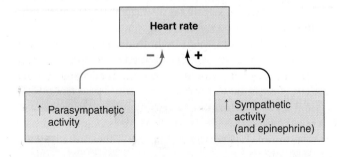

• Parasympathetic influence on the AV node decreases the node's excitability, prolonging transmission of impulses to the ventricles even longer than the usual AV nodal delay.
• Parasympathetic stimulation of the atrial contractile cells shortens the action potential, reducing the slow inward current carried by Ca^{2+}; that is, the plateau phase is shortened. As a result, atrial contraction is weakened.
• The parasympathetic system has little effect on ventricular contraction, because of the sparseness of parasympathetic innervation to the ventricles.

Thus the heart is more "leisurely" under parasympathetic influence—it beats less rapidly, the time between atrial and ventricular contraction is stretched out, and atrial contraction is weaker. These actions are appropriate, considering that the parasympathetic system controls heart action in quiet, relaxed situations when the body is not demanding an enhanced cardiac output.

EFFECT OF SYMPATHETIC STIMULATION ON THE HEART

• In contrast, the sympathetic nervous system, which controls heart action in emergency or exercise situations, when there is a need for greater blood flow, speeds up the heart rate through its effect on the pacemaker tissue. The main effect of sympathetic stimulation on the SA node is to speed up depolarization so that threshold is reached more rapidly. This swifter drift to threshold under sympathetic influence permits more frequent action potentials and a correspondingly faster heart rate (● Figure 9-18 and ▲ Table 9-3).

• Sympathetic stimulation of the AV node reduces the AV nodal delay by increasing conduction velocity.

• Similarly, sympathetic stimulation speeds up spread of the action potential throughout the specialized conduction pathway.

• In the atrial and ventricular contractile cells, both of which have lots of sympathetic nerve endings, sympathetic stimulation increases contractile strength so the heart beats more forcefully and squeezes out more blood. This effect is produced by increasing Ca^{2+} permeability, which enhances the slow Ca^{2+} influx and intensifies Ca^{2+} participation in excitation–contraction coupling.

The overall effect of sympathetic stimulation on the heart, therefore, is to improve its effectiveness as a pump by increasing heart rate, decreasing the delay between atrial and ventricular contraction, decreasing conduction time throughout the heart, and increasing the force of contraction; that is, sympathetic stimulation "revs up" the heart.

CONTROL OF HEART RATE

Thus, as is typical of the autonomic nervous system, parasympathetic and sympathetic effects on heart rate are antagonistic (oppose each other). At any given moment, heart rate is determined largely by the balance between inhibition of the SA node by the vagus nerve and stimulation by the cardiac sympathetic nerves. Under resting conditions, parasympathetic discharge dominates. In fact, if all autonomic nerves to the heart were blocked, the resting heart rate would increase from its average value of 70 beats per minute to about 100 beats per minute, which is the inherent rate of the SA node's spontaneous discharge when not subjected to any nervous influence. (We use 70 beats per minute as the normal rate of SA node discharge because this is the average rate under normal resting conditions when parasympathetic ac-

tivity dominates.) The heart rate can be altered beyond this resting level in either direction by shifting the balance of autonomic nervous stimulation. Heart rate is speeded up by simultaneously increasing sympathetic and decreasing parasympathetic activity; heart rate is slowed by a concurrent rise in parasympathetic activity and decline in sympathetic activity. The relative level of activity in these two autonomic branches to the heart in turn is primarily coordinated by the *cardiovascular control center* in the brain stem.

Although autonomic innervation is the primary means by which heart rate is regulated, other factors affect it as well. The most important is epinephrine, a hormone that on sympathetic stimulation is secreted into the blood from the adrenal medulla and that acts on the heart in a manner similar to norepinephrine (the sympathetic neurotransmitter) to increase the heart rate. Epinephrine therefore reinforces the direct effect that the sympathetic nervous system has on the heart.

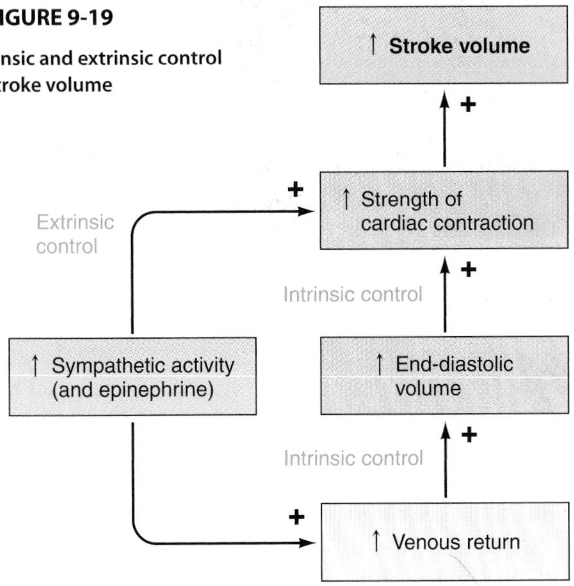

Stroke volume is determined by the extent of venous return and by sympathetic activity.

The other component besides heart rate that determines cardiac output is stroke volume, the amount of blood pumped out by each ventricle during each beat. Two types of controls influence stroke volume: (1) *intrinsic control* related to the extent of venous return and (2) *extrinsic control* related to the extent of sympathetic stimulation of the heart. Both factors increase stroke volume by increasing the strength of heart contraction (● Figure 9-19). Let us examine each factor in detail to see how they influence stroke volume.

Increased end-diastolic volume results in increased stroke volume.

Intrinsic control of stroke volume, which refers to the heart's inherent ability to vary stroke volume, depends on the direct correlation between end-diastolic volume and stroke volume. As more blood returns to the heart, the heart pumps out more blood, but the relationship is not quite as simple as might seem, because the heart does not eject all the blood it contains. This intrinsic control depends on the length–tension relationship of cardiac muscle, which is similar to that of skeletal muscle. For skeletal muscle, the resting muscle length is approximately the optimal length at which maximal tension can be developed during a subsequent contraction. When the skeletal muscle is longer or shorter than this optimal length, the subsequent contraction is weaker (see ● Figure 8-15, p. 216). For cardiac muscle, the resting cardiac muscle fiber length is less than optimal length. Therefore, the length of cardiac muscle fibers normally varies along the ascending limb of the length–tension curve. An increase in cardiac muscle fiber length, by moving closer to the optimal length, increases the contractile tension of the heart on the following systole (● Figure 9-20).

Unlike skeletal muscle, the length–tension curve of cardiac muscle normally does not operate at lengths that fall within the region of the descending limb. That is, within phys-

● **FIGURE 9-19**

Intrinsic and extrinsic control of stroke volume

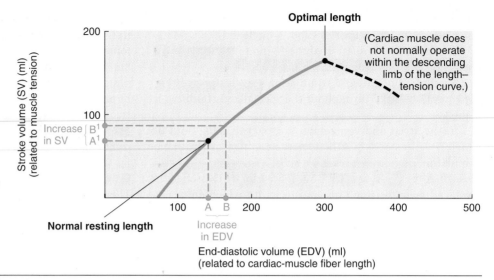

Intrinsic control of stroke volume (Frank-Starling curve). The cardiac muscle fiber's length, which is determined by the extent of venous filling, is normally less than the optimal length for developing maximal tension. Therefore, an increase in end-diastolic volume (that is, an increase in venous return), by moving the cardiac muscle fiber length closer to optimal length, increases the contractile tension of the fibers on the next systole. A stronger contraction squeezes out more blood. Thus, as more blood is returned to the heart and the end-diastolic volume increases, the heart automatically pumps out a correspondingly larger stroke volume.

iologic limits cardiac muscle does not get stretched beyond its optimal length to the point that contractile strength diminishes with further stretching.

FRANK-STARLING LAW OF THE HEART

What causes cardiac muscle fibers to vary in length before contraction? Skeletal muscle length can vary before contraction because of the positioning of the skeletal parts to which the muscle is attached, but cardiac muscle is not attached to any bones. The main determinant of cardiac muscle fiber length is degree of diastolic filling. An analogy is a balloon filled with water—the more water you put in, the larger the balloon becomes, and the more it is stretched. Likewise, the greater the diastolic filling, the larger the end-diastolic volume, and the more the heart is stretched. The more the heart is stretched, the longer the initial cardiac-fiber length before contraction. The increased length results in a greater force on the subsequent cardiac contraction and thus in a greater stroke volume. This intrinsic relationship between end-diastolic volume and stroke volume is known as the **Frank-Starling law of the heart.** Stated simply, the law says that the heart normally pumps out during systole the volume of blood returned to it during diastole; increased venous return results in increased stroke volume. In ● Figure 9-20, assume that the end-diastolic volume increases from point A to point B. You can see this increase in end-diastolic volume is accompanied by a corresponding increase in stroke volume from point A^1 to point B^1.

ADVANTAGES OF THE CARDIAC LENGTH–TENSION RELATIONSHIP

The built-in relationship matching stroke volume with venous return has two important advantages. First, one of the most important functions of this intrinsic mechanism is equalizing output between the right and left sides of the heart, so that blood pumped out by the heart is equally distributed between the pulmonary and systemic circulation. If, for example, the right side of the heart ejects a larger stroke volume,

more blood enters the pulmonary circulation, so venous return to the left side of the heart increases accordingly. The increased end-diastolic volume of the left side of the heart causes it to contract more forcefully, so it too pumps out a larger stroke volume. In this way, output of the two ventricular chambers is kept equal. If such equalization did not happen, too much blood would be dammed up in the venous system before the ventricle with the lower output.

Second, when a larger cardiac output is needed, such as during exercise, venous return is increased through action of the sympathetic nervous system and other mechanisms to be described in the next chapter. The resulting increase in end-diastolic volume automatically increases stroke volume correspondingly. Because exercise also increases heart rate, these two factors act together to increase the cardiac output so more blood can be delivered to the exercising muscles.

▮ Sympathetic stimulation increases the contractility of the heart.

In addition to intrinsic control, stroke volume is also subject to **extrinsic control** by factors originating outside the heart, the most important of which are actions of the cardiac sympathetic nerves and epinephrine (▲ Table 9-3). Sympathetic stimulation and epinephrine enhance the heart's **contractility,** which is the strength of contraction at any given end-diastolic volume. In other words, on sympathetic stimulation the heart contracts more forcefully and squeezes out a greater percentage of the blood it contains, leading to more complete ejection. This increased contractility is due to the increased Ca^{2+} influx triggered by norepinephrine and epinephrine. The extra cytosolic Ca^{2+} lets the myocardial fibers generate more force through greater cross-bridge cycling than they would without sympathetic influence. Normally, the end-diastolic volume is 135 ml and the end-systolic volume is 65 ml for a stroke volume of 70 ml (● Figure 9-21a). Under sympathetic influence, for the same end-diastolic volume of 135 ml, the end-systolic volume might be 35 ml and the stroke volume 100 ml (● Figure 9-21b). In effect, sympathetic

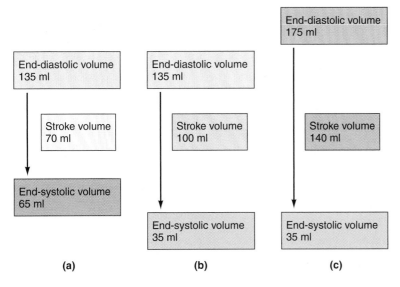

(a)   **(b)**   **(c)**

● **FIGURE 9-21**

Effect of sympathetic stimulation on stroke volume. (a) Normal stroke volume. (b) Stroke volume during sympathetic stimulation. (c) Stroke volume with combination of sympathetic stimulation and increased end-diastolic volume.

stimulation shifts the Frank-Starling curve to the left (● Figure 9-22). Depending on the extent of sympathetic stimulation, the curve can be shifted to varying degrees, up to a maximal increase in contractile strength of about 100% greater than normal.

Sympathetic stimulation increases stroke volume not only by strengthening cardiac contractility but also by enhancing venous return (● Figure 9-21c). Sympathetic stimulation constricts the veins, which squeezes more blood forward from the veins to the heart, increasing the end-diastolic

● **FIGURE 9-22**

Shift of the Frank-Starling curve to the left by sympathetic stimulation. For the same end-diastolic volume (point A), there is a larger stroke volume (from point B to point C) on sympathetic stimulation as a result of increased contractility of the heart. The Frank-Starling curve shifts to the left by variable degrees, depending on the extent of sympathetic stimulation.

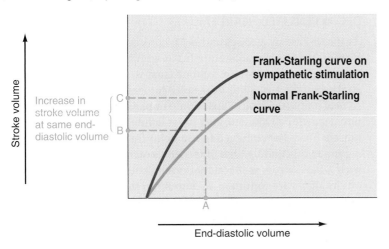

volume and subsequently increasing the stroke volume even further.

SUMMARY OF FACTORS AFFECTING STROKE VOLUME AND CARDIAC OUTPUT

The strength of cardiac muscle contraction and accordingly the stroke volume can thus be graded by (1) varying the initial length of the muscle fibers, which in turn depends on the degree of ventricular filling before contraction (intrinsic control); and (2) varying the extent of sympathetic stimulation (extrinsic control) (● Figure 9-19). This is in contrast to gradation of skeletal muscle, in which twitch summation and recruitment of motor units produce variable strength of muscle contraction. These mechanisms do not apply to cardiac muscle. Twitch summation is impossible because of the long refractory period. Recruitment of motor units is not possible because the heart muscle cells are arranged into functional syncytia where all contractile cells become excited and contract with every beat, instead of into distinct motor units that can be discretely activated.

All the factors that determine cardiac output by influencing heart rate or stroke volume are summarized in ● Figure 9-23. Note that sympathetic stimulation increases cardiac output by increasing both heart rate and stroke volume. Sympathetic activity to the heart increases, for example, dur-

● **FIGURE 9-23**

Control of cardiac output. Because cardiac output equals heart rate times stroke volume, this figure is a composite of ● Figure 9-18 (control of heart rate) and ● Figure 9-19 (control of stroke volume).

PhysioEdge   For an interaction related to this figure, see Media Exercise 9.3: The Heart: Cardiac Output on the CD-ROM.

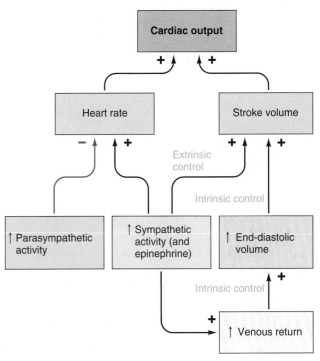

ing exercise when the working skeletal muscles need increased delivery of O_2-laden blood to support their high rate of ATP consumption.

We are next going to examine how a failing heart cannot pump out enough blood, before we turn to the final section of this chapter, on how the heart muscle is nourished.

In heart failure the contractility of the heart decreases.

Clinical Note **Heart failure** is the inability of the cardiac output to keep pace with the body's demands for supplies and removal of wastes. Either one or both ventricles may progressively weaken and fail. When a failing ventricle cannot pump out all the blood returned to it, the veins behind the failing ventricle become congested with blood. Heart failure may occur for a variety of reasons, but the two most common are (1) damage to the heart muscle as a result of a heart attack or impaired circulation to the cardiac muscle and (2) prolonged pumping against a chronically elevated blood pressure. Heart failure presently affects about 5 million Americans, nearly 50% of whom will die within four years of diagnosis.

PRIME DEFECT IN HEART FAILURE

The prime defect in heart failure is a decrease in cardiac contractility; that is, the cardiac muscle cells contract less effectively. The intrinsic ability of the heart to develop pressure and eject a stroke volume is reduced so that the heart operates on a lower length–tension curve (● Figure 9-24a). The Frank-Starling curve shifts downward and to the right such that for a given end-diastolic volume, a failing heart pumps out a smaller stroke volume than a normal healthy heart.

COMPENSATORY MEASURES FOR HEART FAILURE

In the early stages of heart failure, two major compensatory measures help restore the stroke volume to normal. First, sympathetic activity to the heart is reflexively increased, which increases the contractility of the heart toward normal (● Figure 9-24b). Sympathetic stimulation can help compensate only for a limited period of time, however, because the heart becomes less responsive to prolonged sympathetic stimulation. Second, when cardiac output is reduced, the kidneys, in a compensatory attempt to improve their reduced blood flow, retain extra salt and water in the body during urine formation, to expand the blood volume. The increase in circulating blood volume increases the end-diastolic volume. The resultant stretching of the cardiac muscle fibers enables the weakened heart to pump out a normal stroke volume (● Figure 9-24b). The heart is now pumping out the blood returned to it but is operating at a greater cardiac muscle fiber length.

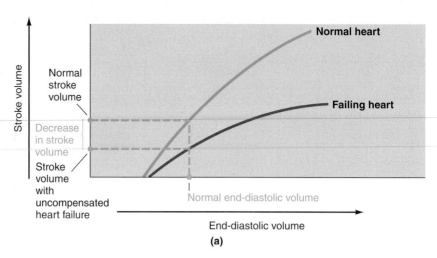

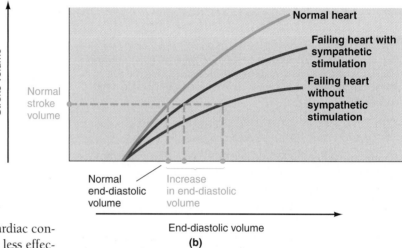

(a)

(b)

● FIGURE 9-24

Compensated heart failure. (a) Shift of the Frank-Starling curve downward and to the right in a failing heart. Because its contractility is decreased, the failing heart pumps out a smaller stroke volume at the same end-diastolic volume than a normal heart does. (b) Compensations for heart failure. Reflex sympathetic stimulation shifts the Frank-Starling curve of a failing heart to the left, increasing the contractility of the heart toward normal. A compensatory increase in end-diastolic volume as a result of blood volume expansion further increases the strength of contraction of the failing heart. Operating at a longer cardiac muscle fiber length, a compensated failing heart is able to eject a normal stroke volume.

DECOMPENSATED HEART FAILURE

As the disease progresses and heart contractility deteriorates further, the heart reaches a point at which it can no longer pump out a normal stroke volume (that is, cannot pump out all the blood returned to it) despite compensatory measures. At this point, the heart slips from compensated heart failure into a state of decompensated heart failure. Now the cardiac muscle fibers are stretched to the point that they are operating in the descending limb of the length–tension curve. *Backward failure* occurs as blood that cannot enter and be pumped out by the heart continues to dam up in the venous system. *Forward failure* occurs simultaneously as the heart fails to pump an adequate amount of blood forward to the tissues because the stroke volume becomes progressively smaller. The congestion in the venous system is why this condition is sometimes termed **congestive heart failure.**

Left-sided failure has more serious consequences than right-sided failure. Backward failure of the left side leads to pulmonary edema (excess tissue fluid in the lungs) because blood dams up in the lungs. This fluid accumulation in the lungs reduces exchange of O_2 and CO_2 between the air and blood in the lungs, reducing arterial oxygenation and elevating levels of acid-forming CO_2 in the blood. In addition, one of the more serious consequences of left-sided forward failure is an inadequate blood flow to the kidneys, which causes a twofold problem. First, vital kidney function is depressed; and second, the kidneys retain even more salt and water in the body during urine formation as they try to expand the plasma volume even further to improve their reduced blood flow. Excessive fluid retention further exacerbates the already existing problems of venous congestion.

Treatment of congestive heart failure therefore includes measures that reduce salt and water retention and increase urinary output as well as drugs that enhance the contractile ability of the weakened heart—digitalis, for example.

 Click on the Media Exercises menu of the CD-ROM and work Media Exercise 9.3: The Heart: Cardiac Output to test your understanding of the previous section.

NOURISHING THE HEART MUSCLE

The heart depends on O_2 delivery and aerobic metabolism to generate the energy necessary for contraction (see p. 30).

The heart receives most of its own blood supply through the coronary circulation during diastole.

Although all the blood passes through the heart, the heart muscle cannot extract O_2 or nutrients from the blood within its chambers for two reasons. First, the watertight endocardial lining does not permit blood to pass from the chamber into the myocardium. Second, the heart walls are too thick to permit diffusion of O_2 and other supplies from the blood in the chamber to the individual cardiac cells. Therefore, like other tissues of the body, heart muscle must receive blood through blood vessels, specifically via the **coronary circulation.** The coronary arteries branch from the aorta just beyond the aortic valve (see ● Figure 9-27, p. 268), and the coronary veins empty into the right atrium.

The heart muscle receives most of its blood supply during diastole. Blood flow to the heart muscle cells is substantially reduced during systole for two reasons. First, the contracting myocardium compresses the major branches of the coronary arteries, and, second, the open aortic valve partially blocks the entrance to the coronary vessels. Thus most coronary arterial flow (about 70%) occurs during diastole, driven by the aortic blood pressure, with flow declining as aortic pressure drops. This limited time for coronary blood flow be-

comes especially important during rapid heart rates, when diastolic time is much reduced. Just when increased demands are placed on the heart to pump more rapidly, it has less time to provide O_2 and nourishment to its own musculature to accomplish the increased workload.

Nevertheless, under normal circumstances the heart muscle does receive adequate blood flow to support its activities, even during exercise, when the rate of coronary blood flow increases up to five times its resting rate. Extra blood is delivered to the cardiac cells primarily by vasodilation, or enlargement, of the coronary vessels, which lets more blood flow through them, especially during diastole. Coronary blood flow is adjusted primarily in response to changes in the heart's O_2 requirements. When cardiac activity increases and the heart thus needs more O_2, local chemical changes induce dilation of the coronary blood vessels, allowing more O_2-rich blood to flow to the more active cardiac cells to meet their increased O_2 demand. This matching of O_2 delivery with O_2 needs is crucial, because the heart muscle depends on oxidative processes to generate energy. The heart cannot get enough ATP through anaerobic metabolism.

Atherosclerotic coronary artery disease can deprive the heart of essential oxygen.

Clinical Note Adequacy of coronary blood flow is relative to the heart's O_2 demands at any given moment. With coronary artery disease, coronary blood flow may not be able to keep pace with rising O_2 needs. The term **coronary artery disease (CAD)** refers to pathologic changes, within the coronary artery walls, that diminish blood flow through these vessels. A given rate of coronary blood flow may be adequate at rest but insufficient in physical exertion or other stressful situations.

Complications of CAD, including heart attacks, make it the single leading cause of death in the United States. CAD is the underlying cause of about 50% of all deaths in this country.

CAD can cause myocardial ischemia and possibly lead to acute myocardial infarction by three mechanisms: (1) profound vascular spasm of the coronary arteries, (2) formation of atherosclerotic plaques, and (3) thromboembolism. We will discuss each in turn.

VASCULAR SPASM

Vascular spasm is an abnormal spastic constriction that transiently narrows the coronary vessels. Vascular spasms are associated with the early stages of CAD and are most often triggered by exposure to cold, physical exertion, or anxiety. The condition is reversible and usually does not last long enough to damage the cardiac muscle.

When too little O_2 is available in the coronary vessels, the endothelium (blood vessel lining) releases *platelet-activating factor* (PAF). PAF, which exerts a variety of actions, was named for its first discovered effect, activating platelets. Among its other effects, PAF, once released from the endothelium, diffuses to the underlying vascular smooth muscle and causes it to contract, bringing about vascular spasm.

DEVELOPMENT OF ATHEROSCLEROSIS

Atherosclerosis is a progressive, degenerative arterial disease that leads to occlusion (gradual blockage) of affected vessels, reducing blood flow through them. Atherosclerosis is characterized by plaques forming beneath the vessel lining within arterial walls. An **atherosclerotic plaque** consists of a lipid-rich core covered by an abnormal overgrowth of smooth muscle cells, topped off by a collagen-rich connective tissue cap. As plaque forms, it bulges into the vessel lumen (● Figure 9-25).

Although not all the contributing factors have been identified, in recent years investigators have sorted out the following complex sequence of events in the gradual development of atherosclerosis:

1. Atherosclerosis is believed to start with injury to the blood vessel wall, which triggers an *inflammatory response* that sets

● **FIGURE 9-25**

Atherosclerotic plaque. (a) Schematic representation of the parts of a plaque. (b) Photomicrograph of a severe atherosclerotic plaque in a coronary vessel.

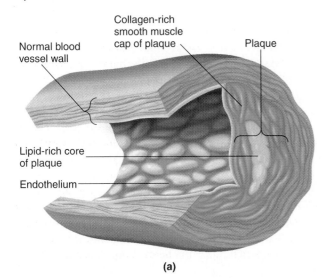

Collagen-rich smooth muscle cap of plaque

Plaque

Normal blood vessel wall

Lipid-rich core of plaque

Endothelium

(a)

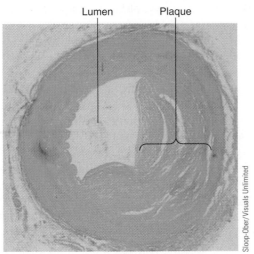

Lumen   Plaque

Sloop-Ober/Visuals Unlimited

(b)

the stage for plaque buildup. Normally inflammation is a protective response that fights infection and promotes repair of damaged tissue (see p. 330). However, when the cause of the injury persists within the vessel wall, the sustained, low-grade inflammatory response over a course of decades can insidiously lead to arterial plaque formation and heart disease. Plaque formation likely has many causes. Suspected artery-abusing agents that may set off the vascular inflammatory response included oxidized cholesterol, free radicals, high blood pressure, homocysteine, chemicals released from fat cells, or even bacteria and viruses that damage blood vessel walls. The most common triggering agent appears to be oxidized cholesterol. (For a further discussion of the role of cholesterol and other factors in the development of atherosclerosis, see the boxed feature, ▶ Beyond the Basics, on pp. 268–269.)

2. Typically, the initial stage of atherosclerosis is characterized by the accumulation beneath the endothelium of excessive amounts of *low-density lipoprotein (LDL)*, the so-called "bad" cholesterol, in combination with a protein carrier. As LDL accumulates within the vessel wall, this cholesterol product becomes oxidized, primarily by oxidative wastes produced by the blood vessel cells. These wastes are *free radicals*, very unstable electron-deficient particles that are highly reactive. Antioxidant vitamins that prevent LDL oxidation, such as *vitamin E*, *vitamin C*, and *beta-carotene*, have been shown to slow plaque deposition.

3. In response to the presence of oxidized LDL and/or other irritants, the endothelial cells produce chemicals that attract *monocytes*, a type of white blood cell, to the site. These immune cells trigger a local inflammatory response.

4. Once they leave the blood and enter the vessel wall, monocytes settle down permanently, enlarge, and become large phagocytic cells called *macrophages*. Macrophages voraciously phagocytize (see p. 25) the oxidized LDL until these cells become so packed with fatty droplets that they appear foamy under a microscope. Now called *foam cells*, these greatly engorged macrophages accumulate beneath the vessel lining and form a visible *fatty streak*, the earliest form of an atherosclerotic plaque.

5. Thus the earliest stage of plaque formation is characterized by the accumulation beneath the endothelium of a cholesterol-rich deposit. The disease progresses as smooth muscle cells within the blood vessel wall migrate from the muscular layer of the blood vessel to a position on top of the lipid accumulation, just beneath the endothelium. This migration is triggered by chemicals released at the inflammatory site. At their new location, the smooth muscle cells continue to divide and enlarge, producing *atheromas*, which are benign (noncancerous) tumors of smooth muscle cells within the blood vessel walls. Together the lipid-rich core and overlying smooth muscle form a maturing plaque.

6. As it continues to develop, the plaque progressively bulges into the lumen of the vessel. The protruding plaque narrows the opening through which blood can flow.

7. Further contributing to vessel narrowing, oxidized LDL inhibits release of *nitric oxide* from the endothelial cells. Nitric oxide is a local chemical messenger that relaxes the under-

lying layer of normal smooth-muscle cells within the vessel wall. Relaxation of these smooth muscle cells dilates the vessel. Because of reduced nitric oxide release, vessels damaged by developing plaques cannot dilate as readily as normal.

8. A thickening plaque also interferes with nutrient exchange for cells located within the involved arterial wall, leading to degeneration of the wall in the vicinity of the plaque. The damaged area is invaded by *fibroblasts* (scar-forming cells), which form a connective tissue cap over the plaque. (The term *sclerosis* means excessive growth of fibrous connective tissue, hence the term *atherosclerosis* for this condition characterized by atheromas and sclerosis, along with abnormal lipid accumulation.)

9. In the later stages of the disease, Ca^{2+} often precipitates in the plaque. A vessel so afflicted becomes hard and cannot distend easily.

THROMBOEMBOLISM AND OTHER COMPLICATIONS OF ATHEROSCLEROSIS

Atherosclerosis attacks arteries throughout the body, but the most serious consequences involve damage to the vessels of the brain and heart. In the brain, atherosclerosis is the prime cause of strokes, whereas in the heart it brings about myocardial ischemia and its complications. The following are potential complications of coronary atherosclerosis:

- *Angina pectoris.* Gradual enlargement of protruding plaque continues to narrow the vessel lumen and progressively diminishes coronary blood flow, triggering increasingly frequent bouts of transient myocardial ischemia as the ability to match blood flow with cardiac O_2 needs becomes more limited. Although the heart cannot normally be "felt," pain is associated with myocardial ischemia. Such cardiac pain, known as **angina pectoris** ("pain of the chest"), can be felt beneath the sternum and is often referred to (appears to come from) the left shoulder and down the left arm. The symptoms of angina pectoris recur whenever cardiac O_2 demands become too great in relation to the coronary blood flow—for example, during exertion or emotional stress. The ischemia associated with the characteristically brief angina attacks is usually temporary and reversible and can be relieved by rest, taking vasodilator drugs such as *nitroglycerin*, or both. Nitroglycerin brings about coronary vasodilation by being metabolically converted to nitric oxide, which in turn relaxes the vascular smooth muscle.

- *Thromboembolism.* The enlarging atherosclerotic plaque can break through the weakened endothelial lining that covers it, exposing blood to the underlying collagen in the collagen-rich connective tissue cap of the plaque. Foam cells release chemicals that can weaken the fibrous cap of a plaque by breaking down the connective tissue fibers. Plaques with thick fibrous caps are considered stable, because they are not likely to rupture. However, plaques with thinner fibrous caps are unstable, because they are likely to rupture and trigger clot formation.

Blood platelets (formed elements of the blood involved in plugging vessel defects and in clot formation) normally do not adhere to smooth, healthy vessel linings. However, when platelets contact collagen at the site of vessel damage, they stick to the site and help promote the formation of a blood clot. Furthermore, foam cells produce a potent clot promoter. Such an abnormal clot attached to a vessel wall is called a **thrombus.** The thrombus may enlarge gradually until it completely blocks the vessel at that site, or the continued flow of blood past the thrombus may break it loose. As it heads downstream, such a freely floating clot, or **embolus,** may completely plug a smaller vessel (● Figure 9-26). Thus, through **thromboembolism** atherosclerosis can result in a gradual or sudden occlusion of a coronary vessel (or any other vessel).

- *Heart attack.* When a coronary vessel is completely plugged, the cardiac tissue served by the vessel soon dies from O_2 deprivation, and a heart attack occurs, unless the area can be supplied with blood from nearby vessels.

Sometimes a deprived area is lucky enough to receive blood from more than one pathway. **Collateral circulation** exists when small terminal branches from adjacent blood vessels nourish the same area. These accessory vessels cannot develop suddenly after an abrupt blockage but may be lifesaving if already developed. Such alternate vascular pathways often develop over a period of time when an atherosclerotic constriction progresses slowly, or they may be induced by sustained demands on the heart through regular aerobic exercise.

In the absence of collateral circulation, the extent of the damaged area during a heart attack depends on the size of

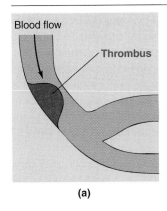

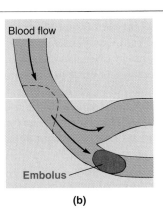

(a)　　　　(b)　　　　(c)

● **FIGURE 9-26**

Results of thromboembolism. (a) A thrombus may enlarge gradually until it completely occludes the vessel at that site. (b) A thrombus may break loose from its attachment, forming an embolus that may completely block a smaller vessel downstream. (c) Scanning electron micrograph of a vessel completely occluded by a thromboembolic lesion.

Lennart Nilsson/Bonnier Alba AB

Atherosclerosis: Cholesterol and Beyond

The cause of atherosclerosis is still not entirely clear. Certain high-risk factors have been associated with an increased incidence of atherosclerosis and coronary heart disease. Included among them are genetic predisposition, obesity, advanced age, smoking, hypertension, lack of exercise, high blood concentrations of C-reactive protein, elevated levels of homocysteine, infectious agents, and, most notoriously, elevated cholesterol levels in the blood.

Sources of Cholesterol

There are two sources of cholesterol for the body: (1) dietary intake of cholesterol, with animal products such as egg yolk, red meats, and butter being especially rich in this lipid (animal fats contain cholesterol, whereas plant fats typically do not); and (2) manufacture of cholesterol by cells, especially liver cells.

"Good" versus "Bad" Cholesterol

Actually, it is not the total blood-cholesterol level but the amount of cholesterol bound to various plasma protein carriers that is most important with regard to the risk of developing atherosclerotic heart disease. Because cholesterol is a lipid, it is not very soluble in blood. Most cholesterol in the blood is attached to specific plasma-protein carriers in the form of lipoprotein complexes, which are soluble in blood. The three major lipoproteins are named for their density of protein as compared to lipid: (1) **high-density lipoproteins (HDL),** which contain the most protein and least cholesterol; (2) **low-density lipoproteins (LDL),** which have less protein and more cholesterol; and (3) **very-low-density lipoproteins (VLDL),** which have the least protein and most

lipid, but the lipid they carry is neutral fat, not cholesterol.

Cholesterol carried in LDL complexes has been termed "bad" cholesterol, because cholesterol is transported *to* the cells, including those lining the blood vessel walls, by LDL. The propensity toward developing atherosclerosis substantially increases with elevated levels of LDL. The presence of oxidized LDL within an arterial wall is a major trigger for the inflammatory process that leads to the development of atherosclerotic plaques (see p. 266).

In contrast, cholesterol carried in HDL complexes has been dubbed "good" cholesterol, because HDL removes cholesterol *from* the cells and transports it to the liver for partial elimination from the body. Not only does HDL help remove excess cholesterol from the tissues, but in addition, it protects by inhibiting oxidation of LDL. The risk of atherosclerosis is inversely related to the concentration of HDL in the blood; that is, elevated levels of HDL are associated with a low incidence of atherosclerotic heart disease.

Some other factors known to influence atherosclerotic risk can be related to HDL levels; for example, cigarette smoking lowers HDL, whereas regular exercise raises HDL.

Cholesterol Uptake by Cells

Unlike most lipids, cholesterol is not used as metabolic fuel by cells. Instead, it serves as an essential component of plasma membranes. In addition, a few special cell types use cholesterol as a precursor for the synthesis of secretory products, such as steroid hormones and bile salts. Although most cells can synthesize some of the cholesterol needed for their own

plasma membranes, they cannot manufacture sufficient amounts and therefore must rely on supplemental cholesterol being delivered by the blood.

Cells accomplish cholesterol uptake from the blood by synthesizing receptor proteins specifically capable of binding LDL and inserting these receptors into the plasma membrane. When an LDL particle binds to one of the membrane receptors, the cell engulfs the particle by receptor-mediated endocytosis, receptor and all (see p. 25). Within the cell, lysosomal enzymes break down the LDL to free the cholesterol, making it available to the cell for synthesis of new cellular membrane. The LDL receptor, which is also freed within the cell, is recycled back to the surface membrane.

If too much free cholesterol accumulates in the cell, there is a shutdown of both the synthesis of LDL receptor proteins (so that less cholesterol is taken up) and the cell's own cholesterol synthesis (so that less new cholesterol is made). Faced with a cholesterol shortage, in contrast, the cell makes more LDL receptors so that it can engulf more cholesterol from the blood.

Maintenance of Blood Cholesterol Level and Cholesterol Metabolism

The maintenance of a blood-borne cholesterol supply to the cells involves an interaction between dietary cholesterol and the synthesis of cholesterol by the liver. When the amount of dietary cholesterol is increased, hepatic (liver) synthesis of cholesterol is turned off, because cholesterol in the blood directly inhibits a hepatic enzyme essential for cholesterol

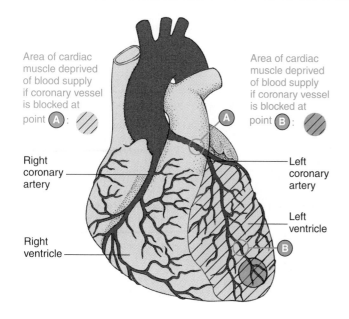

Area of cardiac muscle deprived of blood supply if coronary vessel is blocked at point (A) :

Area of cardiac muscle deprived of blood supply if coronary vessel is blocked at point (B) :

Right coronary artery

Left coronary artery

Left ventricle

Right ventricle

the blocked vessel. The larger the vessel occluded, the greater the area deprived of blood supply. As ● Figure 9-27 illustrates, a blockage at point A in the coronary circulation would cause more extensive damage than would a blockage at point B. Because there are only two major coronary arteries, complete blockage of either one of these main branches results in extensive myocardial damage. Left coronary-artery blockage is most devastating because this vessel supplies blood to 85% of the cardiac tissue.

A heart attack has four possible outcomes: immediate death, delayed death from complications, full functional recovery, or recovery with impaired function (▲ Table 9-4).

● FIGURE 9-27

Extent of myocardial damage as a function of the size of the occluded vessel

synthesis. Thus as more cholesterol is ingested, less is produced by the liver. Conversely, when cholesterol intake from food is reduced, the liver synthesizes more of this lipid because the inhibitory effect of cholesterol on the crucial hepatic enzyme is removed. In this way, the blood concentration of cholesterol is maintained at a fairly constant level despite changes in cholesterol intake; thus it is difficult to significantly reduce cholesterol levels in the blood by decreasing cholesterol intake.

HDL transports cholesterol to the liver. The liver secretes cholesterol as well as cholesterol-derived bile salts into the bile. Bile enters the intestinal tract, where bile salts participate in the digestive process. Most of the secreted cholesterol and bile salts are subsequently reabsorbed from the intestine into the blood to be recycled to the liver. However, the cholesterol and bile salts not reclaimed by absorption are eliminated in the feces and lost from the body.

Thus the liver has a primary role in determining total blood-cholesterol levels, and the interplay between LDL and HDL determines the traffic flow of cholesterol between the liver and the other cells. Whenever these mechanisms are altered, blood cholesterol levels may be affected in such a way as to influence the individual's predisposition to atherosclerosis.

Varying the intake of dietary fatty acids may alter total blood-cholesterol levels by influencing one or more of the mechanisms involving cholesterol balance. The blood cholesterol level tends to be raised by ingesting saturated fatty acids found predominantly in animal fats, because these fatty acids stimulate cholesterol synthesis and inhibit its conversion to bile salts. In contrast, ingesting polyunsaturated fatty acids, the predominant fatty acids of most plants, tends to reduce blood cholesterol levels by enhancing elimination of both cholesterol and cholesterol-derived bile salts in the feces.

Other Risk Factors Besides Cholesterol

Despite the strong links between cholesterol and heart disease, over half of all patients with heart attacks have a normal cholesterol profile and no other well-established risk factors. Clearly, other factors are involved in the development of coronary artery disease in these people. These same factors may also contribute to development of atherosclerosis in people with unfavorable cholesterol levels. Following are among the leading other possible risk factors:

- Elevated blood levels of the amino acid **homocysteine** have recently been implicated as a strong predictor for heart disease, independent of the person's cholesterol/lipid profile. Homocysteine is formed as an intermediate product during metabolism of the essential dietary amino acid *methionine*. Investigators believe homocysteine contributes to atherosclerosis by promoting proliferation of vascular smooth muscle cells, an early step in development of this artery-clogging condition. Furthermore, homocysteine appears to damage endothelial cells and may cause oxidation of LDL, both of which can contribute to plaque formation. Three B vitamins—*folic acid, vitamin B$_{12}$,* and *vitamin B$_6$*—all play key roles in pathways that clear homocysteine from the blood. Therefore, these B vitamins are all needed to keep blood homocysteine at safe levels.

- Indicating the role of inflammation in forming atherosclerotic plaques, people with elevated levels of **C-reactive protein,** a blood-borne marker of inflammation, have a higher risk for developing coronary artery disease. In one study, people with a high level of C-reactive protein in their blood were three times more likely to have a heart attack over the next 10 years than those with a low level of this inflammatory protein. Because inflammation plays a crucial role in the development of atherosclerosis, anti-inflammatory drugs, such as aspirin, help prevent heart attacks. Furthermore, aspirin protects against heart attacks through its role in inhibiting clot formation.

- Accumulating data suggest that an infectious agent may be the underlying culprit in a significant number of cases of atherosclerotic disease. Among the leading suspects are respiratory infection–causing *Chlamydia pneumoniae,* cold sore–causing *herpes virus,* and gum disease–causing bacteria. Importantly, if a link between infections and coronary artery disease can be confirmed, antibiotics may be added to the regimen of heart disease prevention strategies.

As you can see, the relationship between atherosclerosis, cholesterol, and other factors is far from clear. Much research on this complex disease is currently in progress, because the incidence of atherosclerosis is so high and its consequences are potentially fatal.

▲ **TABLE 9-4**

Possible Outcomes of Acute Myocardial Infarction (Heart Attack)

IMMEDIATE DEATH	DELAYED DEATH FROM COMPLICATIONS	FULL FUNCTIONAL RECOVERY	RECOVERY WITH IMPAIRED FUNCTION
Acute cardiac failure because heart is too weak to pump effectively to support body tissues	Fatal rupture of dead, degenerating area of heart wall	Replacement of damaged area with a strong scar, accompanied by enlargement of remaining normal contractile tissue to compensate for lost cardiac musculature	Persistence of permanent functional defects, such as bradycardia or conduction blocks, caused by destruction of irreplaceable autorhythmic or conductive tissues
Fatal ventricular fibrillation from damage to specialized conducting tissue or from O$_2$ deprivation	Slowly progressing congestive heart failure because weakened heart cannot pump out all the blood returned to it		

CHAPTER IN PERSPECTIVE: FOCUS ON HOMEOSTASIS

Survival depends on continual delivery of needed supplies to all body cells and on ongoing removal of wastes generated by the cells. Furthermore, regulatory chemical messengers, such as hormones, must be transported from their production site to their action site, where they control a variety of activities, most of which are directed toward maintaining a stable internal environment.

The circulatory system contributes to homeostasis by serving as the body's transport system. It provides a way to rapidly move materials from one part of the body to another. Without the circulatory system, materials would not get where they need to go to support life-sustaining activities nearly rapidly enough. For example, O_2 would take months to years to diffuse from the body surface to internal organs, yet through the heart's swift pumping action the blood can pick up and deliver O_2 and other substances to all the cells in a few seconds.

The heart serves as a dual pump to continuously circulate blood between the lungs, where O_2 is picked up, and the other body tissues, which use O_2 to support their energy-generating chemical reactions. As blood is pumped through the various tissues, other substances besides O_2 are also exchanged between the blood and tissues. For example, the blood picks up nutrients as it flows through the digestive organs, and other tissues remove nutrients from the blood as it flows through them.

Although all the body tissues constantly depend on the life-supporting blood flow provided to them by the heart, the heart itself is quite an independent organ. It can take care of many of its own needs without any outside influence. Contraction of this magnificent muscle is self-generated through a carefully orchestrated interplay of changing ionic permeabilities. Local mechanisms within the heart ensure that blood flow to the cardiac muscle normally meets the heart's need for O_2. In addition, the heart has built-in capabilities to vary its strength of contraction, depending on the amount of blood returned to it. The heart does not act entirely autonomously, however. It is innervated by the autonomic nervous system and is influenced by the hormone epinephrine, both of which can vary heart rate and contractility, depending on the body's needs for blood delivery. Furthermore, as with all tissues, the cells that make up the heart depend on the other body systems to maintain a stable internal environment in which they can survive and function.

CHAPTER SUMMARY

Introduction (pp. 241–242)

▊ The circulatory system is the transport system of the body.

▊ The three basic components of the circulatory system are the heart (the pump), the blood vessels (the passageways), and the blood (the transport medium).

Anatomy of the Heart (pp. 242–246)

▊ The heart is basically a dual pump that provides the driving pressure for blood flow through the pulmonary and systemic circulations. *(Review Figures 9-1 and 9-2.)*

▊ The heart has four chambers: Each half of the heart consists of an atrium, or venous input chamber, and a ventricle, or arterial output chamber. *(Review Figures 9-2 and 9-4.)*

▊ Four heart valves direct the blood in the right direction and keep it from flowing in the other direction. *(Review Figures 9-3 and 9-4.)*

▊ Contraction of the spirally arranged cardiac muscle fibers produces a wringing effect important for efficient pumping. Also important for efficient pumping is that the muscle fibers in each chamber act as a functional syncytium, contracting as a coordinated unit. *(Review Figure 9-5.)*

Electrical Activity of the Heart (pp. 246–255)

▊ The heart is self-excitable, initiating its own rhythmic contractions.

▊ Autorhythmic cells are 1% of the cardiac muscle cells; they do not contract but are specialized to initiate and conduct action potentials. Autorhythmic cells display a pacemaker potential, a slow drift to threshold potential. *(Review Figure 9-6.)* The other 99% of cardiac cells are contractile cells that contract in response to the spread of an action potential initiated by autorhythmic cells.

▊ The cardiac impulse originates at the SA node, the pacemaker of the heart, which has the fastest rate of spontaneous depolarization to threshold. *(Review Table 9-1 and Figures 9-7 and 9-8.)*

▊ Once initiated, the action potential spreads throughout the right and left atria, partially facilitated by specialized conduction pathways but mostly by cell-to-cell spread of the impulse through gap junctions. *(Review Figure 9-9.)*

▊ The impulse passes from the atria into the ventricles through the AV node, the only point of electrical contact between these chambers. The action potential is delayed briefly at the AV node, ensuring that atrial contraction precedes ventricular contraction to allow complete ventricular filling. *(Review Figures 9-7 and 9-9.)*

▊ The impulse then travels rapidly down the interventricular septum via the bundle of His and rapidly disperses throughout the myocardium by means of the Purkinje fibers. The rest of the ventricular cells are activated by cell-to-cell spread of the impulse through gap junctions. *(Review Figures 9-7 and 9-9.)*

▊ Thus the atria contract as a single unit, followed after a brief delay by a synchronized ventricular contraction.

▊ The action potentials of cardiac contractile cells exhibit a prolonged positive phase, or plateau, accompanied by a prolonged period of contraction, which ensures adequate ejection time. This plateau is primarily due to activation of slow Ca^{2+} channels. *(Review Figure 9-10.)*

▊ Because a long refractory period occurs in conjunction with this prolonged plateau phase, summation and tetanus of cardiac muscle are impossible, ensuring the alternate periods of contraction and relaxation essential for pumping of blood. *(Review Figure 9-12.)*

▊ The spread of electrical activity throughout the heart can be recorded from the body surface. This record, the ECG, can provide useful information about the status of the heart. *(Review Figures 9-13, 9-14, and 9-15.)*

Mechanical Events of the Cardiac Cycle (pp. 255–259)

▪ The cardiac cycle consists of three important events:
1. The generation of electrical activity as the heart autorhythmically depolarizes and repolarizes
2. Mechanical activity consisting of alternate periods of systole (contraction and emptying) and diastole (relaxation and filling), which are initiated by the rhythmic electrical cycle
3. Directional flow of blood through the heart chambers, guided by valve opening and closing induced by pressure changes that are generated by mechanical activity

▪ Valve closing gives rise to two normal heart sounds. The first heart sound is caused by closing of the atrioventricular (AV) valves and signals the onset of ventricular systole. The second heart sound is due to closing of the aortic and pulmonary valves at the onset of diastole.

▪ The atrial pressure curve remains low throughout the entire cardiac cycle, with only minor fluctuations (normally varying between 0 and 8 mm Hg). The aortic pressure curve remains high the entire time, with moderate fluctuations (normally varying between a systolic pressure of 120 mm Hg and a diastolic pressure of 80 mm Hg). The ventricular pressure curve fluctuates dramatically, because ventricular pressure must be below the low atrial pressure during diastole to allow the AV valve to open for filling, and to force the aortic valve open to allow emptying, it must be above the high aortic pressure during systole. Therefore, ventricular pressure normally varies from 0 mm Hg during diastole to slightly more than 120 mm Hg during systole. *(Review Figure 9-16.)*

▪ The end-diastolic volume is the volume of blood in the ventricle when filling is complete at the end of diastole. The end-systolic volume is the volume of blood remaining in the ventri-cle when ejection is complete at the end of systole. The stroke volume is the volume of blood pumped out by each ventricle each beat.

Cardiac Output and Its Control (pp. 259–265)

▪ Cardiac output, the volume of blood ejected by each ventricle each minute, is determined by heart rate times stroke volume. *(Review Figures 9-18, 9-19, and 9-23.)*

▪ Heart rate is varied by altering the balance of parasympathetic and sympathetic influence on the SA node. Parasympathetic stimulation slows the heart rate, and sympathetic stimulation speeds it up. *(Review Table 9-3 and Figure 9-18.)*

▪ Stroke volume depends on (1) extent of ventricular filling, with an increased end-diastolic volume resulting in a larger stroke volume by means of the length–tension relationship (Frank-Starling law of the heart, a form of intrinsic control), and (2) extent of sympathetic stimulation, with increased sympathetic stimulation resulting in increased contractility of the heart, that is, increased strength of contraction and increased stroke volume at a given end-diastolic volume (extrinsic control). *(Review Figures 9-19 through 9-22.)*

Nourishing the Heart Muscle (pp. 265–268)

▪ Cardiac muscle is supplied with oxygen and nutrients by blood delivered to it by the coronary circulation, not by blood within the heart chambers.

▪ Most coronary blood flow occurs during diastole, because during systole the contracting heart muscle compresses the coronary vessels.

▪ Coronary blood flow is normally varied to keep pace with cardiac oxygen needs.

▪ Coronary blood flow may be compromised by development of atherosclerotic plaques, which can lead to ischemic heart disease ranging in severity from mild chest pain on exertion, to fatal heart attacks. *(Review Table 9-4 and Figures 9-25 through 9-27.)*

REVIEW EXERCISES

Objective Questions (Answers on p. A-43)

1. Adjacent cardiac muscle cells are joined end to end at specialized structures known as _____, which contain two types of membrane junctions: _____ and _____.

2. _____ is an abnormally slow heart rate, whereas _____ is a rapid heart rate.

3. The left ventricle is a stronger pump than the right ventricle because more blood is needed to supply the body tissues than to supply the lungs. *(True or false?)*

4. The heart lies in the left half of the thoracic cavity. *(True or false?)*

5. The only point of electrical contact between the atria and ventricles is the fibrous tissue that surrounds and supports the heart valves. *(True or false?)*

6. The atria and ventricles each act as a functional syncytium. *(True or false?)*

7. Which of the following is the proper sequence of cardiac excitation?
   a. SA node → AV node → atrial myocardium → bundle of His → Purkinje fibers → ventricular myocardium.
   b. SA node → atrial myocardium → AV node → bundle of His → ventricular myocardium → Purkinje fibers.
   c. SA node → atrial myocardium → ventricular myocardium → AV node → bundle of His → Purkinje fibers.
   d. SA node → atrial myocardium → AV node → bundle of His → Purkinje fibers → ventricular myocardium.

8. What percentage of ventricular filling is normally accomplished before atrial contraction begins?
   a. 0%
   b. 20%
   c. 50%
   d. 80%
   e. 100%

9. Sympathetic stimulation of the heart _____.
   a. increases the heart rate.
   b. increases the contractility of the heart muscle.
   c. shifts the Frank-Starling curve to the left.
   d. Both (a) and (b) above are correct.
   e. All of the above are correct.

10. Circle the correct choice in each instance to complete the statements: During ventricular filling, ventricular pressure must be *(greater than/less than)* atrial pressure, whereas during ventricular ejection ventricular pressure must be *(greater than/less than)* aortic pressure. Atrial pressure is always *(greater than/*

less than) aortic pressure. During isovolumetric ventricular contraction and relaxation, ventricular pressure is *(greater than/ less than)* atrial pressure and *(greater than/less than)* aortic pressure.

11. Circle the correct choice in each instance to complete the statement: The first heart sound is associated with closing of the *(AV/semilunar)* valves and signals the onset of *(systole/diastole),* whereas the second heart sound is associated with closing of the *(AV/semilunar)* valves and signals the onset of *(systole/ diastole).*

12. Match the following:

_____ 1. receives O_2-poor blood from the venae cavae
_____ 2. prevent backflow of blood from the ventricles to the atria
_____ 3. pumps O_2-rich blood into the aorta
_____ 4. prevent backflow of blood from the arteries into the ventricles
_____ 5. pumps O_2-poor blood into the pulmonary artery
_____ 6. receives O_2-rich blood from the pulmonary veins

(a) AV valves
(b) semilunar valves
(c) left atrium
(d) left ventricle
(e) right atrium
(f) right ventricle

Essay Questions
1. What are the three basic components of the circulatory system?
2. Trace a drop of blood through one complete circuit of the circulatory system.

3. Describe the location and function of the four heart valves. What keeps each of these valves from everting?
4. What are the three layers of the heart wall? Describe the distinguishing features of the structure and arrangement of cardiac muscle cells. What are the two specialized types of cardiac muscle cells?
5. Why is the SA node the pacemaker of the heart?
6. What is the significance of the AV nodal delay? Why is the ventricular conduction system important?
7. Compare the changes in membrane potential associated with an action potential in a nodal pacemaker cell with those in a myocardial contractile cell. What is responsible for the plateau phase?
8. Why is tetanus of cardiac muscle impossible? Why is this inability advantageous?
9. Draw and label the waveforms of a normal ECG. What electrical event does each component of the ECG represent?
10. Describe the mechanical events (that is, pressure changes, volume changes, valve activity, and heart sounds) of the cardiac cycle. Correlate the mechanical events of the cardiac cycle with the changes in electrical activity.
11. Distinguish between a stenotic and an insufficient valve.
12. Define the following: *end-diastolic volume, end-systolic volume, stroke volume, heart rate, cardiac output,* and *cardiac reserve.*
13. Discuss autonomic nervous system control of heart rate.
14. Describe the intrinsic and extrinsic control of stroke volume.
15. How is the heart muscle provided with blood? Why does the heart receive most of its own blood supply during diastole?
16. What are the pathologic changes and consequences of coronary artery disease?

POINTS TO PONDER

(Explanations on p. A-43)

1. The stroke volume ejected on the next heartbeat after a premature ventricular contraction (PVC) is usually larger than normal. Can you explain why? (*Hint:* At a given heart rate, the interval between a PVC and the next normal beat is longer than the interval between two normal beats.)

2. Trained athletes usually have lower resting heart rates than normal (for example, 50 beats/min in an athlete compared to 70 beats/ min in a sedentary individual). Considering that the resting cardiac output is 5000 ml/min in both trained athletes and sedentary people, what is responsible for the bradycardia of trained athletes?

3. During fetal life, because of the tremendous resistance offered by the collapsed, nonfunctioning lungs, the pressures in the right half of the heart and pulmonary circulation are higher than in the left half of the heart and systemic circulation, a situation that reverses after birth. Also in the fetus, a vessel called the **ductus arteriosus** connects the pulmonary artery and aorta as these major vessels both leave the heart. The blood pumped out by the heart into the pulmonary circulation is shunted from the pulmonary artery into the aorta through the ductus arteriosus, bypassing the nonfunctional lungs. What force is driving blood to flow in this direction through the ductus arteriosus?

At birth, the ductus arteriosus normally collapses and eventually degenerates into a thin, ligamentous strand. On occasion, this fetal bypass fails to close properly at birth, leading to a *patent* (open) *ductus arteriosus.* In what direction would blood flow through a patent ductus arteriosus? What possible outcomes would you predict might occur as a result of this blood flow?

4. Through what regulatory mechanisms can a transplanted heart, which does not have any innervation, adjust cardiac output to meet the body's changing needs?

5. There are two branches of the bundle of His, the right and left bundle branches, each of which travels down its respective side of the ventricular septum (see ● Figure 9-7, p. 247). Occasionally conduction through one of these branches becomes blocked (so-called *bundle-branch block*). In this case, the wave of excitation spreads out from the terminals of the intact branch and eventually depolarizes the whole ventricle, but the normally stimulated ventricle completely depolarizes a considerable time before the ventricle on the side of the defective bundle branch. For example, if the left bundle branch is blocked, the right ventricle will be completely depolarized two to three times more rapidly than the left ventricle. How would this defect affect the heart sounds?

CLINICAL CONSIDERATION

(Explanation on p. A-44)

In a physical exam, Rachel B.'s heart rate was rapid and very irregular. Furthermore, her heart rate, determined directly by listening to her heart with a stethoscope, exceeded the pulse rate taken concurrently at her wrist. Such a difference in heart rate and pulse rate is called a **pulse deficit.** No definite P waves could be detected on Rachel's ECG. The QRS complexes were normal in shape but occurred sporadically. Given these findings, what is the most likely diagnosis of Rachel's condition? Explain why the condition is characterized by a rapid, irregular heartbeat. Would cardiac output be seriously impaired by this condition? Why or why not? What accounts for the pulse deficit?

PHYSIOEDGE RESOURCES

 PhysioEdge CD-ROM

PhysioEdge, the CD-ROM packaged with your text, focuses on the concepts students find most difficult to learn. Figures marked with this icon have associated activities on the CD. For a visual review of concepts in this chapter, check out the following:

Tutorial: Cardiovascular Physiology

Media Exercise 9.1: The Electrocardiogram

Media Exercise 9.2: The Heart: A Dual Pump

Media Exercise 9.3: The Heart: Cardiac Output

PhysioEdge Website

The website for this book contains a wealth of helpful study aids, as well as many ideas for further reading and research. Log on to: **http://www.brookscole.com/hpfundamentals3**

Select Chapter 9 from the drop-down menu or click on one of the many resource areas, including Case Histories, which introduce clinical aspects of human physiology. For this chapter check out: Case Histories 7: Why Am I So Tired?, 9: Endocarditis, 10: Blue Baby, and 11: Congestive Heart Failure.

For Suggested Readings, consult InfoTrac College Edition/ Research on the PhysioEdge website or go directly to InfoTrac College Edition, your online research library, at: **http://infotrac.thomsonlearning.com**

Cardiovascular System
(Blood Vessels)

Body systems maintain homeostasis

Homeostasis
The circulatory system contributes to homeostasis by transporting O_2, CO_2, wastes, electrolytes, and hormones from one part of the body to another.

Homeostasis is essential for survival of cells

Cells
Cells need to have O_2 and nutrients continuously delivered and CO_2 constantly removed by the circulatory system in order to generate the energy needed to power life-sustaining cellular activities by the following chemical reaction:

Food + $O_2 \rightarrow CO_2$ + H_2O + Energy

Cells make up body systems

The **circulatory system** contributes to homeostasis by serving as the body's transport system. The blood vessels transport and distribute blood pumped through them by the heart to meet the body's needs for O_2 and nutrient delivery, waste removal, and hormonal signaling. The highly elastic **arteries** transport blood from the heart to the organs and serve as a pressure reservoir to continue driving blood forward when the heart is relaxing and filling. The **mean arterial blood pressure** is closely regulated to ensure adequate blood delivery to the organs. The amount of blood that flows through a given organ depends on the caliber (internal diameter) of the highly muscular **arterioles** that supply the organ. Arteriolar caliber is subject to control so that cardiac output can be constantly readjusted to best serve the body's needs at the moment. The thin-walled, pore-lined **capillaries** are the actual site of exchange between blood and the surrounding tissue cells. The highly distensible veins return blood from the organs to the heart and also serve as a blood reservoir.

The Blood Vessels and Blood Pressure

INTRODUCTION

Most body cells are not in direct contact with the external environment, yet these cells must make exchanges with this environment, such as picking up O_2 and nutrients and eliminating wastes. Furthermore, chemical messengers must be transported between cells to accomplish integrated activity. To achieve these long-distance exchanges, the cells are linked with each other and with the external environment by vascular (blood vessel) highways. Blood is transported to all parts of the body through a system of vessels that brings fresh supplies to the vicinity of all cells while removing their wastes.

To review, all blood pumped by the right side of the heart passes through the pulmonary circulation to the lungs for O_2 pickup and CO_2 removal. The blood pumped by the left side of the heart into the systemic circulation is distributed in various proportions to the systemic organs through a parallel arrangement of vessels that branch from the aorta (● Figure 10-1) (also see p. 243). This arrangement ensures that all organs receive blood of the same composition; that is, one organ does not receive "leftover" blood that has passed through another organ. Because of this parallel arrangement, blood flow through each systemic organ can be independently adjusted as needed.

In this chapter, we will first examine some general principles regarding blood flow patterns and the physics of blood flow. Then we will turn our attention to the roles of the various types of blood vessels through which blood flows. We will end by discussing how blood pressure is regulated to ensure adequate delivery of blood to the tissues.

▮ To maintain homeostasis, reconditioning organs receive blood flow in excess of their own needs.

Blood is constantly "reconditioned" so that its composition remains relatively constant despite an on-

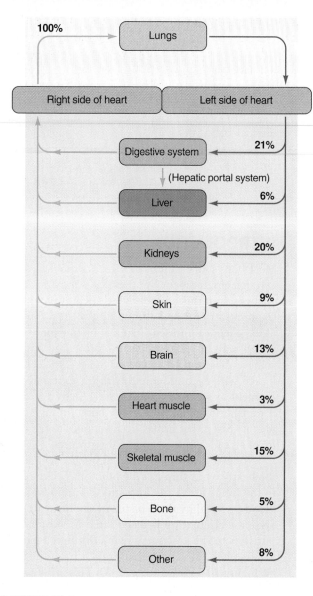

100%

| Lungs |

| Right side of heart | | Left side of heart |

| Digestive system | 21%

(Hepatic portal system)

| Liver | 6%

| Kidneys | 20%

| Skin | 9%

| Brain | 13%

| Heart muscle | 3%

| Skeletal muscle | 15%

| Bone | 5%

| Other | 8%

● **FIGURE 10-1**

Distribution of cardiac output at rest. The lungs receive all the blood pumped out by the right side of the heart, whereas the systemic organs each receive some of the blood pumped out by the left side of the heart. The percentage of pumped blood received by the various organs under resting conditions is indicated. This distribution of cardiac output can be adjusted as needed.

going drain of supplies to support metabolic activities and despite the continual addition of wastes from the tissues. The organs that recondition the blood normally receive much more blood than is necessary to meet their basic metabolic needs, so they can adjust the extra blood to achieve homeostasis. For example, large percentages of the cardiac output are distributed to the digestive tract (to pick up nutrient supplies), to the kidneys (to eliminate metabolic wastes and adjust water and electrolyte composition), and to the skin (to eliminate heat). Blood flow to the other organs—heart, skeletal muscles, and so on—is solely for filling these tissues' metabolic needs and can be adjusted according to their level of activity. For example, during exercise additional blood is

delivered to the active muscles to meet their increased metabolic needs.

Because reconditioning organs—digestive organs, kidneys, and skin—receive blood flow in excess of their own needs, they can withstand temporary reductions in blood flow much better than can other organs that do not have this extra margin of blood supply. The brain in particular suffers irreparable damage when transiently deprived of blood supply. After only four minutes without O_2, permanent brain damage occurs. Therefore, a high priority in the overall operation of the circulatory system is the constant delivery of adequate blood to the brain, which can least tolerate disrupted blood supply.

In contrast, the reconditioning organs can tolerate significant reductions in blood flow for quite a long time, and often do. For example, during exercise some of the blood that normally flows through the digestive organs and kidneys is diverted instead to the skeletal muscles. Likewise, to conserve body heat blood flow through the skin is markedly restricted during exposure to cold.

Later in the chapter, you will see how distribution of cardiac output is adjusted according to the body's current needs. For now, we are going to concentrate on the factors that influence blood flow through a given blood vessel.

▌ Blood flow through vessels depends on the pressure gradient and vascular resistance.

The **flow rate** of blood through a vessel (that is, the volume of blood passing through per unit of time) is directly proportional to the pressure gradient and inversely proportional to vascular resistance:

$$F = \frac{\Delta P}{R}$$

where

F = flow rate of blood through a vessel
ΔP = pressure gradient
R = resistance of blood vessels

PRESSURE GRADIENT

The **pressure gradient** is the difference in pressure between the beginning and end of a vessel. Blood flows from an area of higher pressure to an area of lower pressure down a pressure gradient. Contraction of the heart imparts pressure to the blood, which is the main driving force for flow through a vessel. Because of frictional losses (resistance), the pressure drops as it flows throughout the vessel's length. Accordingly, pressure is higher at the beginning than at the end of the vessel, establishing a pressure gradient for forward flow of blood through the vessel. The greater the pressure gradient forcing blood through a vessel, the greater the rate of flow through that vessel (● Figure 10-2a). Think of a garden hose attached to a faucet. If you turn on the faucet slightly, a small stream of water will flow out of the end of the hose, because the pressure is slightly greater at the beginning than at the end of the hose. If you open the faucet all the way, the pressure gradient

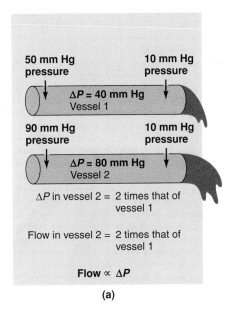

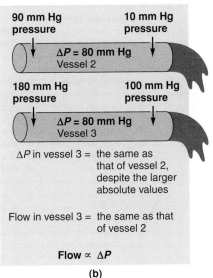

● **FIGURE 10-2**

Relationship of flow to the pressure gradient in a vessel. (a) As the difference in pressure (ΔP) between the two ends of a vessel increases, the flow rate increases proportionately. (b) Flow rate is determined by the *difference* in pressure between the two ends of a vessel, not the magnitude of the pressures at each end.

PhysioEdge For an interaction related to this figure, see Media Exercise 10.1: Blood Flow and Total Cross-Sectional Area on the CD-ROM.

increases tremendously, so that water flows through the hose much faster and spurts from the end of the hose. Note that the *difference* in pressure between the two ends of a vessel, not the absolute pressures within the vessel, determines flow rate (● Figure 10-2b).

RESISTANCE

The other factor influencing flow rate through a vessel is the **resistance,** which is a measure of the hindrance or opposition to blood flow through a vessel, caused by friction be-

tween the moving fluid and the stationary vascular walls. As resistance to flow increases, it is more difficult for blood to pass through the vessel, so flow decreases (as long as the pressure gradient remains unchanged). When resistance increases, the pressure gradient must increase correspondingly to maintain the same flow rate. Accordingly, when the vessels offer more resistance to flow, the heart must work harder to maintain adequate circulation.

Resistance to blood flow depends on three factors: (1) viscosity of the blood, (2) vessel length, and (3) vessel radius, which is by far the most important. The term **viscosity** (designated as η) refers to the friction developed between the molecules of a fluid as they slide over each other during flow of the fluid. The greater the viscosity, the greater the resistance to flow. In general, the thicker a liquid, the more viscous it is. For example, molasses flows more slowly than water, because molasses has greater viscosity. Blood viscosity is determined primarily by the number of circulating red blood cells. Normally, this factor is relatively constant and thus not important in controlling resistance. Occasionally, however, blood viscosity and, accordingly, resistance to flow are altered by an abnormal number of red blood cells. When excessive red blood cells are present, blood flow is more sluggish than normal.

Because blood "rubs" against the lining of the vessels as it flows past, the greater the vessel surface area in contact with the blood, the greater the resistance to flow. Surface area is determined by both the length (L) and radius (r) of the vessel. At a constant radius, the longer the vessel the greater the surface area and the greater the resistance to flow. Because vessel length remains constant in the body, it is not a variable factor in the control of vascular resistance.

Therefore, the major determinant of resistance to flow is the vessel's radius. Fluid passes more readily through a large vessel than through a smaller vessel. The reason is that a given volume of blood comes into contact with much more of the surface area of a small-radius vessel than of a larger-radius vessel, resulting in greater resistance (● Figure 10-3a).

Furthermore, a slight change in the radius of a vessel brings about a notable change in flow, because the resistance is inversely proportional to the fourth power of the radius (multiplying the radius by itself four times):

$$R \propto \frac{1}{r^4}$$

Thus doubling the radius reduces the resistance to 1/16 its original value ($r^4 = 2 \times 2 \times 2 \times 2 = 16$; $R \propto 1/16$) and therefore increases flow through the vessel 16-fold (at the same pressure gradient) (● Figure 10-3b). The converse is also true. Only 1/16th as much blood flows through a vessel at the same driving pressure when its radius is halved. Importantly, the radius of arterioles can be regulated and is the most important factor in controlling resistance to blood flow throughout the vascular circuit.

The factors that affect the flow rate through a vessel are integrated in **Poiseuille's law** as follows:

$$\text{Flow rate} = \frac{\pi P r^4}{8 \eta L}$$

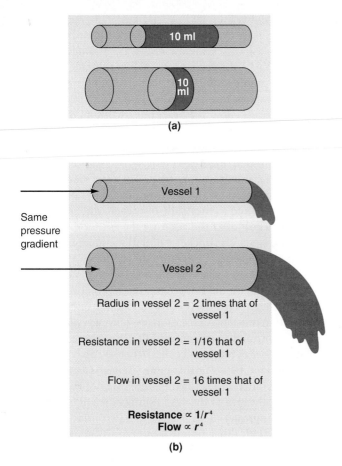

(a)

Vessel 1

Same pressure gradient

Vessel 2

Radius in vessel 2 = 2 times that of vessel 1

Resistance in vessel 2 = 1/16 that of vessel 1

Flow in vessel 2 = 16 times that of vessel 1

Resistance ∝ 1/r^4
Flow ∝ r^4

(b)

● **FIGURE 10-3**

Relationship of resistance and flow to the vessel radius. (a) The same volume of blood comes into contact with a greater surface area of a small-radius vessel compared to a larger-radius vessel. Accordingly, the smaller-radius vessel offers more resistance to blood flow, because the blood "rubs" against a larger surface area. (b) Doubling the radius decreases the resistance to 1/16 and increases the flow 16 times, because the resistance is inversely proportional to the fourth power of the radius.

 PhysioEdge For an interaction related to this figure, see Media Exercise 10.1: Blood Flow and Total Cross-Sectional Area on the CD-ROM.

The significance of the relationships among flow, pressure, and resistance, as largely determined by vessel radius, will become even more apparent as we embark on a voyage through the vessels in the next section.

▎ **The vascular tree consists of arteries, arterioles, capillaries, venules, and veins.**

The systemic and pulmonary circulations each consist of a closed system of vessels (● Figure 10-4). (For the history leading up to the conclusion that blood vessels form a closed system, see the accompanying boxed feature, ▶ Beyond the Basics.) These vascular loops each consist of a continuum of different blood vessel types that begins and ends with the heart, as follows. Looking specifically at the systemic circulation, **arteries**, which carry blood from the heart to the organs, branch into a "tree" of progressively smaller vessels,

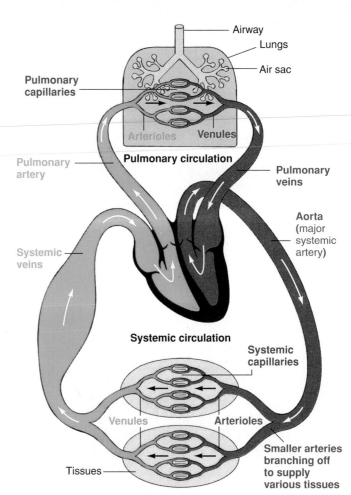

For simplicity, only two capillary beds within two organs are illustrated.

● **FIGURE 10-4**

Basic organization of the cardiovascular system. Arteries progressively branch as they carry blood from the heart to the organs. A separate small arterial branch delivers blood to each of the various organs. As a small artery enters the organ it is supplying, it branches into arterioles, which further branch into an extensive network of capillaries. The capillaries rejoin to form venules, which further unite to form small veins that leave the organ. The small veins progressively merge as they carry blood back to the heart.

with the various branches delivering blood to different regions of the body. When a small artery reaches the organ it is supplying, it branches into numerous **arterioles**. The volume of blood flowing through an organ can be adjusted by regulating the caliber (internal diameter) of the organ's arterioles. Arterioles branch further within the organs into **capillaries**, the smallest of vessels, across which all exchanges are made with surrounding cells. Capillary exchange is the entire purpose of the circulatory system; all other activities of the system are directed toward ensuring an adequate distribution of replenished blood to capillaries for exchange with all cells. Capillaries rejoin to form small **venules**, which further merge to form small **veins** that leave the organs. The small veins progressively unite to form larger veins that eventually empty

Today even grade school children know that blood is pumped by the heart and continually circulates throughout the body in a system of blood vessels. Furthermore, people accept without question that blood picks up O_2 in the lungs from the air we breathe and delivers it to the various organs. This common knowledge was unknown for most of human history, however. Even though the function of blood was described as early as the fifth century B.C., our modern concept of circulation did not develop until 1628, more than 2000 years later, when William Harvey published his now classical study on the circulatory system.

Ancient Greeks believed everything material in the universe consisted of just four elements: earth, air, fire, and water. Extending this view to the human body, they thought these four elements took the form of four "humors": *black bile* (representing earth), *blood* (representing air), *yellow bile* (representing fire), and *phlegm* (representing water). According to the Greeks, disease resulted when one humor was out of normal balance with the rest. The "cure" was logical: Drain off whichever humor was in excess, to restore normal balance. Because the easiest humor to drain off was the blood, bloodletting became standard procedure for treating many illnesses—a practice that persisted well into the Renaissance (which began in the 1300s and extended into the 1600s).

Although the ancient Greeks' notion of the four humors was erroneous, their concept of the necessity of balance within the body was remarkably accurate. As we now know, life depends on homeostasis, maintenance of the proper balance among all elements of the internal environment.

Aristotle (384–322 B.C.), a biologist as well as a philosopher, was among the first to correctly describe the heart at the center of a system of blood vessels. However, he thought the heart was both the seat of intellect (the brain was not identified as the seat of intellect until over a century later) and a furnace that heated the blood. He considered this warmth the vital force of life, because the body cools quickly at death. Aristotle also erroneously theorized that breathing ventilated the "furnace," with air serving as a cooling agent. Aristotle could observe with his eyes the arteries and veins in cadavers but did not have a microscope with which to observe capillaries. (The microscope was not invented until the 17th century.) Thus he did not think arteries and veins were directly connected.

In the third century B.C., Erasistratus, a Greek many considered the first "physiologist," proposed that the liver used food to make blood, which the veins delivered to the other organs. He believed the arteries contained air, not blood. According to his view, *pneuma* ("air"), a living force, was taken in by the lungs, which transferred it to the heart. The heart transformed the air into a "vital spirit" that the arteries carried to the other organs.

Galen (A.D. 130–206), a prolific, outspoken, dogmatic Roman physician, philosopher, and scholar, expanded on the work of Erasistratus and others who had preceded him. Galen further elaborated on the pneumatic theory. He proposed three fundamental members in the body, from lowest to highest: liver, heart, and brain. Each was dominated by a special *pneuma*, or "spirit." (In Greek, *pneuma* encompassed the related ideas of "air," "breath," and "spirit.") Like Erasistratus, Galen believed that the liver made blood from food, taking on a "natural" or "physical" spirit (*pneuma physicon*) in the process. The newly formed blood then proceeded through veins to organs. The natural spirit, which Galen considered a vapor rising from the blood, controlled the functions of nutrition, growth, and reproduction. Once its spirit supply was depleted, the blood moved in the opposite direction through the same venous pathways, returning to the liver to be replenished. When the natural spirit was carried in the venous blood to the heart, it mixed with air that was breathed in and transferred from the lungs to the heart. Contact with air in the heart transformed the natural spirit into a higher-level spirit, the "vital" spirit (*pneuma zotikon*). The vital spirit, which was carried by the arteries, conveyed heat and life throughout the body. The vital spirit was transformed further into a yet higher "animal" or "psychical" spirit (*pneuma psychikon*) in the brain. This ultimate spirit regulated the brain, nerves, feelings, and so on. Thus, according to Galenic theory, the veins and arteries were conduits for carrying different levels of pneuma, and there was no direct connection between the veins and arteries. The heart was not involved in moving blood but instead was the site where blood and air mixed. (We now know that blood and air meet in the lungs for the exchange of O_2 and CO_2.)

Galen was one of the first to understand the need for experimentation, but unfortunately, his impatience and his craving for philosophical and literary fame led him to expound comprehensive theories that were not always based on the time-consuming collection of evidence. Even though his assumptions about bodily structure and functions often were incorrect, his theories were convincing because they seemed a logical way of pulling together what was known at the time. Furthermore, the sheer quantity of his writings helped establish him as an authority. In fact, his writings remained the anatomic and physiologic "truth" for nearly 15 centuries, throughout the Middle Ages and well into the Renaissance. So firmly entrenched was Galenic doctrine that people who challenged its accuracy risked their lives by being declared secular heretics.

Not until the Renaissance and the revival of classical learning did independent-minded European investigators begin to challenge Galen's theories. Most notably, the English physician William Harvey (1578–1657) revolutionized the view of the roles played by the heart, blood vessels, and blood. Through careful observations, experimentation, and deductive reasoning, Harvey was the first to correctly identify the heart as a pump that repeatedly moves a small volume of blood forward in one fixed direction in a circular path through a closed system of blood vessels (the *circulatory system*). He also correctly proposed that blood travels to the lungs to mix with air (instead of air traveling to the heart to mix with blood). Even though he could not see physical connections between arteries and veins, he speculated on their existence. Not until the discovery of the microscope later in the century was the existence of these connections, the capillaries, confirmed, by Marcello Malpighi (1628–1694).

---

into the heart. The arterioles, capillaries, and venules are collectively referred to as the **microcirculation**, because they are only visible through a microscope. The pulmonary circulation consists of the same vessel types, except that all the blood in this loop goes between the heart and lungs. In discussing the vessel types in this chapter, we will refer to their roles in the systemic circulation, starting with systemic arteries.

PhysioEdge

Click on the Media Exercises menu of the CD-ROM and work Media Exercise 10.1: Blood Flow and Total Cross-Sectional Area to test your understanding of the previous section.

ARTERIES

The consecutive segments of the vascular tree are specialized to perform specific tasks (▲ Table 10-1).

▌ Arteries serve as rapid-transit passageways to the organs and as a pressure reservoir.

Arteries are specialized (1) to serve as rapid-transit passageways for blood from the heart to the organs (because of their large radius, arteries offer little resistance to blood flow) and (2) to act as a **pressure reservoir** to provide the driving force for blood when the heart is relaxing.

Let us expand on the role of the arteries as a pressure reservoir. The heart alternately contracts to pump blood into the arteries and then relaxes to refill from the veins. When the heart is relaxing and refilling, no blood is pumped out. However, capillary flow does not fluctuate between cardiac systole and diastole; that is, blood flow is continuous through the capillaries supplying the organs. The driving force for the continued flow of blood to the organs during cardiac relaxation is provided by the elastic properties of the arterial walls.

All vessels are lined with a thin layer of smooth, flat endothelial cells that are continuous with the endothelial lining of the heart. A thick wall made up of smooth muscle and connective tissue surrounds the arteries' endothelial lining. Arterial connective tissue contains an abundance of two types of connective tissue fibers; *collagen fibers,* which provide tensile strength against the high driving pressure of blood ejected from the heart, and *elastin fibers,* which give the arterial walls elasticity so that they behave much like a balloon.

As the heart pumps blood into the arteries during ventricular systole, a greater volume of blood enters the arteries from the heart than leaves them to flow into smaller vessels downstream, because the smaller vessels have a greater resistance to flow. The arteries' elasticity enables them to expand to temporarily hold this excess volume of ejected blood, storing some of the pressure energy imparted by cardiac contraction in their stretched walls—just as a balloon expands to accommodate the extra volume of air you blow into it (● Figure 10-5a). When the heart relaxes and ceases pumping blood into the arteries, the stretched arterial walls passively recoil, like an inflated balloon that is released. This recoil pushes the excess blood contained in the arteries into the vessels downstream, ensuring continued blood flow to the organs when the heart is relaxing and not pumping blood into the system (● Figure 10-5b).

▌ Arterial pressure fluctuates in relation to ventricular systole and diastole.

Blood pressure, the force exerted by the blood against a vessel wall, depends on the volume of blood contained within the vessel and the **compliance**, or **distensibility**, of the vessel walls (how easily they can be stretched). If the volume of blood entering the arteries were equal to the volume of blood leaving the arteries during the same period, arterial blood pressure would remain constant. This is not the case, however. During ventricular systole, a stroke volume of blood enters the arteries from the ventricle while only about one third as much blood leaves the arteries to enter the arterioles. During diastole, no blood enters the arteries, while blood continues to leave, driven by elastic recoil. The maximum pressure exerted in the arteries when blood is ejected into them during systole, the **systolic pressure**, averages 120 mm Hg. The minimum pressure within the arteries when blood is draining off into the rest of the vessels during diastole, the **diastolic pressure**, averages 80 mm Hg. Although ventricular

---

▲ **TABLE 10-1**

Features of Blood Vessels

FEATURE	VESSEL TYPE			
	Arteries	**Arterioles**	**Capillaries**	**Veins**
Number	Several hundred*	Half a million	Ten billion	Several hundred*
Special Features	Thick, highly elastic, walls; large radii*	Highly muscular, well-innervated walls; small radii	Thin walled; large total cross-sectional area	Thin walled; highly distensible; large radii*
Functions	Passageway from heart to organs; serve as pressure reservoir	Primary resistance vessels; determine distribution of cardiac output	Site of exchange; determine distribution of extracellular fluid between plasma and interstitial fluid	Passageway to heart from organs; serve as blood reservoir

*These numbers and special features refer to the large arteries and veins, not to the smaller arterial branches or venules.

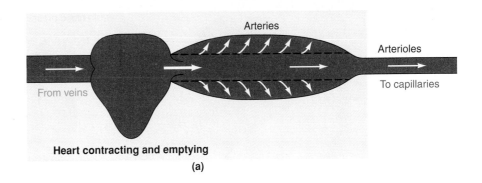

Heart contracting and emptying

(a)

● **FIGURE 10-5**

Arteries as a pressure reservoir. Because of their elasticity, arteries act as a pressure reservoir. (a) The elastic arteries distend during cardiac systole as more blood is ejected into them than drains off into the narrow, high-resistance arterioles downstream. (b) The elastic recoil of arteries during cardiac diastole continues driving the blood forward when the heart is not pumping.

For an animation of this figure, click the Arteries tab in the Cardiovascular Physiology tutorial on the CD-ROM.

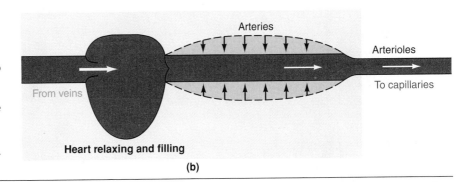

Heart relaxing and filling

(b)

pressure falls to 0 mm Hg during diastole, arterial pressure does not fall to 0 mm Hg, because the next cardiac contraction occurs and refills the arteries before all the blood drains off (● Figure 10-6; also see ● Figure 9-16, p. 256).

▌ Blood pressure can be measured indirectly by using a sphygmomanometer.

Clinical Note The changes in arterial pressure throughout the cardiac cycle can be measured directly by connecting a pressure-measuring device to a needle inserted in an artery. However, it is more convenient and reasonably accurate to measure the pressure indirectly with a **sphygmo-**

● **FIGURE 10-6**

Arterial blood pressure. The systolic pressure is the peak pressure exerted in the arteries when blood is pumped into them during ventricular systole. The diastolic pressure is the lowest pressure exerted in the arteries when blood is draining off into the vessels downstream during ventricular diastole. The pulse pressure is the difference between systolic and diastolic pressure. The mean pressure is the average pressure throughout the cardiac cycle.

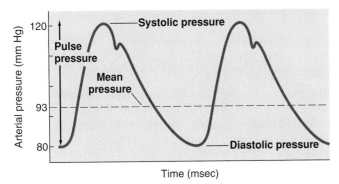

manometer, an externally applied inflatable cuff attached to a pressure gauge. When the cuff is wrapped around the upper arm and then inflated with air, the pressure of the cuff is transmitted through the tissues to the underlying brachial artery, the main vessel carrying blood to the forearm (● Figure 10-7). The technique involves balancing the pressure in the cuff against the pressure in the artery. When cuff pressure is greater than the pressure in the vessel, the vessel is pinched closed so that no blood flows through it. When blood pressure is greater than cuff pressure, the vessel is open and blood flows through.

DETERMINATION OF SYSTOLIC AND DIASTOLIC PRESSURE

During the determination of blood pressure, a stethoscope is placed over the brachial artery at the inside bend of the elbow just below the cuff. No sound can be detected either when blood is not flowing through the vessel or when blood is flowing in the normal, smooth laminar flow (see p. 258). Turbulent blood flow, in contrast, creates vibrations that can be heard. At the onset of a blood pressure determination, the cuff is inflated to a pressure greater than systolic blood pressure so that the brachial artery collapses. Because the externally applied pressure is greater than the peak internal pressure, the artery remains completely pinched closed throughout the entire cardiac cycle; no sound can be heard, because no blood is passing through (point ① in ● Figure 10-7b). As air in the cuff is slowly released, the pressure in the cuff is gradually reduced. When the cuff pressure falls to just below the peak systolic pressure, the artery transiently opens a bit when the blood pressure reaches this peak. Blood escapes through the partially occluded artery for a brief interval before the arterial pressure falls below the cuff pressure and the artery col-

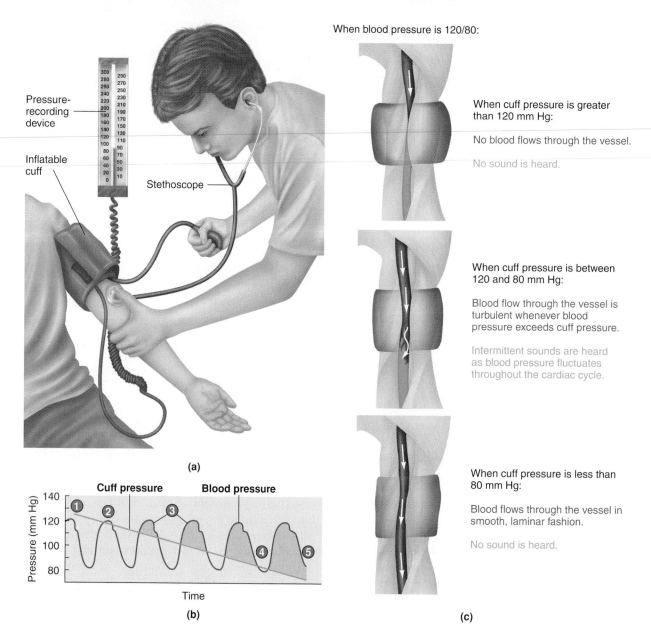

When blood pressure is 120/80:

When cuff pressure is greater than 120 mm Hg:

No blood flows through the vessel.

No sound is heard.

When cuff pressure is between 120 and 80 mm Hg:

Blood flow through the vessel is turbulent whenever blood pressure exceeds cuff pressure.

Intermittent sounds are heard as blood pressure fluctuates throughout the cardiac cycle.

When cuff pressure is less than 80 mm Hg:

Blood flows through the vessel in smooth, laminar fashion.

No sound is heard.

(a)

(b)

(c)

● **FIGURE 10-7**

Sphygmomanometry. (a) Use of a sphygmomanometer in determining blood pressure. The pressure in the inflatable cuff can be varied to prevent or permit blood flow in the underlying brachial artery. Turbulent blood flow can be detected with a stethoscope, whereas smooth laminar flow and no flow are inaudible. (b) Pattern of sounds in relation to cuff pressure compared with blood pressure. The numbers on the illustration refer to the following key points during a blood pressure determination: ① Cuff pressure exceeds blood pressure throughout the cardiac cycle. No sound is heard. ② The first sound is heard at peak systolic pressure. ③ Intermittent sounds are heard as blood pressure cyclically exceeds cuff pressure. ④ The last sound is heard at minimum diastolic pressure. ⑤ Blood pressure exceeds cuff pressure throughout the cardiac cycle. No sound is heard. (c) Blood flow through the brachial artery in relation to cuff pressure and sounds.

lapses once again. This spurt of blood is turbulent, so it can be heard. Thus the highest cuff pressure at which the *first sound* can be heard indicates the *systolic pressure* (point ②). As the cuff pressure continues to fall, blood intermittently spurts through the artery and produces a sound with each subsequent cardiac cycle whenever the arterial pressure exceeds the cuff pressure (point ③).

When the cuff pressure finally falls below diastolic pressure, the brachial artery is no longer pinched closed during any part of the cardiac cycle, and blood can flow uninterrupted through the vessel (point ⑤). With the return of nonturbulent blood flow, no further sounds can be heard. Therefore, the highest cuff pressure at which the *last sound* can be detected indicates the *diastolic pressure* (point ④).

In clinical practice, arterial blood pressure is expressed as systolic pressure over diastolic pressure, with the cutoff for desirable blood pressure being less than 120/80 (120 over 80) mm Hg.

PULSE PRESSURE

The pulse that can be felt in an artery lying close to the surface of the skin is due to the difference between systolic and diastolic pressures. This pressure difference is known as the **pulse pressure**. When blood pressure is 120/80, pulse pressure is 40 mm Hg (120 mm Hg − 80 mm Hg).

❚ Mean arterial pressure is the main driving force for blood flow.

The **mean arterial pressure** is the *average pressure* driving blood forward into the tissues throughout the cardiac cycle. Contrary to what you might expect, mean arterial pressure is not the halfway value between systolic and diastolic pressure (for example, with a blood pressure of 120/80, mean pressure is not 100 mm Hg). The reason is that arterial pressure remains closer to diastolic than to systolic pressure for a longer portion of each cardiac cycle. At resting heart rate, about two thirds of the cardiac cycle is spent in diastole and only one third in systole. As an analogy, if a race car traveled 80 miles per hour (mph) for 40 minutes and 120 mph for 20 minutes, its average speed would be 93 mph, not the halfway value of 100 mph.

Similarly, a good approximation of the mean arterial pressure can be determined using the following formula:

$$\text{Mean arterial pressure} = \text{diastolic pressure} + 1/3 \text{ pulse pressure}$$

$$\text{At } 120/80, \text{ mean arterial pressure} = 80 \text{ mm Hg} + (1/3)40 \text{ mm Hg} = 93 \text{ mm Hg}$$

The mean arterial pressure, not the systolic or diastolic pressure, is monitored and regulated by blood pressure reflexes described later in the chapter.

Because arteries offer little resistance to flow, only a negligible amount of pressure energy is lost in them because of friction. Therefore, arterial pressure—systolic, diastolic, pulse, or mean—is essentially the same throughout the arterial tree (● Figure 10-8).

Blood pressure exists throughout the entire vascular tree, but when discussing a person's "blood pressure" without qualifying which blood vessel type is being referred to, the term is tacitly understood to mean the pressure in the arteries.

ARTERIOLES

When an artery reaches the organ it is supplying, it branches into numerous arterioles within the organ.

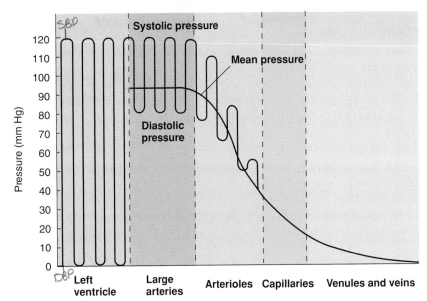

● FIGURE 10-8

Pressures throughout the systemic circulation. Left ventricular pressure swings between a low pressure of 0 mm Hg during diastole to a high pressure of 120 mm Hg during systole. Arterial blood pressure, which fluctuates between a peak systolic pressure of 120 mm Hg and a low diastolic pressure of 80 mm Hg each cardiac cycle, is of the same magnitude throughout the large arteries. Because of the arterioles' high resistance, the pressure drops precipitously and the systolic-to-diastolic swings in pressure are converted to a nonpulsatile pressure when blood flows through the arterioles. The pressure continues to decline but at a slower rate as blood flows through the capillaries and venous system.

❚ Arterioles are the major resistance vessels.

Arterioles are the major resistance vessels in the vascular tree because their radius is small enough to offer considerable resistance to flow. (Even though the capillaries have a smaller radius than the arterioles, you will see later how collectively the capillaries do not offer as much resistance to flow as the arteriolar level of the vascular tree does.) In contrast to the low resistance of the arteries, the high degree of arteriolar resistance causes a marked drop in mean pressure as blood flows through these vessels. On average, the pressure falls from 93 mm Hg, the mean arterial pressure (the pressure of the blood entering the arterioles), to 37 mm Hg, the pressure of the blood leaving the arterioles and entering the capillaries (● Figure 10-8). This decline in pressure helps establish the pressure differential that encourages the flow of blood from the heart to the various organs downstream. If no pressure drop occurred in the arterioles, the pressure at the end of the arterioles (that is, at the beginning of the capillaries) would be equal to the mean arterial pressure. No pressure gradient would exist to drive the blood from the heart to the tissue capillary beds.

Arteriolar resistance also converts the pulsatile systolic-to-diastolic pressure swings in the arteries into the nonfluctuating pressure present in the capillaries.

The radius (and, accordingly, the resistances) of arterioles supplying individual organs can be adjusted independently to accomplish two functions: (1) to variably distribute the cardiac output among the systemic organs, depending on the body's momentary needs, and (2) to help regulate arterial blood pressure. Before considering how such adjustments are important in accomplishing these two functions, we will discuss the mechanisms involved in adjusting arteriolar resistance.

VASOCONSTRICTION AND VASODILATION

Unlike arteries, arteriolar walls contain very little elastic connective tissue. However, they do have a thick layer of smooth muscle that is richly innervated by sympathetic nerve fibers. The smooth muscle is also sensitive to many local chemical changes and to a few circulating hormones. The smooth muscle layer runs circularly around the arteriole (● Figure 10-9a), so when it contracts the vessel's circumference (and its ra-

dius) becomes smaller, increasing resistance and decreasing the flow through that vessel. **Vasoconstriction** is the term applied to such narrowing of a vessel (● Figure 10-9c). The term **vasodilation** refers to enlargement in the circumference and radius of a vessel as a result of its smooth muscle layer relaxing (● Figure 10-9d). Vasodilation leads to decreased resistance and increased flow through that vessel.

VASCULAR TONE

Arteriolar smooth muscle normally displays a state of partial constriction known as **vascular tone**, which establishes a baseline of arteriolar resistance (● Figure 10-9b) (also see p. 232). Two factors are responsible for vascular tone. First, arteriolar smooth muscle has considerable myogenic activity; that is, it shows self-induced contractile activity independent of any neural or hormonal influences (see p. 232). Second, the sympathetic fibers supplying most arterioles continually release norepinephrine, which further enhances muscle tone.

This ongoing tonic activity makes it possible to either increase or decrease the level of contractile activity to accomplish vasoconstriction or vasodilation, respectively. Were it not for tone, it would be impossible to reduce the tension in an arteriolar wall to accomplish vasodilation; only varying degrees of vasoconstriction would be possible.

A variety of factors can influence the level of contractile activity in arteriolar smooth muscle, thereby substantially changing resistance to flow in these vessels. These factors fall into two categories: local (intrinsic) controls, which are important in determining the distribution of cardiac output; and extrinsic controls, which are important in blood pressure regulation. We will look at each of these controls in turn.

▌ **Local control of arteriolar radius is important in determining the distribution of cardiac output.**

The fraction of the total cardiac output delivered to each organ is not always constant; it varies, depending on the demands for blood at the time. The amount of the cardiac output received by each organ is determined by the number and caliber of the arterioles supplying that area. Recall that $F = \Delta P/R$. Because blood is delivered to all organs at the same mean arterial pressure, the driving force for flow is identical for each organ. Therefore, differences in flow to various organs are completely determined by differences in the extent of vascularization and by differences in resistance offered by the arterioles supplying each organ. On a moment-to-moment basis, the distribu-

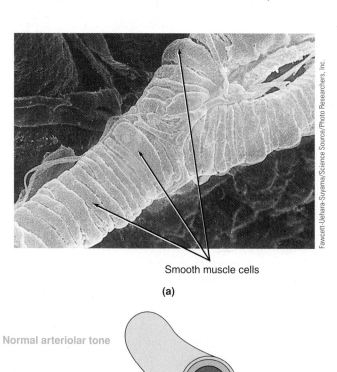

Smooth muscle cells

(a)

Normal arteriolar tone

Cross section
of arteriole

(b)

Vasoconstriction
(increased contraction
of circular smooth
muscle in the arteriolar
wall, which leads to
increased resistance
and decreased
flow through the vessel)

Caused by:
↑ Myogenic activity
↑ Oxygen (O₂)
↓ Carbon dioxide (CO₂)
  and other metabolites
↑ Endothelin
↑ Sympathetic stimulation
  Vasopressin; angiotensin II
  Cold

(c)

Vasodilation
(decreased contraction
of circular smooth
muscle in the arteriolar
wall, which leads to
decreased resistance
and increased flow
through the vessel)

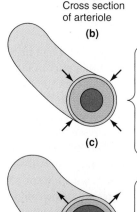

Caused by:
↓ Myogenic activity
↓ O₂
↑ CO₂ and other metabolites
↑ Nitric oxide
↓ Sympathetic stimulation
  Histamine release
  Heat

(d)

● **FIGURE 10-9**

Arteriolar vasoconstriction and vasodilation. (a) A scanning electron micrograph of an arteriole showing how the smooth muscle cells run circularly around the vessel wall. (b) Schematic representation of an arteriole in cross section showing normal arteriolar tone. (c) Outcome of and factors causing arteriolar vasoconstriction. (d) Outcome of and factors causing arteriolar vasodilation.

Fawcett-Uehara-Suyama/Science Source/Photo Researchers, Inc.

tion of cardiac output can be varied by differentially adjusting arteriolar resistance in the various vascular beds.

As an analogy, consider a pipe carrying water, with a number of adjustable valves located throughout its length (● Figure 10-10). Assuming that water pressure in the pipe is constant, differences in the amount of water flowing into a beaker under each valve depend entirely on which valves are open and to what extent. No water enters beakers under closed valves (high resistance), and more water flows into beakers under valves that are opened completely (low resistance) than into beakers under valves that are only partially opened (moderate resistance).

Similarly, more blood flows to areas whose arterioles offer the least resistance to its passage. During exercise, for example, not only is cardiac output increased, but also, because of vasodilation in skeletal muscle and in the heart, a greater percentage of the pumped blood is diverted to these organs to support their increased metabolic activity. Simultaneously, blood flow to the digestive tract and kidneys is reduced, as a result of arteriolar vasoconstriction in these organs (● Figure 10-11). Only the blood supply to the brain remains remarkably constant no matter what the person is doing, be it vigorous physical activity, intense mental concentration, or sleep. Although the total blood flow to the brain remains constant, new imaging techniques demonstrate that regional blood flow varies within the brain in close correlation with local neural activity patterns (see p. 118).

Local (intrinsic) controls are changes within an organ that alter the radius of the vessels and hence adjust blood flow through the organ by directly affecting the smooth muscle of the organ's arterioles. Local influences may be either chemical or physical. Local chemical influences on arteriolar radius include (1) local metabolic changes and (2) histamine release. Local physical influences include (1) local application of heat or cold and (2) myogenic response to stretch. Let's examine the role and mechanism of each of these local influences.

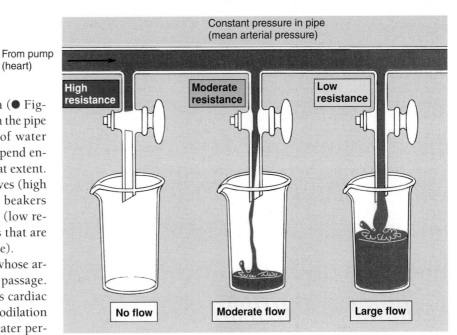

From pump (heart)

Constant pressure in pipe (mean arterial pressure)

High resistance — No flow

Moderate resistance — Moderate flow

Low resistance — Large flow

Control valves = Arterioles

● **FIGURE 10-10**

Flow rate as a function of resistance

Local metabolic influences on arteriolar radius help match blood flow with the organs' needs.

The most important local chemical influences on arteriolar smooth muscle are related to metabolic changes within a given organ. The influence of these local changes on arteriolar radius is important in matching blood flow through an organ with the organ's metabolic needs. Local metabolic controls are especially important in skeletal muscle and in the heart, the organs whose metabolic activity and need for blood supply normally vary most.

ACTIVE HYPEREMIA

Arterioles lie within the organ they are supplying and can be acted on by local factors within the organ. During increased metabolic activity, such as when a skeletal muscle is contracting during exercise, local concentrations of a number of the organ's chemicals change. For example, the local O_2 concentration decreases as the actively metabolizing cells use up more O_2 to support oxidative phosphorylation for ATP production (see p. 29). This and other local chemical changes (such as increased CO_2 and other metabolites) produce local arteriolar dilation by triggering relaxation of the arteriolar smooth muscle in the vicinity. Local arteriolar vasodilation then increases blood flow to that particular area, a response called **active hyperemia** (*hyper* means "above normal"; *emia* means "blood"). When cells are more active metabolically, they need more blood to bring in O_2 and nutrients and to remove metabolic wastes. The increased blood flow meets these increased local needs.

Conversely, when an organ, such as a relaxed muscle, is less active metabolically and thus has reduced needs for blood delivery, the resultant local chemical changes (for example, increased local O_2 concentration) bring about local arteriolar vasoconstriction and a subsequent reduction in blood flow to the area. Local metabolic changes can thus adjust blood flow as needed without involving nerves or hormones.

LOCAL VASOACTIVE MEDIATORS

The local chemical changes that bring about these "selfish" local adjustments in arteriolar caliber to match an organ's blood flow with its needs do not act directly on vascular smooth muscle to change its contractile state. Instead, **endothelial cells**, the single layer of specialized epithelial cells that line the lumen of all blood vessels, release chemical mediators that play a key role in locally regulating arteriolar caliber. Until recently, scientists regarded endothelial cells as little more than

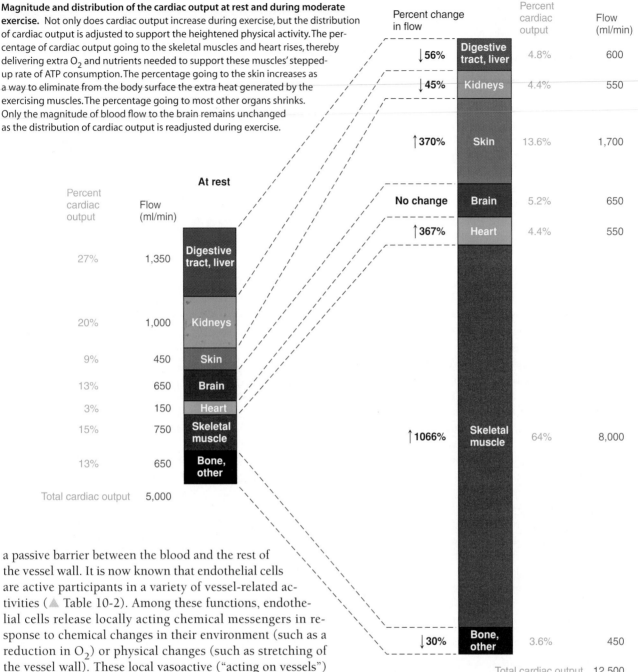

● **FIGURE 10-11**

Magnitude and distribution of the cardiac output at rest and during moderate exercise. Not only does cardiac output increase during exercise, but the distribution of cardiac output is adjusted to support the heightened physical activity. The percentage of cardiac output going to the skeletal muscles and heart rises, thereby delivering extra O_2 and nutrients needed to support these muscles' stepped-up rate of ATP consumption. The percentage going to the skin increases as a way to eliminate from the body surface the extra heat generated by the exercising muscles. The percentage going to most other organs shrinks. Only the magnitude of blood flow to the brain remains unchanged as the distribution of cardiac output is readjusted during exercise.

a passive barrier between the blood and the rest of the vessel wall. It is now known that endothelial cells are active participants in a variety of vessel-related activities (▲ Table 10-2). Among these functions, endothelial cells release locally acting chemical messengers in response to chemical changes in their environment (such as a reduction in O_2) or physical changes (such as stretching of the vessel wall). These local vasoactive ("acting on vessels") mediators act on the underlying smooth muscle to alter its state of contraction.

Among the best studied of these local vasoactive mediators is **nitric oxide (NO)**, which causes local arteriolar vasodilation by inducing relaxation of arteriolar smooth muscle in the vicinity. It does so by inhibiting the entry of contraction-inducing Ca^{2+} into these smooth muscle cells (see p. 230). NO is a small, highly reactive, short-lived gas molecule that once was known primarily as a toxic air pollutant. Yet studies have revealed an astonishing number of biological roles for NO, which is produced in many other tissues besides endothelial cells. In fact, it appears that NO serves as one of the body's most important messenger molecules, as

shown by the range of functions identified for this chemical and listed in ▲ Table 10-3. As you can see, most areas of the body are influenced by this versatile intercellular messenger molecule.

Endothelial cells release other important chemicals besides NO. **Endothelin**, another endothelial vasoactive substance, causes arteriolar smooth-muscle contraction and is one of the most potent vasoconstrictors yet identified. Still other chemicals, released from the endothelium in response to chronic changes in blood flow to an organ, trigger long-

TABLE 10-2

Functions of Endothelial Cells

- Line blood vessels and heart chambers; serve as physical barrier between blood and rest of the vessel wall.
- Secrete vasoactive substances in response to local chemical and physical changes; these substances cause relaxation (vasodilation) or contraction (vasoconstriction) of underlying smooth muscle.
- Secrete substances that stimulate new vessel growth and proliferation of smooth muscle cells in vessel walls.
- Participate in exchange of materials between blood and surrounding tissue cells across capillaries through vesicular transport (see p. 59).
- Influence formation of platelet plugs, clotting, and clot dissolution (see Chapter 11).
- Help determine capillary permeability by contracting to vary size of pores between adjacent endothelial cells.

TABLE 10-3

Functions of Nitric Oxide (NO)

- Causes relaxation of arteriolar smooth muscle. By means of this action, NO plays an important role in controlling blood flow through the tissues and in maintaining mean arterial blood pressure.
- Dilates the arterioles of the penis and clitoris, thus serving as the direct mediator of erection of these reproductive organs. Erection is accomplished by rapid engorgement of these organs with blood.
- Used as chemical warfare against bacteria and cancer cells by macrophages, large phagocytic cells of the immune system.
- Interferes with platelet function and blood clotting at sites of vessel damage.
- Serves as a novel type of neurotransmitter in the brain and elsewhere.
- Plays a role in the changes underlying memory.
- By promoting relaxation of digestive-tract smooth muscle, helps regulate peristalsis, a type of contraction that pushes digestive tract contents forward.
- Relaxes the smooth muscle cells in the airways of the lungs, helping keep these passages open to facilitate movement of air in and out of the lungs.
- May play a role in relaxation of skeletal muscle.

term vascular changes that permanently influence blood flow to a region. Some chemicals, for example, stimulate new vessel growth, a process known as **angiogenesis**.

Local histamine release pathologically dilates arterioles.

Histamine is another local chemical mediator that influences arteriolar smooth muscle, but it is not released in response to local metabolic changes and is not derived from endothelial cells. Although histamine normally does not participate in controlling blood flow, it is important in certain pathological conditions.

Clinical Note Histamine is synthesized and stored within special connective tissue cells in many organs and in certain types of circulating white blood cells. When organs are injured or during allergic reactions, histamine is released and acts as a paracrine in the damaged region (see p. 93). By promoting relaxation of arteriolar smooth muscle, histamine is the major cause of vasodilation in an injured area. The resultant increase in blood flow into the area produces the redness and contributes to the swelling seen with inflammatory responses (see Chapter 11 for further details).

Local physical influences on arteriolar radius include temperature changes and stretch.

Among the physical influences on arteriolar smooth muscle, the effect of temperature changes is exploited clinically but the myogenic response to stretch is most important physiologically. Let's examine each of these effects.

LOCAL HEAT OR COLD APPLICATION

Clinical Note Heat application, by causing localized arteriolar vasodilation, is a useful therapeutic agent for promoting increased blood flow to an area. Conversely, applying ice packs to an inflamed area produces vasoconstriction, which reduces swelling by counteracting histamine-induced vasodilation.

MYOGENIC RESPONSES TO STRETCH

Arteriolar smooth muscle responds to being passively stretched by myogenically increasing its tone, thereby acting to resist the initial passive stretch. Conversely, a decrease in arteriolar stretching induces a reduction in myogenic vessel tone. Endothelial-derived vasoactive substances may also contribute to these mechanically induced responses. The extent of passive stretch varies with the volume of blood delivered to the arterioles from the arteries. For example, when mean arterial pressure falls (for example, because of hemorrhage or a weakened heart), the driving force is reduced, so blood flow to organs decreases. The resultant changes in local metabolites and the reduced stretch in the arterioles collectively bring about arteriolar dilation to help restore tissue blood flow to normal despite the reduced driving pressure.

This completes our discussion of the local control of arteriolar radius. Now let's shift attention to extrinsic control of arteriolar radius.

▌ Extrinsic sympathetic control of arteriolar radius is important in regulating blood pressure.

Extrinsic control of arteriolar radius includes both neural and hormonal influences, the effects of the sympathetic nervous system being the most important. Sympathetic nerve fibers supply arteriolar smooth muscle everywhere in the systemic circulation except in the brain. Recall that a certain level of ongoing sympathetic activity contributes to vascular tone. Increased sympathetic activity produces generalized arteriolar vasoconstriction, whereas decreased sympathetic activity leads to generalized arteriolar vasodilation. These widespread changes in arteriolar resistance bring about changes in mean arterial pressure because of their influence on total peripheral resistance, as follows.

INFLUENCE OF TOTAL PERIPHERAL RESISTANCE ON MEAN ARTERIAL PRESSURE

To find the effect of changes in arteriolar resistance on mean arterial pressure, the formula $F = \Delta P/R$ applies to the entire circulation as well as to a single vessel:

- F: Looking at the circulatory system as a whole, flow (F) through all the vessels in either the systemic or pulmonary circulation is equal to the cardiac output.
- ΔP: The pressure gradient (ΔP) for the entire systemic circulation is the mean arterial pressure. (ΔP equals the difference in pressure between the beginning and the end of the systemic circulatory system. The beginning pressure is the mean arterial pressure as the blood leaves the left ventricle at an average of 93 mm Hg. The end pressure in the right atrium is 0 mm Hg. Therefore, ΔP = 93 mm Hg minus 0 mm Hg = 93 mm Hg, which is equivalent to the mean arterial pressure.)
- R: The total resistance (R) offered by all the systemic peripheral vessels together is the **total peripheral resistance**. By far the greatest percentage of the total peripheral resistance is due to arteriolar resistance, because arterioles are the primary resistance vessels.

Therefore, for the entire systemic circulation, rearranging

$$F = \Delta P/R$$

to

$$\Delta P = F \times R$$

gives us the equation

Mean arterial pressure =
cardiac output × total peripheral resistance

Thus the extent of total peripheral resistance offered collectively by all the systemic arterioles influences the mean arterial pressure immensely. A dam provides an analogy to this relationship. At the same time a dam restricts the flow of water downstream, it increases the pressure upstream by elevating

the water level in the reservoir behind the dam. Similarly, generalized, sympathetically induced vasoconstriction reflexly reduces blood flow downstream to the organs while elevating the upstream mean arterial pressure, thereby increasing the main driving force for blood flow to all the organs.

These effects seem counterproductive. Why increase the driving force for flow to the organs by increasing arterial blood pressure while reducing flow to the organs by narrowing the vessels supplying them? In effect, the sympathetically induced arteriolar responses help maintain the appropriate driving pressure head (that is, the mean arterial pressure) to all organs. The extent to which each organ actually receives blood flow is determined by local arteriolar adjustments that override the sympathetic constrictor effect. If all arterioles were dilated, blood pressure would fall substantially, so there would not be an adequate driving force for blood flow. An analogy is the pressure head for water in the pipes in your home. If the water pressure is adequate, you can selectively obtain satisfactory water flow at any of the faucets by turning the appropriate handle to the open position. If the water pressure in the pipes is too low, however, you cannot obtain satisfactory flow at any faucet, even if you turn the handle to the maximally open position. Tonic sympathetic activity thus constricts most vessels (with the exception of those in the brain) to help maintain a pressure head on which organs can draw as needed through local mechanisms that control arteriolar radius.

No vasoconstriction occurs in the brain. It is important that cerebral arterioles are not reflexly constricted by neural influences, because brain blood flow must remain constant to meet the brain's continual need for O_2, no matter what is going on elsewhere in the body.

Thus sympathetic activity contributes in an important way to maintaining mean arterial pressure, assuring an adequate driving force for blood flow to the brain at the expense of organs that can better withstand reduced blood flow. Other organs that really need additional blood, such as active muscles, obtain it through local controls that override the sympathetic effect.

LOCAL CONTROLS OVERRIDING SYMPATHETIC VASOCONSTRICTION

Skeletal muscles have the most powerful local control mechanisms with which to override generalized sympathetic vasoconstriction. For example, if you are pedaling a bicycle the increased activity in the skeletal muscles of your legs brings about an overriding local, metabolically induced vasodilation in those particular muscles, despite the generalized sympathetic vasoconstriction that accompanies exercise. As a result, more blood flows through your leg muscles but not through your inactive arm muscles.

NO PARASYMPATHETIC INNERVATION TO ARTERIOLES

There is no significant parasympathetic innervation to arterioles, with the exception of the abundant parasympathetic vasodilator supply to the arterioles of the penis and clitoris. The rapid, profuse vasodilation induced by parasympathetic stimulation in these organs (by means of promoting release of NO) is largely responsible for accomplishing erection. Vasodilation elsewhere is produced by decreasing sympathetic vasoconstrictor activity below its tonic level. When mean arterial pressure rises above normal, reflex reduction in sympathetic vasoconstrictor activity accomplishes generalized arteriolar vasodilation to help restore the driving pressure down toward normal.

■ The medullary cardiovascular control center and several hormones regulate blood pressure.

The main region of the brain that adjusts sympathetic output to the arterioles is the **cardiovascular control center** in the medulla of the brain stem. This is the integrating center for blood pressure regulation (described in further detail later in this chapter). Several other brain regions also influence blood distribution, the most notable being the hypothalamus, which, as part of its temperature-regulating function, controls blood flow to the skin to adjust heat loss to the environment.

In addition to neural reflex activity, several hormones also extrinsically influence arteriolar radius. These hormones include the major adrenal medullary hormone epinephrine, which generally reinforces the sympathetic nervous system in most organs, as well as vasopressin and angiotensin II, which are important in controlling fluid balance. Vasopressin is primarily involved in maintaining water balance by regulating the amount of water the kidneys retain for the body during urine formation. Angiotensin II is part of a hormonal pathway, the *renin-angiotensin-aldosterone pathway,* which is important in regulating the body's salt balance. This pathway promotes salt conservation during urine formation and also leads to water retention, because salt exerts a water-holding osmotic effect in the ECF. Thus both these hormones play important roles in maintaining the body's fluid balance, which in turn is an important determinant of plasma volume and blood pressure.

In addition, both vasopressin and angiotensin II are potent vasoconstrictors. Their role in this regard is especially crucial during hemorrhage. A sudden loss of blood reduces the plasma volume, which triggers increased secretion of both these hormones to help restore plasma volume. Their vasoconstrictor effect also helps maintain blood pressure despite abrupt loss of plasma volume. (The functions and control of these hormones are discussed more thoroughly in later chapters.)

This completes our discussion of the various factors that affect total peripheral resistance, the most important of which are controlled adjustments in arteriolar radius. These factors are summarized in ● Figure 10-12.

We are now going to turn our attention to the next vessels in the vascular tree, the capillaries.

CAPILLARIES

Capillaries, the sites for exchange of materials between blood and tissue cells, branch extensively to bring blood within the reach of every cell.

■ Capillaries are ideally suited to serve as sites of exchange.

There are no carrier-mediated transport systems across capillaries, with the exception of those in the brain that play a role in the blood–brain barrier (see p. 113). Materials are exchanged across capillary walls mainly by diffusion.

FACTORS THAT ENHANCE DIFFUSION ACROSS CAPILLARIES

Capillaries are ideally suited to enhance diffusion, in accordance with Fick's law of diffusion (see p. 52). They minimize diffusion distances while maximizing surface area and time available for exchange, as follows:

1. Diffusing molecules have only a short distance to travel between blood and surrounding cells because of the thin capillary wall and small capillary diameter, coupled with the close proximity of every cell to a capillary. This short distance is important because the rate of diffusion slows down as the diffusion distance increases.
    a. Capillary walls are very thin (1 μm in thickness; in contrast, the diameter of a human hair is 100 μm). Capillaries consist of only a single layer of flat endothelial cells—essentially the lining of the other vessel types. No smooth muscle or connective tissue is present.
    b. Each capillary is so narrow (7 μm average diameter) that red blood cells (8 μm diameter) have to squeeze through single file (● Figure 10-13). Consequently, plasma contents are either in direct contact with the inside of the capillary wall or are only a short diffusing distance from it.
    c. Researchers estimate that because of extensive capillary branching, no cell is farther than 0.01 cm (4/1000 inch) from a capillary.
2. Because capillaries are distributed in such incredible numbers (estimates range from 10 to 40 billion capillaries), a tremendous total surface area is available for exchange (an estimated 600 m^2). Despite this large number of capillaries, at any point in time they contain only 5% of the total blood volume (250 ml out of a total of 5000 ml). As a result, a small volume of blood is exposed to an extensive surface area. If all the capillary surfaces were stretched out in a flat sheet and the volume of blood contained within the capillaries were spread over the top, this would be roughly equivalent to spreading a half pint of paint over the floor of a high school gymnasium. Imagine how thin the paint layer would be!
3. Blood flows more slowly in the capillaries than elsewhere in the circulatory system. The extensive capillary branching is responsible for this slow velocity of blood flow through the capillaries. Let's see why blood slows down in the capillaries.

SLOW VELOCITY OF FLOW THROUGH CAPILLARIES

First, we need to clarify a potentially confusing point. The term *flow* can be used in two different contexts—flow rate and velocity of flow. The *flow rate* refers to the *volume* of blood per unit of time flowing through a given segment of the circulatory system (this is the flow we have been talking about in relation to the pressure gradient and resistance). The *velocity of flow* is the linear *speed*, or distance per unit of time, with which blood flows forward through a given segment of the circulatory system. Because the circulatory system is a closed system, the volume of blood flowing through any level of the system must equal the cardiac output. For example, if the heart pumps out 5 liters of blood per minute, and 5 liters/min return to the heart, then 5 liters/min must flow through the arteries, arterioles, capillaries, and veins. Therefore, the flow rate is the same at all levels of the circulatory system.

However, the velocity with which blood flows through the different segments of the vascular tree varies, because velocity of flow is inversely proportional to the total cross-section area of all the vessels at any given level of the circulatory system. Even though the cross-section area of each capillary is extremely small compared to that of the large aorta, the total cross-section area of all the capillaries added together is about 1300 times greater than the cross-section area of the aorta because there are so many capillaries. Accordingly, blood slows considerably as it passes through the capillaries (● Figure 10-14). This slow velocity allows adequate time for ex-

● **FIGURE 10-12**

Factors affecting total peripheral resistance. The primary determinant of total peripheral resistance is the adjustable arteriolar radius. Two major categories of factors influence arteriolar radius: (1) local (intrinsic) control, which is primarily important in matching blood flow through an organ with the organ's metabolic needs and is mediated by local factors acting on the arteriolar smooth muscle in the vicinity, and (2) extrinsic control, which is important in regulating blood pressure and is mediated primarily by sympathetic influence on arteriolar smooth muscle.

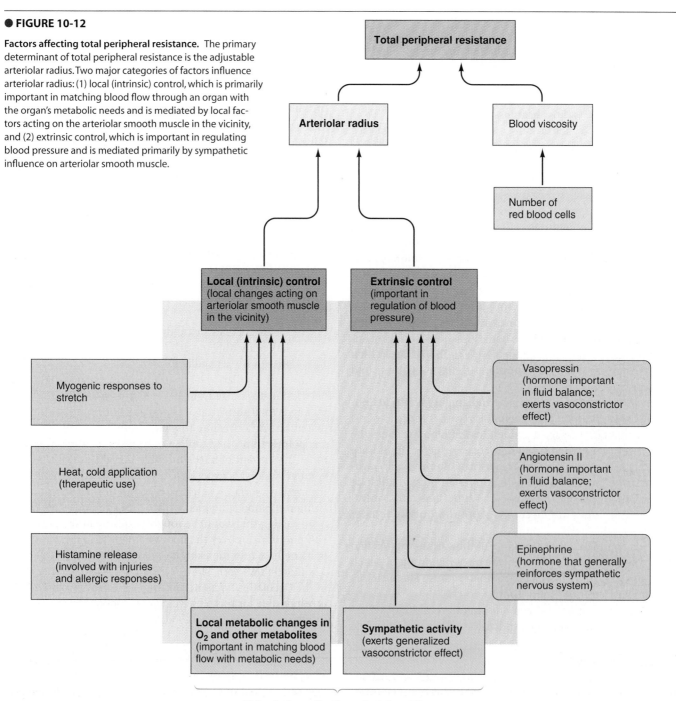

Major factors affecting arteriolar radius

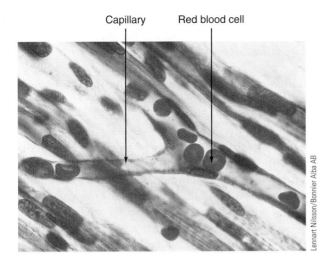

Capillary   Red blood cell

Lennart Nilsson/Bonnier Alba AB

● **FIGURE 10-13**

Capillary bed. Photograph of a capillary bed. The capillaries are so narrow that red blood cells must pass through single file.

change of nutrients and metabolic end products between blood and tissue cells, which is the sole purpose of the entire circulatory system. As the capillaries rejoin to form veins, the total cross-section area is once again reduced, and the velocity of blood flow increases as blood returns to the heart.

As an analogy, consider a river (the arterial system) that widens into a lake (the capillaries), then narrows into a river again (the venous system) (● Figure 10-15). The flow rate is the same throughout the length of this body of water; that is, identical volumes of water are flowing past all the points along the bank of the river and lake. However, the velocity of flow is slower in the wide lake than in the narrow river because the identical volume of water, now spread out over a larger cross-section area, moves forward a much shorter distance in the wide lake than in the narrow river during a given period of time. You could readily observe the forward movement of water in the swift-flowing river, but the forward motion of water in the lake would be unnoticeable.

Also, because of the capillaries' tremendous total cross-section area, the resistance offered by all the capillaries is much lower than that offered by all the arterioles, even though each capillary has a smaller radius than each arteriole. For this reason, the arterioles contribute more to total peripheral resistance. Furthermore, arteriolar caliber (and, accordingly resistance) is subject to control, whereas capillary caliber cannot be adjusted.

▌ Water-filled capillary pores permit passage of small, water-soluble substances.

Diffusion across capillary walls also depends on the walls' permeability to the materials being exchanged. The endothelial cells forming the capillary walls fit together in jigsaw-puzzle fashion, but the closeness of the fit varies considerably between organs. In most capillaries, narrow, water-filled gaps, or **pores**, lie at the junctions between the cells (● Fig-

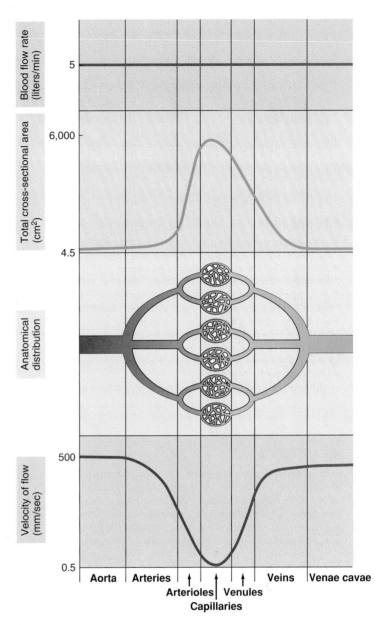

● **FIGURE 10-14**

Comparison of blood flow rate and velocity of flow in relation to total cross-section area. The blood flow rate (*red curve*) is identical through all levels of the circulatory system and is equal to the cardiac output (5 liters/min at rest). The velocity of flow (*purple curve*) varies throughout the vascular tree and is inversely proportional to the total cross-section area (*green curve*) of all the vessels at a given level. Note that the velocity of flow is slowest in the capillaries, which have the largest total cross-section area.

For an interaction related to this figure, see Media Exercise 10.2: Arteriolar Resistance and Flow in Capillaries on the CD-ROM.

ure 10-16). These pores permit passage of water-soluble substances. Lipid-soluble substances, such as O_2 and CO_2, can readily pass through the endothelial cells themselves by dissolving in the lipid bilayer barrier. Large, non–lipid-soluble materials that cannot fit through the pores, such as plasma proteins, are kept from passing.

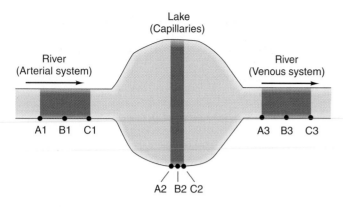

● FIGURE 10-15

Relationship between total cross-section area and velocity of flow.
The three dark blue areas represent equal volumes of water. During one minute, this volume of water moves forward from points A to points C. Therefore, an identical volume of water flows past points B1, B2, and B3 during this minute; that is, the flow rate is the same at all points along the length of this body of water. However, during that minute the identical volume of water moves forward a much shorter distance in the wide lake (A2 to C2) than in the much narrower river (A1 to C1 and A3 to C3). Thus velocity of flow is much slower in the lake than in the river. Similarly, velocity of flow is much slower in the capillaries than in the arterial and venous systems.

Scientists traditionally considered the capillary wall a passive sieve, like a brick wall with permanent gaps in the mortar acting as pores. Recent studies, however, suggest that endothelial cells can actively change to regulate capillary permeability; that is, in response to appropriate signals, the "bricks" can readjust themselves to vary the size of the holes. Thus the degree of leakiness does not necessarily remain constant for a given capillary bed. For example, histamine increases capillary permeability by triggering contractile responses in endothelial cells to widen the intercellular gaps. This is not a muscular contraction, because no smooth muscle cells are present in capillaries. It is due to an actin-myosin contractile apparatus in the nonmuscular capillary endothelial cells. Because of these enlarged pores, the affected capillary wall is leakier. As a result, normally retained plasma proteins escape into the surrounding tissue, where they exert an osmotic effect. Along with histamine-induced vasodilation, the resulting additional local fluid retention contributes to inflammatory swelling.

Vesicular transport also plays a limited role in the passage of materials across the capillary wall. Large non–lipid-soluble molecules such as protein hormones that must be exchanged between blood and surrounding tissues are transported from one side of the capillary wall to the other in endocytotic-exocytotic vesicles (see p. 60).

● FIGURE 10-16

Exchanges across the capillary wall. (a) Slitlike gaps between adjacent endothelial cells form pores within the capillary wall. (b) As depicted in this schematic representation of a cross section of a capillary wall, small water-soluble substances are exchanged between the plasma and the interstitial fluid by passing through the water-filled pores, whereas lipid-soluble substances are exchanged across the capillary wall by passing through the endothelial cells. Proteins to be moved across are exchanged by vesicular transport. Plasma proteins generally cannot escape from the plasma across the capillary wall.

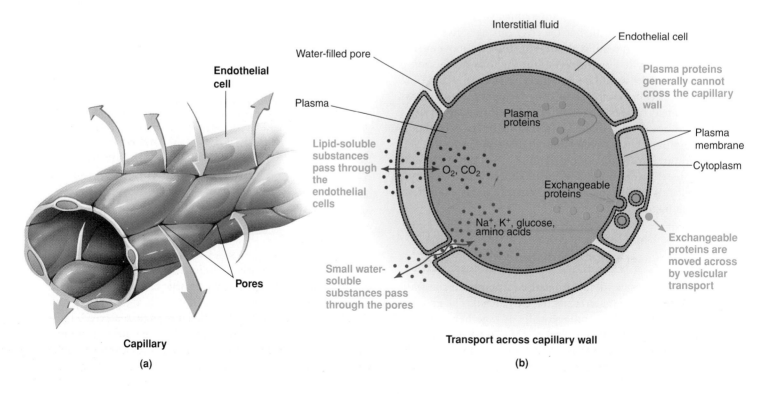

Interstitial fluid is a passive intermediary between blood and cells.

Exchanges between blood and tissue cells are not made directly. Interstitial fluid, the true internal environment in immediate contact with the cells, acts as the go-between. Only 20% of the ECF circulates as plasma. The remaining 80% consists of interstitial fluid, which bathes all the cells in the body. Cells exchange materials directly with interstitial fluid, with the type and extent of exchange being governed by the properties of cellular plasma membranes. Movement across the plasma membrane may be either passive (that is, by diffusion down electrochemical gradients or by facilitated diffusion) or active (that is, by active carrier-mediated transport or by vesicular transport) (see ▲ Table 3-2, p. 60).

In contrast, exchanges across the capillary wall between plasma and interstitial fluid are largely passive. The only transport across this barrier that requires energy is the limited vesicular transport. Because capillary walls are highly permeable, exchange is so thorough that the interstitial fluid takes on essentially the same composition as incoming arterial blood, with the exception of the large plasma proteins that usually do not escape from the blood. Therefore, when we speak of exchanges between blood and tissue cells, we tacitly include interstitial fluid as a passive intermediary.

Exchanges between blood and surrounding tissues across the capillary walls are made in two ways: (1) passive diffusion down concentration gradients, the primary mechanism for exchanging individual solutes; and (2) bulk flow, a process that fills the totally different function of determining the distribution of the ECF volume between the vascular and interstitial fluid compartments. Now let's examine each of these mechanisms in more detail, starting with diffusion.

Diffusion across the capillary walls is important in solute exchange.

Because there are no carrier-mediated transport systems in most capillary walls, solutes cross primarily by diffusion down concentration gradients. The chemical composition of arterial blood is carefully regulated to maintain the concentrations of individual solutes at levels that will promote each solute's movement in the appropriate direction across the capillary walls. The reconditioning organs continuously add nutrients and O_2 and remove CO_2 and other wastes as blood passes through them. Meanwhile, cells constantly use up supplies and generate metabolic wastes. As cells use up O_2 and glucose, the blood constantly brings in fresh supplies of these vital materials, maintaining concentration gradients that favor the net diffusion of these substances from blood to cells. Simultaneously, ongoing net diffusion of CO_2 and other metabolic wastes from cells to blood is maintained by the continual production of these wastes at the cell level and by their constant removal by the circulating blood (● Figure 10-17).

Because the capillary wall does not limit the passage of any constituent except plasma proteins, the extent of ex-

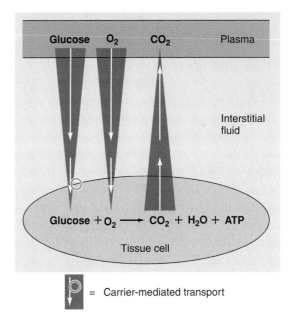

= Carrier-mediated transport

● **FIGURE 10-17**

Independent exchange of individual solutes down their own concentration gradients across the capillary wall

changes for each solute is independently determined by the magnitude of its concentration gradient between blood and surrounding cells. As cells increase their level of activity, they use up more O_2 and produce more CO_2, among other things. This creates larger concentration gradients for O_2 and CO_2 between cells and blood, so more O_2 diffuses out of the blood into the cells and more CO_2 proceeds in the opposite direction to help support the increased metabolic activity.

Bulk flow across the capillary walls is important in extracellular fluid distribution.

The second means by which exchange is accomplished across capillary walls is bulk flow. A volume of protein-free plasma actually filters out of the capillary, mixes with the surrounding interstitial fluid, and then is reabsorbed. This process is called **bulk flow,** because the various constituents of the fluid are moving together in bulk, or as a unit, in contrast to the discrete diffusion of individual solutes down concentration gradients.

The capillary wall acts like a sieve, with fluid moving through its water-filled pores. When pressure inside the capillary exceeds pressure on the outside, fluid is pushed out through the pores in a process known as **ultrafiltration.** Most plasma proteins are retained on the inside during this process because of the pores' filtering effect, although a few do escape. Because all other constituents in plasma are dragged along as a unit with the volume of fluid leaving the capillary, the filtrate is essentially a protein-free plasma. When inward-driving pressures exceed outward pressures across the capillary wall, net inward movement of fluid from the interstitial fluid compartment into the capillaries takes place through the pores, a process known as **reabsorption.**

FORCES INFLUENCING BULK FLOW

Bulk flow occurs because of differences in the hydrostatic and colloid osmotic pressures between plasma and interstitial fluid. Even though pressure differences exist between plasma and surrounding fluid elsewhere in the circulatory system, only the capillaries have pores that let fluids pass through. Four forces influence fluid movement across the capillary wall (● Figure 10-18):

1. **Capillary blood pressure (P_C)** is the fluid or hydrostatic pressure exerted on the inside of the capillary walls by blood. This pressure tends to force fluid *out of* the capillaries into the interstitial fluid. By the level of the capillaries, blood pressure has dropped substantially because of frictional losses in pressure in the high-resistance arterioles upstream. On average, the hydrostatic pressure is 37 mm Hg at the arteriolar end of a tissue capillary (compared to a mean arterial pressure of 93 mm Hg). It declines even further, to 17 mm Hg, at the capillary's venular end because of further frictional loss coupled with the exit of fluid through ultrafiltration along the capillary's length (see ● Figure 10-8, p. 283).

2. **Plasma-colloid osmotic pressure (π_P)**, also known as *oncotic pressure,* is a force caused by colloidal dispersion of plasma proteins (see p. A-10); it encourages fluid movement *into* the capillaries. Because plasma proteins remain in the plasma rather than entering the interstitial fluid, a protein concentration difference exists between plasma and interstitial fluid. Accordingly, there is also a water concentration difference between these two regions. Plasma has a higher protein concentration and a lower water concentration than interstitial fluid does. This difference exerts an osmotic effect that tends to move water from the area of higher water concentra-

tion in interstitial fluid to the area of lower water concentration (or higher protein concentration) in plasma (see p. 53). The other plasma constituents do not exert an osmotic effect, because they readily pass through the capillary wall, so their concentrations are equal in plasma and interstitial fluid. Plasma-colloid osmotic pressure averages 25 mm Hg.

3. **Interstitial fluid hydrostatic pressure (P_{IF})** is the fluid pressure exerted on the outside of the capillary wall by interstitial fluid. This pressure, which tends to force fluid *into* the capillaries, is low, averaging about 1 mm Hg.

4. **Interstitial fluid–colloid osmotic pressure (π_{IF})** is another force that does not normally contribute significantly to bulk flow. The small fraction of plasma proteins that leak across the capillary walls into the interstitial spaces are normally returned to the blood by means of the lymphatic system. Therefore, the protein concentration in the interstitial fluid is extremely low, and the interstitial fluid–colloid osmotic pressure is very close to zero. If plasma proteins pathologically leak into the interstitial fluid, however, as they do when histamine widens the capillary pores during tissue injury, the leaked proteins exert an osmotic effect that tends to promote movement of fluid *out of* the capillaries into the interstitial fluid.

Therefore, the two pressures that tend to force fluid out of the capillary are capillary blood pressure and interstitial fluid–colloid osmotic pressure. The two opposing pressures that tend to force fluid into the capillary are plasma-colloid osmotic pressure and interstitial fluid hydrostatic pressure. Now let's analyze the fluid movement that occurs across a capillary wall because of imbalances in these opposing physical forces (● Figure 10-18).

NET EXCHANGE OF FLUID ACROSS THE CAPILLARY WALL

Net exchange at a given point across the capillary wall can be calculated using the following equation:

● **FIGURE 10-18**

Bulk flow across the capillary wall. Schematic representation of ultrafiltration and reabsorption as a result of imbalances in the forces acting across the capillary wall.

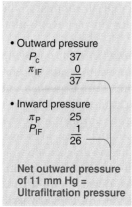

Forces at arteriolar end of capillary

• Outward pressure
P_C	37
π_{IF}	0
	37

• Inward pressure
π_P	25
P_{IF}	1
	26

Net outward pressure of 11 mm Hg = Ultrafiltration pressure

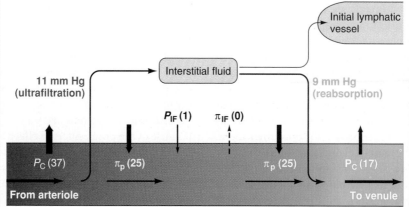

11 mm Hg (ultrafiltration)

Interstitial fluid

Initial lymphatic vessel

9 mm Hg (reabsorption)

P_{IF} (1) π_{IF} (0)

P_C (37) π_P (25) π_P (25) P_C (17)

From arteriole   **To venule**

Blood capillary

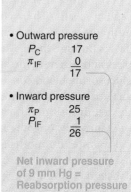

Forces at venular end of capillary

• Outward pressure
P_C	17
π_{IF}	0
	17

• Inward pressure
π_P	25
P_{IF}	1
	26

Net inward pressure of 9 mm Hg = Reabsorption pressure

All values are given in mm Hg.

$$\text{Net exchange pressure} = \underset{\text{(outward pressure)}}{(P_C + \pi_{IF})} - \underset{\text{(inward pressure)}}{(\pi_P + P_{IF})}$$

A positive net exchange pressure (when the outward pressure exceeds the inward pressure) represents an ultrafiltration pressure. A negative net exchange pressure (when the inward pressure exceeds the outward pressure) represents a reabsorption pressure.

At the arteriolar end of the capillary, the outward pressure totals 37 mm Hg, whereas the inward pressure totals 26 mm Hg, for a net outward pressure of 11 mm Hg. Ultrafiltration takes place at the beginning of the capillary as this outward pressure gradient forces a protein-free filtrate through the capillary pores.

By the time the venular end of the capillary is reached, the capillary blood pressure has dropped, but the other pressures have remained essentially constant. At this point the outward pressure has fallen to a total of 17 mm Hg, whereas the total inward pressure is still 26 mm Hg, for a net inward pressure of 9 mm Hg. Reabsorption of fluid takes place as this inward pressure gradient forces fluid back into the capillary at its venular end.

Ultrafiltration and reabsorption, collectively known as *bulk flow*, are thus due to a shift in the balance between the passive physical forces acting across the capillary wall. No active forces or local energy expenditures are involved in the bulk exchange of fluid between the plasma and surrounding interstitial fluid. With only minor contributions from the interstitial fluid forces, ultrafiltration occurs at the beginning of the capillary because capillary blood pressure exceeds plasma-colloid osmotic pressure, whereas by the end of the capillary, reabsorption takes place because blood pressure has fallen below osmotic pressure.

It is important to realize that we have taken "snapshots" at two points—at the beginning and at the end—in a hypothetical capillary. Actually, blood pressure gradually diminishes along the length of the capillary, so that progressively diminishing quantities of fluid are filtered out in the first half of the vessel and progressively increasing quantities of fluid are reabsorbed in the last half (● Figure 10-19).

ROLE OF BULK FLOW

Bulk flow does not play an important role in the exchange of individual solutes between blood and tissues, because the quantity of solutes moved across the capillary wall by bulk flow is extremely small compared to the much larger transfer of solutes by diffusion. Thus ultrafiltration and reabsorption are not important in the exchange of nutrients and wastes. Bulk flow is extremely important, however, in regulating distribution of ECF between plasma and interstitial fluid. Maintenance of proper arterial blood pressure depends in part on an appropriate volume of circulating blood. If plasma volume is reduced (for example, by hemorrhage), blood pressure falls. The resultant lowering of capillary blood pressure alters the balance of forces across the capillary walls. Because the net outward pressure is decreased while the net inward pressure remains unchanged, extra fluid is shifted from the interstitial compartment into the plasma as a result of reduced filtration

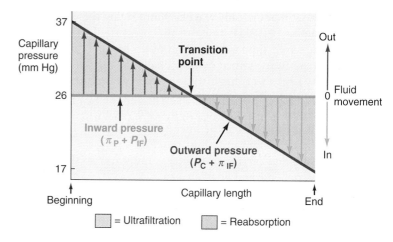

● **FIGURE 10-19**

Net filtration and net reabsorption along the vessel length. The inward pressure ($\pi_P + P_{IF}$) remains constant throughout the length of the capillary whereas the outward pressure ($P_C + \pi_{IF}$) progressively declines throughout the capillary's length. In the first half of the vessel, where the declining outward pressure still exceeds the constant inward pressure, progressively diminishing quantities of fluid are filtered out *(upward red arrows)*. In the last half of the vessel, progressively increasing quantities of fluid are reabsorbed *(downward blue arrows)* as the declining outward pressure falls farther below the constant inward pressure.

and increased reabsorption. The extra fluid soaked up from the interstitial fluid provides additional fluid for the plasma, temporarily compensating for the loss of blood. Meanwhile, reflex mechanisms acting on the heart and blood vessels (described later) also come into play to help maintain blood pressure until long-term mechanisms, such as thirst (and its satisfaction) and reduction of urinary output, can restore the fluid volume to completely compensate for the loss.

Conversely, if the plasma volume becomes overexpanded, as with excessive fluid intake, the resulting rise in capillary blood pressure forces extra fluid from the capillaries into the interstitial fluid, temporarily relieving the expanded plasma volume until the excess fluid can be eliminated from the body by long-term measures, such as increased urinary output.

These internal fluid shifts between the two ECF compartments occur automatically and immediately whenever the balance of forces acting across the capillary walls is changed; they provide a temporary mechanism to help keep plasma volume fairly constant. In the process of restoring plasma volume to an appropriate level, interstitial fluid volume fluctuates, but it is much more important that plasma volume be kept constant, to ensure that the circulatory system functions effectively.

▌ **The lymphatic system is an accessory route by which interstitial fluid can be returned to the blood.**

Even under normal circumstances, slightly more fluid is filtered out of the capillaries into the interstitial fluid than is reabsorbed from the interstitial fluid back into the plasma. On average, the net ultrafiltration pressure starts at 11 mm Hg at the beginning of the capillary, whereas the net reabsorption pressure only reaches 9 mm Hg by the vessel's end (● Figure 10-18). Because of this pressure differential, on average more fluid is filtered out of the first half of the capillary than is reabsorbed in its last half. The extra fluid filtered out as a result of this filtration–reabsorption imbalance is picked up by the **lymphatic system**. This extensive network of one-way vessels provides an accessory route by which fluid can be returned from the interstitial fluid to the blood. The lymphatic system functions much like a storm sewer that picks up and carries away excess rainwater so that it does not accumulate and flood an area.

PICKUP AND FLOW OF LYMPH

Small, blind-ended terminal lymph vessels known as **initial lymphatics** permeate almost every tissue of the body (● Figure 10-20a). The endothelial cells forming the walls of initial lymphatics slightly overlap like shingles on a roof, with their overlapping edges being free instead of attached to the surrounding cells. This arrangement creates one-way, valvelike openings in the vessel wall. Fluid pressure on the outside of the vessel pushes the innermost edge of a pair of overlapping edges inward, creating a gap between the edges (that is, opening the valve). This opening permits interstitial fluid to enter (● Figure 10-20b). Once interstitial fluid enters a lymphatic vessel, it is called **lymph**. Fluid pressure on the inside forces the overlapping edges together, closing the valves so that lymph does not escape. These lymphatic valvelike openings are much larger than the pores in blood capillaries. Consequently, large particles in the interstitial fluid, such as escaped plasma proteins and bacteria, can gain access to initial lymphatics but are excluded from blood capillaries.

Initial lymphatics converge to form larger and larger **lymph vessels**, which eventually empty into the venous system near where the blood enters the right atrium (● Figure 10-21a). Because there is no "lymphatic heart" to provide driving pressure, you may wonder how lymph is directed from the tissues toward the venous system in the thoracic cavity. Lymph flow is accomplished by two mechanisms. First, lymph vessels beyond the initial lymphatics are surrounded by smooth muscle, which contracts rhythmically as a result of myogenic activity. When this muscle is stretched because the vessel is distended with lymph, the muscle inherently contracts more forcefully, pushing the lymph through the vessel. This intrinsic "lymph pump" is the major force for propelling lymph. Stimulation of lymphatic smooth muscle by the sympathetic nervous system further increases the pumping activity of the lymph vessels. Second, because lymph vessels lie between skeletal muscles, contraction of these muscles squeezes the lymph out of the vessels. One-way valves spaced at intervals within the lymph vessels direct the flow of lymph toward its venous outlet in the chest.

FUNCTIONS OF THE LYMPHATIC SYSTEM

Here are the most important functions of the lymphatic system:

- *Return of excess filtered fluid.* Normally, capillary filtration exceeds reabsorption by about 3 liters per day (20 liters filtered, 17 liters reabsorbed) (● Figure 10-21b). Yet the entire

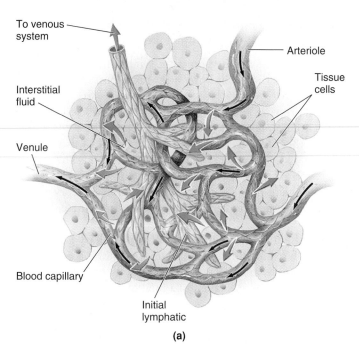

(a)

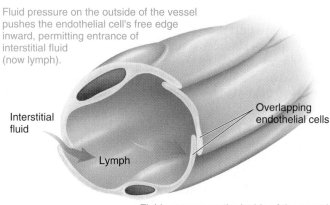

Fluid pressure on the outside of the vessel pushes the endothelial cell's free edge inward, permitting entrance of interstitial fluid (now lymph).

Fluid pressure on the inside of the vessel forces the overlapping edges together so that lymph cannot escape.

(b)

● FIGURE 10-20

Initial lymphatics. (a) Relationship between initial lymphatics and blood capillaries. Blind-ended initial lymphatics pick up excess fluid filtered by blood capillaries and return it to the venous system in the chest. (b) Arrangement of endothelial cells in an initial lymphatic. Note that the overlapping edges of the endothelial cells create valvelike openings in the vessel wall.

blood volume is only 5 liters, and only 2.75 liters of that is plasma. (Blood cells make up the rest of the blood volume.) With an average cardiac output, 7200 liters of blood pass through the capillaries daily under resting conditions (more when cardiac output increases). Even though only a small fraction of the filtered fluid is not reabsorbed by the blood capillaries, the cumulative effect of this process being repeated with every heartbeat results in the equivalent of more than the entire plasma volume being left behind in the interstitial fluid each day. Obviously, this fluid must be returned to the circulating plasma, and this task is done by the lymph vessels. The average rate of flow through the lymph vessels is 3 liters per day, compared with 7200 liters per day through the circulatory system.

- *Defense against disease.* The lymph percolates through **lymph nodes** located en route within the lymphatic system. Passage of this fluid through the lymph nodes is an important aspect of the body's defense mechanism against disease. For example, bacteria picked up from the interstitial fluid are destroyed by special phagocytes within the lymph nodes (see Chapter 11).

- *Transport of absorbed fat.* The lymphatic system is important in the absorption of fat from the digestive tract. The end products of the digestion of dietary fats are packaged by cells lining the digestive tract into fatty particles that are too large to gain access to the blood capillaries but can easily enter the initial lymphatics (see Chapter 15).

- *Return of filtered protein.* Most capillaries permit leakage of some plasma proteins during filtration. These proteins cannot readily be reabsorbed back into the blood capillaries but can easily gain access to the initial lymphatics. If the proteins were allowed to accumulate in the interstitial fluid rather than being returned to the circulation via the lymphatics, the interstitial fluid–colloid osmotic pressure (an outward pressure) would progressively increase while the plasma-colloid osmotic pressure (an inward pressure) would progressively fall. As a result, filtration forces would gradually increase and reabsorption forces would gradually decrease, resulting in progressive accumulation of fluid in the interstitial spaces at the expense of loss of plasma volume.

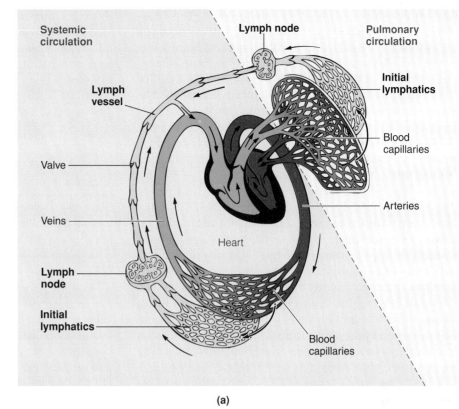

(a)

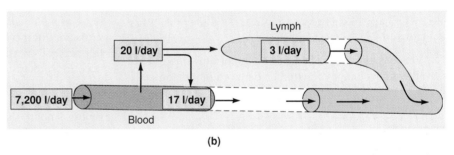

(b)

● **FIGURE 10-21**

Lymphatic system. (a) Lymph empties into the venous system near its entrance to the right atrium. (b) Lymph flow averages 3 liters per day, whereas blood flow averages 7200 liters per day.

▌ Edema occurs when too much interstitial fluid accumulates.

Clinical Note Occasionally, excessive interstitial fluid does accumulate when one of the physical forces acting across the capillary walls becomes abnormal for some reason. Swelling of the tissues because of excess interstitial fluid is known as **edema**. The causes of edema can be grouped into four general categories:

1. *A reduced concentration of plasma proteins* decreases plasma-colloid osmotic pressure. Such a drop in the major inward pressure lets excess fluid filter out, whereas less-than-normal amounts of fluid are reabsorbed; hence extra fluid remains in the interstitial spaces. Edema can be caused by a decreased concentration of plasma proteins in several different ways: excessive loss of plasma proteins in the urine, from kidney disease; reduced synthesis of plasma proteins, from liver disease (the liver synthesizes almost all plasma proteins); a diet deficient in protein; or significant loss of plasma proteins from large burned surfaces.

2. *Increased permeability of the capillary walls* allows more plasma proteins than usual to pass from plasma into surrounding interstitial fluid—for example, via histamine-induced widening of the capillary pores during tissue injury or allergic reactions. The resultant fall in plasma-colloid osmotic pressure decreases the effective inward pressure, whereas the resultant rise in interstitial fluid-colloid osmotic pressure caused by excess protein in the interstitial fluid increases the effective outward force. This imbalance contributes in part to the localized edema associated with injuries (for example, blisters) and allergic responses (for example, hives).

3. *Increased venous pressure,* as when blood dams up in the veins, is accompanied by an increased capillary blood pres-

sure, because the capillaries drain into the veins. This elevation in outward pressure across the capillary walls is largely responsible for the edema seen with congestive heart failure (see p. 265). Regional edema can also occur because of localized restriction of venous return. An example is the swelling often occurring in the legs and feet during pregnancy. The enlarged uterus compresses the major veins that drain the lower extremities as these vessels enter the abdominal cavity. The resultant damming of blood in these veins raises blood pressure in the capillaries of the legs and feet, which promotes regional edema of the lower extremities.

4. *Blockage of lymph vessels* produces edema because the excess filtered fluid is retained in the interstitial fluid rather than being returned to the blood through the lymphatics. Protein accumulation in the interstitial fluid compounds the problem through its osmotic effect. Local lymph blockage can occur, for example, in the arms of women whose major lymphatic drainage channels from the arm have been blocked as a result of lymph node removal during surgery for breast cancer. More widespread lymph blockage occurs with *filariasis*, a mosquito-borne parasitic disease found predominantly in tropical coastal regions. In this condition, small, threadlike filaria worms infect the lymph vessels, where their presence prevents proper lymph drainage. The affected body parts, particularly the scrotum and extremities, become grossly edematous. The condition is often called *elephantiasis* because of the elephant-like appearance of the swollen extremities (● Figure 10-22).

Whatever the cause of edema, an important consequence is a reduced exchange of materials between blood and cells.

● FIGURE 10-22

Elephantiasis. This tropical condition is caused by a mosquito-borne parasitic worm that invades the lymph vessels. As a result of the interference with lymph drainage, the affected body parts, usually the extremities, become grossly edematous, appearing elephant-like.

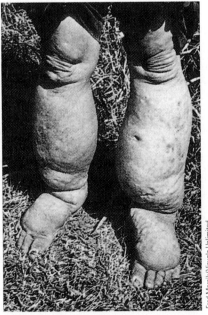

Fred Marsik/Visuals Unlimited

As excess interstitial fluid accumulates, the distance between blood and cells across which nutrients, O_2, and wastes must diffuse increases, so the rate of diffusion decreases. Therefore, cells within edematous tissues may not be adequately supplied.

 Click on the Media Exercises menu of the CD-ROM and work Media Exercise 10.2: Arteriolar Resistance and Flow in Capillaries to test your understanding of the previous section.

VEINS

The venous system completes the circulatory circuit. Blood leaving the capillary beds enters the venous system for transport back to the heart.

■ Veins serve as a blood reservoir as well as passageways back to the heart.

Veins have a large radius, so they offer little resistance to flow. Furthermore, because the total cross-section area of the venous system gradually decreases as smaller veins converge into progressively fewer but larger vessels, blood flow speeds up as blood approaches the heart.

In addition to serving as low-resistance passageways to return blood from the tissues to the heart, systemic veins also serve as a *blood reservoir*. Because of their storage capacity, veins are often called **capacitance vessels**. Veins have much thinner walls with less smooth muscle than arteries do. Also, in contrast to arteries, veins have very little elasticity, because venous connective tissue contains considerably more collagen fibers than elastin fibers. Unlike arteriolar smooth muscle, venous smooth muscle has little inherent myogenic tone. Because of these features, veins are highly distensible, or stretchable, and have little elastic recoil. They easily distend to accommodate additional volumes of blood with only a small increase in venous pressure. Arteries stretched by an excess volume of blood recoil because of the elastic fibers in their walls, driving the blood forward. Veins containing an extra volume of blood simply stretch to accommodate the additional blood without tending to recoil. In this way veins serve as a **blood reservoir**; that is, when demands for blood are low, the veins can store extra blood in reserve because of their passive distensibility. Under resting conditions, the veins contain more than 60% of the total blood volume (● Figure 10-23).

When the stored blood is needed, such as during exercise, extrinsic factors (soon to be described) reduce the capacity of the venous reservoir and drive the extra blood from the veins to the heart so it can be pumped to the tissues. Increased venous return leads to an increased cardiac stroke volume, in accordance with the Frank-Starling law of the heart (see p. 262). In contrast, if too much blood pools in the veins instead of being returned to the heart, cardiac output is abnormally diminished. Thus a delicate balance exists between the capacity of the veins, the extent of venous return, and the cardiac output. We will now turn our attention to the factors that affect venous capacity and contribute to venous return.

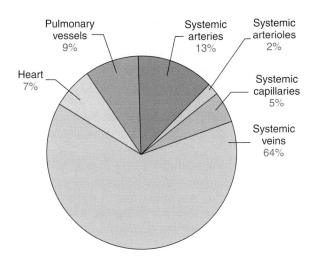

● **FIGURE 10-23**

Percentage of total blood volume in different parts of the circulatory system

▌Venous return is enhanced by a number of extrinsic factors.

Venous capacity (the volume of blood that the veins can accommodate) depends on the distensibility of the vein walls (how much they can stretch to hold blood) and the influence of any externally applied pressure squeezing inwardly on the veins. At a constant blood volume, as venous capacity increases, more blood remains in the veins instead of being returned to the heart. Such venous storage decreases the effective circulating volume, the volume of blood being returned to and pumped out of the heart. Conversely, when venous capacity decreases, more blood is returned to the heart and is subsequently pumped out. Thus changes in venous capacity directly influence the magnitude of venous return, which in turn is an important (although not the only) determinant of effective circulating blood volume. The effective circulating blood volume is also influenced on a short-term basis by passive shifts in bulk flow between the vascular and interstitial fluid compartments and on a long-term basis by factors that control total ECF volume, such as salt and water balance.

The term **venous return** refers to the volume of blood entering each atrium per minute from the veins. Recall that the magnitude of flow through a vessel is directly proportional to the pressure gradient. Much of the driving pressure imparted to the blood by cardiac contraction has been lost by the time the blood reaches the venous system, because of frictional losses along the way, especially during passage through the high-resistance arterioles. By the time the blood enters the venous system, blood pressure averages only 17 mm Hg (● Figure 10-8, p. 283). However, because atrial pressure is near 0 mm Hg, a small but adequate driving pressure still exists to promote the flow of blood through the large-radius, low-resistance veins. If atrial pressure becomes pathologically elevated, as in the presence of a leaky AV valve, the venous-to-atrial pressure gradient is decreased, reducing venous re-

turn and causing blood to dam up in the venous system. Elevated atrial pressure is thus one cause of congestive heart failure (see p. 265).

In addition to the driving pressure imparted by cardiac contraction, five other factors enhance venous return: sympathetically induced venous vasoconstriction, skeletal muscle activity, the effect of venous valves, respiratory activity, and the effect of cardiac suction (● Figure 10-24). Most of these secondary factors affect venous return by influencing the pressure gradient between the veins and the heart. We will examine each in turn.

EFFECT OF SYMPATHETIC ACTIVITY ON VENOUS RETURN

Veins are not very muscular and have little inherent tone, but venous smooth muscle is abundantly supplied with sympathetic nerve fibers. Sympathetic stimulation produces venous vasoconstriction, which modestly elevates venous pressure; this, in turn, increases the pressure gradient to drive more of the stored blood from the veins into the right atrium, thus enhancing venous return. The veins normally have such a large radius that the moderate vasoconstriction from sympathetic stimulation has little effect on resistance to flow. Even when constricted, the veins still have a relatively large radius and are still low-resistance vessels.

It is important to recognize the different outcomes of vasoconstriction in arterioles and veins. Arteriolar vasoconstriction immediately *reduces* flow through these vessels because of their increased resistance (less blood can enter and flow through a narrowed arteriole), whereas venous vasoconstriction immediately *increases* flow through these vessels because of their decreased capacity (narrowing of veins squeezes out more of the blood already in the veins, increasing blood flow through these vessels).

EFFECT OF SKELETAL MUSCLE ACTIVITY ON VENOUS RETURN

Many of the large veins in the extremities lie between skeletal muscles, so muscle contraction compresses the veins. This external venous compression decreases venous capacity and increases venous pressure, in effect squeezing fluid in the veins forward toward the heart (● Figure 10-25). This pumping action, known as the **skeletal muscle pump**, is one way extra blood stored in the veins is returned to the heart during exercise. Increased muscular activity pushes more blood out of the veins and into the heart. Increased sympathetic activity and the resultant venous vasoconstriction also accompany exercise, further enhancing venous return.

The skeletal muscle pump also counters the effect of gravity on the venous system. Let's see how.

COUNTERING THE EFFECTS OF GRAVITY ON THE VENOUS SYSTEM

The average pressures mentioned thus far for various regions of the vascular tree are for a person in the horizontal position. When a person is lying down, the force of gravity is uniformly applied, so it need not be considered. When a person stands up, however, gravitational effects are not uniform. In addition to the usual pressure from cardiac contraction, vessels below heart level are subject to pressure from the weight of the column of blood extending from the heart to the level of the vessel (● Figure 10-26).

There are two important consequences of this increased pressure. First, the distensible veins yield under the increased hydrostatic pressure, further expanding so their capacity is increased. Even though the arteries are subject to the same gravitational effects, they are not nearly as distensible and do not expand like the veins. Much of the blood entering from the capillaries tends to pool in the expanded lower-leg veins instead of returning to the heart. Because venous return is reduced, cardiac output decreases and the effective circulating volume shrinks. Second, the marked increase in capillary blood pressure from gravity causes excessive fluid to filter out of capillary beds in the lower extremities, producing localized edema (that is, swollen feet and ankles).

Two compensatory measures normally counteract these gravitational effects. First, the resultant fall in mean arterial pressure that occurs when a person moves from a lying-down to an upright position triggers sympathetically induced venous vasoconstriction, which drives some of the pooled blood forward. Second, the skeletal muscle pump "interrupts" the column of blood by completely emptying given vein segments intermittently so that a particular portion of a vein is not subjected to the weight of the entire venous column from the heart to that portion's level (● Figure 10-27). Reflex venous vasoconstriction cannot completely compensate for gravitational effects without skeletal muscle activity. When a person stands still for a long time, therefore, blood flow to the brain is reduced because of the decline in effective circulating volume, despite reflexes aimed at maintaining mean arterial pressure. Reduced flow of blood to the brain, in turn, leads to fainting, which returns the person to a horizontal position, eliminating the gravitational effects on the vascular system and restoring effective circulation. For this reason, it is counterproductive to try to hold upright someone who has fainted. Fainting is a remedy to the problem, not the problem itself.

Because the skeletal muscle pump facilitates venous return and helps counteract the detrimental effects of gravity on the circulatory system, when you are working at a desk it's a good idea to get up periodically and when you are on your feet, to move around. The mild muscular activity "gets the blood moving."

EFFECT OF VENOUS VALVES ON VENOUS RETURN

Venous vasoconstriction and external venous compression both drive blood toward the heart. Yet if you squeeze a fluid-filled tube in the middle, fluid is pushed in both directions from the point of constriction (● Figure 10-28a). Then why isn't blood driven backward as well as forward by venous vasoconstriction and the skeletal muscle pump? Blood can only

● **FIGURE 10-24**

Factors that facilitate venous return.

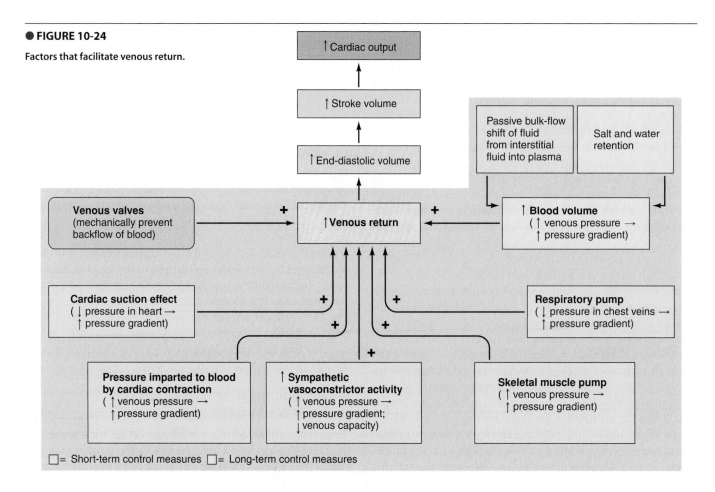

□ = Short-term control measures □ = Long-term control measures

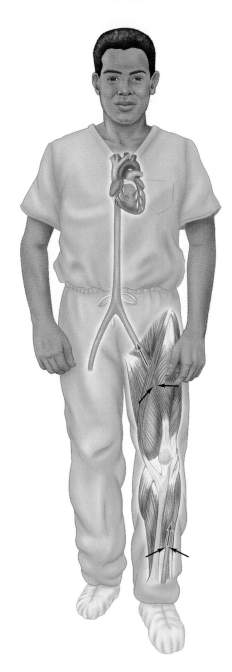

● **FIGURE 10-25**

Skeletal muscle pump enhancing venous return

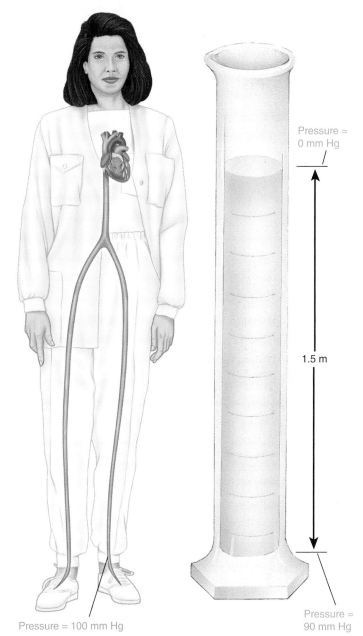

Pressure = 0 mm Hg

1.5 m

Pressure = 90 mm Hg

Pressure = 100 mm Hg

90 mm Hg caused by gravitational effect

10 mm Hg caused by pressure imparted by cardiac contraction

● **FIGURE 10-26**

Effect of gravity on venous pressure. In an upright adult, the blood in the vessels extending between the heart and foot is equivalent to a 1.5-m column of blood. The pressure exerted by this column of blood as a result of the effect of gravity is 90 mm Hg. The pressure imparted to the blood by the heart has declined to about 10 mm Hg in the lower-leg veins because of frictional losses in preceding vessels. Together these pressures produce a venous pressure of 100 mm Hg in the ankle and foot veins. Similarly, the capillaries in the region are subjected to these same gravitational effects.

be driven forward because the large veins are equipped with one-way valves spaced at 2- to 4-cm intervals; these valves let blood move forward toward the heart but keep it from moving back toward the tissues (● Figure 10-28b).

Clinical Note   **Varicose veins** occur when the venous valves become incompetent and can no longer support the column of blood above them. People predisposed to this condition usually have inherited an overdistensibility and weakness of their vein walls. Aggravated by frequent, prolonged standing, the veins become so distended as blood pools in them that the edges of the valves can no longer meet to form a seal. Varicosed superficial leg veins be-

come visibly overdistended and tortuous. Contrary to what might be expected, the chronic pooling of blood in the pathologically distended veins does not reduce cardiac output, because there is a compensatory increase in total circulating

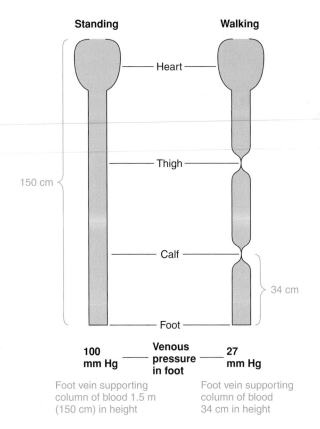

Standing   **Walking**

Heart

Thigh

150 cm

Calf

34 cm

Foot

| 100 mm Hg | Venous pressure in foot | 27 mm Hg |

Foot vein supporting column of blood 1.5 m (150 cm) in height

Foot vein supporting column of blood 34 cm in height

● **FIGURE 10-27**

Effect of contracting the skeletal muscles of the legs in counteracting the effects of gravity. Contraction of skeletal muscles (as in walking) completely empties given vein segments, interrupting the column of blood that the lower veins must support.

(*Source:* Adapted from *Physiology of the Heart and Circulation,* Fourth Edition, by R. C. Little and W. C. Little. Copyright ©1989 Year Book Medical Publishers, Inc. Reprinted by permission of the author and Mosby-Year Book, Inc.)

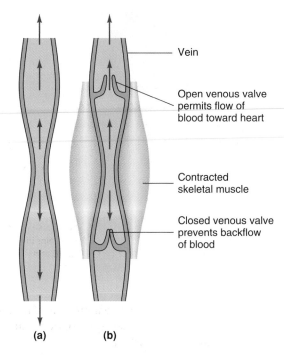

Vein

Open venous valve permits flow of blood toward heart

Contracted skeletal muscle

Closed venous valve prevents backflow of blood

(a)   (b)

● **FIGURE 10-28**

Function of venous valves. (a) When a tube is squeezed in the middle, fluid is pushed in both directions. (b) Venous valves permit the flow of blood only toward the heart.

---

blood volume. Instead, the most serious consequence of varicose veins is the possibility of abnormal clot formation in the sluggish, pooled blood. Particularly dangerous is the risk that these clots may break loose and block small vessels elsewhere, especially the pulmonary capillaries.

EFFECT OF RESPIRATORY ACTIVITY ON VENOUS RETURN

As a result of respiratory activity, the pressure within the chest cavity averages 5 mm Hg less than atmospheric pressure. As the venous system returns blood to the heart from the lower regions of the body, it travels through the chest cavity, where it is exposed to this subatmospheric pressure. Because the venous system in the limbs and abdomen is subject to normal atmospheric pressure, an externally applied pressure gradient exists between the lower veins (at atmospheric pressure) and the chest veins (at 5 mm Hg less than atmospheric pressure). This pressure difference squeezes blood from the lower veins to the chest veins, promoting increased venous return (● Figure 10-29). This mechanism of facilitating ve-

nous return is called the **respiratory pump**, because it results from respiratory activity. Increased respiratory activity as well as the effects of the skeletal muscle pump and venous vasoconstriction all enhance venous return during exercise.

EFFECT OF CARDIAC SUCTION ON VENOUS RETURN

The extent of cardiac filling does not depend entirely on factors affecting the veins. The heart plays a role in its own filling. During ventricular contraction, the AV valves are drawn downward, enlarging the atrial cavities. As a result, the atrial pressure transiently drops below 0 mm Hg, thus increasing the vein-to-atria pressure gradient so that venous return is enhanced. In addition, the rapid expansion of the ventricular chambers during ventricular relaxation creates a transient negative pressure in the ventricles so that blood is "sucked in" from the atria and veins; that is, the negative ventricular pressure increases the vein-to-atria-to-ventricle pressure gradient, further enhancing venous return. Thus the heart functions as a "suction pump" to facilitate cardiac filling.

BLOOD PRESSURE

Mean arterial pressure is the blood pressure that is monitored and regulated in the body, not the arterial systolic or diastolic or pulse pressure nor the pressure in any other part of the vascular tree. Routine blood pressure measurements record the arterial systolic and diastolic pressures, which can be used

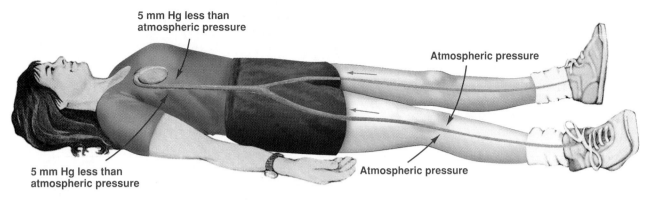

5 mm Hg less than atmospheric pressure

Atmospheric pressure

5 mm Hg less than atmospheric pressure

Atmospheric pressure

● **FIGURE 10-29**

Respiratory pump enhancing venous return. As a result of respiratory activity, the pressure surrounding the chest veins is lower than the pressure surrounding the veins in the extremities and abdomen. This establishes an externally applied pressure gradient on the veins that drives blood toward the heart.

as a yardstick for assessing mean arterial pressure. The cutoff for normal blood pressure has recently been designated by the National Institutes of Health (NIH) as being less than 120/80 mm Hg.

Blood pressure is regulated by controlling cardiac output, total peripheral resistance, and blood volume.

Mean arterial pressure is the main driving force for propelling blood to the tissues. This pressure must be closely regulated for two reasons. First, it must be high enough to ensure sufficient driving pressure; without this pressure, the brain and other organs will not receive adequate flow, no matter what local adjustments are made in the resistance of the arterioles supplying them. Second, the pressure must not be so high that it creates extra work for the heart and increases the risk of vascular damage and possible rupture of small blood vessels.

DETERMINANTS OF MEAN ARTERIAL PRESSURE

Elaborate mechanisms involving the integrated action of the various components of the circulatory system and other body systems are vital in regulating this all-important mean arterial pressure (● Figure 10-30). Remember that the two determinants of mean arterial pressure are cardiac output and total peripheral resistance:

$$\text{Mean arterial pressure} = \text{cardiac output} \times \text{total peripheral resistance}$$

(Do not confuse this equation, which indicates the *determinants* of mean arterial pressure, namely the magnitude of both the cardiac output and total peripheral resistance, with the equation used to *calculate* mean arterial pressure, namely mean arterial pressure = diastolic pressure + 1/3 pulse pressure.)

Recall that a number of factors, in turn, determine cardiac output (● Figure 9-23, p. 263) and total peripheral resistance (● Figure 10-12, p. 290). Thus you can quickly appreciate the complexity of blood pressure regulation. Let's work through ● Figure 10-30, reviewing all the factors that affect mean arterial pressure. Even though we've covered all these factors before, it is useful to pull them all together. The circled numbers in the text correspond to the numbers in the figure.

- Mean arterial pressure depends on cardiac output and total peripheral resistance (1 on ● Figure 10-30).
- Cardiac output depends on heart rate and stroke volume 2 .
- Heart rate depends on the relative balance of parasympathetic activity 3 , which decreases heart rate, and sympathetic activity (tacitly including epinephrine throughout this discussion) 4 , which increases heart rate.
- Stroke volume increases in response to sympathetic activity 5 (extrinsic control of stroke volume).
- Stroke volume also increases as venous return increases 6 (intrinsic control of stroke volume according to the Frank-Starling law of the heart).
- Venous return is enhanced by sympathetically induced venous vasoconstriction 7 , the skeletal muscle pump 8 , the respiratory pump 9 , and cardiac suction 10 .
- The effective circulating blood volume also influences how much blood is returned to the heart 11 . The blood volume depends in the short term on the size of passive bulk-flow fluid shifts between plasma and interstitial fluid across the capillary walls 12 . In the long term, the blood volume depends on salt and water balance 13 , which are hormonally controlled by the renin-angiotensin-aldosterone system and vasopressin, respectively 14 .
- The other major determinant of mean arterial blood pressure, total peripheral resistance, depends on the radius of all arterioles as well as blood viscosity 15 . The main factor determining blood viscosity is the number of red blood cells 16 . However, arteriolar radius is the more important factor determining total peripheral resistance.
- Arteriolar radius is influenced by local (intrinsic) metabolic controls that match blood flow with metabolic needs

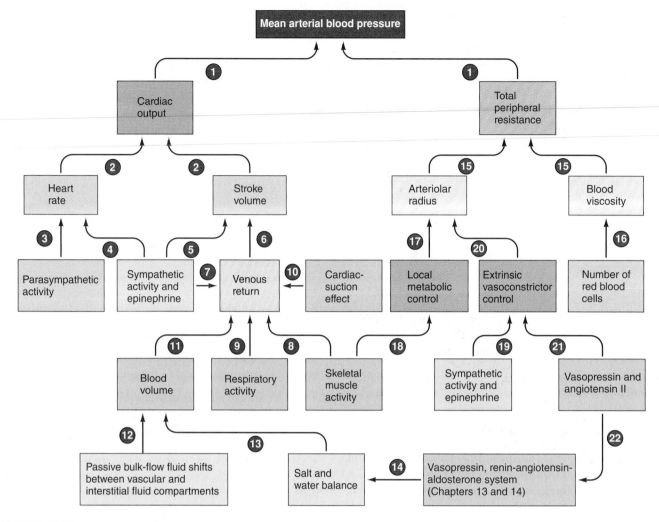

● **FIGURE 10-30**

Determinants of mean arterial blood pressure. Note that this figure is basically a composite of ● Figure 9-23, p. 263, "Control of cardiac output"; ● Figure 10-12, p. 290, "Factors affecting total peripheral resistance"; and ● Figure 10-24, p. 300, "Factors that facilitate venous return." See the text for a discussion of the circled numbers.

 For an interaction related to this figure, see Media Exercise 10.3: Circulation on the CD-ROM.

17. For example, local changes that take place in active skeletal muscles cause local arteriolar vasodilation and increased blood flow to these muscles 18.

• Arteriolar radius is also influenced by sympathetic activity 19, an extrinsic control mechanism that causes arteriolar vasoconstriction 20 to increase total peripheral resistance and mean arterial blood pressure.

• Arteriolar radius is also extrinsically controlled by the hormones vasopressin and angiotensin II, which are potent vasoconstrictors 21 as well as being important in salt and water balance 22.

Altering any of the pertinent factors that influence blood pressure will change blood pressure, unless a compensatory change in another variable keeps the blood pressure constant. Blood flow to any given organ depends on the driving force

of the mean arterial pressure and on the degree of vasoconstriction of the organ's arterioles. Because mean arterial pressure depends on the cardiac output and the degree of arteriolar vasoconstriction, if the arterioles in one organ dilate, the arterioles in other organs must constrict to maintain an adequate arterial blood pressure. An adequate pressure is needed to provide a driving force to push blood not only to the vasodilated organ but also to the brain, which depends on a constant blood supply. Thus the cardiovascular variables must be continuously juggled to maintain a constant blood pressure despite organs' varying needs for blood.

SHORT-TERM AND LONG-TERM CONTROL MEASURES

Mean arterial pressure is constantly monitored by **baroreceptors** (pressure sensors) within the circulatory system. When deviations from normal are detected, multiple reflex responses are initiated to return mean arterial pressure to its normal value. *Short-term* (within seconds) adjustments are made by alterations in cardiac output and total peripheral resistance, mediated by means of autonomic nervous system influences on the heart, veins, and arterioles. *Long-term* (requiring minutes to days) control involves adjusting total blood volume by restoring normal salt and water balance through mechanisms that regulate urine output and thirst (Chapter 14). The size of the total blood volume, in turn, has a profound effect on cardiac output and mean arterial pressure. Let us now turn our attention to the short-term mechanisms involved in ongoing regulation of this pressure.

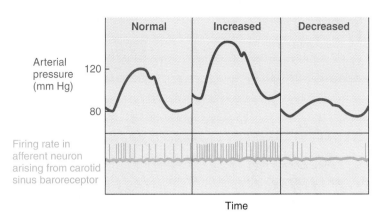

● **FIGURE 10-32**

Firing rate in the afferent neuron from the carotid sinus baroreceptor in relation to the magnitude of mean arterial pressure

---

▌ The baroreceptor reflex is an important short-term mechanism for regulating blood pressure.

Any change in mean arterial pressure triggers an autonomically mediated **baroreceptor reflex** that influences the heart and blood vessels to adjust cardiac output and total peripheral resistance in an attempt to restore blood pressure to normal. Like any reflex, the baroreceptor reflex includes a receptor, an afferent pathway, an integrating center, an efferent pathway, and effector organs.

The most important receptors involved in moment-to-moment regulation of blood pressure, the **carotid sinus** and **aortic arch baroreceptors**, are mechanoreceptors sensitive to

changes in mean arterial pressure. These baroreceptors are strategically located (● Figure 10-31) to provide critical information about arterial blood pressure in the vessels leading to the brain (the carotid sinus baroreceptor) and in the major arterial trunk before it gives off branches that supply the rest of the body (the aortic arch baroreceptor).

The baroreceptors constantly provide information about mean arterial pressure; in other words, they continuously generate action potentials in response to the ongoing pressure within the arteries. When mean arterial pressure increases, the receptor potential of these baroreceptors increases, thus increasing the rate of firing in the corresponding afferent neurons. Conversely, a decrease in the mean arterial pressure slows the rate of firing generated in the afferent neurons by the baroreceptors (● Figure 10-32).

The integrating center that receives the afferent impulses about the state of mean arterial pressure is the **cardiovascular control center**, located in the medulla within the brain stem.

The efferent pathway is the autonomic nervous system. The cardiovascular control center alters the ratio between sympathetic and parasympathetic activity to the effector organs (the heart and blood vessels). To review how autonomic changes alter arterial blood pressure, ● Figure 10-33 summarizes the major effects of parasympathetic and sympathetic stimulation on the heart and blood vessels.

Let's fit all the pieces of the baroreceptor reflex together now by tracing the reflex activity that compensates for an elevation or fall in blood pressure. If for any reason mean arterial pressure rises above normal (● Figure 10-34a), the carotid sinus and aortic arch baroreceptors increase the rate of firing in their

---

● **FIGURE 10-31**

Location of the arterial baroreceptors. The arterial baroreceptors are strategically located to monitor the mean arterial blood pressure in the arteries that supply blood to the brain (carotid sinus baroreceptor) and to the rest of the body (aortic arch baroreceptor).

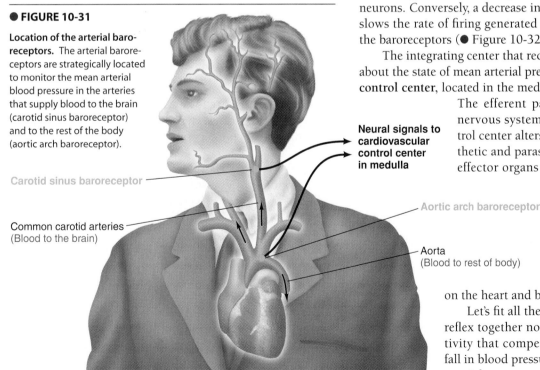

Carotid sinus baroreceptor

Common carotid arteries
(Blood to the brain)

Neural signals to cardiovascular control center in medulla

Aortic arch baroreceptor

Aorta
(Blood to rest of body)

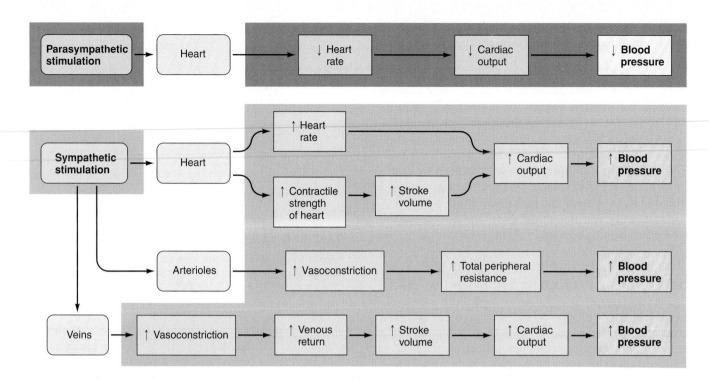

● **FIGURE 10-33**

Summary of the effects of the parasympathetic and sympathetic nervous systems on factors that influence mean arterial blood pressure

respective afferent neurons. On being informed by increased afferent firing that the blood pressure has become too high, the cardiovascular control center responds by decreasing sympathetic and increasing parasympathetic activity to the cardiovascular system. These efferent signals decrease heart rate, decrease stroke volume, and produce arteriolar and venous vasodilation, which in turn lead to a decrease in cardiac output and a decrease in total peripheral resistance, with a subsequent fall in blood pressure back toward normal.

Conversely, when blood pressure falls below normal (● Figure 10-34b) baroreceptor activity decreases, inducing the cardiovascular center to increase sympathetic cardiac and vasoconstrictor nerve activity while decreasing its parasympathetic output. This efferent pattern of activity leads to an increase in heart rate and stroke volume, coupled with arteriolar and venous vasoconstriction. These changes increase both cardiac output and total peripheral resistance, raising blood pressure back toward normal.

Despite these control measures, sometimes blood pressure is not maintained at the appropriate level. We will next examine blood pressure abnormalities.

▌Hypertension is a serious national public-health problem, but its causes are largely unknown.

Sometimes blood-pressure control mechanisms do not function properly or are unable to completely compensate for

Clinical Note changes that have taken place. Blood pressure may be too high (**hypertension** if above 140/90 mm Hg) or too low (**hypotension** if below 100/60 mm Hg). Hypotension in its extreme form is *circulatory shock.* We will first examine hypertension, which is by far the most common of blood pressure abnormalities, and then conclude this chapter with a discussion of hypotension and shock.

There are two broad classes of hypertension, secondary hypertension and primary hypertension, depending on the cause. A definite cause for hypertension can be established in only 10% of the cases. Hypertension that occurs secondary to another known primary problem is called *secondary hypertension.* For example, when the elasticity of the arteries is reduced by the loss of elastin fibers or the presence of calcified atherosclerotic plaques (see p. 267), arterial pressure is chronically increased by this "hardening of the arteries." Likewise, hypertension occurs if the kidneys are diseased and unable to eliminate the normal salt load. Salt retention induces water retention, which expands the plasma volume and leads to hypertension.

The underlying cause is unknown in the remaining 90% of hypertension cases. Such hypertension is known as *primary* (*essential* or *idiopathic*) hypertension. Primary hypertension is a catchall category for blood pressure elevated by a variety of unknown causes rather than by a single disease entity. People show a strong genetic tendency to develop primary hypertension, which can be hastened or worsened by contributing factors such as obesity, stress, smoking, or dietary habits.

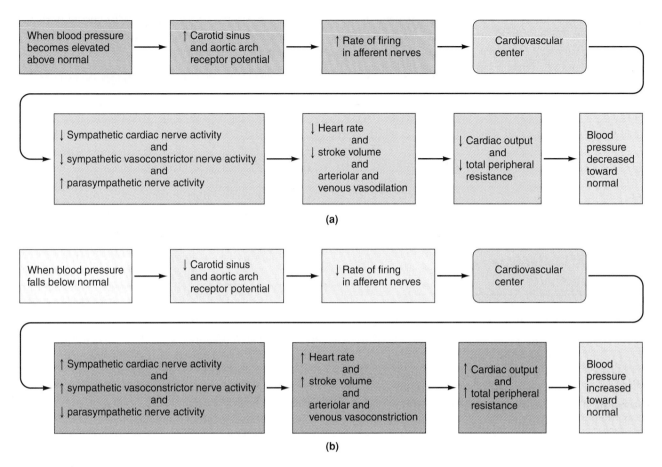

● FIGURE 10-34

Baroreceptor reflexes to restore the blood pressure to normal. (a) Baroreceptor reflex in response to an elevation in blood pressure. (b) Baroreceptor reflex in response to a fall in blood pressure.

Whatever the underlying defect, once initiated, hypertension appears to be self-perpetuating. Constant exposure to elevated blood pressure predisposes vessel walls to the development of atherosclerosis, which further raises blood pressure.

ADAPTATION OF BARORECEPTORS DURING HYPERTENSION

The baroreceptors do not respond to bring the blood pressure back to normal during hypertension because they adapt, or are "reset," to operate at a higher level. In the presence of chronically elevated blood pressure, the baroreceptors still function to regulate blood pressure, but they maintain it at a higher mean pressure.

COMPLICATIONS OF HYPERTENSION

Hypertension imposes stresses on both the heart and the blood vessels. The heart has an increased workload because it is pumping against an increased total peripheral resistance, whereas blood vessels may be damaged by the high internal pressure, particularly when the vessel wall is weakened by the degenerative process of atherosclerosis. Complications of hypertension include congestive heart failure caused by the heart's inability to pump continuously against a sustained elevation in arterial pressure, strokes caused by rupture of brain vessels, and heart attacks caused by rupture of coronary vessels. Spontaneous hemorrhage caused by bursting of small vessels elsewhere in the body may also occur but with less serious consequences; an example is the rupture of blood vessels in the nose, resulting in nosebleeds. Another serious complication of hypertension is renal failure caused by progressive impairment of blood flow through damaged renal blood vessels. Furthermore, retinal damage from changes in the blood vessels supplying the eyes may result in progressive loss of vision.

Until complications occur, hypertension is symptomless, because the tissues are adequately supplied with blood. Therefore, unless blood pressure measurements are made on a routine basis the condition can go undetected until a precipitous complicating event. When you become aware of these potential complications of hypertension and consider that 25% of all adults in America are estimated to have chronic elevated blood pressure, you can appreciate the magnitude of this national health problem.

In its recent guidelines, NIH identified **prehypertension** as a new category for blood pressures in the range between normal and hypertension (between 120/80 to 139/89). Blood pressures in the prehypertension range can usually be reduced by appropriate dietary and exercise measures, whereas those in the hypertension range typically must be treated with blood pressure medication in addition to changing health habits. The goal in managing blood pressures in the prehypertension range is to take action before the pressure climbs into the hypertension range, where serious complications may develop.

We are now going to examine the other extreme, hypotension, looking first at transient orthostatic hypotension, then at the (more serious) circulatory shock.

▌Orthostatic hypotension results from transient inadequate sympathetic activity.

Hypotension, or low blood pressure, occurs either when there is a disproportion between vascular capacity and blood volume (in essence, too little blood to fill the vessels) or when the heart is too weak to drive the blood.

The most common situation in which hypotension occurs transiently is orthostatic hypotension. **Orthostatic (postural) hypotension** is a transient hypotensive condition resulting from insufficient compensatory responses to the gravitational shifts in blood when a person moves from a horizontal to a vertical position, especially after prolonged bed rest. When a person moves from lying down to standing up, pooling of blood in the leg veins from gravity reduces venous return, decreasing stroke volume and thus lowering cardiac output and blood pressure. This fall in blood pressure is normally detected by the baroreceptors, which initiate immediate compensatory responses to restore blood pressure to its proper level. When a long-bedridden patient first starts to rise, however, these reflex compensatory adjustments are temporarily lost or reduced because of disuse. Sympathetic control of the leg veins is inadequate, so when the patient first stands up blood pools in the lower extremities. The resultant orthostatic hypotension and decrease in blood flow to the brain cause dizziness or actual fainting.

▌Circulatory shock can become irreversible.

Clinical Note When blood pressure falls so low that adequate blood flow to the tissues can no longer be maintained, the condition known as **circulatory shock** occurs. Circulatory shock may result from (1) extensive loss of blood volume as through hemorrhage; (2) failure of a weakened heart to pump blood adequately; (3) widespread arteriolar vasodilation triggered by vasodilator substances (such as extensive histamine release in severe allergic reactions); or (4) neurally defective vasoconstrictor tone.

We will now examine the consequences of and compensations for shock, using hemorrhage as an example (● Figure 10-35). This figure may look intimidating, but we will work through it step by step. It is an important example that pulls together many of the principles discussed in this chapter. As before, the circled numbers in the text correspond to the numbers in the figure.

CONSEQUENCES AND COMPENSATIONS OF SHOCK

- Following severe loss of blood, the resultant reduction in circulating blood volume leads to a decrease in venous return ①and a subsequent fall in cardiac output and arterial blood pressure. (Note the blue boxes, which indicate consequences of hemorrhage.)
- Compensatory measures immediately attempt to maintain adequate blood flow to the brain. (Note the pink boxes, which indicate compensations for hemorrhage.)
- The baroreceptor reflex response to the fall in blood pressure brings about increased sympathetic and decreased parasympathetic activity to the heart ②. The result is an increase in heart rate ③to offset the reduced stroke volume ④brought about by the loss of blood volume. With severe fluid loss, the pulse is weak because of the reduced stroke volume but rapid because of the increased heart rate.
- Increased sympathetic activity to the veins produces generalized venous vasoconstriction ⑤, increasing venous return by means of the Frank-Starling mechanism ⑥.
- Simultaneously, sympathetic stimulation of the heart increases the heart's contractility ⑦so that it beats more forcefully and ejects a greater volume of blood, likewise increasing the stroke volume.
- The increase in heart rate and in stroke volume collectively increase cardiac output ⑧.
- Sympathetically induced generalized arteriolar vasoconstriction ⑨leads to an increase in total peripheral resistance ⑩.
- Together, the increase in cardiac output and total peripheral resistance bring about a compensatory increase in arterial pressure ⑪.
- The original fall in arterial pressure is also accompanied by a fall in capillary blood pressure ⑫, which results in fluid shifts from the interstitial fluid into the capillaries to expand the plasma volume ⑬. This response is sometimes termed *autotransfusion*, because it restores the plasma volume as a transfusion does.
- This ECF fluid shift is enhanced by plasma protein synthesis by the liver during the next few days following hemorrhage ⑭. The plasma proteins exert a colloid osmotic pressure that helps retain extra fluid in the plasma.
- Urinary output is reduced, thereby conserving water that normally would have been lost from the body ⑮. This additional fluid retention helps expand the reduced plasma volume ⑯. Expansion of plasma volume further augments the increase in cardiac output brought about by the baroreceptor reflex ⑰. Reduction in urinary output results from decreased renal blood flow caused by compensatory renal arteriolar vasoconstriction ⑱. The reduced plasma volume also triggers increased secretion of the hormone vasopressin and activation of the salt- and water-conserving renin-angiotensin-aldosterone hormonal pathway, which further reduces urinary output ⑲.

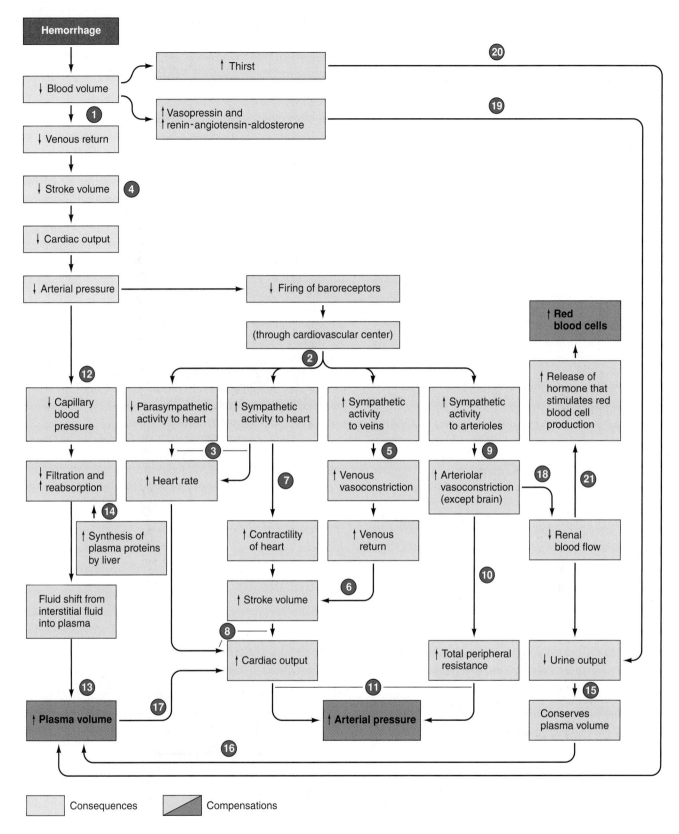

● FIGURE 10-35

Consequences and compensations of hemorrhage. The reduction in blood volume resulting from hemorrhage leads to a fall in arterial pressure. (Note the *blue boxes,* representing consequences of hemorrhage.) A series of compensations ensue *(light pink boxes)* that ultimately restore plasma volume, arterial pressure, and the number of red blood cells toward normal *(dark pink boxes).* Refer to the text (pp. 308 and 310) for an explanation of the circled numbers and a detailed discussion of the compensations.

• Increased thirst is also stimulated by a fall in plasma volume [20]. The resultant increased fluid intake helps restore plasma volume.

• Over a longer course of time (a week or more), lost red blood cells are replaced through increased red blood cell production triggered by a reduction in O_2 delivery to the kidneys [21].

IRREVERSIBLE SHOCK

These compensatory mechanisms are often not enough to counteract substantial fluid loss. Even if they can maintain an adequate blood pressure level, the short-term measures cannot continue indefinitely. Ultimately, fluid volume must be replaced from the outside through drinking, transfusion, or a combination of both. Blood supply to kidneys, digestive tract, skin, and other organs can be compromised to maintain blood flow to the brain only so long before organ damage begins to occur. A point may be reached at which blood pressure continues to drop rapidly because of tissue damage, despite vigorous therapy. This condition is frequently termed *irreversible shock*, in contrast to *reversible shock*, which can be corrected by compensatory mechanisms and effective therapy.

Click on the Media Exercises menu of the CD-ROM and work Media Exercise 10.3: Circulation to test your understanding of the previous section.

CHAPTER IN PERSPECTIVE: FOCUS ON HOMEOSTASIS

Homeostatically, the blood vessels serve as passageways to transport blood to and from the cells for O_2 and nutrient delivery, waste removal, distribution of fluid and electrolytes, and hormonal signaling, among other things. Cells soon die if deprived of their blood supply; brain cells succumb within four minutes. Blood is constantly recycled and reconditioned as it travels through the various organs via the vascular highways. Hence the body needs only a very small volume of blood to maintain the appropriate chemical composition of the entire internal fluid environment on which the cells depend for their survival. For example, O_2 is continually picked up by blood in the lungs and constantly delivered to all the body cells.

The smallest blood vessels, the capillaries, are the actual site of exchange between the blood and surrounding cells. Capillaries bring homeostatically maintained blood within 0.01 cm of every cell in the body; this proximity is critical, because beyond a few centimeters materials cannot diffuse rapidly enough to support life-sustaining activities. Oxygen that would take months to years to diffuse from the lungs to all the cells of the body is continuously delivered at the "doorstep" of every cell, where diffusion can efficiently accomplish short local exchanges between the capillaries and surrounding cells. Likewise, hormones must be rapidly transported through the circulatory system from their sites of production in endocrine glands to their sites of action in other parts of the body. These chemical messengers could not diffuse nearly rapidly enough to their target organs to effectively exert their controlling effects, many of which are aimed toward maintaining homeostasis.

The rest of the circulatory system is designed to transport blood to and from the capillaries. The arteries and arterioles distribute blood pumped by the heart to the capillaries for life-sustaining exchanges to take place, and the venules and veins collect blood from the capillaries and return it to the heart, where the process is repeated.

CHAPTER SUMMARY

Introduction (pp. 275–279)

▪ Materials can be exchanged between various parts of the body and with the external environment by means of the blood vessel network that transports blood to and from all organs. *(Review Figure 10-1.)*

▪ Organs that replenish nutrient supplies and remove metabolic wastes from the blood receive a greater percentage of the cardiac output than is warranted by their metabolic needs. These "reconditioning" organs can better tolerate reductions in blood supply than can organs that receive blood solely for meeting their own metabolic needs.

▪ The brain is especially vulnerable to reductions in its blood supply. Therefore, maintaining adequate flow to this vulnerable organ is a high priority in circulatory function.

▪ The flow rate of blood through a vessel is directly proportional to the pressure gradient and inversely proportional to the resistance. The higher pressure at the beginning of a vessel is established by the pressure imparted to the blood by cardiac contraction. The lower pressure at the end is due to frictional losses as flowing blood rubs against the vessel wall. *(Review Figure 10-2.)*

▪ Resistance, the hindrance to blood flow through a vessel, is influenced most by the vessel's radius. Resistance is inversely proportional to the fourth power of the radius, so small changes in radius profoundly influence flow. As the radius increases, resistance decreases and flow increases. *(Review Figure 10-3.)*

▪ Blood flows in a closed loop between the heart and the organs. The arteries transport blood from the heart throughout the body. The arterioles regulate the amount of blood that flows through each organ. The capillaries are the actual site where materials are exchanged between blood and surrounding tissue cells. The veins return blood from the tissue level back to the heart. *(Review Figure 10-4 and Table 10-1.)*

Arteries (pp. 280–283)

▪ Arteries are large-radius, low-resistance passageways from the heart to the organs.

▪ They also serve as a pressure reservoir. Because of their elasticity, arteries expand to accommodate the extra volume of blood pumped into them by cardiac contraction and then recoil to continue driving the blood forward when the heart is relaxing. *(Review Figure 10-5.)*

▪ Systolic pressure is the peak pressure exerted by the ejected blood against the vessel walls during cardiac systole. Diastolic pressure is the minimum pressure in the arteries when blood is draining off into the vessels downstream during cardiac diastole. (Review Figures 10-6 and 10-7.)

- The average driving pressure throughout the cardiac cycle is the mean arterial pressure, which can be estimated using the following formula: mean arterial pressure = diastolic pressure + 1/3 pulse pressure. (Review Figure 10-8.)

Arterioles (pp. 283–289)

- Arterioles are the major resistance vessels. Their high resistance produces a large drop in mean pressure between the arteries and capillaries. This decline enhances blood flow by contributing to the pressure differential between the heart and organs. (Review Figure 10-8.)
- Tone, a baseline of contractile activity, is maintained in arterioles at all times.
- Arteriolar vasodilation, an expansion of arteriolar caliber above tonic level, decreases resistance and increases blood flow through the vessel, whereas vasoconstriction, a narrowing of the vessel, increases resistance and decreases flow. (Review Figure 10-9.)
- Arteriolar caliber is subject to two types of control mechanisms: local (intrinsic) controls and extrinsic controls.
- Local controls primarily involve local chemical changes associated with changes in the level of metabolic activity in an organ. These changes in local metabolic factors cause the release of vasoactive mediators from the endothelium in the vicinity. These vasoactive mediators act on the underlying arteriolar smooth muscle to bring about an appropriate change in the caliber of the arterioles supplying the organ. By adjusting the resistance to blood flow in this manner, the local control mechanism adjusts blood flow to the organ to match the momentary metabolic needs of the organ. (Review Figure 10-10 and Table 10-2.)
- Arteriolar caliber can be adjusted independently in different organs by local control factors. Such adjustments are important in variably distributing cardiac output. (Review Figure 10-11.)
- Extrinsic control is accomplished primarily by sympathetic nerve influence and to a lesser extent by hormonal influence over arteriolar smooth muscle. Extrinsic controls are important in maintaining mean arterial pressure. Arterioles are richly supplied with sympathetic nerve fibers, whose increased activity produces generalized vasoconstriction and a subsequent increase in mean arterial pressure. Decreased sympathetic activity produces generalized arteriolar vasodilation, which lowers mean arterial pressure. These extrinsically controlled adjustments of arteriolar caliber help maintain the appropriate pressure head for driving blood forward to the tissues. (Review Figure 10-12.)

Capillaries (pp. 289–298)

- The thin-walled, small-radius, extensively branched capillaries are ideally suited to serve as sites of exchange between the blood and surrounding tissue cells. Anatomically, the surface area for exchange is maximized and diffusion distance is minimized in the capillaries. Furthermore, because of their large total cross-section area, the velocity of blood flow through capillaries is relatively slow, providing adequate time for exchanges to take place. (Review Figures 10-13 through 10-15.)
- Two types of passive exchanges—diffusion and bulk flow—take place across capillary walls.
- Individual solutes are exchanged primarily by diffusion down concentration gradients. Lipid-soluble substances pass directly through the single layer of endothelial cells lining a capillary, whereas water-soluble substances pass through water-filled pores between the endothelial cells. Plasma proteins generally do not escape. (Review Figures 10-16 and 10-17.)

- Imbalances in physical pressures acting across capillary walls are responsible for bulk flow of fluid through the pores back and forth between plasma and interstitial fluid. (1) Fluid is forced out of the first portion of the capillary (ultrafiltration), where outward pressures (mainly capillary blood pressure) exceed inward pressures (mainly plasma-colloid osmotic pressure). (2) Fluid is returned to the capillary along its last half, when outward pressures fall below inward pressures. The reason for the shift in balance down the length of the capillary is the continuous decline in capillary blood pressure while the plasma-colloid osmotic pressure remains constant. (Review Figures 10-8, 10-18, and 10-19.)
- Bulk flow is responsible for the distribution of extracellular fluid between plasma and interstitial fluid.
- Normally, slightly more fluid is filtered than is reabsorbed. The extra fluid, any leaked proteins, and bacteria in the tissue are picked up by the lymphatic system. Bacteria are destroyed as lymph passes through the lymph nodes en route to being returned to the venous system. (Review Figures 10-20 and 10-21.)

Veins (pp. 298–302)

- Veins are large-radius, low-resistance passageways through which blood returns from the organs to the heart.
- In addition, veins can accommodate variable volumes of blood and therefore act as a blood reservoir. The capacity of veins to hold blood can change markedly with little change in venous pressure. Veins are thin-walled, highly distensible vessels that can passively stretch to store a larger volume of blood. (Review Figure 10-23.)
- The primary force that produces venous flow is the pressure gradient between the veins and atrium (that is, what remains of the driving pressure imparted to the blood by cardiac contraction). (Review Figures 10-8 and 10-24.)
- Venous return is enhanced by sympathetically induced venous vasoconstriction and by external compression of the veins from contraction of surrounding skeletal muscles, both of which drive blood out of the veins. (Review Figures 10-24 through 10-27.)
- One-way venous valves ensure that blood is driven toward the heart and kept from flowing back toward the tissues. (Review Figure 10-28.)
- Venous return is also enhanced by the respiratory pump and the cardiac suction effect. Respiratory activity produces a less-than-atmospheric pressure in the chest cavity, thus establishing an external pressure gradient that encourages flow from the lower veins that are exposed to atmospheric pressure to the chest veins that empty into the heart. In addition, slightly negative pressures created within the atria during ventricular systole and within the ventricles during ventricular diastole exert a suctioning effect that further enhances venous return and facilitates cardiac filling. (Review Figures 10-24 and 10-29.)

Blood Pressure (pp. 302–310)

- Regulation of mean arterial pressure depends on control of its two main determinants, cardiac output and total peripheral resistance. (Review Figure 10-30.)
- Control of cardiac output in turn depends on regulation of heart rate and stroke volume, whereas total peripheral resistance is determined primarily by the degree of arteriolar vasoconstriction. (Review Figures 9-23 and 10-12.)
- Short-term regulation of blood pressure is accomplished mainly by the baroreceptor reflex. Carotid sinus and aortic arch baroreceptors continuously monitor mean arterial pressure. When they detect a deviation from normal, they signal the medullary cardiovascular center, which responds by adjusting autonomic out-

- put to the heart and blood vessels to restore the blood pressure to normal. *(Review Figures 10-31 through 10-34.)*
- Long-term control of blood pressure involves maintaining proper plasma volume through the kidneys' control of salt and water balance.

- Blood pressure can be abnormally high (hypertension) or abnormally low (hypotension). Severe, sustained hypotension resulting in generalized inadequate blood delivery to the tissues is known as *circulatory shock. (Review Figure 10-35.)*

REVIEW EXERCISES

Objective Questions (Answers on p. A-44)

1. In general, the parallel arrangement of the vascular system enables each organ to receive its own separate arterial blood supply. *(True or false?)*
2. More blood flows through the capillaries during cardiac systole than during diastole. *(True or false?)*
3. The capillaries contain only 5% of the total blood volume at any point in time. *(True or false?)*
4. The same volume of blood passes through the capillaries in a minute as passes through the aorta, even though blood flow is much slower in the capillaries. *(True or false?)*
5. Because of gravitational effects, venous pressure in the lower extremities is greater when a person is standing up than when the person is lying down. *(True or false?)*
6. Which of the following functions is (are) attributable to arterioles?
   a. producing a significant decline in mean pressure, which helps establish the driving pressure gradient between the heart and organs
   b. serving as site of exchange of materials between blood and surrounding tissue cells
   c. acting as main determinant of total peripheral resistance
   d. determining the pattern of distribution of cardiac output
   e. helping regulate mean arterial blood pressure
   f. converting the pulsatile nature of arterial blood pressure into a smooth, nonfluctuating pressure in the vessels further downstream
   g. acting as a pressure reservoir
7. Using the answer code on the right, indicate whether the following factors increase or decrease venous return:
   ____ 1. sympathetically induced venous vasoconstriction
   ____ 2. skeletal muscle activity
   ____ 3. gravitational effects on the venous system
   ____ 4. respiratory activity
   ____ 5. increased atrial pressure associated with a leaky AV valve
   ____ 6. ventricular pressure change associated with diastolic recoil

   (a) increases venous return
   (b) decreases venous return
   (c) has no effect on venous return

8. Using the following answer code, indicate what kind of compensatory changes occur in the factors in question to restore blood pressure to normal in response to hypotension resulting from severe hemorrhage:
   (a) increased
   (b) decreased
   (c) no effect

   ____ 1. rate of afferent firing generated by the carotid sinus and aortic arch baroreceptors
   ____ 2. sympathetic output by the cardiovascular center
   ____ 3. parasympathetic output by the cardiovascular center
   ____ 4. heart rate
   ____ 5. stroke volume
   ____ 6. cardiac output
   ____ 7. arteriolar radius
   ____ 8. total peripheral resistance
   ____ 9. venous radius
   ____ 10. venous return
   ____ 11. urinary output
   ____ 12. fluid retention within the body
   ____ 13. fluid movement from interstitial fluid into plasma across the capillaries

Essay Questions

1. Compare blood flow through reconditioning organs and through organs that do not recondition the blood.
2. Discuss the relationships among flow rate, pressure gradient, and vascular resistance. What is the major determinant of resistance to flow?
3. Describe the structure and major functions of each segment of the vascular tree.
4. How do the arteries serve as a pressure reservoir?
5. Describe the indirect technique of measuring arterial blood pressure by means of a sphygmomanometer.
6. Define *vasoconstriction* and *vasodilation.*
7. Discuss the local and extrinsic controls that regulate arteriolar resistance.
8. What is the primary means by which individual solutes are exchanged across the capillary walls? What forces produce bulk flow across the capillary walls? Of what importance is bulk flow?
9. How is lymph formed? What are the functions of the lymphatic system?
10. Define *edema,* and discuss its possible causes.
11. How do veins serve as a blood reservoir?
12. Compare the effect of vasoconstriction on the rate of blood flow in arterioles and veins.
13. Discuss the factors that determine mean arterial pressure.
14. Review the effects on the cardiovascular system of parasympathetic and sympathetic stimulation.
15. Differentiate between secondary hypertension and primary hypertension. What are the potential consequences of hypertension?

POINTS TO PONDER

(Explanations on p. A-44)

1. During coronary bypass surgery, a piece of vein is often removed from the patient's leg and surgically attached within the coronary circulatory system so that through it blood detours around an occluded coronary artery segment. Why must the patient wear, for an extended period of time after surgery, an elastic support stocking on the limb from which the vein was removed?

2. Assume a person has a blood pressure recording of 125/77:
   a. What is the systolic pressure?
   b. What is the diastolic pressure?
   c. What is the pulse pressure?
   d. What is the mean arterial pressure?
   e. Would any sound be heard when the pressure in an external cuff around the arm was 130 mm Hg? *(Yes or no?)*
   f. Would any sound be heard when cuff pressure was 118 mm Hg?
   g. Would any sound be heard when cuff pressure was 75 mm Hg?

3. A classmate who has been standing still for several hours working on a laboratory experiment suddenly faints. What is the probable explanation? What would you do if the person next to him tried to get him up?

4. A drug applied to a piece of excised arteriole causes the vessel to relax, but an isolated piece of arteriolar muscle stripped from the other layers of the vessel fails to respond to the same drug. What is the probable explanation?

5. Explain how each of the following antihypertensive drugs would lower arterial blood pressure:
   a. drugs that block alpha-1-adrenergic receptors (for example, *phentolamine*) (*Hint:* Review *adrenergic receptors* on p. 191.)
   b. drugs that block beta-1-adrenergic receptors (for example, *metoprolol*)
   c. drugs that directly relax arteriolar smooth muscle (for example, *hydralazine*)
   d. diuretic drugs that increase urinary output (for example, *furosemide*)
   e. drugs that block release of norepinephrine from sympathetic endings (for example, *guanethidine*)
   f. drugs that act on the brain to reduce sympathetic output (for example, *clonidine*)
   g. drugs that block Ca^{2+} channels (for example, *verapamil*)
   h. drugs that interfere with the production of angiotensin II (for example, *captopril*)

CLINICAL CONSIDERATION

(Explanation on p. A-45)

Li-Ying C. has just been diagnosed as having hypertension secondary to a *pheochromocytoma,* a tumor of the adrenal medulla that secretes excessive epinephrine. Explain how this condition leads to secondary hypertension, by describing the effect that excessive epinephrine would have on various factors that determine arterial blood pressure.

PHYSIOEDGE RESOURCES

PhysioEdge CD-ROM

PhysioEdge, the CD-ROM packaged with your text, focuses on the concepts students find most difficult to learn. Figures marked with this icon have associated activities on the CD. For a visual review of concepts in this chapter, check out the following:

Tutorial: Cardiovascular Physiology—Arteries tab

Media Exercise 10.1: Blood Flow and Total Cross-Sectional Area

Media Exercise 10.2: Arteriolar Resistance and Flow in Capillaries

Media Exercise 10.3: Circulation

PhysioEdge Website

The website for this book contains a wealth of helpful study aids, as well as many ideas for further reading and research. Log on to:

http://www.brookscole.com/hpfundamentals3

Select Chapter 10 from the drop-down menu or click on one of the many resource areas, including Case Histories, which introduce clinical aspects of human physiology. For this chapter check out. Case History 8: Toxic Shock Syndrome.

For Suggested Readings, consult **InfoTrac College Edition/ Research** on the PhysioEdge website or go directly to InfoTrac College Edition, your online research library, at: **http://infotrac.thomsonlearning.com**

**Blood; Immune System;
Integumentary System (Skin)**

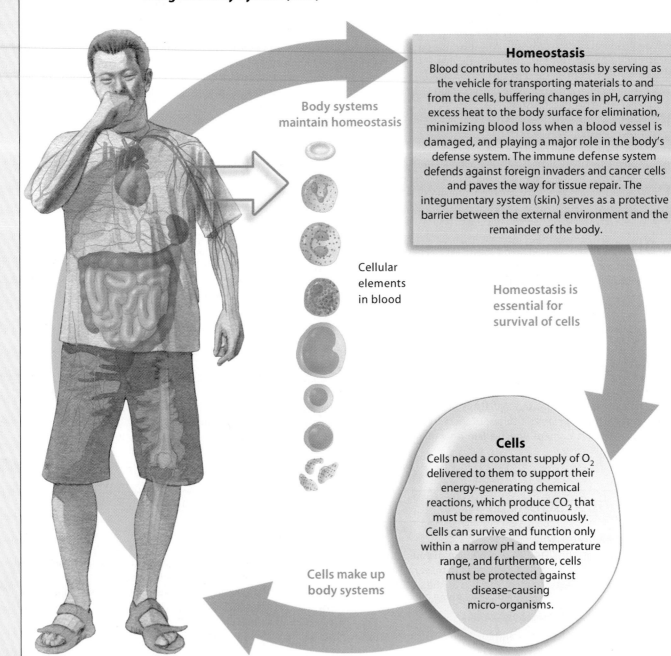

Body systems
maintain homeostasis

Cellular
elements
in blood

Homeostasis

Blood contributes to homeostasis by serving as the vehicle for transporting materials to and from the cells, buffering changes in pH, carrying excess heat to the body surface for elimination, minimizing blood loss when a blood vessel is damaged, and playing a major role in the body's defense system. The immune defense system defends against foreign invaders and cancer cells and paves the way for tissue repair. The integumentary system (skin) serves as a protective barrier between the external environment and the remainder of the body.

Homeostasis is
essential for
survival of cells

Cells

Cells need a constant supply of O_2 delivered to them to support their energy-generating chemical reactions, which produce CO_2 that must be removed continuously. Cells can survive and function only within a narrow pH and temperature range, and furthermore, cells must be protected against disease-causing micro-organisms.

Cells make up
body systems

Blood is the vehicle for long-distance, mass transport of materials between the cells and the external environment or between the cells themselves. Such transport is essential for maintaining homeostasis. Blood consists of a complex liquid **plasma** in which the cellular elements are suspended. Among the blood's cellular elements, **erythrocytes** transport O_2 in the blood. **Platelets** are important in **hemostasis**, the stopping of bleeding from an injured vessel. **Leukocytes,** as part of a complex, multifaceted internal defense system, the **immune system,** are transported in the blood to sites of injury or of invasion by disease-causing micro-organisms. The **integumentary system** (skin) serves as an outer protective barrier that prevents the loss of internal fluids and resists penetration by external agents.

The Blood and Body Defenses

INTRODUCTION

Blood represents about 8% of total body weight and has an average volume of 5 liters in women and 5.5 liters in men. It consists of three types of specialized cellular elements, *erythrocytes (red blood cells)*, *leukocytes (white blood cells)*, and *platelets (thrombocytes)*, suspended in the complex liquid *plasma* (▲ Table 11-1). Erythrocytes and leukocytes are both whole cells, whereas platelets are cell fragments. For convenience, we will refer collectively to all these blood cellular elements as "*blood cells.*"

The constant movement of blood as it flows through the blood vessels keeps the blood cells rather evenly dispersed within the plasma. However, if you put a sample of whole blood in a test tube and treat it to prevent clotting, the heavier cells slowly settle to the bottom and the lighter plasma rises to the top. This process can be speeded up by centrifuging, which quickly packs the cells in the bottom of the tube (● Figure 11-1). Because over 99% of the cells are erythrocytes, the **hematocrit**, or **packed cell volume**, essentially represents the percentage of erythrocytes in the total blood volume. The hematocrit averages 42% for women and slightly higher, 45%, for men. Plasma accounts for the remaining volume. Accordingly, the average volume of plasma in the blood is 58% for women and 55% for men. White blood cells and platelets, which are colorless and less dense than red cells, are packed in a thin, cream-colored layer, the "*buffy coat,*" on top of the packed red cell column. They are less than 1% of the total blood volume.

Let's first consider the properties of the largest portion of the blood, the plasma, before turning our attention to the cellular elements.

PLASMA

Plasma, being a liquid, consists of 90% water.

TABLE 11-1

Blood Constituents and Their Functions

CONSTITUENT	FUNCTIONS
Plasma	
Water	Transport medium; carries heat
Electrolytes	Membrane excitability; osmotic distribution of fluid between ECF and ICF; buffering of pH changes
Nutrients, wastes, gases, hormones	Transported in blood; the blood gas CO_2 plays a role in acid–base balance
Plasma proteins	Exert an osmotic effect that is important in the distribution of ECF between the vascular and interstitial compartments; buffer pH changes; transport many substances; clotting factors; inactive precursor molecules; antibodies
Cellular Elements	
Erythrocytes	Transport O_2 and CO_2 (mainly O_2)
Leukocytes	
Neutrophils	Phagocytes that engulf bacteria and debris
Eosinophils	Attack parasitic worms; important in allergic reactions
Basophils	Release histamine, which is important in allergic reactions, and heparin, which helps clear fat from the blood
Monocytes	In transit to become tissue macrophages
Lymphocytes	
B lymphocytes	Produce antibodies
T lymphocytes	Cell-mediated immune responses
Platelets	Hemostasis

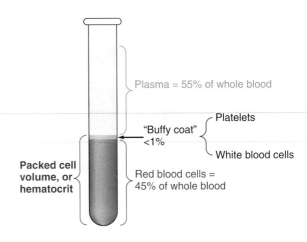

● FIGURE 11-1

Hematocrit. The values given are for men. The average hematocrit for women is 42%, with plasma occupying 58% of the blood volume.

 For an interaction related to this figure, see Media Exercise 11.1: Blood on the CD-ROM.

about 1% of plasma weight. The most abundant electrolytes (ions) in the plasma are Na^+ and Cl^-, the components of common salt. There are smaller amounts of HCO_3^-, K^+, Ca^{2+}, and others. The most notable functions of these ions are their roles in membrane excitability, osmotic distribution of fluid between the ECF and cells, and buffering of pH changes; these functions are discussed elsewhere.

The most plentiful organic constituents by weight are the plasma proteins, which compose 6 to 8% of plasma's total weight. We will examine them more thoroughly in the next section. The remaining small percentage of plasma consists of other organic substances, including nutrients (such as glucose, amino acids, lipids, and vitamins), waste products (creatinine, bilirubin, and nitrogenous substances such as urea); dissolved gases (O_2 and CO_2), and hormones. Most of these substances are merely being transported in the plasma. For example, endocrine glands secrete hormones into the plasma, which transports these chemical messengers to their sites of action.

Many of the functions of plasma are carried out by plasma proteins.

Plasma proteins are the one group of plasma constituents that are not along just for the ride. These important components normally stay in the plasma, where they perform many valuable functions. Here are the most important of these functions, which are elaborated on elsewhere in the text:

1. Unlike other plasma constituents that are dissolved in the plasma water, plasma proteins are dispersed as a colloid (see p. A-10). Furthermore, because they are the largest of the plasma constituents, plasma proteins usually do not exit through the narrow pores in the capillary walls to enter the interstitial fluid. By their presence as a colloidal dispersion in the plasma and their absence in the interstitial fluid, plasma proteins establish an osmotic gradient between blood and

Plasma water is a transport medium for many inorganic and organic substances.

Plasma water serves as a medium for materials being carried in the blood. Also, because water has a high capacity to hold heat, plasma can absorb and distribute much of the heat generated metabolically within tissues, whereas the temperature of the blood itself undergoes only small changes. As blood travels close to the surface of the skin, heat energy not needed to maintain body temperature is eliminated to the environment.

A large number of inorganic and organic substances are dissolved in the plasma. Inorganic constituents account for

interstitial fluid. This colloid osmotic pressure is the primary force preventing excessive loss of plasma from the capillaries into the interstitial fluid and thus helps maintain plasma volume (see p. 294).

2. Plasma proteins are partially responsible for plasma's capacity to buffer changes in pH (see p. 294).

3. Some plasma proteins bind substances that are poorly soluble in plasma (for example, thyroid hormone, cholesterol, and iron) for transport in the plasma.

4. Many of the factors involved in the blood-clotting process are plasma proteins.

5. Some plasma proteins are inactive, circulating precursor molecules, which are activated as needed by specific regulatory inputs. For example, the plasma protein *angiotensinogen* is activated to *angiotensin,* which plays an important role in regulating salt balance in the body (see p. 417).

6. One specific group of plasma proteins, the *gamma globulins,* are *immunoglobulins* (antibodies), which are crucial to the body's defense mechanism.

Plasma proteins are synthesized by the liver, with the exception of gamma globulins, which are produced by lymphocytes, one of the types of white blood cells.

ERYTHROCYTES

Each milliliter of blood contains about 5 billion **erythrocytes** (**red blood cells** or **RBCs**) on average, commonly reported clinically in a **red blood cell count** as 5 million cells per cubic millimeter (mm^3).

▌ The structure of erythrocytes is well suited to their main function of O_2 transport in the blood.

The shape and content of erythrocytes are ideally suited to carry out their primary function, namely transporting O_2 and to a lesser extent CO_2 and hydrogen ion in the blood.

ERYTHROCYTE SHAPE

Erythrocytes are flat, disc-shaped cells indented in the middle on both sides, like a doughnut with a flattened center instead of a hole (that is, they are biconcave discs) (● Figure 11-2). This unique shape contributes in two ways to the efficiency with which red blood cells perform their main function of O_2 transport in the blood. (1) The biconcave shape provides a larger surface area for diffusion of O_2 across the membrane than would a spherical cell of the same volume. (2) The thinness of the cell enables O_2 to diffuse rapidly between the exterior and innermost regions of the cell.

The most important anatomic feature that enables RBCs to transport O_2 is the hemoglobin they contain. Let's look at this unique molecule in more detail.

PRESENCE OF HEMOGLOBIN

Hemoglobin is found only in red blood cells. A **hemoglobin** molecule has two parts: (1) the **globin** portion, a protein made

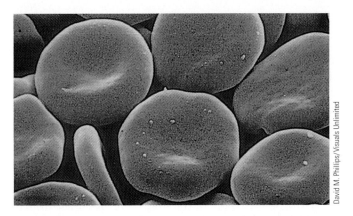

● **FIGURE 11-2**

Anatomic characteristics of erythrocytes. Appearance of erythrocytes under a scanning electron microscope. Note their biconcave shape.

up of four highly folded polypeptide chains, and (2) four iron-containing, nonprotein groups known as **heme groups,** each of which is bound to one of the polypeptides (● Figure 11-3). Each of the four iron atoms can combine reversibly with one molecule of O_2; thus each hemoglobin molecule can pick up four O_2 passengers in the lungs. Because O_2 is poorly soluble in the plasma, 98.5% of the O_2 carried in the blood is bound to hemoglobin (see p. 387).

Hemoglobin is a pigment (that is, it is naturally colored). Because of its iron content, it appears reddish when combined with O_2 and bluish when deoxygenated. Thus fully oxygenated arterial blood is red in color, and venous blood, which has lost some of its O_2 load at the tissue level, has a bluish cast.

● **FIGURE 11-3**

Hemoglobin molecule. A hemoglobin molecule consists of four highly folded polypeptide chains (the globin portion) and four iron-containing heme groups.

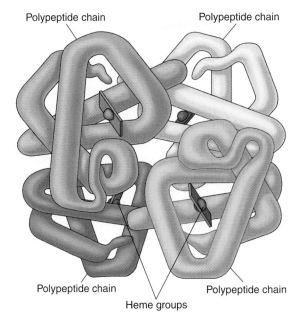

Polypeptide chain   Polypeptide chain

Polypeptide chain   Polypeptide chain

Heme groups

In addition to carrying O_2, hemoglobin can also combine with the following:

1. *Carbon dioxide.* Hemoglobin helps transport this gas from the tissue cells back to the lungs (see p. 391).
2. *The acidic hydrogen-ion portion (H^+) of ionized carbonic acid,* which is generated at the tissue level from CO_2. Hemoglobin buffers this acid so that it minimally alters the pH of the blood (see p. 457).
3. *Carbon monoxide (CO).* This gas is not normally in the blood, but if inhaled it preferentially occupies the O_2-binding sites on hemoglobin, causing CO poisoning (see p. 391).
4. *Nitric oxide (NO).* In the lungs, the vasodilator nitric oxide binds to hemoglobin. This nitric oxide is released at the tissues, where it relaxes and dilates the local arterioles (see p. 286). This vasodilation helps ensure that the O_2-rich blood can make its vital rounds and also helps stabilize blood pressure.

Therefore, hemoglobin plays the key role in O_2 transport while contributing significantly to CO_2 transport and the pH-buffering capacity of blood. Furthermore, by toting along its own vasodilator, hemoglobin helps deliver the O_2 it is carrying.

LACK OF NUCLEUS AND ORGANELLES

To maximize its hemoglobin content, a single erythrocyte is stuffed with more than 250 million hemoglobin molecules, excluding almost everything else. (That means each RBC can carry over a billion O_2 molecules!) Red blood cells contain no nucleus, organelles, or ribosomes. During the cell's development these structures are extruded to make room for more hemoglobin. Thus an RBC is mainly a plasma membrane–enclosed sac full of hemoglobin.

Ironically, even though erythrocytes are the vehicles for transporting O_2 to all other tissues of the body, for energy production they themselves cannot use the O_2 they are carrying. Lacking the mitochondria that house the enzymes for oxidative phosphorylation, erythrocytes must rely entirely on glycolysis for ATP formation (see p. 27).

▌ The bone marrow continuously replaces worn-out erythrocytes.

Each of us has a total of 25 to 30 trillion RBCs streaming through our blood vessels at any given time (100,000 times more in number than the number of the entire U.S. population)! Yet these vital gas-transport vehicles are short-lived and must be replaced at the average rate of 2 to 3 million cells per second.

ERYTHROCYTES' SHORT LIFE SPAN

The price erythrocytes pay for their generous content of hemoglobin, to the exclusion of the usual specialized intracellular machinery, is a short life span. Without DNA and RNA, red blood cells cannot synthesize proteins for cellular repair, growth, and division or for renewing enzyme supplies. Equipped only with initial supplies synthesized before they extrude their nucleus, organelles, and ribosomes, RBCs survive an average of only 120 days, in contrast to nerve and muscle cells, which last a person's entire life. During its short life span of four months, each erythrocyte travels about 700 miles as it circulates through the vasculature.

As a red blood cell ages, its nonreparable plasma membrane becomes fragile and prone to rupture as the cell squeezes through tight spots in the vascular system. Most old RBCs meet their final demise in the **spleen**, because this organ's narrow, winding capillary network is a tight fit for these fragile cells. The spleen lies in the upper left part of the abdomen. In addition to removing most of the old erythrocytes from circulation, the spleen has a limited ability to store healthy erythrocytes in its pulpy interior, serves as a reservoir site for platelets, and contains an abundance of lymphocytes, a type of white blood cell.

ERYTHROPOIESIS

Because erythrocytes cannot divide to replenish their own numbers, the old ruptured cells must be replaced by new cells produced in an erythrocyte factory—the **bone marrow**—which is the soft, highly cellular tissue that fills the internal cavities of bones. The bone marrow normally generates new red blood cells, a process known as **erythropoiesis**, at a rate to keep pace with the demolition of old cells.

In children, most bones are filled with **red bone marrow** that is capable of blood cell production. As a person matures, however, fatty **yellow bone marrow** that is incapable of erythropoiesis gradually replaces red marrow, which remains only in a few isolated places, such as the sternum (breastbone), ribs, and upper ends of the long limb bones.

Red marrow not only produces RBCs but also is the ultimate source for leukocytes and platelets as well. Undifferentiated **pluripotent stem cells** reside in the red marrow, where they continuously divide and differentiate to give rise to each of the types of blood cells (see p. 4 and p. 327). These stem cells, the source of all blood cells, have now been isolated. The search has been difficult, because stem cells are less than 0.1% of all cells in the bone marrow. Although a great deal of research remains to be done, this recent discovery may hold a key to a cure for a host of blood and immune disorders as well as some types of cancer and genetic diseases.

The different types of immature blood cells, along with the stem cells, are intermingled in the red marrow at various stages of development. Once mature, the blood cells are released into the rich supply of capillaries that permeate the red marrow. Regulatory factors act on the *hemopoietic* ("blood-producing") red marrow to govern the type and number of cells generated and discharged into the blood. Of the blood cells, the mechanism for regulating RBC production is the best understood. We will consider it next.

▌ Erythropoiesis is controlled by erythropoietin from the kidneys.

Because O_2 transport in the blood is the erythrocytes' primary function, you might logically suspect that the primary stimulus for increased erythrocyte production would be reduced

O_2 delivery to the tissues. You would be correct, but low O_2 levels do not stimulate erythropoiesis by acting directly on the red bone marrow. Instead, reduced O_2 delivery to the kidneys stimulates them to secrete the hormone **erythropoietin** into the blood, and this hormone in turn stimulates erythropoiesis by the bone marrow (● Figure 11-4). This increased erythropoietic activity elevates the number of circulating RBCs, thereby increasing O_2-carrying capacity of the blood and restoring O_2 delivery to the tissues to normal. Once normal O_2 delivery to the kidneys is achieved, erythropoietin secretion is turned down until needed again. In this way, erythrocyte production is normally balanced against destruction or loss of these cells so that O_2-carrying capacity in the blood stays fairly constant.

When you donate blood, your circulating erythrocyte supply is replenished in less than a week. In severe loss of RBCs, as in hemorrhage or abnormal destruction of young circulating erythrocytes, the rate of erythropoiesis can be increased to more than six times the normal level.

SYNTHETIC ERYTHROPOIETIN

Clinical Note Researchers have identified the gene that directs erythropoietin synthesis, so this hormone can now be produced in a laboratory. Laboratory-produced erythropoietin has become biotechnology's current single biggest moneymaker, with sales exceeding $1 billion annually. This hormone is often used to boost RBC production in patients with suppressed erythropoietic activity, such as those undergoing chemotherapy for cancer. (Chemotherapy drugs interfere with the rapid cell division characteristic of both cancer cells and developing RBCs.) Furthermore, the ready availability of this hormone has diminished the need for blood transfusions. For example, transfusion of a surgical patient's own previously collected blood, coupled with erythropoietin to stimulate further RBC production, has reduced the use of donor blood by as much as 50% in some hospitals. (For a discussion of erythropoietin abuse by some athletes, see the boxed feature, ▶ Beyond the Basics, on p. 320.)

Anemia can be caused by a variety of disorders.

Clinical Note Despite control measures, O_2-carrying capacity cannot always be maintained to meet tissue needs. The term **anemia** refers to a below-normal O_2-carrying capacity of the blood and is characterized by a low hematocrit. It can be brought about by a decreased rate of erythropoiesis, excessive losses of erythrocytes, or a deficiency in the hemoglobin content of erythrocytes. The various causes of anemia can be grouped into six categories:

1. **Nutritional anemia** is caused by a dietary deficiency of a factor needed for erythropoiesis. The production of RBCs depends on an adequate supply of essential raw ingredients, some of which are not synthesized in the body but must be provided by dietary intake. For example, *iron deficiency anemia* occurs when not enough iron is available for synthesis of hemoglobin.

2. **Pernicious anemia** is caused by an inability to absorb enough ingested vitamin B_{12} from the digestive tract. Vitamin B_{12} is essential for normal RBC production and maturation. It is abundant in a variety of foods. The problem is a deficiency of *intrinsic factor*, a special substance secreted by the lining of the stomach (see p. 482). Vitamin B_{12} can be absorbed from the intestinal tract only when this nutrient is bound to intrinsic factor. When intrinsic factor is deficient, not enough ingested vitamin B_{12} is absorbed. The resulting impairment of RBC production and maturation leads to anemia.

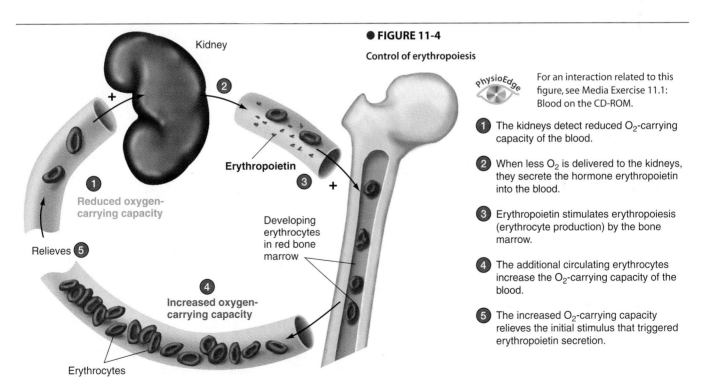

● **FIGURE 11-4**

Control of erythropoiesis

PhysioEdge For an interaction related to this figure, see Media Exercise 11.1: Blood on the CD-ROM.

1 The kidneys detect reduced O_2-carrying capacity of the blood.

2 When less O_2 is delivered to the kidneys, they secrete the hormone erythropoietin into the blood.

3 Erythropoietin stimulates erythropoiesis (erythrocyte production) by the bone marrow.

4 The additional circulating erythrocytes increase the O_2-carrying capacity of the blood.

5 The increased O_2-carrying capacity relieves the initial stimulus that triggered erythropoietin secretion.

Kidney

Erythropoietin

1 Reduced oxygen-carrying capacity

Relieves 5

Developing erythrocytes in red bone marrow

4 Increased oxygen-carrying capacity

Erythrocytes

Blood Doping: Is More of a Good Thing Better?

Exercising muscles require a continual supply of O_2 for generating energy to sustain endurance activities (see p. 219). **Blood doping** is a technique designed to temporarily increase the O_2-carrying capacity of blood in an attempt to gain a competitive advantage. Blood doping involves removing blood from an athlete, then promptly reinfusing the plasma but freezing the RBCs for reinfusion one to seven days before a competitive event. One to four units of blood (one unit equals 450 ml) are usually withdrawn at three- to eight-week intervals before the competition. In the periods between blood withdrawals, increased erythropoietic activity restores the RBC count to a normal level.

Reinfusion of the stored RBCs temporarily increases the red blood cell count and hemoglobin level above normal. Theoretically, blood doping would benefit endurance athletes by improving the blood's O_2-carrying capacity. If too many red cells were infused, however, performance could suffer because the increased blood viscosity would decrease blood flow.

Research indicates that, in a standard laboratory exercise test, athletes who have used blood doping may realize a 5 to 13% increase in aerobic capacity; a reduction in heart rate during exercise compared to the rate during the same exercise in the absence of blood doping; improved performance; and reduced lactic acid levels in the blood. (Lactic acid is produced when muscles resort to less efficient anaerobic glycolysis for energy production—see page 220.)

Blood doping, although probably effective, is illegal in both collegiate athletics and Olympic competition for ethical and medical reasons. Of concern, as with the use of any banned performance-enhancing product, is the loss of fair competition. Furthermore, the practice has even been implicated in the deaths of some athletes. The prohibitive regulations are very difficult to enforce, however. Current procedures permit only the testing of urine, not blood, for the use of illegal drugs among athletes. Blood doping cannot be detected by analyzing a urine sample. The only way of exposing the practice of blood doping is through witnesses or self-admission.

The recent development of synthetic erythropoietin exacerbates the problem of blood doping. Injection of this product stimulates RBC production and thus temporarily increases the O_2-carrying capacity of the blood. Rigorous studies have demonstrated that synthetic erythropoietin may improve an endurance athlete's performance by 7 to 10%. Using erythropoietin is more convenient for the athlete than blood doping and minimizes the number of witnesses involved, thereby reducing the probability of disclosure. Although formally banned, erythropoietin is not detectable using the technology that sports officials currently have available to catch drug cheats.

Erythropoietin is now widely used among competitors in cycling, cross-country skiing, and long-distance running and swimming. This practice is ill advised, however, not only because of legal and ethical implications but because of the dangers of increasing blood viscosity. Synthetic erythropoietin is believed responsible for the deaths of 20 European cyclists since 1987. Unfortunately, too many athletes are willing to take the risks.

---

3. **Aplastic anemia** is caused by failure of the bone marrow to produce enough RBCs, even though all ingredients necessary for erythropoiesis are available. Reduced erythropoietic capability can be caused by destruction of red bone marrow by toxic chemicals (such as benzene), heavy exposure to radiation, invasion of the marrow by cancer cells, or chemotherapy for cancer. The anemia's severity depends on the extent to which erythropoietic tissue is destroyed; severe losses are fatal.

4. **Renal anemia** may result from kidney disease. Because erythropoietin from the kidneys is the primary stimulus for promoting erythropoiesis, inadequate erythropoietin secretion by diseased kidneys leads to insufficient RBC production.

5. **Hemorrhagic anemia** is caused by losing a lot of blood. The loss can be either acute, such as a bleeding wound, or chronic, such as excessive menstrual flow.

6. **Hemolytic anemia** is caused by the rupture of too many circulating erythrocytes. **Hemolysis,** the rupture of RBCs, occurs either because otherwise normal cells are induced to rupture by external factors, as in the invasion of RBCs by *malaria* parasites, or because the cells are defective, as in sickle cell disease. **Sickle cell disease** is the best-known example among various hereditary abnormalities of erythrocytes that make these cells very fragile. It affects about 1 in 650 African Americans. In this condition, a defective type of hemoglobin joins together to form rigid chains that make the RBC stiff and unnaturally shaped, like a crescent or sickle (● Figure 11-5). These deformed RBCs tend to clump together and can block blood flow through small vessels, leading to tissue damage.

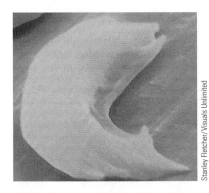

Stanley Fletcher/Visuals Unlimited

● **FIGURE 11-5**

Sickle-shaped red blood cell

Furthermore, the defective erythrocytes are fragile and prone to rupture, even as young cells, as they travel through the narrow splenic capillaries. Despite an accelerated rate of erythropoiesis triggered by the constant excessive loss of RBCs, production may not be able to keep pace with the rate of destruction, and anemia may result.

▌ Polycythemia is an excess of circulating erythrocytes.

Polycythemia, in contrast to anemia, is characterized by too many circulating RBCs. There are two general types of polycythemia, depending on the circumstances triggering the excess RBC production: primary polycythemia and secondary polycythemia.

Primary polycythemia is caused by a tumorlike condition of the bone marrow in which erythropoiesis proceeds at an excessive, uncontrolled rate instead of being subject to the normal erythropoietin regulatory mechanism. The RBC count may reach 11 million cells/mm³ (normal is 5 million cells/mm³), and the hematocrit may be as high as 70 to 80% (normal is 42 to 45%). No benefit is derived from the extra O_2-carrying capacity of the blood, because O_2 delivery is more than adequate with normal RBC numbers. Inappropriate polycythemia has harmful effects, however. The excessive number of red cells increases blood's viscosity up to five to seven times normal (that is, makes the blood "thicker"), causing the blood to flow very sluggishly, which may actually reduce O_2 delivery to the tissues (see p. 277). The increased viscosity also increases the total peripheral resistance, which may elevate the blood pressure, thus increasing the workload of the heart, unless blood-pressure control mechanisms can compensate (see ● Figure 10-12, p. 290).

Secondary polycythemia, in contrast, is an appropriate erythropoietin-induced adaptive mechanism to improve blood's O_2-carrying capacity in response to a prolonged reduction in O_2 delivery to the tissues. It occurs normally in people living at high altitudes, where less O_2 is available in the atmospheric air, or people in whom O_2 delivery to the tissues is impaired by chronic lung disease or cardiac failure. The price paid for improved O_2 delivery is an increased viscosity of the blood.

We are next going to examine platelets, another type of cellular element present in the blood.

PLATELETS AND HEMOSTASIS

An average of 250,000,000 platelets are normally present in each milliliter of blood (range of 150,000 to 350,000/mm³).

▌Platelets are cell fragments shed from megakaryocytes.

Platelets (thrombocytes) are not whole cells but small cell fragments that are shed off the outer edges of extraordinarily large bone marrow–bound cells known as **megakaryocytes** (● Figure 11-6). A single megakaryocyte typically produces about 1000 platelets. Megakaryocytes are derived from the same undifferentiated stem cells that give rise to the erythrocytic and leukocytic cell lines. Platelets are essentially detached vesicles containing pieces of megakaryocyte cytoplasm wrapped in plasma membrane.

Platelets remain functional for an average of 10 days, at which time they are removed from circulation by the tissue macrophages, especially those in the spleen and liver, and are replaced by newly released platelets from the bone marrow. The hormone **thrombopoietin**, produced by the liver, increases the number of megakaryocytes in the bone marrow and stimulates each megakaryocyte to produce more platelets as needed.

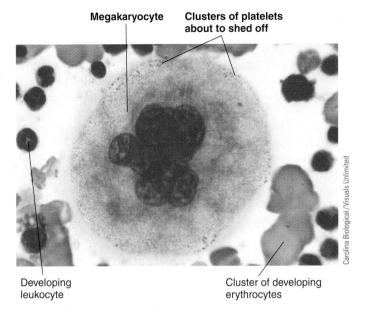

Megakaryocyte   Clusters of platelets about to shed off

Developing leukocyte   Cluster of developing erythrocytes

● FIGURE 11-6

Photomicrograph of a megakaryocyte forming platelets

 For an interaction related to this figure, see Media Exercise 11.3: Thrombocytes and Hemostasis on the CD-ROM.

Because platelets are cell fragments, they lack nuclei. However, they have organelles and cytosolic enzyme systems for generating energy and synthesizing secretory products. Furthermore, platelets contain high concentrations of actin and myosin, which enable them to contract. Their secretory and contractile abilities are important in hemostasis, a topic to which we now turn.

▌Hemostasis prevents blood loss from damaged small vessels.

Hemostasis is the arrest of bleeding from a broken blood vessel—that is, the stopping of hemorrhage (*hemo* means "blood"; *stasis* means "standing"). (Be sure not to confuse this with the term *homeostasis*.) For bleeding to take place from a vessel, there must be a break in the vessel wall, and the pressure inside the vessel must be greater than the pressure outside it to force blood out through the defect. The body's inherent hemostatic mechanisms normally are adequate to seal defects and stop blood loss through small damaged capillaries, arterioles, and venules. These small vessels are often ruptured by minor traumas of everyday life; such traumas are the most common source of bleeding, although we often are not even aware that any damage has taken place. The hemostatic mechanisms normally minimize blood loss from these minor vascular traumas.

The much rarer occurrence of bleeding from medium- to large-size vessels usually cannot be stopped by the body's hemostatic mechanisms alone. Bleeding from a severed artery is more profuse and therefore more dangerous than venous bleeding, because the outward driving pressure is greater in the arteries (that is, arterial blood pressure is much higher than venous pressure).

First-aid measures for a severed artery include applying to the wound external pressure that is greater than the arterial blood pressure, to temporarily halt the bleeding until the torn vessel can be surgically closed. Hemorrhage from a traumatized vein can often be stopped simply by elevating the bleeding body part to reduce gravity's effects on pressure in the vein (see p. 300). If the accompanying drop in venous pressure is not enough to stop the bleeding, mild external compression is usually adequate.

Hemostasis involves three major steps: (1) *vascular spasm*, (2) *formation of a platelet plug*, and (3) *blood coagulation (clotting)*. Platelets play a pivotal role in hemostasis. They obviously play a major part in forming a platelet plug, but they also contribute significantly to the other two steps .

▌ Vascular spasm reduces blood flow through an injured vessel.

A cut or torn blood vessel immediately constricts. The underlying mechanism is unclear but is thought to be an intrinsic response triggered by a paracrine released locally from the inner lining (endothelium) of the injured vessel (see p. 93). This constriction, or **vascular spasm**, slows blood flow through the defect and thus minimizes blood loss. Also, as the opposing endothelial surfaces of the vessel are pressed together by this initial vascular spasm, they become sticky and adhere to each other, further sealing off the damaged vessel. These physical measures alone cannot completely prevent further blood loss, but they minimize blood flow through the break in the vessel until the other hemostatic measures can actually plug up the hole.

▌ Platelets aggregate to form a plug at a vessel tear or cut.

Platelets normally do not stick to the smooth endothelial surface of blood vessels, but when this lining is disrupted because of vessel injury, platelets become activated by the exposed collagen, which is a fibrous protein in the underlying connective tissue. When activated, platelets quickly adhere to the collagen and form a hemostatic **platelet plug** at the site of the defect. Once platelets start aggregating, they release several important chemicals, such as *adenosine diphosphate (ADP)*. ADP makes the surface of nearby circulating platelets sticky, so that they adhere to the first layer of aggregated platelets. These newly aggregated platelets release more ADP, which causes more platelets to pile on, and so on; thus a plug of platelets is rapidly built up at the defect site, in a positive-feedback fashion (● Figure 11-7).

Given the self-perpetuating nature of platelet aggregation, why doesn't the platelet plug continue to develop and expand over the surface of the adjacent normal vessel lining? A key

---

● **FIGURE 11-7**

Formation of a platelet plug. Platelets aggregate at a vessel defect through a positive-feedback mechanism involving the release of adenosine diphosphate (ADP) from platelets, which stick to exposed collagen at the site of the injury. Platelets are prevented from aggregating at the adjacent normal vessel lining by the release of prostacyclin and nitric oxide from the undamaged endothelial cells.

 For an interaction related to this figure, see Media Exercise 11.3: Thrombocytes and Hemostasis on the CD-ROM.

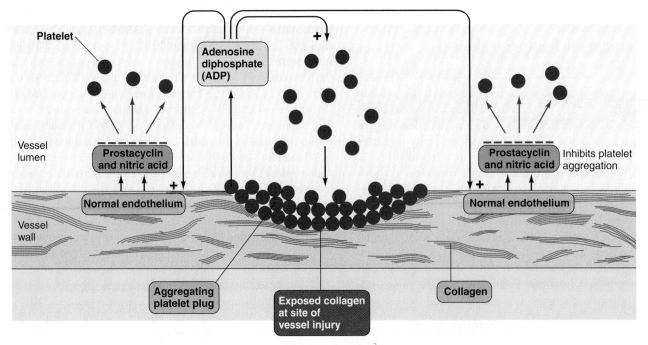

reason is that ADP and other chemicals released by the activated platelets stimulate the release of *prostacyclin* and *nitric oxide* from the adjacent normal endothelium. Both these chemicals profoundly inhibit platelet aggregation. Thus the platelet plug is limited to the defect and does not spread to the nearby undamaged vascular tissue (● Figure 11-7).

The aggregated platelet plug not only physically seals the break in the vessel but also performs three other important roles. (1) The actin-myosin complex within the aggregated platelets contracts to compact and strengthen what was originally a fairly loose plug. (2) The chemicals released from the platelet plug include several powerful vasoconstrictors, which induce profound constriction of the affected vessel to reinforce the initial vascular spasm. (3) The platelet plug releases other chemicals that enhance blood coagulation, the next step of hemostasis. Although the platelet-plugging mechanism alone is often enough to seal the myriad minute tears in capillaries and other small vessels that occur many times daily, larger holes in vessels require the formation of a blood clot to completely stop the bleeding.

▌ Clot formation results from a triggered chain reaction involving plasma clotting factors.

Blood coagulation, or **clotting**, is the transformation of blood from a liquid into a solid gel. Formation of a clot on top of the platelet plug strengthens and supports the plug, reinforcing the seal over a break in a vessel. Furthermore, as blood in the vicinity of the vessel defect solidifies, it can no longer flow. Clotting is the body's most powerful hemostatic mechanism. It is required to stop bleeding from all but the most minute defects.

CLOT FORMATION

The ultimate step in clot formation is the conversion of **fibrinogen**, a large, soluble plasma protein produced by the liver and normally always present in the plasma, into **fibrin**, an insoluble, threadlike molecule. This conversion into fibrin is catalyzed by the enzyme **thrombin** at the site of the injury. Fibrin molecules adhere to the damaged vessel surface, forming a loose, netlike meshwork that traps blood cells, including aggregating platelets. The resulting mass, or **clot**, typically appears red because of the abundance of trapped RBCs, but the foundation of the clot is formed of fibrin derived from the plasma (● Figure 11-8). Except for platelets, which play an important role in ultimately bringing about the conversion of fibrinogen to fibrin, clotting can take place in the absence of all other blood cells.

Because thrombin's action converts the ever-present fibrinogen molecules in the plasma into a blood-stanching clot, thrombin must normally be absent from the plasma except in the vicinity of vessel damage. Otherwise, blood would always be coagulated—a situation incompatible with life. How can thrombin normally be absent from the plasma, yet be readily available to trigger fibrin formation when a vessel is injured? The solution lies in thrombin's existence in the plasma in the

● **FIGURE 11-8**

Erythrocytes trapped in the fibrin meshwork of a clot

 For an interaction related to this figure, see Media Exercise 11.2: Abnormal Hemostasis on the CD-ROM.

form of an inactive precursor called **prothrombin.** What converts prothrombin into thrombin when blood clotting is desirable? This conversion involves the clotting cascade.

THE CLOTTING CASCADE

Yet another activated plasma clotting factor, **factor X**, converts prothrombin to thrombin; factor X itself is normally present in the blood in inactive form and must be converted into its active form by still another activated factor, and so on. Altogether, 12 plasma clotting factors participate in essential steps that lead to the final conversion of fibrinogen into a stabilized fibrin mesh (● Figure 11-9). These factors are designated by roman numerals in the order in which the factors were discovered, not the order in which they participate in the clotting process.[1] Most of these clotting factors are plasma proteins synthesized by the liver. Normally, they are always present in the plasma in an inactive form, like fibrinogen and prothrombin. In contrast to fibrinogen, which is converted into insoluble fibrin strands, prothrombin and the other precursors, when converted to their active form, act as proteolytic (protein-splitting) enzymes. These enzymes activate another specific factor in the clotting sequence. Once the first factor in the sequence is activated, it in turn activates the next factor, and so on, in a series of sequential reactions known as the **clotting cascade**, until thrombin catalyzes the final con-

---

[1]The term *factor VI* is no longer used. What researchers once considered a separate factor VI has now been determined to be an activated form of factor V.

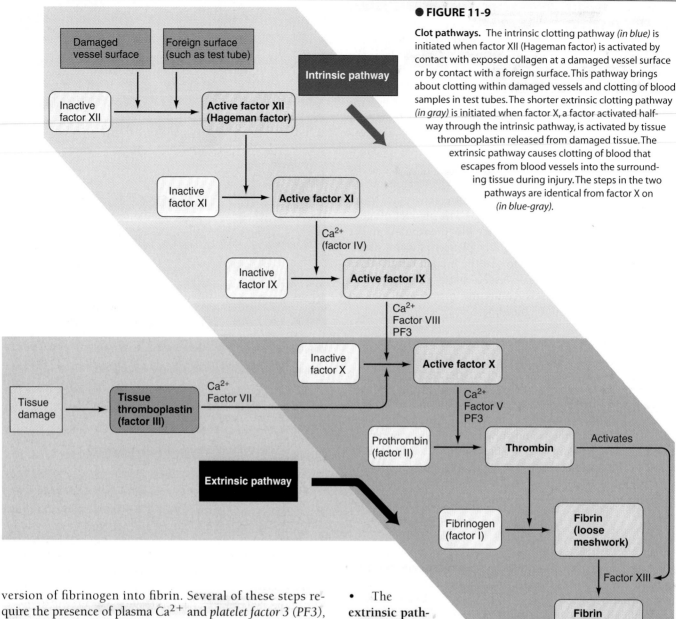

● **FIGURE 11-9**

Clot pathways. The intrinsic clotting pathway *(in blue)* is initiated when factor XII (Hageman factor) is activated by contact with exposed collagen at a damaged vessel surface or by contact with a foreign surface. This pathway brings about clotting within damaged vessels and clotting of blood samples in test tubes. The shorter extrinsic clotting pathway *(in gray)* is initiated when factor X, a factor activated halfway through the intrinsic pathway, is activated by tissue thromboplastin released from damaged tissue. The extrinsic pathway causes clotting of blood that escapes from blood vessels into the surrounding tissue during injury. The steps in the two pathways are identical from factor X on *(in blue-gray)*.

version of fibrinogen into fibrin. Several of these steps require the presence of plasma Ca^{2+} and *platelet factor 3 (PF3)*, a chemical secreted by the aggregated platelet plug. Thus, platelets also contribute to clot formation.

INTRINSIC AND EXTRINSIC PATHWAYS

The clotting cascade may be triggered by the *intrinsic pathway* or the *extrinsic pathway*:

• The **intrinsic pathway** precipitates clotting within damaged vessels as well as clotting of blood samples in test tubes. All elements necessary to bring about clotting by means of the intrinsic pathway are present in the blood. This pathway, which involves seven separate steps (shown in blue in ● Figure 11-9), is set off when **factor XII (Hageman factor)** is activated by coming into contact with either exposed collagen in an injured vessel or a foreign surface such as a glass test tube. Remember that exposed collagen also initiates platelet aggregation. Thus formation of a platelet plug and the chain reaction leading to clot formation are simultaneously set in motion when a vessel is damaged.

• The **extrinsic pathway** takes a shortcut and requires only four steps (shown in gray in ● Figure 11-9). This pathway, which requires contact with tissue factors external to the blood, initiates clotting of blood that has escaped into the tissues. When a tissue is traumatized, it releases a protein complex known as **tissue thromboplastin.** Tissue thromboplastin directly activates factor X, thereby bypassing all preceding steps of the intrinsic pathway. From this point on, the two pathways are identical.

The intrinsic and extrinsic mechanisms usually operate simultaneously. When tissue injury involves rupture of vessels, the intrinsic mechanism stops blood in the injured vessel, whereas the extrinsic mechanism clots blood that escaped into the tissue before the vessel was sealed off. Typically, clots are fully formed in three to six minutes.

CLOT RETRACTION

Once a clot is formed, contraction of the platelets trapped within the clot shrinks the fibrin mesh, pulling the edges of the damaged vessel closer together. During **clot retraction**, fluid is squeezed from the clot. This fluid, which is essentially plasma minus fibrinogen and other clotting precursors that have been removed during the clotting process, is called **serum.**

▌ Fibrinolytic plasmin dissolves clots.

A clot is not meant to be a permanent solution to vessel injury. It is a transient device to stop bleeding until the vessel can be repaired.

VESSEL REPAIR

The aggregated platelets secrete a chemical that helps promote the invasion of fibroblasts ("fiber formers") from the surrounding connective tissue into the wounded area of the vessel. Fibroblasts form a scar at the vessel defect.

CLOT DISSOLUTION

Simultaneous with the healing process, the clot, which is no longer needed to prevent hemorrhage, is slowly dissolved by a fibrinolytic (fibrin-splitting) enzyme called **plasmin**. If clots were not removed after they performed their hemostatic function, the vessels, especially the small ones that endure tiny ruptures on a regular basis, would eventually become obstructed by clots.

Plasmin, like the clotting factors, is a plasma protein produced by the liver and present in the blood in an inactive precursor form, **plasminogen**. Plasmin is activated in a fast cascade of reactions involving many factors, among them factor XII (Hageman factor), which also triggers the chain reaction leading to clot formation (● Figure 11-10). When a clot is rapidly being formed, activated plasmin becomes trapped in the clot and later dissolves it by slowly breaking down the fibrin meshwork.

Phagocytic white blood cells gradually remove the products of clot dissolution. You have observed the slow removal of blood that has clotted after escaping into the tissue layers of your skin following an injury. The black-and-blue marks of such bruised skin result from deoxygenated clotted blood within the skin; this blood is eventually cleared by plasmin action, followed by the phagocytic cleanup crew.

PREVENTING INAPPROPRIATE CLOT FORMATION

In addition to removing clots that are no longer needed, plasmin functions continually to prevent clots from forming inappropriately. Throughout the vasculature, small amounts of fibrinogen are constantly being converted into fibrin, triggered by unknown mechanisms. Clots do not develop, however, because the fibrin is quickly disposed of by plasmin activated by **tissue plasminogen activator (tPA)** from the tissues, especially the lungs. Normally, the low level of fibrin formation is counterbalanced by a low level of fibrinolytic activity, so inappropriate clotting does not occur. Only when a vessel is damaged

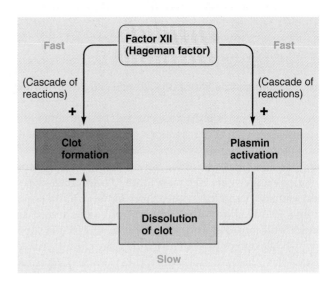

● **FIGURE 11-10**

Role of factor XII in clot formation and dissolution. Activation of factor XII (Hageman factor) simultaneously initiates a fast cascade of reactions that result in clot formation and a fast cascade of reactions that result in plasmin activation. Plasmin, which is trapped in the clot, subsequently slowly dissolves the clot. This action removes the clot when it is no longer needed after the vessel has been repaired.

do additional factors precipitate the explosive chain reaction that leads to more extensive fibrin formation and results in local clotting at the site of injury.

Clinical Note Genetically engineered tPA and other similar chemicals that trigger clot dissolution are frequently used to limit damage to cardiac muscle during heart attacks. Administering a clot-busting drug within the first hours after a clot has blocked a coronary (heart) vessel often dissolves the clot in time to restore blood flow to the cardiac muscle supplied by the blocked vessel before the muscle dies of O_2 deprivation. In recent years, tPA and related drugs have also been used successfully to promptly dissolve a stroke-causing clot within a cerebral (brain) blood vessel, minimizing loss of irreplaceable brain tissue after a stroke (see p. 114).

▌ Inappropriate clotting produces thromboembolism.

Clinical Note Despite protective measures, clots occasionally form in intact vessels. Abnormal or excessive clot formation within blood vessels—what has been dubbed "hemostasis in the wrong place"—can compromise blood flow to vital organs. The body's clotting and anticlotting systems normally function in a check-and-balance manner. Acting in concert, they permit prompt formation of "good" blood clots, thus minimizing blood loss from damaged vessels, while preventing "bad" clots from forming and blocking blood flow in intact vessels. An abnormal intravascular clot attached to a vessel wall is known as a **thrombus,** and freely floating clots are called **emboli.** An enlarging thrombus narrows and can eventually completely occlude the vessel in which it forms. By entering and completely plugging a smaller vessel, a circulating embolus can suddenly block blood flow.

Several factors, acting independently or simultaneously, can cause *thromboembolism*: (1) Roughened vessel surfaces associated with atherosclerosis can lead to thrombus formation (see p. 267). (2) Imbalances in the clotting–anticlotting systems can trigger clot formation. (3) Slow-moving blood is more apt to clot, probably because small quantities of fibrin accumulate in the stagnant blood, for example, in blood pooled in varicosed leg veins (see p. 301). (4) Widespread clotting is occasionally triggered by the release of tissue thromboplastin into the blood from large amounts of traumatized tissue. Similar widespread clotting can occur in **septicemic shock**, in which bacteria or their toxins initiate the clotting cascade.

▌ Hemophilia is the primary condition that produces excessive bleeding.

 In contrast to inappropriate clot formation in intact vessels, the opposite hemostatic disorder is failure of clots to form promptly in injured vessels, resulting in life-threatening hemorrhage from even relatively mild traumas. The most common cause of excessive bleeding is **hemophilia**, which is caused by a deficiency of one of the factors in the clotting cascade. Although a deficiency of any of the clotting factors could block the clotting process, 80% of all hemophiliacs lack the genetic ability to synthesize factor VIII.

In contrast to the more profuse bleeding that accompanies defects in the clotting mechanism, people with a platelet deficiency continuously develop hundreds of small, confined hemorrhagic areas throughout the body tissues as blood leaks from tiny breaks in the small blood vessels before coagulation takes place. Platelets normally are the primary sealers of these ever-occurring minute ruptures. In the skin of a platelet-deficient person, the diffuse capillary hemorrhages are visible as small, purplish blotches, giving rise to the term **thrombocytopenia purpura** ("the purple of thrombocyte deficiency") for this condition. (Recall that *thrombocytes* is another name for platelets.)

Vitamin K deficiency can also cause a bleeding tendency. Vitamin K, commonly known as the blood-clotting vitamin, is essential for normal clot formation.

Now let's focus on the remaining type of blood cell, the leukocytes.

PhysioEdge Click on the Media Exercises menu of the CD-ROM and work Media Exercises 11.2: Abnormal Hemostasis and 11.3: Thrombocytes and Hemostasis to test your understanding of the previous section.

LEUKOCYTES

Leukocytes (**white blood cells** or **WBCs**) are the mobile units of the body's immune defense system. **Immunity** is the body's ability to resist or eliminate potentially harmful foreign materials or abnormal cells. The leukocytes and their derivatives, along with a variety of plasma proteins, make up the **immune system**, an internal defense system that plays a key role in recognizing and either destroying or neutralizing materials within the body that are foreign to the "normal self." Specifically, the immune system

1. Defends against invading **pathogens** (disease-producing micro-organisms such as bacteria and viruses).
2. Functions as a "cleanup crew" that removes worn-out cells (such as aged red blood cells) and tissue debris (for example, tissue damaged by trauma or disease). The latter is essential for wound healing and tissue repair.
3. Identifies and destroys abnormal or mutant cells that have originated in the body. This function, termed *immune surveillance*, is the primary internal-defense mechanism against cancer.

To carry out their functions, the leukocytes largely use a "seek out and attack" strategy; that is, they go to sites of invasion or tissue damage. The main reason WBCs are in the blood is to be rapidly transported from their site of production or storage to wherever they are needed.

▌ Pathogenic bacteria and viruses are the major targets of the immune system.

The primary foreign enemies against which the immune system defends are bacteria and viruses. **Bacteria** are nonnucleated, single-celled micro-organisms self-equipped with all machinery essential for their own survival and reproduction. Pathogenic bacteria that invade the body cause tissue damage and produce disease largely by releasing enzymes or toxins that physically injure or functionally disrupt affected cells and organs. The disease-producing power of a pathogen is known as its **virulence.**

In contrast to bacteria, **viruses** are not self-sustaining cellular entities. They consist only of nucleic acids (genetic material—DNA or RNA) enclosed by a protein coat. Because they lack cellular machinery for energy production and protein synthesis, viruses cannot carry out metabolism and reproduce unless they invade a **host cell** (a body cell of the infected individual) and take over the cell's biochemical facilities for their own uses. Not only do viruses sap the host cell's energy resources, but the viral nucleic acids also direct the host cell to synthesize proteins needed for viral replication.

When a virus becomes incorporated into a host cell, the body's own defense mechanisms may destroy the cell, because they no longer recognize it as a "normal self" cell. Other ways in which viruses can lead to cell damage or death are by depleting essential cell components, dictating that the cell produce substances that are toxic to the cell, or transforming the cell into a cancer cell.

▌ There are five types of leukocytes.

Leukocytes lack hemoglobin (in contrast to erythrocytes) so they are colorless (that is, "white") unless specifically stained for microscopic visibility. Unlike erythrocytes, which are of uniform structure, identical function, and constant number, leukocytes vary in structure, function, and number. There are

Leukocytes					Erythrocyte	Platelets
Polymorphonuclear granulocytes			Mononuclear agranulocytes			
Neutrophil	Eosinophil	Basophil	Monocyte	Lymphocyte		

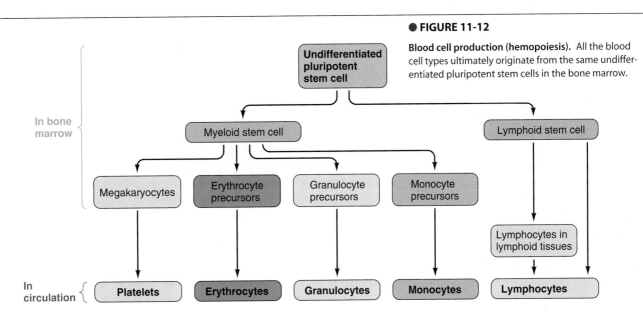

From left to right: Cabisco/Visuals Unlimited; Science VU/Visuals Unlimited; Cabisco/Visuals Unlimited; Science VU/Visuals Unlimited; Williams & Wilkins, Co.; Biophoto Associates/Photo Researchers, Inc.; Williams & Wilkins Co.

● **FIGURE 11-11**

Normal blood cellular elements

For an interaction related to this figure, see Media Exercise 11.1: Blood on the CD-ROM.

five different types of circulating leukocytes—neutrophils, eosinophils, basophils, monocytes, and lymphocytes each with a characteristic structure and function. They are all somewhat larger than erythrocytes.

The five types of leukocytes fall into two main categories, depending on the appearance of their nuclei and the presence or absence of granules in their cytoplasm when viewed microscopically (● Figure 11-11). Neutrophils, eosinophils, and basophils are categorized as **polymorphonuclear** ("many-shaped nucleus") **granulocytes** ("granule-containing cells"). Their nuclei are segmented into several lobes of varying shapes, and their cytoplasm contains an abundance of membrane-enclosed granules. The three types of granulocytes are distinguished on the basis of the varying affinity of their granules for dyes: *eosinophils* have an affinity for the red dye eosin, *basophils* preferentially take up a basic blue dye, and *neutrophils* are neutral, showing no dye preference. Monocytes and lymphocytes are known as **mononuclear** ("single nucleus") **agranulocytes** ("cells lacking granules"). Both have a single, large, nonsegmented nucleus and few granules. Monocytes are the larger of the two and have an oval or kidney-shaped

nucleus. *Lymphocytes,* the smallest of the leukocytes, characteristically have a large spherical nucleus that occupies most of the cell.

Leukocytes are produced at varying rates depending on the changing defense needs of the body.

All leukocytes ultimately originate from the same undifferentiated pluripotent stem cells in the red bone marrow that also give rise to erythrocytes and platelets (● Figure 11-12). The cells destined to become leukocytes eventually differentiate into various committed cell lines and proliferate under the influence of appropriate stimulating factors. Granulocytes and monocytes are produced only in the bone marrow, which releases these mature leukocytes into the blood. Lymphocytes are originally derived from precursor cells in the bone marrow, but most new lymphocytes are actually produced by lymphocytes already in the **lymphoid** (lymphocyte-containing) **tissues,** such as the lymph nodes and tonsils

● **FIGURE 11-12**

Blood cell production (hemopoiesis). All the blood cell types ultimately originate from the same undifferentiated pluripotent stem cells in the bone marrow.

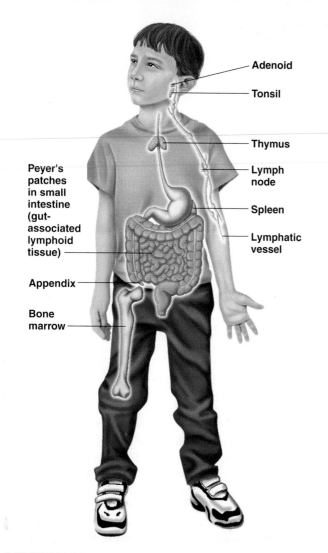

| Adenoid |
| Tonsil |
| Thymus |
| Lymph node |
| Spleen |
| Lymphatic vessel |

Peyer's patches in small intestine (gut-associated lymphoid tissue)

Appendix

Bone marrow

● **FIGURE 11-13**

Lymphoid tissues. The lymphoid tissues, which are dispersed throughout the body, produce, store, or process lymphocytes.

For an interaction related to this figure, see Media Exercise 11.4: The Body's Defenses on the CD-ROM.

(● Figure 11-13). Lymphoid tissues produce, store, and/or process lymphocytes.

The total number of leukocytes normally ranges from 5 to 10 million cells per milliliter of blood, with an average of 7 million cells/ml, expressed as an average **white blood cell count** of 7000/mm³. Leukocytes are the least numerous of the blood cells (about 1 white blood cell for every 700 red blood cells), not because fewer are produced but because they are merely in transit while in the blood. Normally, about two thirds of the circulating leukocytes are granulocytes, mostly neutrophils, whereas one third are agranulocytes, predominantly lymphocytes (▲ Table 11-2). However, the total number of white cells and the percentage of each type may vary considerably to meet changing defense needs. Depending on the type and extent of assault the body is combating, different types of leukocytes are selectively produced at varying rates. Chemical messengers arising from invaded or damaged

▲ TABLE 11-2

Typical Human Blood Cell Count

Total Erythrocytes = 5,000,000,000 cells/ml blood

Red blood cell (RBC) count = 5,000,000/mm³

Total Leukocytes = 7,000,000 cells/ml blood

White blood cell (WBC) count = 7,000/mm³

Differential White Blood Cell Count (percentage distribution of types of leukocytes)

Polymorphonuclear granulocytes		*Mononuclear agranulocytes*	
Neutrophils	60–70%	Lymphocytes	25–33%
Eosinophils	1–4%	Monocytes	2–6%
Basophils	0.25–0.5%		

Total Platelets = 250,000,000/ml blood

Platelet count = 250,000/mm³

tissues or from activated leukocytes themselves govern the rates of production of the various leukocytes. Specific hormones analogous to erythropoietin direct the differentiation and proliferation of each cell type. Some of these hormones have been identified and can be produced in the laboratory; an example is **granulocyte colony–stimulating factor**, which stimulates increased replication and release of granulocytes, especially neutrophils, from the bone marrow. This achievement has opened up the ability to administer these hormones as a powerful new therapeutic tool to bolster a person's normal defense against infection or cancer.

FUNCTIONS AND LIFE SPANS OF LEUKOCYTES

Among the granulocytes, **neutrophils** are phagocytic specialists. Furthermore, scientists recently discovered that neutrophils release a web of extracellular fibers dubbed *neutrophil extracellular traps (NETs)*. These fibers contain bacteria-killing chemicals, enabling NETs to trap, then destroy bacteria extracellularly. Thus neutrophils can destroy bacteria both intracellularly by phagocytosis and extracellularly via the NETs they release. Neutrophils invariably are the first defenders on the scene of bacterial invasion and, accordingly, are very important in inflammatory responses. Furthermore, they scavenge to clean up debris. As might be expected in view of these functions, an increase in circulating neutrophils typically accompanies acute bacterial infections.

Eosinophils are specialists of another type. An increase in circulating eosinophils is associated with allergic conditions (such as asthma and hay fever) and with internal parasite infestations (for example, worms). Eosinophils obviously cannot engulf a much larger parasitic worm, but they do attach to the worm and secrete substances that kill it.

Basophils are the least numerous and most poorly understood of the leukocytes. They are quite similar structurally and functionally to **mast cells**, which never circulate in the blood but instead are dispersed in connective tissue throughout the body. Scientists once believed that basophils became mast cells by migrating from the circulatory system, but researchers have shown that basophils arise from the bone marrow, whereas mast cells are derived from precursor cells in connective tissue. Both basophils and mast cells synthesize and store *histamine* and *heparin,* powerful chemical substances that can be released on appropriate stimulation. Histamine release is important in allergic reactions, whereas heparin speeds up removal of fat particles from the blood after a fatty meal. Heparin can also prevent clotting of blood samples drawn for clinical analysis, but whether it plays a physiologic role as an anticoagulant is still debated.

Once released into the blood from the bone marrow, a granulocyte usually stays in transit in the blood for less than a day before leaving the blood vessels to enter the tissues, where it survives another three to four days unless it dies sooner in the line of duty.

Among the agranulocytes, **monocytes**, like neutrophils, become professional phagocytes. They emerge from the bone marrow while still immature and circulate for only a day or two before settling down in various tissues throughout the body. At their new residences, monocytes continue to mature and greatly enlarge, becoming the large tissue phagocytes known as **macrophages** (*macro* means "large"; *phage* means "eater"). A macrophage's life span may range from months to years unless it is destroyed sooner while performing its phagocytic activity. A phagocytic cell can ingest only a limited amount of foreign material before it succumbs.

Lymphocytes provide immune defense against targets for which they are specifically programmed. There are two types of lymphocytes, B lymphocytes and T lymphocytes (B and T cells). **B lymphocytes** produce *antibodies*, which circulate in the blood. An antibody binds with and marks for destruction (by phagocytosis or other means) the specific kinds of foreign matter, such as bacteria, that induced production of the antibody. **T lymphocytes** do not produce antibodies; instead, they directly destroy their specific target cells by releasing chemicals that punch holes in the victim cell. The target cells of T cells include body cells invaded by viruses and cancer cells. Lymphocytes live for about 100 to 300 days. During this period, most continually recycle among the lymphoid tissues, lymph, and blood, spending only a few hours at a time in the blood. Therefore, only a small part of the total lymphocytes are in transit in the blood at any given moment.

ABNORMALITIES IN LEUKOCYTE PRODUCTION

Clinical Note Even though levels of circulating leukocytes may vary, changes in these levels are normally controlled and adjusted according to the body's needs. However, abnormalities in leukocyte production can occur that are not subject to control; that is, either too few or too many WBCs can be produced. The bone marrow can greatly slow down or even stop production of white blood cells when it is exposed to certain toxic chemical agents (such as benzene and anticancer drugs) or to excessive radiation. The most serious consequence is the reduction in professional phagocytes (neutrophils and macrophages), which greatly reduces the body's defense capabilities against invading micro-organisms. The only defense still available when the bone marrow fails is the immune capabilities of the lymphocytes produced by the lymphoid organs.

In **infectious mononucleosis,** not only does the number of lymphocytes (but not other leukocytes) in the blood increase, but also many of the lymphocytes are atypical in structure. This condition, which is caused by the Epstein-Barr virus, is characterized by pronounced fatigue, a mild sore throat, and low-grade fever. Full recovery usually requires a month or more.

Surprisingly, one of the major consequences of **leukemia,** a cancerous condition that involves uncontrolled proliferation of WBCs, is inadequate defense capabilities against foreign invasion. In leukemia, the WBC count may reach as high as $500,000/mm^3$, compared with the normal $7000/mm^3$, but because most of these cells are abnormal or immature, they cannot perform their normal defense functions. Another devastating consequence of leukemia is displacement of the other blood cell lines in the bone marrow. This results in anemia because of a reduction in erythropoiesis and in internal bleeding because of a deficit of platelets. Consequently, overwhelming infections or hemorrhage are the most common causes of death in leukemic patients.

We are now going to turn our attention to the two major components of the immune system's response to foreign invaders and other targets—innate and adaptive immune responses. In the process, we will further examine the roles of each type of leukocyte.

❙ Immune responses may be either innate and nonspecific, or adaptive and specific.

Protective immunity is conferred by the complementary actions of two separate but interdependent components of the immune system: the *innate immune system* and the *adaptive* or *acquired immune system*. The responses of these two systems differ in timing and in the selectivity of the defense mechanisms.

The **innate immune system** encompasses the body's *nonspecific* immune responses that come into play immediately on exposure to a threatening agent. These nonspecific responses are inherent (innate or built-in) defense mechanisms that nonselectively defend against foreign or abnormal material of any type, even on initial exposure to it. Such responses provide a first line of defense against a wide range of threats, including infectious agents, chemical irritants, and tissue injury from mechanical trauma and burns. Everyone is born with essentially the same innate immune-response mechanisms, although there are some subtle genetic differences.

The components of the innate system are always on guard, ready to unleash a limited, rather crude, repertoire of defense mechanisms at any and every invader. Of the immune effector cells, the neutrophils and macrophages—both phagocytic specialists—are especially important in innate defense. Several groups of plasma proteins also play key roles, as you will see shortly.

The various nonspecific immune responses are set in motion in response to generic molecular patterns associated with threatening agents, such as the carbohydrates typically found in bacterial cell walls but not found in human cells. The responding phagocytic cells are studded with a recently discovered type of plasma membrane protein known as **Toll-like receptors (TLRs)**. TLRs have been dubbed the "eyes of the innate immune system" because these immune sensors recognize and bind with the telltale bacterial markers, allowing the effector cells of the innate system to "see" pathogens as distinct from self-cells. A TLR's recognition of a pathogen triggers the phagocyte to engulf and destroy the infectious micro-organism. Moreover, activation of the TLR induces the phagocytic cell to secrete chemicals, some of which contribute to inflammation, an important innate response to microbial invasion.

The innate mechanisms give us all a rapid but limited and nonselective response to unfriendly challenges of all kinds, much like medieval guardsmen lashing out with brute force weapons at any enemy approaching the walls of the castle they are defending. Innate immunity largely contains and limits the spread of infection. These nonspecific responses are important for keeping the foe at bay until the adaptive immune system, with its highly selective weapons, can be prepared to take over and mount strategies to eliminate the villain.

The **adaptive** or **acquired immune system** relies on *specific* immune responses selectively targeted against a particular foreign material to which the body has already been exposed and has had an opportunity to prepare for an attack aimed discriminatingly at the enemy. The responses of the adaptive immune system are mediated by the B and T lymphocytes. Each B and T cell can recognize and defend against only one particular type of foreign material, such as one kind of bacterium. Among the millions of B and T cells in the body, only the ones specifically equipped to recognize the unique molecular features of a particular infectious agent are called into action to discriminatingly defend against this agent, similar to modern, specially trained military personnel called into active duty to accomplish a very specific task. The chosen lymphocytes multiply, expanding the pool of specialists that can launch a highly targeted attack against the invader. The adaptive immune system thus takes considerably more time to mount and takes on one specific foe.

The adaptive immune system is your ultimate weapon against most pathogens. The repertoire of activated and expanded B and T cells is constantly changing in response to the various pathogens encountered. Thus the adaptive or acquired immune system adapts to wage battle against the specific pathogens in your environment. The targets of the adaptive immune system vary among people, depending on the types of immune assaults each individual meets. Furthermore, this system acquires an ability to more efficiently eradicate a particular foe when rechallenged by the same pathogen in the future. It does so by establishing a pool of memory cells as a result of an encounter with a given pathogen, so that when later exposed to the same agent, it can more swiftly defend against the invader.

The innate and adaptive immune systems work in harmony to contain, then eliminate, harmful agents. We will first examine in more detail the innate immune responses before looking more closely at adaptive immunity.

 Click on the Media Exercises menu of the CD-ROM and work Media Exercise 11.1: Blood to test your understanding of the previous sections.

INNATE IMMUNITY

Innate defenses include the following:

1. *Inflammation,* a nonspecific response to tissue injury in which the phagocytic specialists—neutrophils and macrophages—play a major role, along with supportive input from other immune-cell types.
2. *Interferon,* a family of proteins that nonspecifically defend against viral infection.
3. *Natural killer cells,* a special class of lymphocyte-like cells that spontaneously and nonspecifically lyse (rupture) and thereby destroy virus-infected host cells and cancer cells.
4. The *complement system,* a group of inactive plasma proteins that, when sequentially activated, bring about destruction of foreign cells by attacking their plasma membranes.

We will discuss each of these in turn, beginning with inflammation.

▌ Inflammation is a nonspecific response to foreign invasion or tissue damage.

The term **inflammation** refers to an innate, nonspecific series of highly interrelated events that are set into motion in response to foreign invasion, tissue damage, or both. The ultimate goal of inflammation is to bring to the invaded or injured area phagocytes and plasma proteins that can (1) isolate, destroy, or inactivate the invaders; (2) remove debris; and (3) prepare for subsequent healing and repair. The overall inflammatory response is remarkably similar no matter what the triggering event (bacterial invasion, chemical injury, or mechanical trauma), although some subtle differences may be evident, depending on the injurious agent or the site of damage. The following sequence of events typically occurs during the inflammatory response. As an example we will use bacterial entry into a break in the skin.

DEFENSE BY RESIDENT TISSUE MACROPHAGES

When bacteria invade through a break in the external barrier of skin, the macrophages already in the area immediately begin phagocytizing the foreign microbes. Although usually not enough resident macrophages are present to meet the challenge alone, they defend against infection during the first hour or so, before other mechanisms can be mobilized

LOCALIZED VASODILATION

Almost immediately on microbial invasion, arterioles within the area dilate, increasing blood flow to the site of injury. This localized vasodilation is mainly induced by histamine released from mast cells in the area of tissue damage, (the connective tissue–bound "cousins" of circulating basophils). Increased local delivery of blood brings to the site more phagocytic leukocytes and plasma proteins, both crucial to the defense response.

INCREASED CAPILLARY PERMEABILITY

Released histamine also increases the capillaries' permeability by enlarging the capillary pores (the spaces between the endothelial cells), so plasma proteins that normally are prevented from leaving the blood can escape into the inflamed tissue (see p. 292).

LOCALIZED EDEMA

Accumulation of leaked plasma proteins in the interstitial fluid raises the local interstitial fluid–colloid osmotic pressure. Furthermore, increased capillary blood pressure accompanies the increased local blood flow. Because both these pressures tend to move fluid out of the capillaries, these changes favor enhanced ultrafiltration and reduced reabsorption of fluid across the involved capillaries. The end result of this shift in fluid balance is localized edema (see p. 297). Thus the familiar swelling that accompanies inflammation is due to histamine-induced vascular changes. Likewise, the other well-known gross manifestations of inflammation such as redness and heat are largely caused by the enhanced flow of warm arterial blood to the damaged tissue. Pain is caused both by local distension within the swollen tissue and by the direct effect of locally produced substances on the receptor endings of afferent neurons that supply the area. These observable characteristics of the inflammatory process (swelling, redness, heat, and pain) are coincidental to the primary purpose of the vascular changes in the injured area— to increase the number of leukocytic phagocytes and crucial plasma proteins in the area (● Figure 11-14).

WALLING OFF THE INFLAMED AREA

The leaked plasma proteins most critical to the immune response are those in the complement system as well as clotting and anticlotting factors. On exposure to tissue thromboplastin in the injured tissue and to specific chemicals secreted by phagocytes on the scene, fibrinogen is converted into fibrin. Fibrin forms interstitial fluid clots in the spaces around the bacterial invaders and dam-

aged cells. This walling off of the injured region from the surrounding tissues prevents or at least delays the spread of bacterial invaders and their toxic products. Later, the more slowly activated anticlotting factors gradually dissolve the clots after they are no longer needed.

EMIGRATION OF LEUKOCYTES

Within an hour after injury, the area is teeming with leukocytes that have left the vessels. Neutrophils arrive first, followed during the next 8 to 12 hours by the slower-moving monocytes. The latter swell and mature into macrophages during another 8- to 12-hour period. Once neutrophils or monocytes leave the bloodstream, they never recycle back to the blood.

Leukocytes can emigrate from the blood into the tissues by behaving like amoebas (see p. 36) and wriggling through the capillary pores, than crawling toward the injured area

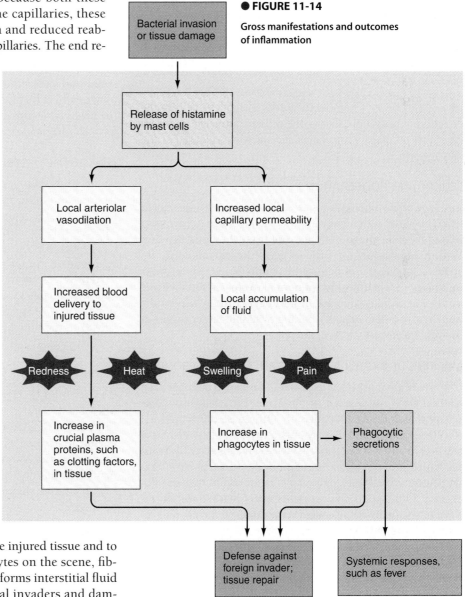

● FIGURE 11-14

Gross manifestations and outcomes of inflammation

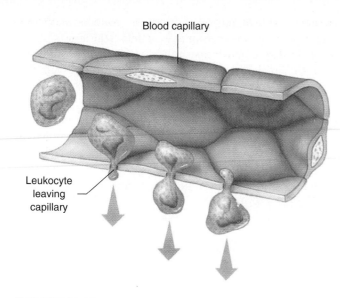

Blood capillary

Leukocyte leaving capillary

(● Figure 11-15). **Chemotaxis** guides phagocytic cells in the direction of migration; that is, the cells are attracted to certain chemical mediators known as *chemotaxins*, released at the damage site.

LEUKOCYTE PROLIFERATION

Resident tissue macrophages as well as leukocytes that exited from the blood and migrated to the inflammatory site are soon joined by new phagocytic recruits from the bone marrow. Within a few hours after the onset of the inflammatory response, the number of neutrophils in the blood may increase up to four to five times that of normal. A slower-commencing but longer-lasting increase in monocyte production by the bone marrow also occurs, making available larger numbers of these macrophage precursor cells.

MARKING OF BACTERIA FOR DESTRUCTION BY OPSONINS

Obviously, phagocytes must be able to distinguish between normal cells and foreign or abnormal cells before accomplishing their destructive mission. Otherwise they could not selectively engulf and destroy only unwanted materials. First, as you already learned, phagocytes, by means of their TLRs, recognize and subsequently engulf infiltrators that have standard bacterial cell wall components not found in human cells. Second, foreign particles are deliberately marked for phagocytic ingestion by being coated with chemical mediators generated by the immune system. Such body-produced chemicals that make bacteria more susceptible to phagocytosis are known as **opsonins**. The most important opsonins are antibodies and one of the activated proteins of the complement system.

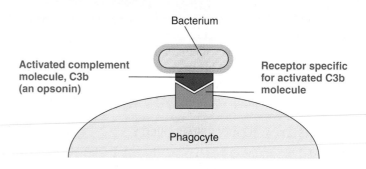

Bacterium

Activated complement molecule, C3b (an opsonin)

Receptor specific for activated C3b molecule

Phagocyte

Structures are not drawn to scale.

● **FIGURE 11-16**

Mechanism of opsonin action. One of the activated complement molecules, C3b, links a foreign cell, such as a bacterium, and a phagocytic cell by nonspecifically binding with the foreign cell and specifically binding with a receptor on the phagocyte. This link ensures that the foreign victim does not escape before it can be engulfed by the phagocyte.

An opsonin enhances phagocytosis by linking the foreign cell to a phagocytic cell (● Figure 11-16). One portion of an opsonin molecule binds nonspecifically to the surface of an invading bacterium, whereas another portion of the opsonin molecule binds to receptor sites specific for it on the phagocytic cell's plasma membrane. This link ensures that the bacterial victim does not have a chance to "get away" before the phagocyte can perform its lethal attack.

LEUKOCYTIC DESTRUCTION OF BACTERIA

Neutrophils and macrophages clear the inflamed area of infectious and toxic agents as well as tissue debris; this clearing action is the main function of the inflammatory response. They do this by both phagocytic and nonphagocytic means.

Phagocytes eventually die from the accumulation of toxic by-products from foreign particle degradation or from inadvertent release of destructive lysosomal chemicals into the cytosol. The **pus** that forms in an infected wound is a collection of these phagocytic cells, both living and dead; necrotic (dead) tissue liquefied by lysosomal enzymes released from the phagocytes; and bacteria.

MEDIATION OF THE INFLAMMATORY RESPONSE BY PHAGOCYTE-SECRETED CHEMICALS

Microbe-stimulated phagocytes release many chemicals that function as mediators of the inflammatory response. These chemical mediators induce a broad range of interrelated immune activities, varying from local responses to the systemic manifestations that accompany microbe invasion. The following are among the most important functions of phagocytic secretions:

1. Some of the chemicals, which are very destructive, directly kill microbes by nonphagocytic means. For example, macrophages secrete *nitric oxide,* a multipurpose chemical that is toxic to nearby microbes (see p. 286). As a more subtle means of destruction, neutrophils secrete **lactoferrin**, a

protein that tightly binds with iron, making it unavailable for use by invading bacteria. Bacterial multiplication depends on high concentrations of available iron.

2. Phagocytic secretions stimulate the release of *histamine* from mast cells in the vicinity. Histamine, in turn, induces the local vasodilation and increased vascular permeability of inflammation.

3. Some phagocytic chemical mediators trigger both the *clotting* and *anticlotting systems* to first enhance the walling-off process and then facilitate gradual dissolution of the fibrous clot after it is no longer needed.

4. One secretion released from neutrophils activates specific plasma proteins known as **kinins** that augment a variety of inflammatory events. For example, kinins activate nearby pain receptors and thus partially produce the soreness associated with inflammation. They also act as powerful chemotaxins to induce phagocyte migration into the affected area.

5. One chemical released by macrophages, **endogenous pyrogen (EP)**, induces the development of fever (*endogenous* means "from within the body"; *pyro* means "fire" or "heat"; *gen* means "production"). This response occurs especially when the invading organisms have spread into the blood. Endogenous pyrogen causes release within the hypothalamus of *prostaglandins,* locally acting chemical messengers that "turn up" the hypothalamic "thermostat" that regulates body temperature. The function of the resulting elevation in body temperature in fighting infection remains unclear. The fact that fever is such a common systemic manifestation of inflammation suggests the raised temperature plays an important beneficial role in the overall inflammatory response, as supported by recent evidence. For example, higher temperatures augment phagocytosis and increase the rate of the many enzyme-dependent inflammatory activities. Furthermore, an elevated body temperature may interfere with bacterial multiplication by increasing bacterial requirements for iron. Resolving the controversial issue of whether a fever can be beneficial is extremely important, given the widespread use of drugs that suppress fever.

6. Secreted chemicals also stimulate synthesis and release of *neutrophils* and *lymphocytes.* This effect is especially prominent in response to bacterial infections.

This list of events augmented by chemicals that phagocytes secrete is not complete, but it illustrates the diversity and complexity of responses these mediators elicit. Thus the effect that phagocytes, especially macrophages, ultimately have on microbial invaders far exceeds their "engulf and destroy" tactics.

TISSUE REPAIR

The ultimate purpose of the inflammatory process is to isolate and destroy injurious agents and to clear the area for tissue repair. In some tissues (for example, skin, bone, and liver), the healthy organ-specific cells surrounding the injured area undergo cell division to replace the lost cells, often repairing the wound perfectly. In typically nonregenerative tissues such as nerve and muscle, however, lost cells are replaced by **scar**

tissue. Fibroblasts, a type of connective tissue cell, start to divide rapidly in the vicinity and secrete large quantities of the protein collagen, which fills in the region vacated by the lost cells and results in the formation of scar tissue (see p. 48). Even in a tissue as readily replaceable as skin, scars sometimes form when complex underlying structures, such as hair follicles and sweat glands, are permanently destroyed by deep wounds.

▌NSAIDs and glucocorticoid drugs suppress the inflammatory response.

Clinical Note Many drugs can suppress the inflammatory process; the most effective are the *nonsteroidal anti-inflammatory drugs,* or *NSAIDs* (aspirin, ibuprofen, and related compounds) and *glucocorticoids* (drugs similar to the steroid hormone cortisol, which is secreted by the adrenal cortex; see p. 554). For example, aspirin interferes with the inflammatory response by decreasing histamine release, thus reducing pain, swelling, and redness. Furthermore, aspirin reduces fever by inhibiting production of prostaglandins, the local mediators of endogenous pyrogen-induced fever.

Glucocorticoids, which are potent anti-inflammatory drugs, suppress almost every aspect of the inflammatory response. In addition, they destroy lymphocytes within lymphoid tissue and reduce antibody production. These therapeutic agents are useful for treating undesirable immune responses, such as allergic reactions (for example, poison ivy rash and asthma) and the inflammation associated with arthritis. However, by suppressing inflammatory and other immune responses that localize and eliminate bacteria, such therapy also reduces the body's ability to resist infection. For this reason, glucocorticoids should be used discriminatingly.

Now let's shift attention from inflammation to interferon, another component of innate immunity.

▌Interferon transiently inhibits multiplication of viruses in most cells.

Besides the inflammatory response, another innate defense mechanism is the release of **interferon** from virus-infected cells. Interferon briefly provides nonspecific resistance to viral infections by transiently interfering with replication of the same or unrelated viruses in other host cells. In fact, interferon was named for its ability to "interfere" with viral replication.

ANTIVIRAL EFFECT OF INTERFERON

When a virus invades a cell, in response to being exposed to viral nucleic acid the cell synthesizes and secretes interferon. Once released into the ECF from a virus-infected cell, interferon binds with receptors on the plasma membranes of healthy neighboring cells or even distant cells that it reaches through the blood, signaling these cells to prepare for possible viral attack. Interferon thus acts as a "whistle-blower," forewarning healthy cells of potential viral attack and helping them prepare to resist. Interferon does not have a direct antiviral effect; instead, it triggers the production of virus-blocking enzymes by potential host cells. When interferon binds with these other cells, they synthesize enzymes that can break down viral messenger RNA (see p. A-22) and inhibit protein synthesis. Both these processes are essential for viral replication. Although viruses are still able to invade these forewarned cells, these pathogens cannot govern cellular protein synthesis for their own replication (● Figure 11-17).

The newly synthesized inhibitory enzymes remain inactive within the tipped-off potential host cell unless it is actually invaded by a virus, at which time the enzymes are activated by the presence of viral nucleic acid. This activation requirement protects the cell's own messenger RNA and protein-synthesizing machinery from unnecessary inhibition by these enzymes should viral invasion not occur. Because activation can take place only during a limited time span, this is a short-term defense mechanism.

Interferon is released nonspecifically from any cell infected by any virus and, in turn, can induce temporary self-protective activity against many different viruses in any other cells that it reaches. Thus it provides a general, rapidly responding defense strategy against viral invasion until more specific but slower-responding immune mechanisms come into play.

ANTICANCER EFFECTS OF INTERFERON

Interferon also exerts anticancer as well as antiviral effects. It markedly enhances the actions of cell-killing cells—the natural killer cells and a special type of T lymphocyte, *cytotoxic T cells*—which attack and destroy both virus-infected cells and cancer cells. Furthermore, interferon itself slows cell division and suppresses tumor growth.

▌ Natural killer cells destroy virus-infected cells and cancer cells on first exposure to them.

Natural killer (NK) cells are naturally occurring, lymphocyte-like cells that nonspecifically destroy virus-infected cells and cancer cells by directly lysing the membranes of such cells on first exposure to them. Their mode of action and major targets are similar to those of cytotoxic T cells, but the latter can fatally attack only the specific types of virus-infected cells and cancer cells to which they have been previously exposed. Furthermore, after exposure cytotoxic T cells require a maturation period before they can launch their lethal assault. The NK cells provide an immediate, nonspecific defense against virus-invaded cells and cancer cells before the more specific and more abundant cytotoxic T cells become functional.

▌ The complement system punches holes in micro-organisms.

The **complement system** is another defense mechanism brought into play nonspecifically in response to invading organisms. This system can be activated in two ways:

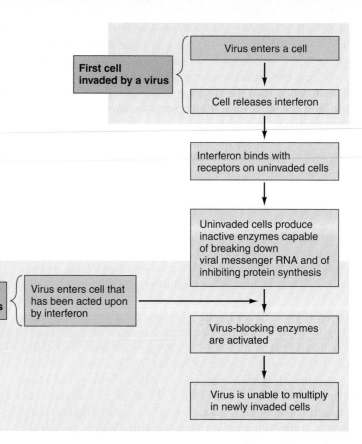

● FIGURE 11-17

Mechanism of action of interferon in preventing viral replication. Interferon, which is released from virus-infected cells, binds with other uninvaded host cells and induces these cells to produce inactive enzymes capable of blocking viral replication. These inactive enzymes are activated only if a virus subsequently invades one of these prepared cells.

1. By exposure to particular carbohydrate chains present on the surfaces of micro-organisms but not found on human cells, a nonspecific innate immune response
2. By exposure to antibodies produced against a specific foreign invader, an adaptive immune response

In fact, the system derives its name from the fact that it "complements" the action of antibodies; it is the primary mechanism activated by antibodies to kill foreign cells. It does so by forming membrane attack complexes that punch holes in the victim cells. In addition to bringing about direct lysis of the invader, the powerful complement cascade reinforces other general inflammatory tactics.

FORMATION OF THE MEMBRANE ATTACK COMPLEX

In the same mode as the clotting and anticlotting systems, the complement system consists of plasma proteins that are produced by the liver and circulate in the blood in inactive form. Once the first component, C1, is activated, it activates the next component, C2, and so on, in a cascade sequence of activation reactions. The five final components, C5 through C9, assemble into a large, doughnut-shaped protein complex, the **membrane attack complex (MAC)**, which attacks the surface membrane of nearby micro-organisms by embedding itself so that a large channel is created through the microbial surface membrane (● Figure 11-18). In other words, the parts make a hole. This hole-punching technique makes the membrane extremely leaky; the resulting osmotic flux of water into the victim cell causes it to swell and burst. This complement-induced lysis is the major means of directly killing microbes without phagocytizing them.

AUGMENTING INFLAMMATION

Unlike the other cascade systems, in which the sole function of the various components is activation of the next precursor in the sequence, several activated proteins in the complement cascade perform additional important functions on their own. Besides the direct destruction of foreign cells by the membrane attack complex, various other activated complement components augment the inflammatory process by

- *Serving as chemotaxins,* which attract and guide professional phagocytes to the site of complement activation (that is, the site of microbial invasion)
- *Acting as opsonins* by binding with microbes and thereby enhancing their phagocytosis
- *Promoting vasodilation and increased vascular permeability,* thus increasing blood flow to the invaded areas
- *Stimulating the release of histamine* from mast cells in the vicinity, which in turn enhances the local vascular changes characteristic of inflammation
- *Activating kinins,* which further reinforce inflammatory reactions

● FIGURE 11-18

Membrane attack complex (MAC) of the complement system. Activated complement proteins C5, C6, C7, C8, and a number of C9s aggregate to form a porelike channel in the plasma membrane of the target cell. The resulting leakage leads to destruction of the cell.

(*Source:* Adapted from the illustration by Dana Burns-Pizer in "How Killer Cells Kill," by John Ding-E Young and Zanvil A. Cohn. Copyright ©1988 Scientific American, Scientific American, Inc. All rights reserved.)

 For an interaction related to this figure, see Media Exercise 11.4: The Body's Defenses on the CD-ROM.

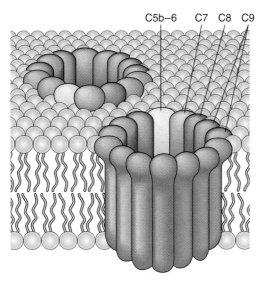

C5b–6   C7   C8   C9

Several activated components in the cascade are very unstable. Because these unstable components can carry out the sequence only in the immediate area in which they are activated before they decompose, the complement attack is confined to the surface membrane of the microbe whose presence initiated activation of the system. Nearby host cells are thus spared from lytic attack.

We have now completed our discussion of innate immunity and are going to turn our attention to adaptive immunity.

 Click on the Media Exercises menu of the CD-ROM and work Media Exercises 11.4: The Body's Defenses and 11.5: Inflammation to test your understanding of the previous section.

ADAPTIVE IMMUNITY: GENERAL CONCEPTS

A specific adaptive immune response is a selective attack aimed at limiting or neutralizing a particular offending target for which the body has been specially prepared after exposure to it.

▌Adaptive immune responses include antibody-mediated immunity and cell-mediated immunity.

There are two classes of adaptive immune responses: **antibody-mediated immunity**, involving production of antibodies by B lymphocyte derivatives known as *plasma cells*; and **cell-mediated immunity**, involving production of *activated T lymphocytes,* which directly attack unwanted cells.

Lymphocytes can specifically recognize and selectively respond to an almost limitless variety of foreign agents as well as cancer cells. The recognition and response processes are different in B and in T cells. In general, B cells recognize free-existing foreign invaders such as bacteria and their toxins and a few viruses, which they combat by secreting antibodies specific for the invaders. T cells specialize in recognizing and destroying body cells gone awry, including virus-infected cells and cancer cells. We will examine each of these processes in detail in the upcoming sections. For now, we are going to explore the different life histories of B and T cells.

ORIGINS OF B AND T CELLS

Both types of lymphocytes, like all blood cells, are derived from common stem cells in the bone marrow. Whether a lymphocyte and all its progeny are destined to be B or T cells depends on the site of final differentiation and maturation of the original cell in the lineage (● Figure 11-19). B cells differentiate and mature in the bone marrow. As for T cells, during fetal life and early childhood, some of the immature lymphocytes from the bone marrow migrate through the blood to the thymus, where they undergo further processing to become T lymphocytes (named for their site of maturation). The **thymus** is a lymphoid tissue located midline within the chest cavity above the heart in the space between the lungs (see ● Figure 11-13, p. 328).

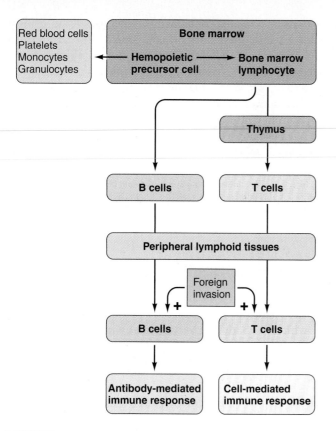

● **FIGURE 11-19**

Origins of B and T cells. B cells are derived from lymphocytes that matured and differentiated in the bone marrow, whereas T cells are derived from lymphocytes that originated in the bone marrow but matured and differentiated in the thymus. New B and T cells are produced by colonies of B and T cells in peripheral lymphoid tissues.

 For an interaction related to this figure, see Media Exercise 11.6: Basics of Specific Immunity on the CD-ROM.

On being released into the blood from either the bone marrow or the thymus, mature B and T cells take up residence and establish lymphocyte colonies in the peripheral lymphoid tissues. Here, on appropriate stimulation, they undergo cell division to produce new generations of either B or T cells, depending on their ancestry. After early childhood, most new lymphocytes are derived from these peripheral lymphocyte colonies rather than from the bone marrow.

Each of us has an estimated 2 trillion lymphocytes, which, if aggregated in a mass, would be about the size of the brain. At any one time, most of these lymphocytes are concentrated in the various strategically located lymphoid tissues, but both B and T cells continually circulate among the lymph, blood, and body tissues, where they remain on constant surveillance.

ROLE OF THYMOSIN

Because most of the migration and differentiation of T cells occurs early in development, the thymus gradually atrophies and becomes less important as the person matures. It does, however, continue to produce **thymosin,** a hormone important in maintaining the T cell lineage. Thymosin enhances proliferation of new T cells within the peripheral lymphoid tissues

and augments the immune capabilities of existing T cells. Secretion of thymosin decreases after about 30 to 40 years of age. This decline has been suggested as a contributing factor in aging. Scientists further speculate that diminishing T cell capacity with advancing age may be linked to increased susceptibility to viral infections and cancer, because T cells play an especially important role in defense against viruses and cancer.

Let's now see how lymphocytes detect their selected target.

❚ An antigen induces an immune response against itself.

Both B and T cells must be able to specifically recognize unwanted cells and other material to be destroyed or neutralized, as being distinct from the body's own normal cells. The presence of antigens enables lymphocytes to make this distinction. An **antigen** is a large, complex, unique molecule that triggers a specific immune response against itself when it gains entry into the body. Foreign proteins are the most common antigens. Antigens may exist as isolated molecules, such as bacterial toxins, or they may be an integral part of a multimolecular structure, as when they are on the surface of an invading foreign microbe.

We will first see how B cells respond to their targeted antigen, and then we'll look at T cells' response to their antigen.

B LYMPHOCYTES: ANTIBODY-MEDIATED IMMUNITY

Each B and T cell has receptors on its surface for binding with one particular type of the multitude of possible antigens. These receptors are the "eyes of the adaptive immune system," although a given lymphocyte can "see" only one unique antigen. This is in contrast to the TLRs of the innate effector cells, which recognize generic "trademarks" characteristic of all microbial invaders. Furthermore, lymphocytes cannot respond directly to new incoming antigen. It must first be processed and presented to them by *antigen-presenting cells,* an activity we will describe in detail later.

❚ Antigens stimulate B cells to convert into plasma cells that produce antibodies.

On binding with processed and presented antigen, most B cells differentiate into active *plasma cells* while others become dormant *memory cells.* We will first examine the role of plasma cells and later turn our attention to memory cells. A **plasma cell** produces **antibodies** that can combine with the specific type of antigen that stimulated activation of the plasma cell. During differentiation into a plasma cell, a B cell swells as the rough endoplasmic reticulum (the site for synthesis of proteins to be exported) greatly expands (● Figure 11-20). Because antibodies are proteins, plasma cells essentially become prolific protein factories, producing up to 2000 antibody molecules per second. So great is the commitment of a plasma cell's protein-synthesizing machinery to antibody production

Unactivated B cell

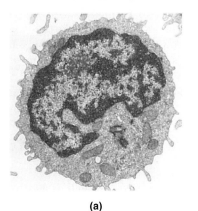

(a)

● **FIGURE 11-20**

Comparison of an unactivated B cell and a plasma cell. Electron micrograph of (a) an unactivated B cell, or small lymphocyte, and (b) a plasma cell. A plasma cell is an activated B cell. It is filled with an abundance of rough endoplasmic reticulum distended with antibody molecules.

Plasma cell

Endoplasmic reticulum

Dr. Dorothea Zucker-Franklin, New York University Medical Center

(b)

that it cannot maintain protein synthesis for its own viability and growth. Consequently, it dies after a brief (five- to seven-day), highly productive life span.

Antibodies are secreted into the blood or lymph, depending on the location of the activated plasma cells, but all antibodies eventually gain access to the blood, where they are known as **gamma globulins**, or **immunoglobulins.**

Antibodies are grouped into five subclasses based on differences in their biological activity. For example, **IgG antibodies** produce most specific immune responses against bacterial invaders and a few types of viruses, whereas **IgE** is the antibody mediator for common allergic responses, such as hay fever, asthma, and hives, and also helps defend against parasitic worms. Note that this classification is based on different ways in which antibodies function. It does not imply there are only five different antibodies. Within each functional subclass are millions of different antibodies, each able to bind only with a specific antigen.

▌ Antibodies are Y-shaped and classified according to properties of their tail portion.

Antibodies of all five subclasses are arranged in the shape of a Y (● Figure 11-21). Characteristics of the arm regions of the Y determine the *specificity* of the antibody (that is, with what antigen the antibody can bind). Properties of the tail portion of the antibody determine the *functional properties* of the antibody (what the antibody does once it binds with antigen).

An antibody has two identical antigen-binding sites, one at the tip of each arm. These **antigen-binding fragments (Fab)** are unique for each different antibody, so that each antibody can interact only with an antigen that specifically matches it, much like a lock and key. The tremendous variation in the antigen-binding fragments of different antibodies leads to the

extremely large number of unique antibodies that can bind specifically with millions of different antigens.

In contrast to these variable Fab regions at the arm tips, the tail portion of every antibody within each immunoglobulin subclass is identical. The tail, the antibody's so-called **constant (Fc) region**, contains binding sites for particular mediators of antibody-induced activities, which vary among the different subclasses. In fact, differences in the constant region are the basis for distinguishing between the different immunoglobulin subclasses. For example, the constant tail region of IgG antibodies, when activated by antigen binding in the Fab region, binds with phagocytic cells and serves as an opsonin to enhance phagocytosis. In comparison, the constant tail region of IgE antibodies attaches to mast cells and basophils, even in the absence of antigen. When the appropriate antigen gains entry to the body and binds with the attached antibodies, this trig-

● **FIGURE 11-21**

Antibody structure. An antibody is Y-shaped. It is able to bind only with the specific antigen that "fits" its antigen-binding sites (Fab) on the arm tips. The tail region (Fc) binds with particular mediators of antibody-induced activities.

PhysioEdge For an interaction related to this figure, see Media Exercise 11.6: Basics of Specific Immunity on the CD-ROM.

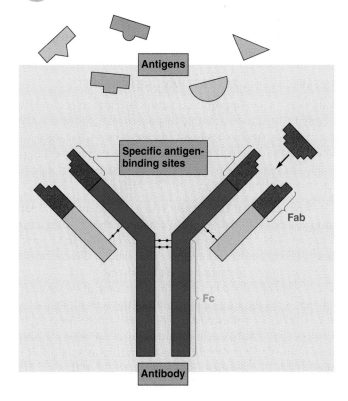

gers the release of histamine from the affected mast cells and basophils. Histamine, in turn, induces the allergic manifestations that follow.

▌ Antibodies largely amplify innate immune responses to promote antigen destruction.

Immunoglobulins cannot directly destroy foreign organisms or other unwanted materials on binding with antigens on their surfaces. Instead, antibodies exert their protective influence by physically hindering antigens or, more commonly, by amplifying innate immune responses (● Figure 11-22).

NEUTRALIZATION AND AGGLUTINATION

Antibodies can physically hinder some antigens from exerting their detrimental effects. For example, by combining with bacterial toxins, antibodies can prevent these harmful chemicals from interacting with susceptible cells. This process is known as **neutralization**. Sometimes multiple antibody molecules can cross-link numerous antigen molecules into chains or lattices of antigen-antibody complexes. The process in which foreign cells, such as bacteria or mismatched transfused red blood cells, bind together in such a clump is known as **agglutination**. When linked antigen-antibody complexes involve soluble antigens, such as tetanus toxin, the lattice can become so large that it precipitates out of solution. (**Precipitation** is the process in which a substance separates from a solution.) Within the body, these physical hindrance mechanisms play only a minor protective role against invading agents.

Clinical Note However, the tendency for certain antigens to agglutinate or precipitate on forming large complexes with antibodies specific for them is useful clinically and experimentally for detecting the presence of particular antigens or antibodies. Pregnancy diagnosis tests, for example, use this principle to detect, in urine, the presence of a hormone secreted soon after conception.

AMPLIFICATION OF INNATE IMMUNE RESPONSES

Antibodies' most important function by far is to profoundly augment the innate immune responses already initiated by the invaders. Antibodies mark foreign material as targets for actual destruction by the complement system, phagocytes, or killer cells while enhancing the activity of these other defense systems as follows:

1. *Activation of the complement system.* When an appropriate antigen binds with an antibody, receptors on the tail portion of the antibody binds with and activates C1, the first component of the complement system. This sets off the cascade of events leading to formation of the membrane attack complex, which is specifically directed at the membrane of the invading cell that bears the antigen that initiated the activation process. In fact, antibody is the most powerful activator of the complement system. The biochemical attack subsequently unleashed against the invader's membrane is the most important mechanism by which antibodies exert their protective influence. The same complement system is activated by an antigen-antibody complex regardless of the type of antigen. Although the binding of antigen to antibody is highly specific, the outcome, which is determined by the antibody's constant tail region, is identical for all activated antibodies within a given subclass; for example, all IgG antibodies activate the same complement system.

2. *Enhancement of phagocytosis.* Antibodies, especially IgG, act as opsonins. The tail portion of an antigen-bound IgG antibody binds with a receptor on the surface of a phagocyte and subsequently promotes phagocytosis of the antigen-containing victim attached to the antibody.

3. *Stimulation of killer (K) cells.* Binding of antibody to antigen also induces attack of the antigen-bearing cell by a **killer (K) cell**. K cells are similar to NK cells except that K cells require the target cell to be coated with antibodies before they can destroy it by lysing its plasma membrane. K cells have receptors for the constant tail portion of antibodies.

In these ways, antibodies, although unable to directly destroy invading bacteria or other undesirable material, bring about destruction of the antigens to which they are specifically attached, by amplifying other nonspecific lethal defense mechanisms.

IMMUNE COMPLEX DISEASE

Clinical Note Occasionally an overzealous antigen-antibody response inadvertently causes damage to normal cells as well as to invading foreign cells. Typically, antigen-antibody complexes, formed in response to foreign invasion, are removed by phagocytic cells after having revved up nonspecific defense strategies. If large numbers of these complexes are continuously produced, however, the phagocytes cannot clear away all the immune complexes formed. Antigen-antibody complexes that are not removed continue to activate the complement system, among other things. Excessive amounts of activated complement and other inflammatory agents may "spill over," damaging the surrounding normal cells as well as the unwanted cells. Furthermore, destruction is not necessarily restricted to the initial site of inflammation. Antigen-antibody complexes may circulate freely and become trapped in the kidneys, joints, brain, small vessels of the skin, and elsewhere, causing widespread inflammation and tissue damage. Such damage produced by immune complexes is referred to as an **immune complex disease**, which can be a complicating outcome of bacterial, viral, or parasitic infection.

More insidiously, immune complex disease can also stem from overzealous inflammatory activity prompted by immune complexes formed by "self-antigens" (proteins synthesized by the person's own body) and antibodies erroneously produced against them. *Rheumatoid arthritis* develops in this way.

▌ Clonal selection accounts for the specificity of antibody production.

Consider the diversity of foreign molecules a person can potentially encounter during a lifetime. Yet each B cell is preprogrammed to respond to only one of these millions of dif-

Neutralization

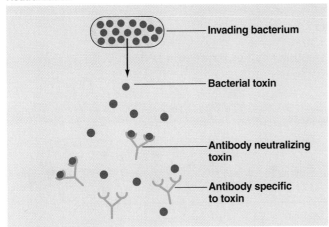

Invading bacterium

Bacterial toxin

Antibody neutralizing toxin

Antibody specific to toxin

Agglutination (clumping of antigenic cells) and **precipitation** (if soluble antigen-antibody complex is too large to stay in solution)

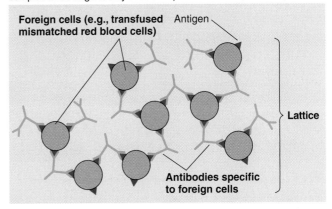

Foreign cells (e.g., transfused mismatched red blood cells)   Antigen

Lattice

Antibodies specific to foreign cells

Activation of complement system

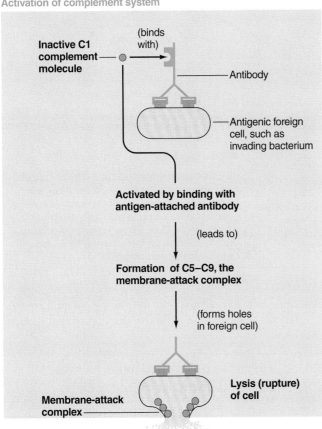

Inactive C1 complement molecule

(binds with)

Antibody

Antigenic foreign cell, such as invading bacterium

Activated by binding with antigen-attached antibody

(leads to)

Formation of C5–C9, the membrane-attack complex

(forms holes in foreign cell)

Lysis (rupture) of cell

Membrane-attack complex

Enhancement of phagocytosis (opsonization)

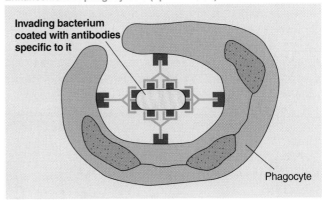

Invading bacterium coated with antibodies specific to it

Phagocyte

Stimulation of killer cells

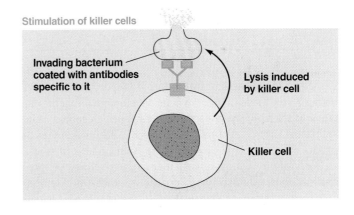

Invading bacterium coated with antibodies specific to it

Lysis induced by killer cell

Killer cell

Structures are not drawn to scale.

● **FIGURE 11-22**

How antibodies help eliminate invading microbes. Antibodies physically hinder antigens through (1) neutralization or (2) agglutination and precipitation. Antibodies amplify innate immune responses by (1) activating the complement system, (2) enhancing phagocytosis by acting as opsonins, and (3) stimulating killer cells.

ferent antigens. Other antigens cannot combine with the same B cell and induce it to secrete different antibodies. The astonishing implication is that each of us is equipped with millions of different preformed B lymphocytes, at least one for every possible antigen that we might ever encounter—including those specific for synthetic substances that do not exist in nature. The clonal selection theory proposes how a "matching" B cell responds to its antigen.

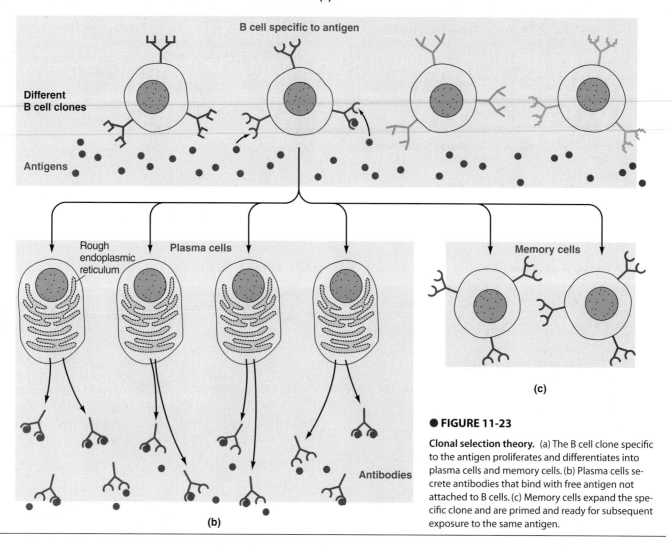

(a)

B cell specific to antigen

Different
B cell clones

Antigens

Rough
endoplasmic
reticulum

Plasma cells

Memory cells

Antibodies

(b)

(c)

● **FIGURE 11-23**

Clonal selection theory. (a) The B cell clone specific to the antigen proliferates and differentiates into plasma cells and memory cells. (b) Plasma cells secrete antibodies that bind with free antigen not attached to B cells. (c) Memory cells expand the specific clone and are primed and ready for subsequent exposure to the same antigen.

Early researchers in immunologic theory believed antibodies were "made to order" whenever a foreign antigen gained entry to the body. In contrast, the currently accepted **clonal selection theory** proposes that diverse B lymphocytes are produced during fetal development, each capable of synthesizing antibody against a particular antigen before ever being exposed to it. All offspring of a particular ancestral B lymphocyte form a family of identical cells, or a **clone**, that is committed to producing the same specific antibody. B cells remain dormant, not actually secreting their particular antibody product nor undergoing rapid division until (or unless) they come into contact with the appropriate antigen. When an antigen gains entry to the body, it "selects" (that is, activates) the particular clone of B cells that bear receptors on their surface uniquely specific for that antigen, hence the term *clonal selection theory* (● Figure 11-23).

The first antibodies produced by a newly formed B cell are inserted into the cell's plasma membrane rather than being secreted. Here they serve as receptor sites for binding with a specific kind of antigen, almost like "advertisements" for the kind of antibody the cell can produce. Binding of the appropriate antigen to a B cell amounts to "placing an order" for the manufacture and secretion of large quantities of that particular antibody.

Selected clones differentiate into active plasma cells and dormant memory cells.

Antigen binding causes the activated B-cell clone to multiply and differentiate into two cell types—plasma cells and memory cells. Most progeny are transformed into plasma cells, which are prolific producers of customized antibodies that contain the same antigen-binding sites as the surface receptors. In the blood, the secreted antibodies combine with invading free (not bound to lymphocytes) antigen, marking it for destruction by the complement system, phagocytic ingestion, or other means.

Not all the new B lymphocytes produced by the specifically activated clone differentiate into antibody-secreting plasma cells. A small proportion become **memory cells**, which do not participate in the current immune attack against the antigen but instead remain dormant and expand the specific clone. If the person is ever re-exposed to the same antigen, these memory cells are primed and ready for even more immediate action than were the original lymphocytes in the clone.

Even though each of us has essentially the same original pool of different B-lymphocyte clones, the pool gradually becomes appropriately biased to respond most efficiently to each person's particular antigenic environment. Those clones specific for antigens to which a person is never exposed remain dormant for life, whereas those specific for antigens in the individual's environment typically become expanded and enhanced by forming highly responsive memory cells.

PRIMARY AND SECONDARY RESPONSES

During initial contact with a microbial antigen, the antibody response is delayed for several days until plasma cells are formed and does not reach its peak for a couple of weeks (● Figure 11-24). This response is known as the **primary response**. Meanwhile, symptoms characteristic of the particular microbial invasion persist until either the invader succumbs to the mounting specific immune attack against it or the infected person dies. After reaching the peak, the antibody levels gradually decline over a period of time. If the same antigen ever reappears, the long-lived memory cells launch a more rapid, more potent, and longer-lasting **secondary response** than occurred during the primary response. This swifter, more powerful immune attack is frequently adequate to prevent or minimize overt infection on subsequent exposures to the same microbe, forming the basis of long-term immunity against a specific disease.

The original antigenic exposure that induces the formation of memory cells can occur through the person either actually having the disease or being vaccinated (● Figure 11-25). Vaccination deliberately exposes the person to a pathogen that has been stripped of its disease-inducing capability but that can still induce antibody formation against itself. (For the early history of vaccination development, see the boxed feature, ▶ Beyond the Basics, on p. 343.)

Memory cells are not formed for some diseases, so lasting immunity is not conferred by an initial exposure, as in the case of "strep throat." The course and severity of the disease are the same each time a person is reinfected with a microbe that the immune system does not "remember," regardless of the number of prior exposures.

▌ Blood types are a form of natural immunity.

Scientists once thought certain antibodies occurred naturally in the blood. Antibodies associated with blood types are the classic example of "natural antibodies," although natural immunity is actually a special case of actively acquired immunity. Let's see how.

ABO BLOOD TYPES

The surface membranes of human erythrocytes contain inherited antigens that vary depending on blood type. With the major blood group system, the **ABO system**, the erythrocytes of people with type A blood contain A antigens, those with type B blood contain B antigens, those with type AB blood have both A and B antigens, and those with type O blood do not have any A or B red blood cell surface antigens.

Antibodies against erythrocyte antigens not present on the body's own erythrocytes begin to appear in human plasma after a baby is about 6 months of age. Accordingly, the plasma of type A blood contains anti-B antibodies, type B blood contains anti-A antibodies, no antibodies related to the ABO system are present in type AB blood, and both anti-A and anti-B antibodies are present in type O blood. Typically, one would expect antibody production against A or B antigen to be induced only if blood containing the alien antigen were injected into the body. However, high levels of these antibodies are found in the plasma of people who have never been exposed to a different type of blood. Consequently, these were considered naturally occurring antibodies, that is, produced without any known exposure to the antigen. Scientists now know that people are unknowingly exposed at an early age to small amounts of A- and B-like antigens associated with common intestinal bacteria. Antibodies produced against these foreign antigens coincidentally also interact with a nearly identical antigen for a foreign blood group, even on first exposure to it.

TRANSFUSION REACTION

Clinical Note If a person is given blood of an incompatible type, two different antigen-antibody interactions take place. By far the more serious consequences arise from the effect of the antibodies in the recipient's

● FIGURE 11-24

Primary and secondary immune responses. (a) Primary response on first exposure to a microbial antigen. (b) Secondary response on subsequent exposure to the same microbial antigen. The primary response does not peak for a couple of weeks, whereas the secondary response peaks in a week. The magnitude of the secondary response is 100 times that of the primary response. (The relative antibody response is in the logarithmic scale.)

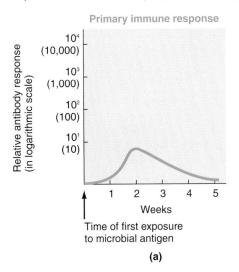
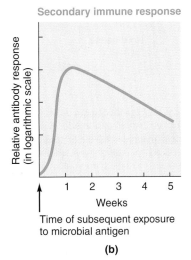

(a)

(b)

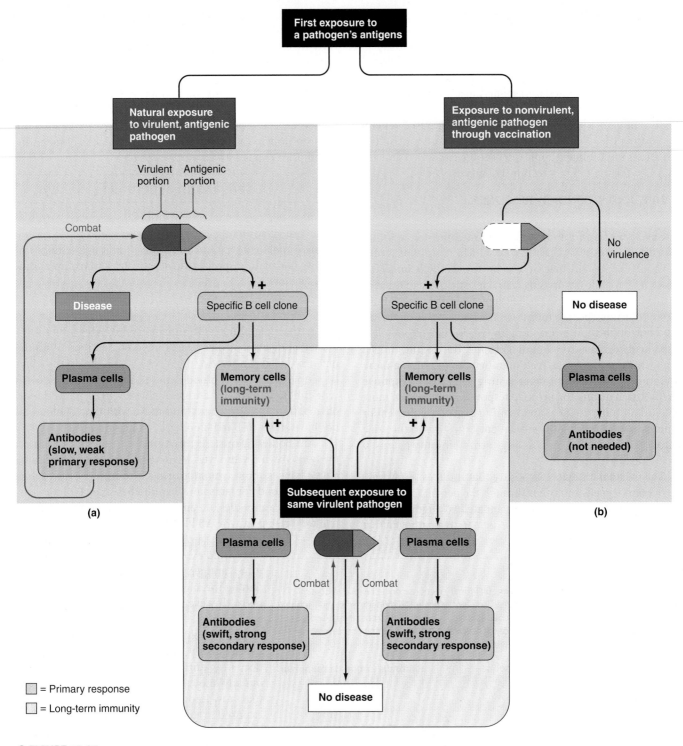

● FIGURE 11-25

Means of acquiring long-term immunity. Long-term immunity against a pathogen can be acquired through having the disease or being vaccinated against it. (a) Exposure to a virulent (disease-producing) pathogen. (b) Vaccination with a modified pathogen that is no longer virulent (that is, can no longer produce disease) but is still antigenic. In both cases, long-term memory cells are produced that mount a swift, secondary response that prevents or minimizes symptoms on a subsequent natural exposure to the same virulent pathogen.

plasma on the incoming donor erythrocytes. The effect of the donor's antibodies on the recipient's erythrocyte-bound antigens is less important unless a large amount of blood is transfused, because the donor's antibodies are so diluted by the recipient's plasma that little red blood cell damage takes place in the recipient.

Antibody interaction with erythrocyte-bound antigen may result in agglutination (clumping) or hemolysis (rupture) of

Vaccination: A Victory over Many Dreaded Diseases

Modern society has come to hope and even expect that vaccines can be developed to protect us from almost any dreaded infectious disease. This expectation has been brought into sharp focus by our current frustration over the inability to date to develop a successful vaccine against HIV, the virus that causes AIDS.

Nearly 2500 years ago, our ancestors were aware of the existence of immune protection. Writing about a plague that struck Athens in 430 B.C., Thucydides observed that the same person was never attacked twice by this disease. However, the ancients did not understand the basis of this protection, so they could not manipulate it to their advantage.

Early attempts to deliberately acquire life-long protection against smallpox, a dreaded disease that was highly infectious and frequently fatal (up to 40% of the sick died), consisted of intentionally exposing oneself by coming into direct contact with a person suffering from a milder form of the disease. The hope was to protect against a future fatal bout of smallpox by deliberately inducing a mild case of the disease. By the beginning of the

17th century, this technique had evolved into using a needle to extract small amounts of pus from active smallpox pustules (the fluid-filled bumps on the skin, which leave a characteristic depressed scar or "pock" mark after healing) and introducing this infectious material into healthy individuals. This inoculation process was done by applying the pus directly to slight cuts in the skin or by inhaling dried pus.

Edward Jenner, an English physician, was the first to demonstrate that immunity against cowpox, a disease similar to but less serious than smallpox, could also protect humans against smallpox. Having observed that milk-maids who got cowpox seemed to be protected from smallpox, Jenner in 1796 inoculated a healthy boy with pus he had extracted from cowpox boils. After the boy recovered, Jenner (not being restricted by modern ethical standards of research on human subjects) deliberately inoculated him with what was considered a normally fatal dose of smallpox infectious material. The boy survived.

Jenner's results were not taken seriously, however, until a century later when, in the

1880s, Louis Pasteur, the first great experimental immunologist, extended Jenner's technique. Pasteur demonstrated that the disease-inducing capability of organisms could be greatly reduced (attenuated) so they could no longer produce disease but would still induce antibody formation when introduced into the body—the basic principle of modern vaccines. His first vaccine was against anthrax, a deadly disease of sheep and cows. Pasteur isolated and heated anthrax bacteria, then injected these attenuated organisms into a group of healthy sheep. A few weeks later at a gathering of fellow scientists, Pasteur injected these vaccinated sheep as well as a group of unvaccinated sheep with fully potent anthrax bacteria. The result was dramatic—all the vaccinated sheep survived, but all the unvaccinated sheep died. Pasteur's notorious public demonstrations such as this, coupled with his charismatic personality, caught the attention of physicians and scientists of the time, sparking the development of modern immunology.

---

the attacked red blood cells. Agglutination and hemolysis of donor red blood cells by antibodies in the recipient's plasma can lead to a sometimes fatal **transfusion reaction.** Agglutinated clumps of incoming donor cells can plug small blood vessels. In addition, one of the most lethal consequences of mismatched transfusions is acute kidney failure caused by the release of large amounts of hemoglobin from damaged donor erythrocytes. If the free hemoglobin in the plasma rises above a critical level, it will precipitate in the kidneys and block the urine-forming structures, leading to acute kidney shutdown.

UNIVERSAL BLOOD DONORS AND RECIPIENTS

Because type O individuals have no A or B antigens, their erythrocytes will not be attacked by either anti-A or anti-B antibodies, so they are considered **universal donors.** Their blood can be transfused into people of any blood type. However, type O individuals can receive only type O blood, because the anti-A and anti-B antibodies in their plasma will attack either A or B antigens in incoming blood. In contrast, type AB individuals are called **universal recipients.** Lacking both anti-A and anti-B antibodies, they can accept donor blood of any type, although they can donate blood only to other AB people. Because their erythrocytes have both A and B antigens, their cells would be attacked if transfused into individuals with antibodies against either of these antigens.

The terms *universal donor* and *universal recipient* are somewhat misleading, however. In addition to the ABO system, many other erythrocyte antigens and plasma antibodies can

cause transfusion reactions, the most important of which is the Rh factor.

OTHER BLOOD-GROUP SYSTEMS

People who have the **Rh factor** (an erythrocyte antigen first observed in rhesus monkeys, hence the designation Rh) are said to have *Rh-positive* blood, whereas those lacking the Rh factor are considered *Rh-negative*. In contrast to the ABO system, no naturally occurring antibodies develop against the Rh factor.

Clinical Note Anti-Rh antibodies are produced only by Rh-negative individuals when (and if) such people are first exposed to the foreign Rh antigen present in Rh-positive blood. A subsequent transfusion of Rh-positive blood could produce a transfusion reaction in such a sensitized Rh-negative person. Rh-positive individuals, in contrast, never produce antibodies against the Rh factor that they themselves possess. Therefore, Rh-negative people should be given only Rh-negative blood, whereas Rh-positive people can safely receive either Rh-negative or Rh-positive blood. The Rh factor is of particular medical importance when an Rh-negative mother develops antibodies against the erythrocytes of an Rh-positive fetus she is carrying, a condition known as **erythroblastosis fetalis,** or **hemolytic disease of the newborn** (see p. 362).

Except in extreme emergencies, it is safest to individually cross-match blood before a transfusion is undertaken even though the ABO and Rh typing is already known, because there are approximately 12 other minor human-erythrocyte antigen systems. Compatibility is determined by mixing the red blood cells from the potential donor with plasma from the recipient. If no clumping occurs, the blood is considered an adequate match for transfusion.

In addition to being an important consideration in transfusions, the various blood-group systems are also of legal importance in disputed paternity cases, because the erythrocyte antigens are inherited. In recent years, however, DNA "fingerprinting" has become a more definitive test.

▌ Lymphocytes respond only to antigens presented to them by antigen-presenting cells.

B cells typically cannot perform their task of antibody production without assistance from macrophages or other antigen-presenting cells

and, in most cases, from T cells as well (● Figure 11-26). Relevant B-cell clones cannot recognize and produce antibodies in response to "raw" foreign antigens entering the body; before reacting to it, a B cell clone must be formally "introduced" to the antigen.

ANTIGEN PRESENTATION

Using macrophages as an example of an **antigen-presenting cell**, invading micro-organisms or other antigens are first en-

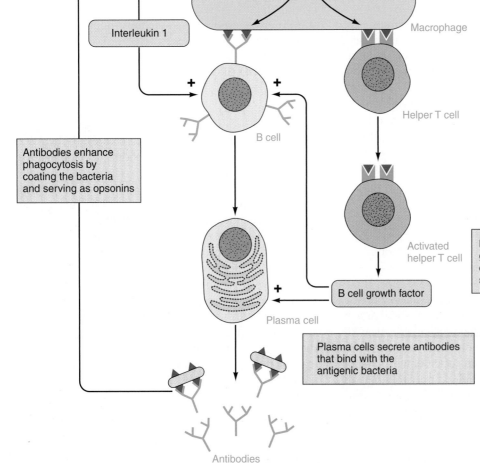

● FIGURE 11-26

Synergistic interactions among macrophages, B cells, and helper T cells. B and T cells cannot react to a newly entering foreign antigen until the antigen has been processed and presented to them by macrophages or other antigen-presenting cells. Macrophages also secrete interleukin 1, which stimulates proliferation of the activated B cells. These B cells are transformed into plasma cells, which produce antibodies against the antigen. Activated helper T cells secrete B cell growth factor, which further stimulates B cell proliferation and antibody production. The antibodies not only lead to the demise of the foreign antigen but also serve as opsonins to enhance phagocytosis by the macrophages.

Macrophages secrete interleukin 1, which enhances B cell proliferation and antibody secretion

Invading bacteria

Macrophages "process and present" bacterial antigen to B and T lymphocyte clones specific to the antigen

Interleukin 1

Macrophage

Antibodies enhance phagocytosis by coating the bacteria and serving as opsonins

B cell

Helper T cell

Activated helper T cell

Helper T cells secrete B cell growth factor that enhances B cell proliferation and antibody secretion

B cell growth factor

Plasma cell

Plasma cells secrete antibodies that bind with the antigenic bacteria

Antibodies

gulfed by macrophages. These large phagocytes cluster around the appropriate B-cell clone and handle the formal introduction. During phagocytosis, the macrophage processes the raw antigen intracellularly and then "presents" the processed antigen by exposing it on the outer surface of the macrophage's plasma membrane in such a way that the adjacent B cells can recognize and be activated by it.

In addition, these antigen-presenting macrophages secrete **interleukin 1**, a multipurpose chemical mediator that enhances differentiation and proliferation of the now-activated B cell clone. Interleukin 1, which is identical or closely related to endogenous pyrogen, is also largely responsible for the fever and malaise accompanying many infections. In collaborative fashion, activated lymphocytes secrete antibodies that, among other things, enhance further phagocytic activity.

Many antigens are similarly presented to T cells. One specialized class of T lymphocytes, called helper T cells, help B cells on being activated by macrophage-presented antigen. The helper T cells secrete a chemical mediator, **B-cell growth factor**, which further contributes to B cell function in concert with the interleukin 1 secreted by macrophages. Therefore, mutually supportive interactions among macrophages, B cells, and helper T cells synergistically reinforce the phagocyte-antibody immune attack against the foreign intruder. ▲ Table 11-3 summarizes the innate and adaptive immune strategies that defend against bacterial invasion.

We are now going to turn our attention to the other roles of T cells besides enhancing B cell activity.

T LYMPHOCYTES: CELL-MEDIATED IMMUNITY

As important as B lymphocytes and their antibody products are in specific defense against invading bacteria and other foreign material, they represent only half of the body's specific immune defenses. The T lymphocytes are equally important in defense against most viral infections and also play an important regulatory role in immune mechanisms. ▲ Table 11-4 compares the properties of these two adaptive effector cells, summarizing what you have already learned about B cells and previewing features you are about to learn about T cells.

▌T cells bind directly with their targets.

Whereas B cells and antibodies defend against conspicuous invaders in the ECF, T cells defend against covert invaders that hide out inside cells where antibodies and the complement system cannot reach them. Unlike B cells, which secrete antibodies that can attack antigen at long distances, T cells do not secrete antibodies. Instead, they must directly contact their targets, a process known as *cell-mediated immunity*. T cells

---

▲ **TABLE 11-3**

Innate and Adaptive Immune Responses to Bacterial Invasion

INNATE IMMUNE MECHANISMS	ADAPTIVE IMMUNE MECHANISMS
Inflammation	Macrophages process and present bacterial antigen to B cells specific to the antigen.
Resident tissue macrophages engulf invading bacteria.	The activated B cell clone proliferates and differentiates into plasma cells and memory cells.
Histamine-induced vascular responses increase blood flow to the area, bringing in additional immune-effector cells and plasma proteins.	Plasma cells secrete customized antibodies, which specifically bind to invading bacteria. Plasma-cell activity is enhanced by:
A fibrin clot walls off the invaded area.	Interleukin 1 secreted by macrophages.
Neutrophils and monocytes/macrophages emigrate to the area to engulf and destroy foreign invaders and to remove cell debris.	Helper T cells, which have been activated by the same bacterial antigen processed and presented to them by macrophages.
Phagocytic cells secrete chemical mediators, which enhance both innate and adaptive immune responses and induce local and systemic symptoms associated with an infection.	Antibodies bind to invading bacteria and enhance innate mechanisms that lead to the bacteria's destruction. Specifically, antibodies:
Nonspecific activation of the complement system	Act as opsonins to enhance phagocytic activity.
Complement components form a hole-punching membrane attack complex that lyses bacterial cells.	Activate the lethal complement system.
Complement components enhance many steps of inflammation.	Stimulate killer cells, which directly lyse bacteria.
	Memory cells persist that are capable of responding more rapidly and more forcefully should the same bacteria be encountered again.

B versus T Lymphocytes

CHARACTERISTIC	B LYMPHOCYTES	T LYMPHOCYTES
Ancestral Origin	Bone marrow	Bone marrow
Site of Maturational Processing	Bone marrow	Thymus
Receptors for Antigen	Antibodies inserted in the plasma membrane serve as surface receptors; highly specific	Surface receptors present but differing from antibodies; highly specific
Bind with	Extracellular antigens such as bacteria, free viruses, and other circulating foreign material	Foreign antigen in association with self-antigen, such as virus-infected cells
Antigen Must Be Processed and Presented by Macrophages	Yes	Yes
Types of Active Cells	Plasma cells	Cytotoxic T cells, helper T cells
Formation of Memory Cells	Yes	Yes
Type of Immunity	Antibody-mediated immunity	Cell-mediated immunity
Secretory Product	Antibodies	Cytokines
Function	Help eliminate free foreign invaders by enhancing innate immune responses against them; provide immunity against most bacteria and a few viruses	Lyse virus-infected cells and cancer cells; provide immunity against most viruses and a few bacteria; aid B cells in antibody production
Life Span	Short	Long

of the killer type release chemicals that destroy targeted cells that they contact, such as virus-infected cells and cancer cells.

Like B cells, T cells are clonal and exquisitely antigen specific. On its plasma membrane, each T cell bears unique receptor proteins, similar although not identical to the surface receptors on B cells. Immature lymphocytes acquire their T cell receptors in the thymus during their differentiation into T cells. Unlike B cells, T cells are activated by foreign antigen only when it is on the surface of a cell that also carries a marker of the individual's own identity; that is, both foreign antigens and **self-antigens** must be on a cell's surface before a T cell can bind with it. During thymic education, T cells learn to recognize foreign antigens only in combination with the person's own tissue antigens—a lesson passed on to all T cells' future progeny. The importance of this dual antigen requirement and the nature of the self-antigens will be described shortly.

A delay of a few days generally follows exposure to the appropriate antigen before **sensitized**, or **activated**, T cells are prepared to launch a cell-mediated immune attack. When exposed to a specific antigen combination, cells of the complementary T-cell clone proliferate and differentiate for several days, yielding large numbers of activated effector T cells that carry out various cell-mediated responses. Like B cells, T cells form a memory pool and display both primary and secondary responses.

▌ The two types of T cells are cytotoxic T cells and helper T cells.

There are two main subpopulations of T cells, depending on their roles when activated by antigen:

1. **Cytotoxic T cells** destroy host cells harboring anything foreign and thus bearing foreign antigen, such as body cells invaded by viruses, cancer cells that have mutated proteins resulting from malignant transformations, and transplanted cells.
2. **Helper T cells** enhance the development of antigen-stimulated B cells into antibody-secreting plasma cells, enhance activity of the appropriate cytotoxic cells, and activate macrophages. Helper T cells do not directly participate in immune destruction of invading pathogens. Instead, they modulate activities of other immune cells.

Helper T cells are by far the most numerous T cells, making up 60 to 80% of circulating T cells. Because of the important role these cells play in "turning on" the full power of all the

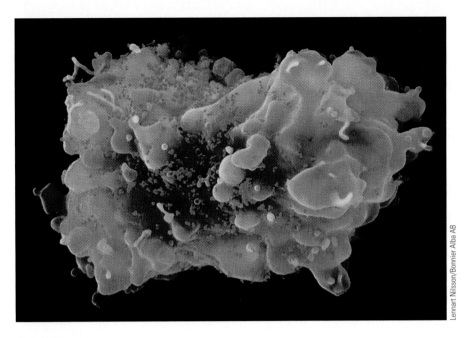

● FIGURE 11-27

AIDS virus. Human immunodeficiency virus (HIV) *(in gray)*, the AIDS-causing virus, on a helper T lymphocyte, HIV's primary target.

other activated lymphocytes and macrophages, helper T cells constitute the immune system's "master switch."

This is why **acquired immunodeficiency syndrome (AIDS)**, caused by the **human immunodeficiency virus (HIV)**, is so devastating to the immune defense system. The AIDS virus selectively invades helper T cells, destroying or incapacitating the cells that normally orchestrate much of the immune response (● Figure 11-27). The virus also invades macrophages, further crippling the immune system, and sometimes enters brain cells as well, leading to the dementia (severe impairment of intellectual capacity) noted in some AIDS victims.

We are next going to examine these two types of T cells in further detail.

▌ Cytotoxic T cells secrete chemicals that destroy target cells.

Cytotoxic T cells are microscopic "hit men." The targets of these destructive cells most frequently are host cells infected with viruses. When a virus invades a body cell, as it must to survive, the cell breaks down the envelope of proteins surrounding the virus and loads a fragment of this viral antigen piggyback onto a newly synthesized self-antigen. This self-antigen and viral antigen complex is inserted into the host cell's surface membrane, where it serves as a red flag indicating the cell is harboring the invader (● Figure 11-28, steps 1 and 2). To attack the intracellular virus, cytotoxic T cells must destroy the infected host cell in the process. Cytotoxic T cells of the clone specific for this particular virus recognize and bind to the viral antigens and self-antigens on the surface of the infected cell (● Figure 11-28, step 3). Thus sen-

● FIGURE 11-28

A cytotoxic T cell lysing a virus-invaded cell

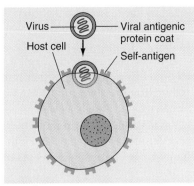

step **1** A virus invades a host cell.

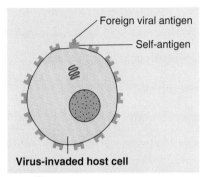

Virus-invaded host cell

step **2** The viral antigen is displayed on the surface of the host cell alongside the cell's self-antigen.

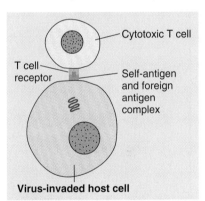

Virus-invaded host cell

step **3** The cytotoxic T cell recognizes and binds with a specific foreign antigen (viral antigen) in association with the self-antigen.

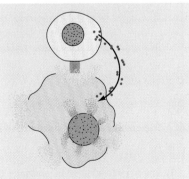

step **4** The cytotoxic T cell releases chemicals that destroy the attacked cell before the virus can enter the nucleus and start to replicate.

The Blood and Body Defenses   **347**

sitized by viral antigen, a cytotoxic T cell can kill the infected cell by either direct or indirect means:

• An activated cytotoxic T cell may directly kill the victim cell by releasing chemicals that lyse the attacked cell before viral replication can begin (● Figure 11-28, step 4). Specifically, cytotoxic T cells as well as NK cells destroy a targeted cell by releasing **perforin** molecules, which penetrate the target cell's surface membrane and join together to form porelike channels (● Figure 11-29). This technique of killing a cell

by punching holes in its membrane is similar to the method employed by the membrane attack complex of the complement cascade. This contact-dependent mechanism of killing has been nicknamed the "kiss of death."

• A cytotoxic T cell can also indirectly bring about death of an infected host cell by releasing **granzymes**, which are enzymes similar to digestive enzymes. Granzymes enter the target cell through the perforin channels. Once inside, these chemicals trigger the virus-infected cell to self-destruct through apoptosis (cell suicide; see pp. 100–101).

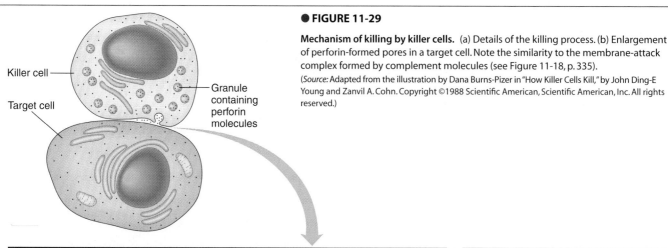

● FIGURE 11-29

Mechanism of killing by killer cells. (a) Details of the killing process. (b) Enlargement of perforin-formed pores in a target cell. Note the similarity to the membrane-attack complex formed by complement molecules (see Figure 11-18, p. 335).

(*Source:* Adapted from the illustration by Dana Burns-Pizer in "How Killer Cells Kill," by John Ding-E Young and Zanvil A. Cohn. Copyright ©1988 Scientific American, Scientific American, Inc. All rights reserved.)

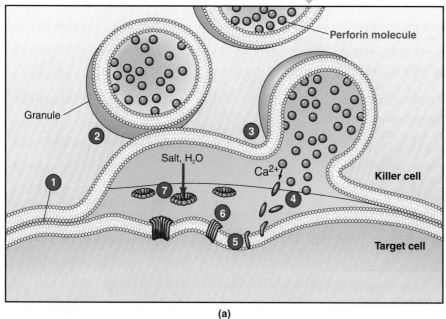

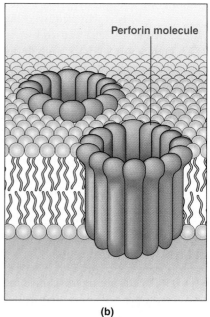

(a)

(b)

1 The killer cell binds to its target cell.

2 As a result of this binding, the killer cell's perforin-containing granules fuse with the plasma membrane.

3 The granules disgorge their perforin by exocytosis into a small pocket of intercellular space between the killer cell and its target.

4 On exposure to Ca^{2+} in this ECF space, the individual perforin molecules change from a spherical to a cylindrical shape.

5 The remodeled perforin molecules bind to the target cell membrane and insert into it.

6 Individual perforin molecules group together like staves of a barrel to form pores.

7 The pores admit salt and H_2O, causing the target cell to swell and burst.

The virus released on destruction of the host cell by either of these methods is directly destroyed in the ECF by phagocytic cells, neutralizing antibodies, and the complement system. Meanwhile the cytotoxic T cell, which has not been harmed in the process, can move on to kill other infected host cells.

The surrounding healthy cells replace the lost cells by means of cell division. Usually, to halt a viral infection only some of the host cells must be destroyed. If the virus has had a chance to multiply, however, with replicated virus leaving the original cell and spreading to other host cells, the cytotoxic T-cell defense mechanism may sacrifice so many of the host cells that serious malfunction may ensue.

Recall that other nonspecific defense mechanisms also come into play to combat viral infections, most notably NK cells, interferon, macrophages, and the complement system. As usual, an intricate web of interplay exists among the immune defenses that are launched against viral invaders (▲ Table 11-5).

❚ Helper T cells secrete chemicals that amplify the activity of other immune cells.

In contrast to cytotoxic T cells, helper T cells are not killer cells. Instead, helper T cells secrete chemicals classified as cytokines that "help," or augment, nearly all aspects of the immune response.

CYTOKINES

Exposure to antigen frequently activates both the B and T cell mechanisms simultaneously. Just as helper T cells can modulate the secretion of antibody by B cells, antibodies may influence the ability of cytotoxic T cells to destroy a victim cell. Most effects that lymphocytes exert on other immune cells are mediated by the secretion of chemical messengers. All chemicals other than antibodies that leukocytes secrete are collectively called **cytokines**, most of which are produced by helper T cells. Unlike antibodies, cytokines do not interact directly with the antigen that induces their production. Instead, cytokines spur other immune cells into action to help ward off the invader. The following are among the best known of the helper T-cell cytokines:

1. As noted earlier, helper T cells secrete *B-cell growth factor*, which enhances the antibody-secreting ability of the activated B-cell clone. Antibody secretion is greatly reduced in the absence of helper T cells.
2. Helper T cells similarly secrete **T-cell growth factor**, also known as **interleukin 2 (IL-2)**, which augments the activity of cytotoxic T cells and even of other helper T cells responsive to the invading antigen. (The 16 known interleukins that mediate interactions between various leukocytes—*interleukin* means "between leukocytes"—were numbered in the order of their discovery.)
3. Some chemicals secreted by T cells act as *chemotaxins* to lure more neutrophils and macrophages-to-be to the invaded area.

▲ **TABLE 11-5**

Defenses against Viral Invasion

When the virus is free in the ECF,

Macrophages

Destroy the free virus by phagocytosis.

Process and present the viral antigen to both B and T cells.

Secrete interleukin 1, which activates B and T cell clones specific to the viral antigen.

Plasma Cells Derived from B Cells Specific to the Viral Antigen Secrete Antibodies That

Neutralize the virus to prevent its entry into a host cell.

Activate the complement cascade that directly destroys the free virus and enhances phagocytosis of the virus by acting as an opsonin.

When the virus has entered a host cell (which it must do to survive and multiply, with the replicated viruses leaving the original host cell to enter the ECF in search of other host cells),

Interferon

Is secreted by virus-infected cells.

Binds with and prevents viral replication in other host cells.

Enhances the killing power of macrophages, natural killer cells, and cytotoxic T cells.

Natural Killer Cells

Nonspecifically lyse virus-infected host cells.

Cytotoxic T Cells

Are specifically sensitized by the viral antigen and lyse the infected host cells before the virus has a chance to replicate.

Helper T Cells

Secrete cytokines, which enhance cytotoxic T-cell activity and B-cell antibody production.

When a virus-infected cell is destroyed, the free virus is released into the ECF, where it is attacked directly by macrophages, antibodies, and the activated complement components.

4. Once macrophages are attracted to the area, **macrophage-migration inhibition factor**, another important cytokine released from helper T cells, keeps these large phagocytic cells in the region by inhibiting their outward migration. As a result, a great number of chemotactically attracted macrophages accumulate in the infected area. This factor also confers greater phagocytic power on the gathered macrophages. These so-called **angry macrophages** have more powerful destructive ability.

5. Some cytokines secreted by helper T cells activate *eosinophils* and promote the development of *IgE antibodies* for defense against parasitic worms.

What normally prevents the adaptive immune system from unleashing its powerful defense capabilities against the body's own self-antigens? We will examine this issue next.

▌ The immune system is normally tolerant of self-antigens.

The term **tolerance** in this context refers to the phenomenon of preventing the immune system from attacking the person's own tissues.

Presumably, during embryonic development some B and T cells are by chance formed that could react against the body's own tissue antigens. If these lymphocyte clones were allowed to function, they would destroy the individual's own body. Fortunately, the immune system normally does not produce antibodies or activated T cells against the body's own self-antigens but instead directs its destructive tactics only at foreign antigens.

At least five different mechanisms are involved in tolerance:

1. *Clonal deletion.* In response to continuous exposure to body antigens early in development, lymphocyte clones specifically capable of attacking these self-antigens in most cases are permanently destroyed. This **clonal deletion** is done by triggering apoptosis of immature cells that would react with the body's own proteins. This physical elimination is the major mechanism by which tolerance is developed.

2. *Clonal anergy.* The premise of **clonal anergy** is that a lymphocyte must receive two specific simultaneous signals to be activated (turned on), one from its compatible antigen and a stimulatory cosignal molecule found only on the surface of an antigen-presenting cell. Both signals are present for foreign antigens, which are introduced to lymphocytes by antigen-presenting cells. Once a B or T cell is turned on by finding its matching antigen in accompaniment with the cosignal, the cell no longer needs the cosignal to interact with other cells. For example, an activated cytotoxic T cell can destroy any virus-invaded cell that bears the viral antigen even though the infected cell does not have the cosignal. In contrast, these dual signals—antigen plus cosignal—never are present for self-antigens because these antigens are not handled by cosignal-bearing antigen-presenting cells. The first exposure to a single signal from a self-antigen turns *off* the compatible T cell, rendering the cell unresponsive to further exposure to the antigen instead of spurring the cell to proliferate. This reaction is referred to as *clonal anergy* (*anergy* means "lack of energy"), because T cells are being inactivated (that is, "become lazy") rather than activated by their antigens. Clonal anergy is a backup to clonal deletion. Anergized lymphocyte clones survive but they can't function.

3. *Receptor editing.* A newly identified means of ridding the body of self-reactive B cells is **receptor editing.** With this mechanism, once a B cell that bears a receptor for one of the body's own antigens encounters the self-antigen, the B cell escapes death or a lifetime of anergy by swiftly changing its antigen receptor to a nonself version. In this way, an originally self-reactive B cell survives but is "rehabilitated" so that it will never target the body's own tissues again.

4. *Antigen sequestering.* Some self-molecules are normally hidden from the immune system, because they never come into direct contact with the ECF in which the immune cells and their products circulate. An example of such a segregated antigen is thyroglobulin, a complex protein sequestered within the hormone-secreting structures of the thyroid gland.

5. *Immune privilege.* A few tissues, most notably the testes and the eyes, have **immune privilege,** because they escape immune attack even when they are transplanted in an unrelated individual. Scientists recently discovered that the cellular plasma membranes in these immune-privileged tissues have a specific molecule that triggers apoptosis of approaching activated lymphocytes that could attack the tissues.

Clinical Note Occasionally the immune system fails to distinguish between self-antigens and foreign antigens and unleashes its deadly powers against one or more of the body's own tissues. A condition in which the immune system fails to recognize and tolerate self-antigens associated with particular tissues is known as an **autoimmune disease.** Autoimmunity underlies more than 80 diseases, many of which are well known. Examples include multiple sclerosis, rheumatoid arthritis, and Type I diabetes mellitus. About 50 million Americans suffer from some type of autoimmune disease.

What is the nature of the self-antigens that the immune system learns to recognize as markers of a person's own cells? That is the topic of the next section.

▌ The major histocompatibility complex is the code for self-antigens.

Self-antigens are plasma membrane–bound glycoproteins (proteins with sugar attached) known as **MHC molecules** because their synthesis is directed by a group of genes called the **major histocompatibility complex** or **MHC.** The MHC genes are the most variable ones in humans. More than 100 different MHC molecules have been identified in human tissue, but each individual has a code for only 3 to 6 of these possible antigens. Because of the tremendous number of different combinations possible, the exact pattern of MHC molecules varies from one individual to another, much like a "biochemical fingerprint" or "molecular identification card," except in identical twins, who have the same MHC self-antigens.

The major histocompatibility (*histo* means "tissue"; *compatibility* means "ability to get along") complex was so named because these genes and the self-antigens they encode were first discerned in relation to tissue typing (similar to blood typing), which is done to obtain the most compatible matches for tissue grafting and transplantation. However, the transfer of tissue from one individual to another does not normally occur in nature. The natural function of MHC antigens lies in their ability to direct the responses of T cells, not in their artificial role in rejecting transplanted tissue.

MHC molecules alone on a cell surface signal to immune cells, "Leave me alone, I'm one of you." T cells typically bind with MHC self-antigens only when associated with a foreign antigen, such as a viral protein, also displayed on the cell surface in a groove on the top of the MHC molecule. Thus T cell receptors bind only with body cells making the statement—by bearing both self- and non–self-antigens on their surface—"I, one of your own kind, have been invaded. Here's a description of the enemy I am housing within." Only T cells that specifically match up with both the self- and foreign antigen can bind with the infected cell.

It would be futile for T cells to bind with free, extracellular antigen—they cannot defend against foreign material unless it is intracellular. To carry out their role of dealing with pathogens that have invaded host cells, it is appropriate that cytotoxic T cells bind only with cells of the organism's own body that viruses have infected—that is, with foreign antigen in association with self-antigen. The outcome of this binding is destruction of the infected body cell. Because cytotoxic T cells do not bind to MHC self-antigens in the absence of foreign antigen, normal body cells are protected from lethal immune attack.

TRANSPLANT REJECTION

Clinical Note T cells do bind with MHC antigens present on the surface of **transplanted cells** in the absence of foreign viral antigen. The ensuing destruction of the transplanted cells triggers the rejection of transplanted or grafted tissues. Presumably, some of the recipient's T cells "mistake" the MHC antigens of the donor cells for a closely resembling combination of a conventional viral foreign antigen complexed with the recipient's MHC self-antigens.

To minimize the rejection phenomenon, technicians match the tissues of donor and recipient according to MHC antigens as closely as possible. Therapeutic procedures to suppress the immune system then follow.

Let's now look in more detail at the role of T cells in defending against cancer.

▌ Immune surveillance against cancer cells involves an interplay among immune cells and interferon.

Besides destroying virus-infected host cells, another important function of the T cell system is recognizing and destroying newly arisen, potentially cancerous tumor cells before they have a chance to multiply and spread, a process known as **immune surveillance.** Any normal cell may be transformed into a cancer cell if mutations occur within its genes that govern cell division and growth. Such mutations may occur by chance alone or, more frequently, by exposure to **carcinogenic** (cancer-causing) factors such as ionizing radiation, certain environmental chemicals, certain viruses, or physical irritants. A few cancers are caused by tumor viruses, which permanently alter particular DNA sequences of the cells they invade. Presumably the immune system recognizes cancer cells because they bear new and different surface antigens alongside the cell's normal self-antigens, because of either genetic mutation or in-

vasion by a tumor virus. At least once a day on average, your immune system destroys a mutated cell that could potentially become cancerous.

BENIGN AND MALIGNANT TUMORS

Clinical Note Cell multiplication and growth are normally under strict control, but the regulatory mechanisms are largely unknown. Cell multiplication in an adult is generally restricted to replacing lost cells. Furthermore, cells normally respect their own place and space in the body's society of cells. If a cell that has transformed into a tumor cell manages to escape immune destruction, however, it defies the normal controls on its proliferation and position. Unrestricted multiplication of a single tumor cell results in a **tumor** that consists of a clone of cells identical to the original mutated cell.

If the mass is slow growing, stays put in its original location, and does not infiltrate the surrounding tissue, it is considered a **benign tumor.** In contrast, the transformed cell may multiply rapidly and form an invasive mass that lacks the "altruistic" behavior characteristic of normal cells. Such invasive tumors are **malignant tumors**, or **cancer.** Malignant tumor cells usually do not adhere well to the neighboring normal cells, so often some of the cancer cells break away from the parent tumor. These "emigrant" cancer cells are transported through the blood to new territories, where they continue to proliferate, forming multiple malignant tumors. The term **metastasis** is applied to this spreading of cancer to other parts of the body.

If a malignant tumor is detected early, before it has metastasized, it can be removed surgically. Once cancer cells have dispersed and seeded multiple cancerous sites, surgical elimination of the malignancy is impossible. In this case, agents that interfere with rapidly dividing and growing cells, such as certain chemotherapeutic drugs, are used in an attempt to destroy the malignant cells. Unfortunately, these agents also harm normal body cells, especially rapidly proliferating cells such as blood cells and the cells lining the digestive tract.

Untreated cancer is eventually fatal in most cases, for several interrelated reasons. The uncontrollably growing malignant mass crowds out normal cells by vigorously competing with them for space and nutrients, yet the cancer cells cannot take over the functions of the cells they are destroying. Cancer cells typically remain immature and do not become specialized, often resembling embryonic cells instead (● Figure 11-30). Such poorly differentiated malignant cells lack the ability to perform the specialized functions of the normal cell type from which they mutated. Affected organs gradually become disrupted to the point that they can no longer perform their life-sustaining functions, and the person dies.

EFFECTORS OF IMMUNE SURVEILLANCE

Immune surveillance against cancer depends on an interplay among three types of immune cells—*cytotoxic T cells, NK cells,* and *macrophages*—as well as *interferon.* Not only can all three of these immune cell types attack and destroy cancer cells directly, but all three also secrete interferon. Interferon in turn inhibits multiplication of cancer cells and increases the killing ability of the immune cells (● Figure 11-31).

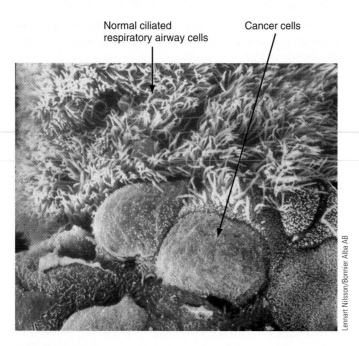

Normal ciliated respiratory airway cells

Cancer cells

Lennart Nilsson/Bonnier Alba AB

● **FIGURE 11-30**

Comparison of normal and cancerous cells in the large respiratory airways. The normal cells display specialized cilia, which constantly contract in whiplike motion to sweep debris and micro-organisms from the respiratory airways so they do not gain entrance to the deeper portions of the lungs. The cancerous cells are not ciliated, so they are unable to perform this specialized defense task.

Because NK cells do not require prior exposure and sensitization to a cancer cell before being able to launch a lethal attack, they are the first line of defense against cancer. In addition, cytotoxic T cells take aim at abnormal cancer cells bearing mutated cell proteins on their surface in conjunction with the MHC molecules. On contacting a cancer cell, both these killer cells release perforin and other toxic chemicals that destroy the targeted mutant cell. Macrophages, in addition to clearing away the remains of the dead victim cell, can engulf and destroy cancer cells intracellularly.

The fact that cancer does sometimes occur means that cancer cells occasionally escape these immune mechanisms. Some cancer cells are believed to survive by evading immune detection, for example, by failing to display identifying antigens on their surface or by being surrounded by counterproductive **blocking antibodies** that interfere with T cell function. Although B cells and antibodies are not believed to play a direct role in cancer defense, B cells, on viewing a mutant cancer cell as an alien to normal self, may produce antibodies against it. These antibodies, for unknown reasons, do not activate the complement system, which could destroy the cancer cells. Instead, the antibodies bind with the antigenic sites on the cancer cell, "hiding" these sites from recognition by cytotoxic T cells. The coating of a tumor cell by blocking antibodies thus protects the harmful cell from attack by deadly T cells. A new finding

reveals that still other successful cancer cells thwart immune attack by turning on their pursuers. They induce the T cells that bind with them to commit suicide.

PhysioEdge

Click on the Media Exercises menu of the CD-ROM and work Media Exercise 11.6: Basics of Specific Immunity to test your understanding of the previous section.

IMMUNE DISEASES

Abnormal functioning of the immune system can lead to immune diseases in two general ways: **immunodeficiency diseases** (too little immune response) and **inappropriate immune attacks** (too much or mistargeted immune response).

▌ Immunodeficiency diseases result from insufficient immune responses.

Clinical Note Immunodeficiency diseases occur when the immune system fails to respond adequately to foreign invasion. The condition may be congenital (present at birth) or acquired (nonhereditary), and it may specifically involve impairment of antibody-mediated immunity, of cell-mediated immunity, or both. In a rare hereditary condition known as **severe combined immunodeficiency**, both B and T cells are lacking. Such people have extremely limited defenses against pathogenic organisms and usually die in infancy unless maintained in a germ-free environment (that is, live in a "bubble"). Acquired (nonhereditary) immunodeficiency

● **FIGURE 11-31**

Immune surveillance against cancer. Anticancer interactions of cytotoxic T cells, natural killer cells, macrophages, and interferon.

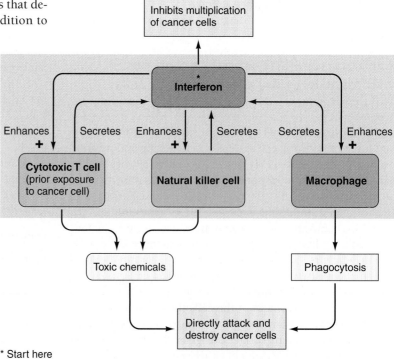

Inhibits multiplication of cancer cells

Interferon

Enhances + | Secretes | Enhances + | Secretes | Secretes | Enhances +

Cytotoxic T cell (prior exposure to cancer cell) | Natural killer cell | Macrophage

Toxic chemicals | Phagocytosis

Directly attack and destroy cancer cells

* Start here

states can arise from inadvertent destruction of lymphoid tissue during prolonged therapy with anti-inflammatory agents, such as cortisol derivatives, or from cancer therapy aimed at destroying rapidly dividing cells (which unfortunately include lymphocytes as well as cancer cells). The most recent and tragically the most common acquired immunodeficiency disease is AIDS, which, as described earlier, is caused by HIV, a virus that invades and incapacitates the critical helper T cells.

Let's now look at inappropriate immune attacks.

▌ Allergies are inappropriate immune attacks against harmless environmental substances.

Clinical Note Inappropriate adaptive immune attacks cause reactions harmful to the body. These include (1) *autoimmune responses,* in which the immune system turns against one of the body's own tissues; (2) *immune complex diseases,* which involve overexuberant antibody responses that "spill over" and damage normal tissue; and (3) allergies. The first two conditions have been described earlier in this chapter, so we will now concentrate on allergies.

An **allergy** is the acquisition of an inappropriate specific immune reactivity, or hypersensitivity, to a normally harmless environmental substance, such as dust or pollen. The offending agent is known as an **allergen.** Subsequent re-exposure of a sensitized individual to the same allergen elicits an immune attack, which may vary from a mild, annoying reaction to a severe, body-damaging reaction that may even be fatal.

Allergic responses are classified into two categories: immediate hypersensitivity and delayed hypersensitivity (▲ Table 11-6). In **immediate hypersensitivity,** the allergic response appears within about 20 minutes after a sensitized person is exposed to an allergen. In **delayed hypersensitivity,** the reaction does not generally show up until a day or so following exposure. The difference in timing is due to the different mediators involved. A particular allergen may activate either a B cell or a T cell response. Immediate allergic reactions involve B cells and are elicited by antibody interactions with an allergen; delayed reactions involve T cells and the more slowly responding process of cell-mediated immunity against the allergen. Let us examine the causes and consequences of each of these reactions in more detail.

TRIGGERS FOR IMMEDIATE HYPERSENSITIVITY

In immediate hypersensitivity, the antibodies involved and the events that ensue on exposure to an allergen differ from the typical antibody-mediated response to bacteria. The most common allergens that provoke immediate hypersensitivities are pollen grains, bee stings, penicillin, certain foods, molds, dust, feathers, and animal fur. For unclear reasons, these allergens bind to and elicit the synthesis of IgE antibodies rather than the IgG antibodies associated with bacterial antigens. When a person with an allergic tendency is first exposed to a particular allergen, compatible helper T cells secrete a cytokine that prods compatible B cells to synthesize IgE antibodies specific for the allergen. During this initial **sensitization period** no symptoms are evoked, but memory cells form that are primed for a more powerful response on subsequent re-exposure to the same allergen.

In contrast to the antibody-mediated response elicited by bacterial antigens, IgE antibodies do not freely circulate. Instead, their tail portions attach to mast cells and basophils, both of which produce and store an arsenal of potent inflammatory chemicals, such as histamine, in preformed granules. Mast cells are most plentiful in regions that come into contact with the external environment, such as the skin, the outer surface of the eyes, and the linings of the respiratory system and digestive tract. Binding of an appropriate allergen with the outreached arm regions of the IgE antibodies that are lodged tail first in a mast cell or basophil triggers the rupture of the cell's granules. As a result, histamine and other chemical mediators spew forth into the surrounding tissue.

▲ **TABLE 11-6**

Immediate versus Delayed Hypersensitivity Reactions

CHARACTERISTIC	IMMEDIATE HYPERSENSITIVITY REACTION	DELAYED HYPERSENSITIVITY REACTION
Time of Onset of Symptoms After Exposure to the Allergen	Within 20 minutes	Within one to three days
Type of Immune Response	Antibody-mediated immunity against the allergen	Cell-mediated immunity against the allergen
Immune Effectors Involved	B cells, IgE antibodies, mast cells, basophils, histamine, slow-reactive substance of anaphylaxis, eosinophil chemotactic factor	T cells
Allergies Commonly Involved	Hay fever, asthma, hives, anaphylactic shock in extreme cases	Contact allergies such as allergies to poison ivy, cosmetics, and household cleaning agents

A single mast cell (or basophil) may be coated with a number of different IgE antibodies, each able to bind with a different allergen. Thus the mast cell can be triggered to release its chemical products by any one of a number of different allergens (● Figure 11-32).

CHEMICAL MEDIATORS OF IMMEDIATE HYPERSENSITIVITY

These released chemicals cause the reactions that characterize immediate hypersensitivity. The following are among the most important chemicals released during immediate allergic reactions:

1. *Histamine,* which brings about vasodilation and increased capillary permeability as well as increased mucus production.
2. **Slow-reactive substance of anaphylaxis (SRS-A)**, which induces prolonged and profound contraction of smooth muscle, especially of the small respiratory airways. SRS-A is a locally acting mediator similar to prostaglandins (see p. 595).
3. **Eosinophil chemotactic factor**, which specifically attracts eosinophils to the area. Interestingly, eosinophils release enzymes that inactivate SRS-A and may also inhibit histamine, perhaps serving as an "off switch" to limit the allergic response.

SYMPTOMS OF IMMEDIATE HYPERSENSITIVITY

Symptoms of immediate hypersensitivity vary depending on the site, allergen, and mediators involved. Most frequently, the reaction is localized to the body site in which the IgE-bearing cells first come into contact with the allergen. If the reaction is limited to the upper respiratory passages after a person inhales an allergen such as ragweed pollen, the released chemicals bring about the symptoms characteristic of **hay fever**—for example, nasal congestion caused by histamine-induced localized edema and sneezing and runny nose caused by increased mucus secretion. If the reaction is concentrated primarily within the bronchioles (the small respiratory airways that lead to the tiny air sacs within the lungs), **asthma** results. Contraction of the smooth muscle in the walls of the bronchioles in response to SRS-A narrows or constricts these passageways, making breathing difficult. Localized swelling in the skin because of allergy-induced histamine release causes **hives.** An allergic reaction in the digestive tract in response to an ingested allergen can lead to diarrhea.

ANAPHYLACTIC SHOCK

A life-threatening systemic reaction can occur if the allergen becomes blood-borne or if very large amounts of chemicals are released from the localized site into the circulation. When large amounts of these chemical mediators gain access to the blood, the extremely serious systemic (involving the entire

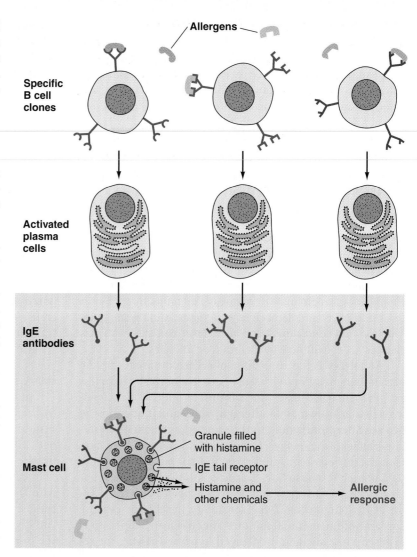

● **FIGURE 11-32**

Role of IgE antibodies and mast cells in immediate hypersensitivity. B cell clones are converted into plasma cells, which secrete IgE antibodies on contact with the allergen for which they are specific. The Fc tail portion of all IgE antibodies, regardless of the specificity of their Fab arm regions, binds to receptor proteins specific for IgE tails on mast cells and basophils. Unlike B cells, each mast cell bears a variety of antibody surface receptors for binding different allergens. When an allergen combines with the IgE receptor specific for it on the surface of a mast cell, the mast cell releases histamine and other chemicals by exocytosis. These chemicals elicit the allergic response.

body) reaction known as **anaphylactic shock** occurs. Severe hypotension that can lead to circulatory shock (see p. 308) results from widespread vasodilation and a massive shift of plasma fluid into the interstitial spaces as a result of a generalized increase in capillary permeability. Concurrently, pronounced bronchiolar constriction occurs and can lead to respiratory failure. The person may suffocate from an inability to move air through the narrowed airways. Unless countermeasures, such as injecting a vasoconstrictor-bronchodilator drug, are undertaken immediately, anaphylactic shock is often fatal. This reaction is why even a single bee sting or a single dose of penicillin can be so dangerous in people sensitized to these allergens.

IMMEDIATE HYPERSENSITIVITY AND ABSENCE OF PARASITIC WORMS

Although the immediate hypersensitivity response differs considerably from the typical IgG antibody response to bacterial infections, it is strikingly similar to the immune response elicited by parasitic worms. Shared characteristics of the immune reactions to allergens and parasitic worms include the production of IgE antibodies and increased basophil and eosinophil activity. This finding has led to the proposal that harmless allergens somehow trigger an immune response designed to fight worms. Mast cells are concentrated in areas where parasitic worms (and allergens) could contact the body. Parasitic worms can penetrate the skin or digestive tract or can attach to the digestive tract lining. Some worms migrate through the lungs during a part of their life cycle. Scientists suspect the IgE response helps ward off these invaders as follows. The inflammatory response in the skin could wall off parasitic worms attempting to burrow in. Coughing and sneezing could expel worms that migrated to the lungs. Diarrhea could help flush out worms before they could penetrate or attach to the digestive tract lining. Interestingly, epidemiologic studies suggest that the incidence of allergies in a country rises as the presence of parasites decreases. Thus, superfluous immediate hypersensitivity responses to normally harmless allergens may represent a pointless marshaling of a honed immune-response system "with nothing better to do" in the absence of parasitic worms.

DELAYED HYPERSENSITIVITY

Some allergens invoke delayed hypersensitivity, a T cell–mediated immune response, rather than an immediate, B cell–IgE antibody response. Among these allergens are poison ivy toxin and certain chemicals to which the skin is frequently exposed, such as cosmetics and household cleaning agents. Most commonly, the response is characterized by a delayed skin eruption that reaches its peak intensity one to three days after contact with an allergen to which the T system has previously been sensitized. To illustrate, poison ivy toxin does not harm the skin on contact, but it activates T cells specific for the toxin, including formation of a memory component. On subsequent exposure to the toxin, activated T cells diffuse into the skin within a day or two, combining with the poison ivy toxin that is present. The resulting interaction gives rise to the tissue damage and discomfort typical of the condition.

▲ Table 11-6 summarizes the distinctions between immediate and delayed hypersensitivities. This completes our discussion of the internal immune defense system. We are now going to turn our attention to external defenses that thwart entry of foreign invaders as a first line of defense.

EXTERNAL DEFENSES

The body's defenses against foreign microbes are not limited to the intricate, interrelated immune mechanisms that destroy micro-organisms that have actually invaded the body. In addition to the internal immune defense system, the body is equipped with external defense mechanisms designed to prevent microbial penetration wherever body tissues are exposed to the external environment. The most obvious external defense is the **skin**, or **integument**, which covers the outside of the body (*integere* means "to cover").

❚ The skin consists of an outer protective epidermis and an inner, connective tissue dermis.

The skin, which is the largest organ of the body, not only serves as a mechanical barrier between the external environment and the underlying tissues but is dynamically involved in defense mechanisms and other important functions as well. The skin consists of two layers, an outer *epidermis* and an inner *dermis* (● Figure 11-33).

EPIDERMIS

The **epidermis** consists of numerous layers of epithelial cells. The inner epidermal layers are composed of cube-shaped cells that are living and rapidly dividing, whereas the cells in the outer layers are dead and flattened. The epidermis has no direct blood supply. Its cells are nourished only by diffusion of nutrients from a rich vascular network in the underlying dermis. The newly forming cells in the inner layers constantly push the older cells closer to the surface, farther and farther from their nutrient supply. This, coupled with the fact that the outer layers are continuously subjected to pressure and "wear and tear," causes these older cells to die and become flattened. Epidermal cells are riveted tightly together by spot desmosomes (see p. 48), which interconnect with intracellular keratin filaments (see p. 37) to form a strong, cohesive covering. During maturation of a keratin-producing cell, keratin filaments progressively accumulate and cross-link with each other within the cytoplasm. As the outer cells die, this fibrous keratin core remains, forming flattened, hardened scales that provide a tough, protective **keratinized layer.** As the scales of the outermost keratinized layer slough or flake off through abrasion, they are continuously replaced by means of cell division in the deeper epidermal layers. The rate of cell division, and consequently the thickness of this keratinized layer, varies in different regions of the body. It is thickest in the areas where the skin is subjected to the most pressure, such as the bottom of the feet. The keratinized layer is airtight, fairly waterproof, and impervious to most substances. It resists anything passing in either direction between the body and the external environment. For example, it minimizes loss of water and other vital constituents from the body and prevents most foreign material from penetrating into the body.

Clinical Note This protective layer's value in holding in body fluids becomes obvious in severe burns. Not only can bacterial infections occur in the unprotected underlying tissue, but even more serious are the systemic consequences of loss of body water and plasma proteins, which escape from the exposed, burned surface. The resulting circulatory disturbances can be life threatening.

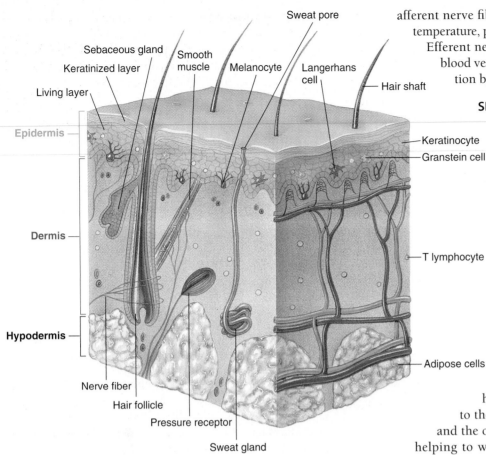

Sweat pore

Sebaceous gland
Keratinized layer
Smooth muscle
Melanocyte
Langerhans cell
Living layer
Hair shaft

Epidermis

Keratinocyte
Granstein cell

Dermis

T lymphocyte

Hypodermis

Adipose cells

Nerve fiber
Hair follicle
Pressure receptor
Sweat gland

● FIGURE 11-33

Anatomy of the skin. The skin consists of two layers, a keratinized outer epidermis and a richly vascularized inner connective tissue dermis. Special infoldings of the epidermis form the sweat glands, sebaceous glands, and hair follicles. The epidermis contains four types of cells: keratinocytes, melanocytes, Langerhans cells, and Granstein cells. The skin is anchored to underlying muscle or bone by the hypodermis, a loose, fat-containing layer of connective tissue.

afferent nerve fibers in the dermis detect pressure, temperature, pain, and other somatosensory input. Efferent nerve endings in the dermis control blood vessel caliber, hair erection, and secretion by the skin's exocrine glands.

SKIN'S EXOCRINE GLANDS AND HAIR FOLLICLES

Special infoldings of the epidermis into the underlying dermis form the skin's exocrine glands—the sweat glands and sebaceous glands—as well as the hair follicles. **Sweat glands**, which are distributed over most of the body, release a dilute salt solution through small openings, the sweat pores, onto the surface of the body. Evaporation of this sweat cools the skin and is important in regulating temperature.

The cells of the **sebaceous glands** produce **sebum**, an oily secretion released into adjacent hair follicles. From there sebum flows to the skin surface, oiling both the hairs and the outer keratinized layers of the skin, helping to waterproof them and prevent them from drying and cracking. Chapped hands or lips indicate insufficient protection by sebum. The sebaceous glands are particularly active during adolescence, causing the oily skin common among teenagers.

Each **hair follicle** is lined by special keratin-producing cells, which secrete keratin and other proteins that form the hair shaft. Hairs increase the sensitivity of the skin's surface to tactile (touch) stimuli. In some other species, this function is more finely tuned. For example, the whiskers on a cat are exquisitely sensitive in this regard. An even more important role of hair in hairier species is heat conservation, but this function is not significant in us relatively hairless humans. Like hair, the **nails** are another special keratinized product derived from living epidermal structures, the nail beds.

HYPODERMIS

The skin is anchored to the underlying tissue (muscle or bone) by the **hypodermis** (*hypo* means "below"), also known as **subcutaneous tissue** (*sub* means "under," *cutaneous* means "skin"), a loose layer of connective tissue. Most fat cells are housed within the hypodermis. These fat deposits throughout the body are collectively referred to as **adipose tissue.**

▌ Specialized cells in the epidermis produce keratin and melanin and participate in immune defense.

The epidermis contains four distinct resident cell types—*melanocytes, keratinocytes, Langerhans cells,* and *Granstein cells*—plus transient T lymphocytes that are scattered through-

Likewise, the skin barrier impedes passage into the body of most materials that come into contact with the body surface, including bacteria and toxic chemicals. Some materials, however, especially lipid-soluble substances, can penetrate intact skin through the lipid bilayers of the plasma membranes of the epidermal cells. Drugs that can be absorbed by the skin, such as nicotine or estrogen, are sometimes used in the form of a cutaneous "patch" impregnated with the drug.

DERMIS

Under the epidermis is the **dermis,** a connective tissue layer that contains many elastin fibers (for stretch) and collagen fibers (for strength), as well as an abundance of blood vessels and specialized nerve endings. The dermal blood vessels not only supply both the dermis and epidermis but also play a major role in temperature regulation. The caliber of these vessels, and hence the volume of blood flowing through them, is subject to control to vary the amount of heat exchange between these skin surface vessels and the external environment (Chapter 16). Receptors at the peripheral endings of

out the epidermis and dermis. Each of these resident cell types performs specialized functions.

Melanocytes produce the pigment **melanin**, which they disperse to surrounding skin cells. The amount and type of melanin, which can vary among black, brown, yellow, and red pigments, are responsible for the different shades of skin color of the various races. In addition to hereditary determination of melanin content, the amount of this pigment can be increased transiently in response to exposure to ultraviolet (UV) light rays from the sun. This additional melanin, the outward appearance of which constitutes a "tan," performs the protective function of absorbing harmful UV rays.

The most abundant epidermal cells are the **keratinocytes**, which, as the name implies, are specialists in keratin production. As they die, they form the outer protective keratinized layer. They also generate hair and nails. A surprising, recently discovered function is that keratinocytes are also important immunologically. They secrete interleukin 1 (a product also secreted by macrophages), which influences the maturation of T cells that tend to localize in the skin.

The two other epidermal cell types also play a role in immunity. **Langerhans cells**, which migrate to the skin from the bone marrow, serve as antigen-presenting cells. Thus the skin not only is a mechanical barrier but actually alerts regulatory lymphocytes if the barrier is breached by invading microorganisms. Langerhans cells present antigen to helper T cells, facilitating their responsiveness to skin-associated antigens. In contrast, **Granstein cells** seem to act as a "brake" on skin-activated immune responses. These cells are the most recently discovered and least understood of the skin's immune cells. Significantly, Langerhans cells are more susceptible to damage by UV radiation (as from the sun) than Granstein cells are. Losing Langerhans cells as a result of exposure to UV radiation can detrimentally lead to a predominant suppressor signal rather than the normally dominant helper signal, leaving the skin more vulnerable to microbial invasion and cancer cells.

▌ Protective measures within body cavities discourage pathogen invasion into the body.

The human body's defense system must guard against entry of potential pathogens not only through the outer surface of the body but also through the internal cavities that communicate directly with the external environment—namely, the digestive system, the genitourinary system, and the respiratory system. These systems use various tactics to destroy microorganisms entering through these routes.

DEFENSES OF THE DIGESTIVE SYSTEM

Saliva secreted into the mouth at the entrance of the digestive system contains an enzyme that lyses certain ingested bacteria. "Friendly" bacteria that live on the back of the tongue convert food-derived nitrate into nitrite, which is swallowed. Acidification of nitrite on reaching the highly acidic stomach generates nitric oxide, which is toxic to a variety of microorganisms. Furthermore, many of the surviving bacteria that are swallowed are killed directly by the strongly acidic gastric

juice in the stomach. Farther down the tract, the intestinal lining is endowed with gut-associated lymphoid tissue. These defensive mechanisms are not 100% effective, however. Some bacteria do manage to survive and reach the large intestine (the last portion of the digestive tract), where they continue to flourish. Surprisingly, this normal microbial population provides a natural barrier against infection within the lower intestine. These harmless resident microbes competitively suppress the growth of potential pathogens that have managed to escape the antimicrobial measures of earlier parts of the digestive tract. On occasion, orally administered antibiotic therapy against an infection elsewhere within the body may actually induce an intestinal infection. By knocking out some of the normal intestinal bacteria, an antibiotic may permit an antibiotic-resistant pathogenic species to overgrow in the intestine.

DEFENSES OF THE GENITOURINARY SYSTEMS

Within the genitourinary (reproductive and urinary) system, would-be invaders encounter hostile conditions in the acidic urine and acidic vaginal secretions. The genitourinary organs also produce a sticky mucus, which, like flypaper, entraps small invading particles. Subsequently, the particles are either engulfed by phagocytes or are swept out as the organ empties (for example, they are flushed out with urine flow).

DEFENSES OF THE RESPIRATORY SYSTEM

The respiratory system is likewise equipped with several important defense mechanisms against inhaled particulate matter. The respiratory system is the largest surface of the body that comes into direct contact with the increasingly polluted external environment. The surface area of the respiratory system that is exposed to the air is 30 times that of the skin. Larger airborne particles are filtered out of the inhaled air by hairs at the entrance of the nasal passages. Lymphoid tissues, the *tonsils* and *adenoids,* provide immunological protection against inhaled pathogens near the beginning of the respiratory system. Farther down in the respiratory airways, millions of tiny hairlike projections known as *cilia* constantly beat in an outward direction (see p. 36). The respiratory airways are coated with a layer of thick, sticky mucus secreted by epithelial cells within the airway lining. This mucus sheet, laden with any inspired particulate debris (such as dust) that adheres to it, is constantly moved upward to the throat by ciliary action. This moving "staircase" of mucus is known as the **mucus escalator.** The dirty mucus is either expectorated (spit out) or in most cases swallowed without the person even being aware of it; any indigestible foreign particulate matter is later eliminated in the feces. Besides keeping the lungs clean, this mechanism is an important defense against bacterial infection, because many bacteria enter the body on dust particles. Also contributing to defense against respiratory infections are antibodies secreted in the mucus. In addition, an abundance of phagocytic specialists called the **alveolar macrophages** scavenge within the air sacs (alveoli) of the lungs. Further respiratory defenses include coughs and sneezes. These commonly experienced reflex mechanisms involve forceful outward expulsion of material in an attempt to remove irritants from the trachea (*coughs*) or nose (*sneezes*).

Clinical Note Cigarette smoking suppresses these normal respiratory defenses. The smoke from a single cigarette can paralyze the cilia for several hours, with repeated exposure eventually leading to ciliary destruction. Failure of ciliary activity to sweep out a constant stream of particulate-laden mucus enables inhaled carcinogens to remain in contact with the respiratory airways for prolonged periods. Furthermore, cigarette smoke incapacitates alveolar macrophages. In addition, noxious agents in tobacco smoke irritate the mucous linings of the respiratory tract, resulting in excess mucus production, which may partially obstruct the airways. "Smoker's cough" is an attempt to dislodge this excess stationary mucus. These and other direct toxic effects on lung tissue lead to the increased incidence of lung cancer and chronic respiratory diseases associated with cigarette smoking. Air pollutants include some of the same substances found in cigarette smoke and can similarly affect the respiratory system.

We will examine the respiratory system in greater detail in the next chapter.

CHAPTER IN PERSPECTIVE: FOCUS ON HOMEOSTASIS

Blood contributes to homeostasis in a variety of ways. First, the composition of interstitial fluid, the true internal environment that surrounds and directly exchanges materials with the cells, depends on the composition of blood plasma. Because of the thorough exchange that occurs between the interstitial and vascular compartments, interstitial fluid has the same composition as plasma with the exception of plasma proteins, which cannot escape through the capillary walls. Thus blood serves as the vehicle for rapid, long-distance, mass transport of materials to and from the cells, and interstitial fluid serves as the go-between.

Homeostasis depends on the blood carrying materials such as O_2 and nutrients to the cells as rapidly as the cells consume these supplies and carrying materials such as metabolic wastes away from the cells as rapidly as the cells produce these products. It also depends on the blood carrying hormonal messengers from their site of production to their distant site of action. Once a substance enters the blood, it can be transported throughout the body within seconds, whereas diffusion of the substance over long distances in a large multicellular organism such as a human would take months to years—a situation incompatible with life. Diffusion can, however, effectively accomplish short local exchanges of materials between blood and surrounding cells through the intervening interstitial fluid.

Blood has special transport capabilities that enable it to move its cargo efficiently throughout the body. For example, life-sustaining O_2 is poorly soluble in water, but blood is equipped with O_2-carrying specialists, the erythrocytes (red blood cells),

which are stuffed full of hemoglobin, a complex molecule that transports O_2. Likewise, homeostatically important water-insoluble hormonal messengers are shuttled in the blood by plasma protein carriers.

Specific components of the blood perform the following additional homeostatic activities that are unrelated to blood's transport function:

- Blood helps maintain the proper pH in the internal environment by buffering changes in the body's acid–base load.
- Blood helps maintain body temperature by absorbing heat produced by heat-generating tissues such as contracting skeletal muscles and distributing it throughout the body. Excess heat is carried by the blood to the body surface for elimination to the external environment.
- Electrolytes in the plasma are important in membrane excitability, which is critical for nerve and muscle function.
- Electrolytes in the plasma are important in osmotic distribution of fluid between the extracellular and intracellular fluid. Plasma proteins play a critical role in distributing extracellular fluid between the plasma and interstitial fluid.
- Through their hemostatic functions, the platelets and clotting factors minimize the loss of life-sustaining blood after vessel injury.
- The leukocytes (white blood cells), their secretory products, and certain types of plasma proteins, such as antibodies, constitute the immune defense system. This system defends the body against invading disease-causing agents, destroys cancer cells, and paves the way for wound healing and tissue repair by clearing away debris from dead or injured cells. These actions indirectly contribute to homeostasis by helping the organs that directly maintain homeostasis stay healthy. We could not survive beyond early infancy were it not for the body's defense mechanisms.

The skin contributes indirectly to homeostasis by serving as a protective barrier between the external environment and the rest of the body cells. It helps prevent harmful foreign agents such as pathogens and toxic chemicals from entering the body and helps prevent the loss of precious internal fluids from the body. The skin also contributes directly to homeostasis by helping maintain body temperature by means of the sweat glands and adjustments in skin blood flow. The amount of heat carried to the body surface for dissipation to the external environment is determined by the volume of warmed blood flowing through the skin.

Other systems that have internal cavities in contact with the external environment, such as the digestive, genitourinary, and respiratory systems, also have defense capabilities to prevent harmful external agents from entering the body through these avenues.

CHAPTER SUMMARY

Introduction (p. 315)

▪ Blood consists of three types of cellular elements—erythrocytes (red blood cells), leukocytes (white blood cells), and platelets (thrombocytes)—suspended in the liquid plasma. *(Review Table 11-1.)*

▪ The 5- to 5.5-liter volume of blood in an adult consists of 42 to 45% erythrocytes, less than 1% leukocytes and platelets, and 55 to 58% plasma. The percentage of whole-blood volume occupied by erythrocytes is the hematocrit. *(Review Figure 11-1.)*

Plasma (pp. 315–317)

▪ Plasma is a complex liquid consisting of 90% water that serves as a transport medium for substances being carried in the blood.

▪ The most abundant inorganic constituents in plasma are Na^+ and Cl^-. The most plentiful organic constituents in plasma are plasma proteins.

Erythrocytes (pp. 317–321)

▪ Erythrocytes are specialized for their primary function of O_2 transport in the blood. They do not contain a nucleus, organelles, or ribosomes but instead are packed full of hemoglobin, an iron-containing molecule that can loosely and reversibly bind with O_2. Because O_2 is poorly soluble in blood, hemoglobin is indispensable for O_2 transport. *(Review Figure 11-3.)*

▪ Unable to replace cell components, erythrocytes are destined to a short life span of about 120 days.

▪ Undifferentiated pluripotent stem cells in the red bone marrow give rise to all cellular elements of the blood. *(Review Figure 11-12.)* Erythrocyte production (erythropoiesis) by the marrow normally keeps pace with the rate of erythrocyte loss, keeping the red cell count constant. Erythropoiesis is stimulated by erythropoietin, a hormone secreted by the kidneys in response to reduced O_2 delivery. *(Review Figure 11-4.)*

Platelets and Hemostasis (pp. 321–326)

▪ Platelets are cell fragments derived from large megakaryocytes in the bone marrow. *(Review Figure 11-6.)*

▪ Platelets play a role in hemostasis, the arrest of bleeding from an injured vessel. The three main steps in hemostasis are (1) vascular spasm, (2) platelet plugging, and (3) clot formation.

▪ Vascular spasm reduces blood flow through an injured vessel.

▪ Aggregation of platelets at the site of vessel injury quickly plugs the defect. Platelets start to aggregate on contact with exposed collagen in the damaged vessel wall. *(Review Figure 11-7.)*

▪ Clot formation reinforces the platelet plug and converts blood in the vicinity of a vessel injury into a nonflowing gel.

▪ Most factors necessary for clotting are always present in the plasma in inactive precursor form. When a vessel is damaged, exposed collagen initiates a cascade of reactions involving successive activation of these clotting factors, ultimately converting fibrinogen into fibrin. *(Review Figure 11-9.)*

▪ Fibrin, an insoluble threadlike molecule, is laid down as the meshwork of the clot; the meshwork in turn entangles blood cellular elements to complete clot formation. *(Review Figure 11-8.)*

▪ Blood that has escaped into the tissues clots on exposure to tissue thromboplastin, which sets the clotting process into motion.

▪ When no longer needed, clots are dissolved by plasmin, a fibrinolytic factor also activated by exposed collagen. *(Review Figure 11-10.)*

Leukocytes (pp. 326–330)

▪ Leukocytes are the defense corps of the body. They attack foreign invaders, destroy abnormal cells that arise in the body, and clean up cellular debris. Leukocytes as well as certain plasma proteins make up the immune system.

▪ The most common invaders are bacteria and viruses.

▪ Each of the five types of leukocytes has a different task: (1) Neutrophils, the phagocytic specialists, are important in engulfing bacteria and debris. (2) Eosinophils specialize in attacking parasitic worms and play a role in allergic responses. (3) Basophils release two chemicals: histamine, which is also important in allergic responses, and heparin, which helps clear fat particles from the blood. (4) Monocytes, on leaving the blood, set up residence in the tissues and greatly enlarge to become the large tissue phagocytes known as macrophages. (5) Lymphocytes are primarily responsible for the specific adaptive immune defenses of the body. *(Review Figure 11-11.)*

▪ Leukocytes are present in the blood only while in transit from their site of production and storage in the bone marrow (and also in the lymphoid tissues in the case of the lymphocytes) to their site of action in the tissues. *(Review Figure 11-13.)* At any given time, most leukocytes are out in the tissues on surveillance missions or performing actual combat missions.

▪ Immunity, the body's ability to resist or eliminate potentially harmful foreign invaders and newly arisen cancer cells, includes both innate and adaptive immune responses. Innate immune responses are nonspecific responses that nonselectively defend against foreign material even on initial exposure to it. Adaptive immune responses are specific responses that selectively target particular invaders for which the body has been specially prepared after a prior exposure. *(Review Table 11-3, p. 345.)*

Innate Immunity (pp. 330–335)

▪ Innate immune responses, which form a first line of defense against atypical cells (foreign, mutant, or injured cells) within the body, include inflammation, interferon, natural killer cells, and the complement system.

▪ Inflammation is a nonspecific response to foreign invasion or tissue damage mediated largely by the professional phagocytes (neutrophils and monocytes-turned-macrophages) and their secretions.

▪ The phagocytic cells destroy foreign and damaged cells both by phagocytosis and by the release of lethal chemicals.

▪ Histamine-induced vasodilation and increased permeability of local vessels at the site of invasion or injury permit enhanced delivery of more phagocytic leukocytes and inactive plasma protein precursors crucial to the inflammatory process, such as clotting factors and components of the complement system. These vascular changes also largely produce the observable local manifestations of inflammation—swelling, redness, heat, and pain. *(Review Figure 11-14.)*

▪ Interferon is nonspecifically released by virus-infected cells and transiently inhibits viral multiplication in other cells to which it binds. *(Review Figure 11-17.)* Interferon further exerts anticancer effects by slowing division and growth of tumor cells as well as by enhancing the power of killer cells.

▪ Natural killer (NK) cells nonspecifically lyse and destroy virus-infected cells and cancer cells on first exposure to them.

- On being activated by microbes themselves at the site of invasion or by antibodies produced against the microbes, the complement system directly destroys the foreign invaders by forming a hole-punching membrane attack complex that inserts into the victim cell's membrane, leading to osmotic rupture of the cell. (*Review Figure 11-18.*)

Adaptive Immunity: General Concepts (pp. 335–336)

- Not only is the adaptive immune system able to recognize foreign molecules as different from self-molecules—so that destructive immune reactions are not unleashed against the body itself—but it can also distinguish between millions of different foreign molecules. Lymphocytes, the effector cells of adaptive immunity, are each uniquely equipped with surface membrane receptors that can bind lock-and-key fashion with only one specific complex foreign molecule, known as an *antigen.*
- There are two broad classes of adaptive immune responses: antibody-mediated immunity involving B cells and cell-mediated immunity involving T cells. (*Review Figure 11-19 and Table 11-4, p. 346.*)

B Lymphocytes: Antibody-Mediated Immunity (pp. 336–345)

- Each B cell recognizes specific free extracellular antigen that is not associated with cell-bound self-antigens, such as that found on the surface of bacteria.
- After being activated by binding with its specific antigen, a B cell rapidly proliferates, producing a clone of its own kind that can specifically wage battle against the invader. Most lymphocytes in the expanded B-cell clone become plasma cells that participate in the primary response against the invader. (*Review Figures 11-20 and 11-23.*)
- Plasma cells are specialized to secrete freely circulating antibodies that besiege the freely existing invading bacteria (or other foreign substance) that induced their production.
- Antibodies are Y-shaped molecules. The antigen-binding sites on the tips of each arm of the antibody determine with what specific antigen the antibody can bind. Properties of the antibody's tail portion determine what the antibody does once it binds with antigen. (*Review Figure 11-21.*)
- Antibodies do not directly destroy material. Instead, they intensify lethal innate immune responses already called into play by the foreign invasion. (*Review Figure 11-22.*)
- Some of the newly developed lymphocytes in the activated B-cell clone do not participate in the attack but become memory cells that lie in wait, ready to launch a swifter and more forceful secondary response should the same foreigner ever invade the body again. (*Review Figures 11-23 through 11-25.*)
- Both B and T cells can recognize and bind with antigen only when it has been processed and presented to them by antigen-presenting cells, such as macrophages. (*Review Figure 11-26.*)

T Lymphocytes: Cell-Mediated Immunity (pp. 345–352)

- T cells accomplish cell-mediated immunity by being in direct contact with their targets.

- There are two types of T cells: cytotoxic T cells and helper T cells.
- The targets of cytotoxic T cells are virally invaded cells and cancer cells, which they destroy by releasing perforin molecules that form a lethal hole-punching complex that inserts into the membrane of the victim cell or by releasing granzymes that trigger the victim cell to undergo apoptosis. (*Review Figures 11-28 and 11-29 and Table 11-5.*)
- Helper T cells bind with other immune cells and release chemicals that augment the activity of these other cells. Chemicals other than antibodies released by leukocytes are known as *cytokines,* most of which are secreted by helper T cells.
- Those lymphocytes produced that can attack the body's own antigen-bearing cells are eliminated or suppressed so that they can't function. In this way, the body is able to "tolerate" (not attack) its own antigens.
- In the process of *immune surveillance,* natural killer cells, cytotoxic T cells, macrophages, and the interferon they collectively secrete normally eradicate newly arisen cancer cells before they have a chance to spread. (*Review Figure 11-31.*)

Immune Diseases (pp. 352–355)

- Immune diseases are of two types: immunodeficiency diseases (insufficient immune responses) or inappropriate immune attacks (excessive or mistargeted immune responses).
- With immunodeficiency diseases, the immune system fails to defend normally against bacterial or viral infections through a deficit of B or T cells, respectively.
- With inappropriate immune attacks, the immune system becomes overzealous. There are three categories of inappropriate attacks:
  1. In autoimmune disease, the immune system erroneously turns against one of the person's own tissues that the system no longer recognizes and tolerates as self.
  2. With immune complex diseases, body tissues are inadvertently destroyed as an overabundance of antigen-antibody complexes activates excessive lethal complement, which destroys surrounding normal cells as well as the antigen.
  3. Allergies, or hypersensitivities, occur when the immune system inappropriately launches a symptom-producing, body-damaging attack against an allergen, a normally harmless environmental antigen. (a) Immediate hypersensitivities involve the production of IgE antibodies by B cells that trigger the release of powerful inflammatory chemicals from mast cells and basophils to bring about a swift response to the allergen. (b) Delayed hypersensitivities involve a more slowly responding cell-mediated, symptom-producing response by T cells against the allergen. (*Review Table 11-6 and Figure 11-32.*)

External Defenses (pp. 355–358)

- The body surfaces exposed to the outside environment—both the outer covering of skin and the linings of internal cavities that communicate with the external environment—serve not only as mechanical barriers to deter would-be pathogenic invaders but also play an active role in thwarting entry of bacteria and other unwanted materials. (*Review Figure 11-33.*)

REVIEW EXERCISES

Objective Questions (Answers on p. A-45)

1. Hemoglobin can carry only O_2. *(True or false?)*
2. Erythrocytes originate from the same undifferentiated pluripotent stem cells as leukocytes and platelets do. *(True or false?)*
3. White blood cells spend the majority of their time in the blood. *(True or false?)*
4. The complement system can only be activated by antibodies. *(True or false?)*
5. Specific adaptive immune responses are accomplished by neutrophils. *(True or false?)*
6. Active immunity against a particular disease can be acquired only by actually having the disease. *(True or false?)*
7. A secondary response has a more rapid onset, is more potent, and has a longer duration than a primary response. *(True or false?)*
8. _____, collectively, are all the chemical messengers other than antibodies secreted by lymphocytes.
9. Which of the following is *not* a function served by the plasma proteins?
   a. facilitating retention of fluid in the blood vessels
   b. playing an important role in blood clotting
   c. binding and transporting certain hormones in the blood
   d. transporting O_2 in the blood
   e. serving as antibodies
   f. contributing to the buffering capacity of the blood
10. Which of the following is *not* directly triggered by exposed collagen in an injured vessel?
    a. initial vascular spasm
    b. platelet aggregation
    c. activation of the clotting cascade
    d. activation of plasminogen
11. Which of the following statements concerning leukocytes is/are *incorrect*?
    a. Monocytes are transformed into macrophages.
    b. T lymphocytes are transformed into plasma cells that secrete antibodies.
    c. Neutrophils are highly mobile phagocytic specialists.
    d. Basophils release histamine.
    e. Lymphocytes arise in large part from lymphoid tissues.
12. Match the following blood abnormalities with their causes:
    ____ 1. deficiency of intrinsic factor
    ____ 2. insufficient amount of iron to synthesize adequate hemoglobin
    ____ 3. destruction of bone marrow
    ____ 4. abnormal loss of blood
    ____ 5. tumorlike condition of bone marrow
    ____ 6. inadequate erythropoietin secretion
    ____ 7. excessive rupture of circulating erythrocytes
    ____ 8. associated with living at high altitudes

    (a) hemolytic anemia
    (b) aplastic anemia
    (c) nutritional anemia
    (d) hemorrhagic anemia
    (e) pernicious anemia
    (f) renal anemia
    (g) primary polycythemia
    (h) secondary polycythemia

13. Match the following:
    (a) complement system
    (b) natural killer cells
    (c) interferon
    (d) inflammation

    ____ 1. a family of proteins that nonspecifically defend against viral infection
    ____ 2. a response to tissue injury in which neutrophils and macrophages play a major role
    ____ 3. a group of plasma proteins that, when activated, bring about destruction of foreign cells by attacking their plasma membranes
    ____ 4. lymphocyte-like entities that spontaneously lyse tumor cells and virus-infected host cells

14. Using the answer code on the right, indicate whether the numbered characteristics of the adaptive immune system apply to antibody-mediated immunity or cell-mediated immunity (or both):
    ____ 1. involves secretion of antibodies
    ____ 2. mediated by B cells
    ____ 3. mediated by T cells
    ____ 4. accomplished by thymus-educated lymphocytes
    ____ 5. triggered by the binding of specific antigens to complementary lymphocyte receptors
    ____ 6. involves formation of memory cells in response to initial exposure to an antigen
    ____ 7. primarily aimed against virus-infected host cells
    ____ 8. protects primarily against bacterial invaders
    ____ 9. directly destroys targeted cells
    ____ 10. involved in rejection of transplanted tissue
    ____ 11. requires binding of a lymphocyte to a free extracellular antigen
    ____ 12. requires dual binding of a lymphocyte with both foreign antigen and self-antigens present on the surface of a host cell

    (a) antibody-mediated immunity
    (b) cell-mediated immunity
    (c) both antibody-mediated and cell-mediated immunity

Essay Questions

1. What is the average blood volume in women and in men?
2. What is the normal percentage of blood occupied by erythrocytes and by plasma? What is the hematocrit? What is the buffy coat?
3. What is the composition of plasma?
4. Describe the structure and functions of erythrocytes.

5. Why can erythrocytes survive for only about 120 days?
6. Describe the process and control of erythropoiesis.
7. Discuss the derivation of platelets.
8. Describe the three steps of hemostasis, including a comparison of the intrinsic and extrinsic pathways by which the clotting cascade is triggered.
9. Compare the structure and functions of the five types of leukocytes.
10. Distinguish between bacteria and viruses.
11. Distinguish between innate and adaptive immune responses.
12. Compare the life history of B cells and of T cells.
13. What is an antigen?
14. Describe the structure of an antibody.
15. In what ways do antibodies exert their effect?
16. Describe the clonal selection theory.
17. Compare the functions of B cells and T cells. What are the roles of the two types of T cells?
18. What mechanisms are involved in tolerance?
19. Describe the factors that contribute to immune surveillance against cancer cells.
20. Distinguish among immunodeficiency disease, autoimmune disease, immune complex disease, immediate hypersensitivity, and delayed hypersensitivity.
21. What are the immune functions of the skin?

POINTS TO PONDER

(Explanations on p. A-45)

1. There are different forms of hemoglobin. *Hemoglobin A* is normal adult hemoglobin. The abnormal form *hemoglobin S* causes RBCs to warp into fragile, sickle-shaped cells. Fetal RBCs contain *hemoglobin F*, the production of which stops soon after birth. Now researchers are trying to goad the genes that direct hemoglobin F synthesis back into action as a means of treating sickle cell anemia. Explain how turning on these fetal genes could be a useful remedy. (Indeed, the first effective drug therapy recently approved for treating sickle cell anemia, *hydroxyurea,* acts on the bone marrow to boost production of fetal hemoglobin.)

2. Low on the list of popular animals are vampire bats, leeches, and ticks, yet these animals may someday indirectly save your life.

Scientists are currently examining the "saliva" of these blood-sucking creatures in search of new chemicals that might limit cardiac muscle damage in heart attack victims. What do you suspect the nature of these sought-after chemicals is?

3. Why does the frequent mutation of HIV (the AIDS virus) make it difficult to develop a vaccine against this virus?

4. What impact would failure of the thymus to develop embryonically have on the immune system after birth?

5. Medical researchers are currently working on ways to "teach" the immune system to view foreign tissue as "self." What useful clinical application will the technique have?

CLINICAL CONSIDERATION

(Explanation on p. A-45)

Heather L., who has Rh-negative blood, has just given birth to her first child, who has Rh-positive blood. Both mother and baby are fine, but the doctor administers an Rh immunoglobulin preparation so that any future Rh-positive babies Heather has will not suffer from hemolytic disease of the newborn (see p. 343). During gestation (pregnancy), fetal and maternal blood do not mix. Instead, materials are exchanged between these two circulatory systems across the placenta, a special organ that develops during gestation from both maternal and fetal structures (see p. 615). Red blood cells are unable to cross the placenta, but antibodies can cross. During the birthing process, a small amount of the infant's blood may enter the maternal circulation.

1. Why did Heather's first-born child not have hemolytic disease of the newborn; that is, why didn't maternal antibodies against the Rh factor attack the fetal Rh-positive red blood cells during gestation?

2. Why would any subsequent Rh-positive babies Heather might carry be likely to develop hemolytic disease of the newborn if she were not treated with Rh immunoglobulin?

3. How would administering Rh immunoglobulin immediately following Heather's first pregnancy with an Rh-positive child prevent hemolytic disease of the newborn in a later pregnancy with another Rh-positive child? Similarly, why must Rh immunoglobulin be given to Heather after the birth of each Rh-positive child she bears?

4. Suppose Heather were not treated with Rh immunoglobulin after the birth of her first Rh-positive child, and a second Rh-positive child developed hemolytic disease of the newborn. Would administering Rh immunoglobulin to Heather immediately after the second birth prevent this condition in a third Rh-positive child? Why or why not?

PHYSIOEDGE RESOURCES

 PhysioEdge CD-ROM

PhysioEdge, the CD-ROM packaged with your text, focuses on the concepts students find most difficult to learn. Figures marked with this icon have associated activities on the CD. For a visual review of concepts in this chapter, check out the following:

Media Exercise 11.1: Blood

Media Exercise 11.2: Abnormal Hemostasis

Media Exercise 11.3: Thrombocytes and Hemostasis

Media Exercise 11.4: The Body's Defenses

Media Exercise 11.5: Inflammation

Media Exercise 11.6: Basics of Specific Immunity

PhysioEdge Website

The website for this book contains a wealth of helpful study aids, as well as many ideas for further reading and research. Log on to: **http://www.brookscole.com/hpfundamentals3**

Select Chapter 11 from the drop-down menu, or click on one of the many resource areas, including Case Histories, which introduce clinical aspects of human physiology. For this chapter check out Case History 17: Pneumococcal Bacteremia, Case History 18: "It's Nothing—Just a Bee Sting," Case History 19: A Close Call, Case History 20: David-Life in a Germ-Free World, Case History 21: AIDS, Case History 22: Acne Can Be Controlled.

For Suggested Readings, consult **InfoTrac College Edition/ Research** on the PhysioEdge website or go directly to InfoTrac College Edition, your online research library, at: **http://infotrac.thomsonlearning.com**

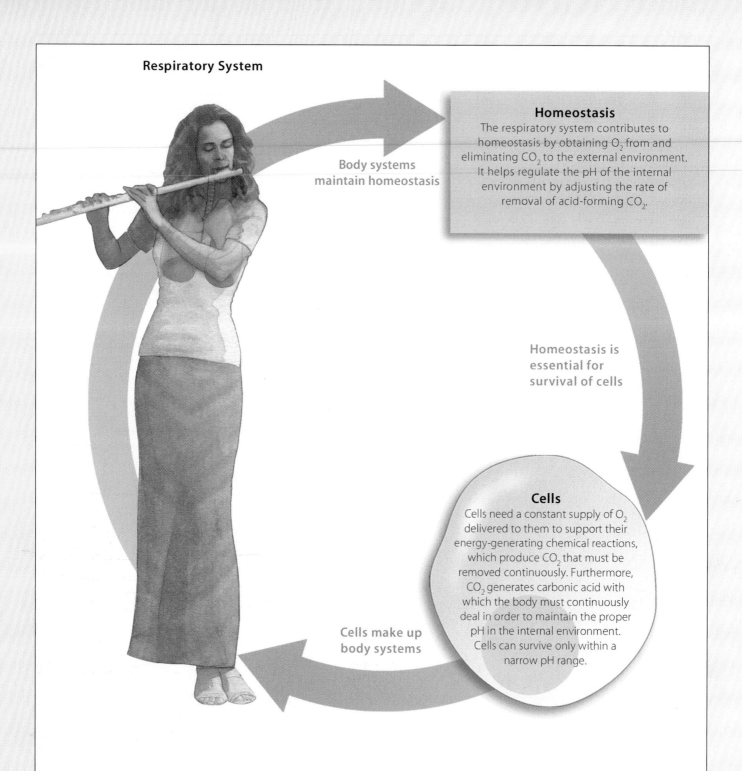

Respiratory System

Homeostasis
The respiratory system contributes to homeostasis by obtaining O_2 from and eliminating CO_2 to the external environment. It helps regulate the pH of the internal environment by adjusting the rate of removal of acid-forming CO_2.

Body systems maintain homeostasis

Homeostasis is essential for survival of cells

Cells
Cells need a constant supply of O_2 delivered to them to support their energy-generating chemical reactions, which produce CO_2 that must be removed continuously. Furthermore, CO_2 generates carbonic acid with which the body must continuously deal in order to maintain the proper pH in the internal environment. Cells can survive only within a narrow pH range.

Cells make up body systems

Energy is essential for sustaining life-supporting cellular activities, such as protein synthesis and active transport across plasma membranes. The cells of the body need a continual supply of O_2 to support their energy-generating chemical reactions. The CO_2 produced during these reactions must be eliminated from the body at the same rate as produced, to prevent dangerous fluctuations in pH (that is, to maintain the acid–base balance), because CO_2 generates carbonic acid.

Respiration involves the sum of the processes that accomplish ongoing passive movement of O_2 from the atmosphere to the tissues to support cell metabolism, as well as the continual passive movement of metabolically produced CO_2 from the tissues to the atmosphere. The **respiratory system** contributes to homeostasis by exchanging O_2 and CO_2 between the atmosphere and blood. The blood transports O_2 and CO_2 between the respiratory system and tissues.

The Respiratory System

4 steps of external resp. :

 PhysioEdge Click on the Tutorials menu of the CD-ROM for a tutorial on The Respiratory System.

INTRODUCTION

The primary function of respiration is to obtain O_2 for use by the body's cells and to eliminate the CO_2 the cells produce.

▮ **The respiratory system does not participate in all steps in respiration.**

Most people think of respiration as the process of breathing in and breathing out. In physiology, however, respiration has a much broader meaning. Respiration encompasses two separate but related processes: internal respiration and external respiration.

INTERNAL AND EXTERNAL RESPIRATION

The term **internal** or **cellular respiration** refers to the intracellular metabolic processes carried out within the mitochondria, which use O_2 and produce CO_2 while deriving energy from nutrient molecules (see p. 29). The term **external respiration** refers to the entire sequence of events in the exchange of O_2 and CO_2 between the external environment and the cells of the body. External respiration, the topic of this chapter, encompasses four steps (● Figure 12-1):

1. Air is alternately moved in and out of the lungs so that air can be exchanged between the atmosphere (external environment) and air sacs (*alveoli*) of the lungs. This exchange is accomplished by the mechanical act of **breathing, or ventilation.** The rate of ventilation is regulated to adjust the flow of air between the atmosphere and alveoli according to the body's metabolic needs for O_2 uptake and CO_2 removal.

2. Oxygen and CO_2 are exchanged between air in the alveoli and blood within the pulmonary (*pulmonary* means "lung") capillaries by the process of diffusion.

3. The blood transports O_2 and CO_2 between the lungs and tissues.

4. Oxygen and CO_2 are exchanged between the tissues and blood by the process of diffusion across the systemic (tissue) capillaries.

The respiratory system does not accomplish all the steps of respiration; it is involved only with ventilation and the exchange of O_2 and CO_2 between the lungs and blood (steps 1 and 2). The circulatory system carries out the remaining steps.

NONRESPIRATORY FUNCTIONS OF THE RESPIRATORY SYSTEM

The respiratory system also fills these nonrespiratory functions:

- It provides a route for water loss and heat elimination. Inspired (inhaled) atmospheric air is humidified and warmed by the respiratory airways before it is expired. Moistening of inspired air is essential to prevent the alveolar linings from drying out. Oxygen and CO_2 cannot diffuse through dry membranes.
- It enhances venous return (see the "respiratory pump," p. 302).
- It helps maintain normal acid–base balance by altering the amount of H^+-generating CO_2 exhaled (see p. 458).
- It enables speech, singing, and other vocalization.
- It defends against inhaled foreign matter (see p. 357).
- It removes, modifies, activates, or inactivates various materials passing through the pulmonary circulation. All blood returning to the heart from the tissues must pass through the lungs before being returned to the systemic circulation. The lungs, therefore, are uniquely situated to act on specific materials that have been added to the blood at the tissue level before they have a chance to reach other parts of the body by means of the arterial system. For example, prostaglandins, a collection of chemical messengers released in many tissues to mediate particular local responses (see p. 595), may spill into the blood, but they are inactivated during passage through the lungs so that they cannot exert systemic effects.
- The nose, a part of the respiratory system, serves as the organ of smell (see p. 178).

■ The respiratory airways conduct air between the atmosphere and alveoli.

The **respiratory system** includes the respiratory airways leading into the lungs, the lungs themselves, and the structures of the thorax (chest) involved in producing movement of air through the airways into and out of the lungs. The **respiratory airways** are tubes that carry air between the atmosphere and air sacs, the latter being the only site where gases can be exchanged between air and blood. The airways begin with the **nasal passages (nose)** (● Figure 12-2a). The nasal passages open into the **pharynx (throat)**, which serves as a common passageway for both the respiratory and digestive systems. Two tubes lead from the pharynx—the **trachea (windpipe)**, through which air is conducted to the lungs, and the **esophagus**, the tube through which food passes to the stomach. Air normally enters the pharynx through the nose, but it can enter by the mouth as well when the nasal passages are congested;

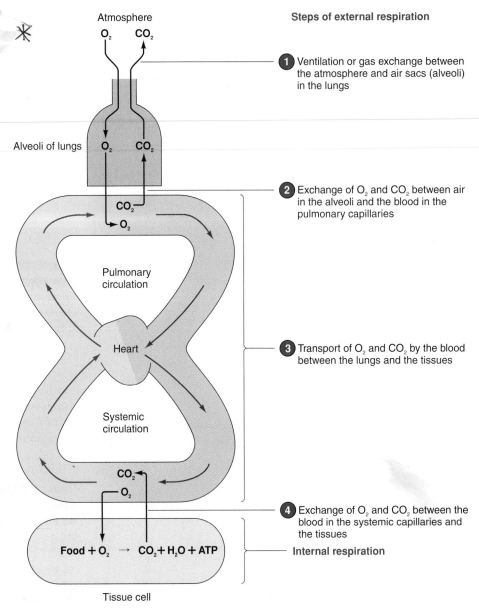

Steps of external respiration

1 Ventilation or gas exchange between the atmosphere and air sacs (alveoli) in the lungs

2 Exchange of O_2 and CO_2 between air in the alveoli and the blood in the pulmonary capillaries

3 Transport of O_2 and CO_2 by the blood between the lungs and the tissues

4 Exchange of O_2 and CO_2 between the blood in the systemic capillaries and the tissues

Internal respiration

● **FIGURE 12-1**

External and internal respiration. External respiration encompasses the steps involved in the exchange of O_2 and CO_2 between the external environment and tissue cells (steps 1 through 4). Internal respiration encompasses the intracellular metabolic reactions involving the use of O_2 to derive energy (ATP) from food, producing CO_2 as a by-product.

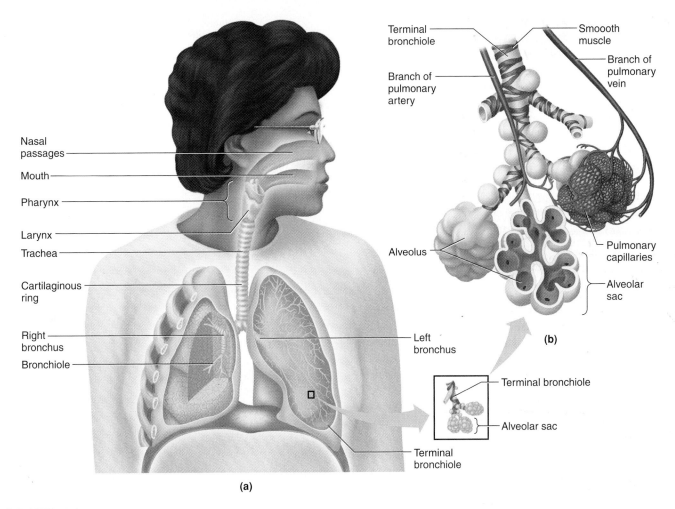

(a)

(b)

● **FIGURE 12-2**

Anatomy of the respiratory system. (a) The respiratory airways. (b) Enlargement of the alveoli (air sacs) at the terminal end of the airways. Most alveoli are clustered in grapelike arrangements at the end of the terminal bronchioles.

(*Source:* Part b adapted from Cecie Starr and Ralph Taggart, *Biology: The Unity and Diversity of Life,* Eighth Edition, Fig. 41.10a, p. 696. Copyright © 1998 Wadsworth.)

 For an interaction related to this figure, see Media Exercise 12.1: Anatomy of the Respiratory System on the CD-ROM.

that is, you can breathe through your mouth when you have a cold. Because the pharynx serves as a common passageway for food and air, reflex mechanisms close off the trachea during swallowing so that food enters the esophagus and not the airways. The esophagus stays closed except during swallowing to keep air from entering the stomach during breathing.

The **larynx,** or **voice box,** is located at the entrance of the trachea. The anterior protrusion of the larynx forms the "Adam's apple." The **vocal folds,** two bands of elastic tissue that lie across the opening of the larynx, can be stretched and positioned in different shapes by laryngeal muscles. As air is moved past the taut vocal folds, they vibrate to produce the many different sounds of speech. The lips, tongue, and soft palate modify the sounds into recognizable sound patterns. During swallowing, the vocal folds assume a function not related to speech; they are brought into tight apposition to each other to close off the entrance to the trachea.

Beyond the larynx, the trachea divides into two main branches, the right and left **bronchi,** which enter the right and left lungs, respectively. Within each lung, the bronchus continues to branch into progressively narrower, shorter, and more numerous airways, much like the branching of a tree. The smaller branches are known as **bronchioles.** Clustered at the ends of the terminal bronchioles are the **alveoli,** the tiny air sacs where gases are exchanged between air and blood (● Figure 12-2b).

■ **The gas-exchanging alveoli are thin-walled, inflatable air sacs encircled by pulmonary capillaries.**

The lungs are ideally structured for gas exchange. According to Fick's law of diffusion, the shorter the distance through which diffusion must take place, the greater the rate of diffusion. Also, the greater the surface area across which diffusion can take place, the greater the rate of diffusion (see p. 52).

The alveoli are clusters of thin-walled, inflatable, grapelike sacs at the terminal branches of the conducting airways. The alveolar walls consist of a single layer of flattened **Type I alveolar cells** (● Figure 12-3a). The walls of the dense network of pulmonary capillaries encircling each alveolus are also only one cell thick. The interstitial space between an alveolus and the surrounding capillary network forms an extremely thin barrier, with only 0.5 μm separating air in the alveoli from blood in the pulmonary capillaries. (A sheet of tracing paper is about 50 times thicker than this air-to-blood barrier.) The thinness of this barrier facilitates gas exchange.

Furthermore, the alveolar air–blood interface presents a tremendous surface area for exchange. The lungs contain about 300 million alveoli, each about 300 μm in diameter. So dense are the pulmonary capillary networks that each alveolus is encircled by an almost continuous sheet of blood (● Figure 12-3b). The total surface area thus exposed between alveolar air and pulmonary capillary blood is about 75 m² (about the size of a tennis court). In contrast, if the lungs consisted of a single hollow chamber of the same dimensions instead of being divided into a myriad alveolar units, the total surface area would be only about 0.01 m².

In addition to the thin, wall-forming Type I cells, the alveolar epithelium also contains **Type II alveolar cells** (● Figure 12-3a). These cells secrete *pulmonary surfactant*, a chemical complex that facilitates lung expansion (described later). Furthermore, defensive alveolar macrophages stand guard within the lumen of the air sacs (see p. 357).

▎The lungs occupy much of the thoracic cavity.

There are two **lungs**, each divided into several lobes and each supplied by one of the bronchi. The lung tissue itself consists of the series of highly branched airways, the alveoli, the pulmonary blood vessels, and large quantities of elastic connective tissue. The only muscle within the lungs is the smooth muscle in the walls of the arterioles and the walls of the bronchioles, both of which are subject to control. No muscle is present within the alveolar walls to cause them to inflate and deflate during the breathing process. Instead, changes in lung volume (and accompanying changes in alveolar volume) are brought about through changes in the dimensions of the thoracic cavity. You will learn about this mechanism after we complete our discussion of respiratory anatomy.

The lungs occupy most of the volume of the **thoracic (chest) cavity**, the only other structures in the chest being the heart and associated vessels, the esophagus, the thymus, and some nerves. The outer chest wall (**thorax**) is formed by 12 pairs of curved **ribs**, which join the **sternum** (breastbone) anteriorly and the **thoracic vertebrae** (backbone) posteriorly.

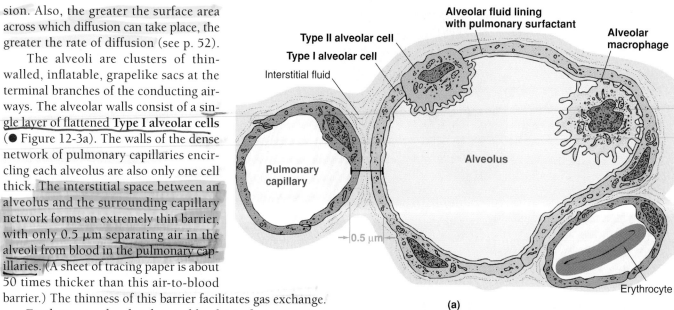

(a)

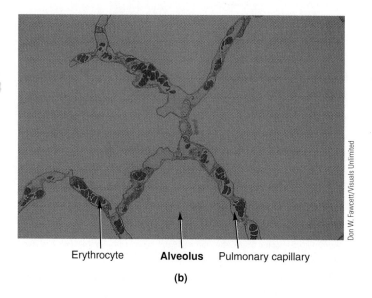

(b)

● **FIGURE 12-3**

Alveolus and associated pulmonary capillaries. (a) A schematic representation of a detailed electron microscope view of an alveolus and surrounding capillaries. A single layer of flattened Type I alveolar cells forms the alveolar walls. Type II alveolar cells embedded within the alveolar wall secrete pulmonary surfactant. Wandering alveolar macrophages are found within the alveolar lumen. (The size of the cells and respiratory membrane is exaggerated compared to the size of the alveolar and pulmonary capillary lumens. The diameter of an alveolus is actually about 600 times larger than the intervening space between air and blood.) (b) A transmission electron micrograph showing several alveoli and the close relationship of the capillaries surrounding them.

The rib cage provides bony protection for the lungs and heart. The **diaphragm**, which forms the floor of the thoracic cavity, is a large, dome-shaped sheet of skeletal muscle that completely separates the thoracic cavity from the abdominal cavity. It is penetrated only by the esophagus and blood vessels traversing the thoracic and abdominal cavities. At the neck, muscles and connective tissue enclose the thoracic cavity.

The only communication between the thorax and the atmosphere is through the respiratory airways into the alveoli.

▌ A pleural sac separates each lung from the thoracic wall.

A double-walled, closed sac called the **pleural sac** separates each lung from the thoracic wall and other surrounding structures (● Figure 12-4). The interior of the pleural sac is known as the **pleural cavity**. In the illustration, the dimensions of the pleural cavity are greatly exaggerated to aid visualization; in reality the layers of the pleural sac are in close contact with one another. The surfaces of the pleura secrete a thin **intrapleural fluid** (*intra* means "within"), which lubricates the pleural surfaces as they slide past each other during respiratory movements.

Pleurisy, an inflammation of the pleural sac, is accompanied by painful breathing, because each inflation and each deflation of the lungs cause a "friction rub."

Click on the Media Exercises menu of the CD-ROM and work Media Exercise 12.1: Anatomy of the Respiratory System to test your understanding of the previous section.

RESPIRATORY MECHANICS

Air tends to move from a region of higher pressure to a region of lower pressure, that is, down a **pressure gradient**.

▌ Interrelationships among pressures inside and outside the lungs are important in ventilation.

Air flows in and out of the lungs during the act of breathing by moving down alternately reversing pressure gradients established between the alveoli and atmosphere by cyclic respiratory muscle activity. Three different pressure considerations are important in ventilation (● Figure 12-5):

1. **Atmospheric (barometric) pressure** is the pressure exerted by the weight of the air in the atmosphere on objects on Earth's surface. At sea level it equals 760 mm Hg (● Figure 12-6). Atmospheric pressure diminishes with increasing altitude above sea level as the layer of air above Earth's surface correspondingly decreases in thickness. Minor fluctuations in atmospheric pressure occur at any height because of changing weather conditions (that is, when barometric pressure is rising or falling).
2. **Intra-alveolar pressure** is the pressure within the alveoli. Because the alveoli communicate with the atmosphere through the conducting airways, air quickly flows down its pressure gradient any time intra-alveolar pressure differs from atmospheric pressure; airflow continues until the two pressures equilibrate (become equal).

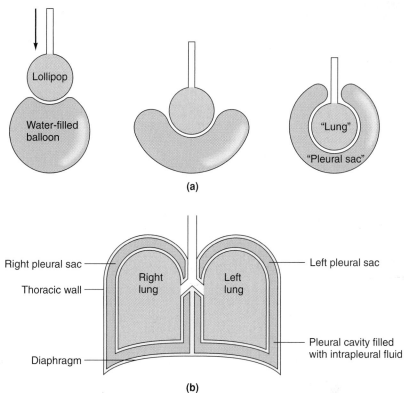

(a)

(b)

● **FIGURE 12-4**

Pleural sac. (a) Pushing a lollipop into a water-filled balloon produces a relationship analogous to that between each double-walled, closed pleural sac and the lung that it surrounds and separates from the thoracic wall. (b) Schematic representation of the relationship of the pleural sac to the lungs and thorax. One layer of the pleural sac closely adheres to the surface of the lung, then reflects back on itself to form another layer, which lines the interior surface of the thoracic wall. The relative size of the pleural cavity between these two layers is grossly exaggerated for the purpose of visualization.

For an interaction related to this figure, see Media Exercise 12.2: Mechanics of Ventilation on the CD-ROM.

3. **Intrapleural pressure** is the pressure within the pleural sac. It is the pressure exerted outside the lungs within the thoracic cavity. The intrapleural pressure is usually less than atmospheric pressure, averaging 756 mm Hg at rest. Just as blood pressure is recorded using atmospheric pressure as a reference point (that is, a systolic blood pressure of 120 mm Hg is 120 mm Hg greater than the atmospheric pressure of 760 mm Hg or in reality 880 mm Hg), 756 mm Hg is sometimes referred to as a pressure of −4 mm Hg. However, there is really no such thing as an absolute negative pressure. A pressure of −4 mm Hg is just negative when compared with the normal atmospheric pressure of 760 mm Hg. To avoid confusion, we will use absolute positive values throughout our discussion of respiration.

Intrapleural pressure does not equilibrate with atmospheric or intra-alveolar pressure, because there is no direct communication between the pleural cavity and either the atmosphere or the lungs. Because the pleural sac is a closed sac with no openings, air cannot enter or leave despite any pres-

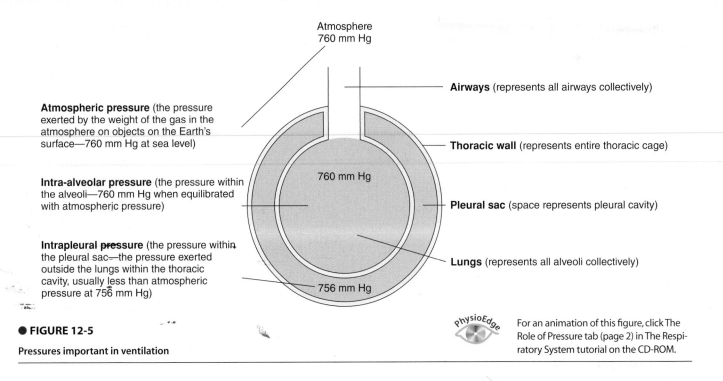

● FIGURE 12-5

Pressures important in ventilation

PhysioEdge For an animation of this figure, click The Role of Pressure tab (page 2) in The Respiratory System tutorial on the CD-ROM.

Atmosphere
760 mm Hg

Airways (represents all airways collectively)

Thoracic wall (represents entire thoracic cage)

760 mm Hg

Pleural sac (space represents pleural cavity)

Lungs (represents all alveoli collectively)

756 mm Hg

Atmospheric pressure (the pressure exerted by the weight of the gas in the atmosphere on objects on the Earth's surface—760 mm Hg at sea level)

Intra-alveolar pressure (the pressure within the alveoli—760 mm Hg when equilibrated with atmospheric pressure)

Intrapleural pressure (the pressure within the pleural sac—the pressure exerted outside the lungs within the thoracic cavity, usually less than atmospheric pressure at 756 mm Hg)

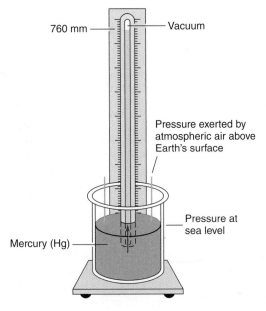

760 mm

Vacuum

Pressure exerted by atmospheric air above Earth's surface

Pressure at sea level

Mercury (Hg)

● FIGURE 12-6

Atmospheric pressure. The pressure exerted on objects by the atmospheric air above Earth's surface at sea level can push a column of mercury to a height of 760 mm. Therefore, atmospheric pressure at sea level is 760 mm Hg.

sure gradients that might exist between it and surrounding regions.

▌ The lungs are normally stretched to fill the larger thorax.

The thoracic cavity is larger than the unstretched lungs, because the thoracic wall grows more rapidly than the lungs during development. However, two forces—the *intrapleural*

fluid's cohesiveness and the *transmural pressure gradient*—hold the thoracic wall and lungs in close apposition, stretching the lungs to fill the larger thoracic cavity.

INTRAPLEURAL FLUID'S COHESIVENESS

The water molecules in the intrapleural fluid resist being pulled apart because they are polar and attracted to each other (see p. A-7). The resultant cohesiveness of the intrapleural fluid tends to hold the pleural surfaces together. Thus the intrapleural fluid can be considered very loosely as a "stickiness" or "glue" between the lining of the thoracic wall and the lung. Have you ever tried to pull apart two smooth surfaces held together by a thin layer of liquid, such as two wet glass slides? If so, you know that the two surfaces act as if they were stuck together by the thin layer of water. Even though you can easily slip the slides back and forth relative to each other (just as the intrapleural fluid facilitates movement of the lungs against the interior surface of the chest wall), you can pull the slides apart only with great difficulty, because the molecules within the intervening liquid resist being separated. This relationship is partly responsible for the fact that changes in thoracic dimension are always accompanied by corresponding changes in lung dimension; that is, when the thorax expands, the lungs—being stuck to the thoracic wall by the intrapleural fluid's cohesiveness—do likewise. An even more important reason that the lungs follow the movements of the chest wall is the transmural pressure gradient that exists across the lung wall.

TRANSMURAL PRESSURE GRADIENT

The intra-alveolar pressure, equilibrated with atmospheric pressure at 760 mm Hg, is greater than the intrapleural pressure of 756 mm Hg, so a greater pressure is pushing outward than is pushing inward across the lung wall. This net outward

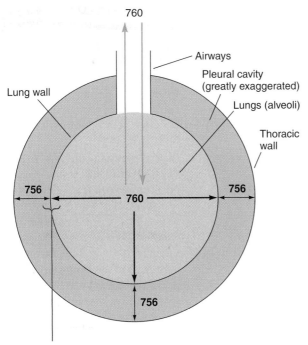

Numbers are mm Hg pressure.

Transmural pressure gradient
across lung wall = intra-alveolar
pressure minus intrapleural
pressure

● **FIGURE 12-7**

Transmural pressure gradient. Across the lung wall, the intra-alveolar pressure of 760 mm Hg pushes outward, while the intrapleural pressure of 756 mm Hg pushes inward. This 4-mm Hg difference in pressure constitutes a transmural pressure gradient that pushes out on the lungs, stretching them to fill the larger thoracic cavity.

pressure differential, the **transmural pressure gradient**, pushes out on the lungs, stretching, or distending them (*trans* means "across"; *mural* means "wall") (● Figure 12-7). Because of this pressure gradient, the lungs are always forced to expand to fill the thoracic cavity.

PNEUMOTHORAX

Clinical Note   Normally, air does not enter the pleural cavity, because there is no communication between the cavity and either the atmosphere or the alveoli. However, if the chest wall is punctured (for example, by a stab wound or a broken rib), air flows down its pressure gradient from the higher atmospheric pressure and rushes into the pleural space (● Figure 12-8a). The abnormal condition of air entering the pleural cavity is known as **pneumothorax** ("air in the chest"). Intrapleural and intra-alveolar pressure are now both equilibrated with atmospheric pressure, so a transmural pressure gradient no longer exists across the lung wall. With no force present to stretch the lung, it collapses to its unstretched size (● Figure 12-8b). (The intrapleural fluid's cohesiveness cannot hold

the lungs and thoracic wall in apposition in the absence of the transmural pressure gradient.) Similarly, pneumothorax and lung collapse can occur if air enters the pleural cavity through a hole in the lung produced, for example, by a disease process (● Figure 12-8c).

▌ Flow of air into and out of the lungs occurs because of cyclic changes in intra-alveolar pressure.

Because air flows down a pressure gradient, the intra-alveolar pressure must be less than atmospheric pressure for air to flow into the lungs during inspiration. Similarly, the intra-alveolar pressure must be greater than atmospheric pressure for air to flow out of the lungs during expiration. Intra-alveolar pressure can be changed by altering the volume of the lungs, in accordance with Boyle's law. **Boyle's law** states that at any constant temperature, the pressure exerted by a gas varies inversely with the volume of the gas (● Figure 12-9); that is, as the volume of a gas increases, the pressure exerted by the gas decreases proportionately. Conversely, the pressure increases proportionately as the volume decreases. Changes in lung volume, and accordingly intra-alveolar pressure, are brought about indirectly by respiratory muscle activity.

● **FIGURE 12-8**

Pneumothorax. (a) Traumatic pneumothorax. A puncture in the chest wall permits air from the atmosphere to flow down its pressure gradient and enter the pleural cavity, abolishing the transmural pressure gradient. (b) When the transmural pressure gradient is abolished, the lung collapses to its unstretched size. (c) Spontaneous pneumothorax. A hole in the lung wall permits air to move down its pressure gradient and enter the pleural cavity from the lungs, abolishing the transmural pressure gradient. As with traumatic pneumothorax, the lung collapses to its unstretched size.

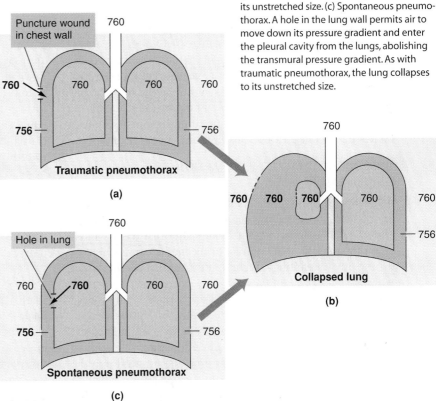

Numbers are mm Hg pressure.

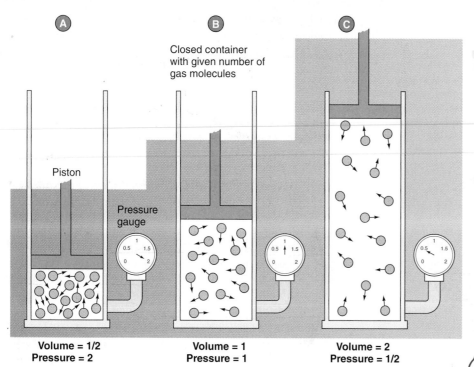

● FIGURE 12-9

Boyle's law. Each container has the same number of gas molecules. Given the random motion of gas molecules, the likelihood of a gas molecule striking the interior wall of the container and exerting pressure varies inversely with the volume of the container at any constant temperature. The gas in container B exerts more pressure than the same gas in larger container C but less pressure than the same gas in smaller container A. This relationship is stated as Boyle's law: $P_1V_1 = P_2V_2$. As the volume of a gas increases, the pressure of the gas decreases proportionately; conversely, the pressure increases proportionately as the volume decreases.

The respiratory muscles that accomplish breathing do not act directly on the lungs to change their volume. Instead, these muscles change the volume of the thoracic cavity, causing a corresponding change in lung volume because the thoracic wall and lungs are linked together by the intrapleural fluid's cohesiveness and the transmural pressure gradient.

Let's follow the changes that occur during one respiratory cycle—that is, one breath in (**inspiration**) and out (**expiration**).

ONSET OF INSPIRATION: CONTRACTION OF INSPIRATORY MUSCLES

Before the beginning of inspiration, the respiratory muscles are relaxed, no air is flowing, and intra-alveolar pressure is equal to atmospheric pressure. The major **inspiratory muscles**—the muscles that contract to accomplish an inspiration during quiet breathing—include the *diaphragm* and *external intercostal muscles* (● Figure 12-10). At the onset of inspiration, these muscles are stimulated to contract, enlarging the thoracic cavity. The major inspiratory muscle is the **diaphragm**, a sheet of skeletal muscle that forms the floor of the thoracic cavity and is innervated by the **phrenic nerve**. The relaxed diaphragm has a dome shape that protrudes upward into the thoracic cavity. When the diaphragm contracts (on stimulation by the phrenic nerve), it descends downward, enlarging the volume of the thoracic cavity by increasing its vertical (top-to-bottom) dimension (● Figure 12-11a). The abdominal wall,

if relaxed, bulges outward during inspiration as the descending diaphragm pushes the abdominal contents downward and forward. Seventy-five percent of the enlargement of the thoracic cavity during quiet inspiration is done by contraction of the diaphragm.

Two sets of **intercostal muscles** lie between the ribs (*inter* means "between"; *costa* means "rib"). The *external intercostal muscles* lie on top of the *internal intercostal muscles*. Contraction of the **external intercostal muscles**, whose fibers run downward and forward between adjacent ribs, enlarges the thoracic cavity in both the lateral (side-to-side) and anteroposterior (front-to-back) dimensions. When the external intercostals contract, they elevate the ribs and subsequently the sternum upward and outward (● Figure 12-11a).

Before inspiration, at the end of the preceding expiration, intra-alveolar pressure is equal to atmospheric pressure, so no air is flowing into or out of the lungs (● Figure 12-12a). As the thoracic cavity enlarges, the lungs are also forced to expand to fill the larger thoracic cavity. As the lungs enlarge, the intra-alveolar pressure drops because the same number of air molecules now occupy a larger lung volume. In a typical inspiratory excursion, the intra-alveolar pressure drops 1 mm Hg to 759 mm Hg (● Figure 12-12b). Because the intra-alveolar pressure is now less than atmospheric pressure, air flows into the lungs down the pressure gradient from higher to lower pressure. Air continues to enter the lungs until no further gradient exists—that is, until intra-alveolar pressure equals atmospheric pressure. Thus lung expansion is not caused by movement of air into the lungs; instead, air flows into the lungs because of the fall in intra-alveolar pressure brought about by lung expansion.

During inspiration, the intrapleural pressure falls to 754 mm Hg as a result of expansion of the thorax. The resultant increase in the transmural pressure gradient during inspiration ensures that the lungs are stretched to fill the expanded thoracic cavity.

ROLE OF ACCESSORY INSPIRATORY MUSCLES

Deeper inspirations (more air breathed in) can be accomplished by contracting the diaphragm and external intercostal muscles more forcefully and by bringing the **accessory inspiratory muscles** into play to further enlarge the thoracic cavity. Contracting these accessory muscles, which are in the neck (● Figure 12-10), raises the sternum and elevates the first two ribs, enlarging the upper portion of the thoracic cavity. As the thoracic cavity increases even further in volume than under resting conditions, the lungs likewise expand even

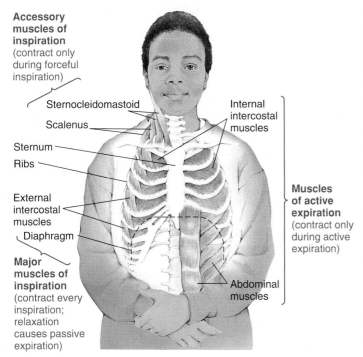

Accessory
muscles of
inspiration
(contract only
during forceful
inspiration)

Sternocleidomastoid

Scalenus

Sternum

Ribs

External
intercostal
muscles

Diaphragm

Major
muscles of
inspiration
(contract every
inspiration;
relaxation
causes passive
expiration)

Internal
intercostal
muscles

Muscles
of active
expiration
(contract only
during active
expiration)

Abdominal
muscles

● **FIGURE 12-10**

Anatomy of the respiratory muscles

PhysioEdge

For an animation of this figure, click the Ventilation tab
(page 1) in The Respiratory System tutorial on the CD-ROM.
For an interaction related to this figure, see Media Exercise 12.2:
Mechanics of Ventilation.

more, dropping the intra-alveolar pressure even further. Consequently, a larger inward flow of air occurs before equilibration with atmospheric pressure is achieved; that is, a deeper breath occurs.

ONSET OF EXPIRATION: RELAXATION OF INSPIRATORY MUSCLES

At the end of inspiration, the inspiratory muscles relax. The diaphragm assumes its original dome-shaped position when it relaxes. The elevated rib cage falls because of gravity when the external intercostals relax. With no forces causing expansion of the chest wall (and accordingly expansion of the lungs), the chest wall and stretched lungs recoil to their preinspiratory size because of their elastic properties, much as a stretched balloon would on release (● Figure 12-11b). As the lungs recoil and become smaller in volume, the intra-alveolar pressure rises, because the greater number of air molecules contained within the larger lung volume at the end of inspiration are now compressed into a smaller volume. In a resting expiration, the intra-alveolar pressure increases about 1 mm Hg above atmospheric level to 761 mm Hg (● Figure 12-12c). Air now leaves the lungs down its pressure gradient from high intra-alveolar pressure to lower atmospheric pressure. Outward flow of air ceases when intra-alveolar pressure becomes equal to atmospheric pressure and a pressure gradient no longer exists. ● Figure 12-13 summarizes the intra-alveolar and intrapleural pressure changes that take place during one respiratory cycle.

FORCED EXPIRATION: CONTRACTION OF EXPIRATORY MUSCLES

During quiet breathing, expiration is normally a passive process, because it is accomplished by elastic recoil of the lungs on relaxation of the inspiratory muscles, with no muscular exertion or energy expenditure required. In contrast, inspiration is *always* active, because it is brought about only by contraction of inspiratory muscles at the expense of energy use. To empty the lungs more completely and more rapidly than is accomplished during quiet breathing, as during the deeper breaths accompanying exercise, expiration does become active. The intra-alveolar pressure must be increased even further above atmospheric pressure than can be accomplished by simple relaxation of the inspiratory muscles and elastic recoil of the lungs. To produce such a **forced**, or **active**, **expiration**, expiratory muscles must contract to further reduce the volume of the thoracic cavity and lungs. The most important **expiratory muscles** are (unbelievable as it may seem at first) the *muscles of the abdominal wall* (● Figure 12-10). As the abdominal muscles contract, the resultant increase in intra-abdominal pressure exerts an upward force on the diaphragm, pushing it further up into the thoracic cavity than its relaxed position, thus decreasing the vertical dimension of the thoracic cavity even more. The other expiratory muscles are the **internal intercostal muscles,** whose contraction pulls the ribs downward and inward, flattening the chest wall and further decreasing the size of the thoracic cavity; this action is just the opposite of that of the external intercostal muscles (● Figure 12-11c).

As active contraction of the expiratory muscles further reduces the volume of the thoracic cavity, the lungs also become further reduced in volume because they do not have to be stretched as much to fill the smaller thoracic cavity; that is, they are permitted to recoil to an even smaller volume. The intra-alveolar pressure increases further as the air in the lungs is confined within this smaller volume. The differential between intra-alveolar and atmospheric pressure is even greater now than during passive expiration, so more air leaves down the pressure gradient before equilibration is achieved. In this way, the lungs are emptied more completely during forceful, active expiration than during quiet, passive expiration.

▌ Airway resistance influences airflow rates.

Thus far we have discussed airflow in and out of the lungs as a function of the magnitude of the pressure gradient between the alveoli and the atmosphere. However, just as flow of blood through the blood vessels depends not only on the pressure gradient but also on the resistance to the flow offered by the vessels, so it is with airflow:

$$F = \frac{\Delta P}{R}$$

where

F = airflow rate

ΔP = difference between atmospheric and intra-alveolar pressure (pressure gradient)

R = resistance of airways, determined by their radius

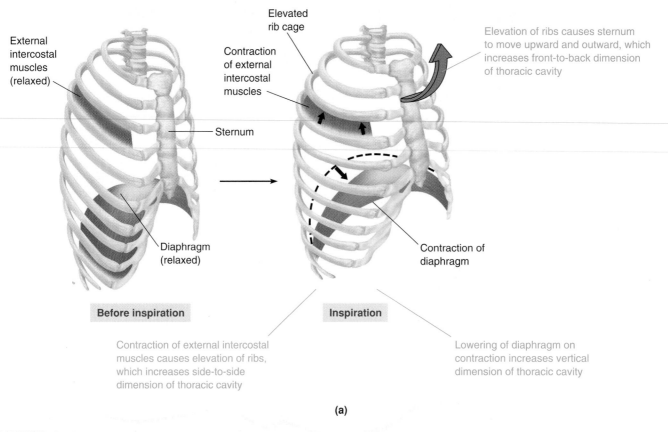

Elevated
rib cage

External
intercostal
muscles
(relaxed)

Contraction
of external
intercostal
muscles

Elevation of ribs causes sternum
to move upward and outward, which
increases front-to-back dimension
of thoracic cavity

Sternum

Diaphragm
(relaxed)

Contraction of
diaphragm

Before inspiration

Inspiration

Contraction of external intercostal
muscles causes elevation of ribs,
which increases side-to-side
dimension of thoracic cavity

Lowering of diaphragm on
contraction increases vertical
dimension of thoracic cavity

(a)

● **FIGURE 12-11**

Respiratory muscle activity during inspiration and expiration. (a) *Inspiration,* during which the diaphragm
descends on contraction, increasing the vertical dimension of the thoracic cavity. Contraction of the external intercostal muscles elevates the ribs and subsequently the sternum to enlarge the thoracic cavity from front to back
and from side to side. (b) *Quiet passive expiration,* during which the diaphragm relaxes, reducing   *(continued)*

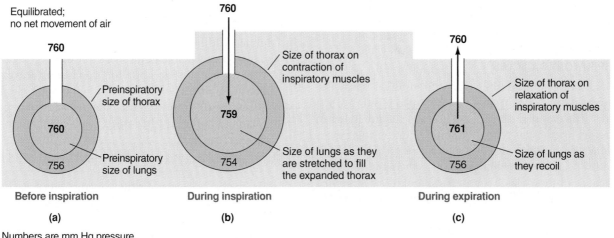

Equilibrated;
no net movement of air

760

760

760

Preinspiratory
size of thorax

Size of thorax on
contraction of
inspiratory muscles

Size of thorax on
relaxation of
inspiratory muscles

760

759

761

756

Preinspiratory
size of lungs

754

Size of lungs as they
are stretched to fill
the expanded thorax

756

Size of lungs as
they recoil

Before inspiration

During inspiration

During expiration

(a)

(b)

(c)

Numbers are mm Hg pressure.

● **FIGURE 12-12**

Changes in lung volume and intra-alveolar pressure during inspiration and expiration. (a) *Before inspiration,* at the end of the preceding expiration. Intra-alveolar pressure is equilibrated with atmospheric pressure
and no air is flowing. (b) *Inspiration.* As the lungs increase in volume during inspiration, the intra-alveolar
pressure decreases, establishing a pressure gradient that favors the flow of air into the alveoli from the
atmosphere; that is, an inspiration occurs. (c) *Expiration.* As the lungs recoil to their preinspiratory size on
relaxation of the inspiratory muscles, the intra-alveolar pressure increases, establishing a pressure gradient
that favors the flow of air out of the alveoli into the atmosphere; that is, an expiration occurs.

PhysioEdge

For an animation of this figure, click The Role of Pressure tab in The Respiratory System
tutorial on the CD-ROM.

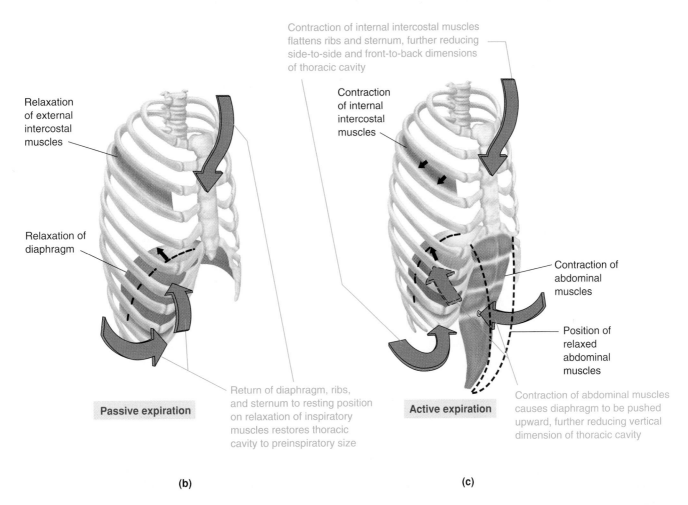

Contraction of internal intercostal muscles flattens ribs and sternum, further reducing side-to-side and front-to-back dimensions of thoracic cavity

Relaxation of external intercostal muscles

Contraction of internal intercostal muscles

Relaxation of diaphragm

Contraction of abdominal muscles

Position of relaxed abdominal muscles

Return of diaphragm, ribs, and sternum to resting position on relaxation of inspiratory muscles restores thoracic cavity to preinspiratory size

Contraction of abdominal muscles causes diaphragm to be pushed upward, further reducing vertical dimension of thoracic cavity

Passive expiration

Active expiration

(b)

(c)

● **FIGURE 12-11** *(continued)*

the volume of the thoracic cavity from its peak inspiratory size. As the external intercostal muscles relax, the elevated rib cage falls because of the force of gravity. This also reduces the volume of the thoracic cavity. (c) *Active expiration,* during which contraction of the abdominal muscles increases the intra-abdominal pressure, exerting an upward force on the diaphragm. This reduces the vertical dimension of the thoracic cavity further than it is reduced during quiet passive expiration. Contraction of the internal intercostal muscles decreases the front-to-back and side-to-side dimensions by flattening the ribs and sternum.

 For an animation of this figure, click the Ventilation tab in The Respiratory System tutorial on the CD-ROM.

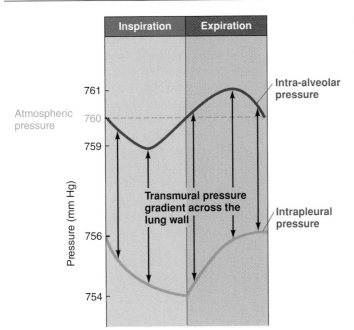

- During inspiration, intra-alveolar pressure is < atmospheric pressure.

- During expiration, intra-alveolar pressure is > atmospheric pressure.

- At the end of both inspiration and expiration, intra-alveolar pressure is equal to atmospheric pressure, because the alveoli are in direct communication with the atmosphere and air continues to flow down its pressure gradient until the two pressures equilibrate.

- Throughout the respiratory cycle, intrapleural pressure is < intra-alveolar pressure. Thus a transmural pressure gradient always exists, and the lung is always stretched to some degree, even during expiration.

● **FIGURE 12-13**

Intra-alveolar and intrapleural pressure changes throughout the respiratory cycle

 For an interaction related to this figure, see Media Exercise 12.2: Mechanics of Ventilation on the CD-ROM.

The primary determinant of resistance to airflow is the radius of the conducting airways. We ignored airway resistance in our preceding discussion of pressure gradient–induced airflow rates because in a healthy respiratory system, the radius of the conducting system is large enough that resistance remains extremely low. Therefore, the pressure gradient between the alveoli and the atmosphere is usually the main factor determining the airflow rate. Indeed, the airways normally offer such low resistance that only very small pressure gradients of 1 to 2 mm Hg need be created to achieve adequate rates of airflow in and out of the lungs. (By comparison, it would take a pressure gradient 250 times greater to move air through a smoker's pipe than through the respiratory airways at the same flow rate.)

Normally, modest adjustments in airway size can be accomplished by autonomic nervous system regulation to suit the body's needs. Parasympathetic stimulation, which occurs in quiet, relaxed situations when the demand for airflow is low, promotes bronchiolar smooth muscle contraction, which increases airway resistance by producing **bronchoconstriction** (a reduction in bronchiolar caliber). In contrast, sympathetic stimulation and to a greater extent its associated hormone, epinephrine, bring about **bronchodilation** (an increase in bronchiolar caliber) and decreased airway resistance by promoting bronchiolar smooth-muscle relaxation. Thus during periods of sympathetic domination, when increased demands for O_2 uptake are actually or potentially placed on the body, bronchodilation ensures that the pressure gradients established by respiratory muscle activity can achieve maximum airflow rates with minimum resistance. Because of this bronchodilator action, epinephrine or similar drugs are useful therapeutic tools to counteract airway constriction in patients with bronchial spasms.

Resistance becomes an extremely important impediment to airflow when airway lumens become abnormally narrowed by disease. We have all transiently experienced the effect that increased airway resistance has on breathing when we have a cold. We know how difficult it is to produce an adequate airflow rate through a "stuffy nose" when the nasal passages are narrowed by swelling and mucus accumulation. More serious is chronic obstructive pulmonary disease, to which we now turn our attention.

▌ Airway resistance is abnormally increased with chronic obstructive pulmonary disease.

Clinical Note **Chronic obstructive pulmonary disease (COPD)** is a group of lung diseases characterized by increased airway resistance resulting from narrowing of the lumen of the lower airways. When airway resistance increases, a larger pressure gradient must be established to maintain even a normal airflow rate. For example, if resistance is doubled by narrowing of airway lumens, ΔP must be doubled through increased respiratory muscle exertion to induce the same flow rate of air in and out of the lungs as a normal individual accomplishes during quiet breathing. Accordingly, patients with COPD must work harder to breathe.

Chronic obstructive pulmonary disease encompasses three chronic (long-term) diseases: *chronic bronchitis, asthma*, and *emphysema*.

CHRONIC BRONCHITIS

Chronic bronchitis is a long-term inflammatory condition of the lower respiratory airways, generally triggered by frequent exposure to irritating cigarette smoke, polluted air, or allergens. In response to the chronic irritation, the airways become narrowed by prolonged edematous thickening of the airway linings, coupled with overproduction of thick mucus. Despite frequent coughing associated with the chronic irritation, the plugged mucus often cannot be satisfactorily removed, especially because the irritants immobilize the ciliary mucus escalator (see p. 357). Pulmonary bacterial infections frequently occur, because the accumulated mucus serves as an excellent medium for bacterial growth.

ASTHMA

In **asthma**, airway obstruction is due to (1) thickening of airway walls, brought about by inflammation and histamine-induced edema (see p. 331); (2) plugging of the airways by excessive secretion of very thick mucus; and (3) airway hyperresponsiveness, characterized by profound constriction of the smaller airways caused by trigger-induced spasm of the smooth muscle in the walls of these airways (see p. 354). Triggers that lead to these inflammatory changes and the exaggerated bronchoconstrictor response include repeated exposure to allergens (such as dust mites or pollen), irritants (as in cigarette smoke), and infections. A growing number of studies suggest that long-term infections with *Chlamydia pneumoniae*, a common cause of lung infections, may underlie up to half of the adult cases of asthma. In severe asthmatic attacks, pronounced clogging and narrowing of the airways can cut off all airflow, leading to death. An estimated 15 million people in the United States have asthma, with the number steadily climbing. Asthma is the most common chronic childhood disease.

EMPHYSEMA

Emphysema is characterized by (1) collapse of the smaller airways and (2) breakdown of alveolar walls. This irreversible condition can arise in two different ways. Most commonly, emphysema results from excessive release of destructive enzymes such as *trypsin* from alveolar macrophages as a defense mechanism in response to chronic exposure to inhaled cigarette smoke or other irritants. The lungs are normally protected from damage by these enzymes by α_1-*antitrypsin*, a protein that inhibits trypsin. Excessive secretion of these destructive enzymes in response to chronic irritation, however, can overwhelm the protective capability of α_1-antitrypsin so that these enzymes destroy not only foreign materials but lung tissue as well. Loss of lung tissue leads to breakdown of alveolar walls and collapse of small airways, the characteristics of emphysema.

Less frequently, emphysema arises from a genetic inability to produce α_1-antitrypsin so that the lung tissue has no protection from trypsin. The unprotected lung tissue gradually disintegrates under the influence of even small amounts of macrophage-released enzymes, in the absence of chronic exposure to inhaled irritants.

DIFFICULTY IN EXPIRING

When COPD of any type increases airway resistance, expiration is more difficult than inspiration. The smaller airways, lacking the cartilaginous rings that hold the larger airways open, are held open by the same transmural pressure gradient that distends the alveoli. Expansion of the thoracic cavity during inspiration indirectly dilates the airways even further than their expiratory dimensions, like alveolar expansion, so airway resistance is lower during inspiration than during expiration. In a normal individual, the airway resistance is always so low that the slight variation between inspiration and expiration is not noticeable. When airway resistance has substantially increased, however, as during an asthmatic attack, the difference is quite noticeable. Thus an asthmatic has more difficulty expiring than inspiring, giving rise to the characteristic "wheeze" as air is forced out through the narrowed airways.

▌ Elastic behavior of the lungs is due to elastic connective tissue and alveolar surface tension.

During the respiratory cycle, the lungs alternately expand during inspiration and recoil during expiration. What properties of the lungs enable them to behave like balloons, being stretchable and then snapping back to their resting position when the stretching forces are removed? Two interrelated concepts are involved in pulmonary elasticity: *compliance* and *elastic recoil.*

The term **compliance** refers to how much effort is required to stretch or distend the lungs; it is analogous to how hard you have to work to blow up a balloon. (By comparison, 100 times more distending pressure is required to inflate a child's toy balloon than to inflate the lungs.) Specifically, compliance is a measure of how much change in lung volume results from a given change in the transmural pressure gradient, the force that stretches the lungs. A highly compliant lung stretches farther for a given increase in the pressure difference than a less compliant lung does. Stated another way, the lower the compliance of the lungs, the larger the transmural pressure gradient that must be created during inspiration to produce normal lung expansion. In turn, a greater-than-normal transmural pressure gradient during inspiration can be achieved only by making the intrapleural pressure more subatmospheric than usual. This is accomplished by greater expansion of the thorax through more vigorous contraction of the inspiratory muscles. Therefore, the less compliant the lungs, the more work required to produce a given degree of inflation. A poorly compliant lung is referred to as a "stiff" lung, because it lacks normal stretchability.

 Respiratory compliance can be decreased by a number of factors, as in *pulmonary fibrosis,* where normal lung tissue is replaced with scar-forming fibrous connective tissue as a result of chronically breathing in asbestos fibers or similar irritants.

The term **elastic recoil** refers to how readily the lungs rebound after having been stretched. It is responsible for the lungs returning to their preinspiratory volume when the inspiratory muscles relax at the end of inspiration.

Pulmonary elastic behavior depends mainly on two factors: *highly elastic connective tissue* in the lungs and *alveolar surface tension.*

PULMONARY ELASTIC CONNECTIVE TISSUE

Pulmonary connective tissue contains large quantities of elastin fibers (see p. 48). Not only do these fibers have elastic properties themselves, but they also are arranged into a meshwork that amplifies their elastic behavior, much like threads in stretch-knit fabric. The entire piece of fabric (or lung) is stretchier and tends to bounce back to its original shape more than the individual threads (elastin fibers) do.

ALVEOLAR SURFACE TENSION

An even more important factor influencing elastic behavior of the lungs is the **alveolar surface tension** displayed by the thin liquid film that lines each alveolus. At an air–water interface, the water molecules at the surface are more strongly attracted to other surrounding water molecules than to the air above the surface. This unequal attraction produces a force known as *surface tension* at the surface of the liquid. Surface tension has a twofold effect. First, the liquid layer resists any force that increases its surface area; that is, it opposes expansion of the alveolus, because the surface water molecules oppose being pulled apart. Accordingly, the greater the surface tension, the less compliant the lungs. Second, the liquid surface area tends to shrink as small as possible, because the surface water molecules, being preferentially attracted to each other, try to get as close together as possible. Thus the surface tension of the liquid lining an alveolus tends to reduce alveolus size, squeezing in on the air inside. This property, along with the rebound of the stretched elastin fibers, produces the lungs' elastic recoil back to their preinspiratory size when inspiration is over.

▌ Pulmonary surfactant decreases surface tension and contributes to lung stability.

The cohesive forces between water molecules are so strong that if the alveoli were lined with water alone, surface tension would be so great that the lungs would collapse. The recoil force attributable to the elastin fibers and high surface tension would exceed the opposing stretching force of the transmural pressure gradient. Furthermore, the lungs would be very poorly compliant, so exhausting muscular efforts would be required to accomplish stretching and inflation of the alveoli. The tremendous surface tension of pure water is normally counteracted by pulmonary surfactant.

PULMONARY SURFACTANT

Pulmonary surfactant is a complex mixture of lipids and proteins secreted by the Type II alveolar cells (● Figure 12-3a,

TABLE 12-1	
Opposing Forces Acting on the Lung	
FORCES KEEPING THE ALVEOLI OPEN	**FORCES PROMOTING ALVEOLAR COLLAPSE**
Transmural pressure gradient	Elasticity of stretched pulmonary connective tissue fibers
Pulmonary surfactant (which opposes alveolar surface tension)	Alveolar surface tension

p. 368). It intersperses between the water molecules in the fluid lining the alveoli and lowers alveolar surface tension, because the cohesive force between a water molecule and an adjacent pulmonary surfactant molecule is very low. By lowering alveolar surface tension, pulmonary surfactant provides two important benefits: (1) it increases pulmonary compliance, reducing the work of inflating the lungs; and (2) it reduces the lungs' tendency to recoil, so they do not collapse as readily.

The opposing forces acting on the lung (that is, the forces keeping the alveoli open and the countering forces that promote alveolar collapse) are summarized in ▲ Table 12-1.

NEWBORN RESPIRATORY DISTRESS SYNDROME

Clinical Note Developing fetal lungs normally cannot synthesize pulmonary surfactant until late in pregnancy. Especially in an infant born prematurely, not enough pulmonary surfactant may be produced to reduce the alveolar surface tension to manageable levels. The resulting collection of symptoms is termed **newborn respiratory distress syndrome**. The infant must make very strenuous inspiratory efforts to overcome the high surface tension in an attempt to inflate the poorly compliant lungs. Moreover, the work of breathing is further increased because the alveoli, in the absence of surfactant, tend to collapse almost completely during each expiration. It is more difficult (requires a greater transmural pressure differential) to expand a collapsed alveolus by a given volume than to increase an already partially expanded alveolus by the same volume. The situation is analogous to blowing up a new balloon. It takes more effort to blow in that first breath of air when starting to blow up a new balloon than to blow additional breaths into the already partially expanded balloon. With newborn respiratory distress syndrome, it is as though with every breath the infant must start blowing up a new balloon. Lung expansion may require transmural pressure gradients of 20 to 30 mm Hg (compared to the normal 4 to 6 mm Hg) to overcome the tendency

of surfactant-deprived alveoli to collapse. Worse yet, the newborn's muscles are still weak. The respiratory distress from surfactant deficiency may soon lead to death if breathing efforts become exhausting or inadequate to support sufficient gas exchange.

This life-threatening condition affects 30,000 to 50,000 newborns, primarily premature infants, each year in the United States. Until the surfactant-secreting cells mature sufficiently,

---

● **FIGURE 12-14**

Variations in lung volume. (a) Normal range and extremes of lung volume in a healthy young adult male. (b) Normal spirogram of a healthy young adult male. (The residual volume cannot be measured with a spirometer but must be determined by another means.)

 PhysioEdge For an interaction related to this figure, see Media Exercise 12.4: Control of Ventilation, Lung Volumes and Terms on the CD-ROM.

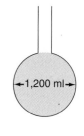

Total lung capacity at maximum inflation

Variation in lung volume with normal, quiet breathing

Minimal lung volume (residual volume) at maximal deflation

←5,700 ml→

←2,200 ml→
←2,700 ml→

Volume of lungs at end of normal inspiration (average 2,700 ml)

Volume of lungs at end of normal expiration (average 2,200 ml)

←1,200 ml→

Difference between end-expiratory and end-inspiratory volume equals tidal volume (average 500 ml)

(a)

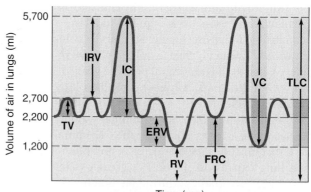

Volume of air in lungs (ml)

Time (sec)

TV = Tidal volume (500 ml)
IRV = Inspiratory reserve volume (3,000 ml)
IC = Inspiratory capacity (3,500 ml)
ERV = Expiratory reserve volume (1,000 ml)
RV = Residual volume (1,200 ml)
FRC = Functional residual capacity (2,200 ml)
VC = Vital capacity (4,500 ml)
TLC = Total lung capacity (5,700 ml)

(b)

Values are average for a healthy young adult male; values for females are somewhat lower.

the condition is treated by surfactant replacement. In addition, drugs can hasten the maturation process.

▌ The lungs normally operate at about "half full."

On average, in healthy young adults, the maximum air that the lungs can hold is about 5.7 liters in males (4.2 liters in females). Anatomic build, age, the distensibility of the lungs, and the presence or absence of respiratory disease affect this total lung capacity. Normally, during quiet breathing the lungs are nowhere near maximally inflated nor are they deflated to their minimum volume. Thus the lungs normally remain moderately inflated throughout the respiratory cycle. At the end of a normal quiet expiration, the lungs still contain about 2200 ml of air. During each typical breath under resting conditions, about 500 ml of air are inspired and the same quantity is expired, so during quiet breathing the lung volume varies between 2200 ml at the end of expiration to 2700 ml at the end of inspiration (● Figure 12-14a). During maximal expiration, lung volume can decrease to 1200 ml in males (1000 ml in females), but the lungs can never be completely deflated, because the small airways collapse during forced expirations at low lung volumes, blocking further outflow.

An important outcome of not being able to empty the lungs completely is that even during maximal expiratory efforts, gas exchange can still continue between blood flowing through the lungs and the remaining alveolar air. As a result, the gas content of the blood leaving the lungs for delivery to the tissues normally remains remarkably constant throughout the respiratory cycle. By contrast, if the lungs completely filled and emptied with each breath, the amount of O_2 taken up and CO_2 dumped off by the blood would fluctuate widely. Another advantage of the lungs not completely emptying with each breath is a reduction in the work of breathing. Recall that it takes less effort to inflate a partially inflated alveolus than a totally collapsed one.

The changes in lung volume that occur with different respiratory efforts can be measured using a spirometer. Let's see what this device is and the various lung volumes and capacities it measures.

LUNG VOLUMES AND CAPACITIES

Basically, a **spirometer** consists of an air-filled drum floating in a water-filled chamber. As the person breathes air in and out of the drum through a tube connecting the mouth to the air chamber, the drum rises and falls in the water chamber (● Figure 12-15). This rise and fall can be recorded as a **spirogram**, which is calibrated to volume changes. The pen records inspiration as an upward deflection and expiration as a downward deflection.

● Figure 12-14b is a hypothetical example of a spirogram in a healthy young adult male. Generally, the values are lower for females. The following lung volumes and lung capacities

● **FIGURE 12-15**

A spirometer. A spirometer is a device that measures the volume of air breathed in and out; it consists of an air-filled drum floating in a water-filled chamber. As a person breathes air in and out of the drum through a connecting tube, the resultant rise and fall of the drum are recorded as a spirogram, which is calibrated to the magnitude of the volume change.

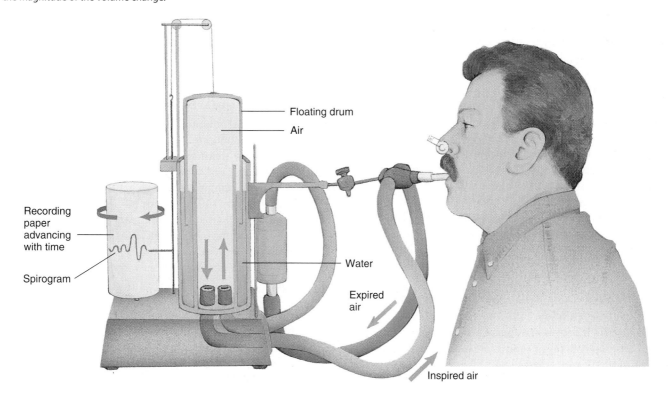

Floating drum

Air

Recording paper advancing with time

Spirogram

Water

Expired air

Inspired air

(a lung capacity is the sum of two or more lung volumes) can be determined.

- **Tidal volume (TV).** The volume of air entering or leaving the lungs during a single breath. Average value under resting conditions = 500 ml.
- **Inspiratory reserve volume (IRV).** The extra volume of air that can be maximally inspired over and above the typical resting tidal volume. The IRV is accomplished by maximal contraction of the diaphragm, external intercostal muscles, and accessory inspiratory muscles. Average value = 3000 ml.
- **Inspiratory capacity (IC).** The maximum volume of air that can be inspired at the end of a normal quiet expiration (IC = IRV + TV). Average value = 3500 ml.
- **Expiratory reserve volume (ERV).** The extra volume of air that can be actively expired by maximally contracting the expiratory muscles beyond that normally passively expired at the end of a typical resting tidal volume. Average value = 1000 ml.
- **Residual volume (RV).** The minimum volume of air remaining in the lungs even after a maximal expiration. Average value = 1200 ml. The residual volume cannot be measured directly with a spirometer, because this volume of air does not move in and out of the lungs. It can be determined indirectly, however, through gas dilution techniques involving inspiration of a known quantity of a harmless tracer gas such as helium.
- **Functional residual capacity (FRC).** The volume of air in the lungs at the end of a normal passive expiration (FRC = ERV + RV). Average value = 2200 ml.
- **Vital capacity (VC).** The maximum volume of air that can be moved out during a single breath following a maximal inspiration. The subject first inspires maximally, then expires maximally (VC = IRV + TV + ERV). The VC represents the maximum volume change possible within the lungs (● Figure 12-16). It is rarely used, because the maximal muscle contractions involved become exhausting, but it is useful in ascertaining the functional capacity of the lungs. Average value = 4500 ml.
- **Total lung capacity (TLC).** The maximum volume of air that the lungs can hold (TLC = VC + RV). Average value = 5700 ml.
- **Forced expiratory volume in 1 second (FEV_1).** The volume of air that can be expired during the first second of expiration in a VC determination. Usually, FEV_1 is about 80% of VC; that is, normally 80% of the air that can be forcibly expired from maximally inflated lungs can be expired within 1 second. This measurement indicates the maximal airflow rate that is possible from the lungs.

RESPIRATORY DYSFUNCTION

Clinical Note Measurement of the lungs' various volumes and capacities is of more than pure academic interest, because such determinations are useful to the diagnostician. Two general categories of respiratory dysfunction yield abnormal results during spirometry—*obstructive lung*

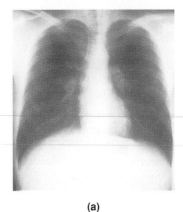 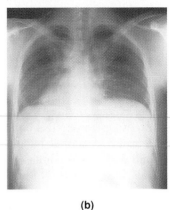

(a)   (b)

SIU/Visuals Unlimited

● **FIGURE 12-16**

X rays of lungs showing maximum volume change. (a) Maximum volume of the lungs at maximum inspiration. (b) Minimum volume of the lungs at maximum expiration. The difference between these two volumes is the vital capacity, which is the maximum volume of air that can be moved out during a single breath following a maximum inspiration.

disease and *restrictive lung disease* (● Figure 12-17). However, these are not the only categories of respiratory dysfunction, nor is spirometry the only pulmonary function test. For example, other conditions affecting respiratory function include diseases impairing diffusion of O_2 and CO_2 across the pulmonary membranes; reduced ventilation because of mechanical failure, as with neuromuscular disorders affecting the respiratory muscles; or failure of adequate pulmonary blood flow.

▌ **Alveolar ventilation is less than pulmonary ventilation because of dead space.**

Various changes in lung volume represent only one factor in determining **pulmonary ventilation**, which is the volume of air breathed in and out in one minute. The other important factor is **respiratory rate**, which averages 12 breaths per minute.

$$\underset{\text{(ml/min)}}{\text{Pulmonary ventilation}} = \underset{\text{(ml/breath)}}{\text{tidal volume}} \times \underset{\text{(breaths/min)}}{\text{respiratory rate}}$$

At an average tidal volume of 500 ml/breath and a respiratory rate of 12 breaths/minute, pulmonary ventilation is 6000 ml, or 6 liters, of air breathed in and out in one minute under resting conditions. For a brief period of time, a healthy young adult male can voluntarily increase his total pulmonary ventilation 25-fold, to 150 liters/min. To increase pulmonary ventilation, both tidal volume and respiratory rate increase, but depth of breathing increases more than frequency of breathing. It is usually more advantageous to have a greater increase in tidal volume than in respiratory rate, because of anatomic dead space.

ANATOMIC DEAD SPACE

Not all the inspired air gets down to the site of gas exchange in the alveoli. Part remains in the conducting airways, where it is not available for gas exchange. The volume of the con-

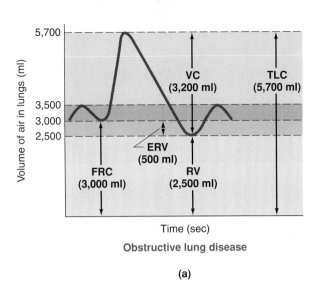

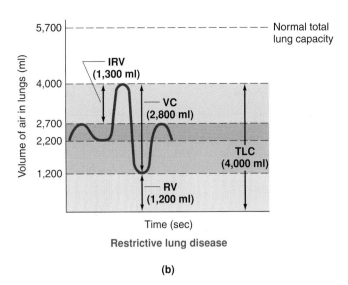

Obstructive lung disease

(a)

Restrictive lung disease

(b)

● **FIGURE 12-17**

Abnormal spirograms associated with obstructive and restrictive lung diseases. (a) *Spirogram in obstructive lung disease.* Because a patient with obstructive lung disease experiences more difficulty emptying the lungs than filling them, the total lung capacity (TLC) is essentially normal, but the functional residual capacity (FRC) and the residual volume (RV) are elevated as a result of the additional air trapped in the lungs following expiration. Because the RV is increased, the vital capacity (VC) is reduced. With more air remaining in the lungs, less of the TLC is available to be used in exchanging air with the atmosphere. Another common finding is a markedly reduced forced expiratory volume in 1 second (FEV_1) because the airflow rate is reduced by the airway obstruction. Even though both the VC and the FEV_1 are reduced, the FEV_1 is reduced more markedly than the VC is. As a result, the FEV_1/VC% is much lower than the normal 80%; that is, much less than 80% of the reduced VC can be blown out during the first second. (b) *Spirogram in restrictive lung disease.* In this disease the lungs are less compliant than normal. Total lung capacity, inspiratory capacity, and VC are reduced, because the lungs cannot be expanded as normal. The percentage of the VC that can be exhaled within 1 second is the normal 80% or an even higher percentage, because air can flow freely in the airways. Therefore, the FEV_1/VC% is particularly useful in distinguishing between obstructive and restrictive lung disease. Also, in contrast to obstructive lung disease, the RV is usually normal in restrictive lung disease.

---

ducting passages in an adult averages about 150 ml. This volume is considered **anatomic dead space**, because air within these conducting airways is useless for exchange. Anatomic dead space greatly affects the efficiency of pulmonary ventilation. In effect, even though 500 ml of air are moved in and out with each breath, only 350 ml are actually exchanged between the atmosphere and alveoli, because of the 150 ml occupying the anatomic dead space.

Looking at ● Figure 12-18, note that at the end of inspiration, the respiratory airways are filled with 150 ml of fresh atmospheric air from the inspiration. During the subsequent expiration, 500 ml of air are expired to the atmosphere. The first 150 ml expired are the fresh air that was retained in the airways and never used. The remaining 350 ml expired are "old" alveolar air that has participated in gas exchange with the blood. During the same expiration, 500 ml of gas also leave the alveoli. The first 350 ml are expired to the atmosphere; the other 150 ml of old alveolar air never reach the outside but remain in the conducting airways.

On the next inspiration, 500 ml of gas enter the alveoli. The first 150 ml to enter the alveoli are the old alveolar air that remained in the dead space during the preceding expiration. The other 350 ml entering the alveoli are fresh air inspired from the atmosphere. Simultaneously, 500 ml of air enter from

the atmosphere. The first 350 ml of atmospheric air reach the alveoli; the other 150 ml remain in the conducting airways to be expired without benefit of being exchanged with the blood, as the cycle repeats itself.

ALVEOLAR VENTILATION

Because the amount of atmospheric air that reaches the alveoli and is actually available for exchange with blood is more important than the total amount breathed in and out, **alveolar ventilation**—the volume of air exchanged between the atmosphere and alveoli per minute—is more important than pulmonary ventilation. In determining alveolar ventilation, the amount of wasted air moved in and out through the anatomic dead space must be taken into account, as follows:

$$\text{Alveolar ventilation} = (\text{tidal volume} - \text{dead space volume}) \times \text{respiratory rate}$$

With average resting values,

$$\text{Alveolar ventilation} = (500 \text{ ml/breath} - 150 \text{ ml dead space volume}) \times 12 \text{ breaths/min}$$
$$= 4200 \text{ ml/min}$$

Thus with quiet breathing, alveolar ventilation is 4200 ml/min, whereas pulmonary ventilation is 6000 ml/min.

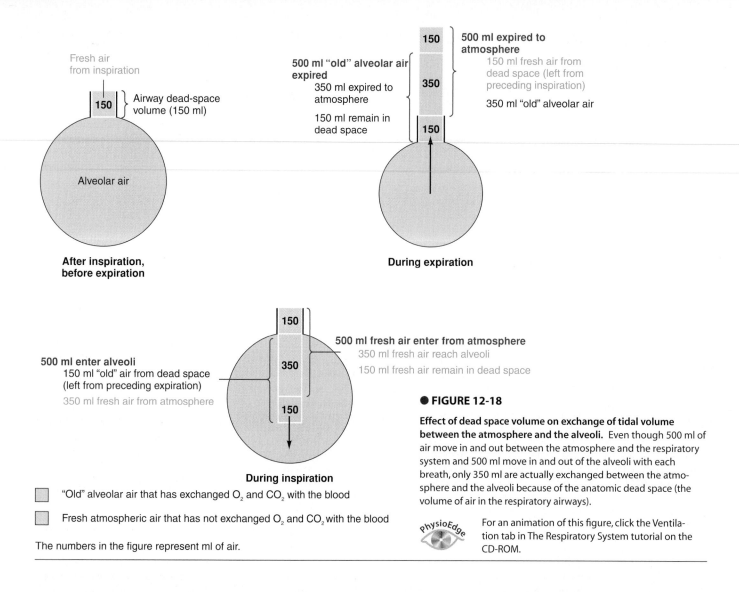

After inspiration, before expiration

Fresh air from inspiration

150 } Airway dead-space volume (150 ml)

Alveolar air

During expiration

500 ml "old" alveolar air expired
350 ml expired to atmosphere
150 ml remain in dead space

150

350

150

150

500 ml expired to atmosphere
150 ml fresh air from dead space (left from preceding inspiration)
350 ml "old" alveolar air

During inspiration

150

350

150

500 ml enter alveoli
150 ml "old" air from dead space (left from preceding expiration)
350 ml fresh air from atmosphere

500 ml fresh air enter from atmosphere
350 ml fresh air reach alveoli
150 ml fresh air remain in dead space

☐ "Old" alveolar air that has exchanged O_2 and CO_2 with the blood

☐ Fresh atmospheric air that has not exchanged O_2 and CO_2 with the blood

The numbers in the figure represent ml of air.

● **FIGURE 12-18**

Effect of dead space volume on exchange of tidal volume between the atmosphere and the alveoli. Even though 500 ml of air move in and out between the atmosphere and the respiratory system and 500 ml move in and out of the alveoli with each breath, only 350 ml are actually exchanged between the atmosphere and the alveoli because of the anatomic dead space (the volume of air in the respiratory airways).

PhysioEdge For an animation of this figure, click the Ventilation tab in The Respiratory System tutorial on the CD-ROM.

EFFECT OF BREATHING PATTERNS ON ALVEOLAR VENTILATION

To understand how important dead space volume is in determining the magnitude of alveolar ventilation, examine the effect of various breathing patterns on alveolar ventilation, as shown in ▲ Table 12-2. If a person deliberately breathes deeply (for example, a tidal volume of 1200 ml) and slowly (for example, a respiratory rate of 5 breaths/min), pulmonary ventilation is 6000 ml/min, the same as during quiet breathing at rest, but alveolar ventilation increases to 5250 ml/min compared to the resting rate of 4200 ml/min. In contrast, if a person deliberately breathes shallowly (for example, a tidal volume of 150 ml) and rapidly (a frequency of 40 breaths/min), pulmonary ventilation would still be 6,000 ml/min; however, alveolar ventilation would be 0 ml/min. In effect, the person would only be drawing air in and out of the anatomic dead space without any atmospheric air being exchanged with the alveoli, where it could be useful. The individual could voluntarily maintain such a breathing pattern for only a few minutes before losing consciousness, at which time normal breathing would resume.

The value of reflexly bringing about a larger increase in depth of breathing than in rate of breathing when pulmonary ventilation increases during exercise, should now be apparent. It is the most efficient means of elevating alveolar ventilation. When tidal volume is increased, the entire increase goes toward elevating alveolar ventilation, whereas an increase in respiratory rate does not go entirely toward increasing alveolar ventilation. When respiratory rate is increased, the frequency with which air is wasted in the dead space also increases, because a portion of *each breath* must move in and out of the dead space. As needs vary, ventilation is normally adjusted to a tidal volume and respiratory rate that meet those needs most efficiently in terms of energy cost.

ALVEOLAR DEAD SPACE

We have assumed that all the atmospheric air entering the alveoli participates in exchanges of O_2 and CO_2 with pulmonary blood. However, the match between air and blood is not always perfect, because not all alveoli are equally ventilated with air and perfused with blood. Any ventilated alveoli that do not participate in gas exchange with blood because

▲ TABLE 12-2

Effect of Different Breathing Patterns on Alveolar Ventilation

BREATHING PATTERN	TIDAL VOLUME (ml/breath)	RESPIRATORY RATE (breaths/min)	DEAD SPACE VOLUME (ml)	PULMONARY VENTILATION (ml/min)*	ALVEOLAR VENTILATION (ml/min)**
Quiet breathing at rest	500	12	150	6,000	4,200
Deep, slow breathing	1,200	5	150	6,000	5,250
Shallow, rapid breathing	150	40	150	6,000	0

*Equals tidal volume × respiratory rate.
**Equals (tidal volume − dead space volume) × respiratory rate.

they are inadequately perfused are considered **alveolar dead space**. In normal people, alveolar dead space is quite small and of little importance, but it can increase to even lethal levels in several types of pulmonary disease.

We have now completed our discussion of respiratory mechanics—all the factors involved in ventilation. We are now going to examine gas exchange between alveolar air and the blood and then between the blood and systemic tissues.

 Click on the Media Exercises menu of the CD-ROM and work Media Exercise 12.2: Mechanics of Ventilation to test your understanding of the previous section.

GAS EXCHANGE

The ultimate purpose of breathing is to provide a continual supply of fresh O_2 for pickup by the blood and to constantly remove CO_2 unloaded from the blood. Blood acts as a transport system for O_2 and CO_2 between the lungs and tissues, with the tissue cells extracting O_2 from the blood and eliminating CO_2 into it.

▌ Gases move down partial pressure gradients.

Gas exchange at both the pulmonary capillary and the tissue capillary levels involves simple passive diffusion of O_2 and CO_2 down *partial pressure gradients*. No active transport mechanisms exist for these gases. Let us see what partial pressure gradients are and how they are established.

PARTIAL PRESSURES

Atmospheric air is a mixture of gases; typical dry air contains about 79% nitrogen (N_2) and 21% O_2, with almost negligible percentages of CO_2, H_2O vapor, other gases, and pollutants. Altogether, these gases exert a total atmospheric pressure of 760 mm Hg at sea level. This total pressure is equal to the sum of the pressures that each gas in the mixture partially contributes. The pressure exerted by a particular gas is directly proportional to the percentage of that gas in the total air mix-

ture. Every gas molecule, no matter what its size, exerts the same amount of pressure; for example, a N_2 molecule exerts the same pressure as an O_2 molecule. Because 79% of the air consists of N_2 molecules, 79% of the 760 mm Hg atmospheric pressure, or 600 mm Hg, is exerted by the N_2 molecules. Similarly, because O_2 represents 21% of the atmosphere, 21% of the 760 mm Hg atmospheric pressure, or 160 mm Hg, is exerted by O_2 (● Figure 12-19). The individual pressure exerted independently by a particular gas within a mixture of gases is known as its **partial pressure**, designated by P_{gas}. Thus the partial pressure of O_2 in atmospheric air, P_{O_2}, is normally 160 mm Hg. The atmospheric partial pressure of CO_2, P_{CO_2}, is negligible at 0.23 mm Hg.

● FIGURE 12-19

Concept of partial pressures. The partial pressure exerted by each gas in a mixture equals the total pressure times the fractional composition of the gas in the mixture.

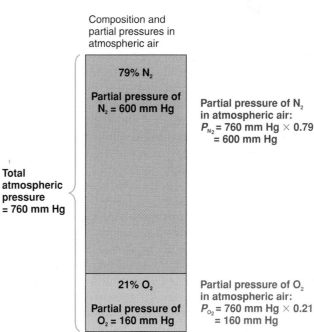

Composition and partial pressures in atmospheric air

79% N_2

Partial pressure of N_2 = 600 mm Hg

Partial pressure of N_2 in atmospheric air:
P_{N_2} = 760 mm Hg × 0.79 = 600 mm Hg

Total atmospheric pressure = 760 mm Hg

21% O_2

Partial pressure of O_2 = 160 mm Hg

Partial pressure of O_2 in atmospheric air:
P_{O_2} = 760 mm Hg × 0.21 = 160 mm Hg

Gases dissolved in a liquid such as blood or another body fluid also exert a partial pressure. The greater the partial pressure of a gas in a liquid, the more of that gas dissolved.

PARTIAL PRESSURE GRADIENTS

A difference in partial pressure between capillary blood and surrounding structures is known as a **partial pressure gradient**. Partial pressure gradients exist between the alveolar air and pulmonary capillary blood. Similarly, partial pressure gradients exist between systemic capillary blood and surrounding tissues. A gas always diffuses down its partial pressure gradient from the area of higher partial pressure to the area of lower partial pressure, similar to diffusion down a concentration gradient.

■ Oxygen enters and CO_2 leaves the blood in the lungs passively down partial pressure gradients.

We are first going to consider the magnitude of alveolar P_{O_2} and P_{CO_2} and then look at the partial pressure gradients that move these two gases between the alveoli and incoming pulmonary capillary blood.

ALVEOLAR P_{O_2} AND P_{CO_2}

Alveolar air is not of the same composition as inspired atmospheric air, for two reasons. First, as soon as atmospheric air enters the respiratory passages, exposure to the moist airways saturates it with H_2O. Like any other gas, water vapor exerts a partial pressure. Humidification of inspired air in effect "dilutes" the partial pressure of the inspired gases, because the sum of the partial pressures must total the atmospheric pressure. Second, alveolar P_{O_2} is also lower than atmospheric P_{O_2} because fresh inspired air is mixed with the large volume of old air that remained in the lungs and dead space at the end of the preceding expiration (the functional residual capacity). At the end of inspiration, less than 15% of the air in the alveoli is fresh air. As a result of humidification and the small turnover of alveolar air, the average alveolar P_{O_2} is 100 mm Hg, compared to the atmospheric P_{O_2} of 160 mm Hg.

It is logical to think that alveolar P_{O_2} would increase during inspiration with the arrival of fresh air and would decrease during expiration. Only small fluctuations of a few mm Hg occur, however, for two reasons. First, only a small proportion of the total alveolar air is exchanged with each breath. The relatively small volume of inspired, high-P_{O_2} air is quickly mixed with the much larger volume of retained alveolar air, which has a lower P_{O_2}. Thus the O_2 in the inspired air can only slightly elevate the level of the total alveolar P_{O_2}. Even this potentially small elevation of P_{O_2} is diminished for another reason. Oxygen is continually moving by passive diffusion down its partial pressure gradient from the alveoli into the blood. The O_2 arriving in the alveoli in the newly inspired air simply replaces the O_2 diffusing out of the alveoli into the pulmonary capillaries. Therefore, alveolar P_{O_2} remains relatively constant at about 100 mm Hg throughout the respiratory cycle. Because pulmonary blood P_{O_2} equilibrates with alveolar P_{O_2}, the P_{O_2} of the blood leaving the lungs likewise remains fairly constant at this same value. Accordingly, the amount of O_2 in the blood available to the tissues varies only slightly during the respiratory cycle.

A similar situation exists in reverse for CO_2. Carbon dioxide, which is continually produced by the body tissues as a metabolic waste product, is constantly added to the blood at the level of the systemic capillaries. In the pulmonary capillaries, CO_2 diffuses down its partial pressure gradient from the blood into the alveoli and is subsequently removed from the body during expiration. As with O_2, alveolar P_{CO_2} remains fairly constant throughout the respiratory cycle but at a lower value of 40 mm Hg.

P_{O_2} AND P_{CO_2} GRADIENTS ACROSS THE PULMONARY CAPILLARIES

As blood passes through the lungs, it picks up O_2 and gives up CO_2 simply by diffusion down partial pressure gradients that exist between the blood and alveoli. Ventilation constantly replenishes alveolar O_2 and removes CO_2, thus maintaining the appropriate partial pressure gradients between the blood and alveoli. The blood entering the pulmonary capillaries is systemic venous blood pumped to the lungs through the pulmonary arteries. This blood, having just returned from the body tissues, is relatively low in O_2, with a P_{O_2} of 40 mm Hg, and is relatively high in CO_2, with a P_{CO_2} of 46 mm Hg. As this blood flows through the pulmonary capillaries, it is exposed to alveolar air (● Figure 12-20). Because the alveolar P_{O_2} at 100 mm Hg is higher than the P_{O_2} of 40 mm Hg in the blood entering the lungs, O_2 diffuses down its partial pressure gradient from the alveoli into the blood until no further gradient exists. As the blood leaves the pulmonary capillaries, it has a P_{O_2} equal to alveolar P_{O_2} at 100 mm Hg.

The partial pressure gradient for CO_2 is in the opposite direction. Blood entering the pulmonary capillaries has a P_{CO_2} of 46 mm Hg, whereas alveolar P_{CO_2} is only 40 mm Hg. Carbon dioxide diffuses from the blood into the alveoli until blood P_{CO_2} equilibrates with alveolar P_{CO_2}. Thus the blood leaving the pulmonary capillaries has a P_{CO_2} of 40 mm Hg. After leaving the lungs, the blood, which now has a P_{O_2} of 100 mm Hg and a P_{CO_2} of 40 mm Hg, is returned to the heart, then pumped out to the body tissues as systemic arterial blood.

Note that blood returning to the lungs from the tissues still contains O_2 (P_{O_2} of systemic venous blood = 40 mm Hg) and that blood leaving the lungs still contains CO_2 (P_{CO_2} of systemic arterial blood = 40 mm Hg). The extra O_2 carried in the blood beyond that normally given up to the tissues represents an immediately available O_2 reserve that can be tapped by the tissue cells whenever their O_2 demands increase. The CO_2 remaining in the blood even after passage through the lungs plays an important role in the acid–base balance of the body, because CO_2 generates carbonic acid. Furthermore, arterial P_{CO_2} is important in driving respiration. This mechanism will be described later.

The amount of O_2 picked up in the lungs matches the amount extracted and used by the tissues. When the tissues metabolize more actively (for example, during exercise), they extract more O_2 from the blood, reducing the systemic venous P_{O_2} even lower than 40 mm Hg—for example, to a P_{O_2}

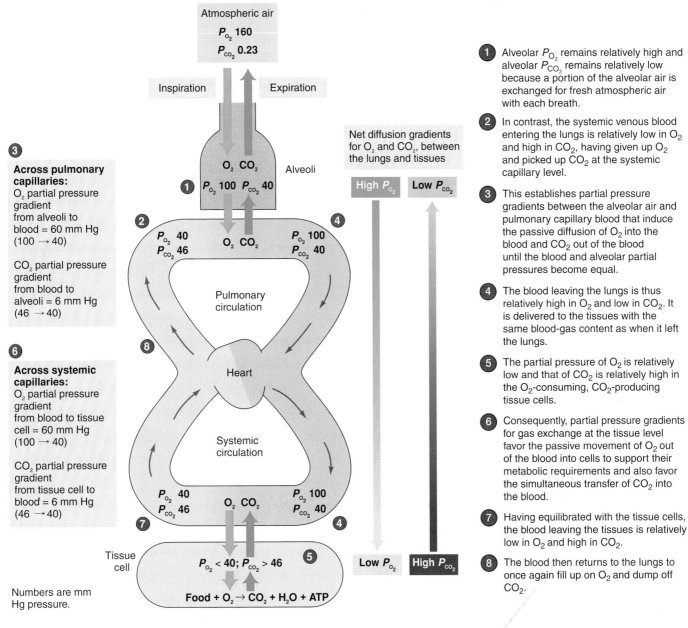

① Alveolar P_{O_2} remains relatively high and alveolar P_{CO_2} remains relatively low because a portion of the alveolar air is exchanged for fresh atmospheric air with each breath.

② In contrast, the systemic venous blood entering the lungs is relatively low in O_2 and high in CO_2, having given up O_2 and picked up CO_2 at the systemic capillary level.

③ This establishes partial pressure gradients between the alveolar air and pulmonary capillary blood that induce the passive diffusion of O_2 into the blood and CO_2 out of the blood until the blood and alveolar partial pressures become equal.

④ The blood leaving the lungs is thus relatively high in O_2 and low in CO_2. It is delivered to the tissues with the same blood-gas content as when it left the lungs.

⑤ The partial pressure of O_2 is relatively low and that of CO_2 is relatively high in the O_2-consuming, CO_2-producing tissue cells.

⑥ Consequently, partial pressure gradients for gas exchange at the tissue level favor the passive movement of O_2 out of the blood into cells to support their metabolic requirements and also favor the simultaneous transfer of CO_2 into the blood.

⑦ Having equilibrated with the tissue cells, the blood leaving the tissues is relatively low in O_2 and high in CO_2.

⑧ The blood then returns to the lungs to once again fill up on O_2 and dump off CO_2.

③ Across pulmonary capillaries:
O_2 partial pressure gradient from alveoli to blood = 60 mm Hg (100 → 40)

CO_2 partial pressure gradient from blood to alveoli = 6 mm Hg (46 → 40)

⑥ Across systemic capillaries:
O_2 partial pressure gradient from blood to tissue cell = 60 mm Hg (100 → 40)

CO_2 partial pressure gradient from tissue cell to blood = 6 mm Hg (46 → 40)

Numbers are mm Hg pressure.

● **FIGURE 12-20**

Oxygen and CO_2 exchange across pulmonary and systemic capillaries caused by partial pressure gradients

PhysioEdge For an animation of this figure, click the Gas Exchange tab (page 2) in The Respiratory System tutorial on the CD-ROM.

of 30 mm Hg. When this blood returns to the lungs, a larger-than-normal P_{O_2} gradient exists between the newly entering blood and alveolar air. The difference in P_{O_2} between the alveoli and blood is now 70 mm Hg (alveolar P_{O_2} of 100 mm Hg and blood P_{O_2} of 30 mm Hg), compared to the normal P_{O_2} gradient of 60 mm Hg (alveolar P_{O_2} of 100 mm Hg and blood P_{O_2} of 40 mm Hg). Therefore, more O_2 diffuses from the alveoli into the blood down the larger partial pressure gradient before blood P_{O_2} equals alveolar P_{O_2}. This additional transfer of O_2 into the blood replaces the increased amount of O_2 consumed, so O_2 uptake matches O_2 use even when O_2 consumption increases. At the same time that more O_2 is dif-

fusing from the alveoli into the blood because of the increased partial pressure gradient, ventilation is stimulated so that O_2 enters the alveoli more rapidly from the atmosphere to replace the O_2 diffusing into the blood. Similarly, the amount of CO_2 given up to the alveoli from the blood matches the amount of CO_2 picked up at the tissues.

▌ Factors other than the partial pressure gradient influence the rate of gas transfer.

We have been discussing diffusion of O_2 and CO_2 between the alveoli and blood as if these gases' partial pressure gradients were the sole determinants of their rates of diffusion. According to Fick's law of diffusion, the diffusion rate of a gas through a sheet of tissue also depends on the surface area and thickness of the membrane through which the gas is diffusing, with the rate of diffusion decreasing as surface area decreases or thickness increases. Because these other factors are relatively constant under resting conditions in a healthy lung, changes in the rate of gas exchange normally are determined primarily by changes in partial pressure gradients between blood and alveoli.

Clinical Note However, several pathologic conditions can markedly reduce pulmonary surface area and, in turn, decrease the rate of gas exchange. Most notably, in *emphysema* surface area is reduced because many alveolar walls are lost, resulting in larger but fewer chambers (● Figure 12-21).

Inadequate gas exchange can also occur when the thickness of the barrier separating the air and blood is pathologically increased, because a gas takes longer to diffuse through the greater thickness. Thickness increases in (1) *pulmonary edema*, an excess accumulation of interstitial fluid between the alveoli and pulmonary capillaries caused by pulmonary in-flammation or left-sided congestive heart failure (see p. 265); (2) *pulmonary fibrosis* involving replacement of delicate lung tissue with thick fibrous tissue in response to certain chronic irritants; and (3) *pneumonia,* which is characterized by inflammatory fluid accumulation within or around the alveoli.

▌ Gas exchange across the systemic capillaries also occurs down partial pressure gradients.

Just as they do at the pulmonary capillaries, O_2 and CO_2 move between the systemic capillary blood and the tissue cells by simple passive diffusion down partial pressure gradients. Refer again to ● Figure 12-20. The arterial blood that reaches the systemic capillaries is essentially the same blood that left the lungs by means of the pulmonary veins, because the only two places in the entire circulatory system at which gas exchange can take place are the pulmonary capillaries and the systemic capillaries. The arterial P_{O_2} is 100 mm Hg, and the arterial P_{CO_2} is 40 mm Hg, the same as alveolar P_{O_2} and P_{CO_2}.

P_{O_2} AND P_{CO_2} GRADIENTS ACROSS THE SYSTEMIC CAPILLARIES

Cells constantly consume O_2 and produce CO_2 through oxidative metabolism. Cellular P_{O_2} averages about 40 mm Hg and P_{CO_2} about 46 mm Hg, although these values are highly variable, depending on the level of cellular metabolic activity. Oxygen moves by diffusion down its partial pressure gradient from the entering systemic capillary blood (P_{O_2} = 100 mm Hg) into the adjacent cells (P_{O_2} = 40 mm Hg) until equilibrium is reached. Therefore, the P_{O_2} of venous blood leaving the systemic capillaries is equal to the tissue P_{O_2} at an average of 40 mm Hg.

The reverse situation exists for CO_2. Carbon dioxide rapidly diffuses out of the cells (P_{CO_2} = 46 mm Hg) into the entering capillary blood (P_{CO_2} = 40 mm Hg) down the partial pressure gradient created by the ongoing production of CO_2. Transfer of CO_2 continues until blood P_{CO_2} equilibrates with tissue P_{CO_2}.[1] Accordingly, blood leaving the systemic capillaries has an average P_{CO_2} of 46 mm Hg. This systemic venous blood, which is relatively low in O_2 (P_{O_2} = 40 mm Hg) and relatively high in CO_2 (P_{CO_2} = 46 mm Hg), returns to the heart and is subsequently pumped to the lungs as the cycle repeats itself.

The more actively a tissue is metabolizing, the lower the cellular P_{O_2} falls and the higher the cellular P_{CO_2} rises. As a consequence of the larger blood-to-cell partial pressure gradients, more O_2 diffuses from the blood into the cells, and more C_{O_2} moves in the opposite direction before blood P_{O_2} and P_{CO_2} achieve equilibrium with the surrounding cells. Thus the amount of O_2 transferred to the cells and the amount of CO_2 carried away from the cells both depend on the rate of cellular metabolism.

---

[1] Actually, the partial pressures of the systemic blood gases never completely equilibrate with tissue P_{O_2} and P_{CO_2}. Because the cells are constantly consuming O_2 and producing CO_2, the tissue P_{O_2} is always slightly less than the P_{O_2} of the blood leaving the systemic capillaries, and the tissue P_{CO_2} always slightly exceeds the systemic venous P_{CO_2}.

● **FIGURE 12-21**

Comparison of normal and emphysematous lung tissue. (a) Photomicrograph of lung tissue from a normal individual. Each of the smallest clear spaces is an alveolar lumen. (b) Photomicrograph of lung tissue from a patient with emphysema. Note the loss of alveolar walls in the emphysematous lung tissue, resulting in larger but fewer alveolar chambers.

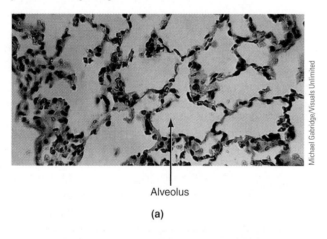

Alveolus

(a)

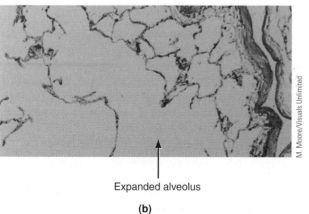

Expanded alveolus

(b)

NET DIFFUSION OF O_2 AND CO_2 BETWEEN THE ALVEOLI AND TISSUES

Net diffusion of O_2 occurs first between the alveoli and blood and then between the blood and tissues, because of the O_2 partial pressure gradients created by continuous use of O_2 in the cells and continuous replenishment of fresh alveolar O_2 provided by alveolar ventilation. Net diffusion of CO_2 occurs in the reverse direction, first between the tissues and blood and then between the blood and alveoli, because of the CO_2 partial pressure gradients created by continuous production of CO_2 in the cells and continuous removal of alveolar CO_2 through the process of alveolar ventilation (● Figure 12-20).

Now let's see how O_2 and CO_2 are transported in the blood between the alveoli and tissues.

GAS TRANSPORT

Oxygen picked up by the blood at the lungs must be transported to the tissues for cell use. Conversely, CO_2 produced at the cell level must be transported to the lungs for elimination.

Most O_2 in the blood is transported bound to hemoglobin.

Oxygen is present in the blood in two forms: physically dissolved and chemically bound to hemoglobin (▲ Table 12-3).

PHYSICALLY DISSOLVED O_2

Very little O_2 physically dissolves in plasma water, because O_2 is poorly soluble in body fluids. The amount dissolved is directly proportional to the P_{O_2} of the blood; the higher the P_{O_2}, the more O_2 dissolved. At a normal arterial P_{O_2} of 100 mm Hg, only 3 ml of O_2 can dissolve in 1 liter of blood. Thus only 15 ml of O_2/min can dissolve in the normal pulmonary blood flow of 5 liters/min (the resting cardiac output). Even under resting conditions, the cells consume 250 ml of O_2/min, and consumption may increase up to 25-fold during strenuous exercise. To deliver the O_2 required by the tissues even at rest, the cardiac output would have to be 83.3 liters/min if O_2 could only be transported in dissolved form. Obviously, there

must be an additional mechanism for transporting O_2 to the tissues. This mechanism is *hemoglobin (Hb)*. Only 1.5% of the O_2 in the blood is dissolved; the remaining 98.5% is transported in combination with Hb. *The O_2 bound to Hb does not contribute to the P_{O_2} of the blood;* thus blood P_{O_2} is not a measure of the total O_2 content of the blood but only of the dissolved portion of O_2.

OXYGEN BOUND TO HEMOGLOBIN

Hemoglobin, an iron-bearing protein molecule contained within the red blood cells, can form a loose, easily reversible combination with O_2 (see p. 317). When not combined with O_2, Hb is referred to as **reduced hemoglobin**, or **deoxyhemoglobin**; when combined with O_2, it is called **oxyhemoglobin** (HbO_2):

$$\underset{\text{reduced hemoglobin}}{Hb + O_2} \quad \rightleftarrows \quad \underset{\text{oxyhemoglobin}}{HbO_2}$$

We need to answer several questions about the role of Hb in O_2 transport. What determines whether O_2 and Hb are combined or dissociated (separated)? Why does Hb combine with O_2 in the lungs and release O_2 at the tissues? How can a variable amount of O_2 be released at the tissues, depending on the level of tissue activity? How can we talk about O_2 transfer between blood and surrounding tissues in terms of O_2 partial pressure gradients when 98.5% of the O_2 is bound to Hb and thus does not contribute to the P_{O_2} of the blood at all?

▌ The P_{O_2} is the primary factor determining the percent hemoglobin saturation.

Each of the four atoms of iron within the heme portions of a hemoglobin molecule can combine with an O_2 molecule, so each Hb molecule can carry up to four molecules of O_2. Hemoglobin is considered *fully saturated* when all the Hb present is carrying its maximum O_2 load. The **percent hemoglobin (% Hb) saturation**, a measure of the extent to which the Hb present is combined with O_2, can vary from 0% to 100%.

The most important factor determining the % Hb saturation is the P_{O_2} of the blood, which in turn is related to the concentration of O_2 physically dissolved in the blood. According to the **law of mass action**, if the concentration of one substance involved in a reversible reaction is increased, the reaction is driven toward the opposite side. Conversely, if the concentration of one substance is decreased, the reaction is driven toward that side. Applying this law to the reversible reaction involving Hb and O_2 ($Hb + O_2 \rightleftarrows HbO_2$), when blood P_{O_2} increases, as in the pulmonary capillaries, the reaction is driven toward the right side of the equation, increasing formation of HbO_2 (increased % Hb saturation). When blood P_{O_2} decreases, as in the systemic capillaries, the reaction is driven toward the left side of the equation and oxygen is released from Hb as HbO_2 dissociates (decreased % Hb saturation). Thus because of the difference in P_{O_2} at the lungs and other tissues, Hb automatically "loads up" on O_2 in the lungs, where ventilation is continually providing fresh supplies of O_2, and "unloads" it in the tissues, which are constantly using up O_2.

▲ **TABLE 12-3**

Methods of Gas Transport in the Blood

GAS	METHOD OF TRANSPORT IN BLOOD	PERCENTAGE CARRIED IN THIS FORM
O_2	Physically dissolved	1.5
	Bound to hemoglobin	98.5
CO_2	Physically dissolved	10
	Bound to hemoglobin	30
	As bicarbonate (HCO_3^-)	60

O₂–Hb DISSOCIATION CURVE

The relationship between blood P_{O_2} and % Hb saturation is not linear, however, a point that is very important physiologically. Doubling the partial pressure does not double the % Hb saturation. Rather, the relationship between these variables follows an S-shaped curve, the O₂–Hb dissociation (or **saturation**) **curve** (● Figure 12-22). At the upper end, between a blood P_{O_2} of 60 and 100 mm Hg, the curve flattens off, or plateaus. Within this pressure range, a rise in P_{O_2} produces only a small increase in the extent to which Hb is bound with O₂. In the P_{O_2} range of 0 to 60 mm Hg, in contrast, a small change in P_{O_2} results in a large change in the extent to which Hb is combined with O₂, as shown by the steep lower part of the curve. Both the upper plateau and lower steep portion of the curve have physiological significance.

SIGNIFICANCE OF THE PLATEAU PORTION OF THE O₂–Hb CURVE

The plateau portion of the curve is in the blood P_{O_2} range that exists at the pulmonary capillaries where O₂ is being

---

● FIGURE 12-22

Oxygen–hemoglobin (O₂–Hb) dissociation (saturation) curve. The percent hemoglobin saturation (the scale on the left side of the graph) depends on the P_{O_2} of the blood. The relationship between these two variables is depicted by an S-shaped curve with a plateau region between a blood P_{O_2} of 60 and 100 mm Hg and a steep portion between 0 and 60 mm Hg. Another way to express the effect of blood P_{O_2} on the amount of O₂ bound with hemoglobin is as the volume percent of O₂ in the blood (ml of O₂ bound with hemoglobin in each 100 ml of blood). That relationship is represented by the scale on the right side of the graph.

For an animation of this figure, click the Gas Transport tab (page 4) in The Respiratory System tutorial on the CD-ROM. For an interaction related to this figure, see Media Exercise 12.3: Gas Transport and Exchange.

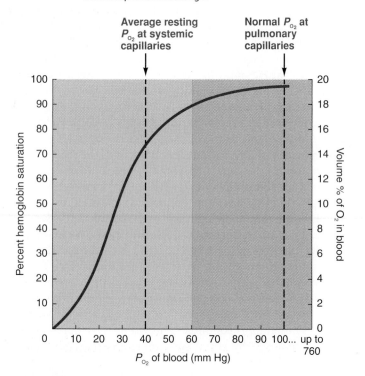

---

loaded onto Hb. The systemic arterial blood leaving the lungs, having equilibrated with alveolar P_{O_2}, normally has a P_{O_2} of 100 mm Hg. Looking at the O₂–Hb curve, note that at a blood P_{O_2} of 100 mm Hg, Hb is 97.5% saturated. Therefore, the Hb in the systemic arterial blood normally is almost fully saturated.

If the alveolar P_{O_2} and consequently the arterial P_{O_2} fall below normal, there is little reduction in the total amount of O₂ transported by the blood until the P_{O_2} falls below 60 mm Hg, because of the plateau region of the curve. If the arterial P_{O_2} falls 40%, from 100 to 60 mm Hg, the concentration of dissolved O₂ as reflected by the P_{O_2} is likewise reduced 40%. At a blood P_{O_2} of 60 mm Hg, however, the % Hb saturation is still remarkably high at 90%. Accordingly, the total O₂ content of the blood is only slightly decreased despite the 40% reduction in P_{O_2}, because Hb is still carrying an almost full load of O₂, and, as mentioned before, the vast majority of O₂ is transported by Hb rather than being dissolved. However, even if the blood P_{O_2} is greatly increased—say, to 600 mm Hg—by breathing pure O₂, very little additional O₂ is added to the blood. A small extra amount of O₂ dissolves, but the % Hb saturation can be maximally increased by only another 2.5%, to 100% saturation. Therefore, in the P_{O_2} range between 60 and 600 mm Hg or even higher, there is only a 10% difference in the amount of O₂ carried by Hb. Thus the plateau portion of the O₂–Hb curve provides a good margin of safety in O₂-carrying capacity of the blood.

Arterial P_{O_2} may be reduced by pulmonary diseases accompanied by inadequate ventilation or defective gas exchange or by circulatory disorders that result in inadequate blood flow to the lungs. It may also fall in healthy people under two circumstances: (1) at high altitudes, where the total atmospheric pressure and hence the P_{O_2} of the inspired air are reduced, or (2) in O₂-deprived environments at sea level, such as if someone were accidentally locked in a vault. Unless the arterial P_{O_2} becomes markedly reduced (falls below 60 mm Hg) in either pathologic conditions or abnormal circumstances, near-normal amounts of O₂ can still be carried to the tissues.

SIGNIFICANCE OF THE STEEP PORTION OF THE O₂–Hb CURVE

The steep portion of the curve between 0 and 60 mm Hg is in the blood P_{O_2} range that exists at the systemic capillaries, where O₂ is unloaded from Hb. In the systemic capillaries, the blood equilibrates with the surrounding tissue cells at an average P_{O_2} of 40 mm Hg. Note on ● Figure 12-22 that at a P_{O_2} of 40 mm Hg, the % Hb saturation is 75%. The blood arrives in the tissue capillaries at a P_{O_2} of 100 mm Hg with 97.5% Hb saturation. Because Hb can only be 75% saturated at the P_{O_2} of 40 mm Hg in the systemic capillaries, nearly 25% of the HbO₂ must dissociate, yielding reduced Hb and O₂. This released O₂ is free to diffuse down its partial pressure gradient from the red blood cells through the plasma and interstitial fluid into the tissue cells.

The Hb in the venous blood returning to the lungs is still normally 75% saturated. If the tissue cells are metabolizing more actively, the P_{O_2} of the systemic capillary blood falls (for example, from 40 to 20 mm Hg) because the cells are consuming O₂ more rapidly. Note on the curve that this drop of

20 mm Hg in P_{O_2} decreases the % Hb saturation from 75% to 30%; that is, about 45% more of the total HbO_2 than normal gives up its O_2 for tissue use. The normal 60 mm Hg drop in P_{O_2} from 100 to 40 mm Hg in the systemic capillaries causes about 25% of the total HbO_2 to unload its O_2. In comparison, a further drop in P_{O_2} of only 20 mm Hg results in an additional 45% of the total HbO_2 unloading its O_2, because the O_2 partial pressures in this range are operating in the steep portion of the curve. In this range, only a small drop in systemic capillary P_{O_2} can automatically make large amounts of O_2 immediately available to meet the O_2 needs of more actively metabolizing tissues. As much as 85% of the Hb may give up its O_2 to actively metabolizing cells during strenuous exercise. In addition to this more thorough withdrawal of O_2 from the blood, even more O_2 is made available to actively metabolizing cells, such as exercising muscles, by circulatory and respiratory adjustments that increase the flow rate of oxygenated blood through the active tissues.

▌ Hemoglobin promotes the net transfer of O_2 at both the alveolar and tissue levels.

We still have not really clarified the role of Hb in gas exchange. Because blood P_{O_2} depends entirely on the concentration of *dissolved* O_2, we could ignore the O_2 bound to Hb in our earlier discussion of O_2 being driven from the alveoli to the blood by a P_{O_2} gradient. However, Hb does play a crucial role in permitting the transfer of large quantities of O_2 before blood P_{O_2} equilibrates with the surrounding tissues (● Figure 12-23).

ROLE OF HEMOGLOBIN AT THE ALVEOLAR LEVEL

Hemoglobin acts as a "storage depot" for O_2, removing O_2 from solution as soon as it enters the blood from the alveoli. Because only dissolved O_2 contributes to P_{O_2}, the O_2 stored in Hb cannot contribute to blood P_{O_2}. When systemic venous blood enters the pulmonary capillaries, its P_{O_2} is considerably lower than alveolar P_{O_2}, so O_2 immediately diffuses into the blood, raising blood P_{O_2}. As soon as the blood P_{O_2} increases, the percentage of Hb that can bind with O_2 likewise increases, as indicated by the O_2–Hb curve. Consequently, most of the O_2 that has diffused into the blood combines with Hb and no longer contributes to blood P_{O_2}. As O_2 is removed from solution by combining with Hb, blood P_{O_2} falls to about the same level it was when the blood entered the lungs, even though the total quantity of O_2 in the blood actually has increased. Because blood P_{O_2} is once again considerably below alveolar P_{O_2}, more O_2 diffuses from the alveoli into the blood, only to be soaked up by Hb again.

Even though we have considered this process stepwise for clarity, net diffusion of O_2 from alveoli to blood occurs con-

---

● FIGURE 12-23

Hemoglobin facilitating a large net transfer of O_2 by acting as a storage depot to keep P_{O_2} low. (a) In the hypothetical situation in which no hemoglobin is present in the blood, the alveolar P_{O_2} and the pulmonary capillary blood P_{O_2} are at equilibrium. (b) Hemoglobin has been added to the pulmonary capillary blood. As the Hb starts to bind with O_2, it removes O_2 from solution. Because only dissolved O_2 contributes to blood P_{O_2}, the blood P_{O_2} falls below that of the alveoli, even though the same number of O_2 molecules are present in the blood as in part (a). By "soaking up" some of the dissolved O_2, Hb favors the net diffusion of more O_2 down its partial pressure gradient from the alveoli to the blood. (c) Hemoglobin is fully saturated with O_2, and the alveolar and blood P_{O_2} are at equilibrium again. The blood P_{O_2} resulting from dissolved O_2 is equal to the alveolar P_{O_2}, despite the fact that the total O_2 content in the blood is much greater than in part (a) when blood P_{O_2} was equal to alveolar P_{O_2} in the absence of Hb.

 For an animation of this figure, click the Gas Transport tab (page 1) in The Respiratory System tutorial on the CD-ROM.

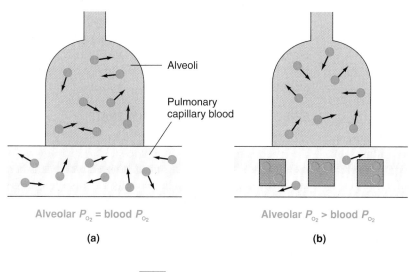
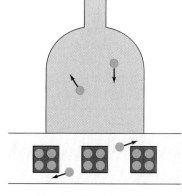

Alveoli

Pulmonary capillary blood

Alveolar P_{O_2} = blood P_{O_2}

(a)

Alveolar P_{O_2} > blood P_{O_2}

(b)

Alveolar P_{O_2} = blood P_{O_2}

(c)

 = O_2 molecule   = Partially saturated hemoglobin molecule   = Fully saturated hemoglobin molecule

tinuously until Hb becomes as completely saturated with O_2 as it can be at that particular P_{O_2}. At a normal P_{O_2} of 100 mm Hg, Hb is 97.5% saturated. Thus by soaking up O_2, Hb keeps blood P_{O_2} low and prolongs the existence of a partial pressure gradient so that a large net transfer of O_2 into the blood can take place. Not until Hb can store no more O_2 (that is, Hb is maximally saturated for that P_{O_2}) does all the O_2 transferred into the blood remain dissolved and directly contribute to the P_{O_2}. Only now does blood P_{O_2} rapidly equilibrate with alveolar P_{O_2} and bring further O_2 transfer to a halt, but this point is not reached until Hb is already loaded to the maximum extent possible. Once blood P_{O_2} equilibrates with alveolar P_{O_2}, no further O_2 transfer can take place, no matter how little or how much total O_2 has already been transferred.

ROLE OF HEMOGLOBIN AT THE TISSUE LEVEL

The reverse situation occurs at the tissue level. Because the P_{O_2} of blood entering the systemic capillaries is considerably higher than the P_{O_2} of the surrounding tissue, O_2 immediately diffuses from the blood into the tissues, lowering blood P_{O_2}. When blood P_{O_2} falls, Hb must unload some stored O_2, because the % Hb saturation is reduced. As the O_2 released from Hb dissolves in the blood, blood P_{O_2} increases and once again exceeds the P_{O_2} of the surrounding tissues. This favors further movement of O_2 out of the blood, although the total quantity of O_2 in the blood has already fallen. Only when Hb can no longer release any more O_2 into solution (when Hb is unloaded to the greatest extent possible for the P_{O_2} existing at the systemic capillaries) can blood P_{O_2} fall as low as in surrounding tissue. At this time, further transfer of O_2 stops. Hemoglobin, because it stores a large quantity of O_2 that can be freed by a slight reduction in P_{O_2} at the systemic capillary level, permits the transfer of tremendously much more O_2 from the blood into the cells than would be possible in its absence.

Thus Hb plays an important role in the *total quantity* of O_2 that the blood can pick up in the lungs and drop off in the tissues. If Hb levels fall to one half of normal, as in a severely anemic patient (see p. 319), the O_2-carrying capacity of the blood falls by 50% even though the arterial P_{O_2} is the normal 100 mm Hg with 97.5% Hb saturation. Only half as much Hb is available to be saturated, emphasizing once again how critical Hb is in determining how much O_2 can be picked up at the lungs and made available to tissues.

▌ **Factors at the tissue level promote the unloading of O_2 from hemoglobin.**

Even though the main factor determining the % Hb saturation is the P_{O_2} of the blood, other factors can affect the affinity, or bond strength, between Hb and O_2 and, accordingly, can shift the O_2–Hb curve (that is, change the % Hb saturation at a given P_{O_2}). These other factors are CO_2, acidity, temperature, and 2,3-bisphosphoglycerate, which we will examine separately. The O_2–Hb dissociation curve with which you are already familiar (● Figure 12-22) is a typical curve at normal arterial CO_2 and acidity levels, normal body temperature, and normal 2,3-bisphosphoglycerate concentration.

EFFECT OF CO_2 ON % Hb SATURATION

An increase in P_{CO_2} shifts the O_2–Hb curve to the right (● Figure 12-24). The % Hb saturation still depends on the P_{O_2}, but for any given P_{O_2} less O_2 and Hb can be combined. This effect is important, because the P_{CO_2} of the blood increases in the systemic capillaries as CO_2 diffuses down its gradient from the cells into the blood. The presence of this additional CO_2 in the blood in effect decreases the affinity of Hb for O_2, so Hb unloads even more O_2 at the tissue level than it would if the reduction in P_{O_2} in the systemic capillaries were the only factor affecting % Hb saturation.

EFFECT OF ACID ON % Hb SATURATION

An increase in acidity also shifts the curve to the right. Because CO_2 generates carbonic acid (H_2CO_3), the blood becomes more acidic at the systemic capillary level as it picks up CO_2 from the tissues. The resulting reduction in Hb affinity for O_2 in the presence of increased acidity aids in releasing even more O_2 at the tissue level for a given P_{O_2}. In actively metabolizing cells, such as exercising muscles, not only is more carbonic acid–generating CO_2 produced, but lactic acid also may be produced if the cells resort to anaerobic metabolism (see p. 29 and p. 220). The resultant local elevation of acid in the working muscles facilitates further unloading of O_2 in the very tissues that need the most O_2.

EFFECT OF TEMPERATURE ON % Hb SATURATION

In a similar manner, a rise in temperature shifts the O_2–Hb curve to the right, resulting in more unloading of O_2 at a given P_{O_2}. An exercising muscle or other actively metabolizing cell produces heat. The resulting local rise in temperature enhances O_2 release from Hb for use by more active tissues.

COMPARISON OF THESE FACTORS AT THE TISSUE AND PULMONARY LEVELS

As you just learned, increases in CO_2, acidity, and temperature at the tissue level, all of which are associated with increased cellular metabolism and increased O_2 consumption, enhance the effect of a drop in P_{O_2} in facilitating the release of O_2 from Hb. These effects are largely reversed at the pulmonary level, where the extra acid-forming CO_2 is blown off and the local environment is cooler. Appropriately, therefore, Hb has a higher affinity for O_2 in the pulmonary capillary environment, enhancing the effect of raised P_{O_2} in loading O_2 onto Hb.

EFFECT OF 2,3-BISPHOSPHOGLYCERATE ON % Hb SATURATION

The preceding changes take place in the *environment* of the red blood cells, but a factor *inside* the red blood cells can also affect the degree of O_2–Hb binding: **2,3-bisphosphoglycerate (BPG)**. This erythrocyte constituent, which is produced during red blood cell metabolism, can bind reversibly with Hb and reduce its affinity for O_2, just as CO_2 and H^+ do. Thus an increased level of BPG, like the other factors, shifts the O_2–Hb curve to the right, enhancing O_2 unloading as the blood flows through the tissues.

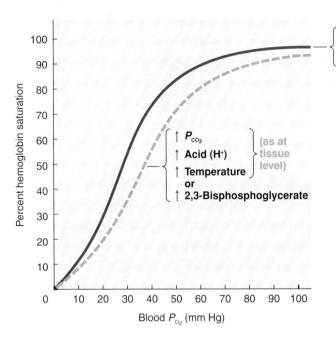

● FIGURE 12-24

Effect of increased P_{CO_2}, H^+, temperature, and 2,3-bisphosphoglycerate on the O_2–Hb curve. Increased P_{CO_2}, acid, temperature, and 2,3-bisphosphoglycerate, as found at the tissue level, shift the O_2–Hb curve to the right. As a result, less O_2 and Hb can be combined at a given P_{O_2}, so that more O_2 is unloaded from Hb for use by the tissues.

PhysioEdge For an interaction related to this figure, see Media Exercise 12.3: Gas Transport and Exchange on the CD-ROM.

BPG production by red blood cells gradually increases whenever Hb in the arterial blood is chronically undersaturated—that is, when arterial HbO_2 is below normal. This condition may occur in people living at high altitudes or in those suffering from certain types of circulatory or respiratory diseases or anemia. By helping unload O_2 from Hb at the tissue level, increased BPG helps maintain O_2 availability for tissue use even though arterial O_2 supply is chronically reduced.

▌ Hemoglobin has a much higher affinity for carbon monoxide than for O_2.

Clinical Note **Carbon monoxide (CO)** and O_2 compete for the same binding sites on Hb, but Hb's affinity for CO is 240 times that of its affinity for O_2. The combination of CO and Hb is known as **carboxyhemoglobin (HbCO)**. Because Hb preferentially latches onto CO, even small amounts of CO can tie up a disproportionately large share of Hb, making Hb unavailable for O_2 transport. Even though the Hb concentration and P_{O_2} are normal, the O_2 content of the blood is seriously reduced.

Fortunately, CO is not a normal constituent of inspired air. It is a poisonous gas produced during the incomplete combustion (burning) of carbon products such as automobile gasoline, coal, wood, and tobacco. Carbon monoxide is especially dangerous because it is so insidious. If CO is being produced in a closed environment so that its concentration con-

tinues to increase (for example, in a parked car with the motor running and windows closed), it can reach lethal levels without the victim ever being aware of the danger. Because it is odorless, colorless, tasteless, and nonirritating, CO is not detectable. Furthermore, for reasons described later the victim has no sensation of breathlessness and makes no attempt to increase ventilation, even though the cells are O_2-starved.

▌ Most CO_2 is transported in the blood as bicarbonate.

When arterial blood flows through the tissue capillaries, CO_2 diffuses down its partial pressure gradient from the tissue cells into the blood. Carbon dioxide is transported in the blood in three ways (● Figure 12-25 and ▲ Table 12-3, p. 387):

1. *Physically dissolved.* As with dissolved O_2, the amount of CO_2 physically dissolved in the blood depends on the P_{CO_2}. Because CO_2 is more soluble than O_2 in plasma water, a greater proportion of the total CO_2 in the blood is physically dissolved compared to O_2. Even so, only 10% of the blood's total CO_2 content is carried this way at the normal systemic venous P_{CO_2} level.
2. *Bound to hemoglobin.* Another 30% of the CO_2 combines with Hb to form **carbamino hemoglobin ($HbCO_2$)**. Carbon dioxide binds with the globin portion of Hb in contrast to O_2, which combines with the heme portions. Reduced Hb has a greater affinity for CO_2 than HbO_2 does. The unloading of O_2 from Hb in the tissue capillaries therefore facilitates the picking up of CO_2 by Hb.
3. *As bicarbonate.* By far the most important means of CO_2 transport is as **bicarbonate (HCO_3^-)**, with 60% of the CO_2 being converted into HCO_3^- by the following chemical reaction, which takes place within the red blood cells:

$$CO_2 + H_2O \overset{\text{carbonic}}{\underset{\text{anhydrase}}{\rightleftarrows}} H_2CO_3 \rightleftarrows H^+ + HCO_3^-$$

In the first step, CO_2 combines with H_2O to form **carbonic acid (H_2CO_3)**. This reaction can occur very slowly in plasma, but it proceeds swiftly within the red blood cells because of the presence of the erythrocyte enzyme **carbonic anhydrase**, which catalyzes (speeds up) the reaction. As is characteristic of acids, some of the carbonic acid molecules spontaneously dissociate into hydrogen ions (H^+) and bicarbonate ions (HCO_3^-). The one carbon and two oxygen atoms of the original CO_2 molecule are thus present in the blood as an integral part of HCO_3^-. This is beneficial because HCO_3^- is more soluble in the blood than CO_2 is.

After the dissociation of H_2CO_3, most of the accumulated H^+ within the erythrocytes becomes bound to Hb. As with CO_2, reduced Hb has a greater affinity for H^+ than HbO_2 does. Therefore, the unloading of O_2 facilitates Hb pickup of CO_2-generated H^+. Because only free, dissolved H^+ contributes to the acidity of a solution, the venous blood would

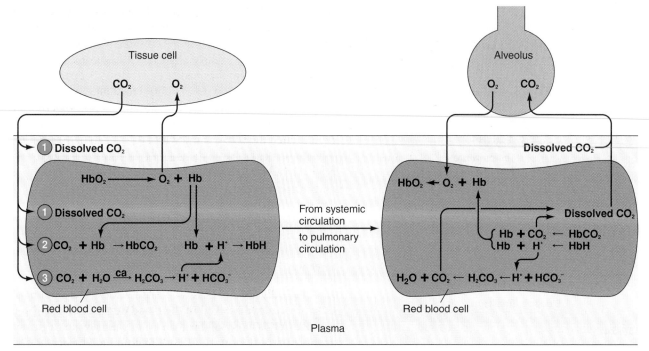

ca = Carbonic anhydrase

● **FIGURE 12-25**

Carbon dioxide transport in the blood. Carbon dioxide (CO_2) picked up at the tissue level is transported in the blood to the lungs in three ways: 1 physically dissolved, 2 bound to hemoglobin (Hb), and 3 as bicarbonate ion (HCO_3^-). Hemoglobin is present only in the red blood cells, as is carbonic anhydrase, the enzyme that catalyzes the production of HCO_3^-. The H^+ generated during the production of HCO_3^- also binds to Hb.

The reactions that occur at the tissue level are reversed at the pulmonary level, where CO_2 diffuses out of the blood to enter the alveoli.

 For an animation of this figure, click the Gas Transport tab (page 2) in The Respiratory System tutorial on the CD-ROM. For an interaction related to this figure, see Media Exercise 12.3: Gas Transport and Exchange.

be considerably more acidic than the arterial blood if Hb did not mop up most of the H^+ generated at the tissue level.

Note how hemoglobin's unloading of O_2 and its uptake of CO_2 and CO_2-generated H^+ at the tissue level work in synchrony. Increased CO_2 and H^+ cause increased O_2 release from Hb, and increased O_2 release from Hb in turn causes increased CO_2 and H^+ uptake by Hb. The entire process is very efficient. Reduced Hb must be carried back to the lungs to refill on O_2 anyway. Meanwhile, after O_2 is released Hb picks up new passengers—CO_2 and H^+—that are going in the same direction to the lungs.

The reactions at the tissue level as CO_2 enters the blood from the tissues are reversed once the blood reaches the lungs and CO_2 leaves the blood to enter the alveoli (● Figure 12-25).

▌ **Various respiratory states are characterized by abnormal blood-gas levels.**

▲ Table 12-4 is a glossary of terms used to describe various states associated with respiratory abnormalities.

ABNORMALITIES IN ARTERIAL P_{O_2}

Clinical Note The term **hypoxia** refers to the condition of having insufficient O_2 at the cell level. There are four general categories of hypoxia:

1. *Hypoxic hypoxia* is characterized by a low arterial-blood P_{O_2} accompanied by inadequate Hb saturation. It is caused by (a) a respiratory malfunction involving inadequate gas exchange, typified by a normal alveolar P_{O_2} but a reduced arterial P_{O_2}, or (b) exposure to high altitude or to a suffocating environment where atmospheric P_{O_2} is reduced so that alveolar and arterial P_{O_2} are likewise reduced.

2. *Anemic hypoxia* is a reduced O_2-carrying capacity of the blood. It can result from (a) a decrease in circulating red blood cells, (b) an inadequate amount of Hb within the red blood cells, or (c) CO poisoning. In all cases of anemic hypoxia, the arterial P_{O_2} is normal, but the O_2 content of the arterial blood is lower than normal because of the reduction in available Hb.

3. *Circulatory hypoxia* arises when too little oxygenated blood is delivered to the tissues. The arterial P_{O_2} and O_2 content are

Miniglossary of Clinically Important Respiratory States

Apnea   Transient cessation of breathing

Asphyxia O_2 starvation of tissues, caused by a lack of O_2 in the air, respiratory impairment, or inability of the tissues to use O_2

Cyanosis   Blueness of the skin resulting from insufficiently oxygenated blood in the arteries

Dyspnea   Difficult or labored breathing

Eupnea   Normal breathing

Hypercapnia   Excess CO_2 in the arterial blood

Hyperpnea   Increased pulmonary ventilation that matches increased metabolic demands, as in exercise

Hyperventilation   Increased pulmonary ventilation in excess of metabolic requirements, resulting in decreased P_{CO_2} and respiratory alkalosis

Hypocapnia   Below-normal CO_2 in the arterial blood

Hypoventilation   Underventilation in relation to metabolic requirements, resulting in increased P_{CO_2} and respiratory acidosis

Hypoxia   Insufficient O_2 at the cellular level

  Anemic hypoxia   Reduced O_2-carrying capacity of the blood

  Circulatory hypoxia   Too little oxygenated blood delivered to the tissues; also known as *stagnant hypoxia*

  Histotoxic hypoxia   Inability of the cells to use O_2 available to them

  Hypoxic hypoxia   Low arterial blood P_{O_2} accompanied by inadequate Hb saturation

Respiratory arrest   Permanent cessation of breathing (unless clinically corrected)

Suffocation O_2 deprivation as a result of an inability to breathe oxygenated air

typically normal, but too little oxygenated blood reaches the cells.

4.   In *histotoxic hypoxia*, O_2 delivery to the tissues is normal, but the cells cannot use the O_2 available to them. The classic example is *cyanide poisoning*. Cyanide blocks cellular enzymes essential for internal respiration.

Hyperoxia, an above-normal arterial P_{O_2}, cannot occur when a person is breathing atmospheric air at sea level. However, breathing supplemental O_2 can increase alveolar and consequently arterial P_{O_2}. Because more of the inspired air is O_2, more of the total pressure of the inspired air is attributable to the O_2 partial pressure, so more O_2 dissolves in the blood before arterial P_{O_2} equilibrates with alveolar P_{O_2}. Even though arterial P_{O_2} increases, the total blood O_2 content does not significantly increase, because Hb is nearly fully saturated at the normal arterial P_{O_2}. In certain pulmonary diseases associated with a reduced arterial P_{O_2}, however, breathing supplemental O_2 can help establish a larger alveoli-to-blood driving gradient, improving arterial P_{O_2}. Far from being advantageous, in contrast, a markedly elevated arterial P_{O_2} can be dangerous. If the arterial P_{O_2} is too high, **oxygen toxicity** can occur. Even though the total O_2 content of the blood is only slightly increased, exposure to a high P_{O_2} can damage some cells. In particular, brain damage and blindness-causing damage to the retina are associated with O_2 toxicity. Therefore, O_2 therapy must be administered cautiously.

ABNORMALITIES IN ARTERIAL P_{CO_2}

Clinical Note   The term **hypercapnia** refers to the condition of having excess CO_2 in arterial blood; it is caused by **hypoventilation** (ventilation inadequate to meet metabolic needs for O_2 delivery and CO_2 removal). With most lung diseases, CO_2 accumulates in arterial blood concurrently with an O_2 deficit.

Hypocapnia, below-normal arterial P_{CO_2} levels, is brought about by hyperventilation. **Hyperventilation** occurs when a person "overbreathes," that is, when the rate of ventilation exceeds the body's metabolic needs for CO_2 removal. As a result, CO_2 is blown off to the atmosphere more rapidly than it is produced in the tissues, and arterial P_{CO_2} falls. Hyperventilation can be triggered by anxiety states, fever, and aspirin poisoning. Alveolar P_{O_2} increases during hyperventilation as more fresh O_2 is delivered to the alveoli from the atmosphere than the blood extracts from the alveoli for tissue consumption, and arterial P_{O_2} increases correspondingly. However, because Hb is almost fully saturated at the normal arterial P_{O_2}, very little additional O_2 is added to the blood. Except for the small extra amount of dissolved O_2, blood O_2 content remains essentially unchanged during hyperventilation.

Increased ventilation is not synonymous with hyperventilation. Increased ventilation that matches an increased metabolic demand, such as the increased need for O_2 delivery and CO_2 elimination during exercise, is termed **hyperpnea**. During exercise, alveolar P_{O_2} and P_{CO_2} remain constant, with the increased atmospheric exchange just keeping pace with the increased O_2 consumption and CO_2 production.

CONSEQUENCES OF ABNORMALITIES IN ARTERIAL BLOOD GASES

Clinical Note   The consequences of reduced O_2 availability to the tissues during hypoxia are apparent. The cells need adequate O_2 supplies to sustain energy-generating metabolic activities. The consequences of abnormal blood CO_2 levels are less obvious. Changes in blood CO_2 concentration primarily affect acid–base balance. Hypercapnia elevates production of carbonic acid. The subsequent generation of excess H^+ produces an acidic condition termed *respiratory acidosis*. Conversely, less-than-normal amounts of H^+ are generated through carbonic acid formation in con-

Effects of Heights and Depths on the Body

Our bodies are optimally equipped for existence at normal atmospheric pressure. Ascent into mountains high above sea level or descent into the depths of the ocean can have adverse effects on the body.

Effects of High Altitude on the Body

The atmospheric pressure progressively declines as altitude increases. At 18,000 ft above sea level, the atmospheric pressure is only 380 mm Hg—half of its normal sea level value. Because the proportion of O_2 and N_2 in the air remains the same, the P_{O_2} of inspired air at this altitude is 21% of 380 mm Hg, or 80 mm Hg, with alveolar P_{O_2} being even lower at 45 mm Hg. At any altitude above 10,000 ft, the arterial P_{O_2} falls into the steep portion of the O_2–Hb curve, below the safety range of the plateau region. As a result, the % Hb saturation in the arterial blood declines precipitously with further increases in altitude.

People who rapidly ascend to altitudes of 10,000 ft or more experience symptoms of **acute mountain sickness** attributable to hypoxic hypoxia and the resultant hypocapnia-induced alkalosis. The increased ventilatory drive to obtain more O_2 causes respiratory

alkalosis, because acid-forming CO_2 is blown off more rapidly than it is produced. Symptoms of mountain sickness include fatigue, nausea, loss of appetite, labored breathing, rapid heart rate (triggered by hypoxia as a compensatory measure to increase circulatory delivery of available O_2 to the tissues), and nerve dysfunc-

F. Westmorland/Photo Researchers, Inc.

tion characterized by poor judgment, dizziness, and incoordination.

Despite these acute responses to high altitude, millions of people live at elevations above 10,000 ft, with some villagers even residing in the Andes at altitudes higher than 16,000 ft. How do they live and function normally? They do so through the process of **acclimatization.** When a person remains at high altitude, the acute compensatory responses of increased ventilation and increased cardiac output are gradually replaced over a period of days by more slowly developing compensatory measures that permit adequate oxygenation of the tissues and restoration of normal acid–base balance. Red blood cell (RBC) production is increased, stimulated by erythropoietin in response to reduced O_2 delivery to the kidneys (see p. 319). The rise in the number of RBCs increases the O_2-carrying capacity of the blood. Hypoxia also promotes the synthesis of BPG within the RBCs, so that O_2 is unloaded from Hb more easily at the tissues (see p. 390). The number of capillaries within the tissues is increased, reducing the distance that O_2 must diffuse from the blood to reach the cells. Furthermore, acclimatized cells can use O_2 more efficiently through an increase in the

junction with hypocapnia. The resultant alkalotic (less acidic than normal) condition is called *respiratory alkalosis* (Chapter 14). (To learn about the effects of mountain climbing and deep sea diving on blood gases, see the accompanying boxed feature, ▶ Beyond the Basics.)

PhysioEdge

Click on the Media Exercises menu of the CD-ROM and work Media Exercise 12.3: Gas Transport and Exchange to test your understanding of the previous section.

CONTROL OF RESPIRATION

Like the heartbeat, breathing must occur in a continuous, cyclic pattern to sustain life processes. Cardiac muscle must rhythmically contract and relax to alternately empty blood from the heart and fill it again. Similarly, inspiratory muscles must rhythmically contract and relax to alternately fill the lungs with air and empty them. Both these activities are accomplished automatically, without conscious effort. However, the underlying mechanisms and control of these two systems are remarkably different.

▌ **Respiratory centers in the brain stem establish a rhythmic breathing pattern.**

Whereas the heart can generate its own rhythm by means of its intrinsic pacemaker activity, the respiratory muscles, being skeletal muscles, require nervous stimulation to induce contraction. The rhythmic pattern of breathing is established by cyclic neural activity to the respiratory muscles. In other words, the pacemaker activity that establishes breathing rhythm resides in the respiratory control centers in the brain, not in the lungs or respiratory muscles themselves. The nerve supply to the heart, not being needed to initiate the heartbeat, only modifies the rate and strength of cardiac contraction. In contrast, the nerve supply to the respiratory system is absolutely essential in maintaining breathing and in reflexly adjusting the level of ventilation to match changing needs for O_2 uptake and CO_2 removal. Furthermore, unlike cardiac activity, which is not subject to voluntary control, respiratory activity can be voluntarily modified to accomplish speaking, singing, whistling, playing a wind instrument, or holding one's breath while swimming.

number of mitochondria, the energy organelles (see p. 26). The kidneys restore the arterial pH to nearly normal by conserving acid that normally would have been lost in the urine.

These compensatory measures are not without undesirable tradeoffs. For example, the greater number of RBCs increases blood viscosity (makes the blood "thicker"), thereby increasing resistance to blood flow. As a result, the heart has to work harder to pump blood through the vessels (see p. 321).

Effects of Deep-Sea Diving on the Body

When a deep-sea diver descends underwater, the body is exposed to greater than atmospheric pressure. Pressure rapidly increases with sea depth as a result of the weight of the water. Pressure is already doubled by about 30 ft below sea level. The air provided by scuba equipment is delivered to the lungs at these high pressures. Recall that (1) the amount of a gas in solution is directly proportional to the partial pressure of the gas and (2) air is composed of 79% N_2. Nitrogen is poorly soluble in body tissues, but the high P_{N_2} that occurs during deep-sea diving causes more of this gas than normal to dissolve in the body tissues. The small amount of N_2 dissolved in the tissues at sea level has no known effect, but as more N_2 dissolves at greater depths, **nitrogen narcosis,** or **"rapture of the deep,"** ensues. Nitrogen narcosis is believed to result from a reduction in the excitability of neurons because of the highly lipid-soluble N_2 dissolving in their lipid membranes. At 150 ft under water, divers experience euphoria and become drowsy, as if they had had a few cocktails. At lower depths, divers become weak and clumsy, and at 350 to 400 ft they lose consciousness. Oxygen toxicity resulting from the high P_{O_2} is another possible harmful effect of descending deep under water.

Another problem associated with deep-sea diving occurs during ascent. If a diver who has been submerged long enough for a significant amount of N_2 to dissolve in the tissues suddenly rises to the surface, the rapid reduction in P_{N_2} causes N_2 to quickly come out of solution and form bubbles of gaseous N_2 in the body. The consequences depend on the amount and location of the bubble formation. This condition is called **decompression sickness** or "**the bends,**" because the victim often bends over in pain. Decompression sickness can be prevented by the diver slowly ascending to the surface or by gradually decompressing in a decompression tank so that the excess N_2 can slowly escape through the lungs without bubble formation.

COMPONENTS OF NEURAL CONTROL OF RESPIRATION

Neural control of respiration involves three distinct components: (1) factors that generate the alternating inspiration/expiration rhythm, (2) factors that regulate the magnitude of ventilation (that is, the rate and depth of breathing) to match body needs, and (3) factors that modify respiratory activity to serve other purposes. The latter modifications may be either voluntary, as in the breath control required for speech, or involuntary, as in the respiratory maneuvers involved in a cough or sneeze.

Respiratory control centers housed in the brain stem generate the rhythmic pattern of breathing. The primary respiratory control center, the *medullary respiratory center,* consists of several aggregations of neuronal cell bodies within the medulla that provide output to the respiratory muscles. In addition, two other respiratory centers lie higher in the brain stem in the pons—the *apneustic center* and *pneumotaxic center.* These pontine centers influence output from the medullary respiratory center (● Figure 12-26). Here is a description of how these various regions interact to establish respiratory rhythmicity.

INSPIRATORY AND EXPIRATORY NEURONS IN THE MEDULLARY CENTER

We rhythmically breathe in and out during quiet breathing because of alternate contraction and relaxation of the inspiratory muscles, namely the diaphragm and external intercostal muscles, supplied by the phrenic nerve and intercostal nerves, respectively. The cell bodies for the neuronal fibers composing these nerves are located in the spinal cord. Impulses originating in the medullary center end on these motor-neuron cell bodies. When these motor neurons are activated, they in turn stimulate the inspiratory muscles, leading to inspiration; when these neurons are not firing, the inspiratory muscles relax, and expiration takes place.

The **medullary respiratory center** consists of two neuronal clusters known as the *dorsal respiratory group* and the *ventral respiratory group* (● Figure 12-26).

• The **dorsal respiratory group (DRG)** consists mostly of *inspiratory neurons* whose descending fibers terminate on the motor neurons that supply the inspiratory muscles. When the DRG inspiratory neurons fire, inspiration takes place; when

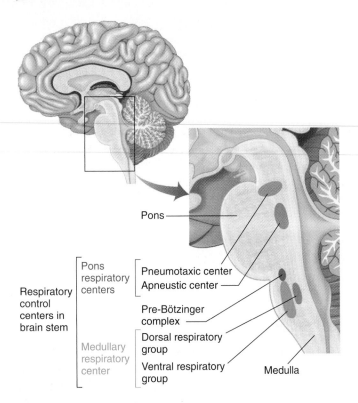

Pons

Pneumotaxic center
Apneustic center

Pre-Bötzinger
complex

Dorsal respiratory
group

Ventral respiratory
group

Medulla

Pons
respiratory
centers

Medullary
respiratory
center

Respiratory
control
centers in
brain stem

● FIGURE 12-26

Respiratory control centers in the brain stem

 For an interaction related to this figure, see Media Exercise 12.4:
Control of Ventilation, Lung Volumes and Terms on the CD-ROM.

they cease firing, expiration occurs. Expiration is brought to an end as the inspiratory neurons once again reach threshold and fire. The DRG has important interconnections with the ventral respiratory group.

• The **ventral respiratory group (VRG)** is composed of *inspiratory neurons* and *expiratory neurons,* both of which remain inactive during normal quiet breathing. This region is called into play by the DRG as an "overdrive" mechanism during periods when demands for ventilation are increased. It is especially important in active expiration. No impulses are generated in the descending pathways from the expiratory neurons during quiet breathing. Only during active expiration do the expiratory neurons stimulate the motor neurons supplying the expiratory muscles (the abdominal and internal intercostal muscles). Furthermore, the VRG inspiratory neurons, when stimulated by the DRG, rev up inspiratory activity when demands for ventilation are high.

GENERATION OF RESPIRATORY RHYTHM

Until recently, the DRG was generally thought to generate the basic rhythm of ventilation. However, generation of respiratory rhythm is now widely believed to lie in the **pre-Bötzinger complex,** a region located near the upper (head) end of the medullary respiratory center (● Figure 12-26). A network of neurons in this region display pacemaker activity, undergoing self-induced action potentials similar to the SA node of the heart. Scientists believe the rate at which the DRG inspi-

ratory neurons rhythmically fire is driven by synaptic input from this complex.

INFLUENCES FROM THE PNEUMOTAXIC AND APNEUSTIC CENTERS

The pontine centers exert "fine-tuning" influences over the medullary center to help produce normal, smooth inspirations and expirations. The **pneumotaxic center** sends impulses to the DRG that help "switch off" the inspiratory neurons, limiting the duration of inspiration. In contrast, the **apneustic center** prevents the inspiratory neurons from being switched off, thus providing an extra boost to the inspiratory drive. In this check-and-balance system, the pneumotaxic center dominates over the apneustic center, helping halt inspiration and letting expiration occur normally. Without the pneumotaxic brakes, the breathing pattern consists of prolonged inspiratory gasps abruptly interrupted by very brief expirations. This abnormal pattern of breathing is known as **apneusis**; hence, the center that promotes this type of breathing is the apneustic center. Apneusis occurs in certain types of severe brain damage.

HERING-BREUER REFLEX

When the tidal volume is large (greater than 1 liter), as during exercise, the **Hering-Breuer reflex** is triggered to prevent overinflation of the lungs. **Pulmonary stretch receptors** within the smooth muscle layer of the airways are activated by stretching of the lungs at large tidal volumes. Action potentials from these stretch receptors travel through afferent nerve fibers to the medullary center and inhibit the inspiratory neurons. This negative feedback from the highly stretched lungs helps cut inspiration short before the lungs become overinflated.

▌ The magnitude of ventilation is adjusted in response to three chemical factors: P_{O_2}, P_{CO_2}, and H^+.

No matter how much O_2 is extracted from the blood or how much CO_2 is added to it at the tissue level, the P_{O_2} and P_{CO_2} of the systemic arterial blood leaving the lungs are held remarkably constant, indicating that arterial blood gas content is precisely regulated. Arterial blood gases are maintained within the normal range by varying the magnitude of ventilation (rate and depth of breathing) to match the body's needs for O_2 uptake and CO_2 removal. If more O_2 is extracted from the alveoli and more CO_2 is dropped off by the blood because the tissues are metabolizing more actively, ventilation increases correspondingly to bring in more fresh O_2 and blow off more CO_2.

The medullary respiratory center receives inputs that provide information about the body's needs for gas exchange. It responds by sending appropriate signals to the motor neurons supplying the respiratory muscles, to adjust the rate and depth of ventilation to meet those needs. The two most obvious signals to increase ventilation are a decreased arterial P_{O_2} or an increased arterial P_{CO_2}. Intuitively, you would suspect that if O_2 levels in the arterial blood declined or if CO_2 accumulated, ventilation would be stimulated to obtain more O_2 or to eliminate excess CO_2. These two factors do indeed in-

▲ TABLE 12-5

Influence of Chemical Factors on Respiration

CHEMICAL FACTOR	EFFECT ON THE PERIPHERAL CHEMORECEPTORS	EFFECT ON THE CENTRAL CHEMORECEPTORS
↓ P_{O_2} in the Arterial Blood	Stimulates only when the arterial P_{O_2} has fallen to the point of being life-threatening (<60 mm Hg); an emergency mechanism	Directly depresses the central chemoreceptors and the respiratory center itself when <60mm Hg
↑ P_{CO_2} in the Arterial Blood (↑ H^+ in the Brain ECF)	Weakly stimulates	Strongly stimulates; is the dominant control of ventilation (Levels >70–80 mm Hg directly depress the respiratory center and central chemoreceptors)
↑ H^+ in the Arterial Blood	Stimulates; important in acid–base balance	Does not affect; cannot penetrate the blood–brain barrier

fluence the magnitude of ventilation, but not to the same degree nor through the same pathway. Also, a third chemical factor, H^+, notably influences the level of respiratory activity. We will examine the role of each of these important chemical factors in the control of ventilation (▲ Table 12-5).

▌ Decreased arterial P_{O_2} increases ventilation only as an emergency mechanism.

Arterial P_{O_2} is monitored by **peripheral chemoreceptors** known as the **carotid bodies** and **aortic bodies**, which lie at the fork of the common carotid arteries on both the right and left sides and in the arch of the aorta, respectively (● Figure 12-27). These chemoreceptors respond to specific changes in the chemical content of the arterial blood that bathes them. They are distinctly different from the carotid sinus and aortic arch baroreceptors located in the same vicinity. The latter monitor pressure changes rather than chemical changes and are important in regulating systemic arterial blood pressure (see p. 305).

EFFECT OF A LARGE DECREASE IN P_{O_2} ON THE PERIPHERAL CHEMORECEPTORS

The peripheral chemoreceptors are not sensitive to modest reductions in arterial P_{O_2}. The arterial P_{O_2} must fall below 60 mm Hg (>40% reduction) before the peripheral chemoreceptors respond by sending afferent impulses to the medullary inspiratory neurons, thereby reflexly increasing ventilation. Because arterial P_{O_2} only falls below 60 mm Hg in the unusual circumstances of severe pulmonary disease or reduced atmospheric P_{O_2}, it does not play a role in the normal ongoing regulation of respiration. This fact may seem surprising at first, because a primary function of ventilation is to provide enough O_2 for uptake by the blood. However, there is no need to increase ventilation until the arterial P_{O_2} falls below 60 mm Hg, because of the safety margin in % Hb saturation afforded by the plateau portion of the O_2–Hb curve. Hemoglobin is still 90%

saturated at an arterial P_{O_2} of 60 mm Hg, but the % Hb saturation drops precipitously when the P_{O_2} falls below this level. Therefore, reflex stimulation of respiration by the peripheral chemoreceptors serves as an important emergency mechanism in dangerously low arterial P_{O_2} states. Indeed, this reflex mechanism is a lifesaver, because a low arterial P_{O_2} tends to directly depress the respiratory center, as it does all the rest of the brain.

Because the peripheral chemoreceptors respond to the P_{O_2} of the blood, *not* the total O_2 content of the blood, O_2 con-

● **FIGURE 12-27**

Location of the peripheral chemoreceptors. The carotid bodies are located in the carotid sinus, and the aortic bodies are located in the aortic arch.

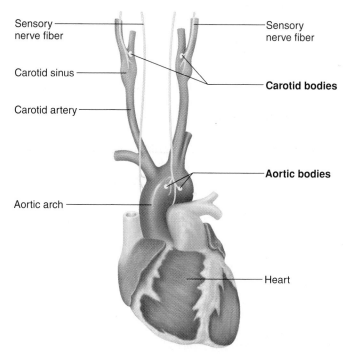

Sensory nerve fiber

Sensory nerve fiber

Carotid sinus

Carotid bodies

Carotid artery

Aortic bodies

Aortic arch

Heart

tent in the arterial blood can fall to dangerously low or even fatal levels without the peripheral chemoreceptors ever responding to reflexly stimulate respiration. Remember that only physically dissolved O_2 contributes to blood P_{O_2}. The total O_2 content in the arterial blood can be reduced in anemic states, in which O_2-carrying Hb is reduced, or in CO poisoning, when Hb is preferentially bound to this molecule rather than to O_2. In both cases, arterial P_{O_2} is normal, so respiration is not stimulated, even though O_2 delivery to the tissues may be so reduced that the person dies from cellular O_2 deprivation.

DIRECT EFFECT OF A LARGE DECREASE IN P_{O_2} ON THE RESPIRATORY CENTER

Except for the peripheral chemoreceptors, the level of activity in all nervous tissue falls in O_2 deprivation. Were it not for stimulatory intervention of the peripheral chemoreceptors when the arterial P_{O_2} falls threateningly low, a vicious cycle ending in cessation of breathing would ensue. Direct depression of the respiratory center by the markedly low arterial P_{O_2} would further reduce ventilation, leading to an even greater fall in arterial P_{O_2}, which would even further depress the respiratory center until ventilation ceased and death occurred.

▌ **Carbon dioxide-generated H$^+$ in the brain is normally the main regulator of ventilation.**

In contrast to arterial P_{O_2}, which does not contribute to the minute-to-minute regulation of respiration, arterial P_{CO_2} is the most important input regulating the magnitude of ventilation under resting conditions. This role is appropriate, because changes in alveolar ventilation have an immediate and pronounced effect on arterial P_{CO_2}. By contrast, changes in ventilation have little effect on % Hb saturation and O_2 availability to the tissues until the arterial P_{O_2} falls by more than 40%. Even slight alterations from normal in arterial P_{CO_2} induce a significant reflex effect on ventilation. An increase in arterial P_{CO_2} reflexly stimulates the respiratory center, with the resultant increase in ventilation promoting elimination of the excess CO_2 to the atmosphere. Conversely, a fall in arterial P_{CO_2} reflexly reduces the respiratory drive. The subsequent decrease in ventilation lets metabolically produced CO_2 accumulate so that P_{CO_2} can be returned to normal.

EFFECT OF INCREASED P_{CO_2} ON THE CENTRAL CHEMORECEPTORS

Surprisingly, given the key role of arterial P_{CO_2} in regulating respiration, no important receptors monitor arterial P_{CO_2} per se. The carotid and aortic bodies are only weakly responsive to changes in arterial P_{CO_2}, so they play only a minor role in reflexly stimulating ventilation in response to an elevation in arterial P_{CO_2}. More important in linking changes in arterial P_{CO_2} to compensatory adjustments in ventilation are the **central chemoreceptors**, located in the medulla near the respiratory center. These central chemoreceptors do not monitor CO_2

itself; however, they are sensitive to changes in CO_2-induced H$^+$ concentration in the brain extracellular fluid (ECF) that bathes them.

Movement of materials across the brain capillaries is restricted by the blood–brain barrier (see p. 113). Because this barrier is readily permeable to CO_2, any increase in arterial P_{CO_2} causes a similar rise in brain ECF P_{CO_2} as CO_2 diffuses down its pressure gradient from the cerebral blood vessels into the brain ECF. The increased P_{CO_2} within the brain ECF correspondingly raises the concentration of H$^+$ according to the law of mass action as it applies to this reaction: $CO_2 + H_2O \rightleftarrows H_2CO_3 \rightleftarrows H^+ + HCO_3^-$. An elevation in H$^+$ concentration in the brain ECF directly stimulates the central chemoreceptors, which in turn increase ventilation by stimulating the respiratory center through synaptic connections (● Figure 12-28). As the excess CO_2 is subsequently blown off, the arterial P_{CO_2} and the P_{CO_2} and H$^+$ concentration of the brain ECF return to normal. Conversely, a decline in arterial P_{CO_2} below normal is paralleled by a fall in P_{CO_2} and H$^+$ in the brain ECF, the result of which is a central chemoreceptor-mediated decrease in ventilation. As CO_2 produced by cell metabolism is consequently allowed to accumulate, arterial

---

● **FIGURE 12-28**

Effect of increased arterial P_{CO_2} on ventilation

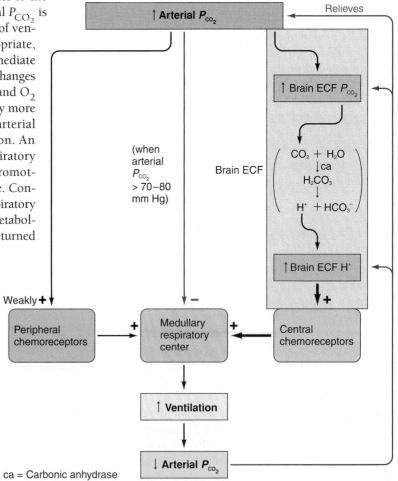

ca = Carbonic anhydrase

P_{CO_2} and P_{CO_2} and H^+ of the brain ECF are restored toward normal.

Unlike CO_2, H^+ cannot readily permeate the blood–brain barrier, so H^+ in the plasma cannot gain access to the central chemoreceptors. Accordingly, the central chemoreceptors respond only to H^+ generated within the brain ECF itself as a result of CO_2 entry. Thus the major mechanism controlling ventilation under resting conditions is specifically aimed at regulating the brain ECF H^+ concentration, which in turn directly reflects the arterial P_{CO_2}. Unless there are extenuating circumstances such as reduced availability of O_2 in the inspired air, arterial P_{O_2} is coincidentally also maintained at its normal value by the brain ECF H^+ ventilatory driving mechanism.

The powerful influence of the central chemoreceptors on the respiratory center is responsible for your inability to deliberately hold your breath for more than about a minute. While you hold your breath, metabolically produced CO_2 continues to accumulate in your blood and then to build up the H^+ concentration in your brain ECF. Finally, the increased P_{CO_2}–H^+ stimulant to respiration becomes so powerful that central-chemoreceptor excitatory input overrides voluntary inhibitory input to respiration, so breathing resumes despite deliberate attempts to prevent it. Breathing resumes long before arterial P_{O_2} falls to the threateningly low levels that trigger the peripheral chemoreceptors. Therefore, you cannot deliberately hold your breath long enough to create a dangerously high level of CO_2 or low level of O_2 in the arterial blood.

DIRECT EFFECT OF A LARGE INCREASE IN P_{CO_2} ON THE RESPIRATORY CENTER

In contrast to the normal reflex stimulatory effect of the increased P_{CO_2}–H^+ mechanism on respiratory activity, very high levels of CO_2 directly depress the entire brain, including the respiratory center, just as very low levels of O_2 do. Up to a P_{CO_2} of 70 to 80 mm Hg, progressively higher P_{CO_2} levels induce correspondingly more vigorous respiratory efforts in an attempt to blow off the excess CO_2. A further increase in P_{CO_2} beyond 70 to 80 mm Hg, however, does not further increase ventilation but actually depresses the respiratory neurons. For this reason, CO_2 must be removed and O_2 supplied in closed environments such as closed-system anesthesia machines, submarines, or space capsules. Otherwise, CO_2 could reach lethal levels, not only because it depresses respiration but also because it produces severe respiratory acidosis.

▌ Adjustments in ventilation in response to changes in arterial H^+ are important in acid–base balance.

Changes in arterial H^+ concentration cannot influence the central chemoreceptors, because H^+ does not readily cross the blood–brain barrier. However, the aortic and carotid body peripheral chemoreceptors are highly responsive to fluctuations in arterial H^+ concentration, in contrast to their weak sensitivity to deviations in arterial P_{CO_2} and their unresponsiveness to arterial P_{O_2} until it falls 40% below normal. In many situations, even though P_{CO_2} is normal, arterial H^+ concentration is changed by the addition or loss of noncarbonic acid from the body. For example, arterial H^+ concentration increases during diabetes mellitus because excess H^+-generating keto acids are abnormally produced and added to the blood. A rise in arterial H^+ concentration reflexly stimulates ventilation by means of the peripheral chemoreceptors. Conversely, the peripheral chemoreceptors reflexly suppress respiratory activity in response to a fall in arterial H^+ concentration resulting from nonrespiratory causes. Changes in ventilation by this mechanism are extremely important in regulating the body's acid–base balance. Changing the magnitude of ventilation can vary the amount of H^+-generating CO_2 eliminated. The resulting adjustment in the amount of H^+ added to the blood from CO_2 can compensate for the nonrespiratory-induced abnormality in arterial H^+ concentration that first elicited the respiratory response. (See Chapter 14 for further details.)

▌ During apnea, a person "forgets to breathe"; during dyspnea, a person feels "short of breath."

Apnea is the transient interruption of ventilation, with breathing resuming spontaneously. If breathing does not resume, the condition is called **respiratory arrest**. Because ventilation is normally decreased and the central chemoreceptors are less sensitive to the arterial P_{CO_2} drive during sleep, especially paradoxical sleep (see p. 134), apnea is most likely to occur during this time. Victims of **sleep apnea** may stop breathing for a few seconds or up to one or two minutes as many as 500 times a night. Mild sleep apnea is not dangerous unless the sufferer has pulmonary or circulatory disease, which can be exacerbated by recurrent bouts of apnea.

SUDDEN INFANT DEATH SYNDROME

In exaggerated cases of sleep apnea, the victim may be unable to recover from an apneic period, and death results. This is the case in **sudden infant death syndrome (SIDS)**, or "crib death." With this tragic form of sleep apnea, a previously healthy 2- to 4-month-old infant is found dead in his or her crib for no apparent reason. The underlying cause of SIDS is the subject of intense investigation. Most evidence suggests that the baby "forgets to breathe" because the respiratory control mechanisms are immature, either in the brain stem or in the chemoreceptors that monitor the body's respiratory status. For example, on autopsy more than half the victims have poorly developed carotid bodies, the more important of the peripheral chemoreceptors. Alternatively, some researchers believe the condition may be triggered by an initial cardiovascular failure rather than by an initial cessation of breathing. Still other investigators propose that some cases may be due to aspiration of gastric (stomach) juice containing the bacterium *Helicobacter pylori*. In one study, this micro-organism was present in 88% of infants who died of SIDS. Scientists speculate that *H. pylori* may lead to the production of ammonia, which can be lethal if it gains access to the blood from the lungs.

Whatever the underlying cause, certain risk factors make babies more vulnerable to SIDS. Among them are sleeping position (an almost 40% higher incidence of SIDS is associated with sleeping on the abdomen rather than on the back or side) and exposure to nicotine during fetal life or after birth. Infants whose mothers smoked during pregnancy or who breathe cigarette smoke in the home are three times more likely to die of SIDS than those not exposed to smoke.

DYSPNEA

Clinical Note People who have dyspnea have the subjective sensation that they are not getting enough air; that is, they feel "short of breath." **Dyspnea** is the mental anguish associated with the unsatiated desire for more adequate ventilation. It often accompanies the labored breathing characteristic of obstructive lung disease or the pulmonary edema associated with congestive heart failure. In contrast, during exercise a person can breathe very hard without experiencing dyspnea, because such exertion is not accompanied by a sense of anxiety over the adequacy of ventilation. Surprisingly, dyspnea is not directly related to chronic elevation of arterial P_{CO_2} or reduction of P_{O_2}. The subjective feeling of air hunger may occur even when alveolar ventilation and the blood gases are normal. Some people experience dyspnea when they *perceive* that they are short of air even though this is not actually the case, such as in a crowded elevator.

PhysioEdge Click on the Media Exercises menu of the CD-ROM and work Media Exercise 12.4: Control of Ventilation, Lung Volumes and Terms to test your understanding of the previous section.

CHAPTER IN PERSPECTIVE: FOCUS ON HOMEOSTASIS

The respiratory system contributes to homeostasis by obtaining O_2 from and eliminating CO_2 to the external environment. All body cells ultimately need an adequate supply of O_2 to use in oxidizing nutrient molecules to generate ATP. Brain cells, which are especially dependent on a continual supply of O_2, die if deprived of O_2 for more than four minutes. Even cells that can resort to anaerobic ("without O_2") metabolism for energy production, such as strenuously exercising muscles, can do so only transiently by incurring an O_2 deficit that ultimately must be made up during the period of excess postexercise O_2 consumption (see p. 220).

As a result of these energy-yielding metabolic reactions, the body produces large quantities of CO_2 that must be eliminated. Because CO_2 and H_2O form carbonic acid, adjustments in the rate of CO_2 elimination by the respiratory system are important in regulating acid–base balance in the internal environment. Cells can survive only within a narrow pH range.

CHAPTER SUMMARY

Introduction (pp. 365–369)

▮ Internal respiration encompasses the intracellular metabolic reactions that use O_2 and produce CO_2 during energy-yielding oxidation of nutrient molecules.

▮ External respiration encompasses the various steps in the transfer of O_2 and CO_2 between the external environment and tissue cells. The respiratory and circulatory systems function together to accomplish external respiration. *(Review Figure 12-1.)*

▮ The respiratory system exchanges air between the atmosphere and lungs through the process of ventilation.

▮ Respiratory airways conduct air from the atmosphere to the alveoli, the gas-exchanging portion of the lungs. *(Review Figure 12-2.)*

▮ Exchange of O_2 and CO_2 between the air in the lungs and the blood in the pulmonary capillaries takes place across the extremely thin walls of the air sacs, or alveoli. The alveolar walls are formed by Type I alveolar cells. Type II alveolar cells secrete pulmonary surfactant. *(Review Figure 12-3.)*

▮ The lungs are housed within the closed compartment of the thorax, the volume of which can be changed by contractile activity of surrounding respiratory muscles.

▮ Each lung is surrounded by a double-walled, closed sac, the pleural sac. *(Review Figure 12-4.)*

Respiratory Mechanics (pp. 369–383)

▮ Ventilation, or breathing, is the process of cyclically moving air in and out of the lungs, so that old alveolar air that has already participated in exchanging O_2 and CO_2 with the pulmonary capillary blood can be exchanged for fresh atmospheric air.

▮ Ventilation is mechanically accomplished by alternately shifting the direction of the pressure gradient for airflow between the atmosphere and alveoli through the cyclical expansion and recoil of the lungs. When intra-alveolar pressure decreases as a result of lung expansion during inspiration, air flows into the lungs from the higher atmospheric pressure. When intra-alveolar pressure increases as a result of lung recoil during expiration, air flows out of the lungs toward the lower atmospheric pressure. *(Review Figures 12-5, 12-6, 12-9, 12-12, and 12-13.)*

▮ Alternate contraction and relaxation of the inspiratory muscles (primarily the diaphragm) indirectly produce periodic inflation and deflation of the lungs by cyclically expanding and compressing the thoracic cavity, with the lungs passively following its movements. *(Review Figures 12-10 and 12-11.)*

▮ The lungs follow the movements of the thoracic cavity by virtue of the intrapleural fluid's cohesiveness and the transmural pressure gradient across the lung wall. The transmural pressure gradient exists because the intrapleural pressure is subatmospheric and thus less than the intra-alveolar pressure. *(Review Figures 12-7 and 12-13.)*

▮ Because energy is required for contracting the inspiratory muscles, inspiration is an active process, but expiration is passive during quiet breathing because it is accomplished by elastic recoil of the lungs on relaxing inspiratory muscles, at no energy expense.

▮ For more forceful active expiration, contraction of the expiratory muscles (namely, the abdominal muscles) further decreases the size of the thoracic cavity and lungs, which further increases the intra-alveolar-to-atmospheric pressure gradient. *(Review Figure 12-11.)*

▮ The larger the gradient between the alveoli and atmosphere in either direction, the larger the airflow rate, because air continues to flow until the intra-alveolar pressure equilibrates with atmospheric pressure.

- Besides being directly proportional to the pressure gradient, airflow rate is also inversely proportional to airway resistance. Because airway resistance, which depends on the caliber of the conducting airways, is normally very low, airflow rate usually depends primarily on the pressure gradient between the alveoli and atmosphere.
- If airway resistance is pathologically increased by chronic obstructive pulmonary disease, the pressure gradient must be correspondingly increased by more vigorous respiratory muscle activity to maintain a normal airflow rate.
- The lungs can be stretched to varying degrees during inspiration and then recoil to their preinspiratory size during expiration because of their elastic behavior.
    1. The term *pulmonary compliance* refers to the distensibility of the lungs—how much they stretch in response to a given change in the transmural pressure gradient, the stretching force exerted across the lung wall.
    2. The term *elastic recoil* refers to the phenomenon of the lungs snapping back to their resting position during expiration.
- Pulmonary elastic behavior depends on the elastic connective tissue meshwork within the lungs and on alveolar surface tension—pulmonary surfactant interaction. Alveolar surface tension, which is due to the attractive forces between the surface water molecules in the liquid film lining each alveolus, tends to resist the alveolus being stretched on inflation (decreases compliance) and tends to return it back to a smaller surface area during deflation (increases lung rebound). (*Review Table 12-1.*)
- If the alveoli were lined by water alone, the surface tension would be so great that the lungs would be poorly compliant and would tend to collapse. Pulmonary surfactant intersperses between the water molecules and lowers the alveolar surface tension, thereby increasing the compliance of the lungs and counteracting the tendency for alveoli to collapse.
- The lungs can be filled to over 5.5 liters on maximal inspiratory effort or emptied to about one liter on maximal expiratory effort. Normally, however, the lungs operate at "half full." The lung volume typically varies from about 2 to 2.5 liters as an average tidal volume of 500 ml of air is moved in and out with each breath. (*Review Figure 12-14.*)
- The amount of air moved in and out of the lungs in one minute, the pulmonary ventilation, is equal to tidal volume times respiratory rate.
- Not all the air moved in and out is available for O_2 and CO_2 exchange with the blood, because part occupies the conducting airways, known as the *anatomic dead space*. Alveolar ventilation, the volume of air exchanged between the atmosphere and alveoli in one minute, is a measure of the air actually available for gas exchange with the blood. Alveolar ventilation equals (tidal volume minus the dead space volume) times respiratory rate. (*Review Figure 12-18 and Table 12-2.*)

Gas Exchange (pp. 383–387)

- Oxygen and CO_2 move across body membranes by passive diffusion down partial pressure gradients.
- The partial pressure of a gas in air is that portion of the total atmospheric pressure contributed by this individual gas, which in turn is directly proportional to the percentage of this gas in the air. The partial pressure of a gas in blood depends on the amount of this particular gas dissolved in the blood. (*Review Figure 12-19.*)
- Net diffusion of O_2 occurs first between the alveoli and blood and then between the blood and tissues as a result of the O_2 partial pressure gradients created by continuous use of O_2 in the cells and continuous replenishment of fresh alveolar O_2 provided by ventilation. (*Review Figure 12-20.*)

- Net diffusion of CO_2 occurs in the reverse direction, first between the tissues and blood and then between the blood and alveoli, as a result of the CO_2 partial pressure gradients created by continuous production of CO_2 in the cells and continuous removal of alveolar CO_2 through ventilation. (*Review Figure 12-20.*)
- Factors other than the partial pressure gradient that influence the rate of gas exchange are the surface area and thickness of the membrane across which the gas is diffusing, according to Fick's law of diffusion.

Gas Transport (pp. 387–394)

- Because O_2 and CO_2 are not very soluble in blood, they must be transported primarily by mechanisms other than simply being physically dissolved. (*Review Table 12-3.*)
- Only 1.5% of the O_2 is physically dissolved in the blood, with 98.5% chemically bound to hemoglobin (Hb). (*Review Figure 12-23.*)
- The primary factor that determines the extent to which Hb and O_2 are combined (the % Hb saturation) is the P_{O_2} of the blood, depicted by an S-shaped curve known as the O_2–Hb dissociation curve. (*Review Figure 12-22.*)
    1. The relationship between blood P_{O_2} and % Hb saturation is such that in the P_{O_2} range of the pulmonary capillaries (the plateau portion of the curve), Hb is still almost fully saturated even if the blood P_{O_2} falls as much as 40%. This provides a margin of safety by ensuring near-normal O_2 delivery to the tissues despite a substantial reduction in arterial P_{O_2}.
    2. In the P_{O_2} range in the systemic capillaries (the steep portion of the curve), Hb unloading increases greatly in response to a small local decline in blood P_{O_2} associated with increased cellular metabolism. In this way, more O_2 is provided to match the increased tissue needs.
- Carbon dioxide picked up at the systemic capillaries is transported in the blood by three methods: (1) 10% is physically dissolved, (2) 30% is bound to Hb, and (3) 60% takes the form of bicarbonate (HCO_3^-). (*Review Table 12-3.*)
- The erythrocyte enzyme carbonic anhydrase catalyzes conversion of CO_2 to HCO_3^- according to the reaction $CO_2 + H_2O \rightleftarrows H_2CO_3 \rightleftarrows H^+ + HCO_3^-$. The carbon and oxygen originally present in CO_2 are now part of the bicarbonate ion. The generated H^+ binds to Hb. These reactions are all reversed in the lungs as CO_2 is eliminated to the alveoli. (*Review Figure 12-25.*)

Control of Respiration (pp. 394–400)

- Ventilation involves two distinct aspects, both subject to neural control: (1) rhythmic cycling between inspiration and expiration and (2) regulation of ventilation magnitude, which in turn depends on control of respiratory rate and depth of tidal volume.
- Respiratory rhythm is established by a complex neuronal network, the pre-Bötzinger complex, that displays pacemaker activity and drives the inspiratory neurons located in the dorsal respiratory group (DRG) of the respiratory control center in the medulla of the brain stem. When these inspiratory neurons fire, impulses ultimately reach the inspiratory muscles to bring about inspiration. (*Review Figure 12-26.*)
- When the inspiratory neurons stop firing, the inspiratory muscles relax and expiration takes place. If active expiration is to occur, the expiratory muscles are activated by output at this time from the medullary expiratory neurons in the ventral respiratory group (VRG) of the medullary respiratory control center.
- This basic rhythm is smoothed out by a balance of activity in the apneustic and pneumotaxic centers located higher in the brain stem, in the pons. The apneustic center prolongs inspiration, whereas the more powerful pneumotaxic center limits inspiration.

- Three chemical factors play a role in determining the magnitude of ventilation: P_{CO_2}, P_{O_2}, and H^+ concentration of the arterial blood. *(Review Table 12-5.)*
- The dominant factor in the minute-to-minute regulation of ventilation is the arterial P_{CO_2}. An increase in arterial P_{CO_2} is the most potent chemical stimulus for increasing ventilation. Changes in arterial P_{CO_2} alter ventilation primarily by bringing about corresponding changes in the brain ECF H^+ concentration, to which the central chemoreceptors are exquisitely sensitive. *(Review Figure 12-28.)*
- The peripheral chemoreceptors are responsive to an increase in arterial H^+ concentration, which likewise reflexly brings about increased ventilation. The resulting adjustment in arterial H^+-generating CO_2 is important in maintaining the acid–base balance of the body. *(Review Figure 12-27.)*
- The peripheral chemoreceptors also reflexly stimulate the respiratory center in response to a marked reduction in arterial P_{O_2} (<60 mm Hg). This response serves as an emergency mechanism to increase respiration when the arterial P_{O_2} levels fall below the safety range provided by the plateau portion of the O_2–Hb curve.

REVIEW EXERCISES

Objective Questions (Answers on p. A-45)

1. Breathing is accomplished by alternate contraction and relaxation of muscles within the lung tissue. *(True or false?)*
2. Normally the alveoli empty completely during maximal expiratory efforts. *(True or false?)*
3. Hemoglobin has a higher affinity for O_2 than for any other substance. *(True or false?)*
4. Rhythmicity of breathing is brought about by pacemaker activity displayed by the respiratory muscles. *(True or false?)*
5. The expiratory neurons send impulses to the motor neurons controlling the expiratory muscles during normal quiet breathing. *(True or false?)*
6. The two forces that tend to keep the alveoli open are _____ and _____.
7. The two forces that promote alveolar collapse are _____ and _____.
8. _____ is a measure of the magnitude of change in lung volume accomplished by a given change in the transmural pressure gradient.
9. _____ is the phenomenon of the lungs snapping back to their resting size after having been stretched.
10. _____ is the erythrocytic enzyme that catalyzes the conversion of CO_2 into HCO_3^-.
11. Which of the following reactions take(s) place at the pulmonary capillaries?
    a. $Hb + O_2 \rightarrow HbO_2$
    b. $CO_2 + H_2O \rightarrow H_2CO_3 \rightarrow H^+ + HCO_3^-$
    c. $Hb + CO_2 \rightarrow HbCO_2$
    d. $HbH \rightarrow Hb + H^+$
12. Using the following answer code, indicate which chemoreceptors are being described:
    (a) peripheral chemoreceptors
    (b) central chemoreceptors
    (c) both peripheral and central chemoreceptors
    (d) neither peripheral nor central chemoreceptors
    _____ 1. stimulated by an arterial P_{O_2} of 80 mm Hg
    _____ 2. stimulated by an arterial P_{O_2} of 55 mm Hg
    _____ 3. directly depressed by an arterial P_{O_2} of 55 mm Hg
    _____ 4. weakly stimulated by an elevated arterial P_{CO_2}
    _____ 5. strongly stimulated by an elevated brain ECF H^+ concentration induced by an elevated arterial P_{CO_2}
    _____ 6. stimulated by an elevated arterial H^+ concentration
13. Indicate the O_2 and CO_2 partial-pressure relationships that are important in gas exchange by circling (greater than) >, (less than) <, or = (equal to) as appropriate in each of the following statements:;
    a. P_{O_2} in blood entering the pulmonary capillaries is (>, <, or =) P_{O_2} in the alveoli.
    b. P_{CO_2} in blood entering the pulmonary capillaries is (>, <, or =) P_{CO_2} in the alveoli.
    c. P_{O_2} in the alveoli is (>, <, or =) P_{O_2} in blood leaving the pulmonary capillaries.
    d. P_{CO_2} in the alveoli is (>, <, or =) P_{CO_2} in blood leaving the pulmonary capillaries.
    e. P_{O_2} in blood leaving the pulmonary capillaries is (>, <, or =) P_{O_2} in blood entering the systemic capillaries.
    f. P_{CO_2} in blood leaving the pulmonary capillaries is (>, <, or =) P_{CO_2} in blood entering the systemic capillaries.
    g. P_{O_2} in blood entering the systemic capillaries is (>, <, or =) P_{O_2} in the tissue cells.
    h. P_{CO_2} in blood entering the systemic capillaries is (>, <, or =) P_{CO_2} in the tissue cells.
    i. P_{O_2} in the tissue cells is (>, <, or =) or approximately =) P_{O_2} in blood leaving the systemic capillaries.
    j. P_{CO_2} in the tissue cells is (>, <, or approximately =) P_{CO_2} in blood leaving the systemic capillaries.
    k. P_{O_2} in blood leaving the systemic capillaries is (>, <, or =) P_{O_2} in blood entering the pulmonary capillaries.
    l. P_{CO_2} in blood leaving the systemic capillaries is (>, <, or =) P_{CO_2} in blood entering the pulmonary capillaries.

Essay Questions

1. Distinguish between internal and external respiration. List the steps in external respiration.
2. Describe the components of the respiratory system. What is the site of gas exchange?
3. Compare atmospheric, intra-alveolar, and intrapleural pressures.
4. Why are the lungs normally stretched even during expiration?
5. Explain why air enters the lungs during inspiration and leaves during expiration.
6. Why is inspiration normally active and expiration normally passive?
7. Why does airway resistance become an important determinant of airflow rates in chronic obstructive pulmonary disease?
8. Explain pulmonary elasticity in terms of compliance and elastic recoil.
9. What are the source and function of pulmonary surfactant?
10. Define the various lung volumes and capacities.
11. Compare pulmonary ventilation and alveolar ventilation. What is the consequence of anatomic and alveolar dead space?
12. What determines the partial pressures of a gas in air and in blood?
13. List the methods of O_2 and CO_2 transport in the blood.
14. What is the primary factor that determines the percent hemoglobin saturation? What is the significance of the plateau and the steep portions of the O_2–Hb dissociation curve?
15. How does hemoglobin promote the net transfer of O_2 from the alveoli to the blood?

16. Define the following: *hypoxic hypoxia, anemic hypoxia, circulatory hypoxia, histotoxic hypoxia, hypercapnia, hypocapnia, hyperventilation, hypoventilation, hyperpnea, apnea,* and *dyspnea.*

17. What are the locations and functions of the three respiratory control centers? Distinguish between the DRG and the VRG.
18. What brain region establishes the rhythmicity of breathing?

POINTS TO PONDER

(Explanations on p. A-46)

1. Why is it important that airplane interiors are pressurized (that is, the pressure is maintained at sea-level atmospheric pressure even though the atmospheric pressure surrounding the plane is substantially lower)? Explain the physiological value of using O_2 masks if the pressure in the airplane interior cannot be maintained.

2. Would hypercapnia accompany the hypoxia produced in each of the following situations? Explain why or why not.
   a. cyanide poisoning
   b. pulmonary edema
   c. restrictive lung disease
   d. high altitude
   e. severe anemia
   f. congestive heart failure
   g. obstructive lung disease

3. At body temperature, the partial pressure of water vapor is 47 mm Hg. If a person lives one mile above sea level at Denver, Colorado, where the atmospheric pressure is 630 mm Hg, what would the P_{O_2} of the inspired air be once it is humidified in the respiratory airways before it reaches the alveoli?

4. Based on what you know about the control of respiration, explain why it is dangerous to voluntarily hyperventilate to lower the arterial P_{CO_2} before going underwater. The purpose of the hyperventilation is to stay under longer before P_{CO_2} rises above normal and drives the swimmer to surface for a breath of air.

5. If a person whose alveolar membranes are thickened by disease has an alveolar P_{O_2} of 100 mm Hg and an alveolar P_{CO_2} of 40 mm Hg, which of the following values of systemic arterial blood gases are most likely to exist?
   a. P_{O_2} = 105 mm Hg, P_{CO_2} = 35 mm Hg
   b. P_{O_2} = 100 mm Hg, P_{CO_2} = 40 mm Hg
   c. P_{O_2} = 90 mm Hg, P_{CO_2} = 45 mm Hg

If the person is administered 100% O_2, will the arterial P_{O_2} increase, decrease, or remain the same? Will the arterial P_{CO_2} increase, decrease, or remain the same?

CLINICAL CONSIDERATION

(Explanation on p. A-46)

Keith M., a former heavy cigarette smoker, has severe emphysema. How does this condition affect his airway resistance? How does this change in airway resistance influence Keith's inspiratory and expiratory efforts? Describe how his respiratory muscle activity and intraalveolar pressure changes compare to normal to accomplish a normal tidal volume. How would his spirogram compare to normal? What influence would Keith's condition have on gas exchange in his lungs? What blood gas abnormalities are likely to be present? Some patients like Keith who have severe chronic lung disease lose their sensitivity to an elevated arterial P_{CO_2}. In a prolonged increase in H^+ generation in the brain ECF, from long-standing CO_2 retention, enough HCO_3^- may cross the blood–brain barrier to buffer, or "neutralize," the excess H^+. The additional HCO_3^- combines with the excess H^+, removing it from solution so that it no longer contributes to free H^+ concentration. When brain ECF HCO_3^- concentration rises, brain ECF H^+ concentration returns to normal although arterial P_{CO_2} and brain ECF P_{CO_2} remain high. The central chemoreceptors are no longer aware of the elevated P_{CO_2}, because the brain ECF H^+ is normal. Because the central chemoreceptors no longer reflexly stimulate the respiratory center in response to the elevated P_{CO_2}, the drive to eliminate CO_2 is blunted in such patients; that is, their level of ventilation is abnormally low considering their high arterial P_{CO_2}. In these patients, the hypoxic drive to ventilation becomes their primary respiratory stimulus, in contrast to normal individuals, in whom the arterial P_{CO_2} level is the dominant factor governing the magnitude of ventilation. Given this situation, would it be appropriate to administer O_2 to Keith to relieve his hypoxic condition?

PHYSIOEDGE RESOURCES

PhysioEdge CD-ROM

PhysioEdge, the CD-ROM packaged with your text, focuses on the concepts students find most difficult to learn. Figures marked with this icon have associated activities on the CD. For a visual review of concepts in this chapter, check out the following:

Tutorial: The Respiratory System

Media Exercise 12.1: Anatomy of the Respiratory System

Media Exercise 12.2: Mechanics of Ventilation

Media Exercise 12.3: Gas Transport and Exchange

Media Exercise 12.4: Control of Ventilation, Lung Volumes and Terms

PhysioEdge Website

The website for this book contains a wealth of helpful study aids, as well as many ideas for further reading and research. Log on to: **http://www.brookscole.com/hpfundamentals3**

Select Chapter 12 from the drop-down menu or click on one of the many resource areas, including Case Histories, which introduce clinical aspects of human physiology. For this chapter check out Case History 1: Fighting for Every Breath, Case History 2: Asthma and Influenza: A Dangerous Combination, Case History 3: When Help Becomes Harm.

For Suggested Readings, consult **InfoTrac College Edition/ Research** on the PhysioEdge website or go directly to InfoTrac College Edition, your online research library, at: **http://infotrac.thomsonlearning.com**

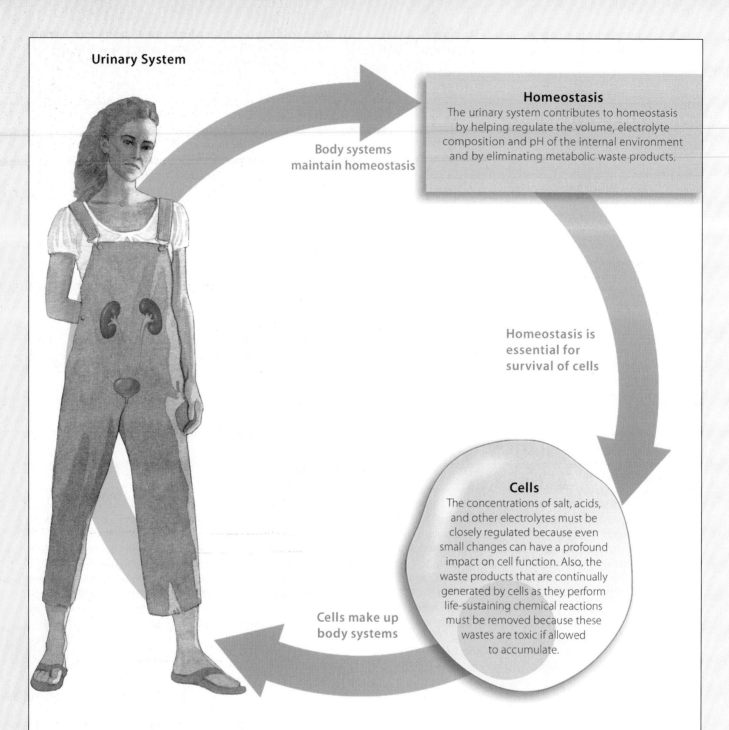

Urinary System

Homeostasis

The urinary system contributes to homeostasis by helping regulate the volume, electrolyte composition and pH of the internal environment and by eliminating metabolic waste products.

Body systems maintain homeostasis

Homeostasis is essential for survival of cells

Cells

The concentrations of salt, acids, and other electrolytes must be closely regulated because even small changes can have a profound impact on cell function. Also, the waste products that are continually generated by cells as they perform life-sustaining chemical reactions must be removed because these wastes are toxic if allowed to accumulate.

Cells make up body systems

The survival and proper functioning of cells depend on the maintenance of stable concentrations of salt, acids, and other electrolytes in the internal fluid environment. Cell survival also depends on continual removal of toxic metabolic wastes that cells produce as they perform life-sustaining chemical reactions. The kidneys play a major role in maintaining homeostasis by regulating the concentration of many plasma constituents, especially electrolytes and water, and by eliminating all metabolic wastes (except CO_2, which is removed by the lungs). As plasma repeatedly filters through the kidneys, they retain constituents of value for the body and eliminate undesirable or excess materials in urine. Of special importance is the kidneys' ability to regulate the volume and osmolarity (solute concentration) of the internal fluid environment by controlling salt and water balance. Also crucial is their ability to help regulate pH by controlling elimination of acid and base in urine.

The Urinary System

 Click on the Tutorials menu of the CD-ROM for a tutorial on Renal Physiology.

INTRODUCTION

Exchanges between the cells and the ECF could notably alter the composition of the small, private internal fluid environment if mechanisms did not keep it stable.

▮ The kidneys perform a variety of functions aimed at maintaining homeostasis.

The kidneys, in concert with hormonal and neural inputs that control their function, are the organs primarily responsible for maintaining the stability of ECF volume, electrolyte composition, and osmolarity (solute concentration). By adjusting the quantity of water and various plasma constituents that are either conserved for the body or eliminated in the urine, the kidneys can maintain water and electrolyte balance within the very narrow range compatible with life, despite a wide range of intake and losses of these constituents through other avenues. The kidneys not only adjust for wide variations in ingestion of water (H_2O), salt, and other electrolytes, but they also adjust urinary output of these ECF constituents to compensate for abnormal losses through heavy sweating, vomiting, diarrhea, or hemorrhage. Thus as the kidneys do what they can to maintain homeostasis, urine composition varies widely.

When the ECF has a surplus of water or a particular electrolyte such as salt (NaCl), the kidneys can eliminate the excess in urine. If there is a deficit, the kidneys cannot provide additional quantities of the depleted constituent, but they can limit urinary losses of the material in short supply and thus conserve it until the person can take in more of the depleted substance. Accordingly, the kidneys can compensate more efficiently for excesses than for deficits. In fact, in some instances the kidneys cannot completely halt the loss of a valuable substance in the urine, even though the substance may be in short supply. A prime example is the case of a H_2O deficit. Even if a person is not consuming any H_2O, the kidneys must put out about half a liter of H_2O in the urine each day to fill another major role as the body's cleaners.

In addition to the kidneys' important regulatory role in maintaining fluid and electrolyte balance, they are the main route for eliminating potentially toxic metabolic wastes and foreign compounds from the body. These wastes cannot be eliminated as solids; they must be excreted in solution, thus obligating the kidneys to produce a minimum volume of around 500 ml of waste-filled urine per day. Because H_2O eliminated in urine is derived from the blood plasma, a person stranded without H_2O eventually urinates him- or herself to death: The plasma volume falls to a fatal level as H_2O is inexorably removed to accompany the wastes.

OVERVIEW OF KIDNEY FUNCTIONS

The kidneys perform the following specific functions, most of which help preserve the constancy of the internal fluid environment:

1. *Maintaining H_2O balance in the body.*
2. *Maintaining the proper osmolarity of body fluids, primarily through regulating H_2O balance.* This function is important to prevent osmotic fluxes into or out of the cells, which could lead to detrimental swelling or shrinking of the cells, respectively (see Chapter 14).
3. *Regulating the quantity and concentration of most ECF ions,* including sodium (Na^+), chloride (Cl^-), potassium (K^+), calcium (Ca^{2+}), hydrogen ion (H^+), bicarbonate (HCO_3^-), phosphate (PO_4^{3-}), sulfate (SO_4^{2-}), and magnesium (Mg^{2+}). Even minor fluctuations in the ECF concentrations of some of these electrolytes can have profound influences. For example, changes in the ECF concentration of K^+ can potentially lead to fatal cardiac dysfunction.
4. *Maintaining proper plasma volume,* which is important in the long-term regulation of arterial blood pressure. This function is accomplished through the kidneys' regulatory role in salt (Na^+ and Cl^-) and H_2O balance (Chapter 14).
5. *Helping maintain the proper acid–base balance* of the body by adjusting urinary output of H^+ and HCO_3^- (Chapter 14).
6. *Excreting (eliminating) the end products (wastes) of bodily metabolism* such as urea, uric acid, and creatinine. If allowed to accumulate, these wastes are toxic, especially to the brain.
7. *Excreting many foreign compounds* such as drugs, food additives, pesticides, and other exogenous nonnutritive materials that have entered the body.
8. *Producing erythropoietin,* a hormone that stimulates red blood cell production (Chapter 11).
9. *Producing renin,* an enzymatic hormone that triggers a chain reaction important in salt conservation by the kidneys.
10. *Converting vitamin D into its active form* (Chapter 17).

▌ The kidneys form the urine; the rest of the urinary system carries the urine to the outside.

The **urinary system** consists of the urine-forming organs— the **kidneys**—and the structures that carry the urine from the kidneys to the outside for elimination from the body (● Fig-

ure 13-1a). The kidneys are a pair of bean-shaped organs that lie in the back of the abdominal cavity, one on each side of the vertebral column, slightly above the waistline. Each kidney is supplied by a **renal artery** and a **renal vein**, which, respectively, enter and leave the kidney at the medial indentation that gives this organ its beanlike form. The kidney acts on the plasma flowing through it to produce urine, conserving materials to be retained in the body and eliminating unwanted materials into the urine.

After urine is formed, it drains into a central collecting cavity, the **renal pelvis,** located at the medial inner core of each kidney (● Figure 13-1b). From there urine is channeled into the **ureter,** a smooth muscle–walled duct that exits at the medial border in close proximity to the renal artery and vein. There are two ureters, one carrying urine from each kidney to the single urinary bladder.

The **urinary bladder,** which temporarily stores urine, is a hollow, distensible, smooth muscle-walled sac. Periodically, urine is emptied from the bladder to the outside through another tube, the **urethra,** as a result of bladder contraction. The urethra in females is straight and short, passing directly from the neck of the bladder to the outside (see ● Figure 18-2, p. 586). In males the urethra is much longer and follows a curving course from the bladder to the outside, passing through both the prostate gland and the penis (● Figure 13-1a; see also ● Figure 18-1, p. 585). The male urethra serves the dual function of providing both a route for eliminating urine from the bladder and a passageway for semen from the reproductive organs. The prostate gland lies below the neck of the bladder and completely encircles the urethra. Prostatic enlargement, which often occurs during middle to older age, can partially or completely occlude the urethra, impeding the flow of urine.

The parts of the urinary system beyond the kidneys merely serve as ductwork to transport urine to the outside. Once formed by the kidneys, urine is not altered in composition or volume as it moves downstream through the rest of the tract.

▌ The nephron is the functional unit of the kidney.

Each kidney consists of about 1 million microscopic functional units known as **nephrons,** which are bound together by connective tissue. Recall that a functional unit is the smallest unit within an organ capable of performing all of that organ's functions. Because the main function of the kidneys is to produce urine and, in so doing, maintain constancy in the ECF composition, a nephron is the smallest unit capable of forming urine.

The arrangement of nephrons within the kidneys gives rise to two distinct regions—an outer region called the **renal cortex,** which looks granular, and an inner region, the **renal medulla,** which is made up of striated triangles, the **renal pyramids** (● Figure 13-1b).

Knowledge of the structural arrangement of an individual nephron is essential for understanding the distinction between the cortical and medullary regions of the kidney and, more importantly, for understanding renal function. Each nephron consists of a *vascular component* and a *tubular com-*

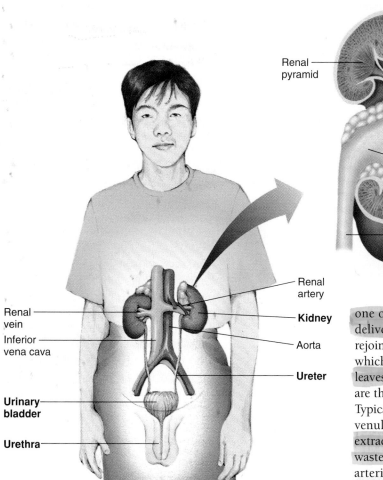

(b)

(a)

FIGURE 13-1

The urinary system. (a) *The components of the urinary system.* The pair of kidneys form the urine, which the ureters carry to the urinary bladder. Urine is stored in the bladder and periodically emptied to the exterior through the urethra. (b) *Longitudinal section of a kidney.* The kidney consists of an outer, granular-appearing renal cortex and an inner, striated-appearing renal medulla. The renal pelvis at the medial inner core of the kidney collects urine after it is formed.

(*Source:* Part b adapted from Ann Stalheim-Smith and Greg K. Fitch, *Understanding Human Anatomy and Physiology,* Fig. 23.4, p. 888. Copyright ©1993 West Publishing Company.)

 For an animation of this figure, click the Anatomy of the Kidney tab in the Renal Physiology tutorial on the CD-ROM. For an interaction related to this figure, see Media Exercise 13.2: The Kidney.

ponent, both of which are intimately related structurally and functionally (● Figure 13-2).

VASCULAR COMPONENT OF THE NEPHRON

The dominant part of the nephron's vascular component is the **glomerulus**, a ball-like tuft of capillaries through which part of the water and solutes is filtered from blood passing through. This filtered fluid, which is almost identical in composition to plasma, then passes through the nephron's tubular component, where various transport processes convert it into urine.

On entering the kidney, the renal artery subdivides to ultimately form many small vessels known as **afferent arterioles,**

one of which supplies each nephron. The afferent arteriole delivers blood to the glomerulus. The glomerular capillaries rejoin to form another arteriole, the **efferent arteriole,** through which blood that was not filtered into the tubular component leaves the glomerulus (● Figure 13-3). The efferent arterioles are the only arterioles in the body that drain from capillaries. Typically, arterioles break up into capillaries that rejoin to form venules. At the glomerular capillaries, no O_2 or nutrients are extracted from the blood for use by the kidney tissues nor are waste products picked up from the surrounding tissue. Thus arterial blood enters the glomerular capillaries through the afferent arteriole, and arterial blood leaves the glomerulus through the efferent arteriole.

The efferent arteriole quickly subdivides into a second set of capillaries, the **peritubular capillaries**, which supply the renal tissue with blood and are important in exchanges between the tubular system and blood during conversion of the filtered fluid into urine. These peritubular capillaries, as their name implies, are intertwined around the tubular system (*peri* means "around"). The peritubular capillaries rejoin to form venules that ultimately drain into the renal vein, by which blood leaves the kidney.

TUBULAR COMPONENT OF THE NEPHRON

The nephron's tubular component is a hollow, fluid-filled tube formed by a single layer of epithelial cells. Even though the tubule is continuous from its beginning near the glomerulus to its ending at the renal pelvis, it is arbitrarily divided into various segments based on differences in structure and function along its length (● Figure 13-2). The tubular component begins with **Bowman's capsule,** an expanded, double-walled invagination that cups around the glomerulus to collect fluid filtered from the glomerular capillaries. The presence of all glomeruli and associated Bowman's capsules in the cortex produces this region's granular appearance.

From Bowman's capsule, the filtered fluid passes into the **proximal tubule,** which lies entirely within the cortex and is highly coiled or convoluted throughout much of its course. The next segment, the **loop of Henle**, forms a sharp U-shaped

● **FIGURE 13-2**

A nephron

PhysioEdge For an interaction related to this figure, see Media Exercise 13.3: The Nephron on the CD-ROM.

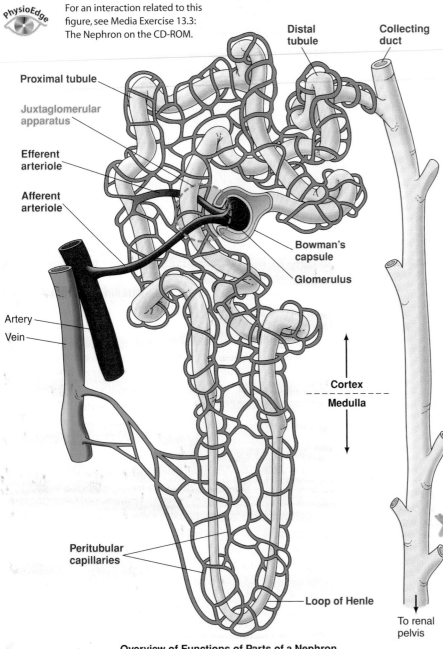

Overview of Functions of Parts of a Nephron

Vascular component
- Afferent arteriole—carries blood to the glomerulus
- Glomerulus—a tuft of capillaries that filters a protein-free plasma into the tubular component
- Efferent arteriole—carries blood from the glomerulus
- Peritubular capillaries—supply the renal tissue; involved in exchanges with the fluid in the tubular lumen

Combined vascular/tubular component
- Juxtaglomerular apparatus—produces substances involved in the control of kidney function

Tubular component
- Bowman's capsule—collects the glomerular filtrate
- Proximal tubule—uncontrolled reabsorption and secretion of selected substances occur here
- Loop of Henle—establishes an osmotic gradient in the renal medulla that is important in the kidney's ability to produce urine of varying concentration
- Distal tubule and collecting duct—variable, controlled reabsorption of Na^+ and H_2O and secretion of K^+ and H^+ occur here; fluid leaving the collecting duct is urine, which enters the renal pelvis

or hairpin loop that dips into the renal medulla. The *descending limb* of Henle's loop plunges from the cortex into the medulla; the *ascending limb* traverses back up into the cortex. The ascending limb returns to the glomerular region of its own nephron, where it passes through the fork formed by the afferent and efferent arterioles. Both the tubular and vascular cells at this point are specialized to form the **juxtaglomerular apparatus**, a structure that lies next to the glomerulus (*juxta* means "next to"). This specialized region plays an important role in regulating kidney function. Beyond the juxtaglomerular apparatus, the tubule once again coils tightly to form the **distal tubule**, which also lies entirely within the cortex. The distal tubule empties into a **collecting duct** or **tubule**, with each collecting duct draining fluid from up to eight separate nephrons. Each collecting duct plunges down through the medulla to empty its fluid contents (now converted into urine) into the renal pelvis. The parallel arrangement of long loops of Henle and collecting ducts in the medulla creates this region's striated appearance.

▌ **The three basic renal processes are glomerular filtration, tubular reabsorption, and tubular secretion.**

Three basic processes are involved in forming urine: *glomerular filtration, tubular reabsorption,* and *tubular secretion.* To aid in visualizing the relationships among these renal processes, it is useful to unwind the nephron schematically, as in ● Figure 13-4.

GLOMERULAR FILTRATION

As blood flows through the glomerulus, protein-free plasma filters through the glomerular capillaries into Bowman's capsule. Normally, about 20% of the plasma that enters the glomerulus is filtered. This process, known as **glomerular filtration**, is the first step in urine formation. On average, 125 ml of glomerular filtrate (filtered fluid) are formed collectively through all the glomeruli each minute. This amounts to 180 liters (about 47.5 gallons) each day. Considering that the average plasma volume in an adult

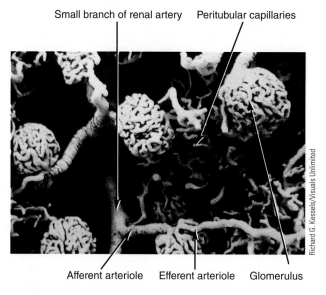

Small branch of renal artery   Peritubular capillaries

Richard G. Kessels/Visuals Unlimited

Afferent arteriole   Efferent arteriole   Glomerulus

● FIGURE 13-3

Scanning electron micrograph of a glomerulus and associated arterioles

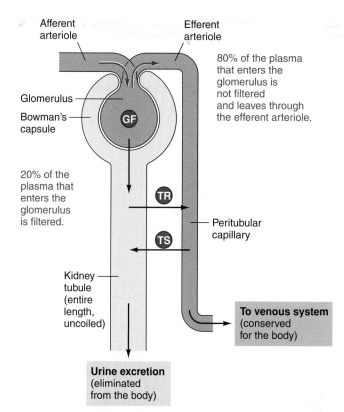

Afferent arteriole   Efferent arteriole

80% of the plasma that enters the glomerulus is not filtered and leaves through the efferent arteriole.

Glomerulus

Bowman's capsule

GF

20% of the plasma that enters the glomerulus is filtered.

TR

TS

Peritubular capillary

Kidney tubule (entire length, uncoiled)

To venous system (conserved for the body)

Urine excretion (eliminated from the body)

 = Glomerular filtration—nondiscriminant filtration of a protein-free plasma from the glomerulus into Bowman's capsule

TR = Tubular reabsorption—selective movement of filtered substances from the tubular lumen into the peritubular capillaries

TS = Tubular secretion—selective movement of nonfiltered substances from the peritubular capillaries into the tubular lumen

● FIGURE 13-4

Basic renal processes. Anything filtered or secreted but not reabsorbed is excreted in the urine and lost from the body. Anything filtered and subsequently reabsorbed, or not filtered at all, enters the venous blood and is saved for the body.

PhysioEdge For an animation of this figure, click the Urine Formation Process tab in the Renal Physiology tutorial on the CD-ROM.

is 2.75 liters, this means that the kidneys filter the entire plasma volume about 65 times per day. If everything filtered passed out in the urine, the total plasma volume would be urinated in less than half an hour! This does not happen, however, because the kidney tubules and peritubular capillaries are intimately related throughout their lengths, so that materials can be transferred between the fluid inside the tubules and the blood within the peritubular capillaries.

TUBULAR REABSORPTION

As the filtrate flows through the tubules, substances of value to the body are returned to the peritubular capillary plasma. This selective movement of substances from inside the tubule (the tubular lumen) into the blood is called **tubular reabsorption.** Reabsorbed substances are not lost from the body in the urine but instead are carried by the peritubular capillaries to the venous system and then to the heart to be recirculated. Of the 180 liters of plasma filtered per day, 178.5 liters on average are reabsorbed. The remaining 1.5 liters left in the tubules passes into the renal pelvis to be eliminated as urine. In general, substances the body needs to conserve are selectively reabsorbed, whereas unwanted substances that must be eliminated stay in the urine.

TUBULAR SECRETION

The third renal process, **tubular secretion,** is the selective transfer of substances from the peritubular capillary blood into the tubular lumen. It provides a second route for substances to enter the renal tubules from the blood, the first being by glomerular filtration. Only about 20% of the plasma flowing through the glomerular capillaries is filtered into Bowman's capsule; the remaining 80% flows on through the efferent arteriole into the peritubular capillaries. A few substances may be discriminatingly transferred by tubular secretion from the plasma in the peritubular capillaries into the tubular lumen.

Tubular secretion provides a mechanism for more rapidly eliminating selected substances from the plasma by extracting an additional quantity of a particular substance from the 80% of unfiltered plasma in the peritubular capillaries and by adding it to the quantity of the substance already present in the tubule as a result of filtration.

URINE EXCRETION

Urine excretion is the elimination of substances from the body in the urine. It is not really a separate process but rather is the result of the first three processes. All plasma constituents that are filtered or secreted but are not reabsorbed remain in the tubules and pass into the renal pelvis to be excreted as urine and eliminated from the body (● Figure 13-4). (Do not confuse excretion with secretion.) Note that anything filtered and subsequently reabsorbed, or not filtered at all, enters the venous blood from the peritubular capillaries and thus is conserved for the body instead of being excreted in urine, despite passing through the kidneys.

THE BIG PICTURE OF THE BASIC RENAL PROCESSES

Glomerular filtration is largely an indiscriminate process. With the exception of blood cells and plasma proteins, all constituents within the blood—H_2O, nutrients, electrolytes, wastes, and so on—nonselectively enter the tubular lumen as a bulk unit during filtration. That is, of the 20% of the plasma that is filtered at the glomerulus, everything in that part of the plasma enters Bowman's capsule except for the plasma proteins. The highly discriminating tubular processes then work on the filtrate to return to the blood a fluid of the composition and volume necessary to maintain the constancy of the internal fluid environment. The unwanted filtered material is left behind in the tubular fluid to be excreted as urine. Glomerular filtration can be thought of as pushing a part of the plasma, with all its essential components as well as those that need to be eliminated from the body, onto a tubular conveyor belt that terminates at the renal pelvis, which is the collecting point for urine within the kidney. All plasma constituents that enter this conveyor belt and are not subsequently returned to the plasma by the end of the line are spilled out of the kidney as urine. It is up to the tubular system to salvage by reabsorption the filtered materials that need to be preserved for the body, while leaving behind substances that must be excreted. In addition, some substances are not only filtered but are also secreted onto the tubular conveyor belt, so that the amounts of these substances excreted in the urine are greater than the amounts that were filtered. For many substances, these renal processes are subject to physiologic control. Thus the kidneys handle each constituent in the plasma in a characteristic manner by a particular combination of filtration, reabsorption, and secretion.

The kidneys act only on the plasma, yet the ECF consists of both plasma and interstitial fluid. The interstitial fluid is actually the true internal fluid environment of the body, because it is the only component of the ECF that comes into direct contact with the cells. However, because of the free exchange between plasma and interstitial fluid across the capillary walls (with the exception of plasma proteins), interstitial fluid composition reflects the composition of plasma. Thus, by performing their regulatory and excretory roles on the plasma, the kidneys maintain the proper interstitial fluid environment for optimal cell function. Most of the rest of this chapter will be devoted to considering how the basic renal

● FIGURE 13-5

Layers of the glomerular membrane

 For an interaction related to this figure, see Media Exercise 13.1: The Urinary System on the CD-ROM.

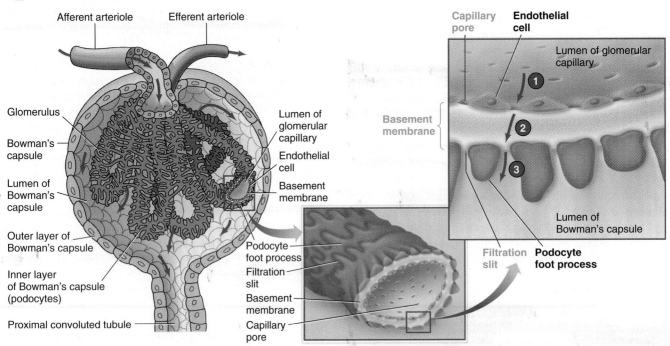

To be filtered, a substance must pass through

1 the pores between the endothelial cells of the glomerular capillary

2 an acellular basement membrane

3 the filtration slits between the foot processes of the podocytes of the inner layer of Bowman's capsule

processes are accomplished and the mechanisms by which they are carefully regulated to help maintain homeostasis.

GLOMERULAR FILTRATION

Fluid filtered from the glomerulus into Bowman's capsule must pass through the following three layers that make up the **glomerular membrane** (● Figure 13-5):

1. The *wall of the glomerular capillaries,* which is a single layer of flattened endothelial cells. It is perforated by many large pores that make it over 100 times more permeable to H_2O and solutes than capillaries elsewhere in the body.
2. The *basement membrane,* which is an acellular (lacking cells) gelatinous layer.
3. The *inner layer of Bowman's capsule,* which consists of **podocytes,** octopus-like cells that encircle the glomerular tuft. Each podocyte bears many elongated foot processes (*podo* means "foot"; a *process* is a projection or appendage) that interdigitate with foot processes of adjacent podocytes, much as you interlace your fingers between each other when you cup your hands around a ball (● Figure 13-6). The narrow slits between adjacent foot processes, known as **filtration slits,** provide a pathway through which fluid leaving the glomerular capillaries can enter the lumen of Bowman's capsule.

Collectively, these layers function as a fine molecular sieve that retains the blood cells and plasma proteins but permits H_2O and solutes of small molecular dimension to filter through. Thus the route that filtered substances take across the glomer-

● FIGURE 13-6

Bowman's capsule podocytes with foot processes and filtration slits. Note the filtration slits between adjacent foot processes on this scanning electron micrograph. The podocytes and their foot processes encircle the glomerular capillaries.

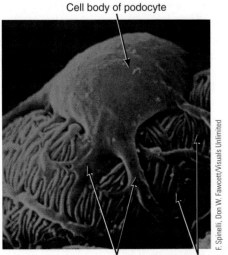

Cell body of podocyte

Foot processes   Filtration slits

F. Spinelli, Don W. Fawcett/Visuals Unlimited

ular membrane is completely extracellular—first through capillary pores, then through the acellular basement membrane, and finally through capsular filtration slits (● Figure 13-5).

■ Glomerular capillary blood pressure is the major force that induces glomerular filtration.

To accomplish glomerular filtration, a force must drive a part of the plasma in the glomerulus through the openings in the glomerular membrane. No active transport mechanisms or local energy expenditures are involved in moving fluid from the plasma across the glomerular membrane into Bowman's capsule. Passive physical forces similar to those acting across capillaries elsewhere accomplish glomerular filtration. Because the glomerulus is a tuft of capillaries, the same principles of fluid dynamics apply here that cause ultrafiltration across other capillaries (see p. 293), except for two important differences: (1) The glomerular capillaries are much more permeable than capillaries elsewhere, so more fluid is filtered for a given filtration pressure; and (2) the balance of forces across the glomerular membrane is such that filtration occurs throughout the entire length of the capillaries. In contrast, the balance of forces in other capillaries shifts, so that filtration occurs in the beginning part of the vessel but reabsorption occurs toward the vessel's end (see ● Figure 10-19, p. 295).

FORCES INVOLVED IN GLOMERULAR FILTRATION

Three physical forces are involved in glomerular filtration (▲ Table 13-1): glomerular capillary blood pressure, plasma-colloid osmotic pressure, and Bowman's capsule hydrostatic pressure. Let's examine the role of each.

1. *Glomerular capillary blood pressure* is the fluid pressure exerted by the blood within the glomerular capillaries. It ultimately depends on contraction of the heart (the source of energy that produces glomerular filtration) and the resistance to blood flow offered by the afferent and efferent arterioles. Glomerular capillary blood pressure, at an estimated average value of 55 mm Hg, is higher than capillary blood pressure elsewhere. The reason for the higher pressure in the glomerular capillaries is the larger diameter of the afferent arteriole compared to the efferent arteriole. Because blood can more readily enter the glomerulus through the wide afferent arteriole than it can leave through the narrower efferent arteriole, glomerular capillary blood pressure is maintained high as a result of blood damming up in the glomerular capillaries. Furthermore, because of the high resistance offered by the efferent arterioles, blood pressure does not have the same tendency to fall along the length of the glomerular capillaries as it does along other capillaries. This raised, nondecremental glomerular blood pressure tends to push fluid out of the glomerulus into Bowman's capsule along the glomerular capillaries' entire length, and it is the major force producing glomerular filtration.

Whereas glomerular capillary blood pressure *favors* filtration, the two other forces acting across the glomerular membrane (plasma-colloid osmotic pressure and Bowman's capsule hydrostatic pressure) *oppose* filtration.

2. *Plasma-colloid osmotic pressure* is caused by the unequal distribution of plasma proteins across the glomerular membrane. Because plasma proteins cannot be filtered, they are in the glomerular capillaries but not in Bowman's capsule. Accordingly, the concentration of H_2O is higher in Bowman's capsule than in the glomerular capillaries. The resulting tendency for H_2O to move by osmosis down its own concentration gradient from Bowman's capsule into the glomerulus opposes glomerular filtration. This opposing osmotic force averages 30 mm Hg, which is slightly higher than across other capillaries. It is higher because much more H_2O is filtered out of the glomerular blood, so the concentration of plasma proteins is higher than elsewhere.

3. *Bowman's capsule hydrostatic pressure,* the pressure exerted by the fluid in this initial part of the tubule, is estimated to be about 15 mm Hg. This pressure, which tends to push fluid out of Bowman's capsule, opposes the filtration of fluid from the glomerulus into Bowman's capsule.

GLOMERULAR FILTRATION RATE

As can be seen in ▲ Table 13-1, the forces acting across the glomerular membrane are not in balance. The total force favoring filtration is attributable to the glomerular capillary blood pressure at 55 mm Hg. The total of the two forces opposing filtration is 45 mm Hg. The net difference favoring filtration (10 mm Hg of pressure) is called the net filtration pressure. This modest pressure forces large volumes of fluid from the blood through the highly permeable glomerular membrane.

Normally, about 20% of the plasma that enters the glomerulus is filtered at the net filtration pressure of 10 mm Hg, producing collectively, through all glomeruli, 180 liters of glomerular filtrate each day for an average **glomerular filtration rate (GFR)** of 125 ml/min.

▌Changes in the GFR primarily result from changes in glomerular capillary blood pressure.

Because the net filtration pressure that accomplishes glomerular filtration is simply due to an imbalance of opposing physical forces between the glomerular capillary plasma and Bowman's capsule fluid, alterations in any of these physical forces can affect the GFR. We will examine the effect that changes in each of these physical forces have on the GFR.

UNREGULATED INFLUENCES ON THE GFR

Plasma-colloid osmotic pressure and Bowman's capsule hydrostatic pressure are not subject to regulation and under normal conditions do not vary much.

Clinical Note However, they can change pathologically and thus inadvertently affect the GFR. Because plasma-colloid osmotic pressure opposes filtration, a decrease in plasma protein concentration, by reducing this pressure, leads to an increase in the GFR. An uncontrollable reduction in plasma protein concentration might occur, for example, in severely burned patients who lose a large quantity of protein-rich, plasma-derived fluid through the exposed burned surface of their skin.

Bowman's capsule hydrostatic pressure can become uncontrollably elevated, and filtration subsequently can decrease, given a urinary tract obstruction, such as a kidney stone or prostatic enlargement. The damming up of fluid behind the obstruction elevates capsular hydrostatic pressure.

CONTROLLED ADJUSTMENTS IN THE GFR

Unlike plasma-colloid osmotic pressure and Bowman's capsule hydrostatic pressure—which may be uncontrollably altered in various disease states and, thereby inappropriately alter the GFR—glomerular capillary blood pressure can be controlled to adjust the GFR to suit the body's needs. Assuming that all other factors stay constant, as the glomerular capillary blood pressure goes up the net filtration pressure increases and the GFR increases correspondingly. The magnitude of the glomerular capillary blood pressure depends on the rate of blood flow within each of the glomeruli. The amount of blood flowing into a glomerulus per minute is determined largely by the resistance offered by the afferent arterioles. If resistance increases in the afferent arteriole, less blood flows into the glomerulus, decreasing the GFR. Controlled changes in the GFR are brought about by the sympathetic nervous system, which adjusts glomerular blood flow by regulating the caliber

▲ **TABLE 13-1**

Forces Involved in Glomerular Filtration

FORCE	EFFECT	MAGNITUDE (mm Hg)
Glomerular Capillary Blood Pressure	Favors filtration	55
Plasma-Colloid Osmotic Pressure	Opposes filtration	30
Bowman's Capsule Hydrostatic Pressure	Opposes filtration	15
Net Filtration Pressure (difference between force favoring filtration and forces opposing filtration)	Favors filtration	10

$$55 - (30 + 15) = 10$$

of the afferent arterioles. The parasympathetic nervous system does not have any influence on the kidneys.

Sympathetic control of the GFR is aimed at long-term regulation of arterial blood pressure. If plasma volume is decreased—for example, by hemorrhage—the resulting fall in arterial blood pressure is detected by the arterial carotid sinus and aortic arch baroreceptors, which initiate neural reflexes to raise blood pressure toward normal (see p. 305). These reflex responses are coordinated by the cardiovascular control center in the brain stem and are mediated primarily through increased sympathetic activity to the heart and blood vessels. Although the resulting increase in both cardiac output and total peripheral resistance helps raise blood pressure toward normal, plasma volume is still reduced. In the long term, plasma volume must be restored to normal. One compensation for a depleted plasma volume is reduced urine output, so that more fluid than normal is conserved for the body. Urine output is reduced in part by reducing the GFR; if less fluid is filtered, less is available to excrete.

No new mechanism is needed to decrease the GFR. It is reduced by the baroreceptor reflex response to a fall in blood pressure (● Figure 13-7). During this reflex, sympathetically induced vasoconstriction occurs in most arterioles throughout the body (including the afferent arterioles) as a compensatory mechanism to increase total peripheral resistance. When the afferent arterioles carrying blood to the glomeruli constrict from increased sympathetic activity, less blood flows into the glomeruli than normal, lowering glomerular capillary blood pressure (● Figure 13-8a). The resulting decrease in GFR in turn reduces urine volume. In this way, some of the H_2O and salt that would otherwise have been lost in urine are saved for the body, helping in the long term to restore plasma volume to normal so that short-term cardiovascular adjustments that have been made are no longer necessary. Other mechanisms, such as increased tubular reabsorption of H_2O and salt as well as increased thirst (described more thoroughly elsewhere), also contribute to long-term maintenance of blood pressure,

● **FIGURE 13-7**

Baroreceptor reflex influence on the GFR in long-term regulation of arterial blood pressure

● **FIGURE 13-8**

Adjustments of afferent arteriole caliber to alter the GFR. (a) Arteriolar adjustment to reduce the GFR. (b) Arteriolar adjustment to increase the GFR.

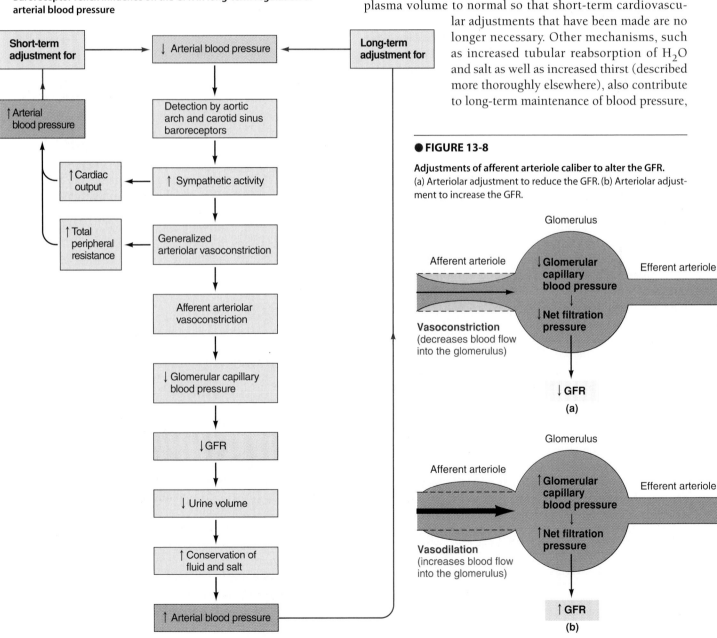

despite a loss of plasma volume, by helping to restore plasma volume.

Conversely, if blood pressure is elevated (for example, because of an expansion of plasma volume following ingestion of excessive fluid), the opposite responses occur. When the baroreceptors detect a rise in blood pressure, sympathetic vasoconstrictor activity to the arterioles, including the renal afferent arterioles, is reflexly reduced, allowing afferent arteriolar vasodilation to occur. As more blood enters the glomeruli through the dilated afferent arterioles, glomerular capillary blood pressure rises, increasing the GFR (● Figure 13-8b). As more fluid is filtered, more fluid is available to be eliminated in urine. Contributing to the increase in urine volume is a hormonally adjusted reduction in the tubular reabsorption of H_2O and salt. These two renal mechanisms—increased glomerular filtration and decreased tubular reabsorption of H_2O and salt— increase urine volume and eliminate the excess fluid from the body. Reduced thirst and fluid intake also help restore an elevated blood pressure to normal.

Before turning our attention to the process of tubular reabsorption, we are first going to examine the percentage of the cardiac output that goes to the kidneys. This will further reinforce the concept of how much blood flows through the kidneys and how much of that fluid is filtered and subsequently acted on by the tubules.

▌ The kidneys normally receive 20 to 25% of the cardiac output.

At the average net filtration pressure, 20% of the plasma that enters the kidneys is converted into glomerular filtrate. That means at an average GFR of 125 ml/min, the total renal plasma flow must average about 625 ml/min. Because 55% of whole blood consists of plasma (that is, hematocrit = 45; see p. 315), the total flow of blood through the kidneys averages 1140 ml/min. This quantity is about 22% of the total cardiac output of 5 liters (5000 ml)/min, although the kidneys compose less than 1% of total body weight.

The kidneys need to receive such a seemingly disproportionate share of the cardiac output because they must continuously perform their regulatory and excretory functions on the huge volumes of plasma delivered to them to maintain stability in the internal fluid environment. Most of the blood goes to the kidneys not to supply the renal tissue but to be adjusted and purified by the kidneys. On average, 20 to 25% of the blood pumped out by the heart each minute "goes to the cleaners" instead of serving its normal purpose of exchanging materials with the tissues. Only by continuously processing such a large proportion of the blood can the kidneys precisely regulate the volume and electrolyte composition of the internal environment and adequately eliminate the large quantities of metabolic waste products that are constantly produced.

TUBULAR REABSORPTION

All plasma constituents except the proteins are indiscriminately filtered together through the glomerular capillaries. In addition to waste products and excess materials that the body must eliminate, the filtered fluid also contains nutrients, electrolytes, and other substances that the body cannot afford to lose in the urine. Indeed, through ongoing glomerular filtration greater quantities of these materials are filtered per day than are even present in the entire body. It is important that the essential materials that are filtered be returned to the blood by *tubular reabsorption,* the discrete transfer of substances from the tubular lumen into the peritubular capillaries.

▌ Tubular reabsorption is tremendous, highly selective, and variable.

Tubular reabsorption is a highly selective process. All constituents except plasma proteins are at the same concentration in the glomerular filtrate as in plasma. In most cases, the quantity of each material that is reabsorbed is the amount required to maintain the proper composition and volume of the internal fluid environment. In general, the tubules have a high reabsorptive capacity for substances needed by the body and little or no reabsorptive capacity for substances of no value (▲ Table 13-2). Accordingly, only a small percentage, if any, of filtered plasma constituents that are useful to the body are present in the urine, most having been reabsorbed and returned to the blood. Only excess amounts of essential materials such as electrolytes are excreted in the urine. For the essential plasma constituents regulated by the kidneys, absorptive capacity may vary depending on the body's needs. In contrast, a large percentage of filtered waste products is present in the urine. These wastes, which are useless or even potentially harmful to the body if allowed to accumulate, are not reabsorbed to any extent. Instead, they stay in the tubules to be eliminated in the urine. As H_2O and other valuable constituents are reabsorbed, the waste products remaining in the tubular fluid become highly concentrated.

Of the 125 ml/min filtered, typically 124 ml/min are reabsorbed. Considering the magnitude of glomerular filtration, the extent of tubular reabsorption is tremendous: The tubules typically reabsorb 99% of the filtered H_2O (47 gallons/day),

▲ **TABLE 13-2**

Fate of Various Substances Filtered by the Kidneys

SUBSTANCE	AVERAGE PERCENTAGE OF FILTERED SUBSTANCE REABSORBED	AVERAGE PERCENTAGE OF FILTERED SUBSTANCE EXCRETED
Water	99	1
Sodium	99.5	0.5
Glucose	100	0
Urea (a waste product)	50	50
Phenol (a waste product)	0	100

100% of the filtered sugar (2.5 pounds/day), and 99.5% of the filtered salt (0.36 pounds/day).

▍Tubular reabsorption involves transepithelial transport.

Throughout its entire length, the tubule wall is one cell thick and is in close proximity to a surrounding peritubular capillary (● Figure 13-9). Adjacent tubular cells do not come into contact with each other except where they are joined by tight junctions (see p. 48) at their lateral edges near their *luminal membranes*, which face the tubular lumen. Interstitial fluid lies in the gaps between adjacent cells—the **lateral spaces**—as well as between the tubules and capillaries. The *basolateral membrane* faces the interstitial fluid at the base and lateral edges of the cell. The tight junctions largely prevent substances from moving *between* the cells, so materials must pass *through* the cells to leave the tubular lumen and gain entry to the blood.

STEPS OF TRANSEPITHELIAL TRANSPORT

To be reabsorbed, a substance must traverse five distinct barriers (● Figure 13-9):

- *Step 1.* It must leave the tubular fluid by crossing the luminal membrane of the tubular cell.
- *Step 2.* It must pass through the cytosol from one side of the tubular cell to the other.
- *Step 3.* It must traverse the basolateral membrane of the tubular cell to enter the interstitial fluid.
- *Step 4.* It must diffuse through the interstitial fluid.
- *Step 5.* It must penetrate the capillary wall to enter the blood plasma.

This entire sequence of steps is known as **transepithelial** ("across the epithelium") **transport**.

PASSIVE VERSUS ACTIVE REABSORPTION

The two types of tubular reabsorption—*passive reabsorption* and *active reabsorption*—depend on whether local energy expenditure is needed for reabsorbing a particular substance. In **passive reabsorption**, all steps in the transepithelial transport of a substance from the tubular lumen to the plasma are passive; that is, no energy is spent for the substance's net movement, which occurs down electrochemical or osmotic gradients (see p. 52). In contrast, **active reabsorption** takes place if any one of the steps in the transepithelial transport of a substance requires energy, even if the four other steps are passive. With active reabsorption, net movement of the substance from the tubular lumen to the plasma occurs *against* an electrochemical gradient. Substances that are actively reabsorbed are of particular importance to the body, such as glucose, amino acids, and other organic nutrients, as well as Na^+ and other electrolytes such as PO_4^{3-}.

Rather than specifically describing the reabsorption process for each of the many filtered substances that are returned to the plasma, we will provide illustrative examples of the general mechanisms involved, after first highlighting the unique and important case of Na^+ reabsorption.

▍An active Na^+–K^+ ATPase pump in the basolateral membrane is essential for Na^+ reabsorption.

Sodium reabsorption is unique and complex. Of the total energy requirement of the kidneys, 80% is used for Na^+ trans-

● FIGURE 13-9

Steps of transepithelial transport

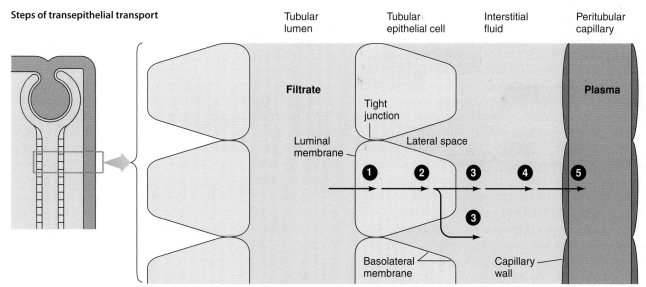

To be reabsorbed (move from the filtrate to the plasma), a substance must traverse five distinct barriers:

1 the luminal cell membrane   **3** the basolateral cell membrane   **5** the capillary wall

2 the cytosol   **4** the interstital fluid

port, indicating the importance of this process. Unlike most filtered solutes, Na$^+$ is reabsorbed throughout most of the tubule, but to varying extents in different regions. Of the Na$^+$ filtered, 99.5% is normally reabsorbed. Of the Na$^+$ reabsorbed, on average 67% is reabsorbed in the proximal tubule, 25% in the loop of Henle, and 8% in the distal and collecting tubules. Sodium reabsorption plays different important roles in each of these segments, as will become apparent as our discussion continues. Here is a preview of these roles.

• Sodium reabsorption in the *proximal tubule* plays a pivotal role in reabsorbing glucose, amino acids, H$_2$O, Cl$^-$, and urea.

• Sodium reabsorption in the ascending limb of the *loop of Henle*, along with Cl$^-$ reabsorption, plays a critical role in the kidneys' ability to produce urine of varying concentrations and volumes, depending on the body's need to conserve or eliminate H$_2$O.

• Sodium reabsorption in the *distal and collecting tubules* is variable and subject to hormonal control. It plays a key role in regulating ECF volume, which is important in long-term control of arterial blood pressure, and is also linked in part to K$^+$ secretion and H$^+$ secretion.

Sodium is reabsorbed throughout the tubule with the exception of the descending limb of the loop of Henle. You will learn about the significance of this exception later. Throughout all Na$^+$-reabsorbing segments of the tubule, the active step in Na$^+$ reabsorption involves the energy-dependent Na$^+$–K$^+$ ATPase carrier located in the tubular cells basolateral membrane (● Figure 13-10). This carrier is the same one that is in all cells and actively extrudes Na$^+$ from the cell (see p. 58). As this basolateral pump transports Na$^+$ out of the tubular cell into the lateral space, it keeps the intracellular Na$^+$ concentration low while it simultaneously builds up the concentration of Na$^+$ in the lateral space; that is, it moves Na$^+$ against a concentration gradient. Because the intracellular Na$^+$ concentration is kept low by basolateral pump activity, a concentration gradient is established that favors the diffusion of Na$^+$ from its higher concentration in the tubular lumen across the luminal border into the tubular cell. The nature of the luminal Na$^+$ channels and/or transport carriers that permit movement of Na$^+$ from the lumen into the cell varies for different parts of the tubule, but in each case, movement of Na$^+$ across the luminal membrane is always a passive step. For example, in the proximal tubule, Na$^+$ crosses the luminal border by a cotransport carrier that simultaneously moves Na$^+$ and an organic nutrient such as glucose from the lumen into the cell. You will learn more about this cotransport process shortly. By contrast, in the collecting duct, Na$^+$ crosses the luminal border through a Na$^+$ channel. Once Na$^+$ enters the cell across the luminal border by whatever means, it is actively extruded to the lateral space by the basolateral

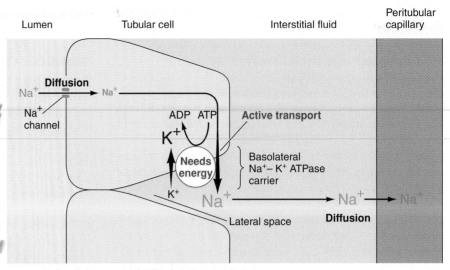

● **FIGURE 13-10**

Sodium reabsorption. The basolateral Na$^+$–K$^+$ ATPase carrier actively transports Na$^+$ from the tubular cell into the interstitial fluid within the lateral space. This process establishes a concentration gradient for diffusion of Na$^+$ from the lumen into the tubular cell and from the lateral space into the peritubular capillary, accomplishing net transport of Na$^+$ from the tubular lumen into the blood at the expense of energy.

Na$^+$–K$^+$ pump. This step is the same throughout the tubule. Sodium continues to diffuse down a concentration gradient from its high concentration in the lateral space into the surrounding interstitial fluid and finally into the peritubular capillary blood. Thus net transport of Na$^+$ from the tubular lumen into the blood occurs at the expense of energy.

First let's consider the importance of regulating Na$^+$ reabsorption in the distal portion of the nephron and examine how this control is accomplished. Later we will explore in further detail the roles of Na$^+$ reabsorption in the proximal tubule and in the loop of Henle.

▌ Aldosterone stimulates Na$^+$ reabsorption in the distal and collecting tubules.

In the proximal tubule and loop of Henle, a constant percentage of the filtered Na$^+$ is reabsorbed regardless of the **Na$^+$ load** (*total amount* of Na$^+$ in the body fluids, *not the concentration* of Na$^+$ in the body fluids). In the distal part of the tubule, the reabsorption of a small percentage of the filtered Na$^+$ is subject to hormonal control. The extent of this controlled, discretionary reabsorption is inversely related to the magnitude of the Na$^+$ load in the body. If there is too much Na$^+$, little of this controlled Na$^+$ is reabsorbed; instead, it is lost in urine, thereby removing excess Na$^+$ from the body. If Na$^+$ is depleted, however, most or all of this controlled Na$^+$ is reabsorbed, conserving for the body Na$^+$ that otherwise would be lost in urine.

The Na$^+$ load in the body is reflected by the ECF volume. Sodium and its accompanying anion Cl$^-$ account for more than 90% of the ECF's osmotic activity. (NaCl is common table salt.) Recall that osmotic pressure can be thought of loosely as a force that attracts and holds H$_2$O (see p. 55).

When the Na$^+$ load is above normal and the ECF's osmotic activity is therefore increased, the extra Na$^+$ holds extra H$_2$O, expanding the ECF volume. Conversely, when the Na$^+$ load is below normal, thereby decreasing ECF osmotic activity, less H$_2$O than normal can be held in the ECF, so the ECF volume is reduced. Because plasma is part of the ECF, the most important result of a change in ECF volume is the matching change in blood pressure with expansion ($\uparrow$ blood pressure) or reduction ($\downarrow$ blood pressure) of the plasma volume. Thus long-term control of arterial blood pressure ultimately depends on Na$^+$-regulating mechanisms. We will now turn our attention to these mechanisms.

ACTIVATION OF THE RENIN-ANGIOTENSIN-ALDOSTERONE SYSTEM

The most important and best known hormonal system involved in regulating Na$^+$ is the **renin-angiotensin-aldosterone system** (**RAAS**). The juxtaglomerular apparatus ($\bullet$ Figure 13-2) secretes a hormone, **renin**, into the blood in response to a fall in NaCl/ECF volume/blood pressure. These interrelated signals for increased renin secretion all indicate the need to expand the plasma volume to increase the arterial pressure to normal on a long-term basis. Through a complex series of events, increased renin secretion brings about increased Na$^+$ reabsorption by the distal and collecting tubules. Chloride always passively follows Na$^+$ down the electrical gradient established by sodium's active movement. The ultimate benefit of this salt retention is that it osmotically promotes H$_2$O retention, which helps restore the plasma volume, thus being important in the long-term control of blood pressure.

Let's examine in further detail the RAAS mechanism by which renin secretion ultimately leads to increased Na$^+$ reabsorption ($\bullet$ Figure 13-11). Once secreted into the blood, renin acts as an enzyme to activate **angiotensinogen** into **angiotensin I**. Angiotensinogen is a plasma protein synthesized by the liver and always present in the plasma in high concentration. On passing through the lungs via the pulmonary cir-

$\bullet$ **FIGURE 13-11**

Renin-angiotensin-aldosterone system. The kidneys secrete the hormone renin in response to reduced NaCl, ECF volume, and arterial blood pressure. Renin activates angiotensinogen, a plasma protein produced by the liver, into angiotensin I. Angiotensin I is converted into angiotensin II by angiotensin-converting enzyme (ACE) produced in the lungs. Angiotensin II stimulates the adrenal cortex to secrete the hormone aldosterone, which stimulates Na$^+$ reabsorption by the kidneys. The resulting retention of Na$^+$ exerts an osmotic effect that holds more H$_2$O in the ECF. Together, the conserved Na$^+$ and H$_2$O help correct the original stimuli that activated this hormonal system. Angiotensin II also promotes arteriolar vasoconstriction to help rectify the original stimuli.

 For an interaction related to this figure, see Media Exercise 13.1: The Urinary System on the CD-ROM.

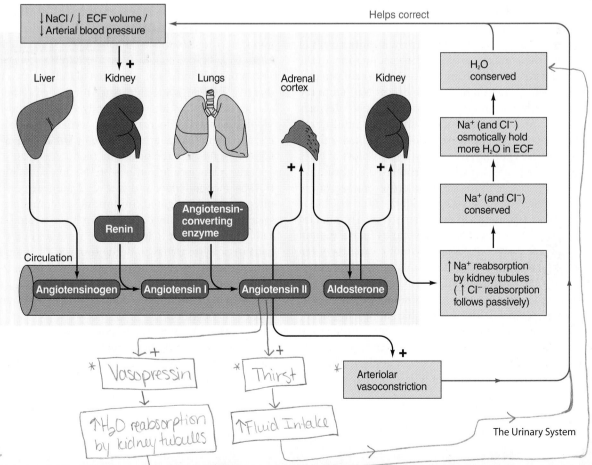

culation, angiotensin I is converted into **angiotensin II** by **angiotensin-converting enzyme (ACE)**, which is abundant in the pulmonary capillaries. Angiotensin II is the main stimulus for secretion of the hormone *aldosterone* from the adrenal cortex. The *adrenal cortex* is an endocrine gland that produces several different hormones, each secreted in response to different stimuli.

FUNCTIONS OF THE RENIN-ANGIOTENSIN-ALDOSTERONE SYSTEM

Among its actions, **aldosterone** increases Na^+ reabsorption by the distal and collecting tubules. It does so by promoting the insertion of additional Na^+ channels into the luminal membranes and additional Na^+–K^+ ATPase carriers into the basolateral membranes of the distal and collecting tubular cells. The net result is a greater passive inward flux of Na^+ into the tubular cells from the lumen and increased active pumping of Na^+ out of the cells into the plasma—that is, an increase in Na^+ reabsorption, with Cl^- following passively. RAAS thus promotes salt retention and a resulting H_2O retention and rise in arterial blood pressure. Acting in negative-feedback fashion, this system alleviates the factors that triggered the initial release of renin—namely, salt depletion, plasma volume reduction, and decreased arterial blood pressure (● Figure 13-11).

In addition to stimulating aldosterone secretion, angiotensin II is also a potent constrictor of the systemic arterioles, directly increasing blood pressure by increasing total peripheral resistance (see p. 290).

The opposite situation exists when the Na^+ load, ECF and plasma volume, and arterial blood pressure are above normal. Under these circumstances, renin secretion is inhibited. Therefore, because angiotensinogen is not activated to angiotensin I and II, aldosterone secretion is not stimulated. Without aldosterone, the small aldosterone-dependent part of Na^+ reabsorption in the distal segments of the tubule does not occur. Instead, this nonreabsorbed Na^+ is lost in urine. In the absence of aldosterone, the ongoing loss of this small percentage of filtered Na^+ can rapidly remove excess Na^+ from the body. Even though only about 8% of the filtered Na^+ depends on aldosterone for reabsorption, this small loss, multiplied many times as the entire plasma volume is filtered through the kidneys many times per day, can lead to a sizable loss of Na^+.

In the complete absence of aldosterone, 20 g of salt may be excreted per day. With maximum aldosterone secretion, all the filtered Na^+ (and, accordingly, all the filtered Cl^-) is reabsorbed, so salt excretion in the urine is zero. The amount of aldosterone secreted, and consequently the relative amount of salt conserved versus salt excreted, usually varies between these extremes, depending on the body's needs. For example, an average salt consumer typically excretes about 10 g of salt per day in the urine, a heavy salt consumer excretes more, and someone who has lost considerable salt during heavy sweating excretes less. By varying the amount of renin and aldosterone secreted in accordance with the salt-determined fluid load in the body, the kidneys can finely adjust the amount of salt conserved or eliminated. In doing so, they maintain the salt load and ECF volume/arterial blood pressure at a relatively constant level despite wide variations in salt consumption and abnormal losses of salt-laden fluid. It should not be surprising that some cases of hypertension (high blood pressure) are due to abnormal increases in RAAS activity.

Clinical Note Many **diuretics**, therapeutic agents that cause *diuresis* (increased urinary output) and thus promote loss of fluid from the body, function by inhibiting tubular reabsorption of Na^+. As more Na^+ is excreted, more H_2O is also lost from the body, helping remove the excess ECF. Diuretics are often beneficial in treating congestive heart failure (see p. 265) as well as certain cases of hypertension.

ACE inhibitor drugs, which block the action of angiotensin-converting enzyme (ACE), are also beneficial in treating these conditions. By blocking the generation of angiotensin II, ACE inhibitors halt the ultimate salt- and fluid-conserving actions and arteriolar constrictor effects of RAAS.

❚ Atrial natriuretic peptide inhibits Na^+ reabsorption.

Whereas RAAS exerts the most powerful influence on the renal handling of Na^+, this Na^+-retaining, blood pressure–raising system is opposed by a Na^+-losing, blood pressure–lowering system that involves the hormone **atrial natriuretic peptide (ANP)** (*natriuretic* means "inducing excretion of large amounts of sodium in the urine"). The heart, in addition to its pump action, produces ANP, which is stored in specialized cardiac atrial muscle cells and is released from the atria when the heart is mechanically stretched by expansion of the ECF volume, including the circulating plasma volume. This expansion, which occurs as a result of Na^+ and H_2O retention, increases arterial blood pressure. In turn, the main action of ANP is to inhibit Na^+ reabsorption in the distal parts of the nephron, thus increasing Na^+ excretion in the urine. This natriuresis brings about an accompanying diuresis. As a result, more salt and water are excreted in the urine. Besides indirectly lowering blood pressure by reducing the Na^+ load and hence the fluid load in the body, ANP also directly lowers blood pressure by decreasing the cardiac output and reducing peripheral vascular resistance by inhibiting sympathetic nervous activity to the heart and blood vessels.

The relative contributions of ANP in maintaining salt and H_2O balance and blood pressure regulation are presently being intensively investigated. Importantly, derangements of this system could logically contribute to hypertension. In fact, recent studies suggest that a deficiency of the counterbalancing natriuretic system may underlie some cases of long-term hypertension by leaving the powerful Na^+-conserving system unopposed. The resulting salt retention, especially in association with high salt intake, could expand ECF volume and elevate blood pressure.

We are now going to shift our attention to the reabsorption of other filtered solutes. However, we will still continue to discuss Na^+ reabsorption, because the reabsorption of many other solutes is linked in some way to Na^+ reabsorption.

Glucose and amino acids are reabsorbed by Na^+-dependent secondary active transport.

Large quantities of nutritionally important organic molecules such as glucose and amino acids are filtered each day. Because these substances normally are completely reabsorbed back into the blood by energy- and Na^+-dependent mechanisms located in the proximal tubule, none of these materials are usually excreted in the urine. This rapid and thorough reabsorption early in the tubules protects against the loss of these important organic nutrients.

Even though glucose and amino acids are moved uphill against their concentration gradients from the tubular lumen into the blood until their concentration in the tubular fluid is virtually zero, no energy is directly used to operate the glucose or amino acid carriers. Glucose and amino acids are transferred by **secondary active transport.** With this process, specialized *cotransport carriers* located in the proximal tubule simultaneously transfer both Na^+ and the specific organic molecule from the lumen into the cell. This luminal cotransport carrier is the means by which Na^+ passively crosses the luminal membrane in the proximal tubule. The lumen-to-cell Na^+ concentration gradient maintained by the energy-consuming basolateral Na^+-K^+ pump drives this cotransport system and pulls the organic molecule against its concentration gradient without the direct expenditure of energy. Specifically, the Na^+ gradient, not ATP, is directly responsible for the cotransport carrier picking up its passengers from the lumen, changing shape, and dropping them off inside the cell. The movement of Na^+ into the cell by this cotransport carrier is downhill because the intracellular Na^+ concentration is low (due to the Na^+-K^+ pump), but the movement of glucose (or amino acid) is uphill because glucose becomes concentrated in the cell. Because the overall process of glucose and amino acid reabsorption depends on the use of energy, these organic molecules are considered to be actively reabsorbed, even though energy is not used directly to transport them across the membrane. In essence, glucose and amino acids get a "free ride" at the expense of energy already used in the reabsorption of Na^+. Once transported into the tubular cells, glucose and amino acids passively diffuse down their concentration gradients across the basolateral membrane into the plasma, facilitated by a carrier that is not dependent on energy.

In general, actively reabsorbed substances exhibit a tubular maximum.

All actively reabsorbed substances bind with plasma membrane carriers that transfer them across the membrane against a concentration gradient. Each carrier is specific for the types of substances it can transport; for example, the glucose cotransport carrier cannot transport amino acids, or vice versa. Because a limited number of each carrier type are present in the cells lining the tubules, there is an upper limit on how much of a particular substance that can be actively transported from the tubular fluid in a given period of time. The maximum reabsorption rate is reached when all the carriers specific for a particular substance are fully occupied or saturated, so they cannot handle any additional passengers at that time (see p. 56). This transport maximum is designated as the **tubular maximum,** or T_m. Any quantity of a substance filtered beyond its T_m is not reabsorbed, and escapes instead into the urine. With the exception of Na^+, all actively reabsorbed substances have a tubular maximum. (Even though individual Na^+ transport carriers can become saturated, the tubules as a whole do not display a tubular maximum for Na^+, because aldosterone promotes the synthesis of more active Na^+-K^+ carriers in the distal and collecting tubular cells as needed.)

The plasma concentrations of some but not all substances that display carrier-limited reabsorption are regulated by the kidneys. How can the kidneys regulate some actively reabsorbed substances but not others, when the renal tubules limit the quantity of each of these substances that can be reabsorbed and returned to the plasma? We will compare glucose, a substance that has a T_m but *is not regulated* by the kidneys, with phosphate, a T_m-limited substance that *is regulated* by the kidneys.

Glucose is an example of an actively reabsorbed substance that is not regulated by the kidneys.

The normal plasma concentration of glucose is 100 mg of glucose/100 ml of plasma. Because glucose is freely filterable at the glomerulus, it passes into Bowman's capsule at the same concentration it has in the plasma. Accordingly, 100 mg of glucose are present in every 100 ml of plasma filtered. With 125 ml of plasma normally being filtered each minute (average GFR = 125 ml/min), 125 mg of glucose pass into Bowman's capsule with this filtrate every minute. The quantity of any substance filtered per minute, known as its **filtered load,** can be calculated as follows:

$$\text{Filtered load of a substance} = \text{plasma concentration} \times \text{GFR of the substance}$$

$$\text{Filtered load of glucose} = 100 \text{ mg}/100 \text{ ml} \times 125 \text{ ml/min}$$

$$= 125 \text{ mg/min}$$

At a constant GFR, the filtered load of glucose is directly proportional to the plasma glucose concentration. Doubling the plasma glucose concentration to 200 mg/100 ml doubles the filtered load of glucose to 250 mg/min, and so on (● Figure 13-12).

TUBULAR MAXIMUM FOR GLUCOSE

The T_m for glucose averages 375 mg/min; that is, the glucose carrier mechanism is capable of actively reabsorbing up to 375 mg of glucose per minute before it reaches its maximum transport capacity. At a normal plasma glucose concentration of 100 mg/100 ml, the 125 mg of glucose filtered per minute can readily be reabsorbed by the glucose carrier mechanism, because the filtered load is well below the T_m for glucose.

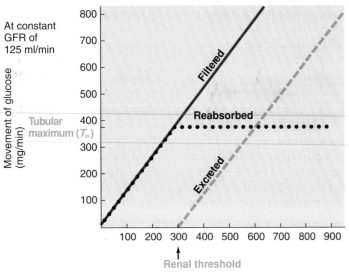

At constant GFR of 125 ml/min

Movement of glucose (mg/min)

Tubular maximum (T_m)

Renal threshold

Plasma concentration of glucose (mg/100 ml)

● **FIGURE 13-12**

Renal handling of glucose as a function of plasma glucose concentration. At a constant GFR, the quantity of glucose filtered per minute is directly proportional to the plasma concentration of glucose. All the filtered glucose can be reabsorbed up to the tubular maximum (T_m). If the amount of glucose filtered per minute exceeds the T_m, the maximum amount of glucose is reabsorbed (a T_m worth), and the rest stays in the filtrate to be excreted in urine. The renal threshold is the plasma concentration at which the T_m is reached and glucose first starts appearing in the urine.

Ordinarily, therefore, no glucose appears in the urine. Not until the filtered load of glucose exceeds 375 mg/min is the T_m reached. When more glucose is filtered per minute than can be reabsorbed because the T_m is exceeded, the maximum amount is reabsorbed, while the rest stays in the filtrate to be excreted. Accordingly, the plasma glucose concentration must be greater than 300 mg/100 ml—more than three times the normal value—before glucose starts spilling into the urine.

RENAL THRESHOLD FOR GLUCOSE

The plasma concentration at which the T_m of a particular substance is reached and the substance first starts appearing in urine is called the **renal threshold**. At the average T_m of 375 mg/min and GFR of 125 ml/min, the renal threshold for glucose is 300 mg/100 ml.[1] Beyond the T_m, reabsorption stays constant at its maximum rate, and any further increase in the filtered load leads to a directly proportional increase in the amount of the substance excreted. For example, at a plasma

[1]This is an idealized situation. In reality, glucose often starts spilling into the urine at glucose plasma concentrations of 180 mg/100 ml and above. Glucose is often excreted before the average renal threshold of 300 mg/100 ml is reached for two reasons. First, not all nephrons have the same T_m, so that some nephrons may have exceeded their T_m and be excreting glucose while others have not yet reached their T_m. Second, the efficiency of the glucose transport mechanism varies depending on the filtered load of glucose. The glucose cotransport carrier may not be working at its maximum capacity at elevated values less than the true T_m so that some of the filtered glucose may fail to be reabsorbed and spill into the urine even though the average renal threshold has not been reached.

glucose concentration of 400 mg/100 ml, the filtered load of glucose is 500 mg/min, 375 mg/min of which can be reabsorbed (a T_m worth) and 125 mg/min of which are excreted in urine. At a plasma glucose concentration of 500 mg/100 ml, the filtered load is 625 mg/min, still only 375 mg/min can be reabsorbed, and 250 mg/min spill into urine (● Figure 13-12).

The plasma glucose concentration can become extremely high in *diabetes mellitus*, an endocrine disorder involving inadequate insulin action. Insulin is a pancreatic hormone that facilitates transport of glucose into many body cells. When cellular glucose uptake is impaired, glucose that cannot gain entry into cells stays in the plasma, elevating the plasma glucose concentration. Consequently, although glucose does not normally appear in urine, it is found in the urine of people with diabetes when the plasma glucose concentration exceeds the renal threshold, even though renal function has not changed.

What happens when plasma glucose concentration falls below normal? The renal tubules, of course, reabsorb all the filtered glucose, because the glucose reabsorptive capacity is far from being exceeded. The kidneys cannot do anything to raise a low plasma glucose level to normal. They simply return all the filtered glucose to the plasma.

REASON WHY THE KIDNEYS DO NOT REGULATE GLUCOSE

The kidneys do not influence plasma glucose concentration over a wide range of values from abnormally low levels up to three times the normal level. Because the T_m for glucose is well above the normal filtered load, the kidneys usually conserve all the glucose, thereby protecting against loss of this important nutrient in urine. The kidneys do not regulate glucose, because they do not maintain glucose at some specific plasma concentration. Instead, this concentration is normally regulated by endocrine and liver mechanisms, with the kidneys merely maintaining whatever plasma glucose concentration is set by these other mechanisms (except when excessively high levels overwhelm the kidneys' reabsorptive capacity). The same general principle holds true for other organic plasma nutrients, such as amino acids and water-soluble vitamins.

▌ Phosphate is an example of an actively reabsorbed substance that is regulated by the kidneys.

The kidneys do directly contribute to the regulation of many electrolytes, such as phosphate (PO_4^{3-}) and calcium (Ca^{2+}), because the renal thresholds of these inorganic ions equal their normal plasma concentrations. We will use PO_4^{3-} as an example. Our diets are generally rich in PO_4^{3-}, but because the tubules can reabsorb up to the normal plasma concentration's worth of PO_4^{3-} and no more, the excess ingested PO_4^{3-} is quickly spilled into the urine, restoring the plasma concentration to normal. The greater the amount of PO_4^{3-} ingested beyond the body's needs, the greater the amount excreted. In this way, the kidneys maintain the desired plasma PO_4^{3-} concentration while eliminating any excess PO_4^{3-} ingested.

Unlike the reabsorption of organic nutrients, the reabsorption of PO_4^{3-} and Ca^{2+} is also subject to hormonal control. Parathyroid hormone can alter the renal thresholds for PO_4^{3-} and Ca^{2+}, thus adjusting the quantity of these electrolytes conserved, depending on the body's momentary needs (Chapter 17).

▌ Active Na^+ reabsorption is responsible for the passive reabsorption of Cl^-, H_2O, and urea.

Not only is secondary active reabsorption of glucose and amino acids linked to the basolateral Na^+–K^+ pump, but passive reabsorption of Cl^-, H_2O, and urea also depends on this active Na^+ reabsorption mechanism.

CHLORIDE REABSORPTION

The negatively charged chloride ions are passively reabsorbed down the electrical gradient created by the active reabsorption of the positively charged sodium ions. The amount of Cl^- reabsorbed is determined by the rate of active Na^+ reabsorption, instead of being directly controlled by the kidneys.

WATER REABSORPTION

Water is passively reabsorbed throughout the length of the tubule as H_2O osmotically follows Na^+ that is actively reabsorbed. Of the H_2O filtered, 65%—117 liters per day—is passively reabsorbed by the end of the proximal tubule. Neither the proximal tubule nor indeed any other part of the tubule directly requires energy for this tremendous reabsorption of H_2O. Another 15% of the filtered H_2O is obligatorily reabsorbed from the loop of Henle. This 80% of the filtered H_2O is reabsorbed in the proximal tubule and Henle's loop regardless of the H_2O load in the body and is not subject to regulation. Variable amounts of the remaining 20% are reabsorbed in the distal portions of the tubule; the extent of reabsorption in the distal and collecting tubules is under direct hormonal control, depending on the body's state of hydration.

During reabsorption, H_2O passes through **aquaporins**, or **water channels**, formed by specific plasma membrane proteins in the tubular cells. Different types of water channels are present in various parts of the nephron. The water channels in the proximal tubule are always open, accounting for the high H_2O permeability of this region. The channels in the distal parts of the nephron, in contrast, are regulated by the hormone *vasopressin*, accounting for the variable H_2O reabsorption in this region. The mechanisms of H_2O reabsorption beyond the proximal tubule will be described later.

UREA REABSORPTION

In addition to Cl^- and H_2O, passive reabsorption of urea is also indirectly linked to active Na^+ reabsorption. Urea is a waste product from the breakdown of protein. The osmotically induced reabsorption of H_2O in the proximal tubule secondary to active Na^+ reabsorption produces a concentration gradient for urea that favors passive reabsorption of this waste, as follows (● Figure 13-13). Extensive reabsorption of H_2O in

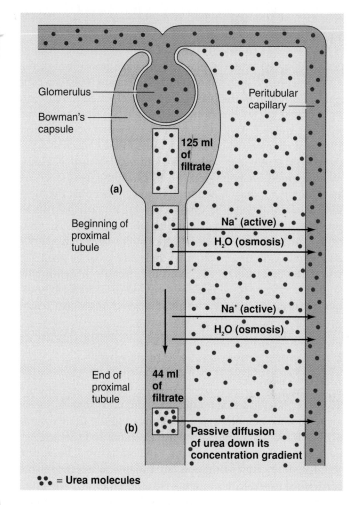

● **FIGURE 13-13**

Passive reabsorption of urea at the end of the proximal tubule. (a) In Bowman's capsule and at the beginning of the proximal tubule, urea is at the same concentration as in the plasma and surrounding interstitial fluid. (b) By the end of the proximal tubule, 65% of the original filtrate has been reabsorbed, concentrating the filtered urea in the remaining filtrate. This establishes a concentration gradient favoring passive reabsorption of urea.

the proximal tubule gradually reduces the original 125 ml/min of filtrate until only 44 ml/min of fluid remain in the lumen by the end of the proximal tubule (65% of the H_2O in the original filtrate, or 81 ml/min, has been reabsorbed). Substances that have been filtered but not reabsorbed become progressively more concentrated in the tubular fluid as H_2O is reabsorbed while they are left behind. Urea is one such substance. Urea's concentration as it is filtered at the glomerulus is identical to its concentration in the plasma entering the peritubular capillaries. The quantity of urea present in the 125 ml of filtered fluid at the beginning of the proximal tubule, however, is concentrated almost threefold in the 44 ml left at the end of the proximal tubule. As a result, the urea concentration within the tubular fluid becomes much greater than the urea concentration in the adjacent capillaries. Therefore, a concentration gradient is created for urea to passively diffuse from the tubular lumen into the peritubular capillary plasma. Because the walls of the proximal tubules are only somewhat permeable to urea, only about 50% of the filtered urea is passively reabsorbed by this means.

Clinical Note Even though only half of the filtered urea is eliminated from the plasma with each pass through the nephrons, this removal rate is adequate. The urea concentration in the plasma becomes elevated only in impaired kidney function, when much less than half of the urea is removed. An elevated urea level was one of the first chemical characteristics to be identified in the plasma of patients with severe renal failure. Accordingly, clinical measurement of **blood urea nitrogen (BUN)** came into use as a crude assessment of kidney function. It is now known that the most serious consequences of renal failure are not attributable to the retention of urea, which itself is not especially toxic, but rather to the accumulation of other substances that are not adequately excreted because of their failure to be properly secreted—most notably H^+ and K^+. Health professionals still often refer to renal failure as **uremia** (urea in the blood), indicating excess urea in the blood, even though urea retention is not this condition's major threat.

In general, unwanted waste products are not reabsorbed.

The other filtered waste products besides urea, such as *phenol* and *creatinine*, are likewise concentrated in the tubular fluid as H_2O leaves the filtrate to enter the plasma, but they are not passively reabsorbed, as urea is. Urea molecules, being the smallest of the waste products, are the only wastes passively reabsorbed by this concentrating effect. Even though the other wastes are also concentrated in the tubular fluid, they cannot leave the lumen down their concentration gradients to be passively reabsorbed, because they cannot permeate the tubular wall. Therefore, the waste products, not being reabsorbed, generally remain in the tubules and are excreted in the urine in highly concentrated form. This excretion of metabolic wastes is not subject to physiologic control. When renal function is normal, however, the excretory processes proceed at a satisfactory rate even though they are not controlled.

We have now completed our discussion of tubular reabsorption and are going to shift our attention to the other basic renal process carried out by the tubules—tubular secretion.

TUBULAR SECRETION

Like tubular reabsorption, tubular secretion involves transepithelial transport, but now the steps are reversed. By providing a second route of entry into the tubules for selected substances, *tubular secretion*, the discrete transfer of substances from the peritubular capillaries into the tubular lumen, is a supplemental mechanism that hastens elimination of these compounds from the body. Anything that gains entry to the tubular fluid, whether by glomerular filtration or tubular secretion, and fails to be reabsorbed is eliminated in the urine.

The most important substances secreted by the tubules are *hydrogen ion (H^+)*, *potassium (K^+)*, and *organic anions and cations*, many of which are compounds foreign to the body.

Hydrogen ion secretion is important in acid–base balance.

Renal H^+ secretion is extremely important in regulating acid–base balance in the body, as you will learn in the next chapter. Hydrogen ion can be added to the filtered fluid by being secreted by the proximal, distal, and collecting tubules, with the extent of H^+ secretion depending on the acidity of the body fluids.

Potassium secretion is controlled by aldosterone.

Potassium ion is selectively moved in opposite directions in different parts of the tubule; it is actively reabsorbed in the proximal tubule and actively secreted in the distal and collecting tubules. Early in the tubule potassium is reabsorbed in a constant, unregulated fashion, whereas K^+ secretion later in the tubule is variable and subject to regulation.

During K^+ depletion, K^+ secretion in the distal parts of the nephron is reduced to a minimum, so only the small percentage of filtered K^+ that escapes reabsorption in the proximal tubule is excreted in the urine. In this way, K^+ that normally would have been lost in urine is conserved for the body. Conversely, when plasma K^+ levels are elevated, K^+ secretion is adjusted so that just enough K^+ is added to the filtrate for elimination to reduce the plasma K^+ concentration to normal. Thus K^+ secretion, not the filtration or reabsorption of K^+, is varied in a controlled fashion to regulate the rate of K^+ excretion and maintain the desired plasma K^+ concentration.

MECHANISM OF K^+ SECRETION

Potassium secretion in the distal and collecting tubules is coupled to Na^+ reabsorption by the energy-dependent basolateral Na^+–K^+ pump (● Figure 13-14). This pump not only moves Na^+ out of the cell into the lateral space but also transports K^+ from the lateral space into the tubular cells. The resulting high intracellular K^+ concentration favors net diffusion of K^+ from the cells into the tubular lumen. Movement across the luminal membrane occurs passively through the large number of K^+ channels in this barrier in the distal and collecting tubules. By keeping the interstitial fluid concentration of K^+ low as it transports K^+ into the tubular cells from the surrounding interstitial fluid, the basolateral pump encourages passive diffusion of K^+ out of the peritubular capillary plasma into the interstitial fluid. Potassium leaving the plasma in this manner is later pumped into the cells, from which it diffuses into the lumen. In this way, the basolateral pump actively induces the net secretion of K^+ from the peritubular capillary plasma into the tubular lumen.

Because K^+ secretion is linked with Na^+ reabsorption by the Na^+–K^+ pump, why isn't K^+ secreted throughout the Na^+-reabsorbing segments of the tubule instead of taking place only in the distal parts of the nephron? The answer lies in the location of the passive K^+ channels. In the distal and collecting tubules, the K^+ channels are concentrated in the luminal membrane, providing a route for K^+ pumped into

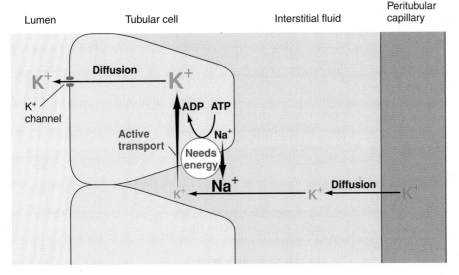

● FIGURE 13-14

Potassium secretion. The basolateral pump simultaneously transports Na$^+$ into the lateral space and K$^+$ into the tubular cell. In the parts of the tubule that secrete K$^+$, this ion leaves the cell through channels located in the luminal border, thus being secreted. (In the parts of the tubule that do not secrete K$^+$, the K$^+$ pumped into the cell during Na$^+$ reabsorption leaves the cell through channels located in the basolateral border, thus being retained in the body.)

the cell to exit into the lumen, thus being secreted. In the other tubular segments, the K$^+$ channels are located primarily in the basolateral membrane. As a result, K$^+$ pumped into the cell from the lateral space by the Na$^+$–K$^+$ pump simply diffuses back out into the lateral space through these channels. This K$^+$ recycling permits the ongoing operation of the Na$^+$–K$^+$ pump to accomplish Na$^+$ reabsorption with no local net effect on K$^+$.

CONTROL OF K$^+$ SECRETION

Several factors can alter the rate of K$^+$ secretion, the most important being aldosterone. This hormone stimulates K$^+$ secretion by the tubular cells late in the nephron simultaneous to enhancing these cells' reabsorption of Na$^+$. A rise in plasma K$^+$ concentration directly stimulates the adrenal cortex to increase its output of aldosterone, which in turn promotes the secretion and ultimate urinary excretion and elimination of excess K$^+$. Conversely, a decline in plasma K$^+$ concentration causes a reduction in aldosterone secretion and a corresponding decrease in aldosterone-stimulated renal K$^+$ secretion.

Note that a rise in plasma K$^+$ concentration directly stimulates aldosterone secretion by the adrenal cortex, whereas a fall in plasma Na$^+$ concentration stimulates aldosterone secretion by means of the complex RAAS pathway. Thus aldosterone secretion can be stimulated by two separate pathways (● Figure 13-15).

The kidneys usually exert a fine degree of control over plasma K$^+$ concentration. This is extremely important, because even minor fluctuations in plasma K$^+$ concentration can detrimentally influence the membrane electrical activity of excitable tissues, adversely affecting their performance.

For example, a rise in ECF K$^+$ concentration causes cardiac overexcitability, which can lead to a rapid heart rate and even fatal cardiac arrhythmias.

❙ **Organic anion and cation secretion helps efficiently eliminate foreign compounds from the body.**

The proximal tubule contains two distinct types of secretory carriers, one for the secretion of organic anions and a separate system for secretion of organic cations.

FUNCTIONS OF ORGANIC ION SECRETORY SYSTEMS

These systems serve two major functions. First, by adding more of a particular type of organic ion to the quantity that has already gained entry to the tubular fluid by glomerular filtration, these organic secretory pathways facilitate excretion of these substances. Included among these organic ions are certain blood-borne chemical messengers such as prostaglandins, histamine, and norepinephrine, that, having served their purpose, must be rapidly removed from the blood so that their biological activity is not unduly prolonged.

● FIGURE 13-15

Dual control of aldosterone secretion of K$^+$ and Na$^+$

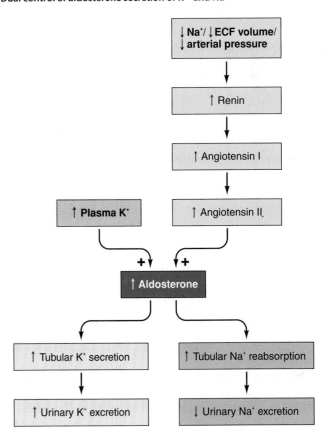

Second, the proximal tubule organic-ion secretory systems play a key role in eliminating many foreign compounds from the body. These systems can secrete a large number of different organic ions, both those produced endogenously (within the body) and those foreign organic ions that have gained access to the body fluids. This nonselectivity permits these organic-ion secretory systems to hasten removal of many foreign organic chemicals, including food additives, environmental pollutants (for example, pesticides), drugs, and other nonnutritive organic substances that have entered the body. Even though this mechanism helps rid the body of potentially harmful foreign compounds, it is not subject to physiologic adjustments. The carriers cannot pick up their secretory pace when confronting an elevated load of these organic ions.

Clinical Note Many drugs, such as penicillin and nonsteroidal anti-inflammatory drugs (NSAIDs), are eliminated from the body by the organic-ion secretory systems. To keep the plasma concentration of these drugs at effective levels, the dosage must be repeated on a regular, frequent basis to keep pace with the rapid removal of these compounds in the urine.

SUMMARY OF REABSORPTIVE AND SECRETORY PROCESSES

This completes our discussion of the reabsorptive and secretory processes that occur across the proximal and distal portions of the nephron. These processes are summarized in ▲ Table 13-3. To generalize, the proximal tubule does most of the reabsorbing. This mass reabsorber transfers much of the filtered water and needed solutes back into the blood in unregulated fashion. Similarly, the proximal tubule is the major site of secretion, with the exception of K^+ secretion. The distal and collecting tubules then determine the final amounts of H_2O, Na^+, K^+, and H^+ that are excreted in the urine and thus eliminated from the body. They do so by fine-tuning the amount of Na^+ and H_2O reabsorbed and the amount of K^+ and H^+ secreted. These processes in the distal part of the nephron are all subject to control, depending on the body's momentary needs. The unwanted filtered waste products are left behind to be eliminated in urine, along with excess amounts of filtered or secreted nonwaste products that fail to be reabsorbed.

We will next focus on the end result of the basic renal processes—what's left in the tubules to be excreted in urine, and, as a consequence, what has been cleared from plasma.

URINE EXCRETION AND PLASMA CLEARANCE

Of the 125 ml of plasma filtered per minute, typically 124 ml/min are reabsorbed, so the final quantity of urine formed averages 1 ml/min. Thus, of the 180 liters filtered per day, 1.5 liters of urine are excreted.

Urine contains high concentrations of various waste products plus variable amounts of the substances regulated by the kidneys, with any excess quantities having spilled into the

▲ **TABLE 13-3**

Summary of Transport across Proximal and Distal Portions of the Nephron

PROXIMAL TUBULE	
Reabsorption	**Secretion**
67% of filtered Na^+ actively reabsorbed; not subject to control; Cl^- follows passively	Variable H^+ secretion depending on acid–base status of body
All filtered glucose and amino acids reabsorbed by secondary active transport; not subject to control	Organic ion secretion; not subject to control
Variable amounts of filtered PO_4^{3-} and other electrolytes reabsorbed; subject to control	
65% of filtered H_2O osmotically reabsorbed; not subject to control	
50% of filtered urea passively reabsorbed; not subject to control	
Almost all filtered K^+ reabsorbed; not subject to control	

DISTAL TUBULE AND COLLECTING DUCT	
Reabsorption	**Secretion**
Variable Na^+ reabsorption, controlled by aldosterone; Cl^- follows passively	Variable H^+ secretion, depending on acid–base status of body
Variable H_2O reabsorption, controlled by vasopressin	Variable K^+ secretion, controlled by aldosterone

urine. Useful substances are conserved by reabsorption, so they do not appear in the urine.

A relatively small change in the quantity of filtrate reabsorbed can bring about a large change in the volume of urine formed. For example, a reduction of less than 1% in the total reabsorption rate, from 124 to 123 ml/min, increases the urinary excretion rate by 100%, from 1 to 2 ml/min.

▌ **Plasma clearance is the volume of plasma cleared of a particular substance per minute.**

Compared to plasma entering the kidneys through the renal arteries, plasma leaving the kidneys through the renal veins lacks the materials that were left behind to be eliminated in the urine. By excreting substances in the urine, the kidneys clean or "clear" the plasma flowing through them of these substances. The **plasma clearance** of any substance is defined as the volume of plasma completely cleared of that substance by the kidneys per minute.[2] It refers, not to the *amount of the substance* removed, but to the *volume of plasma* from which that amount was removed. Plasma clearance is actually a more useful measure than urine excretion; it is more important to know what effect urine excretion has on removing materials from body fluids than to know the volume and composition of discarded urine. Plasma clearance expresses the kidneys' effectiveness in removing various substances from the internal fluid environment. The plasma clearance rate varies for different substances, depending on how the kidneys handle each substance.

▌ If a substance is filtered but not reabsorbed or secreted, its plasma clearance rate equals the GFR.

Assume that a plasma constituent, substance X, is freely filterable at the glomerulus but is not reabsorbed or secreted. As 125 ml/min of plasma are filtered and subsequently reabsorbed, the quantity of substance X originally contained within the 125 ml is left behind in the tubules to be excreted. Thus 125 ml of plasma are cleared of substance X each minute (● Figure 13-16a). (Of the 125 ml/min of plasma filtered, 124 ml/min of the filtered fluid are returned, through reabsorption, to the plasma minus substance X, thus clearing this 124 ml/min of substance X. In addition, the 1 ml/min of fluid lost in urine is eventually replaced by an equivalent volume of ingested H_2O that is already clear of substance X. Therefore, 125 ml of plasma cleared of substance X are, in effect, returned to the plasma for every 125 ml of plasma filtered per minute.)

There is no endogenous chemical with the characteristics of substance X. All substances naturally present in the plasma, even wastes, are reabsorbed or secreted to some extent. However, **inulin** (do not confuse with insulin), a harmless foreign carbohydrate produced by Jerusalem artichokes, is freely filtered and not reabsorbed or secreted—an ideal substance X. Inulin can be injected and its plasma clearance determined as a clinical means of ascertaining the GFR. Because all glomerular filtrate formed is cleared of inulin, the volume of plasma cleared of inulin per minute equals the volume of plasma filtered per minute—that is, the GFR.

▌ If a substance is filtered and reabsorbed but not secreted, its plasma clearance rate is always less than the GFR.

Some or all of a reabsorbable substance that has been filtered is returned to the plasma. Because less than the filtered volume of plasma will have been cleared of the substance, the plasma clearance rate of a reabsorbable substance is always less than the GFR. For example, the plasma clearance for glucose is normally zero. All the filtered glucose is reabsorbed along with the rest of the returning filtrate, so none of the plasma is cleared of glucose (● Figure 13-16b).

For a substance that is partially reabsorbed, such as urea, only part of the filtered plasma is cleared of that substance. With about 50% of the filtered urea being passively reabsorbed, only half of the filtered plasma, or 62.5 ml, is cleared of urea each minute (● Figure 13-16c).

▌ If a substance is filtered and secreted but not reabsorbed, its plasma clearance rate is always greater than the GFR.

Tubular secretion allows the kidneys to clear certain materials from the plasma more efficiently. Only 20% of the plasma entering the kidneys is filtered. The remaining 80% passes unfiltered into the peritubular capillaries. The only means by which this unfiltered plasma can be cleared of any substance during this trip through the kidneys before being returned to the general circulation is by secretion. An example is H^+. Not only is filtered plasma cleared of nonreabsorbable H^+, but the plasma from which H^+ is secreted is also cleared of H^+. For example, if the quantity of H^+ secreted is equivalent to the quantity of H^+ present in 25 ml of plasma, the clearance rate for H^+ will be 150 ml/min at the normal GFR of 125 ml/min. Every minute 125 ml of plasma will lose its H^+ through filtration and failure of reabsorption, and 25 more ml of plasma will lose its H^+ through secretion. The plasma clearance for a secreted substance is always greater than the GFR (● Figure 13-16d).

Just as inulin can be used clinically to determine the GFR, plasma clearance of another foreign compound, the organic anion **para-aminohippuric acid (PAH)**, can be used to measure renal plasma flow. Like inulin, PAH is freely filterable and nonreabsorbable. It differs, however, in that all the PAH in the plasma that escapes filtration is secreted from the peritubular capillaries by the organic-anion secretory pathway in the proximal tubule. Thus PAH is removed from all the plasma that flows through the kidneys—both from plasma that is filtered and subsequently reabsorbed without its PAH, and from unfiltered plasma that continues on in the peritubular capillaries and loses its PAH by active secretion into the tubules. Because all the plasma that flows through the kidneys is cleared of PAH, the plasma clearance for PAH is a reasonable estimate of the rate of plasma flow through the kidneys. Typically, renal plasma flow averages 625 ml/min, for a renal blood flow (plasma plus blood cells) of 1140 ml/min—over 20% of the cardiac output.

FILTRATION FRACTION

If you know PAH clearance (renal plasma flow) and inulin clearance (GFR), you can easily determine the **filtration fraction,** the fraction of the plasma flowing through the glomeruli that is filtered into the tubules.

---

[2] Actually, plasma clearance is an artificial concept, because when a particular substance is excreted in the urine that substance's concentration in the plasma as a whole is uniformly decreased by thorough mixing in the circulatory system. However, it is useful for comparative purposes to consider clearance in effect as the volume of plasma that would have contained the total quantity of the substance (at the substance's concentration prior to excretion) that the kidneys excreted in one minute; that is, the hypothetical volume of plasma completely cleared of that substance per minute.

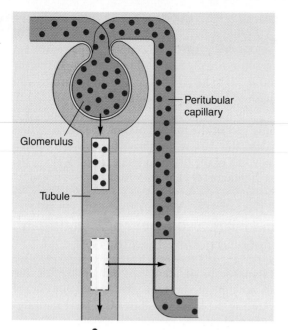

Peritubular
capillary

Glomerulus

Tubule

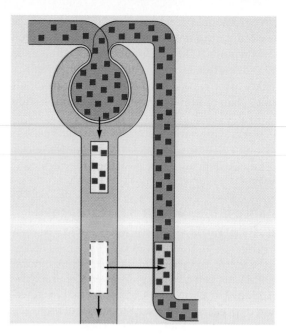

 In urine

For a substance filtered and not reabsorbed or secreted, such as inulin, all of the filtered plasma is cleared of the substance.

(a)

For a substance filtered, not secreted, and completely reabsorbed, such as glucose, none of the filtered plasma is cleared of the substance.

(b)

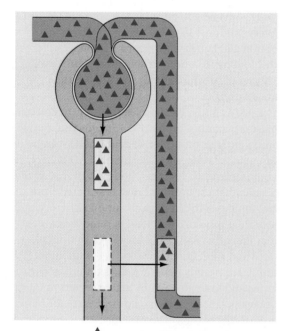

 For a substance filtered, not secreted, and partially reabsorbed, such as urea, only a portion of the filtered plasma is cleared of the substance.

(c)

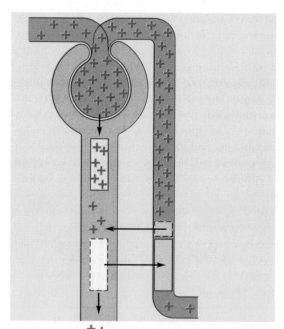

 For a substance filtered and secreted but not reabsorbed, such as hydrogen ion, all of the filtered plasma is cleared of the substance, and the peritubular plasma from which the substance is secreted is also cleared.

(d)

● **FIGURE 13-16**

Plasma clearance for substances handled in different ways by the kidneys

PhysioEdge For an animation of this figure, click the Urine Formation Process tab in the Renal Physiology tutorial on the CD-ROM.

$$\text{Filtration fraction} = \frac{\text{GFR (plasma inulin clearance)}}{\text{renal plasma flow (plasma PAH clearance)}}$$

$$= \frac{125 \text{ ml/min}}{625 \text{ ml/min}} = 20\%$$

Thus 20% of the plasma that enters the glomeruli is typically filtered.

▮ The kidneys can excrete urine of varying concentrations depending on the body's state of hydration.

Having considered how the kidneys deal with a variety of solutes in the plasma, we will now concentrate on renal handling of plasma H_2O. The ECF osmolarity (solute concentration) depends on the relative amount of H_2O compared to solute. At normal fluid balance and solute concentration, the body fluids are **isotonic** at an osmolarity of 300 milliosmols/liter (mosm/liter) (see pp. 55 and A-10). If too much H_2O is present relative to the solute load, the body fluids are **hypotonic**, which means they are too dilute at an osmolarity less than 300 mosm/liter. However, if a H_2O deficit exists relative to the solute load, the body fluids are too concentrated or are **hypertonic**, having an osmolarity greater than 300 mosm/liter.

Knowing that the driving force for H_2O reabsorption throughout the entire length of the tubules is an osmotic gradient between the tubular lumen and surrounding interstitial fluid, you would expect, given osmotic considerations, that the kidneys could not excrete urine more or less concentrated than the body fluids. Indeed, this would be the case if the interstitial fluid surrounding the tubules in the kidneys were identical in osmolarity to the remaining body fluids. Water reabsorption would proceed only until the tubular fluid equilibrated osmotically with the interstitial fluid, and the body would have no way to eliminate excess H_2O when the body fluids were hypotonic, or to conserve H_2O in the presence of hypertonicity.

Fortunately, a large **vertical osmotic gradient** is uniquely maintained in the interstitial fluid of the medulla of each kidney. The concentration of the interstitial fluid progressively increases from the cortical boundary down through the depth of the renal medulla until it reaches a maximum of 1200 mosm/liter in humans at the junction with the renal pelvis (● Figure 13-17).

By a mechanism described shortly, this gradient enables the kidneys to produce urine that ranges in concentration from 100 to 1200 mosm/liter, depending on the body's state of hydration. When the body is in ideal fluid balance, 1 ml/min of isotonic urine is formed. When the body is overhydrated (too much H_2O), the kidneys can produce a large volume of dilute urine (up to 25 ml/min and hypotonic at 100 mosm/liter), eliminating the excess H_2O in the urine. Conversely, the kidneys can put out a small volume of concentrated urine (down to 0.3 ml/min and hypertonic at 1200 mosm/liter) when the body is dehydrated (too little H_2O), conserving H_2O for the body.

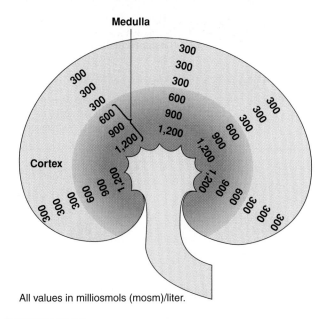

All values in milliosmols (mosm)/liter.

● **FIGURE 13-17**

Vertical osmotic gradient in the renal medulla. Schematic representation of the kidney rotated 90° from its normal position in an upright person for better visualization of the vertical osmotic gradient in the renal medulla. The osmolarity of the interstitial fluid throughout the renal cortex is isotonic at 300 mosm/liter, but the osmolarity of the interstitial fluid in the renal medulla increases progressively from 300 mosm/liter at the boundary with the cortex to a maximum of 1200 mosm/liter at the junction with the renal pelvis.

Unique anatomic arrangements and complex functional interactions between the various nephron components in the renal medulla establish and use the vertical osmotic gradient. In most nephrons the hairpin loop of Henle dips only slightly into the medulla, but in about 20% of nephrons the loop plunges through the entire depth of the medulla so that the tip of the loop lies near the renal pelvis (● Figure 13-18). Flow in the long loops of Henle is considered countercurrent, because the flow in the two adjacent limbs of the loop moves in opposite directions. Also running through the medulla in the descending direction only, on their way to the renal pelvis, are the collecting ducts that serve both types of nephrons. This arrangement, coupled with the permeability and transport characteristics of these tubular segments, plays a key role in the kidneys' ability to produce urine of varying concentrations, depending on the body's needs for water conservation or elimination. Briefly, the long loops of Henle *establish* the vertical osmotic gradient and the collecting ducts of all nephrons *use* the gradient, in conjunction with the hormone vasopressin, to produce urine of varying concentrations. Collectively, this entire functional organization is known as the **medullary countercurrent system.** We will examine each of its facets in greater detail.

▮ The medullary vertical osmotic gradient is established by countercurrent multiplication.

We will follow the filtrate through a long-looped nephron to see how this structure establishes a vertical osmotic gradient

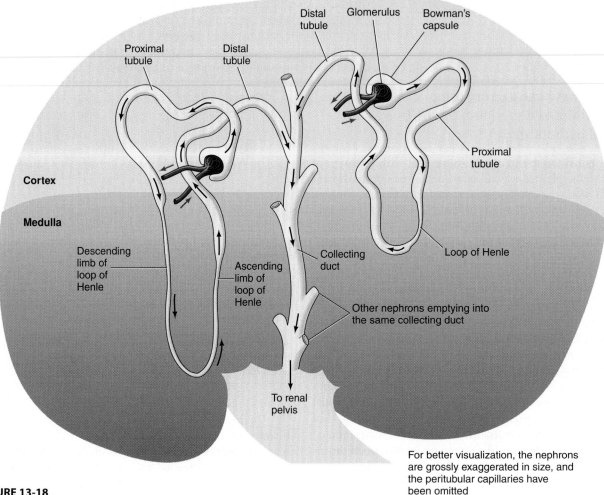

Long-looped nephron important in establishing the medullary vertical osmotic gradient

Most abundant type of nephron

Proximal tubule

Distal tubule

Distal tubule

Glomerulus

Bowman's capsule

Proximal tubule

Cortex

Medulla

Descending limb of loop of Henle

Ascending limb of loop of Henle

Collecting duct

Loop of Henle

Other nephrons emptying into the same collecting duct

To renal pelvis

For better visualization, the nephrons are grossly exaggerated in size, and the peritubular capillaries have been omitted

● **FIGURE 13-18**

Schematic representation of the two types of nephrons. Note that the loop of Henle of long-looped nephrons plunges deep into the medulla.

---

in the medulla. Immediately after the filtrate is formed, uncontrolled osmotic reabsorption of filtered H_2O occurs in the proximal tubule secondary to active Na^+ reabsorption. As a result, by the end of the proximal tubule about 65% of the filtrate has been reabsorbed, but the 35% remaining in the tubular lumen still has the same osmolarity as the body fluids. Therefore the fluid entering the loop of Henle is still isotonic. An additional 15% of the filtered H_2O is obligatorily reabsorbed from the loop of Henle during the establishment and maintenance of the vertical osmotic gradient, with the osmolarity of the tubular fluid being altered in the process.

PROPERTIES OF THE DESCENDING AND ASCENDING LIMBS OF A LONG HENLE'S LOOP

The following functional distinctions between the descending limb of a long Henle's loop (which carries fluid from the proximal tubule down into the depths of the medulla) and the ascending limb (which carries fluid up and out of the medulla

into the distal tubule) are crucial for establishing the incremental osmotic gradient in the medullary interstitial fluid.

The *descending limb*

1. is highly permeable to H_2O.
2. does not actively extrude Na^+. (That is, it does not reabsorb Na^+. It is the only segment of the entire tubule that does not do so.)

The *ascending limb*

1. actively transports NaCl out of the tubular lumen into the surrounding interstitial fluid.
2. is always impermeable to H_2O, so salt leaves the tubular fluid without H_2O osmotically following along.

MECHANISM OF COUNTERCURRENT MULTIPLICATION

The close proximity and countercurrent flow of the two limbs allow important interactions between them. Even though the flow of fluids is continuous through the loop of Henle, we will visualize what happens step by step, much like an animated film run so slowly that each frame can be viewed.

- *Initial scene* (● Figure 13-19a). Before the vertical osmotic gradient is established, the medullary interstitial fluid concentration is uniformly 300 mosm/liter, as is the rest of the body fluids.
- *Step 1* (● Figure 13-19b). The active salt pump in the ascending limb can transport NaCl out of the lumen until the surrounding interstitial fluid is 200 mosm/liter more concentrated than the tubular fluid in this limb. When the ascending limb pump starts actively extruding salt, the medullary interstitial fluid becomes hypertonic. Water cannot follow osmotically from the ascending limb, because this limb is impermeable to H_2O. However, net diffusion of H_2O does occur from the descending limb into the interstitial fluid. The tubular fluid entering the descending limb from the proximal tubule is isotonic. Because the descending limb is highly permeable to H_2O, net diffusion of H_2O occurs by osmosis out of the descending limb into the more concentrated interstitial fluid. The passive movement of H_2O out of the descending limb continues until the osmolarities of the fluid in the descending limb and interstitial fluid become equilibrated. Thus the tubular fluid entering the loop of Henle immediately starts to become more concentrated as it loses H_2O. At equilibrium, the osmolarity of the ascending limb fluid is 200 mosm/liter and the osmolarities of the interstitial fluid and descending limb fluid are equal at 400 mosm/liter.
- *Step 2* (● Figure 13-19c). If we now advance the entire column of fluid in the loop of Henle several frames, a mass of 200-mosm/liter fluid exits from the top of the ascending limb into the distal tubule, and a new mass of isotonic fluid at 300 mosm/liter enters the top of the descending limb from the proximal tubule. At the bottom of the loop, a comparable mass of 400-mosm/liter fluid from the descending limb moves forward around the tip into the ascending limb, placing it opposite a 400-mosm/liter region in the descending limb. Note that the 200-mosm/liter concentration difference has been lost at both the top and the bottom of the loop.
- *Step 3* (● Figure 13-19d). The ascending limb pump again transports NaCl out while H_2O passively leaves the descending limb until a 200-mosm/liter difference is re-established between the ascending limb and both the interstitial fluid and descending limb at each horizontal level. Note, however, that the concentration of tubular fluid is progressively increasing in the descending limb and progressively decreasing in the ascending limb.
- *Step 4* (● Figure 13-19e). As the tubular fluid is advanced still further, the 200-mosm/liter concentration gradient is disrupted once again at all horizontal levels.
- *Step 5* (● Figure 13-19f). Again, active extrusion of NaCl from the ascending limb, coupled with the net diffusion of H_2O out of the descending limb, re-establishes the 200-mosm/liter gradient at each horizontal level.
- Step 6 and on (● Figure 13-19g). As the fluid flows slightly forward again and this stepwise process continues, the fluid in the descending limb becomes progressively more hyper-

tonic until it reaches a maximum concentration of 1200 mosm/liter at the bottom of the loop, four times the normal concentration of body fluids. Because the interstitial fluid always achieves equilibrium with the descending limb, an incremental vertical concentration gradient ranging from 300 to 1200 mosm/liter is likewise established in the medullary interstitial fluid. In contrast, the concentration of the tubular fluid progressively decreases in the ascending limb as salt is pumped out but H_2O is unable to follow. In fact, the tubular fluid even becomes hypotonic before leaving the ascending limb to enter the distal tubule at a concentration of 100 mosm/liter, one third the normal concentration of body fluids.

Note that although a gradient of only 200 mosm/liter exists between the ascending limb and the surrounding fluids at each medullary horizontal level, a much larger vertical gradient exists from the top to the bottom of the medulla. Even though the ascending limb pump can generate a gradient of only 200 mosm/liter, this effect is multiplied into a large vertical gradient because of the countercurrent flow within the loop. This concentrating mechanism accomplished by the loop of Henle is known as **countercurrent multiplication**.

We have artificially described countercurrent multiplication in a stop-and-flow, stepwise fashion to facilitate understanding. It is important to realize that once the incremental medullary gradient is established, it stays constant, because of the continuous flow of fluid coupled with the ongoing ascending-limb active transport and accompanying descending-limb passive fluxes.

BENEFITS OF COUNTERCURRENT MULTIPLICATION

If you consider only what happens to the tubular fluid as it flows through the loop of Henle, the whole process seems an exercise in futility. The isotonic fluid that enters the loop becomes progressively more concentrated as it flows down the descending limb, achieving a maximum concentration of 1200 mosm/liter, only to become progressively more dilute as it flows up the ascending limb, finally leaving the loop at a minimum concentration of 100 mosm/liter. What is the point of concentrating the fluid fourfold and then turning around and diluting it until it leaves at one third the concentration at which it entered? Such a mechanism offers two benefits. First, it establishes a vertical osmotic gradient in the medullary interstitial fluid. This gradient, in turn, is used by the collecting ducts to concentrate the tubular fluid so that a urine *more concentrated* than normal body fluids can be excreted. Second, the fact that the fluid is hypotonic as it enters the distal parts of the tubule enables the kidneys to excrete a urine *more dilute* than normal body fluids. Let's see how.

❚ Vasopressin-controlled, variable H_2O reabsorption occurs in the final tubular segments.

After obligatory H_2O reabsorption from the proximal tubule (65% of the filtered H_2O) and loop of Henle (15% of the filtered H_2O), 20% of the filtered H_2O remains in the lumen to

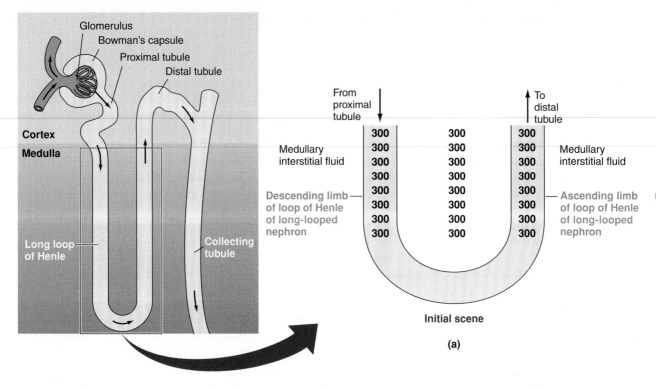

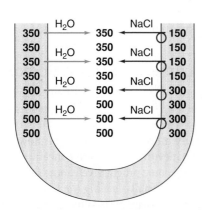

Step 3: The ascending limb pump and descending limb passive fluxes re-establish the 200 mosm/liter gradient at each horizontal level.

(d)

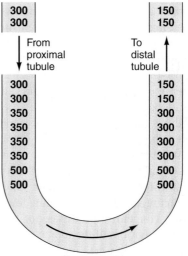

Step 4: Let the fluid flow forward several "frames" once again.

(e)

● **FIGURE 13-19**

Countercurrent multiplication in the renal medulla

 For an animation of this figure, click the Vertical Osmotic Gradient and Countercurrent Multiplication tab in the Renal Physiology tutorial on the CD-ROM.

enter the distal and collecting tubules for variable reabsorption that is under hormonal control. This is still a large volume of filtered H_2O subject to regulated reabsorption; 20% × GFR (180 liters/day) = 36 liters per day to be reabsorbed to varying extents, depending on the body's state of hydration. This is more than 13 times the amount of plasma H_2O in the entire circulatory system.

The fluid leaving the loop of Henle enters the distal tubule at 100 mosm/liter, so it is hypotonic to the surrounding isotonic (300 mosm/liter) interstitial fluid of the renal cortex through which the distal tubule passes. The distal tubule then empties into the collecting duct, which is bathed by progressively increasing concentrations (300 to 1200 mosm/liter) of surrounding interstitial fluid as it descends through the medulla.

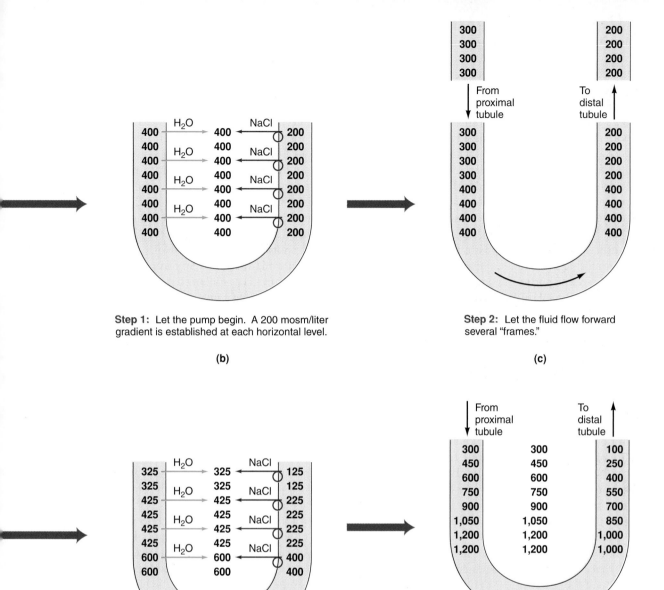

Step 1: Let the pump begin. A 200 mosm/liter gradient is established at each horizontal level.

(b)

Step 2: Let the fluid flow forward several "frames."

(c)

Step 5: The 200 mosm/liter gradient at each horizontal level is established once again.

(f)

Step 6 and on: The final vertical osmotic gradient is established and maintained by the ongoing countercurrent multiplication of the long loops of Henle.

(g)

ROLE OF VASOPRESSIN

For H_2O absorption to occur across a segment of the tubule, two criteria must be met: (1) An osmotic gradient must exist across the tubule, and (2) the tubular segment must be permeable to H_2O. The distal and collecting tubules are *impermeable* to H_2O except in the presence of **vasopressin**, also known as **antidiuretic hormone** (*anti* means "against"; *diuretic*

means "increased urine output"),[3] which increases their permeability to H_2O. Vasopressin is produced by several specific neuronal cell bodies in the *hypothalamus,* part of the brain,

---

[3]Even though textbooks traditionally have tended to use the name *antidiuretic hormone* for this hormone, especially when discussing its actions on the kidney, investigators in the field now prefer *vasopressin.*

then stored in the *posterior pituitary gland,* which is attached to the hypothalamus by a thin stalk. The hypothalamus controls release of vasopressin from the posterior pituitary into the blood. In negative-feedback fashion, vasopressin secretion is stimulated by a H_2O deficit, when the ECF is too concentrated (that is, hypertonic) and H_2O must be conserved for the body, and inhibited by a H_2O excess, when the ECF is too dilute (that is, hypotonic) and surplus H_2O must be eliminated in urine.

Vasopressin reaches the basolateral membrane of the tubular cells lining the distal and collecting tubules through the circulatory system. Here it binds with receptors specific for it. This binding activates the cyclic AMP (cAMP) second-messenger system within the tubular cells (see p. 98), which ultimately increases permeability of the opposite luminal membrane to H_2O by promoting insertion of aquaporins in this membrane. Without these aquaporins, the luminal membrane is impermeable to H_2O. Once H_2O enters the tubular cells from the filtrate through these vasopressin-regulated luminal water channels, it passively leaves the cells down the osmotic gradient across the cells' basolateral membrane (which is always permeable to H_2O) to enter the interstitial fluid. By permitting more H_2O to permeate from the lumen into the tubular cells, these additional luminal channels thus increase H_2O reabsorption from the filtrate into the interstitial fluid. The tubular response to vasopressin is graded; the more vasopressin present, the more water channels inserted, and the greater the permeability of the distal and collecting tubules to H_2O. The increase in luminal membrane water channels is not permanent, however. The channels are retrieved when vasopressin secretion decreases and cAMP activity is similarly decreased. Accordingly, H_2O permeability is reduced when vasopressin secretion decreases.

Vasopressin influences H_2O permeability only in the distal part of the nephron, especially the collecting ducts. It has no influence over the 80% of the filtered H_2O that is obligatorily reabsorbed without control in the proximal tubule and loop of Henle. The ascending limb of Henle's loop is always impermeable to H_2O, even in the presence of vasopressin.

REGULATION OF H$_2$O REABSORPTION IN RESPONSE TO A H$_2$O DEFICIT

When vasopressin secretion increases in response to a H_2O deficit and the permeability of the distal and collecting tubules to H_2O accordingly increases, the hypotonic tubular fluid entering the distal part of the nephron can lose progressively more H_2O by osmosis into the interstitial fluid as the tubular fluid first flows through the isotonic cortex and then is exposed to the ever-increasing osmolarity of the medullary interstitial fluid as it plunges toward the renal pelvis (● Figure 13-20a). As the 100 mosm/liter tubular fluid enters the distal tubule and is exposed to a surrounding interstitial fluid of 300 mosm/ liter, H_2O leaves the tubular fluid by osmosis across the now permeable tubular cells until the tubular fluid reaches a maximum concentration of 300 mosm/liter by the end of the distal tubule. As this 300-mosm/liter tubular fluid progresses farther into the collecting duct, it is exposed to even higher osmo-

larity in the surrounding medullary interstitial fluid. Consequently, the tubular fluid loses more H_2O by osmosis and becomes further concentrated, only to move farther forward and be exposed to an even higher interstitial fluid osmolarity and lose even more H_2O, and so on.

Under the influence of maximum levels of vasopressin, it is possible to concentrate the tubular fluid up to 1200 mosm/ liter by the end of the collecting ducts. No further modification of the tubular fluid occurs beyond the collecting duct, so what remains in the tubules at this point is urine. As a result of this extensive vasopressin-promoted reabsorption of H_2O in the late segments of the tubule, a small volume of urine concentrated up to 1200 mosm/liter can be excreted. As little as 0.3 ml of urine may be formed each minute, less than one third the normal urine flow rate of 1 ml/min. The reabsorbed H_2O entering the medullary interstitial fluid is picked up by the peritubular capillaries and returned to the general circulation, thus being conserved for the body.

Although vasopressin promotes H_2O conservation by the body, it cannot completely halt urine production, even when a person is not taking in any H_2O, because a minimum volume of H_2O must be excreted with the solute wastes. Collectively, the waste products and other constituents eliminated in the urine average 600 mosm each day. Because the maximum urine concentration is 1200 mosm/liter, the minimum volume of urine that is required to excrete these wastes is 500 ml/day (600 mosm of wastes/day ÷ 1200 mosm/liter of urine = 0.5 liter, or 500 ml/day, or 0.3 ml/min). Thus, under maximal vasopressin influence, 99.7% of the 180 liters of plasma H_2O filtered per day is returned to the blood, with an obligatory H_2O loss of half a liter.

The kidneys' ability to tremendously concentrate urine to minimize H_2O loss when necessary is possible only because of the presence of the vertical osmotic gradient in the medulla. If this gradient did not exist, the kidneys could not produce a urine more concentrated than the body fluids no matter how much vasopressin was secreted, because the only driving force for H_2O reabsorption is a concentration differential between the tubular fluid and the interstitial fluid.

REGULATION OF H$_2$O REABSORPTION IN RESPONSE TO A H$_2$O EXCESS

Conversely, when a person consumes large quantities of H_2O, the excess H_2O must be removed from the body without simultaneously losing solutes that are critical for maintaining homeostasis. Under these circumstances, no vasopressin is secreted, so the distal and collecting tubules remain impermeable to H_2O. The tubular fluid entering the distal tubule is hypotonic (100 mosm/liter), having lost salt without an accompanying loss of H_2O in the ascending limb of Henle's loop. As this hypotonic fluid passes through the distal and collecting tubules (● Figure 13-20b), the medullary osmotic gradient cannot exert any influence because of the late tubular segment's impermeability to H_2O. In other words, none of the H_2O remaining in the tubules can leave the lumen to be reabsorbed, even though the tubular fluid is less concentrated than the surrounding interstitial fluid. Thus in the absence

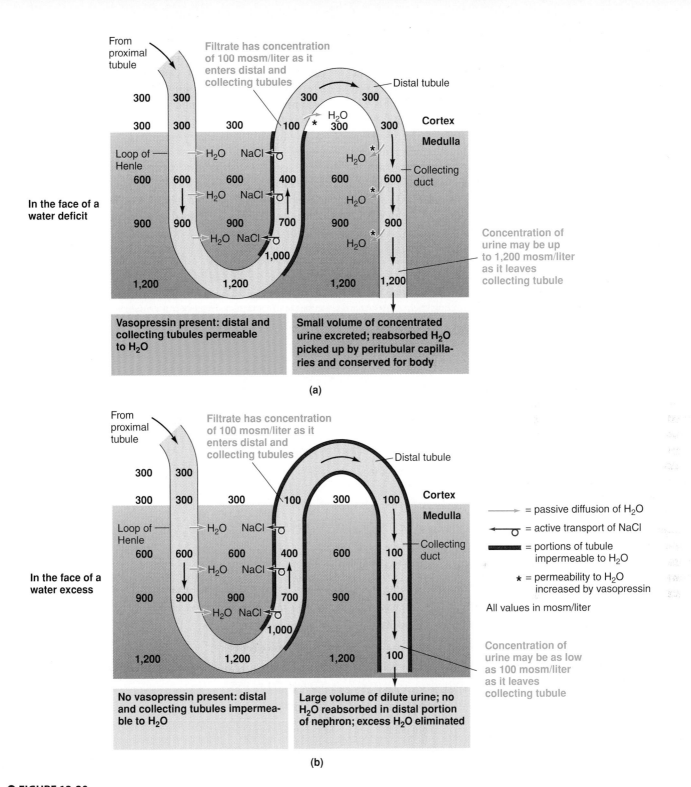

● FIGURE 13-20

Excretion of urine of varying concentration depending on the body's needs

 For an animation of this figure, click the Reabsorption and Urine Concentration tab in the Renal Physiology tutorial on the CD-ROM.

---

of vasopressin, the 20% of the filtered fluid that reaches the distal tubule is not reabsorbed. Meanwhile, excretion of wastes and other urinary solutes remains constant. The net result is a large volume of dilute urine, which helps rid the body of excess H_2O. Urine osmolarity may be as low as 100 mosm/liter, the same as in the fluid entering the distal tubule. Urine flow may be increased up to 25 ml/min in the absence of vasopressin, compared to the normal urine production of 1 ml/min.

The ability to produce urine less concentrated than the body fluids depends on the fact that the tubular fluid is hypotonic as it enters the distal part of the nephron. This dilution is accomplished in the ascending limb as NaCl is actively extruded but H_2O cannot follow. Therefore, the loop of Henle, by simultaneously establishing the medullary osmotic gradient and diluting the tubular fluid before it enters the distal segments, plays a key role in allowing the kidneys to excrete urine that ranges in concentration from 100 to 1200 mosm/liter.

Note that through the combined effects of the medullary vertical osmotic gradient and vasopressin-controlled variability in permeability to H_2O in the distal parts of the nephron, the body is able to retain or lose **free H_2O** (that is, H_2O not accompanied by solutes.) Thus, free H_2O can be reabsorbed without comparable solute reabsorption to correct for hypertonicity of the body fluids. Conversely, to rid the body of excess pure H_2O a large quantity of free H_2O can be excreted unaccompanied by comparable solute excretion (that is, **water diuresis**), thus correcting for hypotonicity of the body fluids. Water diuresis is normally a compensation for ingesting too much H_2O.

Excessive water diuresis follows alcohol ingestion. Because alcohol inhibits vasopressin secretion, the kidneys inappropriately lose too much H_2O. Typically, more fluid is lost in the urine than is consumed in the alcoholic beverage, so the body becomes dehydrated despite substantial fluid ingestion.

▲ **TABLE 13-4**

Potential Ramifications of Renal Failure

Uremic toxicity caused by retention of waste products

    Nausea, vomiting, diarrhea, and ulcers caused by a toxic effect on the digestive system

    Bleeding tendency arising from a toxic effect on platelet function

    Mental changes—such as reduced alertness, insomnia, and shortened attention span, progressing to convulsions and coma—caused by toxic effects on the central nervous system

    Abnormal sensory and motor activity caused by a toxic effect on the peripheral nerves

Metabolic acidosis* caused by the inability of the kidneys to adequately secrete H^+ that is continually being added to the body fluids as a result of metabolic activity

    Altered enzyme activity caused by the action of too much acid on enzymes

    Depression of the central nervous system caused by the action of too much acid interfering with neuronal excitability

Potassium retention* resulting from inadequate tubular secretion of K^+

    Altered cardiac and neural excitability as a result of changing the resting membrane potential of excitable cells

Sodium imbalances caused by the inability of the kidneys to adjust Na^+ excretion to balance the changes in Na^+ consumption

    Elevated blood pressure, generalized edema, and congestive heart failure if too much Na^+ is consumed

    Hypotension and, if severe enough, circulatory shock if too little Na^+ is consumed

Phosphate and calcium imbalances arising from impaired reabsorption of these electrolytes

    Disturbances in skeletal structures caused by abnormalities in deposition of calcium phosphate crystals, which harden bone

Loss of plasma proteins as a result of increased "leakiness" of the glomerular membrane

    Edema caused by a reduction in plasma-colloid osmotic pressure

Inability to vary urine concentration as a result of impairment of the countercurrent system

    Hypotonicity of body fluids if too much H_2O is ingested

    Hypertonicity of body fluids if too little H_2O is ingested

Hypertension arising from the combined effects of salt and fluid retention and vasoconstrictor action of excess angiotensin II

Anemia caused by inadequate erythropoietin production

Depression of the immune system, most likely caused by toxic levels of wastes and acids

    Increased susceptibility to infections

---

*Among the most life-threatening consequences of renal failure.

Dialysis: Cellophane Tubing or Abdominal Lining as an Artificial Kidney

Because chronic renal failure is irreversible and eventually fatal, treatment is aimed at maintaining renal function by alternative methods, such as dialysis and kidney transplant. The process of dialysis bypasses the kidneys to maintain normal fluid and electrolyte balance and remove wastes artificially. In the original method of dialysis, **hemodialysis**, a patient's blood is pumped through cellophane tubing that is surrounded by a large volume of fluid similar in composition to normal plasma. After dialysis, the blood is returned to the patient's circulatory system. Like capillaries, cellophane is highly permeable to most plasma constituents but is impermeable to plasma proteins. As blood flows through the tubing, solutes move across the cellophane down their individual concentration gradients; plasma proteins, however, stay in the blood. Urea and other wastes, which are absent in the dialysis fluid, diffuse out of the plasma into the surrounding fluid, cleaning the blood of these wastes. Plasma constituents that are not regulated by the kidneys and are at normal concentration, such as glucose, do not move across the cellophane into the dialysis fluid, because there is no driving force to produce their movement. (The dialysis fluid's glucose concentration is the same as normal plasma glucose concentration.) Electrolytes, such as K^+ and PO_4^{3-}, which are higher than their normal plasma concentrations because the diseased kidneys cannot eliminate excess quantities of these substances, move out of the plasma until equilibrium is achieved between the plasma and the dialysis fluid. Because the dialysis fluid's solute concentrations are maintained at normal plasma values, the solute concentration of the blood returned to the patient after dialysis is essentially normal. Hemodialysis is repeated as often as necessary to maintain the plasma composition within an acceptable level. Typically, it is done three times per week for several hours at each session.

In a more recent method of dialysis, **continuous ambulatory peritoneal dialysis (CAPD),** the peritoneal membrane (the lining of the abdominal cavity) is used as the dialysis membrane. With this method, 2 liters of dialysis fluid are inserted into the patient's abdominal cavity through a permanently implanted catheter. Urea, K^+, and other wastes and excess electrolytes diffuse from the plasma across the peritoneal membrane into the dialysis fluid, which is drained off and replaced several times a day. The CAPD method offers several advantages: The patient can self-administer it, the patient's blood is continuously purified and adjusted, and the patient can engage in normal activities while dialysis is being accomplished. One drawback is the increased risk of peritoneal infections.

Although dialysis can remove metabolic wastes and foreign compounds and help maintain fluid and electrolyte balance within acceptable limits, this plasma-cleansing technique cannot make up for the failing kidneys' reduced ability to produce hormones (erythropoietin and renin) and to activate vitamin D. One new experimental technique incorporates living kidney cells derived from pigs within a dialysis-like machine. Standard ultrafiltration technology like that used in hemodialysis purifies and adjusts the plasma as usual. Importantly, the living cells not only help maintain even better control of plasma constituents, especially K^+, but also add the deficient renal hormones to the plasma passing through the machine and activate vitamin D. This promising new technology has not yet been tested in large-scale clinical trials.

For now, transplanting a healthy kidney from a donor is another option for treating chronic renal failure. A kidney is one of the few transplants that can be provided by a living donor. Because 25% of the total kidney tissue can maintain the body, both the donor and the recipient have ample renal function with only one kidney each. The biggest problem with transplantation is the possibility that the patient's immune system will reject the organ. Risk of rejection can be minimized by matching the tissue types of the donor and the recipient as closely as possible (the best donor choice is usually a close relative), coupled with immunosuppressive drugs.

▌ Renal failure has wide-ranging consequences.

Clinical Note Urine excretion and the resulting clearance of wastes and excess electrolytes from the plasma are crucial for maintaining homeostasis. When the functions of both kidneys are so disrupted that they cannot perform their regulatory and excretory functions sufficiently to maintain homeostasis, **renal failure** has set in. Renal failure can manifest itself either as *acute renal failure,* characterized by a sudden onset with rapidly reduced urine formation until less than the essential minimum of around 500 ml of urine is being produced per day, or *chronic renal failure,* characterized by slow, progressive, insidious loss of renal function. A person may die from acute renal failure, or the condition may be reversible and lead to full recovery. Chronic renal failure, in contrast, is not reversible. Gradual, permanent destruction of renal tissue eventually proves fatal. Chronic renal failure is insidious, because up to 75% of the kidney tissue can be destroyed before the loss of kidney function is even noticeable. Because of the abundant reserve of kidney function, only 25% of the kidney tissue is needed to adequately maintain all the essential renal excretory and regulatory functions. With less than 25% of functional kidney tissue remaining, however, renal insufficiency becomes apparent. *End-stage renal failure* ensues when 90% of kidney function has been lost.

We will not sort out the different stages and symptoms associated with various renal disorders, but ▲ Table 13-4, which summarizes the potential consequences of renal failure, gives you an idea of the broad effects that kidney impairment can have. The extent of these effects should not be surprising, considering the central role the kidneys play in maintaining homeostasis. When the kidneys cannot maintain a normal internal environment, widespread disruption of cell activities can bring about abnormal function in other organ systems, as well. By the time end-stage renal failure occurs, literally every body system has become impaired to some extent.

Because chronic renal failure is irreversible and eventually fatal, treatment is aimed at maintaining renal function by alternative methods, such as dialysis and kidney transplant. (For further explanation of these procedures, see the accompanying boxed feature, ▶ Beyond the Basics.)

This finishes our discussion of kidney function. For the rest of the chapter, we will focus on the plumbing that stores and carries the urine formed by the kidneys to the outside.

▌ Urine is temporarily stored in the bladder, from which it is emptied by micturition.

Once urine has been formed by the kidneys, it is transmitted through the ureters to the urinary bladder. Urine does not flow through the ureters by gravitational pull alone. Peristaltic contractions of the smooth muscle within the ureteral wall propel the urine forward from the kidneys to the bladder. The ureters penetrate the wall of the bladder obliquely, coursing through the wall several centimeters before they open into the bladder cavity. This anatomic arrangement prevents backflow of urine from the bladder to the kidneys when pressure builds up in the bladder. As the bladder fills, the ureteral ends within its wall are compressed closed. Urine can still enter, however, because ureteral contractions generate enough pressure to overcome the resistance and push urine through the occluded ends.

ROLE OF THE BLADDER

The bladder can accommodate large fluctuations in urine volume. The bladder wall consists of smooth muscle, which can stretch tremendously without building up bladder wall tension (see p. 234). In addition, the highly folded bladder wall flattens out during filling to increase bladder storage capacity. Because the kidneys are continuously forming urine, the bladder must have enough storage capacity to preclude the need to continually get rid of the urine.

The bladder smooth muscle is richly supplied by parasympathetic fibers, stimulation of which causes bladder contraction. If the passageway through the urethra to the outside is open, bladder contraction empties urine from the bladder. The exit from the bladder, however, is guarded by two sphincters, the *internal urethral sphincter* and the *external urethral sphincter.*

ROLE OF THE URETHRAL SPHINCTERS

A **sphincter** is a ring of muscle that, when contracted, closes off passage through an opening. The **internal urethral sphincter**—which is smooth muscle and, accordingly, is under involuntary control—is not really a separate muscle but instead consists of the last part of the bladder. Although it is not a true sphincter, it performs the same function as a sphincter. When the bladder is relaxed, the anatomic arrangement of the internal urethral sphincter region closes the outlet of the bladder.

Farther down the passageway, the urethra is encircled by a layer of skeletal muscle, the **external urethral sphincter.** This sphincter is reinforced by the entire **pelvic diaphragm**, a skeletal muscle sheet that forms the floor of the pelvis and helps support the pelvic organs. The motor neurons that supply the external sphincter and pelvic diaphragm fire continu-

ously at a moderate rate unless they are inhibited, keeping these muscles tonically contracted so they prevent urine from escaping through the urethra. Normally, when the bladder is relaxed and filling, closure of both the internal and external urethral sphincters keeps urine from dribbling out. Furthermore, because they are skeletal muscles the external sphincter and pelvic diaphragm are under voluntary control. The person can deliberately tighten them to prevent urination from occurring even when the bladder is contracting and the internal sphincter is open.

MICTURITION REFLEX

Micturition, or **urination**, the process of bladder emptying, is governed by two mechanisms: the micturition reflex and voluntary control. The **micturition reflex** is initiated when stretch receptors within the bladder wall are stimulated (● Figure 13-21). The bladder in an adult can accommodate up to 250 to 400 ml of urine before the tension within its walls begins to rise sufficiently to activate the stretch receptors. The greater the distension beyond this, the greater the extent of receptor activation. Afferent fibers from the stretch receptors carry impulses into the spinal cord and eventually, by interneurons, stimulate the parasympathetic supply to the bladder and inhibit the motor neuron supply to the external sphincter. Parasympathetic stimulation of the bladder causes it to contract. No special mechanism is required to open the internal sphincter; changes in the shape of the bladder during contraction mechanically pull the internal sphincter open. Simultaneously, the external sphincter relaxes as its motor neuron supply is inhibited. Now both sphincters are open, and urine is expelled through the urethra by the force of bladder contraction. This micturition reflex, which is entirely a spinal reflex, governs bladder emptying in infants. As soon as the bladder fills enough to trigger the reflex, the baby automatically wets.

VOLUNTARY CONTROL OF MICTURITION

In addition to triggering the micturition reflex, bladder filling also gives rise to the conscious urge to urinate. The perception of bladder fullness appears before the external sphincter reflexly relaxes, warning that micturition is imminent. As a result, voluntary control of micturition, learned during toilet training in early childhood, can override the micturition reflex so that bladder emptying can take place at the person's convenience rather than when bladder filling first activates the stretch receptors. If the time when the micturition reflex is initiated is inopportune for urination, the person can voluntarily prevent bladder emptying by deliberately tightening the external sphincter and pelvic diaphragm. Voluntary excitatory impulses from the cerebral cortex override the reflex inhibitory input from the stretch receptors to the involved motor neurons (the relative balance of EPSPs and IPSPs), keeping these muscles contracted so that no urine is expelled (see p. 90).

Urination cannot be delayed indefinitely. As the bladder continues to fill, reflex input from the stretch receptors increases with time. Finally, reflex inhibitory input to the external-

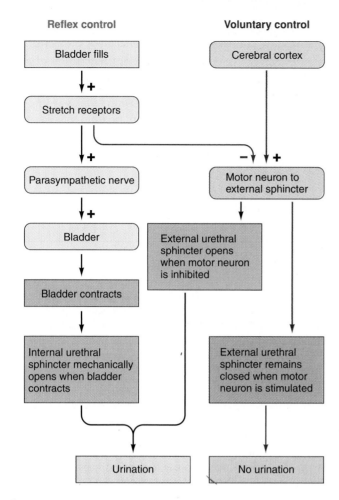

Reflex control

Bladder fills

↓ +

Stretch receptors

↓ +

Parasympathetic nerve

↓ +

Bladder

↓

Bladder contracts

↓

Internal urethral sphincter mechanically opens when bladder contracts

↓

Urination

Voluntary control

Cerebral cortex

− ↓ ↓ +

Motor neuron to external sphincter

External urethral sphincter opens when motor neuron is inhibited

External urethral sphincter remains closed when motor neuron is stimulated

↓

No urination

● **FIGURE 13-21**

Reflex and voluntary control of micturition

sphincter motor neuron becomes so powerful that it can no longer be overridden by voluntary excitatory input, so the sphincter relaxes and the bladder uncontrollably empties.

Micturition can also be deliberately initiated, even though the bladder is not distended, by voluntarily relaxing the external sphincter and pelvic diaphragm. Lowering of the pelvic floor allows the bladder to drop downward, which simultaneously pulls open the internal urethral sphincter and stretches the bladder wall. The subsequent activation of the stretch receptors brings about bladder contraction by the micturition reflex. Voluntary bladder emptying may be further assisted by contracting the abdominal wall and respiratory diaphragm. The resulting increase in intra-abdominal pressure squeezes down on the bladder to facilitate its emptying.

URINARY INCONTINENCE

Clinical Note **Urinary incontinence,** or inability to prevent discharge of urine, occurs when descending pathways in the spinal cord that mediate voluntary control of the external sphincter and pelvic diaphragm are disrupted, as in spinal cord injury. Because the components of the micturition reflex arc are still intact in the lower spinal cord, bladder emptying is governed by an uncontrollable spinal reflex, as in infants. A lesser degree of incontinence characterized by urine escaping when the bladder pressure suddenly increases transiently, such as during coughing or sneezing, can result from impaired sphincter function. This is common in women who have borne children or in men whose sphincters have been injured during prostate surgery.

CHAPTER IN PERSPECTIVE: FOCUS ON HOMEOSTASIS

The kidneys contribute to homeostasis more extensively than any other single organ. They regulate the electrolyte composition, volume, osmolarity, and pH of the internal environment and eliminate all the waste products of bodily metabolism except for respiration-removed CO_2. They fill these regulatory functions by eliminating in the urine substances the body doesn't need, such as metabolic wastes and excess quantities of ingested salt or water, while conserving useful substances. The kidneys can maintain the plasma constituents they regulate within the narrow range compatible with life, despite wide variations in intake and losses of these substances through other avenues. Illustrating the magnitude of the kidneys' task, about a quarter of the blood pumped into the systemic circulation goes to the kidneys to be adjusted and purified, with only three quarters of the blood being used to supply all the other tissues.

The kidneys contribute to homeostasis in the following specific ways:

Regulatory Functions

- The kidneys regulate the quantity and concentration of most ECF electrolytes, including those important in maintaining proper neuromuscular excitability.
- They help maintain proper pH by eliminating excess H^+ (acid) or HCO_3^- (base) in the urine.
- They help maintain proper plasma volume, which is important in long-term regulation of arterial blood pressure, by controlling salt balance in the body. The ECF volume, including plasma volume, reflects total salt load in the ECF, because Na^+ and its attendant anion Cl^- are responsible for over 90% of the ECF's osmotic (water-holding) activity.
- The kidneys maintain water balance in the body, which is important in maintaining proper ECF osmolarity (concentration of solutes). This role is important in maintaining stability of cell volume by keeping water from osmotically moving into or out of the cells, thus preventing them from swelling or shrinking, respectively.

Excretory Functions

- The kidneys excrete the end products of metabolism in urine. If allowed to accumulate, these wastes are toxic to cells.
- The kidneys also excrete many foreign compounds that enter the body.

Hormonal Functions

- The kidneys produce erythropoietin, the hormone that stimulates bone marrow to produce red blood cells. This action contributes to homeostasis by helping maintain the optimal O_2 content of blood. Over 98% of O_2 in blood is bound to hemoglobin within red blood cells.
- They also produce renin, the hormone that initiates the renin-angiotensin-aldosterone pathway for controlling renal tubular Na^+ reabsorption, which is important in long-term maintenance of plasma volume and arterial blood pressure.

Metabolic Functions

- The kidneys help convert vitamin D into its active form. Vitamin D is essential for Ca^{2+} absorption from the digestive tract. Calcium, in turn, exerts a wide variety of homeostatic functions.

CHAPTER SUMMARY

Introduction (pp. 405–411)

- The kidneys eliminate unwanted plasma constituents in the urine while conserving materials of value to the body.
- The urine-forming functional unit of the kidneys is the nephron, which is composed of interrelated vascular and tubular components. *(Review Figure 13-2.)*
- The vascular component consists of two capillary networks in series, the first being the glomerulus, a tuft of capillaries that filters large volumes of protein-free plasma into the tubular component. The second capillary network consists of the peritubular capillaries, which nourish the renal tissue and participate in exchanges between the tubular fluid and plasma. *(Review Figure 13-3.)*
- The tubular component begins with Bowman's capsule, which cups around the glomerulus to catch the filtrate, then continues a specific tortuous course to ultimately empty into the renal pelvis. *(Review Figure 13-2.)* As the filtrate passes through various regions of the tubule, cells lining the tubules modify it, returning to the plasma only those materials necessary for maintaining proper ECF composition and volume. What is left behind in the tubules is excreted as urine.
- The kidneys perform three basic processes in carrying out their regulatory and excretory functions: (1) glomerular filtration, the nondiscriminating movement of protein-free plasma from the blood into the tubules; (2) tubular reabsorption, the selective transfer of specific constituents in the filtrate back into the blood of the peritubular capillaries; and (3) tubular secretion, the highly specific movement of selected substances from peritubular capillary blood into the tubular fluid. Everything filtered or secreted but not reabsorbed is excreted as urine. *(Review Figure 13-4.)*

Glomerular Filtration (pp. 411–414)

- Glomerular filtrate is produced as part of the plasma flowing through each glomerulus is passively forced under pressure through the glomerular membrane into the lumen of the underlying Bowman's capsule. *(Review Figure 13-5.)*
- The net filtration pressure that induces filtration is caused by an imbalance in physical forces acting across the glomerular membrane. A high glomerular capillary blood pressure favoring filtration outweighs the combined opposing forces of plasma-colloid osmotic pressure and Bowman's capsule hydrostatic pressure. *(Review Table 13-1.)*
- Typically, 20 to 25% of the cardiac output is delivered to the kidneys to be acted on by renal regulatory and excretory processes.
- Of the plasma flowing through the kidneys, normally 20% is filtered through the glomeruli, producing an average glomerular filtration rate (GFR) of 125 ml/min.

- The GFR can be deliberately altered by changing the glomerular capillary blood pressure via sympathetic influence on the afferent arterioles as part of the baroreceptor reflex response that compensates for changed arterial blood pressure. Specifically, sympathetically induced afferent arteriolar vasoconstriction decreases the GFR whereas vasodilation of these vessels increases the GFR. As the GFR is altered, the amount of fluid lost in urine changes correspondingly, adjusting plasma volume as needed to help restore blood pressure to normal on a long-term basis. *(Review Figures 13-7 and 13-8.)*

Tubular Reabsorption (pp. 414–422)

- After a protein-free plasma is filtered through the glomerulus, the tubules handle each substance discretely, so that even though the concentrations of all constituents in the initial glomerular filtrate are identical to their concentrations in the plasma (with the exception of plasma proteins), the concentrations of different constituents are variously altered as the filtered fluid flows through the tubular system. *(Review Tables 13-2 and 13-3, p. 424.)*
- The reabsorptive capacity of the tubular system is tremendous. On average, 124 ml out of the 125 ml filtered per minute are reabsorbed.
- Tubular reabsorption involves transepithelial transport from the tubular lumen into the peritubular capillary plasma. This process may be active (requiring energy) or passive (using no energy). *(Review Figure 13-9.)*
- The pivotal event to which most reabsorptive processes are linked in some way is the active reabsorption of Na^+, driven by an energy-dependent Na^+–K^+ ATPase carrier located in the basolateral membrane of almost all tubular cells. *(Review Figure 13-10.)*
- Most Na^+ reabsorption takes place early in the nephron in constant unregulated fashion, but in the distal and collecting tubules, the reabsorption of a small percentage of the filtered Na^+ is variable and subject to control, depending primarily on the complex renin-angiotensin-aldosterone system.
- Because Na^+ and its attendant anion, Cl^-, are the major osmotically active ions in the ECF, the ECF volume is determined by the Na^+ load in the body. In turn, the plasma volume, which reflects the total ECF volume, is important in the long-term determination of arterial blood pressure. Whenever the Na^+ load, ECF volume, plasma volume, and arterial blood pressure are below normal, the kidneys secrete renin, an enzymatic hormone that triggers a series of events ultimately leading to increased secretion of aldosterone from the adrenal cortex. Aldosterone increases Na^+ reabsorption from the distal portions of the tubule, thus correcting for the original reduction in Na^+, ECF volume, and blood pressure. *(Review Figure 13-11.)*

- By contrast, sodium reabsorption is inhibited by atrial natriuretic peptide, a hormone released from the cardiac atria in response to expansion of the ECF volume and a subsequent increase in blood pressure.
- In addition to driving the reabsorption of Na^+, the energy used to supply the Na^+–K^+ ATPase carrier is also ultimately responsible for the reabsorption of organic nutrient molecules from the proximal tubule by secondary active transport.
- The other electrolytes besides Na^+ that are actively reabsorbed by the tubules, such as PO_4^{3-} and Ca^{2+}, have their own independently functioning carrier systems within the proximal tubule.
- Because these carriers, as well as the organic-nutrient cotransport carriers, can become saturated, each exhibits a maximal carrier-limited transport capacity, or T_m. (Review Figure 13-12.)
- Active Na^+ reabsorption also drives the passive reabsorption of Cl^- (via an electrical gradient), H_2O (by osmosis), and urea (down a urea concentration gradient created as a result of extensive osmotic-driven H_2O reabsorption). The small urea molecules are the only waste products that can passively permeate the tubular membranes. Accordingly, urea is the only waste product partially reabsorbed as a result of this concentration effect. (Review Figure 13-13.)
- The other waste products, which are not reabsorbed, remain in the urine in highly concentrated form.

Tubular Secretion (pp. 422–424)
- Tubular secretion also involves transepithelial transport, in this case from the peritubular capillary plasma into the tubular lumen.
- By tubular secretion, the kidney tubules can selectively add some substances to the quantity already filtered. Secretion of substances hastens their excretion in the urine.
- The most important secretory systems are for (1) H^+, which is important in regulating acid–base balance; (2) K^+, which keeps the plasma K^+ concentration at an appropriate level to maintain normal membrane excitability in muscles and nerves; and (3) organic ions, which accomplishes more efficient elimination of foreign organic compounds from the body. (Review Figures 13-14 and 13-15 and Table 13-3.)

Urine Excretion and Plasma Clearance (pp. 424–437)
- Of the 125 ml/min filtered in the glomeruli, normally only 1 ml/min remains in the tubules to be excreted as urine.
- Only wastes and excess electrolytes not wanted by the body are left behind, dissolved in a given volume of H_2O to be eliminated in the urine.
- Because the excreted material is removed or "cleared" from the plasma, the term *plasma clearance* refers to the volume of plasma cleared of a particular substance each minute by renal activity. (Review Figure 13-16.)
- The kidneys can excrete urine of varying volumes and concentrations to either conserve or eliminate H_2O, depending on whether the body has a H_2O deficit or excess, respectively.
- This variable reabsorption is made possible by a vertical osmotic gradient in the medullary interstitial fluid, established by the long loops of Henle via countercurrent multiplication. (Review Figures 13-17 through 13-19.) This vertical osmotic gradient provides a passive driving force for progressive reabsorption of H_2O from the tubular fluid, but the actual extent of H_2O reabsorption depends on the amount of vasopressin (antidiuretic hormone) secreted.
- Vasopressin increases the permeability of the distal and collecting tubules to H_2O; they are impermeable to H_2O in its absence. Vasopressin secretion increases in response to a H_2O deficit, and H_2O reabsorption increases accordingly. Vasopressin secretion is inhibited in response to a H_2O excess, reducing H_2O reabsorption. (Review Figure 13-20.)
- Once formed, urine is propelled through the ureters from the kidneys to the urinary bladder for temporary storage.
- The bladder can accommodate up to 250 to 400 ml of urine before stretch receptors within its wall initiate the micturition reflex. This reflex causes involuntary emptying of the bladder by simultaneous bladder contraction and opening of both the internal and external urethral sphincters. Micturition can transiently be voluntarily prevented until a more opportune time by deliberate tightening of the external sphincter and surrounding pelvic diaphragm. (Review Figure 13-21.)

REVIEW EXERCISES

Objective Questions (Answers on p. A-46)
1. Part of the kidneys' energy supply is used to accomplish glomerular filtration. (True or false?)
2. Sodium reabsorption is under hormonal control throughout the length of the tubule. (True or false?)
3. Glucose and amino acids are reabsorbed by secondary active transport. (True or false?)
4. Water excretion can occur without comparable solute excretion. (True or false?)
5. The functional unit of the kidneys is the _____.
6. _____ is the only ion actively reabsorbed in the proximal tubule and actively secreted in the distal and collecting tubules.
7. The daily minimum volume of obligatory H_2O loss that must accompany excretion of wastes is _____ ml.
8. Which of the following filtered substances is normally not present in the urine at all?
   a. Na^+
   b. PO_4^{3-}
   c. urea
   d. H^+
   e. glucose

9. Reabsorption of which of the following substances is not linked in some way to active Na^+ reabsorption?
   a. glucose
   b. PO_4^{3-}
   c. H_2O
   d. urea
   e. Cl^-

In Questions 10 through 12, indicate the proper sequence through which fluid flows as it traverses the structures in question by writing the identifying letters in the proper order in the blanks.

10. a. ureter   ___ ___ ___ ___ ___
    b. kidney
    c. urethra
    d. bladder
    e. renal pelvis

11. a. efferent arteriole   ___ ___ ___ ___ ___ ___
    b. peritubular capillaries
    c. renal artery
    d. glomerulus
    e. afferent arteriole
    f. renal vein

12. a. loop of Henle _____ _____ _____ _____ _____ _____ _____
    b. collecting duct
    c. Bowman's capsule
    d. proximal tubule
    e. renal pelvis
    f. distal tubule
    g. glomerulus

13. Using the following answer code, indicate what the osmolarity of the tubular fluid is at each of the designated points in a nephron:

    (a) isotonic (300 mosm/liter)
    (b) hypotonic (100 mosm/liter)
    (c) hypertonic (1200 mosm/liter)
    (d) ranging from hypotonic to hypertonic (100 mosm/liter to 1200 mosm/liter)

    _____ 1. Bowman's capsule
    _____ 2. end of proximal tubule
    _____ 3. tip of Henle's loop of long-looped nephron (at the bottom of the U-turn)
    _____ 4. end of Henle's loop of long-looped nephron (before entry into distal tubule)
    _____ 5. end of collecting duct

Essay Questions

1. List the functions of the kidneys.
2. Describe the anatomy of the urinary system. Describe the components of a nephron.
3. Describe the three basic renal processes; indicate how they relate to urine excretion.
4. Distinguish between *secretion* and *excretion*.
5. Discuss the forces involved in glomerular filtration. What is the average GFR?
6. How is GFR regulated as part of the baroreceptor reflex?
7. Why do the kidneys receive a seemingly disproportionate share of the cardiac output? What percentage of renal blood flow is normally filtered?
8. List the steps in transepithelial transport.
9. Distinguish between active and passive reabsorption.
10. Describe all the tubular transport processes that are linked to the basolateral $Na^+–K^+$ ATPase carrier.
11. Describe the renin-angiotensin-aldosterone system. What are the source and function of atrial natriuretic peptide?
12. To what do the terms *tubular maximum* (T_m) and *renal threshold* refer? Compare two substances that display a T_m, one substance that *is* and one that *is not* regulated by the kidneys.
13. What is the importance of tubular secretion? What are the most important secretory processes?
14. What is the average rate of urine formation?
16. Define *plasma clearance*.
16. What establishes a vertical osmotic gradient in the medullary interstitial fluid? Of what importance is this gradient?
17. Discuss the function of vasopressin.
18. Describe the transfer of urine to, the storage of urine in, and the emptying of urine from the bladder.

POINTS TO PONDER

(Explanations on p. A-46)

1. The long-looped nephrons of animals adapted to survive with minimal water consumption, such as desert rats, have relatively much longer loops of Henle than humans have. Of what benefit would these longer loops be?
2. If the plasma concentration of substance X is 200 mg/100 ml and the GFR is 125 ml/min, the filtered load of this substance is

    _____ .

    If the T_m for substance X is 200 mg/min, how much of the substance will be reabsorbed at a plasma concentration of 200 mg/100 ml and a GFR of 125 ml/min? _____ How much of substance X will be excreted? _____
3. *Conn's syndrome* is an endocrine disorder brought about by a tumor of the adrenal cortex that secretes excessive aldosterone in uncontrolled fashion. Given what you know about the functions of aldosterone, describe what the most prominent features of this condition would be.
4. Because of a mutation, a child was born with an ascending limb of Henle that was water permeable. What would be the minimum/maximum urine osmolarities (in units of mosm/liter) the child could produce?
    a. 100/300
    b. 300/1200
    c. 100/100
    d. 1200/1200
    e. 300/300
5. An accident victim suffers permanent damage of the lower spinal cord and is paralyzed from the waist down. Describe what governs bladder emptying in this individual.

CLINICAL CONSIDERATION

(Explanation on p. A-47)

Marcus T. has noted a gradual decrease in his urine flow rate and is now experiencing difficulty in initiating micturition. He needs to urinate frequently, and often he feels as if his bladder is not empty even though he has just urinated. Analysis of Marcus's urine reveals no abnormalities. Are his urinary tract symptoms most likely caused by kidney disease, a bladder infection, or prostate enlargement?

PHYSIOEDGE RESOURCES

 PhysioEdge CD-ROM

PhysioEdge, the CD-ROM packaged with your text, focuses on the concepts students find most difficult to learn. Figures marked with this icon have associated activities on the CD. For a visual review of concepts in this chapter, check out the following:

Tutorial: Renal Physiology

Media Exercise 13.1: The Urinary System

Media Exercise 13.2: The Kidney

Media Exercise 13.3: The Nephron

PhysioEdge Website

The website for this book contains a wealth of helpful study aids, as well as many ideas for further reading and research. Log on to: **http://www.brookscole.com/hpfundamentals3**

Select Chapter 13 from the drop-down menu or click on one of the many resource areas, including Case Histories, which introduce clinical aspects of human physiology. For this chapter check out the following: Case History 11: Congestive Heart Failure, Case History 12: Urinary Tract Infection.

For Suggested Readings, consult **InfoTrac College Edition/ Research** on the PhysioEdge website or go directly to InfoTrac College Edition, your online research library, at: **http://infotrac.thomsonlearning.com**

Systems of Major Importance in Maintaining Fluid and Acid-Base Balance

Body systems maintain homeostasis

Homeostasis
The kidneys, in conjunction with hormones involved in salt and water balance, are responsible for maintaining the volume and osmolarity of the extracellular fluid (internal environment). The kidneys, along with the respiratory system and chemical buffer systems in the body fluids, also contribute to homeostasis by maintaining the proper pH in the internal environment.

$$CO_2 + H_2O \overset{ca}{\rightleftharpoons} H_2CO_3 \rightleftharpoons H^+ + HCO_3^-$$

Chemical buffer systems in body fluids

Homeostasis is essential for survival of cells

Cells
The volume of circulating blood must be maintained to help ensure adequate pressure to drive life-sustaining blood to the cells. The osmolarity of fluid surrounding the cells must be closely regulated to prevent detrimental osmotic movement of water between the cells and ECF. The pH of the internal environment must remain stable because pH changes alter neuromuscular excitability and enzyme activity, among other serious consequences.

Cells make up body systems

Homeostasis depends on maintaining a balance between the input and the output of all constituents present in the internal fluid environment. Regulation of **fluid balance** involves two separate components: *control of ECF volume,* of which circulating plasma volume is a part, and *control of ECF osmolarity* (solute concentration). The kidneys control ECF volume by maintaining **salt balance** and control ECF osmolarity by maintaining **water balance.** The kidneys maintain this balance by adjusting the output of salt and water in the urine as needed to compensate for variable input and abnormal losses of these constituents.

Similarly, the kidneys help maintain **acid–base balance** by adjusting the urinary output of hydrogen ion (acid) and bicarbonate ion (base) as needed. Also contributing to acid–base balance are the lungs, which can adjust the rate at which they excrete hydrogen ion–generating CO_2, and the chemical buffer systems in the body fluids.

Fluid and Acid–Base Balance

BALANCE CONCEPT

The cells of complex multicellular organisms are able to survive and function only within a very narrow range of composition of the extracellular fluid (ECF), the internal fluid environment that bathes them.

▌ The internal pool of a substance is the amount of that substance in the ECF.

The quantity of any particular substance in the ECF is considered a readily available internal **pool**. The amount of the substance in the pool may be increased either by transferring more in from the external environment (most commonly by ingestion) or by metabolically producing it within the body (● Figure 14-1). Substances may be removed from the body by being excreted to the outside or by being used up in a metabolic reaction. If the quantity of a substance is to remain stable within the body, its **input** through ingestion or metabolic production must be balanced by an equal **output** through excretion or metabolic consumption. This relationship, known as the **balance concept**, is extremely important in maintaining homeostasis. Not all input and output pathways are applicable for every body fluid constituent. For example, salt is not synthesized or consumed by the body, so the stability of salt concentration in the body fluids depends entirely on a balance between salt ingestion and salt excretion.

The ECF pool can further be altered by transferring a particular ECF constituent into storage within the body. If the body as a whole has a surplus or deficit of a particular stored substance, the storage site can be expanded or partially depleted to maintain the ECF concentration of the substance within homeostatically prescribed limits. For example, after absorption of a meal, when more glucose is entering the plasma than is being consumed by the cells, the extra glucose can be temporarily stored in muscle and liver cells, in the form of glycogen. This storage depot

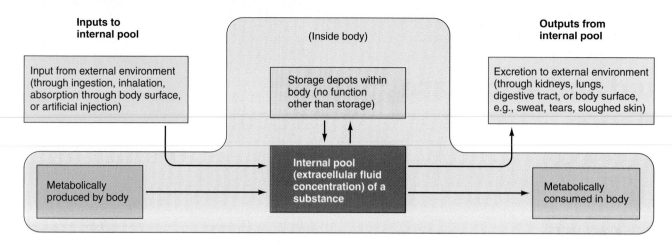

● FIGURE 14-1

Inputs to and outputs from the internal pool of a body constituent

can then be tapped between meals as needed to maintain the plasma glucose level when no new nutrients are being added to the blood by eating. This internal storage capacity is limited, however. Although an internal exchange between the ECF and a storage depot can temporarily restore the plasma concentration of a particular substance to normal, in the long run any excess or deficit of that constituent must be compensated for by appropriate adjustments in total body input or output.

▌ To maintain stable balance of an ECF constituent, its input must equal its output.

When total body input of a particular substance equals its total body output, a **stable balance** exists. When the gains via input for a substance exceed its losses via output, a **positive balance** exists. The result is an increase in the total amount of the substance in the body. In contrast, when losses for a substance exceed its gains, a **negative balance** exists and the total amount of the substance in the body decreases.

Changing the magnitude of any of the input or output pathways for a given substance can alter its plasma concentration. To maintain homeostasis, any change in input must be balanced by a corresponding change in output (for example, increased salt intake must be matched by a corresponding increase in salt output in the urine), and, conversely, increased losses must be compensated for by increased intake. Thus maintaining a stable balance requires control. However, not all input and output pathways are regulated to maintain balance. Generally, input of various plasma constituents is poorly controlled or not controlled at all. We frequently ingest salt and H_2O, for example, not because we need them but because we want them, so the intake of salt and H_2O is highly variable. Likewise, hydrogen ion (H^+) is uncontrollably generated internally and added to the body fluids. Salt, H_2O, and H^+ can also be lost to the external environment to varying degrees through the digestive tract (vomiting), skin (sweating), and elsewhere without regard for salt, H_2O, or H^+ balance in the body. Compensatory adjustments in the urinary excretion of these substances maintain the body fluids' volume and salt and acid composition within the extremely narrow homeostatic range compatible with life despite the wide variations in input and unregulated losses of these plasma constituents.

The rest of this chapter is devoted to discussing the regulation of fluid balance (maintaining salt and H_2O balance) and acid–base balance (maintaining H^+ balance).

FLUID BALANCE

Water is by far the most abundant component of the human body, constituting 60% of body weight on average but ranging from 40 to 80%. The H_2O content of an individual remains fairly constant, largely because the kidneys efficiently regulate H_2O balance, but the percentage of body H_2O varies from person to person. The reason for the wide range in body H_2O among individuals is the variability in the amount of their adipose tissue (fat). Adipose tissue has a low H_2O percentage compared to other tissues. Plasma, as you might suspect, is over 90% H_2O. Even the soft tissues such as skin, muscles, and internal organs consist of 70 to 80% H_2O. The relatively drier skeleton is only 22% H_2O. Fat, however, is the driest tissue of all, having only 10% H_2O content. Accordingly, a high body H_2O percentage is associated with leanness and a low body H_2O percentage with obesity, because a larger proportion of the overweight body consists of relatively dry fat.

▌ Body water is distributed between the ICF and ECF compartments.

Body H_2O is distributed between two major fluid compartments: fluid within the cells, **intracellular fluid (ICF)**, and fluid surrounding the cells, **extracellular fluid (ECF)** (▲ Table 14-1). (The terms "H_2O" and "fluid" are commonly used interchangeably. Although this usage is not entirely accurate,

TABLE 14-1
Classification of Body Fluid

COMPARTMENT	VOLUME OF FLUID (in liters)	PERCENTAGE OF BODY FLUID	PERCENTAGE OF BODY WEIGHT
Total Body Fluid	42	100%	60%
Intracellular Fluid (ICF)	28	67	40
Extracellular Fluid (ECF)	14	33	20
Plasma	2.8	6.6 (20% of ECF)	4
Interstitial fluid	11.2	26.4 (80% of ECF)	16

because it ignores the solutes in body fluids, it is acceptable when discussing total volume of fluids, because the major proportion of these fluids consists of H_2O.)

PROPORTION OF H_2O IN THE MAJOR FLUID COMPARTMENTS

The ICF compartment comprises about two thirds of the total body H_2O. Even though each cell contains its own unique mixture of constituents, these trillions of minute fluid compartments are similar enough to be considered collectively as one large fluid compartment.

The remaining third of the body H_2O found in the ECF compartment is further subdivided into plasma and interstitial fluid. The **plasma**, which makes up about a fifth of the ECF volume, is the fluid portion of blood. The **interstitial fluid**, which represents the other four fifths of the ECF compartment, is the fluid in the spaces between cells. It bathes and makes exchanges with tissue cells.

MINOR ECF COMPARTMENTS

Two other minor categories are included in the ECF compartment: lymph and transcellular fluid. **Lymph** is the fluid being returned from the interstitial fluid to the plasma by means of the lymphatic system, where it is filtered through lymph nodes for immune defense purposes (see p. 296). **Transcellular fluid** consists of a number of small specialized fluid volumes, all of which are secreted by specific cells into a particular body cavity to perform some specialized function. An example of transcellular fluid is *cerebrospinal fluid,* which surrounds, cushions, and nourishes the brain and spinal cord. Although transcellular fluids are extremely important functionally, they represent an insignificant fraction of total body H_2O and can usually be ignored when dealing with problems of fluid balance.

▌ The plasma and interstitial fluid are similar in composition, but the ECF and ICF differ markedly.

Several barriers separate the body fluid compartments, limiting the movement of H_2O and solutes between the various compartments to differing degrees.

THE BARRIER BETWEEN PLASMA AND INTERSTITIAL FLUID: BLOOD VESSEL WALLS

The two components of the ECF—plasma and interstitial fluid—are separated by the walls of the blood vessels. However, H_2O and all plasma constituents except for plasma proteins are continuously and freely exchanged between plasma and interstitial fluid by passive means across the thin, pore-lined capillary walls. Accordingly, plasma and interstitial fluid are nearly identical in composition, except that interstitial fluid lacks plasma proteins. Any change in one of these ECF compartments is quickly reflected in the other compartment, because they are constantly mixing.

THE BARRIER BETWEEN THE ECF AND ICF: CELLULAR PLASMA MEMBRANES

In contrast to the very similar composition of vascular and interstitial fluid compartments, the composition of the ECF differs a lot from that of the ICF (● Figure 14-2). Each cell is surrounded by a highly selective plasma membrane that permits passage of certain materials while excluding others. Movement through the membrane barrier occurs by both passive and active means and may be highly discriminating. Among the major differences between the ECF and ICF are (1) the presence of cell proteins in the ICF that cannot permeate the enveloping membranes to leave the cells and (2) the unequal distribution of Na^+ and K^+ and their attendant anions as a result of the action of the membrane-bound Na^+–K^+ ATPase pump that is in all cells. This pump actively transports Na^+ out of and K^+ into cells; therefore Na^+ is the primary ECF cation, and K^+ is primarily found in the ICF.

Except for the extremely small, electrically unbalanced portion of the total intracellular and extracellular ions that are involved in membrane potential, most ECF and ICF ions are electrically balanced. In the ECF, Na^+ is accompanied primarily by the anion Cl^- (chloride) and to a lesser extent by HCO_3^- (bicarbonate). The major intracellular anions are PO_4^{3-} (phosphate) and the negatively charged proteins trapped within the cell.

▌ Fluid balance is maintained by regulating ECF volume and osmolarity.

Extracellular fluid serves as an intermediary between the cells and external environment. All exchanges of H_2O and other constituents between the ICF and external world must occur through the ECF. Water added to the body fluids always enters the ECF compartment first, and fluid always leaves the body via the ECF.

Plasma is the only fluid that can be acted on directly to control its volume and composition. This fluid circulates through all the reconditioning organs that perform homeostatic adjustments (see p. 276). However, because of the free exchange across the capillary walls, if the volume and composition of the plasma are regulated the volume and composition of the interstitial fluid bathing the cells are likewise regulated. Thus any control mechanism that operates on plasma in effect regulates the entire ECF. The ICF in turn is influenced by changes in the ECF to the extent permitted by the permeability of membrane barriers surrounding cells.

Two factors are regulated to maintain **fluid balance** in the body: ECF volume and ECF osmolarity. Although regulation of these two factors is closely interrelated, both being dependent on the relative NaCl and H_2O load in the body, the reasons why they are closely controlled are significantly different:

1. *ECF volume* must be closely regulated to help *maintain blood pressure*. Maintaining *salt balance* is of primary importance in the long-term regulation of ECF volume.
2. *ECF osmolarity* must be closely regulated to *prevent swelling or shrinking of cells.* Maintaining *water balance* is of primary importance in regulating ECF osmolarity.

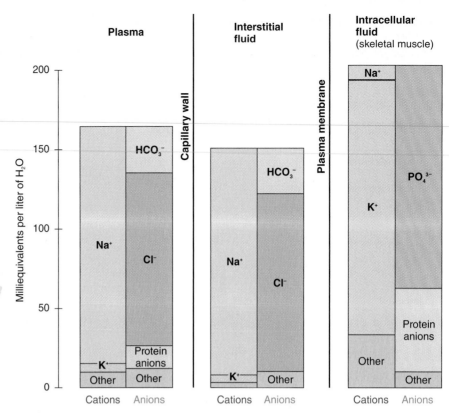

● **FIGURE 14-2**

Ionic composition of the major body-fluid compartments

 For an interaction related to this figure, see Media Exercise 14.1: Fluid and Electrolyte Balance on the CD-ROM.

We will examine each of these factors in more detail.

▌ Controlling ECF volume is important in the long-term regulation of blood pressure.

A reduction in ECF volume causes a fall in arterial blood pressure by decreasing plasma volume. Conversely, expanding ECF volume raises arterial blood pressure by increasing plasma volume. Two compensatory measures come into play to transiently adjust blood pressure until the ECF volume can be restored to normal. Let's review them.

REVIEW OF SHORT-TERM CONTROL MEASURES TO MAINTAIN BLOOD PRESSURE

1. *The baroreceptor reflex alters both cardiac output and total peripheral resistance* to adjust blood pressure in the proper direction through autonomic nervous system effects on the heart and blood vessels (see p. 305). Cardiac output and total peripheral resistance are both increased to raise blood pressure when it falls too low, and conversely, both are decreased to reduce blood pressure when it rises too high.
2. *Fluid shifts occur temporarily and automatically between the plasma and interstitial fluid* as a result of changes in the balance of hydrostatic and osmotic forces acting across the capillary walls that arise when plasma volume deviates from normal (see p. 294). A reduction in plasma volume is partially compensated for by a shift of fluid out of the interstitial compartment into the blood vessels, expanding the circulating plasma volume at the expense of the interstitial compartment. Conversely, when the plasma volume is too large much of the excess fluid shifts into the interstitial compartments.

These two measures provide temporary relief to help keep blood pressure fairly constant, but they are not long-term solutions. Furthermore, these short-term compensatory measures have a limited ability to minimize a change in blood pressure. For example, if the plasma volume is too inadequate, blood pressure remains too low no matter how vigorous the pump action of the heart, how constricted the resistance vessels, or what proportion of interstitial fluid shifts into the blood vessels.

LONG-TERM CONTROL MEASURES TO MAINTAIN BLOOD PRESSURE

It is important, therefore, that other compensatory measures come into play in the long run to restore the ECF volume to normal. Long-term regulation of blood pressure rests with the kidneys and the thirst mechanism, which control urinary output and fluid intake, respectively. In so doing, they make needed fluid exchanges between the ECF and external environment to regulate the body's total fluid volume. Accordingly, they have an important long-term influence on arterial blood pressure. Of these measures, control of urinary output by the kidneys is the most crucial for maintaining blood pressure. You will see why as we discuss these long-term mechanisms in more detail.

Controlling salt balance is primarily important in regulating ECF volume.

By way of review, sodium and its attendant anions account for more than 90% of the ECF's osmotic activity. As the kidneys conserve salt, they automatically conserve H_2O, because H_2O follows Na^+ osmotically. This retained salt solution is isotonic (see p. 55). The more salt in the ECF, the more H_2O in the ECF. The concentration of salt is not changed by changing the amount of salt, because H_2O always follows salt to maintain osmotic equilibrium—that is, to maintain the normal concentration of salt. The total mass of Na^+ salts in the ECF (that is, the *Na$^+$ load*) therefore determines the ECF's volume, and, appropriately, regulation of ECF volume depends primarily on controlling salt balance.

To maintain salt balance at a set level, salt input must equal salt output, thus preventing salt accumulation or deficit in the body. Let's look at the avenues and control of salt input and output.

POOR CONTROL OF SALT INTAKE

The only avenue for salt input is ingestion, which typically is well in excess of the body's need for replacing obligatory salt losses. In our example of a typical daily salt balance (▲ Table 14-2), salt intake is 10.5 g per day. (The average American salt intake is 10 to 15 g per day, although many people are consciously reducing their salt intake.) Yet half a gram of salt per day is adequate to replace the small amounts of salt usually lost in the feces and sweat.

Because humans typically consume salt in excess of our needs, obviously our salt intake is not well controlled. Carnivores (meat eaters) and omnivores (eaters of meat and plants, like humans), which naturally get enough salt in fresh meat

▲ TABLE 14-2
Daily Salt Balance

SALT INPUT		SALT OUTPUT	
Avenue	Amount (g/day)	Avenue	Amount (g/day)
Ingestion	10.5	Obligatory loss in sweat and feces	0.5
		Controlled excretion in urine	10.0
Total input	10.5	Total output	10.5

(meat contains an abundance of salt-rich ECF), normally do not show a physiologic appetite to seek additional salt. In contrast, herbivores (plant eaters), which lack salt naturally in their diets, develop salt hunger and will travel miles to a salt lick. Humans generally have a hedonistic rather than a regulatory appetite for salt; we consume salt because we like it rather than because we have a physiologic need, except in the unusual circumstance of severe salt depletion caused by a deficiency of aldosterone, the salt-conserving hormone.

PRECISE CONTROL OF SALT OUTPUT IN THE URINE

To maintain salt balance, excess ingested salt must be excreted in the urine. The three avenues for salt output are obligatory loss of salt in *sweat* and *feces* and controlled excretion of salt in *urine* (▲ Table 14-2). The total amount of sweat produced is unrelated to salt balance, being determined instead by factors that control body temperature. The small salt loss in feces is not subject to control. Except when sweating heavily or during diarrhea, normally the body uncontrollably loses only about 0.5 g of salt per day.

Because salt consumption is typically far more than the meager amount needed to compensate for uncontrolled losses, the kidneys precisely excrete the excess salt in the urine to maintain salt balance. In our example, 10 g of salt are eliminated in the urine per day so that total salt output exactly equals salt input. By regulating the rate of urinary salt excretion (that is, by regulating the rate of Na^+ excretion, with Cl^- following along), the kidneys normally keep the total Na^+ mass in the ECF constant despite any notable changes in dietary intake of salt or unusual losses through sweating or diarrhea. As a reflection of keeping the total Na^+ mass in the ECF constant, the ECF volume in turn is maintained within the narrowly prescribed limits essential for normal circulatory function.

Deviations in the ECF volume accompanying changes in the salt load trigger renal compensatory responses that quickly bring the Na^+ load and ECF volume back into line. Sodium is freely filtered at the glomerulus and actively reabsorbed, but it is not secreted by the tubules, so the amount of Na^+ excreted in the urine represents the amount of Na^+ that is filtered but is not subsequently reabsorbed:

$$Na^+ \text{ excreted} = Na^+ \text{ filtered} - Na^+ \text{ reabsorbed}$$

The kidneys accordingly adjust the amount of salt excreted by controlling two processes: (1) the glomerular filtration rate (GFR) and (2) more importantly, the tubular reabsorption of Na^+. You have already learned about these regulatory mechanisms, but we are pulling them together here as they relate to the long-term control of ECF volume and blood pressure.

• *The amount of Na^+ filtered is controlled by regulating the GFR.* The amount of Na^+ filtered is equal to the plasma Na^+ concentration times the GFR. At any given plasma Na^+ concentration, any change in the GFR will correspondingly change the amount of Na^+ and accompanying fluid that are filtered. Thus control of the GFR can adjust the amount of Na^+ filtered each minute. Recall that the GFR is deliberately changed to alter the amount of salt and fluid filtered, as part of the general baroreceptor reflex response to change in blood pressure (see ● Figure 13-7, p. 413). Note that the amount of salt filtered is adjusted as part of the general blood pressure-regulating reflexes. Changes in Na⁺ load in the body are not sensed as such; instead, they are monitored indirectly through the effect that Na⁺ ultimately has on blood pressure via its role in determining the ECF volume. Fittingly, baroreceptors that monitor fluctuations in blood pressure bring about adjustments in the amounts of Na⁺ filtered and eventually excreted.

• *The amount of Na⁺ reabsorbed is controlled through the renin-angiotensin-aldosterone system.* The amount of Na⁺ reabsorbed also depends on regulatory systems that play an important role in controlling blood pressure. Although Na⁺ is reabsorbed throughout most of the tubule's length, only its reabsorption in the distal parts of the tubule is subject to control. The main factor controlling the extent of Na⁺ reabsorption in the distal and collecting tubules is the powerful renin-angiotensin-aldosterone system, which promotes Na⁺ reabsorption and thereby Na⁺ retention. Sodium retention in turn promotes osmotic retention of H₂O and subsequent expansion of plasma volume and elevation of arterial blood pressure. Appropriately, this Na⁺-conserving system is activated by a reduction in NaCl, ECF volume, and arterial blood pressure (see ● Figure 13-11, p. 417).

● **FIGURE 14-3**

Dual effect of a fall in arterial blood pressure on renal handling of Na⁺

 For an interaction related to this figure, see Media Exercise 14.1: Fluid and Electrolyte Balance on the CD-ROM.

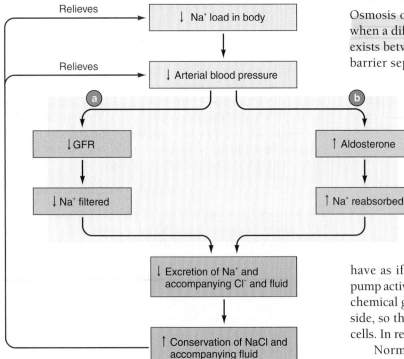

a See Figure 13-7 for details of mechanism.

b See Figure 13-11 for details of mechanism.

Thus control of GFR and Na⁺ reabsorption are closely interrelated, and both are intimately tied in with long-term regulation of ECF volume as reflected by blood pressure. For example, a fall in arterial blood pressure brings about (1) a reflex reduction in the GFR to decrease the amount of Na⁺ filtered and (2) a hormonally adjusted increase in the amount of Na⁺ reabsorbed (● Figure 14-3). Together these effects reduce the amount of Na⁺ excreted, thereby conserving for the body the Na⁺ and accompanying H₂O needed to compensate for the fall in arterial pressure. (To look at how exercising muscles and cooling mechanisms compete for an inadequate plasma volume, see the accompanying boxed feature, ❱ Beyond the Basics.)

❚ Controlling ECF osmolarity prevents changes in ICF volume.

Maintaining fluid balance depends on regulating both ECF volume and ECF osmolarity. Whereas regulating ECF volume is important in long-term control of blood pressure, regulating ECF osmolarity is important in preventing changes in cell volume. The **osmolarity** of a fluid is a measure of the concentration of the individual solute particles dissolved in it. The higher the osmolarity, the higher the concentration of solutes or, to look at it differently, the lower the concentration of H₂O. Recall that water tends to move by osmosis down its own concentration gradient from an area of lower solute (higher H₂O) concentration to an area of higher solute (lower H₂O) concentration.

IONS RESPONSIBLE FOR ECF AND ICF OSMOLARITY

Osmosis occurs across the cellular plasma membranes only when a difference in concentration of nonpenetrating solutes exists between the ECF and ICF. Solutes that can penetrate a barrier separating two fluid compartments quickly become equally distributed between the two compartments and thus do not contribute to osmotic differences.

Sodium and its attendant anions, being by far the most abundant solutes in the ECF in terms of numbers of particles, account for the vast majority of the ECF's osmotic activity. In contrast, K⁺ and its accompanying intracellular anions are responsible for the ICF's osmotic activity. Even though small amounts of Na⁺ and K⁺ passively diffuse across the plasma membrane all the time, these ions behave as if they were nonpenetrating, because of Na⁺–K⁺ pump activity. Any Na⁺ that passively diffuses down its electrochemical gradient into the cell is promptly pumped back outside, so the result is the same as if Na⁺ were barred from the cells. In reverse, K⁺ in effect remains trapped within the cells.

Normally, the osmolarities of the ECF and ICF are the same, because the total concentration of K⁺ and other effectively nonpenetrating solutes inside the cells is equal to the total concentration of Na⁺ and other effectively nonpenetrating solutes in the fluid surrounding the cells. Even though

A Potentially Fatal Clash: When Exercising Muscles and Cooling Mechanisms Compete for an Inadequate Plasma Volume

An increasing number of people of all ages are participating in walking or jogging programs to improve their level of physical fitness and decrease their risk of cardiovascular disease. For people living in environments that undergo seasonal temperature changes, fluid loss can make exercising outdoors dangerous during the transition from the cool days of spring to the hot, humid days of summer. If exercise intensity is not modified until the participant gradually adjusts to the hotter environmental conditions, dehydration and salt loss can indirectly lead to heat cramps, heat exhaustion, or ultimately heat stroke and death.

The term **acclimatization** refers to the gradual adaptations the body makes to maintain long-term homeostasis in response to a prolonged physical change in the surrounding environment, such as a change in temperature. When a person exercises in the heat without gradually adapting to the hotter environment, the body faces a terrible dilemma. During exercise, large amounts of blood must be delivered to the muscles to supply O_2 and nutrients and to remove the wastes that accumulate from their high rate of activity. Exercising muscles also produce heat. To maintain the body temperature in the face of this extra heat, blood flow to the skin is increased, so that heat from the warmed blood can be lost through the skin to the surrounding environment. If the environmental temperature is hotter than the body temperature, heat can

not be lost from the blood to the surrounding environment despite maximal skin vasodilation. Instead, the body gains heat from its warmer surroundings, further adding to the dilemma. Because extra blood is diverted to both the muscles and the skin when a person exercises in the heat, less blood is returned to the heart, and the heart pumps less blood per beat in accordance with the Frank-Starling mechanism (see p. 262). Therefore the heart must beat faster than it would in a cool environment to deliver the same amount of blood per minute. The increased rate of cardiac pumping further contributes to heat production.

The sweat rate also increases so that evaporative cooling can take place to help maintain the body temperature during periods of excessive heat gain. In an unacclimatized person, maximal sweat rate is about 1.5 liters per hour. During sweating, water-retaining salt as well as water is lost. The resulting loss of plasma volume through sweating further depletes the blood supply available for muscular exercise and for cooling through skin vasodilation.

The heart has a maximum rate at which it can pump. If exercise continues at a high intensity and this maximal rate is reached, the exercising muscles win the contest for blood supply. The body responds by constricting the skin arterioles, sacrificing cooling to maintain cardiac output and blood pressure. If exercise continues, body heat continues to rise, and heat exhaustion (rapid, weak pulse; hypoten-

sion, profuse sweating; and disorientation) or heat stroke (failure of the temperature control center in the hypothalamus; hot, dry skin; extreme confusion or unconsciousness; and possibly death) can occur. In fact, every year people die of heat stroke in marathons run during hot, humid weather.

By contrast, if a person exercises in the heat for two weeks at reduced, safe intensities, the body makes the following adaptations so that after acclimatization the person can do the same amount of work as was possible in a cool environment. (1) The plasma volume is increased by as much as 12%. Expansion of the plasma volume provides enough blood to both supply the exercising muscles and direct blood to the skin for cooling. (2) The person begins sweating at a lower temperature so that the body does not get so hot before cooling begins. (3) The sweat rate increases as much as three times, to 4 liters per hour, with a more even distribution over the body. This increase in evaporative cooling reduces the need for cooling by skin vasodilation. (4) The sweat becomes more dilute so that less salt is lost in the sweat. The retained salt exerts an osmotic effect to hold water in the body and help maintain circulating plasma volume. These adaptations take 14 days and occur only if the person exercises in the heat. Being patient until these changes take place can enable the person to exercise safely throughout the summer months.

---

nonpenetrating solutes in the ECF and ICF differ, their concentrations are normally identical, and the number (not the nature) of the unequally distributed particles per volume determines the fluid's osmolarity. Because the osmolarities of the ECF and ICF are normally equal, no net movement of H_2O usually occurs into or out of the cells. Therefore, cell volume normally remains constant.

IMPORTANCE OF REGULATING ECF OSMOLARITY

Any circumstance that results in a loss or gain of *free H_2O* (that is, loss or gain of H_2O that is not accompanied by comparable solute deficit or excess) leads to changes in ECF osmolarity. If there is a deficit of free H_2O in the ECF, the solutes become too concentrated, and ECF osmolarity becomes abnormally high (that is, becomes *hypertonic*) (see p. 55). If there is excess free H_2O in the ECF, the solutes become too dilute, and ECF osmolarity becomes abnormally low (that is, becomes *hypotonic*). When ECF osmolarity changes with respect to ICF osmolarity, osmosis takes place, with H_2O either leaving or entering the cells, depending, re-

spectively, on whether the ECF is more or less concentrated than the ICF.

The osmolarity of the ECF must therefore be regulated to prevent these undesirable shifts of H_2O out of or into the cells. As far as the ECF itself is concerned, the concentration of its solutes does not really matter. However, it is crucial that ECF osmolarity be maintained within very narrow limits to prevent the cells from shrinking or swelling.

Let's examine the fluid shifts that occur between the ECF and ICF when ECF osmolarity becomes hypertonic or hypotonic relative to the ICF. Then we will consider how water balance and consequently ECF osmolarity are normally maintained to minimize harmful changes in cell volume.

▌ **During ECF hypertonicity, the cells shrink as H_2O leaves them.**

 Hypertonicity of the ECF, the excessive concentration of ECF solutes, is usually associated with **dehydration**, or a negative free H_2O balance.

CAUSES OF HYPERTONICITY (DEHYDRATION)

Dehydration with accompanying hypertonicity can be brought about in three major ways:

1. *Insufficient H_2O intake,* such as might occur during desert travel or might accompany difficulty in swallowing
2. *Excessive H_2O loss,* such as might occur during heavy sweating, vomiting, or diarrhea
3. *Diabetes insipidus*

Diabetes insipidus is a disease characterized by a lack of vasopressin. In the absence of vasopressin, the kidneys cannot conserve H_2O because they cannot reabsorb H_2O from the distal parts of the nephron. Such patients typically produce up to 20 liters of very dilute urine daily, compared to the normal average of 1.5 liters per day. Unless H_2O intake keeps pace with this tremendous loss of H_2O in the urine, the person quickly dehydrates.

DIRECTION AND RESULTING SYMPTOMS OF WATER MOVEMENT DURING HYPERTONICITY

Whenever the ECF compartment becomes hypertonic, H_2O moves out of the cells by osmosis into the more concentrated ECF until the ICF osmolarity equilibrates with the ECF. As H_2O leaves them, the cells shrink. Of particular concern is that considerable shrinking of brain neurons disturbs brain function, which can be manifested as mental confusion and irrationality in moderate cases and as possible delirium, convulsions, or coma in more severe hypertonic conditions.

Rivaling the neural symptoms in seriousness are circulatory disturbances that arise from a reduction in plasma volume in association with dehydration. Circulatory problems may range from a slight lowering of blood pressure to circulatory shock and death.

Other more common symptoms become apparent even in mild cases of dehydration. For example, dry skin and sunken eyeballs indicate loss of H_2O from the underlying soft tissues, and the tongue becomes dry and parched because salivary secretion is suppressed.

▌ During ECF hypotonicity, the cells swell as H_2O enters them.

Clinical Note **Hypotonicity** of the ECF is usually associated with **overhydration**; that is, excess free H_2O. When a positive free H_2O balance exists, the ECF is less concentrated (more dilute) than normal.

CAUSES OF HYPOTONICITY (OVERHYDRATION)

Usually, any surplus free H_2O is promptly excreted in the urine, so hypotonicity generally does not occur. However, hypotonicity can arise in three ways:

1. Patients with *renal failure* who cannot excrete a dilute urine become hypotonic when they consume relatively more H_2O than solutes.
2. Hypotonicity can occur transiently in healthy people *if H_2O is rapidly ingested* to such an excess that the kidneys can't respond quickly enough to eliminate the extra H_2O.

3. Hypotonicity can occur when excess H_2O without solute is retained in the body as a result of *inappropriate secretion of vasopressin.*

Vasopressin is normally secreted in response to a H_2O deficit, which is relieved by increasing H_2O reabsorption in the distal part of the nephrons. However, vasopressin secretion, and therefore hormonally controlled tubular H_2O reabsorption, can be increased in response to pain, acute infections, trauma, and other stressful situations, even when the body has no H_2O deficit. The increased vasopressin secretion and resulting H_2O retention elicited by stress are appropriate in anticipation of potential blood loss in the stressful situation. The extra retained H_2O could minimize the effect a loss of blood volume would have on blood pressure. However, because modern-day stressful situations generally do not involve blood loss, the increased vasopressin secretion is inappropriate as far as the body's fluid balance is concerned. The reabsorption and retention of too much H_2O dilute the body's solutes.

DIRECTION AND RESULTING SYMPTOMS OF WATER MOVEMENT DURING HYPOTONICITY

Whichever way it is brought about, excess free H_2O retention first dilutes the ECF compartment, making it hypotonic. The resulting difference in osmotic activity between the ECF and ICF induces H_2O to move by osmosis from the more dilute ECF into the cells, with the cells swelling as H_2O moves into them. Like the shrinking of cerebral neurons, pronounced swelling of brain cells also leads to brain dysfunction. Symptoms include confusion, irritability, lethargy, headache, dizziness, vomiting, drowsiness, and, in severe cases, even convulsions, coma, and death.

Nonneural symptoms of overhydration include weakness, caused by swelling of muscle cells, and circulatory disturbances, including hypertension and edema, caused by expansion of plasma volume.

The condition of overhydration, hypotonicity, and cellular swelling resulting from excess free H_2O retention is known as **water intoxication.** It should not be confused with the fluid retention that occurs with excess salt retention. In the latter case, the ECF is still isotonic because the increase in salt is matched by a corresponding increase in H_2O. Because the interstitial fluid is still isotonic, no osmotic gradient exists to drive the extra H_2O into the cells. The excess salt and H_2O burden is therefore confined to the ECF compartment, with circulatory consequences being the most important concern. In water intoxication, in addition to any circulatory disturbances, symptoms caused by cellular swelling become a problem.

Now let's look at how free H_2O balance is normally maintained.

▌ Controlling water balance by means of vasopressin is important in regulating ECF osmolarity.

Controlling free H_2O balance is crucial for regulating ECF osmolarity. Because increases in free H_2O cause the ECF to become too dilute and deficits of free H_2O cause the ECF to become too concentrated, the osmolarity of the ECF must be immediately corrected by restoring stable free H_2O balance to avoid harmful osmotic fluid shifts into or out of the cells.

To maintain a stable H_2O balance, H_2O input must equal H_2O output.

SOURCES OF H₂O INPUT

• In a person's typical daily H_2O balance (▲ Table 14-3), a little more than a liter of H_2O is added to the body by *drinking liquids*.
• Surprisingly, an amount almost equal to that is obtained from *eating solid food*. Recall that muscles consist of about 75% H_2O; meat is therefore 75% H_2O, because it is animal muscle. Likewise, fruits and vegetables consist of 60 to 90% H_2O. Therefore, people normally get almost as much H_2O from solid foods as from the liquids they drink.
• The third source of H_2O input is *metabolically produced H_2O*. Chemical reactions within the cells convert food and O_2 into energy, producing CO_2 and H_2O in the process. This **metabolic H_2O** produced during cell metabolism and released into the ECF averages about 350 ml/day.

The average H_2O intake from these three sources totals 2600 ml/day. Another source of H_2O often employed therapeutically is intravenous infusion of fluid.

SOURCES OF H₂O OUTPUT

• On the output side of the H_2O balance tally, the body loses close to a liter of H_2O daily without being aware of it. This so-called **insensible loss** (loss of which the person has no sensory awareness) occurs from the *lungs* and *nonsweating skin*. During respiration, inspired air becomes saturated with H_2O within the airways. This H_2O is lost when the moistened air is subsequently expired. Normally, we are not aware of this H_2O loss, but can recognize it on cold days, when H_2O vapor condenses so that we can "see our breath." The other insensible loss is continual loss of H_2O from the skin even in the absence of sweating. Water molecules can diffuse through skin cells and evaporate without being noticed. Fortunately, the skin is fairly waterproofed by its keratinized exterior layer, which protects against a much greater loss of H_2O by this avenue (see p. 37). When this protective surface layer is lost, such as when a person has extensive burns, increased fluid loss from the burned surface can cause serious problems with fluid balance.
• Sensible loss (loss of which the person is aware) of H_2O from the skin occurs through *sweating,* which represents another avenue of H_2O output. At an air temperature of 68°F, an average of 100 ml of H_2O is lost daily through sweating. Loss of water from sweating can vary substantially, of course, depending on the environmental temperature and humidity and the degree of physical activity; it may range from zero up to as much as several liters per hour in very hot weather.
• Another passageway for H_2O loss from the body is through the *feces*. Normally, only about 100 ml of H_2O are lost this way each day. During fecal formation in the large intestine, most H_2O is absorbed out of the digestive tract lumen into the blood, thereby conserving fluid and solidifying the digestive tract's contents for elimination. Additional H_2O can be lost from the digestive tract through vomiting or diarrhea.
• By far the most important output mechanism is *urine excretion*, with 1500 ml (1.5 l) of urine being produced daily on average.

The total H_2O output is 2600 ml/day, the same as the volume of H_2O input in our example. This balance is not by chance. Normally, H_2O input matches H_2O output so that the H_2O in the body remains in balance.

FACTORS REGULATED TO MAINTAIN WATER BALANCE

Of the many sources of H_2O input and output, only two can be regulated to maintain H_2O balance. On the intake side, thirst influences the amount of fluid ingested, and on the output side the kidneys can adjust how much urine is formed. Controlling H_2O output in the urine is the most important mechanism in controlling H_2O balance.

Some of the other factors are regulated, but not for maintaining H_2O balance. Food intake is subject to regulation to maintain energy balance, and control of sweating is important in maintaining body temperature. Metabolic H_2O production and insensible losses are completely unregulated.

CONTROL OF WATER OUTPUT IN THE URINE BY VASOPRESSIN

Fluctuations in ECF osmolarity caused by imbalances between H_2O input and output are quickly compensated for by adjusting urinary excretion of H_2O without changing the usual excretion of salt. That is, H_2O reabsorption and excretion are partially dissociated from solute reabsorption and excretion, so the amount of free H_2O retained or eliminated can be varied to quickly restore ECF osmolarity to normal. Free H_2O reabsorption and excretion are adjusted through changes in vasopressin secretion (see p. 432). Throughout most of the nephron, H_2O reabsorption is important in regulating ECF

▲ **TABLE 14-3**

Daily Water Balance

WATER INPUT		WATER OUTPUT	
Avenue	Quantity (ml/day)	Avenue	Quantity (ml/day)
Fluid intake	1,250	Insensible loss (from lungs and nonsweating skin)	900
H_2O in food intake	1,000		
Metabolically produced H_2O	350	Sweat	100
		Feces	100
		Urine	1,500
Total input	2,600	Total input	2,600

volume, because salt reabsorption is accompanied by comparable H_2O reabsorption. In the distal and collecting tubules, however, variable free H_2O reabsorption can take place without comparable salt reabsorption, because of the vertical osmotic gradient in the renal medulla to which this part of the tubule is exposed. Vasopressin increases the permeability of this late part of the tubule to H_2O. Depending on the amount of vasopressin present, the amount of free H_2O reabsorbed can be adjusted as necessary to restore ECF osmolarity to normal.

CONTROL OF WATER INPUT BY THIRST

Thirst is the subjective sensation that drives you to ingest H_2O. A **thirst center** is located in the hypothalamus in close proximity to the vasopressin-secreting cells. Thirst increases H_2O input, whereas vasopressin decreases H_2O output by reducing urine production.

We are now going to elaborate on the mechanisms that regulate vasopressin secretion and thirst.

▌ Vasopressin secretion and thirst are largely triggered simultaneously.

The hypothalamic control centers that regulate vasopressin secretion (and thus urinary output) and thirst (and thus drinking) act in concert. Vasopressin secretion and thirst are both stimulated by a free H_2O deficit and suppressed by a free H_2O excess. Thus, appropriately, the same circumstances that call for reducing urinary output to conserve body H_2O also give rise to the sensation of thirst to replenish body H_2O.

ROLE OF HYPOTHALAMIC OSMORECEPTORS

The predominant excitatory input for both vasopressin secretion and thirst comes from **hypothalamic osmoreceptors** located near the vasopressin-secreting cells and thirst center. These osmoreceptors monitor the osmolarity of fluid surrounding them, which in turn reflects the concentration of the entire internal fluid environment. As the osmolarity increases (too little H_2O) and the need for H_2O conservation increases, vasopressin secretion and thirst are both stimulated (● Figure 14-4). As a result, reabsorption of H_2O in the distal and collecting tubules is increased so that urinary output is reduced and H_2O is conserved, while H_2O intake is simultaneously encouraged. These actions restore depleted H_2O stores, thus relieving the hypertonic condition by diluting the solutes to normal concentration. In contrast, H_2O excess, manifested

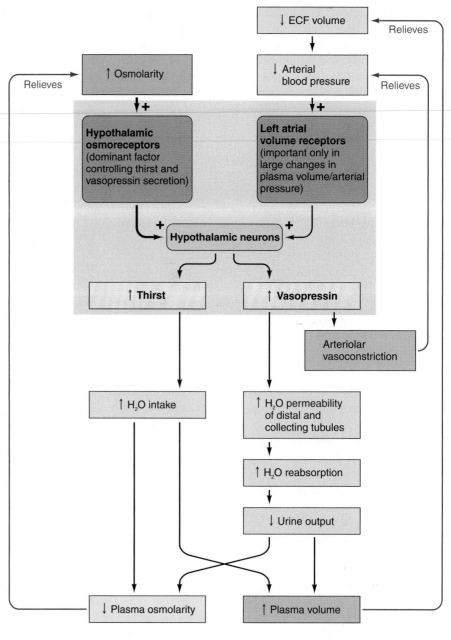

● **FIGURE 14-4**

Control of increased vasopressin secretion and thirst during a H_2O deficit

by reduced ECF osmolarity, prompts increased urinary output (through decreased vasopressin release) and suppresses thirst, which together reduce the water load in the body.

ROLE OF LEFT ATRIAL BARORECEPTORS

Even though the major stimulus for vasopressin secretion and thirst is an increase in ECF osmolarity, the vasopressin-secreting cells and thirst center are both influenced to a moderate extent by changes in ECF volume mediated by input from the **left atrial volume receptors**. Located in the left atrium, these volume receptors monitor the pressure of blood flowing through, which reflects the ECF volume. In response to a major reduction in ECF volume and accordingly in arterial pressure, as during hemorrhage, the left atrial volume receptors reflexly stimulate both vasopressin secretion and thirst. The outpouring of vasopressin and the increased thirst lead to decreased urine output and increased fluid intake, respectively. Furthermore, vasopressin, at the circulating levels elicited by a large decline in ECF volume and arterial pressure, exerts a potent vasoconstrictor effect on arterioles. Both by helping expand ECF and plasma volume and by increasing total peripheral resistance, vasopressin helps relieve the low blood pressure that elicited vasopressin secretion. Conversely, vasopressin and thirst are both inhibited when ECF/plasma volume and arterial blood pressure are elevated.

Recall that low ECF/plasma volume and low arterial blood pressure also reflexly increase aldosterone secretion. The resulting increase in Na^+ reabsorption ultimately leads to osmotic retention of H_2O, expansion of ECF volume, and an increase in arterial blood pressure. In fact, aldosterone-controlled Na^+ reabsorption is the most important factor in regulating ECF volume, with the vasopressin and thirst mechanism playing only a supportive role.

NONPHYSIOLOGIC INFLUENCES ON FLUID INTAKE

Even though the thirst mechanism exists to control H_2O intake, fluid consumption by humans is often influenced more by habit and sociological factors than by the need to regulate H_2O balance. Thus even though H_2O intake is critical in maintaining fluid balance, it is not precisely controlled in humans, who err especially on the side of excess H_2O consumption. We usually drink when we are thirsty, but we often drink even when we are not thirsty because, for example, we are on a coffee break.

With H_2O intake being inadequately controlled and indeed even contributing to H_2O imbalances in the body, the primary factor involved in maintaining H_2O balance is urinary output regulated by the kidneys. Accordingly, *vasopressin-controlled H_2O reabsorption is of primary importance in regulating ECF osmolarity.*

Before shifting attention to acid–base balance, ▲ Table 14-4 summarizes the regulation of ECF volume and osmolarity, the two factors important in maintaining fluid balance.

 Click on the Media Exercises menu of the CD-ROM and work Media Exercise 14.1: Fluid and Electrolyte Balance to test your understanding of the previous section.

ACID–BASE BALANCE

Note: The scope of this topic has been limited to what can be understood by students with no college chemistry background. Students who have the background and need a more chemistry-oriented coverage of acid–base balance should refer to Appendix D, p. A-31, which covers the chemistry in more detail.

The term **acid–base balance** refers to the precise regulation of **free** (that is, unbound) **hydrogen-ion (H^+) concentration** in the body fluids. To indicate the concentration of a chemical, its symbol is enclosed in square brackets, [ ]. Thus [H^+] designates H^+ concentration.

▌ Acids liberate free hydrogen ions, whereas bases accept them.

Acids are a special group of hydrogen-containing substances that *dissociate,* or separate, when in solution to liberate free H^+ and anions (negatively charged ions). Many other substances (for example, carbohydrates) also contain hydrogen, but they are not classified as acids, because the hydrogen is tightly bound within their molecular structure and is never liberated as free H^+.

▲ **TABLE 14-4**

Summary of the Regulation of ECF Volume and Osmolarity

REGULATED VARIABLE	NEED TO REGULATE THE VARIABLE	OUTCOMES IF THE VARIABLE IS NOT NORMAL	MECHANISM FOR REGULATING THE VARIABLE
ECF Volume	Important in the long-term control of arterial blood pressure	↓ECF volume→ ↓arterial blood pressure ↑ECF volume→ ↑arterial blood pressure	Maintenance of salt balance; salt osmotically "holds" H_2O, so the Na^+ load determines the ECF volume. Accomplished primarily by aldosterone-controlled adjustments in urinary Na^+ excretion
ECF Osmolarity	Important to prevent detrimental osmotic movement of H_2O between the ECF and ICF	↓ECF osmolarity (hypotonicity)→ H_2O enters the cell → cells swell ↑ECF osmolarity (hypertonicity)→ H_2O leaves the cells → cells shrink	Maintenance of free H_2O balance. Accomplished primarily by vasopressin-controlled adjustments in excretion of H_2O in the urine

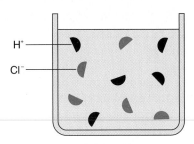

Strong acid (HCl)

(a)

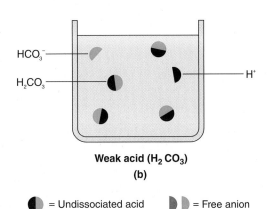

Weak acid (H₂ CO₃)

(b)

= Undissociated acid   = Free anion

= Free H⁺

● **FIGURE 14-5**

Comparison of a strong and a weak acid.
(a) Five molecules of a strong acid. A strong acid
such as HCl (hydrochloric acid) completely dissociates into free H^+ and anions in solution.
(b) Five molecules of a weak acid. A weak acid
such as H_2CO_3 (carbonic acid) only partially
dissociates into free H^+ and anions in solution.

A strong acid has a greater tendency
to dissociate in solution than a weak acid
does; that is, a greater percentage of a
strong acid's molecules separates into
free H^+ and anions. Hydrochloric acid
(HCl) is an example of a strong acid;
every HCl molecule dissociates into free
H^+ and Cl^- (chloride) when dissolved
in H_2O. With a weaker acid such as carbonic acid (H_2CO_3), only a portion of
the molecules dissociate in solution into
H^+ and HCO_3^- (bicarbonate anions).
The remaining H_2CO_3 molecules remain
intact. Because only the free hydrogen
ions contribute to the acidity of a solution, H_2CO_3 is a weaker acid than HCl
because H_2CO_3 does not yield as many
free hydrogen ions per number of acid
molecules present in solution (● Figure 14-5).

A **base** is a substance that can combine with a free H^+ and thus remove it

from solution. A strong base can bind H^+ more readily than
a weak base can.

The pH designation is used to express [H⁺].

The concept of pH has been developed to express $[H^+]$ more
conveniently. Specifically, **pH** equals the logarithm (log) to the
base 10 of the reciprocal of the hydrogen ion concentration:

$$pH = \log \frac{1}{[H^+]}$$

This formula may seem intimidating, but you only need to
glean one important point from it: Because $[H^+]$ is in the denominator, *a high [H⁺] corresponds to a low pH, and a low [H⁺]
corresponds to a high pH*. The greater the $[H^+]$, the larger the
number by which 1 must be divided, and the lower the pH.
You do not need to understand what a logarithm is to recognize this relationship.

● **FIGURE 14-6**

pH considerations in chemistry and physiology. (a) Relationship of pH to
the relative concentrations of H^+ and base (OH^-) under chemically neutral,
acidic, and alkaline conditions. (b) Plasma pH range under normal, acidosis,
and alkalosis conditions.

PhysioEdge   For an interaction related to this figure, see Media Exercise 14.2:
Basics of Acid–Base Balance on the CD-ROM.

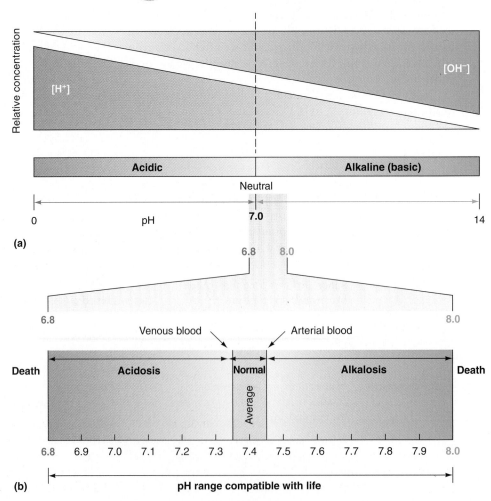

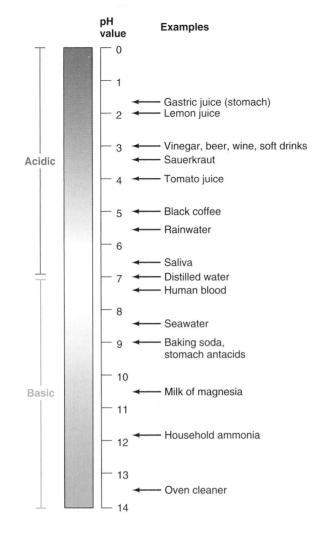

pH value | Examples

0

1

2 ← Gastric juice (stomach)
  ← Lemon juice

3 ← Vinegar, beer, wine, soft drinks
  ← Sauerkraut

Acidic

4 ← Tomato juice

5 ← Black coffee

  ← Rainwater

6

  ← Saliva

7 ← Distilled water
  ← Human blood

8

  ← Seawater

9 ← Baking soda,
    stomach antacids

10

Basic

  ← Milk of magnesia

11

12 ← Household ammonia

13

  ← Oven cleaner

14

● **FIGURE 14-7**

Comparison of pH values of common solutions

The pH of pure H_2O is 7.0, which is considered chemically neutral. Solutions having a pH less than 7.0 contain a higher $[H^+]$ than pure H_2O and are considered **acidic**. Conversely, solutions having a pH value greater than 7.0 have a lower $[H^+]$ and are considered **basic** or **alkaline** (● Figure 14-6a). ● Figure 14-7 compares the pH values of common solutions.

The pH of arterial blood is normally 7.45, and the pH of venous blood is 7.35, for an average blood pH of 7.4. The pH of venous blood is slightly lower (more acidic) than that of arterial blood, because H^+ is generated by the formation of H_2CO_3 from CO_2 picked up at the tissue capillaries. **Acidosis** exists whenever blood pH falls below 7.35, whereas **alkalosis** occurs when blood pH is above 7.45 (● Figure 14-6b). Note that the reference point for determining the body's acid–base status is not the chemically neutral pH of 7.0 but the normal plasma pH of 7.4. Thus a plasma pH of 7.2 is considered acidotic even though in chemistry a pH of 7.2 is considered basic.

An arterial pH of less than 6.8 or greater than 8.0 is not compatible with life. Because death occurs if arterial pH falls outside the range of 6.8 to 8.0 for more than a few seconds, $[H^+]$ in the body fluids must be carefully regulated.

▌ Fluctuations in $[H^+]$ alter nerve, enzyme, and K^+ activity.

Only a narrow pH range is compatible with life, because even small changes in $[H^+]$ have dramatic effects on normal cell function. The main consequences of fluctuations in $[H^+]$ include the following:

1. *Changes in excitability of nerve and muscle cells* are among the major clinical manifestations of pH abnormalities.
    *   The major clinical effect of increased $[H^+]$ (acidosis) is depression of the central nervous system (CNS). Acidotic patients become disoriented and, in more severe cases, eventually die in a state of coma.
    *   In contrast, the major clinical effect of decreased $[H^+]$ (alkalosis) is overexcitability of the nervous system, first the peripheral nervous system and later the CNS. Peripheral nerves become so excitable that they fire even in the absence of normal stimuli. Such overexcitability of the afferent (sensory) nerves gives rise to abnormal "pins-and-needles" tingling sensations. Overexcitability of efferent (motor) nerves brings about muscle twitches and, in more pronounced cases, severe muscle spasms. Death may occur in extreme alkalosis, because spasm of the respiratory muscles seriously impairs breathing. Alternatively, severely alkalotic patients may die of convulsions resulting from overexcitability of the CNS. In less serious situations, CNS overexcitability is manifested as extreme nervousness.
2. Hydrogen ion concentration exerts a *marked influence on enzyme activity*. Even slight deviations in $[H^+]$ alter the shape and activity of protein molecules. Because enzymes are proteins, a shift in the body's acid–base balance disturbs the normal pattern of metabolic activity catalyzed by these enzymes. Some cellular chemical reactions are accelerated; others are depressed.
3. Changes in $[H^+]$ *influence K^+ levels* in the body. When reabsorbing Na^+ from the filtrate, the renal tubular cells secrete either K^+ or H^+ in exchange. Normally, they secrete a preponderance of K^+ compared to H^+. Because of the intimate relationship between secretion of H^+ and K^+ by the kidneys, an increased rate of secretion of one of these ions is accompanied by a decreased rate of secretion of the other. For example, if more H^+ than normal is eliminated by the kidneys, as occurs when the body fluids become acidotic, less K^+ than usual can be excreted. The resulting K^+ retention can affect cardiac function, among other detrimental consequences.

▌ Hydrogen ions are continually added to the body fluids as a result of metabolic activities.

As with any other constituent, to maintain a constant $[H^+]$ in the body fluids, input of hydrogen ions must be balanced by an equal output. On the input side, only a small amount of acid capable of dissociating to release H^+ is taken in with food, such as the weak citric acid found in oranges. Most H^+ in the body fluids is generated internally from metabolic activities.

SOURCES OF H⁺ IN THE BODY

Normally, H⁺ is continually being added to the body fluids from the three following sources:

1. *Carbonic acid formation*. The major source of H⁺ is through H_2CO_3 formation from metabolically produced CO_2. Cellular oxidation of nutrients yields energy, with CO_2 and H_2O as end products. Catalyzed by the enzyme *carbonic anhydrase (ca)*, CO_2 and H_2O form H_2CO_3, which then partially dissociates to liberate free H⁺ and HCO_3^-.

$$CO_2 + H_2O \overset{ca}{\rightleftarrows} H_2CO_3 \rightleftarrows H^+ + HCO_3^-$$

2. *Acids produced during the breakdown of nutrients*. Dietary proteins found abundantly in meat contain a large quantity of sulfur and phosphorus. When these nutrient molecules are broken down, sulfuric acid and phosphoric acid are produced as by-products. Being moderately strong acids, these two acids largely dissociate, liberating free H⁺ into the body fluids.

3. *Acids resulting from intermediary metabolism*. Numerous acids are produced during normal intermediary metabolism. For example, lactic acid is produced by muscles during heavy exercise. These acids partially dissociate to yield free H⁺.

Hydrogen ion generation therefore normally goes on continuously, as a result of ongoing metabolic activities. Further-more, in certain disease states additional acids may be produced that further contribute to the total body pool of H⁺. For example, in diabetes mellitus large quantities of keto acids may be produced by abnormal fat metabolism. Thus input of H⁺ is unceasing, highly variable, and essentially unregulated.

THREE LINES OF DEFENSE AGAINST CHANGES IN [H⁺]

The key to H⁺ balance is maintaining normal alkalinity of the ECF (pH 7.4) despite this constant onslaught of acid. The generated free H⁺ must be largely removed from solution while in the body and ultimately must be eliminated so that the pH of body fluids can remain within the narrow range compatible with life. Mechanisms must also exist to compensate rapidly for the occasional situation in which the ECF becomes too alkaline.

Three lines of defense against changes in [H⁺] operate to maintain [H⁺] of body fluids at a nearly constant level despite unregulated input: (1) the *chemical buffer systems*, (2) the *respiratory mechanism of pH control*, and (3) the *renal mechanism of pH control*. We will look at each of these methods.

▌Chemical buffer systems minimize changes in pH by binding with or yielding free H⁺.

A **chemical buffer system** is a mixture in a solution of two chemical compounds that minimize pH changes when either an acid or a base is added to or removed from the solution. A buffer system consists of a pair of substances involved in a reversible reaction—one substance that can yield free H⁺ as [H⁺] starts to fall and another that can bind with free H⁺ (thus removing it from solution) when [H⁺] starts to rise. A reversible reaction can proceed in either direction depending on the concentrations of the substances involved, as dictated by the *law of mass action* (see p. 387).

An important example of such a buffer system is the carbonic acid:bicarbonate (H_2CO_3:HCO_3^-) buffer pair, which is involved in the following reversible reaction:

$$H_2CO_3 \rightleftarrows H^+ + HCO_3^-$$

When a strong acid such as HCl is added to an unbuffered solution, all the dissociated H⁺ remains free in the solution (● Figure 14-8a). In contrast, when HCl is added to a solution containing the H_2CO_3:HCO_3^- buffer pair, the HCO_3^- immediately binds with the free H⁺ to form H_2CO_3 (● Figure 14-8b). This weak

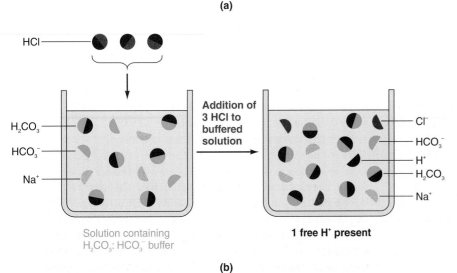

Unbuffered NaCl solution — 3 free H⁺ present

(a)

Solution containing H_2CO_3: HCO_3^- buffer — 1 free H⁺ present

(b)

● **FIGURE 14-8**

Action of chemical buffers. (a) Addition of HCl to an unbuffered solution. All the added hydrogen ions (H⁺) remain free and contribute to the acidity of the solution. (b) Addition of HCl to a buffered solution. Bicarbonate ions (HCO_3^-), the basic member of the buffer pair, bind with some of the added H⁺ and remove them from solution so that they do not contribute to its acidity.

H_2CO_3 dissociates only slightly compared to the marked reduction in pH that occurred when the buffer system was not present and the additional H^+ remained unbound. In the opposite case, when the pH of the solution starts to rise from the addition of base or loss of acid, the H^+-yielding member of the buffer pair, H_2CO_3, releases H^+ to minimize the rise in pH.

The body has four buffer systems: (1) the H_2CO_3:HCO_3^- buffer system, (2) the protein buffer system, (3) the hemoglobin buffer system, and (4) the phosphate buffer system. Each serves a different important role, as you will learn as we examine each in turn (▲ Table 14-5).

▌ The H_2CO_3:HCO_3^- buffer pair is the primary ECF buffer for noncarbonic acids.

The H_2CO_3:HCO_3^- buffer pair is the most important buffer system in the ECF for buffering pH changes brought about by causes other than fluctuations in CO_2-generated H_2CO_3. It is a very effective ECF buffer system for two reasons. First, H_2CO_3 and HCO_3^- are abundant in the ECF, so this system is readily available to resist changes in pH. Second, and more importantly, each component of this buffer pair is closely regulated. The kidneys regulate HCO_3^-, and the respiratory system regulates CO_2, which generates H_2CO_3.

▌ The protein buffer system is primarily important intracellularly.

The most plentiful buffers of the body fluids are the proteins, including the intracellular proteins and the plasma proteins. Proteins are excellent buffers, because they contain both acidic and basic groups that can give up or take up H^+. Quantitatively, the protein system is most important in buffering changes in $[H^+]$ in the ICF, because of the sheer abundance of the intracellular proteins. A more limited number of plasma proteins reinforces the H_2CO_3:HCO_3^- system in extracellular buffering.

▲ **TABLE 14-5**

Chemical Buffers and Their Primary Roles

BUFFER SYSTEM	MAJOR FUNCTIONS
Carbonic Acid: Bicarbonate Buffer System	Primary ECF buffer against noncarbonic-acid changes
Protein Buffer System	Primary ICF buffer; also buffers ECF
Hemoglobin Buffer System	Primary buffer against carbonic acid changes
Phosphate Buffer System	Important urinary buffer; also buffers ICF

▌ The hemoglobin buffer system buffers H^+ generated from carbonic acid.

Hemoglobin (Hb) buffers the H^+ generated from metabolically produced CO_2 in transit between the tissues and lungs. At the systemic capillary level, CO_2 continuously diffuses into the blood from the tissue cells where it is being produced. The greatest percentage of this CO_2 forms H_2CO_3, which partially dissociates into H^+ and HCO_3^-. Most H^+ generated from CO_2 at the tissue level becomes bound to reduced Hb and no longer contributes to acidity of body fluids. Were it not for Hb, blood would become much too acidic after picking up CO_2 at the tissues. With the tremendous buffering capacity of the Hb system, venous blood is only slightly more acidic than arterial blood despite the large volume of H^+-generating CO_2 carried in venous blood. At the lungs, the reactions are reversed and the resulting CO_2 is exhaled.

▌ The phosphate buffer system is an important urinary buffer.

The phosphate buffer system consists of an acid phosphate salt that can donate a free H^+ when $[H^+]$ falls and a basic phosphate salt that can accept a free H^+ when $[H^+]$ rises. Because phosphates are most abundant within the cells, this system contributes significantly to intracellular buffering, being rivaled only by the more plentiful intracellular proteins.

Even more importantly, the phosphate system serves as an excellent urinary buffer. Humans normally consume more phosphate than needed. The excess phosphate filtered through the kidneys is not reabsorbed but remains in the tubular fluid to be excreted (because the renal threshold for phosphate is exceeded). This excreted phosphate buffers urine as it is being formed by removing from solution the H^+ secreted into the tubular fluid. None of the other body-fluid buffer systems are present in the tubular fluid to play a role in buffering urine during its formation. Most or all of the filtered HCO_3^- and CO_2 (alias H_2CO_3) are reabsorbed, whereas Hb and plasma proteins are not even filtered.

▌ Chemical buffer systems act as the first line of defense against changes in $[H^+]$.

All chemical buffer systems act immediately, within fractions of a second, to minimize changes in pH. When $[H^+]$ is altered, the involved buffer systems' reversible chemical reactions shift at once to compensate for the change in $[H^+]$. Accordingly, the buffer systems are the *first line of defense* against changes in $[H^+]$, because they are the first mechanism to respond.

Through the mechanism of buffering, most hydrogen ions seem to disappear from the body fluids between the times of their generation and their elimination. It must be emphasized, however, that none of the chemical buffer systems actually eliminates H^+ from the body. These ions are merely removed from solution by being incorporated within one member of the buffer pair, thus preventing the hydrogen ions from contributing to body fluid acidity. Because each buffer system has a limited capacity to soak up H^+, the H^+ that is unceasingly produced must ultimately be removed from the body. If H^+ were not eventually eliminated, soon all the body fluid buffers would already be bound with H^+ and there would be no further buffering ability.

The respiratory and renal mechanisms of pH control actually eliminate acid from the body instead of merely suppressing it, but they respond more slowly than chemical buffer systems. We will now turn our attention to these other defenses against changes in acid–base balance.

▌ The respiratory system regulates [H^+] by controlling the rate of CO_2 removal.

The respiratory system plays an important role in acid–base balance through its ability to alter pulmonary ventilation and consequently to alter excretion of H^+-generating CO_2. The level of respiratory activity is governed in part by arterial [H^+], as follows (▲ Table 14-6):

• When arterial [H^+] increases as the result of a *nonrespiratory* (or *metabolic*) cause, the respiratory center in the brain stem is reflexly stimulated to increase pulmonary ventilation (the rate at which gas is exchanged between the lungs and the atmosphere) (see p. 399). As the rate and depth of breathing increase, more CO_2 than usual is blown off, so less H_2CO_3 than normal is added to the body fluids. Because CO_2 forms acid, removal of CO_2 in essence removes acid from this source from the body, offsetting extra acid present from a nonrespiratory source.
• Conversely, when arterial [H^+] falls, pulmonary ventilation is reduced. As a result of slower, shallower breathing, metabolically produced CO_2 diffuses from the cells into the blood faster than it is removed from the blood by the lungs, so higher-than-usual amounts of acid-forming CO_2 accumulate in the blood, thus restoring [H^+] toward normal.

The lungs are extremely important in maintaining [H^+]. Every day they remove from body fluids what amounts to 100 times more H^+ derived from carbonic acid than the kidneys remove from sources other than carbonic acid. Furthermore, the respiratory system, through its ability to regulate arterial [CO_2], can adjust the amount of H^+ added to body fluids from this source as needed to restore pH toward normal when fluctuations occur in [H^+] from sources other than carbonic acid.

▌ The respiratory system serves as the second line of defense against changes in [H^+].

Respiratory regulation acts at a moderate speed, coming into play only when chemical buffer systems alone cannot minimize [H^+] changes. When deviations in [H^+] occur, the buffer systems respond immediately, whereas adjustments in ventilation require a few minutes to be initiated. If a deviation in [H^+] is not swiftly and completely corrected by the buffer systems, the respiratory system comes into action a few minutes later, thus serving as the *second line of defense* against changes in [H^+].

Of course, when changes in [H^+] stem from [CO_2] fluctuations that arise from respiratory abnormalities, the respiratory mechanism cannot contribute at all to pH control. For example, if acidosis exists because of CO_2 accumulation caused by lung disease, the impaired lungs cannot possibly compensate for acidosis by increasing the rate of CO_2 removal. The buffer systems (other than the H_2CO_3:HCO_3^- pair) plus renal regulation are the only mechanisms available for defending against respiratory-induced acid–base abnormalities.

Let's now see how the kidneys help maintain acid–base balance.

▌ The kidneys help maintain acid–base balance by adjusting their rate of H^+ excretion, HCO_3^- excretion, and NH_3 secretion.

The kidneys control the pH of body fluids by adjusting three interrelated factors: (1) H^+ excretion, (2) HCO_3^- excretion, and (3) ammonia (NH_3) secretion.

▲ **TABLE 14-6**

Respiratory Adjustments to Acidosis and Alkalosis Induced by Nonrespiratory Causes

RESPIRATORY COMPENSATIONS	ACID–BASE STATUS		
	Normal (pH 7.4)	Nonrespiratory (metabolic) Acidosis (pH 7.1)	Nonrespiratory (metabolic) Alkalosis (pH 7.7)
Ventilation	Normal	↑	↓
Rate of CO_2 Removal	Normal	↑	↓
Rate of H_2CO_3 Formation	Normal	↓	↑
Rate of H^+ Generation from CO_2	Normal	↓	↑

RENAL H$^+$ EXCRETION

Acids are continuously being added to body fluids as a result of metabolic activities, yet the generated H$^+$ must not be allowed to accumulate. Although the body's buffer systems can resist changes in pH by removing H$^+$ from solution, the persistent production of acidic metabolic products would eventually overwhelm the limits of this buffering capacity. Therefore the constantly generated H$^+$ must ultimately be eliminated from the body. The lungs can remove only carbonic acid by eliminating CO$_2$. The task of eliminating H$^+$ derived from sulfuric, phosphoric, lactic, and other acids rests with the kidneys. Furthermore, the kidneys can also eliminate extra H$^+$ derived from carbonic acid.

All the filtered H$^+$ is excreted, but most of the excreted H$^+$ enters the urine via secretion (see p. 422). Energy-dependent carriers in the membrane of the tubular cells secrete the extra H$^+$ from the peritubular capillary plasma into the tubular fluid. Because the kidneys normally excrete H$^+$, urine is usually acidic, having an average pH of 6.0.

The magnitude of H$^+$ secretion depends primarily on a direct effect of the plasma's acid–base status on the kidneys' tubular cells (● Figure 14-9). No neural or hormonal control is involved.

• When the [H$^+$] of the plasma passing through the peritubular capillaries is elevated above normal, the tubular cells respond by secreting greater-than-usual amounts of H$^+$ from the plasma into the tubular fluid to be excreted in the urine.

Conversely, when plasma [H$^+$] is lower than normal, the kidneys conserve H$^+$ by reducing its secretion and subsequent excretion in the urine. The kidneys cannot raise plasma [H$^+$] by reabsorbing more of the filtered H$^+$, because there are no reabsorptive mechanisms for H$^+$. The only way the kidneys can reduce H$^+$ excretion is by secreting less H$^+$.

RENAL HCO$_3^-$ EXCRETION

Before being eliminated by the kidneys, H$^+$ generated from noncarbonic acids is buffered to a large extent by plasma HCO$_3^-$. Appropriately, therefore, renal handling of acid–base balance also involves adjustment of HCO$_3^-$ excretion, depending on the H$^+$ load in plasma (● Figure 14-9).

• When plasma [H$^+$] is higher than normal, the kidneys reabsorb more HCO$_3^-$ than usual, making it available to help buffer the extra H$^+$ load in the body, rather than excreting it in the urine.

• When plasma [H$^+$] is below normal, a smaller proportion of the HCO$_3^-$ pool than usual is tied up buffering H$^+$, so plasma [HCO$_3^-$] is elevated above normal. The kidneys compensate by reabsorbing less HCO$_3^-$, thus excreting more HCO$_3^-$ in the urine and removing the excess, unused HCO$_3^-$ from the plasma.

Note that to compensate for acidosis, the kidneys acidify the urine (by getting rid of extra H$^+$) and alkalinize the plasma (by conserving HCO$_3^-$) to bring pH to normal. In the opposite case—alkalosis—the kidneys make the urine alkaline (by eliminating excess HCO$_3^-$) while acidifying the plasma (by conserving H$^+$) (▲ Table 14-7).

RENAL NH$_3$ SECRETION

The energy-dependent H$^+$ carriers in the tubular cells can secrete H$^+$ against a concentration gradient until the tubular fluid (urine) becomes 800 times more acidic than the plasma. At this point, further H$^+$ secretion stops, because the gradient becomes too great for the secretory process to continue. The kidneys cannot acidify urine beyond a gradient-limited urinary pH of 4.5. If left unbuffered as free H$^+$, only about 1% of the excess H$^+$ typically excreted daily would produce a urinary pH of this magnitude at normal urine flow rates, and elimination of the other 99% of the usually secreted H$^+$ load would be prevented—a situation that would be intolerable. For H$^+$ secretion to proceed, most secreted H$^+$ must be buffered in the tubular fluid so that it does not exist as free H$^+$ and, accordingly, does not contribute to tubular acidity.

Bicarbonate cannot buffer urinary H$^+$ as it does the ECF, because HCO$_3^-$ is not excreted in the urine simultaneously with H$^+$. (Whichever of these substances is in excess in the plasma is excreted in the urine.) There are, however, two important urinary buffers: (1) filtered phosphate buffers and (2) secreted ammonia.

• *Filtered phosphate as a urinary buffer.* Normally, secreted H$^+$ is first buffered by the phosphate buffer system, which is in the tubular fluid because excess ingested phosphate has been filtered but not reabsorbed. The basic member of the phosphate buffer pair binds with secreted H$^+$. Basic phosphate is present in the tubular fluid because of dietary excess, not because of any specific mechanism for buffering secreted H$^+$. When H$^+$ secretion is high, the buffering capacity of urinary phosphates is exceeded, but the kidneys cannot respond by excreting more basic phosphate. Only the quantity of phosphate reabsorbed, not the quantity excreted, is subject to control. As soon as all the basic phosphate ions that are coincidentally excreted have soaked up H$^+$, the acidity of the tubular fluid quickly rises as more H$^+$ ions are secreted. Without additional buffering capacity from another source, H$^+$ secretion would soon halt abruptly as the free [H$^+$] in the tubular fluid quickly rose to the critical limiting level.

● FIGURE 14-9

Control of the rate of tubular H$^+$ secretion

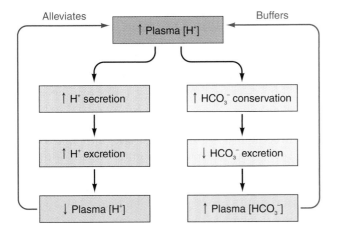

Summary of Renal Responses to Acidosis and Alkalosis

ACID–BASE ABNORMALITY	H^+ SECRETION	H^+ EXCRETION	HCO_3^- REABSORPTION	HCO_3^- EXCRETION	pH OF URINE	COMPENSATORY CHANGE IN PLASMA pH
Acidosis	↑	↑	↑	↓	Acidic	Alkalinization toward normal
Alkalosis	↓	↓	↓	↑	Alkaline	Acidification toward normal

- *Secreted NH_3 as a urinary buffer.* When acidosis exists, the tubular cells secrete **ammonia (NH_3)** into the tubular fluid once the normal urinary phosphate buffers are saturated. This NH_3 enables the kidneys to continue secreting additional H^+ ions, because NH_3 combines with free H^+ in the tubular fluid to form **ammonium ion (NH_4^+)** as follows:

$$NH_3 + H^+ \rightarrow NH_4^+$$

The ammonium ions remain in the tubular fluid and are lost in urine, each one taking a H^+ with it. Thus NH_3 secreted during acidosis buffers excess H^+ in the tubular fluid, so that large amounts of H^+ can be secreted into the urine before the pH falls to the limiting value of 4.5.

▌ The kidneys are a powerful third line of defense against changes in [H^+].

The kidneys need hours to days to compensate for changes in body fluid pH, compared to the immediate responses of the buffer systems and the few minutes of delay before the respiratory system responds. Therefore, they are the *third line of defense* against [H^+] changes in body fluids. However, the kidneys are the most potent acid–base regulatory mechanism; not only can they vary removal of H^+ from any source, but they can also variably conserve or eliminate HCO_3^- depending on the acid–base status of the body.

▌ Acid–base imbalances can arise from either respiratory dysfunction or metabolic disturbances.

Clinical Note Deviations from normal acid–base status are divided into four general categories, depending on the source and direction of the abnormal change in [H^+]. These categories are *respiratory acidosis, respiratory alkalosis, metabolic acidosis,* and *metabolic alkalosis.* An acid–base imbalance that has a respiratory cause is associated with an abnormal [CO_2], giving rise to a change in H_2CO_3–generated H^+. A deviation in pH resulting from any cause other than an abnormal [CO_2] is considered a metabolic acid–base imbalance.

- **Respiratory acidosis** is the result of abnormal CO_2 retention arising from *hypoventilation* (see p. 393). As less-than-normal amounts of CO_2 are lost through the lungs, the resulting increase in H_2CO_3 formation and dissociation leads to an elevated [H^+]. Possible causes include lung disease, depression of the respiratory center by drugs or disease, nerve or muscle disorders that reduce respiratory muscle ability, or (transiently) even the simple act of holding one's breath.

- **Respiratory alkalosis** occurs when excessive CO_2 is lost from the body as a result of *hyperventilation* (see p. 393). When pulmonary ventilation increases out of proportion to the rate of CO_2 production, less H_2CO_3 is formed and [H^+] decreases. Possible causes of respiratory alkalosis include fever, anxiety, and aspirin poisoning, all of which excessively stimulate ventilation without regard to the status of O_2, CO_2, or H^+ in the body fluids.

- **Metabolic acidosis** encompasses all types of acidosis besides that caused by excess CO_2 in body fluids. This is the type of acid–base disorder most frequently encountered. Here are its most common causes:
  1. *Severe diarrhea.* During digestion, a HCO_3^--rich digestive juice is normally secreted into the digestive tract and is later reabsorbed back into the plasma when digestion is completed. During diarrhea, this HCO_3^- is lost from the body rather than being reabsorbed. Because of the loss of HCO_3^-, less HCO_3^- is available to buffer H^+, leading to more free H^+ in the body fluids.
  2. *Diabetes mellitus.* Abnormal fat metabolism resulting from the inability of cells to preferentially use glucose due to inadequate insulin action leads to formation of excess keto acids whose dissociation increases plasma [H^+].
  3. *Strenuous exercise.* When muscles resort to anaerobic glycolysis during strenuous exercise, excess lactic acid is produced, raising plasma [H^+] (see p. 220).
  4. *Uremic acidosis.* In severe renal failure (uremia), the kidneys cannot rid the body of even the normal amounts of H^+ generated from noncarbonic acids formed by ongoing metabolic processes, so H^+ starts to accumulate in body fluids. Also, the kidneys cannot conserve an adequate amount of HCO_3^- for buffering the normal acid load.

- **Metabolic alkalosis** is a reduction in plasma [H^+] caused by a relative deficiency of noncarbonic acids. This condition arises most commonly from the following:

1. *Vomiting* causes abnormal loss of H^+ from the body as a result of lost acidic gastric juices. Hydrochloric acid is secreted into the stomach lumen during digestion and is usually later reabsorbed back into the plasma.

2. *Ingestion* of alkaline drugs can produce alkalosis, such as when baking soda ($NaHCO_3$, which dissociates in solution into Na^+ and HCO_3^-) is used as a self-administered remedy for treating gastric hyperacidity. By neutralizing excess acid in the stomach, HCO_3^- relieves the symptoms of stomach irritation and heartburn, but when more HCO_3^- than needed is ingested the extra HCO_3^- is absorbed from the digestive tract and increases plasma [HCO_3^-]. The extra HCO_3^- binds with some of the free H^+ normally present in plasma from noncarbonic-acid sources, reducing free [H^+]. (In contrast, commercial alkaline products for treating gastric hyperacidity are not absorbed from the digestive tract to any extent and therefore do not alter the body's acid–base status.)

 Click on the Media Exercises menu of the CD-ROM and work Media Exercise 14.2: Basics of Acid–Base Balance to test your understanding of the previous section.

CHAPTER IN PERSPECTIVE: FOCUS ON HOMEOSTASIS

Homeostasis depends on maintaining a balance between the input and output of all constituents present in the internal fluid environment. Regulation of fluid balance involves two separate components: control of salt balance and control of H_2O balance. Control of salt balance is primarily important in the long-term regulation of arterial blood pressure, because the body's salt load affects the osmotic determination of the ECF volume, of which plasma volume is a part. An increased salt load in the ECF leads to an expansion in ECF volume, including plasma volume, which in turn causes a rise in blood pressure. Conversely, a reduction in the ECF salt load brings about a fall in blood pressure. Salt balance is maintained by constantly adjusting salt output in the urine to match unregulated, variable salt intake.

Control of H_2O balance is important in preventing changes in ECF osmolarity, which would induce detrimental osmotic shifts of H_2O between the cells and the ECF. Such shifts of H_2O into or out of the cells would cause the cells to swell or shrink, respectively. Cells, especially brain neurons, do not function normally when swollen or shrunken. Water balance is largely maintained by controlling the volume of H_2O lost in urine to compensate for uncontrolled losses of variable volumes of H_2O from other avenues, such as through sweating or diarrhea, and for poorly regulated H_2O intake. Even though a thirst mechanism exists to control H_2O intake based on need, the amount a person drinks is often influenced by social custom and habit instead of thirst alone.

A balance between input and output of H^+ is critical to maintaining the body's acid–base balance within the narrow limits compatible with life. Deviations in the internal fluid environment's pH lead to altered neuromuscular excitability, to changes in enzymatically controlled metabolic activity, and to K^+ imbalances, which can cause cardiac arrhythmias. These effects are fatal if the pH falls outside the range of 6.8 to 8.0.

Hydrogen ions are uncontrollably and continually added to the body fluids as a result of ongoing metabolic activities, yet the ECF's pH must be kept constant at a slightly alkaline level of 7.4 for optimal body function. Like salt and H_2O balance, control of H^+ output by the kidneys is the main regulatory factor in achieving H^+ balance. The lungs, which can adjust their rate of excretion of H^+-generating CO_2, help the kidneys eliminate H^+. Furthermore, chemical buffer systems can take up or liberate H^+, transiently keeping its concentration constant within the body until its output can be brought into line with its input. Such a mechanism is not available for salt or H_2O balance.

CHAPTER SUMMARY

Balance Concept (pp. 443–444)

▪ The internal pool of a substance is the quantity of that substance in the ECF.

▪ The inputs to the pool are by way of ingestion or metabolic production of the substance. The outputs from the pool are by way of excretion or metabolic consumption of the substance. (Review Figure 14-1.)

▪ Input must equal output to maintain a stable balance of the substance.

Fluid Balance (pp. 444–453)

▪ On average, the body fluids compose 60% of total body weight. This figure varies among individuals, depending on how much fat (a tissue that has a low H_2O content) a person has.

▪ Two thirds of the body H_2O is found in the intracellular fluid (ICF). The remaining third present in the extracellular fluid (ECF) is distributed between plasma (20% of ECF) and interstitial fluid (80% of ECF). (Review Table 14-1.)

▪ Because all plasma constituents are freely exchanged across the capillary walls, the plasma and interstitial fluid are nearly identical in composition, except for the lack of plasma proteins in the interstitial fluid. In contrast, the ECF and ICF have markedly different compositions, because the plasma membrane barriers are highly selective as to what materials are transported into or out of the cells. (Review Figure 14-2.)

▪ The essential components of fluid balance are control of ECF volume by maintaining salt balance and control of ECF osmolarity by maintaining water balance. (Review Tables 14-2, 14-3, and 14-4.)

▪ Because of the osmotic holding power of Na^+, the major ECF cation, a change in the body's total Na^+ content brings about a corresponding change in ECF volume, including plasma volume, which, in turn, alters arterial blood pressure in the same direction. Appropriately, in the long run Na^+-regulating mechanisms compensate for changes in ECF volume and arterial blood pressure. (Review Table 14-4.)

▪ Salt intake is not controlled in humans, but control of salt output in urine is closely regulated. Blood pressure–regulating mechanisms can vary the GFR, and accordingly the amount of Na^+ filtered, by adjusting the caliber of the afferent arterioles supplying the glomeruli. Simultaneously, blood pressure–regulating mechanisms can vary the secretion of aldosterone to adjust Na^+ reabsorption by the renal tubules. Varying Na^+ filtration and

Na$^+$ reabsorption can adjust how much Na$^+$ is excreted in the urine, to regulate plasma volume and consequently arterial blood pressure in the long term. (*Review Figure 14-3.*)

■ Changes in ECF osmolarity are primarily detected and corrected by the systems that maintain H$_2$O balance.

■ ECF osmolarity must be closely regulated to prevent osmotic shifts of H$_2$O between the ECF and ICF, because cell swelling or shrinking is harmful, especially to brain neurons. Excess free H$_2$O in the ECF dilutes ECF solutes; the resulting ECF hypotonicity drives H$_2$O into the cells. An ECF free-H$_2$O deficit, by contrast, concentrates ECF solutes, so H$_2$O leaves the cells to enter the hypertonic ECF. (*Review Table 14-4.*)

■ To prevent these harmful fluxes, free H$_2$O balance is regulated largely by vasopressin and, to a lesser degree, by thirst.

■ Changes in vasopressin secretion and thirst are both governed primarily by hypothalamic osmoreceptors, which monitor ECF osmolarity. The amount of vasopressin secreted determines the extent of free H$_2$O reabsorption by distal portions of the nephrons, thereby determining the volume of urinary output. (*Review Figure 14-4.*)

■ Simultaneously, intensity of thirst controls the volume of fluid intake. However, because the volume of fluid drunk is often not directly correlated with the intensity of thirst, control of urinary output by vasopressin is the most important regulatory mechanism for maintaining H$_2$O balance.

Acid–Base Balance (pp. 453–461)

■ Acids liberate free hydrogen ions (H$^+$) into solution; bases bind with free hydrogen ions and remove them from solution.

■ *Acid–base balance* refers to regulation of H$^+$ concentration ([H$^+$]) in the body fluids.

■ Hydrogen ion concentration is often expressed in terms of pH, which is the logarithm of 1/[H$^+$].

■ The normal pH of the plasma is 7.4, slightly alkaline compared to neutral H$_2$O, which has a pH of 7.0. A pH lower than normal (higher [H$^+$] than normal) indicates a state of acidosis. A pH higher than normal (lower [H$^+$] than normal) characterizes a state of alkalosis. (*Review Figure 14-6.*)

■ Fluctuations in [H$^+$] have profound effects on body chemistry, most notably (1) changes in neuromuscular excitability, (2) disruption of normal metabolic reactions by altering the structure and function of all enzymes, and (3) alterations in plasma [K$^+$] brought about by H$^+$-induced changes in the rate of K$^+$ elimination by the kidneys.

■ The primary challenge in controlling acid–base balance is maintaining normal plasma alkalinity despite continual addition of H$^+$ to the plasma from ongoing metabolic activity. The major source of H$^+$ is from the dissociation of CO$_2$-generated H$_2$CO$_3$.

■ The three lines of defense for resisting changes in [H$^+$] are (1) the chemical buffer systems, (2) respiratory control of pH, and (3) renal control of pH.

■ Chemical buffer systems, the first line of defense, each consist of a pair of chemicals involved in a reversible reaction, one that can liberate H$^+$ and the other that can bind H$^+$. By acting according to the law of mass action, a buffer pair acts immediately to minimize any changes in pH. (*Review Figure 14-8 and Table 14-5.*)

■ The respiratory system, the second line of defense, normally eliminates the metabolically produced CO$_2$ so that H$_2$CO$_3$ does not accumulate in the body fluids.

■ When chemical buffers alone have been unable to immediately minimize a pH change, the respiratory system responds within a few minutes by altering its rate of CO$_2$ removal. An increase in [H$^+$] from sources other than carbonic acid stimulates respiration so that more H$_2$CO$_3$-forming CO$_2$ is blown off, compensating for acidosis by reducing generation of H$^+$ from H$_2$CO$_3$. Conversely, a fall in [H$^+$] depresses respiratory activity so that CO$_2$ and thus H$^+$-generating H$_2$CO$_3$ can accumulate in the body fluids to compensate for alkalosis. (*Review Table 14-6.*)

■ The kidneys are the third and most powerful line of defense. They require hours to days to compensate for a deviation in body fluid pH. However, they not only eliminate H$^+$, but they can also regulate [HCO$_3^-$] in the body fluids.

■ The kidneys compensate for acidosis by secreting the excess H$^+$ in the urine while adding new HCO$_3^-$ to the plasma to expand the HCO$_3^-$ buffer pool. During alkalosis, the kidneys conserve H$^+$ by reducing its secretion in urine. They also eliminate HCO$_3^-$, which is in excess because less HCO$_3^-$ than usual is tied up buffering H$^+$ when H$^+$ is in short supply. (*Review Table 14-7 and Figure 14-9.*)

■ Secreted H$^+$ must be buffered in the tubular fluid to prevent the H$^+$ concentration gradient from becoming so great that it blocks further H$^+$ secretion. Normally, H$^+$ is buffered by the urinary phosphate buffer pair, which is abundant in the tubular fluid because excess dietary phosphate spills into the urine to be excreted from the body. In acidosis, when all the phosphate buffer is already used up in buffering the extra secreted H$^+$, the kidneys secrete NH$_3$ into the tubular fluid to serve as a buffer so that H$^+$ secretion can continue.

■ The four types of acid–base imbalances are respiratory acidosis, respiratory alkalosis, metabolic acidosis, and metabolic alkalosis. Respiratory acid–base disorders stem from deviations from normal [CO$_2$], whereas metabolic acid–base imbalances include all deviations in pH other than those caused by abnormal [CO$_2$].

REVIEW EXERCISES

Objective Questions (Answers on p. A-47)

1. The only avenue by which materials can be exchanged between the cells and the external environment is the ECF. (*True or false?*)

2. Salt balance in humans is poorly regulated because of our hedonistic salt appetite. (*True or false?*)

3. An unintentional increase in CO$_2$ is a cause of respiratory acidosis, but a deliberate increase in CO$_2$ is a compensation for metabolic alkalosis. (*True or false?*)

4. The largest body fluid compartment is the _____.

5. Of the two members of the H$_2$CO$_3$:HCO$_3^-$ buffer system, _____ is regulated by the lungs, whereas _____ is regulated by the kidneys.

6. Which of the following factors does *not* increase vasopressin secretion?
   a. ECF hypertonicity
   b. an ECF volume deficit following hemorrhage
   c. an increase in arterial blood pressure
   d. stressful situations

7. *Indicate all correct answers:* pH
   a. equals log 1/[H$^+$].
   b. is high in acidosis.
   c. falls lower as [H$^+$] increases.

8. *Indicate all correct answers:* Acidosis
   a. causes overexcitability of the nervous system.
   b. exists when the plasma pH falls below 7.35.

c. occurs when CO_2 is blown off more rapidly than it is being produced by metabolic activities.

d. occurs when excessive HCO_3^- is lost from the body such as in diarrhea.

9. *Indicate all correct answers:* The kidney tubular cells secrete NH_3
   a. when the urinary pH becomes too high.
   b. when the body is in a state of alkalosis.
   c. to enable further renal secretion of H^+ to occur.
   d. to buffer excess filtered HCO_3^-.
   e. when there is excess NH_3 in the body fluids.

10. Match each acid–base imbalance with a possible cause:

    _____ 1. respiratory acidosis   a. vomiting
    _____ 2. respiratory alkalosis   b. diabetes mellitus
    _____ 3. metabolic acidosis   c. pneumonia
    _____ 4. metabolic alkalosis   d. aspirin poisoning

Essay Questions

1. Explain the balance concept.
2. Outline the distribution of body H_2O.
3. Compare the ionic composition of plasma, interstitial fluid, and intracellular fluid.

4. What factors are regulated to maintain the body's fluid balance?
5. Why is regulation of ECF volume important? How is it regulated?
6. Why is regulation of ECF osmolarity important? How is it regulated? What are the causes and consequences of ECF hypertonicity and ECF hypotonicity?
7. Outline the sources of input and output in a daily salt balance and a daily H_2O balance. Which are subject to control to maintain the body's fluid balance?
8. Distinguish between an acid and a base.
9. What is the relationship between $[H^+]$ and pH?
10. What is the normal pH of body fluids? How does this compare to the pH of H_2O? Define acidosis and alkalosis.
11. What are the consequences of fluctuations in $[H^+]$?
12. What are the body's sources of H^+?
13. Describe the three lines of defense against changes in $[H^+]$ in terms of the mechanisms and speed of action.
14. List and indicate the functions of each of the body's chemical buffer systems.
15. What are the causes of the four categories of acid–base imbalances?

POINTS TO PONDER

(Explanations on p. A-47)

1. Alcoholic beverages inhibit vasopressin secretion. Given this fact, predict the effect of alcohol on the rate of urine formation. Predict the actions of alcohol on ECF osmolarity. Explain why a person still feels thirsty after excessive consumption of alcoholic beverages.

2. If a person loses 1500 ml of salt-rich sweat and drinks 1000 ml of water during the same time period, what will happen to vasopressin secretion? Why is it important to replace both the water and the salt?

3. If a solute that can penetrate the plasma membrane, such as dextrose (a type of sugar), is dissolved in sterile water at a concentration equal to that of normal body fluids and then is injected intravenously, what is the impact on the body's fluid balance?

4. Why does respiratory alkalosis occur in a person who quickly ascends to high altitudes where arterial PO_2 falls below 60 mm Hg? (*Hint:* See the role of decreased arterial PO_2 in regulating ventilation, on p. 397.)

5. Which of the following reactions would buffer the acidosis accompanying severe pneumonia?
   a. $H^+ + HCO_3^- \rightarrow H_2CO_3 \rightarrow CO_2 + H_2O$
   b. $CO_2 + H_2O \rightarrow H_2CO_3 \rightarrow H^+ + HCO_3^-$
   c. $H^+ + Hb \rightarrow HHb$
   d. $HHb \rightarrow H^+ + Hb$
   e. $NaH_2PO_4 + Na^+ \rightarrow Na_2HPO_4 + H^+$

CLINICAL CONSIDERATION

(Explanation on p. A-47)

Marilyn Y. has had pronounced diarrhea for over a week as a result of having acquired salmonellosis, a bacterial intestinal infection,

from improperly handled food. What impact has this prolonged diarrhea had on her fluid and acid–base balance? In what ways has Marilyn's body been trying to compensate for these imbalances?

PHYSIOEDGE RESOURCES

 **PhysioEdge CD-ROM**

PhysioEdge, the CD-ROM packaged with your text, focuses on the concepts students find most difficult to learn. Figures marked with this icon have associated activities on the CD. For a visual review of concepts in this chapter, check out

Media Exercise 14.1: Fluid and Electrolyte Balance

Media Exercise 14.2: Basics of Acid–Base Balance

PhysioEdge Website

The website for this book contains a wealth of helpful study aids, as well as many ideas for further reading and research. Log on to:
http://www.brookscole.com/hpfundamentals3

Select Chapter 14 from the drop-down menu or click on one of the many resource areas, including Case Histories, which introduce clinical aspects of human physiology. For this chapter check out the following: Case History 5: Eight Years Later.

 For Suggested Readings, consult **InfoTrac College Edition/ Research** on the PhysioEdge website or go directly to InfoTrac College Edition, your online research library, at:
http://infotrac.thomsonlearning.com

Digestive System

Homeostasis
The digestive system contributes to homeostasis by transferring nutrients, water, and electrolytes from the external environment to the internal environment.

Body systems maintain homeostasis

Homeostasis is essential for survival of cells

Cells
Cells need a constant supply of nutrients to support their energy-generating chemical reactions.

$$Food + O_2 \rightarrow CO_2 + H_2O + Energy$$

Also, proper cell function depends on maintaining the availability of water and various electrolytes.

Cells make up body systems

To maintain homeostasis, nutrient molecules used for energy production must continually be replaced by new, energy-rich nutrients. Similarly, water and electrolytes constantly lost in urine and sweat and through other avenues must be replenished regularly. The **digestive system** contributes to homeostasis by transferring nutrients, water, and electrolytes from the external environment to the internal environment. The digestive system does not directly regulate the concentration of any of these constituents in the internal environment. It does not vary nutrient, water, or electrolyte uptake based on body needs (with few exceptions); rather, it optimizes conditions for digesting and absorbing what is ingested.

The Digestive System

CONTENTS AT A GLANCE

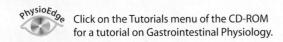

PhysioEdge Click on the Tutorials menu of the CD-ROM for a tutorial on Gastrointestinal Physiology.

INTRODUCTION

The primary function of the **digestive system** is to transfer nutrients, water, and electrolytes from the food we eat into the body's internal environment. Ingested food is essential as an energy source, or fuel, from which the cells can produce ATP to carry out their particular energy-dependent activities, such as active transport, contraction, synthesis, and secretion. Food is also a source of building supplies for the renewal and addition of body tissues.

The act of eating does not automatically make the preformed organic molecules in food available to the body cells as a source of fuel or as building blocks. The food first must be digested, or biochemically broken down, into small, simple molecules that can be absorbed from the digestive tract into the circulatory system for distribution to the cells. Normally, about 95% of the ingested food is made available for the body's use.

We will first provide an overview of the digestive system, examining the common features of the various components of the system, before we begin a detailed tour of the tract from beginning to end.

▌ The digestive system performs four basic digestive processes.

There are four basic digestive processes: *motility, secretion, digestion,* and *absorption.*

MOTILITY

The term **motility** refers to the digestive tract's muscular contractions, of which there are two basic types: propulsive movements and mixing movements. *Propulsive movements* propel or push the contents forward through the digestive tract, with the rate of propulsion varying depending on the functions accomplished by the different regions. That is, the contents are moved forward in a given segment at an appropriate velocity to allow that segment to do its job. For example, transit of food through the esophagus is rapid, which is appropriate because this structure merely serves as a passageway from the mouth to the stomach. In comparison, in the small intestine—the main site of digestion and absorption—the contents are moved forward slowly, allowing time for the breakdown and absorption of food.

Mixing movements serve a twofold function. First, by mixing food with the digestive juices, these movements promote digestion of the food. Second, they facilitate absorption by exposing all parts of the intestinal contents to the absorbing surfaces of the digestive tract.

Contraction of the smooth muscle within the walls of the digestive organs accomplishes movement of material through most of the digestive tract. The exceptions are at the ends of the tract—the mouth through the early part of the esophagus at the beginning and the external anal sphincter at the end—where motility involves skeletal muscle rather than smooth muscle activity. Accordingly, the acts of chewing, swallowing, and defecation have voluntary components, because skeletal muscle is under voluntary control. By contrast, motility accomplished by smooth muscle throughout the rest of the tract is controlled by complex involuntary mechanisms.

SECRETION

A number of digestive juices are secreted into the digestive tract lumen by exocrine glands (see p. 3) along the route, each with its own specific secretory product. Each **digestive secretion** consists of water, electrolytes, and specific organic constituents important in the digestive process, such as enzymes, bile salts, or mucus. The secretory cells extract from the plasma large volumes of water and the raw materials necessary to produce their particular secretion. Secretion of all digestive juices requires energy, both for active transport of some of the raw materials into the cell (others diffuse in passively) and for synthesis of secretory products. On appropriate neural or hormonal stimulation, the secretions are released into the digestive tract lumen. Normally, the digestive secretions are reabsorbed in one form or another back into the blood after their participation in digestion. Failure to do so (because of vomiting or diarrhea, for example) results in loss of this fluid that has been "borrowed" from the plasma.

DIGESTION

Humans consume three different biochemical categories of energy-rich foodstuffs: *carbohydrates, proteins,* and *fats.* These large molecules cannot cross plasma membranes intact to be absorbed from the lumen of the digestive tract into the blood or lymph. The term **digestion** refers to the biochemical breakdown of the structurally complex foodstuffs of the diet into smaller, absorbable units by the enzymes produced within the digestive system as follows:

1. The simplest form of **carbohydrates** is the simple sugars or **monosaccharides** ("one-sugar" molecules), such as **glucose, fructose,** and **galactose,** very few of which are normally found in the diet (see p. A-12). Most ingested carbohydrate is in the form of **polysaccharides** ("many sugar" molecules), which consist of chains of interconnected glucose molecules. The most common polysaccharide consumed is **starch** derived from plant sources. In addition, meat contains **glycogen,** the polysaccharide storage form of glucose in muscle. **Cellulose,** another dietary polysaccharide, found in plant walls, cannot be digested into its constituent monosaccharides by the digestive juices humans secrete; thus it represents the undigested *fiber* or "bulk" of our diets. Besides polysaccharides, a lesser source of dietary carbohydrate is in the form of **disaccharides** ("two-sugar" molecules), including **sucrose** (table sugar, which consists of one glucose and one fructose molecule) and **lactose** (milk sugar made up of one glucose and one galactose molecule). Through the process of digestion, starch, glycogen, and disaccharides are converted into their constituent monosaccharides, principally glucose with small amounts of fructose and galactose. These monosaccharides are the absorbable units for carbohydrates.

2. Dietary **proteins** consist of various combinations of **amino acids** held together by peptide bonds (see p. A-15). Through the process of digestion, proteins are degraded into their constituent amino acids, which are the absorbable units for protein.

3. Most dietary **fats** are in the form of **triglycerides,** which are neutral fats, each consisting of a glycerol with three **fatty acid** molecules attached (*tri* means "three") (see p. A-13). During digestion, two of the fatty acid molecules are split off, leaving a **monoglyceride,** a glycerol molecule with one fatty acid molecule attached (*mono* means "one"). Thus, the end products of fat digestion are monoglycerides and free fatty acids, which are the absorbable units of fat.

Digestion is accomplished by enzymatic **hydrolysis** (breakdown by water; see p. A-17). By adding H_2O at the bond site, enzymes in the digestive secretions break down the bonds that hold the small molecular subunits within the nutrient molecules together, thus setting the small molecules free (● Figure 15-1). The removal of H_2O at the bond sites originally joined these small subunits to form nutrient molecules. Hydrolysis replaces the H_2O and frees the small absorbable units. Digestive enzymes are specific in the bonds they can hydrolyze. As food moves through the digestive tract, it is subjected to various enzymes, each of which breaks down

● **FIGURE 15-1**

An example of hydrolysis. In this example, the disaccharide maltose (the intermediate breakdown product of polysaccharides) is broken down into two glucose molecules by the addition of H_2O at the bond site.

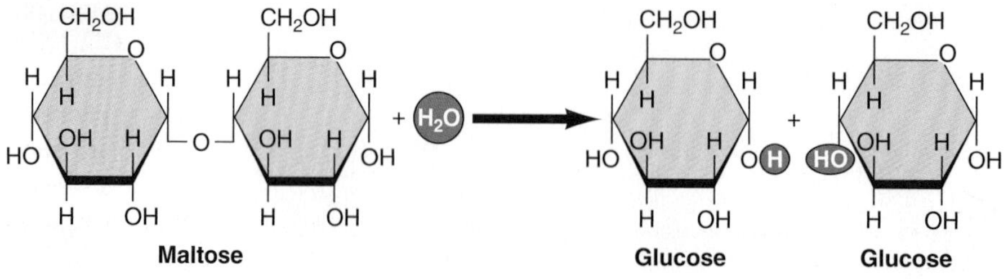

Maltose   Glucose   Glucose

the food molecules even further. In this way, large food molecules are converted to simple absorbable units in a progressive, stepwise fashion, like an assembly line in reverse, as the digestive tract contents are propelled forward.

ABSORPTION

In the small intestine, digestion is completed and most absorption occurs. Through the process of **absorption**, the small absorbable units that result from digestion, along with water, vitamins, and electrolytes, are transferred from the digestive tract lumen into the blood or lymph.

As we examine the digestive tract from beginning to end, we will discuss the four processes of motility, secretion, digestion, and absorption as they take place within each digestive organ (▲ Table 15-1).

▌ The digestive tract and accessory digestive organs make up the digestive system.

The digestive system consists of the *digestive* (or *gastrointestinal*) *tract* plus the accessory digestive organs (*gastro* means "stomach"). The **accessory digestive organs** include the *salivary glands*, the *exocrine pancreas*, and the *biliary system*, which is composed of the *liver* and *gallbladder*. These exocrine organs lie outside the digestive tract and empty their secretions through ducts into the digestive tract lumen.

The **digestive tract** is essentially a tube about 4.5 m (15 feet) in length in its normal contractile state.[1] Running through the middle of the body, the digestive tract includes the following organs (▲ Table 15-1): *mouth; pharynx* (throat); *esophagus; stomach; small intestine* (consisting of the *duodenum, jejunum,* and *ileum*); *large intestine* (the *cecum, appendix, colon,* and *rectum*); and *anus*. Although these organs are continuous with each other, they are considered as separate entities because of their regional modifications, which allow them to specialize in particular digestive activities.

Because the digestive tract is continuous from the mouth to the anus, the lumen of this tube, like the lumen of a straw, is continuous with the external environment. As a result, the contents within the lumen of the digestive tract are technically outside the body, just as the soda you suck through a straw is not a part of the straw. Only after a substance has been absorbed from the lumen across the digestive tract wall is it considered part of the body. This is important, because conditions essential to the digestive process can be tolerated in the digestive tract lumen that could not be tolerated in the body proper. Consider the following examples:

- The pH of the stomach contents falls as low as 2 as a result of the gastric secretion of hydrochloric acid (HCl), yet in the body fluids the range of pH compatible with life is 6.8 to 8.0.

---

[1] Because the uncontracted digestive tract in a cadaver is about twice as long as the contracted tract in a living person, anatomy texts indicate the digestive tract is 30 feet long compared to the length of 15 feet indicated in physiology texts.

- The digestive enzymes that hydrolyze the protein in food could also destroy the body's own tissues that produce them. (Protein is the main structural component of cells.) Therefore, once these enzymes are synthesized in inactive form, they are not activated until they reach the lumen, where they actually attack the food outside the body (that is, within the lumen), thereby protecting the body tissues against self-digestion.
- In the lower part of the intestine live quadrillions of living micro-organisms that are normally harmless and even beneficial, yet if these same micro-organisms enter the body proper (as may happen with a ruptured appendix), they may be extremely harmful or even lethal.

▌ The digestive tract wall has four layers.

The digestive tract wall has the same general structure throughout most of its length from the esophagus to the anus, with some local variations characteristic for each region. A cross section of the digestive tube reveals four major tissue layers (● Figure 15-2, p. 470). From the innermost layer outward they are the *mucosa,* the *submucosa,* the *muscularis externa,* and the *serosa.*

MUCOSA

The **mucosa** lines the luminal surface of the digestive tract. The primary component of the mucosa is a **mucous membrane** that serves as a protective surface as well as being modified in particular areas for secretion and absorption. The mucous membrane contains *exocrine gland cells* for secretion of digestive juices, *endocrine gland cells* for secretion of gastrointestinal hormones, and *epithelial cells* specialized for absorbing digested nutrients.

The mucosal surface is generally highly folded, with many ridges and valleys that greatly increase the surface area available for absorption. The degree of folding varies in different areas of the digestive tract, being most extensive in the small intestine, where maximum absorption occurs, and least extensive in the esophagus, which merely serves as a transit tube.

SUBMUCOSA

The **submucosa** (under the mucosa) is a thick layer of connective tissue that provides the digestive tract with its distensibility and elasticity. It contains the larger blood and lymph vessels, both of which send branches inward to the mucosal layer and outward to the surrounding thick muscle layer. Also, a nerve network known as the *submucous plexus* lies within the submucosa (*plexus* means "network").

MUSCULARIS EXTERNA

The **muscularis externa**, the major smooth-muscle coat of the digestive tube, surrounds the submucosa. In most parts of the tract, the muscularis externa consists of two layers: an *inner circular layer* and an *outer longitudinal layer*. The fibers of the inner smooth-muscle layer (adjacent to the submucosa) run circularly around the circumference of the tube. Contraction of these circular fibers constricts or decreases the diameter of

Anatomy and Functions of Components of the Digestive System

PRINT !

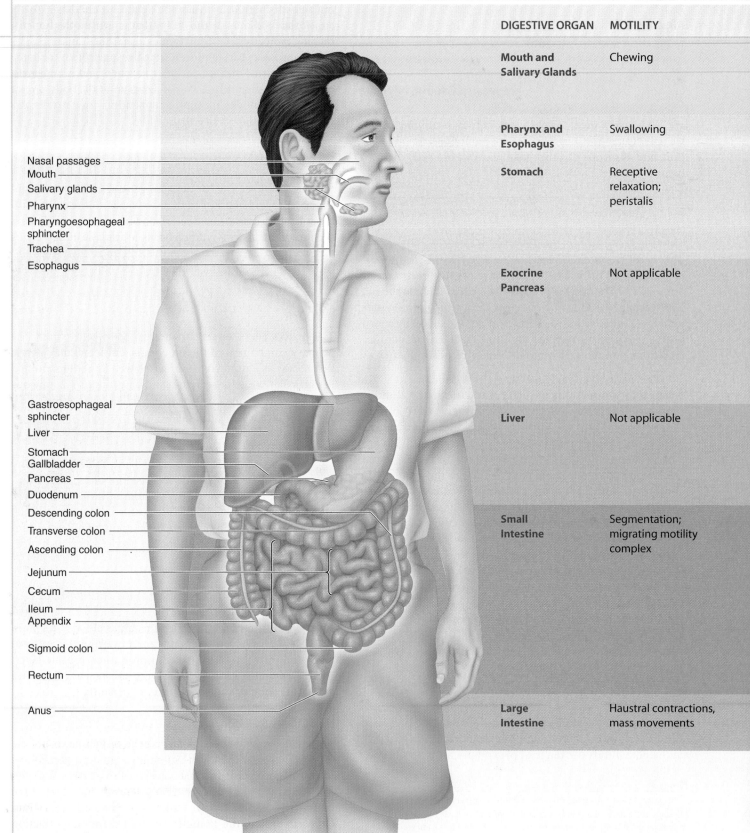

DIGESTIVE ORGAN	MOTILITY
Mouth and Salivary Glands	Chewing
Pharynx and Esophagus	Swallowing
Stomach	Receptive relaxation; peristalis
Exocrine Pancreas	Not applicable
Liver	Not applicable
Small Intestine	Segmentation; migrating motility complex
Large Intestine	Haustral contractions, mass movements

Nasal passages
Mouth
Salivary glands
Pharynx
Pharyngoesophageal sphincter
Trachea
Esophagus

Gastroesophageal sphincter
Liver
Stomach
Gallbladder
Pancreas
Duodenum
Descending colon
Transverse colon
Ascending colon
Jejunum
Cecum
Ileum
Appendix
Sigmoid colon
Rectum
Anus

Contraction of the fibers in the outer layer, which run longitudinally along the length of the tube, shortens the tube. Together, contractile activity of these smooth muscle layers produces the propulsive and mixing movements. Another nerve network, the *myenteric plexus*, lies between the two muscle layers (*myo* means "muscle," *enteric* means "intestine"). Together the submucous and myenteric plexuses help regulate local gut activity.

SEROSA

The outer connective tissue covering of the digestive tract is the **serosa**, which secretes a watery, slippery fluid that lubricates and prevents friction between the digestive organs and surrounding viscera.

Regulation of digestive function is complex and synergistic.

Digestive motility and secretion are carefully regulated to maximize digestion and absorption of ingested food. Four factors are involved in regulating digestive system function: (1) autonomous smooth-muscle function, (2) intrinsic nerve plexuses, (3) extrinsic nerves, and (4) gastrointestinal hormones.

AUTONOMOUS SMOOTH-MUSCLE FUNCTION

Like self-excitable cardiac-muscle cells, some smooth-muscle cells are pacesetter cells that display rhythmic, spontaneous variations in membrane potential. The prominent type of self-induced electrical activity in digestive smooth muscle is **slow-wave potentials** (see p. 232), alternatively referred to as the digestive tract's **basic electrical rhythm (BER)** or **pacesetter potential**. Musclelike but noncontractile cells known as the **interstitial cells of Cajal** are the pacesetter cells that instigate cyclic slow-wave activity. These pacesetter cells lie at the boundary between the longitudinal and circular smooth-muscle layers. Slow waves are not action potentials and do not directly induce muscle contraction; they are rhythmic, wave-like fluctuations in membrane potential that cyclically bring the membrane closer to or farther from threshold potential. If these waves reach threshold at the peaks of depolarization, a volley of action potentials is triggered at each peak, resulting in rhythmic cycles of muscle contraction.

Like cardiac muscle, sheets of smooth muscle cells are connected by gap junctions through which charge-carrying ions can flow (see p. 49). In this way, electrical activity initiated in a digestive-tract pacesetter cell spreads to the adjacent contractile smooth-muscle cells. Thus the whole muscle sheet behaves like a functional syncytium, becoming excited and contracting as a unit when threshold is reached (see p. 231). If threshold is not achieved, the oscillating slow-wave electrical activity continues to sweep across the muscle sheet without being accompanied by contractile activity.

Whether threshold is reached depends on the effect of various mechanical, neural, and hormonal factors that influence the starting point around which the slow-wave rhythm oscillates. If the starting point is nearer the threshold level, as it is when food is present in the digestive tract, the depolar-

SECRETION	DIGESTION	ABSORPTION
Saliva • Amylase • Mucus • Lysozyme *[handwritten: .5% salt & protein]* *[handwritten: 99.5% H₂O]*	Carbohydrate digestion begins	No foodstuffs; a few medications—for example, nitroglycerin
Mucus	None	None
Gastric juice • HCl • Pepsin • Mucus • Intrinsic factor	Carbohydrate digestion continues in body of stomach; protein digestion begins in antrum of stomach	No foodstuffs; a few lipid-soluble substances, such as alcohol and aspirin
Pancreatic digestive enzymes • Trypsin, chymotrypsin, carboxypeptidase • Amylase • Lipase Pancreatic aqueous NaHCO₃ secretion	These pancreatic enzymes accomplish digestion in duodenal lumen	Not applicable
Bile • Bile Salts • Alkaline secretion • Bilirubin	Bile does not digest anything, but bile salts facilitate fat digestion and absorption in duodenal lumen	Not applicable
Succus entericus • Mucus • Salt (Small intestine enzymes—disaccharidases and aminopeptidases—are not secreted but function intracellularly in the brush border)	In lumen, under influence of pancreatic enzymes and bile, carbohydrate and protein digestion continue and fat digestion is completely accomplished; in brush border, carbohydrate and protein digestion completed	All nutrients, most electrolytes, and water
Mucus *[handwritten: Alkaline(NaHCO₃)]*	None	Salt and water, converting contents to feces

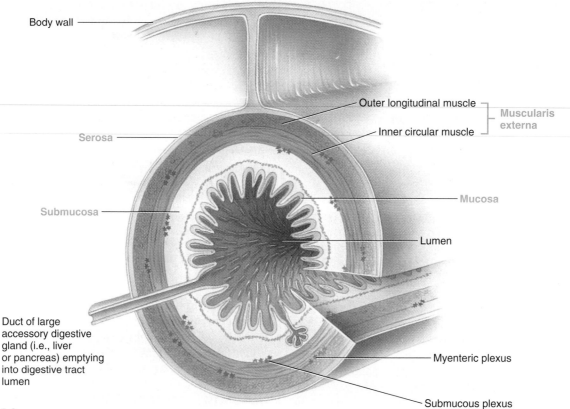

Body wall

Outer longitudinal muscle
Inner circular muscle
Muscularis externa

Serosa

Mucosa

Submucosa

Lumen

Duct of large
accessory digestive
gland (i.e., liver
or pancreas) emptying
into digestive tract
lumen

Myenteric plexus

Submucous plexus

● **FIGURE 15-2**

Layers of the digestive tract wall. The digestive tract wall consists of four major layers: from the innermost out, they are the mucosa, submucosa, muscularis externa, and serosa.

 For an interaction related to this figure, see Media Exercise 15.3: The Intestine and Associated Organs on the CD-ROM.

izing slow-wave peak reaches threshold, so action potential frequency and its accompanying contractile activity increase. Conversely, if the starting point is farther from threshold, as when no food is present, there is less likelihood of reaching threshold, so action potential frequency and contractile activity are reduced.

The rate of self-induced rhythmic digestive contractile activities, such as peristalsis in the stomach, segmentation in the small intestine, and haustral contractions in the large intestine, depends on the inherent rate established by the involved pacesetter cells. (Specific details about these rhythmic contractions will be discussed when we examine the organs involved.)

INTRINSIC NERVE PLEXUSES

The **intrinsic nerve plexuses** are the two major networks of nerve fibers—the **submucous plexus** and the **myenteric plexus**—that lie entirely within the digestive tract wall and run its entire length. Thus, unlike any other body system, the digestive tract has its own intramural ("within-wall") nervous system, which contains as many neurons as the spinal cord and endows the tract with a considerable degree of self-regulation. Together, these two plexuses are often termed the **enteric nervous system.**

The intrinsic plexuses influence all facets of digestive tract activity. Various types of neurons are present in the intrinsic

plexuses. Some are sensory neurons, which have receptors that respond to specific local stimuli in the digestive tract. Other local neurons innervate the smooth muscle cells and exocrine and endocrine cells of the digestive tract to directly affect digestive tract motility, secretion of digestive juices, and secretion of gastrointestinal hormones. As with the central nervous system, these input and output neurons of the enteric nervous system are linked by interneurons. Some of the output neurons are excitatory, and some are inhibitory. These intrinsic nerve networks primarily coordinate local activity within the digestive tract. To illustrate, if a large piece of food gets stuck in the esophagus, the intrinsic plexuses coordinate local responses to push the food forward. Intrinsic nerve activity can in turn be influenced by the extrinsic nerves.

EXTRINSIC NERVES

The **extrinsic nerves** are the nerve fibers from both branches of the autonomic nervous system that originate outside the digestive tract and innervate the various digestive organs. The autonomic nerves influence digestive tract motility and secretion either by modifying ongoing activity in the intrinsic plexuses, altering the level of gastrointestinal hormone secretion, or, in some instances, acting directly on the smooth muscle and glands.

Recall that, in general, the sympathetic and parasympathetic nerves supplying any given tissue exert opposing actions on that tissue. The sympathetic system, which dominates in "fight-or-flight" situations, tends to inhibit or slow down digestive tract contraction and secretion. This action is appropriate, considering that digestive processes are not of highest priority when the body faces an emergency. The parasympathetic nervous system, by contrast, dominates in quiet, "rest-and-digest" situations, when general maintenance types of activities such as digestion can proceed optimally. Accordingly, the parasympathetic nerve fibers supplying the digestive tract, which arrive primarily by way of the vagus nerve, tend to increase smooth muscle motility and promote secretion of digestive enzymes and hormones.

In addition to being called into play during generalized sympathetic or parasympathetic discharge, the autonomic nerves, especially the vagus nerve, can be discretely activated to modify only digestive activity. One of the major purposes of specific activation of extrinsic innervation is to coordinate activity between different regions of the digestive system. For example, the act of chewing food reflexly increases not only salivary secretion but also stomach, pancreatic, and liver secretion via vagal reflexes in anticipation of the arrival of food.

GASTROINTESTINAL HORMONES

Tucked within the mucosa of certain regions of the digestive tract are endocrine gland cells that on appropriate stimulation, release hormones into the blood. These **gastrointestinal hormones** are carried through the blood to other areas of the digestive tract, where they exert either excitatory or inhibitory influences on smooth muscle and exocrine gland cells.

▌ **Receptor activation alters digestive activity through neural reflexes and hormonal pathways.**

The digestive tract wall contains three types of sensory receptors that respond to local changes in the digestive tract: (1) *chemoreceptors* sensitive to chemical components within the lumen, (2) *mechanoreceptors* (pressure receptors) sensitive to stretch or tension within the wall, and (3) *osmoreceptors* sensitive to the osmolarity of the luminal contents. Stimulation of these receptors elicits neural reflexes or secretion of hormones, both of which alter the level of activity in the digestive system's effector cells. These effector cells include smooth muscle cells (for modifying motility), exocrine gland cells (for controlling secretion of digestive juices), and endocrine gland cells (for varying secretion of gastrointestinal hormones) (● Figure 15-3).

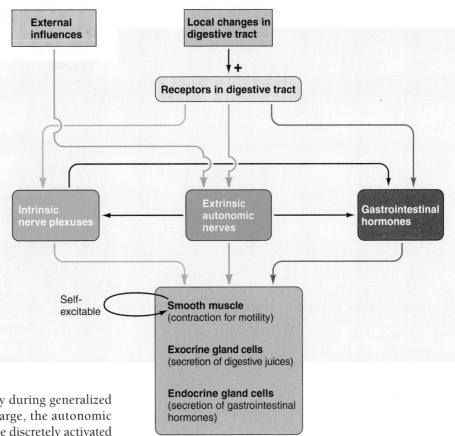

● **FIGURE 15-3**

Summary of pathways controlling digestive system activities

From this overview, you can see that regulation of gastrointestinal function is very complex, being influenced by many synergistic, interrelated pathways designed to ensure that the appropriate responses occur to digest and absorb the ingested food. Nowhere else in the body is so much overlapping control exercised.

We are now going to take a "tour" of the digestive tract, beginning with the mouth and ending with the anus. We will examine the four basic digestive processes of motility, secretion, digestion, and absorption at each digestive organ along the way. ▲ Table 15-1 summarizes these activities and serves as a useful reference throughout the rest of the chapter.

MOUTH

▌ **The oral cavity is the entrance to the digestive tract.**

Entry to the digestive tract is through the **mouth** or **oral cavity**. The opening is formed by the muscular **lips**, which help procure, guide, and contain the food in the mouth. The lips also serve nondigestive functions; they are important in speech (articulation of many sounds depends on a particular lip formation) and as a sensory receptor in interpersonal relationships (for example, as in kissing).

The **palate**, which forms the arched roof of the oral cavity, separates the mouth from the nasal passages. Its presence allows breathing and chewing or sucking to take place simultaneously. Hanging down from the palate in the rear of the throat is a dangling projection, the **uvula**, which plays an important role in sealing off the nasal passages during swallowing. (The uvula is the structure you elevate when you say "ahhh" so that the physician can better see your throat.)

The **tongue**, which forms the floor of the oral cavity, is composed of voluntarily controlled skeletal muscle. Movements of the tongue are important in guiding food within the mouth during chewing and swallowing and also play an important role in speech. Furthermore, the major **taste buds** are embedded in the tongue (see p. 175).

The **pharynx** is the cavity at the rear of the throat. It acts as a common passageway for both the digestive system (by serving as the link between the mouth and esophagus, for food) and the respiratory system (by providing access between the nasal passages and trachea, for air). This arrangement necessitates mechanisms (to be described shortly) to guide food and air into the proper passageways beyond the pharynx. Housed within the side walls of the pharynx are the **tonsils**, lymphoid tissues that are part of the body's defense team.

▌ The teeth do the chewing.

The first step in the digestive process is **chewing**, the motility of the mouth that involves the slicing, tearing, grinding, and mixing of ingested food by the **teeth**. The teeth are firmly embedded in and protrude from the jawbones. The exposed part of a tooth is covered by **enamel**, the hardest structure of the body. Enamel forms before the tooth's eruption, by special cells that are lost as the tooth erupts. Because enamel cannot be regenerated after the tooth has erupted, any defects (**dental caries** or "cavities") that develop in the enamel must be patched by artificial fillings, or else the surface will continue to erode into the underlying living pulp.

The functions of chewing are (1) to grind and break food up into smaller pieces to facilitate swallowing, (2) to mix food with saliva, and (3) to stimulate the taste buds. The latter not only gives rise to the pleasurable subjective sensation of taste but also, in feedforward fashion, reflexly increases salivary, gastric, pancreatic, and bile secretion to prepare for the arrival of food.

The act of chewing can be voluntary, but most chewing during a meal is a rhythmic reflex brought about by activation of the skeletal muscles of the jaws, lips, cheeks, and tongue in response to the pressure of food against the oral tissues.

▌ Saliva begins carbohydrate digestion, is important in oral hygiene, and facilitates speech.

Saliva, the secretion associated with the mouth, is produced largely by three major pairs of salivary glands that lie outside of the oral cavity and discharge saliva through short ducts into the mouth.

Saliva is about 99.5% H_2O and 0.5% electrolytes and protein. The salivary NaCl (salt) concentration is only one seventh of that in the plasma, which is important in perceiving salty tastes. Similarly, discrimination of sweet tastes is enhanced by the absence of glucose in the saliva. The most important salivary proteins are *amylase, mucus,* and *lysozyme.* They contribute to the functions of saliva as follows:

1. Saliva begins digestion of carbohydrate in the mouth through action of **salivary amylase,** an enzyme that breaks polysaccharides down into **maltose,** a disaccharide consisting of two glucose molecules.
2. Saliva facilitates swallowing by moistening food particles, thereby holding them together, and by providing lubrication through the presence of **mucus,** which is thick and slippery.
3. Saliva exerts some antibacterial action by a twofold effect— first by **lysozyme,** an enzyme that lyses, or destroys, certain bacteria by breaking down their cell walls, and second by rinsing away material that may serve as a food source for bacteria.
4. Saliva serves as a solvent for molecules that stimulate the taste buds. Only molecules in solution can react with taste bud receptors. You can demonstrate this for yourself: Dry your tongue and then drop some sugar on it—you cannot taste the sugar until it is moistened.
5. Saliva aids speech by facilitating movements of the lips and tongue. It is difficult to talk when the mouth feels dry.
6. Saliva plays an important role in oral hygiene by helping keep the mouth and teeth clean. The constant flow of saliva helps flush away food residues, shed epithelial cells, and foreign particles. Saliva's contribution in this regard is apparent to anyone who has experienced a foul taste in the mouth when salivation is suppressed for a while, such as during a fever or states of prolonged anxiety.
7. Saliva is rich in bicarbonate buffers, which neutralize acids in food as well as acids produced by bacteria in the mouth, thereby helping prevent dental caries.

Despite these many functions, saliva is not essential for digesting and absorbing foods, because enzymes produced by the pancreas and small intestine can complete food digestion even in the absence of salivary and gastric secretion.

Clinical Note The main problems associated with diminished salivary secretion, a condition known as **xerostomia,** are difficulty in chewing and swallowing, inarticulate speech unless frequent sips of water are taken when talking, and a rampant increase in dental caries unless special precautions are taken.

▌ Salivary secretion is continuous and can be reflexly increased.

On average, about 1 to 2 liters of saliva are secreted per day, ranging from a continuous spontaneous basal rate of 0.5 ml/min to a maximum flow rate of about 5 ml/min in response to a potent stimulus such as sucking on a lemon. The continuous basal secretion of saliva in the absence of apparent stimuli is brought about by constant low-level stimulation by the parasympathetic nerve endings that terminate in the salivary glands. This basal secretion is important in keeping the mouth and throat moist at all times. In addition to this continuous, low-level secretion, salivary secretion may be increased by two types of salivary reflexes, the simple and conditioned salivary reflexes (● Figure 15-4).

SIMPLE AND CONDITIONED SALIVARY REFLEXES

The **simple salivary reflex** occurs when chemoreceptors and pressure receptors within the oral cavity respond to the presence of food. On activation, these receptors initiate impulses in afferent nerve fibers that carry the information to the **salivary center**, which is located in the medulla of the brain stem, as are all the brain centers that control digestive activities. The salivary center in turn sends impulses via the extrinsic autonomic nerves to the salivary glands to promote increased salivation. Dental procedures promote salivary secretion in the absence of food because these manipulations activate pressure receptors in the mouth.

With the **conditioned**, or **acquired, salivary reflex**, salivation occurs without oral stimulation. Just thinking about, seeing, smelling, or hearing the preparation of pleasant food initiates salivation through this reflex. All of us have experienced such "mouth watering" in anticipation of something delicious to eat. This reflex is a learned response based on previous experience. Inputs that arise outside the mouth and are mentally associated with the pleasure of eating act through the cerebral cortex to stimulate the medullary salivary center.

AUTONOMIC INFLUENCE ON SALIVARY SECRETION

The salivary center controls the degree of salivary output by means of the autonomic nerves that supply the salivary glands. Unlike the autonomic nervous system elsewhere in the body, sympathetic and parasympathetic responses in the salivary glands are not antagonistic. Both sympathetic and parasympathetic stimulation increase salivary secretion, but the quantity, characteristics, and mechanisms are different. Parasympathetic stimulation, which exerts the dominant role in salivary secretion, produces a prompt and abundant flow of watery saliva that is rich in enzymes. Sympathetic stimulation, by contrast, produces a much smaller volume of thick saliva that is rich in mucus. Because sympathetic stimulation elicits a smaller volume of saliva, the mouth feels drier than usual during circumstances when the sympathetic system is dominant, such as stress situations. Thus people experience a dry feeling in the mouth when they are nervous about giving a speech.

Salivary secretion is the only digestive secretion entirely under neural control. All other digestive secretions are regulated by both nervous system reflexes and hormones.

▌ Digestion in the mouth is minimal; no absorption of nutrients occurs.

Digestion in the mouth involves the hydrolysis of polysaccharides into disaccharides by amylase. However, most digestion

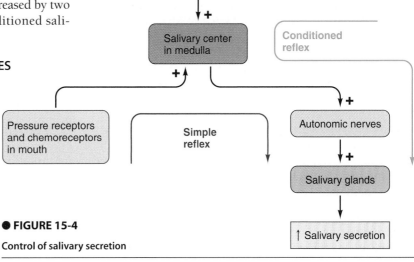

● **FIGURE 15-4**

Control of salivary secretion

by this enzyme is done in the body of the stomach after the food mass and saliva have been swallowed. Acid inactivates amylase, but in the center of the food mass, where stomach acid has not yet reached, this salivary enzyme continues to function for several more hours.

No absorption of foodstuff occurs from the mouth. Importantly, some drugs can be absorbed by the oral mucosa, a prime example being *nitroglycerin,* a vasodilator drug sometimes used by cardiac patients to relieve anginal attacks (see p. 267) associated with myocardial ischemia (see p. 255).

PHARYNX AND ESOPHAGUS

The motility associated with the pharynx and esophagus is **swallowing.** Most of us think of swallowing as the limited act of moving food out of the mouth into the esophagus. However, swallowing actually is the entire process of moving food from the mouth through the esophagus into the stomach.

▌ Swallowing is a sequentially programmed all-or-none reflex.

Swallowing is initiated when a **bolus,** or ball of food, is voluntarily forced by the tongue to the rear of the mouth into the pharynx. The pressure of the bolus stimulates pharyngeal pressure receptors, which send afferent impulses to the **swallowing center** located in the medulla. The swallowing center then reflexly activates in the appropriate sequence the muscles that are involved in swallowing. Swallowing is the most complex reflex in the body. Multiple highly coordinated responses are triggered in a specific all-or-none pattern over a period of time to accomplish the act of swallowing. Swallowing is initiated voluntarily, but once begun it cannot be stopped. Perhaps you have experienced this when a large piece of hard candy inadvertently slipped to the rear of your throat, triggering an unintentional swallow.

During the oropharyngeal stage of swallowing, food is prevented from entering the wrong passageways.

Swallowing is divided into the oropharyngeal stage and the esophageal stage. The **oropharyngeal stage** lasts about 1 second and consists of moving the bolus from the mouth through the pharynx and into the esophagus. When the bolus enters the pharynx, it must be directed into the esophagus and prevented from entering the other openings that communicate with the pharynx. In other words, food must be kept from re-entering the mouth, from entering the nasal passages, and from entering the trachea. All of this is managed by the following coordinated activities (● Figure 15-5):

• The position of the tongue against the hard palate keeps food from re-entering the mouth during swallowing.
• The uvula is elevated and lodges against the back of the throat, sealing off the nasal passage from the pharynx so that food does not enter the nose.
• Food is prevented from entering the trachea primarily by elevation of the larynx and tight closure of the vocal folds across the laryngeal opening, or **glottis**. The first part of the trachea is the *larynx,* or *voice box,* across which are stretched

the *vocal folds* (see p. 367). During swallowing, the vocal folds serve a purpose unrelated to speech. Contraction of laryngeal muscles aligns the vocal folds in tight apposition to each other, thus sealing the glottis entrance. Also, the bolus tilts a small flap of cartilaginous tissue, the **epiglottis** (*epi* means "upon"), backward down over the closed glottis as further protection from food entering the respiratory airways.

• The person does not attempt futile respiratory efforts when the respiratory passages are temporarily sealed off during swallowing, because the swallowing center briefly inhibits the nearby respiratory center.
• With the larynx and trachea sealed off, pharyngeal muscles contract to force the bolus into the esophagus.

The pharyngoesophageal sphincter keeps air from entering the digestive tract during breathing.

The **esophagus** is a fairly straight muscular tube that extends between the pharynx and stomach (see ▲ Table 15-1, p. 468). Lying for the most part in the thoracic cavity, it penetrates the diaphragm and joins the stomach in the abdominal cavity a few centimeters below the diaphragm.

● **FIGURE 15-5**

Oropharyngeal stage of swallowing. (a) Position of the oropharyngeal structures at rest. (b) Changes that occur during the oropharyngeal stage of swallowing to prevent the bolus of food from entering the wrong passageways.

 For an animation of this figure, click the Gastrointestinal Motility tab in the Gastrointestinal Physiology tutorial on the CD-ROM.

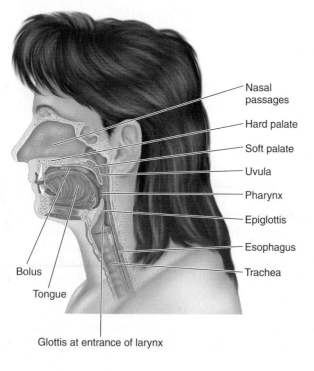

Nasal passages
Hard palate
Soft palate
Uvula
Pharynx
Epiglottis
Esophagus
Trachea
Bolus
Tongue
Glottis at entrance of larynx

(a)

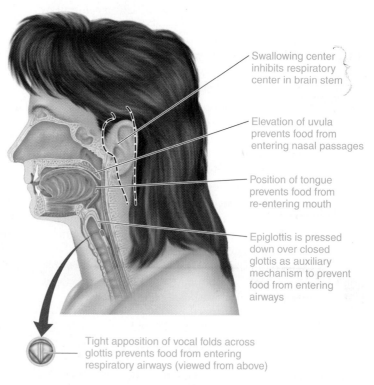

Swallowing center inhibits respiratory center in brain stem

Elevation of uvula prevents food from entering nasal passages

Position of tongue prevents food from re-entering mouth

Epiglottis is pressed down over closed glottis as auxiliary mechanism to prevent food from entering airways

Tight apposition of vocal folds across glottis prevents food from entering respiratory airways (viewed from above)

(b)

The esophagus is guarded at both ends by sphincters. A sphincter is a ringlike muscular structure that, when closed, prevents passage through the tube it guards. The upper esophageal sphincter is the *pharyngoesophageal sphincter,* and the lower sphincter is the *gastroesophageal sphincter.* We will first discuss the role of the pharyngoesophageal sphincter, then the process of esophageal transit of food, and finally the importance of the gastroesophageal sphincter.

Except during a swallow, the **pharyngoesophageal sphincter** keeps the entrance to the esophagus closed to prevent large volumes of air from entering the esophagus and stomach during breathing. Instead, air is directed only into the respiratory airways. Otherwise, the digestive tract would be subjected to large volumes of gas, which would lead to excessive burping. During swallowing, this sphincter opens and allows the bolus to pass into the esophagus. Once the bolus has entered the esophagus, the pharyngoesophageal sphincter closes, the respiratory airways are opened, and breathing resumes. The oropharyngeal stage is complete, and about 1 second has passed since the swallow was first initiated.

▌ Peristaltic waves push food through the esophagus.

The **esophageal stage** of the swallow now begins. The swallowing center triggers a **primary peristaltic wave** that sweeps from the beginning to the end of the esophagus, forcing the bolus ahead of it through the esophagus to the stomach. The term **peristalsis** refers to ringlike contractions of the circular smooth muscle that move progressively forward, pushing the bolus into a relaxed area ahead of the contraction (● Figure 15-6). The peristaltic wave takes about 5 to 9 seconds to reach the lower end of the esophagus.

If a large or sticky swallowed bolus, such as a bite of peanut butter sandwich, fails to be carried along to the stomach by the primary peristaltic wave, the lodged bolus distends the esophagus, stimulating pressure receptors within its walls. As a result, a second, more forceful peristaltic wave is initiated, mediated by the intrinsic nerve plexuses at the level of the distension. These **secondary peristaltic waves** do not involve the swallowing center, nor is the person aware of their occurrence. Distension of the esophagus also reflexly increases salivary secretion. The trapped bolus is eventually dislodged and moved forward through the combination of lubrication by the extra swallowed saliva and the forceful secondary peristaltic waves.

▌ The gastroesophageal sphincter prevents reflux of gastric contents.

Except during swallowing, the **gastroesophageal sphincter** stays contracted to keep a barrier between the stomach and esophagus, reducing the chance of reflux of acidic gastric contents into the esophagus. If gastric contents do flow backward despite the sphincter, the acidity of these contents irritates the esophagus, causing the esophageal discomfort known as **heartburn.** (The heart itself is not involved at all.)

As the peristaltic wave sweeps down the esophagus, the gastroesophageal sphincter relaxes reflexly so that the bolus can pass into the stomach. After the bolus has entered the stomach, the swallow is complete and the gastroesophageal sphincter again contracts.

▌ Esophageal secretion is entirely protective.

Esophageal secretion is entirely mucus. In fact, mucus is secreted throughout the length of the digestive tract. By lubricating the passage of food, esophageal mucus lessens the likelihood that the esophagus will be damaged by any sharp edges in the newly entering food. Furthermore, it protects the esophageal wall from acid and enzymes in gastric juice if gastric reflux occurs.

The entire transit time in the pharynx and esophagus averages a mere 6 to 10 seconds, too short a time for any digestion or absorption in this region. We now move on to our next stop, the stomach.

STOMACH

The **stomach** is a J-shaped saclike chamber lying between the esophagus and small intestine. It is arbitrarily divided into three sections based on anatomical, histological, and functional distinctions (● Figure 15-7). The **fundus** is the part of the stomach that lies above the esophageal opening. The middle or main part of the stomach is the **body.** The smooth muscle layers in the fundus and body are relatively thin, but the lower part of the stomach, the **antrum,** has much heavier musculature. This difference in muscle thickness plays an important role in gastric motility in these two regions, as you will see shortly. There are also glandular differences in the mucosa of these regions, as described later. The end part of the stomach is the **pyloric sphincter,** which acts as a barrier between the stomach and the upper part of the small intestine, the duodenum.

● **FIGURE 15-6**

Peristalsis in the esophagus. As the wave of peristaltic contraction sweeps down the esophagus, it pushes the bolus ahead of it toward the stomach.

For an animation of this figure, click the Gastrointestinal Motility tab in the Gastrointestinal Physiology tutorial on the CD-ROM.

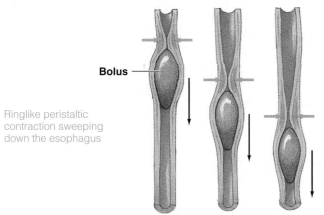

Bolus

Ringlike peristaltic contraction sweeping down the esophagus

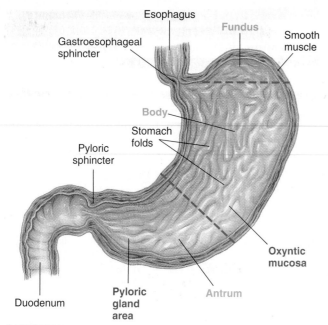

Esophagus

Gastroesophageal sphincter

Fundus

Smooth muscle

Body

Stomach folds

Pyloric sphincter

Oxyntic mucosa

Pyloric gland area

Antrum

Duodenum

● FIGURE 15-7

Anatomy of the stomach. The stomach is divided into three sections based on structural and functional distinctions—the fundus, body, and antrum. The mucosal lining of the stomach is divided into the oxyntic mucosa and the pyloric gland area based on differences in glandular secretion.

PhysioEdge

For an animation of this figure, click the Gastrointestinal Motility tab in the Gastrointestinal Physiology tutorial on the CD-ROM. For an interaction related to this figure, see Media Exercise 15.1: The Stomach.

The stomach stores food and begins protein digestion.

The stomach performs three main functions:

1. The stomach's most important function is to store ingested food until it can be emptied into the small intestine at a rate appropriate for optimal digestion and absorption. It takes hours to digest and absorb a meal that was consumed in only a matter of minutes. Because the small intestine is the primary site for this digestion and absorption, it is important that the stomach store the food and meter it into the duodenum at a rate that does not exceed the small intestine's capacities.
2. The stomach secretes hydrochloric acid (HCl) and enzymes that begin protein digestion.
3. Through the stomach's mixing movements, the ingested food is pulverized and mixed with gastric secretions to produce a thick liquid mixture known as **chyme**. The stomach contents must be converted to chyme before they can be emptied into the duodenum.

We will now discuss how the stomach accomplishes these functions as we examine the four basic digestive processes—motility, secretion, digestion, and absorption—as they relate to the stomach. Starting with motility, gastric motility is complex and subject to multiple regulatory inputs. The four aspects of gastric motility are (1) filling, (2) storage, (3) mixing, and (4) emptying. We begin with gastric filling.

Gastric filling involves receptive relaxation.

When empty, the stomach has a volume of about 50 ml, but it can expand to a capacity of about 1 liter (1000 ml) during a meal. The stomach can accommodate such a 20-fold change in volume with little change in tension in its walls and little rise in intragastric pressure, through the following mechanism. The interior of the stomach is thrown into deep folds. During a meal, the folds get smaller and nearly flatten out as the stomach relaxes slightly with each mouthful, much like the gradual expansion of a collapsed ice bag as it is being filled. This reflex relaxation of the stomach as it is receiving food is called **receptive relaxation**; it enhances the stomach's ability to accommodate the extra volume of food with little rise in stomach pressure. If more than a liter of food is consumed, however, the stomach becomes overdistended and the person experiences discomfort. Receptive relaxation is triggered by the act of eating and is mediated by the vagus nerve.

Gastric storage takes place in the body of the stomach.

A group of pacesetter cells located in the upper fundus region of the stomach generate slow-wave potentials that sweep down the length of the stomach toward the pyloric sphincter at a rate of three per minute. This rhythmic pattern of spontaneous depolarizations—the basic electrical rhythm, or BER, of the stomach—occurs continuously and may or may not be accompanied by contraction of the stomach's circular smooth-muscle layer. Depending on the level of excitability in the smooth muscle, it may be brought to threshold by this flow of current and undergo action potentials, which in turn initiate peristaltic waves that sweep over the stomach in pace with the BER at a rate of three per minute.

Once initiated, the peristaltic wave spreads over the fundus and body to the antrum and pyloric sphincter. Because the muscle layers are thin in the fundus and body, the peristaltic contractions in this region are weak. When the waves reach the antrum, they become much stronger and more vigorous, because the muscle there is much thicker.

Because only feeble mixing movements occur in the body and fundus, food emptied into the stomach from the esophagus is stored in the relatively quiet body without being mixed. The fundic area usually does not store food but contains only a pocket of gas. Food is gradually fed from the body into the antrum, where mixing does take place.

Gastric mixing takes place in the antrum of the stomach.

The strong antral peristaltic contractions mix the food with gastric secretions to produce chyme. Each antral peristaltic wave propels chyme forward toward the pyloric sphincter. Tonic contraction of the pyloric sphincter normally keeps it almost, but not completely, closed. The opening is large enough for water and other fluids to pass through with ease but too small for the thicker chyme to pass through except when a strong antral peristaltic contraction pushes it through. Even then, of the 30 ml of chyme that the antrum can hold, usually only a few milliliters of antral contents are pushed into the duodenum with each peristaltic wave. Before more chyme can be squeezed out, the peristaltic wave reaches the pyloric sphincter and causes it to contract more forcefully, sealing off the exit and blocking further passage into the duodenum. The bulk of the antral chyme that was being propelled forward but failed to be pushed into the duodenum is abruptly halted at the closed sphincter and is tumbled back into the antrum, only to be propelled forward and tumbled back again as the new peristaltic wave advances (● Figure 15-8). This tossing back and forth thoroughly mixes the chyme in the antrum.

▌ Gastric emptying is largely controlled by factors in the duodenum.

In addition to mixing gastric contents, the antral peristaltic contractions are the driving force for gastric emptying. The amount of chyme that escapes into the duodenum with each peristaltic wave before the pyloric sphincter tightly closes depends largely on the strength of peristalsis. The intensity of antral peristalsis can vary markedly under the influence of different signals from both the stomach and duodenum; thus gastric emptying is regulated by both gastric and duodenal factors (▲ Table 15-2). These factors influence the stomach's excitability by slightly depolarizing or hyperpolarizing the gastric smooth muscle. This excitability in turn is a determinant of the degree of antral peristaltic activity. The greater the excitability, the more frequently the BER will generate action potentials, the greater the degree of peristaltic activity in the antrum, and the faster the rate of gastric emptying.

FACTORS IN THE STOMACH THAT INFLUENCE THE RATE OF GASTRIC EMPTYING

The main gastric factor that influences the strength of contraction is the amount of chyme in the stomach. Other things being equal, the stomach empties at a rate proportional to the volume of chyme in it at any given time. Stomach distension triggers increased gastric motility through a direct effect of stretch on the smooth muscle as well as through involvement of the intrinsic plexuses, the vagus nerve, and the stomach hormone *gastrin*. (The source, control, and other functions of this hormone will be described later.)

Furthermore, the degree of fluidity of the chyme in the stomach influences gastric emptying. The stomach contents

---

● **FIGURE 15-8**

Gastric emptying and mixing as a result of antral peristaltic contractions

For an animation of this figure, click the Gastrointestinal Motility tab in the Gastrointestinal Physiology tutorial on the CD-ROM. For an interaction related to this figure, see Media Exercise 15.3: The Intestine and Associated Organs.

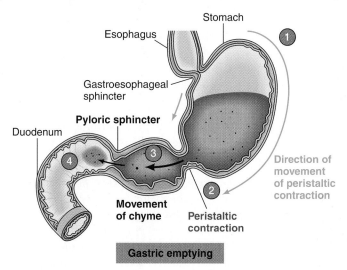

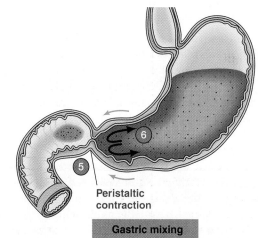

1. A peristaltic contraction originates in the upper fundus and sweeps down toward the pyloric sphincter.

2. The contraction becomes more vigorous as it reaches the thick-muscled antrum.

3. The strong antral peristaltic contraction propels the chyme forward.

4. A small portion of chyme is pushed through the partially open sphincter into the duodenum. The stronger the antral contraction, the more chyme is emptied with each contractile wave.

5. When the peristaltic contraction reaches the pyloric sphincter, the sphincter is tightly closed and no further emptying takes place.

6. When chyme that was being propelled forward hits the closed sphincter, it is tossed back into the antrum. Mixing of chyme is accomplished as chyme is propelled forward and tossed back into the antrum with each peristaltic contraction.

▲ TABLE 15-2

Factors Regulating Gastric Motility and Emptying

FACTORS	MODE OF REGULATION	EFFECTS ON GASTRIC MOTILITY AND EMPTYING
Within the Stomach		
Volume of chyme	Distension has a direct effect on gastric smooth muscle excitability, as well as acting through the intrinsic plexuses, the vagus nerve, and gastrin	Increased volume stimulates motility and emptying
Degree of fluidity	Direct effect; contents must be in a fluid form to be evacuated	Increased fluidity allows more rapid emptying
Within the Duodenum		
Presence of fat, acid, hypertonicity, or distension	Initiates the enterogastric reflex or triggers the release of enterogastrones (cholecystokinin, secretin)	These factors in the duodenum inhibit further gastric motility and emptying until the duodenum has coped with factors already present
Outside the Digestive System		
Emotion	Alters autonomic balance	Stimulates or inhibits motility and emptying
Intense pain	Increases sympathetic activity	Inhibits motility and emptying

must be converted into a finely divided, thick liquid form before emptying. The sooner the appropriate degree of fluidity can be achieved, the more rapidly the contents are ready to be evacuated.

FACTORS IN THE DUODENUM THAT INFLUENCE THE RATE OF GASTRIC EMPTYING

Despite these gastric influences, factors in the duodenum are of primary importance in controlling the rate of gastric emptying. The duodenum must be ready to receive the chyme and can delay gastric emptying by reducing peristaltic activity in the stomach until the duodenum is ready to accommodate more chyme. Even if the stomach is distended and its contents are in a liquid form, it cannot empty until the duodenum is ready to deal with the chyme.

The four most important duodenal factors that influence gastric emptying are *fat, acid, hypertonicity,* and *distension.* The presence of one or more of these stimuli in the duodenum activates appropriate duodenal receptors, triggering either a neural or a hormonal response that puts brakes on gastric motility by reducing the excitability of the gastric smooth muscle. The subsequent reduction in antral peristaltic activity slows down the rate of gastric emptying.

- The *neural response* is mediated through both the intrinsic nerve plexuses and the autonomic nerves. Collectively, these reflexes are called the **enterogastric reflex.**
- The *hormonal response* involves the release from the duodenal mucosa of several hormones collectively known as **enterogastrones.** The blood carries these hormones to the stomach, where they inhibit antral contractions to reduce gas-

tric emptying. The two most important enterogastrones are **secretin** and **cholecystokinin (CCK).** Secretin was the first hormone discovered (in 1902). Because it was a secretory product that entered the blood, it was termed *secretin.* The name *cholecystokinin* derives from the fact that this same hormone also causes contraction of the bile-containing gallbladder (*chole* means "bile," *cysto* means "bladder," and *kinin* means "contraction"). Secretin and CCK are major gastrointestinal hormones that perform other important functions in addition to serving as enterogastrones.

Let's examine why it is important that each of these stimuli in the duodenum (fat, acid, hypertonicity, and distension) delays gastric emptying (acting through the enterogastric reflex or one of the enterogastrones).

- *Fat.* Fat is digested and absorbed more slowly than the other nutrients. Furthermore, fat digestion and absorption take place only within the lumen of the small intestine. Therefore, when fat is already in the duodenum, further gastric emptying of more fatty stomach contents into the duodenum is prevented until the small intestine has processed the fat already there. In fact, fat is the most potent stimulus for inhibition of gastric motility. This is evident when you compare the rate of emptying of a high-fat meal (after six hours some of a bacon-and-eggs meal may still be in the stomach) with that of a protein and carbohydrate meal (a meal of lean meat and potatoes may empty in three hours).
- *Acid.* Because the stomach secretes hydrochloric acid (HCl), highly acidic chyme is emptied into the duodenum, where it is neutralized by sodium bicarbonate ($NaHCO_3$) secreted into the duodenal lumen from the pancreas. Unneutral-

ized acid irritates the duodenal mucosa and inactivates the pancreatic digestive enzymes that are secreted into the duodenal lumen. Appropriately, therefore, unneutralized acid in the duodenum inhibits further emptying of acidic gastric contents until complete neutralization can be accomplished.

- *Hypertonicity.* As molecules of protein and starch are digested in the duodenal lumen, large numbers of amino acid and glucose molecules are released. If absorption of these amino acid and glucose molecules does not keep pace with the rate at which protein and carbohydrate digestion proceeds, these large numbers of molecules remain in the chyme and increase the osmolarity of the duodenal contents. Osmolarity depends on the number of molecules present, not on their size, and one protein molecule may be split into several hundred amino-acid molecules, each of which has the same osmotic activity as the original protein molecule. The same holds true for one large starch molecule, which yields many smaller but equally osmotically active glucose molecules. Because water is freely diffusable across the duodenal wall, it enters the duodenal lumen from the plasma as the duodenal osmolarity rises. Large volumes of water entering the intestine from the plasma lead to intestinal distension, and, more importantly, circulatory disturbances ensue because of the reduction in plasma volume. To prevent these effects, gastric emptying is reflexly inhibited when the osmolarity of the duodenal contents starts to rise. Thus the amount of food entering the duodenum for further digestion into a multitude of additional osmotically active particles is reduced until absorption processes have had an opportunity to catch up.
- *Distension.* Too much chyme in the duodenum inhibits the emptying of even more gastric contents, giving the distended duodenum time to cope with the excess volume of chyme it already contains before it gets any more.

Emotions can influence gastric motility.

Other factors unrelated to digestion, such as emotions, can also alter gastric motility by acting through the autonomic nerves to influence the degree of gastric smooth-muscle excitability. Even though the effect of emotions on gastric motility varies from one person to another and is not always predictable, sadness and fear generally tend to decrease motility, whereas anger and aggression tend to increase it. In addition to emotional influences, intense pain from any part of the body tends to inhibit motility, not just in the stomach but throughout the digestive tract. This response is brought about by increased sympathetic activity.

The stomach does not actively participate in vomiting.

Clinical Note **Vomiting,** or **emesis,** the forceful expulsion of gastric contents out through the mouth, is not accomplished by reverse peristalsis in the stomach, as might be predicted. Actually, the stomach itself does not actively participate in vomiting. The stomach, the esophagus, and associated sphincters are all relaxed during vomiting. The major force for expulsion comes, surprisingly, from contraction of

the respiratory muscles—namely, the diaphragm (the major inspiratory muscle) and the abdominal muscles (the muscles of active expiration).

The complex act of vomiting is coordinated by a **vomiting center** in the medulla. Vomiting begins with a deep inspiration and closure of the glottis. The contracting diaphragm descends downward on the stomach while simultaneous contraction of the abdominal muscles compresses the abdominal cavity, increasing the intra-abdominal pressure and forcing the abdominal viscera upward. As the flaccid stomach is squeezed between the diaphragm from above and the compressed abdominal cavity from below, the gastric contents are forced upward through the relaxed sphincters and esophagus and out through the mouth. The glottis is closed, so vomited material does not enter the respiratory airways. Also, the uvula is raised to close off the nasal cavity. The vomiting cycle may be repeated several times until the stomach is emptied. Vomiting is usually preceded by profuse salivation, sweating, rapid heart rate, and the sensation of nausea, all of which are characteristic of a generalized discharge of the autonomic nervous system.

CAUSES OF VOMITING

Vomiting can be initiated by afferent input to the vomiting center from a number of receptors throughout the body. The causes of vomiting include the following:

- Tactile (touch) stimulation of the back of the throat, which is one of the most potent stimuli. For example, sticking a finger in the back of the throat or even the presence of a tongue depressor or dental instrument in the back of the mouth is enough stimulation to cause gagging and even vomiting in some people.
- Irritation or distension of the stomach and duodenum.
- Elevated intracranial pressure, such as that caused by cerebral hemorrhage. Thus vomiting after a head injury is considered a bad sign; it suggests swelling or bleeding within the cranial cavity.
- Rotation or acceleration of the head producing dizziness, such as in motion sickness.
- Chemical agents, including drugs or noxious substances that initiate vomiting (that is, **emetics**) either by acting in the upper parts of the gastrointestinal tract or by stimulating chemoreceptors in a specialized **chemoreceptor trigger zone** next to the vomiting center in the brain. Activation of this zone triggers the vomiting reflex. For example, chemotherapeutic agents used in treating cancer often cause vomiting by acting on the chemoreceptor trigger zone.
- Psychogenic vomiting induced by emotional factors, including those accompanying nauseating sights and odors and anxiety before taking an examination or in other stressful situations.

EFFECTS OF VOMITING

With excessive vomiting, the body experiences large losses of secreted fluids and acids that normally would be reabsorbed. The resulting reduction in plasma volume can lead to dehydration and circulatory problems, and the loss of acid from the stomach can lead to metabolic alkalosis (see p. 461).

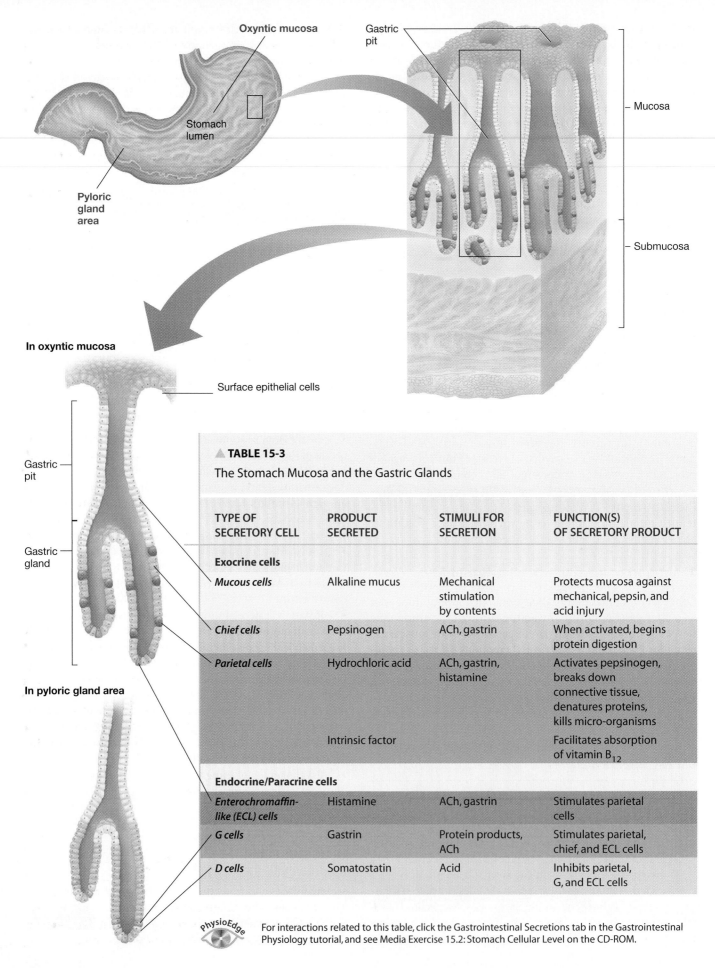

Oxyntic mucosa

Gastric pit

Stomach lumen

Pyloric gland area

Mucosa

Submucosa

In oxyntic mucosa

Surface epithelial cells

Gastric pit

Gastric gland

In pyloric gland area

▲ **TABLE 15-3**

The Stomach Mucosa and the Gastric Glands

TYPE OF SECRETORY CELL	PRODUCT SECRETED	STIMULI FOR SECRETION	FUNCTION(S) OF SECRETORY PRODUCT
Exocrine cells			
Mucous cells	Alkaline mucus	Mechanical stimulation by contents	Protects mucosa against mechanical, pepsin, and acid injury
Chief cells	Pepsinogen	ACh, gastrin	When activated, begins protein digestion
Parietal cells	Hydrochloric acid	ACh, gastrin, histamine	Activates pepsinogen, breaks down connective tissue, denatures proteins, kills micro-organisms
	Intrinsic factor		Facilitates absorption of vitamin B_{12}
Endocrine/Paracrine cells			
Enterochromaffin-like (ECL) cells	Histamine	ACh, gastrin	Stimulates parietal cells
G cells	Gastrin	Protein products, ACh	Stimulates parietal, chief, and ECL cells
D cells	Somatostatin	Acid	Inhibits parietal, G, and ECL cells

PhysioEdge

For interactions related to this table, click the Gastrointestinal Secretions tab in the Gastrointestinal Physiology tutorial, and see Media Exercise 15.2: Stomach Cellular Level on the CD-ROM.

Vomiting is not always harmful, however. Limited vomiting brought about by irritation of the digestive tract can provide a useful service in removing noxious material from the stomach rather than letting it stay and be absorbed. In fact, emetics are sometimes taken after accidental ingestion of a poison to quickly remove the offending substance from the body.

We have now completed our discussion of gastric motility and will shift to gastric secretion.

Gastric digestive juice is secreted by glands located at the base of gastric pits.

Each day the stomach secretes about 2 liters of gastric juice. The cells that secrete gastric juice are in the lining of the stomach, the gastric mucosa, which is divided into two distinct areas: (1) the **oxyntic mucosa**, which lines the body and fundus, and (2) the **pyloric gland area (PGA)**, which lines the antrum. The luminal surface of the stomach is pitted with deep pockets formed by infoldings of the gastric mucosa. The first part of these invaginations are called **gastric pits**, at the base of which lie the **gastric glands**. A variety of secretory cells line these invaginations, some exocrine and some endocrine or paracrine (▲ Table 15-3). Let's look at the gastric exocrine secretory cells first.

Three types of gastric exocrine secretory cells are found in the walls of the pits and glands in the oxyntic mucosa.

- **Mucous cells** line the gastric pits and the entrance of the glands. They secrete a thin, watery *mucus*. (*Mucous* is the adjective; *mucus* is the noun.)
- The deeper parts of the gastric glands are lined by chief and parietal cells. The more numerous **chief cells** secrete the enzyme precursor *pepsinogen*.
- The **parietal cells** secrete *HCl* and *intrinsic factor*.

These exocrine secretions are all released into the gastric lumen. Collectively, they make up the gastric digestive juice.

A few **stem cells** are also found in the gastric pits. These cells rapidly divide and serve as the parent cells of all new cells of the gastric mucosa. The daughter cells that result from cell division either migrate out of the pit to become surface epithelial cells or migrate down deeper to the gastric glands, where they differentiate into chief or parietal cells. Through this activity, the entire stomach mucosa is replaced about every three days.

Between the gastric pits, the gastric mucosa is covered by **surface epithelial cells**, which secrete a thick, viscous, alkaline mucus that forms a visible layer several millimeters thick over the surface of the mucosa.

The gastric glands of the PGA primarily secrete mucus and a small amount of pepsinogen; no acid is secreted in this area, in contrast to the oxyntic mucosa.

Let's consider these exocrine products and their roles in digestion in further detail.

Hydrochloric acid activates pepsinogen.

The parietal cells actively secrete HCl into the lumen of the gastric pits, which in turn empty into the lumen of the stomach.

As a result of this HCl secretion, the pH of the luminal contents falls as low as 2. Hydrogen ion (H^+) and chloride ion (Cl^-) are actively transported by separate pumps in the parietal cells' plasma membrane. Hydrogen ion is actively transported against a tremendous concentration gradient, with the H^+ concentration being as much as 3 million times greater in the lumen than in the blood. Chloride is also actively secreted but against a much smaller concentration gradient of only 1.5 times.

Although HCl does not actually digest anything, it performs several functions that aid digestion. Specifically, HCl

1. Activates the enzyme precursor pepsinogen to an active enzyme, pepsin, and provides an acid medium that is optimal for pepsin activity.
2. Aids in the breakdown of connective tissue and muscle fibers, reducing large food particles into smaller particles.
3. Denatures protein; that is, it uncoils proteins from their highly folded final form, thus exposing more of the peptide bonds for enzymatic attack.
4. Along with salivary lysozyme, kills most of the microorganisms ingested with food, although some do escape and continue to grow and multiply in the large intestine.

Pepsinogen, once activated, begins protein digestion.

The major digestive constituent of gastric secretion is **pepsinogen**, an inactive enzymatic molecule produced by the chief cells. Pepsinogen is stored in the chief cells' cytoplasm within secretory vesicles, from which it is released by exocytosis on appropriate stimulation (see p. 24). When pepsinogen is secreted into the gastric lumen, HCl cleaves off a small fragment of the molecule, converting it to the active form of the enzyme, **pepsin** (● Figure 15-9). Once formed, pepsin acts on other pepsinogen molecules to produce more pepsin. A mechanism such as this, whereby an active form of an enzyme activates other molecules of the same enzyme, is called an **autocatalytic** ("self-activating") **process**.

Pepsin initiates protein digestion by splitting certain amino acid linkages in proteins to yield peptide fragments (small amino-acid chains); it works most effectively in the acid environment provided by HCl. Because pepsin can digest protein, it must be stored and secreted in an inactive form so it does not digest the cells in which it is formed. Therefore, pepsin is maintained in the inactive form of pepsinogen until it reaches the gastric lumen, where it is activated by HCl secreted into the lumen by a different cell type.

Mucus is protective.

The surface of the gastric mucosa is covered by a layer of mucus derived from the surface epithelial cells and mucous cells. This mucus serves as a protective barrier against several forms of potential injury to the gastric mucosa:

- By virtue of its lubricating properties, mucus protects the gastric mucosa against mechanical injury.
- It helps protect the stomach wall from self-digestion, because pepsin is inhibited when it comes in contact with the

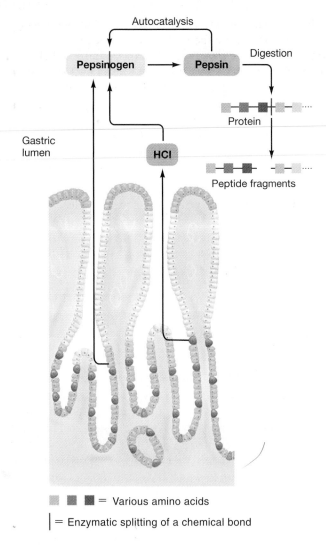

= Various amino acids

| = Enzymatic splitting of a chemical bond

● **FIGURE 15-9**

Pepsinogen activation in the stomach lumen. In the lumen, hydrochloric acid (HCl) activates pepsinogen to its active form, pepsin, by cleaving off a small fragment. Once activated, pepsin autocatalytically activates more pepsinogen and begins protein digestion. Secretion of pepsinogen in the inactive form prevents it from digesting the protein structures of the cells in which it is produced.

mucus layer coating the stomach lining. (However, mucus does not affect pepsin activity in the lumen, where digestion of dietary protein proceeds without interference.)

• Being alkaline, mucus helps protect against acid injury by neutralizing HCl in the vicinity of the gastric lining, but it does not interfere with the function of HCl in the lumen.

Intrinsic factor is essential for absorption of vitamin B$_{12}$.

Intrinsic factor, another secretory product of the parietal cells in addition to HCl, is important in the absorption of vitamin B$_{12}$. This vitamin can be absorbed only when in combination with intrinsic factor. Binding of the intrinsic factor–vitamin B$_{12}$ complex with a special receptor located only in the terminal ileum, the last part of the small intestine, trig-

gers the receptor-mediated endocytosis of the complex at this location (see p. 25).

Vitamin B$_{12}$ is essential for the normal formation of red blood cells. In the absence of intrinsic factor, vitamin B$_{12}$ is not absorbed, so erythrocyte production is defective, and *pernicious anemia* results (see p. 319).

Multiple regulatory pathways influence the parietal and chief cells.

In addition to the gastric exocrine secretory cells, other secretory cells in the gastric glands release endocrine and paracrine regulatory factors instead of products involved in the digestion of nutrients in the gastric lumen (see p. 482). These are as follows (▲ Table 15-3):

• Endocrine cells known as **G cells** found in the gastric pits only in the PGA secrete the hormone *gastrin* into the blood.
• **Enterochromaffin-like (ECL) cells** dispersed among the parietal and chief cells in the gastric glands of the oxyntic mucosa secrete the paracrine *histamine*.
• **D cells,** which are scattered in glands near the pylorus but are more numerous in the duodenum, secrete the paracrine *somatostatin*.

These three regulatory factors from the gastric pits along with the neurotransmitter *acetylcholine (ACh)* primarily control the secretion of gastric digestive juices. Parietal cells have separate receptors for each of these chemical messengers. Three of them—ACh, gastrin, and histamine—increase HCl secretion. The fourth regulatory agent—somatostatin—inhibits HCl secretion. ACh and gastrin also increase pepsinogen secretion through their stimulatory effect on the chief cells. We will now consider each of these chemical messengers in further detail (▲ Table 15-3).

• **Acetylcholine** is a neurotransmitter released from the intrinsic nerve plexuses in response to both local reflexes and vagal stimulation. ACh stimulates both the parietal and chief cells as well as the G cells and ECL cells.
• The G cells secrete the hormone **gastrin** into the blood in response to protein products in the stomach lumen as well in response to ACh. Like secretin and CCK, gastrin is a major gastrointestinal hormone. After being carried by the blood back to the body and fundus of the stomach, gastrin stimulates the parietal and chief cells, promoting secretion of a highly acidic gastric juice. In addition to directly stimulating the parietal cells, gastrin also indirectly promotes HCl secretion by stimulating the ECL cells to release histamine. Gastrin is the main factor that brings about increased HCl secretion during meal digestion.
• **Histamine,** a paracrine, is released from the ECL cells in response to ACh and gastrin. Histamine acts locally on nearby parietal cells to speed up HCl secretion.
• **Somatostatin** is released from the D cells in response to acid. It acts locally as a paracrine in negative-feedback fashion to inhibit secretion by the parietal cells, G cells, and ECL cells, thus turning off the HCl-secreting cells and their most potent stimulatory pathway.

From this list, it is obvious not only that multiple chemical messengers influence the parietal and chief cells but also that these chemicals also influence each other. Next, as we examine the phases of gastric secretion, you will see under what circumstances each of these regulatory agents is released.

Control of gastric secretion involves three phases.

The rate of gastric secretion can be influenced by (1) factors arising before food ever reaches the stomach, (2) factors resulting from the presence of food in the stomach, and (3) factors in the duodenum after food has left the stomach. Accordingly, gastric secretion is divided into three phases—the cephalic, gastric, and intestinal phases.

CEPHALIC PHASE

The *cephalic phase of gastric secretion* refers to the increased secretion of HCl and pepsinogen that occurs in feedforward fashion in response to stimuli acting in the head even before food reaches the stomach (*cephalic* means "head"). Thinking about, tasting, smelling, chewing, and swallowing food increase gastric secretion by vagal nerve activity in two ways. First, vagal stimulation of the intrinsic plexuses promotes increased secretion of ACh, which in turn leads to increased secretion of HCl and pepsinogen by the secretory cells. Second, vagal stimulation of the G cells within the PGA causes the release of gastrin, which in turn further enhances secretion of HCl and pepsinogen, with the effect on HCl being potentiated (made stronger) by gastrin promoting the release of histamine (▲ Table 15-4).

GASTRIC PHASE

The *gastric phase of gastric secretion* begins when food actually reaches the stomach. Stimuli acting in the stomach—namely *protein, distension, caffeine,* and *alcohol*—increase gastric secretion by overlapping efferent pathways. For example, protein in the stomach, the most potent stimulus, stimulates chemoreceptors that activate the intrinsic nerve plexuses, which in turn stimulate the secretory cells. Furthermore, protein brings about activation of the extrinsic vagal fibers to the stomach. Vagal activity further enhances intrinsic nerve stimulation of the secretory cells and triggers the release of gastrin. Protein also directly stimulates the release of gastrin. Gastrin in turn is a powerful stimulus for further HCl and pepsinogen secretion and also calls forth release of histamine, which further increases HCl secretion. Through these synergistic and overlapping pathways, protein induces the secretion of a highly acidic, pepsin-rich gastric juice, which continues the digestion of the protein that first initiated the process (▲ Table 15-4).

When the stomach is distended with protein-rich food that needs to be digested, these secretory responses are appropriate. Caffeine and, to lesser extent, alcohol also stimulate the secretion of a highly acidic gastric juice, even when no food is present. This unnecessary acid can irritate the linings of the stomach and duodenum. For this reason, people with ulcers or gastric hyperacidity should avoid caffeinated and alcoholic beverages.

INTESTINAL PHASE

The *intestinal phase of gastric secretion* encompasses the factors originating in the small intestine that influence gastric secretion. Whereas the other phases are excitatory, this phase is inhibitory. The intestinal phase is important in helping shut off the flow of gastric juices as chyme begins to be emptied into the small intestine, a topic to which we now turn.

Gastric secretion gradually decreases as food empties from the stomach into the intestine.

You now know what factors turn on gastric secretion before and during a meal, but how is the flow of gastric juices shut off when they are no longer needed? Gastric secretion is grad-

▲ **TABLE 15-4**

Stimulation of Gastric Secretion

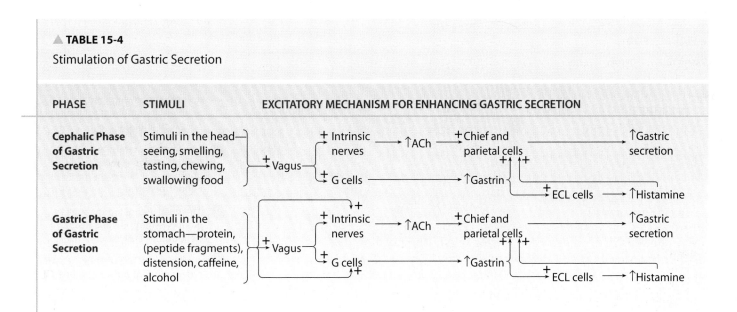

PHASE	STIMULI	EXCITATORY MECHANISM FOR ENHANCING GASTRIC SECRETION
Cephalic Phase of Gastric Secretion	Stimuli in the head—seeing, smelling, tasting, chewing, swallowing food	+ Vagus → + Intrinsic nerves → ↑ACh → + Chief and parietal cells → ↑Gastric secretion; + G cells → ↑Gastrin → + ECL cells → ↑Histamine
Gastric Phase of Gastric Secretion	Stimuli in the stomach—protein, (peptide fragments), distension, caffeine, alcohol	+ Vagus → + Intrinsic nerves → ↑ACh → + Chief and parietal cells → ↑Gastric secretion; + G cells → ↑Gastrin → + ECL cells → ↑Histamine

REGION	STIMULI	INHIBITORY MECHANISM FOR GASTRIC SECRETION		
Body and Antrum	Removal of protein and distension as the stomach empties	→ Intrinsic nerves → Vagus → G cells → ↓Gastrin ↘ ↓Histamine		→ ↓Gastric secretion
Antrum and Duodenum	Accumulation of acid	+ → D cells → ↑Somatostatin	→ Parietal cells → G cells → ECL cells	→ ↓Gastric secretion
Duodenum (intestinal phase of gastric secretion)	Fat Acid Hypertonicity Distension	+ → Enterogastric reflex ↑Enterogastrones (cholecystokinin and secretin)	→ Parietal cells → Chief cells → Smooth muscle cells	→ ↓Gastric secretion and motility

ually reduced in three different ways as the stomach empties (▲ Table 15-5):

- As the meal is gradually emptied into the duodenum, the major stimulus for enhanced gastric secretion—the presence of protein in the stomach—is withdrawn.
- After foods leave the stomach and gastric juices accumulate to such an extent that gastric pH falls very low, somatostatin is released. As a result of somatostatin's inhibitory effects, gastric secretion declines.
- The same stimuli that inhibit gastric motility (fat, acid, hypertonicity, or distension in the duodenum brought about by the emptying of stomach contents into the duodenum) in-

hibit gastric secretion as well. The enterogastric reflex and the enterogastrones suppress the gastric secretory cells while they simultaneously reduce the excitability of the gastric smooth muscle cells. This inhibitory response is the intestinal phase of gastric secretion.

▌ The gastric mucosal barrier protects the stomach lining from gastric secretions.

How can the stomach contain strong acid contents and proteolytic enzymes without destroying itself? You already learned that mucus provides a protective coating. In addition, other barriers to mucosal acid damage are provided by the mucosal

● **FIGURE 15-10**

Gastric mucosal barrier

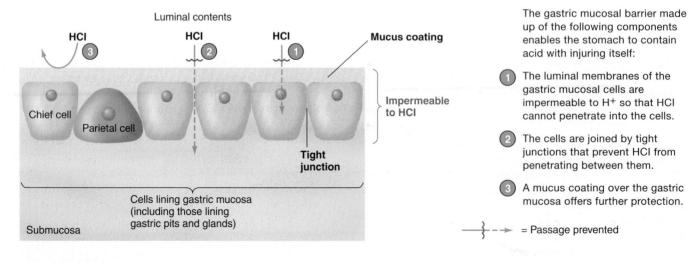

Luminal contents

HCl ③   HCl ②   HCl ①   Mucus coating

Chief cell   Parietal cell

Impermeable to HCl

Tight junction

Cells lining gastric mucosa (including those lining gastric pits and glands)

Submucosa

The gastric mucosal barrier made up of the following components enables the stomach to contain acid with injuring itself:

① The luminal membranes of the gastric mucosal cells are impermeable to H⁺ so that HCl cannot penetrate into the cells.

② The cells are joined by tight junctions that prevent HCl from penetrating between them.

③ A mucus coating over the gastric mucosa offers further protection.

= Passage prevented

Ulcers: When Bugs Break the Barrier

Peptic ulcers are erosions that typically begin in the mucosal lining of the stomach and may penetrate into the deeper layers of the stomach wall. They occur when the gastric mucosal barrier is disrupted and thus pepsin and HCl act on the stomach wall instead of food in the lumen. Frequent backflow of acidic gastric juices into the esophagus or excess unneutralized acid from the stomach in the duodenum can lead to peptic ulcers in these sites as well.

Until recently, the exact cause of ulcers was unknown, but in a surprising discovery in the early 1990s the bacterium *Helicobacter pylori* was pinpointed as the cause of more than 80% of all peptic ulcers. Thirty percent of the population harbors *H. pylori*. Those who have this slow bacterium have a 3 to 12 times greater risk of developing an ulcer within 10 to 20 years of acquiring the infection than those without the bacterium. They are also at increased risk of developing stomach cancer.

For years, scientists had overlooked the possibility that ulcers could be triggered by an infectious agent, because bacteria typically cannot survive in a strongly acidic environment such as the stomach lumen. An exception, *H. pylori* exploits several strategies to survive in this hostile environment. First, these organisms are motile, being equipped with four to six flagella (whiplike appendages; see the accompanying figure), which enable them to tunnel through and take up residence under the stomach's thick layer of alkaline mucus. Here they are protected from the highly acidic gastric contents. Furthermore, *H. pylori* preferentially settles in the antrum, which has no acid-producing parietal cells, although HCl from the upper parts of the stomach does reach the antrum. Also, these bacteria produce *urease*, an enzyme that breaks down urea, an end product of protein metabolism, into ammonia (NH_3) and CO_2. Ammonia serves as a buffer (see p. 460) that neutralizes stomach acid locally in the vicinity of the *H. pylori*.

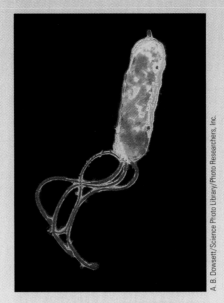

Helicobacter pylori. *Helicobacter pylori*, the bacterium responsible for most cases of peptic ulcers, has flagella that enable it to tunnel beneath the protective mucus layer that coats the stomach lining.

A. B. Dowsett/Science Photo Library/Photo Researchers, Inc.

H. pylori contributes to ulcer formation in part by secreting toxins that cause a persistent inflammation, or *chronic superficial gastritis*, at the site they colonize. *H. pylori* further weakens the gastric mucosal barrier by disrupting the tight junctions between the gastric epithelial cells, thereby making the gastric mucosa leakier than normal.

Alone or in conjunction with this infectious culprit, other factors are known to contribute to ulcer formation. Frequent exposure to some chemicals can break the gastric mucosal barrier; the most important of these are ethyl alcohol and nonsteroidal anti-inflammatory drugs (NSAIDs), such as aspirin, ibuprofen, or more potent medications for the treatment of arthritis or other chronic inflammatory processes.

The barrier frequently breaks in patients with pre-existing debilitating conditions, such as severe injuries or infections. Persistent stressful situations are frequently associated with ulcer formation, presumably because emotional response to stress can stimulate excessive gastric secretion.

When the gastric mucosal barrier is broken, acid and pepsin diffuse into the mucosa and underlying submucosa, with serious pathophysiologic consequences. The surface erosion, or ulcer, progressively enlarges as increasing levels of acid and pepsin continue to damage the stomach wall. Two of the most serious consequences of ulcers are (1) hemorrhage resulting from damage to submucosal capillaries and (2) perforation, or complete erosion through the stomach wall, resulting in the escape of potent gastric contents into the abdominal cavity.

Treatment of ulcers includes antibiotics, H-2 histamine receptor blockers, and proton pump inhibitors. With the discovery of the infectious component of most ulcers, antibiotics are now a treatment of choice. The other drugs are also used alone or in combination with antibiotics.

Two decades before the discovery of *H. pylori*, researchers discovered an antihistamine (*cimetidine*) that specifically blocks H-2 receptors, the type of receptors that bind histamine released from the stomach. These receptors differ from H-1 receptors that bind the histamine involved in allergic respiratory disorders. Accordingly, traditional antihistamines used for respiratory allergies (such as hay fever and asthma) are not effective against ulcers, nor is cimetidine useful for respiratory problems.

Another recent class of drugs used in treating ulcers inhibits acid secretion by directly blocking the pump that transports H^+ into the stomach lumen. These so-called *proton-pump inhibitors* (H^+ is a naked proton without its electron) help reduce the corrosive effect of HCl on the exposed tissue.

---

lining itself. First, the luminal membranes of the gastric mucosal cells are almost impermeable to H^+, so acid cannot penetrate into the cells and damage them. Furthermore, the lateral edges of these cells are joined together near their luminal borders by tight junctions, so acid cannot diffuse between the cells from the lumen into the underlying submucosa (see p. 48). The properties of the gastric mucosa that enable the stomach to contain acid without injuring itself constitute the **gastric mucosal barrier** (● Figure 15-10). These protective mechanisms are further enhanced by the fact that the entire stomach lining is replaced every three days. Because of rapid mucosal turnover, cells are usually replaced before they are exposed to the wear and tear of harsh gastric conditions long enough to suffer damage.

Clinical Note

Despite the protection provided by mucus, by the gastric mucosal barrier, and by the frequent turnover of cells, the barrier occasionally is broken, and the gastric wall is injured by its acidic and enzymatic contents. When this occurs, an erosion, or **peptic ulcer**, of the stomach wall results. Excessive gastric reflux into the esophagus and dumping of excessive acidic gastric contents into the duodenum can lead to peptic ulcers in these locations as well. (For a further discussion of ulcers, see the accompanying boxed feature, ▶ Beyond the Basics.)

We now turn to the remaining two digestive processes in the stomach, gastric digestion and absorption.

Carbohydrate digestion continues in the body of the stomach; protein digestion begins in the antrum.

Two separate digestive processes take place within the stomach. In the body of the stomach, food remains in a semisolid mass, because peristaltic contractions in this region are too weak for mixing to occur. Because food is not mixed with gastric secretions in the body of the stomach, very little protein digestion occurs here. In the interior of the mass, however, carbohydrate digestion continues under the influence of salivary amylase. Even though acid inactivates salivary amylase, the unmixed interior of the food mass is free of acid.

Digestion by the gastric juice itself is done in the antrum of the stomach, where the food is thoroughly mixed with HCl and pepsin, beginning protein digestion.

The stomach absorbs alcohol and aspirin but no food.

No food or water is absorbed into the blood through the stomach mucosa. However, two noteworthy nonnutrient substances are absorbed directly by the stomach—*ethyl alcohol* and *aspirin*. Alcohol is somewhat lipid soluble, so it can diffuse through the lipid membranes of the epithelial cells that line the stomach and can enter the blood through the submucosal capillaries. Yet although alcohol can be absorbed by the gastric mucosa, it can be absorbed even more rapidly by the small-intestine mucosa, because the surface area for absorption in the small intestine is much greater than in the stomach. Thus alcohol absorption occurs more slowly if gastric emptying is delayed so that the alcohol remains in the stomach longer. Because fat is the most potent duodenal stimulus for inhibiting gastric motility, consuming fat-rich foods (for example, whole milk, pizza, or nuts) before or during alcohol ingestion delays gastric emptying and prevents the alcohol from producing its effects as rapidly.

Another category of substances absorbed by the gastric mucosa includes weak acids, most notably *acetylsalicylic acid* (aspirin). In the highly acidic environment of the stomach lumen, weak acids are lipid soluble, so they can be absorbed quickly by crossing the plasma membranes of the epithelial cells that line the stomach. Most other drugs are not absorbed until they reach the small intestine, so they do not begin to take effect as quickly.

Having completed our coverage of the stomach, we will move to the next part of the digestive tract, the small intestine and the accessory digestive organs that release their secretions into the small-intestine lumen.

 Click on the Media Exercises menu of the CD-ROM and work Media Exercises 15.1: The Stomach and 15.2: Stomach: Cellular Level to test your understanding of the previous sections.

PANCREATIC AND BILIARY SECRETIONS

When gastric contents are emptied into the small intestine, they are mixed not only with juice secreted by the small-intestine mucosa but also with the secretions of the exocrine pancreas and liver that are released into the duodenal lumen. We will discuss the roles of each of these accessory digestive organs before we examine the contributions of the small intestine itself.

The pancreas is a mixture of exocrine and endocrine tissue.

The **pancreas** is an elongated gland that lies behind and below the stomach, above the first loop of the duodenum (● Figure 15-11). This mixed gland contains both exocrine and endocrine tissue. The predominant exocrine part consists of grapelike clusters of secretory cells that form sacs known as **acini**, which connect to ducts that eventually empty into the duodenum. The smaller endocrine part consists of isolated islands of endocrine tissue, the **islets of Langerhans**, which are dispersed throughout the pancreas. The most important hormones secreted by the islet cells are insulin and glucagon (Chapter 17). The exocrine and endocrine pancreas have nothing in common except sharing the same location.

The exocrine pancreas secretes digestive enzymes and an aqueous alkaline fluid.

The **exocrine pancreas** secretes a pancreatic juice consisting of two components—(1) *pancreatic enzymes* actively secreted by the *acinar cells* that form the acini and (2) an *aqueous alkaline solution* actively secreted by the *duct cells* that line the pancreatic ducts. The aqueous (watery) alkaline component is rich in sodium bicarbonate ($NaHCO_3$).

Like pepsinogen, pancreatic enzymes are stored within secretory vesicles after being produced, then are released by exocytosis as needed. These pancreatic enzymes are important because they can almost completely digest food in the absence of all other digestive secretions. The acinar cells secrete three different types of pancreatic enzymes capable of digesting all three categories of foodstuffs: (1) **proteolytic enzymes** for protein digestion, (2) **pancreatic amylase** for carbohydrate digestion, and (3) **pancreatic lipase** for fat digestion.

PANCREATIC PROTEOLYTIC ENZYMES

The three major pancreatic proteolytic enzymes are *trypsinogen, chymotrypsinogen,* and *procarboxypeptidase,* each of which is secreted in an inactive form. When **trypsinogen** is secreted into the duodenal lumen, it is activated to its active enzyme form, **trypsin,** by **enterokinase,** an enzyme embedded in the luminal border of the cells that line the duodenal mucosa. Trypsin then autocatalytically activates more trypsinogen. Like pepsinogen, trypsinogen must remain inactive within the pancreas to prevent this proteolytic enzyme from digesting the cells in which it is formed. Trypsinogen remains inactive, therefore, until it reaches the duodenal lumen, where enterokinase triggers the activation process, which then proceeds autocatalytically. As further protection, the pancreas also produces a chemical known as **trypsin inhibitor**, which blocks trypsin's actions if spontaneous activation of trypsinogen inadvertently occurs within the pancreas.

Chymotrypsinogen and **procarboxypeptidase**, the other pancreatic proteolytic enzymes, are converted by trypsin to their active forms, **chymotrypsin** and **carboxypeptidase**, respectively, within the duodenal lumen. Thus, once enterokinase has activated some of the trypsin, trypsin then carries out the rest of the activation process.

Each of these proteolytic enzymes attacks different peptide linkages. The end products that result from this action are a mixture of small peptide chains and amino acids. Mucus secreted by the intestinal cells protects against digestion of the small-intestine wall by the activated proteolytic enzymes.

● **FIGURE 15-11**

Schematic representation of the exocrine and endocrine portions of the pancreas. The exocrine pancreas secretes into the duodenal lumen a digestive juice composed of digestive enzymes secreted by the acinar cells and an aqueous NaHCO₃ solution secreted by the duct cells. The endocrine pancreas secretes the hormones insulin and glucagon into the blood.

PhysioEdge For an animation of this figure, click the Gastrointestinal Secretions tab in the Gastrointestinal Physiology tutorial on the CD-ROM. For an interaction related to this figure, see Media Exercise 15.3: The Intestine and Associated Organs.

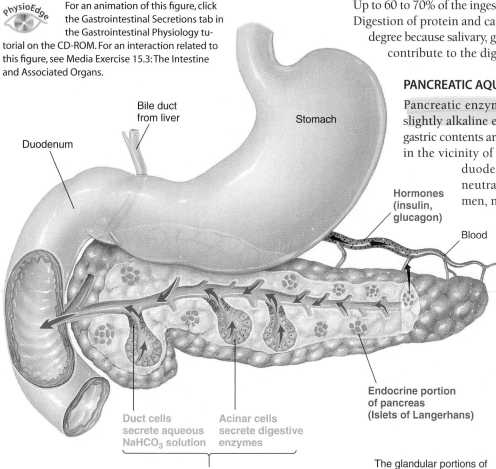

Bile duct from liver

Stomach

Duodenum

Hormones (insulin, glucagon)

Blood

Duct cells secrete aqueous NaHCO₃ solution

Acinar cells secrete digestive enzymes

Endocrine portion of pancreas (Islets of Langerhans)

Exocrine portion of pancreas (Acinar and duct cells)

The glandular portions of the pancreas are grossly exaggerated.

PANCREATIC AMYLASE

Like salivary amylase, pancreatic amylase contributes to carbohydrate digestion by converting polysaccharides into the disaccharide maltose. Amylase is secreted in the pancreatic juice in an active form, because active amylase does not endanger the secretory cells. These cells do not contain any polysaccharides.

PANCREATIC LIPASE

Pancreatic lipase is extremely important because it is the only enzyme secreted throughout the entire digestive system that can digest fat. Pancreatic lipase hydrolyzes dietary triglycerides into monoglycerides and free fatty acids, which are the absorbable units of fat. Like amylase, lipase is secreted in its active form because there is no risk of pancreatic self-digestion by lipase. Triglycerides are not a structural component of pancreatic cells.

PANCREATIC INSUFFICIENCY

Clinical Note When pancreatic enzymes are deficient, digestion of food is incomplete. Because the pancreas is the only significant source of lipase, pancreatic enzyme deficiency results in serious maldigestion of fats. The main clinical manifestation of pancreatic exocrine insufficiency is **steatorrhea**, or excessive undigested fat in the feces. Up to 60 to 70% of the ingested fat may be excreted in the feces. Digestion of protein and carbohydrates is impaired to a lesser degree because salivary, gastric, and small intestinal enzymes contribute to the digestion of these two foodstuffs.

PANCREATIC AQUEOUS ALKALINE SECRETION

Pancreatic enzymes function best in a neutral or slightly alkaline environment, yet the highly acidic gastric contents are emptied into the duodenal lumen in the vicinity of pancreatic enzyme entry into the duodenum. This acidic chyme must be neutralized quickly in the duodenal lumen, not only to allow optimal functioning of the pancreatic enzymes but also to prevent acid damage to the duodenal mucosa. The alkaline (NaHCO₃-rich) fluid secreted by the pancreatic duct cells into the duodenal lumen serves the important function of neutralizing the acidic chyme as the latter is emptied into the duodenum from the stomach. This aqueous NaHCO₃ secretion is by far the largest component of pancreatic secretion. The volume of pancreatic secretion ranges between 1 and 2 liters per day, depending on the types and degree of stimulation.

Pancreatic exocrine secretion is regulated by secretin and CCK.

Pancreatic exocrine secretion is regulated primarily by hormonal mechanisms. A small amount of parasympathetically induced pancreatic secretion occurs during the cephalic phase of digestion, with a further token increase occurring during the gastric phase in response to gastrin. However, the predominant stimulation of pancreatic secretion occurs during the intestinal phase of digestion when chyme is in the small intestine. The release of the two major enterogastrones, secretin and cholecystokinin (CCK), in response to chyme in the duodenum plays the central role in controlling pancreatic secretion (● Figure 15-12).

ROLE OF SECRETIN IN PANCREATIC SECRETION

Of the factors that stimulate enterogastrone release, the primary stimulus specifically for secretin release is acid in the duodenum. Secretin in turn is carried by the blood to the pancreas, where it stimulates the duct cells to markedly increase their secretion of a $NaHCO_3$-rich aqueous fluid into the duodenum. Even though other stimuli may cause the release of secretin, it is appropriate that the most potent stimulus is acid, because secretin promotes the alkaline pancreatic secretion that neutralizes the acid. This mechanism provides a control system for maintaining neutrality of the chyme in the intestine. The amount of secretin released is proportional to the amount of acid that enters the duodenum, so the amount of $NaHCO_3$ secreted parallels the duodenal acidity.

ROLE OF CCK IN PANCREATIC SECRETION

Cholecystokinin is important in regulating pancreatic digestive enzyme secretion. The main stimulus for release of CCK from the duodenal mucosa is the presence of fat and to a lesser extent protein products. The circulatory system transports CCK to the pancreas where it stimulates the pancreatic acinar cells to increase digestive enzyme secretion. Among these enzymes are lipase and the proteolytic enzymes, which appropriately further digest the fat and protein that initiated the response and also help digest carbohydrate. In contrast to fat and protein, carbohydrate does not have any direct influence on pancreatic digestive enzyme secretion.

We will now look at the contributions of the remaining accessory digestive unit, the liver and gallbladder.

The liver performs various important functions including bile production.

Besides pancreatic juice, the other secretory product emptied into the duodenal lumen is **bile**. The **biliary system** includes the *liver,* the *gallbladder,* and associated ducts.

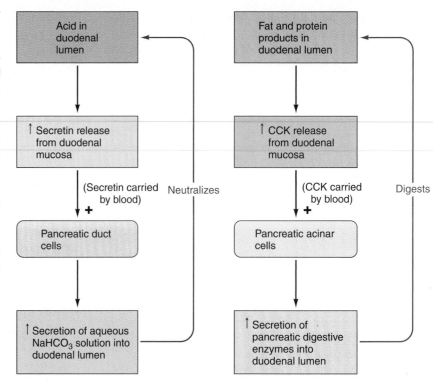

● **FIGURE 15-12**

Hormonal control of pancreatic exocrine secretion

LIVER FUNCTIONS

The **liver** is the largest and most important metabolic organ in the body; it can be viewed as the body's major biochemical factory. Its importance to the digestive system is its secretion of *bile salts,* which aid fat digestion and absorption. The liver also performs a wide variety of functions not related to digestion, including

1. Metabolic processing of the major categories of nutrients (carbohydrates, proteins, and lipids) after their absorption from the digestive tract
2. Detoxifying or degrading body wastes and hormones as well as drugs and other foreign compounds
3. Synthesizing plasma proteins, including those needed for blood clotting and those that transport steroid and thyroid hormones and cholesterol in the blood
4. Storing glycogen, fats, iron, copper, and many vitamins
5. Activating vitamin D, which the liver does in conjunction with the kidneys
6. Removing bacteria and worn-out red blood cells, thanks to its resident macrophages (see p. 329)
7. Excreting cholesterol and bilirubin, the latter being a breakdown product derived from the destruction of worn-out red blood cells

Given this wide range of complex functions, there is amazingly little specialization among cells within the liver. Each liver cell, or **hepatocyte**, performs the same wide variety of metabolic and secretory tasks (*hepato* means "liver"; *cyte* means "cell"). The specialization comes from the highly developed organelles within each hepatocyte. The only liver function not accomplished by the hepatocytes is the phagocytic activity carried out by the resident macrophages.

LIVER BLOOD FLOW

To carry out these wide-ranging tasks, the anatomic organization of the liver permits each hepatocyte to be in direct contact with blood from two sources: arterial blood coming from the aorta and venous blood coming directly from the digestive tract. Like other cells, the hepatocytes receive fresh arterial blood via the hepatic artery, which supplies their oxygen and delivers blood-borne metabolites for hepatic processing. Venous blood also enters the liver by the **hepatic portal system,** a unique and complex vascular connection between the digestive tract and liver (● Figure 15-13). The veins draining the digestive tract do not directly join the inferior vena cava, the large vein that returns blood to the heart. Instead, the veins from the stomach and intestine enter the hepatic portal vein, which carries the products absorbed from the digestive tract directly to the liver for processing, storage, or detoxification before they gain access to the general circulation. Within the liver, the portal vein once again breaks up into a capillary network (the liver *sinusoids*) to permit exchange between the blood and hepatocytes before draining into the hepatic vein, which joins the inferior vena cava.

▮ The liver lobules are delineated by vascular and bile channels.

The liver is organized into functional units known as **lobules,** which are hexagonal arrangements of tissue surrounding a central vein (● Figure 15-14a). At each of the six outer corners of the lobule are three vessels: a branch of the hepatic artery, a branch of the hepatic portal vein, and a bile duct. Blood from the branches of both the hepatic artery and the portal vein flows from the periphery of the lobule into large, expanded capillary spaces called **sinusoids,** which run between rows of liver cells to the central vein like spokes on a bicycle wheel (● Figure 15-14b). The hepatocytes are arranged between the sinusoids in plates two cell layers thick, so that each lateral edge faces a sinusoidal pool of blood. The central veins of all the liver lobules converge to form the hepatic vein, which carries the blood away from the liver. The thin bile-carrying channel, a **bile canaliculus,** runs between the cells within each hepatic plate. Hepatocytes continuously secrete bile into these thin channels, which carry the bile to a bile duct at the periphery of the lobule. The bile ducts from the various lobules converge to eventually form the common bile duct, which transports the bile from the liver to the duodenum. Each hepatocyte is in contact with a sinusoid on one side and a bile canaliculus on the other side.

▮ Bile is continuously secreted by the liver and is diverted to the gallbladder between meals.

The opening of the bile duct into the duodenum is guarded by the **sphincter of Oddi,** which prevents bile from entering the duodenum except during digestion of meals (● Figure 15-15). When this sphincter is closed, most of the bile secreted by the

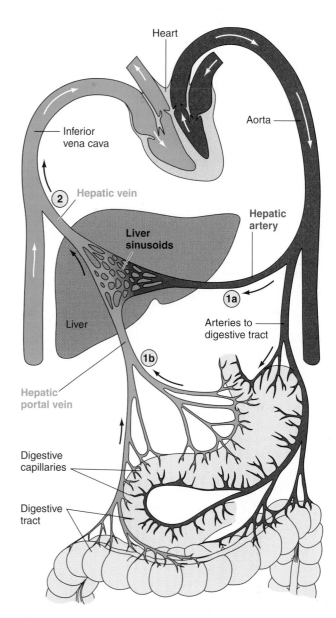

(1) The liver receives blood from two sources:

(a) Arterial blood, which provides the liver's O_2 supply and carries blood-borne metabolites for hepatic processing, is delivered by the **hepatic artery**.

(b) Venous blood draining the digestive tract is carried by the **hepatic portal vein** to the liver for processing and storage of newly absorbed nutrients.

(2) Blood leaves the liver via the **hepatic vein**.

● **FIGURE 15-13**

Schematic representation of liver blood flow

liver is diverted back up into the **gallbladder,** a small, saclike structure tucked beneath but not directly connected to the liver. Thus bile is not transported directly from the liver to the gallbladder. The bile is subsequently stored and concentrated in the gallbladder between meals. After a meal, bile enters the duodenum as a result of the combined effects of gallbladder emptying and increased bile secretion by the liver. The

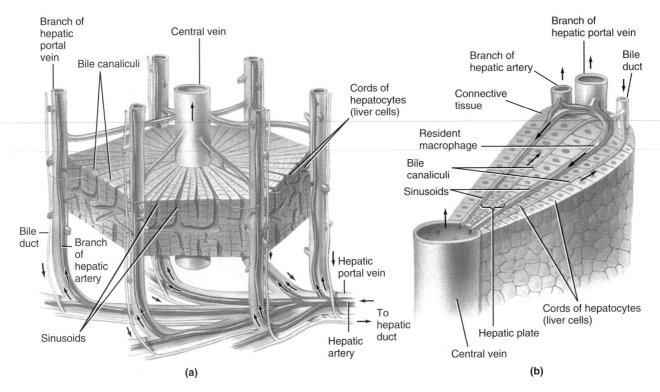

● FIGURE 15-14

Anatomy of the liver. (a) Hepatic lobule. (b) Wedge of a hepatic lobule.

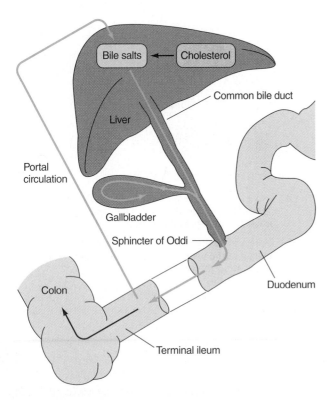

● FIGURE 15-15

Enterohepatic circulation of bile salts. The majority of bile salts are recycled between the liver and small intestine through the enterohepatic circulation *(blue arrows)*. After participating in fat digestion and absorption, most bile salts are reabsorbed by active transport in the terminal ileum and returned through the hepatic portal vein to the liver, which resecretes them in the bile.

amount of bile secreted per day ranges from 250 ml to 1 liter, depending on the degree of stimulation.

▌Bile salts are recycled through the enterohepatic circulation.

Bile contains several organic constituents, namely *bile salts, cholesterol, lecithin,* and *bilirubin* (all derived from hepatocyte activity) in an *aqueous alkaline fluid* (added by the duct cells) similar to the pancreatic NaHCO$_3$ secretion. Even though bile does not contain any digestive enzymes, it is important for the digestion and absorption of fats, primarily through the activity of bile salts.

Bile salts are derivatives of cholesterol. They are actively secreted into the bile and eventually enter the duodenum along with the other biliary constituents. Following their participation in fat digestion and absorption, most bile salts are reabsorbed back into the blood by special active transport mechanisms located in the terminal ileum. From here bile salts are returned by the hepatic portal system to the liver, which resecretes them into the bile. This recycling of bile salts (and some of the other biliary constituents) between the small intestine and liver is called the **enterohepatic circulation** (*entero* means "intestine"; *hepatic* means "liver") (● Figure 15-15).

The total amount of bile salts in the body averages about 3 to 4 g, yet 3 to 15 g of bile salts may be emptied into the duodenum in a single meal. Obviously, bile salts must be recycled many times per day. Usually, only about 5% of the secreted bile escapes into the feces daily. These lost bile salts are replaced by new bile salts synthesized by the liver; thus the size of the pool of bile salts is kept constant.

▌ Bile salts aid fat digestion and absorption.

Bile salts aid fat digestion through their detergent action (emulsification) and facilitate fat absorption by participating in the formation of micelles. Both functions are related to the structure of bile salts. Let's see how.

DETERGENT ACTION OF BILE SALTS

The term **detergent action** refers to bile salts' ability to convert large fat globules into a **lipid emulsion** consisting of many small fat droplets suspended in the aqueous chyme, thus increasing the surface area available for attack by pancreatic lipase. Fat globules, no matter their size, are made up primarily of undigested triglyceride molecules. To digest fat, lipase must come into direct contact with the triglyceride molecule. Because triglycerides are not soluble in water, they tend to aggregate into large droplets in the watery environment of the small-intestine lumen. If bile salts did not emulsify these large droplets, lipase could act on the triglyceride molecules only at the surface of the large droplets, and fat digestion would be greatly prolonged.

Bile salts exert a detergent action similar to that of the detergent you use to break up grease when you wash dishes. A bile salt molecule contains a lipid-soluble part (a steroid derived from cholesterol) plus a negatively charged, water-soluble part. Bile salts *adsorb* on the surface of a fat droplet; that is, the lipid-soluble part of the bile salt dissolves in the fat droplet, leaving the charged water-soluble part projecting from the surface of the droplet (● Figure 15-16a). Intestinal mixing movements break up large fat droplets into smaller ones. These small droplets would quickly recoalesce were it not for bile salts adsorbing on their surface and creating a shell of water-soluble negative charges on the surface of each little droplet. Because like charges repel, these negatively charged groups on the droplet surfaces cause the fat droplets to repel each other (● Figure 15-16b). This electrical repulsion prevents the small droplets from recoalescing into large fat droplets and thus produces

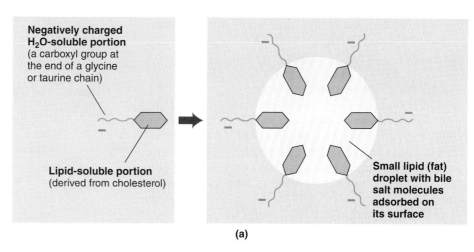

Negatively charged H₂O-soluble portion (a carboxyl group at the end of a glycine or taurine chain)

Lipid-soluble portion (derived from cholesterol)

Small lipid (fat) droplet with bile salt molecules adsorbed on its surface

(a)

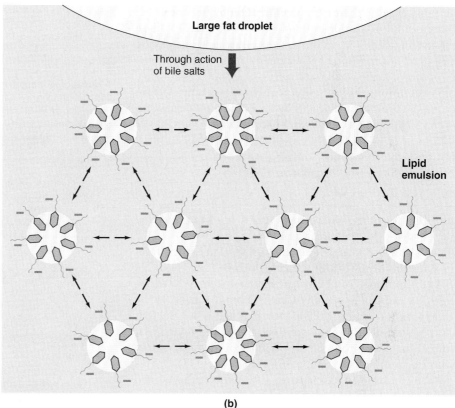

Large fat droplet

Through action of bile salts

Lipid emulsion

(b)

● **FIGURE 15-16**

Schematic structure and function of bile salts. (a) Schematic representation of the structure of bile salts and their adsorption on the surface of a small fat droplet. A bile salt consists of a lipid-soluble part that dissolves in the fat droplet and a negatively charged, water-soluble part that projects from the surface of the droplet. (b) Formation of a lipid emulsion through the action of bile salts. When a large fat droplet is broken up into smaller fat droplets by intestinal contractions, bile salts adsorb on the surface of the small droplets, creating shells of negatively charged, water-soluble bile salt components that cause the fat droplets to repel each other. This action holds the fat droplets apart and prevents them from recoalescing, increasing the surface area of exposed fat available for digestion by pancreatic lipase.

a lipid emulsion that increases the surface area available for lipase action.

MICELLAR FORMATION

Bile salts—along with cholesterol and lecithin, which are also constituents of bile—play an important role in facilitating fat absorption through micellar formation. Like bile salts, lecithin has both a lipid-soluble and a water-soluble part, whereas cholesterol is almost totally insoluble in water. In a **micelle**, the bile salts and lecithin aggregate in small clusters with their fat-soluble parts huddled together in the middle to form a hydrophobic ("water-fearing") core, while their water-soluble parts form an outer hydrophilic ("water-loving") shell (● Figure 15-17). A micelle is about one millionth the size of an emulsified lipid droplet. Micelles, being water soluble by virtue of their hydrophilic shells, can dissolve water-insoluble (and hence lipid-soluble) substances in their lipid-soluble cores. Micelles thus provide a handy vehicle for carrying water-insoluble substances through the watery luminal contents. The most important lipid-soluble substances carried within micelles are the products of fat digestion (monoglycerides and free fatty acids) as well as fat-soluble vitamins, which are all transported to their sites of absorption by this means. If they did not hitch a ride in the water-soluble micelles, these nutrients would float on the surface of the aqueous chyme (just as oil floats on top of water), never reaching the absorptive surfaces of the small intestine.

In addition, cholesterol, a highly water-insoluble substance, dissolves in the micelle's hydrophobic core. This mechanism is important in cholesterol homeostasis.

▌ Bilirubin is a waste product excreted in the bile.

Bilirubin, the other major constituent of bile, does not play a role in digestion at all but instead is a waste product excreted in the bile. Bilirubin is the primary bile pigment derived from the breakdown of worn-out red blood cells. The typical life span of a red blood cell in the circulatory system is 120 days. Worn-out red blood cells are removed from the blood by the macrophages that line the liver sinusoids and reside in other areas in the body. Bilirubin is the end product from degradation of the heme (iron-containing) part of the hemoglobin contained within these old red blood cells (see p. 317). This bilirubin is extracted from the blood by the hepatocytes and is actively excreted into the bile.

Bilirubin is a yellow pigment that gives bile its yellow color. Within the intestinal tract, this pigment is modified by bacterial enzymes, giving rise to the characteristic brown color of feces. When bile secretion does not occur, as when the bile duct is completely obstructed by a gallstone, the feces are grayish white. A small amount of bilirubin is normally reabsorbed by the intestine back into the blood, and when it is eventually excreted in the urine, it is largely responsible for the urine's yellow color. The kidneys cannot excrete bilirubin until after it has been modified during its passage through the liver and intestine.

Clinical Note If bilirubin is formed more rapidly than it can be excreted, it accumulates in the body and causes **jaundice.** Patients with this condition appear yellowish, with this color being seen most easily in the whites of their eyes. Jaundice can be brought about in three different ways:

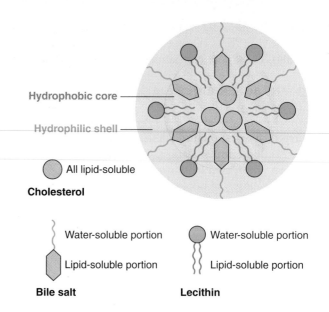

Hydrophobic core

Hydrophilic shell

○ All lipid-soluble
Cholesterol

〜 Water-soluble portion
⬡ Lipid-soluble portion
Bile salt

○ Water-soluble portion
〜 Lipid-soluble portion
Lecithin

● FIGURE 15-17

Schematic representation of a micelle. Bile constituents (bile salts, lecithin, and cholesterol) aggregate to form micelles that consist of a hydrophilic (water-soluble) shell and a hydrophobic (lipid-soluble) core. Because the outer shell of a micelle is water soluble, the products of fat digestion, which are not water soluble, can be carried through the watery luminal contents to the absorptive surface of the small intestine by dissolving in the micelle's lipid-soluble core.

PhysioEdge For an animation of this figure, click the Gastrointestinal Secretions tab in the Gastrointestinal Physiology tutorial on the CD-ROM. For an interaction related to this figure, see Media Exercise 15.3: The Intestine and Associated Organs.

1. *Prehepatic* (the problem occurs "before the liver"), or *hemolytic, jaundice* is due to excessive breakdown (hemolysis) of red blood cells, which results in the liver being presented with more bilirubin than it is capable of excreting.
2. *Hepatic* (the problem is the "liver") *jaundice* occurs when the liver is diseased and cannot deal with even the normal load of bilirubin.
3. *Posthepatic* (the problem occurs "after the liver"), or *obstructive, jaundice* occurs when the bile duct is obstructed, such as by a gallstone, so that bilirubin cannot be eliminated in the feces.

▌ Bile salts are the most potent stimulus for increased bile secretion.

Bile secretion may be increased by chemical, hormonal, and neural mechanisms:

- *Chemical mechanism (bile salts).* Any substance that increases bile secretion by the liver is called a **choleretic**. The most potent choleretic is bile salts themselves. Between meals, bile is stored in the gallbladder, but during a meal bile is emptied into the duodenum as the gallbladder contracts. After bile salts participate in fat digestion and absorption, they are reabsorbed and returned by the enterohepatic circulation to the liver, where they act as potent choleretics to stimulate further bile secretion. Therefore, during a meal, when bile salts are needed and being used, bile secretion by the liver is enhanced.

- *Hormonal mechanism (secretin).* Besides increasing the aqueous $NaHCO_3$ secretion by the pancreas, secretin also stimulates an aqueous alkaline bile secretion by the liver ducts without any corresponding increase in bile salts.
- *Neural mechanism (vagus nerve).* Vagal stimulation of the liver plays a minor role in bile secretion during the cephalic phase of digestion, promoting an increase in liver bile flow before food ever reaches the stomach or intestine.

▌ The gallbladder stores and concentrates bile between meals and empties during meals.

Even though the factors just described increase bile secretion by the liver during and after a meal, bile secretion by the liver occurs continuously. Between meals the secreted bile is shunted into the gallbladder, where it is stored and concentrated. Active transport of salt out of the gallbladder, with water following osmotically, results in a 5 to 10 times concentration of the organic constituents.

Clinical Note Because the gallbladder stores this concentrated bile, it is the primary site for precipitation of concentrated bile constituents into gallstones. Fortunately, the gallbladder does not play an essential digestive role, so its removal as a treatment for gallstones or other gallbladder disease presents no particular problem. The bile secreted between meals is stored instead in the common bile duct, which becomes dilated.

During digestion of a meal, when chyme reaches the small intestine, the presence of food, especially fat products, in the duodenal lumen triggers the release of CCK. This hormone stimulates contraction of the gallbladder and relaxation of the sphincter of Oddi, so bile is discharged into the duodenum, where it appropriately aids in the digestion and absorption of the fat that initiated the release of CCK.

Having looked at the accessory digestive organs that empty their exocrine products into the small-intestine lumen, we are now going to examine the contributions of the small intestine itself.

SMALL INTESTINE

The **small intestine** is the site where most digestion and absorption take place. No further digestion is accomplished after the luminal contents pass beyond the small intestine, and no further absorption of nutrients occurs, although the large intestine does absorb small amounts of salt and water. The small intestine lies coiled within the abdominal cavity, extending between the stomach and large intestine. It is arbitrarily divided into three segments—the **duodenum**, the **jejunum**, and the **ileum**.

As usual, we will examine motility, secretion, digestion, and absorption in the small intestine, in that order. Small-

intestine motility includes *segmentation* and the *migrating motility complex.* Let's consider segmentation first.

▌ Segmentation contractions mix and slowly propel the chyme.

Segmentation, the small intestine's primary method of motility during digestion of a meal, both mixes and slowly propels the chyme. Segmentation consists of oscillating, ringlike contractions of the circular smooth muscle along the small intestine's length; between the contracted segments are relaxed areas containing a small bolus of chyme. The contractile rings occur every few centimeters, dividing the small intestine into segments like a chain of sausages. These contractile rings do not sweep along the length of the intestine as peristaltic waves do. Rather, after a brief period of time, the contracted segments relax, and ringlike contractions appear in the previously relaxed areas (● Figure 15-18). The new contraction forces the chyme in a previously relaxed segment to move in both directions into the now relaxed adjacent segments. A newly relaxed segment therefore receives chyme from both the contracting segment immediately ahead of it and the one immediately behind it. Shortly thereafter, the areas of contraction and relaxation alternate again. In this way, the chyme is chopped, churned, and thoroughly mixed. These contractions can be compared to squeezing a pastry tube with your hands to mix the contents.

INITIATION AND CONTROL OF SEGMENTATION

Segmentation contractions are initiated by the small intestine's pacesetter cells, which produce a basic electrical rhythm

---

● **FIGURE 15-18**

Segmentation. Segmentation consists of ringlike contractions along the length of the small intestine. Within a matter of seconds, the contracted segments relax and the previously relaxed areas contract. These oscillating contractions thoroughly mix the chyme within the small-intestine lumen.

 For an animation of this figure, click the Gastrointestinal Motility tab in the Gastrointestinal Physiology tutorial on the CD-ROM.

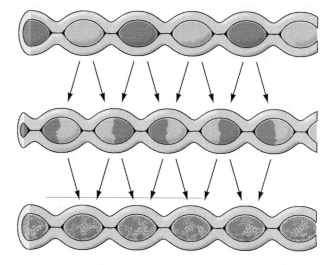

(BER) similar to the gastric BER that governs peristalsis in the stomach. If the small-intestine BER brings the circular smooth-muscle layer to threshold, segmentation contractions are induced, with the frequency of segmentation following the frequency of the BER.

The circular smooth muscle's degree of responsiveness and thus the intensity of segmentation contractions can be influenced by factors that influence the excitability of the small-intestine smooth muscle cells by moving the starting potential around which the BER oscillates closer to or farther from the threshold. Segmentation is slight or absent between meals but becomes very vigorous immediately after a meal. Both the duodenum and the ileum start to segment simultaneously when the meal first enters the small intestine. The duodenum starts to segment primarily in response to local distension caused by the presence of chyme. Segmentation of the empty ileum, in contrast, is brought about by gastrin secreted in response to the presence of chyme in the stomach, a mechanism known as the **gastroileal reflex**. Extrinsic nerves can modify the strength of these contractions. Parasympathetic stimulation enhances segmentation, whereas sympathetic stimulation depresses segmental activity.

FUNCTIONS OF SEGMENTATION

Segmentation not only accomplishes mixing but also slowly moves chyme through the small intestine. How can this be, when each segmental contraction propels chyme both forward and backward? The chyme slowly progresses forward because the frequency of segmentation declines along the length of the small intestine. The pacesetter cells in the duodenum spontaneously depolarize faster than those farther down the tract, with segmentation contractions occurring in the duodenum at a rate of 12 per minute, compared to only 9 per minute in the terminal ileum. Because segmentation occurs with greater frequency in the upper part of the small intestine than in the lower part, more chyme on average is pushed forward than is pushed backward. As a result, chyme is moved very slowly from the upper to the lower part of the small intestine, being shuffled back and forth to accomplish thorough mixing and absorption in the process. This slow propulsive mechanism is advantageous because it allows ample time for the digestive and absorptive processes to take place. The contents usually take 3 to 5 hours to move through the small intestine.

▌ The migrating motility complex sweeps the intestine clean between meals.

When most of the meal has been absorbed, segmentation contractions cease and are replaced between meals by the **migrating motility complex**, or "**intestinal housekeeper**." This between-meal motility consists of weak, repetitive peristaltic waves that move a short distance down the intestine before dying out. The waves start at the stomach and migrate down the intestine; that is, each new peristaltic wave is initiated at a site a little farther down the small intestine. These short peristaltic waves take about 100 to 150 minutes to gradually migrate from the stomach to the end of the small intestine, with each contraction sweeping any remnants of the preced-

ing meal plus mucosal debris and bacteria forward toward the colon, just like a good "intestinal housekeeper." After the end of the small intestine is reached, the cycle begins again and continues to repeat itself until the next meal. The migrating motility complex is believed to be regulated between meals by a speculated hormone *motilin* secreted by endocrine cells of the small-intestine mucosa. When the next meal arrives, segmental activity is triggered again, and the migrating motility complex ceases.

▌ The ileocecal juncture prevents contamination of the small intestine by colonic bacteria.

At the juncture between the small and large intestines, the last part of the ileum empties into the cecum (● Figure 15-19). Two factors contribute to this region's ability to act as a barrier between the small and large intestines. First, the anatomic arrangement is such that valvelike folds of tissue protrude from the ileum into the lumen of the cecum. When the ileal contents are pushed forward, this **ileocecal valve** is easily pushed open, but the folds of tissue are forcibly closed when the cecal contents attempt to move backward. Second, the smooth muscle within the last several centimeters of the ileal wall is thickened, forming a sphincter that is under neural and hormonal control. Most of the time this **ileocecal sphincter** remains at least mildly constricted. Pressure on the cecal side of the sphincter causes it to contract more forcibly; distension of the ileal side causes the sphincter to relax, a reaction mediated by the intrinsic plexuses in the area. In this way, the ileocecal juncture prevents the bacteria-laden contents of the large intestine from contaminating the small intestine and at the same time lets the ileal contents pass into the colon. If the colonic bacteria gained access to the nutrient-rich small intestine, they would multiply rapidly. Relaxation of the sphincter is enhanced by the release of gastrin at the onset of a meal, when increased gastric activity is taking place. This relaxation allows the undigested fibers and unabsorbed solutes from the preceding meal to be moved forward as the new meal enters the tract.

▌ Small-intestine secretions do not contain any digestive enzymes.

Each day the exocrine gland cells in the small-intestine mucosa secrete into the lumen about 1.5 liters of an aqueous salt and mucus solution called **succus entericus** ("juice of intestine"). Secretion increases after a meal in response to local stimulation of the small-intestine mucosa by the presence of chyme.

No digestive enzymes are secreted into this intestinal juice. The small intestine does synthesize digestive enzymes, but they act intracellularly within the borders of the epithelial cells that line the lumen instead of being secreted directly into the lumen.

The mucus in the secretion provides protection and lubrication. Furthermore, this aqueous secretion provides plenty of H_2O to participate in the enzymatic digestion of food. Recall that digestion involves hydrolysis—bond breakage by re-

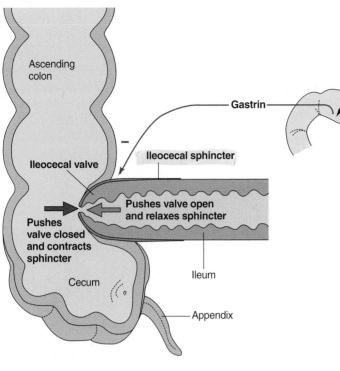

● FIGURE 15-19

Control of the ileocecal valve/sphincter. The juncture between the ileum and large intestine is the ileocecal valve, which is surrounded by thickened smooth muscle, the ileocecal sphincter. Pressure on the cecal side pushes the valve closed and contracts the sphincter, preventing the bacteria-laden colonic contents from contaminating the nutrient-rich small intestine. The valve/sphincter opens and allows ileal contents to enter the large intestine in response to pressure on the ileal side of the valve and to the hormone gastrin secreted as a new meal enters the stomach.

action with H_2O—which proceeds most efficiently when all the reactants are in solution.

The small-intestine enzymes complete digestion intracellularly.

Digestion within the small-intestine lumen is accomplished by the pancreatic enzymes, with fat digestion being enhanced by bile secretion. As a result of pancreatic enzymatic activity, fats are completely reduced to their absorbable units of monoglycerides and free fatty acids, proteins are broken down into small peptide fragments and some amino acids, and carbohydrates are reduced to disaccharides and some monosaccharides. Thus fat digestion is completed within the small-intestine lumen, but carbohydrate and protein digestion have not been brought to completion.

Special hairlike projections on the luminal surface of the small-intestine epithelial cells, the **microvilli**, form the **brush border** (see p. 37). The brush-border plasma membrane contains three categories of membrane-bound enzymes:

1. **Enterokinase**, which activates the pancreatic enzyme trypsinogen.
2. The **disaccharidases (maltase, sucrase, and lactase)**, which complete carbohydrate digestion by hydrolyzing the remaining disaccharides (maltose, sucrose, and lactose, respectively) into their constituent monosaccharides.
3. The **aminopeptidases**, which hydrolyze the small peptide fragments into their amino acid components, thereby completing protein digestion.

Thus carbohydrate and protein digestion are completed intracellularly within the confines of the brush border. (▲ Table 15-6 provides a summary of the digestive processes for the three major categories of nutrients.)

A fairly common disorder, **lactose intolerance**, involves a deficiency of lactase, the disaccharidase specific for the digestion of lactose, or milk sugar. Most children under 4 years of age have adequate lactase, but this may be gradually lost so that in many adults, lactase activity is diminished or absent. When lactose-rich milk or dairy products are consumed by a person with lactase deficiency, the undigested lactose remains in the lumen and has several related consequences. First, accumulation of undigested lactose creates an osmotic gradient that draws H_2O into the intestinal lumen. Second, bacteria living in the large intestine have lactose-splitting ability, so they eagerly attack the lactose as an energy source, producing large quantities of CO_2 and methane gas in the process. Distension of the intestine by both fluid and gas produces pain (cramping) and diarrhea. Infants with lactose intolerance may also suffer from malnutrition.

Finally we are ready to discuss absorption of nutrients. Up to this point, no food, water, or electrolytes have been absorbed.

The small intestine is remarkably well adapted for its primary role in absorption.

All products of carbohydrate, protein, and fat digestion, as well as most of the ingested electrolytes, vitamins, and water, are normally absorbed by the small intestine indiscriminately. Usually only the absorption of calcium and iron is adjusted to the body's needs. Thus, the more food that is consumed, the more that will be digested and absorbed, as people who are trying to control their weight are all too painfully aware.

Most absorption occurs in the duodenum and jejunum; very little occurs in the ileum, not because the ileum does not have absorptive capacity but because most absorption has already been accomplished before the intestinal contents reach the ileum. The small intestine has an abundant reserve absorptive capacity. About 50% of the small intestine can be removed with little interference to absorption—with one exception. If the terminal ileum is removed, vitamin B_{12} and bile salts are not properly absorbed, because the specialized transport mechanisms for these two substances are located only in this region. All other substances can be absorbed throughout the small intestine's length.

The mucous lining of the small intestine is remarkably well adapted for its special absorptive function for two reasons: (1) It has a very large surface area, and (2) the epithelial

Digestive Processes for the Three Major Categories of Nutrients

NUTRIENTS	ENZYMES FOR DIGESTING NUTRIENT	SOURCE OF ENZYMES	SITE OF ACTION OF ENZYMES	ACTION OF ENZYMES	ABSORBABLE UNITS OF NUTRIENTS
Carbohydrate	Amylase	Salivary glands	Mouth and body of stomach	Hydrolyzes polysaccharides to disaccharides	
		Exocrine pancreas	Small-intestine lumen		
	Disaccharidases (maltase, sucrase, lactase)	Small-intestine epithelial cells	Small-intestine brush border	Hydrolyze disaccharides to monosaccharides	Monosaccharides, especially glucose
Protein	Pepsin	Stomach chief cells	Stomach antrum	Hydrolyzes protein to peptide fragments	
	Trypsin, chymotrypsin carboxypeptidase	Exocrine pancreas	Small-intestine lumen	Attack different peptide fragments	
	Aminopeptidases	Small-intestine epithelial cells	Small-intestine brush border	Hydrolyze peptide fragments to amino acids	Amino acids
Fat	Lipase	Exocrine pancreas	Small-intestine lumen	Hydrolyzes triglycerides to fatty acids and monoglycerides	Fatty acids and monoglycerides
	Bile salts (not an enzyme)	Liver	Small-intestine lumen	Emulsify large fat globules for attack by pancreatic lipase	

cells in this lining have a variety of specialized transport mechanisms.

ADAPTATIONS THAT INCREASE THE SMALL INTESTINE'S SURFACE AREA

The following special modifications of the small-intestine mucosa greatly increase the surface area available for absorption (● Figure 15-20):

• The inner surface of the small intestine is thrown into permanent circular folds that are visible to the naked eye and increase the surface area threefold.

• Projecting from this folded surface are microscopic fingerlike projections known as **villi**, which give the lining a velvety appearance and increase the surface area by another 10 times (● Figure 15-21). The surface of each villus is covered by epithelial cells interspersed occasionally with mucous cells.

• Even smaller hairlike projections known as the *brush border* or **microvilli** arise from the luminal surface of these epithelial cells, increasing the surface area another 20-fold. Each epithelial cell has as many as 3000 to 6000 of these micro-

villi, which are visible only with an electron microscope. The small-intestine enzymes perform their functions within the membrane of this brush border.

Altogether, the folds, villi, and microvilli provide the small intestine with a luminal surface area 600 times greater than if it were a tube of the same length and diameter lined by a flat surface. In fact, if the surface area of the small intestine were spread out flat, it would cover an entire tennis court.

● **FIGURE 15-20** ▶

Small-intestine absorptive surface. (a) Gross structure of the small intestine. (b) One of the circular folds of the small-intestine mucosa, which collectively increase the absorptive surface area threefold. (c) Microscopic fingerlike projection known as a *villus*. Collectively, the villi increase the surface area another tenfold. (d) Electron microscope view of a villus epithelial cell, depicting the presence of microvilli on its luminal border; the microvilli increase the surface area another 20-fold. Altogether, these surface modifications increase the small intestine's absorptive surface area 600-fold.

 For an animation of this figure, click the Digestion and Absorption tab (page 1) in the Gastrointestinal Physiology tutorial on the CD-ROM.

Malabsorption (impairment of absorption) may be caused by damage to or reduction of the surface area of the small intestine. One of the most common causes is **gluten enteropathy.** In this condition, the person's small intestine is abnormally sensitive to *gluten,* a protein constituent of wheat and some other grains. Through an unknown mechanism, exposure to gluten damages the intestinal villi: the normally luxuriant array of villi is reduced, the mucosa becomes flattened, and the brush border becomes short and stubby (● Figure 15-22). Because this loss of villi decreases the surface area available for absorption, absorption of all nutrients is impaired. The condition is treated by eliminating gluten from the diet.

STRUCTURE OF A VILLUS

Absorption across the digestive tract wall involves transepithelial transport similar to movement of material across the kidney tubules (see p. 415). Each villus has the following major components (● Figure 15-20c):

- *Epithelial cells that cover the surface of the villus.* The epithelial cells are joined at their lateral borders by tight junctions, which limit passage of luminal contents between the cells, although the tight junctions in the small intestine are leakier than those in the stomach. Within their luminal brush borders, these epithelial cells have carriers for absorption of specific nutrients and electrolytes from the lumen as well as the intracellular digestive enzymes that complete carbohydrate and protein digestion.
- *A connective tissue core.*
- *A capillary network.* Each villus is supplied by an arteriole that breaks up into a capillary network within the villus core. The capillaries rejoin to form a venule that drains away from the villus.
- *A terminal lymphatic vessel.* Each villus is supplied by a single blind-ended lymphatic vessel known as the **central lacteal,** which occupies the center of the villus core.

During the process of absorption, digested substances enter the capillary network or the central lacteal. To be absorbed, a substance must pass completely through the epi-

(a)

Circular fold

Villus

(b)

Epithelial cell

Mucous cell

Capillaries

Central lacteal

(c)

Crypt of Lieberkühn

Arteriole

Venule

Lymphatic vessel

Microvilli

(d)

● **FIGURE 15-21**

Scanning electron micrograph of villi projecting from the small-intestine mucosa

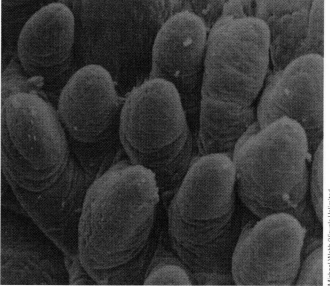

Normal

Brush border

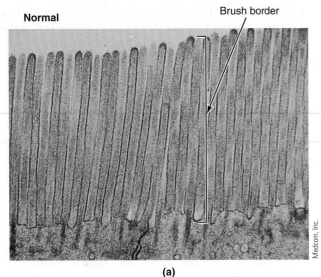

(a)

Gluten enteropathy

Brush border

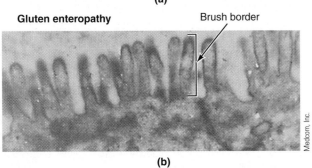

(b)

Medcom, Inc.

● **FIGURE 15-22**

Reduction in the brush border with gluten enteropathy. (a) Electron micrograph of the brush border of a small-intestine epithelial cell in a normal individual. (b) Electron micrograph of the short, stubby brush border of a small-intestine epithelial cell in a patient with gluten enteropathy.

thelial cell, diffuse through the interstitial fluid within the connective tissue core of the villus, and then cross the wall of a capillary or lymph vessel. Like renal transport, intestinal absorption may be active or passive, with active absorption involving energy expenditure during at least one of the trans-epithelial transport steps.

▮ The mucosal lining experiences rapid turnover.

Dipping down into the mucosal surface between the villi are shallow invaginations known as the **crypts of Lieberkühn** (● Figure 15-20c). Unlike the gastric pits, these intestinal crypts do not secrete digestive enzymes, but they do secrete water and electrolytes, which, along with the mucus secreted by the cells on the villus surface, constitute the intestinal juice.

Furthermore, the crypts function as nurseries. The epithelial cells lining the small intestine slough off and are replaced at a rapid rate as a result of high mitotic activity of *stem cells* in the crypts. New cells that are continually being produced in the crypts migrate up the villi and, in the process, push off the older cells at the tips of the villi into the lumen. In this manner, more than 100 million intestinal cells are shed

per minute. The entire trip from crypt to tip averages about three days, so the epithelial lining of the small intestine is replaced approximately every three days. Because of this high rate of cell division, the crypt stem cells are very sensitive to damage by radiation and anticancer drugs, both of which may inhibit cell division.

The new cells undergo several changes as they migrate up the villus. The concentration of brush border enzymes increases and the capacity for absorption improves, so the cells at the tip of the villus have the greatest digestive and absorptive capability. Just at their peak, these cells are pushed off by the newly migrating cells. Thus the luminal contents are constantly exposed to cells that are optimally equipped to complete the digestive and absorptive functions efficiently. Furthermore, just as in the stomach, the rapid turnover of cells in the small intestine is essential because of the harsh luminal conditions. Cells exposed to the abrasive and corrosive luminal contents are easily damaged and cannot live for long, so they must be continually replaced by a fresh supply of newborn cells.

The old cells sloughed off into the lumen are not entirely lost to the body. These cells are digested, with the cell constituents being absorbed into the blood and reclaimed for synthesis of new cells, among other things.

In addition to the stem cells, *defensive cells* are also found in the crypts. These cells safeguard the stem cells by producing lysozyme and other chemicals that thwart bacteria.

We now turn our attention to the mechanisms through which the specific dietary constituents are normally absorbed.

▮ Special mechanisms facilitate absorption of most nutrients.

The epithelial lining of the small intestine is specialized to accomplish absorption of luminal contents.

SALT AND WATER ABSORPTION

Sodium may be absorbed both passively and actively. When the electrochemical gradient favors movement of Na^+ from the lumen to the blood, passive diffusion of Na^+ can occur *between* the intestinal epithelial cells through the leaky tight junctions into the interstitial fluid within the villus. Movement of Na^+ *through* the cells is energy dependent, being driven by the Na^+–K^+ ATPase pumps located at the cells' basolateral borders, similar to the process of Na^+ reabsorption across the kidney tubules (see pp. 416 and 419).

As with the renal tubules in the early part of the nephron, the absorption of Cl^-, H_2O, glucose, and amino acids from the small intestine is linked to this energy-dependent Na^+ absorption. Chloride passively follows down the electrical gradient created by Na^+ absorption and can be actively absorbed as well if needed. Water is reabsorbed passively down the osmotic gradient produced by active reabsorption of Na^+.

CARBOHYDRATE ABSORPTION

Absorption of the digestion end products of both carbohydrates and proteins involves special carrier-mediated transport systems that require energy expenditure and Na^+ cotransport, and both categories of end products are absorbed into the blood.

Dietary carbohydrate is presented to the small intestine for absorption mainly in the forms of the disaccharides maltose (the product of polysaccharide digestion), sucrose, and lactose. The disaccharidases located in the brush borders of the small intestine cells further reduce these disaccharides into the absorbable monosaccharide units of glucose, galactose, and fructose.

Glucose and galactose are both absorbed by secondary active transport, in which cotransport carriers on the luminal border transport both the monosaccharide and Na^+ from the lumen into the interior of the intestinal cell. The operation of these cotransport carriers, which do not directly use energy themselves, depends on the Na^+ concentration gradient established by the energy-consuming basolateral Na^+–K^+ pump (see p. 58). Glucose (or galactose), having been concentrated in the cell by the cotransport carriers, leaves the cell down its concentration gradient by means of a passive carrier in the basolateral border to enter the blood within the villus. Fructose is absorbed into the blood solely by facilitated diffusion (passive carrier-mediated transport; see p. 57).

PROTEIN ABSORPTION

Both ingested proteins and endogenous (within the body) proteins that have entered the digestive tract lumen from the three following sources are digested and absorbed:

1. Digestive enzymes, all of which are proteins, that have been secreted into the lumen
2. Proteins within the cells that are pushed off from the villi into the lumen during the process of mucosal turnover
3. Small amounts of plasma proteins that normally leak from the capillaries into the digestive tract lumen

About 20 to 40 g of endogenous protein enter the lumen each day from these three sources. This quantity can amount to more than the quantity of protein in ingested food. All endogenous proteins must be digested and absorbed along with the dietary proteins to prevent depletion of the body's protein stores. The amino acids absorbed from both food and endogenous protein are used primarily to synthesize new protein in the body.

The protein presented to the small intestine for absorption is primarily in the form of amino acids, which are absorbed across the intestinal cells by secondary active transport, similar to glucose and galactose absorption. Thus glucose, galactose, and amino acids all get a "free ride" in on the energy expended for Na^+ transport. Small peptides are broken down into their constituent amino acids by the aminopeptidases in the brush borders. Like monosaccharides, amino acids enter the capillary network within the villus.

FAT ABSORPTION

Fat absorption is quite different from carbohydrate and protein absorption, because the insolubility of fat in water presents a special problem. Fat must be transferred from the watery chyme through the watery body fluids, even though fat is not water soluble. Therefore, fat must undergo a series of physical and chemical transformations to circumvent this problem during its digestion and absorption (● Figure 15-23).

When the stomach contents are emptied into the duodenum, the ingested fat is aggregated into large, oily triglyceride droplets that float in the chyme. Recall that through the bile salts' detergent action in the small intestine, the large droplets are dispersed into a lipid emulsification of small droplets, exposing a much greater surface area of fat for digestion by pancreatic lipase. The products of lipase digestion (monoglycerides and free fatty acids) are also not very water soluble, so very little of these end products of fat digestion can diffuse through the aqueous chyme to reach the absorptive lining. However, biliary components facilitate absorption of these fatty end products by forming micelles.

Remember that micelles are water-soluble particles that can carry the end products of fat digestion within their lipid-soluble interiors. Once these micelles reach the luminal membranes of the epithelial cells, the monoglycerides and free fatty acids passively diffuse from the micelles through the lipid component of the epithelial cell membranes to enter the interior of these cells. As these fat products leave the micelles and are absorbed across the epithelial cell membranes, the micelles can pick up more monoglycerides and free fatty acids, which have been produced from digestion of other triglyceride molecules in the fat emulsion.

Bile salts continuously repeat their fat-solubilizing function down the length of the small intestine until all fat is absorbed. Then the bile salts themselves are reabsorbed in the terminal ileum by special active transport. This is an efficient process, because relatively small amounts of bile salts can facilitate digestion and absorption of large amounts of fat, with each bile salt performing its ferrying function repeatedly before it is reabsorbed.

Once within the interior of the epithelial cells, the monoglycerides and free fatty acids are resynthesized into triglycerides. These triglycerides conglomerate into droplets and are coated with a layer of lipoprotein (synthesized by the endoplasmic reticulum of the epithelial cell), which makes the fat droplets water soluble. The large, coated fat droplets, known as **chylomicrons**, are extruded by exocytosis from the epithelial cells into the interstitial fluid within the villus. The chylomicrons subsequently enter the central lacteals rather than the capillaries because of the structural differences between these two vessels. Capillaries have a basement membrane (an outer layer of polysaccharides) that prevents chylomicrons from entering, but the lymph vessels do not have this barrier. Thus fat can be absorbed into the lymphatics but not directly into the blood.

The actual absorption or transfer of monoglycerides and free fatty acids from the chyme across the luminal membranes of the intestinal epithelial cells is a passive process, because the lipid-soluble fatty end products merely dissolve in and pass through the lipid part of the membrane. However, the overall sequence of events needed for fat absorption does require

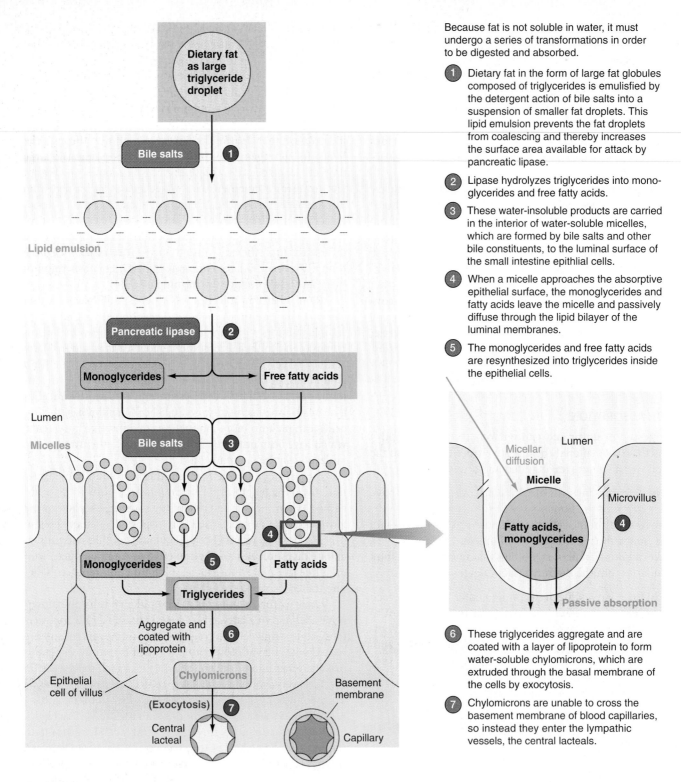

Because fat is not soluble in water, it must undergo a series of transformations in order to be digested and absorbed.

1. Dietary fat in the form of large fat globules composed of triglycerides is emulsified by the detergent action of bile salts into a suspension of smaller fat droplets. This lipid emulsion prevents the fat droplets from coalescing and thereby increases the surface area available for attack by pancreatic lipase.

2. Lipase hydrolyzes triglycerides into monoglycerides and free fatty acids.

3. These water-insoluble products are carried in the interior of water-soluble micelles, which are formed by bile salts and other bile constituents, to the luminal surface of the small intestine epithlial cells.

4. When a micelle approaches the absorptive epithelial surface, the monoglycerides and fatty acids leave the micelle and passively diffuse through the lipid bilayer of the luminal membranes.

5. The monoglycerides and free fatty acids are resynthesized into triglycerides inside the epithelial cells.

6. These triglycerides aggregate and are coated with a layer of lipoprotein to form water-soluble chylomicrons, which are extruded through the basal membrane of the cells by exocytosis.

7. Chylomicrons are unable to cross the basement membrane of blood capillaries, so instead they enter the lympathic vessels, the central lacteals.

● FIGURE 15-23

Fat digestion and absorption

 For an animation of this figure, click the Digestion and Absorption tab (page 2) in the Gastrointestinal Physiology tutorial on the CD-ROM.

energy. For example, bile salts are actively secreted by the liver, and the resynthesis of triglycerides and formation of chylomicrons within the epithelial cells are active processes.

VITAMIN ABSORPTION

Water-soluble vitamins are primarily absorbed passively with water, whereas fat-soluble vitamins are carried in the micelles and absorbed passively with the end products of fat digestion. Some of the vitamins can also be absorbed by carriers, if necessary. Vitamin B_{12} is unique in that it must be in combination with gastric intrinsic factor for absorption by receptor-mediated endocytosis in the terminal ileum.

IRON AND CALCIUM ABSORPTION

In contrast to the almost complete, unregulated absorption of other ingested electrolytes, dietary iron and calcium may not be absorbed completely because their absorption is subject to regulation, depending on the body's needs for these electrolytes. Normally only enough iron and calcium are actively absorbed into the blood to maintain the homeostasis of these electrolytes, with excess ingested quantities being lost in the feces.

❚ Most absorbed nutrients immediately pass through the liver for processing.

The venules that leave the small-intestine villi, along with those from the rest of the digestive tract, empty into the hepatic portal vein, which carries the blood to the liver. Consequently, anything absorbed into the digestive capillaries first must pass through the hepatic biochemical factory before entering the general circulation. Thus the products of carbohydrate and protein digestion are channeled into the liver, where many of these energy-rich products are subjected to immediate metabolic processing. Furthermore, harmful substances that may have been absorbed are detoxified by the liver before gaining access to the general circulation. After passing through the portal circulation, the venous blood from the digestive system empties into the vena cava and returns to the heart to be distributed throughout the body, carrying glucose and amino acids for use by the tissues.

Fat, which cannot penetrate the intestinal capillaries, is picked up by the central lacteal and enters the lymphatic system instead, bypassing the hepatic portal system. Contractions of the villi periodically compress the central lacteal and "milk" the lymph out of this vessel. The lymph vessels eventually converge to form the *thoracic duct*, a large lymph vessel that empties into the venous system within the chest. In this way fat ultimately gains access to the blood. The absorbed fat is carried by the systemic circulation to the liver and to other tissues of the body. Therefore, the liver does have a chance to act on the digested fat, but not until the fat has been diluted by the blood in the general circulatory system. This dilution of fat protects the liver from being inundated with more fat than it can handle at one time.

❚ Extensive absorption by the small intestine keeps pace with secretion.

The small intestine normally absorbs about 9 liters of fluid per day in the form of H_2O and solutes, including the absorbable units of nutrients, vitamins, and electrolytes. How can that be, when humans normally ingest only about 1250 ml of fluid and consume 1250 g of solid food (80% of which is

H_2O) per day (see p. 451)? ▲ Table 15-7 illustrates the tremendous daily absorption performed by the small intestine. Each day about 9500 ml of H_2O and solutes enter the small intestine. Note that of this 9500 ml, only 2500 ml are ingested from the external environment. The remaining 7000 ml (7 liters) of fluid are digestive juices derived from the plasma. Recall that plasma is the ultimate source of digestive secretions, because the secretory cells extract from the plasma the necessary raw materials for their secretory product. Considering that the entire plasma volume is only about 2.75 liters, absorption must closely parallel secretion to keep the plasma volume from falling sharply.

Of the 9500 ml of fluid entering the small-intestine lumen per day, about 95%, or 9000 ml of fluid, is normally absorbed by the small intestine back into the plasma, with only 500 ml of the small-intestine contents passing on into the colon. Thus the body does not lose the digestive juices. After the constituents of the juices are secreted into the digestive tract lumen and perform their function, they are returned to the plasma. The only secretory product that escapes from the body is bilirubin, a waste product that must be eliminated.

❚ Diarrhea results in loss of fluid and electrolytes.

 Clinical Note **Diarrhea** is characterized by passage of a highly fluid fecal matter, often with increased frequency of defecation. The abnormal fluidity of the feces usually occurs because the small intestine is unable to absorb fluid

▲ **TABLE 15-7**

Volumes Absorbed by the Small and Large Intestine per Day

Volume entering small intestine per day				
Sources	Ingested		Food eaten	1,250 g*
			Fluid drunk	1,250 ml
	Secreted from the plasma		Saliva	1,500 ml
			Gastric juice	2,000 ml
			Pancreatic juice	1,500 mil
			Bile	500 ml
			Intestinal juice	1,500 ml
				9,500 ml
Volume absorbed by small intestine per day				9,000 ml
Volume entering colon from small intestine per day				500 ml
Volume absorbed by colon per day				350 ml
Volume of feces eliminated from colon per day				150 g*

*One milliliter of H_2O weighs 1 g. Therefore, because a high percentage of food and feces is H_2O, we can roughly equate grams of food or feces with milliliters of fluid.

as extensively as normal. This extra unabsorbed fluid passes out in the feces. The most common cause of diarrhea is excessive small intestinal motility, which arises either from local irritation of the gut wall by bacterial or viral infection of the small intestine or from emotional stress. Rapid transit of the small-intestine contents does not allow enough time for adequate absorption of fluid. Not only are some of the ingested materials lost, but some of the secreted materials that normally would have been reabsorbed are lost as well. Excessive loss of intestinal contents causes dehydration, loss of unabsorbed nutrients, and metabolic acidosis resulting from the loss of HCO_3^- (see p. 460).

LARGE INTESTINE

The **large intestine** consists of the colon, cecum, appendix, and rectum (● Figure 15-24). The **cecum** forms a blind-ended pouch below the junction of the small and large intestines at the ileocecal valve. The small, finger-like projection at the bottom of the cecum is the **appendix,** a lymphoid tissue that houses lymphocytes (see p. 327). The **colon,** which makes up most of the large intestine, is not coiled like the small intestine but consists of three relatively straight parts—the *ascending colon,* the *transverse colon,* and the *descending colon.* The end part of the descending colon becomes S-shaped, forming the *sigmoid colon* (*sigmoid* means "S-shaped"), then straightens out to form the **rectum** (*rectum* means "straight").

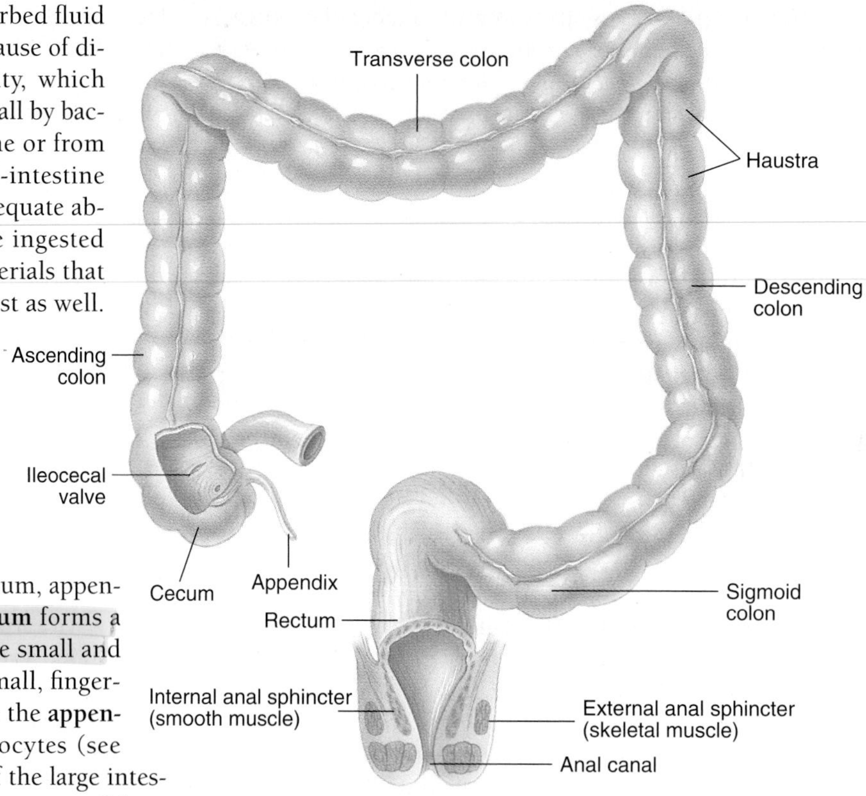

▮ The large intestine is primarily a drying and storage organ.

The colon normally receives about 500 ml of chyme from the small intestine each day. Because most digestion and absorption have been accomplished in the small intestine, the contents delivered to the colon consist of indigestible food residues (such as cellulose), unabsorbed biliary components, and the remaining fluid. The colon extracts more H_2O and salt from the contents. What remains to be eliminated is known as **feces.** The primary function of the large intestine is to store feces before defecation. Cellulose and other indigestible substances in the diet provide bulk and help maintain regular bowel movements by contributing to the volume of the colonic contents.

▮ Haustral contractions slowly shuffle the colonic contents back and forth.

Most of the time, movements of the large intestine are slow and nonpropulsive, as is appropriate for its absorptive and storage functions. The colon's main motility is **haustral contractions** initiated by the autonomous rhythmicity of colonic smooth muscle cells. These contractions, which throw the large intestine into pouches or sacs called **haustra,** are similar to small-intestine segmentations but occur much less fre-

● FIGURE 15-24

Anatomy of the large intestine

quently. Thirty minutes may elapse between haustral contractions, whereas segmentation contractions in the small intestine occur at rates of between 9 and 12 per minute. The location of the haustral sacs gradually changes as a relaxed segment that has formed a sac slowly contracts while a previously contracted area simultaneously relaxes to form a new sac. These movements are nonpropulsive; they slowly shuffle the contents in a back-and-forth mixing movement that exposes the colonic contents to the absorptive mucosa. Haustral contractions are largely controlled by locally mediated reflexes involving the intrinsic plexuses.

▮ Mass movements propel feces long distances.

Three to four times a day, generally after meals, a marked increased in motility takes place during which large segments of the ascending and transverse colon contract simultaneously, driving the feces one third to three fourths of the length of the colon in a few seconds. These massive contractions, appropriately called **mass movements,** drive the colonic contents into the distal part of the large intestine, where material is stored until defecation occurs.

When food enters the stomach, mass movements are triggered in the colon primarily by the **gastrocolic reflex,** which is mediated from the stomach to the colon by gastrin and by the extrinsic autonomic nerves. In many people, this reflex is most evident after the first meal of the day and is often followed by the urge to defecate. Thus when a new meal enters the digestive tract, reflexes are initiated to move the existing contents farther along down the tract to make way for the incoming food. The gastroileal reflex moves the remaining small-intestine contents into the large intestine, and the gastrocolic reflex pushes the colonic contents into the rectum, triggering the defecation reflex.

Feces are eliminated by the defecation reflex.

When mass movements of the colon move feces into the rectum, the resultant distension of the rectum stimulates stretch receptors in the rectal wall, initiating the **defecation reflex**. This reflex causes the **internal anal sphincter** (which is smooth muscle) to relax and the rectum and sigmoid colon to contract more vigorously. If the **external anal sphincter** (which is skeletal muscle) is also relaxed, defecation occurs. Being skeletal muscle, the external anal sphincter is under voluntary control. The initial distension of the rectal wall is accompanied by the conscious urge to defecate. If circumstances are unfavorable for defecation, voluntary tightening of the external anal sphincter can prevent defecation despite the defecation reflex. If defecation is delayed, the distended rectal wall gradually relaxes, and the urge to defecate subsides until the next mass movement propels more feces into the rectum, once again distending the rectum and triggering the defecation reflex. During periods of inactivity, both anal sphincters remain contracted to ensure fecal continence.

When defecation does occur, it is usually assisted by voluntary straining movements that involve simultaneous contraction of the abdominal muscles and a forcible expiration against a closed glottis. This maneuver greatly increases intraabdominal pressure, which helps expel the feces.

Constipation occurs when the feces become too dry.

If defecation is delayed too long, **constipation** may result. When colonic contents are retained for longer periods of time than normal, more than the usual amount of H_2O is absorbed from the feces, so they become hard and dry. Normal variations in frequency of defecation among individuals range from after every meal to up to once a week. When the frequency is delayed beyond what is normal for a particular person, constipation and its attendant symptoms may occur. These symptoms include abdominal discomfort, dull headache, loss of appetite sometimes accompanied by nausea, and mental depression. Contrary to popular belief, these symptoms are not caused by toxins absorbed from the retained fecal material. Although bacterial metabolism produces some potentially toxic substances in the colon, these substances normally pass through the portal system and are removed by the liver before they can reach the systemic circulation. Instead, the symptoms associated with constipation are caused by prolonged distension of the large intestine, particularly the rectum; the symptoms promptly disappear after relief from distension.

Possible causes for delayed defecation that might lead to constipation include (1) ignoring the urge to defecate; (2) decreased colon motility accompanying aging, emotion, or a low-bulk diet; (3) obstruction of fecal movement in the large bowel caused by a local tumor or colonic spasm; and (4) impairment of the defecation reflex, such as through injury of the nerve pathways involved.

Clinical Note If hardened fecal material becomes lodged in the appendix, it may obstruct normal circulation and mucus secretion in this narrow, blind-ended appendage. This blockage leads to inflammation of the appendix, or **appendicitis**. The inflamed appendix often becomes swollen and filled with pus, and the tissue may die as a result of local circulatory interference. If not surgically removed, the diseased appendix may rupture, spewing its infectious contents into the abdominal cavity.

Large-intestine secretion is entirely protective.

The large intestine does not secrete any digestive enzymes. None are needed, because digestion is completed before chyme ever reaches the colon. Colonic secretion consists of an alkaline ($NaHCO_3$) mucus solution, whose function is to protect the large-intestine mucosa from mechanical and chemical injury. The mucus provides lubrication to facilitate passage of the feces, whereas the $NaHCO_3$ neutralizes irritating acids produced by local bacterial fermentation.

No digestion takes place within the large intestine because there are no digestive enzymes. However, the colonic bacteria do digest some of the cellulose and use it for their own metabolism.

The colon contains myriads of beneficial bacteria.

Because of slow colonic movement, bacteria have time to grow and accumulate in the large intestine. In contrast, in the small intestine the contents are normally moved through too rapidly for bacterial growth to occur. Furthermore, the mouth, stomach, and small intestine secrete antibacterial agents, but the colon does not. Not all ingested bacteria are destroyed by lysozyme and HCl, however. The surviving bacteria continue to thrive in the large intestine. About 10 times more bacteria live in the human colon than the human body has cells. An estimated 500 to 1000 different species of bacteria typically live in the colon. These colonic micro-organisms not only are typically harmless but in fact they provide beneficial functions. For example, indigenous bacteria (1) make nutritional contributions, such as synthesizing vitamin K and raising colonic acidity, thereby promoting the absorption of calcium, magnesium, and zinc; (2) enhance intestinal immunity by competing with potentially pathogenic microbes for nutrients and space (see p. 357); (3) promote colonic motility; and (4) help maintain colonic mucosal integrity.

The large intestine absorbs salt and water, converting the luminal contents into feces.

Some absorption takes place within the colon, but not to the same extent as in the small intestine. Because the luminal surface of the colon is fairly smooth, it has considerably less absorptive surface area than the small intestine. Furthermore, the colon does not have the specialized transport mechanisms found in the small intestine for absorbing glucose or amino acids. When excessive small intestine motility delivers the contents to the colon before absorption of nutrients has been completed, the colon cannot absorb these materials, and they are lost in diarrhea.

The colon normally absorbs salt and H_2O. Sodium is actively absorbed, Cl^- follows passively down the electrical gradient, and H_2O follows osmotically. The colon absorbs token amounts of other electrolytes as well as vitamin K synthesized by colonic bacteria.

Through absorption of salt and H_2O, a firm fecal mass is formed. Of the 500 ml of material entering the colon per day from the small intestine, the colon normally absorbs about 350 ml, leaving 150 g of feces to be eliminated from the body each day (see ▲ Table 15-6). This fecal material normally consists of 100 g of H_2O and 50 g of solid, including undigested cellulose, bilirubin, bacteria, and small amounts of salt. Thus, contrary to popular thinking, the digestive tract is not a major excretory passageway for eliminating wastes from the body. The main waste product excreted in the feces is bilirubin. The other fecal constituents are unabsorbed food residues and bacteria, which were never actually a part of the body.

▌ Intestinal gases are absorbed or expelled.

Occasionally, instead of feces passing from the anus, intestinal gas, or **flatus**, passes out. Most gas in the colon is due to bacterial activity, with the quantity and nature of the gas depending on the type of food eaten and the characteristics of the colonic bacteria. Some foods, such as beans, contain types of carbohydrates that humans cannot digest but that can be attacked by gas-producing bacteria. Much of the gas is absorbed through the intestinal mucosa. The rest is expelled through the anus.

To selectively expel gas when feces are also present in the rectum, the person voluntarily contracts the abdominal muscles and external anal sphincter at the same time. When abdominal contraction raises the pressure against the contracted anal sphincter sufficiently, the pressure gradient forces air out at a high velocity through a slitlike anal opening that is too narrow for solid feces to escape through. This passage of air at high velocity causes the edges of the anal opening to vibrate, giving rise to the characteristic low-pitched sound accompanying passage of gas.

 Click on the Media Exercises menu of the CD-ROM and work Media Exercise 15.3: The Intestine and Associated Organs to test your understanding of the previous sections.

OVERVIEW OF THE GASTROINTESTINAL HORMONES

Throughout our discussion of digestion, we have repeatedly mentioned different functions of the three major gastrointestinal hormones: gastrin, secretin, and CCK. Let's now fit all of these functions together so you can appreciate the overall adaptive importance of these interactions. We will also introduce a more recently identified hormone, GIP, the fourth hormone to be confirmed as a gastrointestinal hormone.

GASTRIN

Protein in the stomach stimulates the release of gastrin, which performs the following functions:

1. It acts in multiple ways to increase secretion of HCl and pepsinogen. These two substances in turn are of primary importance in initiating digestion of the protein that promoted their secretion.
2. It enhances gastric motility, stimulates ileal motility, relaxes the ileocecal sphincter, and induces mass movements in the colon—functions that are all aimed at keeping the contents moving through the tract on the arrival of a new meal.

Predictably, gastrin secretion is inhibited by an accumulation of acid in the stomach and by the presence in the duodenal lumen of acid and other constituents that necessitate a delay in gastric secretion.

SECRETIN

As the stomach empties into the duodenum, the presence of acid in the duodenum stimulates the release of secretin, which performs the following interrelated functions:

1. It inhibits gastric emptying to prevent further acid from entering the duodenum until the acid already present is neutralized.
2. It inhibits gastric secretion to reduce the amount of acid being produced.
3. It stimulates the pancreatic duct cells to produce a large volume of aqueous $NaHCO_3$ secretion, which is emptied into the duodenum to neutralize the acid.
4. It stimulates secretion by the liver of a $NaHCO_3$-rich bile, which likewise is emptied into the duodenum to assist in the neutralization process.

Neutralization of the acidic chyme in the duodenum helps prevent damage to the duodenal walls and provides a suitable environment for the optimal functioning of the pancreatic digestive enzymes, which are inhibited by acid.

CCK

As chyme empties from the stomach, fat and other nutrients enter the duodenum. These nutrients, especially fat and to a lesser extent protein products, cause the release of CCK, which performs the following interrelated functions:

1. It inhibits gastric motility and secretion, thereby allowing adequate time for the nutrients already in the duodenum to be digested and absorbed.
2. It stimulates the pancreatic acinar cells to increase secretion of pancreatic enzymes, which continue the digestion of these nutrients in the duodenum (this action is especially important for fat digestion, because pancreatic lipase is the only enzyme that digests fat).

3. It causes contraction of the gallbladder and relaxation of the sphincter of Oddi so that bile is emptied into the duodenum to aid fat digestion and absorption. Bile salts' detergent action is particularly important in enabling pancreatic lipase to perform its digestive task. Once again, the multiple effects of CCK are remarkably well adapted to dealing with the fat and other nutrients whose presence in the duodenum triggered this hormone's release.

4. Besides facilitating the digestion of ingested nutrients, CCK is an important regulator of food intake. It plays a key role in satiety, the sensation of having had enough to eat (see p. 515).

GIP

A more recently recognized hormone released by the duodenum, **glucose-dependent insulinotropic peptide** or **GIP**, helps promote metabolic processing of the nutrients once they are absorbed. Specifically, this hormone stimulates insulin release by the pancreas. Again, this is remarkably adaptive. As soon as the meal is absorbed, the body has to shift its metabolic gears to use and store the newly arriving nutrients. The metabolic activities of this absorptive phase are largely under the control of insulin (see p. 563 and p. 566). Stimulated by the presence of a meal in the digestive tract, GIP initiates the release of insulin in anticipation of absorption of the meal, in a feedforward fashion. Insulin is especially important in promoting the uptake and storage of glucose. Appropriately, glucose in the duodenum has recently been shown to increase GIP secretion.

This overview of the multiple, integrated, adaptive functions of the gastrointestinal hormones provides an excellent example of the remarkable efficiency of the human body.

CHAPTER IN PERSPECTIVE: FOCUS ON HOMEOSTASIS

To maintain constancy in the internal environment, materials that are used up in the body (such as nutrients and O_2) or uncontrollably lost from the body (such as evaporative H_2O loss from the airways or salt loss in sweat) must constantly be replaced by new supplies of these materials from the external environment. All these replacement supplies except O_2 are acquired through the digestive system. Fresh supplies of O_2 are transferred to the internal environment by the respiratory system, but all the nutrients, H_2O, and various electrolytes needed to maintain homeostasis are acquired through the digestive system. The large, complex food that is ingested is broken down by the digestive system into small absorbable units. These small energy-rich nutrient molecules are transferred across the small-intestine epithelium into the blood for delivery to the cells to replace the nutrients constantly used for ATP production and for repair and growth of body tissues. Likewise, ingested H_2O, salt, and other electrolytes are absorbed by the intestine into the blood.

Unlike most body systems, regulation of digestive system activities is not aimed at maintaining homeostasis. The quantity of nutrients and H_2O ingested is subject to control, but the quantity of ingested materials absorbed by the digestive tract is not subject to control, with few exceptions. The hunger mechanism governs food intake to help maintain energy balance (Chapter 16), and the thirst mechanism controls H_2O intake to help maintain H_2O balance (Chapter 14). However, we often do not heed these control mechanisms, and often eat and drink even when we are not hungry or thirsty. Once these materials are in the digestive tract, the digestive system does not vary its rate of nutrient, H_2O, or electrolyte uptake according to body needs (with the exception of iron and calcium); rather, it optimizes conditions for digesting and absorbing what is ingested. Truly, what you eat is what you get. The digestive system is subject to many regulatory processes, but these are not influenced by the nutritional or hydration state of the body. Instead, these control mechanisms are governed by the composition and volume of digestive tract contents so that the rate of motility and secretion of digestive juices are optimal for digestion and absorption of the ingested food.

If excess nutrients are ingested and absorbed, the extra is placed in storage, such as in adipose tissue (fat), so that the blood level of nutrient molecules is kept at a constant level. Excess ingested H_2O and electrolytes are eliminated in the urine to homeostatically maintain the blood levels of these constituents.

CHAPTER SUMMARY

Introduction (pp. 465–471)

▮ The four basic digestive processes are motility, secretion, digestion, and absorption.

▮ The three classes of energy-rich nutrients are digested into absorbable units as follows: (1) Dietary carbohydrates in the form of the polysaccharides starch and glycogen are digested into their absorbable units of monosaccharides, especially glucose. *(Review Figure 15-1.)* (2) Dietary proteins are digested into their absorbable units of amino acids. (3) Dietary fats in the form of triglycerides are digested into their absorbable units of monoglycerides and free fatty acids.

▮ The digestive system consists of the digestive tract and accessory digestive organs (salivary glands, exocrine pancreas, and biliary system.) *(Review Table 15-1.)*

▮ The digestive tract is a continuous tube that runs from the mouth to the anus, with local modifications that reflect regional specializations for carrying out digestive functions.

▮ The lumen of the digestive tract is continuous with the external environment, so its contents are technically outside the body; this arrangement permits digestion of food without self-digestion occurring in the process.

▮ The digestive tract wall has four layers throughout most of its length. From innermost outward, they are the mucosa, submucosa, muscularis externa, and serosa. *(Review Figure 15-2.)*

▮ Digestive activities are carefully regulated by synergistic autonomous, neural (both intrinsic and extrinsic), and hormonal mechanisms to ensure that the ingested food is maximally made available to the body for energy production and as synthetic raw materials. *(Review Figure 15-3.)*

Mouth (pp. 471–473)

▪ **Motility:** Food enters the digestive system through the mouth, where it is chewed and mixed with saliva to facilitate swallowing.

▪ **Secretion:** The salivary enzyme, amylase, begins the digestion of carbohydrates. More important than its minor digestive function, saliva is essential for articulate speech and plays an important role in dental health. Salivary secretion is controlled by a salivary center in the medulla, mediated by autonomic innervation of the salivary glands. *(Review Figure 15-4.)*

▪ **Digestion:** Salivary amylase begins to digest polysaccharides into the disaccharide maltose, a process that continues in the stomach after the food has been swallowed until amylase is eventually inactivated by the acidic gastric juice. *(Review Table 15-6, p. 496.)*

▪ **Absorption:** No absorption of nutrients occurs from the mouth.

Pharynx and Esophagus (pp. 473–475)

▪ **Motility:** Following chewing, the tongue propels the bolus of food to the rear of the throat, which initiates the swallowing reflex. The swallowing center in the medulla coordinates a complex group of activities that result in closure of the respiratory passages and propulsion of food through the pharynx and esophagus into the stomach. *(Review Figures 15-5 and 15-6.)*

▪ **Secretion:** The esophageal secretion, mucus, is protective in nature.

▪ **Digestion and absorption:** No nutrient digestion or absorption occurs in the pharynx or esophagus.

Stomach (pp. 475–486)

▪ The stomach, a saclike structure located between the esophagus and small intestine, stores ingested food for variable periods of time until the small intestine is ready to process it further for final absorption. *(Review Figure 15-7.)*

▪ **Motility:** The four aspects of gastric motility are gastric filling, storage, mixing, and emptying.

▪ Gastric filling is facilitated by vagally mediated receptive relaxation of the stomach muscles.

▪ Gastric storage takes place in the body of the stomach, where peristaltic contractions of the thin muscle walls are too weak to mix the contents.

▪ Gastric mixing in the thick-muscled antrum results from vigorous peristaltic contractions. *(Review Figure 15-8.)*

▪ Gastric emptying is influenced by the following factors in the stomach and duodenum. (1) The volume and fluidity of chyme in the stomach tend to promote emptying of the stomach contents. (2) The duodenal factors, which are the dominant factors controlling gastric emptying, tend to delay gastric emptying until the duodenum is ready to receive and process more chyme. The specific factors in the duodenum that delay gastric emptying are fat, acid, hypertonicity, and distension. They delay gastric emptying by inhibiting stomach peristaltic activity by the enterogastric reflex and the enterogastrones, secretin and cholecystokinin, which are secreted by the duodenal mucosa. *(Review Table 15-2.)*

▪ **Secretion:** Gastric secretions into the stomach lumen include (1) HCl (from the parietal cells), which activates pepsinogen, denatures protein, and kills bacteria; (2) pepsinogen (from the chief cells), which, once activated, initiates protein digestion; (3) mucus (from the mucous cells), which provides a protective coating to supplement the gastric mucosal barrier, enabling the stomach

to contain the harsh luminal contents without self-digestion; and (4) intrinsic factor (from the parietal cells), which plays a vital role in vitamin B_{12} absorption, a constituent essential for normal red blood cell production. *(Review Table 15-3 and Figures 15-9 and 15-10.)*

▪ The stomach also secretes the following endocrine and paracrine regulatory factors: (1) the hormone gastrin (from the G cells), which plays a dominant role in stimulating gastric secretion; (2) the paracrine histamine (from the ECL cells), a potent stimulant of acid secretion by the parietal cells; and (3) the paracrine somatostatin (from the D cells), which inhibits gastric secretion. *(Review Table 15-3.)*

▪ Gastric secretion is under complex control mechanisms. Gastric secretion is increased during the cephalic and gastric phases of gastric secretion before and during a meal by mechanisms involving excitatory vagal and intrinsic nerve responses along with the stimulatory actions of gastrin and histamine. After the meal empties from the stomach, gastric secretion is reduced by the withdrawal of stimulatory factors, the release of inhibitory somatostatin, and the inhibitory actions of the enterogastric reflex and enterogastrones during the intestinal phase of gastric secretion. *(Review Tables 15-4 and 15-5.)*

▪ **Digestion:** Carbohydrate digestion continues in the body of the stomach under the influence of the swallowed salivary amylase. Protein digestion is initiated by pepsin in the antrum of the stomach, where vigorous peristaltic contractions mix the food with gastric secretions, converting it to a thick liquid mixture known as *chyme. (Review Table 15-6, p. 496.)*

▪ **Absorption:** No nutrients are absorbed from the stomach.

Pancreatic and Biliary Secretions (pp. 486–493)

▪ Pancreatic exocrine secretions and bile from the liver both enter the duodenal lumen.

▪ Pancreatic secretions include (1) potent digestive enzymes from the acinar cells, which digest all three categories of foodstuff; and (2) an aqueous $NaHCO_3$ solution from the duct cells, which neutralizes the acidic contents emptied into the duodenum from the stomach. This neutralization is important to protect the duodenum from acid injury and to allow pancreatic enzymes, which are inactivated by acid, to perform their important digestive functions. *(Review Figure 15-11.)*

▪ The pancreatic digestive enzymes include (1) the proteolytic enzymes trypsinogen, chymotrypsinogen, and procarboxypeptidase, which are secreted in inactive form and are activated in the duodenal lumen on exposure to enterokinase and activated trypsin; (2) pancreatic amylase, which continues carbohydrate digestion; and (3) lipase, which accomplishes fat digestion. *(Review Table 15-6, p. 496.)*

▪ Pancreatic secretion is primarily under hormonal control, which matches composition of the pancreatic juice with the needs in the duodenal lumen. Secretin stimulates the pancreatic duct cells and cholecystokinin (CCK) stimulates the acinar cells. *(Review Figure 15-12.)*

▪ The liver, the body's largest and most important metabolic organ, performs many varied functions. Its contribution to digestion is the secretion of bile, which contains bile salts. *(Review Figure 15-14.)*

▪ Bile salts aid fat digestion through their detergent action and facilitate fat absorption by forming water-soluble micelles that can carry the water-insoluble products of fat digestion to their absorption site. *(Review Figures 15-16 and 15-17.)*

▪ Between meals, bile is stored and concentrated in the gallbladder, which is stimulated by cholecystokinin to contract and empty the bile into the duodenum during meal digestion. After participating in fat digestion and absorption, bile salts are reabsorbed and returned via the hepatic portal system to the liver, where they not only are resecreted but also act as a potent choleretic to stimulate the secretion of even more bile. *(Review Figures 15-13 and 15-15.)*

■ Bile also contains bilirubin, a derivative of degraded hemoglobin, which is the major excretory product in the feces.

Small Intestine (pp. 493–502)

■ The small intestine is the main site for digestion and absorption.

■ **Motility:** Segmentation, the small intestine's primary motility during digestion of a meal, thoroughly mixes the food with pancreatic, biliary, and small-intestine juices to facilitate digestion; it also exposes the products of digestion to the absorptive surfaces. *(Review Figure 15-18.)*

■ **Digestion:** The pancreatic enzymes continue carbohydrate and protein digestion in the small-intestine lumen. The small-intestine brush-border enzymes complete the digestion of carbohydrates and protein. Fat is digested entirely in the small-intestine lumen, by pancreatic lipase. *(Review Table 15-6, p. 496.)*

■ **Absorption:** The small-intestine lining is remarkably adapted to its digestive and absorptive function. Its folds bear a rich array of fingerlike projections, the villi, which have a multitude of even smaller hairlike protrusions, the microvilli. Altogether, these surface modifications tremendously increase the area available to house the membrane-bound enzymes and to accomplish both active and passive absorption. *(Review Figures 15-20 through 15-22.)* This impressive lining is replaced about every three days to ensure an optimally healthy and functional presence of epithelial cells despite harsh lumen conditions.

■ The energy-dependent process of Na^+ absorption provides the driving force for Cl^-, water, glucose, and amino acid absorption. All these absorbed products enter the blood.

■ Because they are not soluble in water, the products of fat digestion must undergo a series of transformations that enable them to be passively absorbed, eventually entering the lymph. *(Review Figure 15-23.)*

■ The small intestine absorbs almost everything presented to it, from ingested food to digestive secretions to sloughed epithe-

lial cells. Only a small amount of fluid and indigestible food residue passes on to the large intestine. *(Review Table 15-7.)*

Large Intestine (pp. 502–504) *(Review Figure 15-24.)*

■ The colon serves primarily to concentrate and store undigested food residues (fiber, the indigestible cellulose in plant walls) and bilirubin until they can be eliminated from the body as feces.

■ **Motility:** Haustral contractions slowly shuffle the colonic contents back and forth to mix and facilitate absorption of most of the remaining fluid and electrolytes. Mass movements several times a day, usually after meals, propel the feces long distances. Movement of feces into the rectum triggers the defecation reflex, which the person can voluntarily prevent by contracting the external anal sphincter if the time is inopportune for elimination.

■ **Secretion:** The alkaline mucus secretion of the large intestine is primarily protective in function.

■ **Digestion and absorption:** No secretion of digestive enzymes or absorption of nutrients takes place in the colon, all nutrient digestion and absorption having been completed in the small intestine. Absorption of some of the remaining salt and water converts the colonic contents into feces.

Overview of the Gastrointestinal Hormones (pp. 504–505)

■ The three major gastrointestinal hormones are gastrin from the stomach mucosa and secretin and cholecystokinin from the duodenal mucosa. Each of these hormones performs multiple interrelated functions.

■ Gastrin is released primarily in response to the presence of protein products in the stomach, and its effects promote digestion of protein and the movement of materials through the digestive tract.

■ Secretin is released primarily in response to the presence of acid in the duodenum, and its effects neutralize the acid in the duodenal lumen.

■ Cholecystokinin is released primarily in response to the presence of fat products in the duodenum, and its effects optimize conditions for digesting fat and other nutrients.

REVIEW EXERCISES

Objective Questions (Answers on p. A-47)

1. The extent of nutrient uptake from the digestive tract depends on the body's needs. *(True or false?)*
2. The stomach is relaxed during vomiting. *(True or false?)*
3. Acid cannot normally penetrate into or between the cells lining the stomach, which enables the stomach to contain acid without injuring itself. *(True or false?)*
4. Protein is continually lost from the body through digestive secretions and sloughed epithelial cells, which pass out in the feces. *(True or false?)*
5. Foodstuffs not absorbed by the small intestine are absorbed by the large intestine. *(True or false?)*
6. The endocrine pancreas secretes secretin and CCK. *(True or false?)*

7. When food is mechanically broken down and mixed with gastric secretions, the resultant thick, liquid mixture is known as _____ .
8. The two substances absorbed by specialized transport mechanisms located only in the terminal ileum are _____ and _____ .
9. The most potent choleretic is _____ .
10. Which of the following is *not* a function of saliva?
    a. begins digestion of carbohydrate
    b. facilitates absorption of glucose across the oral mucosa
    c. facilitates speech
    d. exerts an antibacterial effect
    e. plays an important role in oral hygiene

11. Match the following:
_____ 1. prevents re-entry of food into the mouth during swallowing
_____ 2. triggers the swallowing reflex
_____ 3. seals off the nasal passages during swallowing
_____ 4. prevents air from entering the esophagus during breathing
_____ 5. closes off the respiratory airways during swallowing
_____ 6. prevents gastric contents from backing up into the esophagus

(a) closure of the pharyngo-esophageal sphincter
(b) elevation of the uvula
(c) position of the tongue against the hard palate
(d) closure of the gastro-esophageal sphincter
(e) bolus pushed to the rear of the mouth by the tongue
(f) tight apposition of the vocal folds

12. Use the following answer code to identify the characteristics of the listed substances:
(a) pepsin
(b) mucus
(c) HCl
(d) intrinsic factor
(e) histamine
_____ 1. activates pepsinogen
_____ 2. inhibits amylase
_____ 3. essential for vitamin B_{12} absorption
_____ 4. can act autocatalytically
_____ 5. a potent stimulant for acid secretion
_____ 6. breaks down connective tissue and muscle fibers
_____ 7. begins protein digestion
_____ 8. serves as a lubricant
_____ 9. kills ingested bacteria
_____10. is alkaline
_____11. deficient in pernicious anemia
_____12. coats the gastric mucosa

Essay Questions
1. Describe the four basic digestive processes.
2. List the three categories of energy-rich foodstuffs and the absorbable units of each.
3. List the components of the digestive system. Describe the cross-sectional anatomy of the digestive tract.
4. What four general factors are involved in regulating digestive system function? What is the role of each?
5. Describe the types of motility in each component of the digestive tract. What factors control each type of motility?
6. State the composition of the digestive juice secreted by each component of the digestive system. Describe the factors that control each digestive secretion.
7. List the enzymes involved in digesting each category of foodstuff. Indicate the source and control of secretion of each of the enzymes.
8. Why are some digestive enzymes secreted in inactive form? How are they activated?
9. What absorption processes take place within each component of the digestive tract? What special adaptations of the small intestine enhance its absorptive capacity?
10. Describe the absorptive mechanisms for salt, water, carbohydrate, protein, and fat.
11. What are the contributions of the accessory digestive organs? What are the nondigestive functions of the liver?
12. Summarize the functions of each of the three major gastrointestinal hormones.
13. What waste product is excreted in the feces?
14. How is vomiting accomplished? What are the causes and consequences of vomiting, diarrhea, and constipation?
15. Describe the process of mucosal turnover in the stomach and small intestine.

POINTS TO PONDER

(Explanations on p. A-47)
1. Why do patients who have had a large part of their stomachs removed for treatment of stomach cancer or severe peptic ulcer disease have to eat small quantities of food frequently instead of consuming three meals a day?
2. The number of immune cells in the *gut-associated lymphoid tissue (GALT)* housed in the mucosa is estimated to be equal to the total number of these defense cells in the rest of the body. Speculate on the adaptive significance of this extensive defense capability of the digestive system.
3. How would defecation be accomplished in a patient paralyzed from the waist down by a lower spinal cord injury?

4. After bilirubin is extracted from the blood by the liver, it is conjugated (combined) with glycuronic acid by the enzyme glucuronyl transferase within the liver. Only when conjugated can bilirubin be actively excreted into the bile. For the first few days of life, the liver does not make adequate quantities of glucuronyl transferase. Explain how this transient enzyme deficiency leads to the common condition of jaundice in newborns.
5. Explain why removal of either the stomach or the terminal ileum leads to pernicious anemia.

CLINICAL CONSIDERATION

(Explanation on p. A-48)
Thomas W. experiences a sharp pain in his upper right abdomen after eating a high-fat meal. Also, he has noted that his feces are grayish white instead of brown. What is the most likely cause of his symptoms? Explain why each of these symptoms occurs with this condition.

PHYSIOEDGE RESOURCES

PhysioEdge CD-ROM

PhysioEdge, the CD-ROM packaged with your text, focuses on the concepts students find most difficult to learn. Figures marked with this icon have associated activities on the CD. For a visual review of concepts in this chapter, check out the following:

Tutorial: Gastrointestinal Physiology

Media Exercise 15.1: The Stomach

Media Exercise 15.2: Stomach: Cellular Level

Media Exercise 15.3: The Intestine and Associated Organs

PhysioEdge Website

The website for this book contains a wealth of helpful study aids, as well as many ideas for further reading and research. Log on to: **http://www.brookscole.com/hpfundamentals3** Select Chapter 15 from the drop-down menu or click on one of the many resource areas, including Case Histories, which introduce clinical aspects of human physiology. For this chapter check out Case History 6: Just Stress.

For Suggested Readings, consult **InfoTrac College Edition/ Research** on the PhysioEdge website or go directly to InfoTrac College Edition, your online research library, at: **http://infotrac.thomsonlearning.com**

Components Important in Energy Balance and Temperature Regulation

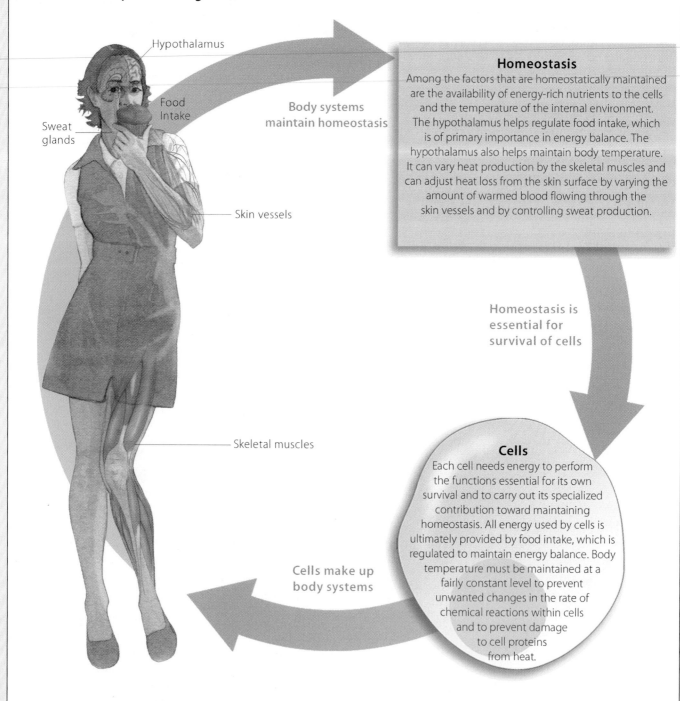

Hypothalamus

Food Intake

Sweat glands

Skin vessels

Skeletal muscles

Body systems maintain homeostasis

Homeostasis
Among the factors that are homeostatically maintained are the availability of energy-rich nutrients to the cells and the temperature of the internal environment. The hypothalamus helps regulate food intake, which is of primary importance in energy balance. The hypothalamus also helps maintain body temperature. It can vary heat production by the skeletal muscles and can adjust heat loss from the skin surface by varying the amount of warmed blood flowing through the skin vessels and by controlling sweat production.

Homeostasis is essential for survival of cells

Cells
Each cell needs energy to perform the functions essential for its own survival and to carry out its specialized contribution toward maintaining homeostasis. All energy used by cells is ultimately provided by food intake, which is regulated to maintain energy balance. Body temperature must be maintained at a fairly constant level to prevent unwanted changes in the rate of chemical reactions within cells and to prevent damage to cell proteins from heat.

Cells make up body systems

Food intake is essential to power cell activities. For body weight to remain constant, the caloric value of food must equal total energy needs. **Energy balance** and thus body weight are maintained by controlling food intake.

Energy expenditure generates heat, which is important in **temperature regulation.** Humans, usually in environments cooler than their bodies, must constantly generate heat to maintain their body temperatures. Also, they must have

mechanisms to cool the body if it gains too much heat from heat-generating skeletal muscle activity or from a hot external environment. Body temperature must be regulated because the rate of cellular chemical reactions depends on temperature, and overheating damages cell proteins.

The hypothalamus is the major integrating center for maintenance of both energy balance and body temperature.

Energy Balance and Temperature Regulation

ENERGY BALANCE

Each cell in the body needs energy to perform the functions essential for the cell's own survival (such as active transport and cellular repair) and to carry out its specialized contributions toward maintenance of homeostasis (such as gland secretion or muscle contraction). All energy used by cells is ultimately provided by food intake.

▐ Most food energy is ultimately converted into heat in the body.

According to the **first law of thermodynamics**, energy can be neither created nor destroyed. Therefore, energy is subject to the same kind of input–output balance as are the chemical components of the body such as H_2O and salt (see p. 443).

ENERGY INPUT AND OUTPUT

The energy in ingested food constitutes *energy input* to the body. Chemical energy locked in the bonds that hold the atoms together in nutrient molecules is released when these molecules are broken down in the body. Cells capture a portion of this nutrient energy in the high-energy phosphate bonds of ATP (see p. 27). Energy harvested from biochemical processing of ingested nutrients either is used immediately to perform biological work or is stored in the body for later use as needed during periods when food is not being digested and absorbed.

Energy output or *expenditure* by the body falls into two categories (● Figure 16-1): external work and internal work. **External work** is the energy expended when skeletal muscles are contracted to move external objects or to move the body in relation to the environment. **Internal work** constitutes all other forms of biological energy expenditure that do not accomplish mechanical work outside the body. Internal work encompasses two types of energy-dependent activities: (1) skeletal muscle activity used for purposes other than external work, such as the contractions associated with postural maintenance and

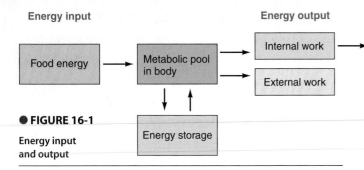

● **FIGURE 16-1**

Energy input and output

shivering; and (2) all the energy-expending activities that must go on all the time just to sustain life. The latter include the work of pumping blood and breathing; the energy required for active transport of critical materials across plasma membranes; and the energy used during synthetic reactions essential for the maintenance, repair, and growth of cellular structures—in short, the "metabolic cost of living."

CONVERSION OF NUTRIENT ENERGY TO HEAT

Not all energy in nutrient molecules can be harnessed to perform biological work. Energy cannot be created or destroyed, but it can be converted from one form to another. The energy in nutrient molecules not used to energize work is transformed into **thermal energy**, or **heat**. During biochemical processing, only about 50% of the energy in nutrient molecules is transferred to ATP; the other 50% of nutrient energy is immediately lost as heat. During ATP expenditure by the cells, another 25% of the energy derived from ingested food becomes heat. Because the body is not a heat engine, it cannot convert heat into work. Therefore, not more than 25% of nutrient energy is available for work, either external or internal. The remaining 75% is lost as heat during the sequential transfer of energy from nutrient molecules to ATP to cellular systems.

Furthermore, of the energy actually captured for use by the body, almost all expended energy eventually becomes heat. To exemplify, energy expended by the heart to pump blood is gradually changed into heat by friction as blood flows through the vessels. Likewise, energy used in synthesizing structural protein eventually appears as heat when that protein is degraded during the normal course of turnover of bodily constituents. Even in performing external work, skeletal muscles convert chemical energy into mechanical energy inefficiently; as much as 75% of the expended energy is lost as heat. Thus all energy liberated from ingested food that is not directly used for moving external objects or stored in fat (adipose tissue) deposits (or, in the case of growth, as protein) eventually becomes body heat. This heat is not entirely wasted energy, however, because much of it is used to maintain body temperature.

▌ The metabolic rate is the rate of energy use.

The rate at which energy is expended by the body during both external and internal work is known as the **metabolic rate**:

$$\text{Metabolic rate} = \frac{\text{energy expenditure}}{\text{unit of time}}$$

Because most of the body's energy expenditure eventually appears as heat, the metabolic rate is normally expressed in terms of the rate of heat production in kilocalories per hour. The basic unit of heat energy is the **calorie**, which is the amount of heat required to raise the temperature of 1 g of H_2O by 1 degree C. This unit is too small to be convenient when discussing the human body because of the magnitude of heat involved, so the **kilocalorie** or **Calorie**, which is equivalent to 1000 calories, is used. When nutritionists speak of "calories" in quantifying the energy content of various foods, they are actually referring to kilocalories. Four kilocalories of heat energy are released when 1 g of glucose is oxidized or "burned," whether the oxidation takes place inside or outside the body.

CONDITIONS FOR MEASURING THE BASAL METABOLIC RATE

The metabolic rate and consequently the amount of heat produced vary depending on a variety of factors, such as exercise, anxiety, shivering, and food intake. Increased skeletal muscle activity is the factor that can increase metabolic rate to the greatest extent. Even slight increases in muscle tone notably elevate the metabolic rate, and various levels of physical activity markedly alter energy expenditure and heat production (▲ Table 16-1). For this reason, a person's metabolic rate is determined under standardized basal conditions established to control as many as possible of the variables that can alter metabolic rate. Specifically, the person should be at physical and mental rest at comfortable room temperature and should not have eaten within the last 12 hours. In this way, the metabolic activity necessary to maintain the basic body functions at rest can be determined. Thus the so-called **basal metabolic rate (BMR)** is a reflection of the body's "idling speed," or the minimal waking rate of internal energy expenditure.

METHODS OF MEASURING THE BASAL METABOLIC RATE

Clinical Note The rate of heat production in BMR determinations can be measured directly or indirectly. With **direct calorimetry**, the person sits in an insulated chamber with water circulating through the walls. The difference in the temperature of the water entering and leaving the chamber reflects the amount of heat liberated by the person and picked up by the water as it passes through the chamber. Even though this method provides a direct measurement of heat production, it is not practical, because a calorimeter chamber is costly and takes up a lot of space. Therefore, a more practical method of indirectly determining the rate of heat production was developed for widespread use. With **indirect calorimetry**, only the person's O_2 uptake per unit of time is measured, which is a simple task using minimal equipment. Recall that

$$\text{Food} + O_2 \rightarrow CO_2 + H_2O + \text{energy (mostly transformed into heat)}$$

Accordingly, a direct relationship exists between the volume of O_2 used and the quantity of heat produced. This relation-

TABLE 16-1

Rate of Energy Expenditure for a 70-kg Person during Different Types of Activity

FORM OF ACTIVITY	ENERGY EXPENDITURE (Kcal/hour)
Sleeping	65
Awake, lying still	77
Sitting at rest	100
Standing relaxed	105
Getting dressed	118
Typewriting	140
Walking slowly on level (2.6 mi/hr)	200
Carpentry, painting a house	240
Sexual intercourse	280
Bicycling on level (5.5 mi/hr)	304
Shoveling snow, sawing wood	480
Swimming	500
Jogging (5.3 mi/hr)	570
Rowing (20 strokes/min)	828
Walking up stairs	1,100

ship also depends on the type of food being oxidized. Although carbohydrates, proteins, and fats require different amounts of O_2 for their oxidation and yield different amounts of kilocalories when oxidized, an average estimate can be made of the quantity of heat produced per liter of O_2 consumed on a typical mixed American diet. This approximate value, known as the **energy equivalent of** O_2, is 4.8 kilocalories of energy liberated per liter of O_2 consumed. Using this method, the metabolic rate of a person consuming 15 liters/hr of O_2 can be estimated as follows:

$$\begin{array}{ll} 15 & \text{liters/hr} & = O_2 \text{ consumption} \\ \times 4.8 & \text{kilocalories/liter} & = \text{energy equivalent of } O_2 \\ \hline 72 & \text{kilocalories/hr} & = \text{estimated metabolic rate} \end{array}$$

In this way, a simple measurement of O_2 consumption can be used to reasonably approximate heat production in determining metabolic rate.

Once the rate of heat production is determined under the prescribed basal conditions, it must be compared with normal values for people of the same sex, age, height, and weight, because these factors all affect the basal rate of energy expenditure. For example, a large man actually has a higher rate of heat production than a smaller man, but, expressed in terms of total surface area (which is a reflection of height and weight), the output in kilocalories per hour per square meter of surface area is normally about the same.

FACTORS INFLUENCING THE BASAL METABOLIC RATE

Thyroid hormone is the primary but not sole determinant of the rate of basal metabolism. As thyroid hormone increases, the BMR increases correspondingly.

Surprisingly, the BMR is not the body's lowest metabolic rate. The rate of energy expenditure during sleep is 10 to 15% lower than the BMR, presumably because of the more complete muscle relaxation that occurs during the paradoxical stage of sleep (see p. 135).

▌ Energy input must equal energy output to maintain a neutral energy balance.

Because energy cannot be created or destroyed, energy input must equal energy output, as follows:

$$\text{Energy input} = \text{energy output}$$

$$\begin{array}{c} \text{Energy in food} = \text{external} + \text{internal heat} \pm \text{stored} \\ \text{consumed} \quad\quad \text{work} \quad\quad \text{production} \quad\quad \text{energy} \end{array}$$

There are three possible states of energy balance:

- *Neutral energy balance.* If the amount of energy in food intake exactly equals the amount of energy expended by the muscles in performing external work plus the basal internal energy expenditure that eventually appears as body heat, then energy input and output are exactly in balance, and body weight remains constant.
- *Positive energy balance.* If the amount of energy in food intake is greater than the amount of energy expended by means of external work and internal functioning, the extra energy taken in but not used is stored in the body, primarily as adipose tissue, so body weight increases.
- *Negative energy balance.* Conversely, if the energy derived from food intake is less than the body's immediate energy requirements, the body must use stored energy to supply energy needs, and body weight decreases accordingly.

To maintain a constant body weight (with the exception of minor fluctuations caused by changes in H_2O content), energy acquired through food intake must equal energy expenditure by the body. Because the average adult maintains a fairly constant weight over long periods of time, this implies that precise homeostatic mechanisms exist to maintain a long-term balance between energy intake and energy expenditure. Theoretically, total body energy content could be maintained at a constant level by regulating the magnitude of food intake, physical activity, or internal work and heat production. Control of food intake to match changing metabolic expenditures is the major means of maintaining a neutral energy balance. The level of physical activity is principally under voluntary control, and mechanisms that alter the degree of internal work and heat production are aimed primarily at regulating body temperature rather than total energy balance.

However, after several weeks of eating less or more than desired, small counteracting changes in metabolism may occur. For example, a compensatory increase in the body's efficiency of energy use in response to underfeeding partially explains why some dieters become stuck at a plateau after having lost the first 10 or so pounds of weight fairly easily. Similarly, a compensatory reduction in the efficiency of energy use in response to overfeeding accounts in part for the difficulty experienced by very thin people who are deliberately trying to gain weight. Despite these modest compensatory changes in metabolism, regulation of food intake is the most important factor in the long-term maintenance of energy balance and body weight.

▌ Food intake is controlled primarily by the hypothalamus.

Even though food intake is adjusted to balance changing energy expenditures over a period of time, there are no calorie receptors per se to monitor energy input, energy output, or total body energy content. Instead, various blood-borne chemical factors that signal the body's nutritional state, such as how much fat is stored or the feeding status, are important in regulating food intake. Control of food intake does not depend on changes in a single signal but is determined by the integration of many inputs that provide information about the body's energy status. Multiple molecular signals together ensure that feeding behavior is synchronized with the body's immediate and long-term energy needs. Some information is used for short-term regulation of food intake, helping to control meal size and frequency. Even so, over a 24-hour period the energy in ingested food rarely matches energy expenditure for that day. The correlation between total caloric intake and total energy output is excellent, however, over long periods of time. As a result, the total energy content of the body— and, consequently, body weight—remains relatively constant on a long-term basis. Thus energy homeostasis, that is, energy balance, is carefully regulated.

ROLE OF THE ARCUATE NUCLEUS: NPY AND MELANOCORTINS

Control of energy balance and food intake is primarily a function of the hypothalamus. The **arcuate nucleus** of the hypothalamus plays a central role in both the long-term control of energy balance and body weight and the short-term control of food intake on a meal-to-meal basis. The arcuate nucleus is an arc-shaped collection of neurons located adjacent to the floor of the third ventricle. Multiple, highly integrated, redundant pathways crisscross into and out of the arcuate nucleus, indicative of the complex systems involved in feeding and satiety. **Feeding**, or **appetite**, **signals** give rise to the sensation of **hunger**, driving us to eat. By contrast, **satiety** is the feeling of being full. **Satiety signals** tell us when we have had enough and suppress the desire to eat.

The arcuate nucleus has two subsets of neurons that function in an opposing manner. One subset releases *neuropeptide Y*, while the other releases *melanocortins*. **Neuropeptide Y (NPY)**, one of the most potent appetite stimulators ever found, leads to increased food intake, thus promoting weight gain. **Melanocortins**, a group of hormones traditionally known to be important in varying the skin color for the purpose of camouflage in some species, have been shown to exert an unexpected role in energy homeostasis. Melanocortins, most notably *melanocyte-stimulating hormone* (see p. 537), suppress appetite, thus leading to reduced food intake and weight loss. Melanocortins do not play a role in determining skin coloration in humans. Their importance in our species lies in part in toning down appetite in response to increased fat stores.

But NPY and melanocortins are likely not the final effectors in appetite control. Recent evidence suggests that these arcuate-nucleus chemical messengers may in turn influence the release of neuropeptides in other parts of the brain that exert more direct control over food intake. Scientists are currently trying to unravel these other factors that act downstream from NPY and melanocortins to regulate appetite.

Based on current evidence, the following regulatory inputs to the arcuate nucleus and beyond are important in the long-term maintenance of energy balance and the short-term control of food intake at meals (● Figure 16-2).

ROLE OF FAT STORES AND LEPTIN

Scientists' notion of fat cells (**adipocytes**) in adipose tissue as merely storage space for triglyceride fat has undergone a dramatic change in the past decade with the discovery of their active role in energy homeostasis. Adipocytes secrete **leptin**, a hormone essential for normal body weight regulation (*leptin* means "thin"). The amount of leptin in the blood is an excellent indicator of the total amount of triglyceride fat stored in adipose tissue: The larger the fat stores, the more leptin released into the blood. This blood-borne signal, discovered in the mid-1990s, was the first molecular satiety signal identified.

The arcuate nucleus is the major site for leptin action. Acting in negative-feedback fashion, increased leptin from burgeoning fat stores serves as a "trim down" signal. Leptin suppresses appetite, thus decreasing food consumption and promoting weight loss, by inhibiting hypothalamic output of appetite-stimulating NPY and stimulating output of appetite-suppressing melanocortins. Conversely, a decrease in fat stores and the resultant decline in leptin secretion bring about an increase in appetite, leading to weight gain. The leptin signal is generally considered responsible for the long-term matching of food intake to energy expenditure so that total body energy content remains balanced and body weight remains constant.

Interestingly, leptin has recently been shown to also be important in reproduction. It is one of the triggers for the onset of puberty, signaling that the female has enough long-term energy (adipose) stores to sustain a pregnancy.

In addition to the importance of leptin and perhaps other signals in the long-term control of body weight, other factors play a role in controlling the timing and size of meals. Several blood-borne messengers from the digestive tract and pancreas are important in regulating how often and how much we eat in a given day as follows.

THE EXTENT OF GHRELIN AND PYY$_{3-36}$ SECRETION

Two peptides important in the short-term control of food intake have recently been identified: *ghrelin* and *PYY$_{3-36}$*, which signify hunger and fullness, respectively. Both are secreted by the digestive tract. **Ghrelin**, the so-called "hunger hormone," is a potent appetite stimulator produced by the stomach and regulated by the feeding status. Secretion of this mealtime stimulator peaks before meals and makes people feel like eating, then falls once food is eaten. Ghrelin stimulates appetite by activating the hypothalamic NPY-secreting neurons.

PYY_{3-36} is a counterpart of ghrelin. The secretion of PYY_{3-36}, which is produced by the small and large intestines, is at its lowest level before a meal but rises during meals and signals satiety. This peptide acts by inhibiting the appetite-stimulating NPY-secreting neurons and stimulating the appetite-suppressing melanocortin-secreting neurons in the arcuate nucleus. By thwarting appetite, PYY_{3-36} is believed to be an important mealtime terminator.

The following other factors are also involved in signaling where the body is on the hunger–satiety scale.

● **FIGURE 16-2**

Factors that influence food intake

*Other chemicals are also released from this area that exert similar functions.

Satiety signals important in short-term control of the timing and size of meals

Adipose tissue-related signals important in long-term matching of food intake to energy expenditure to control body weight

Psychosocial and environmental factors that influence food intake

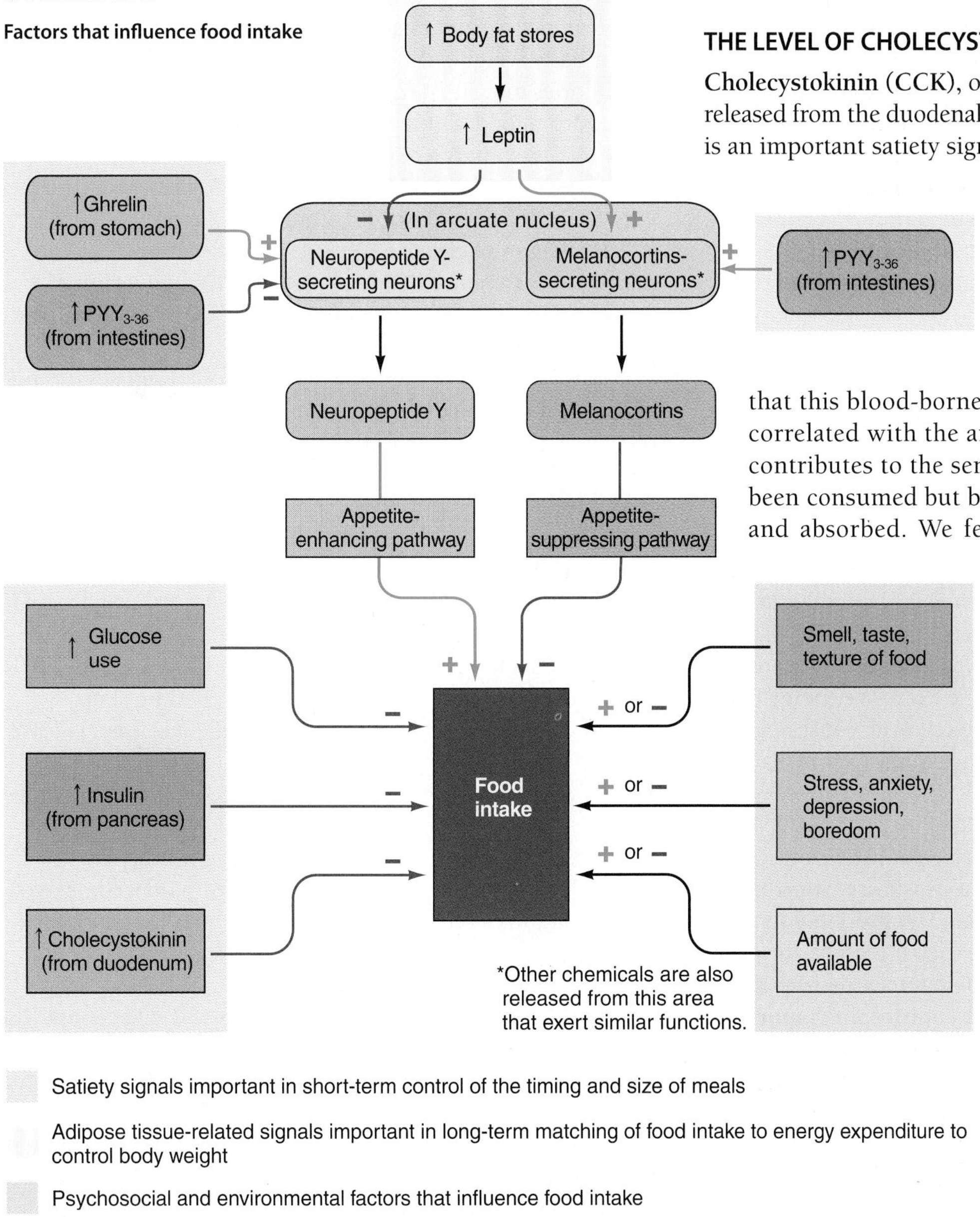

THE EXTENT OF GLUCOSE USE AND INSULIN SECRETION

Satiety is signaled by increased glucose use, such as occurs during a meal when more glucose is available for use because it is being absorbed from the digestive tract. Conversely, after absorption of a meal is complete, and no new glucose is entering the blood, the resultant reduction in the cells' glucose use arouses the sensation of hunger. The extent of glucose use appears more important in determining the timing of meals than in the long-term control of body weight.

In a related mechanism, increased insulin in the blood signals satiety. Insulin, a hormone secreted by the pancreas in response to a rise in the concentration of glucose and other nutrients in the blood following a meal, stimulates cellular uptake, use, and storage of glucose and other nutrients. Thus the increase in insulin secretion that accompanies nutrient abundance and promotes increased glucose use is an appropriate satiety signal.

The exact pathways by which increased glucose use and insulin secretion signal satiety have not been figured out.

THE LEVEL OF CHOLECYSTOKININ SECRETION

Cholecystokinin (CCK), one of the gastrointestinal hormones released from the duodenal mucosa during digestion of a meal, is an important satiety signal for regulating the size of meals. CCK is secreted in response to the presence of nutrients in the small intestine. Through multiple effects on the digestive system, CCK facilitates digestion and absorption of these nutrients (see p. 504). It is appropriate that this blood-borne signal, whose rate of secretion is correlated with the amount of nutrients ingested, also contributes to the sense of being filled after a meal has been consumed but before it has actually been digested and absorbed. We feel satisfied when adequate food to replenish the stores is in the digestive tract, even though the body's energy stores are still low. This explains why we stop eating before the ingested food is made available to meet the body's energy needs. The exact pathways by which CCK signals satiety are unknown.

PSYCHOSOCIAL AND ENVIRONMENTAL INFLUENCES

Thus far we have described involuntary signals that automatically occur to control food intake. However, as with water intake, people's eating habits are also shaped by psychological, social, and environmental factors. Often our decision to eat or stop eating is not determined merely by whether we are hungry or full, respectively. Frequently, we eat out of habit (eating three meals a day on schedule no matter what our status on the hunger–satiety continuum) or because of social custom (food often plays a prime role in entertainment, leisure, and business activities). Even well-intentioned family pressure—"Clean your plate before you can leave the table"—can have an impact on the amount consumed.

Furthermore, the amount of pleasure derived from eating can reinforce feeding behavior. Eating foods with an enjoyable taste, smell, and texture can increase appetite and food intake. This has been demonstrated in an experiment in which rats were offered their choice of highly palatable human foods. They overate by as much as 70 to 80% and became obese. When the rats returned to eating their regular monotonous but nutritionally balanced rat chow, their obesity was rapidly reversed as their food intake was controlled once again by physiologic drives rather than by hedonistic urges for the tastier offerings.

Stress, anxiety, depression, and boredom have also been shown to alter feeding behavior in ways that are unrelated to energy needs in both experimental animals and humans. People often eat to satisfy psychological needs rather than because they are hungry. Furthermore, environmental influences, such as the amount of food available, play an important role in determining extent of food intake. Thus any comprehensive explanation of how food intake is controlled must take into account these voluntary eating acts that can reinforce or override the internal signals governing feeding behavior.

▌ Obesity occurs when more kilocalories are consumed than are burned up.

Clinical Note **Obesity** is defined as excessive fat content in the adipose tissue stores; the arbitrary boundary for obesity is generally considered to be greater than 20% overweight compared to normal standards. Over half of the adults in the United States are overweight, and one third are clinically obese. Obesity occurs when, over a period of time, more kilocalories are ingested in food than are used to support the body's energy needs, with the excessive energy being stored as triglycerides in adipose tissue. The causes of obesity are many, and some remain obscure. Some factors that may be involved include the following:

- *Disturbances in the leptin signaling pathway.* Some cases of obesity have been linked to leptin resistance. Some investigators suggest that the hypothalamic centers involved in maintaining energy homeostasis are "set at a higher level" in obese people. For example, the problem may lie with faulty leptin receptors in the brain that do not respond appropriately to the high levels of circulating leptin from abundant adipose stores. Thus the brain does not detect leptin as a signal to turn down appetite until a higher set point (and accordingly greater fat storage) is achieved. This could explain why overweight people do tend to maintain their weight but at a higher set point than normal.

- *Lack of exercise.* Numerous studies have shown that, on average, fat people do not eat any more than thin people. One possible explanation is that overweight persons do not overeat but "underexercise"—the "couch potato" syndrome. Very low levels of physical activity typically are not accompanied by comparable reductions in food intake.

For this reason, modern technology is partly to blame for the current obesity epidemic. Our ancestors had to exert physical effort to eke out a subsistence. By comparison, we now have machines to replace much manual labor, remote controls to operate our machines with minimal effort, and computers that encourage long hours of sitting. We have to make a conscious effort to exercise.

- *Differences in the "fidget factor."* **Nonexercise activity thermogenesis (NEAT),** or the "fidget factor," might explain some variation in fat storage among people. Those who engage in toe tapping or other types of repetitive, spontaneous physical activity expend a substantial number of kilocalories throughout the day without a conscious effort.

- *Differences in extracting energy from food.* Another reason why lean people and obese people may have dramatically different body weights despite consuming the same number of kilocalories may lie in the efficiency with which each extracts energy from food. Studies suggest that leaner individuals tend to derive less energy from the food they consume, because they convert more of the food's energy into heat than into energy for immediate use or for storage.

- *Hereditary tendencies.* Often differences in the regulatory pathways for energy balance—either those governing food intake or those influencing energy expenditure—arise from genetic variations.

- *Development of an excessive number of fat cells as a result of overfeeding.* One of the problems in fighting obesity is that once fat cells are created, they do not disappear with dieting and weight loss. Even if a dieter loses a large portion of the triglyceride fat stored in these cells, the depleted cells remain, ready to refill. Therefore, rebound weight gain after losing weight is difficult to avoid and discouraging for the dieter.

- *The existence of certain endocrine disorders such as hypothyroidism (see p. 551).* Hypothyroidism involves a deficiency of thyroid hormone, the main factor that bumps up the BMR so that the body burns more calories in its idling state.

- *An abundance of convenient, highly palatable, energy-dense, relatively inexpensive foods.*

- *Emotional disturbances in which overeating replaces other gratifications.*

- *A possible virus link.* One intriguing new proposal links a relatively common cold virus to a propensity to become overweight and may account for a portion of the current obesity epidemic.

Despite this rather lengthy list, our knowledge about the causes and control of obesity is still rather limited, as evidenced by the number of people who are constantly trying to stabilize their weight at a more desirable level. This is important from more than an aesthetic viewpoint. It is known that obesity, especially of the android type, can predispose an individual to illness and premature death from a multitude of

What the Scales Don't Tell You

Body composition is the percentage of body weight that is composed of lean tissue and adipose tissue. Assessing body composition is an important step in evaluating a person's health status. One crude means of assessing body composition is by calculating the **body mass index (BMI)** using the following formula:

$$BMI = \frac{(\text{weight in pounds}) \times 700}{(\text{height in inches})^2}$$

A BMI of 25 or less is considered healthy, whereas a BMI of 30 or higher is considered to place the person at increased risk for various diseases and premature death. BMIs between this range are considered borderline.

BMI determinations and the age–height–weight tables used by insurance companies can be misleading for determining healthy body weight. By these charts, many athletes, for example, would be considered overweight. A football player may be 6 feet 5 inches tall and weigh 300 pounds, but have only 12% body fat. This player's extra weight is muscle, not fat, and therefore is not a detriment to his health. A sedentary person, in contrast, may be normal on the height–weight charts but have 30% body fat. This person should maintain body weight while increasing muscle mass and decreasing fat. Ideally, men should have 15% fat or less and women should have 20% fat or less.

The most accurate method for assessing body composition is underwater weighing. This technique is based on the fact that lean tissue is denser than water and fat tissue is less dense than water. (You can readily demonstrate this for yourself by dropping a piece of lean meat and a piece of fat into a glass of water; the lean meat will sink and the fat will float.) In underwater weighing, the person breathes out as much air as possible and then completely submerges in a tank of water while sitting in a swing that is attached to a scale. The results are used to determine body density using equations that take into consideration the density of water, the difference between the person's weight in air and underwater, and the residual volume of air remaining in the lungs. Because of the difference in density between lean and fat tissue, people who have more fat have a lower density and weigh relatively less underwater than in air compared to their lean counterparts. Body composition is then determined by means of an equation that correlates percentage fat with body density.

Another common way to assess body composition is skinfold thickness. Because approximately half of the body's total fat content is located just beneath the skin, total body fat can be estimated from measurements of skinfold thickness taken at various sites on the body. Skinfold thickness is determined by pinching up a fold of skin at one of the designated sites and measuring its thickness by means of a caliper, a hinged instrument that fits over the fold and is calibrated to measure thickness. Mathematical equations specific for the person's age and sex can be used to predict the percentage of fat from the skinfold thickness scores. A major criticism of skinfold assessments is that accuracy depends on the investigator's skill.

There are different ways to be fat, and one way is more dangerous than the other. Obese patients can be classified into two categories— *android*, a male-type of adipose tissue distribution, and *gynoid*, a female-type distribution—based on the anatomic distribution of adipose tissue measured as the ratio of waist circumference to hip circumference. **Android obesity** is characterized by abdominal fat distribution (people shaped like "apples"), whereas **gynoid obesity** is characterized by fat distribution in the hips and thighs (people shaped like "pears"). Both sexes can display either android or gynoid obesity.

Android obesity is associated with a number of disorders, including insulin resistance, type II (adult-onset) diabetes mellitus, excess blood lipid levels, high blood pressure, coronary heart disease, and stroke. Gynoid obesity is not associated with high risk for these diseases. Because android obesity is associated with increased risk for disease, it is most important for apple-shaped overweight individuals to reduce their fat stores.

Research on the success of weight reduction programs indicates that it is very difficult for people to lose weight, but when weight loss occurs, it is from the areas of increased stores. Because very-low-calorie diets are difficult to maintain, an alternative to severely cutting caloric intake to lose weight is to increase energy expenditure through physical exercise. Exercise physiologists often assess body composition as an aid in prescribing and evaluating exercise programs. Exercise generally reduces the percentage of body fat and, by increasing muscle mass, increases the percentage of lean tissue. An aerobic exercise program further helps reduce the risk of the disorders associated with android obesity.

---

diseases. (To learn about the differences between android and gynoid obesity, see the accompanying boxed feature, ❯ Beyond the Basics.)

❙ People suffering from anorexia nervosa have a pathologic fear of gaining weight.

Clinical Note The converse of obesity is generalized nutritional deficiency. The obvious causes for reduction of food intake below energy needs are lack of availability of food, interference with swallowing or digestion, and impairment of appetite.

One poorly understood disorder in which lack of appetite is a prominent feature is **anorexia nervosa**. Patients with this disorder, most commonly adolescent girls and young women, have a morbid fear of becoming fat. They have a distorted body image, tending to visualize themselves as being much heavier than they actually are. Because they have an aversion to food, they eat very little and consequently lose considerable weight, perhaps even starving themselves to death. Other characteristics of the condition include altered secretion of many hormones, absence of menstrual periods, and low body temperature. It is unclear whether these symptoms occur secondarily as a result of general malnutrition or arise independently of the eating disturbance as a part of a primary hypothalamic malfunction. Many investigators think the underlying problem may be psychological rather than biological. Some experts suspect that anorexics may suffer from addiction to endogenous opiates, self-produced morphinelike substances (see p. 150) that are thought to be released during prolonged starvation.

Click on the Media Exercises menu of the CD-ROM and work Media Exercise 16.1: Basics of Energy Balance to test your understanding of the previous section.

TEMPERATURE REGULATION

Humans are usually in environments cooler than their bodies, but they constantly generate heat internally, which helps maintain body temperature. Heat production ultimately depends on the oxidation of metabolic fuel derived from food.

Changes in body temperature in either direction alter cell activity—an increase in temperature speeds up cellular chemical reactions, whereas a fall in temperature slows down these reactions. Because cell function is sensitive to fluctuations in internal temperature, humans homeostatically maintain body temperature at a level that is optimal for cellular metabolism to proceed in a stable fashion. Overheating is more serious than cooling. Even moderate elevations of body temperature begin to cause nerve malfunction and irreversible protein denaturation. Most people suffer convulsions when the internal body temperature reaches about 106°F (41°C); 110°F (43.3°C) is considered the upper limit compatible with life.

By contrast, most of the body's tissues can transiently withstand substantial cooling. This characteristic is useful during cardiac surgery when the heart must be stopped. The patient's body temperature is deliberately lowered. The cooled tissues need less nourishment than they do at normal body temperature because of their reduced metabolic activity. However, a pronounced, prolonged fall in body temperature slows metabolism to a fatal level.

Internal core temperature is homeostatically maintained at 100°F.

Normal body temperature has traditionally been considered 98.6°F (37°C). However, a recent study indicates that normal body temperature varies among individuals and varies throughout the day, ranging from 96.0°F in the morning to 99.9°F in the evening with an overall average of 98.2°F. These values are considered normal for temperatures taken orally (by mouth). This variation is due to an innate biological rhythm or "biological clock."

Furthermore, there is no one body temperature because the temperature varies from organ to organ. From a thermoregulatory viewpoint, the body may conveniently be viewed as a *central core* surrounded by an *outer shell*. The temperature within the inner core, which consists of the abdominal and thoracic organs, the central nervous system, and the skeletal muscles, generally remains fairly constant. This internal **core temperature** is subject to precise regulation to maintain its homeostatic constancy. The core tissues function best at a relatively constant temperature of around 100°F. The skin and subcutaneous fat constitute the outer shell. In contrast to the constant high temperature in the core, the temperature within the shell is generally cooler and may vary substantially. For example, skin temperature may fluctuate between 68° and 104°F without damage. In fact, as you will see, the temperature of the skin is deliberately varied as a control measure to help maintain the core's thermal constancy.

Several easily accessible sites are used for monitoring body temperature. The oral and axillary (under the armpit) temperatures are comparable, whereas rectal temperature averages about 1 degree F higher. Also recently available is a temperature-monitoring instrument that scans the heat generated by the eardrum and converts this temperature into an oral equivalent. However, none of these measurements is an absolute indication of the internal core temperature, which is a bit higher at 100°F than the monitored sites.

Heat input must balance heat output to maintain a stable core temperature.

The core temperature is a reflection of the body's total heat content. To maintain a constant total heat content and thus a stable core temperature, heat input to the body must balance heat output (● Figure 16-3). *Heat input* occurs by way of heat gain from the external environment and internal heat production, the latter being the most important source of heat for the body. Recall that most of the body's energy expenditure ultimately appears as heat. This heat is important in maintaining core temperature. In fact, usually more heat is generated than required to maintain body temperature at a normal level, so the excess heat must be eliminated from the body. *Heat output* occurs by way of heat loss from exposed body surfaces to the external environment.

Balance between heat input and output is frequently disturbed by (1) changes in internal heat production for purposes unrelated to regulation of body temperature, most notably by exercise, which markedly increases heat production; and (2) changes in the external environmental temperature that influence the degree of heat gain or heat loss that occurs between the body and its surroundings. To maintain body temperature within narrow limits despite changes in metabolic heat production and changes in environmental temperature, compensatory adjustments must take place in heat loss and heat gain mechanisms. If the core temperature starts to fall, heat production is increased and heat loss is minimized so that normal temperature can be restored. Conversely, if the temperature starts to rise above normal, it can be corrected by increasing heat loss while simultaneously reducing heat production.

● **FIGURE 16-3**

Heat input and output

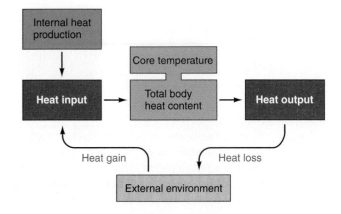

We will now elaborate on the means by which heat gains and losses can be adjusted to maintain body temperature.

▌Heat exchange takes place by radiation, conduction, convection, and evaporation.

All heat loss or heat gain between the body and external environment must take place between the body surface and its surroundings. The same physical laws of nature that govern heat transfer between inanimate objects also control the transfer of heat between the body surface and environment. The temperature of an object may be thought of as a measure of the concentration of heat within the object. Accordingly, heat always moves down its concentration gradient; that is, down a **thermal gradient** from a warmer to a cooler region (*thermo* means "heat").

The body uses four mechanisms of heat transfer: *radiation, conduction, convection,* and *evaporation.*

RADIATION

Radiation is the emission of heat energy from the surface of a warm body in the form of **electromagnetic waves**, or **heat waves**, which travel through space (● Figure 16-4 ①). When radiant energy strikes an object and is absorbed, the energy of the wave motion is transformed into heat within the object. The human body both emits (source of heat loss) and absorbs (source of heat gain) radiant energy. Whether the body loses or gains heat by radiation depends on the difference in temperature between the skin surface and the surfaces of other objects in the body's environment. Because net transfer of heat by radiation is always from warmer objects to cooler ones, the body gains heat by radiation from objects warmer than the skin surface, such as the sun, a radiator, or burning logs. By contrast, the body loses heat by radiation to objects in its environment whose surfaces are cooler than the surface of the skin, such as building walls, furniture, or trees. On average, humans lose close to half of their heat energy through radiation.

CONDUCTION

Conduction is the transfer of heat between objects of differing temperatures that are in direct contact with each other, with heat moving down its thermal gradient from the warmer to the cooler object (● Figure 16-4 ②). When you hold a snowball, for example, your hand becomes cold because heat moves by conduction from your hand to the snowball. Conversely, when you apply a heating pad to a body part, the part is warmed up as heat is transferred directly from the pad to the body.

Similarly, you either lose or gain heat by conduction to the layer of air in direct contact with your body. The direction of heat transfer depends on whether the air is cooler or warmer, respectively, than your skin. Only a small percentage of total heat exchange between the skin and environment takes place by conduction alone, however, because air is not a very good conductor of heat. (This is why swimming pool water at 80°F feels cooler than air at the same temperature; heat is conducted more rapidly from the body surface into the water, which is a good conductor, than into the air, which is a poor conductor.)

● **FIGURE 16-4**

Mechanisms of heat transfer

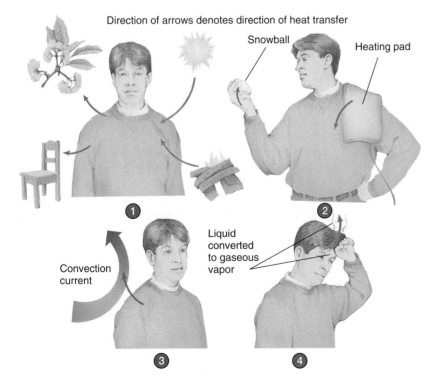

Direction of arrows denotes direction of heat transfer

Snowball

Heating pad

Convection current

Liquid converted to gaseous vapor

① **Radiation**—the transfer of heat energy from a warmer object to a cooler object in the form of electromagnetic waves ("heat waves"), which travel through space.

② **Conduction**—the transfer of heat from a warmer to a cooler object that is in direct contact with the warmer one. The heat is transferred through the movement of thermal energy from molecule to adjacent molecule.

③ **Convection**—the transfer of heat energy by air currents. Cool air warmed by the body through conduction rises and is replaced by more cool air. This process is enhanced by the forced movement of air across the body surface.

④ **Evaporation**—conversion of a liquid such as sweat into a gaseous vapor, a process that requires heat (the heat of vaporization), which is absorbed from the skin .

CONVECTION

The term **convection** refers to the transfer of heat energy by *air* (or *H_2O*) *currents*. As the body loses heat by conduction to the surrounding cooler air, the air in immediate contact with the skin is warmed. Because warm air is lighter (less dense) than cool air, the warmed air rises while cooler air moves in next to the skin to replace the vacating warm air. The process is then repeated (● Figure 16-4 ③). These air movements, known as *convection currents*, help carry heat away from the body. If it were not for convection currents, no further heat could be dissipated from the skin by conduction once the temperature of the layer of air immediately around the body equilibrated with skin temperature.

The combined conduction-convection process of dissipating heat from the body is enhanced by forced movement of air across the body surface, either by external air movements, such as those caused by the wind or a fan, or by movement of the body through the air, as during bicycle riding. Because forced air movement sweeps away the air warmed by conduction and replaces it with cooler air more rapidly, a greater total amount of heat can be carried away from the body over a given time period. Thus wind makes us feel cooler on hot days, and windy days in the winter are more chilling than calm days at the same cold temperature. For this reason, weather forecasters have developed the concept of *wind chill factor*.

EVAPORATION

Evaporation is the final method of heat transfer used by the body. When water evaporates from the skin surface, the heat required to transform water from a liquid to a gaseous state is absorbed from the skin, thereby cooling the body (● Figure 16-4 ④). Evaporative heat loss makes you feel cooler when your bathing suit is wet than when it is dry. Evaporative heat loss occurs continually from the linings of the respiratory airways and from the surface of the skin. Heat is continuously lost through the H_2O vapor in the expired air as a result of the air's humidification during its passage through the respiratory system. Similarly, because the skin is not completely waterproof, H_2O molecules constantly diffuse through the skin and evaporate. This ongoing evaporation from the skin is completely unrelated to the sweat glands. These passive evaporative heat-loss processes are not subject to physiologic control and go on even in very cold weather, when the problem is one of conserving body heat.

Sweating is an active evaporative heat-loss process under sympathetic nervous control. The rate of evaporative heat loss can be deliberately adjusted by varying the extent of sweating, which is an important homeostatic mechanism to eliminate excess heat as needed.

Sweat is a dilute salt solution actively extruded to the surface of the skin by sweat glands dispersed all over the body. The sweat glands can produce up to 4 liters of sweat per hour. Sweat must be evaporated from the skin for heat loss to occur. If sweat merely drips from the surface of skin surface or is wiped away, no heat loss is accomplished. The most important factor determining the extent of evaporation of sweat is the *relative humidity* of the surrounding air (the percentage of H_2O vapor actually present in the air compared to the greatest amount that the air can possibly hold at that temperature; for example, a relative humidity of 70% means that the air contains 70% of the H_2O vapor it is capable of holding). When the relative humidity is high, the air is already almost fully saturated with H_2O, so it has limited ability to take up additional moisture from the skin. Thus little evaporative heat loss can occur on hot, humid days. The sweat glands continue to secrete, but the sweat simply remains on the skin or drips off, instead of evaporating and producing a cooling effect. As a measure of the discomfort associated with combined heat and high humidity, meteorologists have devised the *temperature–humidity index*.

▌ The hypothalamus integrates a multitude of thermosensory inputs.

The hypothalamus serves as the body's thermostat. The home thermostat keeps track of the temperature in a room and triggers a heating mechanism (the furnace) or a cooling mechanism (the air conditioner) as necessary to maintain the room temperature at the indicated setting. Similarly, the hypothalamus, as the body's thermoregulatory integrating center, receives afferent information about the temperature in various regions of the body and initiates extremely complex, coordinated adjustments in heat gain and heat loss mechanisms as necessary to correct any deviations in core temperature from the normal setting. The hypothalamus is far more sensitive than your home thermostat. The hypothalamus can respond to changes in blood temperature as small as 0.01 degree C.

To appropriately adjust the delicate balance between the heat loss mechanisms and the opposing heat-producing and heat-conserving mechanisms, the hypothalamus must be apprised continuously of both the core and the skin temperature by specialized temperature-sensitive receptors called **thermoreceptors.** The core temperature is monitored by *central thermoreceptors,* which are located in the hypothalamus itself as well as elsewhere in the central nervous system and the abdominal organs. *Peripheral thermoreceptors* monitor skin temperature throughout the body and transmit information about changes in surface temperature to the hypothalamus.

Two centers for temperature regulation have been identified in the hypothalamus. The *posterior region* is activated by cold and subsequently triggers reflexes that mediate heat production and heat conservation. The *anterior region,* which is activated by warmth, initiates reflexes that mediate heat loss. Let's examine the means by which the hypothalamus fulfills its thermoregulatory functions.

▌ Shivering is the primary involuntary means of increasing heat production.

The body can gain heat as a result of internal heat production generated by metabolic activity or from the external environment if the latter is warmer than body temperature. Because body temperature usually is higher than environmental temperature, metabolic heat production is the primary source of body heat. In a resting person, most body heat is produced by the thoracic and abdominal organs as a result of ongoing, cost-of-living metabolic activities. Above and beyond this basal level, the rate of metabolic heat production can be variably increased primarily by changes in skeletal muscle activity or to a lesser extent by certain hormonal actions. Thus changes in skeletal muscle activity constitute the major way heat gain is controlled for temperature regulation.

ADJUSTMENTS IN HEAT PRODUCTION BY SKELETAL MUSCLES

In response to a fall in core temperature caused by exposure to cold, the hypothalamus takes advantage of the fact that increased skeletal muscle activity generates more heat. Acting through descending pathways that terminate on the motor neurons controlling the skeletal muscles, the hypothalamus first gradually increases skeletal muscle tone. (Muscle tone is the constant level of tension within the muscles.) Soon shivering begins. **Shivering** consists of rhythmic, oscillating skeletal muscle contractions that occur at a rapid rate of 10 to 20 per second. This mechanism is very effective in increasing heat production; all the energy liberated during these muscle tremors is converted to heat because no external work is accomplished. Within a matter of seconds to minutes, internal heat production may increase two- to fivefold as a result of shivering.

Frequently, these reflex changes in skeletal muscle activity are augmented by increased voluntary, heat-producing actions such as bouncing up and down or hand clapping. Such behavioral responses appear to share neural systems in common with the involuntary physiologic responses. The hypothalamus and limbic system are extensively involved with controlling motivated behavior (see p. 127).

In the opposite situation—a rise in core temperature caused by heat exposure—two mechanisms are employed to reduce heat-producing skeletal muscle activity: Muscle tone is reflexly reduced, and voluntary movement is curtailed. When the air becomes very warm, people often complain it is "too hot even to move."

NONSHIVERING THERMOGENESIS

Although reflex and voluntary changes in muscle activity are the major means of increasing the rate of heat production, **nonshivering (chemical) thermogenesis** also plays a role in thermoregulation. In most experimental animals, chronic cold exposure brings about an increase in metabolic heat production that is independent of muscle contraction, instead being brought about by changes in heat-generating chemical activity. In humans, nonshivering thermogenesis is most important in newborns, because they lack the ability to shiver. Nonshivering thermogenesis is mediated by the hormones epinephrine and thyroid hormone, both of which increase heat production by stimulating fat metabolism. Newborns have deposits of a special type of adipose tissue known as **brown fat**, which is especially capable of converting chemical energy into heat. The role of nonshivering thermogenesis in adults remains controversial.

Having examined the mechanisms for adjusting heat production, we now turn to the other side of the equation, adjustments in heat loss.

▌ The magnitude of heat loss can be adjusted by varying the flow of blood through the skin.

Heat loss mechanisms are also subject to control, again largely by the hypothalamus. When we are hot, we want to increase heat loss to the environment; when we are cold, we want to decrease heat loss. The amount of heat lost to the environment by radiation and conduction-convection is largely determined by the temperature gradient between the skin and external environment. The body's central core is a heat-generating chamber in which the temperature must be maintained at approximately 100°F. Surrounding the core is an insulating shell through which heat exchanges between the body and external environment take place. To maintain a constant core temperature, the insulative capacity and temperature of the shell can be adjusted to vary the temperature gradient between the skin and external environment, thereby influencing the extent of heat loss.

The insulative capacity of the shell can be varied by controlling the amount of blood flowing through the skin. Skin blood flow serves two functions. First, it provides a nutritive blood supply to the skin. Second, as blood is pumped to the skin from the heart, it has been heated in the central core and carries this heat to the skin. Most skin blood flow is for the function of temperature regulation; at normal room temperature, 20 to 30 times more blood flows through the skin than is needed to meet the skin's nutritional needs.

In the process of thermoregulation, skin blood flow can vary tremendously, from 400 ml/min up to 2500 ml/min. The more blood that reaches the skin from the warm core, the closer the skin's temperature is to the core temperature. The skin's blood vessels diminish the effectiveness of the skin as an insulator by carrying heat to the surface, where it can be lost from the body by radiation and conduction-convection. Accordingly, vasodilation of the skin vessels (specifically, the arterioles), which permits increased flow of heated blood through the skin, increases heat loss. Conversely, vasoconstriction of the skin vessels, which reduces skin blood flow, decreases heat loss by keeping the warm blood in the central core, where it is insulated from the external environment. This response conserves heat that otherwise would have been lost.

These skin vasomotor responses are coordinated by the hypothalamus by means of sympathetic nervous system output. Increased sympathetic activity to the skin vessels produces heat-conserving vasoconstriction in response to cold exposure, whereas decreased sympathetic activity produces heat-losing vasodilation of the skin vessels in response to heat exposure.

The Extremes of Heat and Cold Can Be Fatal

Prolonged exposure to temperature extremes in either direction can overtax the body's thermoregulatory mechanisms, leading to disorders and even death if severe enough.

Heat-Related Disorders

Heat exhaustion is a state of collapse, usually manifested by fainting, that is caused by reduced blood pressure brought about as a result of overtaxing the heat loss mechanisms. Extensive sweating reduces cardiac output by depleting the plasma volume, and pronounced skin vasodilation causes a drop in total peripheral resistance. Because blood pressure is determined by cardiac output times total peripheral resistance, blood pressure falls, an insufficient amount of blood is delivered to the brain, and fainting takes place. Thus heat exhaustion is a consequence of overactivity of the heat loss mechanisms rather than a breakdown of these mechanisms. Because the heat loss mechanisms have been very active, body temperature is only mildly elevated in heat exhaustion. By forcing the cessation of activity when the heat loss mechanisms are no longer able to cope with heat gain through exercise or a hot environment, heat exhaustion serves as a safety valve to help prevent the more serious consequences of heat stroke.

Heat stroke is an extremely dangerous situation that arises from the complete breakdown of the hypothalamic thermoregulatory systems. Heat exhaustion may progress into heat stroke if the heat loss mechanisms continue to be overtaxed. Heat stroke is more likely to occur on overexertion during a prolonged exposure to a hot, humid environment. The elderly, in whom thermoregulatory responses are generally slower and less efficient, are particularly vulnerable to heat stroke during prolonged, stifling heat waves. So too are individuals who are taking certain common tranquilizers, such as Valium, because these drugs interfere with the hypothalamic thermoregulatory centers' neurotransmitter activity.

The most striking feature of heat stroke is a lack of compensatory heat loss measures, such as sweating, in the face of a rapidly rising body temperature. No sweating occurs, despite a markedly elevated body temperature, because the hypothalamic thermoregulatory control centers are not functioning properly and cannot initiate heat loss mechanisms. During the development of heat stroke, body temperature starts to climb as the heat loss mechanisms are eventually overwhelmed by prolonged, excessive heat gain. Once the core temperature reaches the point at which the hypothalamic temperature-control centers are damaged by the heat, the body temperature rapidly rises even higher because of the complete shutdown of heat loss mechanisms. Furthermore, as the body temperature increases, the rate of metabolism increases correspondingly, because higher temperatures speed up the rate of all chemical reactions; the result is even greater heat production. This positive-feedback state sends the temperature spiraling upward. Heat stroke is a very dangerous situation that is rapidly fatal if untreated. Even with treatment to halt and reverse the rampant rise in body temperature, there is still a high rate of mortality. The rate of permanent disability in survivors is also high because of irreversible protein denaturation caused by the high internal heat.

Cold-Related Disorders

At the other extreme, the body can be harmed by cold exposure in two ways: frostbite and generalized hypothermia. **Frostbite** involves excessive cooling of a particular part of the body to the point where tissue in that area is damaged. If exposed tissues actually freeze, tissue damage results from disruption of the cells by formation of ice crystals or by lack of liquid water.

Hypothermia, a fall in body temperature, occurs when generalized cooling of the body exceeds the ability of the normal heat-producing and heat-conserving regulatory mechanisms to match the excessive heat loss. As hypothermia sets in, the rate of all metabolic processes slows down because of the declining temperature. Higher cerebral functions are the first affected by body cooling, leading to loss of judgment, and to apathy, disorientation, and tiredness, all of which diminish the cold victim's ability to initiate voluntary mechanisms to reverse the falling body temperature. As body temperature continues to plummet, depression of the respiratory center occurs, reducing the ventilatory drive so that breathing becomes slow and weak. Activity of the cardiovascular system also is gradually reduced. The heart is slowed and cardiac output decreased. Cardiac rhythm is disturbed, eventually leading to ventricular fibrillation and death.

▌ The hypothalamus simultaneously coordinates heat production and heat loss mechanisms.

Let's now pull together the coordinated adjustments in heat production and heat loss in response to exposure to either a cold or a hot environment (▲ Table 16-2).

COORDINATED RESPONSES TO COLD EXPOSURE

In response to cold exposure, the posterior region of the hypothalamus directs increased heat production such as by shivering, while simultaneously decreasing heat loss (that is, conserving heat) by skin vasoconstriction and other measures.

Because there is a limit to the body's ability to reduce skin temperature through vasoconstriction, even maximum vasoconstriction is not sufficient to prevent excessive heat loss when the external temperature falls too low. Accordingly, other measures must be instituted to further reduce heat loss. In animals with dense fur or feathers, the hypothalamus, acting through the sympathetic nervous system, brings about contraction of the tiny muscles at the base of the hair or feather shafts to lift the hair or feathers off the skin surface. This puffing up traps a layer of poorly conductive air between the skin surface and environment, thus increasing the insulating barrier between the core and the cold air and reducing heat loss. Even though the hair shaft muscles contract in humans in response to cold exposure, this heat retention mechanism is ineffective because of the low density and fine texture of most human body hair. The result instead is useless *goosebumps*.

After maximum skin vasoconstriction has been achieved as a result of exposure to cold, further heat dissipation in humans can be prevented only by behavioral adaptations, such as postural changes that reduce as much as possible the exposed surface area from which heat can escape. These postural

▲ **TABLE 16-2**

Coordinated Adjustments in Response to Cold or Heat Exposure

IN RESPONSE TO COLD EXPOSURE (coordinated by the posterior hypothalamus)		IN RESPONSE TO HEAT EXPOSURE (coordinated by the anterior hypothalamus)	
Increased Heat Production	**Decreased Heat Loss (heat conservation)**	**Decreased Heat Production**	**Increased Heat Loss**
Increased muscle tone	Skin vasoconstriction	Decreased muscle tone	Skin vasodilation
Shivering	Postural changes to reduce exposed surface area (hunching shoulders, etc.)*	Decreased voluntary exercise*	Sweating
Increased voluntary exercise*	Warm clothing*		Cool clothing*
Nonshivering thermogenesis			

*Behavioral adaptations.

changes include maneuvers such as hunching over, clasping the arms in front of the chest, or curling up in a ball. Putting on warmer clothing further insulates the body from too much heat loss.

COORDINATED RESPONSES TO HEAT EXPOSURE

Under the opposite circumstance—heat exposure—the anterior part of the hypothalamus reduces heat production by decreasing skeletal muscle activity and promotes increased heat loss by inducing skin vasodilation. When even maximal skin vasodilation is inadequate to rid the body of excess heat, sweating is brought into play to accomplish further heat loss through evaporation. In fact, if the air temperature rises above the temperature of maximally vasodilated skin, the temperature gradient reverses itself so that heat is gained from the environment. Sweating is the only means of heat loss under these conditions.

Humans also employ voluntary measures, such as using fans, wetting the body, drinking cold beverages, and wearing cool clothing, to further enhance heat loss. Contrary to popular belief, wearing light-colored, loose clothing is cooler than being nude. Naked skin absorbs almost all the radiant energy that strikes it, whereas light-colored clothing reflects almost all the radiant energy that falls on it. Thus if light-colored clothing is loose and thin enough to permit convection currents and evaporative heat loss to occur, wearing it is actually cooler than going without any clothes at all. (For a discussion of the effects of extreme heat or cold exposure, see the accompanying boxed feature, ▶ Beyond the Basics.)

▌ **During a fever, the hypothalamic thermostat is "reset" at an elevated temperature.**

Clinical Note The term **fever** refers to an elevation in body temperature as a result of infection or inflammation. In response to microbial invasion, certain phagocytic cells (macrophages) release a chemical known as **endogenous pyrogen**, which, among its many infection-fighting effects (see p. 333), acts on the hypothalamic thermoregulatory center to raise the setting of the thermostat (● Figure 16-5). The hypothalamus now maintains the temperature at the new set level instead of maintaining normal body temperature. If, for example, endogenous pyrogen raises the set point to 102°F,

● **FIGURE 16-5**

Fever production

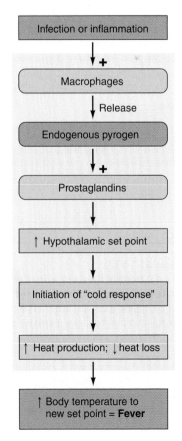

the hypothalamus senses that the normal prefever temperature is too cold, so it initiates the cold response mechanisms to raise the temperature to 102°F. Specifically, the hypothalamus initiates shivering to rapidly increase heat production, and promotes skin vasoconstriction to rapidly reduce heat loss, both of which drive the temperature upward. These events account for the sudden cold chills often experienced at the onset of a fever. Feeling cold, the person may put on more blankets as a voluntary mechanism that helps raise body temperature by conserving body heat. Once the new temperature is achieved, body temperature is regulated as normal in response to cold and heat but at a higher setting. Thus fever production in response to an infection is a deliberate outcome and is not due to a breakdown of thermoregulatory mechanisms. Although the physiological significance of a fever is still unclear, many medical experts believe that a rise in body temperature has a beneficial role in fighting infection. A fever augments the inflammatory response and may interfere with bacterial multiplication.

During fever production, endogenous pyrogen raises the set point of the hypothalamic thermostat by triggering the local release of *prostaglandins*, which are local chemical mediators that act directly on the hypothalamus. Aspirin reduces a fever by inhibiting the synthesis of prostaglandins. Aspirin does not lower the temperature in a nonfebrile person, because in the absence of endogenous pyrogen, prostaglandins are not present in the hypothalamus in appreciable quantities.

The exact molecular cause of a fever "breaking" naturally is unknown, although it presumably results from reduced pyrogen release or decreased prostaglandin synthesis. When the hypothalamic set point is restored to normal, the temperature at 102°F (in this example) is too high. The heat response mechanisms are instituted to cool down the body. Skin vasodilation occurs, and sweating commences. The person feels hot and throws off extra covers. The gearing up of these heat loss mechanisms by the hypothalamus reduces the temperature to normal.

 Click on the Media Exercises menu of the CD-ROM and work Media Exercise 16.2: Body Temperature Regulation to test your understanding of the previous section.

CHAPTER IN PERSPECTIVE: FOCUS ON HOMEOSTASIS

Because energy can be neither created or destroyed, for body weight and body temperature to remain constant, input must equal output in the case of, respectively, the body's total energy balance and its heat energy balance. If total energy input exceeds total energy output, the extra energy is stored in the body, and body weight increases. Similarly, if the input of heat energy exceeds its output, body temperature increases. Conversely, if output exceeds input, body weight decreases or body temperature falls. The hypothalamus is the major integrating center for maintaining both a constant total energy balance (and thus a constant body weight) and a constant heat energy balance (and thus a constant body temperature).

Body temperature, which is one of the homeostatically regulated factors of the internal environment, must be maintained within narrow limits, because the structure and reactivity of the chemicals that compose the body are temperature sensitive. Deviations in body temperature outside a limited range result in protein denaturation and death of the individual if the temperature rises too high, or metabolic slowing and death if the temperature falls too low.

Body weight, in contrast, varies widely among individuals. Only the extremes of imbalances between total energy input and output become incompatible with life. For example, in the face of insufficient energy input in the form of ingested food during prolonged starvation, the body resorts to breaking down muscle protein to meet its needs for energy expenditure once the adipose stores are depleted. Body weight dwindles because of this self-cannibalistic mechanism until death finally occurs as a result of loss of heart muscle, among other things. At the other extreme, when the food energy consumed greatly exceeds the energy expended, the extra energy input is stored as adipose tissue, and body weight increases. The resultant gross obesity can also lead to heart failure. Not only must the heart work harder to pump blood to the excess adipose tissue, but obesity also predisposes the person to atherosclerosis and heart attacks (see p. 265).

CHAPTER SUMMARY

Energy Balance (pp. 511–517)

▮ Energy input to the body in the form of food energy must equal energy output, because energy cannot be created or destroyed.

▮ Energy output or expenditure includes (1) external work, performed by skeletal muscles to move an external object or move the body through the external environment; and (2) internal work, which consists of all other energy-dependent activities that do not accomplish external work, including active transport, smooth and cardiac muscle contraction, glandular secretion, and protein synthesis. *(Review Figure 16-1.)*

▮ Only about 25% of the chemical energy in food is harnessed to do biological work. The rest is immediately converted to heat.

Furthermore, all the energy expended to accomplish internal work is eventually converted into heat, and 75% of the energy expended by working skeletal muscles is lost as heat. Therefore, most of the energy in food ultimately appears as body heat.

▮ The metabolic rate, which is energy expenditure per unit of time, is measured in kilocalories of heat produced per hour.

▮ The basal metabolic rate (BMR) is a measure of the body's minimal waking rate of internal energy expenditure.

▮ For a neutral energy balance, the energy in ingested food must equal energy expended in performing external work and transformed into heat. If more energy is consumed than is expended, the extra energy is stored in the body, primarily as adipose tissue, so body weight increases. By contrast, if more energy is expended than is available in the food, body energy stores are used to support energy expenditure, so body weight decreases.

■ Usually, body weight remains fairly constant over a prolonged period of time (except during growth) because food intake is adjusted to match energy expenditure on a long-term basis.

■ Food intake is controlled primarily by the hypothalamus by means of complex, incompletely understood regulatory mechanisms in which hunger and satiety are important components. Feeding or appetite signals give rise to the sensation of hunger and promote eating, whereas satiety signals lead to the sensation of fullness and suppress eating.

■ The arcuate nucleus of the hypothalamus plays a key role in energy homeostasis by virtue of the two clusters of appetite-regulating neurons it contains: neurons that secrete neuropeptide Y (NPY), which increases appetite and food intake, and neurons that secrete melanocortins, which suppress appetite and food intake. (Review Figure 16-2.)

■ Adipocytes in fat stores secrete the hormone leptin, which reduces appetite and decreases food consumption by inhibiting the NPY-secreting neurons and stimulating the melanocortins-secreting neurons of the arcuate nucleus. This mechanism is primarily important in the long-term matching of energy intake with energy output, thus maintaining body weight over the long term. (Review Figure 16-2.)

■ Short-term control of the timing and size of meals is mediated primarily by the actions of signals arising from the digestive tract and pancreas. Of major importance are two peptides secreted by the digestive tract. (1) Ghrelin, a mealtime initiator, is secreted by the stomach before a meal and signals hunger. Its secretion drops when food is consumed. Ghrelin stimulates appetite and promotes feeding behavior by stimulating the NPY-secreting neurons. (2) PYY_{3-36}, a mealtime terminator, is secreted by the small and large intestines during a meal and signals satiety. Its secretion is lowest before a meal. PYY_{3-36} inhibits the NPY-secreting neurons and stimulates the melanocortins-secreting neurons. (Review Figure 16-2.)

■ Other satiety signals that tell you when to end a meal include (1) increased glucose use, (2) increased insulin, and (3) increased cholecystokinin.

■ Psychosocial and environmental factors can also influence food intake above and beyond the internal signals that govern feeding behavior.

Temperature Regulation (pp. 518–524)

■ The body can be thought of as a heat-generating core (internal organs, CNS, and skeletal muscles) surrounded by a shell of variable insulating capacity (the skin).

■ The skin exchanges heat energy with the external environment, with the direction and amount of heat transfer depending on the environmental temperature and the momentary insulating capacity of the shell. (Review Figure 16-3.)

■ The four physical means by which heat is exchanged between the body and external environment are (1) radiation (net movement of heat energy via electromagnetic waves); (2) conduction (exchange of heat energy by direct contact); (3) convection (transfer of heat energy by means of air currents); and (4) evaporation (extraction of heat energy from the body by the heat-requiring conversion of liquid H_2O to H_2O vapor). Because heat energy moves from warmer to cooler objects, radiation, conduction, and convection can be channels for either heat loss or heat gain, depending on whether surrounding objects are cooler or warmer, respectively, than the body surface. Normally, they are avenues for heat loss, along with evaporation resulting from sweating. (Review Figure 16-4.)

■ To prevent serious cell malfunction, the core temperature must be held constant at about 100°F (equivalent to an average oral temperature of 98.2°F) by continuously balancing heat gain and heat loss despite changes in environmental temperature and variation in internal heat production.

■ This thermoregulatory balance is controlled by the hypothalamus. The hypothalamus is apprised of the skin temperature by peripheral thermoreceptors and of the core temperature by central thermoreceptors, the most important of which are located in the hypothalamus itself.

■ The primary means of heat gain is heat production by metabolic activity, the biggest contributor being skeletal muscle contraction.

■ Heat loss is adjusted by sweating and by controlling to the greatest extent possible the temperature gradient between the skin and surrounding environment. The latter is done by regulating the caliber of the skin's blood vessels. (1) Vasoconstriction of the skin vessels reduces the flow of warmed blood through the skin so that skin temperature falls. The layer of cool skin between the core and environment increases the insulating barrier between the warm core and the external air. (2) Conversely, skin vasodilation brings more warmed blood through the skin so that skin temperature approaches the core temperature, thus reducing the insulative capacity of the skin.

■ On exposure to cool surroundings the core temperature starts to fall as heat loss increases, because of the larger-than-normal skin-to-air temperature gradient. The hypothalamus responds to reduce the heat loss by inducing skin vasoconstriction while simultaneously increasing heat production through heat-generating shivering. (Review Table 16-2.)

■ Conversely, in response to a rise in core temperature (resulting either from excessive internal heat production accompanying exercise or from excessive heat gain on exposure to a hot environment), the hypothalamus triggers heat loss mechanisms, such as skin vasodilation and sweating, while simultaneously decreasing heat production, such as by reducing muscle tone. (Review Table 16-2.)

■ In both cold and heat responses, voluntary behavioral actions also contribute importantly to maintenance of thermal homeostasis.

■ A fever occurs when endogenous pyrogen released from white blood cells in response to infection raises the hypothalamic set point. An elevated core temperature develops as the hypothalamus initiates cold response mechanisms to raise the core temperature to the new set point. (Review Figure 16-5.)

REVIEW EXERCISES

Objective Questions (Answers on p. A-48)

1. If more food energy is consumed than is expended, the excess energy is lost as heat. (True or false?)

2. All the energy within nutrient molecules can be harnessed to perform biological work. (True or false?)

3. Each liter of O_2 contains 4.8 kilocalories of heat energy. (*True or false?*)

4. A body temperature greater than 98.2°F is always indicative of a fever. (*True or false?*)

5. Core temperature is relatively constant, but skin temperature can vary markedly. (*True or false?*)

6. Sweat that drips off the body has no cooling effect. (*True or false?*)

7. Production of "goosebumps" in response to cold exposure has no value in regulating body temperature. (*True or false?*)

8. The posterior region of the hypothalamus triggers shivering and skin vasoconstriction. (*True or false?*)

9. The _____ of the hypothalamus contains two sets of neurons that act in opposition to each other to regulate food intake and energy balance.

10. The primary means of involuntarily increasing heat production is _____.

11. Increased heat production independent of muscle contraction is known as _____.

12. The only means of heat loss when the environmental temperature exceeds the core temperature is _____.

13. Which of the following statements concerning heat exchange between the body and the external environment is *incorrect*?
    a. Heat gain is primarily by means of internal heat production.
    b. Radiation serves as a means of heat gain but not of heat loss.
    c. Heat energy always moves down its concentration gradient from warmer to cooler objects.
    d. The temperature gradient between the skin and the external air is subject to control.
    e. Very little heat is lost from the body by conduction alone.

14. Which of the following statements concerning fever production is *incorrect*?
    a. Endogenous pyrogen is released by white blood cells in response to microbial invasion.

b. The hypothalamic set point is elevated.
c. The hypothalamus initiates cold response mechanisms to increase the body temperature.
d. Prostaglandins appear to mediate the effect.
e. The hypothalamus is not effective in regulating body temperature during a fever.

15. Using the following answer code, indicate which mechanism of heat transfer is being described:
    (a) radiation   (c) convection
    (b) conduction   (d) evaporation
    ____ 1. sitting on a cold metal chair
    ____ 2. sunbathing on the beach
    ____ 3. a gentle breeze
    ____ 4. sitting in front of a fireplace
    ____ 5. sweating
    ____ 6. riding in a car with the windows open
    ____ 7. lying under an electric blanket
    ____ 8. sitting in a wet bathing suit
    ____ 9. fanning yourself
    ____ 10. immersion in cold water

Essay Questions
1. Differentiate between external and internal work.
2. Define *metabolic rate* and *basal metabolic rate*. Explain the process of indirect calorimetry.
3. Describe the three states of energy balance.
4. By what means is energy balance primarily maintained?
5. List the sources of heat input and output for the body.
6. Describe the role of the following in the long-term regulation of energy balance and the short-term control of the timing and size of meals: neuropeptide Y, melanocortins, adipocytes, leptin, ghrelin, PYY_{3-36}, glucose use, insulin, and cholecystokinin.
7. Discuss the compensatory measures that occur in response to a fall in core temperature as a result of cold exposure and in response to a rise in core temperature as a result of heat exposure.

POINTS TO PONDER

(Explanations on p. A-48)
1. Explain how drugs that selectively inhibit CCK increase feeding behavior in experimental animals.
2. What advice would you give an overweight friend who asks for your help in designing a safe, sensible, inexpensive program for losing weight?
3. Why is it dangerous to engage in heavy exercise on a hot, humid day?
4. Describe the avenues for heat loss in a person soaking in a hot bath.
5. Consider the difference between you and a fish in a local pond with regard to control of body temperature. Humans are *thermoregulators;* they can maintain a remarkably constant, rather high inter-

nal body temperature despite the body's exposure to a wide range of environmental temperatures. To maintain thermal homeostasis, humans physiologically manipulate mechanisms within their bodies to adjust heat production, heat conservation, and heat loss. In contrast, fish are *thermoconformers;* their body temperatures conform to the temperature of their surroundings. Thus their body temperatures vary capriciously with changes in the environmental temperature. Even though fish produce heat, they cannot physiologically regulate internal heat production, nor can they control heat exchange with their environment to maintain a constant body temperature when the temperature in their surroundings rises or falls. Knowing this, do you think fish run a fever when they have a systemic infection? Why or why not?

CLINICAL CONSIDERATION

(Explanation on p. A-48)
Michael F., a near-drowning victim, was pulled from the icy water by rescuers 15 minutes after he fell through the thin ice on which he was skating. Michael is now alert and recuperating in the hospital.

How can you explain his "miraculous" survival even though he was submerged for 15 minutes and irreversible brain damage, soon followed by death, normally occurs if the brain is deprived of its critical O_2 supply for more than four or five minutes?

PHYSIOEDGE RESOURCES

 PhysioEdge CD-ROM

PhysioEdge, the CD-ROM packaged with your text, focuses on the concepts students find most difficult to learn. Figures marked with this icon have associated activities on the CD. For a visual review of concepts in this chapter, check out the following:

Media Exercise 16.1: Basics of Energy Balance

Media Exercise 16.2: Body Temperature Regulation

PhysioEdge Website

The website for this book contains a wealth of helpful study aids, as well as many ideas for further reading and research. Log on to:
http://www.brookscole.com/hpfundamentals3
Select Chapter 16 from the drop-down menu or click on one of the many resource areas.

For Suggested Readings, consult **InfoTrac College Edition/ Research** on the PhysioEdge website or go directly to InfoTrac College Edition, your online research library, at:
http://infotrac.thomsonlearning.com

Endocrine System

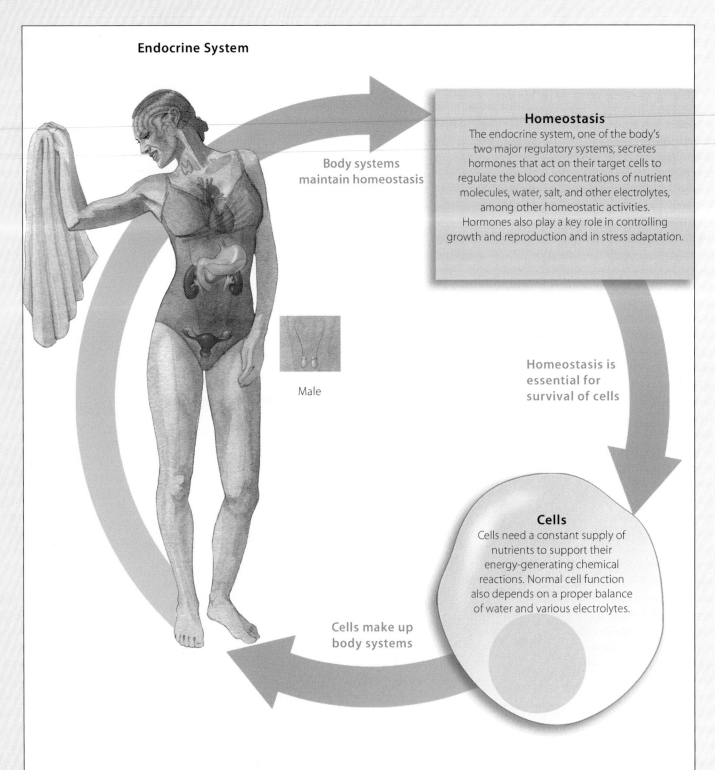

Homeostasis

The endocrine system, one of the body's two major regulatory systems, secretes hormones that act on their target cells to regulate the blood concentrations of nutrient molecules, water, salt, and other electrolytes, among other homeostatic activities. Hormones also play a key role in controlling growth and reproduction and in stress adaptation.

Body systems maintain homeostasis

Male

Homeostasis is essential for survival of cells

Cells

Cells need a constant supply of nutrients to support their energy-generating chemical reactions. Normal cell function also depends on a proper balance of water and various electrolytes.

Cells make up body systems

The **endocrine system,** by means of the blood-borne **hormones** it secretes, generally regulates activities that require duration rather than speed. Most target-cell activities under hormonal control are directed toward maintaining homeostasis. The **pineal gland** secretes a hormone important in establishing the body's biological rhythms. The **hypothalamus** and **posterior pituitary gland** act as a unit to release hormones essential for maintaining water balance and for giving birth and for breast-feeding. The **anterior pituitary gland** secretes hormones that promote growth and control the hormonal output of several other endocrine glands. The **thyroid gland** controls the body's basal metabolic rate. The **adrenal glands** secrete hormones important in metabolizing nutrient molecules, in adapting to stress, and in maintaining salt balance. The **endocrine pancreas** secretes hormones important in metabolizing nutrient molecules. The **parathyroid glands** secrete a hormone important in Ca^{2+} metabolism.

The Endocrine System

GENERAL PRINCIPLES OF ENDOCRINOLOGY

The **endocrine system** consists of the ductless endocrine glands (see p. 5) that are scattered throughout the body (● Figure 17-1). Even though the endocrine glands for the most part are not connected anatomically, they constitute a system in a functional sense. They all accomplish their functions by secreting hormones into the blood, and many functional interactions take place among the various endocrine glands. Once secreted, a hormone travels in the blood to its distant target cells, where it regulates or directs a particular function. Even though the blood distributes hormones throughout the body, only specific target cells can respond to each hormone, because only the target cells have receptors for binding with the particular hormone (see p. 103).

The binding of a hormone with its specific target-cell receptors initiates a chain of events within the target cells to bring about the hormone's final effect. Recall that the means by which a hormone brings about its ultimate physiologic effect depends on whether the hormone is hydrophilic (peptide hormones and catecholamines) or lipophilic (steroid hormones and thyroid hormone). *Peptide hormones*, the most abundant chemical category of hormone, are chains of amino acids of varying length. *Catecholamines*, produced by the adrenal medulla, are derived from the amino acid tyrosine. *Steroid hormones*, produced by the adrenal cortex and reproductive endocrine glands, are neutral lipids derived from cholesterol. *Thyroid hormone*, produced only by the thyroid gland, is an iodinated tyrosine derivative. To review, hydrophilic hormones on binding with surface membrane receptors primarily act through second-messenger systems to alter the activity of preexisting proteins, such as enzymes, within the target cell. Lipophilic hormones, by contrast, activate genes on binding with receptors in the nucleus, thus bringing about formation of new proteins in the target cell that carry out the desired response. Hydrophilic hormones circulate in the blood largely dissolved in the plasma, whereas lipophilic hormones are largely bound to plasma proteins (see pp. 95–102).

Some organs are exclusively endocrine in function (they specialize in hormone secretion alone, the anterior pituitary being an example), whereas other organs of the endocrine system perform nonendocrine functions in addition to secreting hormones. For example, the testes produce sperm and also secrete the male sex hormone testosterone.

Hormones exert a variety of regulatory effects throughout the body.

The endocrine system is one of the body's two major regulatory systems, the other being the nervous system, with which you are already familiar (Chapters 4 through 7). The endocrine system primarily controls activities that require duration rather than speed, including the following:

1. Regulating organic metabolism and H_2O and electrolyte balance, which are important collectively in maintaining a constant internal environment
2. Inducing adaptive changes to help the body cope with stressful situations
3. Promoting smooth, sequential growth and development
4. Controlling reproduction
5. Regulating red blood cell production
6. Along with the autonomic nervous system, controlling and integrating both circulation and the digestion and absorption of food

Some hormones regulate the production and secretion of another hormone. A hormone that has as its primary function the regulation of hormone secretion by another endocrine gland is classified functionally as a **tropic hormone** (*tropic* means "nourishing"). Tropic hormones stimulate and maintain their endocrine target tissues. For example, the tropic hormone thyroid-stimulating hormone (TSH), from the anterior pituitary, stimulates thyroid hormone secretion by the thyroid gland and also maintains the structural integrity of this gland. In the absence of TSH, the thyroid gland atrophies (shrinks) and produces very low levels of its hormone.

The plasma concentration of a hormone is normally regulated by changes in its rate of secretion.

The primary function of most hormones is the regulation of various homeostatic activities. Because hormones' effects are proportional to their concentrations in the plasma, these concentrations are subject to control according to homeostatic

● **FIGURE 17-1**

The endocrine system

PhysioEdge For an interaction related to this figure, see Media Exercises 17.1: Overview of the Classic Endocrine Glands and 17.4: Endocrine Roles by Organs of Mixed or Uncertain Functions on the CD-ROM.

■ Solely endocrine function

■ Mixed function

■ Complete function uncertain

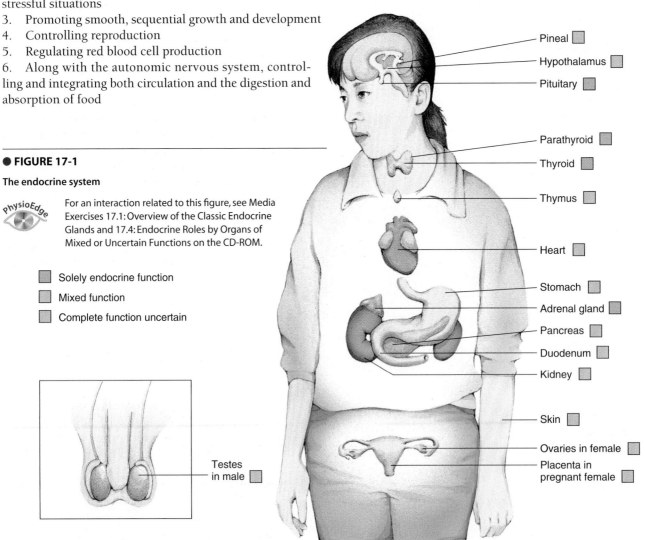

Pineal

Hypothalamus

Pituitary

Parathyroid

Thyroid

Thymus

Heart

Stomach

Adrenal gland

Pancreas

Duodenum

Kidney

Skin

Ovaries in female

Placenta in pregnant female

Testes in male

need. Normally, the plasma concentration of a hormone is regulated by appropriate adjustments in the rate of its secretion. Endocrine glands do not secrete their hormones at a constant rate; the secretion rates of all hormones vary subject to control, often by a combination of several complex mechanisms. The regulatory system for each hormone is considered in detail in later sections. For now, we will address these general mechanisms of controlling secretion that are common to many different hormones: negative-feedback control, neuroendocrine reflexes, and diurnal (circadian) rhythms.

NEGATIVE-FEEDBACK CONTROL

Negative feedback is a prominent feature of hormonal control systems. Stated simply, *negative feedback exists when the output of a system counteracts a change in input,* maintaining a controlled variable within a narrow range around a set level (see p. 13). Negative feedback maintains the plasma concentration of a hormone at a given level, similar to the way in which a home heating system maintains the room temperature at a given set point. Control of hormonal secretion provides some classic physiologic examples of negative feedback. For example, when the plasma concentration of free circulating thyroid hormone falls below a given "set point," the anterior pituitary secretes thyroid-stimulating hormone (TSH), which stimulates the thyroid to increase its secretion of thyroid hormone (● Figure 17-2). Thyroid hormone in turn inhibits further secretion of TSH by the anterior pituitary. Negative feedback ensures that once thyroid gland secretion has been "turned on" by TSH, it will not continue unabated but instead will be "turned off" when the appropriate level of free circulating thyroid hormone has been achieved. Thus the effect of a particular hormone's actions can inhibit its own secretion. The feedback loops often become quite complex.

● FIGURE 17-2

Negative-feedback control

 For an interaction related to this figure, see Media Exercise 17.2: Anterior Pituitary Gland on the CD-ROM.

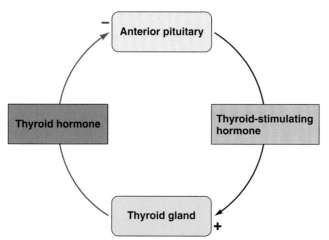

NEUROENDOCRINE REFLEXES

Many endocrine control systems involve **neuroendocrine reflexes,** which include neural as well as hormonal components. The purpose of such reflexes is to produce a sudden increase in hormone secretion (that is, "turn up the thermostat setting") in response to a specific stimulus, frequently a stimulus external to the body. Some endocrine control systems include both feedback control (which maintains a constant basal level of the hormone) and neuroendocrine reflexes (which cause sudden bursts in secretion in response to a sudden increased need for the hormone). An example is the increased secretion of cortisol, the "stress hormone," by the adrenal cortex during a stress response (see ● Figure 17-20, p. 555).

DIURNAL (CIRCADIAN) RHYTHMS

The secretion rates of many hormones rhythmically fluctuate up and down as a function of time. The most common endocrine rhythm is the **diurnal** ("day–night"), or **circadian** ("around a day") **rhythm,** which is characterized by repetitive oscillations in hormone levels that are very regular and cycle once every 24 hours. This rhythmicity is caused by endogenous oscillators similar to the self-paced respiratory neurons in the brain stem that control the rhythmic motions of breathing, except the timekeeping oscillators cycle on a much longer time scale. Furthermore, unlike the rhythmicity of breathing, endocrine rhythms are locked on, or **entrained,** to external cues such as the light–dark cycle. That is, the inherent 24-hour cycles of peak and ebb of hormone secretion are set to "march in step" with cycles of light and dark. For example, cortisol secretion rises during the night, reaching its peak secretion in the morning before a person gets up, then falls throughout the day to its lowest level at bedtime (● Figure 17-3). Inherent hormonal rhythmicity and entrainment are not accomplished by the endocrine glands themselves but result from the central nervous system changing the set point of these glands. We discuss the master biological clock further in a later section. Negative-feedback control mechanisms operate to maintain whatever set point is estab-

● FIGURE 17-3

Diurnal rhythm of cortisol secretion

(*Source:* Adapted from *Clinical Endocrine Physiology,* by George A. Hedge, Howard D. Colby, and Robert L. Goodman, Figure 1-13, p. 28. © 1987, with permission from Elsevier.

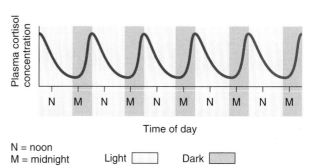

lished for that time of day. Some endocrine cycles operate on time scales other than a circadian rhythm, a well-known example being the monthly menstrual cycle.

▌ The plasma concentration of a hormone is influenced by its rate of excretion.

Even though the plasma concentration of a hormone is normally regulated by adjusting its rate of secretion, alterations in its rate of excretion can also influence the hormone's plasma concentration, sometimes inappropriately.

Hormones and their metabolites are typically eliminated from the blood by urinary excretion. In contrast to the tight controls on hormone secretion, hormone inactivation and excretion are not regulated.

The liver is the most common site for metabolic hormone inactivation, but some hormones are also inactivated in the blood, kidneys, or target cells. The amount of time after a hormone is secreted before it is inactivated, and the means by which this takes place, differ for different classes of hormones. In general, the hydrophilic peptides and catecholamines are easy targets for blood and tissue enzymes, so they remain in the blood only briefly (a few minutes to a few hours) before being enzymatically inactivated. In the case of some peptide hormones, such as insulin, the target cell actually engulfs the bound hormone by endocytosis and degrades it intracellularly. In contrast, binding of lipophilic hormones to plasma proteins makes them less vulnerable to metabolic inactivation and keeps them from escaping into urine. Therefore, lipophilic hormones are removed from plasma much more slowly. They may persist in the blood for hours (steroids) or up to a week (thyroid hormone). Lipophilic hormones typically undergo a series of reactions that reduce their biologic activity and make them more water soluble so they can be freed from their plasma protein carriers and be eliminated in urine.

Clinical Note When liver and kidney function are normal, measuring urinary concentrations of hormones and their metabolites provides a useful, noninvasive way to assess endocrine function, because the rate of excretion of these products in urine directly reflects their rate of secretion by the endocrine glands. Because the liver and kidneys are important in removing hormones from the blood, patients with liver or kidney disease may suffer from excess activity of certain hormones solely because hormone elimination is reduced.

▌ Endocrine disorders result from hormone excess or deficiency or decreased target-cell responsiveness.

Clinical Note Endocrine disorders most commonly result from abnormal plasma concentrations of a hormone caused by inappropriate rates of secretion—that is, too little hormone secreted (**hyposecretion**) or too much hormone secreted (**hypersecretion**). Occasionally endocrine dys-

function arises because target cell responsiveness to the hormone is abnormally low, even though plasma concentration of the hormone is normal. This unresponsiveness may be caused, for example, by an inborn lack of receptors for the hormone, as in *testicular feminization syndrome*. In this condition, receptors for testosterone, a masculinizing hormone produced by the male testes, are not produced because of a specific genetic defect. Although adequate testosterone is available, masculinization does not take place, just as if no testosterone were present. Abnormal responsiveness may also occur if the target cells for a particular hormone lack an enzyme essential to carrying out the response.

▌ The responsiveness of a target cell can be varied by regulating the number of hormone-specific receptors.

In contrast to endocrine dysfunction caused by *unintentional* receptor abnormalities, the target cell receptors for a particular hormone can be *deliberately altered* as a result of physiologic control mechanisms. A target cell's response to a hormone is correlated with the number of the cell's receptors occupied by molecules of that hormone, which in turn depends not only on the plasma concentration of the hormone but also on the number of receptors in the target cell for that hormone. Thus the response of a target cell to a given plasma concentration can be fine-tuned up or down by varying the number of receptors available for hormone binding.

DOWN REGULATION

As an illustration of this fine-tuning, when the plasma concentration of insulin is chronically elevated, the total number of target cell receptors for insulin is reduced as a direct result of the effect an elevated level of insulin has on the insulin receptors. This phenomenon, known as **down regulation**, constitutes an important locally acting negative-feedback mechanism that prevents the target cells from overreacting to the high concentration of insulin; that is, the target cells are *desensitized* to insulin, helping blunt the effect of insulin hypersecretion.

PERMISSIVENESS, SYNERGISM, AND ANTAGONISM

A given hormone's effects are influenced not only by the concentration of the hormone itself but also by the concentrations of other hormones that interact with it. Because hormones are widely distributed through the blood, target cells may be exposed simultaneously to many different hormones, giving rise to numerous complex hormonal interactions on target cells. Hormones frequently alter the receptors for other kinds of hormones as part of their normal physiologic activity. A hormone can influence the activity of another hormone at a given target cell in one of three ways: permissiveness, synergism, and antagonism.

• With **permissiveness**, one hormone must be present in adequate amounts for the full exertion of another hormone's effect. In essence, the first hormone, by enhancing a target cell's responsiveness to another hormone, "permits" this other hormone to exert its full effect. For example, thyroid hormone increases the number of receptors for epinephrine in epinephrine's target cells, increasing the effectiveness of epinephrine. In the absence of thyroid hormone, epinephrine is only marginally effective.

• **Synergism** occurs when the actions of several hormones are complementary and their combined effect is greater than the sum of their separate effects. An example is the synergistic action of follicle-stimulating hormone and testosterone, both of which are required for maintaining the normal rate of sperm production. Synergism results from each hormone's influence on the number or affinity of receptors for the other hormone.

• **Antagonism** occurs when one hormone causes the loss of another hormone's receptors, reducing the effectiveness of the second hormone. To illustrate, progesterone (a hormone secreted during pregnancy that decreases contractions of the uterus) inhibits uterine responsiveness to estrogen (another hormone secreted during pregnancy that increases uterine contractions). By causing loss of estrogen receptors on uterine smooth muscle, progesterone prevents estrogen from exerting its excitatory effects during pregnancy and thus keeps the uterus a quiet (noncontracting) environment suitable for the developing fetus.

Having completed our discussion of the general principles of endocrinology, we now begin an examination of the individual endocrine glands and their hormones. ▲ Table 17-1 summarizes the most important specific functions of the major hormones. Some of these hormones have been introduced elsewhere and are not discussed further; these are the gastrointestinal hormones (Chapter 15), the renal hormones (erythropoietin in Chapter 11 and renin in Chapter 13), thrombopoietin from the liver (Chapter 11), thymosin from the thymus (Chapter 11), and atrial natriuretic peptide from the heart (Chapter 13). The remainder of the hormones are described in greater detail in this and the next chapter. We start with those in the brain itself or in close association with the brain—namely, the pineal gland, the hypothalamus, and the pituitary gland.

 Click on the Media Exercises menu of the CD-ROM and work Media Exercise 17.1: Overview of the Classic Endocrine Glands to test your understanding of the previous section.

▲ **TABLE 17-1**

Summary of the Major Hormones

ENDOCRINE GLAND	HORMONES	TARGET CELLS	MAJOR FUNCTIONS OF HORMONES
Hypothalamus	Releasing and inhibiting hormones (TRH, CRH, GnRH, GHRH, GHIH, PRH, PIH)	Anterior pituitary	Controls release of anterior pituitary hormones
Posterior Pituitary (hormones stored in)	Vasopressin (antidiuretic hormone)	Kidney tubules	Increases H_2O reabsorption
		Arterioles	Produces vasoconstriction
	Oxytocin	Uterus	Increases contractility
		Mammary glands (breasts)	Causes milk ejection
Anterior Pituitary	Thyroid-stimulating hormone (TSH)	Thyroid follicular cells	Stimulates T_3 and T_4 secretion
	Adrenocorticotropic hormone (ACTH)	Adrenal cortex	Stimulates cortisol secretion
	Growth hormone	Bone; soft tissues	Essential but not solely responsible for growth; stimulates growth of bones and soft tissues; metabolic effects include protein anabolism, fat mobilization, and glucose conservation
		Liver	Stimulates somatomedin secretion

(continued)

ENDOCRINE GLAND	HORMONES	TARGET CELLS	MAJOR FUNCTIONS OF HORMONES
Anterior Pituitary *(continued)*	Follicle-stimulating hormone (FSH)	*Females:* ovarian follicles	Promotes follicular growth and development; stimulates estrogen secretion
		Males: seminiferous tubules in testes	Stimulates sperm production
	Luteinizing hormone (LH) (interstitial cell–stimulating hormone—ICSH)	*Females:* ovarian follicle and corpus luteum	Stimulates ovulation, corpus luteum development, and estrogen and progesterone secretion
		Males: interstitial cells of Leydig in testes	Stimulates testosterone secretion
	Prolactin	*Females:* mammary glands	Promotes breast development; stimulates milk secretion
		Males	Uncertain
Thyroid Gland Follicular Cells	Tetraiodothyronine (T_4 or thyroxine); tri-iodothyronine (T_3)	Most cells	Increases the metabolic rate; essential for normal growth and nerve development
Thyroid Gland C Cells	Calcitonin	Bone	Decreases plasma calcium concentration
Adrenal Cortex			
Zona glomerulosa	Aldosterone (mineralocorticoid)	Kidney tubules	Increases Na^+ reabsorption and K^+ secretion
Zona fasciculata and zona reticularis	Cortisol (glucocorticoid)	Most cells	Increases blood glucose at the expense of protein and fat stores; contributes to stress adaption
	Androgens (dehydroepiandrosterone)	*Females:* bone and brain	Responsible for the pubertal growth spurt and sex drive in females
Adrenal Medulla	Epinephrine and norepinephrine	Sympathetic receptor sites throughout the body	Reinforces the sympathetic nervous system; contributes to stress adaptation and blood pressure regulation
Endocrine Pancreas (Islets of Langerhans)	Insulin (β cells)	Most cells	Promotes cellular uptake, use, and storage of absorbed nutrients
	Glucagon (α cells)	Most cells	Important for maintaining nutrient levels in blood during postabsorptive state
	Somatostatin (D cells)	Digestive system	Inhibits digestion and absorption of nutrients
		Pancreatic islet cells	Inhibits secretion of all pancreatic hormones
Parathyroid Gland	Parathyroid hormone (PTH)	Bone, kidneys, intestine	Increases plasma calcium concentration; decreases plasma phosphate concentration; stimulates vitamin D activation

(continued)

ENDOCRINE GLAND	HORMONES	TARGET CELLS	MAJOR FUNCTIONS OF HORMONES
Gonads			
Female: ovaries	Estrogen (estradiol)	Female sex organs; body as a whole	Promotes follicular development; governs development of secondary sexual characteristics: stimulates uterine and breast growth
		Bone	Promotes closure of the epiphyseal plate
	Progesterone	Uterus	Prepares for pregnancy
Male: testes	Testosterone	Male sex organs; body as a whole	Stimulates sperm production; governs development of secondary sexual characteristics; promotes sex drive
		Bone	Enhances pubertal growth spurt; promotes closure of the epiphyseal plate
Testes and ovaries	Inhibin	Anterior pituitary	Inhibits secretion of follicle-stimulating hormone
Pineal Gland	Melatonin	Brain; anterior pituitary; reproductive organs; immune system; possibly others	Entrains body's biological rhythm with external cues; believed to inhibit gonadotropins; initiation of puberty possibly caused by a reduction in melatonin secretion; acts as an antioxidant; enhances immunity
Placenta	Estrogen (estriol); progesterone	Female sex organs	Help maintain pregnancy; prepare breasts for lactation
	Chorionic gonadotropin	Ovarian corpus luteum	Maintains corpus luteum of pregnancy
Kidneys	Renin ($\rightarrow$ angiotensin)	Adrenal cortex (acted on by angiotensin, which is activated by renin)	Stimulates aldosterone secretion
	Erythropoietin	Bone marrow	Stimulates erythrocyte production
Stomach	Gastrin	Digestive-tract exocrine glands and smooth muscles; pancreas; liver; gallbladder	Control of motility and secretion to facilitate digestive and absorptive processes
Duodenum	Secretin; cholecystokinin		
	Glucose-dependent insulinotropic peptide	Endocrine pancreas	Stimulates insulin secretion
Liver	Somatomedins	Bone; soft tissues	Promotes growth
	Thrombopoietin	Bone marrow	Stimulates platelet production
Skin	Vitamin D	Intestine	Increases absorption of ingested calcium and phosphate
Thymus	Thymosin	T lymphocytes	Enhances T lymphocyte proliferation and function
Heart	Atrial natriuretic peptide	Kidney tubules	Inhibits Na^+ reabsorption

PINEAL GLAND AND CIRCADIAN RHYTHMS

The **pineal gland**, a tiny, pinecone-shaped structure located in the center of the brain (see ● Figure 5-5, p. 115, and ● Figure 17-1, p. 530), secretes the hormone **melatonin**. (Do not confuse melatonin with the skin-darkening pigment, *melanin*—see p. 357.) Although melatonin was discovered in 1959, investigators have only recently begun to unravel its many functions. One of melatonin's most widely accepted roles is helping to keep the body's inherent circadian rhythms in synchrony with the light–dark cycle. We will first examine circadian rhythms in general before looking at the role of melatonin in this regard and considering other functions of this hormone.

▌ The suprachiasmatic nucleus is the master biological clock.

Hormone secretion rates are not the only factor in the body that fluctuates cyclically over a 24-hour period. Humans have similar biological clocks for other bodily functions such as temperature regulation (see p. 518). The master biological clock that serves as the pacemaker for the body's circadian rhythms is the **suprachiasmatic nucleus (SCN)**. It consists of a cluster of nerve cell bodies in the hypothalamus above the optic chiasm, the point at which part of the nerve fibers from each eye cross to the opposite half of the brain (*supra* means "above"; *chiasm* means "cross") (see p. 163 and ● Figure 5-5, p. 115). The self-induced rhythmic firing of the SCN neurons plays a major role in establishing many of the body's inherent daily rhythms.

ROLE OF CLOCK PROTEINS

Scientists have now unraveled the underlying molecular mechanisms responsible for the SCN's circadian oscillations. Specific self-starting genes within the nuclei of SCN neurons set in motion a series of events that brings about the synthesis of **clock proteins** in the cytosol surrounding the nucleus. As the day wears on, these clock proteins continue to accumulate, finally reaching a critical mass, at which time they are transported back into the nucleus. Here they block the genetic process responsible for their own production. The level of clock proteins gradually dwindles as they degrade within the nucleus, thus removing their inhibitory influence from the clock-protein genetic machinery. No longer being blocked, these genes once again rev up the production of more clock proteins, as the cycle repeats itself. Each cycle takes about a day. The fluctuating levels of clock proteins bring about cyclic changes in neural output from the SCN that in turn lead to cyclic changes in effector organs throughout the day. An example is the diurnal variation in cortisol secretion (see ● Figure 17-3, p. 531). Circadian rhythms are thus linked to fluctuations in clock proteins, which use a feedback loop to control their own production. In this way, internal timekeeping is a self-sustaining mechanism built into the genetic makeup of the SCN neurons.

SYNCHRONIZATION OF THE BIOLOGICAL CLOCK WITH ENVIRONMENTAL CUES

On its own, this biological clock generally cycles a bit slower than the 24-hour environmental cycle. Without any external cues, the SCN sets up cycles that average about 25 hours. The cycles are consistent for a given individual but vary somewhat among different people. If this master clock were not continually adjusted to keep pace with the world outside, the body's circadian rhythms would become progressively out of sync with the cycles of light (periods of activity) and dark (periods of rest). Thus the SCN must be reset daily by external cues so that the body's biologic rhythms are synchronized with the activity levels driven by the surrounding environment. The effect of not maintaining the internal clock's relevance to the environment is well known by people who experience **jet lag** when their inherent rhythm is out of step with external cues. The SCN works in conjunction with the pineal gland and its hormonal product melatonin to synchronize the various circadian rhythms with the 24-hour day–night cycle.

▌ Melatonin helps keep the body's circadian rhythms in time with the light–dark cycle.

Daily changes in light intensity are the major environmental cue used to adjust the SCN master clock. Special photoreceptors in the retina pick up light signals and transmit them directly to the SCN. These photoreceptors are distinct from the rods and cones used to perceive, or see, light (see p. 160). This is the major way the internal clock is coordinated to a 24-hour day. The eyes cue the pineal gland about the absence or presence of light by means of a neural pathway that passes through the SCN. This pathway is distinct from the neural systems that result in vision perception. Melatonin is the hormone of darkness. Melatonin secretion increases up to 10-fold during the darkness of night and then falls to low levels during the light of day. Fluctuations in melatonin secretion in turn help entrain the body's biologic rhythms with the external light–dark cues.

Other proposed roles of melatonin besides regulating the body's biological clock include the following:

- Melatonin induces a natural sleep without the side effects that accompany hypnotic sedatives.
- Melatonin is believed to inhibit the hormones that stimulate reproductive activity. Puberty may be initiated by a reduction in melatonin secretion.
- Melatonin appears to be a very effective **antioxidant**, a defense tool against biologically damaging free radicals. *Free radicals* are very unstable electron-deficient particles that are highly reactive and destructive. Free radicals have been implicated in several chronic diseases such as coronary artery disease (see p. 266) and cancer and are believed to contribute to the aging process.
- Evidence suggests that melatonin may slow the aging process, perhaps by removing free radicals or by other means.
- Melatonin appears to enhance immunity and has been shown to reverse some of the age-related shrinkage of the

thymus, the source of T lymphocytes (see p. 335), in old experimental animals.

Because of melatonin's many proposed roles, use of supplemental melatonin for a variety of conditions is very promising. However, most researchers are cautious about recommending supplemental melatonin until its effectiveness as a drug is further substantiated. Meanwhile, many people are turning to melatonin as a health food supplement; as such, it is not regulated by the Food and Drug Administration for safety and effectiveness. The two biggest self-prescribed uses of melatonin are to prevent jet lag and as a sleep aid.

We now shift our attention to the other endocrine glands located in or near the brain—the hypothalamus and pituitary.

 Click on the Media Exercises menu of the CD-ROM and work Media Exercise 17.4: Endocrine Roles by Organs of Mixed or Uncertain Functions to test your understanding of the previous section.

HYPOTHALAMUS AND PITUITARY

The **pituitary gland**, or **hypophysis**, is a small endocrine gland located in a bony cavity at the base of the brain just below the hypothalamus (● Figure 17-4). The pituitary is connected to the hypothalamus by a thin connecting stalk. If you point one finger between your eyes and another finger toward one of your ears, the imaginary point where these lines would intersect is about the location of your pituitary.

● **FIGURE 17-4**

Anatomy of the pituitary gland. (a) Relation of the pituitary gland to the hypothalamus and to the rest of the brain. (b) Schematic enlargement of the pituitary gland and its connection to the hypothalamus.

PhysioEdge For an interaction related to this figure, see Media Exercise 17.2: Anterior Pituitary Gland on the CD-ROM.

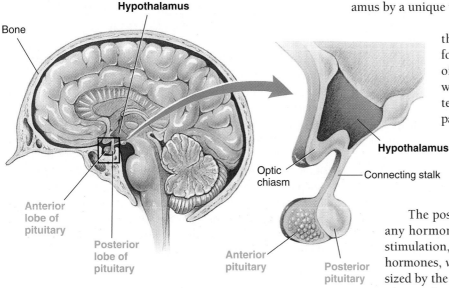

(a)   (b)

The pituitary gland consists of anterior and posterior lobes.

The pituitary has two anatomically and functionally distinct lobes, the **posterior pituitary** and the **anterior pituitary**. The posterior pituitary is composed of nervous tissue and thus is also termed the **neurohypophysis**. The anterior pituitary consists of glandular epithelial tissue and accordingly is also called the **adenohypophysis** (*adeno* means "glandular"). The anterior and posterior pituitary only have their location in common.

In some species, the adenohypophysis also includes a third, well-defined *intermediate lobe,* but humans lack this lobe. In lower vertebrates, the intermediate lobe secretes **melanocyte-stimulating hormone** or **MSH,** which regulates skin coloration by controlling the dispersion of granules containing the pigment **melanin.** By causing variable skin darkening in certain amphibians, reptiles, and fishes, MSH plays a vital role in the camouflage of these species.

In humans, a part of the anterior pituitary that briefly exists as a separate intermediate lobe during fetal development secretes a small amount of MSH. It is not involved in differences in the amount of melanin deposited in the skin of various races, nor with the process of skin tanning, although excessive MSH activity does darken skin. Instead MSH in humans plays a totally different role in helping control food intake (see p. 514). It also appears to influence excitability of the nervous system, perhaps improving memory and learning.

The hypothalamus and posterior pituitary act as a unit to secrete vasopressin and oxytocin.

The release of hormones from both the posterior and the anterior pituitary is directly controlled by the hypothalamus, but the nature of the relationship is entirely different. The posterior pituitary connects to the hypothalamus by a neural pathway, whereas the anterior pituitary connects to the hypothalamus by a unique vascular link.

Looking first at the posterior pituitary, the hypothalamus and posterior pituitary form a neuroendocrine system that consists of a population of neurosecretory neurons whose cell bodies lie in two well-defined clusters in the hypothalamus and whose axons pass down through the thin connecting stalk to terminate on capillaries in the posterior pituitary (● Figure 17-5). Functionally as well as anatomically, the posterior pituitary is simply an extension of the hypothalamus.

The posterior pituitary does not actually produce any hormones. It simply stores and, on appropriate stimulation, releases into the blood two small peptide hormones, *vasopressin* and *oxytocin,* which are synthesized by the neuronal cell bodies in the hypothalamus. The synthesized hormones are packaged in secretory granules that are transported down the cytoplasm of

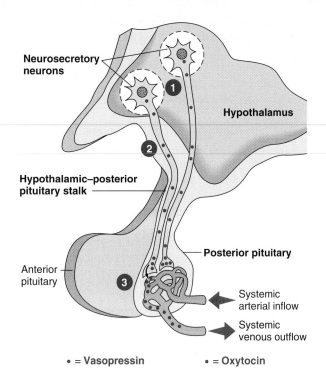

Neurosecretory neurons

Hypothalamus

Hypothalamic–posterior pituitary stalk

Anterior pituitary

Posterior pituitary

Systemic arterial inflow

Systemic venous outflow

• = Vasopressin   • = Oxytocin

1 The cell bodies of specific neurosecretory neurons in the hypothalamus secrete either vasopressin or oxytocin depending on the neuron.

2 The hormone travels down the axon to be stored in the neuronal terminals within the posterior pituitary.

3 On excitation of the neuron, the stored hormone is released from these terminals into the systemic blood for distribution throughout the body.

● **FIGURE 17-5**

Relationship of the hypothalamus and posterior pituitary

PhysioEdge   For an interaction related to this figure, see Media Exercise 17.3: Posterior Pituitary Gland on the CD-ROM.

the axon and stored in the neuronal terminals within the posterior pituitary. Each terminal stores either vasopressin or oxytocin, but not both. Thus these hormones can be released independently as needed. On stimulatory input to the hypothalamus, either vasopressin or oxytocin is released into the systemic blood from the posterior pituitary by exocytosis of the appropriate secretory granules. This hormonal release is triggered in response to action potentials that originate in the hypothalamic cell body and sweep down the axon to the neuronal terminal in the posterior pituitary. As in any other neuron, action potentials are generated in these neurosecretory neurons in response to synaptic input to their cell bodies.

The actions of vasopressin and oxytocin are briefly summarized here to make our endocrine story complete. They are described more thoroughly elsewhere—vasopressin in Chapter 13 and oxytocin in Chapter 18.

VASOPRESSIN

Vasopressin (antidiuretic hormone, ADH) has two major effects that correspond to its two names: (1) it enhances the

retention of H_2O by the kidneys (an antidiuretic effect), and (2) it causes contraction of arteriolar smooth muscle (a vessel pressor effect). The first effect has more physiologic importance. Under normal conditions, vasopressin is the primary endocrine factor that regulates urinary H_2O loss and overall H_2O balance. In contrast, typical levels of vasopressin play only a minor role in regulating blood pressure by means of the hormone's pressor effect.

OXYTOCIN

Oxytocin stimulates contraction of the uterine smooth muscle to help expel the infant during childbirth, and it promotes ejection of the milk from the mammary glands (breasts) during breast-feeding.

▌ Most anterior pituitary hormones are tropic.

Unlike the posterior pituitary, which releases hormones synthesized by the hypothalamus, the anterior pituitary itself synthesizes the hormones it releases into the blood. Different cell populations within the anterior pituitary secrete six major peptide hormones. The actions of each of these hormones is described in detail in later sections. For now, here is a brief statement of their primary effects to provide a rationale for their names (● Figure 17-6):

1. **Growth hormone (GH, somatotropin)**, the primary hormone responsible for regulating overall body growth, is also important in intermediary metabolism.
2. **Thyroid-stimulating hormone (TSH, thyrotropin)** stimulates secretion of thyroid hormone and growth of the thyroid gland.
3. **Adrenocorticotropic hormone (ACTH, adrenocorticotropin)** stimulates cortisol secretion by the adrenal cortex and promotes growth of the adrenal cortex.
4. **Follicle-stimulating hormone (FSH)** has different functions in females and males. In females it stimulates growth and development of ovarian follicles, within which the ova, or eggs, develop. It also promotes secretion of the hormone estrogen by the ovaries. In males FSH is required for sperm production.
5. **Luteinizing hormone (LH)** also functions differently in females and males. In females LH is responsible for ovulation and luteinization (that is, the formation of a hormone-secreting corpus luteum in the ovary following ovulation). LH also regulates ovarian secretion of the female sex hormones, estrogen and progesterone. In males the same hormone stimulates the interstitial cells of Leydig in the testes to secrete the male sex hormone, testosterone, giving rise to its alternate name of **interstitial cell-stimulating hormone (ICSH)**.
6. **Prolactin (PRL)** enhances breast development and milk production in females. Its function in males is uncertain, although evidence indicates that it may induce the production of testicular LH receptors.

TSH, ACTH, FSH, and LH are all tropic hormones, because they each regulate the secretion of another specific endocrine gland. FSH and LH are collectively referred to as **gonado-**

tropins because they control secretion of the sex hormones by the gonads (ovaries and testes). Because growth hormone exerts its growth-promoting effects indirectly by stimulating the release of liver hormones, the *somatomedins*, it too is sometimes categorized as a tropic hormone. Among the anterior pituitary hormones, prolactin is the only one that does not stimulate secretion of another hormone. Of the tropic hormones, FSH, LH, and growth hormone exert effects on nonendocrine target cells in addition to stimulating secretion of other hormones.

● **FIGURE 17-6**

Functions of the anterior pituitary hormones. Five different endocrine cell types produce the six anterior-pituitary hormones—TSH, ACTH, growth hormone, LH and FSH (produced by the same cell type), and prolactin—which exert a wide range of effects throughout the body.

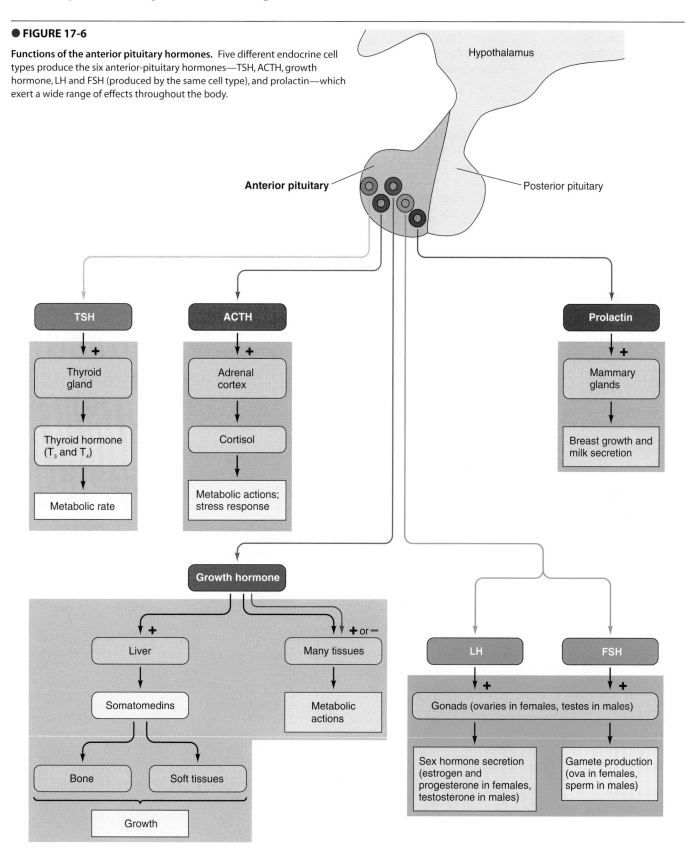

■ Hypothalamic releasing and inhibiting hormones help regulate anterior pituitary hormone secretion.

None of the anterior pituitary hormones are secreted at a constant rate. Even though each of these hormones has a unique control system, there are some common regulatory patterns. The two most important factors that regulate anterior pituitary hormone secretion are (1) hypothalamic hormones and (2) feedback by target gland hormones.

Because the anterior pituitary secretes hormones that control the secretion of various other hormones, it long held the undeserved title of "master gland." Scientists now know that the release of each anterior-pituitary hormone is largely controlled by still other hormones produced by the hypothalamus. The secretion of these regulatory neurohormones in turn is controlled by a variety of neural and hormonal inputs to the hypothalamic neurosecretory cells.

ROLE OF THE HYPOTHALAMIC RELEASING AND INHIBITING HORMONES

The secretion of each anterior-pituitary hormone is stimulated or inhibited by one or more of the seven hypothalamic **hypophysiotropic hormones** (*hypophysis* means "pituitary"; *tropic* means "nourishing"). These peptide hormones are listed in ▲ Table 17-2. Depending on their actions, these hormones are called **releasing hormones** or **inhibiting hormones**. In each

case, the primary action of the hormone is apparent from its name. For example, **thyrotropin-releasing hormone (TRH)** stimulates the release of TSH (alias thyrotropin) from the anterior pituitary, whereas **prolactin-inhibiting hormone**

▲ TABLE 17-2

Major Hypophysiotropic Hormones

HORMONE	EFFECT ON THE ANTERIOR PITUITARY
Thyrotropin-Releasing Hormone (TRH)	Stimulates release of TSH (thyrotropin) and prolactin
Corticotropin-Releasing Hormone (CRH)	Stimulates release of ACTH (corticotropin)
Gonadotropin-Releasing Hormone (GnRH)	Stimulates release of FSH and LH (gonadotropins)
Growth Hormone–Releasing Hormone (GHRH)	Stimulates release of growth hormone
Growth Hormone–Inhibiting Hormone (GHIH)	Inhibits release of growth hormone and TSH
Prolactin-Releasing Hormone (PRH)	Stimulates release of prolactin
Prolactin-Inhibiting Hormone (PIH)	Inhibits release of prolactin

● **FIGURE 17-7**

Hierarchic chain of command and negative feedback in endocrine control. The general pathway involved in the hierarchic chain of command among the hypothalamus, anterior pituitary, and peripheral target endocrine gland is depicted on the left. The pathway on the right leading to cortisol secretion provides a specific example of this endocrine chain of command. The hormone ultimately secreted by the target endocrine gland, such as cortisol, acts in negative feedback fashion to reduce secretion of the regulatory hormones higher in the chain of command.

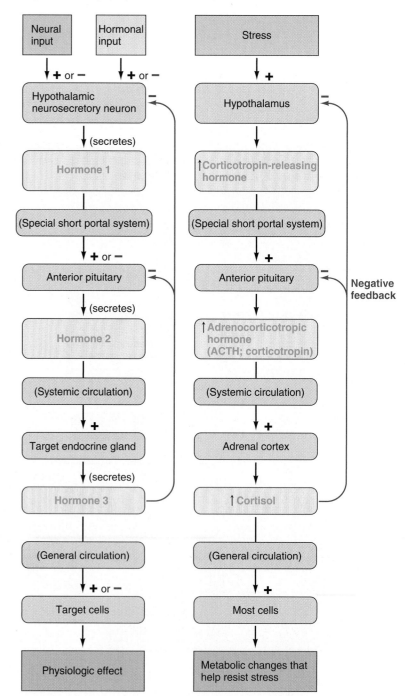

(PIH) inhibits the release of prolactin from the anterior pituitary. Note that hypophysiotropic hormones in most cases are involved in a three-hormone hierarchic chain of command (● Figure 17-7): The hypothalamic hypophysiotropic hormone *(hormone 1)* controls the output of an anterior-pituitary tropic hormone *(hormone 2)*. This tropic hormone in turn regulates secretion of the target endocrine gland's hormone *(hormone 3)*, which exerts the final physiologic effect.

Although endocrinologists originally speculated that there was one hypophysiotropic hormone for each anterior pituitary hormone, many hypothalamic hormones have more than one effect, so their names indicate only the function first identified. Moreover, a single anterior pituitary hormone may be regulated by two or more hypophysiotropic hormones, which may even exert opposing effects. For example, **growth hormone–releasing hormone (GHRH)** stimulates growth hormone secretion, whereas **growth hormone–inhibiting hormone (GHIH)**, also known as *somatostatin*, inhibits it.

ROLE OF THE HYPOTHALAMIC-HYPOPHYSEAL PORTAL SYSTEM

The hypothalamic regulatory hormones reach the anterior pituitary by means of a unique vascular link. In contrast to the direct neural connection between the hypothalamus and posterior pituitary, the anatomic and functional link between the hypothalamus and anterior pituitary is an unusual capillary-to-capillary connection, the **hypothalamic-hypophyseal portal system.** A portal system is a vascular arrangement in which venous blood flows directly from one capillary bed through a connecting vessel to another capillary bed. The largest and best-known portal system is the hepatic portal system, which drains intestinal venous blood directly into the liver for immediate processing of absorbed nutrients (see p. 489). Although much smaller, the hypothalamic-hypophyseal portal system is no less important, because it provides a critical link between the brain and much of the endocrine system. It begins in the base of the hypothalamus with a group of capillaries that recombine into small portal vessels, which pass down through the connecting stalk into the anterior pituitary. Here the portal vessels branch to form most of the anterior pituitary capillaries, which in turn drain into the systemic venous system (● Figure 17-8).

As a result, almost all the blood supplied to the anterior pituitary must first pass through the hypothalamus. Because materials can be exchanged between the blood and surrounding tissue only at the capillary level, the hypothalamic-hypophyseal portal system provides a route where releasing and inhibiting hormones can be picked up at the hypothalamus and delivered immediately and directly to the anterior pituitary at relatively high concentrations, completely bypassing the general circulation. If the portal system did not exist, once the hypophysiotropic hormones were picked up in the hypothalamus they would be returned to the heart by the systemic venous system. From here, they would travel to the lungs and back to the heart through the pulmonary circulation, and finally enter the systemic arterial system for delivery throughout the body, including the anterior pituitary. This process would not only take longer, but the hypophysiotropic hormones would be considerably diluted by the much larger volume of blood flowing through this usual circulatory route.

● FIGURE 17-8

Vascular link between the hypothalamus and anterior pituitary

 PhysioEdge For an interaction related to this figure, see Media Exercise 17.2: Anterior Pituitary Gland on the CD-ROM.

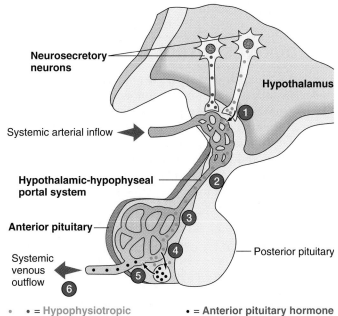

1 Hypophysiotropic hormones (releasing hormones and inhibiting hormones) produced by neurons in the hypothalamus enter the hypothalamic capillaries.

2 These hypothalamic capillaries rejoin to form the hypothalamic-hypophyseal portal system. This vascular link passes to the anterior pituitary.

3 Here it branches into the anterior pituitary capillaries.

4 The hypophysiotropic hormones leave the blood across the anterior pituitary capillaries and control the release of anterior pituitary hormones.

5 On stimulation by the appropriate hypothalamic releasing hormone, a given anterior pituitary hormone is secreted into these capillaries.

6 The anterior pituitary capillaries rejoin to form a vein, through which the anterior pituitary hormones leave for ultimate distribution throughout the body by the systemic circulation.

Labels: Neurosecretory neurons; Hypothalamus; Systemic arterial inflow; Hypothalamic-hypophyseal portal system; Anterior pituitary; Posterior pituitary; Systemic venous outflow

• • = Hypophysiotropic hormones • = Anterior pituitary hormone

The axons of the neurosecretory neurons that produce the hypothalamic regulatory hormones terminate on the capillaries at the origin of the portal system. These hypothalamic neurons secrete their hormones in the same way as the hypothalamic neurons that produce vasopressin and oxytocin. The hormone is synthesized in the cell body and then transported to the axon terminal. It is stored there until its release into an adjacent capillary on appropriate stimulation. The major difference is that the hypophysiotropic hormones are released into the portal vessels, which deliver them to the anterior pituitary where they control the release of anterior pituitary hormones into the general circulation. In contrast, the hypothalamic hormones stored in the posterior pituitary are themselves released into the general circulation.

CONTROL OF HYPOTHALAMIC RELEASING AND INHIBITING HORMONES

What regulates secretion of these hypophysiotropic hormones? Like other neurons, the neurons secreting these regulatory hormones receive abundant input of information (both neural and hormonal and both excitatory and inhibitory) that they must integrate. Studies are still in progress to unravel the complex neural input from many diverse areas of the brain to the hypophysiotropic secretory neurons. Some of these inputs carry information about a variety of environmental conditions. One example is the marked increase in secretion of corticotropin-releasing hormone (CRH) in response to stress (● Figure 17-7). Numerous neural connections also exist between the hypothalamus and the portions of the brain concerned with emotions (the limbic system—see p. 126). Thus emotions greatly influence secretion of hypophysiotropic hormones. The menstrual irregularities sometimes experienced by women who are emotionally upset are a common manifestation of this relationship.

In addition to being regulated by different regions of the brain, the hypophysiotropic neurons are also controlled by various chemical inputs that reach the hypothalamus through the blood. The most common blood-borne factors that influence hypothalamic neurosecretion are the negative-feedback effects of target gland hormones, to which we now turn our attention.

▌ Target gland hormones inhibit hypothalamic and anterior pituitary hormone secretion via negative feedback.

In most cases, hypophysiotropic hormones initiate a three-hormone sequence: (1) hypophysiotropic hormone, (2) anterior-pituitary tropic hormone, and (3) hormone from the peripheral target endocrine gland. Typically, in addition to producing its physiologic effects, the target gland hormone also suppresses secretion of the tropic hormone that is driving it. This negative feedback is accomplished by the target gland hormone acting either directly on the pituitary itself or on the release of hypothalamic hormones, which in turn regulate anterior pituitary function (● Figure 17-7). As an example, consider the CRH-ACTH-cortisol system. Hypothalamic CRH

(corticotropin-releasing hormone) stimulates the anterior pituitary to secrete ACTH (adrenocorticotropic hormone, alias corticotropin), which in turn stimulates the adrenal cortex to secrete cortisol. The final hormone in the system, cortisol, inhibits the hypothalamus to reduce CRH secretion and also reduces the sensitivity of the ACTH-secreting cells to CRH by acting directly on the anterior pituitary. Through this double-barreled approach, cortisol exerts negative-feedback control to stabilize its own plasma concentration. If plasma cortisol levels start to rise above a prescribed set level, cortisol suppresses its own further secretion by its inhibitory actions at the hypothalamus and anterior pituitary. This mechanism ensures that once a hormonal system is activated, its secretion does not continue unabated. If plasma cortisol levels fall below the desired set point, cortisol's inhibitory actions at the hypothalamus and anterior pituitary are reduced, so the driving forces for cortisol secretion (CRH-ACTH) increase accordingly. The other target-gland hormones act by similar negative-feedback loops to maintain their plasma levels relatively constant at the set point.

Diurnal rhythms are superimposed on this type of stabilizing negative-feedback regulation; that is, the set point changes as a function of the time of day. Furthermore, other controlling inputs may break through the negative-feedback control to alter hormone secretion (that is, change the set level) at times of special need. For example, stress raises the set point for cortisol secretion.

The detailed functions and control of all the anterior pituitary hormones except growth hormone are discussed elsewhere in conjunction with the target tissues that they influence; for example, thyroid-stimulating hormone is covered with the discussion of the thyroid gland. Accordingly, growth hormone is the only anterior-pituitary hormone we elaborate on at this time.

Click on the Media Exercises menu of the CD-ROM and work Media Exercises 17.2: Anterior Pituitary Gland and 17.3: Posterior Pituitary Gland to test your understanding of the previous section.

ENDOCRINE CONTROL OF GROWTH

In growing children, continuous net protein synthesis occurs under the influence of growth hormone as the body steadily gets larger. Weight gain alone is not synonymous with growth, because weight gain may occur from retaining excess H_2O or storing fat without true structural growth of tissues. Growth requires net synthesis of proteins and includes lengthening of the long bones (the bones of the extremities) as well as increases in the size and number of cells in the soft tissues.

▌ Growth depends on growth hormone but is influenced by other factors as well.

Although, as the name implies, growth hormone (GH) is absolutely essential for growth, it alone is not wholly responsible for determining the rate and final magnitude of growth in a given individual. The following factors affect growth:

- *Genetic determination* of an individual's maximum growth capacity. Attaining this full growth potential further depends on the other factors listed here.
- *An adequate diet,* including enough total protein and ample essential amino acids to accomplish the protein synthesis necessary for growth. Malnourished children never achieve their full growth potential. By contrast, a person cannot exceed his or her genetically determined maximum by eating a more than adequate diet. The excess food intake produces obesity instead of growth.
- *Freedom from chronic disease and stressful environmental conditions.* Stunted growth under adverse circumstances is due in large part to the prolonged stress-induced secretion of cortisol from the adrenal cortex. Cortisol exerts several potent antigrowth effects, such as promoting protein breakdown, inhibiting growth in the long bones, and blocking the secretion of GH.
- *Normal levels of growth-influencing hormones.* In addition to the absolutely essential GH, other hormones, including thyroid hormone, insulin, and the sex hormones, play secondary roles in promoting growth.

▌ Growth hormone is essential for growth, but it also exerts metabolic effects not related to growth.

GH is the most abundant hormone produced by the anterior pituitary, even in adults in whom growth has already ceased, although GH secretion typically starts to decline after middle age. The continued high secretion of GH beyond the growing period implies that this hormone has important influences other than on growth. In addition to promoting growth, GH has important metabolic effects and enhances the immune system. We will briefly describe GH's metabolic actions before turning our attention to its growth-promoting actions.

METABOLIC ACTIONS OF GH UNRELATED TO GROWTH

GH increases fatty acid levels in the blood by enhancing the breakdown of triglyceride fat stored in adipose tissue, and it increases blood glucose levels by decreasing glucose uptake by muscles. Muscles use the mobilized fatty acids instead of glucose as a metabolic fuel. Thus the overall metabolic effect of GH is to mobilize fat stores as a major energy source while conserving glucose for glucose-dependent tissues such as the brain. The brain can use only glucose as its metabolic fuel, yet nervous tissue cannot store glycogen (stored glucose) to any extent. This metabolic pattern is suitable for maintaining the body during prolonged fasting or other situations when the body's energy needs exceed available glucose stores.

GROWTH-PROMOTING ACTIONS OF GH ON SOFT TISSUES

When tissues are responsive to its growth-promoting effects, GH stimulates growth of both soft tissues and the skeleton. GH promotes growth of soft tissues by (1) increasing the number of cells (**hyperplasia**) and (2) increasing the size of cells (**hypertrophy**). GH increases the number of cells by stimulating cell division and by preventing apoptosis (programmed

cell death; see p. 100). GH increases the size of cells by favoring synthesis of proteins, the main structural component of cells. GH stimulates almost all aspects of protein synthesis while it simultaneously inhibits protein degradation. It promotes the uptake of amino acids (the raw materials for protein synthesis) by cells, decreasing blood amino-acid levels in the process. Furthermore, it stimulates the cellular machinery responsible for accomplishing protein synthesis according to the cell's genetic code.

Growth of the long bones resulting in increased height is the most dramatic effect of GH. Before you can understand the means by which GH stimulates bone growth, you must first become familiar with the structure of bone and how growth of bone is accomplished.

▌ Bone grows in thickness and in length by different mechanisms, both stimulated by growth hormone.

Bone is a living tissue. Being a form of connective tissue, it consists of cells and an extracellular organic matrix produced by the cells. The bone cells that produce the organic matrix are known as **osteoblasts** ("bone formers"). The organic matrix is composed of collagen fibers (see p. 48) in a semisolid gel. This matrix has a rubbery consistency and is responsible for the tensile strength of bone (the resilience of bone to breakage when tension is applied). Bone is made hard by precipitation of *calcium phosphate crystals* within the matrix. These inorganic crystals provide the bone with compressional strength (the ability of bone to hold its shape when squeezed or compressed). If bones consisted entirely of inorganic crystals, they would be brittle, like pieces of chalk. Bones have structural strength approaching that of reinforced concrete, yet they are not brittle and are much lighter in weight, because they have the structural blending of an organic scaffolding hardened by inorganic crystals. **Cartilage** is similar to bone, except that living cartilage is not calcified.

A long bone basically consists of a fairly uniform cylindrical shaft, the **diaphysis**, with a flared articulating knob at either end, an **epiphysis**. In a growing bone, the diaphysis is separated at each end from the epiphysis by a layer of cartilage known as the **epiphyseal plate** (● Figure 17-9). The central cavity of the bone is filled with bone marrow, the site of blood cell production (see p. 318).

BONE GROWTH

Growth in *thickness* of bone is achieved by adding new bone on top of the outer surface of already existing bone. This growth is produced by osteoblasts within the connective tissue sheath that covers the outer bone. As osteoblast activity deposits new bone on the external surface, other cells within the bone, the **osteoclasts** ("bone-breakers"), dissolve the bony tissue on the inner surface next to the marrow cavity. In this way, the marrow cavity enlarges to keep pace with the increased circumference of the bone shaft.

Growth in *length* of long bones is accomplished by a different mechanism. Bones grow in length as a result of activity

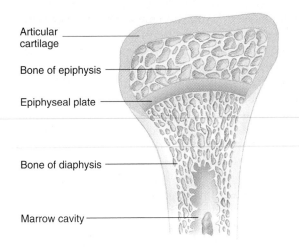

Articular cartilage

Bone of epiphysis

Epiphyseal plate

Bone of diaphysis

Marrow cavity

● **FIGURE 17-9**

Anatomy of long bones

of the cartilage cells in the epiphyseal plates. During growth, cartilage cells on the outer edge of the plate next to the epiphysis divide and multiply, temporarily widening the epiphyseal plate. As new cartilage cells are formed on the epiphyseal border, the older cartilage cells toward the diaphyseal border die and are replaced by osteoblasts, which swarm upward from the diaphysis. These new tenants lay down bone around the persisting remnants of disintegrating cartilage until bone entirely replaces the inner region of cartilage on the diaphyseal side of the plate. When this **ossification** ("bone formation") is complete, the bone on the diaphyseal side has lengthened, and the epiphyseal plate has returned to its original thickness. The cartilage that bone has replaced on the diaphyseal end of the plate is as thick as the new cartilage on the epiphyseal end of the plate. Thus bone growth is made possible by the growth and death of cartilage, which acts like a "spacer" to push the epiphysis farther out while it provides a framework for future bone formation on the end of the diaphysis.

MATURE, NONGROWING BONE

As the extracellular matrix produced by an osteoblast calcifies, the osteoblast becomes entombed by the matrix it has deposited around itself. Osteoblasts trapped within a calcified matrix do not die, because they are supplied by nutrients transported to them through small canals that the osteoblasts themselves form by sending out cytoplasmic extensions around which the bony matrix is deposited. Thus, within the final bony product a network of permeating tunnels radiates from each entrapped osteoblast, serving as a lifeline system for nutrient delivery and waste removal. The entrapped osteoblasts, now called **osteocytes**, retire from active bone-forming duty, because their imprisonment prevents them from laying down new bone. However, they are involved in the hormonally regulated exchange of calcium between bone and the blood. This exchange is under the control of parathyroid hormone (discussed later in the chapter), not GH.

GROWTH-PROMOTING ACTIONS OF GH ON BONES

GH promotes growth of bone in both thickness and length. It stimulates osteoblast activity and also stimulates proliferation of epiphyseal cartilage, thereby making space for more bone formation. GH can promote lengthening of long bones as long as the epiphyseal plate remains cartilaginous, or is "open." At the end of adolescence, under the influence of the sex hormones these plates completely ossify, or "close," so that the bones cannot lengthen any further despite the presence of GH. Thus, after the plates are closed the individual does not grow any taller.

❙ **Growth hormone exerts its growth-promoting effects indirectly by stimulating somatomedins.**

GH does not act directly on its target cells to bring about its growth-producing actions (increased cell division, enhanced protein synthesis, and bone growth). These effects are directly brought about by peptide mediators known as **somatomedins**. These peptides are also referred to as **insulin-like growth factors (IGF)** because they are structurally and functionally similar to insulin.

IGF synthesis is stimulated by GH and mediates most of this hormone's growth-promoting actions. The major source of circulating IGF is the liver, which releases this peptide product into the blood in response to GH stimulation. IGF is also produced by most other tissues, although they do not release it into the blood to any extent. Researchers propose that IGF produced locally in target tissues may act through paracrine means (see p. 93) for at least some of the GH-induced effects. Such a mechanism could explain why blood levels of GH are no higher, and indeed circulating IGF levels are lower, during the first several years of life compared to adult values, even though growth is quite rapid during the postnatal period. Local production of IGF in target tissues may possibly be more important than delivery of blood-borne IGF during this time.

Production of IGF is controlled by a number of factors other than GH, including nutritional status, age, and tissue-specific factors. For example, the gonadotropins and sex hormones stimulate IGF production within reproductive organs such as the testes in males and the ovaries and uterus in females. Thus control of IGF production is complex and subject to a variety of systemic and local factors.

❙ **Growth hormone secretion is regulated by two hypophysiotropic hormones.**

The control of GH secretion is complex, with two hypothalamic hypophysiotropic hormones playing a key role.

GROWTH HORMONE–RELEASING HORMONE AND GROWTH HORMONE–INHIBITING HORMONE

Two antagonistic regulatory hormones from the hypothalamus are involved in controlling growth hormone secretion: growth hormone–releasing hormone (GHRH), which is stimulatory, and growth hormone–inhibiting hormone (GHIH or somatostatin), which is inhibitory (● Figure 17-10). (Note the distinctions among *somatotropin*, alias growth hormone; *somatomedin*, a liver hormone that directly mediates the effects of GH; and *somatostatin*, which inhibits GH secretion.) Any factor that increases GH secretion could theoretically do so either by stimulating GHRH release or by inhibiting GHIH release. Endocrinologists do not know which of these pathways is used in each specific case.

As with the other hypothalamus–anterior pituitary axes, negative-feedback loops participate in the regulation of GH secretion. Both GH and the somatomedins inhibit pituitary secretion of GH.

FACTORS THAT INFLUENCE GH SECRETION

A number of factors influence GH secretion by acting on the hypothalamus. GH secretion displays a well-characterized diurnal rhythm. Through most of the day, GH levels tend to be low and fairly constant. About an hour after the onset of deep sleep, however, GH secretion markedly increases up to five times the daytime value, then rapidly drops over the next several hours.

Superimposed on this diurnal fluctuation in GH secretion are further bursts in secretion that occur in response to exercise, stress, and low blood glucose, the major stimuli for increased secretion. The benefits of increased GH secretion during these situations when energy demands outstrip the body's glucose reserves are presumably that glucose is conserved for the brain and fatty acids are provided as an alternative energy source for muscle.

A rise in blood amino acids after a high-protein meal also enhances GH secretion. In turn GH promotes the use of these amino acids for protein synthesis. GH is also stimulated by a decline in blood fatty acids. Because GH mobilizes fat, such regulation helps maintain fairly constant blood fatty-acid levels.

Note that the known regulatory inputs for GH secretion are aimed at adjusting the levels of glucose, amino acids, and fatty acids in the blood. No known growth-related signals influence GH secretion. The whole issue of what really controls growth is complicated by the fact that GH levels during early childhood, a period of quite rapid linear growth, are similar to those in normal adults. As mentioned earlier, the poorly understood control of IGF activity may be important in this regard. Another related question is, Why aren't adult tissues still responsive to GH's growth-promoting effects? We know we do not grow any taller after adolescence because the epiphyseal plates have closed, but why don't soft tissues continue to grow through hypertrophy and hyperplasia under the influence of GH? Further research is needed to unravel these mysteries.

Abnormal growth-hormone secretion results in aberrant growth patterns.

 Clinical Note Diseases related to both deficiencies and excesses of growth hormone can occur. The effects on the pattern of growth are much more pronounced than the metabolic consequences.

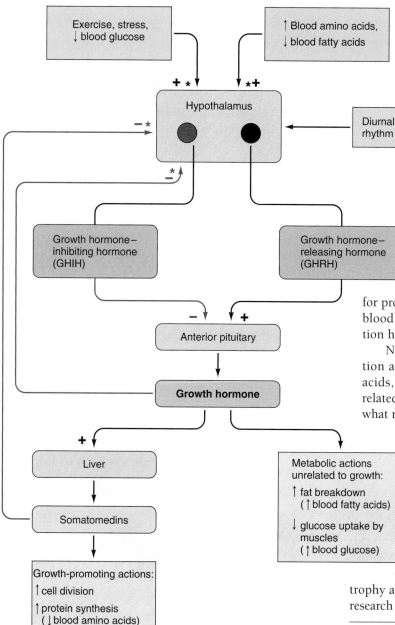

● **FIGURE 17-10**

Control of growth hormone secretion

*These factors all increase growth hormone secretion, but it is unclear whether they do so by stimulating GHRH or inhibiting GHIH, or both.

*These factors inhibit growth hormone secretion in negative-feedback fashion, but it is unclear whether they do so by stimulating GHIH or inhibiting GHRH or inhibiting the anterior pituitary itself.

● **FIGURE 17-11**

Examples of the effect of abnormalities in growth hormone secretion on growth. The man at the left displays pituitary dwarfism resulting from underproduction of growth hormone in childhood. The male at the center of the photograph has gigantism caused by excessive growth hormone secretion in childhood. The woman at the right is of average height.

GROWTH HORMONE DEFICIENCY

GH deficiency may be caused by a pituitary defect (lack of GH) or may occur secondary to hypothalamic dysfunctions (lack of GHRH). Hyposecretion of GH in a child results in **dwarfism**. The predominant feature is short stature caused by retarded skeletal growth (● Figure 17-11). Less obvious characteristics include poorly developed muscles (reduced muscle-protein synthesis) and excess subcutaneous fat (less fat mobilization).

The onset of GH deficiency in adulthood after growth is already complete produces relatively few symptoms. GH-deficient adults tend to have reduced skeletal-muscle mass and strength (less muscle protein) as well as decreased bone density (less osteoblast activity during ongoing bone remodeling). Furthermore, be-

cause GH is essential for maintaining cardiac muscle mass and performance in adulthood, GH-deficient adults may be at increased risk of developing heart failure. (For a discussion of GH therapy, see the accompanying boxed feature, ❯ Beyond the Basics.)

GROWTH HORMONE EXCESS

Hypersecretion of GH is most often caused by a tumor of the GH-producing cells of the anterior pituitary. The symptoms depend on the age of the individual when the abnormal secretion begins. If overproduction of GH begins in childhood before the epiphyseal plates close, the principal manifestation is a rapid growth in height without distortion of body proportions. Appropriately, this condition is known as **gigantism** (● Figure 17-11). If not treated by removal of the tumor or by drugs that block the effect of GH, the person may reach a height of 8 feet or more. All the soft tissues grow correspondingly, so the body is still well proportioned.

If GH hypersecretion occurs after adolescence when the epiphyseal plates have already closed, further growth in height is prevented. Under the influence of excess GH, however, the bones become thicker and the soft tissues, especially connective tissue and skin, proliferate. This disproportionate growth pattern produces a disfiguring condition known as **acromegaly** (*acro* means "extremity"; *megaly* means "large"). Bone thickening is most obvious in the extremities and face. A marked coarsening of the features to an almost apelike appearance gradually develops as the jaws and cheekbones become more prominent because of the thickening of the facial bones and the skin (● Figure 17-12). The hands and feet enlarge, and the fingers and toes become greatly thickened. Peripheral nerve disorders often occur as nerves are entrapped by overgrown connective tissue or bone.

● **FIGURE 17-12**

Progressive development of acromegaly. In this series of photos from childhood to the present, note how the patient's brow bones, cheekbones, and jaw bones are becoming progressively more prominent as a result of ongoing thickening of the bones and skin caused by excessive GH secretion.

Age 13	Age 21	Age 35

Growth and Youth in a Bottle?

Unlike most hormones, growth hormone's structure and activity differ among species. As a result, growth hormone (GH) from animal sources is ineffective for treating GH deficiency in humans. In the past, the only source of human GH was pituitary glands from human cadavers. This supply was never adequate, and it was finally removed from the market by the Food and Drug Administration (FDA) because of fear of viral contamination. Recently, however, the technique of genetic engineering has made available an unlimited supply of human GH. The gene that directs synthesis of GH in humans has been introduced into bacteria, converting the bacteria into "factories" that synthesize human GH.

Even though adequate human GH is now available through genetic engineering, a new problem for the medical community is to determine under which circumstances synthetic GH treatment is appropriate. Currently the FDA has approved treatment only for the following uses: (1) for GH-deficient children, (2) for adults with a pituitary tumor or other disease that causes severe GH deficiency, and (3) for AIDS patients suffering from severe muscle wasting. Although not approved by the FDA for this use, GH therapy is also widely used to promote faster healing of skin in severely burned patients.

Another group whom replacement GH therapy may benefit is the elderly. GH secretion typically peaks during a person's 20s, then in many people may start to dwindle after age 40. This decline may contribute to some of the characteristic signs of aging, such as

- Decreased muscle mass (GH promotes synthesis of proteins, including muscle protein)
- Increased fat deposition (GH promotes leanness by mobilizing fat stores for use as an energy source)
- Reduced bone density (GH promotes the activity of bone-forming cells)
- Thinner, sagging skin (GH promotes proliferation of skin cells)

(However, inactivity is also believed to play a major role in age-associated reductions in muscle mass, bone density, and strength.)

Several studies in the early 1990s suggested that some of these consequences of aging may be counteracted through the use of synthetic GH in people over age 60. Elderly men treated with supplemental GH showed increased muscle mass, reduced fatty tissue, and thickened skin. In similar studies in elderly women, supplemental GH therapy did not increase muscle mass significantly but did reduce fat mass and did protect against bone loss.

Even though these early results were exciting, further studies have been more discouraging. For example, despite increased lean body mass, many treated people surprisingly do not have increased muscle strength or exercise capabilities. Also, when GH is supplemented for an extended time or in large doses, harmful side effects include an increased likelihood of diabetes, kidney stones, high blood pressure, headaches, joint pain, and carpal tunnel syndrome. (*Carpal tunnel syndrome* involves thickening and narrowing of the tunnel in the wrist through which the nerve supply to the hand muscles passes; *carpal* means "wrist.") Furthermore, synthetic GH is costly ($15,000 to $20,000 annually) and must be injected regularly. Also, some scientists worry that sustained administration of synthetic GH may raise the risk of developing cancer by promoting uncontrolled cell proliferation. For these reasons, many investigators no longer view synthetic GH as a potential "fountain of youth." Instead, they hope it can be used in a more limited way to strengthen muscle and bone sufficiently in the many elderly who have GH deficits, to help reduce the incidence of bone-breaking falls that often lead to disability and loss of independence. The National Institute of Aging is currently sponsoring a nationwide series of studies involving GH therapy in the elderly to help sort out potentially legitimate roles of this supplemental hormone.

An ethical dilemma is whether the drug should be used by others who have normal GH levels but want the product's growth-promoting actions for cosmetic or athletic reasons, such as normally growing teenagers who wish to attain even greater height. The drug is already being used illegally by some athletes and bodybuilders. Furthermore, a recent study found that only 4 out of 10 children who are legitimately receiving GH therapy under medical supervision are actually GH deficient. The others are receiving the treatment because parents, physicians, and the children are being swayed by perceived cultural pressures that favor height rather than by medical factors.

Using the drug in children with normal GH levels may be problematic, because synthetic GH is a double-edged product. Although it promotes growth and muscle mass, it also has negative effects, such as potential troubling side effects. Furthermore, a recent British study revealed that supplemental GH therapy in children who do not lack the hormone redistributes the body's fat and protein. The investigators compared two groups of otherwise healthy 6- to 8-year-olds who were among the shortest for their age. One group consisted of children who were receiving GH, the other of children who were not. At the end of six months, the children taking the synthetic hormone had outpaced the untreated group in growth by more than 1.5 inches per year. However, the untreated children added both muscle and fat as they grew, whereas the treated children became unusually muscular and lost up to 76% of their body fat. The loss of fat became especially obvious in their faces and limbs, giving them a rawboned, gangly appearance. It is unclear what long-term effects—either deleterious or desirable—these dramatic changes in body composition might have. Scientists also express concern that these readily observable physical changes may be accompanied by more subtle abnormalities in organs and cells. Thus, the debate about whether to use GH in normal but short children is likely to continue until further investigation demonstrates if it is safe and appropriate.

THYROID GLAND

The **thyroid gland** consists of two lobes of endocrine tissue joined in the middle by a narrow portion of the gland, giving it a bow tie shape (● Figure 17-13a). The gland is even located in the appropriate place for a bow tie, lying over the trachea just below the larynx.

▌ The major cells that secrete thyroid hormone are organized into colloid-filled follicles.

The major thyroid secretory cells, known as **follicular cells**, are arranged into hollow spheres, each of which forms a functional unit called a **follicle**. On a microscopic section (● Figure 17-13b), the follicles appear as rings of follicular cells

contact with the extracellular fluid that surrounds the follicle, similar to an inland lake that is not in direct contact with the oceans that surround a continent.

The chief constituent of the colloid is a large protein molecule known as **thyroglobulin (Tg)**, within which are incorporated the thyroid hormones in their various stages of synthesis. The follicular cells produce two iodine-containing hormones derived from the amino acid tyrosine: **tetraiodothyronine (T_4 or thyroxine)** and **tri-iodothyronine (T_3)**. The prefixes *tetra* and *tri* and the subscripts *4* and *3* denote the number of iodine atoms incorporated into each of these hormones. These two hormones, collectively referred to as **thyroid hormone**, are important regulators of overall basal metabolic rate.

Interspersed in the interstitial spaces between the follicles is another secretory cell type, the **C cells**, so called because they secrete the peptide hormone **calcitonin**. Calcitonin plays a role in calcium metabolism and is not related in any way to the two other major thyroid hormones. We will here discuss T_4 and T_3 and talk about calcitonin later, in a section dealing with endocrine control of calcium balance.

▌ Thyroid hormone is synthesized and stored on the thyroglobulin molecule.

The basic ingredients for thyroid hormone synthesis are tyrosine and iodine, both of which must be taken up from the blood by the follicular cells. The synthesis, storage, and secretion of thyroid hormone involve the following steps:

1. All steps of thyroid hormone synthesis take place on the thyroglobulin molecules within the colloid. Thyroglobulin itself is produced by the endoplasmic reticulum/Golgi complex of the thyroid follicular cells. The amino acid tyrosine becomes incorporated in the much larger thyroglobulin molecules as the latter are being produced. Once produced, tyrosine-containing thyroglobulin is exported from the follicular cells into the colloid by exocytosis (step ① in ● Figure 17-14).

2. The iodine needed for thyroid hormone synthesis must be obtained from dietary intake. The thyroid captures iodine from the blood and transfers it into the colloid by an *iodine pump*—the powerful, energy-requiring carrier proteins in the outer membranes of the follicular cells (step ②). Almost all the iodine in the body is moved against its concentration gradient to become trapped in the thyroid for thyroid hormone synthesis. Iodine serves no other function in the body.

3. Within the colloid, iodine is quickly attached to a tyrosine within the thyroglobulin molecule. Attachment of one iodine to tyrosine yields **monoiodotyrosine (MIT)** (step 3a). Attachment of two iodines to tyrosine yields **di-iodotyrosine (DIT)** (step 3b).

4. Next, a coupling process occurs between the iodinated tyrosine molecules to form the thyroid hormones. Coupling of one MIT (with one iodine) and one DIT (with two iodines) yields **tri-iodothyronine** or T_3 (with three iodines) (step 4a). Coupling of two DITs (each bearing two iodine atoms) yields **tetraiodothyronine** (T_4 or **thyroxine**), the four-iodine form of thyroid hormone (step 4b). Coupling does not occur between two MIT molecules.

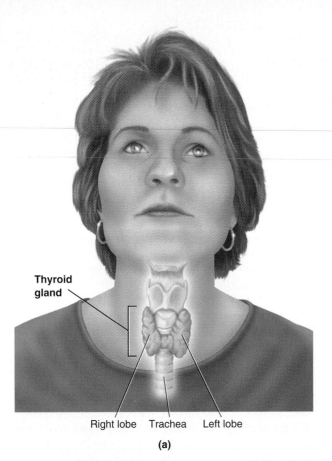

Thyroid gland

Right lobe   Trachea   Left lobe

(a)

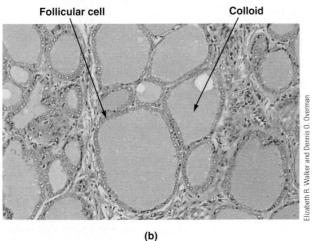

Follicular cell   **Colloid**

(b)

Elizabeth R. Walker and Dennis O. Overman

● FIGURE 17-13

Anatomy of the thyroid gland. (a) Gross anatomy of the thyroid gland, anterior view. The thyroid gland lies over the trachea just below the larynx and consists of two lobes connected by a thin strip. (b) Light-microscope appearance of the thyroid gland. The thyroid gland is composed primarily of colloid-filled spheres enclosed by a single layer of follicular cells.

For an interaction related to this figure, see Media Exercise 17.5: Thyroid Functions on the CD-ROM.

enclosing an inner lumen filled with **colloid**, a substance that serves as an extracellular storage site for thyroid hormone. Note that the colloid within the follicular lumen is extracellular (that is, outside of the thyroid cells), even though it is located within the interior of the follicle. Colloid is not in direct

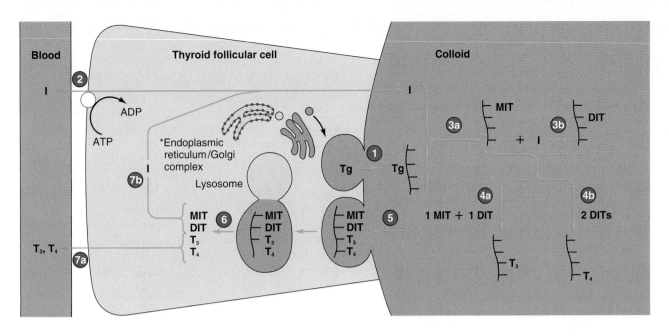

Tg = Thyroglobulin
I = Iodine
MIT = Monoiodotyrosine

DIT = Di-iodotyrosine
T_3 = Tri-iodothyronine
T_4 = Tetraiodothyronine (thyroxine)

* Organelles not drawn to scale. Endoplasmic reticulum/Golgi complex are proportionally too small.

1. Tyrosine-containing Tg produced within the thyroid follicular cells is transported into the colloid by exocytosis.

2. Iodine is actively transported from the blood into the colloid by the follicular cells.

3a. Attachment of one iodine to tyrosine within the Tg molecule yields MIT.

3b. Attachment of two iodines to tyrosine yields DIT.

4a. Coupling of one MIT and one DIT yields T_3.

4b. Coupling of two DITs yields T_4.

5. On appropriate stimulation, the thyroid follicular cells engulf a portion of Tg-containing colloid by phagocytosis.

6. Lysosomes attack the engulfed vesicle and split the iodinated products from Tg.

7a. T_3 and T_4 diffuse into the blood.

7b. MIT and DIT are deiodinated, and the freed iodine is recycled for synthesizing more hormone.

● **FIGURE 17-14**

Synthesis, storage, and secretion of thyroid hormone

All these products remain attached to thyroglobulin. Thyroid hormones remain stored in this form in the colloid until they are split off and secreted. Sufficient thyroid hormone is normally stored to supply the body's needs for several months.

▌ To secrete thyroid hormone, the follicular cells phagocytize thyroglobulin-laden colloid.

The release of thyroid hormone into the systemic circulation is a rather complex process, for two reasons. First, before their release T_3 and T_4 are still bound within the thyroglobulin molecule. Second, these hormones are stored at an inland extracellular site, the follicular lumen, so they must be transported completely across the follicular cells to reach the cap-

illaries that course through the interstitial spaces between the follicles.

The process of thyroid hormone secretion essentially involves the follicular cells "biting off" a piece of colloid, breaking the thyroglobulin molecule down into its component parts, and "spitting out" the freed T_3 and T_4 into the blood. On appropriate stimulation for thyroid hormone secretion, the follicular cells internalize a portion of the thyroglobulin-hormone complex by phagocytizing a piece of colloid (step 5 of ● Figure 17-14). Within the cells, the membrane-enclosed droplets of colloid coalesce with lysosomes, whose enzymes split off the biologically active thyroid hormones, T_3 and T_4, as well as the inactive iodotyrosines, MIT and DIT (step 6). The thyroid hormones, being very lipophilic, pass freely through the outer membranes of the follicular cells and into the blood (step 7a).

The MIT and DIT are of no endocrine value. The follicular cells contain an enzyme that swiftly removes the iodine from MIT and DIT, allowing the freed iodine to be recycled for synthesis of more hormone (step 7b). This highly specific enzyme will remove iodine only from the worthless MIT and DIT, not the valuable T_3 or T_4.

▌ Most of the secreted T_4 is converted into T_3 outside the thyroid.

About 90% of the secretory product released from the thyroid gland is in the form of T_4, yet T_3 is about four times more potent in its biologic activity. However, most of the secreted T_4 is converted into T_3, or *activated,* by being stripped of one of its iodines outside of the thyroid gland, primarily in the liver and kidneys. About 80% of the circulating T_3 is derived from secreted T_4 that has been peripherally stripped. Therefore, T_3 is the major biologically active form of thyroid hormone at the cellular level, even though the thyroid gland secretes mostly T_4.

▌ Thyroid hormone is the main determinant of the basal metabolic rate and exerts other effects as well.

Virtually every tissue in the body is affected either directly or indirectly by thyroid hormone. The effects of T_3 and T_4 can be grouped into several overlapping categories.

EFFECT ON METABOLIC RATE AND HEAT PRODUCTION

Thyroid hormone increases the body's overall basal metabolic rate or "idling speed" (see p. 512). It is the most important regulator of the body's rate of O_2 consumption and energy expenditure under resting conditions.

Closely related to thyroid hormone's overall metabolic effect is its **calorigenic** ("heat-producing") **effect.** Increased metabolic activity results in increased heat production.

SYMPATHOMIMETIC EFFECT

Any action similar to one produced by the sympathetic nervous system is known as a **sympathomimetic** ("sympathetic mimicking") **effect.** Thyroid hormone increases target cell responsiveness to catecholamines (epinephrine and norepinephrine), the chemical messengers used by the sympathetic nervous system and its hormonal reinforcements from the adrenal medulla. Thyroid hormone accomplishes this permissive action by causing a proliferation of specific catecholamine target-cell receptors. Because of this action, many of the effects observed when thyroid hormone secretion is elevated are similar to those that accompany activation of the sympathetic nervous system.

EFFECT ON THE HEART

Through its effect of increasing the heart's responsiveness to circulating catecholamines, thyroid hormone increases heart rate and force of contraction, thus increasing cardiac output.

EFFECT ON GROWTH AND THE NERVOUS SYSTEM

Thyroid hormone is essential for normal growth because of its effects on growth hormone (GH). Thyroid hormone not only stimulates GH secretion but also promotes the effects of GH (or somatomedins) on the synthesis of new structural proteins and on skeletal growth. Thyroid-deficient children have stunted growth that can be reversed by thyroid replacement therapy. Unlike excess GH, however, excess thyroid hormone does not produce excessive growth.

Thyroid hormone plays a crucial role in the normal development of the nervous system, especially the CNS, an effect impeded in children who have thyroid deficiency from birth. Thyroid hormone is also essential for normal CNS activity in adults.

▌ Thyroid hormone is regulated by the hypothalamus-pituitary-thyroid axis.

Thyroid-stimulating hormone (TSH), the thyroid tropic hormone from the anterior pituitary, is the most important physiologic regulator of thyroid hormone secretion (● Figure 17-15). Almost every step of thyroid hormone synthesis and release is stimulated by TSH.

In addition to enhancing thyroid hormone secretion, TSH maintains the structural integrity of the thyroid gland. In the absence of TSH, the thyroid atrophies (decreases in size) and secretes its hormones at a very low rate. Conversely, it undergoes hypertrophy (increase in the size of each follicular cell) and hyperplasia (increase in the number of follicular cells) in response to excess TSH stimulation.

The hypothalamic **thyrotropin-releasing hormone (TRH),** in tropic fashion, "turns on" TSH secretion by the anterior pituitary, whereas thyroid hormone, in negative-feedback fashion, "turns off" TSH secretion by inhibiting the anterior pituitary. Like other negative-feedback loops, the one between thyroid hormone and TSH tends to maintain a stable thyroid-hormone output.

Unlike most other hormonal systems, the hormones in the thyroid axis in an adult normally do not undergo sudden, wide swings in secretion. The only known factor that increases TRH secretion (and, accordingly, TSH and thyroid hormone secretion) is exposure to cold in newborn infants, a highly adaptive mechanism. Scientists think the dramatic increase in heat-producing thyroid-hormone secretion helps maintain body temperature during the abrupt drop in surrounding temperature at birth, as the infant passes from the mother's warm body to the cooler environmental air. A similar TSH response to cold exposure does not occur in adults, although it would make sense physiologically and does occur in many types of experimental animals.

Various types of stress inhibit TSH and thyroid hormone secretion, presumably through neural influences on the hypothalamus, although the adaptive importance of this inhibition is unclear.

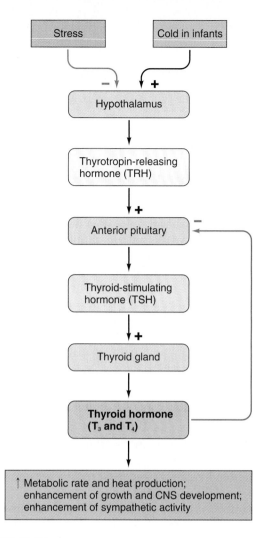

● FIGURE 17-15

Regulation of thyroid hormone secretion

 PhysioEdge   For an interaction related to this figure, see Media Exercise 17.5: Thyroid Functions on the CD-ROM.

▌ Abnormalities of thyroid function include both hypothyroidism and hyperthyroidism.

Clinical Note   Abnormalities of thyroid function are among the most common of all endocrine disorders. They fall into two major categories—**hypothyroidism** and **hyperthyroidism**—reflecting deficient and excess thyroid hormone secretion, respectively. The consequences of too little or too much thyroid hormone secretion are largely predictable, given knowledge of the functions of thyroid hormone.

HYPOTHYROIDISM

Hypothyroidism can result (1) from primary failure of the thyroid gland itself; (2) secondary to a deficiency of TRH, TSH, or both; or (3) from an inadequate dietary supply of iodine.

The symptoms of hypothyroidism are largely caused by a reduction in overall metabolic activity. Among other things, a patient with hypothyroidism has a reduced basal metabolic

rate; displays poor tolerance of cold (lack of the calorigenic effect); has a tendency to gain excessive weight (not burning fuels at a normal rate); is easily fatigued (lower energy production); has a slow, weak pulse (caused by a reduction in the rate and strength of cardiac contraction and a lowered cardiac output); and exhibits slow reflexes and slow mental responsiveness (because of the effect on the nervous system). The mental effects are characterized by diminished alertness, slow speech, and poor memory.

If a person has hypothyroidism from birth, a condition known as **cretinism** develops. Because adequate levels of thyroid hormone are essential for normal growth and CNS development, cretinism is characterized by dwarfism and mental retardation as well as other general symptoms of thyroid deficiency. The mental retardation is preventable if replacement therapy is started promptly, but it is not reversible once it has developed for a few months after birth, even with later treatment with thyroid hormone.

HYPERTHYROIDISM

The most common cause of hyperthyroidism is **Graves' disease.** This is an autoimmune disease in which the body erroneously produces **thyroid-stimulating immunoglobulin (TSI),** an antibody whose target is the TSH receptors on the thyroid cells. TSI stimulates both secretion and growth of the thyroid in a manner similar to TSH. Unlike TSH, however, TSI is not subject to negative-feedback inhibition by thyroid hormone, so thyroid secretion and growth continue unchecked (● Figure 17-16). Less frequently, hyperthyroidism occurs secondary to excess TRH or TSH or in association with a hypersecreting thyroid tumor.

As expected, the hyperthyroid patient has an elevated basal metabolic rate. The resultant increase in heat production leads to excessive perspiration and poor tolerance of heat. Body weight typically falls because the body is burning fuel at

---

● FIGURE 17-16

Role of thyroid-stimulating immunoglobulin in Graves' disease. Thyroid-stimulating immunoglobulin, an antibody erroneously produced in the autoimmune condition of Graves' disease, binds with the TSH receptors on the thyroid gland and continuously stimulates thyroid hormone secretion outside the normal negative-feedback control system.

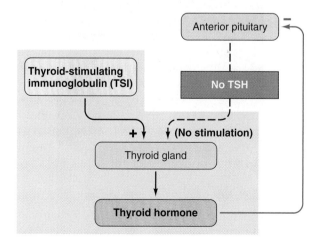

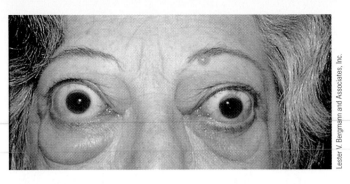

● **FIGURE 17-17**

Patient displaying exophthalmos. Abnormal fluid retention behind the eyeballs causes them to bulge forward.

an abnormally rapid rate. Heart rate and strength of contraction may increase so much the individual has palpitations (an unpleasant awareness of the heart's activity).

A prominent feature of Graves' disease but not of the other types of hyperthyroidism is **exophthalmos** (bulging eyes) (● Figure 17-17). Complex, water-retaining carbohydrates are deposited behind the eyes, although why this happens is still unclear. The resulting fluid retention pushes the eyeballs forward so they bulge from their bony orbit.

▌A goiter develops when the thyroid gland is overstimulated.

Clinical Note A **goiter** is an enlarged thyroid gland. Because the thyroid lies over the trachea, a goiter is readily palpable and usually highly visible (● Figure 17-18). A goiter occurs whenever either TSH or TSI excessively stimu-

● **FIGURE 17-18**

Patient with a goiter

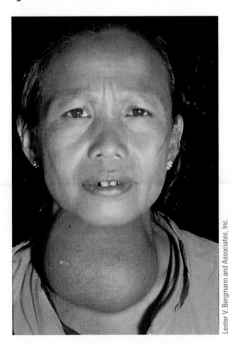

lates the thyroid gland. A goiter may accompany hypothyroidism or hyperthyroidism, but it need not be present in either condition. Knowing the hypothalamus-pituitary-thyroid axis and feedback control, we can predict which types of thyroid dysfunction will produce a goiter. Let's consider hypothyroidism first.

- Hypothyroidism secondary to hypothalamic or anterior pituitary failure will not be accompanied by a goiter, because the thyroid gland is not being adequately stimulated, let alone excessively stimulated.
- With hypothyroidism caused by thyroid gland failure or lack of iodine, a goiter does develop, because the circulating level of thyroid hormone is so low that there is little negative-feedback inhibition on the anterior pituitary, and TSH secretion is therefore elevated. TSH acts on the thyroid to increase the size and number of follicular cells and to increase their rate of secretion. If the thyroid cells cannot secrete hormone because of a lack of a critical enzyme or lack of iodine, no amount of TSH will be able to induce these cells to secrete T_3 and T_4. However, TSH can still promote hypertrophy and hyperplasia of the thyroid, with a consequent paradoxical enlargement of the gland (that is, a goiter), even though the gland is still underproducing.

Similarly, a goiter may or may not accompany hyperthyroidism:

- Excessive TSH secretion resulting from a hypothalamic or anterior pituitary defect would obviously be accompanied by a goiter and excess T_3 and T_4 secretion because of over-stimulation of thyroid growth.
- In Graves' disease, a hypersecreting goiter occurs because TSI promotes growth of the thyroid as well as enhancing secretion of thyroid hormone. Because the high levels of circulating T_3 and T_4 inhibit the anterior pituitary, TSH secretion itself is low.
- Hyperthyroidism resulting from overactivity of the thyroid in the absence of overstimulation, such as caused by an uncontrolled thyroid tumor, is not accompanied by a goiter. The spontaneous secretion of excessive amounts of T_3 and T_4 inhibits TSH, so there is no stimulatory input to promote growth of the thyroid.

 Click on the Media Exercises menu of the CD-ROM and work Media Exercise 17.5: Thyroid Functions to test your understanding of the previous section.

ADRENAL GLANDS

There are two **adrenal glands**, one embedded above each kidney in a capsule of fat (*ad* means "next to"; *renal* means "kidney") (● Figure 17-19a).

▌Each adrenal gland consists of a steroid-secreting cortex and a catecholamine-secreting medulla.

Each adrenal is composed of two endocrine organs, one surrounding the other. The outer layers composing the **adrenal** cortex secrete a variety of steroid hormones; the inner portion, the **adrenal medulla**, secretes catecholamines. Thus the adrenal cortex and medulla secrete hormones belonging to different chemical categories, whose functions, mechanisms of action, and regulation are entirely different. We will first examine the adrenal cortex before turning our attention to the adrenal medulla.

▌ The adrenal cortex secretes mineralocorticoids, glucocorticoids, and sex hormones.

About 80% of the adrenal gland is composed of the cortex, which consists of three layers or zones: the **zona glomerulosa**, the outermost layer; the **zona fasciculata**, the middle and largest portion; and the **zona reticularis**, the innermost zone (● Figure 17-19b). The adrenal cortex produces a number of different **adrenocortical hormones**, all of which are steroids derived from the common precursor molecule, cholesterol. Slight variations in structure confer different functional capabilities on the various adrenocortical hormones. On the basis of their primary actions, the adrenal steroids can be divided into three categories:

1. **Mineralocorticoids**, mainly *aldosterone*, influence mineral (electrolyte) balance, specifically Na^+ and K^+ balance.

2. **Glucocorticoids**, primarily *cortisol*, play a major role in glucose metabolism as well as in protein and lipid metabolism.

3. **Sex hormones** are identical or similar to those produced by the gonads (testes in males, ovaries in females). The most abundant and physiologically important of the adrenocortical sex hormones is *dehydroepiandrosterone*, a "male" sex hormone.

Of the two major adrenocortical hormones, aldosterone is produced exclusively in the zona glomerulosa, whereas cortisol synthesis is limited to the two inner layers of the cortex, with the zona fasciculata being the major source of this glucocorticoid. No other steroidogenic tissues have the capability of producing either mineralocorticoids or glucocorticoids. In contrast, the adrenal sex hormones, also produced by the two inner cortical zones, are produced in far greater abundance in the gonads.

▌ Mineralocorticoids' major effects are on Na^+ and K^+ balance and blood pressure homeostasis.

The actions and regulation of the primary adrenocortical mineralocorticoid, **aldosterone**, are described thoroughly elsewhere (Chapter 13). The principal site of aldosterone action is on the distal and collecting tubules of the kidney, where it promotes Na^+ retention and enhances K^+ elimination during the formation of urine. The promotion of Na^+ retention

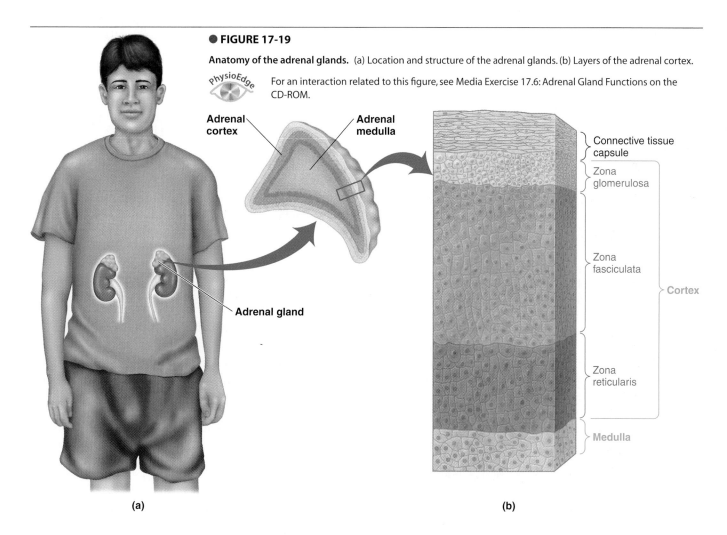

● **FIGURE 17-19**

Anatomy of the adrenal glands. (a) Location and structure of the adrenal glands. (b) Layers of the adrenal cortex.

PhysioEdge For an interaction related to this figure, see Media Exercise 17.6: Adrenal Gland Functions on the CD-ROM.

Adrenal cortex

Adrenal medulla

Adrenal gland

Connective tissue capsule

Zona glomerulosa

Zona fasciculata

Zona reticularis

Cortex

Medulla

(a)

(b)

by aldosterone secondarily induces osmotic retention of H_2O, expanding the ECF volume, which is important in the long-term regulation of blood pressure.

Mineralocorticoids are *essential for life*. Without aldosterone, a person rapidly dies from circulatory shock because of the marked fall in plasma volume caused by excessive losses of H_2O-holding Na^+. With most other hormonal deficiencies, death is not imminent, even though a chronic hormonal deficiency may eventually lead to a premature death.

Aldosterone secretion is increased by (1) activation of the renin-angiotensin-aldosterone system by factors related to a reduction in Na^+ and a fall in blood pressure and (2) direct stimulation of the adrenal cortex by a rise in plasma K^+ concentration (see ● Figure 13-15, p. 423). Adrenocorticotropic hormone (ACTH) from the anterior pituitary primarily promotes the secretion of cortisol, not aldosterone. Thus, unlike cortisol regulation, the regulation of aldosterone secretion is largely independent of anterior pituitary control.

▌ Glucocorticoids exert metabolic effects and play a key role in adaptation to stress.

Cortisol, the primary glucocorticoid, plays an important role in carbohydrate, protein, and fat metabolism; executes significant permissive actions for other hormonal activities; and helps people resist stress.

METABOLIC EFFECTS

The overall effect of cortisol's metabolic actions is to increase the concentration of blood glucose at the expense of protein and fat stores. Specifically, cortisol performs the following functions:

- It stimulates hepatic **gluconeogenesis**, the conversion of noncarbohydrate sources (namely, amino acids) into carbohydrate within the liver (*gluco* means "glucose," *neo* means "new," *genesis* means "production"). Between meals or during periods of fasting, when no new nutrients are being absorbed into the blood for use and storage, the glycogen (stored glucose) in the liver tends to become depleted as it is broken down to release glucose into the blood. Gluconeogenesis is an important factor in replenishing hepatic glycogen stores and thus in maintaining normal blood-glucose levels between meals. This is essential because the brain can use only glucose as its metabolic fuel, yet nervous tissue cannot store glycogen to any extent. The concentration of glucose in the blood must therefore be maintained at an appropriate level to adequately supply the glucose-dependent brain with nutrients.
- It inhibits glucose uptake and use by many tissues, but not the brain, thus sparing glucose for use by the brain, which absolutely requires it as a metabolic fuel. This action contributes to the increase in blood glucose concentration brought about by gluconeogenesis.
- It stimulates protein degradation in many tissues, especially muscle. By breaking down a portion of muscle proteins into their constituent amino acids, cortisol increases the blood amino-acid concentration. These mobilized amino acids are available for use in gluconeogenesis or wherever else they are needed, such as for repair of damaged tissue or synthesis of new cellular structures.
- It facilitates lipolysis, the breakdown of lipid (fat) stores in adipose tissue, thus releasing free fatty acids into the blood (*lysis* means "breakdown"). The mobilized fatty acids are available as an alternative metabolic fuel for tissues that can use this energy source in lieu of glucose, thereby conserving glucose for the brain.

PERMISSIVE ACTIONS

Cortisol is extremely important for its permissiveness (see p. 532). For example, cortisol must be present in adequate amounts to permit the catecholamines to induce vasoconstriction. A person lacking cortisol, if untreated, may go into circulatory shock in a stressful situation that demands immediate widespread vasoconstriction.

ROLE IN ADAPTATION TO STRESS

Cortisol plays a key role in adaptation to stress. **Stress** is the generalized, nonspecific response of the body to any factor that overwhelms, or threatens to overwhelm, the body's compensatory abilities to maintain homeostasis. Contrary to popular usage, the agent inducing the response is correctly called a *stressor,* whereas *stress* refers to the state induced by the stressor. The following types of noxious stimuli illustrate the range of factors that can induce a stress response: *physical* (trauma, surgery, intense heat or cold); *chemical* (reduced O_2 supply, acid–base imbalance); *physiologic* (heavy exercise, hemorrhagic shock, pain); *psychological* or *emotional* (anxiety, fear, sorrow); and *social* (personal conflicts, change in lifestyle). Stress of any kind is one of the major stimuli for increased cortisol secretion.

Although cortisol's precise role in adapting to stress is not known, a speculative but plausible explanation might be as follows. A primitive human or an animal wounded or faced with a life-threatening situation must forgo eating. A cortisol-induced shift away from protein and fat stores in favor of expanded carbohydrate stores and increased availability of blood glucose would help protect the brain from malnutrition during the imposed fasting period. Also, the amino acids liberated by protein degradation would provide a readily available supply of building blocks for tissue repair if physical injury occurred. Thus an increased pool of glucose, amino acids, and fatty acids is available for use as needed.

ANTI-INFLAMMATORY AND IMMUNOSUPPRESSIVE EFFECTS

Clinical Note When cortisol or synthetic cortisol-like compounds are administered to yield higher-than-physiologic concentrations of glucocorticoids (that is, *pharmacologic levels*), not only are all the metabolic effects magnified, but several important new actions not evidenced at normal physiologic levels are seen. The most noteworthy of glucocorticoids' pharmacologic effects are *anti-inflammatory* and *immunosuppressive effects*. Synthetic glucocorticoids have been developed that maximize the anti-inflammatory and immunosuppressive effects of these steroids while minimizing the metabolic effects.

Administering large amounts of glucocorticoid inhibits almost every step of the inflammatory response, making these steroids effective drugs in treating conditions in which the inflammatory response itself has become destructive, such as *rheumatoid arthritis*. Glucocorticoids used in this manner do not affect the underlying disease process; they merely suppress the body's response to the disease. Because glucocorticoids also exert multiple inhibitory effects on the overall immune process, such as "knocking out of commission" the white blood cells responsible for antibody production and destruction of foreign cells, these agents have also proved useful in managing various allergic disorders and in preventing organ transplant rejections.

When these steroids are employed therapeutically, they should be used only when warranted and then only sparingly, for several important reasons. First, because they suppress the normal inflammatory and immune responses that form the backbone of the body's defense system, a glucocorticoid-treated person has limited ability to resist infections. Second, in addition to the anti-inflammatory and immunosuppressive effects readily exhibited at pharmacologic levels, other less desirable effects may also be observed with prolonged exposure to higher-than-normal concentrations of glucocorticoids. These effects include development of gastric ulcers, high blood pressure, atherosclerosis, menstrual irregularities, and bone thinning. Third, high levels of exogenous glucocorticoids act in negative-feedback fashion to suppress the hypothalamus-pituitary axis that drives normal glucocorticoid secretion and maintains the integrity of the adrenal cortex. Prolonged suppression of this axis can lead to irreversible atrophy of the cortisol-secreting cells of the adrenal gland and thus to permanent inability of the body to produce its own cortisol.

▌ Cortisol secretion is regulated by the hypothalamus-pituitary-adrenal cortex axis.

Cortisol secretion by the adrenal cortex is regulated by a negative-feedback system involving the hypothalamus and anterior pituitary (● Figure 17-20). ACTH from the anterior pituitary stimulates the adrenal cortex to secrete cortisol. ACTH is derived from a large precursor molecule, **pro-opiomelanocortin**, produced within the endoplasmic reticulum of the anterior pituitary's ACTH-secreting cells. Prior to secretion, this large precursor is pruned into ACTH and several other biologically active peptides, namely, *melanocyte-stimulating hormone (MSH)* and a morphinelike substance, β-*endorphin* (see p. 150). The possible significance of the fact that these multiple secretory products are developed from a single precursor molecule will be addressed later.

The ACTH-producing cells in turn secrete only at the command of corticotropin-releasing hormone (CRH) from the hypothalamus. The feedback control loop is completed by cortisol's inhibitory actions on CRH and ACTH secretion by the hypothalamus and anterior pituitary, respectively.

The negative-feedback system for cortisol maintains the level of cortisol secretion relatively constant around the set point. Superimposed on the basic negative-feedback control

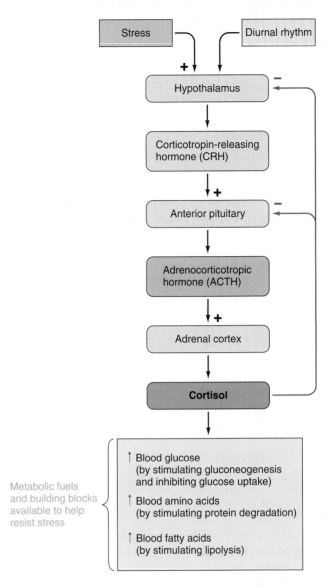

● **FIGURE 17-20**

Control of cortisol secretion

---

system are two additional factors that influence plasma cortisol concentrations by changing the set point: *diurnal rhythm* and *stress,* both of which act on the hypothalamus to vary the secretion rate of CRH.

INFLUENCE OF DIURNAL RHYTHM ON CORTISOL SECRETION

Recall that the plasma cortisol concentration displays a characteristic diurnal rhythm, with the highest level occurring in the morning and the lowest level at night (see ● Figure 17-3, p. 531). This diurnal rhythm, which is intrinsic to the hypothalamus-pituitary control system, is related primarily to the sleep–wake cycle. The peak and low levels are reversed in a person who works at night and sleeps during the day. Such time-dependent variations in secretion are of more than academic interest, because it is important clinically to know at what time of day a blood sample was taken when interpreting the significance of a particular value. Furthermore, the linking of cortisol secretion to day–night activity patterns raises serious questions about the common practice of swing shifts at work (that is, constantly switching day and night shifts among employees).

INFLUENCE OF STRESS ON CORTISOL SECRETION

The other major factor that is independent of, and in fact can override, the stabilizing negative-feedback control is stress. Dramatic increases in cortisol secretion, mediated by the central nervous system through enhanced activity of the CRH-ACTH system, occur in response to all kinds of stressful situations. The magnitude of the increase in plasma cortisol concentration is generally proportional to the intensity of the stressful stimulation; a greater increase in cortisol levels is evoked in response to severe stress than to mild stress.

❚ The adrenal cortex secretes both male and female sex hormones in both sexes.

In both sexes, the adrenal cortex produces both *androgens,* or "male" sex hormones, and *estrogens,* or "female" sex hormones. The main site of production for the sex hormones is the gonads: the testes for androgens and the ovaries for estrogens. Accordingly, males have a preponderance of circulating androgens, whereas in females estrogens predominate. However, no hormones are unique to either males or females (except those from the placenta during pregnancy), because the adrenal cortex in both sexes produces small amounts of the sex hormone of the opposite sex.

Under normal circumstances, the adrenal androgens and estrogens are not sufficiently abundant or powerful to induce masculinizing or feminizing effects, respectively. The only adrenal sex hormone that has any biologic importance is the androgen **dehydroepiandrosterone (DHEA)**. The testes' primary androgen product is the potent testosterone, but the most abundant adrenal androgen is the much weaker DHEA. Adrenal DHEA is overpowered by testicular testosterone in males but is of physiologic significance in females, who otherwise lack androgens. This adrenal androgen governs androgen-dependent processes in the female such as growth of pubic and axillary (armpit) hair, enhancement of the pubertal growth spurt, and development and maintenance of the female sex drive.

A surge in DHEA secretion begins at puberty and peaks between the ages of 25 and 30. After 30, DHEA secretion slowly tapers off until, by the age of 60, the plasma DHEA concentration is less than 15% of its peak level.

Clinical Note Some scientists suspect that the age-related decline of DHEA and other hormones such as GH and melatonin plays a role in some problems of aging. Early studies with DHEA replacement therapy demonstrated some physical improvement, such as an increase in lean muscle mass and a decrease in fat, but the most pronounced effect was a marked increase in psychological well-being and an improved ability to cope with stress. Advocates for DHEA replacement therapy do not suggest that maintaining youthful levels of this hormone is a fountain of youth (that is, it is not going to extend the life span), but they do propose that it may help people feel and act younger as they age. Other scientists caution that evidence supporting DHEA as an anti-aging therapy is still sparse. Also, they are concerned about DHEA supplementation until it has been thoroughly studied for possible harmful side effects. For example, some research suggests a potential increase in the risk of heart disease among women taking DHEA because of an observed reduction in HDL, the "good' cholesterol (see p.268). Also, some experts fear that DHEA supplementation may raise the odds of acquiring ovarian or breast cancer in women and prostate cancer in men. Ironically, although the Food and Drug Administration (FDA) banned sales of DHEA as an over-the-counter drug in 1985 because of concerns about very real risks coupled with little proof of benefits, the product is widely available today as an unregulated food supplement. DHEA can be marketed as a dietary supplement without approval by the FDA as long as the product label makes no specific medical claims.

❚ The adrenal cortex may secrete too much or too little of any of its hormones.

Clinical Note Although uncommon, there are a number of different disorders of adrenocortical function. Excessive secretion may occur with any of the three categories of adrenocortical hormones. Accordingly, three main patterns of symptoms resulting from hyperadrenalism can be distinguished, depending on which hormone type is in excess: aldosterone hypersecretion, cortisol hypersecretion, and adrenal androgen hypersecretion.

ALDOSTERONE HYPERSECRETION

Excess mineralocorticoid secretion may be caused by (1) a hypersecreting adrenal tumor made up of aldosterone-secreting cells (**Conn's syndrome**) or (2) inappropriately high activity of the renin-angiotensin-aldosterone system. The symptoms are related to the exaggerated effects of aldosterone—namely, excessive Na^+ retention (*hypernatremia*) and K^+ depletion (*hypokalemia*). Also, high blood pressure (hypertension) is generally present, at least partially because of excessive Na^+ and fluid retention.

CORTISOL HYPERSECRETION

Excessive cortisol secretion (**Cushing's syndrome**) can be caused by (1) overstimulation of the adrenal cortex by excessive amounts of CRH and/or ACTH or (2) adrenal tumors that uncontrollably secrete cortisol independent of ACTH. The prominent characteristics of this syndrome are related to the exaggerated effects of glucocorticoid, with the main symptoms being reflections of excessive gluconeogenesis. When too many amino acids are converted into glucose, the body suffers from combined glucose excess (high blood glucose) and protein shortage. For example, loss of structural protein within the walls of the small blood vessels leads to easy bruisability. For reasons that are unclear, some of the extra glucose is deposited as body fat in locations characteristic for this disease, typically in the abdomen, above the shoulder blades, and in the face. The abnormal fat distributions in the latter two locations are descriptively called a "buffalo hump"

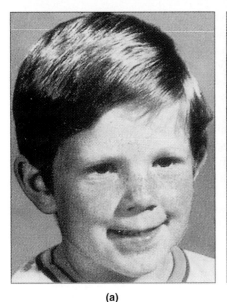
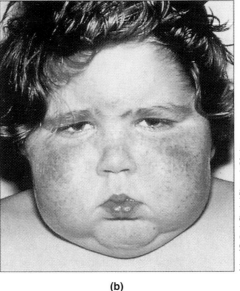

(a) **(b)**

● **FIGURE 17-21**

Patient with Cushing's syndrome. (a) Young boy prior to onset of the condition. (b) Only four months later, the same boy displaying a "moon-face" characteristic of Cushing's syndrome.

and a "moon face," respectively (● Figure 17-21). The appendages, in contrast, remain thin.

ADRENAL ANDROGEN HYPERSECRETION

Excess adrenal androgen secretion, a masculinizing condition, is more common than the extremely rare feminizing condition of excess adrenal estrogen secretion. Either condition is referred to as **adrenogenital syndrome**, emphasizing the pronounced effects that excessive adrenal sex hormones have on the genitalia and associated sexual characteristics. Adrenogenital syndrome is most commonly the result of an enzymatic defect that causes the cortisol-secreting cells to produce androgen instead of cortisol.

The symptoms that result from excess androgen secretion depend on the sex of the individual and the age when the hyperactivity first begins.

• *In adult females.* Because androgens exert masculinizing effects, a woman with this disease tends to develop a male pattern of body hair, a condition referred to as **hirsutism.** She usually also acquires other male secondary sexual characteristics, such as deepening of the voice and more muscular arms and legs. The breasts become smaller, and menstruation may cease as a result of androgen suppression of the woman's hypothalamus-pituitary-ovarian pathway for her own female sex-hormone secretion.

• *In newborn females.* Female infants born with adrenogenital syndrome manifest male-type external genitalia, because excessive androgen secretion occurs early enough during fetal life to induce development of their genitalia along male lines, similar to the development of males under the influence of testicular androgen. The clitoris, which is the female homolog of the male penis, enlarges under androgen influence and takes on a penile appearance so in some cases it is

difficult at first to determine the child's sex. Thus this hormonal abnormality is one of the major causes of **female pseudohermaphroditism**, a condition in which female gonads (ovaries) are present but the external genitalia resemble those of a male. (A true hermaphrodite has the gonads of both sexes.)

• *In prepubertal males.* Excessive adrenal androgen secretion in prepubertal boys causes them to prematurely develop male secondary sexual characteristics—for example, deep voice, beard, enlarged penis, and sex drive. This condition is referred to as **precocious pseudopuberty** to differentiate it from true puberty, which occurs as a result of increased testicular activity. In precocious pseudopuberty, the androgen secretion from the adrenal cortex is not accompanied by sperm production or any other gonadal activity, because the testes are still in their nonfunctional prepubertal state.

• *In adult males.* Overactivity of adrenal androgens in adult males has no apparent effect, because any masculinizing effect induced by the weak DHEA, even when in excess, is unnoticeable in the face of the powerful masculinizing effects of the much more abundant and potent testosterone from the testes.

ADRENOCORTICAL INSUFFICIENCY

If one adrenal gland is nonfunctional or removed, the other healthy organ can take over the function of both through hypertrophy and hyperplasia. Therefore, both glands must be affected before adrenocortical insufficiency occurs.

In **Addison's disease**, all layers of the adrenal cortex are undersecreting. This condition is most commonly caused by autoimmune destruction of the cortex by erroneous production of adrenal cortex–attacking antibodies, in which case both aldosterone and cortisol are deficient. Adrenocortical insufficiency may also occur because of a pituitary or hypothalamic abnormality, resulting in insufficient ACTH secretion. In this case, only cortisol is deficient, because aldosterone secretion does not depend on ACTH stimulation.

The symptoms associated with aldosterone deficiency in Addison's disease are the most threatening. If severe enough, the condition is fatal, because aldosterone is essential for life. However, the loss of adrenal function may develop slowly and insidiously so that aldosterone secretion may be subnormal but not totally lacking. Patients with aldosterone deficiency display K^+ retention (*hyperkalemia*) caused by reduced K^+ loss in the urine, and Na^+ depletion (*hyponatremia*) caused by excessive urinary loss of Na^+. The former disturbs cardiac rhythm. The latter reduces ECF volume, including circulating blood volume, which in turn lowers blood pressure (hypotension).

Symptoms of cortisol deficiency are as would be expected: poor response to stress, hypoglycemia (low blood glucose)

Atlas of Pediatric Physical Diagnosis/Mosby

caused by reduced gluconeogenic activity, and lack of permissive action for many metabolic activities.

We are now going to shift our attention from the adrenal cortex to the adrenal medulla.

The adrenal medulla is a modified sympathetic postganglionic neuron.

The adrenal medulla is actually a modified part of the sympathetic nervous system. A sympathetic pathway consists of two neurons in sequence—a preganglionic neuron originating in the CNS, whose axonal fiber terminates on a second peripherally located postganglionic neuron, which in turn terminates on the effector organ (see p. 186). The neurotransmitter released by sympathetic postganglionic fibers is norepinephrine.

The adrenal medulla consists of modified postganglionic sympathetic neurons. Unlike ordinary postganglionic sympathetic neurons, those in the adrenal medulla do not have axonal fibers that terminate on effector organs. Instead, on stimulation by the preganglionic fiber the ganglionic cell bodies within the adrenal medulla release their chemical transmitter directly into the circulation (see ● Figure 7-4, p. 190).In this case, the transmitter qualifies as a hormone instead of a neurotransmitter. Like sympathetic fibers, the adrenal medulla does release norepinephrine, but its most abundant secretory output is a similar chemical messenger known as **epinephrine**. Both epinephrine and norepinephrine belong to the chemical class of catecholamines, which are derived from the amino acid tyrosine.

Once produced, epinephrine and norepinephrine are stored in **chromaffin granules**, which are similar to the transmitter storage vesicles found in sympathetic nerve endings. Catecholamines are secreted into the blood by exocytosis of chromaffin granules. Their release is analogous to the release mechanism for secretory vesicles that contain stored peptide hormones or the release of norepinephrine at sympathetic postganglionic terminals.

Adrenomedullary norepinephrine is generally secreted in quantities too small to exert significant effects on target cells. Therefore, for practical purposes we can assume that norepinephrine effects are predominantly mediated directly by the sympathetic nervous system and that epinephrine effects are brought about exclusively by the adrenal medulla.

Epinephrine reinforces the sympathetic nervous system and exerts additional metabolic effects.

Together, the sympathetic nervous system and adrenomedullary epinephrine mobilize the body's resources to support peak physical exertion in emergency or stressful situations. The sympathetic and epinephrine actions constitute a fight-or-flight response that prepares the person to combat an enemy or flee from danger (see p. 189). Epinephrine, by circulating in the blood, can reach catecholamine target cells that are not directly innervated by the sympathetic nervous

system, such as the liver and adipose tissue. Accordingly, even though epinephrine and norepinephrine exert similar effects in many tissues, with epinephrine generally reinforcing sympathetic nervous activity, epinephrine can exert some unique effects because it gets into places not supplied by sympathetic fibers. For example, epinephrine prompts the mobilization of stored carbohydrate from the liver and fat from adipose tissue to provide immediately available energy for use as needed to fuel muscular work.

Realize, however, that epinephrine functions only at the bidding of the sympathetic nervous system, which is solely responsible for stimulating its secretion from the adrenal medulla. When the sympathetic system is activated under conditions of fear or stress, it simultaneously triggers a surge of adrenomedullary catecholamine release. The concentration of epinephrine in the blood may increase up to 300 times normal, with the amount of epinephrine released depending on the type and intensity of the stressful stimulus. Thus sympathetic activity indirectly controls actions of epinephrine. By having the more versatile circulating epinephrine at its call, the sympathetic nervous system has a means of reinforcing its own neurotransmitter effects plus a way of executing additional actions on tissues that it does not directly innervate.

Because both components of the adrenal gland play an extensive role in responding to stress, this is an appropriate place to pull together the major factors involved in the stress response.

The stress response is a generalized pattern of reactions to any situation that threatens homeostasis.

Recall that a variety of noxious stimuli can elicit a stress response. Different stressors may produce some specific responses characteristic of that stressor; for example, the body's specific response to cold exposure is shivering and skin vasoconstriction, whereas the specific response to bacterial invasion includes increased phagocytic activity and antibody production. In addition to their specific response, however, all stressors also produce a similar nonspecific, generalized response regardless of the type of stressor (● Figure 17-22). This set of responses common to all noxious stimuli is called the **general adaptation syndrome**. When a stressor is recog-

---

● **FIGURE 17-22**

Action of a stressor on the body

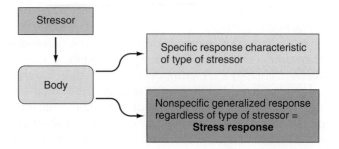

nized, both nervous and hormonal responses bring about defensive measures to cope with the emergency. The result is a state of intense readiness and mobilization of biochemical resources.

To appreciate the value of the multifaceted stress response, imagine a primitive cave dweller who has just seen a large wild beast lurking in the shadows. We will consider both the neural and hormonal responses that would take place in this scenario. The body responds in the same way to modern-day stressors. You are already familiar with all these responses. At this time we are just examining how these responses work together.

ROLES OF THE SYMPATHETIC NERVOUS SYSTEM AND EPINEPHRINE IN STRESS

The major neural response to such a stressful stimulus is generalized activation of the sympathetic nervous system. The resultant increase in cardiac output and ventilation as well as the diversion of blood from vasoconstricted regions of suppressed activity, such as the digestive tract and kidneys, to the more active vasodilated skeletal muscles and heart prepare the body for a fight-or-flight response. Simultaneously, the sympathetic system calls forth hormonal reinforcements in the form of a massive outpouring of epinephrine from the adrenal medulla. Epinephrine strengthens sympathetic responses and reaches places not innervated by the sympathetic system to perform additional functions, such as mobilizing carbohydrate and fat stores.

ROLES OF THE CRH-ACTH-CORTISOL SYSTEM IN STRESS

Besides epinephrine, a number of other hormones are involved in the overall stress response (⚠ Table 17-3). The predominant hormonal response is activation of the CRH-ACTH-cortisol system. Recall that cortisol's role in helping the body cope with stress is presumed to be related to its metabolic effects. Cortisol breaks down fat and protein stores while expanding carbohydrate stores and increasing the availability of blood glucose. A logical assumption is that the increased pool of glucose, amino acids, and fatty acids is available for use as needed, such as to sustain nourishment to the brain and provide building blocks for repair of damaged tissues.

In addition to the effects of cortisol in the hypothalamus-pituitary-adrenal cortex axis, ACTH may also play a role in resisting stress. ACTH is one of several peptides that facilitate learning and behavior. Thus an increase in ACTH during psychosocial stress may help the body cope more readily with similar stressors in the future by facilitating the learning of appropriate behavioral responses. Furthermore, ACTH is not released alone from its anterior pituitary storage vesicles. Pruning of the large pro-opiomelanocortin precursor molecule yields not only ACTH but also morphinelike β-endorphin, which is cosecreted with ACTH on stimulation by CRH during stress. As a potent endogenous opiate, β-endorphin may exert a role in mediating analgesia (reduction of pain perception) if physical injury is inflicted during stress (see p. 150).

⚠ **TABLE 17-3**

Major Hormonal Changes during the Stress Response

HORMONE	CHANGE	PURPOSE SERVED
Epinephrine	↑	Reinforces the sympathetic nervous system to prepare the body for "fight or flight"
		Mobilizes carbohydrate and fat energy stores; increases blood glucose and blood fatty acids
CRH-ACTH-Cortisol	↑	Mobilizes energy stores and metabolic building blocks for use as needed; increases blood glucose, blood amino acids, and blood fatty acids
		ACTH facilitates learning and behavior
		β-Endorphin cosecreted with ACTH may mediate analgesia
Renin-Angiotensin-Aldosterone	↑	Conserve salt and H_2O to expand the plasma volume; help sustain blood pressure when acute loss of plasma volume occurs
Vasopressin	↑	Angiotensin II and vasopressin cause arteriolar vasoconstriction to increase blood pressure
		Vasopressin facilitates learning

ROLE OF BLOOD PRESSURE–SUSTAINING HORMONAL RESPONSES IN STRESS

In addition to the hormonal changes that mobilize energy stores during stress, other hormones are simultaneously called into play to sustain blood volume and blood pressure during the emergency. The sympathetic system and epinephrine play major roles in acting directly on the heart and blood vessels to improve circulatory function. In addition, the renin-angiotensin-aldosterone system is activated as a consequence of a sympathetically induced reduction of blood supply to the kidneys (see p. 417). Vasopressin secretion is also increased during stressful situations (see p. 450). Collectively, these hormones expand the plasma volume by promoting retention of salt and H_2O. Presumably, the enlarged plasma volume serves as a protective measure to help sustain blood pressure should acute loss of plasma fluid occur through hemorrhage or heavy sweating during the impending period of danger. Vasopressin and angiotensin also have direct vasopressor effects, which would be of benefit in maintaining an adequate arterial pressure in the event of acute blood loss (see p. 289). Vasopressin is further believed to facilitate learning, which has implications for future adaptation to stress.

▌ The multifaceted stress response is coordinated by the hypothalamus.

All the individual responses to stress just described are either directly or indirectly influenced by the hypothalamus (● Figure 17-23). The hypothalamus receives input concerning physical and emotional stressors from virtually all areas of the brain and from many receptors throughout the body. In response, the hypothalamus directly activates the sympathetic nervous system, secretes CRH to stimulate ACTH and cortisol release, and triggers the release of vasopressin. Sympathetic stimulation in turn brings about the secretion of epinephrine. Furthermore, vasoconstriction of the renal afferent arterioles by the catecholamines indirectly triggers the secretion of renin by reducing the flow of oxygenated blood through the kidneys. Renin in turn sets in motion the renin-angiotensin-aldosterone mechanism. In this way, the hypothalamus integrates the responses of both the sympathetic nervous system and the endocrine system during stress.

Acceleration of cardiovascular and respiratory activity, retention of salt and H_2O, and mobilization of metabolic fuels and building blocks can be of benefit in response to a physi-

cal stressor, such as an athletic competition. Most of the stressors in our everyday lives are psychosocial in nature, however, yet they induce these same magnified responses. Stressors such as anxiety about an exam, conflicts with loved ones, or impatience while sitting in a traffic jam can elicit a stress response. Although the rapid mobilization of body resources is appropriate in the face of real or threatened physical injury, it is generally inappropriate in response to nonphysical stress. If no extra energy is demanded, no tissue is damaged, and no blood lost, body stores are being broken down and fluid retained needlessly, probably to the detriment of the emotionally stressed individual.

 Click on the Media Exercises menu of the CD-ROM and work Media Exercise 17.6: Adrenal Gland Functions to test your understanding of the previous section.

ENDOCRINE CONTROL OF FUEL METABOLISM

We have just discussed the metabolic changes that are elicited during the stress response. Now we will concentrate on the metabolic patterns that occur in the absence of stress, including the hormonal factors that govern this normal metabolism.

● **FIGURE 17-23**

Integration of the stress response by the hypothalamus

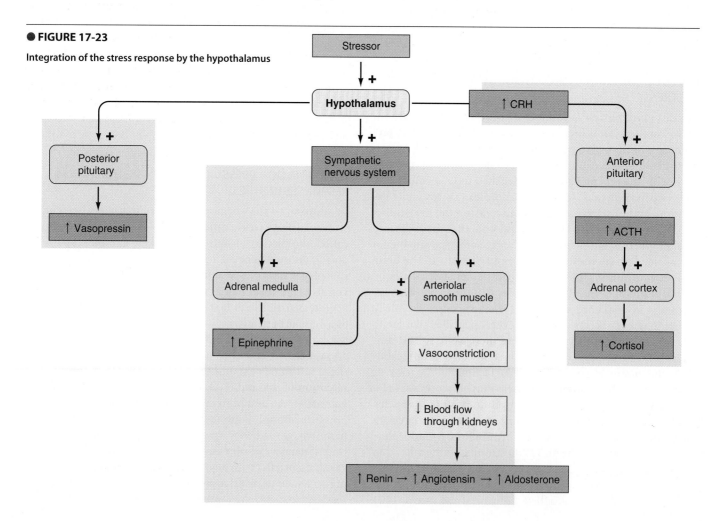

The term **metabolism** refers to all the chemical reactions that occur within the cells of the body. Those reactions involving the degradation, synthesis, and transformation of the three classes of energy-rich organic molecules—protein, carbohydrate, and fat—are collectively known as **intermediary metabolism** or **fuel metabolism** (▲ Table 17-4).

During the process of digestion, large nutrient molecules (**macromolecules**) are broken down into their smaller absorbable subunits as follows: Proteins are converted into amino acids, complex carbohydrates into monosaccharides (mainly glucose), and triglycerides (dietary fats) into monoglycerides and free fatty acids. These absorbable units are transferred from the digestive tract lumen into the blood, either directly or by way of the lymph (Chapter 15).

ANABOLISM AND CATABOLISM

These organic molecules are constantly exchanged between the blood and body cells. The chemical reactions in which the organic molecules participate within the cells are categorized into two metabolic processes: anabolism and catabolism (● Figure 17-24). **Anabolism** is the buildup or synthesis of larger organic macromolecules from the small organic molecular subunits. Anabolic reactions generally require energy input in the form of ATP. These reactions result in either (1) the manufacture of materials needed by the cell, such as cellular structural proteins or secretory products, or (2) storage of excess ingested nutrients not immediately needed for energy production or needed as cellular building blocks. Storage is in the form of glycogen (the storage form of glucose) or fat

reservoirs. **Catabolism** is the breakdown, or degradation, of large, energy-rich organic molecules within cells. Catabolism encompasses two levels of breakdown: (1) hydrolysis (see p. 466) of large cellular organic macromolecules into their smaller subunits, similar to the process of digestion except that the reactions take place within the body cells instead of within the digestive tract lumen (for example, release of glucose by the catabolism of stored glycogen); and (2) oxidation of the smaller subunits, such as glucose, to yield energy for ATP production (see p. 29).

As an alternative to energy production, the smaller, multipotential organic subunits derived from intracellular hydrolysis may be released into the blood. These mobilized glucose, fatty acid, and amino acid molecules can then be used as needed for energy production or cellular synthesis elsewhere in the body.

In an adult, the rates of anabolism and catabolism are generally in balance, so the adult body remains in a dynamic steady state and appears unchanged even though the organic molecules that determine its structure and function are continuously being turned over. During growth, anabolism exceeds catabolism.

INTERCONVERSIONS AMONG ORGANIC MOLECULES

In addition to being able to resynthesize catabolized organic molecules back into the same type of molecules, many cells of the body, especially liver cells, can convert most types of small organic molecules into other types—as in, for example, transforming amino acids into glucose or fatty acids. Because of these interconversions, adequate nourishment can be provided by a wide range of molecules present in different types of foods. There are limits, however. **Essential nutrients**, such as the essential amino acids and vitamins, cannot be formed in the body by conversion from another type of organic molecule.

▲ **TABLE 17-4**

Summary of Reactions in Fuel Metabolism

METABOLIC PROCESS	REACTION	CONSEQUENCE
Glycogenesis	Glucose → glycogen	↓Blood glucose
Glycogenolysis	Glycogen → glucose	↑Blood glucose
Gluconeogenesis	Amino acids → glucose	↑Blood glucose
Protein Synthesis	Amino acids → protein	↓Blood amino acids
Protein Degradation	Protein → amino acids	↑Blood amino acids
Fat Synthesis (lipogenesis, or triglyceride synthesis)	Fatty acids and glycerol → triglycerides	↓Blood fatty acids
Fat Breakdown (lipolysis, or triglyceride degradation)	Triglycerides → fatty acids and glycerol	↑Blood fatty acids

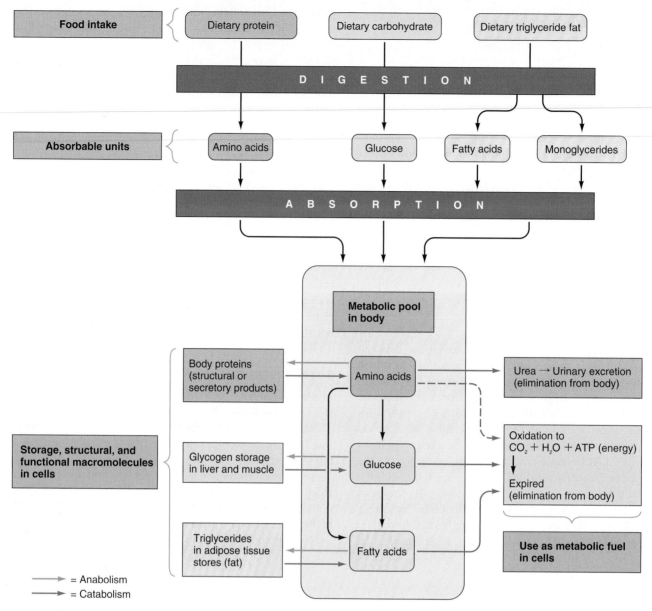

● **FIGURE 17-24**

Summary of the major pathways involving organic nutrient molecules

The major fate of both ingested carbohydrates and fats is catabolism to yield energy. Amino acids are predominantly used for protein synthesis but can be used to supply energy after being converted to carbohydrate or fat. Thus all three categories of foodstuff can be used as fuel, and excesses of any foodstuff can be deposited as stored fuel, as you will see shortly.

At a superficial level, fuel metabolism appears relatively simple: The amount of nutrients in the diet must be sufficient to meet the body's needs for energy production and cellular synthesis. This apparently simple relationship is complicated, however, by two important considerations: (1) nutrients taken in at meals must be stored and then released between meals, and (2) the brain must be continuously supplied with glucose. Let us examine the implications of each of these considerations.

▌ Because food intake is intermittent, nutrients must be stored for use between meals.

Dietary fuel intake is intermittent, not continuous. As a result, excess energy must be absorbed during meals and stored for use during fasting periods between meals, when dietary sources of metabolic fuel are not available (▲ Table 17-5).

• *Excess circulating glucose* is stored in the liver and muscle as *glycogen,* a large molecule consisting of interconnected glucose molecules. Because glycogen is a relatively small energy reservoir, less than a day's energy needs can be stored in this form. Once the liver and muscle glycogen stores are "filled up," additional glucose is transformed into fatty acids and glycerol, which are used to synthesize *triglycerides* (glycerol with three fatty acids attached), primarily in adipose tissue (fat).

Stored Metabolic Fuel in the Body

METABOLIC FUEL	CIRCULATING FORM	STORAGE FORM	MAJOR STORAGE SITE	PERCENTAGE OF TOTAL BODY ENERGY CONTENT (and calories*)	RESERVOIR CAPACITY	ROLE
Carbohydrate	Glucose	Glycogen	Liver, muscle	1% (1,500 calories)	Less than a day's worth of energy	First energy source; essential for the brain
Fat	Free fatty acids	Triglycerides	Adipose tissue	77% (143,000 calories)	About two months' worth of energy	Primary energy reservoir; energy source during a fast
Protein	Amino acids	Body proteins	Muscle	22% (41,000 calories)	Death results long before capacity is fully used because of structural and functional impairment	Source of glucose for the brain during a fast; last resort to meet other energy needs

*Actually refers to kilocalories; see p. 512.

- *Excess circulating fatty acids* derived from dietary intake also become incorporated into triglycerides.
- *Excess circulating amino acids* not needed for protein synthesis are not stored as extra protein but are converted to glucose and fatty acids, which ultimately end up being stored as triglycerides.

Thus the major site of energy storage for excess nutrients of all three classes is adipose tissue. Normally, enough triglyceride is stored to provide energy for about two months, more so in an overweight person. Consequently, during any prolonged period of fasting, the fatty acids released from triglyceride catabolism serve as the primary source of energy for most tissues.

As a third energy reservoir, a substantial amount of energy is stored as *structural protein*, primarily in muscle, the most abundant protein mass in the body. Protein is not the first choice to tap as an energy source, however, because it serves other essential functions; in contrast, the glycogen and triglyceride reservoirs serve solely as energy depots.

▌ The brain must be continuously supplied with glucose.

The second factor complicating fuel metabolism besides intermittent nutrient intake and the resultant necessity of storing nutrients is that the brain normally depends on the delivery of adequate blood glucose as its sole source of energy. Consequently, the blood glucose concentration must be maintained above a critical level. The blood glucose concentration is typically 100 mg glucose/100 ml plasma and is normally kept within the narrow limits of 70 to 110 mg/100 ml. Liver glycogen is an important reservoir for maintaining blood glucose levels during a short fast. However, liver glycogen is depleted relatively rapidly, so during a longer fast other mechanisms must meet the energy requirements of the glucose-dependent brain. First, when no new dietary glucose is entering the blood, tissues not obligated to use glucose shift their metabolic gears to burn fatty acids instead, sparing glucose for the brain. Fatty acids are made available by catabolism of triglyceride stores as an alternative energy source for tissues that are not glucose dependent. Second, amino acids can be converted to glucose by gluconeogenesis, whereas fatty acids cannot. Thus once glycogen stores are depleted despite glucose sparing, new glucose supplies for the brain are provided by the catabolism of body proteins and conversion of the freed amino acids into glucose.

▌ Metabolic fuels are stored during the absorptive state and mobilized during the postabsorptive state.

The preceding discussion should make clear that the disposition of organic molecules depends on the body's metabolic state. The two functional metabolic states—the *absorptive state* and the *postabsorptive state*—are related to eating and fasting cycles, respectively (▲ Table 17-6).

ABSORPTIVE STATE

After a meal, ingested nutrients are being absorbed and entering the blood during the **absorptive**, or **fed**, **state**. During

Comparison of Absorptive and Postabsorptive States

METABOLIC FUEL	ABSORPTIVE STATE	POSTABSORPTIVE STATE
Carbohydrate	Glucose used as major energy source	Glycogen degradation and depletion
	Glycogen synthesis and storage	Glucose sparing to conserve glucose for the brain
	Excess converted and stored as triglyceride fat	Production of new glucose through gluconeogenesis
Fat	Triglyceride synthesis and storage	Triglyceride catabolism
		Fatty acids used as the major energy source for non–glucose dependent tissues
Protein	Protein synthesis	Protein catabolism
	Excess converted and stored as triglyceride fat	Amino acids used for gluconeogenesis

this time, glucose is plentiful and serves as the major energy source. Very little of the absorbed fat and amino acids is used for energy during the absorptive state, because most cells use glucose when it is available. Extra nutrients not immediately used for energy or structural repairs are channeled into storage as glycogen or triglycerides.

POSTABSORPTIVE STATE

The average meal is completely absorbed in about four hours. Therefore, on a typical three-meals-a-day diet, no nutrients are being absorbed from the digestive tract during late morning and late afternoon and throughout the night. These times constitute the **postabsorptive**, or **fasting**, state. During this state, endogenous energy stores are mobilized to provide energy, whereas gluconeogenesis and glucose sparing maintain the blood glucose at an adequate level to nourish the brain. Synthesis of protein and fat is curtailed. Instead, stores of these organic molecules are catabolized for glucose formation and energy production, respectively. Carbohydrate synthesis does occur through gluconeogenesis, but the use of glucose for energy is greatly reduced.

Note that the blood concentration of nutrients does not fluctuate markedly between the absorptive and postabsorptive states. During the absorptive state, the glut of absorbed nutrients is swiftly removed from the blood and placed into storage;

during the postabsorptive state, these stores are catabolized to maintain the blood concentrations at levels necessary to fill tissue energy demands.

ROLES OF KEY TISSUES IN METABOLIC STATES

During these alternating metabolic states, various tissues play different roles as summarized here.

- The *liver* plays the primary role in maintaining normal blood-glucose levels. It stores glycogen when excess glucose is available, releases glucose into the blood when needed, and is the principal site for metabolic interconversions such as gluconeogenesis.
- *Adipose tissue* serves as the primary energy storage site and is important in regulating fatty acid levels in the blood.
- *Muscle* is the primary site of amino acid storage and is the major energy user.
- The *brain* normally can use only glucose as an energy source, yet it does not store glycogen, making it mandatory that blood glucose levels be maintained.

▌ **The pancreatic hormones, insulin and glucagon, are most important in regulating fuel metabolism.**

How does the body "know" when to shift its metabolic gears from a system of net anabolism and nutrient storage to one of net catabolism and glucose sparing? The flow of organic nutrients along metabolic pathways is influenced by a variety of hormones, including insulin, glucagon, epinephrine, cortisol, and growth hormone. Under most circumstances, the pancreatic hormones, insulin and glucagon, are the dominant hormonal regulators that shift the metabolic pathways back and forth from net anabolism to net catabolism and glucose sparing, depending on whether the body is in a state of feasting or fasting, respectively.

The **pancreas** is an organ composed of both exocrine and endocrine tissues. The exocrine portion secretes a watery, alkaline solution and digestive enzymes through the pancreatic duct into the digestive tract lumen. Scattered throughout the pancreas between the exocrine cells are clusters, or "islands," of endocrine cells known as the **islets of Langerhans** (see ● Figure 15-11, p. 487). The most abundant pancreatic endocrine cells are the **β (beta) cells**, the site of *insulin* synthesis and secretion, and the **α (alpha) cells**, which produce *glucagon*. Less common, the **D (delta) cells** are the pancreatic site of *somatostatin* synthesis. We will briefly highlight somatostatin now and then will pay the most attention to insulin and glucagon, the most important hormones in the regulation of fuel metabolism.

Pancreatic **somatostatin** inhibits the digestive system in a variety of ways, the overall effect of which is to inhibit digestion of nutrients and to decrease nutrient absorption. Somatostatin is released from the pancreatic D cells in direct response to an increase in blood glucose and blood amino acids during absorption of a meal. By exerting its inhibitory effects, pancreatic somatostatin acts in negative-feedback fashion to put the brakes on the rate at which the meal is being digested and absorbed, thereby preventing excessive plasma levels of nutrients. Pancreatic somatostatin may also play a paracrine role in regulating pancreatic hormone secretion. The local presence of somatostatin decreases the secretion of insulin, glucagon, and somatostatin itself.

Somatostatin is also produced by cells lining the digestive tract, where it acts locally as a paracrine to inhibit most digestive processes (see p. 482). Furthermore, somatostatin (alias GHIH) is produced by the hypothalamus, where it inhibits the secretion of growth hormone and TSH (see p. 540). We will next consider insulin and then glucagon, followed by a discussion of how insulin and glucagon function as an endocrine unit to shift metabolic gears between the absorptive and postabsorptive states.

▌ Insulin lowers blood glucose, fatty acid, and amino acid levels and promotes their storage.

Insulin has important effects on carbohydrate, fat, and protein metabolism. It lowers the blood levels of glucose, fatty acids, and amino acids and promotes their storage. As these nutrient molecules enter the blood during the absorptive state, insulin promotes their cellular uptake and conversion into glycogen, triglycerides, and protein, respectively. Insulin exerts its many effects either by altering transport of specific blood-borne nutrients into cells or by altering the activity of the enzymes involved in specific metabolic pathways.

ACTIONS ON CARBOHYDRATES

The maintenance of blood glucose homeostasis is a particularly important function of the pancreas. Insulin exerts four effects that lower blood-glucose levels and promote carbohydrate storage:

1. Insulin facilitates glucose transport into most cells. (The mechanism of this increased glucose uptake is explained after insulin's other blood-glucose lowering effects are listed.)
2. Insulin stimulates **glycogenesis**, the production of glycogen from glucose, in both skeletal muscle and the liver.
3. Insulin inhibits **glycogenolysis**, the breakdown of glycogen into glucose. By inhibiting the breakdown of glycogen into glucose, insulin likewise favors carbohydrate storage and decreases glucose output by the liver.
4. Insulin further decreases hepatic glucose output by inhibiting gluconeogenesis, the conversion of amino acids into glucose in the liver

Thus insulin decreases the concentration of blood glucose by promoting the cells' uptake of glucose from the blood for use and storage, while simultaneously blocking the two mechanisms by which the liver releases glucose into the blood (glycogenolysis and gluconeogenesis). Insulin is the only hormone capable of lowering the blood glucose level. Insulin promotes the uptake of glucose by most cells through glucose transporter recruitment, a topic to which we now turn our attention.

Glucose transport between the blood and cells is accomplished by means of a plasma membrane carrier known as a glucose transporter (GLUT). Six forms of glucose transporters have been identified, named in the order they were discovered—GLUT-1, GLUT-2, and so on. These glucose transporters all accomplish passive facilitated diffusion of glucose across the plasma membrane (see p. 57). Each member of the GLUT family performs slightly different functions. For example, *GLUT-1* transports glucose across the blood–brain barrier and *GLUT-3* is the main transporter of glucose into neurons. The glucose transporter responsible for the majority of glucose uptake by most cells of the body is *GLUT-4*, which operates only at the bidding of insulin. Glucose molecules cannot readily penetrate most cell membranes in the absence of insulin, making most tissues highly dependent on insulin for uptake of glucose from the blood and for its subsequent use. GLUT-4 is especially abundant in the tissues that account for the bulk of glucose uptake from the blood during the absorptive state, namely skeletal muscle and adipose tissue cells.

GLUT-4 is the only type of glucose transporter that responds to insulin. Unlike the other types of GLUT molecules, which are always present in the plasma membranes at the sites where they perform their functions, in the absence of insulin GLUT-4 is excluded from the plasma membrane. Insulin promotes glucose uptake by **transporter recruitment.** Insulin-dependent cells maintain a pool of intracellular vesicles containing GLUT-4. Insulin induces these vesicles to move to the plasma membrane and fuse with it, thus inserting GLUT-4 molecules into the plasma membrane. In this way, increased insulin secretion promotes a rapid 10- to 30-fold increase in glucose uptake by insulin-dependent cells. When insulin secretion decreases, these glucose transporters are retrieved from the membrane and returned to the intracellular pool.

Several tissues do not depend on insulin for their glucose uptake—namely the brain, working muscles, and the liver. The brain, which requires a constant supply of glucose for its minute-to-minute energy needs, is freely permeable to glucose at all times by means of GLUT-1 and GLUT-3 molecules. Skeletal muscle cells do not depend on insulin for their glucose uptake during exercise, even though they are dependent at rest. Muscle contraction triggers the insertion of GLUT-4 into the plasma membranes of exercising muscle cells in the absence of insulin. This fact is important in managing diabetes mellitus (insulin deficiency), as described later. The liver also does not depend on insulin for glucose uptake, because it does not use GLUT-4.

Insulin also exerts important actions on fat and protein.

ACTIONS ON FAT

Insulin exerts multiple effects to lower blood fatty acids and promote triglyceride storage:

1. It enhances the entry of fatty acids from the blood into adipose tissue cells.
2. It increases the transport of glucose into adipose tissue cells by means of GLUT-4 recruitment.
3. It promotes chemical reactions that ultimately use fatty acids and glucose derivatives for triglyceride synthesis.

4. It inhibits lipolysis (fat breakdown), reducing the release of fatty acids from adipose tissue into the blood.

Collectively, these actions favor removal of fatty acids and glucose from the blood and promote their storage as triglycerides.

ACTIONS ON PROTEIN

Insulin lowers blood amino-acid levels and enhances protein synthesis through several effects:

1. It promotes the active transport of amino acids from the blood into muscles and other tissues. This effect decreases the circulating amino-acid level and provides the building blocks for protein synthesis within the cells.
2. It increases the rate of amino acid incorporation into protein by stimulating the cells' protein-synthesizing machinery.
3. It inhibits protein degradation.

The collective result of these actions is a protein anabolic effect. For this reason, insulin is essential for normal growth.

SUMMARY OF INSULIN'S ACTIONS

In short, insulin primarily exerts its effects by acting on nonworking skeletal muscle, the liver, and adipose tissue. It stimulates biosynthetic pathways that lead to increased glucose use, increased carbohydrate and fat storage, and increased protein synthesis. In so doing, this hormone lowers the blood glucose, fatty acid, and amino acid levels. This metabolic pattern is characteristic of the absorptive state. Indeed, insulin secretion rises during this state and shifts metabolic pathways to net anabolism.

When insulin secretion is low, the opposite effects occur. The rate of glucose entry into cells is reduced, and net catabolism occurs rather than net synthesis of glycogen, triglycerides, and protein. This pattern is reminiscent of the postabsorptive state; indeed, insulin secretion is reduced during the postabsorptive state. However, the other major pancreatic hormone, glucagon, also plays an important role in shifting from absorptive to postabsorptive metabolic patterns, as described later.

❙ **The primary stimulus for increased insulin secretion is an increase in blood glucose concentration.**

The primary control of insulin secretion is a direct negative-feedback system between the pancreatic β cells and the concentration of glucose in the blood flowing to them. An elevated blood-glucose level, such as during absorption of a meal, directly stimulates the β cells to synthesize and release insulin. The increased insulin in turn reduces the blood glucose to normal and promotes use and storage of this nutrient. Conversely, a fall in blood glucose below normal, such as during fasting, directly inhibits insulin secretion. Lowering the rate of insulin secretion shifts metabolism from the absorptive to the postabsorptive pattern. Thus this simple negative-feedback

system can maintain a relatively constant supply of glucose to the tissues without requiring the participation of nerves or other hormones.

In addition to plasma glucose concentration, other inputs are involved in regulating insulin secretion, as follows (● Figure 17-25):

- An elevated blood amino-acid level, such as after a high-protein meal, directly stimulates the β cells to increase insulin secretion. In negative-feedback fashion, the increased insulin enhances the entry of these amino acids into the cells, lowering the blood amino-acid level while promoting protein synthesis.
- The major gastrointestinal hormones secreted by the digestive tract in response to the presence of food, especially glucose-dependent insulinotropic peptide (see p. 505), stimulate pancreatic insulin secretion in addition to having direct regulatory effects on the digestive system. Through this control, insulin secretion is increased in "feedforward," or anticipatory, fashion even before nutrient absorption increases the blood concentration of glucose and amino acids.
- The autonomic nervous system also directly influences insulin secretion. The islets are richly innervated by both parasympathetic (vagal) and sympathetic nerve fibers. The increase in parasympathetic activity that occurs in response to food in the digestive tract stimulates insulin release. This, too, is a feedforward response in anticipation of nutrient absorption. In contrast, sympathetic stimulation and the concurrent increase in epinephrine both inhibit insulin secretion. The fall in insulin level allows the blood glucose level to rise, an appropriate response to the circumstances under which

---

● **FIGURE 17-25**

Factors controlling insulin secretion

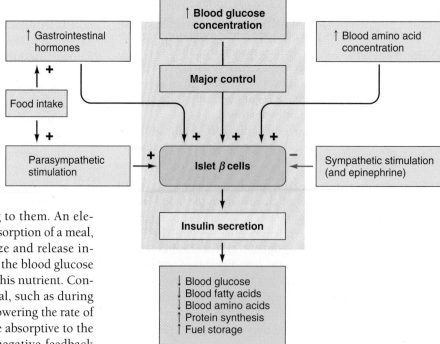

generalized sympathetic activation occurs—namely, stress (fight or flight) and exercise. In both these situations, extra fuel is needed for increased muscle activity.

❚ The symptoms of diabetes mellitus are characteristic of an exaggerated postabsorptive state.

Clinical Note **Diabetes mellitus** is by far the most common of all endocrine disorders. The acute symptoms of diabetes mellitus are attributable to inadequate insulin action. Because insulin is the only hormone capable of lowering blood glucose levels, one of the most prominent features of diabetes mellitus is elevated blood-glucose levels, or *hyperglycemia*. *Diabetes* literally means "syphon" or "running through," a reference to the large urine volume accompanying this condition. A large urine volume occurs both in diabetes mellitus (a result of insulin insufficiency) and in diabetes insipidus (a result of vasopressin deficiency). *Mellitus* means "sweet"; *insipidus* means "tasteless." The urine of patients with diabetes mellitus acquires its sweetness from excess blood glucose that spills into the urine, whereas the urine of patients with diabetes insipidus contains no sugar, so it is tasteless. (Aren't you glad you were not a health professional at the time when these two conditions were distinguished on the basis of the taste of the urine?)

Diabetes mellitus has two major variants, differing in the capacity for pancreatic insulin secretion: *Type I diabetes*, characterized by a lack of insulin secretion, and *Type II diabetes*, characterized by normal or even increased insulin secretion but reduced sensitivity of insulin's target cells to its presence. (For a further discussion of the distinguishing features of these two types of diabetes mellitus, see the boxed feature, ❚ Beyond the Basics, on pp. 568–569.)

The acute consequences of diabetes mellitus can be grouped according to the effects of inadequate insulin action on carbohydrate, fat, and protein metabolism (● Figure 17-26, p. 570). The figure may look overwhelming, but the numbers, which correspond to the numbers in the following discussion, help you work your way through this complex disease step-by-step.

CONSEQUENCES RELATED TO EFFECTS ON CARBOHYDRATE METABOLISM

Because the postabsorptive metabolic pattern is induced by low insulin activity, the changes that occur in diabetes mellitus are an exaggeration of this state, with the exception of hyperglycemia. In the usual fasting state, the blood glucose level is slightly below normal. Hyperglycemia, the hallmark of diabetes mellitus, arises from reduced glucose uptake by cells, coupled with increased output of glucose from the liver (1 in ● Figure 17-26). As the glucose-yielding processes of glycogenolysis and gluconeogenesis proceed unchecked in the absence of insulin, hepatic output of glucose increases. Because many of the body's cells cannot use glucose without the help of insulin, an ironic extracellular glucose excess occurs coincident with an intracellular glucose deficiency— "starvation in the midst of plenty." Even though the non–

insulin-dependent brain is adequately nourished during diabetes mellitus, further consequences of the disease lead to brain dysfunction, as you will see shortly.

When the blood glucose rises to the level where the amount of glucose filtered exceeds the tubular cells' capacity for reabsorption, glucose appears in the urine (*glucosuria*) 2 . Glucose in the urine exerts an osmotic effect that draws H_2O with it, producing an osmotic diuresis characterized by *polyuria* (frequent urination) 3 . The excess fluid lost from the body leads to dehydration 4 , which in turn can ultimately lead to peripheral circulatory failure because of the marked reduction in blood volume 5 . Circulatory failure, if uncorrected, can lead to death because of low cerebral blood flow 6 or secondary renal failure resulting from inadequate filtration pressure 7 . Furthermore, cells lose water as the body becomes dehydrated by an osmotic shift of water from the cells into the hypertonic extracellular fluid 8 . Brain cells are especially sensitive to shrinking, so nervous system malfunction ensues 9 (see p. 450). Another characteristic symptom of diabetes mellitus is *polydipsia* (excessive thirst) 10 , which is actually a compensatory mechanism to counteract the dehydration.

The story is not complete. In intracellular glucose deficiency, appetite is stimulated, leading to *polyphagia* (excessive food intake) 11 . Despite increased food intake, however, progressive weight loss occurs from the effects of insulin deficiency on fat and protein metabolism.

CONSEQUENCES RELATED TO EFFECTS ON FAT METABOLISM

Triglyceride synthesis decreases while lipolysis increases, resulting in large-scale mobilization of fatty acids from triglyceride stores 12 . The increased blood fatty acids are largely used by the cells as an alternative energy source. Increased liver use of fatty acids results in the release of excessive ketone bodies into the blood, causing *ketosis* 13 . Ketone bodies include several different acids, such as acetoacetic acid, that result from incomplete breakdown of fat during hepatic energy production. Therefore, this developing ketosis leads to progressive metabolic acidosis 14 . Acidosis depresses the brain and, if severe enough, can lead to diabetic coma and death 15 .

A compensatory measure for metabolic acidosis is increased ventilation to blow off extra, acid-forming CO_2 16 . Exhalation of one of the ketone bodies, acetone, causes a "fruity" breath odor that smells like a combination of Juicy Fruit gum and nail polish remover. Sometimes, because of this odor, passersby unfortunately mistake a patient collapsed in a diabetic coma for a "wino" passed out in a state of drunkenness. (This situation illustrates the merits of medical alert identification tags.) People with Type I diabetes are much more prone to develop ketosis than are Type II diabetics.

CONSEQUENCES RELATED TO EFFECTS ON PROTEIN METABOLISM

The effects of a lack of insulin on protein metabolism result in a net shift toward protein catabolism. The net breakdown of muscle proteins leads to wasting and weakness of skeletal muscles 17 and, in child diabetics, a reduction in overall growth.

Diabetics and Insulin: Some Have It and Some Don't

There are two distinct types of diabetes mellitus (see the accompanying table). **Type I (insulin-dependent,** or **juvenile-onset) diabetes mellitus,** which accounts for about 10% of all cases of diabetes, is characterized by a lack of insulin secretion. Because their pancreatic β cells secrete no or nearly no insulin, Type I diabetics require exogenous insulin for survival. Hence this form of the disease has the alternative name *insulin-dependent diabetes mellitus.* In **Type II (non–insulin-dependent,** or **maturity-onset) diabetes mellitus,** insulin secretion may be normal or even increased, but insulin's target cells are less sensitive than normal to this hormone. Although either type can first be manifested at any age, Type I is more prevalent in children, whereas Type II more generally arises in adulthood, hence the age-related designations of the two conditions.

Diabetes of both types currently affects 16 million people in the United States, costing this country an estimated $100 billion annually in health care expenses. The diabetes-related death rate in the United States has increased by 30% since 1980, largely because the incidence of the disease has been rising. In 2000, diabetes mellitus afflicted 150 million people worldwide, and this number is projected to climb to 220 million by 2010. Because diabetes is so prevalent and exacts such a huge economic toll, coupled with the fact that it forces a change in the lifestyle of affected individuals and places them at increased risk for developing a variety of troublesome and even life-threatening conditions, intensive research is directed toward better understanding and controlling or preventing both types of the disease.

Underlying Defect in Type I Diabetes

Type I diabetes is an autoimmune process involving the erroneous, selective destruction of the pancreatic β cells by inappropriately activated T lymphocytes (see p. 350). The precise cause of this self-destructive immune attack remains unclear. Some individuals have a genetic susceptibility to acquiring Type I diabetes. Environmental triggers also appear to play an important role.

Underlying Defect in Type II Diabetes

Type II diabetics do secrete insulin, but the affected individuals exhibit *insulin resistance.* That is, the basic problem in Type II diabetes is not lack of insulin but reduced sensitivity of insulin's target cells to its presence. The cause of reduced insulin sensitivity remains elusive despite intense investigation. Various genetic and lifestyle factors appear important in the development of Type II diabetes. Obesity is the biggest risk factor; 90% of Type II diabetics are obese.

Researchers used to think that Type II diabetes developed because of excessive down regulation of insulin receptors (see p. 532), but that turns out not to be the case. Recent research suggests that adipose cells secrete a hormone known as **resistin,** which interferes with insulin action in experimental animals. This could be an important link between obesity and insulin resistance. Resistin is distinct from leptin, the hormone secreted by adipose cells that plays a role in controlling food intake (see p. 514). Still other studies suggest that defects in the insulin signaling pathways may be the underlying culprit in Type II diabetes.

Early in the development of the disease, the resulting decrease in sensitivity to insulin is overcome by secretion of additional insulin. However, the sustained overtaxing of the pancreas eventually exceeds the reserve secretory capacity of the genetically weak β cells. Even though insulin secretion may be normal or somewhat elevated, symptoms of insulin insufficiency develop because the amount of insulin is still inadequate to prevent significant hyperglycemia. The symptoms in Type II diabetes are usually slower in onset and less severe than in Type I diabetes.

Treatment of Diabetes

The conventional treatment for Type I diabetes is a controlled balance of regular insulin injections timed around meals, dietary management of the amounts and types of food consumed, and exercise. Insulin must be administered by injection, because if it were swallowed this peptide hormone would be digested by proteolytic enzymes in the stomach and small intestine. Exercise is also useful in managing both types of diabetes, because working muscles are not insulin dependent. Exercising muscles take up and use some of the excess glucose in the blood, reducing the overall need for insulin.

Whereas Type I diabetics are permanently insulin dependent, dietary control and weight reduction may be all that is necessary to completely reverse the symptoms in Type II diabetics. Therefore, Type II diabetes is alternatively known as *non–insulin dependent diabetes.* Four classes of oral medications are currently available for use if needed for treating Type II diabetes in conjunction with a dietary and exercise regime. These pills help the patient's body use its own insulin more effectively, each by a different mechanism, as follows:

1. By stimulating the β cells to secrete more insulin than they do on their own
2. By suppressing the liver's output of glucose
3. By slowing glucose absorption into the blood from the digestive tract, thus blunting the surge of glucose immediately after a meal
4. By making the cells more receptive to insulin (that is, by reducing insulin resistance)

Because none of these oral drugs deliver new insulin to the body, they cannot replace insulin injections for people with Type I diabetes. Furthermore, sometimes the weakened β cells of Type II diabetics eventually burn out and can no longer produce insulin. In such

Reduced amino-acid uptake coupled with increased protein degradation results in excess amino acids in the blood 18. The increased circulating amino acids can be used for additional gluconeogenesis, which further aggravates the hyperglycemia 19.

As you can readily appreciate from this overview, diabetes mellitus is a complicated disease that can disturb both carbohydrate, fat, and protein metabolism and fluid and acid–base balance. It can also have repercussions on the circulatory system, kidneys, respiratory system, and nervous system.

LONG-TERM COMPLICATIONS

In addition to these potential consequences of untreated diabetes, which can be explained on the basis of insulin's short-term metabolic effects, numerous long-range complications of this disease frequently occur after 15 to 20 years despite treatment to prevent the short-term effects. These chronic

Comparison of Type I and Type II Diabetes Mellitus

CHARACTERISTIC	TYPE I DIABETES	TYPE II DIABETES
Level of Insulin Secretion	None or almost none	May be normal or exceed normal
Typical Age of Onset	Childhood	Adulthood
Percentage of Diabetics	10–20%	80–90%
Basic Defect	Autoimmune destruction of β cells	Reduced sensitivity of insulin's target cells
Treatment	Insulin injections; dietary management; exercise	Dietary control and weight reduction; exercise; sometimes oral hypoglycemic drugs

a case, the previously non–insulin-dependent patient must be placed on insulin therapy for life.

New Approaches to Managing Diabetes
Several new approaches are currently available for insulin-dependent diabetics that preclude the need for the one or two insulin injections daily.

- Implanted insulin pumps can deliver a prescribed amount of insulin on a regular basis, but the recipient must time meals with care to match the automatic insulin delivery.
- Pancreas transplants are also being performed more widely now, with increasing success rates. On the downside, recipients of pancreas transplants must take immuno-suppressive drugs for life to prevent rejection of their donated organs. Also, donor organs are in short supply.

Current research on several fronts may dramatically change the approach to diabetic therapy in the near future. The following new treatments are on the horizon, most of which do away with the dreaded daily injections:

- Some methods under development circumvent the need for insulin injections by using alternate routes of administration that bypass the destructive digestive-tract enzymes. These include the use of inhaled powdered insulin and using ultrasound to force insulin into the skin from an insulin-impregnated patch.
- Some researchers are seeking methods to protect swallowed insulin from destruction by the digestive tract, and others have identified a potential oral substitute for insulin—namely a nonpeptide chemical that binds with the insulin receptors and brings about the same intracellular responses as insulin does. Because this in-

sulin mimic is not a protein, it would not be destroyed by the proteolytic digestive enzymes if taken as a pill.

- Another hope is pancreatic islet transplants. Scientists have developed several types of devices that isolate donor islet cells from the recipient's immune system. Such immunoisolation of islet cells would permit use of grafts from other animals, circumventing the shortage of human donor cells. Pig islet cells would be an especially good source, because pig insulin is nearly identical to human insulin.
- Some researchers hope they will be able to coax stem cells to develop into insulin-secreting cells that can be implanted.
- In a related approach, other scientists are turning to genetic engineering in the hope of developing surrogates for pancreatic β cells. An example is the potential reprogramming of the small-intestine endocrine cells that produce glucose-dependent insulinotropic hormone (GIP). Recall that GIP normally promotes insulin release in feed-forward fashion when food is in the digestive tract. The goal is to cause these non-β cells to cosecrete both insulin and GIP on feeding.
- Another approach under development is an implanted, glucose-detecting, insulin-releasing "artificial pancreas" that would continuously monitor the patient's blood glucose level and deliver insulin in response to need.
- On another front, scientists are hopeful of one day developing immunotherapies that specifically block the attack of the immune system against the β cells, thus curbing or preventing Type I diabetes.
- Still other investigators are scurrying to unravel the defective pathways underlying Type II diabetes with the goal of finding new therapeutic targets to prevent, halt, or reverse this condition.

complications, which account for the shorter life expectancy of diabetics, primarily involve degenerative disorders of the vasculature and nervous system. Cardiovascular lesions are the most common cause of premature death in diabetics. Heart disease and strokes occur with greater incidence than in non-diabetics. Because vascular lesions often develop in the kidneys and retinas of the eyes, diabetes is the leading cause of both kidney failure and blindness in the United States. Impaired delivery of blood to the extremities may cause these

tissues to become gangrenous, and toes or even whole limbs may have to be amputated. In addition to circulatory problems, degenerative lesions in nerves lead to multiple neuropathies that result in dysfunction of the brain, spinal cord, and peripheral nerves. The latter is most often characterized by pain, numbness, and tingling, especially in the extremities.

Regular exposure of tissues to excess blood glucose over a prolonged time leads to tissue alterations responsible for the development of these long-range vascular and neural

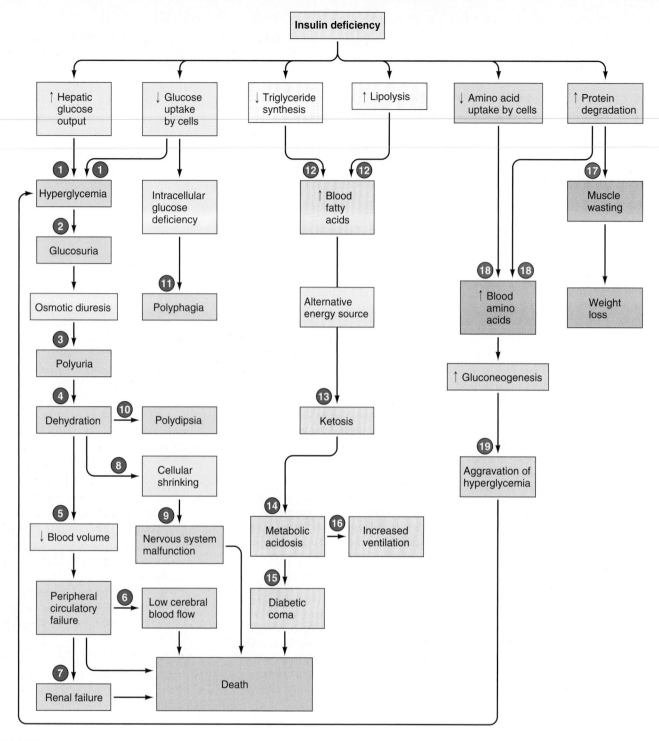

● FIGURE 17-26

Acute effects of diabetes mellitus. The acute consequences of diabetes mellitus can be grouped according to the effects of inadequate insulin action on carbohydrate, fat, and protein metabolism. These effects ultimately cause death through a variety of pathways. See pp. 567–568 for an explanation of the circled numbers.

degenerative complications. Thus the best management for diabetes mellitus is to continuously keep blood glucose levels within normal limits to diminish the incidence of these chronic abnormalities. (See the ❱ Beyond the Basics feature on diabetes on pp. 568–569 for current and potential future treatment strategies for this disorder.)

❚ Insulin excess causes brain-starving hypoglycemia.

Let's now look at the opposite of diabetes mellitus, insulin excess, which is characterized by *hypoglycemia* (low blood glucose) and can arise in two different ways. First, insulin excess can occur in a diabetic patient when too much insulin has been injected for the person's caloric intake and exercise level, resulting in so-called **insulin shock.** Second, blood insulin level may rise abnormally high in a nondiabetic individual who has a β cell tumor or whose β cells are over-responsive to glucose, a condition called **reactive hypoglycemia.** Such β cells "overshoot" and secrete more insulin than necessary, in response to elevated blood glucose after a high-carbohydrate meal. The excess insulin drives too much glucose into the cells, resulting in hypoglycemia.

The consequences of insulin excess are primarily manifestations of the effects of hypoglycemia on the brain. Recall that the brain relies on a continuous supply of blood glucose for its nourishment and that glucose uptake by the brain does not depend on insulin. With insulin excess, more glucose than necessary is driven into the other insulin-dependent cells. The result is a lowering of the blood glucose level, so that not enough glucose is left in the blood to be delivered to the brain. In hypoglycemia, the brain literally starves. The symptoms, therefore, are primarily referable to depressed brain function, which, if severe enough, may rapidly progress to unconsciousness and death. People with over-responsive β cells usually do not become sufficiently hypoglycemic to manifest these more serious consequences, but they do show milder symptoms of depressed CNS activity.

The treatment of hypoglycemia depends on the cause. At the first indication of a hypoglycemic attack with insulin overdose, the diabetic person should eat or drink something sugary. Prompt treatment of severe hypoglycemia is imperative to prevent brain damage.

Ironically, even though reactive hypoglycemia is characterized by a low blood-glucose level, people with this disorder are treated by limiting their intake of sugar and other glucose-yielding carbohydrates to prevent their β cells from over-responding to a high glucose intake. Giving a symptomatic individual with reactive hypoglycemia something sugary temporarily alleviates the symptoms. The blood glucose level is transiently restored to normal so that the brain's energy needs are once again satisfied. However, as soon as the extra glucose triggers further insulin release, the situation is merely aggravated.

▌ Glucagon in general opposes the actions of insulin.

Even though insulin plays a central role in controlling metabolic adjustments between the absorptive and postabsorptive states, the secretory product of the pancreatic-islet α cells, **glucagon,** is also very important. Many physiologists view the insulin-secreting β cells and the glucagon-secreting α cells as a coupled endocrine system whose combined secretory output is a major factor in regulating fuel metabolism.

Glucagon affects many of the same metabolic processes that insulin influences, but in most cases glucagon's actions are opposite to those of insulin. The major site of action of glucagon is the liver, where it exerts a variety of effects on carbohydrate, fat, and protein metabolism.

ACTIONS ON CARBOHYDRATE

The overall effects of glucagon on carbohydrate metabolism result in an increase in hepatic glucose production and release and thus an increase in blood glucose levels. Glucagon exerts its hyperglycemic effects by decreasing glycogen synthesis, promoting glycogenolysis, and stimulating gluconeogenesis.

ACTIONS ON FAT

Glucagon also antagonizes the actions of insulin with regard to fat metabolism by promoting fat breakdown and inhibiting triglyceride synthesis. Thus the blood levels of fatty acids increase under glucagon's influence.

ACTIONS ON PROTEIN

Glucagon inhibits hepatic protein synthesis and promotes degradation of hepatic protein. Stimulation of gluconeogenesis further contributes to glucagon's catabolic effect on hepatic protein metabolism. Glucagon promotes protein catabolism in the liver, but it does not have any significant effect on blood amino-acid levels because it does not affect muscle protein, the major protein store in the body.

▌ Glucagon secretion is increased during the postabsorptive state.

Considering the catabolic effects of glucagon on energy stores, you would be correct in assuming that glucagon secretion increases during the postabsorptive state and decreases during the absorptive state, just the opposite of insulin secretion. In fact, insulin is sometimes referred to as a "hormone of feasting" and glucagon as a "hormone of fasting." Insulin tends to put nutrients in storage when their blood levels are high, such as after a meal, whereas glucagon promotes catabolism of nutrient stores between meals to keep up the blood nutrient levels, especially blood glucose.

Like insulin secretion, the major factor regulating glucagon secretion is a direct effect of the blood glucose concentration on the endocrine pancreas. In this case, the pancreatic α cells increase glucagon secretion in response to a fall in blood glucose. The hyperglycemic actions of this hormone tend to raise the blood glucose level back to normal. Conversely, an increase in blood glucose concentration, such as after a meal, inhibits glucagon secretion, which tends to drop the blood glucose level back to normal.

▌ Insulin and glucagon work as a team to maintain blood glucose and fatty acid levels.

Thus a direct negative-feedback relationship exists between blood glucose concentration and both the β cells' and α cells' rates of secretion, but in opposite directions. An elevated blood-glucose level stimulates insulin secretion but inhibits glucagon secretion, whereas a fall in blood glucose level leads to decreased insulin secretion and increased glucagon secretion (● Figure 17-27). Because insulin lowers and glucagon raises blood glucose, the changes in secretion of these pancreatic hormones in response to deviations in blood glucose work together homeostatically to restore blood glucose levels to normal.

No known clinical abnormalities are caused by glucagon deficiency or excess.

▌ Epinephrine, cortisol, and growth hormone also exert direct metabolic effects.

The pancreatic hormones are the most important regulators of normal fuel metabolism. However, several other hormones exert direct metabolic effects, even though control of their secretion is keyed to factors other than transitions in metabolism between feasting and fasting states (▲ Table 17-7).

The stress hormones, epinephrine and cortisol, both increase blood levels of glucose and fatty acids through a variety of metabolic effects. In addition, cortisol mobilizes amino acids by promoting protein catabolism. Neither hormone plays important roles in regulating fuel metabolism under resting conditions, but both are important for the metabolic responses to stress.

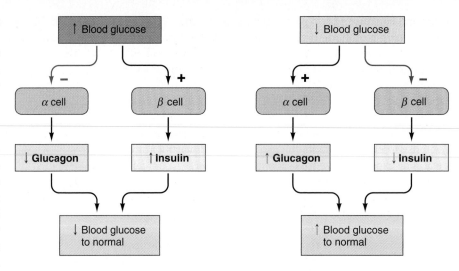

● **FIGURE 17-27**

Complementary interactions of glucagon and insulin. Counteracting actions of glucagon and insulin on blood glucose during absorption of a high-protein meal

Growth hormone has protein anabolic effects in muscle. In fact, this is one of its growth-promoting features. Although GH can elevate the blood levels of glucose and fatty acids, it

▲ **TABLE 17-7**

Summary of Hormonal Control of Fuel Metabolism

HORMONE	MAJOR METABOLIC EFFECTS				CONTROL OF SECRETION	
	Effect on blood glucose	Effect on blood fatty acids	Effect on blood amino acids	Effect on muscle protein	Major stimuli for secretion	Primary role in metabolism
Insulin	↓	↓	↓	↑	↑Blood glucose	Primary regulator of absorptive and postabsorptive cycles
Glucagon	↑	↑	No effect	No effect	↓Blood glucose	Regulation of absorptive and postabsorptive cycles in concert with insulin
Epinephrine	↑	↑	No effect	No effect	Sympathetic stimulation during stress	Provision of energy for emergencies
Cortisol	↑	↑	↑	↓	Stress	Mobilization of metabolic fuels and building blocks during adaptation to stress
Growth hormone	↑	↑	↓	↑	Deep sleep Exercise Stress ↓Blood glucose	Promotion of growth; normally little role in metabolism; mobilization of fuels plus glucose sparing in extenuating circumstances

↑ = increase
↓ = decrease

is normally of little importance to the overall regulation of fuel metabolism. Deep sleep, stress, exercise, and severe hypoglycemia stimulate GH secretion, possibly to provide fatty acids as an energy source and spare glucose for the brain under these circumstances.

Note that, with the exception of the anabolic effects of GH on protein metabolism, all the metabolic actions of these other hormones are opposite to those of insulin. Insulin alone can reduce blood glucose and blood fatty-acid levels, whereas glucagon, epinephrine, cortisol, and GH all increase blood levels of these nutrients. Because of this, these other hormones are considered **insulin antagonists.** Thus the main reason diabetes mellitus has such devastating metabolic consequences is that no other control mechanism is available to pick up the slack to promote anabolism when insulin activity is insufficient, so the catabolic reactions promoted by other hormones proceed unchecked. The only exception is protein anabolism stimulated by GH.

 PhysioEdge Click on the Media Exercises menu of the CD-ROM and work Media Exercise 17.7: Endocrine Pancreas and Fuel Metabolism to test your understanding of the previous section.

ENDOCRINE CONTROL OF CALCIUM METABOLISM

Besides regulating the concentration of organic nutrient molecules in the blood by manipulating anabolic and catabolic pathways, the endocrine system also regulates the plasma concentration of a number of inorganic electrolytes. As you already know, aldosterone controls Na^+ and K^+ concentrations in the ECF. Three other hormones—*parathyroid hormone, calcitonin,* and *vitamin D*—control calcium (Ca^{2+}) and phosphate (PO_4^{3-}) metabolism. These hormonal agents concern themselves with regulating plasma Ca^{2+}, and in the process, plasma PO_4^{3-} is also maintained. Plasma Ca^{2+} concentration is one of the most tightly controlled variables in the body. The need for the precise regulation of plasma Ca^{2+} stems from its critical influence on so many body activities.

▌ Plasma Ca^{2+} must be closely regulated to prevent changes in neuromuscular excitability.

About 99% of the Ca^{2+} in the body is in crystalline form within the skeleton and teeth. Of the remaining 1%, about 0.9% is found intracellularly within the soft tissues; less than 0.1% is present in the ECF. Approximately half of the plasma Ca^{2+} is not free to leave the plasma and participate in chemical reactions, for example by being bound to plasma proteins. The other half of the plasma Ca^{2+} is freely diffusible and can readily pass into the interstitial fluid and interact with the cells. Only this free Ca^{2+} is biologically active and subject to regulation; it constitutes less than one thousandth of the total Ca^{2+} in the body.

This small, freely diffusible fraction of ECF Ca^{2+} plays a vital role in a number of essential activities, the most important of which is its effect on neuromuscular excitability. Even minor variations in the concentration of free ECF Ca^{2+} have a profound and immediate impact on the sensitivity of excitable tissues. A fall in free Ca^{2+} results in overexcitability of nerves and muscles; conversely, a rise in free Ca^{2+} depresses neuromuscular excitability. These effects result from the influence of Ca^{2+} on membrane permeability to Na^+. A decrease in free Ca^{2+} increases Na^+ permeability, with the resultant influx of Na^+ moving the resting potential closer to threshold. Consequently, in the presence of *hypocalcemia* (low blood Ca^{2+}) excitable tissues may be brought to threshold by normally ineffective physiologic stimuli, so that skeletal muscles discharge and contract (go into spasm) "spontaneously" (in the absence of normal stimulation). If severe enough, spastic contraction of the respiratory muscles results in death by asphyxiation. *Hypercalcemia* (elevated blood Ca^{2+}) is also life threatening, because it causes cardiac arrhythmias and generalized depression of neuromuscular excitability.

Because of the profound effects of deviations in free Ca^{2+}, especially on neuromuscular excitability, the plasma concentration of this electrolyte is regulated with extraordinary precision. Let's see how.

Maintaining the proper plasma concentration of free Ca^{2+} differs from the regulation of Na^+ and K^+ in two important ways. Sodium and K^+ homeostasis is maintained primarily by regulating the urinary excretion of these electrolytes so that controlled output matches uncontrolled input. Although urinary excretion of Ca^{2+} is hormonally controlled, in contrast to Na^+ and K^+, not all ingested Ca^{2+} is absorbed from the digestive tract; instead, the extent of absorption is hormonally controlled and depends on the Ca^{2+} status of the body. In addition, bone serves as a large Ca^{2+} reservoir that can be drawn on to maintain the free plasma Ca^{2+} concentration within the narrow limits compatible with life should dietary intake become too low. Exchange of Ca^{2+} between the ECF and bone is also subject to hormonal control. Similar in-house stores are not available for Na^+ and K^+. Thus regulation of Ca^{2+} metabolism depends on hormonal control of exchanges between the ECF and three other compartments: bone, kidneys, and intestine.

▌ Parathyroid hormone raises free plasma Ca^{2+} levels by its effects on bone, kidneys, and intestine.

Parathyroid hormone (PTH), a peptide hormone secreted by the **parathyroid glands,** is the principal regulator of Ca^{2+} metabolism. The four, rice grain–sized parathyroid glands lie on the back surface of the thyroid gland, one in each corner. Like aldosterone, PTH *is essential for life.* The overall effect of PTH is to increase the Ca^{2+} concentration of plasma (and, accordingly, of the entire ECF), thereby preventing hypocalcemia. In the complete absence of PTH, death ensues within a few days, usually because of asphyxiation caused by hypocalcemic spasm of respiratory muscles. By its actions on bone, kidneys, and intestine, PTH raises the plasma Ca^{2+} level when it starts to fall so that hypocalcemia and its effects are normally avoided. This hormone also acts to lower plasma PO_4^{3-} concentration. We will consider each of these mechanisms, beginning with an overview of bone remodeling and PTH's actions on bone.

Bone continuously undergoes remodeling.

Recall that bone is a living tissue composed of an organic extracellular matrix impregnated with calcium phosphate salts, with 99% of the body's Ca^{2+} being found in the skeleton. By mobilizing some of these Ca^{2+} stores in the bone, PTH raises the plasma Ca^{2+} concentration when it starts to fall.

Despite the apparent inanimate nature of bone, its constituents are continually being turned over. **Bone deposition** (formation) and **bone resorption** (removal) normally go on concurrently, so that bone is constantly being remodeled, much as people remodel buildings by tearing down walls and replacing them. Through remodeling, the adult human skeleton is completely regenerated an estimated every 10 years. Bone remodeling serves two purposes: (1) It keeps the skeleton appropriately "engineered" for maximum effectiveness in its mechanical uses, and (2) it helps maintain the plasma Ca^{2+} level. Let us examine in more detail the underlying mechanisms and controlling factors for each of these purposes.

Recall that three types of bone cells are present in bone. The *osteoblasts* secrete the extracellular organic matrix within which the calcium phosphate crystals precipitate. The *osteocytes* are the retired osteoblasts imprisoned within the bony wall they have deposited around themselves. The *osteoclasts* resorb bone in their vicinity by releasing acids that dissolve the calcium phosphate crystals and enzymes that break down the organic matrix. Thus a constant cellular tug-of-war goes on in bone, with bone-forming osteoblasts countering the efforts of the bone-destroying osteoclasts. These construction and demolition crews, working side by side, continuously remodel bone. Throughout most of adult life, the rates of bone formation and bone resorption are about equal, so total bone mass remains fairly constant during this period. As a child grows, the bone builders keep ahead of the bone destroyers under the influence of GH and IGF.

Mechanical stress favors bone deposition

Mechanical stress also tips the balance in favor of bone deposition, causing bone mass to increase and the bones to strengthen. Mechanical factors adjust the strength of bone in response to the demands placed on it. The greater the physical stress and compression to which a bone is subjected, the greater the rate of bone deposition. For example, the bones of athletes are stronger and more massive than those of sedentary people.

By contrast, bone mass diminishes and the bones weaken when bone resorption gains a competitive edge over bone deposition in response to removal of mechanical stress. For example, bone mass decreases in people who undergo prolonged bed confinement or those in space flight. Early astronauts lost up to 20% of their bone mass during their time in orbit. Therapeutic exercises can limit or prevent such loss of bone.

Bone mass also decreases as a person ages. Bone density peaks when a person is in the 30s, then starts to decline after age 40. By 50 to 60 years of age, bone resorption often exceeds bone formation. The result is a reduction in bone mass known as **osteoporosis** (meaning "porous bones") (● Figure 17-28). This bone-thinning condition is characterized by a diminished laying down of organic matrix as a result of reduced osteoblast activity and/or increased osteoclast activity rather than abnormal bone calcification. The underlying cause of osteoporosis is uncertain. Plasma Ca^{2+} and PO_4^{3-} levels are normal, as is PTH. Osteoporosis occurs with greatest frequency in postmenopausal women, suggesting that estrogen withdrawal plays a role. Continued physical activity throughout life appears to retard or prevent bone loss, even in the elderly.

PTH promotes the transfer of Ca^{2+} from bone to plasma.

In addition to the factors geared toward controlling the mechanical effectiveness of bone, throughout life PTH uses bone as a "bank" from which it withdraws Ca^{2+} as needed to maintain the plasma Ca^{2+} level. Parathyroid hormone has two major effects on bone that raise plasma Ca^{2+} concentration:

1. As an immediate effect, PTH promotes rapid movement of Ca^{2+} into the plasma from Ca^{2+}-rich bone fluid. Recall that long cytoplasmic processes extend out from the entombed osteocytes into an extensive network of small canals in the bone to permit exchange of substances between trapped osteocytes and the circulation. **Bone fluid** surrounds these cytoplasmic extensions within the canals. PTH accomplishes the fast exchange of Ca^{2+} between bone fluid and the plasma by activating membrane-bound Ca^{2+} pumps located in the plasma membranes of these cytoplasmic extensions. These pumps promote movement of Ca^{2+} from the bone fluid into the osteocytes, from which the Ca^{2+} is transferred into the plasma circulating through the bone. Through this means, PTH draws Ca^{2+} out of the "quick-cash branch"

● **FIGURE 17-28**

Comparison of normal and osteoporotic bone. Note the reduced density of osteoporotic bone compared to normal bone.

Normal bone

Osteoporotic bone

of the bone bank and rapidly increases the plasma Ca^{2+} level without actually entering the bank (that is, without breaking down mineralized bone itself). Under normal conditions, this exchange is sufficient for maintaining plasma Ca^{2+} concentration.

2. Under conditions of chronic hypocalcemia, such as may occur with dietary Ca^{2+} deficiency, PTH stimulates actual localized dissolution of bone, promoting a slower transfer into the plasma of both Ca^{2+} and PO_4^{3-} from the minerals within the bone itself. It does so by stimulating osteoclasts to gobble up bone and transiently inhibiting the bone-forming activity of the osteoblasts. Bone contains so much Ca^{2+} compared to the plasma (more than 1000 times as much) that even when PTH promotes increased bone resorption, no immediate effects on the skeleton are discernible, because such a tiny amount of bone is affected. Yet the negligible amount of Ca^{2+} "borrowed" from the bone bank can be lifesaving in terms of restoring the free plasma Ca^{2+} level to normal. The borrowed Ca^{2+} is then redeposited in the bone at another time when Ca^{2+} supplies are more abundant. Meanwhile, the plasma Ca^{2+} level has been maintained without sacrificing bone integrity. However, prolonged excess PTH secretion over months or years eventually leads to the formation of cavities throughout the skeleton, which are filled with very large, over-stuffed osteoclasts.

When PTH promotes dissolution of the calcium phosphate crystals in bone to harvest their Ca^{2+} content, both Ca^{2+} and PO_4^{3-} are released into the plasma. An elevation in plasma PO_4^{3-} is undesirable, but PTH deals with this dilemma by its actions on the kidneys, a topic to which we now turn.

▌ PTH acts on the kidneys to conserve Ca^{2+} and eliminate PO_4^{3-}.

PTH stimulates Ca^{2+} conservation and promotes PO_4^{3-} elimination by the kidneys during urine formation. Under the influence of PTH, the kidneys can reabsorb more of the filtered Ca^{2+}, so less Ca^{2+} escapes into urine. This effect increases the plasma Ca^{2+} level and decreases urinary Ca^{2+} losses. (It would be counterproductive to dissolve bone to obtain more Ca^{2+} only to lose it in urine.)

At the same time that it stimulates renal Ca^{2+} reabsorption, PTH decreases PO_4^{3-} reabsorption, thus increasing urinary PO_4^{3-} excretion. As a result, PTH reduces plasma PO_4^{3-} levels at the same time it increases plasma Ca^{2+} concentrations.

This PTH-induced removal of extra PO_4^{3-} from the body fluids is essential for preventing reprecipitation of the Ca^{2+} that has been freed from bone as a result of PTH-induced bone dissolution. Because of the solubility characteristics of calcium phosphate salt, if plasma PO_4^{3-} and plasma Ca^{2+} levels were allowed to increase simultaneously, some of the plasma Ca^{2+} would be forced back into bone through calcium-phosphate crystal formation. This redeposition of Ca^{2+} would lower plasma Ca^{2+}, just the opposite of the needed effect. However, through PTH's action on the kidneys, the plasma concentration of PO_4^{3-} declines even though extra PO_4^{3-} is being re-

leased from bone into the plasma as a result of PTH's action on bone. In this way, the self-defeating redeposition of released Ca^{2+} back into bone is prevented.

The third important action of PTH on the kidneys (besides increasing Ca^{2+} reabsorption and decreasing PO_4^{3-} reabsorption) is to enhance the activation of vitamin D by the kidneys.

▌ PTH indirectly promotes absorption of Ca^{2+} and PO_4^{3-} by the intestine.

Although PTH has no direct effect on the intestine, it indirectly increases both Ca^{2+} and PO_4^{3-} absorption from the small intestine by playing a role in vitamin D activation. This vitamin, in turn, directly increases intestinal absorption of Ca^{2+} and PO_4^{3-}, a topic we will discuss more thoroughly shortly.

▌ The primary regulator of PTH secretion is the plasma concentration of free Ca^{2+}.

All the effects of PTH raise plasma Ca^{2+} levels. Appropriately, PTH secretion is increased in response to a fall in plasma Ca^{2+} concentration and decreased by a rise in plasma Ca^{2+} levels. The secretory cells of the parathyroid glands are directly and exquisitely sensitive to changes in free plasma Ca^{2+}. Because PTH regulates plasma Ca^{2+} concentration, this relationship forms a simple negative-feedback loop for controlling PTH secretion without involving any nervous or other hormonal intervention (● Figure 17-29).

▌ Calcitonin lowers the plasma Ca^{2+} concentration but is not important in the normal control of Ca^{2+} metabolism.

Calcitonin, the hormone produced by the C cells of the thyroid gland, also exerts an influence on plasma Ca^{2+} levels.

● **FIGURE 17-29**

Negative-feedback loops controlling parathyroid hormone (PTH) and calcitonin secretion

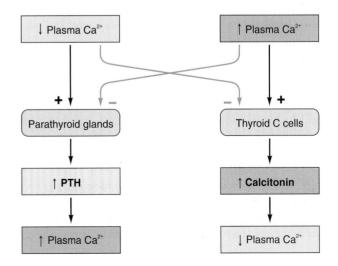

Like PTH, calcitonin has two effects on bone, but in this case both effects *decrease* plasma Ca^{2+} levels. First, on a short-term basis calcitonin decreases Ca^{2+} movement from the bone fluid into the plasma. Second, on a long-term basis calcitonin decreases bone resorption by inhibiting the activity of osteoclasts. The suppression of bone resorption lowers plasma PO_4^{3-} levels as well as reduces plasma Ca^{2+} concentration. The hypocalcemic and hypophosphatemic effects of calcitonin are due entirely to this hormone's actions on bone. It has no effect on the kidneys or intestine.

As with PTH, the primary regulator of calcitonin release is the free plasma Ca^{2+} concentration, but in contrast to its effect on PTH release, an increase in plasma Ca^{2+} stimulates calcitonin secretion and a fall in plasma Ca^{2+} inhibits calcitonin secretion (● Figure 17-29).

Most evidence suggests, however, that calcitonin plays little or no role in the normal control of Ca^{2+} or PO_4^{3-} metabolism. Although calcitonin protects against hypercalcemia, this condition rarely occurs under normal circumstances. Moreover, neither thyroid removal nor calcitonin-secreting tumors alter circulating levels of Ca^{2+} or PO_4^{3-}, implying that this hormone is not normally essential for maintaining Ca^{2+} or PO_4^{3-} homeostasis. Calcitonin may, however, play a role in protecting skeletal integrity when there is a large Ca^{2+} demand, such as during pregnancy or breast-feeding. Furthermore, some experts speculate that calcitonin may hasten the storage of newly absorbed Ca^{2+} following a meal. Gastrointestinal hormones secreted during digestion of a meal have been shown to stimulate the release of calcitonin.

▌ Vitamin D is actually a hormone that increases calcium absorption in the intestine.

The final factor involved in regulating Ca^{2+} metabolism is **cholecalciferol**, or **vitamin D**, a steroidlike compound essential for Ca^{2+} absorption in the intestine. Strictly speaking, vitamin D should be considered a hormone, because the body can produce it in the skin from a precursor related to cholesterol on exposure to sunlight. It is subsequently released into the blood to act at a distant target site, the intestine. The skin, therefore, is actually an endocrine gland and vitamin D a hormone. Traditionally, however, this chemical messenger has been considered a vitamin, for two reasons. First, it was originally discovered and isolated from a dietary source and tagged as a vitamin. Second, even though the skin would be an adequate source of vitamin D if it were exposed to sufficient sunlight, indoor dwelling and clothing in response to cold weather and social customs preclude significant exposure of the skin to sunlight in the United States and many other parts of the world most of the time. At least part of the essential vitamin D must therefore be derived from dietary sources.

ACTIVATION OF VITAMIN D

Regardless of its source, vitamin D is biologically inactive when it first enters the blood from either the skin or the digestive tract. It must be activated by two sequential biochemical alterations. The first of these reactions occurs in the liver

and the second in the kidneys. The kidney enzymes involved in the second step of vitamin D activation are stimulated by PTH in response to a fall in plasma Ca^{2+}. To a lesser extent, a fall in plasma PO_4^{3-} also enhances the activation process.

FUNCTION OF VITAMIN D

The most dramatic and biologically important effect of activated vitamin D is to increase Ca^{2+} absorption in the intestine. Unlike most dietary constituents, dietary Ca^{2+} is not indiscriminately absorbed by the digestive system. In fact, the majority of ingested Ca^{2+} is typically not absorbed but is lost instead in the feces. When needed, more dietary Ca^{2+} is absorbed into the plasma under the influence of vitamin D. Independently of its effects on Ca^{2+} transport, the active form of vitamin D also increases intestinal PO_4^{3-} absorption. Furthermore, vitamin D increases the responsiveness of bone to PTH. Thus vitamin D and PTH are closely interdependent (● Figure 17-30).

▌ Disorders in Ca^{2+} metabolism may arise from abnormal levels of PTH or vitamin D.

 Clinical Note The primary disorders that affect Ca^{2+} metabolism are too much or too little PTH or a deficiency of vitamin D.

PTH HYPERSECRETION

Excess PTH secretion, or **hyperparathyroidism**, which is usually caused by a hypersecreting tumor in one of the parathyroid glands, is characterized by hypercalcemia and hypophosphatemia. The affected person can be asymptomatic or symptoms can be severe, depending on the magnitude of the problem. The following are among the possible consequences:

- Hypercalcemia reduces the excitability of muscle and nervous tissue, leading to muscle weakness and neurologic disorders, including decreased alertness, poor memory, and depression. Cardiac disturbances may also occur.
- Excessive mobilization of Ca^{2+} and PO_4^{3-} from skeletal stores leads to thinning of bone, which may result in skeletal deformities and increased incidence of fractures.
- An increased incidence of Ca^{2+}-containing kidney stones occurs because the excess quantity of Ca^{2+} being filtered through the kidneys may precipitate and form stones. These stones may impair renal function. Passage of the stones through the ureters causes extreme pain. Because of these potential multiple consequences, hyperparathyroidism has been called a disease of "bones, stones, and abdominal groans."
- To further account for the "abdominal groans," hypercalcemia can cause digestive disorders such as peptic ulcers, nausea, and constipation.

PTH HYPOSECRETION

Because of the parathyroid glands' close anatomic relation to the thyroid, the most common cause of deficient PTH secretion, or **hypoparathyroidism**, used to be inadvertent removal of the parathyroid glands (before doctors knew about their existence) during surgical removal of the thyroid gland (to treat thyroid disease). If all the parathyroid tissue was removed, these patients died, of course, because PTH is essential for life. Physicians were puzzled why some patients died soon after thyroid removal even though no surgical complications were apparent. Now that the location and importance of the parathyroid glands have been discovered, surgeons are careful to leave parathyroid tissue during thyroid removal. Rarely, PTH hyposecretion results from an autoimmune attack against the parathyroid glands.

Hypoparathyroidism leads to hypocalcemia and hyperphosphatemia. The symptoms are mainly caused by increased neuromuscular excitability from the reduced level of free plasma Ca^{2+}. In the complete absence of PTH, death is imminent because respiratory muscles go into hypocalcemic spasm. With a relative deficiency rather than a complete absence of PTH, milder symptoms of increased neuromuscular excitability become evident. Muscle cramps and twitches occur from spontaneous activity in the motor nerves, whereas tingling and pins-and-needles sensations result from spontaneous

activity in the sensory nerves. Mental changes include irritability and paranoia.

VITAMIN D DEFICIENCY

The major consequence of vitamin D deficiency is impaired intestinal absorption of Ca^{2+}. In the face of reduced Ca^{2+} uptake, PTH maintains the plasma Ca^{2+} level at the expense of the bones. As a result, the bone matrix is not properly mineralized, because Ca^{2+} salts are not available for deposition. The demineralized bones become soft and deformed, bowing under the pressures of weight bearing, especially in children. This condition is known as **rickets** in children and **osteomalacia** in adults.

 Click on the Media Exercises menu of the CD-ROM and work Media Exercise 17.8: Endocrine Control of Calcium Metabolism to test your understanding of the previous section.

 ## CHAPTER IN PERSPECTIVE: FOCUS ON HOMEOSTASIS

The endocrine system is one of the body's two major regulatory systems, the other being the nervous system. Through its relatively slowly acting hormone messengers, the endocrine system generally regulates activities that require duration rather than speed. Endocrine glands secrete hormones in response to specific stimuli. The hormones in turn exert effects that act in negative-feedback fashion to resist the change that induced their secretion, thus maintaining stability in the internal environment. The specific contributions of the endocrine glands to homeostasis include the following:

- The pineal gland secretes melatonin, which helps entrain the body's circadian rhythm to the environmental cycle of light (period of activity)/dark (period of inactivity).

- The hypothalamus–posterior pituitary unit secretes vasopressin, which acts on the kidneys to help maintain H_2O balance. Control of H_2O balance in turn is essential for maintaining ECF osmolarity and proper cell volume.

- For the most part, the hormones secreted by the anterior pituitary do not directly contribute to homeostasis. Instead, most are tropic; that is, they stimulate the secretion of other hormones.

- Two closely related hormones secreted by the thyroid gland, tetraiodothyronine (T_4) and triiodothyronine (T_3), increase the overall metabolic rate. Not only does this action influence the rate at which cells use nutrient molecules and O_2 within the internal environment, but it also produces heat, which helps maintain body temperature.

- The adrenal cortex secretes three classes of hormones. Aldosterone, the primary mineralocorticoid, is essential for Na^+ and K^+ balance. Because of Na^+'s osmotic effect, Na^+ balance is critical to

● **FIGURE 17-30**

Interactions between PTH and vitamin D in controlling plasma calcium

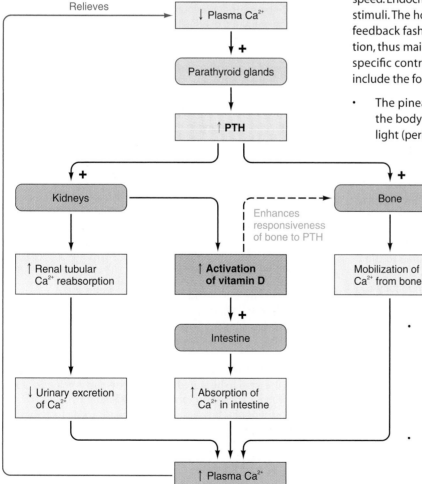

maintaining the proper ECF volume and arterial blood pressure. This action is essential for life. Without aldosterone's Na^+- and H_2O-conserving effect, so much plasma volume would be lost in the urine that death would quickly ensue. Maintaining K^+ balance is essential for homeostasis because changes in extracellular K^+ profoundly impact neuromuscular excitability, jeopardizing normal heart function, among other detrimental effects.

- Cortisol, the primary glucocorticoid secreted by the adrenal cortex, increases the plasma concentrations of glucose, fatty acids, and amino acids above normal. Although these actions destabilize the concentrations of these molecules in the internal environment, they indirectly contribute to homeostasis by making the molecules readily available as energy sources or building blocks for tissue repair to help the body adapt to stressful situations.

- The sex hormones secreted by the adrenal cortex do not contribute to homeostasis.

- The major hormone secreted by the adrenal medulla, epinephrine, generally reinforces activities of the sympathetic nervous system. It contributes to homeostasis directly by its role in blood pressure regulation. Epinephrine also contributes to homeostasis indirectly by helping prepare the body for peak physical responsiveness in fight-or-flight situations. This includes increasing the plasma concentrations of glucose and fatty acids above normal, providing additional energy sources for increased physical activity.

- The two major hormones secreted by the endocrine pancreas, insulin and glucagon, are important in shifting metabolic pathways between the absorptive and postabsorptive states, which maintains the appropriate plasma levels of nutrient molecules.

- Parathyroid hormone from the parathyroid glands is critical to maintaining plasma concentration of Ca^{2+}. PTH is essential for life because of Ca^{2+}'s effect on neuromuscular excitability. In the absence of PTH, death rapidly occurs from asphyxiation caused by pronounced spasms of the respiratory muscles.

Unrelated to homeostasis, hormones direct the growing process and control most aspects of the reproductive system.

CHAPTER SUMMARY

General Principles of Endocrinology (pp. 529–535)

▪ Hormones are long-distance chemical messengers secreted by the ductless endocrine glands into the blood, which transports the hormones to specific target sites where they control a particular function by altering protein activity within target cells. (Review Table 17-1, pp. 533–535.)

▪ Even though hormones can reach all tissues via the blood, they exert their effects only at their target cells, because these cells alone have unique receptors for binding the hormone.

▪ Hormones are grouped into two categories based on differences in their solubility and are further grouped according to their chemical structure—hydrophilic hormones (peptide hormones and catecholamines) and lipophilic hormones (steroid hormones and thyroid hormone).

▪ The endocrine system is especially important in regulating organic metabolism, H_2O and electrolyte balance, growth, and reproduction and in helping the body cope with stress. (Review Figure 17-1.)

▪ Some hormones are tropic, meaning their function is to stimulate and maintain other endocrine glands.

▪ The plasma concentration of each hormone is normally controlled by regulated changes in the rate of hormone secretion. Secretory output of endocrine cells is primarily influenced by two types of direct regulatory inputs: (1) neural input, which increases hormone secretion in response to a specific need and also governs diurnal variations in secretion, and (2) input from another hormone, which involves either stimulatory input from a tropic hormone or inhibitory input from a target cell hormone in negative-feedback fashion. (Review Figures 17-2 and 17-3.)

Pineal Gland and Circadian Rhythms (pp. 536–537)

▪ The suprachiasmatic nucleus (SCN) is the body's master biological clock. Self-induced cyclic variations in the concentration of clock proteins within the SCN bring about cyclic changes in neural discharge from this area. Each cycle takes about a day and drives the body's circadian (daily) rhythms.

▪ The inherent rhythm of this endogenous oscillator is a bit longer than 24 hours. Therefore, each day the body's circadian rhythms must be entrained or adjusted to keep pace with environmental cues so that the internal rhythms are synchronized with the external light–dark cycle.

▪ In the eyes, special photoreceptors that respond to light but are not involved in vision send input to the SCN. Acting through the SCN, the pineal gland's secretion of the hormone melatonin rhythmically fluctuates with the light–dark cycle, decreasing in the light and increasing in the dark. Melatonin in turn is believed to synchronize the body's natural circadian rhythms, such as diurnal (day–night) variations in hormone secretion and body temperature, with external cues such as the light–dark cycle.

Hypothalamus and Pituitary (pp. 537–542)

▪ The pituitary gland consists of two distinct lobes, the posterior pituitary and the anterior pituitary. (Review Figure 17-4.)

▪ The hypothalamus, a portion of the brain, secretes nine peptide hormones; two are stored in the posterior pituitary, and seven are carried through a special vascular link—the hypothalamic-hypophyseal portal system—to the anterior pituitary, where they regulate the release of particular anterior-pituitary hormones. (Review Figures 17-5 and 17-8.)

▪ The posterior pituitary is essentially a neural extension of the hypothalamus. Two small peptide hormones, vasopressin and oxytocin, are synthesized within the cell bodies of neurosecretory neurons located in the hypothalamus, from which they pass down the axon to be stored in nerve terminals within the posterior pituitary. These hormones are independently released from the posterior pituitary into the blood in response to action potentials originating in the hypothalamus. (1) Vasopressin conserves water during urine formation. (2) Oxytocin stimulates uterine contraction during childbirth and milk ejection during breast-feeding. (Review Figure 17-5.)

▪ The anterior pituitary secretes six different peptide hormones that it produces itself. Five anterior-pituitary hormones are tropic.

(1) Thyroid-stimulating hormone (TSH) stimulates secretion of thyroid hormone. (2) Adrenocorticotropic hormone (ACTH) stimulates secretion of cortisol by the adrenal cortex. (3) and (4) The gonadotropic hormones—follicle-stimulating hormone (FSH) and luteinizing hormone (LH)—stimulate production of gametes (eggs and sperm) as well as secretion of sex hormones. (5) Growth hormone (GH) stimulates growth indirectly by stimulating secretion of somatomedins, which in turn promote growth of bone and soft tissues. GH exerts metabolic effects as well. (6) Prolactin stimulates milk secretion and is not tropic to another endocrine gland. *(Review Figure 17-6.)*

▮ The anterior pituitary releases its hormones into the blood at the bidding of releasing and inhibiting hormones from the hypothalamus. The hypothalamus in turn is influenced by a variety of neural and hormonal controlling inputs. *(Review Figures 17-7 and 17-8 and Table 17-2.)*

▮ Both the hypothalamus and the anterior pituitary are inhibited in negative-feedback fashion by the product of the target endocrine gland in the hypothalamus–anterior pituitary–target gland axis. *(Review Figure 17-7.)*

Endocrine Control of Growth (pp. 542–546)

▮ Growth depends not only on growth hormone and other growth-influencing hormones such as thyroid hormone, insulin, and the sex hormones but also on genetic determination, an adequate diet, and freedom from chronic disease or stress.

▮ Growth hormone promotes growth indirectly by stimulating the liver's production of somatomedins, or insulin-like growth factors (IGF), which act directly on bone and soft tissues to bring about most growth-promoting actions.

▮ Growth hormone also directly exerts metabolic effects unrelated to growth, such as conservation of carbohydrates and mobilization of fat stores.

▮ Growth hormone secretion by the anterior pituitary is regulated in negative-feedback fashion by two hypothalamic hormones, growth hormone–releasing hormone and growth hormone–inhibiting hormone. *(Review Figure 17-10.)*

▮ Growth hormone levels are not highly correlated with periods of rapid growth. The primary signals for increased growth-hormone secretion are related to metabolic needs rather than growth, namely, deep sleep, stress, exercise, and low blood-glucose levels.

Thyroid Gland (pp. 547–552)

▮ The thyroid gland contains two types of endocrine secretory cells: (1) follicular cells, which produce the iodine-containing hormones, T_4 (thyroxine or tetraiodothyronine) and T_3 (tri-iodothyronine), collectively known as thyroid hormone, and (2) C cells, which synthesize a Ca^{2+}-regulating hormone, calcitonin. *(Review Figures 17-13 and 17-14.)*

▮ Thyroid hormone is the primary determinant of the overall metabolic rate of the body. By accelerating the metabolic rate of most tissues, it increases heat production. Thyroid hormone also enhances the actions of the chemical mediators of the sympathetic nervous system. Through this and other means, thyroid hormone indirectly increases cardiac output. Finally, thyroid hormone is essential for normal growth as well as the development and function of the nervous system.

▮ Thyroid hormone secretion is regulated by a negative-feedback system between hypothalamic TRH, anterior pituitary TSH, and thyroid gland T_3 and T_4. The feedback loop maintains thyroid hormone levels relatively constant. Cold exposure in newborn infants is the only input to the hypothalamus known to be effective in increasing TRH and thereby thyroid hormone secretion. *(Review Figure 17-15.)*

Adrenal Glands (pp. 552–560)

▮ Each adrenal gland (of the pair) consists of two separate endocrine organs—an outer, steroid-secreting adrenal cortex and an inner, catecholamine-secreting adrenal medulla. *(Review Figure 17-19.)*

▮ The adrenal cortex secretes three different categories of steroid hormones: mineralocorticoids (primarily aldosterone), glucocorticoids (primarily cortisol), and adrenal sex hormones (primarily the weak androgen, dehydroepiandrosterone).

▮ Aldosterone regulates Na^+ and K^+ balance and is important for blood pressure homeostasis, which is achieved secondarily by the osmotic effect of Na^+ in maintaining the plasma volume, a lifesaving effect.

▮ Control of aldosterone secretion is related to Na^+ and K^+ balance and to blood pressure regulation, and is not influenced by ACTH.

▮ Cortisol helps regulate fuel metabolism and is important in stress adaptation. It increases blood levels of glucose, amino acids, and fatty acids and spares glucose for use by the glucose-dependent brain. The mobilized organic molecules are available for use as needed for energy or for repair of injured tissues.

▮ Cortisol secretion is regulated by a negative-feedback loop involving hypothalamic CRH and pituitary ACTH. Stress is the most potent stimulus for increasing activity of the CRH/ACTH/cortisol axis. *(Review Figures 17-7 and 17-20.)*

▮ Dehydroepiandrosterone governs the sex drive and growth of pubertal hair in females.

▮ The adrenal medulla consists of modified sympathetic postganglionic neurons, which secrete the catecholamine epinephrine into the blood in response to sympathetic stimulation. For the most part, epinephrine reinforces the sympathetic system in mounting general systemic "fight-or-flight" responses and in maintaining arterial blood pressure. Epinephrine also exerts important metabolic effects, namely increasing blood glucose and blood fatty acids.

▮ The primary stimulus for increased adrenomedullary secretion is activation of the sympathetic system by stress. *(Review Figure 17-23 and Table 17-3.)*

Endocrine Control of Fuel Metabolism (pp. 560–573)

▮ Intermediary or fuel metabolism is, collectively, the synthesis (anabolism), breakdown (catabolism), and transformations of the three classes of energy-rich organic nutrients—carbohydrate, fat, and protein—within the body. *(Review Figure 17-24 and Table 17-4.)*

▮ Glucose and fatty acids derived respectively from carbohydrates and fats are primarily used as metabolic fuels, whereas amino acids derived from proteins are primarily used for synthesis of structural and enzymatic proteins. *(Review Table 17-5.)*

▮ During the absorptive state following a meal, excess absorbed nutrients not immediately needed for energy production or protein synthesis are stored to a limited extent as glycogen in the liver and muscle but mostly as triglycerides in adipose tissue. *(Review Table 17-6.)*

▮ During the postabsorptive state between meals when no new nutrients are entering the blood, the glycogen and triglyceride stores are catabolized to release nutrient molecules into the blood. If necessary, body proteins are degraded to release amino acids for conversion into glucose. The blood glucose concentration must be maintained above a critical level even during the postabsorptive state, because the brain depends on blood-delivered glucose as its energy source. Tissues not dependent on glucose switch to fatty acids as their metabolic fuel, sparing glucose for the brain. *(Review Table 17-6.)*

These shifts in metabolic pathways between the absorptive and postabsorptive state are hormonally controlled. The most important hormone in this regard is insulin. Insulin is secreted by the β cells of the islets of Langerhans, the endocrine portion of the pancreas. The other major pancreatic hormone, glucagon, is secreted by the α cells of the islets. *(Review Table 17-7, p. 572.)*

Insulin is an anabolic hormone; it promotes the cellular uptake of glucose, fatty acids, and amino acids and enhances their conversion into glycogen, triglycerides, and proteins, respectively. In so doing, it lowers the blood concentrations of these small organic molecules.

Insulin secretion is increased during the absorptive state, primarily by a direct effect of an elevated blood glucose on the β cells, and is largely responsible for directing the organic traffic into cells during this state. *(Review Figures 17-25 and 17-27.)*

Glucagon mobilizes the energy-rich molecules from their stores during the postabsorptive state. Glucagon, which is secreted in response to a direct effect of a fall in blood glucose on the pancreatic α cells, in general opposes the actions of insulin. *(Review Figure 17-27.)*

Endocrine Control of Calcium Metabolism (pp. 573–577)

Changes in the concentration of free, diffusible plasma Ca^{2+}, the biologically active form of this ion, produce profound and life-threatening effects, most notably on neuromuscular excitability. Hypercalcemia reduces excitability, whereas hypocalcemia brings about overexcitability of nerves and muscles. If the overexcitability is severe enough, fatal spastic contractions of respiratory muscles can occur.

Three hormones regulate the plasma concentration of Ca^{2+} (and concurrently regulate PO_4^{3})—parathyroid hormone (PTH), calcitonin, and vitamin D.

PTH, whose secretion is directly increased by a fall in plasma Ca^{2+} concentration, acts on bone, kidneys, and the intestine to raise the plasma Ca^{2+} concentration. In so doing, it is essential for life by preventing the fatal consequences of hypocalcemia. The specific effects of PTH on bone promote Ca^{2+} movement from the bone fluid into the plasma in the short term and promote localized dissolution of bone by enhancing activity of the osteoclasts (bone-dissolving cells) in the long term. *(Review Figures 17-29 and 17-30.)*

Dissolution of the calcium phosphate bone crystals releases PO_4^{3-} as well as Ca^{2+} into the plasma. PTH acts on the kidneys to enhance the reabsorption of filtered Ca^{2+}, thereby reducing the urinary excretion of Ca^{2+} and increasing its plasma concentration. Simultaneously, PTH reduces renal PO_4^{3-} reabsorption, in this way increasing PO_4^{3-} excretion and lowering plasma PO_4^{3-} levels. This is important because a rise in plasma PO_4^{3-} would force the redeposition of some of the plasma Ca^{2+} back into the bone.

Furthermore, PTH facilitates the activation of vitamin D, which in turn stimulates Ca^{2+} and PO_4^{3-} absorption from the intestine. *(Review Figure 17-30.)*

Vitamin D can be synthesized from a cholesterol derivative in the skin when exposed to sunlight, but frequently this endogenous source is inadequate, so vitamin D must be supplemented by dietary intake. From either source, vitamin D must be activated first by the liver and then by the kidneys (the site of PTH regulation of vitamin D activation) before it can exert its effect on the intestine.

Calcitonin, a hormone produced by the C cells of the thyroid gland, is the third factor that regulates Ca^{2+}. In negative-feedback fashion, calcitonin is secreted in response to an increase in plasma Ca^{2+} concentration and acts to lower plasma Ca^{2+} levels by inhibiting activity of bone osteoclasts. Calcitonin is unimportant except during the rare condition of hypercalcemia. *(Review Figure 17-29.)*

REVIEW EXERCISES

Objective Questions (Answers on p. A-48)

1. All endocrine glands are exclusively endocrine in function. *(True or false?)*

2. Growth hormone levels in the blood are no higher during the early childhood growing years than during adulthood. *(True or false?)*

3. "Male" sex hormones are produced in both males and females by the adrenal cortex. *(True or false?)*

4. Excess glucose and amino acids as well as fatty acids can be stored as triglycerides. *(True or false?)*

5. Insulin is the only hormone that can lower blood glucose levels. *(True or false?)*

6. All ingested Ca^{2+} is indiscriminately absorbed in the intestine. *(True or false?)*

7. A hormone that has as its primary function the regulation of another endocrine gland is classified functionally as a _____ hormone.

8. The _____ in the hypothalamus is the body's master biological clock.

9. Activity within the cartilaginous layer of bone known as the _____ brings about the lengthening of long bones.

10. The lumen of the thyroid follicle is filled with _____, the chief constituent of which is a large protein molecule known as _____.

11. _____ is the conversion of glucose into glycogen. _____ is the conversion of glycogen into glucose. _____ is the conversion of amino acids into glucose.

12. The three compartments with which ECF Ca^{2+} is exchanged are _____, _____, and _____.

13. Indicate the relationships among the hormones in the hypothalamic–anterior pituitary–adrenal cortex system by using the following answer code to identify which hormone belongs in each blank:

    (a) cortisol
    (b) ACTH
    (c) CRH

    (1) _____ from the hypothalamus stimulates the secretion of (2) _____ from the anterior pituitary. (3) _____ in turn stimulates the secretion of (4) _____ from the adrenal cortex. In negative-feedback fashion, (5) _____ inhibits secretion of (6) _____ and furthermore reduces the sensitivity of the anterior pituitary to (7) _____.

14. Indicate the primary circulating form and storage form of each of the three classes of organic nutrients:

	Primary Circulating Form	*Primary Storage Form*
Carbohydrate	1. _____	2. _____
Fat	3. _____	4. _____
Protein	5. _____	6. _____

Essay Questions

1. List the overall functions of the endocrine system.
2. How is the plasma concentration of a hormone normally regulated?
3. List and briefly state the functions of the posterior pituitary hormones.
4. List and briefly state the functions of the anterior pituitary hormones.
5. Compare the relationship between the hypothalamus and posterior pituitary with the relationship between the hypothalamus and anterior pituitary. Describe the role of the hypothalamic-hypophyseal portal system and the hypothalamic releasing and inhibiting hormones.
6. Describe the actions of growth hormone that are unrelated to growth. What are growth hormone's growth-promoting actions? What is the role of somatomedins?
7. Discuss the control of growth hormone secretion.
8. Describe the steps of thyroid hormone synthesis.
9. What are the effects of T_3 and T_4? Which is more potent? What is the source of most circulating T_3?
10. Describe the regulation of thyroid hormone.
11. What hormones are secreted by the adrenal cortex? What are the functions and control of each of these hormones?
12. What is the relationship of the adrenal medulla to the sympathetic nervous system? What are the functions of epinephrine? How is epinephrine release controlled?
13. Define *stress*. Describe the neural and hormonal responses to a stressor.
14. Define *fuel metabolism, anabolism,* and *catabolism.*
15. Distinguish between the absorptive and postabsorptive states with regard to the handling of nutrient molecules.
16. Name the two major cell types of the islets of Langerhans, and indicate the hormonal product of each.
17. Compare the functions and control of insulin secretion with those of glucagon secretion.
18. Why must plasma Ca^{2+} be closely regulated?
19. Discuss the contributions of parathyroid hormone, calcitonin, and vitamin D to Ca^{2+} metabolism. Describe the source and control of each of these hormones.

POINTS TO PONDER

(Explanations on p. A-48)

1. A new supervisor at a local hospital decides to rotate the nursing staff to a different shift every week so that one group of employees is not always "stuck" on an undesirable shift. From a physiologic viewpoint, do you think this proposal is advisable?
2. Why would males with testicular feminization syndrome be unusually tall?

3. Gigantism caused by a pituitary tumor is usually treated by surgically removing the pituitary gland. What hormonal replacement therapy would have to be instituted following this procedure?
4. Why do doctors recommend that people who are allergic to bee stings and thus are at risk for anaphylactic shock (see p. 354) carry a vial of epinephrine for immediate injection in case of a sting?
5. Why would an infection tend to raise the blood glucose level of a diabetic individual?

CLINICAL CONSIDERATION

(Explanation on p. A-49)

Najma G. sought medical attention after her menstrual periods ceased and she started growing excessive facial hair. Also, she had been thirstier than usual and urinated more frequently. A clinical evaluation revealed that Najma was hyperglycemic. Her physician told her that she had an endocrine disorder dubbed "diabetes of bearded ladies." Based on her symptoms and your knowledge of the endocrine system, what underlying defect do you think she has?

PHYSIOEDGE RESOURCES

PhysioEdge CD-ROM

PhysioEdge, the CD-ROM packaged with your text, focuses on the concepts students find most difficult to learn. Figures marked with this icon have associated activities on the CD. For a visual review of concepts in this chapter, check out the following:

Media Exercise 17.1: Overview of the Classic Endocrine Glands

Media Exercise 17.2: Anterior Pituitary Gland

Media Exercise 17.3: Posterior Pituitary Gland

Media Exercise 17.4: Endocrine Roles by Organs of Mixed or Uncertain Functions

Media Exercise 17.5: Thyroid Functions

Media Exercise 17.6: Adrenal Gland Functions

Media Exercise 17.7: Endocrine Pancreas and Fuel Metabolism

Media Exercise 17.8: Endocrine Control of Calcium Metabolism

PhysioEdge Website

The website for this book contains a wealth of helpful study aids, as well as many ideas for further reading and research. Log on to:
http://www.brookscole.com/hpfundamentals3
Select Chapter 17 from the drop-down menu or click on one of the many resource areas, including Case Histories, which introduce clinical aspects of human physiology. For this chapter check out: Case History 4: Starvation in the Midst of Plenty, Case History 5: Eight Years Later.

For Suggested Readings, consult **InfoTrac College Edition/Research** on the PhysioEdge website or go directly to InfoTrac College Edition, your online research library, at:
http://infotrac.thomsonlearning.com

Reproductive System

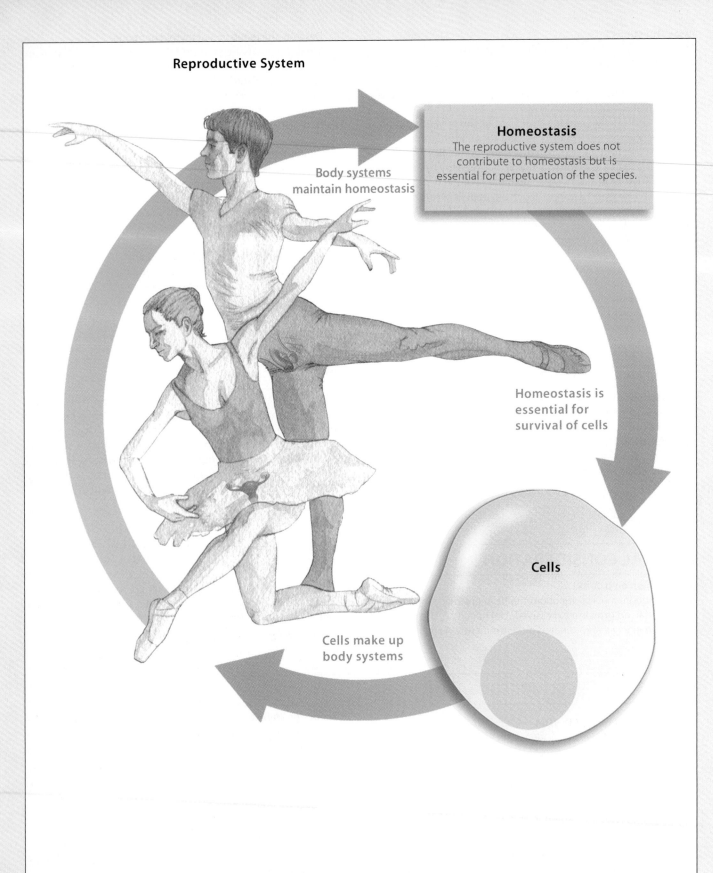

Body systems maintain homeostasis

Homeostasis
The reproductive system does not contribute to homeostasis but is essential for perpetuation of the species.

Homeostasis is essential for survival of cells

Cells

Cells make up body systems

Normal functioning of the **reproductive system** is not aimed toward homeostasis and is not necessary for survival of an individual, but it is essential for survival of the species.

Only through reproduction can the complex genetic blueprint of each species survive beyond the lives of individual members of the species.

...tive System

INTRODUCTION

The central theme of this book has been the physiologic processes aimed at maintaining homeostasis to ensure survival of the individual. We are now going to disembark from this theme to discuss the reproductive system, which primarily serves the purpose of perpetuating the species.

■ The reproductive system includes the gonads, reproductive tract, and accessory sex glands.

Reproduction depends on the union of male and female **gametes** (**reproductive**, or **germ**, **cells**), each with a half set of chromosomes, to form a new individual with a full, unique set of chromosomes. Unlike the other body systems, which are essentially identical in the two sexes, the reproductive systems of males and females are remarkably different, befitting their different roles in the reproductive process. The **male** and **female reproductive systems** are designed to enable union of genetic material from the two sexual partners, and the female system is equipped to house and nourish the offspring to the developmental point at which it can survive independently in the external environment.

The **primary reproductive organs**, or **gonads**, consist of a pair of **testes** in the male and a pair of **ovaries** in the female. In both sexes, the mature gonads perform the dual function of (1) producing gametes (**gametogenesis**), that is, **spermatozoa (sperm)** in the male and **ova (eggs)** in the female, and (2) secreting sex hormones, specifically, **testosterone** in males and **estrogen** and **progesterone** in females.

In addition to the gonads, the reproductive system in each sex includes a **reproductive tract** encompassing a system of ducts that are specialized to transport or house the gametes after they are produced, plus **accessory sex glands** that empty their supportive secretions into these passageways. In females, the *breasts* are also considered accessory reproductive organs. The externally visible portions of the reproductive system are known as **external genitalia**.

SECONDARY SEXUAL CHARACTERISTICS

The **secondary sexual characteristics** are the many external characteristics not directly involved in reproduction that distinguish males and females, such as body configuration and hair distribution. In humans, for example, males have broader shoulders whereas females have curvier hips, and males have beards whereas females do not. Testosterone in the male and estrogen in the female govern the development and maintenance of these characteristics. Progesterone has no influence on secondary sexual characteristics. Even though growth of axillary and pubic hair at puberty is promoted in both sexes by androgens—testosterone in males and adrenocortical dehydroepiandrosterone in females (see p. 556)—this hair growth is not a secondary sexual characteristic, because both sexes display this feature. Thus testosterone and estrogen alone govern the nonreproductive distinguishing features.

In some species, the secondary sexual characteristics are of great importance in courting and mating behavior; for example, the rooster's headdress or comb attracts the female's attention, and the stag's antlers are useful to ward off other males. In humans, the differentiating marks between males and females do serve to attract the opposite sex, but attraction is also strongly influenced by the complexities of human society and cultural behavior.

OVERVIEW OF MALE REPRODUCTIVE FUNCTIONS AND ORGANS

The essential reproductive functions of the male are

1. Production of sperm (*spermatogenesis*)
2. Delivery of sperm to the female

The sperm-producing organs, the testes, are suspended outside the abdominal cavity in a skin-covered sac, the **scrotum**, which lies within the angle between the legs. The male reproductive system is designed to deliver sperm to the female reproductive tract in a liquid vehicle, *semen,* which is conducive to sperm viability. The major **male accessory sex glands,** whose secretions provide the bulk of the semen, are the *seminal vesicles, prostate gland,* and *bulbourethral glands* (● Figure 18-1). The **penis** is the organ used to deposit semen in the female. Sperm exit each testis through the **male reproductive tract**, consisting on each side of an *epididymis* and *ductus (vas) deferens.* These pairs of reproductive tubes empty into a single *urethra,* the canal that runs the length of the penis and empties to the exterior. These parts of the male reproductive system are described more thoroughly later when their functions are discussed.

OVERVIEW OF FEMALE REPRODUCTIVE FUNCTIONS AND ORGANS

The female's role in reproduction is more complicated than the male's. The essential female reproductive functions include

1. Production of ova (*oogenesis*)
2. Reception of sperm
3. Transport of the sperm and ovum to a common site for union (*fertilization,* or *conception*)
4. Maintenance of the developing fetus until it can survive in the outside world (*gestation,* or *pregnancy*), including formation of the *placenta,* the organ of exchange between mother and fetus
5. Giving birth to the baby (*parturition*)
6. Nourishing the infant after birth by milk production (*lactation*)

The product of fertilization is known as an **embryo** during the first two months of intrauterine development when tissue differentiation is taking place. Beyond this time, the developing living being is recognizable as human and is known as a **fetus** during the remainder of gestation. Although no further tissue differentiation takes place during fetal life, it is a time of tremendous tissue growth and maturation.

The ovaries and female reproductive tract lie within the pelvic cavity (● Figure 18-2a and b). The female reproductive tract consists of the following components. Two **oviducts** (**uterine,** or **Fallopian tubes**), which are in close association with the two ovaries, pick up ova on ovulation (ovum release from an ovary) and serve as the site for fertilization. The thick-walled hollow **uterus** is primarily responsible for maintaining the fetus during its development and expelling it at the end of pregnancy. The **vagina** is a muscular, expandable tube that connects the uterus to the external environment. The lowest portion of the uterus, the **cervix,** projects into the vagina and contains a single, small opening, the **cervical canal.** Sperm are deposited in the vagina by the penis during sexual intercourse. The cervical canal serves as a pathway for sperm through the uterus to the site of fertilization in the oviduct, and when greatly dilated during parturition, serves as the passageway for delivery of the baby from the uterus.

The **vaginal opening** is located in the **perineal region** between the urethral opening anteriorly and the anal opening posteriorly (● Figure 18-2c). It is partially covered by a thin mucous membrane, the **hymen,** which typically is physically disrupted by the first sexual intercourse. The vaginal and urethral openings are surrounded laterally by two pairs of skin folds, the **labia minora** and **labia majora.** The smaller labia minora are located medially to the more prominent labia majora. The **clitoris,** a small erotic structure composed of tissue similar to the penis, lies at the anterior end of the folds of the labia minora. The female external genitalia are collectively referred to as the **vulva.**

▌ Reproductive cells each contain a half set of chromosomes.

The DNA molecules that carry the cell's genetic code are not randomly crammed into the nucleus but are precisely organized into **chromosomes** (see p. A-19). Each chromosome consists of a different DNA molecule that contains a unique set of genes. **Somatic (body) cells** contain 46 chromosomes (the **diploid number**), which can be sorted into 23 pairs on the basis of various distinguishing features. Chromosomes composing a matched pair are termed **homologous chromosomes,** one member of each pair having been derived from

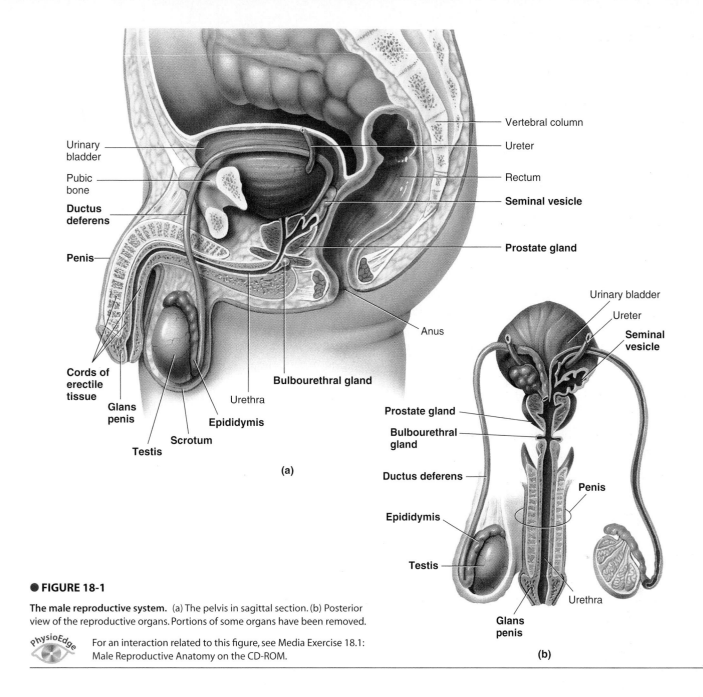

● FIGURE 18-1

The male reproductive system. (a) The pelvis in sagittal section. (b) Posterior view of the reproductive organs. Portions of some organs have been removed.

PhysioEdge For an interaction related to this figure, see Media Exercise 18.1: Male Reproductive Anatomy on the CD-ROM.

the individual's maternal parent and the other member from the paternal parent. Gametes (that is, sperm and eggs) contain only one member of each homologous pair for a total of 23 chromosomes (the **haploid number**).

Gametogenesis is accomplished by meiosis.

Most cells in the human body have the ability to reproduce themselves, a process important in growth, replacement, and repair of tissues. Cell division involves two components: division of the nucleus and division of the cytoplasm. Nuclear division in somatic cells is accomplished by **mitosis**. In mitosis, the chromosomes replicate (make duplicate copies of themselves), then the identical chromosomes are separated so that a complete set of genetic information (that is, a di-

ploid number of chromosomes) is distributed to each of the two new daughter cells. Nuclear division in the specialized case of gametes is accomplished by **meiosis**, in which only a half set of genetic information (that is, a haploid number of chromosomes) is distributed to each of four new daughter cells (see p. A-27).

During meiosis, a specialized diploid germ cell undergoes one chromosome replication followed by two nuclear divisions. In the first meiotic division, the replicated chromosomes do not separate into two individual, identical chromosomes but remain joined together. The doubled chromosomes sort themselves into homologous pairs, and the pairs separate so that each of two daughter cells receives a half set of doubled chromosomes. During the second meiotic division, the doubled chromosomes within each of the two daughter

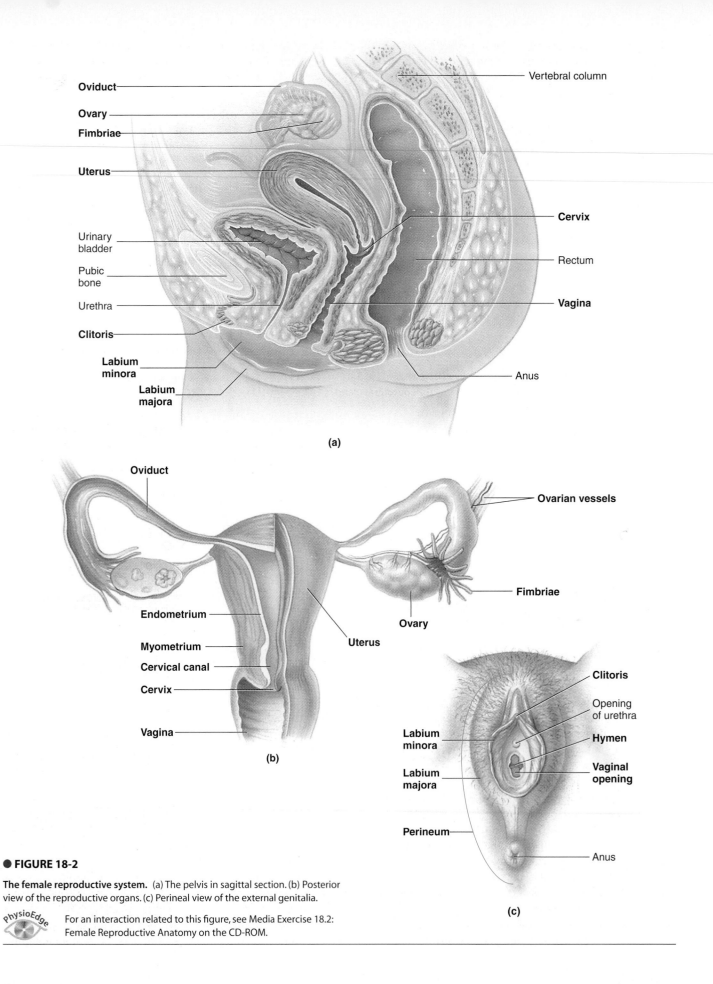

● **FIGURE 18-2**

The female reproductive system. (a) The pelvis in sagittal section. (b) Posterior
view of the reproductive organs. (c) Perineal view of the external genitalia.

PhysioEdge For an interaction related to this figure, see Media Exercise 18.2:
Female Reproductive Anatomy on the CD-ROM.

cells separate and are distributed into two cells, yielding four daughter cells, each containing a half set of chromosomes, a single member of each pair. During this process, the maternally and paternally derived chromosomes of each homologous pair are distributed to the daughter cells in random assortments containing one member of each chromosome pair without regard for its original derivation. That is, not all of the mother-derived chromosomes go to one daughter cell and the father-derived chromosomes to the other cell. This genetic mixing provides novel combinations of chromosomes. Crossing over contributes even further to genetic diversity. *Crossing over* refers to the physical exchange of chromosome material between the homologous pairs prior to their separation during the first meiotic division (see p. A-28).

Thus sperm and ova each have a unique haploid number of chromosomes. When fertilization takes place, a sperm and ovum fuse to form the start of a new individual with 46 chromosomes, one member of each chromosomal pair having been inherited from the mother and the other member from the father.

▌ The sex of an individual is determined by the combination of sex chromosomes.

Whether individuals are destined to be males or females is a genetic phenomenon determined by the sex chromosomes they possess. As the 23 chromosome pairs are separated during meiosis, each sperm or ovum receives only one member of each chromosome pair. Of the chromosome pairs, 22 are **autosomal chromosomes** that code for general human characteristics as well as for specific traits such as eye color. The remaining pair of chromosomes consists of the **sex chromosomes**, of which there are two genetically different types—a larger **X chromosome** and a smaller **Y chromosome**.

Sex determination depends on the combination of sex chromosomes: **Genetic males** have both an X and a Y sex chromosome; **genetic females** have two X sex chromosomes. Thus the genetic difference responsible for all the anatomic and functional distinctions between males and females is the single Y chromosome. Males have it; females do not.

As a result of meiosis during gametogenesis, all chromosome pairs are separated so that each daughter cell contains only one member of each pair, including the sex chromosome pair. When the XY sex chromosome pair separates during sperm formation, half the sperm receive an X chromosome and the other half a Y chromosome. In contrast, during oogenesis, every ovum receives an X chromosome, because separation of the XX sex chromosome pair yields only X chromosomes. During fertilization, combination of an X-bearing sperm with an X-bearing ovum produces a genetic female, XX, whereas union of a Y-bearing sperm with an X-bearing ovum results in a genetic male, XY. Thus genetic sex is determined at the time of conception and depends on which type of sex chromosome is contained within the fertilizing sperm.

▌ Sexual differentiation along male or female lines depends on the presence or absence of masculinizing determinants.

Differences between males and females exist at three levels: genetic, gonadal, and phenotypic (anatomic) sex (● Figure 18-3).

GENETIC AND GONADAL SEX

Genetic sex, which depends on the combination of sex chromosomes at the time of conception, in turn determines **gonadal sex**, that is, whether testes or ovaries develop. The presence

● **FIGURE 18-3**

Sexual differentiation

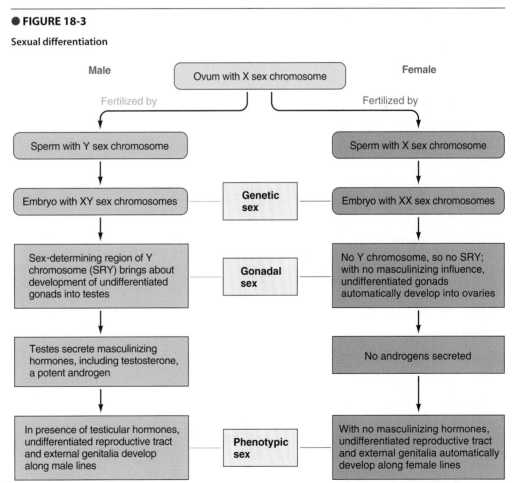

The Reproductive System   **587**

or absence of a Y chromosome determines gonadal differentiation. For the first month and a half of gestation, all embryos have the potential to differentiate along either male or female lines, because the developing reproductive tissues of both sexes are identical and indifferent. Gonadal specificity appears during the seventh week of intrauterine life when the indifferent gonadal tissue of a genetic male begins to differentiate into testes under the influence of the **sex-determining region** of the Y chromosome (**SRY**), the single gene that is responsible for sex determination. This gene triggers a chain of reactions that leads to physical development of a male. SRY "masculinizes" the gonads (induces their development into testes). Because genetic females lack the SRY gene, their gonadal cells never receive a signal for testicular formation, so by default during the ninth week the undifferentiated gonadal tissue starts developing into ovaries instead.

PHENOTYPIC SEX

Phenotypic sex, the apparent anatomic sex of an individual, depends on the genetically determined gonadal sex. The term **sexual differentiation** refers to the embryonic development of the external genitalia and reproductive tract along either male or female lines. As with the undifferentiated gonads, embryos of both sexes have the potential to develop either male or female reproductive tracts and external genitalia. For example, the same embryonic tissue can develop into either a penis or a clitoris. Differentiation into a male-type reproductive system is induced by **androgens,** which are masculinizing hormones secreted by the developing testes. Testosterone is the most potent androgen. The absence of these testicular hormones in female fetuses results in the development of a female-type reproductive system. By 10 to 12 weeks of gestation, the sexes can easily be distinguished by the anatomic appearance of the external genitalia.

Note that the undifferentiated embryonic reproductive tissue passively develops into a female structure unless actively acted on by masculinizing factors. In the absence of male testicular hormones, a female reproductive tract and external genitalia develop regardless of the genetic sex of the individual. For feminization of the fetal genital tissue, ovaries do not even need to be present. Such a control pattern for determining sex differentiation is appropriate, considering that fetuses of both sexes are exposed to high concentrations of female sex hormones throughout gestation. If female sex hormones influenced the development of the reproductive tract and external genitalia, all fetuses would be feminized.

ERRORS IN SEXUAL DIFFERENTIATION

Clinical Note Genetic sex and phenotypic sex are usually compatible; that is, a genetic male anatomically appears to be a male and functions as a male, and the same compatibility holds true for females. Occasionally, however, discrepancies occur between genetic and anatomic sexes because of errors in sexual differentiation, as the following examples illustrate:

• If testes in a genetic male fail to properly differentiate and secrete hormones, the result is the development of an appar-

ent anatomic female in a genetic male, who, of course, will be sterile. Similarly, genetic males whose target cells lack receptors for testosterone are feminized, even though their testes secrete adequate testosterone (see p. 532, testicular feminization syndrome).

• The adrenal gland normally secretes a weak androgen, *dehydroepiandrosterone,* in insufficient quantities to masculinize females. However, pathologically excessive secretion of this hormone in a genetically female fetus during critical developmental stages imposes differentiation of the reproductive tract and genitalia along male lines (see adrenogenital syndrome, p. 557).

Sometimes these discrepancies between genetic sex and apparent sex are not recognized until puberty, when the discovery produces a psychologically traumatic gender identity crisis. For example, a masculinized genetic female with ovaries but with male-type external genitalia may be reared as a boy until puberty, when breast enlargement (caused by estrogen secretion by the awakening ovaries) and lack of beard growth (caused by lack of testosterone secretion in the absence of testes) signal an apparent problem. Therefore, it is important to diagnose any problems in sexual differentiation in infancy. Once a sex has been assigned, it can be reinforced, if necessary, with surgical and hormonal treatment so that psychosexual development can proceed as normally as possible. Less dramatic cases of inappropriate sex differentiation often appear as sterility problems.

 PhysioEdge Click on the Media Exercises menu of the CD-ROM and work Media Exercises 18.1: Male Reproductive Anatomy and 18.2: Female Reproductive Anatomy to test your understanding of the previous section.

MALE REPRODUCTIVE PHYSIOLOGY

In the embryo, the testes develop from the gonadal ridge located at the rear of the abdominal cavity. In the last months of fetal life, they begin a slow descent, passing out of the abdominal cavity through the **inguinal canal** into the scrotum, one testis dropping into each pocket of the scrotal sac. Testosterone from the fetal testes induces descent of the testes into the scrotum. Although the time varies somewhat, descent is usually complete by the seventh month of gestation. As a result, descent is complete in 98% of full-term baby boys.

Clinical Note However, in a substantial percentage of premature male infants the testes are still within the inguinal canal at birth. In most instances of retained testes, descent occurs naturally before puberty or can be encouraged with administration of testosterone. Rarely, a testis remains undescended into adulthood, a condition known as **cryptorchidism** ("hidden testis").

▌ **The scrotal location of the testes provides a cooler environment essential for spermatogenesis.**

The temperature within the scrotum averages several degrees Celsius less than normal body (core) temperature. Descent of the testes into this cooler environment is essential, because spermatogenesis is temperature sensitive and cannot occur at normal body temperature. Therefore, a cryptorchid is unable to produce viable sperm.

The position of the scrotum in relation to the abdominal cavity can be varied by a spinal reflex mechanism that plays an important role in regulating testicular temperature. Reflex contraction of scrotal muscles on exposure to a cold environment raises the scrotal sac to bring the testes closer to the warmer abdomen. Conversely, relaxation of the muscles on exposure to heat permits the scrotal sac to become more pendulous, moving the testes farther from the warm core of the body.

▌ The testicular Leydig cells secrete masculinizing testosterone.

The testes perform the dual function of producing sperm and secreting testosterone. About 80% of the testicular mass consists of highly coiled **seminiferous tubules,** within which spermatogenesis takes place. The endocrine cells that produce testosterone—the **Leydig,** or **interstitial, cells**—lie in the connective tissue (interstitial tissue) between the seminiferous tubules (● Figure 18-4b). Thus the portions of the testes that produce sperm and secrete testosterone are structurally and functionally distinct.

Testosterone is a steroid hormone derived from a cholesterol precursor molecule, as are the female sex hormones, estrogen and progesterone. Once produced, some of the testosterone is secreted into the blood, where it is transported to its target sites of action. A substantial portion of the newly synthesized testosterone goes into the lumen of the seminiferous tubules, where it plays an important role in sperm production.

Most but not all of testosterone's actions ultimately function to ensure delivery of sperm to the female. The effects of testosterone can be grouped into five categories: (1) effects on the reproductive system before birth; (2) effects on sex-specific tissues after birth; (3) other reproduction-related effects; (4) effects on secondary sexual characteristics; and (5) nonreproductive actions (▲ Table 18-1).

EFFECTS ON THE REPRODUCTIVE SYSTEM BEFORE BIRTH

Before birth, testosterone secretion by the fetal testes masculinizes the reproductive tract and external genitalia and promotes descent of the testes into the scrotum, as already described. After birth, testosterone secretion ceases, and the testes and remainder of the reproductive system remain small and nonfunctional until puberty.

EFFECTS ON SEX-SPECIFIC TISSUES AFTER BIRTH

Puberty is the period of arousal and maturation of the previously nonfunctional reproductive system, culminating in sexual maturity and the ability to reproduce. It usually begins sometime between the ages of 10 and 14; on average it begins about two years earlier in females than in males. Usually lasting three to five years, puberty encompasses a complex sequence of endocrine, physical, and behavioral events. **Adolescence** is a broader concept that refers to the entire transition period between childhood and adulthood, not just to sexual maturation.

At puberty, the Leydig cells start secreting testosterone once again. Testosterone is responsible for growth and maturation of the entire male reproductive system. Under the influence of the pubertal surge in testosterone secretion, the testes enlarge and start producing sperm for the first time, the accessory sex glands enlarge and become secretory, and the penis and scrotum enlarge.

Ongoing testosterone secretion is essential for spermatogenesis and for maintaining a mature male reproductive tract throughout adulthood. Once initiated at puberty, testosterone secretion and spermatogenesis occur continuously throughout the male's life. Testicular efficiency gradually declines after 45 to 50 years of age, however, even though men in their 70s and beyond may continue to enjoy an active sex life, and some even father a child at this late age. The gradual reduction in circulating testosterone levels and in sperm production is not caused by a decrease in stimulation of the testes but probably arises instead from degenerative changes associated with aging that occur in the small testicular blood vessels. This gradual decline is often termed "male menopause" or "andropause," although it is not specifically programmed, as is female menopause.

OTHER REPRODUCTION-RELATED EFFECTS

Testosterone governs the development of sexual libido at puberty and helps maintain the sex drive in the adult male. Stimulation of this behavior by testosterone is important for facilitating delivery of sperm to females. In humans, libido is also influenced by many interacting social and emotional factors.

In another reproduction-related function, testosterone participates in the normal negative-feedback control of gonadotropin hormone secretion by the anterior pituitary, a topic covered more thoroughly later.

EFFECTS ON SECONDARY SEXUAL CHARACTERISTICS

All male secondary sexual characteristics depend on testosterone for their development and maintenance. These nonreproductive male characteristics induced by testosterone include (1) the male pattern of hair growth (for example, beard and chest hair and, in genetically predisposed men, baldness); (2) a deep voice caused by enlargement of the larynx and thickening of the vocal folds; (3) thick skin; and (4) the male body configuration (for example, broad shoulders and heavy arm and leg musculature) as a result of protein deposition.

NONREPRODUCTIVE ACTIONS

Testosterone exerts several important effects not related to reproduction. It has a general protein anabolic (synthesis) effect and promotes bone growth, thus contributing to the more muscular physique of males and to the pubertal growth spurt. Ironically, testosterone not only stimulates bone growth but eventually prevents further growth by sealing the growing ends of the long bones (that is, ossifying, or "closing," the epiphyseal plates—see p. 544).

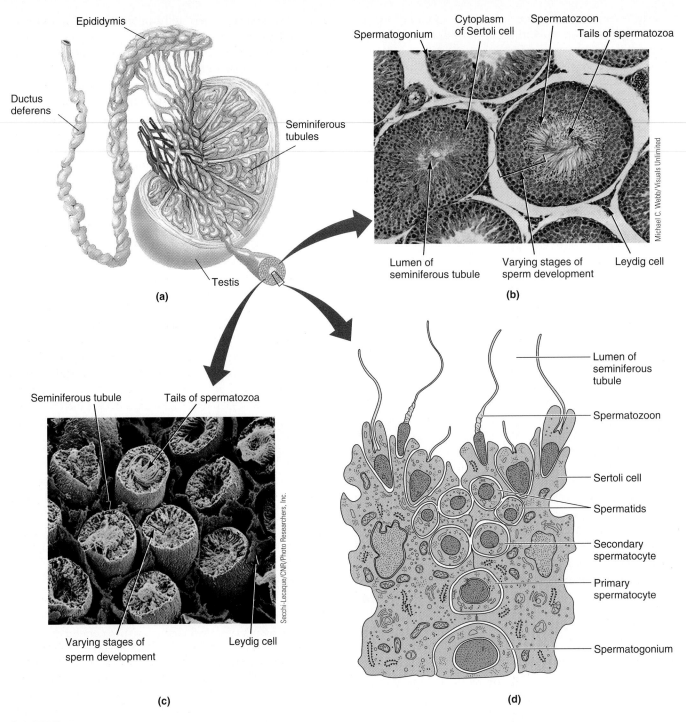

● FIGURE 18-4

Testicular anatomy depicting the site of spermatogenesis. (a) Longitudinal section of a testis showing the location and arrangement of the seminiferous tubules, the sperm-producing portion of the testis. (b) Light micrograph of a cross section of a seminiferous tubule. The undifferentiated germ cells (the spermatogonia) lie in the periphery of the tubule, and the differentiated spermatozoa are in the lumen, with the various stages of sperm development in between. (c) Scanning electron micrograph of a cross section of a seminiferous tubule. (d) Relationship of the Sertoli cells to the developing sperm cells.

 For an interaction related to this figure, see Media Exercise 18.3: Male Reproductive Physiology on the CD-ROM.

In animals, testosterone induces aggressive behavior, but whether it influences human behavior other than in the area of sexual behavior is an unresolved issue. Even though some athletes and bodybuilders who take testosterone-like anabolic androgenic steroids to increase muscle mass have been observed to display more aggressive behavior (see p. 222), it

▲ **TABLE 18-1**

Effects of Testosterone

Effects before Birth

Masculinizes the reproductive tract and external genitalia

Promotes descent of the testes into the scrotum

Effects on Sex-Specific Tissues

Promotes growth and maturation of the reproductive system at puberty

Essential for spermatogenesis

Maintains the reproductive tract throughout adulthood

Other Reproductive Effects

Develops the sex drive at puberty

Controls gonadotropin hormone secretion

Effects on Secondary Sexual Characteristics

Induces the male pattern of hair growth (e.g., beard)

Causes the voice to deepen because vocal folds thicken

Promotes muscle growth responsible for the male body configuration

Nonreproductive Actions

Exerts a protein anabolic effect

Promotes bone growth at puberty and then closure of the epiphyseal plates

May induce aggressive behavior

is unclear to what extent general behavioral differences between the sexes are hormonally induced or result from social conditioning.

We now shift attention from testosterone secretion to the other function of the testes—sperm production.

▌ Spermatogenesis yields an abundance of highly specialized, mobile sperm.

About 250 m (800 feet) of sperm-producing seminiferous tubules are packed within the testes (● Figure 18-4a). Two functionally important cell types are present in these tubules: *germ cells,* most of which are in various stages of sperm development, and *Sertoli cells,* which provide crucial support for spermatogenesis (● Figure 18-4b, c, and d). **Spermatogenesis** is a complex process by which relatively undifferentiated primordial germ cells, the **spermatogonia** (each of which contains a diploid complement of 46 chromosomes), proliferate and are converted into extremely specialized, motile spermatozoa (sperm), each bearing a randomly distributed haploid set of 23 chromosomes.

Microscopic examination of a seminiferous tubule reveals layers of germ cells in an anatomic progression of sperm development, starting with the least differentiated in the outer layer and moving inward through various stages of division to the lumen, where the highly differentiated sperm are ready for exit from the testis (● Figure 18-4b, c, and d). Spermatogenesis takes 64 days for development from a spermatogonium to a mature sperm. Up to several hundred million sperm may reach maturity daily. Spermatogenesis encompasses three major stages: *mitotic proliferation, meiosis,* and *packaging* (● Figure 18-5).

MITOTIC PROLIFERATION

Spermatogonia located in the outermost layer of the tubule continuously divide mitotically, with all new cells bearing the full complement of 46 chromosomes identical to those of the parent cell. Such proliferation provides a continual supply of new germ cells. Following mitotic division of a spermatogonium, one of the daughter cells remains at the outer edge of the tubule as an undifferentiated spermatogonium, thus maintaining the germ cell line. The other daughter cell starts moving toward the lumen while undergoing the various steps required to form sperm, which will be released into the lumen. In humans, the sperm-forming daughter cell divides mitotically twice more to form four identical **primary spermatocytes.** After the last mitotic division, the primary spermatocytes enter a resting phase during which the chromosomes are duplicated and the doubled strands remain together in preparation for the first meiotic division.

MEIOSIS

During meiosis, each primary spermatocyte (with a diploid number of 46 doubled chromosomes) forms two **secondary spermatocytes** (each with a haploid number of 23 doubled chromosomes) during the first meiotic division, finally yielding four **spermatids** (each with 23 single chromosomes) as a result of the second meiotic division.

No further division takes place beyond this stage of spermatogenesis. Each spermatid is remodeled into a single spermatozoon. Because each sperm-producing spermatogonium mitotically produces four primary spermatocytes and each primary spermatocyte meiotically yields four spermatids (spermatozoa-to-be), the spermatogenic sequence in humans can theoretically produce 16 spermatozoa each time a spermatogonium initiates this process. Usually, however, some cells are lost at various stages, so the efficiency of productivity is rarely this high.

PACKAGING

Even after meiosis, spermatids still resemble undifferentiated spermatogonia structurally, except for their half complement of chromosomes. Production of extremely specialized, mobile spermatozoa from spermatids requires extensive remodeling, or **packaging**, of cell elements. Sperm are essentially "stripped-down" cells in which most of the cytosol and any organelles not needed for delivering the sperm's genetic information to an ovum have been extruded. Thus sperm travel lightly, taking with them only the bare essentials to accomplish fertilization.

A **spermatozoon** has four parts (● Figure 18-6): a head, an acrosome, a midpiece, and a tail. The **head** consists pri-

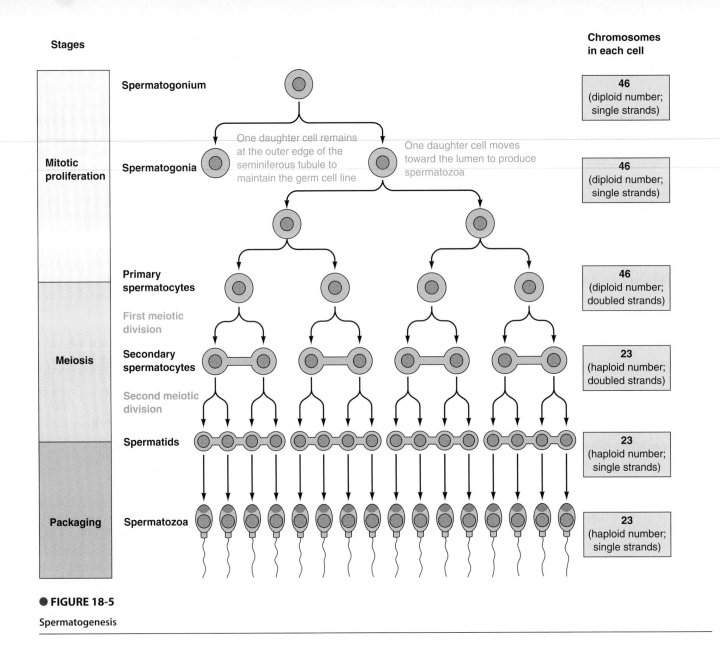

Stages

Chromosomes in each cell

| Spermatogonium | | 46 (diploid number; single strands) |

Mitotic proliferation

Spermatogonia — One daughter cell remains at the outer edge of the seminiferous tubule to maintain the germ cell line — One daughter cell moves toward the lumen to produce spermatozoa — 46 (diploid number; single strands)

Primary spermatocytes — 46 (diploid number; doubled strands)

First meiotic division

Meiosis

Secondary spermatocytes — 23 (haploid number; doubled strands)

Second meiotic division

Spermatids — 23 (haploid number; single strands)

Packaging

Spermatozoa — 23 (haploid number; single strands)

● **FIGURE 18-5**

Spermatogenesis

marily of the nucleus, which contains the sperm's complement of genetic information. The **acrosome**, an enzyme-filled vesicle that caps the tip of the head, is used as an "enzymatic drill" for penetrating the ovum. Mobility for the spermatozoon is provided by a long, whiplike **tail**, movement of which is powered by energy generated by the mitochondria concentrated within the **midpiece** of the sperm.

▌ **Throughout their development, sperm remain intimately associated with Sertoli cells.**

In addition to the spermatogonia and developing sperm cells, the seminiferous tubules also house the **Sertoli cells.** The Sertoli cells lie side-by-side and form a ring that extends from the outer surface of the tubule to the lumen. Each Sertoli cell spans the entire distance from the outer membrane to the fluid-filled lumen. Developing sperm cells are tucked between adjacent Sertoli cells, with spermatogonia lying at the outer

perimeter of the tubule (● Figure 18-4b and d). During spermatogenesis, developing sperm cells arising from spermatogonial mitotic activity migrate toward the lumen in close association with the adjacent Sertoli cells, undergoing their further divisions during this migration. The cytoplasm of the Sertoli cells envelops the migrating sperm cells, which remain buried within these cytoplasmic recesses throughout their development.

Sertoli cells perform the following functions essential for spermatogenesis:

1. The Sertoli cells form a **blood–testes barrier** that prevents blood-borne substances from passing between the cells to gain entry to the lumen of the seminiferous tubule. Because of this barrier, only selected molecules that can pass through the Sertoli cells reach the intratubular fluid. As a result, the composition of the intratubular fluid varies considerably from that of the blood. The unique composition of this fluid that bathes the germ cells is critical for later stages of sperm devel-

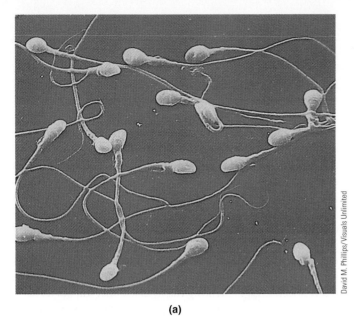

David M. Phillips/Visuals Unlimited

(a)

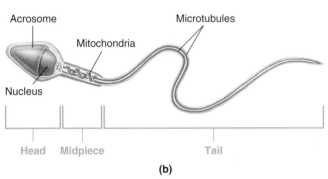

(b)

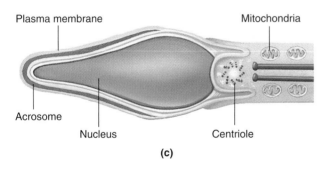

(c)

● **FIGURE 18-6**

Anatomy of a spermatozoon. (a) A phase-contrast photomicrograph of human spermatozoa. (b) Schematic representation of a spermatozoon in "frontal" view. (c) Longitudinal section of the head portion of a spermatozoon in "side" view.

opment. The blood–testes barrier also prevents the antibody-producing cells in the ECF from reaching the tubular sperm factory, thus preventing the formation of antibodies against the highly differentiated spermatozoa.

2.   Because the secluded developing sperm cells do not have direct access to blood-borne nutrients, the Sertoli cells provide nourishment for them.

3.   The Sertoli cells have an important phagocytic function. They engulf the cytoplasm extruded from the spermatids dur-

ing their remodeling, and they destroy defective germ cells that fail to successfully complete all stages of spermatogenesis.

4.   The Sertoli cells secrete into the lumen **seminiferous tubule fluid**, which "flushes" the released sperm from the tubule into the epididymis for storage and further processing.

5.   An important component of this Sertoli secretion is **androgen-binding protein**. As the name implies, this protein binds androgens (that is, testosterone), thus maintaining a very high level of this hormone within the lumen of the seminiferous tubules. This high local concentration of testosterone is essential for sustaining sperm production.

6.   The Sertoli cells are the site of action for control for spermatogenesis by both testosterone and follicle-stimulating hormone (FSH). The Sertoli cells themselves release another hormone, *inhibin*, which acts in negative-feedback fashion to regulate FSH secretion.

▌ LH and FSH from the anterior pituitary control testosterone secretion and spermatogenesis.

The testes are controlled by the two gonadotropic hormones secreted by the anterior pituitary, **luteinizing hormone (LH)** and **follicle-stimulating hormone (FSH)**, which are named for their functions in females (see p. 538).

FEEDBACK CONTROL OF TESTICULAR FUNCTION

LH and FSH act on separate components of the testes (● Figure 18-7). LH acts on the Leydig (interstitial) cells to regulate testosterone secretion, accounting for its alternative name in males—*interstitial cell–stimulating hormone (ICSH)*. FSH acts on the seminiferous tubules, specifically the Sertoli cells, to enhance spermatogenesis. (There is no alternative name for FSH in males.) Secretion of both LH and FSH from the anterior pituitary is stimulated in turn by a single hypothalamic hormone, **gonadotropin-releasing hormone (GnRH)** (see p. 540).

Testosterone, the product of LH stimulation of the Leydig cells, acts in negative-feedback fashion to inhibit LH secretion in two ways. The predominant negative-feedback effect of testosterone is to decrease GnRH release by acting on the hypothalamus, thus indirectly decreasing both LH and FSH release by the anterior pituitary. In addition, testosterone acts directly on the anterior pituitary to reduce the responsiveness of the LH secretory cells to GnRH. The latter action explains why testosterone exerts a greater inhibitory effect on LH secretion than on FSH secretion.

The testicular inhibitory signal specifically directed at controlling FSH secretion is the peptide hormone **inhibin**, which is secreted by the Sertoli cells. Inhibin acts directly on the anterior pituitary to inhibit FSH secretion. This feedback inhibition of FSH by a Sertoli cell product is appropriate, because FSH stimulates spermatogenesis by acting on the Sertoli cells.

ROLES OF TESTOSTERONE AND FSH IN SPERMATOGENESIS

Both testosterone and FSH play critical roles in controlling spermatogenesis, each exerting its effect by acting on the Ser-

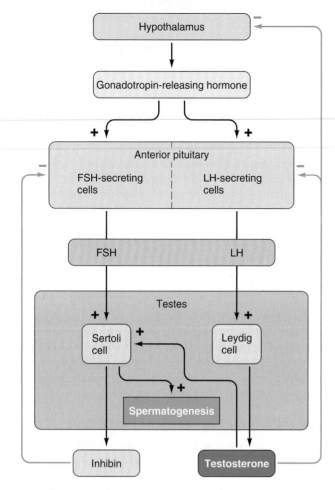

● **FIGURE 18-7**

Control of testicular function

During the prepubertal period, GnRH activity is inhibited. The pubertal process is initiated by an increase in GnRH activity sometime between 8 and 12 years of age. Early in puberty, GnRH secretion occurs only at night, causing a brief nocturnal increase in LH secretion and accordingly, testosterone secretion. The extent of GnRH secretion gradually increases as puberty progresses until the adult pattern of GnRH, FSH, LH, and testosterone secretion is established. Under the influence of the rising levels of testosterone during puberty, the physical changes that encompass the secondary sexual characteristics and reproductive maturation become evident.

The factors responsible for initiating puberty in humans remain unclear. The leading proposal focuses on a potential role for the hormone *melatonin,* which is secreted by the *pineal gland* within the brain (see p. 536). Melatonin, whose secretion decreases during exposure to the light and increases during exposure to the dark, has an antigonadotropic effect in many species. Light striking the eyes inhibits the nerve pathways that stimulate melatonin secretion. In many seasonally breeding species, the overall decrease in melatonin secretion in connection with longer days and shorter nights initiates the mating season. Some researchers suggest that an observed reduction in the overall rate of melatonin secretion at puberty in humans—particularly during the night, when the peaks in GnRH secretion first occur—is the trigger for the onset of puberty.

Having completed our discussion of testicular function, we are now going to shift our attention to the roles of the other components of the male reproductive system.

▌ The reproductive tract stores and concentrates sperm and increases their fertility.

The remainder of the male reproductive system (besides the testes) is designed to deliver sperm to the female reproductive tract. Essentially, it consists of (1) a tortuous pathway of tubes (the reproductive tract), which transports sperm from the testes to outside the body; (2) several accessory sex glands, which contribute secretions that are important to the viability and motility of the sperm; and (3) the penis, which is designed to penetrate and deposit the sperm within the vagina of the female. We will examine each of these parts in greater detail, beginning with the reproductive tract.

COMPONENTS OF THE MALE REPRODUCTIVE TRACT

A comma-shaped **epididymis** is loosely attached to the rear surface of each testis (● Figures 18-1, p. 585, and 18-4a, p. 590). After sperm are produced in the seminiferous tubules, they are swept into the epididymis as a result of the pressure created by the continual secretion of tubular fluid by the Sertoli cells. The epididymal ducts from each testis converge to form a large, thick-walled, muscular duct called the **ductus (vas) deferens**. The ductus deferens from each testis passes up out of the scrotal sac and runs back through the inguinal canal into the abdominal cavity, where it eventually empties into the urethra at the neck of the bladder (● Figure 18-1). The urethra carries sperm out of the penis during ejaculation, the forceful expulsion of semen from the body.

toli cells. Testosterone is essential for both mitosis and meiosis of the germ cells, whereas FSH is required for spermatid remodeling. Testosterone concentration is much higher in the testes than in the blood, because a substantial portion of this hormone produced locally by the Leydig cells is retained in the intratubular fluid complexed with androgen-binding protein secreted by the Sertoli cells. Only this high concentration of testicular testosterone is adequate to sustain sperm production.

▌ Gonadotropin-releasing hormone activity increases at puberty.

Even though the fetal testes secrete testosterone, which directs masculine development of the reproductive system, after birth the testes become dormant until puberty. During the prepubertal period, LH and FSH are not secreted at adequate levels to stimulate any significant testicular activity. The prepubertal delay in the onset of reproductive capability allows time for the individual to mature physically (although not necessarily psychologically) enough to handle child rearing. (This physical maturation is especially important in the female, whose body must support the developing fetus.)

FUNCTIONS OF THE EPIDIDYMIS AND DUCTUS DEFERENS

These ducts perform several important functions. The epididymis and ductus deferens serve as the sperm's exit route from the testis. As they leave the testis, the sperm are capable of neither movement nor fertilization. They gain both capabilities during their passage through the epididymis. This maturational process is stimulated by the testosterone retained within the tubular fluid bound to androgen-binding protein. Sperm's capacity to fertilize is enhanced even further by exposure to secretions of the female reproductive tract. This enhancement of sperm's capacity in the male and female reproductive tracts is known as **capacitation**. The epididymis also concentrates the sperm a hundredfold by absorbing most of the fluid that enters from the seminiferous tubules. The maturing sperm are slowly moved through the epididymis into the ductus deferens by rhythmic contractions of the smooth muscle in the walls of these tubes.

The ductus deferens serves as an important site for sperm storage. Because the tightly packed sperm are relatively inactive and their metabolic needs are accordingly low, they can be stored in the ductus deferens for many days, even though they have no nutrient blood supply and are nourished only by simple sugars present in the tubular secretions.

VASECTOMY

Clinical Note In a **vasectomy**, a common sterilization procedure in males, a small segment of each ductus deferens (alias vas deferens, hence the term *vasectomy*) is surgically removed after it passes from the testis but before it enters the inguinal canal, thus blocking the exit of sperm from the testes. The sperm that build up behind the tied-off testicular end of the severed ductus are removed by phagocytosis. Although this procedure blocks sperm exit, it does not interfere with testosterone activity, because the Leydig cells secrete testosterone into the blood, not through the ductus deferens. Thus testosterone-dependent masculinity or libido should not diminish after a vasectomy.

▌ The accessory sex glands contribute the bulk of the semen.

Several accessory sex glands—the seminal vesicles and prostate—empty their secretions into the duct system before it joins the urethra (● Figure 18-1, p. 585). A pair of saclike *seminal vesicles* empty into the last portion of the two ductus deferens, one on each side. The *prostate* is a large single gland that completely surrounds the urethra. Another pair of accessory sex glands, the *bulbourethral glands,* drain into the urethra after it has passed through the prostate and just before it enters the penis. Numerous mucus-secreting glands also lie along the length of the urethra.

SEMEN

During ejaculation, the accessory sex glands contribute secretions that provide support for the continuing viability of the sperm inside the female reproductive tract. These secretions constitute the bulk of the **semen**, which is a mixture of accessory sex gland secretions, sperm, and mucus. Sperm make up only a small percentage of the total ejaculated fluid.

FUNCTIONS OF THE MALE ACCESSORY SEX GLANDS

Although the accessory sex gland secretions are not absolutely essential for fertilization, they do greatly facilitate the process:

- The **seminal vesicles** (1) supply fructose, which serves as the primary energy source for ejaculated sperm; (2) secrete *prostaglandins*, which stimulate contractions of the smooth muscle in both the male and female reproductive tracts, thereby helping to transport sperm from their storage site in the male to the site of fertilization in the female oviduct; (3) provide more than half the semen, which helps wash the sperm into the urethra and also dilutes the thick mass of sperm, enabling them to become mobile; and (4) secrete fibrinogen, a precursor of fibrin, which forms the meshwork of a clot (see p. 323).
- The **prostate gland** (1) secretes an alkaline fluid that neutralizes the acidic vaginal secretions, an important function because sperm are more viable in a slightly alkaline environment; and (2) provides clotting enzymes and fibrinolysin. The prostatic clotting enzymes act on fibrinogen from the seminal vesicles to produce fibrin, which "clots" the semen, thus helping keep the ejaculated sperm in the female reproductive tract during withdrawal of the penis. Shortly thereafter, the seminal clot is broken down by *fibrinolysin*, a fibrin-degrading enzyme from the prostate, thus releasing mobile sperm within the female tract.
- During sexual arousal, the **bulbourethral glands** secrete a mucuslike substance that provides lubrication for sexual intercourse.

▲ Table 18-2 summarizes the locations and functions of the components of the male reproductive system.

Before turning to the act of delivering sperm to the female (sexual intercourse), we are going to briefly discuss the diverse roles of prostaglandins, which were first discovered in semen but are abundant throughout the body.

▌ Prostaglandins are ubiquitous, locally acting chemical messengers.

Although **prostaglandins** were first identified in the semen and were believed to be of prostate gland origin (hence their name, even though they are actually secreted into the semen by the seminal vesicles), their production and actions are by no means limited to the reproductive system. Being among the most ubiquitous chemical messengers in the body, they are produced in virtually all tissues from arachidonic acid, a fatty acid constituent of the phospholipids within the plasma membrane. On appropriate stimulation, arachidonic acid is split from the plasma membrane by a membrane-bound enzyme and then is converted into the appropriate prostaglandin, which acts as a paracrine locally within or near its site of production (see p. 93). After prostaglandins act, they are rapidly inactivated by local enzymes before they gain access to the blood, or if they do reach the circulatory system, they are

▲ TABLE 18-2

Location and Functions of the Components of the Male Reproductive System

COMPONENT	NUMBER AND LOCATION	FUNCTIONS
Testis	Pair; located in the scrotum, a skin-covered sac suspended within the angle between the legs	Produce sperm
		Secrete testosterone
Epididymis and Ductus Deferens	Pair; one epididymis attached to the rear of each testis; one ductus deferens travels from each epididymis up out of the scrotal sac through the inguinal canal and empties into the urethra at the neck of the bladder	Serve as the sperm's exit route from the testis
		Serve as the site for maturation of the sperm for motility and fertility
		Concentrate and store the sperm
Seminal Vesicle	Pair; both empty into the last portion of the ductus deferens, one on each side	Supply fructose to nourish the ejaculated sperm
		Secrete prostaglandins that stimulate motility to help transport the sperm within the male and female
		Provide the bulk of the semen
		Provide precursors for the clotting of semen
Prostate Gland	Single; completely surrounds the urethra at the neck of the bladder	Secretes an alkaline fluid that neutralizes the acidic vaginal secretions
		Triggers clotting of the semen to keep the sperm in the vagina during penis withdrawal
Bulbourethral Gland	Pair; both empty into the urethra, one on each side, just before the urethra enters the penis	Secrete mucus for lubrication

swiftly degraded on their first pass through the lungs so that they are not dispersed through the systemic arterial system.

Prostaglandins and other closely related arachidonic-acid derivatives—namely, *prostacyclins, thromboxanes,* and *leukotrienes*—are collectively known as **eicosanoids** and are among the most biologically active compounds known. Prostaglandins exert a bewildering variety of effects. Not only are slight variations in prostaglandin structure accompanied by profound differences in biological action, but the same prostaglandin molecule may even exert opposite effects in different tissues. Besides enhancing sperm transport in semen, these abundant chemical messengers are known or suspected to exert other actions in the female reproductive system and in the respiratory, urinary, digestive, nervous, and endocrine systems, in addition to affecting platelet aggregation, fat metabolism, and inflammation (▲ Table 18-3).

As prostaglandins' various actions are better understood, new ways of manipulating them therapeutically are becoming available. A classic example is the use of aspirin, which blocks the conversion of arachidonic acid into prostaglandins, for fever reduction and pain relief. Prostaglandin action is also therapeutically inhibited in the treatment of premenstrual symptoms and menstrual cramping. Furthermore, specific prostaglandins have been medically administered in such diverse situations as inducing labor, treating asthma, and treating gastric ulcers.

Next, before considering the female in greater detail, we will examine the means by which males and females come together to accomplish reproduction.

Click on the Media Exercises menu of the CD-ROM and work Media Exercise 18.3: Male Reproductive Physiology to test your understanding of the previous section.

SEXUAL INTERCOURSE BETWEEN MALES AND FEMALES

Ultimately, union of male and female gametes to accomplish reproduction in humans requires delivery of sperm-laden semen into the female vagina through the **sex act,** also known as **sexual intercourse, coitus,** or **copulation.**

▌ The male sex act is characterized by erection and ejaculation.

The *male sex act* involves two components: (1) **erection,** or hardening of the normally flaccid penis to permit its entry into the vagina, and (2) **ejaculation,** or forceful expulsion of semen into the urethra and out of the penis (▲ Table 18-4). In addition to these strictly reproduction-related components, the **sexual response cycle** also encompasses broader physiologic responses that can be divided into four phases:

▲ TABLE 18-3

Known or Suspected Actions of Prostaglandins

BODY SYSTEM ACTIVITY	ACTIONS OF PROSTAGLANDINS
Reproductive System	Promote sperm transport by action on smooth muscle in the male and female reproductive tracts
	Play a role in ovulation
	Important in menstruation
	Contribute to preparation of the maternal portion of the placenta
	Contribute to parturition
Respiratory System	Some promote bronchodilation, others bronchoconstriction
Urinary System	Increase the renal blood flow
	Increase excretion of water and salt
Digestive System	Inhibit HCl secretion by the stomach
	Stimulate intestinal motility
Nervous System	Influence neurotransmitter release and action
	Act at the hypothalamic "thermostat" to increase body temperature
	Exacerbate sensation of pain
Endocrine System	Enhance cortisol secretion
	Influence tissue responsiveness to hormones in many instances
Circulatory System	Influence platelet aggregation
Fat Metabolism	Inhibit fat breakdown
Defense System	Promote many aspects of inflammation, including development of fever

1. The *excitement phase* includes erection and heightened sexual awareness.
2. The *plateau phase* is characterized by intensification of these responses, plus more generalized body responses, such as steadily increasing heart rate, blood pressure, respiratory rate, and muscle tension.
3. The *orgasmic phase* includes ejaculation as well as other responses that culminate the mounting sexual excitement and are collectively experienced as an intense physical pleasure.
4. The *resolution phase* returns the genitalia and body systems to their prearousal state.

The human sexual response is a multicomponent experience that, in addition to these physiologic phenomena, also

encompasses emotional, psychological, and sociological factors. We will examine only the physiologic aspects of sex.

▌ Erection is accomplished by penis vasocongestion.

Erection is accomplished by engorgement of the penis with blood. The penis consists almost entirely of **erectile tissue** made up of three columns of spongelike vascular spaces extending the length of the organ (● Figure 18-1, p. 585). In the absence of sexual excitation, the erectile tissues contain little blood, because the arterioles that supply these vascular chambers are constricted. As a result, the penis remains small and flaccid. During sexual arousal, these arterioles reflexly dilate and the erectile tissue fills with blood, causing the penis to enlarge both in length and width and to become more rigid. The veins that drain the erectile tissue are mechanically compressed by this engorgement and expansion of the vascular spaces, reducing venous outflow and thereby contributing even further to the buildup of blood, or *vasocongestion*. These local vascular responses transform the penis into a hardened, elongated organ capable of penetrating the vagina.

ERECTION REFLEX

The erection reflex is a spinal reflex triggered by stimulation of highly sensitive mechanoreceptors located in the *glans penis*, which caps the tip of the penis. A recently identified **erection-generating center** lies in the lower spinal cord. Tactile stimulation of the glans reflexly triggers increased parasympathetic vasodilator activity and decreased sympathetic vasoconstrictor activity to the penile arterioles. The result is rapid, pronounced vasodilation of these arterioles and an ensuing erection (● Figure 18-8). As long as this spinal reflex arc remains intact, erection is possible even in men paralyzed by a higher spinal-cord injury.

This parasympathetically induced vasodilation is the major instance of direct parasympathetic control over blood vessel caliber in the body. Parasympathetic stimulation brings about relaxation of penile arteriolar smooth muscle by nitric oxide, which causes arteriolar vasodilation in response to local tissue changes elsewhere in the body (see p. 286). Arterioles are typically supplied only by sympathetic nerves, with increased sympathetic activity producing vasoconstriction and decreased sympathetic activity resulting in vasodilation (see p. 288). Concurrent parasympathetic stimulation and sympathetic inhibition of penile arterioles accomplish vasodilation more rapidly and in greater magnitude than is possible in other arterioles supplied only by sympathetic nerves. Through this efficient means of rapidly increasing blood flow into the penis, the penis can become completely erect in as little as 5 to 10 seconds. At the same time, parasympathetic impulses promote secretion of lubricating mucus from the bulbourethral glands and the urethral glands in preparation for coitus.

A flurry of recent research has led to the discovery of numerous regions throughout the brain that can influence the male sexual response. The erection-influencing brain sites appear extensively interconnected and function as a unified network to either facilitate or inhibit the basic spinal erec-

▲ **TABLE 18-4**

Components of the Male Sex Act

COMPONENTS OF THE MALE SEX ACT	DEFINITION	ACCOMPLISHED BY
Erection	Hardening of the normally flaccid penis to permit its entry into the vagina	Engorgement of the penis erectile tissue with blood as a result of marked parasympathetically induced vasodilation of the penile arterioles and mechanical compression of the veins
Ejaculation		
Emission phase	Emptying of sperm and accessory sex-gland secretions (semen) into the urethra	Sympathetically induced contraction of the smooth muscle in the walls of the ducts and accessory sex glands
Expulsion phase	Forceful expulsion of semen from the penis	Motor neuron–induced contraction of the skeletal muscles at the base of the penis

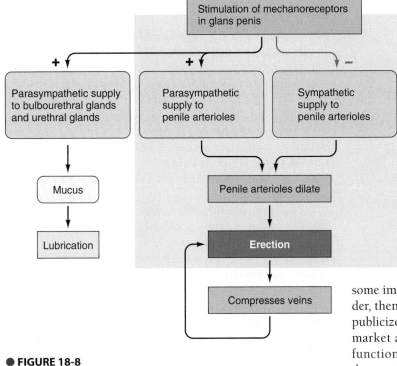

● **FIGURE 18-8**

Erection reflex

tion reflex, depending on the momentary circumstances. As an example of facilitation, psychic stimuli, such as viewing something sexually exciting, can induce an erection in the complete absence of tactile stimulation of the penis. In contrast, failure to achieve an erection despite appropriate stimulation may result from inhibition of the erection reflex by higher brain centers. Let's examine erectile dysfunction in more detail.

ERECTILE DYSFUNCTION

Clinical Note A pattern of failing to achieve or maintain an erection suitable for sexual intercourse—**erectile dysfunction** or **impotence**—may be attributable to psychological or physical factors. An occasional episode of a failed erection does not constitute impotence, but a man who becomes overly anxious about his ability to perform the sex act may well be on his way to chronic failure. Anxiety can lead to erectile dysfunction, which fuels the man's anxiety level and thus perpetuates the problem. Impotence may also arise from physical limitations, including nerve damage, certain medications that interfere with autonomic function, and problems with blood flow through the penis.

Erectile dysfunction is widespread. Over 50% of men between ages 40 and 70 experience some impotence, climbing to nearly 70% by age 70. No wonder, then, that more prescriptions were written for the much-publicized drug *sildenafil (Viagra)* during its first year on the market after its approval in 1998 for treating erectile dysfunction than for any other new drug in history. Sildenafil does not produce an erection, but it amplifies and prolongs an erectile response triggered by usual means of stimulation. Here's how the drug works. Nitric oxide released in response to parasympathetic stimulation activates a membrane-bound enzyme *guanylate cyclase* within nearby arteriolar smooth-muscle cells. This enzyme activates *cyclic guanosine monophosphate (cGMP)*, an intracellular second messenger similar to cAMP (see p. 98). Cyclic GMP in turn leads to relaxation of the penile arteriolar smooth muscle, bringing about pronounced local vasodilation. Under normal circumstances, once cGMP is activated and brings about an erection, this second messenger is broken down by the intracellular enzyme *phos-*

phodiesterase 5 (PDE5). Sildenafil inhibits PDE5. As a result, cGMP remains active longer so that penile arteriolar vasodilation continues and the erection is sustained long enough for a formerly impotent man to accomplish the sex act. Just as pushing a pedal on a piano will not cause a note to be played but will prolong a played note, sildenafil cannot cause the release of nitric oxide and subsequent activation of erection-producing cGMP, but it can prolong the triggered response. The drug has no benefit for those who do not have erectile dysfunction, but its success rate has been high among sufferers of the condition. Side effects have been limited because the drug concentrates in the penis, thus having more impact on this organ than elsewhere in the body.

▌ Ejaculation includes emission and expulsion.

The second component of the male sex act is *ejaculation*. Like erection, ejaculation is a spinal reflex. The same types of tactile and psychic stimuli that induce erection cause ejaculation when the level of excitation intensifies to a critical peak. The overall ejaculatory response occurs in two phases: emission and expulsion.

EMISSION

First, sympathetic impulses cause sequential contraction of smooth muscles in the prostate, reproductive ducts, and seminal vesicles. This contractile activity delivers prostatic fluid, then sperm, and finally seminal vesicle fluid (collectively, semen) into the urethra. This phase of the ejaculatory reflex is called **emission**. During this time, the sphincter at the neck of the bladder is tightly closed to prevent semen from entering the bladder and urine from being expelled along with the ejaculate through the urethra.

EXPULSION

Second, the filling of the urethra with semen triggers nerve impulses that activate a series of skeletal muscles at the base of the penis. Rhythmic contractions of these muscles occur at 0.8-second intervals and increase the pressure within the penis, forcibly expelling the semen through the urethra to the exterior. This is the **expulsion phase** of ejaculation.

ORGASM

The rhythmic contractions that occur during semen expulsion are accompanied by involuntary rhythmic throbbing of pelvic muscles and peak intensity of the overall body responses that were climbing during the earlier phases. Heavy breathing, a heart rate of up to 180 beats per minute, marked generalized skeletal-muscle contraction, and heightened emotions are characteristic. These pelvic and overall systemic responses that culminate the sex act are associated with an intense pleasure characterized by a feeling of release and complete gratification, an experience known as **orgasm**.

RESOLUTION

During the resolution phase following orgasm, sympathetic vasoconstrictor impulses slow the inflow of blood into the penis, causing the erection to subside. A deep relaxation ensues, often accompanied by a feeling of fatigue. Muscle tone returns to normal, while the cardiovascular and respiratory systems return to their prearousal level of activity. Once ejaculation has occurred, a temporary refractory period of variable duration ensues before sexual stimulation can trigger another erection. Males therefore cannot experience multiple orgasms within a matter of minutes, as females sometimes do.

VOLUME AND SPERM CONTENT OF THE EJACULATE

The volume and sperm content of the ejaculate depend on the length of time between ejaculations. The average volume of semen is 2.75 ml, ranging from 2 to 6 ml, the higher volumes following periods of abstinence. An average human ejaculate contains about 180 million sperm (66 million/ml), but some ejaculates contain as many as 400 million sperm.

Clinical Note Both quantity and quality of sperm are important determinants of fertility. A man is considered clinically infertile if his sperm concentration falls below 20 million/ml of semen. Even though only one spermatozoon actually fertilizes the ovum, large numbers of accompanying sperm are needed to provide sufficient acrosomal enzymes to break down the barriers surrounding the ovum until the victorious sperm penetrates into the ovum's cytoplasm. The quality of sperm also must be taken into account when assessing the fertility potential of a semen sample. The presence of substantial numbers of sperm with abnormal motility or structure, such as sperm with distorted tails, reduces the chances of fertilization. (For a discussion of how environmental estrogens may be decreasing sperm counts as well as negatively impacting the male and female reproductive systems in other ways, see the boxed feature, ▶ Beyond the Basics, on p. 600.)

▌ The female sexual cycle is very similar to the male cycle.

Both sexes experience the same four phases of the sexual cycle—excitement, plateau, orgasm, and resolution. Furthermore, the physiologic mechanisms responsible for orgasm are fundamentally the same in males and females.

The excitement phase in females can be initiated by either physical or psychological stimuli. Tactile stimulation of the clitoris and surrounding perineal area is an especially powerful sexual stimulus. These stimuli trigger spinal reflexes that bring about parasympathetically induced vasodilation of arterioles throughout the vagina and external genitalia, especially the clitoris. The resultant inflow of blood becomes evident as swelling of the labia and erection of the clitoris. The latter—like its male homolog, the penis—is composed largely of erectile tissue. Vasocongestion of the vaginal capillaries forces fluid out of the vessels into the vaginal lumen. This fluid, which is the first positive indication of sexual arousal, serves as the primary lubricant for intercourse. Additional lubrication is provided by the mucus secretions from the male and by mucus released during sexual arousal from glands located at the outer opening of the vagina.

Environmental "Estrogens": Bad News for the Reproductive System

Unknowingly, during the last 50 years we humans have been polluting our environment with synthetic endocrine-disrupting chemicals as an unintended side effect of industrialization. Known as **endocrine disrupters,** these hormonelike pollutants bind with the receptor sites normally reserved for the naturally occurring hormones. Depending on how they interact with the receptors, these disrupters can either mimic or block normal hormonal activity. Most endocrine disrupters exert feminizing effects. Many of these environmental contaminants mimic or alter the action of estrogen, the feminizing steroid hormone produced by the female ovaries. Although not yet conclusive, laboratory and field studies suggest that these estrogen disrupters might be responsible for some disturbing trends in reproductive health problems, such as falling sperm counts in males and an increased incidence of breast cancer in females.

Estrogenic pollutants are everywhere. They contaminate our food, drinking water, and air. Proved feminizing synthetic compounds include (1) certain weed killers and insecticides, (2) some detergent breakdown products, (3) petroleum by-products found in car exhaust, (4) a common food preservative used to retard rancidity, and (5) softeners that make plastics flexible. These plastic softeners are commonly found in food packaging and can readily leach into food with which they come in contact, especially during heating. They have also been found to leach into the saliva from some babies' plastic teething toys. They are also in numerous medical products, such as the bags in which blood is stored. Plastic softeners are among the most plentiful industrial contaminants in our environment.

Investigators are only beginning to identify and understand the implications for reproductive health of myriad synthetic chemicals that have become such an integral part of modern societies. An estimated 87,000 synthetic chemicals are already in our environment. Scientists suspect that the estrogen-mimicking chemicals among these may underlie a spectrum of reproductive disorders that have been rising in the past 50 years—the same time period during which large amounts of these pollutants have been introduced into our environment. Here are examples of male reproductive dysfunctions that may be circumstantially linked to exposure to environmental estrogen disrupters:

- *Falling sperm counts.* The average sperm count fell from 113 million sperm/ml of semen in 1940 to 66 million/ml in 1990. Making matters worse, the volume of a single ejaculate has declined from 3.40 ml to 2.75 ml. This means that men on average are now ejaculating less than half the number of sperm as men did 50 years ago—a drop from more than 380 million sperm to about 180 million sperm per ejaculate. Furthermore, the number of motile sperm has also dipped. Importantly, the sperm count has not declined in the less polluted areas of the world during the same time period.
- *Increased incidence of testicular and prostate cancer.* Cases of testicular cancer have tripled since 1940, and the rate continues to climb. Prostate cancer has also been on the rise over the same time period.
- *Rising number of male reproductive tract abnormalities at birth.* The incidence of *cryptorchidism* (undescended testis) nearly doubled from the 1950s to the 1970s. The number of cases of *hypospadia,* a malformation of the penis, more than doubled between the mid-1960s and the mid-1990s. Hypospadia results when the urethral fold fails to fuse closed during the development of a male fetus.
- *Evidence of gender bending in animals.* Some fish and wild animal populations severely exposed to environmental estrogens—such as those living in or near water heavily polluted with hormone-mimicking chemical wastes—display a high rate of grossly impaired reproductive systems. Examples include male fish that are hermaphrodites (having both male and female reproductive parts) and male alligators with abnormally small penises. Similar reproductive abnormalities have been identified in land mammals. Presumably, excessive estrogen exposure is emasculating these populations.
- *Decline in male births.* Many countries are reporting a slight decline in the ratio of baby boys to baby girls being born. Although several plausible explanations have been put forth, many researchers attribute this troubling trend to disruption of normal male fetal development by environmental estrogens. In one compelling piece of circumstantial evidence, people inadvertently exposed to the highest level of an endocrine-disrupting agent during an industrial accident have subsequently had all daughters and no sons, whereas those least exposed have had the normal ratio of girls and boys.

Environmental estrogens are also implicated in the rising incidence of breast cancer in females. Breast cancer is 25 to 30% more prevalent now than in the 1940s. Many of the established risk factors for breast cancer, such as starting to menstruate earlier than usual and undergoing menopause later than usual, are associated with an elevation in the total lifetime exposure to estrogen. Because increased exposure to natural estrogen bumps up the risk for breast cancer, prolonged exposure to environmental estrogens may be contributing to the rising prevalence of this malignancy among women (and men too).

In addition to the estrogen disrupters, scientists recently identified a new class of chemical offenders—androgen disrupters that either mimic or suppress the action of male hormones. For example, studies suggest that bacteria in waste water from pulp mills can convert the sterols in pine pulp into androgens. By contrast, antiandrogen compounds have been found in the fungicides commonly sprayed on vegetable and fruit crops. Yet another cause for concern are the androgens used by the livestock industry to enhance the production of muscle (that is, meat) in feedlot cattle. (Androgens have a protein anabolic effect.) These drugs do not end up in the meat, but they can get into drinking water and other food as hormone-laden feces contaminate rivers and streams.

In response to the growing evidence that has emerged circumstantially linking a wide variety of environmental pollutants to disturbing reproductive abnormalities, the U.S. Congress legally mandated the Environmental Protection Agency (EPA) in 1996 to determine which synthetic chemicals might be endocrine disrupters. In response, the EPA formed an advisory committee, which in 1998 proposed an ambitious plan to begin comprehensive testing of manufactured compounds for their potential to disrupt hormones in humans and wildlife. Although eventually all the 87,000 existing synthetic compounds will be tested, the initial screening is narrowed to evaluating the endocrine-disrupting potential of 15,000 widely used chemicals. Declaring this a national health priority, the government has allocated millions of dollars for this research.

During the plateau phase, the changes initiated during the excitement phase intensify, while systemic responses similar to those in the male (such as increased heart rate, blood pressure, respiratory rate, and muscle tension) occur. Further vasocongestion of the lower third of the vagina during this time reduces its inner capacity so that it tightens around the thrusting penis, heightening tactile sensation for both the female and the male. Simultaneously, the uterus raises upward, lifting the cervix and enlarging the upper two thirds of the vagina. This ballooning, or **tenting effect**, creates a space for ejaculate deposition.

If erotic stimulation continues, the sexual response culminates in orgasm as sympathetic impulses trigger rhythmic contractions of the pelvic musculature at 0.8-second intervals, the same rate as in males. The contractions occur most intensely in the engorged lower third of the vaginal canal. Systemic responses identical to those of the male orgasm also occur. In fact, the orgasmic experience in females parallels that of males with two exceptions. First, there is no female counterpart to ejaculation. Second, females do not become refractory following an orgasm, so they can respond immediately to continued erotic stimulation and achieve multiple orgasms.

During resolution, pelvic vasocongestion and the systemic manifestations gradually subside. As with males, this is a time of great physical relaxation for females.

We will now examine how females fulfill their part of the reproductive process.

FEMALE REPRODUCTIVE PHYSIOLOGY

Female reproductive physiology is much more complex than male reproductive physiology.

▌ Complex cycling characterizes female reproductive physiology.

Unlike the continuous sperm production and essentially constant testosterone secretion characteristic of the male, release of ova is intermittent, and secretion of female sex hormones displays wide cyclic swings. The tissues influenced by these sex hormones also undergo cyclic changes, the most obvious of which is the monthly menstrual cycle. During each cycle, the female reproductive tract is prepared for the fertilization and implantation of an ovum released from the ovary at ovulation. If fertilization does not occur, the cycle repeats. If fertilization does occur, the cycles are interrupted while the female system adapts to nurture and protect the newly conceived human being until it has developed into an individual capable of living outside the maternal environment. Furthermore, the female continues her reproductive functions after birth by producing milk (lactation) for the baby's nourishment. Thus the female reproductive system is characterized by complex cycles that are interrupted only by more complex changes should pregnancy ensue.

The ovaries, as the primary female reproductive organs, perform the dual function of producing ova (oogenesis) and secreting the female sex hormones, estrogen and progesterone. These hormones act together to promote fertilization of the ovum and to prepare the female reproductive system for pregnancy. Estrogen in the female governs many functions similar to those carried out by testosterone in the male, such as maturation and maintenance of the entire female reproductive system and establishment of female secondary sexual characteristics. In general, the actions of estrogen are important to preconception events. Estrogen is essential for ova maturation and release, development of physical characteristics that are sexually attractive to males, and transport of sperm from the vagina to the site of fertilization in the oviduct. Furthermore, estrogen contributes to breast development in anticipation of lactation. The other ovarian steroid, progesterone, is important in preparing a suitable environment for nourishing a developing embryo/fetus and for contributing to the breasts' ability to produce milk.

As in males, reproductive capability begins at puberty in females, but unlike males, who have reproductive potential throughout life, female reproductive potential ceases during middle age at menopause.

▌ The steps of gametogenesis are the same in both sexes, but the timing and outcome differ sharply.

Oogenesis contrasts sharply with spermatogenesis in several important aspects, even though the identical steps of chromosome replication and division take place during gamete production in both sexes. The undifferentiated primordial germ cells in the fetal ovaries, the **oogonia** (comparable to the spermatogonia), divide mitotically to give rise to 6 to 7 million oogonia by the fifth month of gestation, when mitotic proliferation ceases.

FORMATION OF PRIMARY OOCYTES AND PRIMARY FOLLICLES

During the last part of fetal life, the oogonia begin the early steps of the first meiotic division but do not complete it. Known now as **primary oocytes**, they contain the diploid number of 46 replicated chromosomes, which are gathered into homologous pairs but do not separate. The primary oocytes remain in this state of **meiotic arrest** for years until they are prepared for ovulation.

Before birth, each primary oocyte is surrounded by a single layer of **granulosa cells.** Together, an oocyte and surrounding granulosa cells make up a **primary follicle.** Oocytes that are not incorporated into follicles self-destruct by apoptosis (see p. 100). At birth only about 2 million primary follicles remain, each containing a single primary oocyte capable of producing a single ovum. The traditional view is that no new oocytes or follicles appear after birth, with the follicles already present in the ovaries at birth serving as a reservoir from which all ova throughout the reproductive life of a female must arise. However, researchers recently discovered, in mice at least, that new oocytes and follicles are produced after birth from previously unknown ovarian stem cells capable of generating primordial germ cells (oogonia). Despite the potential for similar egg-generating stem cells in humans, the follicular pool gradually dwindles away as a result of processes that "use up" the oocyte-containing follicles.

The pool of primary follicles gives rise to an ongoing trickle of developing follicles. Once it starts to develop, a follicle is destined for one of two fates: It will reach maturity and ovulate, or it will degenerate to form scar tissue, a process known as **atresia.** Of the total pool of follicles, only about 400 will mature and release ova; 99.98% never ovulate but instead undergo atresia at some stage in development. By menopause, which occurs on average in a woman's early 50s, few primary follicles remain, having either already ovulated or become atretic. From this point on, the woman's reproductive capacity ceases.

This limited gamete potential in females is in sharp contrast to the continual process of spermatogenesis in males, who have the potential to produce several hundred million sperm in a single day. Furthermore, considerable chromosome wastage occurs in oogenesis compared with spermatogenesis. Let's see how.

FORMATION OF SECONDARY OOCYTES AND SECONDARY FOLLICLES

The primary oocyte within a primary follicle is still a diploid cell that contains 46 doubled chromosomes. From puberty until menopause, a portion of the resting pool of follicles starts developing into *secondary (antral) follicles* on a cyclic basis. The mechanisms determining which follicles in the pool will develop during a given cycle are unknown. Development of a secondary follicle is characterized by growth of the primary oocyte and by expansion and differentiation of the surrounding cell layers. The oocyte enlarges about a thousandfold. This oocyte enlargement is caused by a buildup of cytoplasmic materials that will be needed by the early embryo.

Just before ovulation, the primary oocyte, whose nucleus has been in meiotic arrest for years, completes its first meiotic division. This division yields two daughter cells, each receiving a haploid set of 23 doubled chromosomes, analogous to the formation of secondary spermatocytes (● Figure 18-9). However, almost all the cytoplasm remains with one of the daughter cells, now called the **secondary oocyte,** which is destined to become the ovum. The chromosomes of the other daughter cell together with a small share of cytoplasm form the **first polar body.** In this way, the ovum-to-be loses half of its chromosomes to form a haploid gamete but retains all of its nutrient-rich cytoplasm. The nutrient-poor polar body soon degenerates.

FORMATION OF A MATURE OVUM

Actually, the secondary oocyte, and not the mature ovum, is ovulated and fertilized, but common usage refers to the developing female gamete as an *ovum* even in its primary and secondary oocyte stages. Sperm entry into the secondary oocyte is needed to trigger the second meiotic division. Oocytes that are not fertilized never complete this final division. During this division, a half set of chromosomes along with a thin layer of cytoplasm is extruded as the **second polar body.** The other half set of 23 unpaired chromosomes remains behind in what is now the **mature ovum.** These 23 maternal chromosomes unite with the 23 paternal chromosomes of the penetrating sperm to complete fertilization. If the first polar body has not already degenerated, it too undergoes the second meiotic division at the same time the fertilized secondary oocyte is dividing its chromosomes.

COMPARISON OF STEPS IN OOGENESIS AND SPERMATOGENESIS

The steps involved in chromosome distribution during oogenesis parallel those of spermatogenesis, except that the cytoplasmic distribution and time span for completion sharply differ. Just as four haploid spermatids are produced by each primary spermatocyte, four haploid daughter cells are produced by each primary oocyte (if the first polar body does not degenerate before it completes the second meiotic division). In spermatogenesis, each daughter cell develops into a highly specialized, motile spermatozoon unencumbered by unessential cytoplasm and organelles, its only destiny being to supply half of the genes for a new individual. In oogenesis, however, of the four daughter cells only the one destined to become the ovum receives cytoplasm. This uneven distribution of cytoplasm is important, because the ovum, in addition to providing half the genes, provides all of the cytoplasmic components needed to support early development of the fertilized ovum. The large, relatively undifferentiated ovum contains numerous nutrients, organelles, and structural and enzymatic proteins. The three other cytoplasm-scarce daughter cells, the polar bodies, rapidly degenerate, their chromosomes being deliberately wasted.

Note also the considerable difference in time to complete spermatogenesis and oogenesis. It takes about two months for a spermatogonium to develop into fully remodeled spermatozoa. In contrast, development of an oogonium (present before birth) to a mature ovum requires anywhere from 11 years (beginning of ovulation at onset of puberty) to 50 years (end of ovulation at onset of menopause). The actual length of the active steps in meiosis is the same in both males and females, but in females the developing eggs remain in meiotic arrest for a variable number of years.

 Clinical Note The older age of ova released by women in their late 30s and 40s is believed to account for the higher incidence of genetic abnormalities, such as Down's syndrome, in children born to women in this age range.

▌ The ovarian cycle consists of alternating follicular and luteal phases.

After the onset of puberty, the ovary constantly alternates between two phases: the **follicular phase,** which is dominated by the presence of *maturing follicles;* and the **luteal phase,** which is characterized by the presence of the *corpus luteum*

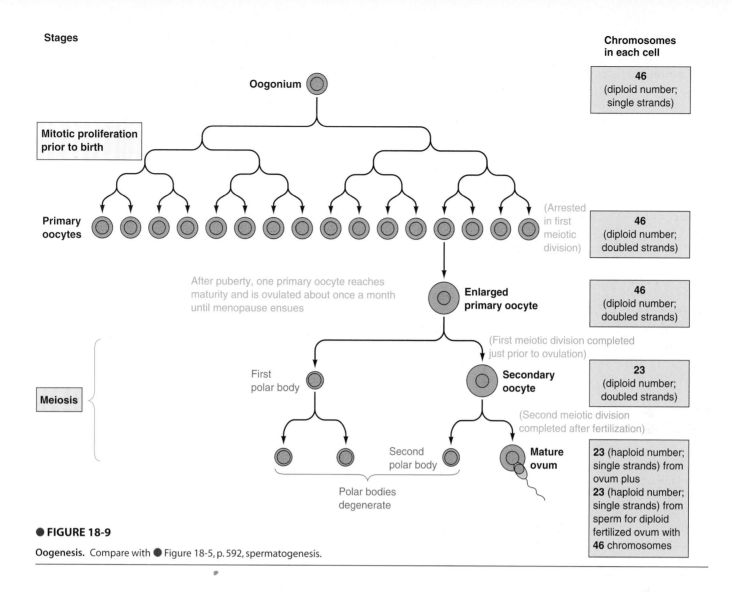

Stages

Oogonium

Mitotic proliferation prior to birth

Primary oocytes

(Arrested in first meiotic division)

After puberty, one primary oocyte reaches maturity and is ovulated about once a month until menopause ensues

Enlarged primary oocyte

(First meiotic division completed just prior to ovulation)

Meiosis

First polar body

Secondary oocyte

(Second meiotic division completed after fertilization)

Second polar body

Polar bodies degenerate

Mature ovum

Chromosomes in each cell

46 (diploid number; single strands)

46 (diploid number; doubled strands)

46 (diploid number; doubled strands)

23 (diploid number; doubled strands)

23 (haploid number; single strands) from ovum plus **23** (haploid number; single strands) from sperm for diploid fertilized ovum with **46** chromosomes

● **FIGURE 18-9**

Oogenesis. Compare with ● Figure 18-5, p. 592, spermatogenesis.

(to be described shortly). This cycle is normally interrupted only by pregnancy and is finally terminated by menopause. The average ovarian cycle lasts 28 days, but this varies among women and among cycles in any particular woman. The follicle operates in the first half of the cycle to produce a mature egg ready for ovulation at midcycle. The corpus luteum takes over during the last 14 days of the cycle to prepare the female reproductive tract for pregnancy if fertilization of the released egg occurs.

▌ The follicular phase is characterized by the development of maturing follicles.

At any given time throughout the cycle, a portion of the primary follicles is starting to develop. However, only those that do so during the follicular phase, when the hormonal environment is right to promote their maturation, continue beyond the early stages of development. The others, lacking hormonal support, undergo atresia. During follicular development, as the primary oocyte is synthesizing and storing materials for future use if fertilized, important changes are taking place

in the cells surrounding the reactivated oocyte in preparation for the egg's release from the ovary (● Figure 18-10a).

PROLIFERATION OF GRANULOSA CELLS AND FORMATION OF THE ZONA PELLUCIDA

First, the single layer of granulosa cells in a primary follicle proliferates to form several layers that surround the oocyte. These granulosa cells secrete a thick, gel-like "rind" that covers the oocyte and separates it from the surrounding granulosa cells. This intervening membrane is known as the **zona pellucida.**

PROLIFERATION OF THECAL CELLS; ESTROGEN SECRETION

At the same time the oocyte is enlarging and the granulosa cells are proliferating, specialized ovarian connective tissue cells in contact with the expanding granulosa cells proliferate and differentiate to form an outer layer of **thecal cells.** The thecal and granulosa cells, collectively known as **follicular cells,** function as a unit to secrete estrogen.

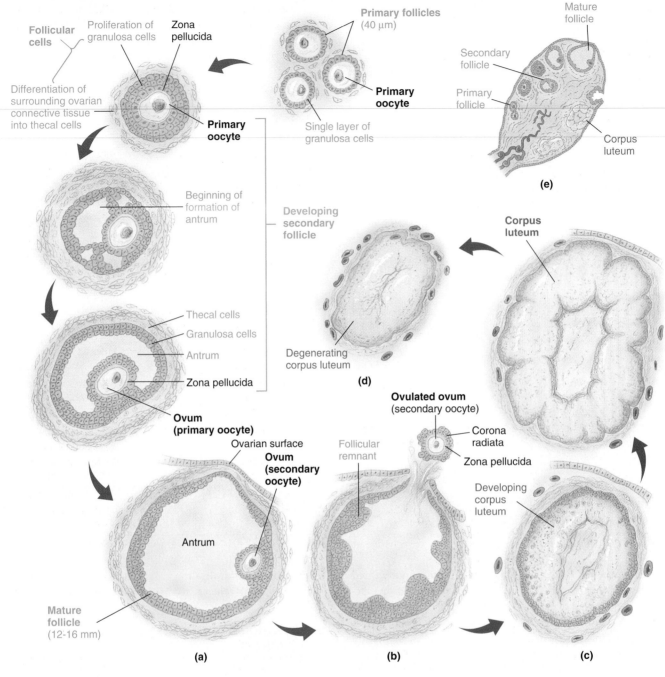

● FIGURE 18-10

Development of the follicle, ovulation, and formation of the corpus luteum. (a) Stages in follicular development from a primary follicle through a mature follicle. (b) Rupture of a mature follicle and release of an ovum (secondary oocyte) at ovulation. (c) Formation of a corpus luteum from the old follicular cells after ovulation. (d) Degeneration of the corpus luteum if the released ovum is not fertilized. (e) Ovary (actual size), showing development of a follicle, ovulation, and formation and degeneration of a corpus luteum.

FORMATION OF THE ANTRUM

The hormonal environment of the follicular phase promotes enlargement and development of the follicular cells' secretory capacity, converting the primary follicle into a **secondary**, or **antral, follicle** capable of estrogen secretion. During this stage of follicular development, a fluid-filled cavity or **antrum** forms in the middle of the granulosa cells (● Figures 18-10a and

18-11). As the follicular cells start producing estrogen, some of this hormone is secreted into the blood for distribution throughout the body. However, a portion of the estrogen collects in the hormone-rich antral fluid.

The oocyte has reached full size by the time the antrum begins to form. The shift to an antral follicle initiates a period of rapid follicular growth. During this time, the follicle

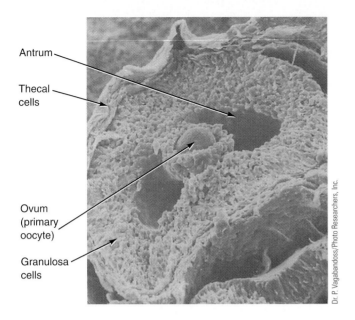

Antrum

Thecal cells

Ovum (primary oocyte)

Granulosa cells

Dr. P. Vagabandos/Photo Researchers, Inc.

● **FIGURE 18-11**

Scanning electron micrograph of a developing secondary follicle

increases in size from a diameter of less than 1 mm to 12 to 16 mm shortly before ovulation. Part of the follicular growth is due to continued proliferation of the granulosa and thecal cells, but most is due to a dramatic expansion of the antrum. As the follicle grows, estrogen is produced in increasing quantities.

FORMATION OF A MATURE FOLLICLE

One of the follicles usually grows more rapidly than the others, developing into a **mature** (**preovulatory, tertiary,** or **Graafian**) **follicle** within about 14 days after the onset of follicular development. The antrum occupies most of the space in a mature follicle. The oocyte, surrounded by the zona pellucida and a single layer of granulosa cells, is displaced asymmetrically at one side of the growing follicle, in a little mound that protrudes into the antrum.

OVULATION

The greatly expanded mature follicle bulges on the ovarian surface, creating a thin area that ruptures to release the oocyte at **ovulation**. Rupture of the follicle is facilitated by the release from the follicular cells of enzymes that digest the connective tissue in the follicular wall. The bulging wall is thus weakened so that it balloons out even farther, to the point that it can no longer contain the rapidly expanding follicular contents.

Just before ovulation, the oocyte completes its first meiotic division. The ovum (secondary oocyte), still surrounded by its tightly adhering zona pellucida and granulosa cells (now called the **corona radiata,** meaning "radiating crown") is swept out of the ruptured follicle into the abdominal cavity by the leaking antral fluid (● Figure 18-10b). The released ovum is quickly drawn into the oviduct, where fertilization may or may not take place.

The other developing follicles that failed to reach maturation and ovulate undergo degeneration, never to be reactivated. Occasionally, two (or perhaps more) follicles reach maturation and ovulate at about the same time. If both are fertilized, **fraternal twins** result. Because fraternal twins arise from separate ova fertilized by separate sperm, they share no more in common than any other two siblings except for the same birth date. **Identical twins,** in contrast, develop from a single fertilized ovum that completely divides into two separate, genetically identical embryos at a very early stage in development.

Rupture of the follicle at ovulation signals the end of the follicular phase and ushers in the luteal phase.

▮ The luteal phase is characterized by the presence of a corpus luteum.

The ruptured follicle left behind in the ovary after release of the ovum changes rapidly.

FORMATION OF THE CORPUS LUTEUM; ESTROGEN AND PROGESTERONE SECRETION

The granulosa and thecal cells remaining in the remnant follicle soon undergo a dramatic structural transformation to form the **corpus luteum,** in a process called **luteinization** (● Figure 18-10c). The follicular-turned-luteal cells enlarge and are converted into very active steroid hormone–producing tissue. Abundant storage of cholesterol, the steroid precursor molecule, in lipid droplets within the corpus luteum gives this tissue a yellowish appearance, hence its name (*corpus* means "body"; *luteum* means "yellow").

The corpus luteum secretes into the blood abundant quantities of progesterone along with smaller amounts of estrogen. Estrogen secretion in the follicular phase followed by progesterone secretion in the luteal phase is essential for preparing the uterus for implantation of a fertilized ovum. The corpus luteum becomes fully functional within four days after ovulation, but it continues to increase in size for another four or five days.

DEGENERATION OF THE CORPUS LUTEUM

If the released ovum is not fertilized and does not implant, the corpus luteum degenerates within 14 days after its formation (● Figure 18-10d). The luteal phase is now over, and one ovarian cycle is complete. A new wave of follicular development, which begins when degeneration of the old corpus luteum is completed, signals the onset of a new follicular phase.

CORPUS LUTEUM OF PREGNANCY

If fertilization and implantation do occur, the corpus luteum continues to grow and produce increasing quantities of progesterone and estrogen instead of degenerating. Now called the *corpus luteum of pregnancy,* this ovarian structure persists until pregnancy ends. It provides the hormones essential for maintaining pregnancy until the developing placenta can take over this crucial function. You will learn more about the role of these structures later.

▌The ovarian cycle is regulated by complex hormonal interactions.

The ovary has two related endocrine units: the estrogen-secreting follicle during the first half of the cycle and the corpus luteum, which secretes both progesterone and estrogen, during the last half of the cycle. These units are sequentially triggered by complex cyclic hormonal relationships among the hypothalamus, anterior pituitary, and these two ovarian endocrine units.

As in the male, gonadal function in the female is directly controlled by the anterior pituitary gonadotropic hormones, follicle-stimulating hormone (FSH) and luteinizing hormone (LH). These hormones, in turn, are regulated by hypothalamic gonadotropin-releasing hormone (GnRH) and feedback actions of gonadal hormones. Differing from the male, however, control of the female gonads is complicated by the cyclic nature of ovarian function. For example, the effects of FSH and LH on the ovaries depend on the stage of the ovarian cycle. Also in contrast to the male, FSH is not strictly responsible for gametogenesis, nor is LH solely responsible for gonadal hormone secretion. We will consider control of follicular function, ovulation, and the corpus luteum separately, using ● Figure 18-12 as a means of integrating the various concurrent and sequential activities that take place throughout the cycle. To facilitate correlation between this rather intimidating figure and the accompanying text description of this complex cycle, the circled numbers in the figure and its legend correspond to the circled numbers in the text description.

CONTROL OF FOLLICULAR FUNCTION

We begin with the follicular phase of the ovarian cycle 1 . The factors that initiate follicular development are poorly understood. The early stages of preantral follicular growth and oocyte maturation do not require gonadotropic stimulation. Hormonal support is required, however, for antrum formation, further follicular development 2 , and estrogen secretion 3 . Estrogen, FSH 4 , and LH 5 are all needed. Antrum formation is induced by FSH. Both FSH and estrogen stimulate proliferation of the granulosa cells. Both LH and FSH are required for synthesis and secretion of estrogen by the follicle. Part of the estrogen produced by the growing follicle is secreted into the blood and is responsible for the steadily increasing plasma estrogen levels during the follicular phase 8 . The remainder of the estrogen remains within the follicle, contributing to the antral fluid and stimulating further granulosa cell proliferation.

The secreted estrogen, in addition to acting on sex-specific tissues such as the uterus, inhibits the hypothalamus and anterior pituitary in negative-feedback fashion (● Figure 18-13). The low but rising levels of estrogen characterizing the follicular phase act directly on the hypothalamus to inhibit GnRH secretion, thus suppressing GnRH-prompted release of FSH and LH from the anterior pituitary. However, estrogen's primary effect is directly on the pituitary itself. Estrogen reduces the sensitivity to GnRH of the cells that produce gonadotropic hormones, especially the FSH-secreting cells.

This differential sensitivity of FSH- and LH-secreting cells induced by estrogen is at least in part responsible for the fact that the plasma FSH level, unlike the plasma LH concentration, declines during the follicular phase as the estrogen level rises 6 . Another contributing factor to the fall in FSH during the follicular phase is secretion of *inhibin* by the follicular cells. Inhibin preferentially inhibits FSH secretion by acting at the anterior pituitary, just as it does in the male. The decline in FSH secretion brings about atresia of all but the single most mature of the developing follicles.

In contrast to FSH, LH secretion continues to rise slowly during the follicular phase 7 despite inhibition of GnRH (and thus, indirectly, LH) secretion. This seeming paradox is due to the fact that estrogen alone cannot completely suppress **tonic** (low-level, ongoing) **LH secretion**; to completely inhibit tonic LH secretion, both estrogen and progesterone are required. Because progesterone does not appear until the luteal phase of the cycle, the basal level of circulating LH slowly increases during the follicular phase under incomplete inhibition by estrogen alone.

CONTROL OF OVULATION

Ovulation and subsequent luteinization of the ruptured follicle are triggered by an abrupt, massive increase in LH secretion 9 . This **LH surge** brings about four major changes in the follicle:

1. It halts estrogen synthesis by the follicular cells 11 .
2. It reinitiates meiosis in the oocyte of the developing follicle.
3. It triggers production of locally acting prostaglandins, which induce ovulation by promoting vascular changes that cause rapid swelling of the follicle while inducing enzymatic digestion of the follicular wall. Together these actions lead to rupture of the weakened wall that covers the bulging follicle 10 .
4. It causes differentiation of follicular cells into luteal cells. Because the LH surge triggers both ovulation and luteinization, formation of the corpus luteum automatically follows ovulation 12 . Thus the midcycle burst in LH secretion is a dramatic point in the cycle; it terminates the follicular phase and initiates the luteal phase 15 .

The two different modes of LH secretion—the tonic secretion of LH 7 responsible for promoting ovarian hormone secretion and the LH surge 9 that causes ovulation—not only occur at different times and produce different effects on the ovaries but also are controlled by different mechanisms. Tonic LH secretion is partially suppressed 7 by the inhibitory action of the low, rising levels of estrogen 3 during the follicular phase and is completely suppressed 17 by the increasing levels of progesterone during the luteal phase 13 . Because tonic LH secretion stimulates both estrogen and progesterone secretion, this is a typical negative-feedback control system.

In contrast, the LH surge is triggered by a *positive-feedback effect*. Whereas the low, rising levels of estrogen early in the follicular phase *inhibit* LH secretion, the high level of estrogen that occurs during peak estrogen secretion late in the follicular phase 8 *stimulates* LH secretion and initiates the LH

● FIGURE 18-12

Correlation between hormonal levels and cyclic ovarian and uterine changes. During the follicular phase (the first half of the ovarian cycle ①, the ovarian follicle ② secretes estrogen ③ under the influence of FSH ④, LH ⑤, and estrogen ③ itself. The low but rising levels of estrogen (1) inhibit FSH secretion, which declines during the last part of the follicular phase ⑥, and (2) incompletely suppress tonic LH secretion, which continues to rise throughout the follicular phase ⑦. When the follicular output of estrogen reaches its peak ⑧, the high level of estrogen triggers a surge in LH secretion at midcycle ⑨. This LH surge brings about ovulation of the mature follicle 10. Estrogen secretion plummets 11 when the follicle meets its demise at ovulation.

The old follicular cells are transformed into the corpus luteum 12, which secretes progesterone 13 as well as estrogen 14 during the luteal phase the last half of the ovarian cycle 15. Progesterone strongly inhibits both FSH 16 and LH 17, which continue to decrease throughout the luteal phase. The corpus luteum degenerates 18 in about two weeks if the released ovum has not been fertilized and implanted in the uterus. Progesterone 19 and estrogen 20 levels sharply decrease when the corpus luteum degenerates, removing the inhibitory influences on FSH and LH. As these anterior pituitary hormone levels start to rise again 21, 22 on the withdrawal of inhibition, they begin to stimulate the development of a new batch of follicles as a new follicular phase is ushered in ①, ②.

Concurrent uterine phases reflect the influences of the ovarian hormones on the uterus. Early in the follicular phase, the highly vascularized, nutrient-rich endometrial lining is sloughed off (the uterine menstrual phase) 23. This sloughing results from the withdrawal of estrogen and progesterone 19, 20 when the old corpus luteum degenerated at the end of the preceding luteal phase 18. Late in the follicular phase, the rising levels of estrogen ③ cause the endometrium to thicken (the uterine proliferative phase) 24. After ovulation 10, progesterone from the corpus luteum 13 brings about vascular and secretory changes in the estrogen-primed endometrium to produce a suitable environment for implantation (the uterine secretory, or progestational, phase) 25. When the corpus luteum degenerates 18, a new ovarian follicular phase ①, ② and uterine menstrual phase 23 begin.

PhysioEdge For an animation of this figure, click the Hormonal Communication tab in the Neural and Hormonal Communication tutorial on the CD-ROM. For an interaction related to this figure, see Media Exercise 18.4: Female Reproductive Physiology.

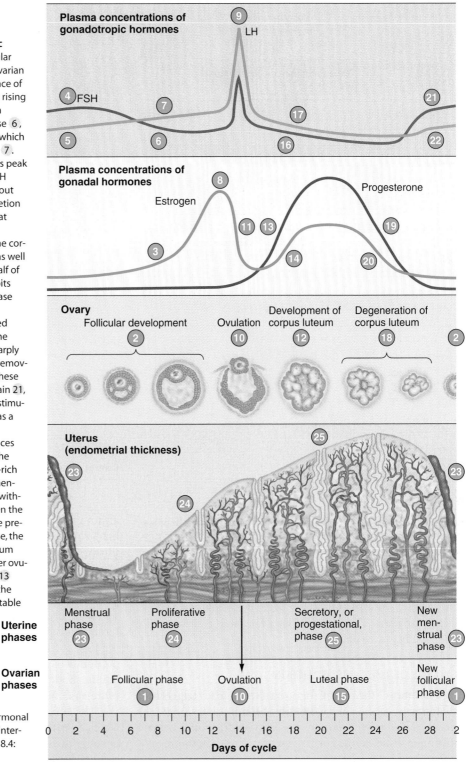

surge (● Figure 18-14). Thus LH enhances estrogen production by the follicle, and the resultant peak estrogen concentration stimulates LH secretion. The high plasma concentration of estrogen acts directly on the hypothalamus to increase GnRH, thereby increasing both LH and FSH secretion. It also acts directly on the anterior pituitary to specifically increase the sensitivity of LH-secreting cells to GnRH. The latter effect accounts in large part for the much greater surge in LH se-

cretion compared to FSH secretion at midcycle ⑨. Also, continued inhibin secretion by the follicular cells preferentially inhibits the FSH-secreting cells, keeping the FSH levels from rising as high as the LH levels. There is no known role for the modest midcycle surge in FSH that accompanies the pronounced and pivotal LH surge. Because only a mature, preovulatory follicle, not follicles in earlier stages of development, can secrete high-enough levels of estrogen to trigger the LH

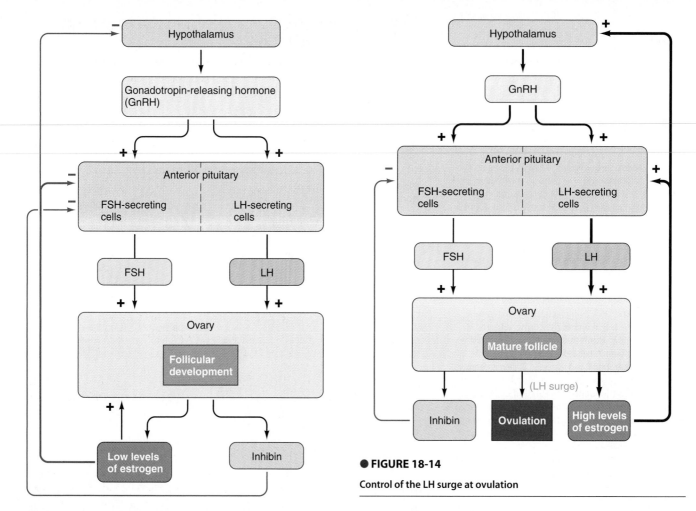

● FIGURE 18-13

Feedback control of FSH and tonic LH secretion during the follicular phase

● FIGURE 18-14

Control of the LH surge at ovulation

surge, ovulation is not induced until a follicle has reached the proper size and degree of maturation. In a way, then, the follicle lets the hypothalamus know when it is ready to be stimulated to ovulate. The LH surge lasts for about a day at midcycle, just before ovulation.

CONTROL OF THE CORPUS LUTEUM

LH "maintains" the corpus luteum; that is, after triggering development of the corpus luteum, LH stimulates ongoing steroid hormone secretion by this ovarian structure. Under the influence of LH, the corpus luteum secretes both progesterone 13 and estrogen 14, with progesterone being its most abundant hormonal product. The plasma progesterone level increases for the first time during the luteal phase. No progesterone is secreted during the follicular phase. Therefore, the follicular phase is dominated by estrogen and the luteal phase by progesterone.

A transitory drop in the level of circulating estrogen occurs at midcycle 11 as the estrogen-secreting follicle meets its demise at ovulation. The estrogen level climbs again during the luteal phase because of the corpus luteum's activity,

although it does not reach the same peak as during the follicular phase. What keeps the moderately high estrogen level during the luteal phase from triggering another LH surge? Progesterone. Even though a high level of estrogen stimulates LH secretion, progesterone, which dominates the luteal phase, powerfully inhibits LH secretion as well as FSH secretion 17, 16 (● Figure 18-15). Inhibition of FSH and LH by progesterone prevents new follicular maturation and ovulation during the luteal phase. Under progesterone's influence, the reproductive system is gearing up to support the just-released ovum, should it be fertilized, instead of preparing other ova for release. No inhibin is secreted by the luteal cells.

The corpus luteum functions for two weeks, then degenerates if fertilization does not occur 18. The mechanisms that govern degeneration of the corpus luteum are not fully understood. The declining level of circulating LH 17, driven down by inhibitory actions of progesterone, undoubtedly contributes to the corpus luteum's downfall. Prostaglandins and estrogen released by the luteal cells themselves may play a role. Demise of the corpus luteum terminates the luteal phase and sets the stage for a new follicular phase. As the corpus luteum degenerates, plasma progesterone 19 and estrogen 20 levels fall rapidly, because these hormones are no longer being produced. Withdrawal of the inhibitory effects of these hormones on the hypothalamus allows FSH 21 and tonic LH 22

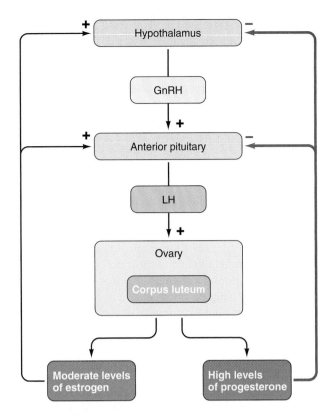

● FIGURE 18-15

Feedback control during the luteal phase

secretion to modestly increase once again. Under the influence of these gonadotropic hormones, another batch of primary follicles ② is induced to mature as a new follicular phase begins ① .

▌Cyclic uterine changes are caused by hormonal changes during the ovarian cycle.

The fluctuations in circulating levels of estrogen and progesterone during the ovarian cycle induce profound changes in the uterus, giving rise to the **menstrual**, or **uterine**, **cycle**. Because it reflects hormonal changes during the ovarian cycle, the menstrual cycle averages 28 days, as does the ovarian cycle, although even normal adults vary considerably from this mean. This variability primarily reflects differing lengths of the follicular phase; the duration of the luteal phase is fairly constant. The outward manifestation of the cyclical changes in the uterus is the menstrual bleeding once during each menstrual cycle (that is, once a month). Less obvious changes take place throughout the cycle, however, as the uterus is prepared for implantation should a released ovum be fertilized, then is stripped clean of its prepared lining (menstruation) if implantation does not occur, only to repair itself and start preparing for the ovum that will be released during the next cycle.

We will briefly examine the influences of estrogen and progesterone on the uterus and then consider the effects of cyclic fluctuations of these hormones on uterine structure and function.

INFLUENCES OF ESTROGEN AND PROGESTERONE ON THE UTERUS

The uterus consists of two main layers: the **myometrium**, the outer smooth-muscle layer; and the **endometrium**, the inner lining that contains numerous blood vessels and glands. Estrogen stimulates growth of both the myometrium and the endometrium. It also induces the synthesis of progesterone receptors in the endometrium. Thus progesterone can exert an effect on the endometrium only after it has been "primed" by estrogen. Progesterone acts on the estrogen-primed endometrium to convert it into a hospitable and nutritious lining suitable for implantation of a fertilized ovum. Under the influence of progesterone, the endometrial connective tissue becomes loose and edematous as a result of an accumulation of electrolytes and water, which facilitates implantation of the fertilized ovum. Progesterone further prepares the endometrium to sustain an early-developing embryo by inducing the endometrial glands to secrete and store large quantities of glycogen and by causing tremendous growth of the endometrial blood vessels. Progesterone also reduces the contractility of the uterus to provide a quiet environment for implantation and embryonic growth.

The menstrual cycle consists of three phases: the *menstrual phase*, the *proliferative phase*, and the *secretory*, or *progestational*, *phase*.

MENSTRUAL PHASE

The **menstrual phase** is the most overt phase, characterized by discharge of blood and endometrial debris from the vagina. By convention, the first day of menstruation is considered the start of a new cycle. It coincides with termination of the ovarian luteal phase and onset of the follicular phase ㉓, ● Figure 18-12. As the corpus luteum degenerates because fertilization and implantation of the ovum released during the preceding cycle did not take place ⑱, circulating levels of estrogen and progesterone drop precipitously ⑲, ㉒. Because the net effect of estrogen and progesterone is to prepare the endometrium for implantation of a fertilized ovum, withdrawal of these steroids deprives the highly vascular, nutrient-rich uterine lining of its hormonal support.

The fall in ovarian hormone levels also stimulates release of a uterine prostaglandin that causes vasoconstriction of the endometrial vessels, disrupting the blood supply to the endometrium. The subsequent reduction in O_2 delivery causes death of the endometrium, including its blood vessels. The resulting bleeding through the disintegrating vessels flushes the dying endometrial tissue into the uterine lumen. The entire uterine lining sloughs during each menstrual period except for a deep, thin layer of epithelial cells and glands, from which the endometrium will regenerate. The same local uterine prostaglandin also stimulates mild rhythmic contractions of the uterine myometrium. These contractions help expel the blood and endometrial debris from the uterine cavity out through the vagina as **menstrual flow**. Excessive uterine contractions caused by prostaglandin overproduction produce the menstrual cramps (**dysmenorrhea**) some women experience.

The average blood loss during a single menstrual period is 50 to 150 ml. In addition to the blood and endometrial debris, large numbers of leukocytes are found in the menstrual flow. These white blood cells play an important defense role in helping the raw endometrium resist infection.

Menstruation typically lasts for about five to seven days after degeneration of the corpus luteum, coinciding in time with the early portion of the ovarian follicular phase 23. Withdrawal of estrogen and progesterone 19, 20 on degeneration of the corpus luteum leads simultaneously to sloughing of the endometrium (menstruation) 23 and development of new follicles in the ovary 1 , 2 under the influence of rising gonadotropic hormone levels 21, 22. The drop in gonadal hormone secretion removes inhibitory influences from the hypothalamus and anterior pituitary, so FSH and LH secretion increases and a new follicular phase begins. After five to seven days under the influence of FSH and LH, the newly growing follicles are secreting enough estrogen 3 to induce repair and growth of the endometrium.

PROLIFERATIVE PHASE

Thus menstrual flow ceases, and the **proliferative phase** of the uterine cycle begins concurrent with the last portion of the ovarian follicular phase as the endometrium starts to repair itself and proliferate 24 under the influence of estrogen from the newly growing follicles. When the menstrual flow ceases, a thin endometrial layer less than 1 mm thick remains. Estrogen stimulates proliferation of epithelial cells, glands, and blood vessels in the endometrium, increasing this lining to a thickness of 3 to 5 mm. The estrogen-dominant proliferative phase lasts from the end of menstruation to ovulation. Peak estrogen levels 8 trigger the LH surge 9 responsible for ovulation 10.

SECRETORY OR PROGESTATIONAL PHASE

After ovulation, when a new corpus luteum is formed 12, the uterus enters the **secretory**, or **progestational**, **phase**, which coincides in time with the ovarian luteal phase 25. The corpus luteum secretes large amounts of progesterone 13 and estrogen 14. Progesterone converts the thickened, estrogen-primed endometrium to a richly vascularized, glycogen-filled tissue. This period is called either the *secretory phase,* because the endometrial glands are actively secreting glycogen, or the *progestational* ("before pregnancy") *phase,* referring to the development of a lush endometrial lining capable of supporting an early embryo. If fertilization and implantation do not occur, the corpus luteum degenerates and a new follicular phase and menstrual phase begin once again.

▌ Pubertal changes in females are similar to those in males.

Regular menstrual cycles are absent in both young and aging females, but for different reasons. The female reproductive system does not become active until puberty. Unlike the fetal testes, the fetal ovaries need not be functional, because in the absence of fetal testosterone secretion in a female the reproductive system is automatically feminized, without requiring the presence of female sex hormones. The female reproductive system remains quiescent from birth until puberty, which occurs at about 12 years of age when hypothalamic GnRH activity increases for the first time. As in the male, the mechanisms that govern the onset of puberty are not clearly understood but are believed to involve the pineal gland and melatonin secretion.

GnRH begins stimulating release of anterior pituitary gonadotropic hormones, which in turn stimulate ovarian activity. The resulting secretion of estrogen by the activated ovaries induces growth and maturation of the female reproductive tract as well as development of the female secondary sexual characteristics. Estrogen's prominent action in the latter regard is to promote fat deposition in strategic locations, such as the breasts, buttocks, and thighs, giving rise to the typical curvaceous female figure. Enlargement of the breasts at puberty is due primarily to fat deposition in the breast tissue, not to functional development of the mammary glands. The pubertal rise in estrogen also closes the epiphyseal plates, halting further growth in height, similar to the effect of testosterone in males. Three other pubertal changes in females—growth of axillary and pubic hair, the pubertal growth spurt, and development of libido—are attributable to a spurt in adrenal androgen secretion at puberty, not to estrogen.

▌ Menopause is unique to females.

The cessation of a woman's menstrual cycles at menopause sometime between the ages of 45 and 55 has traditionally been attributed to the limited supply of ovarian follicles present at birth. According to this proposal, once this reservoir is depleted, ovarian cycles, and hence menstrual cycles, cease. Thus the termination of reproductive potential in a middle-aged woman is "preprogrammed" at her own birth. Recent evidence suggests, however, that a midlife hypothalamic change instead of aging ovaries may trigger the onset of menopause. Evolutionarily, menopause may have developed as a mechanism that prevented pregnancy in women beyond the time that they could likely rear a child before their own death.

Menopause is preceded by a period of progressive ovarian failure characterized by increasingly irregular cycles and dwindling estrogen levels. In addition to the ending of ovarian and menstrual cycles, the loss of ovarian estrogen following menopause brings about many physical and emotional changes. These changes include vaginal dryness, which can cause discomfort during sex, and gradual atrophy of the genital organs. However, postmenopausal women still have a sex drive, because of their adrenal androgens.

Males do not experience complete gonadal failure as females do, for two reasons. First, a male's germ cell supply is unlimited because mitotic activity of the spermatogonia continues. Second, gonadal hormone secretion in males is not inextricably dependent on gametogenesis, as in females. If female sex hormones were produced by separate tissues unrelated to those governing gametogenesis, as are male sex hormones, estrogen and progesterone secretion would not automatically stop when oogenesis stopped.

You have now learned about the events that take place if fertilization does not occur. Because the primary function of the reproductive system is, of course, reproduction, we will next turn our attention to the sequence of events that take place when fertilization does occur.

▌ The oviduct is the site of fertilization.

Fertilization, the union of male and female gametes, normally occurs in the upper third of the oviduct. Thus both the ovum and the sperm must be transported from their gonadal site of production to the upper oviduct (● Figure 18-16).

OVUM TRANSPORT TO THE OVIDUCT

When the ovum is released at ovulation, it is quickly picked up by the oviduct. The dilated end of the oviduct cups around the ovary and contains **fimbriae,** fingerlike projections that contract in a sweeping motion to guide the released ovum into the oviduct (● Figures 18-2b and 18-16). Furthermore, the fimbriae are lined by cilia, fine, hairlike projections that beat in waves toward the interior of the oviduct, further assuring the ovum's passage into the oviduct (see p. 36.)

Conception can take place during a very limited time span in each cycle (the **fertile period**). If not fertilized, the ovum begins to disintegrate within 12 to 24 hours and is subsequently phagocytized by cells that line the reproductive tract. Fertilization must therefore occur within 24 hours after ovulation, when the ovum is still viable. Sperm typically survive about 48 hours but can survive up to five days in the female reproductive tract, so sperm deposited within five days before ovulation to 24 hours after ovulation may be able to fertilize the released ovum, although these times vary considerably.

SPERM TRANSPORT TO THE OVIDUCT

Once sperm are deposited in the vagina at ejaculation, they must travel through the cervical canal, through the uterus, and then up to the egg in the upper third of the oviduct (● Figure 18-16). The first sperm arrive in the oviduct within half an hour after ejaculation. Even though sperm are mobile by means of whiplike contractions of their tails, 30 minutes is much too soon for a sperm's own mobility to transport it to the site of fertilization. To make this formidable journey, sperm need the help of the female reproductive tract.

The first hurdle is passage through the cervical canal. Throughout most of the cycle, the cervical mucus is too thick to permit sperm penetration. The cervical mucus becomes thin and watery enough to permit sperm to penetrate only when estrogen levels are high, as in the presence of a mature follicle about to ovulate. Sperm migrate up the cervical canal under their own power. The canal remains penetrable for only two or three days during each cycle, around the time of ovulation.

Once the sperm have entered the uterus, contractions of the myometrium churn them around in "washing-machine" fashion. This action quickly disperses sperm throughout the uterine cavity. When sperm reach the oviduct, they are propelled to the fertilization site in the upper end of the oviduct by upward contractions of the oviduct smooth muscle. These myometrial and oviduct contractions that facilitate sperm transport are induced by the high estrogen level just before ovulation, aided by seminal prostaglandins. Furthermore, new research indicates ova are not passive partners in conception. Mature eggs release **allurin**, a recently identified chemical that attracts sperm and causes them to propel themselves toward the waiting female gamete.

Even around ovulation time when sperm can penetrate the cervical canal, of the several hundred million sperm deposited in a single ejaculate only a few thousand make it to the oviduct (● Figure 18-16). That only a very small percentage of the deposited sperm ever reach their destination is one reason why sperm concentration must be so high (20 million/ml of semen) for a man to be fertile. The other reason is that the acrosomal enzymes of many sperm are needed to break down the barriers surrounding the ovum (● Figure 18-17).

FERTILIZATION

The tail of the sperm is used to maneuver for final penetration of the ovum. To fertilize an ovum, a sperm must first pass through the corona radiata and zona pellucida surrounding it. The acrosomal enzymes, which are exposed as the acrosomal membrane dis-

● **FIGURE 18-16**

Ovum and sperm transport to the site of fertilization

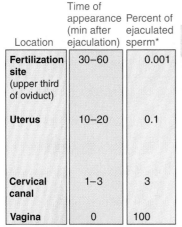

Location	Time of appearance (min after ejaculation)	Percent of ejaculated sperm*
Fertilization site (upper third of oviduct)	30–60	0.001
Uterus	10–20	0.1
Cervical canal	1–3	3
Vagina	0	100

*Based on data from animals. Sperm and ovum enlarged.

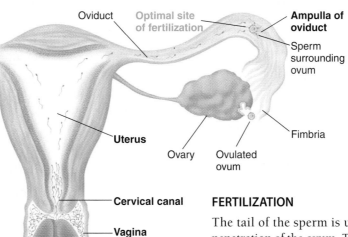

● **FIGURE 18-17** ▶

Scanning electron micrograph of sperm amassed at the surface of an ovum

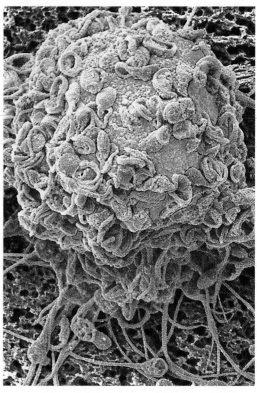

David Scharf

rupts on contact with the corona radiata, enable the sperm to tunnel a path through these protective barriers (● Figure 18-18). The first sperm to reach the ovum itself fuses with the plasma membrane of the ovum (actually a secondary oocyte), triggering a chemical change in the ovum's surrounding membrane that makes this outer layer impenetrable to the entry of any more sperm.

The head of the fused sperm is gradually pulled into the ovum's cytoplasm by a growing cone that engulfs it. The

● **FIGURE 18-18**

Process of fertilization. (a) Schematic representation of sperm tunneling the barriers surrounding the ovum. (b) Scanning electron micrograph of a spermatozoon in which the acrosomal membrane has been disrupted and the acrosomal enzymes *(in red)* are exposed.

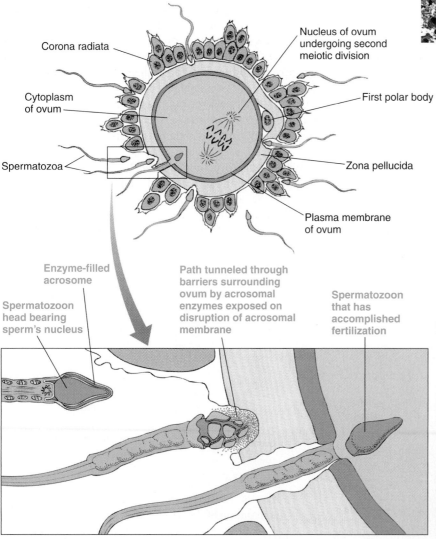

(a)

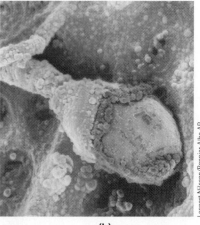

Lennart Nilsson/Bonnier Alba AB

(b)

sperm's tail is frequently lost in this process, but the head carries the crucial genetic information. Recent evidence suggests the sperm releases nitric oxide when it has completely penetrated into the cytoplasm. This nitric oxide promotes the release of stored Ca^{2+} within the egg. This intracellular Ca^{2+} release triggers the final meiotic division of the secondary oocyte. Within an hour, the sperm and egg nuclei fuse, thanks to a molecular complex provided by the sperm that helps the male and female chromosome sets unite. In addition to contributing its half of the chromosomes to the fertilized ovum, now called a **zygote**, the victorious sperm also activates ovum enzymes essential for the early embryonic developmental program.

❚ The blastocyst implants in the endometrium through the action of its trophoblastic enzymes.

During the first three to four days following fertilization, the zygote remains within the oviduct, then descends and floats freely within the uterine cavity for another three to four days before implanting.

THE BEGINNING STEPS

The zygote is not idle during this time, however. It rapidly undergoes a number of mitotic cell divisions to form a solid ball of cells called the *morula* (● Figure 18-19).

During the first six to seven days, while the developing embryo is in transit in the oviduct and floating in the uterine lumen, the newly developing corpus luteum that formed after ovulation is producing progesterone in increasing quantities.

Prior to implantation, nourishment is provided for the developing embryo first by nutrients stored in the cytoplasm of the ovum and then by nutrient-rich endometrial secretions stimulated by the rising levels of progesterone.

Meanwhile, the uterine lining is simultaneously being prepared for implantation under the influence of luteal-phase progesterone. During this time, the uterus is in its secretory or progestational phase, storing up glycogen and becoming richly vascularized.

IMPLANTATION OF THE BLASTOCYST IN THE PREPARED ENDOMETRIUM

By the time the endometrium is suitable for implantation (about a week after ovulation), the morula has descended to the uterus and continued to proliferate and differentiate into a *blastocyst* capable of implantation. The week's delay after fertilization and before implantation allows time for both the endometrium and the developing embryo to prepare for implantation.

A **blastocyst** is a single-layer hollow ball of about 50 cells encircling a fluid-filled cavity, with a dense mass of cells grouped together at one side (● Figure 18-19). This dense mass,

---

● **FIGURE 18-19**

Early stages of development from fertilization to implantation. Note that the fertilized ovum progressively divides and differentiates into a blastocyst as it moves from the site of fertilization in the upper oviduct to the site of implantation in the uterus.

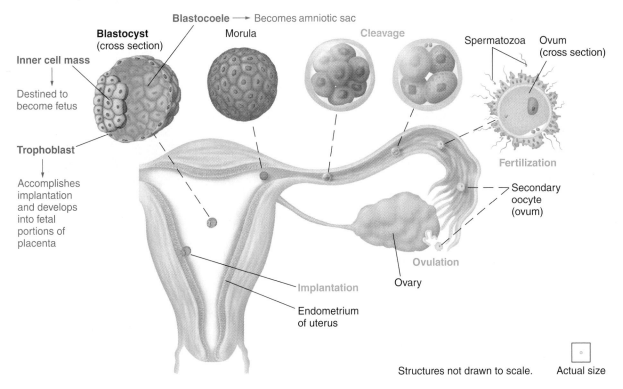

Structures not drawn to scale. Actual size

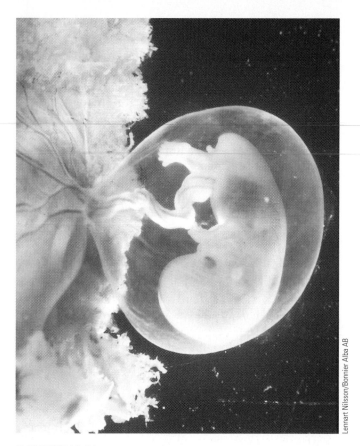

● **FIGURE 18-20**

A human fetus surrounded by the amniotic sac. The fetus is near the end of the first trimester of development.

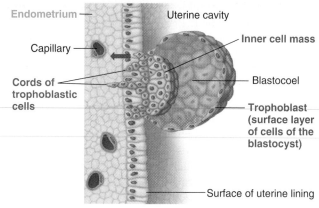

1 When the free-floating blastocyst adheres to the endometrial lining, cords of trophoblastic cells begin to penetrate the endometrium.

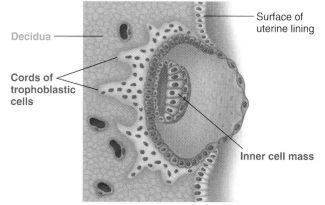

2 Advancing cords of trophoblastic cells tunnel deeper into the endometrium, carving out a hole for the blastocyst. The boundaries between the cells in the advancing trophoblastic tissue disintegrate.

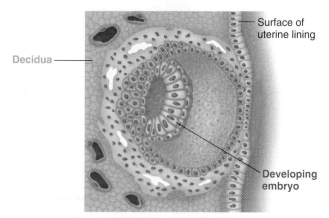

3 When implantation is finished, the blastocyst is completely buried in the endometrium.

● **FIGURE 18-21**

Implantation of the blastocyst

known as the **inner cell mass**, becomes the embryo/fetus itself. The rest of the blastocyst is never incorporated into the fetus, instead serving a supportive role during intrauterine life. The thin outermost layer, the **trophoblast**, accomplishes implantation, after which it develops into the fetal portion of the placenta. The fluid-filled cavity, the **blastocoele**, becomes the fluid within the **amniotic sac**, which surrounds and cushions the fetus throughout gestation (● Figure 18-20).

When the blastocyst is ready to implant, its surface becomes sticky. By this time the endometrium is ready to accept the early embryo. The blastocyst adheres to the uterine lining on the side of its inner–cell mass (● Figure 18-21, step **1**). **Implantation** begins when, on contact with the endometrium, the trophoblastic cells overlying the inner cell mass release protein-digesting enzymes. These enzymes digest pathways between the endometrial cells, permitting fingerlike cords of trophoblastic cells to penetrate into the depths of the endometrium, where they continue to digest uterine cells (step **2**). Through its cannibalistic actions, the trophoblast performs the dual functions of (1) accomplishing implantation as it carves out a hole in the endometrium for the blastocyst and (2) making metabolic fuel and raw materials available for the developing embryo as the advancing trophoblastic projections break down the nutrient-rich endometrial tissue.

Stimulated by the invading trophoblast, the endometrial tissue at the contact site undergoes dramatic changes that enhance its ability to support the implanting embryo, such as increased vascularization and enhanced nutrient storage. The endometrial tissue so modified at the implantation site is called the **decidua**. It is into this super-rich decidual tissue that the blastocyst becomes embedded. After the blastocyst burrows into the decidua by means of trophoblastic activity, a layer of endometrial cells covers over the surface of the hole, completely burying the blastocyst within the uterine lining (● Figure 18-21, step ③). The trophoblastic layer continues to digest the surrounding decidual cells, providing energy for the embryo until the placenta develops.

CONTRACEPTION

Clinical Note Couples wishing to engage in sexual intercourse but avoid pregnancy have available a number of methods of **contraception** ("against conception"). These methods act by blocking one of three major steps in the reproductive process: sperm transport to the ovum, ovulation, or implantation. (See the boxed feature, ▶ Beyond the Basics, on pp. 616–617 for further details on the ways and means of contraception.)

Next let's examine the placenta in further detail.

▌ The placenta is the organ of exchange between maternal and fetal blood.

The glycogen stores in the endometrium are only sufficient to nourish the embryo during its first few weeks. To sustain the growing embryo/fetus for the duration of its intrauterine life, the **placenta**, a specialized organ of exchange between the maternal and fetal blood, rapidly develops (● Figure 18-22, p. 618). The placenta is derived from both trophoblastic and decidual tissue.

FORMATION OF THE PLACENTA

By day 12, the embryo is completely embedded in the decidua. By this time the trophoblastic layer is two cell layers thick and is called the **chorion**. As the chorion continues to release enzymes and expand, it forms an extensive network of cavities within the decidua. As the expanding chorion erodes decidual capillary walls, maternal blood leaks from the capillaries and fills these cavities. Fingerlike projections of chorionic tissue extend into the pools of maternal blood. Soon the developing embryo sends out capillaries into these chorionic projections to form **placental villi.**

Each placental villus contains embryonic (later fetal) capillaries surrounded by a thin layer of chorionic tissue, which separates the embryonic/fetal blood from the maternal blood in the intervillus spaces. Maternal and fetal blood do not actually mingle, but the barrier between them is extremely thin. To visualize this relationship, think of your hands (the fetal capillary blood vessels) in rubber gloves (the chorionic tissue) immersed in water (the pool of maternal blood). Only the rubber gloves separate your hands from the water. In the same way, only the thin chorionic tissue (plus the capillary wall of the fetal vessels) separates the fetal and maternal blood. All exchanges between these two bloodstreams take place across this extremely thin barrier. This entire system of interlocking maternal (decidual) and fetal (chorionic) structures makes up the placenta.

Even though not fully developed, the placenta is well established and operational by five weeks after implantation. By this time, the heart of the developing embryo is pumping blood into the placental villi as well as to the embryonic tissues. Throughout gestation, fetal blood continuously traverses between the placental villi and the circulatory system of the fetus by means of the **umbilical artery** and **umbilical vein**, which are wrapped within the **umbilical cord**, a lifeline between the fetus and the placenta (● Figure 18-22). The maternal blood within the placenta is continuously replaced as fresh blood enters through the uterine arterioles, percolates through the intervillus spaces, where it exchanges substances with fetal blood in the surrounding villi, then exits through the uterine vein.

FUNCTIONS OF THE PLACENTA

During intrauterine life, the placenta performs the functions of the digestive system, the respiratory system, and the kidneys for the "parasitic" fetus. The fetus has these organ systems, but within the uterine environment they cannot (and do not need to) function. Nutrients and O_2 diffuse from the maternal blood across the thin placental barrier into the fetal blood, whereas CO_2 and other metabolic wastes simultaneously diffuse from the fetal blood into the maternal blood. The nutrients and O_2 brought to the fetus in the maternal blood are acquired by the mother's digestive and respiratory systems, and the CO_2 and wastes transferred into the maternal blood are eliminated by the mother's lungs and kidneys, respectively. Thus the mother's digestive tract, respiratory system, and kidneys serve the fetus's needs as well as her own.

Clinical Note Unfortunately, many drugs, environmental pollutants, other chemical agents, and micro-organisms in the mother's bloodstream also can cross the placental barrier, and some of them may harm the developing fetus. For example, newborns who have become "addicted" during gestation by their mother's abuse of a drug such as heroin suffer withdrawal symptoms after birth. Even more common chemical agents such as aspirin, alcohol, and agents in cigarette smoke can reach the fetus and have adverse effects. Likewise, fetuses can acquire AIDS before birth if their mothers are infected with the virus. A pregnant women should therefore be very cautious about potentially harmful exposure from any source.

The placenta assumes yet another important responsibility—it becomes a temporary endocrine organ during pregnancy, a topic to which we now turn.

The Ways and Means of Contraception

The term **contraception** refers to the process of avoiding pregnancy while engaging in sexual intercourse. A number of methods of contraception are available that range in ease of use and effectiveness (see the accompanying table). These methods can be grouped into three categories based on the means by which they prevent pregnancy: blockage of sperm transport to the ovum, prevention of ovulation, or blockage of implantation. After examining the most common ways in which contraception can be accomplished by each of these means, we will take a glimpse at future contraceptive possibilities on the horizon before concluding with a discussion of termination of unwanted pregnancies.

Blockage of Sperm Transport to the Ovum

- *Natural contraception* or the *rhythm method* of birth control relies on abstinence from intercourse during the woman's fertile period. The woman can predict when ovulation is to occur based on keeping careful records of her menstrual cycles. Because of variability in cycles, this technique is only partially effective. The time of ovulation can be determined more precisely by recording body temperature each morning before getting up. Body temperature rises slightly about a day after ovulation has taken place. The temperature rhythm method is not useful in determining when it is safe to engage in intercourse before ovulation, but it can be helpful in determining when it is safe to resume sex after ovulation.

- *Coitus interruptus* involves withdrawal of the penis from the vagina before ejaculation occurs. This method is only moderately effective, however, because timing is difficult, and some sperm may pass out of the urethra prior to ejaculation.

- *Chemical contraceptives,* such as spermicidal ("sperm-killing") jellies, foams, creams, and suppositories, when inserted into the vagina are toxic to sperm for about an hour after application.

- *Barrier methods* mechanically prevent sperm transport to the oviduct. For males, the *condom* is a thin, strong rubber or latex sheath placed over the erect penis before ejaculation to prevent sperm from entering the vagina. For females, the *diaphragm,* which must be fitted by a trained professional, is a flexible rubber dome that is inserted through the vagina and positioned over the cervix to block sperm entry into the cervical canal. Barrier methods are often used in conjunction with spermicidal agents for increased effectiveness.

- *Sterilization,* which involves surgical disruption of either the ductus deferens *(vasectomy)* in men or the oviduct *(tubal ligation)* in women, is considered a permanent method of preventing sperm and ovum from uniting.

Prevention of Ovulation

- *Oral contraceptives,* or *birth control pills,* available only by prescription, prevent ovulation primarily by suppressing gonadotropin secretion. These pills, which contain synthetic estrogen-like and progesterone-like steroids, are taken for three weeks, either in combination or in sequence, and then are withdrawn for one week. These steroids, like the natural steroids produced during the ovarian cycle, inhibit GnRH and thus FSH and LH secretion. As a result, follicle maturation and ovulation do not take place, so conception is impossible. The endometrium responds to the exogenous steroids by thickening and developing secretory capacity, just as it would to the natural hormones. When these synthetic steroids are withdrawn after three weeks, the endometrial lining sloughs and menstruation occurs, as it normally would on degeneration of the corpus luteum.

- Several other contraceptive methods contain synthetic female sex hormones and

Average Failure Rate of Common Contraceptive Techniques

CONTRACEPTIVE METHOD	AVERAGE FAILURE RATE (annual pregnancies/ 100 women)
None	90
Natural (rhythm) methods	20–30
Coitus interruptus	23
Chemical contraceptives	20
Barrier methods	10–15
Oral contraceptives	2–2.5
Intrauterine device	4

■ **Hormones secreted by the placenta play a critical role in maintaining pregnancy.**

The placenta has the remarkable capacity to secrete a number of peptide and steroid hormones essential for maintaining pregnancy. The most important are *human chorionic gonadotropin, estrogen,* and *progesterone.* Serving as the major endocrine organ of pregnancy, the placenta is unique among endocrine tissues in two regards. First, it is a transient tissue. Second, secretion of its hormones is not subject to extrinsic control, in contrast to the stringent, often complex mechanisms that regulate the secretion of other hormones. Instead, the type and rate of placental hormone secretion depend primarily on the stage of pregnancy.

SECRETION OF HUMAN CHORIONIC GONADOTROPIN

One of the first endocrine events is secretion by the developing chorion of **human chorionic gonadotropin (hCG)**, a peptide hormone that prolongs the life span of the corpus luteum. Recall that during the ovarian cycle, the corpus luteum degenerates and the highly prepared, luteal-dependent uterine lining sloughs off if fertilization and implantation do not occur. When fertilization does occur, the implanted blastocyst saves itself from being flushed out in menstrual flow, by

act similarly to birth control pills to prevent ovulation. These include *long-acting subcutaneous* ("under the skin") *implantation* of hormone-containing capsules that gradually release hormones at a nearly steady rate for five years and *birth control patches* impregnated with hormones that are absorbed through the skin.

Blockage of Implantation

Medically, pregnancy is not considered to begin until implantation. According to this view, any mechanism that interferes with implantation is said to prevent pregnancy. Not all hold this view, however. Some consider pregnancy to begin at time of fertilization. To them, any interference with implantation is a form of abortion. Therefore, methods of contraception that rely on blockage of implantation are more controversial than methods that prevent fertilization from taking place.

- Blockage of implantation is most commonly done by a physician inserting a small *intrauterine device (IUD)* into the uterus. The IUD's mechanism of action is not completely understood, although most evidence suggests that the presence of this foreign object in the uterus induces a local inflammatory response that prevents implantation of a fertilized ovum.

- Implantation can also be blocked by so-called *morning-after pills,* also known as *emergency contraception.* The first term is actually a misnomer, because these pills can prevent pregnancy if taken within 72 hours after, not just the morning after, unprotected sexual intercourse. The most common form of emergency contraception is a kit consisting of high doses of birth control pills. These pills, available only by prescription, work in different ways to prevent pregnancy depending on where the woman is in her cycle when she takes the pills. They can either suppress ovulation or cause premature degeneration of the corpus luteum, thus preventing implantation of a fertilized ovum by withdrawing the developing endometrium's hormonal support. These kits are for emergency use only—for instance, if a condom breaks or in the case of rape—and should not be used as a substitute for ongoing contraceptive methods.

Future Possibilities

- A future birth-control technique is *immunocontraception*—the use of vaccines that prod the immune system to produce antibodies targeted against a particular protein critical to the reproductive process. For example, in the testing stage is a vaccine that induces the formation of antibodies against human chorionic gonadotropin so that this essential corpus luteum–supporting hormone is not effective if pregnancy occurs.

- Some researchers are exploring ways to block the union of sperm and egg by interfering with a specific interaction that normally occurs between the male and female gametes. For example, under study are chemicals introduced into the vagina that trigger premature release of the acrosomal enzymes, depriving the sperm of a means to fertilize an ovulated egg.

- Some scientists are seeking ways to manipulate hormones to block spermatogenesis in males without hindering testosterone secretion, although a "male birth-control pill" remains a distant possibility.

- One interesting avenue being explored holds hope for a unisex contraceptive that would stop sperm in their tracks and could be used by either males or females. Based on preliminary findings, the hope is to use Ca^{2+}-blocking drugs to prevent the entry of Ca^{2+} into sperm tails. As in muscle cells, Ca^{2+} switches on the contractile apparatus responsible for the sperm's motility. With no Ca^{2+}, sperm would not be able to maneuver to accomplish fertilization.

Termination of Unwanted Pregnancies

- When contraceptive practices fail or are not used and an unwanted pregnancy results, women often turn to *abortion* to terminate the pregnancy. More than half of the approximately 6.4 million pregnancies in the United States each year are unintended, and about 1.6 million of them end with an abortion. Although surgical removal of an embryo/fetus is legal in the United States, the practice of abortion is fraught with emotional, ethical, and political controversy.

- In late 2000, amid considerable controversy, the "abortion pill," *RU 486,* or *mifepristone,* was approved for use in the United States, even though it has been available in other countries since 1988. This drug terminates an early pregnancy by chemical interference rather than by surgery. RU 486, a progesterone antagonist, binds tightly with the progesterone receptors on the target cells but does not evoke progesterone's usual effects and prevents progesterone from binding and acting. Deprived of progesterone activity, the highly developed endometrial tissue sloughs off, carrying the implanted embryo with it. RU 486 administration is followed in 48 hours by a prostaglandin that induces uterine contractions to help expel the endometrium and embryo.

producing hCG. This hormone, which is functionally similar to LH, stimulates and maintains the corpus luteum so it does not degenerate. Now called the **corpus luteum of pregnancy,** this ovarian endocrine unit grows even larger and produces increasingly greater amounts of estrogen and progesterone for an additional 10 weeks until the placenta takes over secretion of these steroid hormones. Because of the persistence of estrogen and progesterone, the thick, pulpy endometrial tissue is maintained instead of sloughing. Accordingly, menstruation ceases during pregnancy.

Stimulation by hCG is necessary to maintain the corpus luteum of pregnancy because LH, which maintains the corpus luteum during the normal luteal phase of the uterine cycle, is suppressed through feedback inhibition by the high levels of progesterone.

Maintenance of a normal pregnancy depends on high concentrations of progesterone and estrogen. Thus hCG production is critical during the first trimester to maintain ovarian output of these hormones. The secretion rate of hCG increases rapidly during early pregnancy to save the corpus luteum from demise. Peak secretion of hCG occurs about 60 days after the end of the last menstrual period (● Figure 18-23). By the 10th week of pregnancy, hCG output declines to a low rate of secretion that is maintained for the duration of gestation. The fall in hCG occurs at a time when the corpus luteum is no longer needed for its steroid hormone output, because the placenta has begun to secrete substantial quantities of estrogen and progesterone.

Human chorionic gonadotropin is eliminated from the body in the urine. Pregnancy diagnosis tests can detect hCG in urine as early as the first month of pregnancy, about two weeks after the first missed menstrual period. Because this is before the growing embryo can be detected by physical examination, the test permits early confirmation of pregnancy.

A frequent early clinical sign of pregnancy is **morning sickness**, a daily bout of nausea and vomiting that often occurs in the morning but can take place at any time of day. Because this condition usually appears shortly after implantation and coincides with the time of peak hCG production, it is speculated that this early placental hormone may trigger the symptoms, perhaps by acting on the chemoreceptor trigger zone in the vomiting center (see p. 479).

SECRETION OF ESTROGEN AND PROGESTERONE

Why doesn't the developing placenta start producing estrogen and progesterone in the first place instead of secreting hCG, which in turn stimulates the corpus luteum to secrete these two critical hormones? The answer is that, for different reasons, the placenta cannot produce enough estrogen or progesterone in the first trimester of pregnancy. In the case of estrogen, the placenta does not have all the enzymes needed for complete synthesis of this hormone. Estrogen synthesis requires a complex interaction between the placenta and the fetus: The placenta converts the androgen hormone produced by the fetal adrenal cortex, dehydroepiandrosterone (DHEA),

● **FIGURE 18-22**

Placentation. (a) Relationship between the developing fetus and uterus as pregnancy progresses. (b) Schematic representation of interlocking maternal and fetal structures that form the placenta. Fingerlike projections of chorionic (fetal) tissue form the placental villi, which protrude into a pool of maternal blood. Decidual (maternal) capillary walls are broken down by the expanding chorion so that maternal blood oozes through the spaces between the placental villi. Fetal placental capillaries branch off of the umbilical artery and project into the placental villi. Fetal blood flowing through these vessels is separated from the maternal blood by only the thin chorionic layer that forms the placental villi. Maternal blood enters through the maternal arterioles, then percolates through the pool of blood in the intervillus spaces. Here, exchanges are made between the fetal and maternal blood before the fetal blood leaves through the umbilical vein and maternal blood exits through the maternal venules.

(*Source:* Part (a) adapted from Cecie Starr, *Biology: Concepts and Applications,* Fourth Edition, Fig. 38.25b, p. 655. Copyright ©2000 Brooks/Cole)

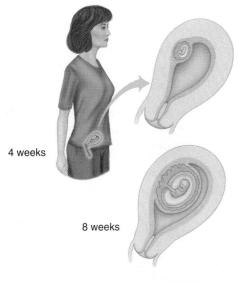

4 weeks

8 weeks

12 weeks

Full term

(a)

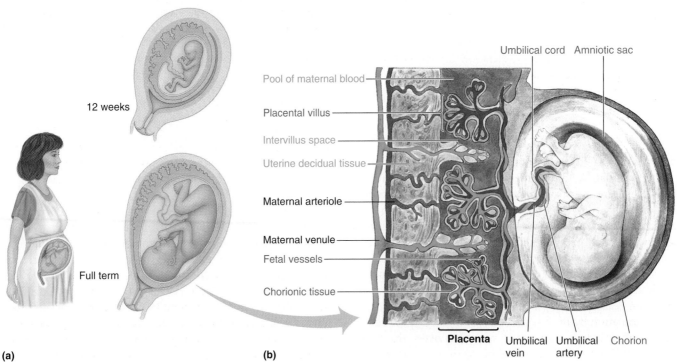

Umbilical cord   Amniotic sac

Pool of maternal blood

Placental villus

Intervillus space

Uterine decidual tissue

Maternal arteriole

Maternal venule

Fetal vessels

Chorionic tissue

Placenta   Umbilical vein   Umbilical artery   Chorion

(b)

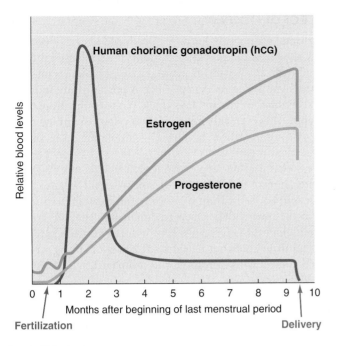

● FIGURE 18-23

Secretion rates of placental hormones

into estrogen. The placenta cannot produce estrogen until the fetus has developed to the point that its adrenal cortex is secreting DHEA into the blood. The placenta extracts DHEA from the fetal blood and converts it into estrogen, which it then secretes into the maternal blood.

In the case of progesterone, the placenta can synthesize this hormone soon after implantation. Even though the early placenta has the enzymes necessary to convert cholesterol extracted from the maternal blood into progesterone, it does not produce much of this hormone, because the amount of progesterone produced is proportional to placental weight. The placenta is simply too small in the first 10 weeks of pregnancy to produce enough progesterone to maintain the endometrial tissue. The notable increase in circulating progesterone in the last seven months of gestation reflects placental growth during this period.

ROLES OF ESTROGEN AND PROGESTERONE DURING PREGNANCY

As noted earlier, high concentrations of estrogen and progesterone are essential to maintain a normal pregnancy. Estrogen stimulates growth of the myometrium, which increases in size throughout pregnancy. The stronger uterine musculature is needed to expel the fetus during labor. Estrogen also promotes development of the ducts within the mammary glands, through which milk will be ejected during lactation.

Progesterone performs various roles throughout pregnancy. Its main function is to prevent miscarriage by suppressing contractions of the uterine myometrium. Progesterone also promotes formation of a mucus plug in the cervical canal, to prevent vaginal contaminants from reaching the uterus.

Finally, placental progesterone stimulates the development of milk glands in the breasts, in preparation for lactation.

▐ Maternal body systems respond to the increased demands of gestation.

The period of **gestation (pregnancy)** is about 38 weeks from conception (40 weeks from the end of the last menstrual period). During gestation, the embryo/fetus develops and grows to the point of being able to leave its maternal life-support system. Meanwhile, a number of physical changes within the mother accommodate the demands of pregnancy. The most obvious change is uterine enlargement. The uterus expands and increases in weight more than 20 times, exclusive of its contents. The breasts enlarge and develop the ability to produce milk. Body systems other than the reproductive system also make needed adjustments. The volume of blood increases by 30%, and the cardiovascular system responds to the increasing demands of the growing placental mass. Weight gain during pregnancy is due only in part to the weight of the fetus. The remainder is mostly from increased weight of the uterus, including the placenta, and increased blood volume. Respiratory activity increases by about 20% to handle the additional fetal requirements for O_2 utilization and CO_2 removal. Urinary output increases, and the kidneys excrete the additional wastes from the fetus.

The increased metabolic demands of the growing fetus increase nutritional requirements for the mother. In general, the fetus takes what it needs from the mother, even if this leaves the mother with a nutritional deficit. The placenta produces a hormone similar to growth hormone that decreases maternal use of glucose and mobilizes fatty acids, making available greater quantities of these nutrients for the fetus. If the mother does not consume enough Ca^{2+} in her diet, yet another placental hormone similar to parathyroid hormone mobilizes Ca^{2+} from the maternal bones to ensure adequate calcification of the fetal bones.

▐ Changes during late gestation prepare for parturition.

Parturition (labor, delivery, or **birth)** requires (1) dilation of the cervical canal to accommodate passage of the fetus from the uterus through the vagina and to the outside and (2) contractions of the uterine myometrium that are sufficiently strong to expel the fetus.

Several changes take place during late gestation in preparation for the onset of parturition. During the first two trimesters of gestation, the uterus remains relatively quiet, because of the inhibitory effect of the high levels of progesterone on the uterine muscle. During the last trimester, however, the uterus becomes progressively more excitable, so that mild contractions (**Braxton-Hicks contractions**) are experienced with increasing strength and frequency. Sometimes these contractions become regular enough to be mistaken for the onset of labor, a phenomenon called "false labor."

Throughout gestation, the exit of the uterus remains sealed by the rigid, tightly closed cervix. As parturition approaches, the cervix begins to soften (or "ripen") as a result of the dissociation of its tough connective tissue (collagen) fibers. Because of this softening, the cervix becomes malleable so that it can gradually yield, dilating the exit, as the fetus is forcefully pushed against it during labor. This cervical softening is caused by **relaxin**, a peptide hormone produced by the corpus luteum of pregnancy and by the placenta. Relaxin also "relaxes" the birth canal by loosening the connective tissue between pelvic bones.

Meanwhile, the fetus shifts downward (the baby "drops") and is normally oriented so that the head is in contact with the cervix in preparation for exiting through the birth canal. In a **breech birth**, any part of the body other than the head approaches the birth canal first.

▌ Scientists are closing in on the factors that trigger the onset of parturition.

Rhythmic, coordinated contractions, usually painless at first, begin at the onset of true labor. As labor progresses, the contractions increase in frequency, intensity, and discomfort. These strong, rhythmic contractions force the fetus against the cervix, dilating the cervix. Then, after having dilated the cervix enough for the fetus to pass through, these contractions force the fetus out through the birth canal.

The exact factors triggering the increase in uterine contractility and thus initiating parturition are not fully established, although much progress has been made in recent years in unraveling the sequence of events. Let's take a look at what is known about this process.

ROLE OF HIGH ESTROGEN LEVELS

During early gestation, maternal estrogen levels are relatively low, but as gestation proceeds placental estrogen secretion continues to rise. In the immediate days before the onset of parturition, soaring levels of estrogen bring about changes in the uterus and cervix to prepare them for labor and delivery (● Figure 18-24). First, high levels of estrogen promote the synthesis of connexons within the uterine smooth-muscle cells. These myometrial cells are not functionally linked to any extent throughout most of gestation. The newly manufactured connexons are inserted in the myometrial plasma membranes to form gap junctions that electrically link together the uterine smooth-muscle cells so they become able to contract as a coordinated unit (see p. 49).

Simultaneously, high levels of estrogen dramatically and progressively increase the concentration of myometrial receptors for oxytocin. Together, these myometrial changes collectively bring about the increased uterine responsiveness to oxytocin that ultimately initiates labor.

In addition to preparing the uterus for labor, the increasing levels of estrogen also promote production of local prostaglandins that contribute to cervical ripening by stimulating cervical enzymes that locally degrade collagen fibers.

ROLE OF OXYTOCIN

Oxytocin is a peptide hormone produced by the hypothalamus, stored in the posterior pituitary, and released into the blood from the posterior pituitary on nervous stimulation by the hypothalamus (see p. 538). A powerful uterine muscle stimulant, oxytocin plays the key role in the progression of labor. However, this hormone was once discounted as the trigger for parturition, because the circulating levels of oxytocin remain constant prior to the onset of labor. The discovery that uterine responsiveness to oxytocin is 100 times greater at term than in nonpregnant women (because of the increased concentration of myometrial oxytocin receptors) led to the now widely accepted conclusion that labor is initiated when the oxytocin receptor concentration reaches a critical threshold that permits the onset of strong, coordinated contractions in response to ordinary levels of circulating oxytocin.

ROLE OF CORTICOTROPIN-RELEASING HORMONE

Until recently, scientists were baffled by the factors that raise levels of placental estrogen secretion. Recent research has shed new light on the probable mechanism. Evidence suggests that *corticotropin-releasing hormone (CRH)* secreted by the fetal portion of the placenta into both the maternal and fetal circulations not only drives the manufacture of placental estrogen, thus ultimately dictating the timing of the onset of labor, but also promotes changes in the fetal lungs needed for breathing air (● Figure 18-24). Recall that CRH is normally secreted by the hypothalamus and regulates the output of ACTH by the anterior pituitary (see p.555). In turn, ACTH stimulates production of both cortisol and DHEA by the adrenal cortex. In the fetus, much of the CRH comes from the placenta rather than solely from the fetal hypothalamus. The additional cortisol secretion summoned by the extra CRH promotes fetal lung maturation. Specifically, cortisol stimulates the synthesis of pulmonary surfactant, which facilitates lung expansion and reduces the work of breathing (see p. 377).

The bumped-up rate of DHEA secretion by the adrenal cortex in response to placental CRH leads to the rising levels of placental estrogen secretion. Recall that the placenta converts DHEA from the fetal adrenal gland into estrogen, which enters the maternal bloodstream. When sufficiently high, this estrogen sets in motion the events that initiate labor. Thus pregnancy duration and delivery timing are determined largely by the placenta's rate of CRH production. That is, a "**placental clock**" ticks out the length of time until parturition. When a critical level of placental CRH is reached, parturition is triggered. This critical CRH level ensures that when labor begins, the infant is ready for life outside the womb. It does so by concurrently increasing the fetal cortisol needed for lung maturation and the estrogen needed for the uterine changes that bring on labor. The remaining unanswered puzzle regarding the placental clock is, What controls placental secretion of CRH?

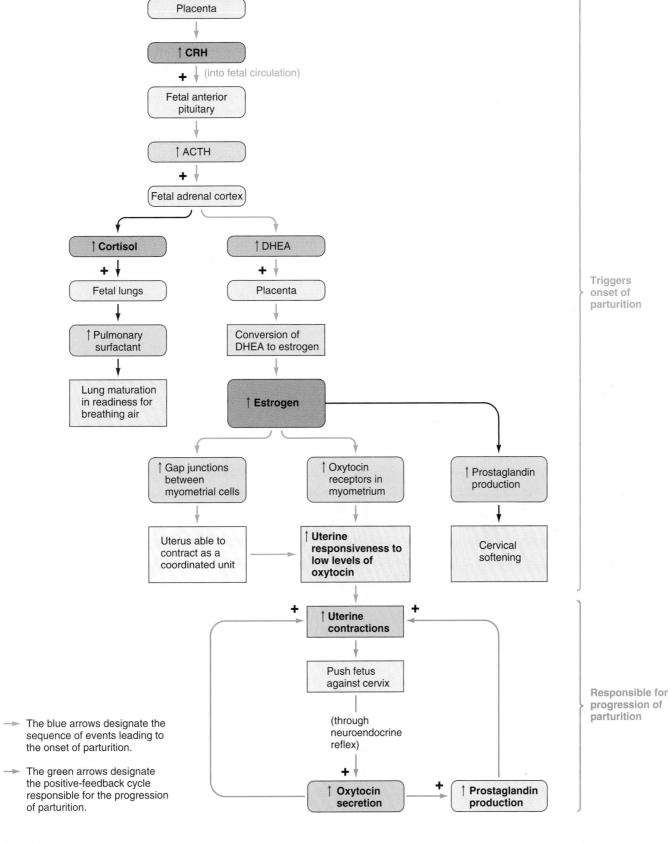

● **FIGURE 18-24**

Initiation and progression of parturition

▌ Parturition is accomplished by a positive-feedback cycle.

Once high levels of estrogen increase uterine responsiveness to oxytocin to a critical level and regular uterine contractions begin, myometrial contractions progressively increase in frequency, strength, and duration throughout labor until they expel the uterine contents. At the beginning of labor, contractions lasting 30 seconds or less occur about every 25 to 30 minutes; by the end, they last 60 to 90 seconds and occur every 2 to 3 minutes.

As labor progresses, a positive-feedback cycle involving oxytocin and prostaglandin ensues, incessantly increasing myometrial contractions (● Figure 18-24). Each uterine contraction begins at the top of the uterus and sweeps downward, forcing the fetus toward the cervix. Pressure of the fetus against the cervix does two things. First, the fetal head pushing against the softened cervix wedges open the cervical canal. Second, cervical stretch stimulates the release of oxytocin through a neuroendocrine reflex. Stimulation of receptors in the cervix in response to fetal pressure sends a neural signal up the spinal cord to the hypothalamus, which in turn triggers oxytocin release from the posterior pituitary. This additional oxytocin promotes more powerful uterine contractions. As a result, the fetus is pushed more forcefully against the cervix, stimulating the release of even more oxytocin, and so on. This cycle is reinforced as oxytocin stimulates prostaglandin production by the decidua. As a powerful myometrial stimulant, prostaglandin further enhances uterine contractions. Oxytocin secretion, prostaglandin production, and uterine contractions continue to increase in positive-feedback fashion throughout labor until delivery relieves the pressure on the cervix.

STAGES OF LABOR

Labor is divided into three stages: (1) cervical dilation, (2) delivery of the baby, and (3) delivery of the placenta (● Figure 18-25). At the onset of labor or sometime during the first

● **FIGURE 18-25**

Stages of labor. (a) Position of the fetus near the end of pregnancy. (b) Stages of labor.

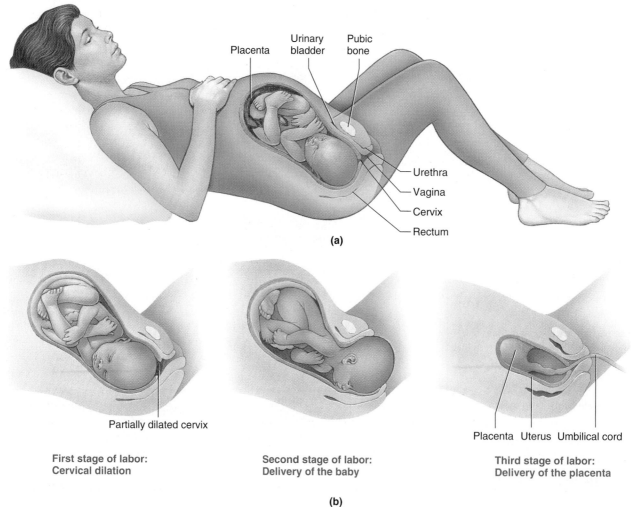

Placenta Urinary bladder Pubic bone

Urethra
Vagina
Cervix
Rectum

(a)

Partially dilated cervix

Placenta Uterus Umbilical cord

First stage of labor:
Cervical dilation

Second stage of labor:
Delivery of the baby

Third stage of labor:
Delivery of the placenta

(b)

stage, the membranes surrounding the amniotic sac, or "bag of waters," rupture. As amniotic fluid escapes out of the vagina, it helps lubricate the birth canal.

- *First stage.* During the first stage, the cervix is forced to dilate to accommodate the diameter of the baby's head, usually to a maximum of 10 cm. This stage is the longest, lasting from several hours to as long as 24 hours in a first pregnancy. If another part of the fetus's body other than the head is oriented against the cervix, it is generally less effective than the head as a wedge. The head has the largest diameter of the baby's body. If the baby approaches the birth canal feet first, the feet may not dilate the cervix enough to let the head pass. In such a case, without medical intervention the baby's head would remain stuck behind the too-narrow cervical opening.
- *Second stage.* The second stage of labor, the actual birth of the baby, begins once cervical dilation is complete. When the infant begins to move through the cervix and vagina, stretch receptors in the vagina activate a neural reflex that triggers contractions of the abdominal wall in synchrony with the uterine contractions. These abdominal contractions greatly increase the force pushing the baby through the birth canal. The mother can help deliver the infant by voluntarily contracting the abdominal muscles at this time in unison with each uterine contraction (that is, by "pushing" with each "labor pain"). Stage 2 is usually much shorter than the first stage and lasts 30 to 90 minutes. The infant is still attached to the placenta by the umbilical cord at birth. The cord is tied and severed, with the stump shriveling up in a few days to form the **umbilicus (navel).**
- *Third stage.* Shortly after delivery of the baby, a second series of uterine contractions separates the placenta from the myometrium and expels it through the vagina. Delivery of the placenta, or **afterbirth,** constitutes the third stage of labor, typically the shortest stage, being completed within 15 to 30 minutes after the baby is born. After the placenta is expelled, continued contractions of the myometrium constrict the uterine blood vessels supplying the site of placental attachment, to prevent hemorrhage.

UTERINE INVOLUTION

After delivery, the uterus shrinks to its pregestational size, a process known as **involution,** which takes four to six weeks to complete. During involution, the remaining endometrial tissue not expelled with the placenta gradually disintegrates and sloughs off, producing a vaginal discharge that continues for three to six weeks following parturition. After this period, the endometrium is restored to its nonpregnant state.

Involution occurs largely because of the precipitous fall in circulating estrogen and progesterone when the placental source of these steroids is lost at delivery. The process is facilitated in mothers who breast-feed their infants, because oxytocin is released in response to suckling. In addition to playing an important role in lactation, this periodic nursing-induced release of oxytocin promotes myometrial contractions that help maintain uterine muscle tone, enhancing in-

volution. Involution is usually complete in about four weeks in nursing mothers but takes about six weeks in those who do not breast-feed.

Lactation requires multiple hormonal inputs.

The female reproductive system supports the new being from the moment of conception through gestation and continues to nourish it during its early life outside the supportive uterine environment. Milk (or its equivalent) is essential for survival of the newborn. Accordingly, during gestation the **mammary glands,** or **breasts,** are prepared for **lactation (milk production).**

The breasts in nonpregnant females consist mostly of adipose tissue and a rudimentary duct system. Breast size is determined by the amount of adipose tissue, which has nothing to do with the ability to produce milk.

PREPARATION OF THE BREASTS FOR LACTATION

Under the hormonal environment present during pregnancy, the mammary glands develop the internal glandular structure and function necessary for milk production. A breast capable of lactating has a network of progressively smaller ducts that branch out from the nipple and terminate in lobules (● Figure 18-26a). Each lobule is made up of a cluster of sac-like epithelial-lined **alveoli** that constitute the milk-producing glands. Milk is synthesized by the epithelial cells, then secreted into the alveolar lumen, which is drained by a milk-collecting duct that transports the milk to the surface of the nipple (● Figure 18-26b).

During pregnancy, the high concentration of *estrogen* promotes extensive duct development, whereas the high level of *progesterone* stimulates abundant alveolar-lobular formation. Elevated concentrations of *prolactin* (an anterior pituitary hormone stimulated by the rising levels of estrogen) also contribute to mammary gland development by inducing the synthesis of enzymes needed for milk production.

PREVENTION OF LACTATION DURING GESTATION

Most of these changes in the breasts occur during the first half of gestation, so the mammary glands are fully capable of producing milk by the middle of pregnancy. However, milk secretion does not occur until parturition. The high estrogen and progesterone concentrations during the last half of pregnancy prevent lactation by blocking prolactin's stimulatory action on milk secretion. Prolactin is the primary stimulant of milk secretion. Thus even though the high levels of placental steroids induce the development of the milk-producing machinery in the breasts, they prevent these glands from becoming operational until the baby is born and milk is needed.

The abrupt decline in estrogen and progesterone that occurs with loss of the placenta at parturition initiates lactation. (We have now completed our discussion of the functions of estrogen and progesterone during gestation and lactation as well as throughout the reproductive life of females. These functions are summarized in ▲ Table 18-5.)

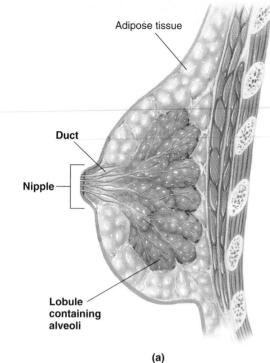

(a)

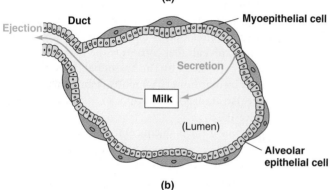

(b)

● FIGURE 18-26

Mammary gland anatomy. (a) Internal structure of the mammary gland, lateral view. (b) Schematic representation of the microscopic structure of an alveolus within the mammary gland. The alveolar epithelial cells secrete milk into the lumen. Contraction of the surrounding myoepithelial cells ejects the secreted milk out through the duct.

STIMULATION OF LACTATION VIA SUCKLING

Once milk production begins after delivery, two hormones are critical for maintaining lactation: (1) *prolactin,* which acts on the alveolar epithelium to promote secretion of milk, and (2) *oxytocin,* which induces **milk ejection.** The latter term refers to the forced expulsion of milk from the lumen of the alveoli out through the ducts. Release of both of these hormones is stimulated by a neuroendocrine reflex triggered by suckling (● Figure 18-27). Let's examine each of these hormones and their roles in further detail.

- *Oxytocin release and milk ejection.* The infant cannot directly suck milk out of the alveolar lumen. Instead, milk must be actively squeezed out of the alveoli into the ducts and hence toward the nipple, by contraction of specialized **myoepithe-**

Actions of Estrogen and Progesterone

ESTROGEN

Effects on Sex-Specific Tissues

Essential for egg maturation and release

Stimulates growth and maintenance of entire female reproductive tract

Stimulates follicle maturation

Thins the cervical mucus to permit sperm penetration

Enhances transport of sperm to the oviduct by stimulating upward contractions of the uterus and oviduct

Stimulates growth of the endometrium and myometrium

Induces synthesis of endometrial progesterone receptors

Triggers onset of parturition by increasing uterine responsiveness to oxytocin during late gestation through a twofold effect: by inducing synthesis of myometrial oxytocin receptors and by increasing myometrial gap junctions so that the uterus can contract as a coordinated unit in response to oxytocin

Other Reproductive Effects

Promotes development of secondary sexual characteristics

Controls GnRH and gonadotropin secretion
    Low levels inhibit secretion
    High levels responsible for triggering LH surge

Stimulates duct development in the breasts during gestation

Inhibits milk-secreting actions of prolactin during gestation

Nonreproductive Effects

Promotes fat deposition

Increases bone density

Closes the epiphyseal plates

PROGESTERONE

Prepares a suitable environment for nourishment of a developing embryo/fetus

Promotes formation of a thick mucus plug in cervical canal

Inhibits hypothalamic GnRH and gonadotropin secretion

Stimulates alveolar development in the breasts during gestation

Inhibits milk-secreting actions of prolactin during gestation

Inhibits uterine contractions during gestation

lial cells (musclelike epithelial cells) that surround each alveolus (● Figure 18-26b). The infant's suckling of the breast stimulates sensory nerve endings in the nipple, initiating action potentials that travel up the spinal cord to the hypothalamus. Thus activated, the hypothalamus triggers a burst of oxytocin release from the posterior pituitary. Oxytocin in

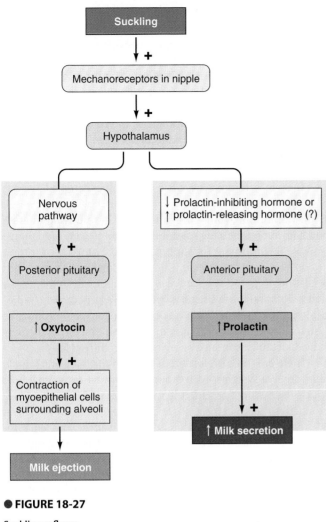

● FIGURE 18-27

Suckling reflexes

hypophyseal portal system to stimulate prolactin secretion by the anterior pituitary (see p. 541). This role of oxytocin is distinct from the roles of oxytocin produced by the hypothalamus and stored in the posterior pituitary.

Throughout most of the female's life, PIH is the dominant influence, so prolactin concentrations normally remain low. During lactation, a burst in prolactin secretion occurs each time the infant suckles. Afferent impulses initiated in the nipple on suckling are carried by the spinal cord to the hypothalamus. This reflex ultimately leads to prolactin release by the anterior pituitary, although it is unclear whether this is from inhibition of PIH or stimulation of PRH secretion or both. Prolactin then acts on the alveolar epithelium to promote secretion of milk to replace the ejected milk (● Figure 18-27).

Concurrent stimulation by suckling of both milk ejection and milk production ensures that the rate of milk synthesis keeps pace with the baby's needs for milk. The more the infant nurses, the more milk is removed by letdown and the more milk is produced for the next feeding.

▍ Breast-feeding is advantageous to both the infant and the mother.

Nutritionally, **milk** is composed of water, triglyceride fat, the carbohydrate lactose (milk sugar), a number of proteins, vitamins, and the minerals calcium and phosphate.

ADVANTAGES OF BREAST-FEEDING FOR THE INFANT

In addition to nutrients, milk contains a host of immune cells, antibodies, and other chemicals that help protect the infant against infection until it can mount an effective immune response on its own a few months after birth. **Colostrum,** the milk produced for the first five days after delivery, contains lower concentrations of fat and lactose but higher concentrations of immunoprotective components. All human babies acquire some passive immunity during gestation by antibodies passing across the placenta from the mother to the fetus. These antibodies are short-lived, however, and often do not persist until the infant can fend for itself immunologically. Breast-fed babies gain additional protection during this vulnerable period.

ADVANTAGES OF BREAST-FEEDING FOR THE MOTHER

Breast-feeding is also advantageous for the mother. Oxytocin release triggered by nursing hastens uterine involution. In addition, suckling suppresses the menstrual cycle by inhibiting LH and FSH secretion, probably by inhibiting GnRH. Lactation, therefore, tends to prevent ovulation, decreasing the likelihood of another pregnancy (although it is not a reliable means of contraception). This mechanism permits all the mother's resources to be directed toward the newborn instead of being shared with a new embryo.

CESSATION OF MILK PRODUCTION AT WEANING

When the infant is weaned, two mechanisms contribute to the cessation of milk production. First, without suckling prolactin secretion is not stimulated, removing the main stimu-

turn stimulates contraction of the myoepithelial cells in the breasts to induce milk ejection, or "milk letdown." Milk letdown continues only as long as the infant continues to nurse. In this way, the milk ejection reflex ensures that the breasts release milk only when and in the amount needed by the baby. Even though the alveoli may be full of milk, the milk cannot be released without oxytocin. The reflex can become conditioned to stimuli other than suckling, however. For example, the infant's cry can trigger milk letdown, causing a spurt of milk to leak from the nipples. In contrast, psychological stress, acting through the hypothalamus, can easily inhibit milk ejection. For this reason, a positive attitude toward breast-feeding and a relaxed environment are essential for successful breast-feeding.

- *Prolactin release and milk secretion.* Suckling not only triggers oxytocin release but also stimulates prolactin secretion. Prolactin output by the anterior pituitary is controlled by two hypothalamic secretions: **prolactin-inhibiting hormone (PIH)** and **prolactin-releasing hormone (PRH).** PIH is now known to be *dopamine,* which also serves as a neurotransmitter in the brain. The chemical nature of PRH has not been identified with certainty, but scientists suspect PRH is oxytocin secreted by the hypothalamus into the hypothalamic-

lus for continued milk synthesis and secretion. Also, because there is no suckling and thus no oxytocin release, milk letdown does not occur. Because milk production does not immediately shut down, milk accumulates in the alveoli, engorging the breasts. The resulting pressure buildup acts directly on the alveolar epithelial cells to suppress further milk production. Cessation of lactation at weaning therefore results from a lack of suckling-induced stimulation of both prolactin and oxytocin secretion.

▌ The end is a new beginning.

Reproduction is an appropriate way to end our discussion of physiology. The single cell resulting from the union of male and female gametes divides mitotically and differentiates into a multicellular individual made up of a number of different body systems that interact cooperatively to maintain homeostasis (that is, stability in the internal environment). All the life-supporting homeostatic processes introduced throughout this book begin all over again at the start of a new life.

 Click on the Media Exercises menu of the CD-ROM and work Media Exercise 18.4: Female Reproductive Physiology to test your understanding of the previous section.

 ## CHAPTER IN PERSPECTIVE: FOCUS ON HOMEOSTASIS

The reproductive system is unique in that it is not essential for homeostasis or for survival of the individual, but it is essential for sustaining the thread of life from generation to generation. Reproduction depends on the union of male and female gametes (reproductive cells), each with a half set of chromosomes, to form a new individual with a full, unique set of chromosomes. Unlike the other body systems, which are essentially identical in the two sexes, the reproductive systems of males and females are remarkably different, befitting their different roles in the reproductive process.

The male system is designed to continuously produce huge numbers of mobile spermatozoa that are delivered to the female during the sex act. Male gametes must be produced in abundance for two reasons: (1) Only a small percentage of them survive the hazardous journey through the female reproductive tract to the site of fertilization; and (2) the cooperative effort of many spermatozoa is required to break down the barriers surrounding the female gamete (ovum or egg) to enable one spermatozoon to penetrate and unite with the ovum.

The female reproductive system undergoes complex changes on a cyclic monthly basis. During the first half of the cycle, a single nonmotile ovum is prepared for release. During the second half, the reproductive system is geared toward preparing a suitable environment for supporting the ovum if fertilization (union with a spermatozoon) occurs. If fertilization does not occur, the prepared supportive environment within the uterus sloughs off, and the cycle starts over again as a new ovum is prepared for release. If fertilization occurs, the female reproductive system adjusts to support growth and development of the new individual until it can survive on its own on the outside.

There are three important parallels in the male and female reproductive systems, even though they differ considerably in structure and function. First, the same set of undifferentiated reproductive tissues in the embryo can develop into either a male or a female system, depending on the presence or absence, respectively, of male-determining factors. Second, the same hormones—namely, hypothalamic GnRH and anterior pituitary FSH and LH—control reproductive function in both sexes. In both cases, gonadal steroids and inhibin act in negative-feedback fashion to control hypothalamic and anterior pituitary output. Third, the same events take place in the developing gamete's nucleus during sperm formation and egg formation, although males produce millions of sperm in one day, whereas females produce only about 400 ova in a lifetime.

CHAPTER SUMMARY

Introduction (pp. 583–588)

- Both sexes produce gametes (reproductive cells), sperm in males and ova (eggs) in females, each of which bears one member of each of the 23 pairs of chromosomes present in human cells. Union of a sperm and an ovum at fertilization results in the beginning of a new individual with 23 complete pairs of chromosomes, half from the father and half from the mother.

- The reproductive system is anatomically and functionally distinct in males and females. Males produce sperm and deliver them into the female. Females produce ova, accept sperm delivery, and provide a suitable environment for supporting development of a fertilized ovum until the new individual can survive on its own in the external world.

- In both sexes, the reproductive system consists of (1) a pair of gonads, testes in males and ovaries in females, which are the primary reproductive organs that produce the gametes and secrete sex hormones; (2) a reproductive tract composed of a system of ducts that transport and/or house the gametes after they are produced; and (3) accessory sex glands that provide supportive secretions for the gametes. The externally visible portions of the reproductive system constitute the external genitalia. *(Review Figures 18-1 and 18-2.)*

- Secondary sexual characteristics are the distinguishing features between males and females not directly related to reproduction.

- Sex determination is a genetic phenomenon dependent on the combination of sex chromosomes at the time of fertilization, an XY combination being a genetic male and an XX combination a genetic female.

- The term *sex differentiation* refers to the embryonic development of the gonads, reproductive tract, and external genitalia along male or female lines, which gives rise to the apparent anatomic sex of the individual. In the presence of masculinizing factors, a male reproductive system develops; in their absence, a female system develops. *(Review Figure 18-3.)*

Male Reproductive Physiology (pp. 588–596)

- The testes are located in the scrotum. The cooler temperature in the scrotum compared to the abdominal cavity is essential for spermatogenesis.
- Spermatogenesis (sperm production) occurs in the testes' highly coiled seminiferous tubules. (*Review Figures 18-4 and 18-5.*)
- Leydig cells in the interstitial spaces between these tubules secrete the male sex hormone testosterone into the blood. (*Review Figure 18-4.*)
- Testosterone is secreted before birth to masculinize the developing reproductive system; then its secretion ceases until puberty, at which time it begins once again and continues throughout life. Testosterone is responsible for maturation and maintenance of the entire male reproductive tract, for development of secondary sexual characteristics, and for stimulating libido. (*Review Table 18-1.*)
- The testes are regulated by the anterior pituitary hormones, luteinizing hormone (LH) and follicle-stimulating hormone (FSH). These gonadotropic hormones in turn are under control of hypothalamic gonadotropin-releasing hormone (GnRH). (*Review Figure 18-7.*)
- Testosterone secretion is regulated by LH stimulation of the Leydig cells, and, in negative-feedback fashion, testosterone inhibits gonadotropin secretion. (*Review Figure 18-7.*)
- Spermatogenesis requires both testosterone and FSH. Testosterone stimulates the mitotic and meiotic divisions required to transform the undifferentiated diploid germ cells, the spermatogonia, into undifferentiated haploid spermatids. FSH stimulates the remodeling of spermatids into highly specialized motile spermatozoa. (*Review Figure 18-5.*)
- A spermatozoon consists only of a DNA-packed head bearing an enzyme-filled acrosome at its tip for penetrating the ovum, a midpiece containing the metabolic machinery for energy production, and a whiplike motile tail. (*Review Figure 18-6.*)
- Also present in the seminiferous tubules are Sertoli cells, which protect, nurse, and enhance the germ cells throughout their development. Sertoli cells also secrete inhibin, a hormone that inhibits FSH secretion, completing the negative-feedback loop. (*Review Figures 18-4 and 18-7.*)
- The still immature sperm are flushed out of the seminiferous tubules into the epididymis by fluid secreted by the Sertoli cells.
- The epididymis and ductus deferens store and concentrate the sperm and increase their motility and fertility prior to ejaculation. (*Review Table 18-2.*)
- During ejaculation, the sperm are mixed with secretions released by the accessory glands. (*Review Table 18-2.*)
- The seminal vesicles supply fructose for energy and prostaglandins, which promote smooth muscle motility in both the male and female reproductive tracts, to enhance sperm transport. The seminal vesicles contribute the bulk of the semen.
- The prostate gland contributes an alkaline fluid for neutralizing the acidic vaginal secretions.
- The bulbourethral glands release lubricating mucus.

Sexual Intercourse between Males and Females (pp. 596–601)

- The male sex act consists of erection and ejaculation, which are part of a much broader systemic, emotional response that typifies the male sexual response cycle. (*Review Table 18-4.*)
- Erection is a hardening of the normally flaccid penis that enables it to penetrate the female vagina. Erection is accomplished by marked vasocongestion of the penis brought about by reflexly induced vasodilation of the arterioles supplying the penile erectile tissue. (*Review Figure 18-8.*)
- When sexual excitation reaches a critical peak, ejaculation occurs. It consists of two stages: (1) emission, the emptying of semen (sperm and accessory sex gland secretions) into the urethra; and (2) expulsion of semen from the penis. The latter is accompanied by a set of characteristic systemic responses and intense pleasure referred to as *orgasm*.
- Females experience a sexual cycle similar to males, with both having excitation, plateau, orgasmic, and resolution phases. The major difference is that women do not ejaculate.
- During the female sexual response, the outer portion of the vagina constricts to grip the penis, whereas the inner part expands to create space for sperm deposition.

Female Reproductive Physiology (pp. 601–626)

- In the nonpregnant state, female reproductive function is controlled by a complex, cyclic negative-feedback control system between the hypothalamus (GnRH), anterior pituitary (FSH and LH), and ovaries (estrogen, progesterone, and inhibin). During pregnancy, placental hormones become the main controlling factors.
- The ovaries perform the dual and interrelated functions of oogenesis (producing ova) and secreting estrogen and progesterone. (*Review Table 18-5, p. 624.*) Two related ovarian endocrine units sequentially accomplish these functions: the follicle and the corpus luteum.
- The same steps in chromosome replication and division take place in oogenesis as in spermatogenesis, but the timing and end result are markedly different. Spermatogenesis is accomplished within two months, whereas the similar steps in oogenesis take anywhere from 12 to 50 years to complete on a cyclical basis from the onset of puberty until menopause. A female is born with a limited, largely nonrenewable supply of germ cells, whereas postpubertal males can produce several hundred million sperm each day. Each primary oocyte yields only one cytoplasm-rich ovum along with three doomed cytoplasm-poor polar bodies that disintegrate, whereas each primary spermatocyte yields four equally viable spermatozoa. (*Review Figure 18-9.*)
- Oogenesis and estrogen secretion take place within an ovarian follicle during the first half of each reproductive cycle (the follicular phase) under the influence of FSH, LH, and estrogen. (*Review Figures 18-10 through 18-13.*)
- At approximately midcycle, the maturing follicle releases a single ovum (ovulation). Ovulation is triggered by an LH surge brought about by the high level of estrogen produced by the mature follicle. (*Review Figures 18-10, 18-12, and 18-14.*)
- Under the influence of LH, the empty follicle is then converted into a corpus luteum, which produces progesterone as well as estrogen during the last half of the cycle (the luteal phase). This endocrine unit prepares the uterus for implantation if the released ovum is fertilized. (*Review Figures 18-10, 18-12, and 18-15.*)
- If fertilization and implantation do not occur, the corpus luteum degenerates. The consequent withdrawal of hormonal support for the highly developed uterine lining causes it to disintegrate and slough, producing menstrual flow. Simultaneously, a new follicular phase is initiated. (*Review Figure 18-12.*)
- Menstruation ceases and the uterine lining (endometrium) repairs itself under the influence of rising estrogen levels from the newly maturing follicle. (*Review Figure 18-12.*)
- If fertilization does take place, it occurs in the oviduct as the released egg and sperm deposited in the vagina are both transported to this site. (*Review Figures 18-16 and 18-18.*)
- The fertilized ovum begins to divide mitotically. Within a week it grows and differentiates into a blastocyst capable of implantation. (*Review Figure 18-19.*)

- Meanwhile, the endometrium has become richly vascularized and stocked with stored glycogen under the influence of luteal-phase progesterone. Into this especially prepared lining the blastocyst implants by means of enzymes released by the trophoblasts, which form the blastocyst's outer layer. These enzymes digest the nutrient-rich endometrial tissue, accomplishing the dual function of carving a hole in the endometrium for implantation of the blastocyst while at the same time releasing nutrients from the endometrial cells for use by the developing embryo. (Review Figure 18-21.)
- After implantation, an interlocking combination of fetal and maternal tissues, the placenta, develops. The placenta is the organ of exchange between the maternal and fetal blood and also acts as a transient, complex endocrine organ that secretes a number of hormones essential for pregnancy. Human chorionic gonadotropin, estrogen, and progesterone are the most important of these hormones. (Review Figure 18-22.)
- Human chorionic gonadotropin maintains the corpus luteum of pregnancy, which secretes estrogen and progesterone during the first trimester of gestation until the placenta takes over this function the last two trimesters. High levels of estrogen and progesterone are essential for maintaining a normal pregnancy. (Review Figure 18-23.)
- At parturition, rhythmic contractions of increasing strength, duration, and frequency accomplish the three stages of labor: dilation of the cervix, birth of the baby, and delivery of the placenta (afterbirth). (Review Figure 18-25.)
- Once the contractions are initiated at the onset of labor, a positive-feedback cycle is established that progressively increases their force. As contractions push the fetus against the cervix, secretion of oxytocin, a powerful uterine muscle stimulant, is reflexly increased. The extra oxytocin causes stronger contractions, giving rise to even more oxytocin release, and so on. This positive-feedback cycle progressively intensifies until cervical dilation and delivery are complete. (Review Figures 18-24 and 18-25.)
- During gestation, the breasts are specially prepared for lactation. The elevated levels of placental estrogen and progesterone, respectively, promote development of the ducts and alveoli in the mammary glands. (Review Figure 18-26.)
- Prolactin stimulates the synthesis of enzymes essential for milk production by the alveolar epithelial cells. However, the high gestational level of estrogen and progesterone prevents prolactin from promoting milk production. Withdrawal of the placental steroids at parturition initiates lactation.
- Lactation is sustained by suckling, which triggers the release of oxytocin and prolactin. Oxytocin causes milk ejection by stimulating the myoepithelial cells surrounding the alveoli to squeeze the secreted milk out through the ducts. Prolactin stimulates the production of more milk to replace the milk ejected as the baby nurses. (Review Figures 18-26 and 18-27.)

REVIEW EXERCISES

Objective Questions (Answers on p. A-49)

1. It is possible for a genetic male to have the anatomic appearance of a female. (True or false?)
2. Testosterone secretion essentially ceases from birth until puberty. (True or false?)
3. Prostaglandins are derived from arachidonic acid found in the plasma membrane. (True or false?)
4. Females do not experience erection. (True or false?)
5. Most of the lubrication for sexual intercourse is provided by the female. (True or false?)
6. If a follicle does not reach maturity during one ovarian cycle, it can finish maturing during the next cycle. (True or false?)
7. Low but rising levels of estrogen inhibit tonic LH secretion, whereas high levels of estrogen stimulate the LH surge. (True or false?)
8. Spermatogenesis takes place within the _____ of the testes, stimulated by the hormones _____ and _____.
9. The source of estrogen and progesterone during the first 10 weeks of gestation is the _____.
10. Detection of _____ in the urine is the basis of pregnancy diagnosis tests.
11. Which of the following statements concerning chromosomal distribution is *incorrect*?
    a. All human somatic cells contain 23 chromosomal pairs for a total diploid number of 46 chromosomes.
    b. Each gamete contains 23 chromosomes, one member of each chromosomal pair.
    c. During meiotic division, the members of the chromosome pairs regroup themselves into the original combinations derived from the individual's mother and father for separation into haploid gametes.
    d. Sex determination depends on the combination of sex chromosomes, an XY combination being a genetic male, XX a genetic female.
    e. The sex chromosome content of the fertilizing sperm determines the sex of the offspring.
12. When the corpus luteum degenerates,
    a. circulating levels of estrogen and progesterone rapidly decline.
    b. FSH and LH secretion start to rise as the inhibitory effects of the gonadal steroids are withdrawn.
    c. the endometrium sloughs off.
    d. Both a and b are correct.
    e. All of the preceding are correct.
13. Match the following:
    ____ 1. secrete(s) prostaglandins
    ____ 2. increase(s) motility and fertility of sperm
    ____ 3. secrete(s) an alkaline fluid
    ____ 4. provide(s) fructose
    ____ 5. storage site for sperm
    ____ 6. concentrate(s) the sperm a hundredfold
    ____ 7. secrete(s) fibrinogen
    ____ 8. provide(s) clotting enzymes
    ____ 9. contain(s) erectile tissue

    (a) epididymis and ductus deferens
    (b) prostate gland
    (c) seminal vesicles
    (d) bulbourethral glands
    (e) penis
14. Using the following answer code, indicate when each event takes place during the ovarian cycle:
    (a) occurs during the follicular phase
    (b) occurs during the luteal phase
    (c) occurs during both the follicular and luteal phases
    ____ 1. development of antral follicles
    ____ 2. secretion of estrogen
    ____ 3. secretion of progesterone
    ____ 4. menstruation
    ____ 5. repair and proliferation of the endometrium
    ____ 6. increased vascularization and glycogen storage in the endometrium

Essay Questions

1. What are the primary reproductive organs, gametes, sex hormones, reproductive tract, accessory sex glands, external genitalia, and secondary sexual characteristics in males and in females?
2. List the essential reproductive functions of the male and of the female.
3. Discuss the differences between males and females with regard to genetic, gonadal, and phenotypic sex.
4. Of what functional significance is the scrotal location of the testes?
5. Discuss the source and functions of testosterone.
6. Describe the three major stages of spermatogenesis. Discuss the functions of each part of a spermatozoon. What are the roles of Sertoli cells?
7. Discuss the control of testicular function.
8. Compare the sex act in males and females.
9. Compare oogenesis with spermatogenesis.
10. Describe the events of the follicular and luteal phases of the ovarian cycle. Correlate the phases of the uterine cycle with those of the ovarian cycle.
11. How are the ovum and spermatozoa transported to the site of fertilization? Describe the process of fertilization.
12. Describe the process of implantation and placenta formation.
13. What are the functions of the placenta? What hormones does the placenta secrete?
14. What is the role of human chorionic gonadotropin?
15. What is the leading proposal for the mechanism that initiates parturition? What are the stages of labor? What is the role of oxytocin?
16. Describe the hormonal factors that play a role in lactation.
17. Summarize the actions of estrogen and progesterone.

POINTS TO PONDER

(Explanations on p. A-49)

1. The hypothalamus releases GnRH in pulsatile bursts once every two to three hours, with no secretion occurring in between. The blood concentration of GnRH depends on the frequency of these bursts of secretion. A promising line of research for a new method of contraception involves administration of GnRH-like drugs. In what way could such drugs act as contraceptives when GnRH is the hypothalamic hormone that triggers the chain of events leading to ovulation? (*Hint:* The anterior pituitary is "programmed" to respond only to the normal pulsatile pattern of GnRH.)

2. Occasionally, testicular tumors composed of interstitial cells of Leydig may secrete up to 100 times the normal amount of testosterone. When such a tumor develops in young children, they grow up much shorter than their genetic potential. Explain why. What other symptoms would be present?

3. What type of sexual dysfunction might arise in men taking drugs that inhibit sympathetic nervous system activity as part of the treatment for high blood pressure?

4. Explain the physiologic basis for administering a posterior pituitary extract to induce or facilitate labor.

5. The symptoms of menopause are sometimes treated with supplemental estrogen and progesterone. Why wouldn't treatment with GnRH or FSH and LH also be effective?

CLINICAL CONSIDERATION

(Explanation on p. A-49)

Maria A., who is in her second month of gestation, has been experiencing severe abdominal cramping. Her physician has diagnosed her condition as a tubal pregnancy: The developing embryo is implanted in the oviduct instead of in the uterine endometrium. Why must this pregnancy be surgically terminated?

PHYSIOEDGE RESOURCES

PhysioEdge CD-ROM

PhysioEdge, the CD-ROM packaged with your text, focuses on the concepts students find most difficult to learn. Figures marked with this icon have associated activities on the CD. For a visual review of concepts in this chapter, check out the following:

Tutorial: Neural and Hormonal Communication—Hormones tab

Media Exercise 18.1: Male Reproductive Anatomy

Media Exercise 18.2: Female Reproductive Anatomy

Media Exercise 18.3: Male Reproductive Physiology

Media Exercise 18.4: Female Reproductive Physiology

PhysioEdge Website

The website for this book contains a wealth of helpful study aids, as well as many ideas for further reading and research. Log on to:
http://www.brookscole.com/hpfundamentals3
Select Chapter 18 from the drop-down menu or click on one of the many resource areas, including Case Histories, which introduce clinical aspects of human physiology. For this chapter check out Case History 23: Pregnancy Tests.

For Suggested Readings, consult **InfoTrac College Edition/ Research** on the PhysioEdge website or go directly to InfoTrac College Edition, your online research library, at:
http://infotrac.thomsonlearning.com

The Metric System

▲ TABLE A-1

Metric Measures and English Equivalents

UNIT	MEASURE	SYMBOL	ENGLISH EQUIVALENT
Linear Measure			
1 kilometer	= 1,000 meters	10^3m km	0.62137 mile
1 meter		10^0m m	39.37 inches
1 decimeter	= 1/10 meter	10^{-1}m dm	3.937 inches
1 centimeter	= 1/100 meter	10^{-2}m cm	0.3937 inch
1 millimeter	= 1/1,000 meter	10^{-3}m mm	Not used
1 micrometer (or micron)	= 1/1,000,000 meter	10^{-6}m μm (or μ)	Not used
1 nanometer	= 1/1,000,000,000 meter	10^{-9}m nm	Not used
Measures of Capacity (for fluids and gases)			
1 liter		l	1.0567 U.S. liquid quarts
1 milliliter	= 1/1,000 liter = volume of 1 g of water at stp*	ml	
Measures of Volume			
1 cubic meter		m^3	
1 cubic decimeter	= 1/1,000 cubic meter = 1 liter (l)	dm^3	
1 cubic centimeter	= 1/1,000,000 cubic meter = 1 milliliter (ml)	cm^3 = ml	
1 cubic millimeter	= 1/100,000,000 cubic meter	mm^3	
Measures of Mass			
1 kilogram	= 1,000 grams	kg	2.2046 pounds
1 gram		g	15.432 grains
1 milligram	= 1/1,000 gram	mg	0.01 grain (about)
1 microgram	= 1/1,000,000 gram	μg (or mcg)	

*stp = standard temperature and pressure

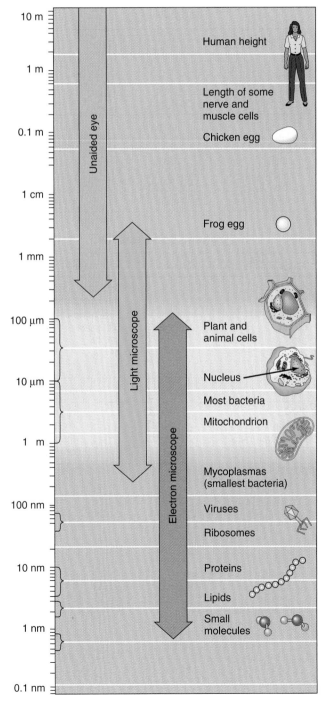

Comparison of human height in meters with the sizes of some biological and molecular structures

Metric–English Conversions

Length

English		Metric	
inch	=	2.54	centimeters
foot	=	0.30	meter
yard	=	0.91	meter
mile (5,280 feet)	=	1.61	kilometer

To convert	multiply by	to obtain
inches	2.54	centimeters
feet	30.00	centimeters
centimeters	0.39	inches
millimeters	0.039	inches

Mass/Weight

English		Metric	
grain	=	64.80	milligrams
ounce	=	28.35	grams
pound	=	453.60	grams
ton (short) (2,000 pounds)	=	0.91	metric ton

To convert	multiply by	to obtain
ounces	28.3	grams
pounds	453.6	grams
pounds	0.45	kilograms
grams	0.035	ounces
kilograms	2.2	pounds

Volume and Capacity

English		Metric	
cubic inch	=	16.39	cubic centimeters
cubic foot	=	0.03	cubic meter
cubic yard	=	0.765	cubic meters
ounce	=	0.03	liter
pint	=	0.47	liter
quart	=	0.95	liter
gallon	=	3.79	liters

To convert	multiply by	to obtain
fluid ounces	30.00	milliliters
quart	0.95	liters
milliliters	0.03	fluid ounces
liters	1.06	quarts

Linear Measurement Comparison

Fahrenheit–Celsius Temperature Comparison

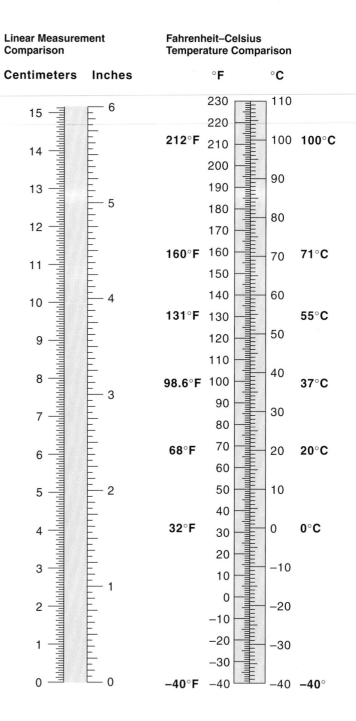

A Review of Chemical Principles

By Spencer Seager, Weber State College, and Lauralee Sherwood

CHEMICAL LEVEL OF ORGANIZATION IN THE BODY

Matter is anything that occupies space and has mass, including all living and nonliving things in the universe. **Mass** is the amount of matter in an object. **Weight**, in contrast, is the effect of gravity on that mass. The more gravity exerted on a mass, the greater the weight of the mass. An astronaut has the same mass whether on Earth or in space but is weightless in the zero gravity of space.

▎Atoms

All matter is made up of tiny particles called **atoms.** These particles are too small to be seen individually, even with the most powerful electron microscopes available today.

Even though extremely small, atoms consist of three types of even smaller subatomic particles. Different types of atoms vary in the number of these various subatomic particles they contain. **Protons** and **neutrons** are particles of nearly identical mass, with protons carrying a positive charge and neutrons having no charge. **Electrons** have a much smaller mass than protons and neutrons and are negatively charged. An atom consists of two regions—a dense, central *nucleus* made of protons and neutrons surrounded by a three-dimensional *electron cloud*, where electrons move rapidly around the nucleus in orbitals (● Figure B-1). The magnitude of the charge of a proton exactly matches that of an electron, but it is opposite in sign, being positive. In all atoms, the number of protons in the nucleus is equal to the number of electrons moving around the nucleus, so their charges balance and the atoms are neutral.

▎Elements and atomic symbols

A pure substance composed of only one type of atom is called an **element**. A pure sample of the element carbon contains only carbon atoms, even though the atoms might be arranged in the form of diamond or in the form of graphite (pencil "lead"). Each element is designated by an **atomic symbol,** a one- or two-letter chemical shorthand for the element's name. Usually these symbols are easy to follow, because they are derived from the English name for the element. Thus H stands for *hydrogen,* C for *carbon,* and O for *oxygen.* In a few cases, the atomic symbol is based on the element's Latin name—for example, Na for *sodium (natrium* in Latin) and K for *potassium (kalium).* Of the 109 known elements, 26 are normally found in the body. Four elements—oxygen, carbon, hydrogen, and nitrogen—compose 96% of the body's mass.

▎Compounds and molecules

Pure substances composed of more than one type of atom are known as **compounds.** Pure water, for example, is a compound that contains atoms of hydrogen and atoms of oxygen in a 2-to-1 ratio, regardless of whether the water is in the form of liquid, solid (ice), or vapor (steam). A **molecule** is the smallest unit of a pure substance that has the properties of that substance and is capable of a stable, independent existence. For example, a molecule of water consists of 2 atoms of hydrogen and 1 atom of oxygen, held together by chemical bonds.

● **FIGURE B-1**

The atom. The atom consists of two regions. The central nucleus contains protons and neutrons and makes up 99.9% of the mass. Surrounding the nucleus is the electron cloud, where the electrons move rapidly around the nucleus. (Figure not drawn to scale.)

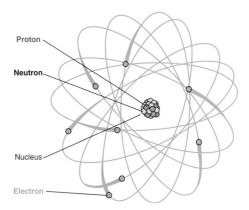

Proton

Neutron

Nucleus

Electron

Atomic number

Exactly what are we talking about when we say a "type" of atom? That is, how do carbon, hydrogen, and oxygen atoms differ? The answer is the number of protons in the nucleus. Regardless of where they are found, all hydrogen atoms have 1 proton in the nucleus, all carbon atoms have 6, and all oxygen atoms have 8. Of course, these numbers also represent the number of electrons moving around each nucleus, because the number of electrons and number of protons in an atom are equal. The number of protons in the nucleus of an atom of an element is called the **atomic number** of the element.

Atomic weight

As expected, tiny atoms have tiny masses. For example, the actual mass of a hydrogen atom is 1.67×10^{-24} g, that of a carbon atom is 1.99×10^{-23} g, and that of an oxygen atom is 2.66×10^{-23} g. These very small numbers are inconvenient to work with in calculations, so a system of relative masses has been developed. These relative masses simply compare the actual masses of the atoms with each other. Suppose the actual masses of two people were determined to be 45.50 kg and 113.75 kg. Their relative masses are determined by dividing each mass by the smaller mass of the two: 45.50/45.50 = 1.00, and 113.75/45.50 = 2.50. Thus the relative masses of the two people are 1.00 and 2.50; these numbers simply express the fact that the mass of the heavier person is 2.50 times that of the other person. The relative masses of atoms are called **atomic masses**, or **atomic weights**, and are given in atomic mass units (*amu*). In this system, hydrogen atoms, the least massive of all atoms, have an atomic weight of 1.01 amu. The atomic weight of carbon atoms is 12.01 amu, and that of oxygen atoms is 16.00 amu. Thus, oxygen atoms have a mass about 16 times that of hydrogen atoms. ▲ Table B-1 gives the atomic weights and some other characteristics of the elements that are most important physiologically.

CHEMICAL BONDS

Because all matter is made up of atoms, atoms must somehow be held together to form matter. The forces holding atoms together are called *chemical bonds*. Not all chemical bonds are formed in the same way, but all involve the electrons of atoms. Whether one atom will bond with another depends on the number and arrangement of its electrons. An atom's electrons are arranged in electron shells, to which we now turn our attention.

Electron shells

Electrons tend to move around the nucleus in a specific pattern. The orbitals, or pathways traveled by electrons around the nucleus, are arranged in an orderly series of concentric layers known as **electron shells**, which consecutively surround the nucleus. Each electron shell can hold a specific number of electrons. The first (innermost) shell closest to

▲ **TABLE B-1**

Characteristics of Selected Elements

NAME AND SYMBOL	NUMBER OF PROTONS	ATOMIC NUMBER	ATOMIC WEIGHT (amu)
Hydrogen (H)	1	1	1.01
Carbon (C)	6	6	12.01
Nitrogen (N)	7	7	14.01
Oxygen (O)	8	8	16.00
Sodium (Na)	11	11	22.99
Magnesium (Mg)	12	12	24.31
Phosphorus (P)	15	15	30.97
Sulfur (S)	16	16	32.06
Chlorine (Cl)	17	17	35.45
Potassium (K)	19	19	39.10
Calcium (Ca)	20	20	40.08

the nucleus can contain a maximum of only 2 electrons, no matter what the element is. The second shell can hold a total of 8 more electrons. The third shell can hold a maximum of 18 electrons. As the number of electrons increases with increasing atomic number, still more electrons occupy successive shells, each at a greater distance from the nucleus. Each successive shell from the nucleus has a higher **energy level**. Because the negatively charged electrons are attracted to the positively charged nucleus, it takes more energy for an electron to overcome the nuclear attraction and orbit farther from the nucleus. Thus the first electron shell has the lowest energy level and the outermost shell of an atom has the highest energy level.

In general, electrons belong to the lowest energy shell possible, up to the maximum capacity of each shell. For example, hydrogen atoms have only 1 electron, so it is in the first shell. Helium atoms have 2 electrons, which are both in the first shell and fill it. Carbon atoms have 6 electrons, 2 in the first shell and 4 in the second shell, whereas the 8 electrons of oxygen are arranged with 2 in the first shell and 6 in the second shell.

Bonding characteristics of an atom; valence

Atoms tend to undergo processes that result in a filled outermost electron shell. Thus the electrons of the outer or higher-energy shell determine the bonding characteristics of an atom and its ability to interact with other atoms. Atoms that have a vacancy in their outermost shell tend to either give up, accept, or share electrons with other atoms (whichever is most favorable energetically) so that all participating atoms have filled outer shells. For example, an atom that has only 1 electron in its outermost shell may empty this shell so its remaining shells are completely full. By contrast, another atom that

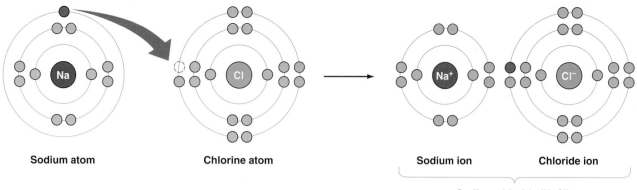

Sodium atom   Chlorine atom   Sodium ion   Chloride ion

Sodium chloride (NaCl)

● **FIGURE B-2**

Ions and ionic bonds. Sodium (Na) and chlorine (Cl) atoms both have partially filled outermost shells. There-fore, sodium tends to give up its lone electron in the outer shell to chlorine, filling chlorine's outer shell. As a result, sodium becomes a positively charged ion, and chlorine becomes a negatively charged ion known as *chloride.* The oppositely charged ions attract each other, forming an ionic bond.

lacks only 1 electron in its outer shell may acquire the defi-cient electron from the first atom to fill all its shells to the max-imum. The number of electrons an atom loses, gains, or shares to achieve a filled outer shell is known as the atom's **valence.** A **chemical bond** is the force of attraction that holds partici-pating atoms together as a result of an interaction between their outermost electrons.

Consider sodium atoms (Na) and chlorine atoms (Cl) (● Figure B-2). Sodium atoms have 11 electrons: 2 in the first shell, 8 in the second shell, and 1 in the third shell. Chlorine atoms have 17 electrons: 2 in the first shell, 8 in the second shell, and 7 in the third shell. Because 8 electrons are required to fill the second and third shells, sodium atoms have 1 elec-tron more than is needed to provide a filled second shell, whereas chlorine atoms have 1 less electron than is needed to fill the third shell. Each sodium atom can lose an electron to a chlorine atom, leaving each sodium with 10 electrons, 8 of which are in the second shell, which is full and is now the outer shell occupied by electrons. By accepting 1 electron, each chlorine atom now has a total of 18 electrons, with 8 of them in the third, or outer, shell, which is now full.

▌ Ions; ionic bonds

Recall that atoms are electrically neutral, because they have an identical number of positively charged protons and nega-tively charged electrons. By giving up and accepting elec-trons, the sodium atoms and chlorine atoms have achieved filled outer shells, but now each atom is unbalanced electri-cally. Although each sodium now has 10 electrons, it still has 11 protons in the nucleus and a net electrical charge, or va-lence, of +1. Similarly, each chlorine now has 18 electrons, but only 17 protons. Thus each chlorine has a −1 charge. Such charged atoms are called **ions.** Positively charged ions are called **cations;** negatively charged ions are called **anions.** As a helpful hint to keep these terms straight, imagine the "t" in *cation* as standing for a "+" sign and the first "n" in *anion* as standing for "negative."

Note that both a cation and anion are formed whenever an electron is transferred from one atom to another. Because opposite charges attract, sodium ions (Na^+) and charged chlorine atoms, now called *chloride* ions (Cl^-), are attracted toward each other. This electrical attraction that holds cations and anions together is known as an **ionic bond.** Ionic bonds hold Na^+ and Cl^- together in the compound **sodium chlo-ride, NaCl,** which is common table salt. A sample of sodium chloride actually contains sodium and chloride ions in a three-dimensional geometric arrangement called a *crystal lattice.* The ions of opposite charge occupy alternate sites within the lattice (● Figure B-3).

● **FIGURE B-3**

Crystal lattice for sodium chloride (table salt)

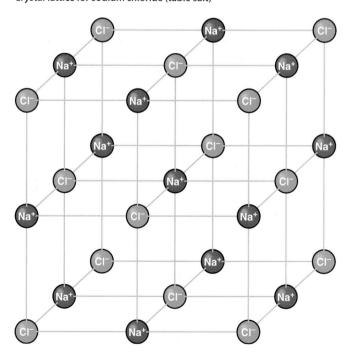

Covalent bonds

It is not favorable, energywise, for an atom to give up or accept more than 3 electrons. Nevertheless, carbon atoms, which have 4 electrons in their outer shell, form compounds. They do so by another bonding mechanism, *covalent bonding*. Atoms that would have to lose or gain 4 or more electrons to achieve outer-shell stability usually bond by *sharing* electrons. Shared electrons actually orbit around *both* atoms. Thus a carbon atom can share its 4 outer electrons with the 4 electrons of four hydrogen atoms, as shown in Equation B-1, where the outer-shell electrons are shown as dots around the symbol of each atom. (The resulting compound is methane, CH_4, a gas made up of individual CH_4 molecules.)

$$\overset{\cdot}{\underset{\cdot}{C}}\cdot + 4\cdot H \rightarrow H\!:\!\overset{\overset{\displaystyle H}{..}}{\underset{\underset{\displaystyle H}{..}}{C}}\!:\!H \qquad \text{Eq. B-1}$$

Shared electron pairs

Shared electron pairs

Each electron that is shared by two atoms is counted toward the number of electrons needed to fill the outer shell of each atom. Thus each carbon atom shares four pairs, or 8 electrons, and so has 8 in its outer shell. Each hydrogen shares one pair, or 2 electrons, and so has a filled outer shell. (Remember, hydrogen atoms need only 2 electrons to complete their outer shell, which is the first shell.) The sharing of a pair of electrons by atoms binds them together by means of a **covalent bond** (● Figure B-4). Covalent bonds are the strongest of chemical bonds; that is, they are the hardest to break.

Covalent bonds also form between some identical atoms. For example, two hydrogen atoms can complete their outer shells by sharing one electron pair made from the single electrons of each atom, as shown in Equation B-2:

$$H\cdot + \cdot H \rightarrow H\!:\!H \qquad \text{Eq. B-2}$$

Thus hydrogen gas consists of individual H_2 molecules (● Figure B-4a). (A subscript after a chemical symbol indicates the number of that type of atom present in the molecule.) Several other nonmetallic elements also exist as molecules, because covalent bonds form between identical atoms; oxygen (O_2) is an example (● Figure B-4b).

Often an atom can form covalent bonds with more than one atom. One of the most familiar examples is water (H_2O), consisting of two hydrogen atoms each forming a single covalent bond with one oxygen atom (● Figure B-4c). Equation B-3 represents the formation of water's covalent bonds:

● **FIGURE B-4**

A covalent bond. A covalent bond is formed when atoms that share a pair of electrons are both attracted toward the shared pair.

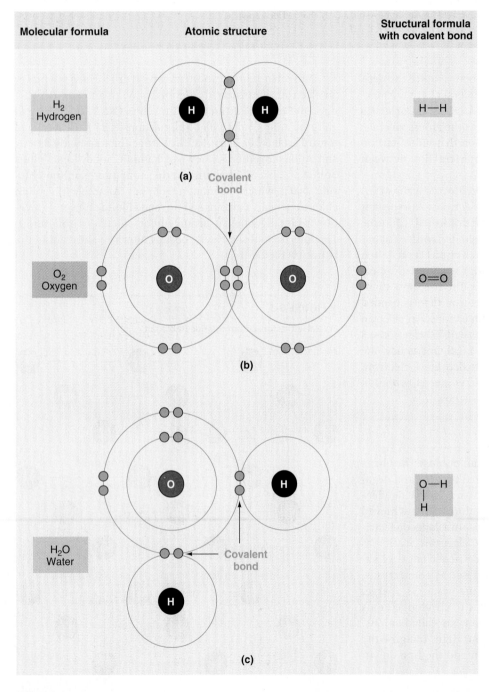

Molecular formula	Atomic structure	Structural formula with covalent bond
H_2 Hydrogen	(a) Covalent bond	H—H
O_2 Oxygen	(b)	O=O
H_2O Water	(c) Covalent bond	O—H | H

$$H\cdot \atop H\cdot} \quad + \cdot \ddot{O}: \rightarrow H:\ddot{O}: \atop H} \qquad \text{Eq. B-3}$$

The water molecule is sometimes represented as

$$H\!\!-\!\!O \atop | \atop H}$$

where the nonshared electron pairs are not shown and the covalent bonds, or shared pairs, are represented by dashes.

Nonpolar and polar molecules

The electrons between two atoms in a covalent bond are not always shared equally. When the atoms sharing an electron pair are identical, such as two oxygen atoms, the electrons are attracted equally by both atoms and so are shared equally. The result is a **nonpolar molecule**. The term *nonpolar* implies no difference at the two ends (two "poles") of the bond. Because both atoms within the molecule exert the same pull on the shared electrons, each shared electron spends the same amount of time orbiting each atom. Thus both atoms remain electrically neutral in a nonpolar molecule such as O_2.

When the sharing atoms are not identical, unequal sharing of electrons occurs, because atoms of different elements do not exert the same pull on shared electrons. For example, an oxygen atom strongly attracts electrons when it is bonded to other atoms. A **polar molecule** results from the unequal sharing of electrons between different types of atoms covalently bonded together. The water molecule is a good example of a polar molecule. The oxygen atom pulls the shared electrons more strongly than do the hydrogen atoms within each of the two covalent bonds. Consequently, the electron of each hydrogen atom tends to spend more time orbiting around the oxygen atom than at home around the hydrogen atom. Because of this nonuniform distribution of electrons, the oxygen side of the water molecule where the shared electrons spend more time is slightly negative, and the two hydrogens that are visited less frequently by the electrons are slightly more positive (● Figure B-5). Note that the entire

---

● FIGURE B-5

A polar molecule. A water molecule is an example of a polar molecule, in which the distribution of shared electrons is not uniform. Because the oxygen atom pulls the shared electrons more strongly than the hydrogen atoms do, the oxygen side of the molecule is slightly negatively charged, and the hydrogen sides are slightly positively charged.

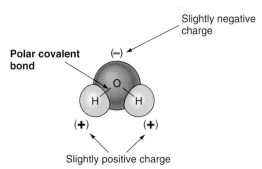

water molecule has the same number of electrons as it has protons, and so as a whole has no net charge. This is unlike ions, which have an electron excess or deficit. Polar molecules have a balanced number of protons and electrons but an unequal distribution of the shared electrons among the atoms making up the molecule.

Hydrogen bonds

Polar molecules are attracted to other polar molecules. In water, for example, an attraction exists between the positive hydrogen ends of some molecules and the negative oxygen ends of others. Hydrogen is not a part of all polar molecules, but when it is covalently bonded to an atom that strongly attracts electrons to form a covalent molecule, the attraction of the positive (hydrogen) end of the polar molecule to the negative end of another polar molecule is called a **hydrogen bond** (● Figure B-6). Thus, the polar attractions of water molecules to each other are an example of hydrogen bonding.

CHEMICAL REACTIONS

Processes in which chemical bonds are broken and/or formed are called **chemical reactions**. Reactions are represented by equations in which the reacting substances (**reactants**) are typically written on the left, the newly produced substances (**products**) are written on the right, and an arrow meaning "yields" points from the reactants to the products. These conventions are illustrated in Equation B-4:

$$A + B \rightarrow C + D \qquad \text{Eq. B-4}$$
$$\text{Reactants} \quad \text{Products}$$

Balanced equations

A chemical equation is a "chemical bookkeeping" ledger that describes what happens in a reaction. By the **law of conservation of mass**, the total mass of all materials entering a re-

---

● FIGURE B-6

A hydrogen bond. A hydrogen bond is formed by the attraction of a positively charged hydrogen end of a polar molecule to the negatively charged end of another polar molecule.

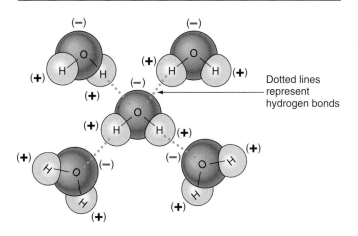

Dotted lines represent hydrogen bonds

action equals the total mass of all the products. Thus the total number of atoms of each element must always be the same on the left and right sides of the equation, because no atoms are lost. Such equations in which the same number of atoms of each type appear on both sides are called **balanced equations**. When writing a balanced equation, the number *preceding* a chemical symbol designates the number of independent (unjoined) atoms, ions, or molecules of that type, whereas a number written as a subscript *following* a chemical symbol denotes the number of a particular atom within a molecule. The absence of a number indicates "one" of that particular chemical. Let's look at a specific example, the oxidation of glucose (the sugar that cells use as fuel), as shown in Equation B-5:

$$C_6H_{12}O_6 + 6O_2 \rightarrow 6CO_2 + 6H_2O$$

glucose   oxygen   carbon   water   Eq. B-5
dioxide

According to this equation, 1 molecule of glucose reacts with 6 molecules of oxygen to produce 6 molecules of carbon dioxide and 6 molecules of water. Note the following balance in this reaction:

•   6 carbon atoms on the left (in 1 glucose molecule) and 6 carbon atoms on the right (in 6 carbon dioxide molecules)
•   12 hydrogen atoms on the left (in 1 glucose molecule) and 12 on the right (in 6 water molecules, each containing 2 hydrogen atoms)
•   18 oxygen atoms on the left (6 in 1 glucose molecule plus 12 more in the 6 oxygen molecules) and 18 on the right (12 in 6 carbon dioxide molecules, each containing 2 oxygen atoms, and 6 more in the 6 water molecules, each containing 1 oxygen atom)

▌ Reversible and irreversible reactions

Under appropriate conditions, the products of a reaction can be changed back to the reactants. For example, carbon dioxide gas dissolves in and reacts with water to form carbonic acid, H_2CO_3:

$$CO_2 + H_2O \rightarrow H_2CO_3 \qquad \text{Eq. B-6}$$

Carbonic acid is not very stable, however, and as soon as some is formed, part of it decomposes to give carbon dioxide and water:

$$H_2CO_3 \rightarrow CO_2 + H_2O \qquad \text{Eq. B-7}$$

Reactions that go in both directions are called **reversible reactions**. They are usually represented by double arrows pointing in both directions:

$$CO_2 + H_2O \rightleftarrows H_2CO_3 \qquad \text{Eq. B-8}$$

Theoretically, every reaction is reversible. Often, however, conditions are such that a reaction, for all practical purposes, goes only in one direction; such a reaction is called **irreversible**. For example, an irreversible reaction takes place when an explosion occurs, because the products do not remain in the vicinity of the reaction site to get together to react.

▌ Catalysts; enzymes

The rates (speeds) of chemical reactions are influenced by a number of factors, of which catalysts are one of the most important. A **catalyst** is a "helper" molecule that speeds up a reaction without being used up in the reaction. Living organisms use catalysts known as **enzymes.** These enzymes exert amazing influence on the rates of chemical reactions that take place in the organisms. Reactions that take weeks or even months to occur under normal laboratory conditions take place in seconds under the influence of enzymes in the body. One of the fastest-acting enzymes is **carbonic anhydrase**, which catalyzes the reaction between carbon dioxide and water to form carbonic acid. This reaction is important in the transport of carbon dioxide from tissue cells, where it is produced metabolically, to the lungs, where it is excreted. The equation for the reaction was shown in Equation B-6. Each molecule of carbonic anhydrase catalyzes the conversion of 36 million CO_2 molecules per minute! Enzymes are important in essentially every chemical reaction that takes place in living organisms.

MOLECULAR AND FORMULA WEIGHT AND THE MOLE

Because molecules are made up of atoms, the relative mass of a molecule is simply the sum of the relative masses (atomic weights) of the atoms found in the molecule. The relative masses of molecules are called **molecular masses** or **molecular weights**. The molecular weight of water, H_2O, is thus the sum of the atomic weights of 2 hydrogen atoms and 1 oxygen atom, or 1.01 amu + 1.01 amu + 16.00 amu = 18.02 amu.

Not all compounds exist in the form of molecules. Ionically bonded substances such as sodium chloride consist of three-dimensional arrangements of sodium ions (Na^+) and chloride ions (Cl^-) in a 1-to-1 ratio. The formulas for ionic compounds reflect only the ratio of the ions in the compound and should not be interpreted in terms of molecules. Thus the formula for sodium chloride, NaCl, indicates that the ions combine in a 1-to-1 ratio. It is convenient to apply the concept of relative masses to ionic compounds even though they do not exist as molecules. The **formula weight** for such compounds is defined as the sum of the atomic weights of the atoms found in the formula. Thus the formula weight of NaCl is equal to the sum of the atomic weights of 1 sodium atom and 1 chlorine atom, or 22.99 amu + 35.45 amu = 58.44 amu.

As you have seen, chemical reactions can be represented by equations and discussed in terms of numbers of molecules, atoms, and ions reacting with each other. To carry out reactions in the laboratory, however, a scientist cannot count out numbers of reactant particles but instead must be able to weigh out the correct amount of each reactant. Using the mole concept makes this task possible. A **mole** (abbreviated *mol*) of a pure element or compound is the amount of material contained in a sample of the pure substance that has a mass in grams equal to the substance's atomic weight (for elements) or the molecular weight or formula weight (for compounds). Thus 1 mole of potassium, K, would be a sample of the element with a mass of 39.10 g. Similarly, a mole of H_2O would have a mass of 18.02 g, and a mole of NaCl would be a sample with a mass of 58.44 g.

The fact that atomic weights, molecular weights, and formula weights are relative masses leads to a fundamental characteristic of moles. One mole of oxygen atoms has a mass of 16.00 g, and 1 mole of hydrogen atoms has a mass of 1.01 g. Thus the ratio of the masses of 1 mole of each element is 16.00/1.01, the same as the ratio of the atomic weights for the two elements. Recall that these atomic weights compare the relative masses of oxygen and hydrogen. Accordingly, the number of oxygen atoms present in 16 grams of oxygen (1 mole of oxygen) is the same as the number of hydrogen atoms present in 1.01 grams of hydrogen. Therefore 1 mole of oxygen contains exactly the same number of oxygen atoms as the number of hydrogen atoms in 1 mole of hydrogen. Thus it is possible and sometimes useful to think of a mole as a specific number of particles. This number, called **Avogadro's number**, is equal to 6.02×10^{23}.

SOLUTIONS, COLLOIDS, AND SUSPENSIONS

In contrast to a compound, a **mixture** consists of two or more types of elements or molecules physically blended together (intermixed) instead of being linked by chemical bonds. A compound has very different properties from the individual elements of which it is composed. For example, the solid, white NaCl (table salt) crystals you use to flavor your food are very different from either sodium (a silvery white metal) or chlorine (a poisonous yellow-green gas found in bleach). By comparison, each component of a mixture retains its own chemical properties. If you mix salt and sugar together, each retains its own distinct taste and other individual properties. The constituents of a compound can only be separated by chemical means—bond breakage. By contrast, the components of a mixture can be separated by physical means, such as filtration or evaporation. The most common mixtures in the body are mixtures of water and various other substances. These mixtures are categorized as *solutions, colloids,* or *suspensions,* depending on the size and nature of the substance mixed with water.

Solutions

Most chemical reactions in the body take place between reactants that have dissolved to form solutions. **Solutions** are homogenous mixtures containing a relatively large amount of one substance called the **solvent** (the dissolving medium) and smaller amounts of one or more substances called **solutes** (the dissolved particles). Salt water, for example, contains mostly water, which is thus the solvent, and a smaller amount of salt, which is the solute. Water is the solvent in most solutions found in the human body.

Electrolytes; nonelectrolytes

When ionic solutes are dissolved in water to form solutions, the resulting solution will conduct electricity. This is not true for most covalently bonded solutes. For example, a salt–water solution conducts electricity, but a sugar–water solution does not. When salt dissolves in water, the solid lattice of Na^+ and Cl^- is broken down, and the individual ions are separated and distributed uniformly throughout the solution. These mobile, charged ions conduct electricity through the solution. Solutes that form ions in solution and conduct electricity are called **electrolytes.** When sugar dissolves, however, individual covalently bonded sugar molecules leave the solid and become uniformly distributed throughout the solution. These uncharged molecules cannot conduct a current. Solutes that do not form conductive solutions are called **nonelectrolytes.**

Measures of concentration

The amount of solute dissolved in a specific amount of solution can vary. For example, a salt–water solution might contain 1 g of salt in 100 ml of solution, or it could contain 10 g of salt in 100 ml of solution. Both solutions are salt–water solutions, but they have different concentrations of solute. The **concentration** of a solution indicates the relationship between the amount of solute and the amount of solution. Concentrations can be given in a number of different units.

MOLARITY

Concentrations given in terms of **molarity (M)** give the number of moles of solute in exactly 1 liter of solution. Thus a half molar (0.5 M) solution of NaCl would contain one-half mole, or 29.22 g, of NaCl in each liter of solution.

NORMALITY

When the solute is an electrolyte, it is sometimes useful to express the concentration of the solution in a unit that gives information about the amount of ionic charge in the solution. This is done by expressing concentration in terms of **normality (N)**. The normality of a solution gives the number of equivalents of solute in exactly 1 liter of solution. An **equivalent** of an electrolyte is the amount that produces 1 mole of positive (or negative) charges when it dissolves. The number of equivalents of an electrolyte can be calculated by multiplying the number of moles of electrolyte by the total number of positive charges produced when one formula unit of the electrolyte dissolves. Consider NaCl and calcium chloride ($CaCl_2$) as examples. The ionization reactions for one formula unit of each solute are as follows:

$$NaCl \rightarrow Na^+ + Cl^- \qquad \text{Eq. B-9}$$
$$CaCl_2 \rightarrow Ca^{2+} + 2Cl^- \qquad \text{Eq. B-10}$$

Thus 1 mole of NaCl produces 1 mole of positive charges (Na^+) and so contains 1 equivalent:

$$(1 \text{ mole NaCl})(1) = 1 \text{ equivalent}$$

where the number 1 used to multiply the 1 mole of NaCl came from the +1 charge on Na^+.

One mole of $CaCl_2$ produces 1 mole of Ca^{2+}, which is 2 moles of positive charge. Thus 1 mole of $CaCl_2$ contains 2 equivalents:

$$(1 \text{ mole } CaCl_2)(2) = 2 \text{ equivalents}$$

where the number 2 used in the multiplication came from the +2 charge on Ca^{2+}.

If two solutions were made such that one contained 1 mole of NaCl per liter and the other contained 1 mole of $CaCl_2$ per liter, the NaCl solution would contain 1 equivalent of solute per liter and would be 1 normal (1 N). The $CaCl_2$ solution would contain 2 equivalents of solute per liter and would be 2 normal (2 N).

OSMOLARITY

Another expression of concentration frequently used in physiology is **osmolarity (osm)**, which indicates the total *number* of solute particles in a liter of solution instead of the relative weights of the specific solutes. The osmolarity of a solution is the product of M and *n,* where *n* is the number of moles of solute particles obtained when 1 mole of solute dissolves. Because nonelectrolytes such as glucose do not dissociate in solution, $n = 1$ and the osmolarity (*n* times M) is equal to the molarity of the solution. For electrolyte solutions, the osmolarity exceeds the molarity by a factor equal to the number of ions produced on dissociation of each molecule in solution. For example, because a NaCl molecule dissociates into two ions, Na^+ and Cl^-, the osmolarity of a 1 M solution of NaCl is 2×1 M = 2 osm.

▌ Colloids and suspensions

In solutions, solute particles are ions or small molecules. By contrast, the particles in colloids and suspensions are much larger than ions or small molecules. In colloids and suspensions, these particles are known as **dispersed-phase particles** instead of *solutes.* When the dispersed-phase particles are no more than about 100 times the size of the largest solute particles found in a solution, the mixture is called a **colloid.** The dispersed-phase particles of colloids generally do not settle out. All dispersed-phase particles of colloids carry electrical charges of the same sign. Thus they repel each other. The constant buffeting from these collisions keeps the particles from settling. The most abundant colloids in the body are small functional proteins dispersed in the body fluids. An example is the colloidal dispersion of the plasma proteins in the blood (see p. 316).

When dispersed-phase particles are larger than those in colloids, if the mixture is left undisturbed the particles will settle out because of the force of gravity. Such mixtures are called **suspensions.** The major example of a suspension in the body is the mixture of blood cells suspended in the plasma. The constant movement of blood as it circulates through the blood vessels keeps the blood cells rather evenly dispersed within the plasma. However, if a blood sample is placed in a test tube and treated to prevent clotting, the heavier blood cells gradually settle to the bottom of the tube.

INORGANIC AND ORGANIC CHEMICALS

Chemicals are commonly classified into two categories: inorganic and organic.

▌ Distinction between inorganic and organic chemicals

The original criterion used for this classification was the origin of the chemicals. Those that came from living or once-living sources were *organic,* and those that came from other sources were *inorganic.* Today the basis for classification is the element carbon. **Organic** chemicals are generally those that contain carbon. All others are classified as **inorganic.** A few carbon-containing chemicals are also classified as inorganic; the most common are pure carbon in the form of diamond and graphite, carbon dioxide (CO_2), carbon monoxide (CO), carbonates such as limestone ($CaCO_3$), and bicarbonates such as baking soda ($NaHCO_3$).

The unique ability of carbon atoms to bond to each other and form networks of carbon atoms results in an interesting fact. Even though organic chemicals all contain carbon, millions of these compounds have been identified. Some were isolated from natural plant or animal sources, and many have been synthesized in laboratories. Inorganic chemicals include all the other 108 elements and their compounds. The number of known inorganic chemicals made up of all these other elements is estimated to be about 250,000, compared to millions of organic compounds made up predominantly of carbon.

▌ Monomers and polymers

Another result of carbon's ability to bond to itself is the large size of some organic molecules. Organic molecules range in size from methane, CH_4, a small, simple molecule with one carbon atom, to molecules such as DNA that contain as many as a million carbon atoms. Organic molecules essential for life are called **biological molecules,** or **biomolecules** for short. Some biomolecules are rather small organic compounds, including *simple sugars, fatty acids, amino acids,* and *nucleotides.* These small, single units, known as **monomers** (meaning "single unit"), are subunits, or building blocks, for the synthesis of larger biomolecules, including *complex carbohydrates, lipids, proteins,* and *nucleic acids,* respectively. These larger organic molecules are called **polymers** (meaning "many units"), reflecting the fact that they are made by the bonding together of a number of smaller monomers. For example, starch is formed by linking many glucose molecules together. Very large organic polymers are often referred to as **macromolecules,** reflecting their large size (*macro* means "large"). Macromolecules include many naturally occurring molecules, such as DNA and structural proteins, as well as many synthetically produced molecules, such as synthetic textiles (for example, nylon) and plastics.

ACIDS, BASES, AND SALTS

Acids, bases, and salts may be inorganic or organic compounds.

▌Acids and bases

Acids and bases are chemical opposites, and salts are produced when acids and bases react with each other. In 1887, Swedish chemist Svante Arrhenius proposed a theory defining acids and bases. He said that an *acid* is any substance that will dissociate, or break apart, when dissolved in water and in the process release a hydrogen ion, H^+. Similarly, *bases* are substances that dissociate when dissolved in water and in the process release a hydroxyl ion, OH^-. Hydrogen chloride (HCl) and sodium hydroxide (NaOH) are examples of Arrhenius acids and bases; their dissociations in water are represented in Equations B-11 and B-12, respectively:

$$HCl \rightarrow H^+ + Cl^- \qquad \text{Eq. B-11}$$

$$NaOH \rightarrow Na^+ + OH^- \qquad \text{Eq. B-12}$$

Note that the hydrogen ion is a bare proton, the nucleus of a hydrogen atom. Also note that both HCl and NaOH would behave as electrolytes.

Arrhenius did not know that free hydrogen ions cannot exist in water. They covalently bond to water molecules to form hydronium ions, as shown in Equation B-13:

$$H^+ + \ddot{\underset{H}{O}}-H \rightarrow \left[H-\underset{H}{\ddot{O}}-H \right]^+ \qquad \text{Eq. B-13}$$

In 1923, Johannes Brønsted in Denmark and Thomas Lowry in England proposed an acid–base theory that took this behavior into account. They defined an **acid** as any hydrogen-containing substance that donates a proton (hydrogen ion) to another substance (an acid is a *proton donor*) and a **base** as any substance that accepts a proton (a base is a *proton acceptor*). According to these definitions, the acidic behavior of HCl given in Equation B-11 is rewritten as shown in Equation B-14:

$$HCl + H_2O \rightleftarrows H_3O^+ + Cl^- \qquad \text{Eq. B-14}$$

Note that this reaction is reversible, and the hydronium ion is represented as H_3O^+. In Equation B-14, the HCl acts as an acid in the forward (left-to-right) reaction, whereas water acts as a base. In the reverse reaction (right-to-left), the hydronium ion gives up a proton and thus is an acid, whereas the chloride ion, Cl^-, accepts the proton and so is a base. It is still common practice to use equations such as B-11 to simplify representation of the dissociation of an acid, even though scientists recognize that equations like B-14 are more correct.

▌Salts; neutralization reactions

At room temperature, **inorganic salts** are crystalline solids that contain the positive ion (cation) of an Arrhenius base such as NaOH and the negative ion (anion) of an acid such as HCl. Salts can be produced by mixing solutions of appropriate acids and bases, allowing a neutralization reaction to occur. In **neutralization reactions**, the acid and base react to form a salt and water. Most salts that form are water soluble and can be recovered by evaporating the water. Equation B-15 is a neutralization reaction:

$$HCl + NaOH \rightarrow NaCl + H_2O \qquad \text{Eq. B-15}$$

When acids or bases are used as solutes in solutions, the concentrations can be expressed as normalities just as they were earlier for salts. An equivalent of acid is the amount that gives up 1 mole of H^+ in solution. Thus, 1 mole of HCl is also 1 equivalent, but 1 mole of H_2SO_4 is 2 equivalents. Bases are described in a similar way, but an equivalent is the amount of base that gives 1 mole of OH^-.

See Chapter 14 for a discussion of acid–base balance in the body.

FUNCTIONAL GROUPS OF ORGANIC MOLECULES

Organic molecules consist of carbon and one or more additional elements covalently bonded to one another in "Tinker Toy" fashion. The simplest organic molecules, hydrocarbons, such as methane and petroleum products, have only hydrogen atoms attached to a carbon backbone of varying lengths. All biomolecules always have additional elements besides hydrogen added to the carbon backbone. The carbon backbone forms the stable portion of most biomolecules. Other atoms covalently bonded to the carbon backbone, either alone or in clusters, form what is termed *functional groups*. All organic compounds can be classified according to the functional group or groups they contain. **Functional groups** are specific combinations of atoms that generally react in the same way, regardless of the number of carbon atoms in the molecule to which they are attached. For example, all *aldehydes* contain a functional group that contains one carbon atom, one oxygen atom, and one hydrogen atom covalently bonded in a specific way:

$$\underset{(-C-H)}{\overset{\displaystyle O}{\overset{\displaystyle \|}{}}}$$

The carbon atom in an aldehyde group forms a single covalent bond with the hydrogen atom and a **double bond** (a bond in which two covalent bonds are formed between the same atoms, designated by a double line between the atoms) with the oxygen atom. The aldehyde group is attached to the rest of the molecule by a single covalent bond extending to the left of the carbon atom. Most reactions of aldehydes involve this group, so most aldehyde reactions are the same regardless of the size and nature of the rest of the molecule to which the aldehyde group is attached. Reactions of physiologic importance often occur between two functional groups or between one functional group and a small molecule such as water.

CARBOHYDRATES

Carbohydrates are organic compounds of tremendous biological and commercial importance. They are widely distributed in nature and include such familiar substances as starch, table sugar, and cellulose. Carbohydrates have five important functions in living organisms: they provide energy, they serve as a stored form of chemical energy, they provide dietary fiber, they supply carbon atoms for the synthesis of cell components, and they form part of the structural elements of some cells.

▌Chemical composition of carbohydrates

Carbohydrates contain carbon, hydrogen, and oxygen. They acquired their name because most of them contain these three elements in an atomic ratio of 1 carbon to 2 hydrogens to 1 oxygen. This ratio suggests that the general formula is CH_2O and that the compounds are simply carbon hydrates ("watered" carbons), or carbohydrates. It is now known that they are not hydrates of carbon, but the name persists. All carbohydrates have a large number of functional groups per molecule. The most common functional groups in carbohydrates are *alcohol, ketone* and *aldehyde*—

$$(-OH), \quad (-\overset{\overset{\textstyle O}{\|}}{C}-), \quad (-\overset{\overset{\textstyle O}{\|}}{C}-H)$$
$$\text{Alcohol} \quad \text{Ketone} \quad \text{Aldehyde}$$

or functional groups formed by reactions between pairs of these three.

▌Types of carbohydrates

The simplest carbohydrates are simple sugars, also called **monosaccharides.** As their name indicates, they consist of single units called saccharides (*mono* means "one"). The molecular structure of *glucose,* an important monosaccharide, is shown in ● Figure B-7a. In solution, most glucose molecules assume the ring form shown in ● Figure B-7b. Other common monosaccharides are *fructose, galactose,* and *ribose.*

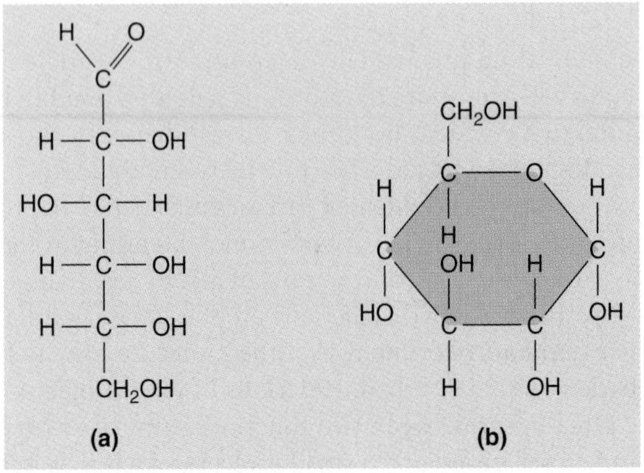

Disaccharides are sugars formed by linking two monosaccharide molecules together through a covalent bond (*di* means "two"). Some common examples of disaccharides are *sucrose* (common table sugar) and *lactose* (milk sugar). Sucrose molecules are formed from one glucose and one fructose molecule. Lactose molecules each contain one glucose and one galactose unit.

Because of the many functional groups on carbohydrate molecules, large numbers of simple carbohydrate molecules can bond together and form long chains and branched networks. The resultant substances, **polysaccharides**, contain many saccharide units (*poly* means "many"). Three common polysaccharides made up entirely of glucose units are glycogen, starch, and cellulose.

- *Glycogen* is a storage carbohydrate found in animals. It is a highly branched polysaccharide that averages a branch every 8 to 12 glucose units. The structure of glycogen is represented in ● Figure B-8, where each circle represents 1 glucose unit.
- *Starch,* a storage carbohydrate of plants, consists of two fractions, amylose and amylopectin. Amylose consists of long, essentially unbranched chains of glucose units. Amylopectin is a highly branched network of glucose units averaging 24 to 30 glucose units per branch. Thus it is less highly branched than glycogen.
- *Cellulose,* a structural carbohydrate of plants, exists in the form of long, unbranched chains of glucose units. The bonding between the glucose units of cellulose is slightly different from the bonding between the glucose units of glycogen and starch. Humans have digestive enzymes that catalyze the breaking (hydrolysis) of the glucose-to-glucose bonds in starch but lack the necessary enzymes to hydrolyze cellulose glucose-to-glucose bonds. Thus starch is a food for humans, but cellulose is not. Cellulose is the indigestible fiber in our diets.

LIPIDS

Lipids are a diverse group of organic molecules made up of substances with widely different compositions and molecular structures. Unlike carbohydrates, which are classified on the

● **FIGURE B-7**

Forms of glucose. (a) Chain. (b) Ring.

(a) **(b)**

● **FIGURE B-8**

A simplified representation of glycogen. Each circle represents a glucose molecule.

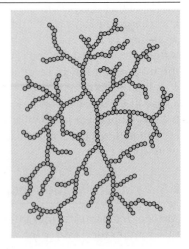

basis of their *molecular structure,* substances are classified as lipids on the basis of their *solubility.* **Lipids** are insoluble in water but soluble in nonpolar solvents such as alcohol. Thus, lipids are the waxy, greasy, or oily compounds found in plants and animals. Lipids repel water, a useful characteristic of the protective wax coatings found on some plants. Fats and oils are energy-rich and have relatively low densities. These properties account for the use of fats and oils as stored energy in plants and animals. Still other lipids occur as structural components, especially in cellular membranes. The oily plasma membrane that surrounds each cell serves as a barrier that separates the intracellular contents from the surrounding extracellular fluid (see p. 2, p. 20, p. 43, and p. 44).

▮ Simple lipids

Simple lipids contain just two types of components, fatty acids and alcohols. **Fatty acid molecules** consist of a hydrocarbon chain with a *carboxylic acid* functional group (—COOH) on the end. The hydrocarbon chain can be of variable length, but natural fatty acids always contain an even number of carbon atoms. The hydrocarbon chain can also contain one or more double bonds between carbon atoms. Fatty acids with no double bonds are called **saturated fatty acids,** whereas those with double bonds are called **unsaturated fatty acids.** The more double bonds present, the higher the degree of unsaturation. Saturated fatty acids predominate in dietary animal products (for example, meat, eggs, and dairy products), whereas unsaturated fatty acids are more prevalent in plant products (for example, grains, vegetables, and fruits). Consumption of a greater proportion of saturated than unsaturated fatty acids is linked with a higher incidence of cardiovascular disease (see p. 269).

The most common alcohol found in simple lipids is **glycerol** (glycerin), a three-carbon alcohol that has three alcohol functional groups (—OH).

Simple lipids called fats and oils are formed by a reaction between the carboxylic acid group of three fatty acids and the three alcohol groups of glycerol. The resulting lipid is an E-shaped molecule called a **triglyceride.** Such lipids are classified as fats or oils on the basis of their melting points.

Fats are solids at room temperature, whereas *oils* are liquids. Their melting points depend on the degree of unsaturation of the fatty acids of the triglyceride. The melting point goes down with increasing degree of unsaturation. Thus oils contain more unsaturated fatty acids than fats do. Examples of the components of fats and oils and a typical triglyceride molecule are shown in ● Figure B-9.

When triglycerides form, a molecule of water is released as each fatty acid reacts with glycerol. Adipose tissue in the body contains triglycerides. When the body uses adipose tissue as an energy source, the triglycerides react with water to release free fatty acids into the blood. The fatty acids can be used as an immediate energy source by many organs. In the liver, free fatty acids are converted into compounds called **ketone bodies.** Two of the ketone bodies are acids, and one is the ketone called *acetone.* Excess ketone bodies are produced during diabetes mellitus, a condition in which most cells resort to using fatty acids as an energy source because the cells cannot take up adequate amounts of glucose in the face of inadequate insulin action (see p. 567).

▮ Complex lipids

Complex lipids have more than two types of components. The different complex lipids usually have three or more of the following components: glycerol, fatty acids, a phosphate group, an alcohol other than glycerol, and a carbohydrate. Those that contain phosphate are called **phospholipids.** ● Figure B-10 shows representations of a few complex lipids; it emphasizes the components but does not give details of the molecular structures.

Steroids are lipids that have a unique structural feature consisting of a fused carbon ring system containing three 6-member rings and a single 5-member ring (● Figure B-11). Different steroids have this characteristic ring structure but have different functional groups and carbon chains attached.

Cholesterol, a steroidal alcohol, is the most abundant steroid in the human body. It is a component of cell membranes and is used by the body to produce other important steroids that include bile salts, male and female sex hormones, and adrenocortical hormones. The structures of cholesterol

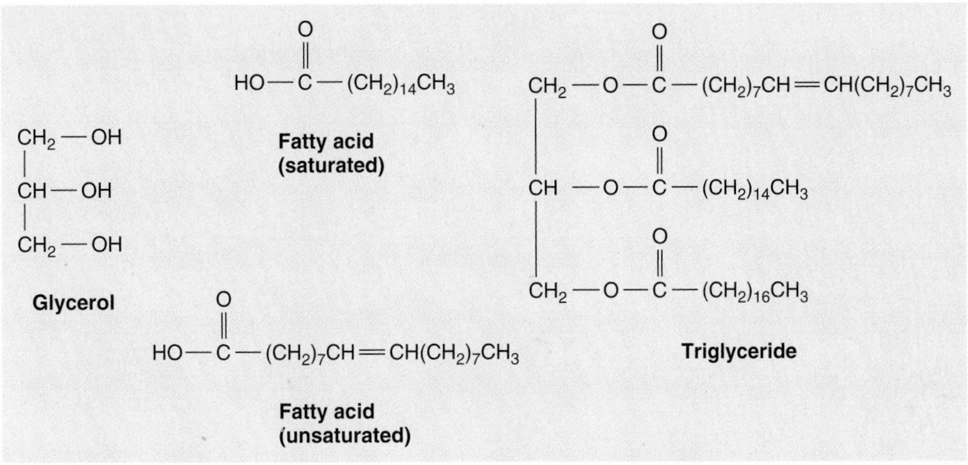

● **FIGURE B-9**

Triglyceride components and structure

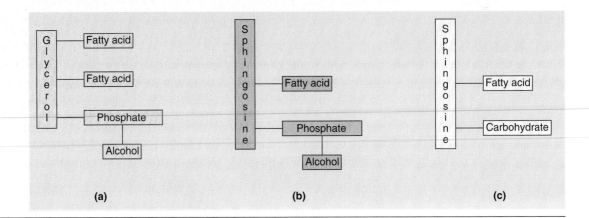

● FIGURE B-10

Examples of complex lipids. (a) A phosphoglyceride. (b) A sphingolipid (sphingosine is an alcohol). (c) A glycolipid.

and cortisol, an important adrenocortical hormone, are shown in ● Figure B-12.

PROTEINS

The name *protein* is derived from the Greek word *proteios,* which means "of first importance." It is certainly an appropriate term for these very important biological compounds. Proteins are indispensable components of all living things, where they play crucial roles in all biological processes. Proteins are the main structural component of cells, and all chemical reactions in the body are catalyzed by enzymes, all of which are proteins.

▌ Chemical composition of proteins

Proteins are macromolecules made up of monomers called **amino acids.** Hundreds of different amino acids, both natural and synthetic, are known, but only 20 are commonly found in natural proteins. From this seemingly limited pool of amino acids, cells build thousands of different types of proteins, each with a distinctly different function, much as composers create a diversity of unique music from a relatively small number of notes. Different proteins are constructed by varying the types and numbers of amino acids used and by varying the order in which they are linked together. Proteins are not built haphazardly, though, by randomly linking together amino acids.

● FIGURE B-11

The steroid ring system. (a) Detailed. (b) Simplified.

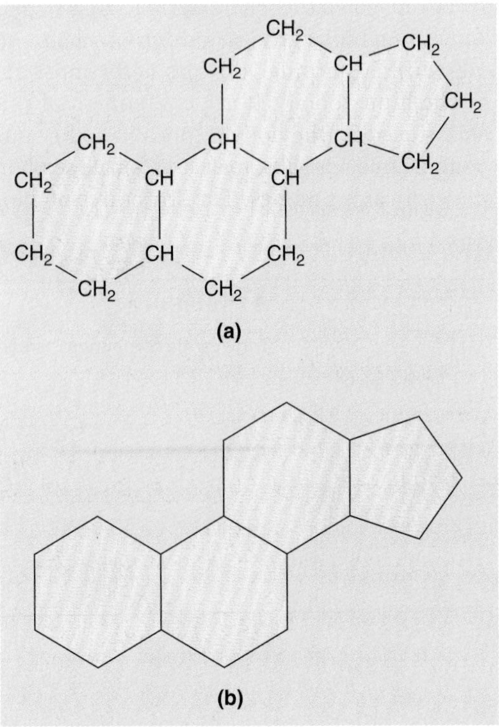

● FIGURE B-12

Examples of steroidal compounds

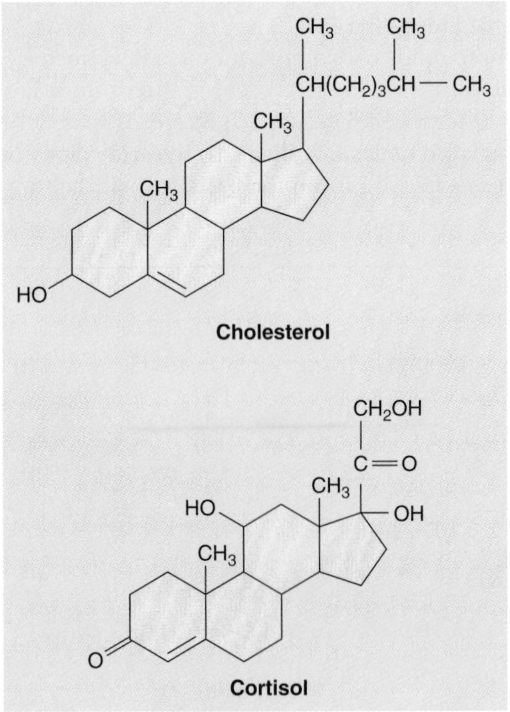

Cholesterol

Cortisol

Every protein in the body is deliberately and precisely synthesized under the direction of the blueprint laid down in the person's genes. Thus amino acids are assembled in a specific pattern to produce a given protein to accomplish a particular structural or functional task in the body. (More information about protein synthesis can be found in Appendix C.)

Peptide bonds

Each amino acid molecule has three important parts: an amino functional group ($-NH_2$), a carboxyl functional group ($-COOH$), and a characteristic side chain or R group. These components are shown in ● Figure B-13. Amino acids form long chains as a result of reactions between the amino group of one amino acid and the carboxyl group of another amino acid. This reaction is illustrated in Equation B-16 in which the components of the carboxyl group are shown in the expanded form for clarity:

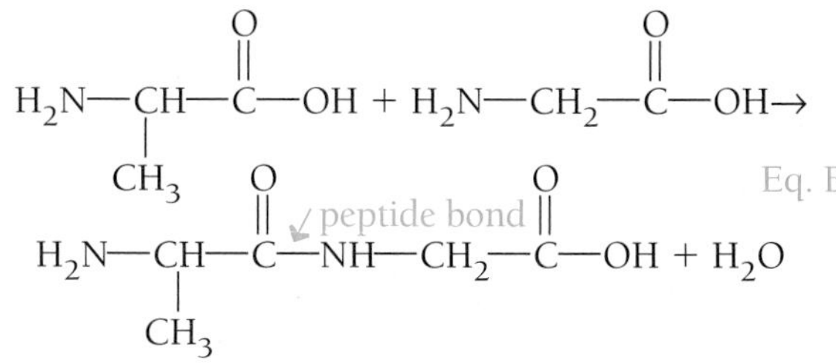

$$H_2N-CH-C(O)-OH + H_2N-CH_2-C(O)-OH \rightarrow$$
$$CH_3$$

peptide bond

$$H_2N-CH-C(O)-NH-CH_2-C(O)-OH + H_2O \qquad \text{Eq. B-16}$$
$$CH_3$$

Notice that after the two molecules react, the ends of the product still have an amino group and a carboxyl group that can react to extend the chain length. The covalent bond formed in the reaction is called a **peptide bond** (● Figure B-14).

On a molecular scale, proteins are immense molecules. Their size can be illustrated by comparing a glucose molecule to a molecule of hemoglobin, a protein. Glucose has a molecular weight of 180 amu and a molecular formula of $C_6H_{12}O_6$. Hemoglobin, a relatively small protein, has a molecular weight of 65,000 amu and a molecular formula of $C_{2952}H_{4664}O_{832}N_{812}S_8Fe_4$.

Levels of protein structure

The many atoms in a protein are not arranged in a random way. In fact, proteins have a high degree of structural organization that plays an important role in their behavior in the body.

● **FIGURE B-13**

The general structure of amino acids

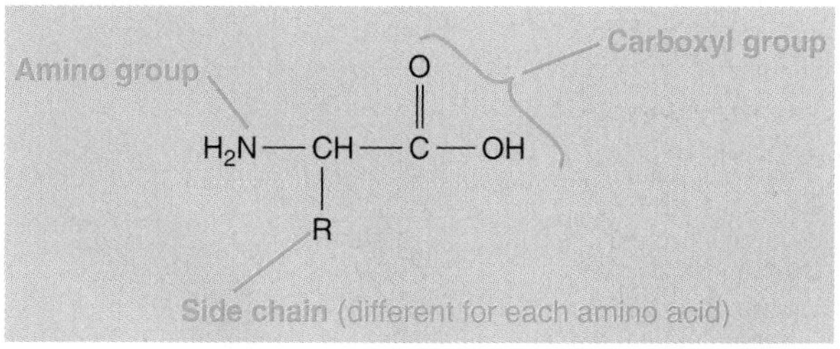

Amino group
Carboxyl group
$$H_2N-CH-C(O)-OH$$
$$R$$
Side chain (different for each amino acid)

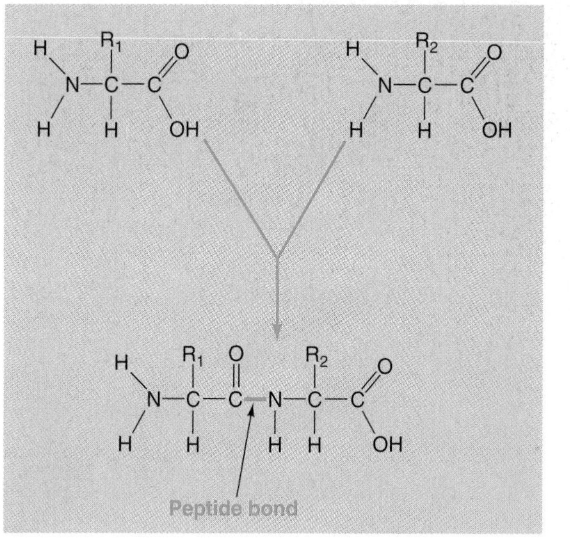

● **FIGURE B-14**

A peptide bond. In forming a peptide bond, the carboxyl group of one amino acid reacts with the amino group of another amino acid.

PRIMARY STRUCTURE

The first level of protein structure is called the **primary structure**. It is simply the order in which amino acids are bonded together to form the protein chain. Amino acids are frequently represented by three-letter abbreviations, such as Gly for glycine and Arg for arginine. In this way the primary structure of a protein can be represented as in ● Figure B-15, which shows part of the primary structure of human insulin, or as in ● Figure B-16a, which depicts a portion of the primary structure of hemoglobin.

SECONDARY STRUCTURE

The second level of protein structure, called the **secondary structure**, results when hydrogen bonding occurs between the amino hydrogen of one amino acid in the primary chain and the carboxyl oxygen

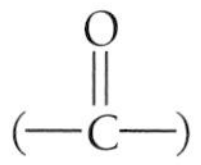

$$(-C(O)-)$$

of another amino acid in the same or another chain. When the hydrogen bonding occurs between amino acids in the same chain, the chain takes a coiled, helical shape called the alpha (α) helix, which is by far the most common secondary structure found in natural proteins (● Figure B-16b). Other secondary structures such as the beta (β) pleated sheet

● **FIGURE B-15**

A portion of the primary protein structure of human insulin

Thr—Lys—Pro—Thr—Tyr—Phe—Phe—Gly—Arg— · · · · ·

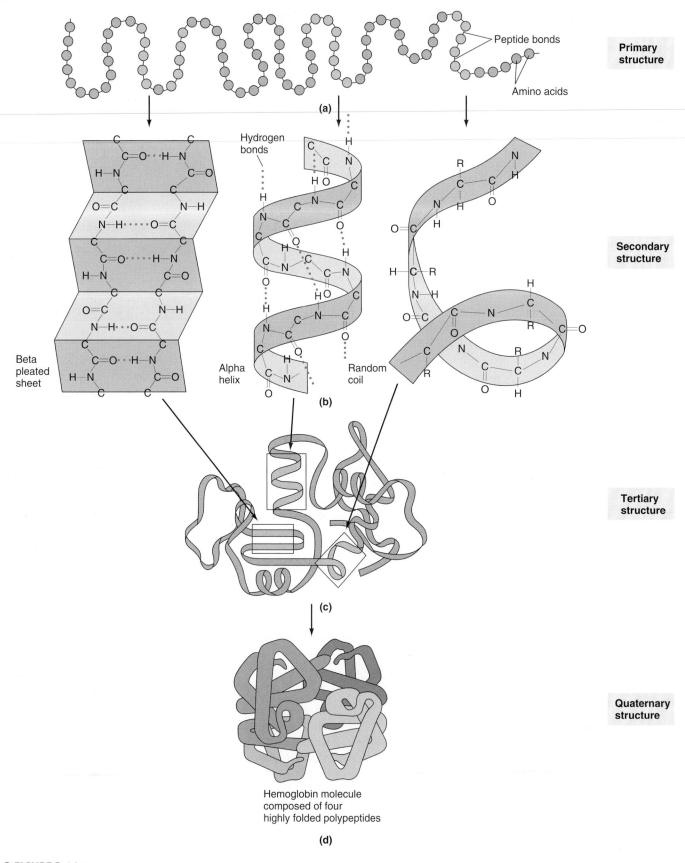

Primary structure

Peptide bonds

Amino acids

(a)

Secondary structure

Hydrogen bonds

Beta pleated sheet

Alpha helix

Random coil

(b)

Tertiary structure

(c)

Quaternary structure

Hemoglobin molecule composed of four highly folded polypeptides

(d)

● **FIGURE B-16**

Levels of protein structure. Proteins can have four levels of structure. (a) The primary structure is a particular sequence of amino acids bonded in a chain. (b) At the secondary level, hydrogen bonding occurs between various amino acids within the chain, causing the chain to assume a particular shape. The most common secondary protein structure in the body is the alpha helix. (c) The tertiary structure is formed by the folding of the secondary structure into a functional three-dimensional configuration. (d) Many proteins form a fourth level of structure composed of several polypeptides, as exemplified by hemoglobin.

and random coils also form as a result of hydrogen bonding between amino acids located in different parts of the same chain.

TERTIARY AND QUATERNARY STRUCTURE

The third level of structure in proteins is the **tertiary structure**. It results when functional groups of the side chains of amino acids in the protein chain react with each other. Several different types of interactions are possible, as shown in ● Figure B-17. Tertiary structures can be visualized by letting a length of wire represent the chain of amino acids in the primary structure of a protein. Next imagine that the wire is wound around a pencil to form a helix, which represents the secondary structure. The pencil is removed, and the helical structure is now folded back on itself or carefully wadded into a ball. Such folded or spherical structures represent the tertiary structure of a protein (● Figure B-16c).

All functional proteins exist in at least a tertiary structure. Sometimes, several polypeptides interact with each other to form a fourth level of protein structure, the **quaternary structure**. For example, hemoglobin contains four highly folded polypeptide chains (the **globin** portion) (● Figure B-16d). Four iron-containing *heme* groups, one tucked within the interior of each of the folded polypeptide subunits, completes the quaternary structure of hemoglobin (see ● Figure 11-3, p. 317).

▌ Hydrolysis and denaturation

One of the important functions of proteins is to serve as enzymes that catalyze the many essential chemical reactions of the body. In addition to catalyzing reactions, proteins can undergo reactions themselves. Two of the most important are hydrolysis and denaturation.

FIGURE B-17

Side chain interactions leading to the tertiary protein structure

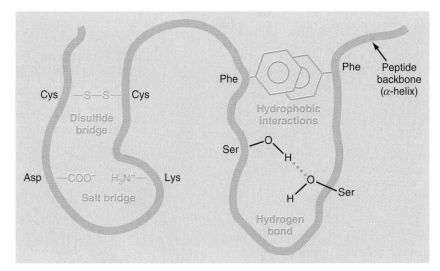

HYDROLYSIS

Notice that according to Equation B-16, the formation of peptide bonds releases water molecules. Under appropriate conditions, it is possible to reverse such reactions by adding water to the peptide bonds and breaking them. **Hydrolysis** ("breakdown by H_2O") reactions of this type convert large proteins into smaller fragments or even into individual amino acids. Hydrolysis is the means by which digestive enzymes break down ingested food into small units that can be absorbed from the digestive tract lumen into the blood. (Dietary proteins, carbohydrates, and fats are all produced through formation of bonds that release H_2O.)

DENATURATION

Denaturation of proteins occurs when the bonds holding a protein chain in its characteristic tertiary or secondary conformation are broken. When this happens, the protein chain takes on a random, disorganized conformation. Denaturation can result when proteins are heated (including when body temperature rises too high; see p. 518), exposed to extremes of pH (see p. 455), or treated with specific chemicals such as alcohol or heavy metal ions. In some instances, denaturation is accompanied by coagulation or precipitation, as illustrated by the changes in the white of an egg as it is fried.

NUCLEIC ACIDS

Nucleic acids are high-molecular-weight macromolecules responsible for storing and using genetic information in living cells and passing it on to future generations. These important biomolecules are classified into two categories: **deoxyribonucleic acids (DNA)** and **ribonucleic acids (RNA)**. DNA is found primarily in the cell's nucleus, and RNA is found primarily in the cytoplasm that surrounds the nucleus.

Both types of nucleic acid are made up of units called **nucleotides**, which in turn are composed of three simpler components. Each nucleotide contains an organic nitrogenous base, a sugar, and a phosphate group. The three components are chemically bonded together with the sugar molecule between the base and the phosphate. In RNA, the sugar is ribose, whereas in DNA it is deoxyribose. When nucleotides bond together to form nucleic acid chains, the bonding is between the phosphate of one nucleotide and the sugar of another. The resulting nucleic acids consist of chains of alternating phosphates and sugar molecules, with a base molecule extending out of the chain from each sugar molecule (see ● Figure C-1, p. A-20).

The chains of nucleic acid have structural features somewhat like those in proteins. DNA takes the form of two chains that mutually coil around one another to form the well-known double helix. Some RNA occurs in essentially straight chains, whereas in other types the chain forms specific loops or helices. See Appendix C for further details.

HIGH-ENERGY BIOMOLECULES

Not all nucleotides are used to construct nucleic acids. One very important nucleotide—**adenosine triphosphate (ATP)**—is used as the body's primary energy carrier. Certain bonds in ATP temporarily store energy that is harnessed during the metabolism of foods and made available to the parts of the cells where it is needed to do specific cellular work (see pp. 27–30). Let's see how ATP functions in this role. Structurally, ATP is a modified RNA (ribose-containing) nucleotide that has adenine as its base and two additional phosphates bonded in sequence to the original nucleotide phosphate. Thus adenosine triphosphate, as the name implies, has a total of three phosphates attached in a string to *adenosine,* the composite of ribose and adenine (● Figure B-18). Attaching these additional phosphates requires considerable energy input. The high-energy input used to create these **high-energy phosphate bonds** is "stored" in the bonds for later use. Most energy transfers in the body involve ATP's terminal phosphate bond. When energy is needed, the third phosphate is cleaved off by hydrolysis, yielding *adenosine diphosphate (ADP)* and an inorganic phosphate (Pᵢ) and releasing energy in the process (Equation B-17):

$$ATP \rightarrow ADP + P_i + \text{energy for use by cell} \qquad \text{Eq. B-17}$$

Why use ATP as an energy currency that cells can cash in by splitting of the high-energy phosphate bonds as needed?

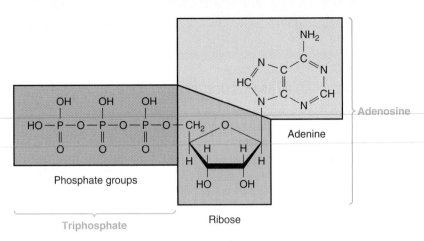

● **FIGURE B-18**

The structure of ATP

Why not just directly use the energy released during the oxidation of nutrient molecules such as glucose? If all the chemical energy stored in glucose were released at once, most of the energy would be squandered, because the cell could not capture much of the energy for immediate use. Instead, the energy trapped within the glucose bonds is gradually released and harnessed as cellular "bite-size pieces" in the form of the high-energy phosphate bonds of ATP.

Under the influence of an enzyme, ATP can be converted to a cyclic form of adenosine monophosphate, which contains only one phosphate group, the other two having been cleaved off. The resultant molecule, called **cyclic AMP** or **cAMP,** serves as an intracellular messenger, affecting the activities of a number of enzymes involved in important reactions in the body (see p. 98).

Storage, Replication, and Expression of Genetic Information

DEOXYRIBONUCLEIC ACID (DNA) AND CHROMOSOMES

The nucleus of the cell houses **deoxyribonucleic acid (DNA)**, the genetic blueprint that is unique for each individual.

Functions of DNA

As genetic material, DNA serves two essential functions. First, it contains "instructions" for assembling the cell's structural and enzymatic proteins. Cellular enzymes in turn control formation of other cell structures and also determine the functional activity of the cell by regulating the rate at which metabolic reactions proceed. The nucleus serves as the cell's control center by directly or indirectly controlling almost all cell activities through the role its DNA plays in governing protein synthesis. Because cells make up the body, the DNA code determines the structure and the function of the body as a whole. The DNA an organism has not only dictates whether the organism is a human, a toad, or a pea but also determines the unique physical and functional characteristics of that individual, all of which ultimately depend on the proteins produced under DNA control.

Second, by replicating (making copies of itself), DNA perpetuates the genetic blueprint within all new cells formed within the body and passes on genetic information from parents to children. We will first examine the structure of DNA and the coding mechanism it uses, then turn our attention to the means by which DNA replicates itself and controls protein synthesis.

Structure of DNA

Deoxyribonucleic acid is a huge molecule, composed (in humans) of millions of nucleotides arranged into two long, paired strands that spiral around each other to form a double helix. Each **nucleotide** has three components: (1) a *nitrogenous base,* a ring-shaped organic molecule containing nitrogen; (2) a 5-carbon ring-shaped sugar molecule, which in DNA is *deoxyribose;* and (3) a phosphate group. Nucleotides are joined end to end by linkages between the sugar of one nucleotide and the phosphate group of the adjacent nucleotide to form a long polynucleotide ("many nucleotide") strand with a sugar–phosphate backbone and bases projecting out one side (● Figure C-1). The four different bases in DNA are the double-ringed bases **adenine (A)** and **guanine (G)** and the single-ringed bases **cytosine (C)** and **thymine (T)**. The two polynucleotide strands within a DNA molecule are wrapped around each other so that their bases all project to the interior of the helix. The strands are held together by weak hydrogen bonds formed between the bases of adjoining strands (● Figure B-6, see p. A-7). Base pairing is highly specific: Adenine pairs only with thymine, and guanine pairs only with cytosine (● Figure C-2).

GENES

The composition of the repetitive sugar–phosphate backbones that form the "sides" of the DNA "ladder" is identical for every molecule of DNA, but the sequence of the linked bases that form the "rungs" varies among different DNA molecules. The particular sequence of bases in a DNA molecule serves as "instructions," or a "code," that dictates the assembly of amino acids into a given order for the synthesis of specific **polypeptides** (chains of amino acids linked by peptide bonds; see p. A-15). A **gene** is a stretch of DNA that codes for the synthesis of a particular polypeptide. Polypeptides, in turn, are folded into a three-dimensional configuration to form a functional protein. Not all portions of a DNA molecule code for structural or enzymatic proteins. Some stretches of DNA code for proteins that regulate genes. Other segments appear important in organizing and packaging DNA within the nucleus. Still other regions are "nonsense" base sequences that have no apparent significance.

Packaging of DNA into chromosomes

The DNA molecules within each human cell, if lined up end to end, would extend more than 2 m (2,000,000 mm), yet these molecules are packed into a nucleus that is only 5 mm in diameter. These molecules are not randomly crammed into the nucleus but are precisely

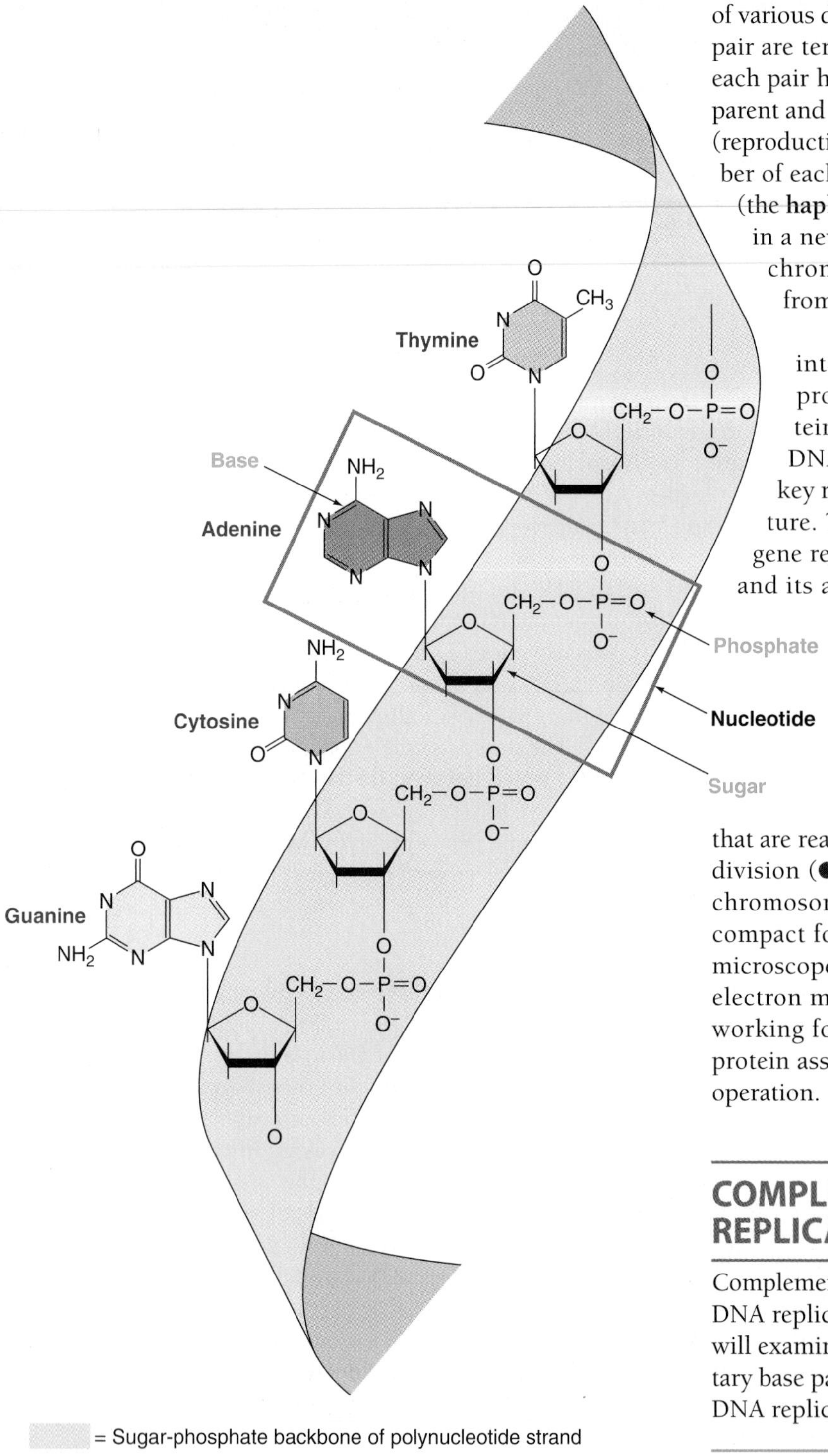

Thymine

Base

Adenine

Cytosine

Guanine

Phosphate

Nucleotide

Sugar

░ = Sugar-phosphate backbone of polynucleotide strand

● **FIGURE C-1**

Polynucleotide strand. Sugar–phosphate bonds link adjacent nucleotides together to form a polynucleotide strand with bases projecting to one side. The sugar–phosphate backbone is identical in all polynucleotides, but the sequence of the bases varies.

organized into **chromosomes**. Each chromosome consists of a different DNA molecule and contains a unique set of genes.

Somatic (body) **cells** contain 46 chromosomes (the **diploid number**), which can be sorted into 23 pairs on the basis

of various distinguishing features. Chromosomes in a matched pair are termed **homologous chromosomes**, one member of each pair having been derived from the individual's maternal parent and the other member from the paternal parent. **Germ** (reproductive) **cells** (sperm and eggs) contain only one member of each homologous pair for a total of 23 chromosomes (the **haploid number**). Union of a sperm and an egg results in a new diploid cell with 46 chromosomes—a set of 23 chromosomes from the mother and another set of 23 from the father.

DNA molecules are packaged and compressed into discrete chromosomal units in part by nuclear proteins associated with DNA. Two classes of proteins—histone and nonhistone proteins—bind with DNA. **Histones** form bead-shaped bodies that play a key role in packaging DNA into its chromosomal structure. The **nonhistones** are believed to be important in gene regulation. The complex formed between the DNA and its associated proteins is known as **chromatin**. The long threads of DNA within a chromosome are wound around histones at regular intervals, thus compressing a given DNA molecule to about one sixth its fully extended length. This "beads-on-a-string" structure is further folded and supercoiled into higher and higher levels of organization to further condense DNA into rodlike chromosomes that are readily visible through a light microscope during cell division (● Figure C-3). When the cell is not dividing, the chromosomes partially "unravel" or decondense to a less compact form of chromatin that is indistinct under a light microscope but appears as thin strands and clumps with an electron microscope. The decondensed form of DNA is its working form; that is, it is the form used as a template for protein assembly. Let's look at this working form of DNA in operation.

COMPLEMENTARY BASE PAIRING, REPLICATION, AND TRANSCRIPTION

Complementary base pairing serves as the foundation for both DNA replication and the initial step of protein synthesis. We will examine the mechanism and significance of complementary base pairing in each of these circumstances, starting with DNA replication.

▌ DNA replication

During DNA replication, the two decondensed DNA strands "unzip" as the weak bonds between the paired bases are enzymatically broken. Then **complementary base pairing** takes place: New nucleotides present within the nucleus pair with the exposed bases from each unzipped strand (● Figure C-4). New adenine-bearing nucleotides pair with exposed thymine-bearing nucleotides in an old strand, and new guanine-bearing nucleotides pair with exposed cytosine-bearing nucleotides in an old strand. This complementary base pairing is initiated at one end of the two old strands and proceeds in an orderly

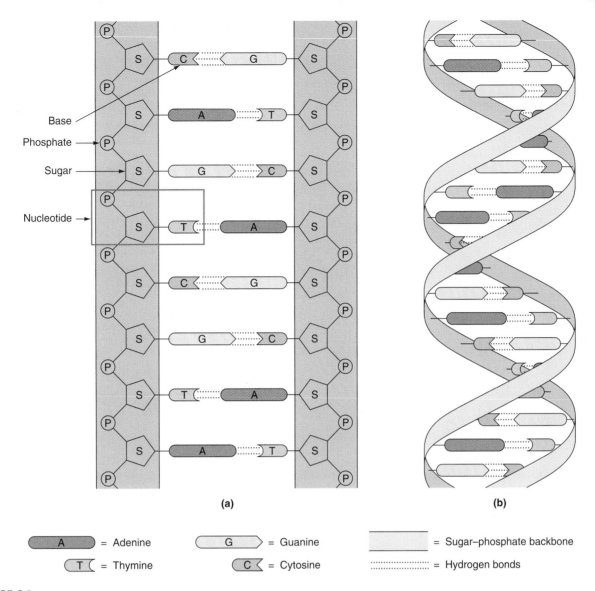

Base
Phosphate
Sugar
Nucleotide

(a)

(b)

$\boxed{A}$ = Adenine	$\boxed{G}$ = Guanine	= Sugar–phosphate backbone
$\boxed{T}$ = Thymine	$\boxed{C}$ = Cytosine	= Hydrogen bonds

● **FIGURE C-2**

Complementary base pairing in DNA. (a) Two polynucleotide strands held together by weak hydrogen bonds formed between the bases of adjoining strands—adenine always paired with thymine and guanine always paired with cytosine. (b) Arrangement of the two bonded polynucleotide strands of a DNA molecule into a double helix.

fashion to the other end. The new nucleotides attracted to and thus aligned in a prescribed order by the old nucleotides are sequentially joined by sugar–phosphate linkages to form two new strands that are complementary to each of the old strands. This replication process results in two complete double-stranded DNA molecules, one strand within each molecule having come from the original DNA molecule and one strand having been newly formed by complementary base pairing. These two DNA molecules are both identical to the original DNA molecule, with the "missing" strand in each of the original separated strands having been produced as a result of the imposed pattern of base pairing. This replication process, which occurs only during cell division, is essential for perpetuating the genetic code in both the new daughter cells. The duplicate copies of DNA are separated and evenly distributed

to the two halves of the cell before it divides. We will cover the topic of cell division in more detail later.

DNA transcription and messenger RNA

At other times, when DNA is not replicating in preparation for cell division, it serves as a blueprint for dictating cell protein synthesis. How is this accomplished when DNA is in the nucleus and protein synthesis is carried out by ribosomes in the cytoplasm? Several types of another nucleic acid, **ribonucleic acid (RNA)**, serve as the "go-between."

STRUCTURE OF RIBONUCLEIC ACID

Ribonucleic acid differs structurally from DNA in three regards: (1) The 5-carbon sugar in RNA is ribose instead of deoxyri-

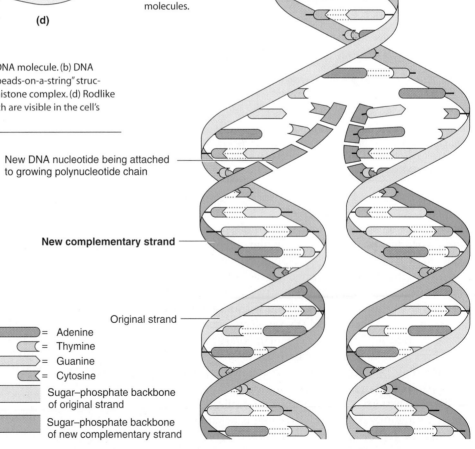

The DNA instructions for assembling a particular protein coded in the base sequence of a given gene are "transcribed" into a molecule of **messenger RNA (mRNA)**. The segment of the DNA molecule to be copied uncoils, and the base pairs separate to expose the particular sequence of bases in the gene. In any given gene, only one of the DNA strands is used as a template for transcribing RNA, with the copied strand varying for different genes along the same DNA molecule. The beginning and end of a gene within a DNA strand are designated by particular base sequences that serve as "start" and "stop" signals.

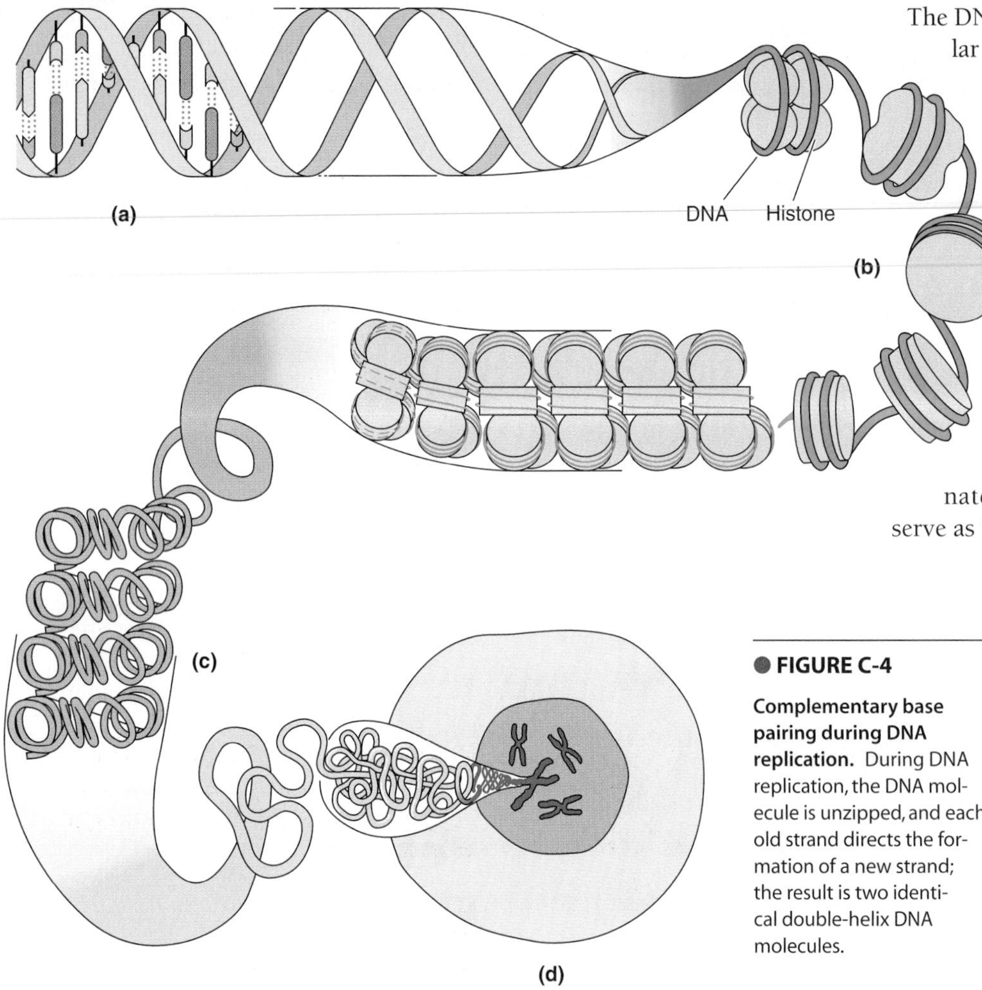

(a)

DNA   Histone

(b)

(c)

(d)

● **FIGURE C-3**

Levels of organization of DNA. (a) Double helix of a DNA molecule. (b) DNA molecule wound around histone proteins, forming a "beads-on-a-string" structure. (c) Further folding and supercoiling of the DNA–histone complex. (d) Rodlike chromosomes, the most condensed form of DNA, which are visible in the cell's nucleus during cell division.

● **FIGURE C-4**

Complementary base pairing during DNA replication. During DNA replication, the DNA molecule is unzipped, and each old strand directs the formation of a new strand; the result is two identical double-helix DNA molecules.

New DNA nucleotide being attached to growing polynucleotide chain

New complementary strand

Original strand

bose, the only difference between them being the presence in ribose of a single oxygen atom that is absent in deoxyribose; (2) RNA contains the closely related base **uracil** instead of thymine, with the three other bases being the same as in DNA; and (3) RNA is single-stranded and not self-replicating.

All RNA molecules are produced in the nucleus using DNA as a template, then exit the nucleus through openings in the nuclear membrane, called *nuclear pores,* which are large enough for passage of RNA molecules but block the much larger DNA molecules.

= Adenine
= Thymine
= Guanine
= Cytosine

Sugar–phosphate backbone of original strand

Sugar–phosphate backbone of new complementary strand

TRANSCRIPTION

Transcription is accomplished by complementary base pairing of free RNA nucleotides with their DNA counterparts in the exposed gene (● Figure C-5). The same pairing rules apply except that uracil, the RNA nucleotide substitute for thymine, pairs with adenine in the exposed DNA nucleotides. As soon as the RNA nucleotides pair with their DNA counterparts, sugar–phosphate bonds are formed to join the nucleotides together into a single-stranded RNA molecule that is released from DNA once transcription is complete. The original conformation of DNA is then restored. The RNA strand is much shorter than a DNA strand, because only a one-gene segment of DNA is transcribed into a single RNA molecule. The length of the finished RNA transcript varies, depending on the size of the gene. Within its nucleotide base sequence, this RNA transcript contains instructions for assembling a particular protein. Note that the message is coded in a base sequence that is *complementary to, not identical to,* the original DNA code.

Messenger RNA delivers the final coded message to the ribosomes for **translation** into a particular amino acid sequence to form a given protein. Thus genetic information flows from DNA (which can replicate itself) through RNA to protein. This is accomplished first by *transcription* of the DNA code into a complementary RNA code, followed by *translation* of the RNA code into a specific protein (● Figure C-6). In the next section, you will learn more about the steps in translation. The structural and functional characteristics of the cell as determined by its protein composition can be varied, subject to control, depending on which genes are "switched on" to produce mRNA.

Free nucleotides present in the nucleus cannot be randomly joined together to form either DNA or RNA strands, because the enzymes required to link together the sugar and phosphate components of nucleotides are active only when bound to DNA. This ensures that DNA, mRNA, and protein assembly occur only according to genetic plan.

TRANSLATION AND PROTEIN SYNTHESIS

Three forms of RNA participate in protein synthesis. Besides messenger RNA, two other forms of RNA are required for translation of the genetic message into cell protein: ribosomal RNA and transfer RNA.

- **Messenger RNA** carries the coded message from nuclear DNA to a cytoplasmic ribosome, where it directs the synthesis of a particular protein.
- **Ribosomal RNA (rRNA)** is an essential component of *ribosomes,* the "workbenches" for protein synthesis (see p. 21). Ribosomal RNA "reads" the base sequence code of mRNA and translates it into the appropriate amino-acid sequence during protein synthesis.

● **FIGURE C-5**

Complementary base pairing during DNA transcription. During DNA transcription, a messenger RNA molecule is formed as RNA nucleotides are assembled by complementary base pairing at a given segment of one strand of an unzipped DNA molecule (that is, a gene).

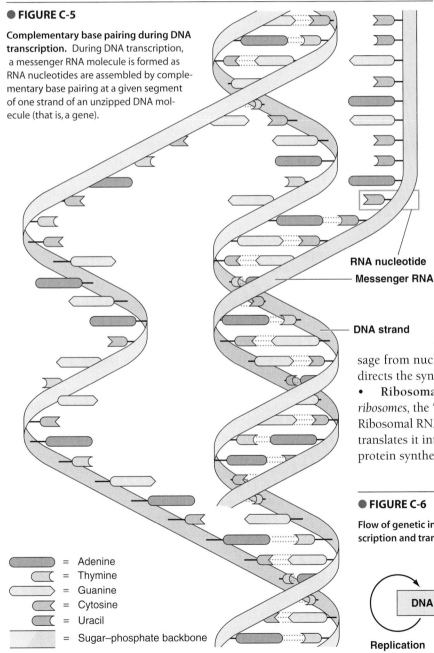

RNA nucleotide
Messenger RNA

DNA strand

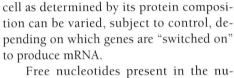

= Adenine
= Thymine
= Guanine
= Cytosine
= Uracil
= Sugar–phosphate backbone

● **FIGURE C-6**

Flow of genetic information from DNA through RNA to protein by transcription and translation

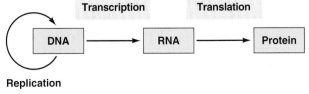

Transcription   Translation

DNA → RNA → Protein

Replication

- **Transfer RNA (tRNA)** transfers the appropriate amino acids in the cytosol to their designated site in the amino acid sequence of the protein under construction.

Triplet code; codon

Twenty different amino acids are used to construct proteins, yet only four different nucleotide bases are used to code for these 20 amino acids. In the "genetic dictionary," each different amino acid is specified by a **triplet code** that consists of a specific sequence of three bases in the DNA nucleotide chain. For example, the DNA sequence ACA (adenine, cytosine, adenine) specifies the amino acid cysteine, whereas the sequence ATA specifies the amino acid tyrosine. Each DNA triplet code is transcribed into mRNA as a complementary code word, or **codon**, consisting of a sequenced order of the three bases that pair with the DNA triplet. For example, the DNA triplet code ATA is transcribed as UAU (uracil, adenine, uracil) in mRNA.

Sixty-four different DNA triplet combinations (and, accordingly, 64 different mRNA codon combinations) are possible using the four different nucleotide bases (4^3). Of these possible combinations, 61 code for specific amino acids and the remaining 3 serve as "stop signals." A stop signal acts as a "period" at the end of a "sentence." The sentence consists of a series of triplet codes that specify the amino acid sequence in a particular protein. When the stop codon is reached, ribosomal RNA releases the finished polypeptide product. Because 61 triplet codes each specify a particular amino acid and there are 20 different amino acids, a given amino acid may be specified by more than one base-triplet combination. For example, tyrosine is specified by the DNA sequence ATG as well as by ATA. In addition, one DNA triplet code, TAC (mRNA codon sequence AUG) functions as a "start signal" in addition to specifying the amino acid methionine. This code marks the place on mRNA where translation is to begin so that the message is started at the correct end and thus reads in the right direction. Interestingly, the same genetic dictionary is used universally; a given three-base code stands for the same amino acid in all living things, including micro-organisms, plants, and animals.

Ribosomes

A **ribosome** brings together all components that participate in protein synthesis—mRNA, tRNA, and amino acids—and provides the enzymes and energy required for linking the amino acids together. The nature of the protein synthesized by a given ribosome is determined by the mRNA message being translated. Each mRNA serves as a code for only one particular polypeptide.

A ribosome is an rRNA-protein structure organized into two subunits of unequal size. These subunits are brought together only when a protein is being synthesized (● Figure C-7, step ①). During assembly of a ribosome, an mRNA molecule attaches to the smaller of the ribosomal subunits by means of a *leader sequence,* a section of mRNA that precedes the start

codon. The small subunit with mRNA attached then binds to a large subunit to form a complete, functional ribosome. When the two subunits unite, a groove is formed that accommodates the mRNA molecule as it is being translated.

Transfer RNA and anticodons

Free amino acids in the cytoplasm cannot "recognize" and bind directly with their specific codons in mRNA. Transfer RNA is required to bring the appropriate amino acid to its proper codon. Even though tRNA is single-stranded, as are all RNA molecules, it is folded back onto itself into a **T** shape with looped ends (● Figure C-8). The open-ended stem portion recognizes and binds to a specific amino acid. There are at least 20 different varieties of tRNA, each able to bind with only one of the 20 different kinds of amino acids. A tRNA is said to be "charged" when it is carrying its passenger amino acid. The loop end of a tRNA opposite the amino-acid binding site contains a sequence of three exposed bases, known as the **anticodon**, which is complementary to the mRNA codon that specifies the amino acid being carried. Through complementary base pairing, a tRNA can bind with mRNA and insert its amino acid into the protein under construction only at the site designated by the codon for the amino acid. For example, the tRNA molecule that binds with tyrosine bears the anticodon AUA, which can pair only with the mRNA codon UAU, which specifies tyrosine. This dual binding function of tRNA molecules ensures that the correct amino acids are delivered to mRNA for assembly in the order specified by the genetic code. Transfer RNA can only bind with mRNA at a ribosome, so protein assembly does not occur except in the confines of a ribosome.

Steps of protein synthesis

The three steps of protein synthesis are initiation, elongation, and termination.

1. *Initiation.* Protein synthesis is initiated when a charged tRNA molecule bearing the anticodon specific for the start codon binds at this site on mRNA (● Figure C-7, step ②).

2. *Elongation.* A second charged tRNA bearing the anticodon specific for the next codon in the mRNA sequence then occupies the site next to the first tRNA (step ③). At any given time, a ribosome can accommodate only two tRNA molecules bound to adjacent codons. Through enzymatic action, a peptide bond is formed between the two amino acids that are linked to the stems of the adjacent tRNA molecules (step ④). The linkage is subsequently broken between the first tRNA and its amino acid passenger, leaving the second tRNA with a chain of two amino acids. The uncharged tRNA molecule (the one minus its amino acid passenger) is released from mRNA (step ⑤). The ribosome then moves along the mRNA molecule by precisely three bases, a distance of one codon (step ⑥), so that the tRNA bearing the chain of two amino acids is moved into the number one ribosomal site for tRNA. Then, an incoming charged tRNA with a complemen-

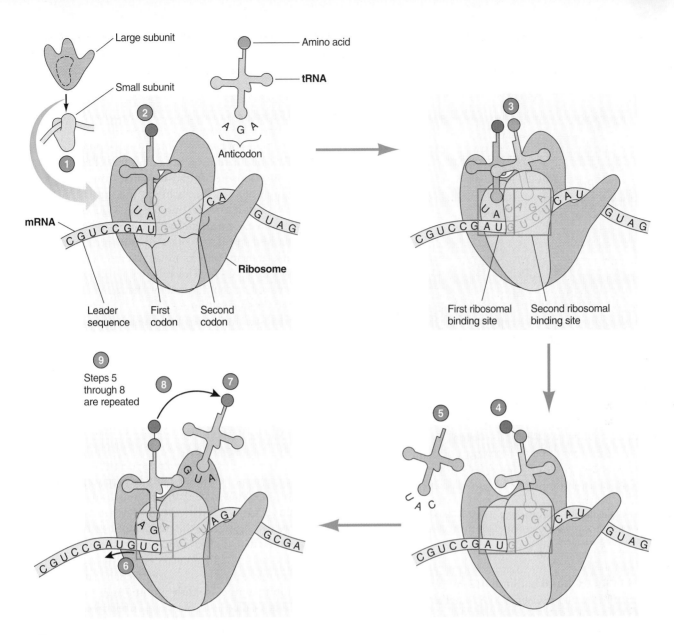

1. On binding with a messenger RNA (mRNA) molecule, the small ribosomal subunit joins with the large subunit to form a functional ribosome.

2. A transfer RNA (tRNA), charged with its specific amino acid passenger, binds to mNRA by means of complementary base pairing between the tRNA anticodon and the first mRNA codon positioned in the first ribosomal binding site.

3. Another tRNA molecule attaches to the next codon on mRNA positioned in the second ribosomal binding site.

4. The amino acid from the first tRNA is linked to the amino acid on the second tRNA.

5. The first tRNA detaches.

6. The mRNA molecule shifts forward one codon (a distance of a three-base sequence).

7. Another charged tRNA moves in to attach with the next codon on mRNA, which has now moved into the second ribosomal binding site.

8. The amino acids from the tRNA in the first ribosomal site are linked with the amino acid in the second site.

9. This process continues (that is, steps 5 through 8 are repeated), with the polypeptide chain continuing to grow, until a stop codon is reached and the polypeptide chain is released.

● **FIGURE C-7**

Ribosomal assembly and protein translation

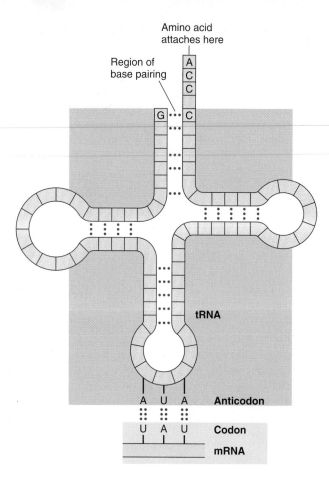

Amino acid attaches here

Region of base pairing

A
C
C

G ••• C

tRNA

A U A   **Anticodon**

U A U   **Codon**

mRNA

● **FIGURE C-8**

Structure of a tRNA molecule. The open end of a tRNA molecule attaches to free amino acids. The anticodon loop of the tRNA molecule attaches to a complementary mRNA codon.

tary anticodon for the third codon in the mRNA sequence occupies the number two ribosomal site that was vacated by the second tRNA (step 7). The chain of two amino acids subsequently binds with and is transferred to the third tRNA to form a chain of three amino acids (step 8). Through repetition of this process, amino acids are subsequently added one at a time to a growing polypeptide chain in the order designated by the codon sequence as the ribosomal translation machinery moves stepwise along the mRNA molecule one codon at a time (step 9). This process is rapid. Up to 10 to 15 amino acids can be added per second.

3. *Termination.* Elongation of the polypeptide chain continues until the ribosome reaches a stop codon in the mRNA molecule, at which time the polypeptide is released. The polypeptide is then folded and modified into a full-fledged protein. The ribosomal subunits dissociate and are free to reassemble into another ribosome for translation of other mRNA molecules.

▌ Energy cost of protein synthesis

Protein synthesis is expensive, in terms of energy. Attachment of each new amino acid to the growing polypeptide chain re-

quires a total investment of splitting four high-energy phosphate bonds—two to charge tRNA with its amino acid, one to bind tRNA to the ribosomal-mRNA complex, and one to move the ribosome forward one codon.

▌ Polyribosomes

A number of copies of a given protein can be produced from a single mRNA molecule before the latter is chemically degraded. As one ribosome moves forward along the mRNA molecule, a new ribosome attaches at the starting point on mRNA and also starts translating the message. Attachment of many ribosomes to a single mRNA molecule results in a *polyribosome*. Multiple copies of the identical protein are produced as each ribosome moves along and translates the same message (● Figure C-9). The released proteins are used within the cytosol, except for the few that move into the nucleus through the nuclear pores.

Recall that, in contrast to the cytosolic polyribosomes, ribosomes directed to bind with the rough endoplasmic reticulum (ER) feed their growing polypeptide chains into the ER lumen (see p. 21). The resultant proteins are subsequently packaged for export out of the cell or for replacing membrane components within the cell.

▌ Control of gene activity and protein transcription

Because each somatic cell in the body has the identical DNA blueprint, you might assume that they would all produce the same proteins. This is not the case, however, because different cell types can transcribe different sets of genes and thus synthesize different sets of structural and enzymatic proteins. For example, only red blood cells can synthesize hemoglobin, even though all body cells carry the DNA instructions for hemoglobin synthesis. Only about 7% of the DNA sequences in a typical cell are ever transcribed into mRNA for ultimate expression as specific proteins.

Control of gene expression involves gene regulatory proteins that activate ("switch on") or repress ("switch off") the genes that code for specific proteins within a given cell. Various DNA segments that do not code for structural and enzy-

● **FIGURE C-9**

A polyribosome. A polyribosome is formed by numerous ribosomes simultaneously translating mRNA.

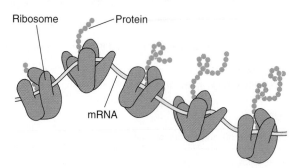

Ribosome   Protein

mRNA

matic proteins code for synthesis of these regulatory proteins. The molecular mechanisms by which these regulatory genes in turn are controlled in human cells are only beginning to be understood. In some instances, regulatory proteins are controlled by **gene-signaling factors** that bring about differential gene activity among various cells to accomplish specialized tasks. The largest group of known gene-signaling factors in humans is the hormones. Some hormones exert their homeostatic effect by selectively altering the transcription rate of the genes that code for enzymes that are in turn responsible for catalyzing the reaction(s) regulated by the hormone. For example, the hormone cortisol promotes the breakdown of fat stores by stimulating synthesis of the enzyme that catalyzes the conversion of stored fat into its component fatty acids. In other cases, gene action appears to be time specific; that is, certain genes are expressed only at a certain developmental stage in the individual. This is especially important during embryonic development.

CELL DIVISION

Most cells in the human body can reproduce themselves, a process important in growth, replacement, and repair of tissues. The rate at which cells divide is highly variable. Cells within the deeper layers of the intestinal lining divide every few days to replace cells that are continually sloughed off the surface of the lining into the digestive tract lumen. In this way, the entire intestinal lining is replaced about every three days (see p. 498). At the other extreme are nerve cells, which permanently lose the ability to divide beyond a certain period of fetal growth and development. Consequently, when nerve cells are lost through trauma or disease, they cannot be replaced (see p. 3). In between these two extremes are cells that divide infrequently except when needed to replace damaged or destroyed tissue. The factors that control the rate of cell division remain obscure.

▌ Mitosis

Cell division involves two components: nuclear division and cytoplasmic division. Nuclear division in somatic cells is accomplished by **mitosis**, in which a complete set of genetic information (that is, a diploid number of chromosomes) is distributed to each of two new daughter cells.

A cell capable of dividing alternates between periods of mitosis and nondivision. The interval of time between cell division is known as **interphase.** Because mitosis takes less than an hour to complete, the vast majority of cells in the body at any given time are in interphase.

Replication of DNA and growth of the cell take place during interphase in preparation for mitosis. Although mitosis is a continuous process, it displays four distinct phases: *prophase, metaphase, anaphase,* and *telophase* (● Figure C-10).

PROPHASE

1.   Chromatin condenses and becomes microscopically visible as chromosomes. The condensed duplicate strands of DNA,

known as *sister chromatids,* remain joined together within the chromosome at a point called the *centromere* (● Figure C-11).
2.   Cells contain a pair of **centrioles**, short cylindrical structures that form the mitotic spindle during cell division (see ● Figure 2-1, p. 21). The centriole pair divides, and the daughter centrioles move to opposite ends of the cell, where they assemble between them a mitotic spindle made up of microtubules (see p. 34).
3.   The membrane surrounding the nucleus starts to break down.

METAPHASE

1.   The nuclear membrane completely disappears.
2.   The 46 chromosomes, each consisting of a pair of sister chromatids, align themselves at the midline, or equator, of the cell. Each chromosome becomes attached to the spindle by means of several spindle fibers that extend from the centriole to the centromere of the chromosome.

ANAPHASE

1.   The centromeres split, converting each pair of sister chromatids into two identical chromosomes, which separate and move toward opposite poles of the spindle.
2.   At the end of anaphase, an identical set of 46 chromosomes is present at each of the poles, for a transient total of 92 chromosomes in the soon-to-be-divided cell.

TELOPHASE

1.   The cytoplasm divides through formation and gradual tightening of an actin contractile ring at the midline of the cell, thus forming two separate daughter cells, each with a full diploid set of chromosomes.
2.   The spindle fibers disassemble.
3.   The chromosomes uncoil to their decondensed chromatin form.
4.   A nuclear membrane reforms in each new cell.

Cell division is complete with the end of telophase. Each of the new cells now enters interphase.

▌ Meiosis

Nuclear division in the specialized case of germ cells is accomplished by **meiosis**, in which only half a set of genetic information (that is, a haploid number of chromosomes) is distributed to each daughter cell. Meiosis differs from mitosis in several important ways (● Figure C-10). Specialized diploid germ cells undergo one chromosome replication followed by two nuclear divisions to produce four haploid germ cells.

MEIOSIS I

1.   During prophase of the first meiotic division, the members of each homologous pair of chromosomes line up side by side to form a **tetrad**, which is a group of four sister chromatids with two identical chromatids within each member of the pair.

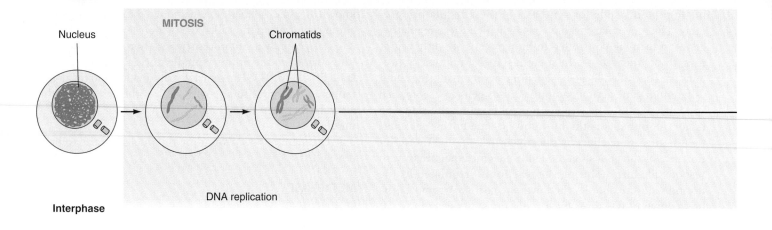

Nucleus

MITOSIS

Chromatids

DNA replication

Interphase

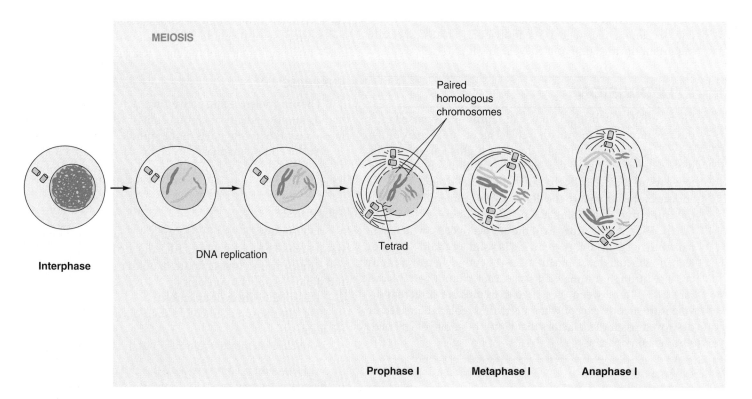

MEIOSIS

Paired
homologous
chromosomes

DNA replication

Tetrad

Interphase

Prophase I   **Metaphase I**   **Anaphase I**

● **FIGURE C-10**

A comparison of events in mitosis and meiosis

2.   The process of crossing over occurs during this period, when the maternal copy and the paternal copy of each chromosome are paired. **Crossing over** involves a physical exchange of chromosome material between nonsister chromatids within a tetrad (● Figure C-12). This process yields new chromosome combinations, thus contributing to genetic diversity.

3.   During metaphase, the 23 tetrads line up at the equator.

4.   At anaphase, homologous chromosomes, each consisting of a pair of sister chromatids joined at the centromere, separate and move toward opposite poles. Maternally and pater-

nally derived chromosomes migrate to opposite poles in random assortments of one member of each chromosome pair without regard for its original derivation. This genetic mixing provides novel new combinations of chromosomes.

5.   During the first telophase, the cell divides into two cells. Each cell contains 23 chromosomes consisting of two sister chromatids.

MEIOSIS II

1.   Following a brief interphase in which no further replication occurs, the 23 unpaired chromosomes line up at the equa-

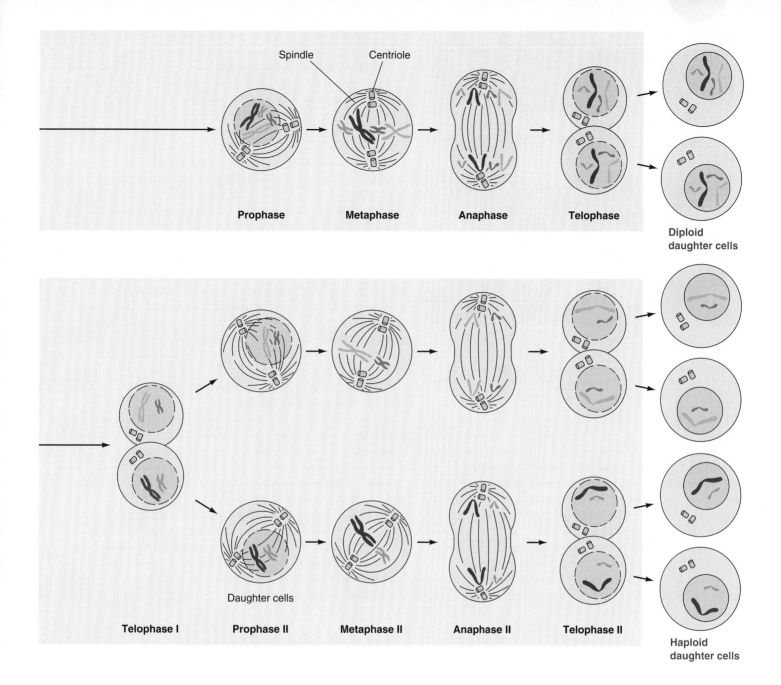

Spindle **Centriole**

Prophase **Metaphase** **Anaphase** **Telophase**

Diploid daughter cells

Telophase I **Prophase II** **Metaphase II** **Anaphase II** **Telophase II**

Daughter cells

Haploid daughter cells

tor, the centromeres split, and the sister chromatids separate for the first time into independent chromosomes that move to opposite poles.

2. During cytoplasmic division, each of the daughter cells derived from the first meiotic division forms two new daughter cells. The end result is four daughter cells, each containing a haploid set of chromosomes.

Union of a haploid sperm and haploid egg results in a zygote (fertilized egg) that contains the diploid number of chromosomes. Through mitosis and cell differentiation, the

zygote develops into a new multicellular individual. Because DNA is normally faithfully replicated in its entirety during each mitotic division, all cells in the body have an identical aggregate of DNA molecules. Structural and functional variations between different cell types result from differential gene expression.

▌ Mutations

An estimated 10^{16} cell divisions take place in the body during a person's lifetime to accomplish growth, repair, and nor-

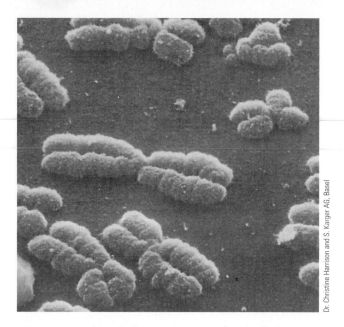

● **FIGURE C-11**

A scanning electron micrograph of human chromosomes from a dividing cell. The replicated chromosomes appear as double structures, with identical sister chromatids joined in the middle at a common centromere.

Dr. Christine Harrison and S. Karger AG, Basel

mal cell turnover. Because more than 3 billion nucleotides must be replicated during each cell division, no wonder "copying errors" occasionally occur. Any change in the DNA sequence is known as a **point (gene) mutation.** A point mutation arises when a base is inadvertently substituted, added, or deleted during the replication process.

When a base is inserted in the wrong position during DNA replication, the mistake can often be corrected by a built-in "proofreading" system. Repair enzymes remove the newly replicated strand back to the defective segment, at which time normal base pairing resumes to resynthesize a corrected strand. Not all mistakes can be corrected, however.

Mutations can arise spontaneously by chance alone or they can be induced by **mutagens,** which are factors that increase the rate at which mutations take place. Mutagens include various chemical agents as well as ionizing radiation such as X rays and atomic radiation. Mutagens promote mutations either by chemically altering the DNA base code through a variety of mechanisms or by interfering with the repair enzymes so that abnormal base segments cannot be cut out.

Depending on the location and nature of a change in the genetic code, a given mutation may (1) have no noticeable effect if it does not alter a critical region of a cell protein; (2) adversely alter cell function if it impairs the function of a crucial protein; (3) be incompatible with the life of the cell, in which case the cell dies and the mutation is lost with it; or (4) in rare cases, prove beneficial if a more efficient structural or enzymatic protein results. If a mutation occurs in a body cell (a **somatic mutation**), the outcome will be reflected as an alteration in all future copies of the cell in the affected individual, but it will not be perpetuated beyond the life of the individual. If, by contrast, a mutation occurs in a sperm- or egg-producing cell (**germ cell mutation**), the genetic alteration may be passed on to succeeding generations.

In most instances, **cancer** results from multiple somatic mutations that occur over a course of time within DNA segments known as **proto-oncogenes.** Proto-oncogenes are normal genes whose coded products are important in the regulation of cell growth and division. These genes have the potential of becoming overzealous **oncogenes** ("cancer genes"), which induce the uncontrolled cell proliferation characteristic of cancer. Proto-oncogenes can become cancer producing as a result of several sequential mutations in the gene itself or by changes in adjacent regions that regulate the proto-oncogenes. Less frequently, tumor viruses become incorporated in the DNA blueprint and act as oncogenes.

● **FIGURE C-12**

Crossing over. (a) During prophase I of meiosis, each homologous pair of chromosomes lines up side by side to form a tetrad. (b) Physical exchange of chromosome material occurs between nonsister chromatids. (c) As a result of this crossing over, new combinations of genetic material are formed within the chromosomes.

Centromere

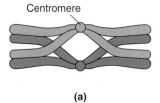

(a)

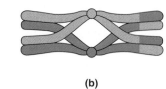

(b)

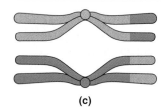

(c)

The Chemistry of Acid–Base Balance

DISSOCIATION CONSTANTS FOR ACIDS

Acids are a special group of hydrogen-containing substances that dissociate, or separate, when in solution to liberate free H^+ and anions (negatively charged ions). The extent of dissociation for a given acid is always constant; that is, when in solution, the same proportion of a particular acid's molecules always separate to liberate free H^+, with the other portion always remaining intact. The constant degree of dissociation for a particular acid (in this example, H_2CO_3) is expressed by its **dissociation constant (K)** as follows:

$$[H^+] [HCO_3^-]/[H_2CO_3] = K$$

where

$[H^+] [HCO_3^-]$ represents the concentration of ions resulting from H_2CO_3 dissociation.

$[H_2CO_3]$ represents the concentration of intact (undissociated) H_2CO_3.

The dissociation constant varies for different acids.

THE LOGARITHMIC NATURE OF pH

Recall that **pH** equals the logarithm (log) to the base 10 of the reciprocal of the hydrogen ion concentration:

$$pH = \log 1/[H^+]$$

Every unit change in pH actually represents a 10-fold change in $[H^+]$ because of the logarithmic relationship. A log to the base 10 indicates how many times 10 must be multiplied to produce a given number. For example, the log of $10 = 1$, whereas the log of $100 = 2$, because 1 ten equals 10 and 2 tens multiplied together equals 100 ($10 \times 10 = 100$). Numbers less than 10 have logs less than 1. Numbers between 10 and 100 have logs between 1 and 2, and so on. Accordingly, each unit of change in pH indicates a 10-fold change in $[H^+]$. For example, a solution with a pH of 7 has a $[H^+]$ 10 times less than that of a solution with a pH of 6 (a 1 pH-unit difference) and 100 times less than that of a solution with a pH of 5 (a 2-pH-unit difference).

CHEMICAL BUFFER SYSTEMS

Recall that a **chemical buffer system** consists of a pair of substances involved in a reversible reaction—one substance that can yield free H^+ as the $[H^+]$ starts to fall and another that can bind with free H^+ (thus removing it from solution) when $[H^+]$ starts to rise. An important example of a buffer system is the **carbonic acid–bicarbonate (H_2CO_3:HCO_3^-) buffer system**, which is the primary ECF buffer for noncarbonic acids and is involved in the following reversible reaction:

$$H_2CO_3 \rightleftarrows H^+ + HCO_3^-$$

Because CO_2 generates H_2CO_3, the H_2CO_3:HCO_3^- buffer system in the body involves CO_2 by means of the following reaction, with which you are already familiar:

$$CO_2 + H_2O \rightleftarrows H_2CO_3 \rightleftarrows H^+ + HCO_3^-$$

Chemical buffer systems function according to the law of mass action, which states that if the concentration of one of the substances involved in a reversible reaction is increased, the reaction is driven toward the opposite side; if the concentration of one of the substances is decreased, the reaction is driven toward that side.

Let's apply the law of mass action to the reversible reaction involving the H_2CO_3:HCO_3^- buffer system. When new H^+ is added to the plasma from any source other than CO_2 (for example through lactic acid released into the ECF from exercising muscles), the preceding reaction is driven toward the left side of the equation. As the extra H^+ binds with HCO_3^-, it no longer contributes to the acidity of body fluids, so the rise in $[H^+]$ abates. In the converse situation, when the plasma $[H^+]$ occasionally falls below normal for some reason other than a change in CO_2 (such as the loss of plasma-derived HCl in the gastric juices during vomiting), the reaction is driven toward the right side of the equation. Dissolved CO_2 and H_2O in the plasma form H_2CO_3, which generates additional H^+ to make up for the H^+ deficit. In so doing, the H_2CO_3:HCO_3^- buffer system resists the fall in $[H^+]$.

This system cannot buffer changes in pH induced by fluctuations in H_2CO_3. A buffer system cannot buffer itself. Consider, for example, the situation in which the plasma $[H^+]$ is elevated by CO_2 retention from a respiratory problem. The rise in CO_2 drives the reaction to the right according to the law of mass action, elevating $[H^+]$. The increase in $[H^+]$ occurs as a result of the reaction being driven to the *right* by an increase in CO_2, so the elevated $[H^+]$ cannot drive the reaction to the *left* to buffer the increase in $[H^+]$. Only if the increase in $[H^+]$ is brought about by some mechanism other than CO_2 accumulation can this buffer system be shifted to the CO_2 side of the equation and effectively reduce $[H^+]$. Likewise, in the opposite situation, the H_2CO_3:HCO_3^- buffer system cannot compensate for a reduction in $[H^+]$ from a deficit of CO_2 by generating more H^+-yielding H_2CO_3 when the problem in the first place is a shortage of H_2CO_3-forming CO_2.

The **hemoglobin buffer system** resists fluctuations in pH caused by changes in CO_2 levels. At the systemic capillary level, CO_2 continuously diffuses into the blood from the tissue cells where it is produced. The greatest percentage of this CO_2 forms H_2CO_3, which partially dissociates into H^+ and HCO_3^-. Simultaneously, some oxyhemoglobin (HbO_2) releases O_2, which diffuses into the tissues. Reduced (unoxygenated) Hb has a greater affinity for H^+ than HbO_2 does. Therefore, most H^+ generated from CO_2 at the tissue level becomes bound to reduced Hb and no longer contributes to the acidity of body fluids:

$$H^+ + Hb \rightleftarrows HHb$$

At the lungs, the reactions are reversed. As Hb picks up O_2 diffusing from the alveoli (air sacs) into the red blood cells, the affinity of Hb for H^+ is decreased, so H^+ is released. This liberated H^+ combines with HCO_3^- to yield H_2CO_3. In turn, H_2CO_3 produces CO_2, which is exhaled. Meanwhile, the hydrogen has been reincorporated into neutral H_2O molecules.

The intracellular proteins and plasma proteins constitute the **protein buffer system.** Their acidic and basic side groups enable them to give up or take up H^+ respectively.

The **phosphate buffer system** is composed of an acidic phosphate salt (NaH_2PO_4) that can donate a free H^+ when the $[H^+]$ falls and a basic phosphate salt (Na_2HPO_4) that can accept a free H^+ when the $[H^+]$ rises. Basically, this buffer pair can alternately switch a H^+ for a Na^+ as demanded by the $[H^+]$.

HENDERSON-HASSELBALCH EQUATION

The relationship between $[H^+]$ and the members of a buffer pair can be expressed according to the **Henderson-Hasselbalch equation,** which, for the H_2CO_3:HCO_3^- buffer system is as follows:

$$pH = pK + \log [HCO_3^-]/[H_2CO_3]$$

Although you do not need to know the mathematical manipulations involved, it is helpful to understand how this formula is derived. Recall that the dissociation constant K for H_2CO_3 is

$$[H^+] [HCO_3^-]/[H_2CO_3] = K$$

and that the relationship between pH and $[H^+]$ is

$$pH = \log 1/[H^+]$$

Then, by solving the dissociation constant formula for $[H^+]$ (that is, $[H^+] = K \times [H_2CO_3]/[HCO_3^-]$) and replacing this value for $[H^+]$ in the pH formula, one comes up with the Henderson-Hasselbalch equation.

Practically speaking, $[H_2CO_3]$ directly reflects the concentration of dissolved CO_2, henceforth referred to as $[CO_2]$, because most of the CO_2 in the plasma is converted into H_2CO_3. (The dissolved CO_2 concentration is equivalent to P_{CO_2}, as described in the chapter on respiration.) Therefore, the equation becomes

$$pH = pK + \log [HCO_3^-]/[CO_2]$$

The pK is the logarithm of 1/K and, like K, always remains a constant for any given acid. For H_2CO_3, the pK is 6.1. Because the pK is always a constant, changes in pH are associated with changes in the ratio between $[HCO_3^-]$ and $[CO_2]$.

Normally, the ratio between $[HCO_3^-]$ and $[CO_2]$ in the ECF is 20 to 1; that is, there is 20 times more HCO_3^- than CO_2. Plugging this ratio into our formula,

$$pH = pK + \log [HCO_3^-]/[CO_2]$$
$$= 6.1 + \log 20/1$$

The log of 20 is 1.3. Therefore pH = 6.1 + 1.3 = 7.4, which is the normal pH of plasma.

- When the ratio of $[HCO_3^-]$ to $[CO_2]$ increases above 20/1, pH increases. Accordingly, either a rise in $[HCO_3^-]$ or a fall in $[CO_2]$, both of which increase the $[HCO_3^-]/[CO_2]$ ratio if the other component remains constant, shifts the acid–base balance toward the alkaline side.
- In contrast, when the $[HCO_3^-]/[CO_2]$ ratio decreases below 20/1, pH decreases toward the acid side. This can occur either if the $[HCO_3^-]$ decreases or if the $[CO_2]$ increases while the other component remains constant.

Because $[HCO_3^-]$ is regulated by the kidneys and $[CO_2]$ by the lungs, plasma pH can be shifted up and down by kidney and lung influences. The kidneys and lungs regulate pH (and thus free $[H^+]$) largely by controlling plasma $[HCO_3^-]$ and $[CO_2]$, respectively, to restore their ratio to normal. Accordingly,

$$pH \propto \frac{[HCO_3^-] \text{ controlled by kidney function}}{[CO_2] \text{ controlled by respiratory function}}$$

Because of this relationship, not only do both the kidneys and lungs normally participate in pH control but renal or respiratory dysfunction can also induce acid–base imbalances by altering the $[HCO_3^-]/[CO_2]$ ratio.

RESPIRATORY REGULATION OF HYDROGEN ION CONCENTRATION

The major source of H^+ in the body fluids is through H_2CO_3 formation from metabolically produced CO_2. Cellular oxidation of nutrients yields energy, with CO_2 and H_2O as end products. Catalyzed by the enzyme *carbonic anhydrase (ca)*, CO_2 and H_2O form H_2CO_3, which then partially dissociates to liberate free H^+ and HCO_3^-.

$$CO_2 + H_2O \overset{ca}{\rightleftarrows} H_2CO_3 \rightleftarrows H^+ + HCO_3^-$$

Within the systemic capillaries, the CO_2 level in the blood increases as metabolically produced CO_2 enters from the tissues. This drives the reaction to the acid side, generating H^+ as well as HCO_3^- in the process. In the lungs, the reaction is reversed: CO_2 diffuses from the blood flowing through the pulmonary capillaries into the alveoli, from which it is expired to the atmosphere. The resultant reduction in blood CO_2 drives the reaction toward the CO_2 side. Hydrogen ion and HCO_3^- form H_2CO_3, which rapidly decomposes into CO_2 and H_2O once again. The CO_2 is exhaled while the hydrogen ions generated at the tissue level are incorporated into H_2O molecules.

The lungs are extremely important in maintaining the $[H^+]$ of plasma. Every day they remove from body fluids what amounts to 100 times more H^+ derived from carbonic acid than the kidneys remove from sources other than carbonic acid. When the respiratory system can keep pace with the rate of metabolism, there is no net gain or loss of H^+ in the body fluids from metabolically produced CO_2. When the rate of CO_2 removal by the lungs does not match the rate of CO_2 production at the tissue level, however, the resulting accumulation or deficit of CO_2 leads to an excess or shortage, respectively, of free H^+ from this source. Therefore, respiratory abnormalities can lead to acid–base imbalances.

On the other hand, the respiratory system, through its ability to regulate arterial $[CO_2]$, can adjust the amount of H^+ added to body fluids from this source as needed to restore pH toward normal when fluctuations occur in $[H^+]$ from sources other than carbonic acid; that is the respiratory system can compensate for nonrespiratory-induced acid–base imbalances.

However, the respiratory system alone can return the pH to only 50% to 75% of the way toward normal. Two reasons contribute to the respiratory system's inability to fully compensate for a nonrespiratory-induced acid–base imbalance.

1. First, during respiratory compensation for a deviation in pH, the peripheral chemoreceptors, which increase ventilation in response to an elevated arterial $[H^+]$, and the central chemoreceptors, which increase ventilation in response to a rise in $[CO_2]$ (by monitoring CO_2-generated H^+ in brain ECF—see p. 398), work at odds. Consider what happens in response to an acidosis arising from a nonrespiratory cause. When the peripheral chemoreceptors detect an increase in arterial $[H^+]$, they reflexly stimulate the respiratory center to step up ventilation, causing more acid-forming CO_2 to be blown off. In response to the resultant fall in CO_2, however, the central chemoreceptors start to inhibit the respiratory center. By opposing the action of the peripheral chemoreceptors, the central chemoreceptors stop the compensatory increase in ventilation short of restoring pH all the way to normal.

2. Second, the driving force for the compensatory increase in ventilation is diminished as the pH moves toward normal. Ventilation is increased by the peripheral chemoreceptors in response to a rise in arterial $[H^+]$, but as the $[H^+]$ is gradually reduced by stepped-up removal of H^+-forming CO_2, the enhanced ventilatory response is also gradually reduced.

RENAL REGULATION OF HYDROGEN ION CONCENTRATION

Renal control of $[H^+]$ is the most potent acid–base regulatory mechanism; not only can the kidneys vary H^+ removal, but they can also variably conserve or eliminate HCO_3^- depending on the acid–base status of the body. For example, during renal compensation for acidosis, not only is extra H^+ excreted in urine, but extra HCO_3^- is added to the plasma to buffer (by means of the $H_2CO_3:HCO_3^-$ system) more H^+ that remains in body fluids. By simultaneously removing acid (H^+) from and adding base (HCO_3^-) to body fluids, the kidneys can restore the pH toward normal more effectively than the lungs, which can adjust only the amount of H^+-forming CO_2 in the body.

Also contributing to the kidneys' acid–base regulatory potency is their ability to return pH almost exactly to normal. In contrast to the respiratory system's inability to fully compensate for a pH abnormality, the kidneys can continue to respond to a change in pH until compensation is essentially complete.

The kidneys control pH of body fluids by adjusting three interrelated factors: (1) H^+ excretion, (2) HCO_3^- excretion, and (3) ammonia (NH_3) secretion. We will examine each of these mechanisms in further detail.

▌ Renal H⁺ excretion

Almost all the excreted H^+ enters the urine via secretion. Recall that the filtration rate of H^+ equals plasma $[H^+]$ times GFR. Because plasma $[H^+]$ is extremely low (less than in pure H_2O except during extreme acidosis, when pH falls below 7.0), the filtration rate of H^+ is likewise extremely low. This minute amount of filtered H^+ is excreted in urine. However, most excreted H^+ gains entry into tubular fluid by being actively secreted.

The H^+ secretory process begins in the tubular cells with CO_2 from three sources: CO_2 diffused into the tubular cells from either the (1) plasma or (2) tubular fluid or (3) CO_2 metabolically produced within the tubular cells. Influenced by carbonic anhydrase, CO_2 and H_2O form H_2CO_3, which dissociates into H^+ and HCO_3^-. An energy-dependent carrier in the luminal membrane then transports H^+ out of the cell into the tubular lumen.

The magnitude of H^+ secretion depends primarily on a direct effect of the plasma's acid–base status on the kidneys' tubular cells. No neural or hormonal control is involved.

- When the $[H^+]$ of plasma passing through the peritubular capillaries is elevated above normal, the tubular cells respond by secreting greater-than-usual amounts of H^+ from the plasma into the tubular fluid to be excreted in urine.
- Conversely, when plasma $[H^+]$ is lower than normal, the kidneys conserve H^+ by reducing its secretion and subsequent excretion in urine. The kidneys cannot raise plasma $[H^+]$ by reabsorbing more of the filtered H^+, because there are no reabsorptive mechanisms for H^+. The only way the kidneys can reduce H^+ excretion is by secreting less H^+.

Because chemical reactions for H^+ secretion begin with CO_2, the rate at which they proceed is influenced by $[CO_2]$.

- When plasma $[CO_2]$ increases, these reactions proceed more rapidly and the rate of H^+ secretion speeds up.
- Conversely, the rate of H^+ secretion slows when plasma $[CO_2]$ falls below normal.

These responses are especially important in renal compensations for acid–base abnormalities involving a change in H_2CO_3 caused by respiratory dysfunction. The kidneys can therefore adjust H^+ excretion to compensate for changes in both carbonic and noncarbonic acids.

▌ Renal handling of HCO_3^-

The kidneys regulate plasma $[HCO_3^-]$ by two interrelated mechanisms: (1) variable reabsorption of filtered HCO_3^- back into the plasma and (2) variable addition of new HCO_3^- to the plasma. Both these mechanisms are inextricably linked with H^+ secretion by the kidney tubules. Every time a H^+ is secreted into the tubular fluid, a HCO_3^- is simultaneously transferred into the peritubular capillary plasma. Whether a filtered HCO_3^- is reabsorbed or a new HCO_3^- is added to the plasma in accompaniment with H^+ secretion depends on whether filtered HCO_3^- is present in the tubular fluid to react with the secreted H^+ as follows:

H^+ SECRETION COUPLED WITH HCO_3^- ABSORPTION

Bicarbonate is freely filtered, but because the luminal membranes of tubular cells are impermeable to filtered HCO_3^-, it cannot diffuse into these cells. Therefore, reabsorption of HCO_3^- must occur indirectly (● Figure D-1). Hydrogen ion secreted into the tubular fluid combines with filtered HCO_3^- to form H_2CO_3. Under the influence of carbonic anhydrase, which is present on the surface of the luminal membrane, H_2CO_3 decomposes into CO_2 and H_2O within the filtrate. Unlike HCO_3^-, CO_2 can easily penetrate tubular cell mem-

branes. Within the cells, CO_2 and H_2O, under the influence of intracellular carbonic anhydrase, form H_2CO_3, which dissociates into H^+ and HCO_3^-. Because HCO_3^- can permeate the tubular cells' basolateral membrane, it passively diffuses out of the cells and into the peritubular capillary plasma. Meanwhile, the generated H^+ is actively secreted. Because the disappearance of a HCO_3^- from the tubular fluid is coupled with the appearance of another HCO_3^- in the plasma, a HCO_3^- has, in effect, been "reabsorbed." Even though the HCO_3^- entering the plasma is not the same HCO_3^- that was filtered, the net result is the same as if HCO_3^- were directly reabsorbed.

Normally, slightly more hydrogen ions are secreted into the tubular fluid than bicarbonate ions are filtered. Accordingly, all the filtered HCO_3^- is usually absorbed, because secreted H^+ is available in the tubular fluid to combine with it to form highly reabsorbable CO_2. The vast majority of the secreted H^+ combines with HCO_3^- and is not excreted, because it is "used up" in HCO_3^- reabsorption. However, the slight excess of secreted H^+ that is not matched by filtered HCO_3^- is excreted in the urine. This normal H^+ excretion rate keeps pace with the normal rate of noncarbonic-acid H^+ production.

H^+ SECRETION AND EXCRETION COUPLED WITH ADDITION OF NEW HCO_3^- TO THE PLASMA

Secretion of H^+ that is *excreted* is coupled with the *addition of new HCO_3^-* to the plasma, in contrast to the secreted H^+ that is coupled with *HCO_3^- reabsorption* and is *not excreted*, instead being incorporated into reabsorbable H_2O molecules. When all the filtered HCO_3^- has been reabsorbed and additional secreted H^+ is generated by dissociation of H_2CO_3, the

---

● FIGURE D-1

Hydrogen ion secretion coupled with bicarbonate reabsorption. Because the disappearance of a filtered HCO_3^- from the tubular fluid is coupled with the appearance of another HCO_3^- in the plasma, HCO_3^- is considered to have been "reabsorbed."

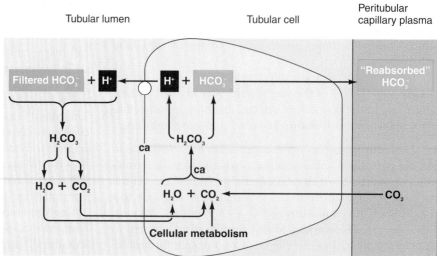

Tubular lumen   Tubular cell   Peritubular capillary plasma

ca = Carbonic anhydrase

HCO_3^- produced by this reaction diffuses into the plasma as a "new" HCO_3^-. It is termed "new" because its appearance in plasma is not associated with reabsorption of filtered HCO_3^- (● Figure D-2). Meanwhile, the secreted H^+ combines with urinary buffers, especially basic phosphate (HPO_4^{2-}) and is excreted.

Renal handling of H^+ and HCO_3^- during acidosis and alkalosis

When plasma $[H^+]$ is elevated during acidosis, more H^+ is secreted than normal. At the same time, less HCO_3^- is filtered than normal because more of the plasma HCO_3^- is used up in buffering the excess H^+ in the ECF. This greater-than-usual inequity between filtered HCO_3^- and secreted H^+ has two consequences. First, more of the secreted H^+ is excreted in the urine, because more hydrogen ions are entering the tubular fluid at a time when fewer are needed to reabsorb the reduced quantities of filtered HCO_3^-. In this way, extra H^+ is eliminated from the body, making the urine more acidic than normal. Second, because excretion of H^+ is linked with the addition of new HCO_3^- to the plasma, more HCO_3^- than usual enters the plasma passing through the kidneys. This additional HCO_3^- is available to buffer excess H^+ present in the body.

In the opposite situation of alkalosis, the rate of H^+ secretion diminishes, while the rate of HCO_3^- filtration increases compared to normal. When plasma $[H^+]$ is below normal, a smaller proportion of the HCO_3^- pool is tied up buffering H^+, so plasma $[HCO_3^-]$ is elevated above normal. As a result, the rate of HCO_3^- filtration correspondingly increases. Not all the filtered HCO_3^- is reabsorbed, because bicarbonate ions are in excess of secreted hydrogen ions in the tubular fluid and HCO_3^- cannot be reabsorbed without first reacting with H^+. Excess HCO_3^- is left in the tubular fluid to be excreted in the urine, thus reducing plasma $[HCO_3^-]$ while making the urine alkaline.

Secreted NH_3 as a urinary buffer

In contrast to the phosphate buffers, which are in the tubular fluid because they have been filtered but not reabsorbed, NH_3 is deliberately synthesized from the amino acid *glutamine* within the tubular cells. Once synthesized, NH_3 readily diffuses passively down its concentration gradient into the tubular fluid; that is, it is secreted. The rate of NH_3 secretion is controlled by a direct effect on the tubular cells of the amount of excess H^+ to be transported in the urine. When someone has been acidotic for more than two or three days, the rate of NH_3 production increases substantially. This extra NH_3 provides additional buffering capacity to allow H^+ secretion to continue after the normal phosphate-buffering capacity is overwhelmed during renal compensation for acidosis.

ACID–BASE IMBALANCES

Because of the relationship between $[H^+]$ and the concentrations of the members of a buffer pair, changes in $[H^+]$ are reflected by changes in the ratio of $[HCO_3^-]$ to $[CO_2]$. Recall that the normal ratio is 20/1. Using the Henderson-Hasselbalch equation and with pK being 6.1 and the log of 20 being 1.3, normal pH = 6.1 + 1.3 = 7.4. Determinations of $[HCO_3^-]$ and $[CO_2]$ provide more meaningful information about the underlying factors responsible for a particular acid–base status than do direct measurements of $[H^+]$ alone. The following rules of thumb apply when examining acid–base imbalances *before any compensations take place*:

1. A change in pH that has a respiratory cause is associated with an abnormal $[CO_2]$, giving rise to a change in carbonic acid–generated H^+. In contrast, a pH deviation of metabolic origin will be associated with an abnormal $[HCO_3^-]$ as a result of the participation of HCO_3^- in buffering abnormal amounts of H^+ generated from noncarbonic acids.
2. Anytime the $[HCO_3^-]/[CO_2]$ ratio falls below 20/1, an acidosis exists. The log of any number lower than 20 is less than 1.3 and, when added to the pK of 6.1, yields an acidotic pH below 7.4. Anytime the ratio exceeds 20/1, an alkalosis exists.

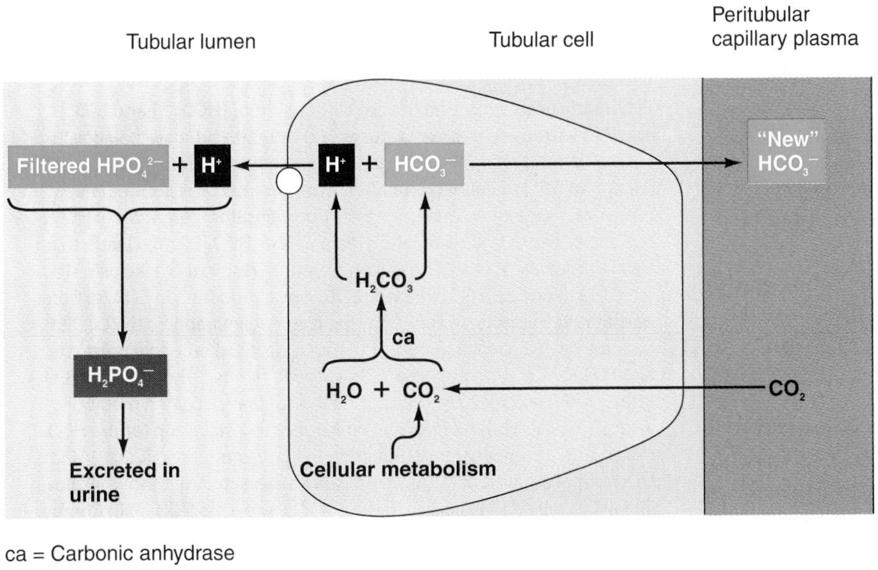

● **FIGURE D-2**

Hydrogen ion secretion and excretion coupled with the addition of new HCO_3^- to the plasma. Secreted H^+ does not combine with filtered HPO_4^{2-} and is not subsequently excreted until all the filtered HCO_3^- has been "reabsorbed," as depicted in ● Figure D-1. Once all the filtered HCO_3^- has combined with secreted H^+, further secreted H^+ is excreted in the urine, primarily in association with urinary buffers such as basic phosphate. Excretion of H^+ is coupled with the appearance of new HCO_3^- in the plasma. The "new" HCO_3^- represents a net gain rather than being merely a replacement for filtered HCO_3^-.

Tubular lumen — Tubular cell — Peritubular capillary plasma

Filtered HPO_4^{2-} + H^+ ← H^+ + HCO_3^- → "New" HCO_3^-

H_2CO_3

ca

H_2O + CO_2 ← CO_2

$H_2PO_4^-$

Excreted in urine

Cellular metabolism

ca = Carbonic anhydrase

The log of any number greater than 20 is more than 1.3 and, when added to the pK of 6.1, yields an alkalotic pH above 7.4.

Putting these two points together,

- *Respiratory acidosis* has a ratio of less than 20/1 arising from an increase in $[CO_2]$.
- *Respiratory alkalosis* has a ratio greater than 20/1 because of a decrease in $[CO_2]$.
- *Metabolic acidosis* has a ratio of less than 20/1 associated with a fall in $[HCO_3^-]$.
- *Metabolic alkalosis* has a ratio greater than 20/1 arising from an elevation in $[HCO_3^-]$.

We will examine each of these categories separately in more detail, paying particular attention to possible causes and the compensations that occur. The "balance beam" concept, presented in ● Figure D-3 in conjunction with the Henderson-Hasselbalch equation, will help you better visualize the contributions of the lungs and kidneys to the causes of and compensations for various acid–base disorders. The normal situation is represented in ● Figure D-3a.

Respiratory acidosis

In uncompensated respiratory acidosis (● Figure D-3b), $[CO_2]$ is elevated because of hypoventilation (in our example, it is doubled) whereas $[HCO_3^-]$ is normal, so the ratio is 20/2 (10/1) and pH is reduced. Let us clarify a potentially confusing point. You might wonder why when $[CO_2]$ is elevated and drives the reaction $CO_2 + H_2O \rightleftarrows H_2CO_3 \rightleftarrows H^+ + HCO_3^-$ to the right, we say that $[H^+]$ becomes elevated but $[HCO_3^-]$ remains normal, although the same quantities of H^+ and HCO_3^- are produced when CO_2-generated H_2CO_3 dissociates. The answer lies in the fact that normally the $[HCO_3^-]$ is 600,000 times the $[H^+]$. For every one hydrogen ion and 600,000 bicarbonate ions present in the ECF, the generation of one additional H^+ and one HCO_3^- doubles the $[H^+]$ (a 100% increase) but only increases the $[HCO_3^-]$ 0.00017% (from 600,000 to 600,001 ions). Therefore, an elevation in $[CO_2]$ brings about a pronounced increase in $[H^+]$, but $[HCO_3^-]$ remains essentially normal.

Compensatory measures act to restore pH to normal.

- The chemical buffer systems immediately take up additional H^+.
- The respiratory mechanism usually cannot respond with compensatory increased ventilation, because impaired respiration is the problem in the first place.
- Thus the kidneys are most important in compensating for respiratory acidosis. They conserve all the filtered HCO_3^- and add new HCO_3^- to the plasma while simultaneously secreting and, accordingly, excreting more H^+.

As a result, HCO_3^- stores in the body become elevated. In our example (● Figure D-3c), the plasma $[HCO_3^-]$ is doubled, so the $[HCO_3^-]/[CO_2]$ ratio is 40/2 rather than 20/2 as it was in the uncompensated state. A ratio of 40/2 is equivalent to a normal 20/1 ratio, so pH is once again the normal 7.4. Enhanced renal conservation of HCO_3^- has fully compensated

for CO_2 accumulation, thus restoring pH to normal, although both $[CO_2]$ and $[HCO_3^-]$ are now distorted. Note that maintenance of normal pH depends on preserving a normal ratio between $[HCO_3^-]$ and $[CO_2]$, no matter what the absolute values of each of these buffer components are. (Compensation is never fully complete because pH can be restored close to but not precisely to normal. In our examples, however, we assume full compensation, for ease in mathematical calculations. Also bear in mind that the values used are only representative. Deviations in pH actually occur over a range, and the degree to which compensation can be accomplished varies.)

Respiratory alkalosis

Looking at the biochemical abnormalities in uncompensated respiratory alkalosis (● Figure D-3d), the increase in pH reflects a reduction in $[CO_2]$ (half the normal value in our example) as a result of hyperventilation, whereas the $[HCO_3^-]$ remains normal. This yields an alkalotic ratio of 20/0.5, which is comparable to 40/1.

Compensatory measures act to shift pH back toward normal.

- The chemical buffer systems liberate H^+ to diminish the severity of the alkalosis.
- As plasma $[CO_2]$ and $[H^+]$ fall below normal because of excessive ventilation, two of the normally potent stimuli for driving ventilation are removed. This effect tends to "put brakes" on the extent to which some nonrespiratory factor such as fever or anxiety can overdrive ventilation. Therefore, hyperventilation does not continue completely unabated.
- If the situation continues for a few days, the kidneys compensate by conserving H^+ and excreting more HCO_3^-.

If, as in our example (● Figure D-3e), the HCO_3^- stores are reduced by half by loss of HCO_3^- in the urine, the $[HCO_3^-]/[CO_2]$ ratio becomes 10/0.5, equivalent to the normal 20/1. Therefore, the pH is restored to normal by reducing the HCO_3^- load to compensate for the CO_2 loss.

● **FIGURE D-3** ▶

Schematic representation of the relationship of $[HCO_3^-]$ and $[CO_2]$ to pH in various acid–base statuses. (a) Normal acid–base balance. The $[HCO_3^-]/[CO_2]$ ratio is 20/1. (b) Uncompensated respiratory acidosis. The $[HCO_3^-]/[CO_2]$ ratio is reduced (20/2), because CO_2 has accumulated. (c) Compensated respiratory acidosis. Compensatory retention of HCO_3^- to balance the CO_2 accumulation restores the $[HCO_3^-]/[CO_2]$ ratio to a normal equivalent (40/2). (d) Uncompensated respiratory alkalosis. The $[HCO_3^-]/[CO_2]$ ratio is increased (20/0.5) by a reduction in CO_2. (e) Compensated respiratory alkalosis. Compensatory elimination of HCO_3^- to balance the CO_2 deficit restores the $[HCO_3^-]/[CO_2]$ ratio to a normal equivalent (10/0.5). (f) Uncompensated metabolic acidosis. The $[HCO_3^-]/[CO_2]$ ratio is reduced (10/1) by a HCO_3^- deficit. (g) Compensated metabolic acidosis. Conservation of HCO_3^-, which partially makes up for the HCO_3^- deficit, and a compensatory reduction in CO_2 restore the $[HCO_3^-]/[CO_2]$ to a normal equivalent (15/0.75). (h) Uncompensated metabolic alkalosis. The $[HCO_3^-]/[CO_2]$ ratio is increased (40/1) by excess HCO_3^-. (i) Compensated metabolic alkalosis. Elimination of some of the extra HCO_3^- and a compensatory increase in CO_2 restore the $[HCO_3^-]/[CO_2]$ ratio to a normal equivalent (25/1.25).

▌ Metabolic acidosis

Uncompensated metabolic acidosis (● Figure D-3f), is always characterized by a reduction in plasma $[HCO_3^-]$ (in our example it is halved), whereas $[CO_2]$ remains normal, producing an acidotic ratio of 10/1. The problem may arise from excessive loss of HCO_3^--rich fluids from the body (as in diarrhea) or from an accumulation of noncarbonic acids (as in diabetes mellitus or uremic acidosis). In the case of accumulation of noncarbonic acids, plasma HCO_3^- is used up in buffering the additional H^+.

Except in uremic acidosis, metabolic acidosis is compensated for by both respiratory and renal mechanisms as well as by chemical buffers.

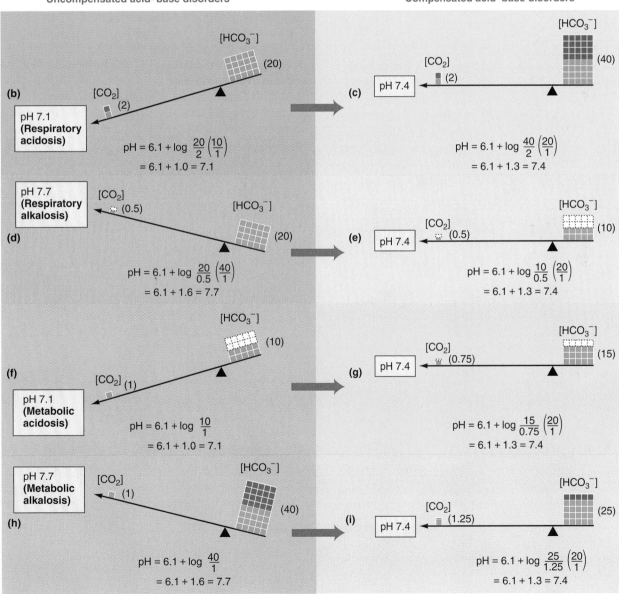

The lengths of the arms of the balance beams are not to scale.

- The chemical buffer systems take up extra H^+.
- The lungs blow off additional H^+-generating CO_2.
- The kidneys excrete more H^+ and conserve more HCO_3^-.

In our example (● Figure D-3g), these compensatory measures restore the ratio to normal by reducing $[CO_2]$ to 75% of normal and by raising $[HCO_3^-]$ halfway back toward normal (up from 50 to 75% of the normal value). This brings the ratio to 15/0.75 (equivalent to 20/1).

Note that in compensating for metabolic acidosis, the lungs deliberately displace $[CO_2]$ from normal in an attempt to restore $[H^+]$ toward normal. Whereas in respiratory-induced acid–base disorders an abnormal $[CO_2]$ is the *cause* of the $[H^+]$ imbalance, in metabolic acid–base disorders $[CO_2]$ is intentionally shifted from normal as an important *compensation* for the $[H^+]$ imbalance.

When kidney disease causes metabolic acidosis, complete compensation is not possible because the renal mechanism is not available for pH regulation. Recall that the respiratory system can compensate only up to 75% of the way toward normal. Uremic acidosis is very serious, because the kidneys cannot help restore pH all the way to normal.

▌Metabolic alkalosis

Uncompensated metabolic alkalosis is associated with an increase in $[HCO_3^-]$, which, in the uncompensated state, is not accompanied by a change in $[CO_2]$. In our example (● Figure D-3h), $[HCO_3^-]$ is doubled, producing an alkalotic ratio of 40/1.

This condition arises most commonly from excess ingestion of alkaline drugs, such as when baking soda ($NaHCO_3$) is used as a self-administered remedy for the treatment of gastric hyperacidity, or from excessive vomiting of acidic gastric juices. Hydrochloric acid is secreted into the stomach lumen during digestion. Bicarbonate is added to the plasma during gastric HCl secretion. This HCO_3^- is neutralized by H^+ as the gastric secretions are eventually reabsorbed back into the plasma, so normally there is no net addition of HCO_3^- to the plasma from this source. However, when this acid is lost from the body during vomiting not only is plasma $[H^+]$ decreased, but also reabsorbed H^+ is no longer available to neutralize the extra HCO_3^- added to the plasma during gastric HCl secretion. Thus loss of HCl in effect increases plasma $[HCO_3^-]$.

The following compensatory measures come into play to restore the pH toward normal:

- The chemical buffer systems immediately liberate H^+.
- Ventilation is reduced so that extra H^+-generating CO_2 is retained in the body fluids.
- If the condition persists for several days, the kidneys conserve H^+ and excrete the excess HCO_3^- in the urine.

The resultant compensatory increase in $[CO_2]$ (up 25% in our example; ● Figure D-3i) and the partial reduction in $[HCO_3^-]$ (75% of the way back down toward normal in our example) together restore the $[HCO_3^-]/[CO_2]$ ratio back to the equivalent of 20/1 at 25/1.25.

Thus an individual's acid–base status cannot be assessed on the basis of pH alone. Even though the pH is essentially normal, determinations of $[HCO_3^-]$ and $[CO_2]$ can reveal compensated acid–base disorders.

Text References to Exercise Physiology

BOXED FEATURES RELATED TO EXERCISE PHYSIOLOGY

EXERCISE REFERENCES BY TOPIC

Answers to End-of-Chapter Objective Questions, Points to Ponder, and Clinical Considerations

CHAPTER 1   HOMEOSTASIS: THE FOUNDATION OF PHYSIOLOGY

▌ Objective Questions

(Questions on p. 16.)

1.e   2. b   3. T   4. F   5. T   6. muscle tissue, nervous tissue, epithelial tissue, connective tissue   7. secretion   8. exocrine, endocrine, hormones   9. intrinsic, extrinsic   10. 1.d, 2.g, 3.a, 4.e, 5.b, 6.j, 7.h, 8.i, 9.c, 10.f

▌ Points to Ponder

(Questions on p. 17.)

1. The respiratory system eliminates internally produced CO_2 to the external environment. A decrease in CO_2 in the internal environment brings about a reduction in respiratory activity (that is, slower, shallower breathing) so that CO_2 produced within the body accumulates instead of being blown off as rapidly as normal to the external environment. The extra CO_2 retained in the body increases the CO_2 levels in the internal environment to normal.
2. (b) (c) (b)
3. b
4. immune defense system
5. When a person is engaged in strenuous exercise, the temperature-regulating center in the brain brings about widening of the blood vessels of the skin. The resultant increased blood flow through the skin carries the extra heat generated by the contracting muscles to the body surface, where it can be lost to the surrounding environment.

▌ Clinical Consideration

(Question on p. 17)

Loss of fluids threatens the maintenance of proper plasma volume and blood pressure. Loss of acidic digestive juices threatens maintenance of the proper pH in the internal fluid environment. The urinary system helps restore the proper plasma volume and pH by reducing the amount of water and acid eliminated in the urine. The respiratory system helps restore the pH by adjusting the rate of removal of acid-forming CO_2. Adjustments are made in the circulatory system to help maintain blood pressure despite fluid loss. Increased thirst encourages increased fluid intake to help restore plasma volume. These compensatory changes in the urinary, respiratory, and circulatory systems, as well as the sensation of thirst, are all regulated by the two regulatory systems, the nervous and endocrine systems. Furthermore, the endocrine system makes internal adjustments to help maintain the concentration of nutrients in the internal environment even though no new nutrients are being absorbed from the digestive system.

CHAPTER 2   CELLULAR PHYSIOLOGY

▌ Objective Questions

(Questions on p. 39.)

1. plasma membrane   2. deoxyribonucleic acid (DNA), nucleus   3. organelles, cytosol, cytoskeleton   4. endoplasmic reticulum, Golgi complex   5. oxidative   6. adenosine triphosphate (ATP)   7. F   8. 1.b, 2.a, 3.b   9. 1.b, 2.c, 3.c, 4.a, 5.b, 6.c, 7.a, 8.c

▌ Points to Ponder

(Questions on p. 40.)

1. 24 moles O_2/day $\times$ 6 moles ATP/mole O_2 = 144 moles ATP/day
    144 moles ATP/day $\times$ 507 g ATP/mole = 73,000 g ATP/day
    1000 g/2.2 lb = 73,000 g/x lb
    1000 x = 160,600
    x = approximately 160 lb
2. The chief cells have an extensive rough endoplasmic reticulum, with this organelle being responsible for synthesizing these cells' protein secretory product, namely pepsinogen. Because the parietal cells do not secrete a protein product to the cells' exterior, they do not need an extensive rough endoplasmic reticulum.
3. With cyanide poisoning, cell activities that depend on ATP expenditure could not continue, such as synthesis of new chemical compounds, membrane transport, and mechanical work. The resultant inability of the heart to pump blood and failure of the respiratory muscles to accomplish breathing would lead to imminent death.
4. ATP is required for muscle contraction. Muscles can store limited supplies of nutrient fuel for use in generating ATP. During anaerobic exercise, muscles generate ATP from these nutrient stores by means of glycolysis, which yields two molecules of ATP per glucose molecule processed. During aerobic exercise, muscles can generate ATP by oxidative phosphorylation, which yields 36 molecules of ATP per glucose molecule processed. Because glycolysis inefficiently generates ATP from nutrient fuels, it rapidly depletes the muscle's limited stores of fuel, and ATP can no longer be produced to sustain the muscle's contractile activity. Aerobic exercise, in contrast, can be sustained for prolonged periods. Not only does oxidative phosphorylation use far less nutrient fuel to generate ATP, but it can be supported by nutrients delivered to the muscle by the blood instead of relying on fuel stored in the muscle. Intense anaerobic exercise outpaces the ability to deliver supplies to the muscle by the blood, so the muscle must rely on stored fuel and inefficient glycolysis, thus limiting anaerobic exercise to brief periods of time before energy sources are depleted.
5. skin. The mutant keratin weakens the skin cells of patients with epidermolysis bullosa so that the skin blisters in response to even a light touch.

▌ Clinical Consideration

(Question on p. 40.)

Some hereditary forms of male sterility involving nonmotile sperm have been traced to defects in the cytoskeletal components of the sperm's flagella. These same individuals usually also have long histories of recurrent respiratory tract disease because the same types of defects are present in their respiratory cilia, which are unable to clear mucus and inhaled particles from the respiratory system.

CHAPTER 3   THE PLASMA MEMBRANE AND MEMBRANE POTENTIAL

▌ Objective Questions

(Questions on p. 67.)

1. T   2. F   3. F   4. negative, positive   5. 1.b, 2.a, 3.b, 4.a, 5.c, 6.b, 7.a, 8.b   6. 1.a, 2.a, 3.b, 4.a, 5.b   7. 1.c, 2.b, 3.a, 4.a, 5.c, 6.b, 7.c, 8.a, 9.b

▌ Points to Ponder

(Questions on p. 68.)

1. c. As Na^+ moves from side 1 to side 2 down its concentration gradient, Cl^- remains on side 1, unable to permeate the membrane. The resultant separation of charges produces a membrane potential, negative on side 1 because of unbalanced chloride ions and positive on side 2 because of unbalanced sodium ions. Sodium does not continue to move to side 2 until its concentration gradient is dissipated because of the development of an opposing electrical gradient.

2. more positive. Because the electrochemical gradient for Na^+ is inward, the membrane potential would become more positive as a result of an increased influx of Na^+ into the cell if the membrane were more permeable to Na^+ than to K^+. (Indeed, this is what happens during the rising phase of an action potential once threshold potential is reached—see Chapter 4).

3. d. active transport. Leveling off of the curve designates saturation of a carrier molecule, so carrier-mediated transport is involved. The graph indicates that active transport is being used instead of facilitated diffusion, because the concentration of the substance in the intracellular fluid is greater than the concentration in the extracellular fluid at all points until after the transport maximum is reached. Thus the substance is being moved *against* a concentration gradient, so active transport must be the transport method used.

4. vesicular transport. The maternal antibodies in the infant's digestive tract lumen are taken up by the intestinal cells by pinocytosis and are extruded on the opposite side of the cell into the interstitial fluid by exocytosis. The antibodies are picked up from the intestinal interstitial fluid by the blood supply to the region.

5. accelerate. During an action potential, Na^+ enters and K^+ leaves the cell. Repeated action potentials would eventually "run down" the Na^+ and K^+ concentration gradients were it not for the Na^+–K^+ pump returning the Na^+ that entered back to the outside and the K^+ that left back to the inside. Indeed, the rate of pump activity is accelerated by the increase in both ICF Na^+ and ECF K^+ concentrations that occurs as a result of action potential activity, thus hastening the restoration of the concentration gradients after the action potential ceases.

▌ Clinical Consideration

(Question on p. 69)

As Cl^- is secreted by the intestinal cells into the intestinal tract lumen, Na^+ follows passively along the established electrical gradient. Water passively accompanies this salt (Na^+ and Cl^-) secretion by osmosis. Increased secretion of Cl^- and the subsequent passively induced secretion of Na^+ and water are responsible for the severe diarrhea that characterizes cholera.

CHAPTER 4   PRINCIPLES OF NEURONAL AND HORMONAL COMMUNICATION

▌ Objective Questions

(Questions on p. 106.)

1. T   2. F   3. F   4. F   5. T   6. F   7. refractory period   8. axon hillock   9. synapse   10. temporal summation   11. spatial summation   12. convergence, divergence   13. G protein   14. 1.b, 2.a, 3.a, 4.b, 5.b, 6.a   15. 1.a, 2.b, 3.a, 4.b, 5.d, 6.b, 7.b, 8.b, 9.a, 10.b, 11.a, 12.c

▌ Points to Ponder

(Questions on p. 107.)

1. c. The action potentials would stop as they met in the middle. As the two action potentials moving toward each other both reached the middle of the axon, the two adjacent patches of membrane in the middle would be in a refractory period so further propagation of either action potential would be impossible.

2. A subthreshold stimulus would transiently depolarize the membrane but not sufficiently to bring the membrane to threshold, so no action potential would occur. Because a threshold stimulus would bring the membrane to threshold, an action potential would occur. An action potential of the same magnitude and duration would occur in response to a suprathreshold stimulus as to a threshold stimulus. Because of the all-or-none law, a stimulus larger than that necessary to bring the membrane to threshold would not produce a larger action potential. (The magnitude of the stimulus is coded in the *frequency* of action potentials generated in the neuron, not the *size* of the action potentials.)

3. The hand could be pulled away from the hot stove by flexion of the elbow accomplished by summation of EPSPs at the cell bodies of the neurons controlling the biceps muscle, thus bringing these neurons to threshold. The subsequent action potentials generated in these neurons would stimulate contraction of the biceps. Simultaneous contraction of the triceps muscle, which would oppose the desired flexion of the elbow, could be prevented by generation of IPSPs at the cell bodies of the neurons controlling this muscle. These IPSPs would keep the triceps neurons from reaching threshold and firing so that the triceps would not be stimulated to contract.

The arm could deliberately be extended despite a painful finger prick by voluntarily generating EPSPs to override the reflex IPSPs at the neuronal cell bodies controlling the triceps while simultaneously generating IPSPs to override the reflex EPSPs at the neuronal cell bodies controlling the biceps.

4. Treatment for Parkinson's disease is aimed toward restoring dopamine activity in the basal nuclei. However, this treatment may lead to excessive dopamine activity in otherwise normal areas of the brain that also use dopamine as a neurotransmitter. Excessive dopamine activity in a particular region of the brain (the limbic system) is believed to be among the causes of schizophrenia. Therefore, symptoms of schizophrenia sometimes occur as a side effect during treatment for Parkinson's disease.

5. An EPSP, being a graded potential, spreads decrementally from its site of initiation in the postsynaptic neuron. If presynaptic neuron A (near the axon hillock of the postsynaptic cell) and presynaptic neuron B (on the opposite side of the postsynaptic cell body) both initiate EPSPs of the same magnitude and frequency, the EPSPs from A will be of greater strength when they reach the axon hillock than will the EPSPs from B. An EPSP from B will decrease more in magnitude as it travels farther before reaching the axon hillock, the region of lowest threshold and thus the site of action potential initiation. Temporal summation of the larger EPSPs from A may bring the axon hillock to threshold and initiate an action potential in the postsynaptic neuron, whereas temporal summation of the weaker EPSPs from B at the axon hillock may not be sufficient to bring this region to threshold. Thus the proximity of a presynaptic neuron to the axon hillock can bias its influence on the postsynaptic cell.

1. T 2. F 3. T 4. T 5. F 6. T 7. F 8. F 9. transduction
10. adequate stimulus 11. 1.d, 2.f, 3.i, 4.g, 5.c, 6.a, 7.h, 8.e, 9.b
12. 1.a, 2.b, 3.c, 4.c, 5.c, 6.a, 7.b, 8.b

Clinical Consideration

(Question on p. 107)

Initiation and propagation of action potentials would not occur in nerve fibers acted on by local anesthetic because blockage of Na^+ channels by the local anesthetic would prevent the massive opening of voltage-gated Na^+ channels at threshold potential. As a result, pain impulses (action potentials in nerve fibers that carry pain signals) would not be initiated and propagated to the brain and reach the level of conscious awareness.

CHAPTER 5   THE CENTRAL NERVOUS SYSTEM

Objective Questions

(Questions on p. 142.)

1. F 2. F 3. F 4. T 5. F 6. consolidation 7. dorsal, ventral
8. 1.d, 2.c, 3.f, 4.e, 5.a, 6.b 9. 1.a, 2.c, 3.a and b, 4.b, 5.a, 6.c, 7.c

Points to Ponder

(Questions on p. 142.)

1. Only the left hemisphere has language ability. When sharing of information between the two hemispheres is prevented as a result of severance of the corpus callosum, visual information presented only to the right hemisphere cannot be verbally identified by the left hemisphere, because the left hemisphere is unaware of the information. However, the information can be recognized by nonverbal means, of which the right hemisphere is capable.
2. c. A severe blow to the back of the head is most likely to traumatize the visual cortex in the occipital lobe.
3. Insulin excess drives too much glucose into insulin-dependent cells so that the blood glucose falls below normal and insufficient glucose is delivered to the non–insulin-dependent brain. Therefore, the brain, which depends on glucose as its energy source, does not receive adequate nourishment.
4. Salivation when seeing or smelling food, striking the appropriate letter on the keyboard when typing, playing the correct note on a musical instrument, and many of the actions involved in driving a car are conditioned reflexes. You undoubtedly will have many other examples.
5. Strokes occur when a portion of the brain is deprived of its vital O_2 and glucose supply because the cerebral blood vessel supplying the area either is blocked by a clot or has ruptured. Although a clot-dissolving drug could be helpful in restoring blood flow through a cerebral vessel blocked by a clot, such a drug would be detrimental in the case of a ruptured cerebral vessel sealed by a clot. Dissolution of a clot sealing a ruptured vessel would lead to renewed hemorrhage through the vessel and exacerbation of the problem.

Clinical Consideration

(Question on p. 143.)

The deficits following the stroke—numbness and partial paralysis on the upper right side of the body and inability to speak—are indicative of damage to the left somatosensory cortex and left primary motor cortex in the regions devoted to the upper part of the body plus Broca's area.

CHAPTER 6   THE PERIPHERAL NERVOUS SYSTEM: AFFERENT DIVISION; SPECIAL SENSES

Objective Questions

(Questions on p. 182.)

Points to Ponder

(Questions on p. 183.)

1. Pain is a conscious warning that tissue damage is occurring or about to occur. A patient unable to feel pain because of a nerve disorder does not consciously take measures to withdraw from painful stimuli and thus prevent more serious tissue damage.
2. Pupillary dilation (mydriasis) can be deliberately induced by ophthalmic instillation of either an adrenergic drug (such as epinephrine or related compound) or a cholinergic blocking drug (such as atropine or related compounds). Adrenergic drugs produce mydriasis by causing contraction of the sympathetically supplied radial (dilator) muscle of the iris. Cholinergic blocking drugs cause pupillary dilation by blocking parasympathetic activity to the circular (constrictor) muscle of the iris so that action of the adrenergically controlled radial muscle of the iris is unopposed.
3. The defect would be in the left optic tract or optic radiation.
4. Fluid accumulation in the middle ear in accompaniment with middle ear infections impedes the normal movement of the tympanic membrane, ossicles, and oval window in response to sound. All these structures vibrate less vigorously in the presence of fluid, causing temporary hearing impairment. Chronic fluid accumulation in the middle ear is sometimes relieved by surgical implantation of drainage tubes in the eardrum. Hearing is restored to normal as the fluid drains to the exterior. Usually, the tube "falls out" as the eardrum heals and pushes out the foreign object.
5. The sense of smell is reduced when you have a cold, even though the cold virus does not directly adversely affect the olfactory receptor cells, because odorants do not reach the receptor cells as readily when the mucous membranes lining the nasal passageways are swollen and excess mucus is present.

Clinical Consideration

(Question on p. 183.)

Syncope most frequently occurs as a result of inadequate delivery of blood carrying sufficient oxygen and glucose supplies to the brain. Possible causes include circulatory disorders such as impaired pumping of the heart or low blood pressure; respiratory disorders resulting in poorly oxygenated blood; anemia, in which the oxygen-carrying capacity of the blood is reduced; or low blood glucose from improper endocrine management of blood glucose levels. Vertigo, in contrast, typically results from a dysfunction of the vestibular apparatus, arising, for example, from viral infection or trauma, or abnormal neural processing of vestibular information, as, for example, with a brain tumor.

CHAPTER 7   THE PERIPHERAL NERVOUS SYSTEM: EFFERENT DIVISION

Objective Questions

(Questions on p. 199.)

1. T 2. F 3. c 4. c 5. sympathetic, parasympathetic 6. adrenal medulla 7. 1.a, 2.b, 3.a, 4.b, 5.a, 6.a, 7.b 8. 1.b, 2.b, 3.a, 4.a, 5.b, 6.b, 7.a

Points to Ponder

(Questions on p. 200.)

1. By promoting arteriolar constriction, epinephrine administered in conjunction with local anesthetics reduces blood flow to the region and thus helps the anesthetic stay in the region instead of being carried away by the blood.

2. No. Atropine blocks the effect of acetylcholine at muscarinic receptors but does not affect nicotinic receptors. Nicotinic receptors are present on the motor end plates of skeletal muscle fibers.

3. The voluntarily controlled external urethral sphincter is composed of skeletal muscle and supplied by the somatic nervous system.

4. By interfering with normal acetylcholine activity at the neuromuscular junction, α bungarotoxin leads to skeletal muscle paralysis, with death ultimately occurring as a result of an inability to contract the diaphragm and breathe.

5. If the motor neurons that control the respiratory muscles, especially the diaphragm, are destroyed by poliovirus or amyotrophic lateral sclerosis, the person is unable to breathe and dies (unless breathing is assisted by artificial means).

▌ Clinical Consideration

(Question on p. 200.)

Drugs that block β_1 receptors are useful for prolonged treatment of angina pectoris because they interfere with sympathetic stimulation of the heart during exercise or emotionally stressful situations. By preventing increased cardiac metabolism and thus an increased need for oxygen delivery to the cardiac muscle during these situations, beta blockers can reduce the frequency and severity of angina attacks.

CHAPTER 8 MUSCLE PHYSIOLOGY

▌ Objective Questions

(Questions on p. 237.)

1. F 2. F 3. F 4. T 5. F 6. T 7. concentric, eccentric 8. alpha, gamma 9. denervation atrophy, disuse atrophy 10. a, b, e 11. b 12. 1.f, 2.d, 3.c, 4.e, 5.b, 6.g, 7.a 13. 1.a, 2.a, 3.a, 4.b, 5.b, 6.b

▌ Points to Ponder

(Questions on p. 238.)

1. By placing increased demands on the heart to sustain increased delivery of O_2 and nutrients to working skeletal muscles, regular aerobic exercise induces changes in cardiac muscle that enable it to use O_2 more efficiently, such as increasing the number of capillaries supplying blood to the heart muscle. Intense exercise of short duration, such as weight training, in contrast, does not induce cardiac efficiency. Because this type of exercise relies on anaerobic glycolysis for ATP formation, no demands are placed on the heart for increased delivery of blood to the working muscles.

2. The length of the thin filaments is represented by the distance between a Z line and the edge of the adjacent H zone. This distance remains the same in a relaxed and contracted myofibril, leading to the conclusion that the thin filaments do not change in length during muscle contraction.

3. Regular bouts of anaerobic, short-duration, high-intensity resistance training would be recommended for competitive downhill skiing. By promoting hypertrophy of the fast glycolytic fibers, such exercise better adapts the muscles to activities that require intense strength for brief periods, such as a swift, powerful descent downhill. In contrast, regular aerobic exercise would be more beneficial for competitive cross-country skiers. Aerobic exercise induces metabolic changes within the oxidative fibers that enable the muscles to use O_2 more efficiently. These changes, which include an increase in mitochondria and capillaries within the oxidative fibers, adapt the muscles to better endure the prolonged activity of cross-country skiing without fatiguing.

4. Because the site of voluntary control to overcome the micturition reflex is at the external urethral sphincter and not the bladder, the external urethral sphincter must be skeletal muscle, which is innervated by the voluntarily controlled somatic nervous system, and the bladder must be smooth muscle, which is innervated by the involuntarily controlled autonomic nervous sys-

tem. The only other type of involuntarily controlled muscle besides smooth muscle is cardiac muscle, which is found only in the heart. Therefore, the bladder must be smooth, not cardiac, muscle.

5.

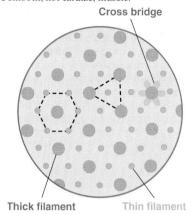

Schematic representation of the geometric relation among thick and thin filaments and cross bridges.

▌ Clinical Consideration

(Question on p. 239.)

The muscles in the immobilized leg have undergone disuse atrophy. The physician or physical therapist can prescribe regular resistance-type exercises that specifically use the atrophied muscles to help restore them to their normal size.

CHAPTER 9 CARDIAC PHYSIOLOGY

▌ Objective Questions

(Questions on p. 271.)

1. intercalated discs, desmosomes, gap junctions 2. bradycardia, tachycardia 3. F 4. F 5. F 6. T 7. d 8. d 9. e 10. less than, greater than, less than, greater than, less than 11. AV, systole, semilunar, diastole 12. 1.e, 2.a, 3.d, 4.b, 5.f, 6.c

▌ Points to Ponder

(Questions on p. 272.)

1. Because, at a given heart rate, the interval between a premature beat and the next normal beat is longer than the interval between two normal beats, the heart fills for a longer period of time following a premature beat before the next period of contraction and emptying begins. Because of the longer filling time, the end-diastolic volume is larger, and, according to the Frank-Starling law of the heart, the subsequent stroke volume will also be correspondingly larger.

2. Trained athletes' hearts are stronger and can pump blood more efficiently so that the resting stroke volume is larger than in an untrained person. For example, if the resting stroke volume of a strong-hearted athlete is 100 ml, a resting heart rate of only 50 beats/minute produces a normal resting cardiac output of 5000 ml/minute. An untrained individual with a resting stroke volume of 70 ml, in contrast, must have a heart rate of about 70 beats/minute to produce a comparable resting cardiac output.

3. In the fetus, blood is shunted from the pulmonary artery into the aorta through the ductus arteriosus, thus bypassing the nonfunctional lungs. Blood flows in this direction down its pressure gradient, with pulmonary artery pressure being greater than aortic pressure during fetal life.

The direction of flow through a patent ductus arteriosus is the reverse of the flow that occurs through this vascular connection during fetal life. With a patent ductus arteriosus, some of the blood present in the aorta is shunted through the still-open ductus arteriosus into the pulmonary artery because, after birth, the aortic pressure is greater than the pulmonary artery pressure.

As a result of this abnormal shunting, not all of the blood pumped out by the left ventricle goes into the systemic circulation, and excessive blood enters the pulmonary circulation. If the condition is not corrrected by tying off the patent ductus arteriosus, the left ventricle compensates by hypertrophying (enlarging and becoming stronger) so that it can pump out even more blood. This extra output provides adequate systemic circulation even through part of the left ventricular output is diverted to the pulmonary circulation. The right ventricle also hypertrophies, enabling it to pump against the elevated pulmonary arterial pressure, which is increased due to the excess volume of blood shunted into the pulmonary circulation. If the condition is not corrected, this extra workload on the heart eventually leads to heart failure and premature death.

4. A transplanted heart that does not have any innervation adjusts the cardiac output to meet the body's changing needs by means of both intrinsic control (the Frank-Starling mechanism) and extrinsic hormonal influences, such as the effect of epinephrine on the rate and strength of cardiac contraction.

5. In left bundle-branch block, the right ventricle becomes completely depolarized more rapidly than the left ventricle. As a result, the right ventricle contracts before the left ventricle, and the right AV valve is forced closed prior to closure of the left AV valve. Because the two AV valves do not close in unison, the first heart sound is "split"; that is, two distinct sounds in close succession can be detected as closure of the left valve lags behind closure of the right valve.

▌ Clinical Consideration

(Question on p. 273.)

The most likely diagnosis is atrial fibrillation. This condition is characterized by rapid, irregular, uncoordinated depolarizations of the atria. Many of these depolarizations reach the AV node at a time when it is not in its refractory period, thus bringing about frequent ventricular depolarizations and a rapid heartbeat. However, because impulses reach the AV node erratically, the ventricular rhythm and thus the heartbeat are also very irregular as well as being rapid.

Ventricular filling is only slightly reduced despite the fact that the fibrillating atria are unable to pump blood because most ventricular filling occurs during diastole prior to atrial contraction. Because of the erratic heartbeat, variable lengths of time are available between ventricular beats for ventricular filling. However, the majority of ventricular filling occurs early in ventricular diastole after the AV valves first open, so even though the filling period may be shortened, the extent of filling may be near normal. Only when the ventricular filling period is very short is ventricular filling substantially reduced.

Cardiac output, which depends on stroke volume and heart rate, usually is not seriously impaired with atrial fibrillation. Because ventricular filling is only slightly reduced during most cardiac cycles, stroke volume, as determined by the Frank-Starling mechanism, is likewise only slightly reduced. Only when the ventricular filling period is very short and the cardiac muscle fibers are operating on the lower end of their length–tension curve is the resultant ventricular contraction weak. When the ventricular contraction becomes too weak, the ventricles eject a small or no stroke volume. During most cardiac cycles, however, the slight reduction in stroke volume is often offset by the increased heart rate, so that cardiac output is usually near normal. Furthermore, if the mean arterial blood pressure falls because the cardiac output does decrease, increased sympathetic stimulation of the heart brought about by the baroreceptor reflex helps restore cardiac output to normal by shifting the Frank-Starling curve to the left.

On those cycles when ventricular contractions are too weak to eject enough blood to produce a palpable wrist pulse, if the heart rate is determined directly, either by the apex beat or via the ECG, and the pulse rate is taken concurrently at the wrist, the heart rate will exceed the pulse rate, producing a pulse deficit.

CHAPTER 10 THE BLOOD VESSELS AND BLOOD PRESSURE

▌ Objective Questions

(Questions on p. 312.)

1. T   2. F   3. T   4. T   5. T   6. a, c, d, e, f   7. 1.a, 2.a, 3.b, 4.a, 5.b, 6.a   8. 1.b, 2.a, 3.b, 4.a, 5.a, 6.a, 7.b, 8.a, 9.b, 10.a, 11.b, 12.a, 13.a

▌ Points to Ponder

(Questions on p. 313.)

1. An elastic support stocking increases external pressure on the remaining veins in the limb to produce a favorable pressure gradient that promotes venous return to the heart and minimizes swelling that would result from fluid retention in the extremity.

2. a. 125 mm Hg
   b. 77 mm Hg
   c. 48 mm Hg; (125 mm Hg − 77 mm = 48 mm Hg)
   d. 93 mm Hg; $[77 + \frac{1}{3}(48) = 77 + 16 = 93$ mm Hg$]$
   e. No; no blood would be able to get through the brachial artery, so no sound would be heard.
   f. Yes; blood would flow through the brachial artery when the arterial pressure was between 118 and 125 mm Hg and would not flow through when the arterial pressure fell below 118 mm Hg. The turbulence created by this intermittent blood flow would produce sounds.
   g. No; blood would flow continuously through the brachial artery in smooth, laminar fashion, so no sound would be heard.

3. The classmate has apparently fainted because of insufficient blood flow to the brain as a result of pooling of blood in the lower extremities brought about by standing still for a prolonged time during the laboratory experiment. When the person faints and assumes a horizontal position, the pooled blood will quickly be returned to the heart, improving cardiac output and blood flow to his brain. Trying to get the person up would be counterproductive, so the classmate trying to get him up should be advised to let him remain lying down until he recovers on his own.

4. The drug is apparently causing the arteriolar smooth muscle to relax by causing the release of a local vasoactive chemical mediator from the endothelial cells that induces relaxation of the underlying smooth muscle.

5. a. Because activation of alpha$_1$-adrenergic receptors in vascular smooth muscle brings about vasoconstriction, blockage of alpha$_1$-adrenergic receptors reduces vasoconstrictor activity, thereby lowering the total peripheral resistance and arterial blood pressure.
   b. Because activation of beta$_1$-adrenergic receptors, which are found primarily in the heart, increases the rate and strength of cardiac contraction, drugs that block beta$_1$-adrenergic receptors reduce cardiac ouput and thus arterial blood pressure by decreasing the rate and strength of the heart beat.
   c. Drugs that directly relax arteriolar smooth muscle lower arterial blood pressure by promoting arteriolar vasodilation and reducing total peripheral resistance.
   d. Diuretic drugs reduce the plasma volume, thereby lowering arterial blood pressure, by increasing urinary output. Salt and water that normally would have been retained in the plasma are excreted in the urine.
   e. Because sympathetic activity promotes generalized arteriolar vasoconstriction, thereby increasing total peripheral resistance and arterial blood pressure, drugs that block the release of norepinephrine from sympathetic endings lower blood pressure by preventing this sympathetic vasoconstrictor effect.
   f. Similarly, drugs that act on the brain to reduce sympathetic output lower blood pressure by preventing the effect of sympathetic activity on promoting arteriolar vasoconstriction and the resultant increase in total peripheral resistance and arterial blood pressure.

g. Drugs that block Ca^{2+} channels reduce the entry of Ca^{2+} into the vascular smooth-muscle cells from the ECF in response to excitatory input. Because the level of contractile activity in vascular smooth-muscle cells depends on their cytosolic Ca^{2+} concentration, drugs that block Ca^{2+} channels reduce the contractile activity of these cells by reducing Ca^{2+} entry and lowering their cytosolic Ca^{2+} concentration. Total peripheral resistance and, accordingly, arterial blood pressure are decreased as a result of reduced arteriolar contractile activity.

h. Drugs that interfere with the production of angiotensin II block activation of the hormonal pathway that promotes salt and water conservation (the renin-angiotensin-aldosterone system). As a result, more salt and water are lost in the urine, and less fluid is retained in the plasma. The resultant reduction in plasma volume lowers the arterial blood pressure.

▌ Clinical Consideration

(Question on p. 313.)

The abnormally elevated levels of epinephrine found with a pheochromocytoma bring about secondary hypertension by (1) increasing the heart rate; (2) increasing cardiac contractility, which increases stroke volume; (3) causing venous vasoconstriction, which increases venous return and subsequently stroke volume by means of the Frank-Starling mechanism; and (4) causing arteriolar vasoconstriction, which increases total peripheral resistance. Increased heart rate and stroke volume both lead to increased cardiac output. Increased cardiac output and increased total peripheral resistance both lead to increased arterial blood pressure.

CHAPTER 11 THE BLOOD AND BODY DEFENSES

▌ Objective Questions

(Questions on p. 361.)

1. F 2. T 3. F 4. F 5. F 6. F 7. T 8. cytokines 9. d 10. a
11. b 12. 1.e, 2.c, 3.b, 4.d, 5.g, 6.f, 7.a, 8.h 13. 1.c, 2.d, 3.a, 4.b 14. 1.a, 2.a, 3.b, 4.b, 5.c, 6.c, 7.b, 8.a, 9.b, 10.b, 11.a, 12.b

▌ Points to Ponder

(Questions on p. 362.)

1. If the genes that direct fetal hemoglobin-F synthesis could be reactivated in a patient with sickle cell anemia, a portion of the abnormal hemoglobin S that causes the erythrocytes to warp into defective sickle-shaped cells would be replaced by "healthy" hemoglobin F, thus sparing a portion of the RBCs from premature rupture. Hemoglobin F would not completely replace hemoglobin S because the gene for synthesis of hemoglobin S would still be active.
2. Most heart-attack deaths are attributable to the formation of abnormal clots that prevent normal blood flow. The sought-after chemicals in the "saliva" of blood-sucking creatures are agents that break up or prevent the formation of these abnormal clots.

Although genetically engineered tissue-plasminogen activator (tPA) is already being used as a clot-busting drug, this agent brings about degradation of fibrinogen as well as fibrin. Thus, even though the life-threatening clot in the coronary circulation is dissolved, the fibrinogen supplies in the blood are depleted for up to 24 hours until new fibrinogen is synthesized by the liver. If the patient sustains a ruptured vessel in the interim, insufficient fibrinogen might be available to form a blood-staunching clot. For example, many patients treated with tPA suffer hemorrhagic strokes within 24 hours of treatment due to incomplete sealing of a ruptured cerebral vessel. Therefore, scientists are searching for better alternatives to combat abnormal clot formation by examining the naturally occurring chemicals produced by blood-

sucking creatures that permit them to suck a victim's blood without the blood clotting.
3. A vaccine against a particular microbe can be effective only if it induces formation of antibodies and/or activated T cells against a stable antigen that is present on all microbes of this type. Because HIV frequently mutates, it has not been possible to produce a reliable vaccine against it. Specific immune responses induced by vaccination against one form of HIV could prove to be ineffective against a slightly modified version of the virus.
4. Failure of the thymus to develop embryonically would lead to an absence of T lymphocytes and no cell-mediated immunity after birth. This outcome would seriously compromise the individual's ability to defend against viral invasion and cancer.
5. Researchers are currently working on ways to "teach" the immune system to view foreign tissue as "self" as a means of preventing the immune systems of organ transplant patients from rejecting the foreign tissue while leaving the patients' immune defense capabilities fully intact. The immunosuppressive drugs currently being used to prevent transplant rejection cripple the recipients' immune defense systems and leave the patients more vulnerable to microbial invasion.

▌ Clinical Consideration

(Question on p. 362.)

Heather's firstborn Rh-positive child did not have hemolytic disease of the newborn, because the fetal and maternal blood did not mix during gestation. Consequently, Heather did not produce any maternal antibodies against the fetus's Rh factor during gestation.

Because a small amount of the infant's blood likely entered the maternal circulation during the birthing process, Heather would produce antibodies against the Rh factor as she was first exposed to it at that time. During any subsequent pregnancies with Rh-positive fetuses, Heather's maternal antibodies against the Rh factor could cross the placental barrier and bring about destruction of fetal erythrocytes.

If, however, any Rh factor that accidentally mixed with the maternal blood during the birthing process were immediately tied up by Rh immunoglobulin administered to the mother, the Rh factor would not be available to induce maternal antibody production. Thus no anti-Rh antibodies would be present in the maternal blood to threaten the RBCs of an Rh-positive fetus in a subsequent pregnancy. (The exogenously administered Rh immunoglobulin, being a passive form of immunity, is short-lived. In contrast, the active immunity that would result if Heather were exposed to Rh factor would be long-lived because of the formation of memory cells.)

Rh immunoglobulin must be administered following the birth of every Rh-positive child Heather bears to sop up any Rh factor before it can induce antibody production. Once an immune attack against Rh factor is launched, subsequent treatment with Rh immunoglobulin will not reverse the situation. Thus if Heather were not treated with Rh immunoglobulin following the birth of a first Rh-positive child, and a second Rh-positive child developed hemolytic disease of the newborn, administration of Rh immunoglobulin following the second birth would not prevent the condition in a third Rh-positive child. Nothing could be done to eliminate the maternal antibodies already present.

CHAPTER 12 THE RESPIRATORY SYSTEM

▌ Objective Questions

(Questions on p. 402.)

1. F 2. F 3. F 4. F 5. F 6. transmural pressure gradient, pulmonary surfactant action 7. pulmonary elasticity, alveolar surface tension 8. compliance 9. elastic recoil 10. carbonic anhydrase 11. a 12. 1.d, 2.a, 3.b, 4.a, 5.b, 6.a 13. a.<, b.>, c. =, d. =, e. = , f. =, g.>, h.,<, i. approximately =, j. approximately =, k. =, l. =

(Questions on p. 403.)

1. Total atmospheric pressure decreases with increasing altitude, yet the percentage of O_2 in the air remains the same. At an altitude of 30,000 feet, the atmospheric pressure is only 226 mm Hg. Because 21% of atmospheric air consists of O_2, the P_{O_2} of inspired air at 30,000 feet is only 47.5 mm Hg, and alveolar P_{O_2} is even lower at about 20 mm Hg. At this low P_{O_2}, hemoglobin is only about 30% saturated with O_2—much too low to sustain tissue needs for O_2.

The P_{O_2} of inspired air can be increased by two means when flying at high altitude. First, by pressurizing the plane's interior to a pressure comparable to that of atmospheric pressure at sea level, the P_{O_2} of inspired air within the plane is 21% of 760 mm Hg, or the normal 160 mm Hg. Accordingly, alveolar and arterial P_{O_2} and percent hemoglobin saturation are likewise normal. In the emergency situation of failure to maintain internal cabin pressure, breathing pure O_2 can raise the P_{O_2} considerably above that accomplished by breathing normal air. When a person is breathing pure O_2, the entire pressure of inspired air is attributable to O_2. For example, with a total atmospheric pressure of 226 mm Hg at an altitude of 30,000 feet, the P_{O_2} of inspired pure O_2 is 226 mm Hg, which is more than adequate to maintain normal arterial hemoglobin saturation.

2. a. Hypercapnia would not accompany the hypoxia associated with cyanide poisoning. In fact, CO_2 levels decline, because oxidative metabolism is blocked by the tissue poisons so that CO_2 is not being produced.

b. Hypercapnia could but may not accompany the hypoxia associated with pulmonary edema. Pulmonary diffusing capacity is reduced in pulmonary edema, but O_2 transfer suffers more than CO_2 transfer because CO_2 is more soluble than O_2 in the body fluids, and thus CO_2 diffuses more rapidly than O_2. Therefore the blood is more likely to have equilibrated with alveolar P_{CO_2} than with alveolar P_{O_2} by the end of the pulmonary capillaries. As a result, hypoxia occurs much more readily than hypercapnia in these circumstances.

c. Hypercapnia would accompany the hypoxia associated with restrictive lung disease because ventilation is inadequate to meet the metabolic needs for both O_2 delivery and CO_2 removal. Both O_2 and CO_2 exchange between the lungs and atmosphere are equally affected.

d. Hypercapnia would not accompany the hypoxia associated with high altitude. In fact, arterial P_{CO_2} levels actually decrease. One of the compensatory responses in acclimatization to high altitudes is reflex stimulation of ventilation as a result of the reduction in arterial P_{O_2}. This compensatory hyperventilation to obtain more O_2 blows off too much CO_2 in the process, so arterial P_{CO_2} levels decline below normal.

e. Hypercapnia would not accompany the hypoxia associated with severe anemia. Reduced O_2-carrying capacity of the blood has no influence on blood CO_2 content, so arterial P_{CO_2} levels are normal.

f. Hypercapnia would exist in accompaniment with circulatory hypoxia associated with congestive heart failure. Just as the diminished blood flow fails to deliver adequate O_2 to the tissues, it also fails to remove sufficient CO_2.

g. Hypercapnia would accompany the hypoxic hypoxia associated with obstructive lung disease because ventilation would be inadequate to meet the metabolic needs for both O_2 delivery and CO_2 removal. Both O_2 and CO_2 exchange between the lungs and atmosphere would be equally affected.

3. $P_{O_2} = 122$ mm Hg
0.21 (atmospheric pressure − partial pressure of H_2O)
= 0.21(630 mm Hg − 47 mm Hg)
= 0.21(583 mm Hg) = 122 mm Hg

4. Voluntarily hyperventilating before going underwater lowers the arterial P_{CO_2} but does not increase the O_2 content in the blood. Because the P_{CO_2} is below normal, the person can hold his or her breath longer than usual before the arterial P_{CO_2} increases to the point that he or she is driven to surface for a breath. Therefore, the person can stay underwater longer. The risk, however, is that the O_2 content of the blood, which was normal, not increased, before going underwater, continues to fall. Therefore, the O_2 level in the blood can fall dangerously low before the CO_2 level builds to the point of driving the per-

son to take a breath. Low arterial P_{O_2} does not stimulate respiratory activity until it has plummeted to 60 mm Hg. Meanwhile, the person may lose consciousness and drown due to inadequate O_2 delivery to the brain. If the person does not hyperventilate so that both the arterial P_{CO_2} and O_2 content are normal before going underwater, the buildup of CO_2 will drive the person to the surface for a breath before the O_2 levels fall to a dangerous point.

5. c. The arterial P_{O_2} will be less than the alveolar P_{O_2}, and the arterial P_{CO_2} will be greater than the alveolar P_{CO_2}. Because pulmonary diffusing capacity is reduced, arterial P_{O_2} and P_{CO_2} do not equilibrate with alveolar P_{O_2} and P_{CO_2}.

If the person is administered 100% O_2, the alveolar P_{O_2} will increase, and the arterial P_{O_2} will increase accordingly. Even though arterial P_{O_2} will not equilibrate with alveolar P_{O_2}, it will be higher than when the person is breathing atmospheric air.

The arterial P_{CO_2} will remain the same whether the person is administered 100% O_2 or is breathing atmospheric air. The alveolar P_{CO_2} and thus the blood-to-alveolar P_{CO_2} gradient are not changed by breathing 100% O_2 because the P_{CO_2} in atmospheric air and 100% O_2 are both essentially zero (P_{CO_2} in atmospheric air = 0.23 mm Hg).

■ Clinical Consideration

(Question on p. 403.)

Emphysema is characterized by a collapse of smaller respiratory airways and a breakdown of alveolar walls. Because of the collapse of smaller airways, airway resistance is increased with emphysema. As with other chronic obstructive pulmonary diseases, expiration is impaired to a greater extent than inspiration because airways are naturally dilated slightly more during inspiration than expiration as a result of the greater transmural pressure gradient during inspiration. Because airway resistance is increased, a patient with emphysema must produce larger-than-normal intra-alveolar pressure changes to accomplish a normal tidal volume. Unlike quiet breathing in a normal person, the accessory inspiratory muscles (neck muscles) and the muscles of active expiration (abdominal muscles and internal intercostal muscles) must be brought into play to inspire and expire a normal tidal volume of air.

The spirogram would be characteristic of chronic obstructive pulmonary disease. Because the patient experiences more difficulty emptying the lungs than filling them, the total lung capacity would be essentially normal, but the functional residual capacity and the residual volume would be elevated as a result of the additional air trapped in the lungs following expiration. Because the residual volume is increased, the inspiratory capacity and vital capacity will be reduced. Also, the FEV_1 will be markedly reduced because the airflow rate is decreased by the airway obstruction. The FEV_1-to-vital capacity ratio will be much lower than the normal 80%.

Because of the reduced surface area for exchange as a result of a breakdown of alveolar walls, gas exchange would be impaired. Therefore, arterial P_{CO_2} would be elevated and arterial P_{O_2} reduced compared to normal.

Ironically, administering O_2 to this patient to relieve his hypoxic condition would markedly depress his drive to breathe by elevating the arterial P_{O_2} and removing the primary driving stimulus for respiration. Because of this danger, O_2 therapy should either not be administered or administered extremely cautiously.

CHAPTER 13   THE URINARY SYSTEM

■ Objective Questions

(Questions on p. 439.)

1. F 2. F 3. T 4. T 5. nephron 6. potassium 7. 500   8. e
9. b 10. b, e, a, d, c   11. c, e, d, a, b, f   12. g, c, d, a, f, b, e   13. 1.a, 2.a, 3.c, 4.b, 5.d

■ Points to Ponder

(Questions on p. 440.)

1. The longer loops of Henle in desert rats (known as *kangaroo rats*) permit a greater magnitude of countercurrent multiplication and thus a larger medullary vertical osmotic gradient. As a result, these rodents can produce urine that is concentrated up to an osmolarity of almost 6000 mosm/liter, which is five times more concentrated than maximally concentrated human urine at 1200 mosm/liter. Because of this tremendous concentrating ability, kangaroo rats never have to drink; the H_2O produced metabolically within their cells during oxidation of foodstuff (food + O_2 yields CO_2 + H_2O + energy) is sufficient for their needs.

2. a. 250 mg/min filtered

   filtered load of substance = plasma concentration of substance × GFR
   filtered load of substance = 200 mg/100 ml × 125 ml/min = 250 mg/min

   b. 200 mg/min reabsorbed

   A T_m's worth of the substance will be reabsorbed.

   c. 50 mg/min excreted

   amount of substance excreted = amount of substance filtered − amount of substance reabsorbed = 250 mg/min − 200 mg/min = 50 mg/min

3. Aldosterone stimulates Na^+ reabsorption and K^+ secretion by the renal tubules. Therefore, the most prominent features of Conn's syndrome (hypersecretion of aldosterone) are hypernatremia (elevated Na^+ levels in the blood) caused by excessive Na^+ reabsorption, hypophosphatemia (below-normal K^+ levels in the blood) caused by excessive K^+ secretion, and hypertension (elevated blood pressure) caused by excessive salt and water retention.

4. e. 300/300. If the ascending limb were permeable to water, it would not be possible to establish a vertical osmotic gradient in the medullary interstitial fluid, nor would the ascending-limb fluid become hypotonic before entering the distal tubule. As the ascending limb pumped NaCl into the interstitial fluid, water would osmotically follow, so both the interstitial fluid and the ascending limb would remain isotonic at 300 mosm/liter. With the tubular fluid entering the distal tubule being 300 mosm/liter instead of the normal 100 mosm/liter, it would not be possible to produce urine with an osmolarity less than 300 mosm/liter. Likewise, in the absence of the medullary vertical osmotic gradient, it would not be possible to produce urine more concentrated than 300 mosm/liter, no matter how much vasopressin was present.

5. Because the descending pathways between the brain and the motor neurons supplying the external urethral sphincter and pelvic diaphragm are no longer intact, the accident victim can no longer voluntarily control micturition. Therefore, bladder emptying in this individual will be governed entirely by the micturition reflex.

❚ Clinical Consideration

(Question on p. 440.)

prostate enlargement

CHAPTER 14   FLUID AND ACID–BASE BALANCE

❚ Objective Questions

(Questions on p. 462.)

1. T   2. F   3. T   4. intracellular fluid   5. $[H_2CO_3]$, $[HCO_3^-]$ 6. c   7. a, c   8. b, d   9. c   10. 1.c, 2.d, 3.b, 4.a

❚ Points to Ponder

(Questions on p. 463.)

1. The rate of urine formation increases when alcohol inhibits vasopressin secretion and the kidneys are unable to reabsorb water from the distal and collecting tubules. Because extra free water that normally would have been reabsorbed from the distal parts of the tubule is lost from the body in the urine, the body becomes dehydrated and the ECF osmolarity increases following alcohol consumption. That is, more fluid is lost in the urine than is consumed in the alcoholic beverage as a result of alcohol's action on vasopressin. Thus the imbibing person experiences a water deficit and still feels thirsty, despite the recent fluid consumption.

2. If a person loses 1500 ml of salt-rich sweat and drinks 1000 ml of water without replacing the salt during the same time period, there will still be a volume deficit of 500 ml, and the body fluids will have become hypotonic (the remaining salt in the body will be diluted by the ingestion of 1000 ml of free H_2O). As a result, the hypothalamic osmoreceptors (the dominant input) will signal the vasopressin-secreting cells to *decrease* vasopressin secretion and thus increase urinary excretion of the extra free water that is making the body fluids too dilute. Simultaneously, the left atrial volume receptors will signal the vasopressin-secreting cells to *increase* vasopressin secretion to conserve water during urine formation and thus help to relieve the volume deficit. These two conflicting inputs to the vasopressin-secreting cells are counterproductive. This is why it is important to replace both water and salt following heavy sweating or abnormal loss of other salt-rich fluids. If salt is replaced along with water intake, the ECF osmolarity remains close to normal, and the vasopressin-secreting cells receive signals only to increase vasopressin secretion to help restore the ECF volume to normal.

3. When a dextrose solution equal in concentration to that of normal body fluids is injected intravenously, the ECF volume is expanded but the ECF and ICF are still osmotically equal. Therefore, no net movement of water occurs between the ECF and ICF. When the dextrose enters the cell and is metabolized, however, the ECF becomes hypotonic as this solute leaves the plasma. If the excess free water is not excreted in the urine rapidly enough, water will move into the cells by osmosis.

4. When arterial P_{O_2} falls below 60 mm Hg, the peripheral chemoreceptors reflexly stimulate ventilation in an attempt to obtain more O_2 for the body. Too much CO_2 is blown off in the process, inadvertently decreasing the production of H_2CO_3 from CO_2 and thereby reducing generation of H^+ from this source. Thus respiratory alkalosis occurs as a result of compensatory physiologic mechanisms at high altitude.

5. c. The hemoglobin buffer system buffers carbonic-acid–generated hydrogen ion. In the case of respiratory acidosis accompanying severe pneumonia, the $H^+ + Hb \rightarrow HHb$ reaction will be shifted toward the HHb side, thus removing some of the extra free H^+ from the blood.

❚ Clinical Consideration

(Question on p. 463.)

The resultant prolonged diarrhea will lead to dehydration and metabolic acidosis due, respectively, to excessive loss in the feces of fluid and $NaHCO_3$ that normally would have been absorbed into the blood.

Compensatory measures for dehydration have included increased vasopressin secretion, resulting in increased water reabsorption by the distal and collecting tubules and a subsequent reduction in urine output. Simultaneously, fluid intake has been encouraged by increased thirst. The metabolic acidosis has been combatted by removal of excess H^+ from the ECF by the HCO_3^- member of the $H_2CO_3{:}HCO_3^-$ buffer system, by increased ventilation to reduce the amount of acid-forming CO_2 in the body fluids, and by the kidneys excreting extra H^+ and conserving HCO_3^-.

CHAPTER 15   THE DIGESTIVE SYSTEM

❚ Objective Questions

(Questions on p. 507.)

1. F   2. T   3. T   4. F   5. F   6. F   7. chyme   8. vitamin B_{12}, bile salts   9. bile salts   10. b   11. 1.c, 2.e, 3.b, 4.a, 5.f, 6.d   12. 1.c, 2.c, 3.d, 4.a, 5.e, 6.c, 7.a, 8.b, 9.c, 10.b, 11.d, 12.b

❚ Points to Ponder

(Questions on p. 508.)

1. Patients who have had their stomachs removed must eat small quantities of food frequently instead of consuming the typical three meals a day because they have lost the ability to store food in the stomach and meter it into the small intestine at an optimal rate. If a person without a stomach consumed a large meal that entered the small intestine all at once, the luminal contents would quickly become too hypertonic as digestion of the large nutrient molecules into a multitude of small, osmotically active, absorbable units outpaced the more slowly acting process of absorption of these units. As a consequence of this increased luminal osmolarity, water would enter the small intestine lumen from the plasma by osmosis, resulting in circulatory disturbances as well as intestinal distension. To prevent this "dumping syndrome" from occurring, the patient must "feed" the small intestine only small amounts of food at a time so that absorption of the digestive end products can keep pace with their rate of production. The person has to consciously take over metering the delivery of food into the small intestine because the stomach is no longer present to assume this responsibility.

2. The gut-associated lymphoid tissue (GALT) launches an immune attack against any pathogenic (disease-causing) micro-organisms that enter the readily accessible digestive tract and escape destruction by salivary lysozyme or gastric HCl. This action defends against entry of these potential pathogens into the body proper. The large number of immune cells in the GALT is adaptive as a first line of defense against foreign invasion when considering that the surface area of the digestive tract lining represents the largest interface between the body proper and the external environment.

3. Defecation would be accomplished entirely by the defecation reflex in a patient paralyzed from the waist down because of lower spinal cord injury. Voluntary control of the external anal sphincter would be impossible because of interruption in the descending pathway between the primary motor cortex and the motor neuron supplying this sphincter.

4. When insufficient glucuronyl transferase is available in the neonate to conjugate all of the bilirubin produced during erythrocyte degradation with glycuronic acid, the extra unconjugated bilirubin cannot be excreted into the bile. Therefore, this extra bilirubin remains in the body, giving rise to mild jaundice in the newborn.

5. Removal of the stomach leads to pernicious anemia because of the resultant lack of intrinsic factor, which is necessary for absorption of vitamin B_{12}. Removal of the terminal ileum leads to pernicious anemia because this is the only site where vitamin B_{12} can be absorbed.

▌ Clinical Consideration

(Question on p. 508.)

A person whose bile duct is blocked by a gallstone experiences a painful "gallbladder attack" after eating a high-fat meal because the ingested fat triggers the release of cholecystokinin, which stimulates gallbladder contraction. As the gallbladder contracts and bile is squeezed into the blocked bile duct, the duct becomes distended prior to the blockage. This distension is painful.

The feces are grayish white because no bilirubin-containing bile enters the digestive tract when the bile duct is blocked. Bilirubin, when acted on by bacterial enzymes, produces the brown color of feces, which are grayish white in its absence.

CHAPTER 16 ENERGY BALANCE AND TEMPERATURE REGULATION

▌ Objective Questions

(Questions on p. 525.)

1. F 2. F 3. F 4. F 5. T 6. T 7. T 8. T 9. arcuate nucleus
10. shivering 11. nonshivering thermogenesis 12. sweating 13. b
14. e 15. 1.b, 2.a, 3.c, 4.a, 5.d, 6.c, 7.b, 8.d, 9.c, 10.b

▌ Points to Ponder

(Questions on p. 526.)

1. Evidence suggests that CCK serves as a satiety signal. It is believed to serve as a signal to stop eating when enough food has been consumed to meet the body's energy needs, even though the food is still in the digestive tract. Therefore, when drugs that inhibit CCK release are administered to experimental animals, the animals overeat because this satiety signal is not released.

2. Don't go on a "crash diet." Be sure to eat a nutritionally balanced diet that provides all essential nutrients, but reduce total caloric intake, especially by cutting down on high-fat foods. Spread out consumption of the food throughout the day instead of just eating several large meals. Avoid bedtime snacks. Burn more calories through a regular exercise program.

3. Engaging in heavy exercise on a hot day is dangerous because of problems arising from trying to eliminate the extra heat generated by the exercising muscles. First, there will be conflicting demands for distribution of the cardiac output—temperature-regulating mechanisms will trigger skin vasodilation to promote heat loss from the skin surface, whereas metabolic changes within the exercising muscles will induce local vasodilation in the muscles to match the increased metabolic needs with increased blood flow. Further exacerbating the problem of conflicting demands for blood flow is the loss of effective circulating plasma volume resulting from the loss of a large volume of fluid through another important cooling mechanism, sweating. Therefore, it is difficult to maintain an effective plasma volume and blood pressure and simultaneously keep the body from overheating when engaging in heavy exercise in the heat, so heat exhaustion is likely to ensue.

4. When a person is soaking in a hot bath, loss of heat by radiation, conduction, convection, and evaporation is limited to the small surface area of the body exposed to the cooler air. Heat is being gained by conduction at the larger skin surface area exposed to the hotter water.

5. The thermoconforming fish would not run a fever when it has a systemic infection because it has no mechanisms for regulating internal heat production or for controlling heat exchange with its environment. The fish's body temperature varies capriciously with the external environment no matter whether it has a systemic infection or not. It cannot maintain body temperature at a "normal" set point or an elevated set point (i.e., a fever).

▌ Clinical Consideration

(Question on p. 526.)

Cooled tissues need less nourishment than they do at normal body temperature because of their pronounced reduction in metabolic activity. The lower O_2 need of cooled tissues accounts for the occasional survival of drowning victims who have been submerged in icy water considerably longer than one could normally survive without O_2.

CHAPTER 17 THE ENDOCRINE SYSTEM

▌ Objective Questions

(Questions on p. 580.)

1. F 2. T 3. T 4. T 5. T 6. F 7. tropic 8. suprachiasmatic nucleus 9. epiphyseal plate 10. colloid, thyroglobulin 11. glycogenesis, glycogenolysis, gluconeogenesis 12. bone, kidneys, digestive tract
13. 1.c, 2.b, 3.b, 4.a, 5.a, 6.c, 7.c 14. 1. glucose, 2. glycogen, 3. free fatty acids, 4. triglycerides, 5. amino acids, 6. body proteins

▌ Points to Ponder

(Questions on p. 581.)

1. It is not advisable to rotate the nursing staff on different shifts every week because such a practice would disrupt the individuals' natural circadian rhythms. Such disruption can affect physical and psychological health and have a negative impact on work performance.

2. Males with testicular feminization syndrome would be unusually tall because of the inability of testosterone to promote closure of the epiphyseal plates of the long bones in the absence of testosterone receptors.

3. Hormonal replacement therapy following pituitary gland removal should include thyroid hormone (the thyroid gland will not produce sufficient thyroid hormone in the absence of TSH) and glucocorticoid (because of the absence of ACTH), especially in stress situations. If indicated, male or female sex hormones can be replaced, even though these hormones are not essential for survival. For example, testosterone in males plays an important role in libido. Growth hormone and prolactin need not be replaced, because their absence will produce no serious consequences in this individual. Vasopressin may have to be replaced if the blood picks up insufficient quantities of this hormone at the hypothalamus in the absence of the posterior pituitary.

4. Anaphylactic shock is an extremely serious allergic reaction brought about by massive release of chemical mediators in response to exposure to a specific allergen—such as one associated with a bee sting—to which the individual has been highly sensitized. These chemical mediators bring about circulatory shock (severe hypotension) through a twofold effect: (1) by relaxing arteriolar smooth muscle, thus causing widespread arteriolar vasodilation and a resultant fall in total peripheral resistance and arterial blood pressure; and (2) by causing a generalized increase in capillary permeability, resulting in a shift of fluid from the plasma into the interstitial fluid. This shift decreases the effective circulating volume, further reducing arterial blood pressure. Additionally, these chemical mediators bring about pronounced bronchoconstriction, making it impossible for the victim to move sufficient air through the narrowed airways. Because these responses take place rapidly and can be fatal, people allergic to bee stings are advised to keep injectable epinephrine in their possession. By promoting arteriolar vasoconstriction through its action on α_1 receptors in arteriolar smooth muscle and promoting bronchodilation through its action on β_2 receptors in bronchiolar smooth muscle (see p. 191), epinephrine counteracts the life-threatening effects of the anaphylactic reaction to the bee sting. Waiting to receive attention from medical personnel is likely to be too late.

5. An infection elicits the stress response, which brings about increased secretion of cortisol and epinephrine, both of which increase the blood glucose level. This can become a problem in managing the blood glucose level of a diabetic patient. When the blood glucose is elevated too high, the patient can reduce it by injecting additional insulin, or, preferably, by reducing carbohydrate intake and/or exercising to use up some of the extra blood glucose. In a normal individual, the check-and-balance system between insulin and the other hormones that oppose insulin's actions helps maintain the blood glucose within reasonable limits during the stress response.

Clinical Consideration

(Question on p. 581.)

"Diabetes of bearded ladies" is descriptive of both excess cortisol and excess adrenal androgen secretion. Excess cortisol secretion causes hyperglycemia and glucosuria. Glucosuria promotes osmotic diuresis, which leads to dehydration and a compensatory increased sensation of thirst. All these symptoms—hyperglycemia, glucosuria, polyuria, and polydipsia—mimic diabetes mellitus. Excess adrenal androgen secretion in females promotes masculinizing characteristics, such as beard growth. Simultaneous hypersecretion of both cortisol and adrenal androgen most likely occurs secondary to excess CRH/ACTH secretion, because ACTH stimulates both cortisol and androgen production by the adrenal cortex.

CHAPTER 18 THE REPRODUCTIVE SYSTEM

Objective Questions

(Questions on p. 628.)

1. T 2. T 3. T 4. F 5. T 6. F 7. T 8. seminiferous tubules, FSH, testosterone 9. corpus luteum of pregnancy 10. human chorionic gonadotropin 11. c 12. e 13. 1.c, 2.a, 3.b, 4.c, 5.a, 6.a, 7.c, 8.b, 9.e 14. 1.a, 2.c, 3.b, 4.a, 5.a, 6.b

Points to Ponder

(Questions on p. 629.)

1. The anterior pituitary responds only to the normal pulsatile pattern of GnRH and does not secrete gonadotropins in response to continuous exposure to GnRH. In the absence of FSH and LH secretion, ovulation and other events of the ovarian cycle do not ensue, so continuous GnRH administration may find use as a contraceptive technique.

2. Testosterone hypersecretion in a young boy causes premature closure of the epiphyseal plates so that he stops growing before he reaches his genetic potential for height. The child would also display signs of precocious pseudopuberty, characterized by premature development of secondary sexual characteristics, such as deep voice, beard, enlarged penis, and sex drive.

3. A potentially troublesome side effect of drugs that inhibit sympathetic nervous system activity as part of the treatment for high blood pressure is males' inability to carry out the sex act. Both divisions of the autonomic nervous system are required for the male sex act. Parasympathetic activity is essential for accomplishing erection, and sympathetic activity is important for ejaculation.

4. Posterior pituitary extract contains an abundance of stored oxytocin, which can be administered to induce or facilitate labor by increasing uterine contractility. Exogenous oxytocin is most successful in inducing labor if the woman is near term, presumably because of the increasing concentration of myometrial oxytocin receptors at that time.

5. GnRH or FSH and LH are not effective in treating the symptoms of menopause because the ovaries are no longer responsive to the gonadotropins. Thus treatment with these hormones would not cause estrogen and progesterone secretion. In fact, GnRH, FSH, and LH levels are already elevated in postmenopausal women because of lack of negative feedback by the ovarian hormones.

Clinical Consideration

(Question on p. 629.)

The first warning of a tubal pregnancy is pain caused by stretching of the oviduct by the growing embryo. A tubal pregnancy must be surgically terminated because the oviduct cannot expand as the uterus does to accommodate the growing embryo. If not removed, the enlarging embryo will rupture the oviduct, causing possibly lethal hemorrhage.

Glossary

A band One of the dark bands that alternate with light (I) bands to create a striated appearance in a skeletal or cardiac muscle fiber when these fibers are viewed with a light microscope

absorptive state The metabolic state following a meal when nutrients are being absorbed and stored; fed state

accessory digestive organs Exocrine organs outside the wall of the digestive tract that empty their secretions through ducts into the digestive tract lumen

accessory sex glands Glands that empty their secretions into the reproductive tract

accommodation The ability to adjust the strength of the lens in the eye so that both near and far sources can be focused on the retina

acetylcholine (ACh) (as´-uh-teal-KŌ-lēn) The neurotransmitter released from all autonomic preganglionic fibers, parasympathetic postganglionic fibers, and motor neurons

acetylcholinesterase (AChE) (as´uh-teal-kō-luh-NES-tuh-rās) An enzyme present in the motor end-plate membrane of a skeletal muscle fiber that inactivates acetylcholine

ACh See *acetylcholine*

AChE See *acetylcholinesterase*

acid A hydrogen-containing substance that yields a free hydrogen ion and anion on dissociation

acidosis (as-i-DŌ-sus) Blood pH of less than 7.35

ACTH See *adrenocorticotropic hormone*

acquired immune responses Responses that are selectively targeted against particular foreign material to which the body has previously been exposed; see also *antibody-mediated immunity* and *cell-mediated immunity*

actin The contractile protein that forms the backbone of the thin filaments in muscle fibers

active expiration Emptying of the lungs more completely than when at rest by contracting the expiratory muscles; also called *forced expiration*

active force A force that requires expenditure of cellular energy (ATP) in the transport of a substance across the plasma membrane

active reabsorption When any one of the five steps in the transepithelial transport of a substance reabsorbed across the kidney tubules requires energy expenditure

active transport Active carrier-mediated transport involving transport of a substance against its concentration gradient across the plasma membrane

acuity Discriminative ability; the ability to discern between two different points of stimulation

adaptation A reduction in receptor potential despite sustained stimulation of the same magnitude

adenosine diphosphate (ADP) (uh-DEN-uh-sēn) The two-phosphate product formed from the splitting of ATP to yield energy for the cell's use

adenosine triphosphate (ATP) The body's common energy "currency," which consists of an adenosine with three phosphate groups attached; splitting of the high-energy, terminal phosphate bond provides energy to power cellular activities

adenylyl cyclase (ah-DEN-il-il sī-klās) The membrane-bound enzyme that is activated by a G protein intermediary in response to binding of an extracellular messenger with a surface membrane receptor and that in turn activates cyclic AMP, an intracellular second messenger

ADH See *vasopressin*

adipose tissue The tissue specialized for storage of triglyceride fat; found under the skin in the hypodermis

ADP See *adenosine diphosphate*

adrenal cortex (uh-DRĒ-nul) The outer portion of the adrenal gland; secretes three classes of steroid hormones: glucocorticoids, mineralocorticoids, and sex hormones

adrenal medulla (muh-DUL-uh) The inner portion of the adrenal gland; an endocrine gland that is a modified sympathetic ganglion that secretes the hormones epinephrine and norepinephrine into the blood in response to sympathetic stimulation

adrenergic fibers (ad´-ruh-NUR-jik) Nerve fibers that release norepinephrine as their neurotransmitter

adrenocorticotropic hormone (ACTH) (ad-rē´-nō-kor´-tuh-kō-TRŌP-ik) An anterior pituitary hormone that stimulates cortisol secretion by the adrenal cortex and promotes growth of the adrenal cortex

aerobic Referring to a condition in which oxygen is available

aerobic exercise Exercise that can be supported by ATP formation accomplished by oxidative phosphorylation because adequate O_2 is available to support the muscle's modest energy demands; also called *endurance-type exercise*

afferent arteriole (AF-er-ent ar-TIR-ē-ōl) The vessel that carries blood into the glomerulus of the kidney's nephron

afferent division The portion of the peripheral nervous system that carries information from the periphery to the central nervous system

afferent neuron Neuron that possesses a sensory receptor at its peripheral ending and carries information to the central nervous system

albumin (al-BEW-min) The smallest and most abundant of the plasma proteins; binds and transports many water-insoluble substances in the blood; contributes extensively to plasma-colloid osmotic pressure

aldosterone (al-dō-steer-OWN) or (al-DOS-tuh-rōn) The adrenocortical hormone that stimulates Na^+ reabsorption by the distal and collecting tubules of the kidney's nephron during urine formation

alkalosis (al´-kuh-LŌ-sus) Blood pH of greater than 7.45

allergy Acquisition of an inappropriate specific immune reactivity to a normally harmless environmental substance

all-or-none law An excitable membrane either responds to a stimulus with a maximal action potential that spreads nondecrementally throughout the membrane or does not respond with an action potential at all

alpha (α) cells The endocrine pancreatic cells that secrete the hormone glucagon

alveolar surface tension (al-VĒ-ō-lur) The surface tension of the fluid lining the alveoli in the lungs; see *surface tension*

alveolar ventilation The volume of air exchanged between the atmosphere and alveoli per minute; equals (tidal volume minus dead space volume) times respiratory rate

alveoli (al-VĒ-ō-lī) The air sacs across which O_2 and CO_2 are exchanged between the blood and air in the lungs

amoeboid movement (uh-MĒ-boid) "Crawling" movement of white blood cells, similar to the means by which amoebas move

anabolism (ah-NAB-ō-li-zum) The buildup, or synthesis, of larger organic molecules from the small organic molecular subunits

anaerobic (an´-uh-RŌ-bik) Referring to a condition in which oxygen is not present

anaerobic exercise High-intensity exercise that can be supported by ATP formation accomplished by anaerobic glycolysis for brief periods of time when O_2 delivery to a muscle is inadequate to support oxidative phosphorylation

analgesic (an-al-JEE-zic) Pain relieving

androgen A masculinizing "male" sex hormone; includes testosterone from the testes and dehydroepiandrosterone from the adrenal cortex

anemia A reduction below normal in O_2-carrying capacity of the blood

anion (AN-ī-on) Negatively charged ion that has gained one or more electrons in its outer shell

ANP See *atrial natriuretic peptide*

antagonism Actions opposing each other; in the case of hormones, when one hormone causes the loss of another hormone's receptors, reducing the effectiveness of the second hormone

anterior pituitary The glandular portion of the pituitary that synthesizes, stores, and secretes six different hormones: growth hormone, TSH, ACTH, FSH, LH, and prolactin

antibody An immunoglobulin produced by a specific activated B lymphocyte (plasma cell) against a particular antigen; binds with the specific antigen against which it is produced and promotes the antigenic invader's destruction by augmenting nonspecific immune responses already initiated against the antigen

antibody-mediated immunity A specific immune response accomplished by antibody production by B cells

antidiuretic hormone (an´-ti-dī´-yū-RET-ik) See *vasopressin*

antigen A large, complex molecule that triggers a specific immune response against itself when it gains entry into the body

antioxidant A substance that helps inactivate biologically damaging free radicals

antrum (of ovary) The fluid-filled cavity formed within a developing ovarian follicle

antrum (of stomach) The lower portion of the stomach

aorta (ā-OR-tah) The large vessel that carries blood from the left ventricle

aortic valve A one-way valve that permits the flow of blood from the left ventricle into the aorta during ventricular emptying but prevents the backflow of blood from the aorta into the left ventricle during ventricular relaxation

apoptosis (ā-pop-TŌ-sis) Programmed cell death; deliberate self-destruction of a cell.

aqueous humor (Ā-kwē-us) The clear, watery fluid in the anterior chamber of the eye; provides nourishment for the cornea and lens

arcuate nucleus The subcortical brain region that houses neurons that secrete appetite-enhancing neuropeptide Y and those that secrete appetite-suppressing melanocortins

arterioles (ar-TIR-ē-ōlz) The highly muscular, high-resistance vessels, the caliber of which can be changed subject to control to determine how much of the cardiac output is distributed to each of the various tissues

artery A vessel that carries blood away from the heart

ascending tract A bundle of nerve fibers of similar function that travels up the spinal cord to transmit signals derived from afferent input to the brain

asthma An obstructive pulmonary disease characterized by profound constriction of the smaller airways caused by allergy-induced spasm of the smooth muscle in the walls of these airways

astrocyte A type of glial cell in the brain; major functions include holding the neurons together in proper spatial relationship and inducing the brain capillaries to form tight junctions important in the blood–brain barrier

atherosclerosis (ath-uh-rō-skluh-RŌ-sus) A progressive, degenerative arterial disease that leads to gradual blockage of affected vessels, reducing blood flow through them

atmospheric pressure The pressure exerted by the weight of the air in the atmosphere on objects on Earth's surface; equals 760 mm Hg at sea level

ATP See *adenosine triphosphate*

ATPase An enzyme that can split ATP

atrial natriuretic peptide (ANP) (Ā-trē-al NĀ-tree-ur-eh´tik) A peptide hormone released from the cardiac atria that promotes urinary loss of Na^+

atrioventricular (AV) node (ā´-trē-ō-ven-TRIK-yuh-lur) A small bundle of specialized cardiac cells at the junction of the atria and ventricles that is the only site of electrical contact between the atria and ventricles

atrioventricular (AV) valve A one-way valve that permits the flow of blood from the atrium to the ventricle during filling of the heart but prevents the backflow of blood from the ventricle to the atrium during emptying of the heart

atrium (atria, plural) (Ā-tree-um) An upper chamber of the heart that receives blood from the veins and transfers it to the ventricle

atrophy (AH-truh-fē) Decrease in mass of an organ

autoimmune disease Disease characterized by erroneous production of antibodies against one of the body's own tissues

autonomic nervous system The portion of the efferent division of the peripheral nervous system that innervates smooth and cardiac muscle and exocrine glands; composed of two subdivisions, the sympathetic nervous system and the parasympathetic nervous system

autorhythmicity The ability of an excitable cell to rhythmically initiate its own action potentials

AV nodal delay The delay in impulse transmission between the atria and ventricles at the AV node to allow enough time for the atria to become completely depolarized and contract, emptying their contents into the ventricles, before ventricular depolarization and contraction occur

AV valve See *atrioventricular valve*

axon A single, elongated tubular extension of a neuron that conducts action potentials away from the cell body; also known as a *nerve fiber*

axon hillock The first portion of a neuronal axon plus the region of the cell body from which the axon leaves; the site of action-potential initiation in most neurons

axon terminals The branched endings of a neuronal axon, which release a neurotransmitter that influences target cells in close association with the axon terminals

baroreceptor reflex An autonomically mediated reflex response that influences the heart and blood vessels to oppose a change in mean arterial blood pressure

baroreceptors Receptors located within the circulatory system that monitor blood pressure

basal metabolic rate (BMR) (BĀ-sul) The minimal waking rate of internal energy expenditure; the body's "idling speed"

basal nuclei Several masses of gray matter located deep within the white matter of the cerebrum of the brain; play an important inhibitory role in motor control

base A substance that can combine with a free hydrogen ion and remove it from solution

basic electrical rhythm (BER) Self-induced electrical activity of the digestive-tract smooth muscle

basilar membrane (BAS-ih-lar) The membrane that forms the floor of the middle compartment of the cochlea and bears the organ of Corti, the sense organ for hearing

basophils (BAY-so-fills) White blood cells that synthesize, store, and release histamine, which is important in allergic responses, and heparin, which hastens the removal of fat particles from the blood

BER See *basic electrical rhythm*

beta (β) cells The endocrine pancreatic cells that secrete the hormone insulin

bicarbonate (HCO_3^-) The anion resulting from dissociation of carbonic acid, H_2CO_3

bile salts Cholesterol derivatives secreted in the bile that facilitate fat digestion through their detergent action and facilitate fat absorption through their micellar formation

biliary system (BIL-ē-air´-ē) The bile-producing system, consisting of the liver, gallbladder, and associated ducts

bilirubin (bill-eh-RŪ-bin) A bile pigment, which is a waste product derived from the degradation of hemoglobin during the breakdown of old red blood cells

blastocyst The developmental stage of the fertilized ovum by the time it is ready to implant; consists of a single-layered sphere of cells encircling a fluid-filled cavity

blood–brain barrier (BBB) Special structural and functional features of the brain capillaries that limit access of materials from the blood into the brain tissue

B lymphocytes (B cells) White blood cells that produce antibodies against specific targets to which they have been exposed

body of the stomach The main, or middle, part of the stomach

body system A collection of organs that perform related functions and interact to accomplish a common activity that is essential for survival of the whole body; for example, the digestive system

bone marrow The soft, highly cellular tissue that fills the internal cavities of bones and is the source of most blood cells

Bowman's capsule The beginning of the tubular component of the kidney's nephron that cups around the glomerulus and collects the glomerular filtrate as it is formed

Boyle's law (boils) At any constant temperature, the pressure exerted by a gas varies inversely with the volume of the gas

brain stem The portion of the brain that is continuous with the spinal cord, serves as an integrating link between the spinal cord and higher brain levels, and controls many life-sustaining processes, such as breathing, circulation, and digestion

bronchioles (BRONG-kē-ōlz) The small, branching airways within the lungs

bronchoconstriction Narrowing of the respiratory airways

bronchodilation Widening of the respiratory airways

brush border The collection of microvilli projecting from the luminal border of epithelial cells lining the digestive tract and kidney tubules

buffer See *chemical buffer system*

bulk flow Movement in bulk of a protein-free plasma across the capillary walls between the blood and surrounding interstitial fluid; encompasses ultrafiltration and reabsorption

bundle of His (hiss) A tract of specialized cardiac cells that rapidly transmits an action potential down the interventricular septum of the heart

calcitonin (kal´-suh-TŌ-nun) A hormone secreted by the thyroid C cells that lowers plasma Ca^{2+} levels

calmodulin (kal´-MA-jew-lin) An intracellular Ca^{2+} binding protein that, on activation by Ca^{2+}, induces a change in structure and function of another intracellular protein; especially important in smooth-muscle excitation–contraction coupling

CAMs See *cell adhesion molecules*

capillaries The thin-walled, pore-lined smallest of blood vessels, across which exchange between the blood and surrounding tissues takes place

carbonic anhydrase (an-HĪ-drās) The enzyme that catalyzes the conversion of CO_2 and H_2O into carbonic acid, H_2CO_3

cardiac cycle One period of systole and diastole

cardiac muscle The specialized muscle found only in the heart

cardiac output (CO) The volume of blood pumped by each ventricle each minute; equals stroke volume times heart rate

cardiovascular control center The integrating center located in the medulla of the brain stem that controls mean arterial blood pressure

carrier-mediated transport Transport of a substance across the plasma membrane facilitated by a carrier molecule

carrier molecules Membrane proteins, which, by undergoing reversible changes in shape so that specific binding sites are alternately exposed at either side of the membrane, can bind with and transfer particular substances unable to cross the plasma membrane on their own

cascade A series of sequential reactions that culminates in a final product, such as a clot

catabolism (kuh-TAB-ō-li-zum) The breakdown, or degradation, of large, energy-rich molecules within cells

catecholamines (kat´-uh-KŌ-luh-means) The chemical classification of the adreno-medullary hormones

cations (KAT-ī-onz) Positively charged ions that have lost one or more electrons from their outer shell

C cells The thyroid cells that secrete calcitonin

cell The smallest unit capable of carrying out the processes associated with life; the basic unit of both structure and function of living organisms

cell adhesion molecules (CAMs) Proteins that protrude from the surface of the plasma membrane and form loops or other appendages that the cells use to grip each other and the surrounding connective-tissue fibers

cell body The portion of a neuron that houses the nucleus and organelles

cell-mediated immunity A specific immune response accomplished by activated T lymphocytes, which directly attack unwanted cells

center A functional collection of cell bodies within the central nervous system

central chemoreceptors (kē-mō-rē-SEP-turz) Receptors located in the medulla near the respiratory center that respond to changes in ECF H^+ concentration resulting from changes in arterial P_{CO_2} and adjust respiration accordingly

central nervous system (CNS) The brain and spinal cord

centrioles (SEN-tree-ōls) A pair of short, cylindrical structures within a cell that form the mitotic spindle during cell division

cerebellum (ser´-uh-BEL-um) The part of the brain attached at the rear of the brain stem and concerned with maintaining proper position of the body in space and subconscious coordination of motor activity

cerebral cortex The outer shell of gray matter in the cerebrum; site of initiation of all voluntary motor output and final perceptual processing of all sensory input as well as integration of most higher neural activity

cerebral hemispheres The cerebrum's two halves, which are connected by a thick band of neuronal axons

cerebrospinal fluid (ser´-uh-brō-SPĪ-nul) or (sah-REE-brō-SPĪ-nul) A special cushioning fluid that is produced by, surrounds, and flows through the central nervous system

cerebrum (SER-uh-brum) or (sah-REE-brum) The division of the brain that consists of the basal nuclei and cerebral cortex

channels Small, water-filled passageways through the plasma membrane; formed by membrane proteins that span the membrane and provide highly selective passage for small water-soluble substances such as ions

chemical bonds The forces holding atoms together

chemical buffer system A mixture in a solution of two or more chemical compounds that minimize pH changes when either an acid or a base is added to or removed from the solution

chemical mediator A chemical that is secreted by a cell and that influences an activity outside of the cell

chemically gated channels Channels in the plasma membrane that open or close in response to the binding of a specific chemical messenger with a membrane receptor site that is in close association with the channel

chemoreceptor (kē-mo-rē-sep´-tur) A sensory receptor sensitive to specific chemicals

chemotaxin (kē-mō-TAK-sin) A chemical released at an inflammatory site that attracts phagocytes to the area

chief cells The cells in the gastric pits that secrete pepsinogen

cholecystokinin (CCK) (kō´-luh-sis-tuh-kī-nun) A hormone released from the duodenal mucosa primarily in response to the presence of fat; inhibits gastric motility and secretion, stimulates pancreatic enzyme secretion, and stimulates gallbladder contraction

cholesterol A type of fat molecule that serves as a precursor for steroid hormones and bile salts and is a stabilizing component of the plasma membrane

cholinergic fibers (kō´-lin-ER-jik) Nerve fibers that release acetylcholine as their neurotransmitter

chronic obstructive pulmonary disease A group of lung diseases characterized by increased airway resistance resulting from narrowing of the lumen of the lower airways; includes asthma, chronic bronchitis, and emphysema

chyme (kīm) A thick liquid mixture of food and digestive juices

cilia (SILL-ee-ah) Motile, hairlike protrusions from the surface of cells lining the respiratory airways and the oviducts

ciliary body The portion of the eye that produces aqueous humor and contains the ciliary muscle

ciliary muscle A circular ring of smooth muscle within the eye whose contraction increases the strength of the lens to accommodate for near vision

circadian rhythm (sir-KĀ-dē-un) Repetitive oscillations in the set point of various body activities, such as hormone levels and body temperature, that are very regular and have a frequency of one cycle every 24 hours, usually linked to light–dark cycles; diurnal rhythm; biological rhythm

circulatory shock When mean arterial blood pressure falls so low that adequate blood flow to the tissues can no longer be maintained

citric acid cycle A cyclical series of biochemical reactions that involves the further processing of intermediate breakdown products of nutrient molecules, resulting in the generation of carbon dioxide and the preparation of hydrogen carrier molecules for entry into the high-energy–yielding electron transport chain

CNS See *central nervous system*

cochlea (KOK-lē-uh) The snail-shaped portion of the inner ear that houses the receptors for sound

collecting tubule The last portion of tubule in the kidney's nephron that empties into the renal pelvis

colloid (KOL-oid) The thyroglobulin-containing substance enclosed within the thyroid follicles

complement system A collection of plasma proteins that are activated in cascade fashion on exposure to invading microorganisms, ultimately producing a membrane attack complex that destroys the invaders

compliance The distensibility of a hollow, elastic structure, such as a blood vessel or the lungs; a measure of how easily the structure can be stretched

concentration gradient A difference in concentration of a particular substance between two adjacent areas

cones The eye's photoreceptors used for color vision in the light

congestive heart failure The inability of the cardiac output to keep pace with the body's needs for blood delivery, with blood damming up in the veins behind the failing heart

connective tissue Tissue that serves to connect, support, and anchor various body parts; distinguished by relatively few cells dispersed within an abundance of extracellular material

contiguous conduction The means by which an action potential is propagated throughout a nonmyelinated nerve fiber; local current flow between an active and adjacent inactive area brings the inactive area to threshold, triggering an action potential in a previously inactive area

contractile proteins Myosin and actin, whose interaction brings about shortening (contraction) of a muscle fiber

controlled variable Some factor that can vary but is controlled to reduce the amount of variability and keep the factor at a relatively steady state

convergence The converging of many presynaptic terminals from thousands of other neurons on a single neuronal cell body and its dendrites so that activity in the single neuron is influenced by the activity in many other neurons

core temperature The temperature within the inner core of the body (abdominal and thoracic organs, central nervous system, and skeletal muscles) that is homeostatically maintained at about 100°F

cornea (KOR-nee-ah) The clear, anterior-most outer layer of the eye through which light rays pass to the interior of the eye

coronary artery disease Atherosclerotic plaque formation and narrowing of the coronary arteries that supply the heart muscle

coronary circulation The blood vessels that supply the heart muscle

corpus luteum (LOO-tē-um) The ovarian structure that develops from a ruptured follicle after ovulation

cortisol (KORT-uh-sol) The adrenocortical hormone that plays an important role in carbohydrate, protein, and fat metabolism and helps the body resist stress

cranial nerves The 12 pairs of peripheral nerves, the majority of which arise from the brain stem

cross bridges The myosin molecules' globular heads that protrude from a thick filament within a muscle fiber and interact with the actin molecules in the thin filaments to bring about shortening of the muscle fiber during contraction

cyclic adenosine monophosphate (cyclic AMP or cAMP) An intracellular second messenger derived from adenosine triphosphate (ATP)

cyclic AMP See *cyclic adenosine monophosphate*

cytokines All chemicals other than antibodies that are secreted by lymphocytes

cytoplasm (SĪ-tō-plaz´-um) The portion of the cell interior not occupied by the nucleus

cytoskeleton A complex intracellular protein network that acts as the "bone and muscle" of the cell

cytosol (SĪ-tuh-sol´) The semiliquid portion of the cytoplasm not occupied by organelles

cytotoxic T cells (sī-tō-TOK-sik) The population of T cells that destroys host cells bearing foreign antigen, such as body cells invaded by viruses or cancer cells

dead-space volume The volume of air that occupies the respiratory airways as air is moved in and out and that is not available to participate in exchange of O_2 and CO_2 between the alveoli and atmosphere

dehydration A water deficit in the body

dehydroepiandrosterone (DHEA) (dē-HĪ-drō-ep-i-and-row-steer-own) The androgen (masculinizing hormone) secreted by the adrenal cortex in both sexes

dendrites Projections from the surface of a neuron's cell body that carry signals toward the cell body

deoxyribonucleic acid (DNA) (dē-OK-sē-rī-bō-new-klā-ik) The cell's genetic material, which is found within the nucleus and which provides codes for protein synthesis and serves as a blueprint for cell replication

depolarization (de´-pō-luh-ruh-ZĀ-shun) A reduction in membrane potential from resting potential; movement of the potential from resting toward 0 mV

dermis The connective tissue layer that lies under the epidermis in the skin; contains the skin's blood vessels and nerves

descending tract A bundle of nerve fibers of similar function that travels down the spinal cord to relay messages from the brain to efferent neurons

desmosome (dez´-muh-sōm) An adhering junction between two adjacent but non-touching cells formed by the extension of filaments between the cells' plasma membranes; most abundant in tissues that are subject to considerable stretching

DHEA See *dehydroepiandrosterone*

diabetes insipidus (in-SIP´-ud-us) An endocrine disorder characterized by a deficiency of vasopressin

diabetes mellitus (muh-LĪ-tus) An endocrine disorder characterized by inadequate insulin action

diaphragm (DIE-uh-fram) A dome-shaped sheet of skeletal muscle that forms the floor of the thoracic cavity; the major inspiratory muscle

diastole (dī-AS-tō-lē) The period of cardiac relaxation and filling

diencephalon (dī´-un-SEF-uh-lan) The division of the brain that consists of the thalamus and hypothalamus

diffusion Random collisions and intermingling of molecules as a result of their continuous thermally induced random motion

digestion The breaking-down process whereby the structurally complex foodstuffs of the diet are converted into smaller absorbable units by the enzymes produced within the digestive system

diploid number (DIP-loid) A complete set of 46 chromosomes (23 pairs), as found in all human somatic cells

distal tubule A highly convoluted tubule that extends between the loop of Henle and the collecting duct in the kidney's nephron

diurnal rhythm (dī-URN´-ul) Repetitive oscillations in hormone levels that are very regular and have a frequency of one cycle every 24 hours, usually linked to the light–dark cycle; circadian rhythm; biological rhythm

divergence The diverging, or branching, of a neuron's axon terminals, so that activity in this single neuron influences the many other cells with which its terminals synapse

DNA See *deoxyribonucleic acid*

dorsal root ganglion A cluster of afferent neuronal cell bodies located adjacent to the spinal cord

ECG See *electrocardiogram*

edema (i-DĒ-muh) Swelling of tissues as a result of excess interstitial fluid

EDV See *end-diastolic volume*

EEG See *electroencephalogram*

effector organs The muscles or glands that are innervated by the nervous system and that carry out the nervous system's orders to bring about a desired effect, such as a particular movement or secretion

efferent division (EF-er-ent) The portion of the peripheral nervous system that carries instructions from the central nervous system to effector organs

efferent neuron Neuron that carries information from the central nervous system to an effector organ

efflux (Ē-flux) Movement out of the cell

elastic recoil Rebound of the lungs after having been stretched

electrical gradient A difference in charge between two adjacent areas

electrocardiogram (ECG) The graphic record of the electrical activity that reaches the surface of the body as a result of cardiac depolarization and repolarization

electrochemical gradient The simultaneous existence of an electrical gradient and concentration (chemical) gradient for a particular ion

electroencephalogram (EEG) (i-lek´-trō-in-SEF-uh-luh-gram´) A graphic record of the collective postsynaptic potential activity in the cell bodies and dendrites located in the cortical layers under a recording electrode

electrolytes Solutes that form ions in solution and conduct electricity

embolus (EM-bō-lus)   A freely floating clot

emphysema (em´-fuh-ZĒ-muh)   A pulmonary disease characterized by collapse of the smaller airways and a breakdown of alveolar walls

end-diastolic volume (EDV)   The volume of blood in the ventricle at the end of diastole, when filling is complete

endocrine glands   Ductless glands that secrete hormones into the blood

endocytosis (en´-dō-sī-TŌ-sis)   Internalization of extracellular material within a cell as a result of the plasma membrane forming a pouch that contains the extracellular material, then sealing at the surface of the pouch to form a small, intracellular, membrane-enclosed vesicle with the contents of the pouch trapped inside

endogenous opiates (en-DAJ´-eh-nus ō´-pē-ātz)   Endorphins and enkephalins, which bind with opiate receptors and are important in the body's natural analgesic system

endogenous pyrogen (pī´-ruh-jun)   A chemical released from macrophages during inflammation that acts by means of local prostaglandins to raise the set point of the hypothalamic thermostat to produce a fever

endometrium (en´-dō-MĒ-trē-um)   The lining of the uterus

endoplasmic reticulum (ER) (en´-dō-PLAZ-mik ri-TIK-yuh-lum)   An organelle consisting of a continuous membranous network of fluid-filled tubules and flattened sacs, partially studded with ribosomes; synthesizes proteins and lipids for formation of new cell membrane and other cell components and manufactures products for secretion

endothelium (en´-dō-THĒ-lē-um)   The thin, single-celled layer of epithelial cells that lines the entire circulatory system

end-plate potential (EPP)   The graded receptor potential that occurs at the motor end plate of a skeletal muscle fiber in response to binding with acetylcholine

end-systolic volume (ESV)   The volume of blood in the ventricle at the end of systole, when emptying is complete

endurance-type exercise   See *aerobic exercise*

enterogastrones (ent´-uh-rō-GAS-trōnz)   Hormones secreted by the duodenal mucosa that inhibit gastric motility and secretion; include secretin, cholecystokinin, and gastric inhibitory peptide

enterohepatic circulation (en´-tur-ō-hi-PAT-ik)   The recycling of bile salts and other bile constituents between the small intestine and liver by means of the hepatic portal vein

enzyme   A special protein molecule that speeds up a particular chemical reaction in the body

eosinophils (ē´-uh-SIN-uh-fils)   White blood cells that are important in allergic responses and in combating internal parasite infestations

epidermis (ep´-uh-DER-mus)   The outer layer of the skin, consisting of numerous layers of epithelial cells, with the outermost layers being dead and flattened

epinephrine (ep´-uh-NEF-rin)   The primary hormone secreted by the adrenal medulla; important in preparing the body for "fight-or-flight" responses and in regulating arterial blood pressure; adrenaline

epiphyseal plate (eh-pif-i-SEE-al)   A layer of cartilage that separates the diaphysis (shaft) of a long bone from the epiphysis (flared end); the site of growth of bones in length before the cartilage ossifies (turns into bone)

epithelial tissue (ep´-uh-THĒ-lē-ul)   A functional grouping of cells specialized in the exchange of materials between the cell and its environment; lines and covers various body surfaces and cavities and forms secretory glands

EPSP   See *excitatory postsynaptic potential*

equilibrium potential (E_K+)   The potential that exists when the concentration gradient and opposing electrical gradient for a given ion exactly counterbalance each other so there is no net movement of the ion

erythrocytes (i-RITH-ruh-sīts)   Red blood cells, which are plasma membrane-enclosed bags of hemoglobin that transport O_2 and to a lesser extent CO_2 and H^+ in the blood

erythropoiesis (i-rith´-rō-poi-Ē-sus)   Erythrocyte production by the bone marrow

erythropoietin   The hormone released from the kidneys in response to a reduction in O_2 delivery to the kidneys; stimulates the bone marrow to increase erythrocyte production

esophagus (i-SOF-uh-gus)   A straight muscular tube that extends between the pharynx and stomach

estrogen   Feminizing "female" sex hormone

ESV   See *end-systolic volume*

excitable tissue   Tissue capable of producing electrical signals when excited; includes nervous and muscle tissue

excitation–contraction coupling   The series of events linking muscle excitation (the presence of an action potential) to muscle contraction (filament sliding and sarcomere shortening)

excitatory postsynaptic potential (EPSP) (pōst´-si-NAP-tik)   A small depolarization of the postsynaptic membrane in response to neurotransmitter binding, bringing the membrane closer to threshold

excitatory synapse (SIN-aps´)   Synapse in which the postsynaptic neuron's response to neurotransmitter release is a small depolarization of the postsynaptic membrane, bringing the membrane closer to threshold

exercise physiology   The study of both the functional changes that occur in response to a single session of exercise and the adaptations that result from regular, repeated exercise sessions

exocrine glands   Glands that secrete through ducts to the outside of the body or into a cavity that communicates with the outside

exocytosis (eks´-ō-sī-TŌ-sis)   Fusion of a membrane-enclosed intracellular vesicle with the plasma membrane, followed by the opening of the vesicle and the emptying of its contents to the outside

expiration   A breath out

expiratory muscles   The skeletal muscles whose contraction reduces the size of the thoracic cavity and lets the lungs recoil to a smaller size, bringing about movement of air from the lungs to the atmosphere

external intercostal muscles   Inspiratory muscles whose contraction elevates the ribs, thereby enlarging the thoracic cavity

external environment   The environment that surrounds the body

external work   Energy expended by contracting skeletal muscles to move external objects or to move the body in relation to the environment

extracellular fluid   All the body's fluid outside the cells; consists of interstitial fluid and plasma

extracellular matrix   An intricate meshwork of fibrous proteins embedded in a watery, gel-like substance; secreted by local cells

extrinsic controls   Regulatory mechanisms initiated outside of an organ that alter the activity of the organ; accomplished by the nervous and endocrine systems

extrinsic nerves   The nerves that originate outside of the digestive tract and innervate the various digestive organs

facilitated diffusion   Passive carrier-mediated transport involving transport of a substance down its concentration gradient across the plasma membrane

fatigue Inability to maintain muscle tension at a given level despite sustained stimulation

feedforward mechanism A response designed to prevent an anticipated change in a controlled variable

fibrinogen (fī-BRIN-uh-jun) A large, soluble plasma protein that is converted into an insoluble, threadlike molecule that forms the meshwork of a clot during blood coagulation

Fick's law of diffusion The rate of net diffusion of a substance across a membrane is directly proportional to the substance's concentration gradient, the membrane's permeability to the substance, and the surface area of the membrane and inversely proportional to the substance's molecular weight and the diffusion distance

fight-or-flight response The changes in activity of the various organs innervated by the autonomic nervous system in response to sympathetic stimulation, which collectively prepare the body for strenuous physical activity in the face of an emergency or stressful situation, such as a physical threat from the outside environment

fire When an excitable cell undergoes an action potential

first messenger An extracellular messenger, such as a hormone, that binds with a surface membrane receptor and activates an intracellular second messenger to carry out the desired cellular response

flagellum (fluh-JEL-um) The single, long, whiplike appendage that serves as the tail of a spermatozoon

follicle (of ovary) A developing ovum and the surrounding specialized cells

follicle-stimulating hormone (FSH) An anterior pituitary hormone that stimulates ovarian follicular development and estrogen secretion in females and stimulates sperm production in males

follicular cells (of ovary) (fah-LIK-you-lar) Collectively, the granulosa and thecal cells

follicular cells (of thyroid gland) The cells that form the walls of the colloid-filled follicles in the thyroid gland and secrete thyroid hormone

follicular phase The phase of the ovarian cycle dominated by the presence of maturing follicles prior to ovulation

Frank-Starling law of the heart Intrinsic control of the heart such that increased venous return resulting in increased end-diastolic volume leads to an increased strength of contraction and increased stroke volume; that is, the heart normally pumps out all the blood returned to it

free radicals Very unstable electron-deficient particles that are highly reactive and destructive

frontal lobes The lobes of the cerebral cortex that lie at the top of the brain in front of the central sulcus and that are responsible for voluntary motor output, speaking ability, and elaboration of thought

FSH See *follicle-stimulating hormone*

fuel metabolism See *intermediary metabolism*

functional syncytium (sin-sish´-ē-um) A group of smooth or cardiac muscle cells that are interconnected by gap junctions and function electrically and mechanically as a single unit

functional unit The smallest component of an organ that can perform all the functions of the organ

gametes (GAM-ētz) Reproductive, or germ, cells, each containing a haploid set of chromosomes; sperm and ova

ganglion (GAN-glē-un) A collection of neuronal cell bodies located outside the central nervous system

ganglion cells The nerve cells in the outermost layer of the retina and whose axons form the optic nerve

gap junction A communicating junction formed between adjacent cells by small connecting tunnels that permit passage of charge-carrying ions between the cells so that electrical activity in one cell is spread to the adjacent cell

gastrin A hormone secreted by the pyloric gland area of the stomach that stimulates the parietal and chief cells to secrete a highly acidic gastric juice

gestation Pregnancy

glands Epithelial tissue derivatives that are specialized for secretion

glial cells (glē-ul) Serve as the connective tissue of the CNS and help support the neurons both physically and metabolically; include astrocytes, oligodendrocytes, ependymal cells, and microglia

glomerular filtration (glow-MER-yu-lur) Filtration of a protein-free plasma from the glomerular capillaries into the tubular component of the kidney's nephron as the first step in urine formation

glomerular filtration rate (GFR) The rate at which glomerular filtrate is formed

glomerulus (glow-MER-yu-lus) A ball-like tuft of capillaries in the kidney's nephron that filters water and solute from the blood as the first step in urine formation

glucagon (GLOO-kuh-gon) The pancreatic hormone that raises blood glucose and blood fatty-acid levels

glucocorticoids (gloo´-kō-KOR-ti-koidz) The adrenocortical hormones that are important in intermediary metabolism and in helping the body resist stress; primarily cortisol

gluconeogenesis (gloo´-kō-nē-ō-JEN-uh-sus) The conversion of amino acids into glucose

glycogen (GLĪ-kō-jen) The storage form of glucose in the liver and muscle

glycogenesis (glī´-kō-JEN-i-sus) The conversion of glucose into glycogen

glycogenolysis (glī´-kō-juh-NOL-i-sus) The conversion of glycogen to glucose

glycolysis (glī-KOL-uh-sus) A biochemical process that takes place in the cell's cytosol and involves the breakdown of glucose into two pyruvic acid molecules

GnRH See *gonadotropin-releasing hormone*

Golgi complex (GOL-jē) An organelle consisting of sets of stacked, flattened membranous sacs; processes raw materials transported to it from the endoplasmic reticulum into finished products and sorts and directs the finished products to their final destination

gonadotropin-releasing hormone (GnRH) (gō-nad´-uh-TRŌ-pin) The hypothalamic hormone that stimulates the release of FSH and LH from the anterior pituitary

gonadotropins FSH and LH; hormones that are tropic to the gonads

gonads (GŌ-nadz) The primary reproductive organs, which produce the gametes and secrete the sex hormones; testes and ovaries

G protein A membrane-bound intermediary, which, when activated on binding of an extracellular first messenger to a surface receptor, activates the enzyme adenylyl cyclase on the intracellular side of the membrane in the cAMP second-messenger system

gradation of contraction Variable magnitudes of tension produced in a single whole muscle

graded potential A local change in membrane potential that occurs in varying grades of magnitude; serves as a short-distance signal in excitable tissues

gray matter The portion of the central nervous system composed primarily of densely packaged neuronal cell bodies and dendrites

growth hormone (GH) An anterior pituitary hormone that is primarily responsible for regulating overall body growth and is also important in intermediary metabolism; somatotropin

H+ See *hydrogen ion*

haploid number (HAP-loid) The number of chromosomes found in gametes; a half set of chromosomes, one member of each pair, for a total of 23 chromosomes in humans

Hb See *hemoglobin*

hCG See *human chorionic gonadotropin*

heart failure An inability of the cardiac output to keep pace with the body's demands for supplies and for removal of wastes

helper T cells The population of T cells that enhances the activity of other immune-response effector cells

hematocrit (hi-mat′-uh-krit) The percentage of blood volume occupied by erythrocytes as they are packed down in a centrifuged blood sample

hemoglobin (HĒ-muh-glō′-bun) A large iron-bearing protein molecule found within erythrocytes that binds with and transports most O_2 in the blood; also carries some of the CO_2 and H^+ in the blood

hemolysis (hē-MOL-uh-sus) Rupture of red blood cells

hemostasis (hē′-mō-STĀ-sus) The stopping of bleeding from an injured vessel

hepatic portal system (hi-PAT-ik) A complex vascular connection between the digestive tract and liver such that venous blood from the digestive system drains into the liver for processing of absorbed nutrients before being returned to the heart

hippocampus (hip-oh-CAM-pus) The elongated, medial portion of the temporal lobe that is a part of the limbic system and is especially crucial for forming long-term memories

histamine A chemical released from mast cells or basophils that brings about vasodilation and increased capillary permeability; important in allergic responses

homeostasis (hō′-mē-ō-STĀ-sus) Maintenance by the highly coordinated, regulated actions of the body systems of relatively stable chemical and physical conditions in the internal fluid environment that bathes the body's cells

hormone A long-distance chemical mediator that is secreted by an endocrine gland into the blood, which transports it to its target cells

hormone response element (HRE) The specific attachment site on DNA for a given steroid hormone and its nuclear receptor

host cell A body cell infected by a virus

HRE See *hormone response element*

human chorionic gonadotropin (hCG) (kō-rē-ON-ik gō-nad′-uh-TRŌ-pin) A hormone secreted by the developing placenta that stimulates and maintains the corpus luteum of pregnancy

hydrogen ion (H+) The cationic portion of a dissociated acid

hydrolysis (hī-DROL-uh-sis) The digestion of a nutrient molecule by the addition of water at a bond site

hydrostatic (fluid) pressure (hī-dro-STAT-ik) The pressure exerted by fluid on the walls that contain it

hyperglycemia (hī′-pur-glī-SĒ-mē-uh) Elevated blood-glucose concentration

hyperplasia (hī-pur-PLĀ-zē-uh) An increase in the number of cells

hyperpolarization An increase in membrane potential from resting potential; potential becomes even more negative than at resting potential

hypersecretion Too much of a particular hormone secreted

hypertension (hī′-pur-TEN-shun) Sustained, above-normal mean arterial blood pressure

hypertonic (hī′-pur-TON-ik) **solution** A solution having an osmolarity greater than normal body fluids; more concentrated than normal

hypertrophy (hī-PUR-truh-fē) Increase in the size of an organ as a result of an increase in the size of its cells

hyperventilation Overbreathing; when the rate of ventilation is in excess of the body's metabolic needs for CO_2 removal

hypophysiotropic hormones (hi-PŌ-fiz-ē-oh-TRO-pik) Hormones secreted by the hypothalamus that regulate the secretion of anterior pituitary hormones; see also *releasing hormone* and *inhibiting hormone*

hyposecretion Too little of a particular hormone secreted

hypotension (hi-po-TEN-chun) Sustained, below-normal mean arterial blood pressure

hypothalamic hypophyseal portal system (hī-pō-thuh-LAM-ik hī-pō-FIZ-ē-ul) The vascular connection between the hypothalamus and anterior pituitary gland used for the pickup and delivery of hypophysiotropic hormones

hypothalamus (hī′-pō-THAL-uh-mus) The brain region located beneath the thalamus that is concerned with regulating many aspects of the internal fluid environment, such as water and salt balance and food intake; serves as an important link between the autonomic nervous system and endocrine system

hypotonic (hī′-pō-TON-ik) **solution** A solution having an osmolarity less than normal body fluids; more dilute than normal

hypoventilation Underbreathing; ventilation inadequate to meet the metabolic needs for O_2 delivery and CO_2 removal

hypoxia (hī-POK-sē-uh) Insufficient O_2 at the cellular level

I band One of the light bands that alternate with dark (A) bands to create a striated appearance in a skeletal or cardiac muscle fiber when these fibers are viewed with a light microscope

IGF See *insulin-like growth factor*

immune surveillance Recognition and destruction of newly arisen cancer cells by the immune system

immunity The body's ability to resist or eliminate potentially harmful foreign materials or abnormal cells

immunoglobulins (im′-ū-nō-GLOB-yū-lunz) Antibodies; gamma globulins

impermeable Prohibiting passage of a particular substance through the plasma membrane

implantation The burrowing of a blastocyst into the endometrial lining

inflammation An innate, nonspecific series of highly interrelated events, especially involving neutrophils, macrophages, and local vascular changes, that are set into motion in response to foreign invasion or tissue damage

influx Movement into the cell

inhibin (in-HIB-un) A hormone secreted by the Sertoli cells of the testes or by the ovarian follicles that inhibits FSH secretion

inhibiting hormone A hypothalamic hormone that inhibits the secretion of a particular anterior pituitary hormone

inhibitory postsynaptic potential (IPSP) (pōst′-si-NAP-tik) A small hyperpolarization of the postsynaptic membrane in response to neurotransmitter binding, thereby moving the membrane farther from threshold

inhibitory synapse (SIN-aps′) Synapse in which the postsynaptic neuron's response to neurotransmitter release is a small hyperpolarization of the postsynaptic membrane, moving the membrane farther from threshold

innate immune responses Inherent defense responses that nonselectively defend against foreign or abnormal material, even on initial exposure to it; see also *inflammation, interferon, natural killer cells,* and *complement system*

inorganic Referring to substances that do not contain carbon; from nonliving sources

inspiration A breath in

inspiratory muscles The skeletal muscles whose contraction enlarges the thoracic cavity, bringing about lung expansion and movement of air into the lungs from the atmosphere

insulin (IN-suh-lin) The pancreatic hormone that lowers blood levels of glucose, fatty acids, and amino acids and promotes their storage

insulin-like growth factor (IGF) Synonymous with somatomedins

integrating center A region that determines efferent output based on processing of afferent input

integument (in-TEG-yuh-munt) The skin and underlying connective tissue

intercostal muscles (int-ur-KOS-tul) The muscles that lie between the ribs; see also *external intercostal muscles* and *internal intercostal muscles*

interferon (in´-tur-FĒR-on) A chemical released from virus-invaded cells that provides nonspecific resistance to viral infections by transiently interfering with replication of the same or unrelated viruses in other host cells

intermediary metabolism The collective set of intracellular chemical reactions that involve the degradation, synthesis, and transformation of small nutrient molecules; also known as *fuel metabolism*

intermediate filaments Threadlike cytoskeletal elements that play a structural role in parts of the cells subject to mechanical stress

internal environment The body's aqueous extracellular environment, which consists of the plasma and interstitial fluid and which must be homeostatically maintained for the cells to make life-sustaining exchanges with it

internal intercostal muscles Expiratory muscles whose contraction pulls the ribs downward and inward, thereby reducing the size of the thoracic cavity

internal respiration The intracellular metabolic processes carried out within the mitochondria that use O_2 and produce CO_2 during the derivation of energy from nutrient molecules

internal work All forms of biological energy expenditure that do not accomplish mechanical work outside of the body

interneuron Neuron that lies entirely within the central nervous system and is important for integrating peripheral responses

to peripheral information as well as for the abstract phenomena associated with the "mind"

interstitial fluid (in´-tur-STISH-ul) The portion of the extracellular fluid that surrounds and bathes all the body's cells

intra-alveolar pressure (in´-truh-al-VĒ-uh-lur) The pressure within the alveoli

intracellular fluid The fluid collectively contained within all the body's cells

intrapleural pressure (in´-truh-PLOOR-ul) The pressure within the pleural sac

intrinsic controls Local control mechanisms inherent to an organ

intrinsic factor A special substance secreted by the parietal cells of the stomach that must be combined with vitamin B_{12} for this vitamin to be absorbed by the intestine; deficiency produces pernicious anemia

intrinsic nerve plexuses Interconnecting networks of nerve fibers within the digestive tract wall

ion An atom that has gained or lost one or more of its electrons, so that it is not electrically balanced

IPSP See *inhibitory postsynaptic potential*

iris A pigmented smooth muscle that forms the colored portion of the eye and controls pupillary size

islets of Langerhans (LAHNG-er-honz) The endocrine portion of the pancreas that secretes the hormones insulin and glucagon into the blood

isometric contraction (ī´-sō-MET-rik) A muscle contraction in which the development of tension occurs at constant muscle length

isotonic (ī´-sō-TON-ik) **solution** A solution having an osmolarity equal to normal body fluids

isotonic contraction A muscle contraction in which muscle tension remains constant as the muscle fiber changes length

juxtaglomerular apparatus (juks´-tuh-glō-MER-yū-lur) A cluster of specialized vascular and tubular cells at a point where the ascending limb of the loop of Henle passes through the fork formed by the afferent and efferent arterioles of the same nephron in the kidney

keratin (CARE-uh-tin) The protein found in the intermediate filaments in skin cells that give the skin strength and help form a waterproof outer layer

killer (K) cells Cells that destroy a target cell that has been coated with antibodies by lysing its membrane

lactation Milk production by the mammary glands

lactic acid An end product formed from pyruvic acid during the anaerobic process of glycolysis

larynx (LARE-inks) The "voice box" at the entrance of the trachea; contains the vocal cords

lateral sacs The expanded saclike regions of a muscle fiber's sarcoplasmic reticulum; store and release calcium, which plays a key role in triggering muscle contraction

law of mass action If the concentration of one of the substances involved in a reversible reaction is increased, the reaction is driven toward the opposite side, and if the concentration of one of the substances is decreased, the reaction is driven toward that side

left ventricle The heart chamber that pumps blood into the systemic circulation

length–tension relationship The relationship between the length of a muscle fiber at the onset of contraction and the tension the fiber can achieve on a subsequent tetanic contraction

lens A transparent, biconvex structure of the eye that refracts (bends) light rays and whose strength can be adjusted to accommodate for vision at different distances

leptin A hormone released from adipose tissue that plays a key role in long-term regulation of body weight by acting on the hypothalamus to suppress appetite

leukocytes (LOO-kuh-sīts) White blood cells, which are the immune system's mobile defense units

Leydig cells (LĪ-dig) The interstitial cells of the testes that secrete testosterone

LH See *luteinizing hormone*

LH surge The burst in LH secretion that occurs at midcycle of the ovarian cycle and triggers ovulation

limbic system (LIM-bik) A functionally interconnected ring of forebrain structures that surrounds the brain stem and is concerned with emotions, basic survival and sociosexual behavioral patterns, motivation, and learning

lipid emulsion A suspension of small fat droplets held apart as a result of adsorption of bile salts on their surface

loop of Henle (HEN-lē) A hairpin loop that extends between the proximal and distal tubule of the kidney's nephron

lumen (LOO-men) The interior space of a hollow organ or tube

luteal phase (LOO-tē-ul)   The phase of the ovarian cycle dominated by the presence of a corpus luteum

luteinization (loot´-ē-un-uh-ZĀ-shun)   Formation of a postovulatory corpus luteum in the ovary

luteinizing hormone (LH)   An anterior pituitary hormone that stimulates ovulation, luteinization, and secretion of estrogen and progesterone in females and stimulates testosterone secretion in males

lymph   Interstitial fluid that is picked up by the lymphatic vessels and returned to the venous system, meanwhile passing through the lymph nodes for defense purposes

lymphocytes   White blood cells that provide immune defense against targets for which they are specifically programmed

lymphoid tissues   Tissues that produce and store lymphocytes, such as lymph nodes and tonsils

lysosomes (LĪ-sō-sōmz)   Organelles consisting of membrane-enclosed sacs containing powerful hydrolytic enzymes that destroy unwanted material within the cell, such as internalized foreign material or cellular debris

macrophages (MAK-ruh-fājs)   Large, tissue-bound phagocytes

mast cells   Cells located within connective tissue that synthesize, store, and release histamine, as during allergic responses

mean arterial blood pressure   The average pressure responsible for driving blood forward through the arteries into the tissues throughout the cardiac cycle; equals cardiac output times total peripheral resistance

mechanically gated channels   Channels that open or close in response to stretching or other mechanical deformation

mechanoreceptor (meh-CAN-oh-rē-SEP-tur) or (mek´-uh-nō-rē-SEP-tur)   A sensory receptor sensitive to mechanical energy, such as stretching or bending

medullary respiratory center (MED-you-LAIR-ē)   Several aggregations of neuronal cell bodies within the medulla that provide output to the respiratory muscles and receive input important for regulating the magnitude of ventilation

meiosis (mī-ō-sis)   Cell division in which the chromosomes replicate followed by two nuclear divisions so that only a half set of chromosomes is distributed to each of four new daughter cells

melanocyte-stimulating hormone (MSH) (mel-AH-nō-sīt)   A hormone produced by the anterior pituitary in humans and by the intermediate lobe of the pituitary in lower vertebrates; regulates skin coloration by controlling the dispersion of melanin granules in lower vertebrates; involved with control of food intake and possibly memory and learning in humans

melatonin (mel-uh-TŌ-nin)   A hormone secreted by the pineal gland during darkness that helps entrain the body's biological rhythms with the external light/dark cues

membrane attack complex   A collection of the five final activated components of the complement system that aggregate to form a porelike channel in the plasma membrane of an invading microorganism, with the resultant leakage leading to destruction of the invader

membrane potential   A separation of charges across the membrane; a slight excess of negative charges lined up along the inside of the plasma membrane and separated from a slight excess of positive charges on the outside

memory cells   B or T cells that are newly produced in response to a microbial invader but that do not participate in the current immune response against the invader: instead they remain dormant, ready to launch a swift, powerful attack should the same micro-organism invade again in the future

menstrual cycle (men´-stroo-ul)   The cyclical changes in the uterus that accompany the hormonal changes in the ovarian cycle

menstrual phase   The phase of the menstrual cycle characterized by sloughing of endometrial debris and blood out through the vagina

messenger RNA   Carries the transcribed genetic blueprint for synthesis of a particular protein from nuclear DNA to the cytoplasmic ribosomes where the protein synthesis takes place

metabolic acidosis (met-uh-bol´-ik)   Acidosis resulting from any cause other than excess accumulation of carbonic acid in the body

metabolic alkalosis (al´-kuh-LŌ-sus)   Alkalosis caused by a relative deficiency of noncarbonic acid

metabolic rate   Energy expenditure per unit of time

micelle (mī-SEL)   A water-soluble aggregation of bile salts, lecithin, and cholesterol that has a hydrophilic shell and a hydrophobic core; carries the water-insoluble products of fat digestion to their site of absorption

microfilaments   Cytoskeletal elements made of actin molecules (as well as myosin molecules in muscle cells); play a major role in various cellular contractile systems and serve as a mechanical stiffener for microvilli

microtubules   Cytoskeletal elements made of tubulin molecules arranged into long, slender, unbranched tubes that help maintain asymmetric cell shapes and coordinate complex cell movements

microvilli (mī´-krō-VIL-ī)   Actin-stiffened, nonmotile, hairlike projections from the luminal surface of epithelial cells lining the digestive tract and kidney tubules; tremendously increase the surface area of the cell exposed to the lumen

micturition (mik-too-RISH-un) or (mik-chuh-RISH-un)   The process of bladder emptying; urination

milk ejection   The squeezing out of milk produced and stored in the alveoli of the breasts by means of contraction of the myoepithelial cells that surround each alveolus

mineralocorticoids (min-uh-rul-ō-KOR-ti-koidz)   The adrenocortical hormones that are important in Na^+ and K^+ balance; primarily aldosterone

mitochondria (mī-tō-KON-drē-uh)   The energy organelles, which contain the enzymes for oxidative phosphorylation

mitosis (mī-TŌ-sis)   Cell division in which the chromosomes replicate before nuclear division, so that each of the two daughter cells receives a full set of chromosomes

mitotic spindle   The system of microtubules assembled during mitosis along which the replicated chromosomes are directed away from each other toward opposite sides of the cell prior to cell division

molecule   A chemical substance formed by the linking of atoms together; the smallest unit of a given chemical substance

monocytes (MAH-nō-sīts)   White blood cells that emigrate from the blood, enlarge, and become macrophages, large-tissue phagocytes

monosaccharides (mah´-nō-SAK-uh-rīdz)   Simple sugars, such as glucose; the absorbable unit of digested carbohydrates

motor activity   Movement of the body accomplished by contraction of skeletal muscles

motor end plate   The specialized portion of a skeletal muscle fiber that lies immediately underneath the terminal button of the motor neuron and possesses receptor sites for binding acetylcholine released from the terminal button

motor neurons   The neurons that innervate skeletal muscle and whose axons constitute the somatic nervous system

motor unit One motor neuron plus all the muscle fibers it innervates

motor unit recruitment The progressive activation of a muscle fiber's motor units to accomplish increasing gradations of contractile strength

mucosa (mew-KŌ-sah) The innermost layer of the digestive tract that lines the lumen

multiunit smooth muscle A smooth muscle mass that consists of multiple discrete units that function independently of each other and that must be separately stimulated by autonomic nerves to contract

muscarinic receptor Type of cholinergic receptor found at the effector organs of all parasympathetic postganglionic fibers

muscle fiber A single muscle cell, which is relatively long and cylindrical in shape

muscle tension See *tension*

muscle tissue A functional grouping of cells specialized for contraction and force generation

myelin (MĪ-uh-lun) An insulative lipid covering that surrounds myelinated nerve fibers at regular intervals along the axon's length; each patch of myelin is formed by a separate myelin forming cell that wraps itself jelly-roll fashion around the neuronal axon

myelinated fibers Neuronal axons covered at regular intervals with insulative myelin

myocardial ischemia (mī-ō-KAR-dē-ul is-KĒ-mē-uh) Inadequate blood supply to the heart tissue

myocardium (mī-ō-KAR-dē-um) The cardiac muscle within the heart wall

myofibril (mī-ō-FĪB-rul) A specialized intracellular structure of muscle cells that contains the contractile apparatus

myometrium (mī-ō-mē-TRĒ-um) The smooth muscle layer of the uterus

myosin (MĪ-uh-sun) The contractile protein that forms the thick filaments in muscle fibers

Na⁺–K⁺ pump A carrier that actively transports Na^+ out of the cell and K^+ into the cell

natural killer cells Naturally occurring, lymphocyte-like cells that nonspecifically destroy virus-infected cells and cancer cells by directly lysing their membranes on first exposure to them

negative balance Situation in which the losses for a substance exceed its gains so that the total amount of the substance in the body decreases

negative feedback A regulatory mechanism in which a change in a controlled variable triggers a response that opposes the change, thus maintaining a relatively steady set point for the regulated factor

nephron (NEF-ron´) The functional unit of the kidney; consisting of an interrelated vascular and tubular component, it is the smallest unit that can form urine

nerve A bundle of peripheral neuronal axons, some afferent and some efferent, enclosed by a connective tissue covering and following the same pathway

nervous system One of the two major regulatory systems of the body; in general, coordinates rapid activities of the body, especially those involving interactions with the external environment

nervous tissue A functional grouping of cells specialized for initiation and transmission of electrical signals

net diffusion The difference between two opposing movements

net filtration pressure The net difference in the hydrostatic and osmotic forces acting across the glomerular membrane that favors the filtration of a protein-free plasma into Bowman's capsule

neuroglia See *glial cells*

neurohormones Hormones released into the blood by neurosecretory neurons

neuromodulators (ner´ō-MA-jew-lā´-torz) Chemical messengers that bind to neuronal receptors at nonsynaptic sites (that is, not at the subsynaptic membrane) and bring about long-term changes that subtly depress or enhance synaptic effectiveness

neuromuscular junction The juncture between a motor neuron and a skeletal muscle fiber

neuron (NER-on) A nerve cell, typically consisting of a cell body, dendrites, and an axon and specialized to initiate, propagate, and transmit electrical signals

neuropeptides Large, slowly acting peptide molecules released from axon terminals along with classical neurotransmitters; most neuropeptides function as neuromodulators

neurotransmitter The chemical messenger that is released from the axon terminal of a neuron in response to an action potential and influences another neuron or an effector with which the neuron is anatomically linked

neutrophils (new´-truh-filz) White blood cells that are phagocytic specialists and important in inflammatory responses and defense against bacterial invasion

nicotinic receptor (nick´-o-TIN-ik) Type of cholinergic receptor found at all autonomic ganglia and the motor end plates of skeletal muscle fibers

nitric oxide A recently identified local chemical mediator released from endothelial cells and other tissues; exerts a wide array of effects, ranging from causing local arteriolar vasodilation to acting as a toxic agent against foreign invaders to serving as a unique type of neurotransmitter

nociceptor (nō-sē-SEP-tur) A pain receptor, sensitive to tissue damage

nodes of Ranvier (RAN-vē-ā) The portions of a myelinated neuronal axon between the segments of insulative myelin; the axonal regions where the axonal membrane is exposed to the ECF and membrane potential exists

norepinephrine (nor´-ep-uh-NEF-run) The neurotransmitter released from sympathetic postganglionic fibers; noradrenaline

nucleus (of brain) (NŪ-klē-us) A functional aggregation of neuronal cell bodies within the brain

nucleus (of cells) A distinct spherical or oval structure that is usually located near the center of a cell and that contains the cell's genetic material, deoxyribonucleic acid (DNA)

occipital lobes (ok-SIP´-ut-ul) The lobes of the cerebral cortex that are located posteriorly and that initially process visual input

O₂–Hb dissociation curve A graphic depiction of the relationship between arterial P_{O_2} and percent hemoglobin saturation

oogenesis (ō´-ō-JEN-uh-sus) Egg production

opsonin (OP´-suh-nun) Body-produced chemical that links bacteria to macrophages, thereby making the bacteria more susceptible to phagocytosis

optic nerve The bundle of nerve fibers that leave the retina, relaying information about visual input

optimal length The length before the onset of contraction of a muscle fiber at which maximal force can be developed on a subsequent tetanic contraction

organ A distinct structural unit composed of two or more types of primary tissue organized to perform one or more particular functions; for example, the stomach

organelles (or´-gan-ELZ) Distinct, highly organized, membrane-bound intracellular compartments, each containing a specific set of chemicals for carrying out a particular cellular function

organic Referring to substances that contain carbon; from living or once-living sources

organism A living entity, either single-celled (unicellular) or made up of many cells (multicellular)

organ of Corti (KOR-tē) The sense organ of hearing within the inner ear that contains hair cells whose hairs are bent in response to sound waves, setting up action potentials in the auditory nerve

osmolarity (oz´-mō-LAR-ut-ē) A measure of the concentration of solute molecules in a solution

osmosis (os-MŌ-sis) Movement of water across a membrane down its own concentration gradient toward the area of higher solute concentration

osteoblasts (OS-tē-ō-blasts´) Bone cells that produce the organic matrix of bone

osteoclasts Bone cells that dissolve bone in their vicinity

otolith organs (ŌT´-ul-ith) Sense organs in the inner ear that provide information about rotational changes in head movement; include the utricle and saccule

oval window The membrane-covered opening that separates the air-filled middle ear from the upper compartment of the fluid-filled cochlea in the inner ear

overhydration Water excess in the body

ovulation (ov´-yuh-LĀ-shun) Release of an ovum from a mature ovarian follicle

oxidative phosphorylation (fos´-fōr-i-LĀ-shun) The entire sequence of mitochondrial biochemical reactions that uses oxygen to extract energy from the nutrients in food and transforms it into ATP, producing CO_2 and H_2O in the process

oxyhemoglobin (ok-si-HĒ-muh-glō-bun) Hemoglobin combined with O_2

oxyntic mucosa (ok-SIN-tic) The mucosa that lines the body and fundus of the stomach; contains gastric pits that lead to the gastric glands lined by mucous neck cells, parietal cells, and chief cells

oxytocin (ok´-sē-TŌ-sun) A hypothalamic hormone that is stored in the posterior pituitary and stimulates uterine contraction and milk ejection

pacemaker activity Self-excitable activity of an excitable cell in which its membrane potential gradually depolarizes to threshold on its own

pancreas (PAN-krē-us) A mixed gland composed of an exocrine portion that secretes digestive enzymes and an aqueous alkaline secretion into the duodenal lumen and an endocrine portion that secretes the hormones insulin and glucagon into the blood

paracrine (PEAR-uh-krin) A local chemical messenger whose effect is exerted only on neighboring cells in the immediate vicinity of its site of secretion

parasympathetic nervous system (pear´-uh-sim-puh-THET-ik) The subdivision of the autonomic nervous system that dominates in quiet, relaxed situations and promotes body maintenance activities such as digestion and emptying of the urinary bladder

parathyroid glands (pear´-uh-THĪ-roid) Four small glands located on the posterior surface of the thyroid gland that secrete parathyroid hormone

parathyroid hormone (PTH) A hormone that raises plasma Ca^{2+} levels

parietal cells (puh-RĪ-ut-ul) The stomach cells that secrete hydrochloric acid and intrinsic factor

parietal lobes The lobes of the cerebral cortex that lie at the top of the brain behind the central sulcus and contain the somatosensory cortex

partial pressure The individual pressure exerted independently by a particular gas within a mixture of gases

partial pressure gradient A difference in the partial pressure of a gas between two regions that promotes the movement of the gas from the region of higher partial pressure to the region of lower partial pressure

parturition (par´-too-RISH-un) Delivery of a baby

passive expiration Expiration accomplished during quiet breathing as a result of elastic recoil of the lungs on relaxation of the inspiratory muscles, with no energy expenditure required

passive force A force that does not require expenditure of cellular energy to accomplish transport of a substance across the plasma membrane

passive reabsorption When none of the steps in the transepithelial transport of a substance reabsorbed across the kidney tubules requires energy expenditure

pathogens (PATH-uh-junz) Disease-causing micro-organisms, such as bacteria or viruses

pathophysiology (path´-ō-fiz-UL-uh-jē) Abnormal functioning of the body associated with disease

pepsin; pepsinogen (pep-SIN-uh-jun) An enzyme secreted in inactive form by the stomach that, once activated, begins protein digestion

peptide hormones Hormones that consist of a chain of specific amino acids of varying length

percent hemoglobin saturation A measure of the extent to which the hemoglobin present is combined with O_2

perception The conscious interpretation of the external world as created by the brain from a pattern of nerve impulses delivered to it from sensory receptors

peripheral chemoreceptors (kē´-mō-rē-SEP-turz) The carotid and aortic bodies, which respond to changes in arterial P_{O_2}, P_{CO_2}, and H^+ and adjust respiration accordingly

peripheral nervous system (PNS) Nerve fibers that carry information between the central nervous system and other parts of the body

peristalsis (per´-uh-STOL-sus) Ringlike contractions of the circular smooth muscle of a tubular organ that move progressively forward with a stripping motion, pushing the contents of the organ ahead of the contraction

peritubular capillaries (per´-i-TŪ-bū-lur) Capillaries that intertwine around the tubules of the kidney's nephron; they supply the renal tissue and participate in exchanges between the tubular fluid and blood during the formation of urine

permeable Permitting passage of a particular substance

permissiveness When one hormone must be present in adequate amounts for the full exertion of another hormone's effect

pernicious anemia (per-NEE-shus) The anemia produced as a result of intrinsic factor deficiency

peroxisomes (puh-ROK´-suh-sōmz) Organelles consisting of membrane-bound sacs that contain powerful oxidative enzymes that detoxify various wastes produced within the cell or foreign compounds that have entered the cell

pH The logarithm to the base 10 of the reciprocal of the hydrogen ion concentration; $pH = \log 1/[H^+]$ or $pH = -\log[H^+]$

phagocytosis (fag´-oh-sī-TŌ-sus) A type of endocytosis in which large, multimolecular, solid particles are engulfed by a cell

pharynx (FARE-inks) The back of the throat, which serves as a common passageway for the digestive and respiratory systems

phosphorylation (fos´-fōr-i-LĀ-shun) Addition of a phosphate group to a molecule

photoreceptor A sensory receptor responsive to light

phototransduction The mechanism of converting light stimuli into electrical activity by the rods and cones of the eye

physiology (fiz-ē-OL-ō-gē) The study of body functions

pineal gland (PIN-ē-ul)   A small endocrine gland located in the center of the brain that secretes the hormone melatonin

pinocytosis (pin-oh-cī-TŌ-sus)   Type of endocytosis in which the cell internalizes fluid

pituitary gland (pih-TWO-ih-tair-ee)   A small endocrine gland connected by a stalk to the hypothalamus; consists of the anterior pituitary and posterior pituitary

placenta (plah-SEN-tah)   The organ of exchange between the maternal and fetal blood; also secretes hormones that support the pregnancy

plaque   A deposit of cholesterol and other lipids, perhaps calcified, in thickened, abnormal smooth-muscle cells within blood vessels as a result of atherosclerosis

plasma   The liquid portion of the blood

plasma cell   An antibody-producing derivative of an activated B lymphocyte

plasma clearance   The volume of plasma that is completely cleared of a given substance by the kidneys per minute

plasma-colloid osmotic pressure (KOL-oid os-MOT-ik)   The force caused by the unequal distribution of plasma proteins between the blood and surrounding fluid that encourages fluid movement into the capillaries

plasma membrane   A protein-studded lipid bilayer that encloses each cell, separating it from the extracellular fluid

plasma proteins   The proteins that remain within the plasma, where they perform a number of important functions; include albumins, globulins, and fibrinogen

plasticity (plas-TIS-uh-tē)   The ability of portions of the brain to assume new responsibilities in response to the demands placed on it

platelets (PLĀT-lets)   Specialized cell fragments in the blood that participate in hemostasis by forming a plug at a vessel defect

pleural sac (PLOOR-ul)   A double-walled, closed sac that separates each lung from the thoracic wall

pluripotent stem cells   Precursor cells that reside in the bone marrow and continuously divide and differentiate to give rise to each of the types of blood cells

polycythemia (pol-i-sī-THĒ-mē-uh)   Excess circulating erythrocytes, accompanied by an elevated hematocrit

polysaccharides (pol´-ī-SAK-uh-rīdz)   Complex carbohydrates, consisting of chains of interconnected glucose molecules

positive balance   Situation in which the gains via input for a substance exceed its losses via output, so that the total amount of the substance in the body increases

positive feedback   A regulatory mechanism in which the input and the output in a control system continue to enhance each other so that the controlled variable is progressively moved farther from a steady state

postabsorptive state   The metabolic state after a meal is absorbed during which endogenous energy stores must be mobilized and glucose must be spared for the glucose-dependent brain; fasting state

posterior pituitary   The neural portion of the pituitary that stores and releases into the blood on hypothalamic stimulation two hormones produced by the hypothalamus, vasopressin and oxytocin

postganglionic fiber (pōst´-gan-glē-ON-ik)   The second neuron in the two-neuron autonomic nerve pathway; originates in an autonomic ganglion and terminates on an effector organ

postsynaptic neuron (pōst´-si-NAP-tik)   The neuron that conducts its action potentials away from a synapse

preganglionic fiber   The first neuron in the two-neuron autonomic nerve pathway; originates in the central nervous system and terminates on an autonomic ganglion

pressure gradient   A difference in pressure between two regions that drives the movement of blood or air from the region of higher pressure to the region of lower pressure

presynaptic neuron (prē-si-NAP-tik)   The neuron that conducts its action potentials toward a synapse

primary active transport   A carrier-mediated transport system in which energy is directly required to operate the carrier and move the transported substance against its concentration gradient

primary motor cortex   The portion of the cerebral cortex that lies anterior to the central sulcus and is responsible for voluntary motor output

progestational phase   See *secretory phase*

prolactin (PRL) (prō-LAK-tun)   An anterior pituitary hormone that stimulates breast development and milk production in females

proliferative phase   The phase of the menstrual cycle during which the endometrium is repairing itself and thickening following menstruation; lasts from the end of the menstrual phase until ovulation

pro-opiomelanocortin (prō-op´Ē-ō-ma-LAN-oh-kor´-tin)   A large precursor molecule produced by the anterior pituitary that is cleaved into adrenocorticotropic hormone, melanocyte-stimulating hormone, and endorphin

proprioception (prō´-prē-ō-SEP-shun)   Awareness of position of body parts in relation to each other and to surroundings

prostaglandins (pros´-tuh-GLAN-dins)   Local chemical mediators that are derived from a component of the plasma membrane, arachidonic acid

prostate gland   A male accessory sex gland that secretes an alkaline fluid, which neutralizes acidic vaginal secretions

protein kinase (KĪ-nase)   An enzyme that phosphorylates and thereby induces a change in the shape and function of a particular intracellular protein

proteolytic enzymes (prōt´-ē-uh-LIT-ik)   Enzymes that digest protein

proximal tubule (PROKS-uh-mul)   A highly convoluted tubule that extends between Bowman's capsule and the loop of Henle in the kidney's nephron

PTH   See *parathyroid hormone*

pulmonary artery (PULL-mah-nair-ē)   The large vessel that carries blood from the right ventricle to the lungs

pulmonary circulation   The closed loop of blood vessels carrying blood between the heart and lungs

pulmonary surfactant (sur-FAK-tunt)   A phospholipoprotein complex secreted by the Type II alveolar cells that intersperses between the water molecules that line the alveoli, thereby lowering the surface tension within the lungs

pulmonary valve   A one-way valve that permits the flow of blood from the right ventricle into the pulmonary artery during ventricular emptying but prevents the backflow of blood from the pulmonary artery into the right ventricle during ventricular relaxation

pulmonary veins   The large vessels that carry blood from the lungs to the heart

pulmonary ventilation   The volume of air breathed in and out in one minute; equals tidal volume times respiratory rate

pupil   An adjustable round opening in the center of the iris through which light passes to the interior portions of the eye

Purkinje fibers (pur-KIN-jē)   Small terminal fibers that extend from the bundle of His and rapidly transmit an action potential throughout the ventricular myocardium

pyloric gland area (PGA) (pī-LŌR-ik)   The specialized region of the mucosa in the antrum of the stomach that secretes gastrin

pyloric sphincter (pī-lōr´-ik SFINGK-tur) The juncture between the stomach and duodenum

reabsorption The net movement of interstitial fluid into the capillary

receptor See *sensory receptor* or *receptor site*

receptor potential The graded potential change that occurs in a sensory receptor in response to a stimulus; generates action potentials in the afferent neuron fiber

receptor site Membrane protein that binds with a specific extracellular chemical messenger, thereby bringing about a series of membrane and intracellular events that alter the activity of the particular cell

reduced hemoglobin Hemoglobin that is not combined with O_2

reflex Any response that occurs automatically without conscious effort; the components of a reflex arc include a receptor, afferent pathway, integrating center, efferent pathway, and effector

refraction Bending of a light ray

refractory period (rē-FRAK-tuh-rē) The time period when a recently activated patch of membrane is refractory (unresponsive) to further stimulation, preventing the action potential from spreading backward into the area through which it has just passed, thereby ensuring the unidirectional propagation of the action potential away from the initial site of activation

regulatory proteins Troponin and tropomyosin, which play a role in regulating muscle contraction by either covering or exposing the sites of interaction between the contractile proteins

releasing hormone A hypothalamic hormone that stimulates the secretion of a particular anterior pituitary hormone

renal cortex An outer granular-appearing region of the kidney

renal medulla (RĒ-nul muh-DUL-uh) An inner striated-appearing region of the kidney

renal threshold The plasma concentration at which the T_m of a particular substance is reached and the substance first starts appearing in the urine .

renin (RĒ-nin) An enzymatic hormone released from the kidneys in response to a decrease in NaCl/ECF volume/arterial blood pressure; activates angiotensinogen

renin-angiotensin-aldosterone system (RAAS) (an´jē-ō-TEN-sun al-dō-steer-OWN) The salt-conserving system triggered by the release of renin from the kidneys, which activates angiotensin, which stimulates al-

dosterone secretion, which stimulates Na^+ reabsorption by the kidney tubules during the formation of urine

repolarization (rē´-pō-luh-ruh-ZĀ-shun) Return of membrane potential to resting potential following a depolarization

reproductive tract The system of ducts that are specialized to transport or house the gametes after they are produced

residual volume The minimum volume of air remaining in the lungs even after a maximal expiration

resistance Hindrance of flow of blood or air through a passageway (blood vessel or respiratory airway, respectively)

respiration The sum of processes that accomplish ongoing passive movement of O_2 from the atmosphere to the tissues, as well as the continual passive movement of metabolically produced CO_2 from the tissues to the atmosphere

respiratory acidosis (as-i-DŌ-sus) Acidosis resulting from abnormal retention of CO_2 arising from hypoventilation

respiratory airways The system of tubes that conducts air between the atmosphere and the alveoli of the lungs

respiratory alkalosis (al´-kuh-LŌ-sus) Alkalosis caused by excessive loss of CO_2 from the body as a result of hyperventilation

respiratory rate Breaths per minute

resting membrane potential The membrane potential that exists when an excitable cell is not displaying an electrical signal

reticular activating system (RAS) (ri-TIK-ū-lur) Ascending fibers that originate in the reticular formation and carry signals upward to arouse and activate the cerebral cortex

reticular formation A network of interconnected neurons that runs throughout the brain stem and initially receives and integrates all synaptic input to the brain

retina The innermost layer in the posterior region of the eye that contains the eye's photoreceptors, the rods and cones

ribonucleic acid (RNA) (rī-bō-new-KLĀ-ik) A nucleic acid that exists in three forms (messenger RNA, ribosomal RNA, and transfer RNA), which participate in gene transcription and protein synthesis

ribosomes (RĪ-bō-sōms) Special ribosomal RNA-protein complexes that synthesize proteins under the direction of nuclear DNA

right atrium (Ā-trē-um) The heart chamber that receives venous blood from the systemic circulation

right ventricle The heart chamber that pumps blood into the pulmonary circulation

RNA See *ribonucleic acid*

rods The eye's photoreceptors used for night vision

round window The membrane-covered opening that separates the lower chamber of the cochlea in the inner ear from the middle ear

salivary amylase (AM-uh-lās´) An enzyme produced by the salivary glands that begins carbohydrate digestion in the mouth and continues in the body of the stomach after the food and saliva have been swallowed

saltatory conduction (SAL-tuh-tōr´-ē) The means by which an action potential is propagated throughout a myelinated fiber, with the impulse jumping over the myelinated regions from one node of Ranvier to the next

SA node See *sinoatrial node*

sarcomere (SAR-kō-mir) The functional unit of skeletal muscle; the area between two Z lines within a myofibril

sarcoplasmic reticulum (ri-TIK-yuh-lum) A fine meshwork of interconnected tubules that surrounds a muscle fiber's myofibrils; contains expanded lateral sacs, which store calcium that is released into the cytosol in response to a local action potential

saturation When all the binding sites on a carrier molecule are occupied

secondary active transport A transport mechanism in which a carrier molecule for glucose or an amino acid is driven by a Na^+ concentration gradient established by the energy-dependent Na^+ pump to transfer the glucose or amino acid uphill without directly expending energy to operate the carrier

secondary sexual characteristics The many external characteristics that are not directly involved in reproduction but that distinguish males and females

second messenger An intracellular chemical that is activated by binding of an extracellular first messenger to a surface receptor site and that triggers a preprogrammed series of biochemical events, which result in altered activity of intracellular proteins to control a particular cellular activity

secretin (si-KRĒT-´n) A hormone released from the duodenal mucosa primarily in response to the presence of acid; inhibits gastric motility and secretion and stimulates secretion of a $NaHCO_3$ solution from the pancreas and from the liver

secretion Release to a cell's exterior, on appropriate stimulation, of substances that have been produced by the cell

secretory phase The phase of the menstrual cycle characterized by the development of a lush endometrial lining capable of supporting a fertilized ovum; also known as the *progestational phase*

secretory vesicles (VES-i-kuls)   Membrane-enclosed sacs containing proteins that have been synthesized and processed by the endoplasmic reticulum/Golgi complex of the cell and which will be released to the cell's exterior by exocytosis on appropriate stimulation

segmentation   The small intestine's primary method of motility; consists of oscillating, ringlike contractions of the circular smooth muscle along the small intestine's length

self-antigens   Antigens that are characteristic of a person's own cells

semen (SĒ-men)   A mixture of accessory sex-gland secretions and sperm

semicircular canal   Sense organ in the inner ear that detects rotational or angular acceleration or deceleration of the head

semilunar valves (sem´-ī-LEW-nur)   The aortic and pulmonary valves

seminal vesicles (VES-i-kuls)   Male accessory sex glands that supply fructose to ejaculated sperm and secrete prostaglandins

seminiferous tubules (sem´-uh-NIF-uh-rus)   The highly coiled tubules within the testes that produce spermatozoa

sensory afferent   Pathway coming into the central nervous system that carries information that reaches the level of consciousness

sensory input   Includes somatic sensation and special senses

sensory receptor   An afferent neuron's peripheral ending, which is specialized to respond to a particular stimulus in its environment

Sertoli cells (sur-TŌ-lē)   Cells located in the seminiferous tubules that support spermatozoa during their development

set point   The desired level at which homeostatic control mechanisms maintain a controlled variable

signal transduction   The sequence of events in which incoming signals (instructions from extracellular chemical messengers such as hormones) are conveyed to the cell's interior for execution

single-unit smooth muscle   The most abundant type of smooth muscle; made up of muscle fibers that are interconnected by gap junctions so that they become excited and contract as a unit; also known as *visceral smooth muscle*

sinoatrial (SA) node (sī-nō-Ā-trē-ul)   A small specialized autorhythmic region in the right atrial wall of the heart that has the fastest rate of spontaneous depolarizations and serves as the normal pacemaker of the heart

skeletal muscle   Striated muscle, which is attached to the skeleton and is responsible for movement of the bones in purposeful relation to one another; innervated by the somatic nervous system and under voluntary control

slow-wave potentials   Self-excitable activity of an excitable cell in which its membrane potential undergoes gradually alternating depolarizing and hyperpolarizing swings

smooth muscle   Involuntary muscle innervated by the autonomic nervous system and found in the walls of hollow organs and tubes

somatic cells (sō-MAT-ik)   Body cells, as contrasted with reproductive cells

somatic nervous system   The portion of the efferent division of the peripheral nervous system that innervates skeletal muscles; consists of the axonal fibers of the alpha motor neurons

somatic sensation   Sensory information arising from the body surface, including somesthetic sensation and proprioception

somatomedins (sō´-mat-uh-MĒ-dinz)   Hormones secreted by the liver or other tissues, in response to growth hormone, that act directly on the target cells to promote growth

somatosensory cortex   The region of the parietal lobe immediately behind the central sulcus; the site of initial processing of somesthetic and proprioceptive input

somesthetic sensations (SŌ-mes-THEH-tik)   Awareness of sensory input such as touch, pressure, temperature, and pain from the body's surface

sound waves   Traveling vibrations of air that consist of regions of high pressure caused by compression of air molecules alternating with regions of low pressure caused by rarefaction of the molecules

spatial summation   The summing of several postsynaptic potentials arising from the simultaneous activation of several excitatory (or several inhibitory) synapses

special senses   Vision, hearing, taste, and smell

specificity   Ability of carrier molecules to transport only specific substances across the plasma membrane

spermatogenesis (spur´-mat-uh-JEN-uh-sus)   Sperm production

sphincter (sfink-tur)   A voluntarily controlled ring of skeletal muscle that controls passage of contents through an opening into or out of a hollow organ or tube

spinal reflex   A reflex that is integrated by the spinal cord

spleen   A lymphoid tissue in the upper left part of the abdomen that stores lymphocytes and platelets and destroys old red blood cells

state of equilibrium   No net change in a system is occurring

stem cells   Relatively undifferentiated precursor cells that give rise to highly differentiated, specialized cells

steroids (STEER-oidz)   Hormones derived from cholesterol

stimulus   A detectable physical or chemical change in the environment of a sensory receptor

stress   The generalized, nonspecific response of the body to any factor that overwhelms, or threatens to overwhelm, the body's compensatory abilities to maintain homeostasis

stretch reflex   A monosynaptic reflex in which an afferent neuron originating at a stretch-detecting receptor in a skeletal muscle terminates directly on the efferent neuron supplying the same muscle to cause it to contract and counteract the stretch

stroke volume (SV)   The volume of blood pumped out of each ventricle with each contraction, or beat, of the heart

subcortical regions   The brain regions that lie under the cerebral cortex, including the basal nuclei, thalamus, and hypothalamus

submucosa   The connective tissue layer of the digestive tract that lies under the mucosa and contains the larger blood and lymph vessels and a nerve network

substance P   The neurotransmitter released from pain fibers

subsynaptic membrane (sub-sih-NAP-tik)   The portion of the postsynaptic cell membrane that lies immediately underneath a synapse and contains receptor sites for the synapse's neurotransmitter

suprachiasmatic nucleus (soup´-ra-kī-as-MAT-ik)   A cluster of nerve cell bodies in the hypothalamus that serves as the master biological clock, acting as the pacemaker that establishes many of the body's circadian rhythms

surface tension   The force at the liquid surface of an air–water interface resulting from the greater attraction of water molecules to the surrounding water molecules than to the air above the surface; a force that tends to decrease the area of a liquid surface and resists stretching of the surface

sympathetic nervous system   The subdivision of the autonomic nervous system that dominates in emergency ("fight-or-flight") or stressful situations and prepares the body for strenuous physical activity

synapse (SIN-aps´) The specialized junction between two neurons where an action potential in the presynaptic neuron influences the membrane potential of the postsynaptic neuron by means of the release of a chemical messenger that diffuses across the small cleft that separates the two neurons

synergism (SIN-er-jiz´-um) When several actions are complementary, so that their combined effect is greater than the sum of their separate effects

systemic circulation (sis-TEM-ik) The closed loop of blood vessels carrying blood between the heart and body systems

systole (SIS-tō-lē) The period of cardiac contraction and emptying

T₃ See *tri-iodothyronine*

T₄ See *thyroxine*

Tₘ See *transport maximum* and *tubular maximum*

tactile (TACK-til) Referring to touch

target cell receptors Receptors located on a target cell that are specific for a particular chemical mediator

target cells The cells that a particular extracellular chemical messenger, such as a hormone or a neurotransmitter, influences

temporal lobes The lobes of the cerebral cortex that are located laterally and that initially process auditory input

temporal summation The summing of several postsynaptic potentials occurring very close together in time because of successive firing of a single presynaptic neuron

tension The force produced during muscle contraction by shortening of the sarcomeres, resulting in stretching and tightening of the muscle's elastic connective tissue and tendon, which transmit the tension to the bone to which the muscle is attached

terminal button A motor neuron's enlarged knoblike ending that terminates near a skeletal muscle fiber and releases acetylcholine in response to an action potential in the neuron

testosterone (tes-TOS-tuh-rōn) The male sex hormone, secreted by the Leydig cells of the testes

tetanus (TET´-n-us) A smooth, maximal muscle contraction that occurs when the fiber is stimulated so rapidly that it does not have a chance to relax at all between stimuli

thalamus (THAL-uh-mus) The brain region that serves as a synaptic integrating center for preliminary processing of all sensory input on its way to the cerebral cortex

thick filaments Specialized cytoskeletal structures within skeletal muscle that are made up of myosin molecules and interact with the thin filaments to shorten the fiber during muscle contraction

thin filaments Specialized cytoskeletal structures within skeletal muscle that are made up of actin, tropomyosin, and troponin molecules and interact with the thick filaments to shorten the fiber during muscle contraction

thoracic cavity (thō-RAS-ik) Chest cavity

threshold potential The critical potential that must be reached before an action potential is initiated in an excitable cell

thrombus An abnormal clot attached to the inner lining of a blood vessel

thymus (THIGH-mus) A lymphoid gland located midline in the chest cavity that processes T lymphocytes and produces the hormone thymosin, which maintains the T cell lineage

thyroglobulin (thī´-rō-GLOB-yuh-lun) A large, complex molecule on which all steps of thyroid hormone synthesis and storage take place

thyroid gland A bilobed endocrine gland that lies over the trachea and secretes three hormones: thyroxine and tri-iodothyronine, which regulate overall basal metabolic rate, and calcitonin, which contributes to control of calcium balance

thyroid hormone Collectively, the hormones secreted by the thyroid follicular cells, namely, thyroxine and tri-iodothyronine

thyroid-stimulating hormone (TSH) An anterior pituitary hormone that stimulates secretion of thyroid hormone and promotes growth of the thyroid gland; thyrotropin

thyroxine (thī-ROCKS-in) The most abundant hormone secreted by the thyroid gland; important in the regulation of overall metabolic rate; also known as *tetraiodothyronine* or T_4

tidal volume The volume of air entering or leaving the lungs during a single breath

tight junction An impermeable junction between two adjacent epithelial cells formed by the sealing together of the cells' lateral edges near their luminal borders; prevents passage of substances between the cells

tissue (1) A functional aggregation of cells of a single specialized type, such as nerve cells forming nervous tissue; (2) the aggregate of various cellular and extracellular components that make up a particular organ, such as lung tissue

T lymphocytes (T cells) White blood cells that accomplish cell-mediated immune responses against targets to which they have

been previously exposed; see also *cytotoxic T cells* and *helper T cells*

tone The ongoing baseline of activity in a given system or structure, as in muscle tone, sympathetic tone, or vascular tone

total peripheral resistance The resistance offered by all the peripheral blood vessels, with arteriolar resistance contributing most extensively

trachea (TRĀ-kē-uh) The "windpipe"; the conducting airway that extends from the pharynx and branches into two bronchi, each entering a lung

tract A bundle of nerve fibers (axons of long interneurons) with a similar function within the spinal cord

transduction Conversion of stimuli into action potentials by sensory receptors

transepithelial transport (tranz-ep-i-THĒ-lē-al) The entire sequence of steps involved in the transfer of a substance across the epithelium between either the renal tubular lumen or digestive tract lumen and the blood

transmural pressure gradient The pressure difference across the lung wall (intra-alveolar pressure is greater than intrapleural pressure) that stretches the lungs to fill the thoracic cavity, which is larger than the unstretched lungs

transporter recruitment The phenomenon of inserting additional transporters (carriers) for a particular substance into the plasma membrane, thereby increasing membrane permeability to the substance, in response to an appropriate stimulus

transport maximum (T_m) The maximum rate of a substance's carrier-mediated transport across the membrane when the carrier is saturated; known as *tubular maximum* in the kidney tubules

transverse tubule (T tubule) A perpendicular infolding of the surface membrane of a muscle fiber; rapidly spreads surface electric activity into the central portions of the muscle fiber

triglycerides (trī-GLIS-uh-rīdz) Neutral fats composed of one glycerol molecule with three fatty acid molecules attached

tri-iodothyronine (T_3) (trī-ī-ō-dō-THĪ-rō-nēn) The most potent hormone secreted by the thyroid follicular cells; important in the regulation of overall metabolic rate

trophoblast (TRŌF-uh-blast´) The outer layer of cells in a blastocyst that is responsible for accomplishing implantation and developing the fetal portion of the placenta

tropic hormone (TRŌ-pik) A hormone that regulates the secretion of another hormone

tropomyosin (trōp´-uh-MĪ-uh-sun) One of the regulatory proteins in the thin filaments of muscle fibers

troponin (tro-PŌ-nun) One of the regulatory proteins found in the thin filaments of muscle fibers

TSH See *thyroid-stimulating hormone*

T tubule See *transverse tubule*

tubular maximum (T_m) The maximum amount of a substance that the renal tubular cells can actively transport within a given time period; the kidney cells' equivalent of transport maximum

tubular reabsorption The selective transfer of substances from tubular fluid into peritubular capillaries during urine formation

tubular secretion The selective transfer of substances from peritubular capillaries into the tubular lumen during urine formation

twitch A brief, weak contraction that occurs in response to a single action potential in a muscle fiber

twitch summation The addition of two or more muscle twitches as a result of rapidly repetitive stimulation, resulting in greater tension in the fiber than that produced by a single action potential

tympanic membrane (tim-PAN-ik) The eardrum, which is stretched across the entrance to the middle ear and which vibrates when struck by sound waves funneled down the external ear canal

Type I alveolar cells (al-VĒ-ō-lur) The single layer of flattened epithelial cells that forms the wall of the alveoli within the lungs

Type II alveolar cells The cells within the alveolar walls that secrete pulmonary surfactant

ultrafiltration The net movement of a protein-free plasma out of the capillary into the surrounding interstitial fluid

ureter (yū-RĒ-tur) A duct that transmits urine from the kidney to the bladder

urethra (yū-RĒ-thruh) A tube that carries urine from the bladder to outside the body

urine excretion The elimination of substances from the body in the urine; anything filtered or secreted and not reabsorbed is excreted

vagus nerve (VĀ-gus) The tenth cranial nerve, which serves as the major parasympathetic nerve

vasoconstriction (vā´-zō-kun-STRIK-shun) The narrowing of a blood vessel lumen as a result of contraction of the vascular circular smooth muscle

vasodilation The enlargement of a blood vessel lumen as a result of relaxation of the vascular circular smooth muscle

vasopressin (vā-zō-PRES-sin) A hormone secreted by the hypothalamus, then stored and released from the posterior pituitary; increases the permeability of the distal and collecting tubules of the kidneys to water and promotes arteriolar vasoconstriction; also known as *antidiuretic hormone (ADH)*

vaults Recently discovered organelles shaped like octagonal barrels; believed to serve as transporters for messenger RNA and/or the ribosomal subunits from the nucleus to sites of protein synthesis

vein A vessel that carries blood toward the heart

vena cava (VĒ-nah CĀV-ah), (venae cavae, plural; VĒ-nē cāv-ē) A large vein that empties blood into the right atrium

venous return (VĒ-nus) The volume of blood returned to each atrium per minute from the veins

ventilation The mechanical act of moving air in and out of the lungs; breathing

ventricle (VEN-tri-kul) A lower chamber of the heart that pumps blood into the arteries

vesicle (VES-i-kul) A small, intracellular, fluid-filled, membrane-enclosed sac

vesicular transport Movement of large molecules or multimolecular materials into or out of the cell by means of being enclosed in a vesicle, as in endocytosis or exocytosis

vestibular apparatus (veh-STIB-yuh-lur) The component of the inner ear that provides information essential for the sense of equilibrium and for coordinating head movements with eye and postural movements; consists of the semicircular canals, utricle, and saccule

villus (**villi**, plural) (VIL-us) Microscopic fingerlike projections from the inner surface of the small intestine

virulence (VIR-you-lentz) The disease-producing power of a pathogen

visceral afferent A pathway coming into the central nervous system that carries subconscious information derived from the internal viscera

visceral smooth muscle (VIS-uh-rul) See *single-unit smooth muscle*

viscosity (vis-KOS-i-tē) The friction developed between molecules of a fluid as they slide over each other during flow of the fluid; the greater the viscosity, the greater the resistance to flow

vital capacity The maximum volume of air that can be moved out during a single breath following a maximal inspiration

voltage-gated channels Channels in the plasma membrane that open or close in response to changes in membrane potential

white matter The portion of the central nervous system composed of myelinated nerve fibers

Z line A flattened disclike cytoskeletal protein that connects the thin filaments of two adjoining sarcomeres

Credits

This page constitutes an extension of the copyright page. We have made every effort to trace the ownership of all copyrighted material and to secure permission from copyright holders. In the event of any question arising as to the use of any material, we will be pleased to make the necessary corrections in future printings. Thanks are due to the following authors, publishers, and agents for permission to use the material indicated.

Illustrations

Chapter 1. 6 Adapted from Cecie Starr and Ralph Taggart, *Biology: The Unity and Diversity of Life,* Eighth Edition, Fig. 33.11, p. 552–553. Copyright © 1998 Wadsworth Publishing Company.

Chapter 3. 44 Adapted from Cecie Starr and Ralph Taggart, *Biology: The Unity and Diversity of Life,* Eighth Edition, Fig. 4-2c, p. 56. Copyright © 1998 Wadsworth Publishing Company.

Chapter 4. 101 Adapted with permission from George A. Hedge, Howard D. Colby, and Robert L. Goodman, *Clinical Endocrine Physiology* (Philadelphia: W. B. Saunders Company, 1987), Figure 1-9, p. 20.

Chapter 6. 173 Adapted from Cecie Starr and Ralph Taggart, *Biology: The Unity and Diversity of Life,* Eighth Edition, Fig. 36.10b, p. 595. Copyright © 1998 Wadsworth Publishing Company.

Chapter 10. 302 Adapted from *Physiology of the Heart and Circulation,* Fourth Edition, by R. C. Little and W. C. Little. Copyright © 1989 Year Book Medical Publishers, Inc. Reprinted by permission of the author and Mosby-Yearbook, Inc.

Chapter 11. 335, 348 © Dana Burns-Pizer. Used with permission.

Chapter 13. 407 Adapted from Ann Atalheim-Smith and Greg K. Fitch, *Understanding Human Anatomy and Physiology,* Fig 23.4, p. 888. Copyright © 1993 West Publishing Company.

Chapter 17. 531 Adapted from *Clinical Endocrine Physiology,* by George A. Hedge, Howard D. Colby, and Robert L. Goodman, Figure 1-13, p. 28, ©1987, with permission from Elsevier.

Chapter 18. 618 Adapted from Cecie Starr, *Biology: Concepts and Applications,* Fourth Edition, Fig 38.25, p. 655. Copyright © 2000 Brooks/Cole.

Photos

Chapter 2. 22 Fig. 2-2: (a–b, bottom) Don W. Fawcett/Visuals Unlimited. 23 Fig. 2-4: David M. Phillips/Visuals Unlimited. 24 Fig. 2-5: (b) Dr. Birgit Satir, Albert Einstein College of Medicine. 24 Fig. 2-6: (right) Don W. Fawcett/Photo Researchers, Inc. 25 Fig. 2-7: (a) Don W. Fawcett/Photo Researchers, Inc. 26 Fig. 2-8: (b) Bill Longcore/Photo Researchers, Inc. 33 Fig. 2-14: (c) Dr. Leonard H. Rome/UCLA School of Medicine. 34 Fig. 2-15: (a, b) Elizabeth R. Walker, Ph.D., Department of Anatomy, School of Medicine, West Virginia University and Dennis O. Overman, Ph.D., Department of Anatomy, School of Medicine, West Virginia. 36 Fig. 2-17: PIR-CNRI/Science Photo Library/Photo Researchers, Inc. 36 Fig. 2-18: M. Abbey/Visuals Unlimited. 37 Fig. 2-19: Reproduced from: R. G. Kessel and R. H. Kardon, *Tissues and Organs: A Text Atlas of Scanning Electron Microscopy,* W. H. Freeman, 1979, all rights reserved.

Chapter 3. 44 Fig. 3-1: Don W. Fawcett/Visuals Unlimited.

Chapter 4. 79 Fig. 4-9: (b) David M. Phillips/Visuals Unlimited. 84 Fig. 4-13: (c) C. Raines/Visuals Unlimited. 86 Fig. 4-14: (b) E. R. Lewis, T. E. Everhart, and Y. Y. Zevi, University of California/Visuals Unlimited.

Chapter 5. 112 Fig. 5-4: Nancy Kedersha, Ph.D., Research Scientist-Cell Biology, Harvard Medical School. 115 Fig. 5-5: (a, b) © Mark Nielsen. 118 Fig. 5-7: (b) Courtesy Washington University School of Medicine, St. Louis. 124 Fig. 5-12: (a, b) © Mark Nielsen.

Chapter 6. 154 Fig. 6-11: (b) Bill Beatty/Visuals Unlimited. 156 Fig. 6-15: (b) Patricia N. Farnsworth, Ph.D. Professor of Physiology and Opthamology, University of Medicine and Dentistry of New Jersey, New Jersey Medical School. 159 Fig. 6-18: A. L. Blum/Visuals Unlimited. 172 Fig. 6-32: (a, b) R. S. Preston and J. E. Hawkins, Kresge Hearing Institute, University of Michigan. 173 Fig. 6-33: (c) Dean Hillman, Ph.D., Professor of Otolaryngology, Physiology and Neuroscience: New York University Medical School.

Chapter 7. 195 Fig. 7-5: Eric Grave/Photo Researchers, Inc.

Chapter 8. 206 Fig. 8-3: (a) Reprinted with permission from Sydney Schochet Jr., M.D., Professor, Department of Pathology, School of Medicine, West Virginia University: *Diagnostic Pathology of Skeletal Muscle and Nerve* (Stamford, Connecticut: Appleton & Lange, 1986, Figure 1-13); (b) M. Abbey/Science Source/Photo Researchers, Inc. 230 Fig. 8-21: (a) Dr. Brian Eyde/Science Source/Photo Researchers, Inc.; (b) Dr. Brenda Russell, Professor of Physiology, University of Illinois.

Chapter 9. 246 Fig. 9-5: John Cunningham/Visuals Unlimited. 266 Fig. 9-25: (b) Sloop-Ober/Visuals Unlimited. 267 Fig. 9-26: (c) Copyright Boehringer

Ingelheim International GmbH, photo Lennart Nilsson/Bonnier Alba AB.

Chapter 10. **284** Fig. 10-9: (a) Fawcett-Uehara-Suyama/ Science Source/Photo Researchers, Inc. **291** Fig. 10-13: From *Behold Man* (Boston, Little, Brown and Company, 1974: 63). Photo Lennart Nilsson/Bonnier Alba AB. **298** Fig. 10-22: Fred Marsik/ Visuals Unlimited.

Chapter 11. **317** Fig. 11-2: David M. Phillips/ Visuals Unlimited. **320** Fig. 11-5: Stanley Fletcher/Visuals Unlimited. **321** Fig. 11-6: Carolina Biological/Visuals Unlimited. **323** Fig. 11-8: Copyright Boehringer Ingelheim International GmbH, photo Lennart Nilsson/Bonnier Alba AB. **327** Fig. 11-11: (from left to right) © Cabisco/Visuals Unlimited; © Science VU/Visuals Unlimited; © Stanley Fletcher/Visuals Unlimited; © Stanley Fletcher/Visuals Unlimited; From *A Color Atlas and Instruction Manual of Peripheral Blood Cell Morphology* by Barbara O'Connor (Baltimore, M.D., Williams & Wilkins Co., 1984). Reprinted with permission of the author and publisher; © Biophoto Associates/Photo Researchers, Inc.; From Barbara O'Connor, *A Color Atlas and Instruction Manual of Peripheral Blood Cell Morphology* (Baltimore, MD: Williams & Wilkins Co., 1984). Reprinted with permission of the author and publisher. **337** Fig. 11-20:

(a, b) Contributed by Dr. Dorothea Zucker-Franklin, New York University Medical Center. **347** Fig. 11-27: Copyright Boehringer Ingelheim International GmbH, photo Lennart Nilsson/Bonnier Alba AB. **352** Fig. 11-30: Copyright Boehringer Ingelheim International GmbH, photo Lennart Nilsson/ Bonnier Alba AB.

Chapter 12. **368** (b) Don W. Fawcett/Visuals Unlimited. **380** Fig. 12-16: (a, b) SIU/Visuals Unlimited. **386** Fig. 12-21: (a) Michael Gabridge/Visuals Unlimited; (b) © M. Moore/ Visuals Unlimited. **394** F. Westmorland/ Photo Researchers, Inc. **395** Dave B. Fleetham/Visuals Unlimited.

Chapter 13. **409** Fig. 13-3: Richard G. Kessel/ Visuals Unlimited. **411** Fig. 13-6: F. Spinelli, Don W. Fawcett/Visuals Unlimited.

Chapter 15. **485** A. B. Dowsett/Science Photo Library/Photo Researchers, Inc. **497** Fig. 15-21: Michael C. Webb/Visuals Unlimited. **498** Fig. 15-22: (a, b) Thomas W. Sheehy, M.D.; Robert L. Slaughter, M.D.: *The Malabsorption Syndrome* by Medcom, Inc. Reproduced by permission of Medcom, Inc.

Chapter 17. **546** Fig. 17-11: Science Photo Library/Photo Researchers, Inc. **546** Fig. 17-12: (all) © Dean Warchak @ www .acromegalysupport.com. **548** Fig. 17-13: (b) Elizabeth R. Walker, Ph.D., Associate Professor, Department of Anatomy, School

of Medicine, West Virginia University, and Dennis O. Overman, Ph.D. Associate Professor, Department of Anatomy, School of Medicine, West Virginia University. **552** Fig. 17-17, Fig. 17-18: © Lester V. Bergmann and Associates, Inc. **557** Fig. 17-21: (a, b) From Zitelli, et al.; *ATLAS of PEDIATRIC PHYSICAL DIAGNOSIS,* 4/ed, © 2002 Mosby. **574** Fig. 17-28: (both) D. P. Motta, Department of Anatomy, University of "La Sapienza" Rome/SPL/Photo Researchers, Inc.

Chapter 18. **590** Fig. 18-4: (b) Michael C. Webb/Visuals Unlimited; (c) Secchi-Lecaque/ CNR/Science Photo Library/Photo Researchers, Inc. **593** Fig. 18-6: (a) David M. Phillips/Visuals Unlimited. **605** Fig. 18-11: Dr. P. Bagavandoss, Department of Biological Sciences, Kent State University/Photo Researchers, Inc. **612** Fig. 18-17: © David Scharf. **612** Fig. 18-18: (b) Photo Lennart Nilsson/Bonnier Alba AB, *A Child is Born.* Copyright © 1966, 1977 Dell Publishing Company, Inc. **614** Fig. 18-20: Photo Lennart Nilsson/Bonnier Alba AB, *A Child is Born.* Copyright © 1966, 1977 Dell Publishing Company, Inc.

Appendix C. **A-30** Fig. C-11: From Christine J. Harrison et al.: *Cytogenics and Cell Genetics,* 35:21–27 (1983), Figure 3B. Reprinted with permission from Dr. Christine J. Harrison and S. Karger AG, Basel.

Index

psychoactive drugs, 127
skin, absorption through, 356
synaptic transmission and, 91–92
vomiting and, 479
Duct cells, 4, 486
Ductus (vas) deferens, 584, 594
functions of, 595, 596
location of, 594, 596
vasectomy, 616
Duodenum
absorption in, 495
bile entering, 489
in digestive tract, 467
emptying of stomach and, 477, 478–479
hormones secreted by, 478
segmentation in, 494
summary of hormones, 535
Dwarfism, 546
cretinism and, 551
Laron dwarfism, 100
Dynorphin, 150
Dyslexia, 122
Dysmenorrhea, 609
Dyspnea, 393, 399–400, 400
Dystonias, 198

Ear canal, 167, 176
Ears, 165–175. *See also* Sound
anatomy of, 165
cochlea, 169–172
deafness, 172
equilibrium and, 165, 172–175
external ear, 167–168
functions of major components of, 176
hearing, 165–172
inner ear, 169–175
middle ear, 168–169
otolith organs, 172, 173–175
semicircular canals, 172–173
temperature, eardrum, 518
transmission of sound waves, 170
tympanic membrane, 167–169
vestibular apparatus, 172–175
Earwax, 167
Eating. *See* Foods
Eccentric contractions, 218
ECF (extracellular fluid). *See also* Balance
concept; Fluid balance; Kidneys;
Membrane potential
barrier between ICF (intracellular fluid)
and, 445
bulk flow and distribution of, 295
calcium (Ca^{2+}) in, 573
in cardiac muscle, 235
hypertonicity, 449–550
hypotonicity, 450
internal pool and, 443–444
as isotonic, 55
osmolarity, regulation of, 448–453
smooth muscle contraction and, 230–231
sodium (Na^+) and, 416–417, 448–449
vesicular transport, 59–61
volume, regulation of, 446–448

Ectopic focus, 249
Edema, 297–298
inflammation and, 331
pulmonary edema, 386
Effectors, 13
muscarinic receptors and, 191
organs, 110
reflex arc, 138
Efferent arterioles, 407
Efferent axons, 111
Efferent division of peripheral nervous
system (PNS), 184–201
Efferent fibers, 111
in spinal nerves, 136–137
Efferent nervous system, 110
Efferent neurons, 110–111, 138
comparison of types, 194
Eggs. *See* Ova
Eicosanoids, 596
Ejaculation, 594, 596, 599
sperm content of ejaculate, 599
spinal reflexes and, 139
volume of ejaculate, 599
Elasticity of lungs, 377
Elastic recoil of lungs, 377
Elastin, 5, 48
in arteries, 280
in dermis, 356
in pulmonary connective tissue, 377
Electrical defibrillation, 249
Electrical gradient, 52
Electrocardiogram (ECG), 252–253
diagnostic use of, 254–255
leads for, 253
specific events, recording, 253–254
waveforms of, 253–254
Electrocerebral silence, 123
Electrochemical gradients, 52
Electroencephalogram (EEG), 123
sleep and, 134
Electrolytes, A-9
urine, excretion in, 414
Electromagnetic spectrum, 153
Electromagnetic waves, 153, 519
Electron cloud, A-3
Electron microscopes, 20
Electron transport chain, 29
Electrons, A-3
covalent bonds, A-6–A-7
shells, A-4
Elements, A-3
in balanced equations, A-8
characteristics of, A-4
Elephantiasis, 298
Elimination. *See also* Kidneys; Waste
products
cells performing, 3
homeostasis and, 8
of hormones, 532
Emboli, 267, 325–326
Embryo, 584
Embryonic stem cells, 6
Emergency contraception, 617

Emetics, 479, 481
Emission phase of ejaculation, 599
Emmetropia, 157–158
Emotions
constipation and, 503
diarrhea and, 502
gastric motility and, 479
hypothalamus and, 125, 542
limbic system and, 126
menopause and, 610
neurotransmitters and, 127
obesity and, 516
pain receptors and, 149
peptic ulcers and, 485
pheromones and, 180
vomiting and, 479
Emphysema
airway obstruction and, 376–377
gas exchange and, 52, 386
Enamel of teeth, 472
End-diastolic volume (EDV), 255
Endocrine disrupters, 600
Endocrine pancreas, 486, 528, 564–572
autonomic nervous system, effects of, 189
schematic representation of, 487
summary of hormones, 534
Endocrine system, 4, 5, 10, 528. *See also*
Hormones; specific endocrine glands;
specific hormones
anatomy of, 530
autonomic nervous system, effects of,
189
disorders of, 532
digestive mucosa, endocrine gland cells
in, 467
functional interaction with nervous
system, 103–104
homeostasis and, 12
negative feedback in, 540
nervous system compared, 102–104
neuroendocrine reflexes, 531
obesity and, 516
placenta as endocrine organ, 615–619
principles of, 529–535
prostaglandins, actions of, 597
specificity of action, 103
tropic hormones, 530
as wireless, 103
Endocytosis, 59–60
forms of, 25
Endocytotic vesicles, 35
Endogenous opiates, 150
anorexia nervosa and, 517
Endogenous pyrogen (EP), 333
in fevers, 523–524
Endometrium, 609
blastocyst, implantation of, 613–615
Endoplasmic reticulum (ER), 20, 21–22
cystic fibrosis and, 47
lumen, 21–22
secretion process for, 23
Endorphins, 150. *See also* β-endorphins
acupuncture endorphin hypothesis, 151

pain and, 150
space flight and, 224
Exocrine glands, 3–4
  autonomic nervous system, effects of, 189
  mucosa, cells in, 467
  of skin, 356
  stomach, cells in, 5
Exocrine pancreas, 467
  aqueous alkaline solution, 486, 487
  autonomic nervous system, effects of, 189
  cholecystokinin (CCK) affecting, 488
  digestive enzymes from, 486–487
  insufficiency, 487
  schematic representation of, 487
  secretin affecting, 488
Exocytosis, 24, 59, 60–61
Exophthalmos, 552
Expiration, 373, 374
  chronic obstructive pulmonary disease (COPD) and, 377
  forced expiration, 373, 375
Expiratory muscles, 373
Expiratory neurons, 396
Expiratory reserve volume (ERV), 380
Expression, 121
Expulsion phase of ejaculation, 599
Extension of muscles, 218
External anal sphincter, 503
External auditory meatus, 167, 176
External ear, 165, 167–168, 176
External eye muscles, 165
External genitalia, 583
External intercostal muscles, 372
External respiration, 365–366
External urethral sphincter, 436
External work, 511–512
Extracellular chemical messengers, 93. *See also* Hormones; Neurotransmitters; Paracrines
  chemically-gated channels, opening, 94
  neurohormones, 93, 94
  second-messenger pathways, activation of, 94–95
  signal transduction and, 94
Extracellular fluid (ECF), 8, 20, 44
Extracellular matrix (ECM), 48
Extrafusal fibers, 225
Extrinsic control, 12
  of stroke volume, 262–263
Extrinsic nerves of digestive tract, 470–471
Extrinsic pathway for blood clotting, 324
Eyelashes, 151
Eyelids, 150–151
Eyes, 150–165. *See also* Photoreceptors
  accommodation by lens, 155–158
  amblyopia, 109
  autonomic nervous system, effects of, 189
  diabetes mellitus and, 569
  disorders of, 156–158
  equilibrium and, 174–175
  exophthalmos, 552
  in fight-or-flight response, 190
  functions of components of, 164

Graves' disease and, 552
immune privilege of, 350
protective mechanisms for, 150–151
refraction of light in, 153–155
sensitivity of, 161–162
structure of, 151–152
thalamus and, 163–165
vestibular nuclei and, 174–175
vestibulocerebellum and, 130
visual field, 162–163

Facilitated diffusion, 57
Factor X, 323–324
Factor XII, 324, 325
FAD (flavine adenine dinucleotide), 29
$FADH_2$ in electron transport chain, 29
Fainting, 300
Fallopian tubes, 584
Farsightedness, 157–158
Fasting, 564
  glucagon in, 571
Fast muscle fibers, 220–221
Fast pain pathway, 150
Fatigue, 214
  glycolysis and, 220
  hypothyroidism and, 551
Fats, 466, 496, A-12–A-14. *See also* Fatty acids; Triglycerides
  bile salts, role in digestion of, 491, 499
  breakdown, 561
  catabolism of, 562
  central lacteal picking up, 501
  cortisol and degradation of, 554
  cytosol and, 33
  diabetes mellitus and, 567
  diagram of digestion of, 500
  digestion by lipase, 487
  emptying of stomach, effect on, 478
  energy and, 31
  food intake and, 514
  glucagon and, 571
  insulin and, 565–566
  lymphatic system and, 297
  pancreatic lipase for digesting, 487
  in plasma membrane, 44
  prostaglandins, actions of, 597
  simple lipids, A-13
  small intestine, absorption by, 499–501
  storage of, 563
  synthesis, 561
Fatty acids, 466, A-13
  as absorbable units of fat digestion, 466, 487
  glucagon and, 571
  glucose transformed to, 562
  growth hormone affecting, 543, 545
  insulin and, 565–566
  triglycerides and, 466, 487, 563
Fatty streak, 266
FDA (Food and Drug Administration)
  Botox, 198
  DHEA (dehydroepiandrosterone) sales, 556

growth hormone treatments, 547
on melatonin, 537
Feces. *See also* Diarrhea
  bilirubin and, 492
  constipation, 503
  elimination of, 503
  mass movement of, 502–503
  salt (NaCl) absorption and, 504
  steatorrhea, 487
  water and, 451, 504
Feedback, 12–13. *See also* Negative feedback; Positive feedback
Feedforward, 12–13, 14–15
  insulin secretion as, 566
Feeding signals, 514
Female pseudohermaphroditism, 557
Female reproductive system, 583, 601–626. *See also* Pregnancy
  anatomy of, 586
  cycling in, 601
  gametogenesis in, 601–602
  hormonal control of, 606–609
  menopause, 610–611
  menstrual cycle, 609–610
  oogenesis, 601–602
  ovarian cycle, 602–609
  overview of, 584
  ovulation, 605
  prostaglandins, role in, 596, 597
  puberty and, 610
  sexual intercourse and, 599–601
Feminization, 588
Fertility
  ejaculate, volume and sperm content of, 599
  period, fertile, 611
Fertilization, 584
  process of, 612
  sex chromosomes and, 587
Fetus, 584
  photograph of, 614
Fever, 523–524
  interleukin 1 and, 345
Fibrillation, 249
Fibrin, 323
Fibrinogen, 323
  in semen, 595
Fibrinolysin, 595
Fibroblasts, 48, 267, 325
  tissue scar tissue and, 333
Fibronectin, 48
Fick's law of diffusion, 52
  capillaries and, 289
  gas exchange and, 386
  lungs and, 367–368
Fidget factor, 516
Fight-or-flight response, 189–190, 559
Filariasis, 298
Filtered load, 419
Filtration fraction, 425, 427
Filtration slits, 411
Final common pathway, 192
Firing of action potential, 76

glomerular filtration rate (GFR) and, 425–427
of inulin, 425
of para-aminohippuric acid (PAH), 425
for reabsorbed substances, 425
for secreted substances, 425
Plasma-colloid osmotic pressure, 294, 412
Plasma membrane, 2, 18, 20, 35, 43. *See also* Membrane potential; Membrane transport
as barrier between ECF and ICF, 445
cAMP (cyclic adenosine monophosphate) and, 99
carbohydrates in, 46–47
cell-to-cell adhesions and, 47–50
cholesterol in, 45, 268
cystic fibrosis and, 47
endocytosis and, 25–26, 59–60
exocytosis and, 24, 60–61
fluid mosaic model of, 45
lipid bilayer of, 44–46
phospholipids in, 44–46
proteins in, 46
as trilaminar structure, 43–44
Plasma proteins, 316
in complement system, 334–335
functions of, 316–317
gamma globulins, 317
kinins, 333
plasmin, 325
renal failure and loss of, 434
Plasmin, 325
Plasminogen, 325
Plasticity of brain, 121
Plateau phase of action potential in cardiac contractile cells, 250
Plateau phase of sexual response, 250–251
in females, 599, 601
in males, 597
Plateau portion of O_2-Hb saturation curve, 388
Platelet-activating factor (PAF), 265
Platelet plug, 322–323
Platelets, 314, 316, 321–326. *See also* Blood clotting; Hemostasis
formation of plug, 322
hemophilia and, 326
leukemia and, 329
megakaryocytes producing, 321
nitric oxide (NO) affecting, 287
number of, 328
red bone marrow and, 318
Pleural cavity, 369
Pleural sac, 369
intrapleural pressure, 369–370
pneumothorax, 371
Pleurisy, 369
Pluripotent stem cells, 318
PMS (premenstrual syndrome), 596
Pneumonia
Fick's law of diffusion and, 52
gas exchange and, 386
Pneumotaxic center, 395, 396

Pneumothorax, 371
Podocytes, 411
Point (gene) mutations, A-30
Poiseuille's law, 277–278
Poison ivy, 355
Poisons. *See* Toxins
Polarization, 72
Polar molecules, A-7
Poliovirus, 192
Pollution, estrogenic, 600
Polycythemia, 320–321
Polymers, A-10
Polymodal nociceptors, 149
Polymorphonuclear granulocytes, 327, 328
Polynucleotide strand, A-19–A-20
Polypeptides, A-19
Polyphagia, 567
Polyribosomes, A-26
Polysaccharides, A-12
digestion of, 466
Polysynaptic reflexes, 139
Pons, 115, 131–133
Positive balance, 444
Positive feedback, 13–14
parturition and, 622–623
Postabsorptive state, 564, 566, 570–571
Posterior parietal cortex, 118, 121
Posterior pituitary gland, 528, 537
summary of hormones, 533
vasopressin and, 432, 537–538
Postganglionic fiber, 186, 192
single-unit smooth muscle and, 233
Posthepatic jaundice, 492
Postsynaptic neurons, 85
excitatory synapses, 86–87
grand postsynaptic potential, 88–89
inhibitory synapses, 87
integration, importance of, 90
spatial summation, 89, 90
synaptic delay, 87–88
temporal summation, 89–90
Postsynaptic potentials, 75
Potassium (K^+). *See also* Na^+–K^+ APTase carriers; Na^+–K^+ pump; Voltage-gated channels
astrocytes and, 112
basolateral pump, 422–423
blood–brain barrier (BBB) and, 114
channels and, 46
end-plate potential (EPP), 195
equilibrium potential for, 63–64
excitatory synapses and, 86–87
homeostasis and, 9
hyperkalemia, 557
inhibitory synapses and, 87
K^+ equilibrium potential, 63–64
kidneys regulating, 406
membrane potential and, 62
pH fluctuations and, 455
retention of, 434
solubility of, 50
tubules secreting, 422–423

Potassium (K^+) channels
action potentials and, 77–78
in cardiac cells, 250–251
Pre-Bötzinger complex, 396
Precipitation, 338, 339
Precocious pseudopuberty, 557
Prefrontal association cortex, 118, 122
autonomic output and, 191
working memory and, 130
Preganglionic fiber, 186, 192
Pregnancy. *See also* Parturition; Placenta
antigens and diagnosis of, 338
blastocyst, implantation of, 613–615
contraception, 615, 616–617
corpus luteum of, 605, 617
edema in, 298
estrogen, role of, 619, 624
fertilization, 611–613
gestation period, 619
human chorionic gonadotropin (hCG), 616–618
inner cell mass, 614
lactation and, 623
maternal body in, 619
progesterone, actions of, 619, 624
Rh factor and, 343
Prehepatic jaundice, 492
Prehypertension guidelines, 308
Premature ventricular contraction (PVC), 249
Prematurity
newborn respiratory distress syndrome, 378
testes and, 588
Premotor cortex, 118, 121
Preovulatory follicle, 605
Preprohormones, 96
Presbyopia, 156
Pressure gradients
and air flow, 369–370
and blood flow, 276–277
partial pressure gradients, 383–384
Pressure reservoir, 280, 281
Presynaptic neurons, 85
PRH (prolactin-releasing hormone), 540
Primary auditory cortex, 118, 169
Primary follicles, 601–602
Primary hypertension, 306
Primary immune response, 341
Primary motor cortex, 118, 120
motor neuron output and, 223, 224
Primary olfactory cortex, 179
Primary oocytes, 601–602
Primary peristaltic wave, 475
Primary polycythemia, 321
Primary spermatocytes, 591
Primary visual cortex, 118
PRL. *See* Prolactin
Procarboxypeptidase, 486–487
Procedural memories, 130
cerebrocerebellum and, 131
Products, A-7
Progestational phase of menstrual cycle, 610

ANATOMICAL TERMS USED TO INDICATE DIRECTION AND ORIENTATION

anterior	situated in front of or in the front part of
posterior	situated behind or toward the rear
ventral	toward the belly or front surface of the body; synonymous with anterior
dorsal	toward the back surface of the body, synonymous with posterior
medial	denoting a position nearer the midline of the body or a body structure
lateral	denoting a position toward the side or farther from the midline of the body or a body structure
superior	toward the head
inferior	away from the head
proximal	closer to a reference point
distal	farther from a reference point
sagittal section	a vertical plane that divides the body or a body structure into right and left sides
longitudinal section	a plane that lies parallel to the length of the body or a body structure
cross section	a plane that runs perpendicular to the length of the body or a body structure
frontal or coronal section	a plane parallel to and facing the front part of the body

WORD DERIVATIVES COMMONLY USED IN PHYSIOLOGY

a; an-	absence or lack	*epi-*	above; over	*oto-*	ear
ad-; af-	toward	*erythro-*	red	*para-*	near
adeno-	glandular	*gastr-*	stomach	*pariet-*	wall
angi-	vessel	*-gen; -genic*	produce	*peri-*	around
anti-	against	*gluc-; glyc-*	sweet	*phago-*	eat
archi-	old	*hemo-*	blood	*pod*	footlike
-ase	splitter	*hemi-*	half	*-poiesis*	formation
auto-	self	*hepat-*	liver	*poly-*	many
bi-	two; double	*homeo-*	sameness	*post-*	behind; after
-blast	former	*hyper-*	above; excess	*pre-*	ahead of; before
brady-	slow	*hypo-*	below; deficient	*pro-*	before
cardi-	heart	*inter-*	between	*pseudo-*	false
cephal-	head	*intra-*	within	*pulmon-*	lung
cerebr-	brain	*kal-*	potassium	*rect-*	straight
-cide	kill; destroy	*leuko-*	white	*ren-*	kidney
chondr-	cartilage	*lip-*	fat	*reticul-*	network
contra-	against	*macro-*	large	*retro-*	backward
cost-	rib	*mamm-*	breast	*sacchar-*	sugar
crani-	skull	*mening-*	membrane	*sarc-*	muscle
-crine	secretion	*micro-*	small	*semi-*	half
crypt-	hidden	*mono-*	single	*-some*	body
cutan-	skin	*multi-*	many	*sub-*	under
-cyte	cell	*myo-*	muscle	*supra-*	upon; above
de-	lack of	*natr-*	sodium	*tachy-*	rapid
di-	two; double	*neo-*	new	*therm-*	temperature
dys-	difficult; faulty	*nephr-*	kidney	*-tion*	act or process of
ef-	away from	*neuro-*	nerve	*trans-*	across
-elle	tiny; miniature	*oculo-*	eye	*tri-*	three
encephalo-	brain	*-oid*	resembling	*vaso-*	vessel
endo-	within; inside	*ophthalmo-*	eye	*-uria*	urine
ecto-; exo-; extra-	outside; away from	*oral-*	mouth		
-emia	blood	*osteo-*	bone		